FOURTEENTH EDITION

Earth Science

Edward J. Tarbuck

Frederick K. Lutgens

Illustrated by

Dennis Tasa

PEARSON

Boston   Columbus   Indianapolis
New York   San Francisco   Upper Saddle River
Amsterdam   Cape Town   Dubai   London
Madrid   Munich   Paris   Montréal   Toronto
Delhi   Mexico City   São Paulo   Sydney
Hong Kong   Seoul   Singapore   Taipei   Tokyo

About Our Sustainability Initiatives

Pearson recognizes the environmental challenges facing this planet and acknowledges our responsibility in making a difference. This book has been carefully crafted to minimize environmental impact. The binding, cover, and paper come from facilities that minimize waste, energy consumption, and the use of harmful chemicals. Pearson closes the loop by recycling every out-of-date text returned to our warehouse.

Along with developing and exploring digital solutions to our market's needs, Pearson has a strong commitment to achieving carbon neutrality. As of 2009, Pearson became the first carbon- and climate-neutral publishing company. Since then, Pearson remains strongly committed to measuring, reducing, and offsetting our carbon footprint.

The future holds great promise for reducing our impact on Earth's environment, and Pearson is proud to be leading the way. We strive to publish the best books with the most up-to-date and accurate content, and to do so in ways that minimize our impact on Earth. To learn more about our initiatives, please visit **www.pearson.com/responsibility**.

PEARSON

Acquisitions Editor: Andrew Dunaway
Senior Marketing Manager: Maureen McLaughlin
Project Manager: Crissy Dudonis
Project Management Team Lead: Gina M. Cheselka
Executive Development Editor: Jonathan Cheney
Director of Development: Jennifer Hart
Content Producer: Timothy Hainley
Project Manager, Instructor Media: Eddie Lee
Editorial Assistant: Sarah Shefveland
Senior Marketing Assistant: Nicola Houston
Full Service/Composition: Cenveo® Publisher Services

Project Manager, Full Service: Heidi Allgair
Photo Manager: Maya Melenchuk
Photo Researcher: Kristin Piljay
Text Permissions Manager: Alison Bruckner
Design Manager: Derek Bacchus
Interior Design: Elise Lansdon Design
Cover Design: Derek Bacchus
Photo and Illustration Support: International Mapping
Operations Specialist: Christy Hall
Cover Image Credit: Michael Collier

Credits and acknowledgments borrowed from other sources and reproduced, with permission, in this textbook appear on the appropriate page within text or are listed below.

Page 9: From J. Bronowski, The Common Sense of Science, p. 148. © 1953 Harvard University Press.
Page 12: From L. Pasteur, Lecture, University of Lille (7 December 1854). **Page 215:** From R.T. Chamberlain, "Some of the Objections to Wegener's Theory," In: THEORY OF CONTINENTAL DRIFT: A SYMPOSIUM, University of Chicago Press, pp. 83-87, 1928. **Page 264:** W. Mooney, USGS Seismologist.
Page 349: From J. Hutton, Theory of Earth, 1700; From J. Hutton, Transactions of the Royal Society of Edinburgh, 1788. **Page 488:** From A.J. Herbertson, "Outlines of Physiography," 1901. **Page 566:** Sir Francis Bacon. **Page 644:** Copernicus, De Revolutionibus, Orbium Coelestium (On the Revolution of the Heavenly Spheres). **Page 648:** Joseph Louis Lagrange, Oeuvres de Lagrange, 1867.

Library of Congress Cataloging–in–Publication Data
Tarbuck, Edward J.
  Earth science / Edward J. Tarbuck, Frederick K. Lutgens; illustrated by Dennis Tasa. – 14th ed.
    p. cm.
  Includes index.
  ISBN 978-0-321-92809-2 – ISBN 0-321-92809-1
1. Earth sciences – Textbooks. I. Lutgens, Frederick K. II. Title.
QE26.3.T38 2015
550–dc23
                    2013012995

5   16

www.pearsonhighered.com   ISBN-10: 0-321-92809-1; ISBN-13: 978-0-321-92809-2

BRIEF CONTENTS

FIND SMART FIGURES AND MOBILE FIELD TRIP FIGURES

In addition to the many informative and colorful illustrations and photos throughout this text, you will find two kinds of special figures that offer additional learning opportunities. These figures contain QR codes which the student can scan with a smart phone to explore exciting expanded online learning materials.

 Find **SmartFigures** *where you see this icon.*

 Find **Mobile Field Trip Figures** *where you see this icon.*

CONTENTS

UNIT TWO | SCULPTING EARTH'S SURFACE 94

4 Weathering, Soil, and Mass Wasting 95

5 Running Water and Groundwater 131

6 Glaciers, Deserts, and Wind 171

UNIT THREE | FORCES WITHIN 208

7 Plate Tectonics: A Scientific Revolution Unfolds 209

14 Ocean Water and Ocean Life 433

15 The Dynamic Ocean 453

UNIT SIX | EARTH'S DYNAMIC ATMOSPHERE 484

16 The Atmosphere: Composition, Structure, and Temperature 485

GEO GRAPHICS | **Acid Precipitation 491**

17 Moisture, Clouds, and Precipitation 517

18 Air Pressure and Wind 551

UNIT SEVEN | EARTH'S PLACE IN THE UNIVERSE 638

PREFACE

Earth Science, 14th edition, is a college-level text designed for an introductory course in Earth science. It consists of seven units that emphasize broad and up-to-date coverage of basic topics and principles in geology, oceanography, meteorology, and astronomy. The textbook is intended to be a meaningful, nontechnical survey for undergraduate students who have little background in science. Usually these students are taking an Earth science class to meet a portion of their college's or university's general requirements.

In addition to being informative and up-to-date, *Earth Science,* 14th edition, strives to meet the need of beginning students for a readable and user-friendly text and a highly usable tool for learning basic Earth science principles and concepts.

NEW TO THIS EDITION

- **SmartFigures—art that teaches.** Inside every chapter are several *SmartFigures. Earth Science,* 14th edition, has more than 100 of these figures. Just use your mobile device to scan the Quick Response (QR) code next to a SmartFigure, and the art comes alive. Each 3- to 5-minute feature, prepared and narrated by Professor Callan Bentley, is a mini-lesson that examines and explains the concepts illustrated by the figure. It is truly *art that teaches.*
- **Mobile Field Trips.** Scattered through this new edition of Earth Science are thirteen Mobile Field Trips. On each trip, you will accompany geologist–pilot–photographer Michael Collier in the air and on the ground to see and learn about landscapes that relate to discussions in the chapter. These extraordinary field trips are accessed in the same way as SmartFigures. You will scan a QR code that accompanies a figure in the chapter—usually one of Michael's outstanding photos.
- **New and expanded active learning path.** *Earth Science,* 14th edition, is designed for learning. Every chapter begins with *Focus on Concepts.* Each numbered learning objective corresponds to a major section in the chapter. The statements identify the knowledge and skills students should master by the end of the chapter, helping students prioritize key concepts. Within the chapter, each major section concludes with *Concept Checks* that allow students to check their understanding and comprehension of important ideas and terms before moving on to the next section. Chapters conclude with sections called *Give It Some Thought* and *Examining the Earth System.* The questions and problems in these sections challenge learners by involving them in activities that require higher-order thinking skills such as application, analysis, and synthesis of material in the chapter. The questions and problems in Examining the Earth System are intended to develop an awareness of and appreciation for some of the Earth system's many interrelationships.
- **Concepts in Review.** This all-new end-of-chapter feature is an important part of the text's revised active learning path. Each review is coordinated with the *Focus on Concepts* at the beginning of the chapter and with the numbered sections within the chapter. It is a readable and concise overview of key ideas, which makes it a valuable review tool for students. Photos, diagrams, and questions also help students focus on important ideas and test their understanding.

- **Eye on Earth.** Within every chapter are two or three images, often aerial or satellite views, that challenge students to apply their understanding of basic facts and principles. A brief explanation of each image is followed by questions that help focus students on visual analysis and critical thinking.
- **GEOgraphics.** As you turn the pages of each chapter, you will encounter striking visual features that we call GEOgraphics. They are engaging magazine-style "geo-essays" that explore topics that promote greater understanding and add interest to the story each chapter is telling.
- **An unparalleled visual program.** In addition to more than 200 new, high-quality photos and satellite images, dozens of figures are new or have been redrawn by renowned geoscience illustrator Dennis Tasa. Maps and diagrams are frequently paired with photographs for greater effectiveness. Further, many new and revised figures have additional labels that narrate the process being illustrated and guide students as they examine the figures. The result is a visual program that is clear and easy to understand.
- **MasteringGeology™.** MasteringGeology delivers engaging, dynamic learning opportunities—focused on course objectives and responsive to each student's progress—that are proven to help students absorb course material and understand difficult concepts. Assignable activities in MasteringGeology include Encounter Earth activities using Google Earth™, SmartFigure activities, Mobile Field Trips, GeoTutor activities, GigaPan® activities, Geoscience Animation activities, GEODe tutorial activities, and more. MasteringGeology also includes all instructor resources and a robust Study Area with resources for students.
- **Significant updating and revision of content.** A basic function of a college science text book is to provide clear, understandable presentations that are accurate, engaging, and up-to-date. Our number-one goal is to keep *Earth Science* current, relevant, and highly readable for beginning students. Every part of this text has been examined carefully with this goal in mind. Many discussions, case studies, and examples have been revised. This 14th edition represents perhaps the *most extensive and thorough revision* in the long history of this textbook.
- **Learning Catalytics™.** Learning Catalytics is a "bring your own device" student engagement, assessment, and classroom intelligence system. Learning Catalytics is a technology that has grown out of twenty years of cutting edge research, innovation, and implementation of interactive teaching and peer instruction. Available integrated with MasteringGeology.

DISTINGUISHING FEATURES

Readability

The language of this textbook is straightforward and *written to be understood.* Clear, readable discussions with a minimum of technical language are the rule. The frequent headings and subheadings help students follow discussions and identify the important ideas presented in each chapter. In this 14th edition, we have continued to improve readability by examining chapter organization and flow and by writing in a more personal style. Significant portions of several chapters have been substantially rewritten in an effort to make the material easier to understand.

Focus on Basic Principles

Although many topical issues are treated in this 14th edition of *Earth Science*, it should be emphasized that the main focus of this new edition remains the same as the focus of each of its predecessors: to promote student understanding of basic Earth science principles. As much as possible, we have attempted to provide the reader with a sense of the observational techniques and reasoning processes that constitute the Earth sciences.

A Strong Visual Component

Earth science is highly visual, and art and photographs play a critical role in an introductory textbook. As in all previous editions, Dennis Tasa, a gifted artist and respected geoscience illustrator, has worked closely with the authors to plan and produce the diagrams, maps, graphs, and sketches that are so basic to student understanding. The result is art that is clearer and easier to understand than ever before.

Our aim is to get *maximum effectiveness* from the visual component of the text. Michael Collier, an award-winning geologist–photographer aided greatly in this quest. As you read through this text, you will see dozens of his extraordinary aerial photographs. His contribution truly helps bring geology alive for the reader.

FOR THE INSTRUCTOR

Pearson continues to improve the instructor resources for this text, with the goal of saving you time in preparing for your classes.

MasteringGeology from Pearson is an online homework, tutorial, and assessment system designed to improve results by helping students quickly master concepts. Students using MasteringGeology benefit from self-paced tutorials that feature specific wrong-answer feedback and hints to keep them engaged and on track. MasteringGeology™ offers:

- Assignable activities, including Encounter Earth activities using Google Earth™, SmartFigure activities, GeoTutor activities, GigaPan® activities, Geoscience Animation activities, GEODe tutorial activities, and more
- Additional Give It Some Thought questions, Test Bank questions, and Reading Quizzes
- A student Study Area with Geoscience Animations, GEODe: Earth Science activities, SmartFigures, Video Field Trips *In the News* RSS feeds, Self Study Quizzes, Web Links, Glossary, and Flashcards
- Pearson eText for *Earth Science*, 14th edition, which gives students access to the text whenever and wherever they can access the Internet and includes powerful interactive and customization functions See www.masteringgeology.com

Learning Catalytics

Learning Catalytics™ is a "bring your own device" student engagement, assessment, and classroom intelligence system. With Learning Catalytics you can:

- Assess students in real time, using open-ended tasks to probe student understanding.
- Understand immediately where students are and adjust your lecture accordingly.
- Improve your students' critical-thinking skills.
- Access rich analytics to understand student performance.

- Add your own questions to make Learning Catalytics fit your course exactly.
- Manage student interactions with intelligent grouping and timing.

Learning Catalytics is a technology that has grown out of twenty years of cutting edge research, innovation, and implementation of interactive teaching and peer instruction. Available integrated with MasteringGeology. www.learningcatalytics.com

Instructor's Resource DVD

The Instructor's Resource DVD puts all your lecture resources in one easy-to-reach place:

- Three PowerPoint® presentations for each chapter
- The Geoscience Animation Library
- All the line art, tables, and photos from the text, in .jpg files
- "Images of Earth" photo gallery
- Instructor's Manual in Microsoft Word
- Test Bank in Microsoft Word
- TestGen test-generation and management software

PowerPoints®

The Instructor's Resource DVD provides three PowerPoint files for each chapter to cut down on your preparation time, no matter what your lecture needs:

- **Art.** All the line art, tables, and photos from the text have been preloaded into PowerPoint slides for easy integration into your presentations.
- **Lecture outline.** This set averages 35 slides per chapter and includes customizable lecture outlines with supporting art.
- **Classroom Response System (CRS) questions.** These questions have been authored for use in conjunction with any classroom response system. You can electronically poll your class for responses to questions, pop quizzes, attendance, and more.

Animations and "Images of Earth"

The Pearson Prentice Hall Geoscience Animation Library includes more than 100 animations illustrating many difficult-to-visualize topics in Earth science. Created through a unique collaboration among five of Pearson Prentice Hall's leading geoscience authors, these animations represent a significant step forward in lecture presentation aids. They are provided both as Flash files and, for your convenience, preloaded into PowerPoint slides.

"Images of Earth" allows you to supplement your personal and text-specific slides with an amazing collection of more than 300 geologic photos contributed by Marli Miller (University of Oregon) and other professionals in the field. The photos are available on the Instructor's Resource DVD.

Instructor's Manual with Test Bank

The *Instructor's Manual* contains learning objectives, chapter outlines, answers to end-of-chapter questions, and suggested short demonstrations to spice up your lecture. The Test Bank incorporates art and averages 75 multiple-choice, true/false, short-answer, and critical thinking questions per chapter.

TestGen

Use this electronic version of the Test Bank to customize and manage your tests. Create multiple versions, add or edit questions, add illustrations, and so on. This powerful software easily addresses your customization needs.

Course Management

Pearson Prentice Hall offers instructor and student media for the 14th edition of *Earth Science* in formats compatible with Blackboard and other course management platforms. Contact your local Pearson representative for more information.

FOR THE STUDENT

The student resources to accompany *Earth Science*, 14th edition, have been further refined, with the goal of focusing the students' efforts and improving their understanding of Earth science concepts.

MasteringGeology from Pearson is an online homework, tutorial, and assessment system designed to improve results by helping students quickly master concepts. Students using MasteringGeology benefit from self-paced tutorials that feature specific wrong-answer feedback and hints to keep them engaged and on track. MasteringGeology™ also offers students the Study Area, which contains:

- **Geoscience Animation Library.** More than 100 animations illustrating many difficult to understand Earth science concepts.
- **GEODe: Earth Science.** An interactive visual walkthrough of each chapter's content.
- *In the News* **RSS Feeds.** Current Earth science events and news articles are pulled into the site, with assessment.
- **SmartFigures and Mobile Field Trips**
- **Pearson eText**
- **Optional Self Study Quizzes**
- **Web Links**
- **Glossary**
- **Flashcards**

FOR THE LABORATORY

Applications and Investigations in Earth Science, 8th edition, was written by Ed Tarbuck, Fred Lutgens, and Ken Pinzke. This full-color laboratory manual contains 23 exercises that provide students with hands-on experience in geology, oceanography, meteorology, astronomy, and Earth science skills. The lab manual is available at a discount when purchased with the text; please contact your local Pearson representative for more details.

ACKNOWLEDGMENTS

Writing a college textbook requires the talents and cooperation of many people. It is truly a team effort, and the authors are fortunate to be part of an extraordinary team at Pearson Education. In addition to being great people to work with, all are committed to producing the best textbooks possible. Special thanks to our geology editor, Andy Dunaway, who invested a great deal of time, energy, and effort in this project. We appreciate his enthusiasm, hard work, and quest for excellence. We also appreciate our conscientious project manager, Crissy Dudonis, whose job it was to keep track of all that was going on—and a lot was going on. The text's new design and striking cover resulted from the creative talents of Derek Bacchus and his team. We think it is a job well done. As always, our marketing manager, Maureen McLaughlin, provided helpful advice and many good ideas. *Earth Science,* 14th edition, was truly improved with the help of our developmental editor, Jonathan Cheney. Many thanks. The production team was led by Gina Cheselka at Pearson Education and by Heidi Allgair at Cenveo® Publisher Services. It was their job to make this text into a finished product. The talents of copy editor Kitty Wilson, compositor Annamarie Boley, and photo researcher Kristin Piljay were an important part of the production process. We think they all did a great job. They are true professionals, with whom we are very fortunate to be associated.

The authors owe a special thanks to three people who were a very important part of this project:

- Working with Dennis Tasa, who is responsible for all of the text's outstanding illustrations, is always special for us. He has been a part of our team for more than 30 years. We not only value his artistic talents, hard work, patience, and imagination but his friendship as well.
- As you read this text, you will see dozens of extraordinary photographs by Michael Collier, an award-winning geologist, author, and photographer. Most are aerial shots taken from his nearly 60-year-old Cessna 180. Michael was also responsible for preparing the remarkable Mobile Field Trips that are scattered through the text. Among his many awards is the American Geological Institute Award for Outstanding Contribution to the Public Understanding of the Geosciences. We think that Michael's photographs and field trips are the next best thing to being there. We were very fortunate to have had Michael's assistance on *Earth Science*, 14th edition. Thanks, Michael.
- Callan Bentley has been an important addition to the *Earth Science* team. Callan is an assistant professor of geology at Northern Virginia Community College in Annandale, where he has been honored many times as an outstanding teacher. He is a frequent contributor to *Earth* magazine and is author of the popular geology blog Mountain Beltway. Callan was responsible for preparing the *SmartFigures* that appear throughout *Earth Science*'s 24 chapters. As you take advantage of these outstanding learning aids, you will hear his voice explaining the ideas. Callan also helped with the preparation of the Concepts in Review feature found at the end of each chapter. We appreciate Callan's contributions to this new edition of *Earth Science*.

Great thanks also go to our colleagues who prepared in-depth reviews. Their critical comments and thoughtful input helped guide our work and clearly strengthened the text. Special thanks to:

Patricia Anderson, *California State University—San Marcos*
J. Bret Bennington, *Hofstra University*
Nahid Brown, *Northeastern Illinois University*
Brett Burkett, *Collin College*
Barry Cameron, *University of Wisconsin—Milwaukee*
Haluk Cetin, *Murray State University*
Natasha Cleveland, *Frederick Community College*
Adam Davis, *Vincennes University*
Anne Egger, *Central Washington University*
Joseph Galewsky, *The University of New Mexico*
Leslie Kanat, *Johnson State College*
Mustapha Kane, *Florida Gateway College at Lake City, FL*
Alyson Lighthart, *Portland Community College*
Rob Martin, *Florida State College at Jacksonville*
Ron Metzger, *Southwestern Oregon Community College*
Sadredin (Dean) Moosavi, *Rochester Community and Technical College*
Carol Mueller, *Harford Community College*
Jessica Olney, *Hillsborough Community College*
David Pitts, *University of Houston—Clear Lake*
Steven Schimmrich, *SUNY Ulster*
Xiaoming Zhai, *College of Lake County*

Last, but certainly not least, we gratefully acknowledge the support and encouragement of our wives, Joanne Bannon and Nancy Lutgens. Preparation of *Earth Science*, 14th edition, would have been far more difficult without their patience and understanding.

Ed Tarbuck
Fred Lutgens

New learning path helps students master the concepts

The new edition is designed to support a new four-part learning path, an innovative structure which facilitates active learning and easily allows students to focus on important ideas as they pause to assess their progress at frequent intervals.

The chapter-opening Focus on Concepts lists the learning objectives for each chapter. Each section of the chapter is tied to a specific learning objective, providing students with a clear learning path to the chapter content.

UNIT THREE | FORCES WITHIN

7 Plate Tectonics: A Scientific Revolution Unfolds

FOCUS ON CONCEPTS

Each statement represents the primary **LEARNING OBJECTIVE** for the corresponding major heading within the chapter. After you complete the chapter, you should be able to:

7.1 Discuss the view that most geologists held prior to the 1960s regarding the geographic positions of the ocean basins and continents.

7.2 List and explain the evidence Wegener presented to support his continental drift hypothesis.

7.3 Discuss the two main objections to the continental drift hypothesis.

7.4 List the major differences between Earth's lithosphere and asthenosphere and explain the importance of each in the plate tectonics theory.

7.5 Sketch and describe the movement along a divergent plate boundary that results in the formation of new oceanic lithosphere.

7.6 Compare and contrast the three types of convergent plate boundaries and name a location where each type can be found.

7.7 Describe the relative motion along a transform plate boundary and locate several examples on a plate boundary map.

7.8 Explain why plates such as the African and Antarctic plates are getting larger, while the Pacific plate is getting smaller.

7.9 List and explain the evidence used to support the plate tectonics theory.

7.10 Describe two methods researchers use to measure relative plate motion.

7.11 Summarize what is meant by plate–mantle convection and explain two of the primary driving forces of plate motion.

Climber ascending Chang Zheng Peak near Mount Everest.
(Photo by Stock Connection/SuperStock)

209

Each chapter section concludes with Concept Checks, a feature that lists questions tied to the section's learning objective, allowing students to monitor their grasp of significant facts and ideas.

7.4 CONCEPT CHECKS

1 What major ocean floor feature did oceanographers discover after World War II?

2 Compare and contrast the lithosphere and the asthenosphere.

3 List the seven largest lithospheric plates.

4 List the three types of plate boundaries and describe the relative motion at each of them.

FOCUS ON CONCEPTS

Each statement represents the primary **LEARNING OBJECTIVE** for the corresponding major heading within the chapter. After you complete the chapter, you should be able to:

7.1 Discuss the view that most geologists held prior to the 1960s regarding the geographic positions of the ocean basins and continents.

7.2 List and explain the evidence Wegener presented to support his continental drift hypothesis.

7.3 Discuss the two main objections to the continental drift hypothesis.

7.4 List the major differences between Earth's lithosphere and asthenosphere and explain the importance of each in the plate tectonics theory.

7.5 Sketch and describe the movement along a divergent plate boundary that results in the formation of new oceanic lithosphere.

Concepts in Review, a fresh approach to the typical end-of-chapter material, provides students with a structured and highly visual review of the chapter.

Key Terms   Section Title   Learning Objective   Review Statements

7 CONCEPTS IN REVIEW

Plate Tectonics: A Scientific Revolution Unfolds

Consistent with the Focus on Concepts and Concept Checks, the Concepts in Review is structured around the section title and the corresponding learning objective for each section.

7.1 FROM CONTINENTAL DRIFT TO PLATE TECTONICS

Discuss the view that most geologists held prior to the 1960s regarding the geographic positions of the ocean basins and continents.

- Fifty years ago, most geologists thought that ocean basins were very old and that continents were fixed in place. Those ideas were discarded with a scientific revolution that revitalized geology: the theory of plate tectonics. Supported by multiple kinds of evidence, plate tectonics is the foundation of modern Earth science.

7.2 CONTINENTAL DRIFT: AN IDEA BEFORE ITS TIME

List and explain the evidence Wegener presented to support his continental drift hypothesis.

KEY TERMS continental drift, supercontinent, Pangaea

- German meteorologist Alfred Wegener formulated the idea of continental drift in 1917. He suggested that Earth's continents are not fixed in place but have moved slowly over geologic time.
- Wegener reconstructed a super-continent called Pangaea that existed about 200 million years ago, during the late Paleozoic and early Mesozoic.
- Wegener's evidence that Pangaea existed but later broke into pieces that drifted apart included (1) the shape of the continents, (2) continental fossil organisms that matched across oceans, (3) matching rock types and modern mountain belts, and (4) sedimentary rocks that recorded ancient climates, including glaciers on the southern portion of Pangaea.

Q Why did Wegener choose organisms such as *Glossopteris* and *Mesosaurus* as evidence for continental drift, as opposed to other fossil organisms such as sharks or jellyfish?

John Cancalosi/AGE Fotostock

7.3 THE GREAT DEBATE

Discuss the two main objections to the continental drift hypothesis.

- Wegener's hypothesis suffered from two flaws: It proposed tidal forces as the mechanism for the motion of continents, and it implied that the continents would have plowed their way through weaker oceanic crust, like a boat cutting through a thin layer of sea ice. Geologists rejected the idea of continental drift when Wegener

7.4 THE THEORY OF PLATE TECTONICS

List the major differences between Earth's lithosphere and asthenosphere and explain the tectonics theory.

KEY TERMS ocean ridge system, theory of plate tectonics, lithosphere, asthenosphere, lithospheric plate (plate)

- Research conducted during World War II led to new insights that helped revive Wegener's hypothesis of con revealed previously unknown features, including an extremely long mid-ocean ridge system. Sampling of th young relative to the continents.
- The lithosphere is Earth's outermost rocky layer that is broken into plates. It is relatively stiff and deforms b is the asthenosphere, a relatively weak layer that deforms by flowing. The lithosphere consists both of crust upper mantle.
- There are seven large plates, another seven intermediate-size plates, and numerous relatively small m that may either be divergent (moving apart from each other), convergent (moving toward each other), each other).

Give It Some Thought (GIST) is found at the end of each chapter and consists of questions and problems asking students to analyze, synthesize, and think critically about Earth science. GIST questions relate back to the chapter's learning objectives, and can easily be assigned using MasteringGeology™.

GIVE IT SOME THOUGHT

1. After referring to the section in the Introduction titled "The Nature of Scientific Inquiry," answer the following:
   a. What observation led Alfred Wegener to develop his continental drift hypothesis?
   b. Why did the majority of the scientific community reject the continental drift hypothesis?
   c. Do you think Wegener followed the basic principles of scientific inquiry? Support your answer.
2. Referring to the accompanying diagrams that illustrate the three types of convergent plate boundaries, complete the following:
   a. Identify each type of convergent boundary.
   b. On what type of crust do volcanic island arcs develop?
   c. Why are volcanoes largely absent where two continental blocks collide?
   d. Describe two ways that oceanic–oceanic convergent boundaries are different from oceanic–continental boundaries. How are they similar?

A.   B.   C.

3. Some predict that California will sink into the ocean. Is this idea consistent with the theory of plate tectonics? Explain.
4. Refer to the accompanying hypothetical plate map to answer the following questions:
   a. How many portions of plates are shown?
   b. Are continents A, B, and C moving toward or away from each other? How did you determine your answer?
   c. Explain why active volcanoes are more likely to be found on continents A and B than on continent C.
   d. Provide at least one scenario in which volcanic activity might be triggered on continent C.

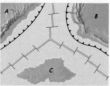

Oceanic ridge   Subduction zone

5. Volcanoes, such as the Hawaiian chain, that form over mantle plumes are some of the largest shield volcanoes on Earth. However, several shield volcanoes on Mars are gigantic compared to those on Earth. What does this difference tell us about the role of plate motion in shaping the Martian surface?

6. Imagine that you are studying seafloor spreading along two different oceanic ridges. Using data from a magnetometer, you produced the two accompanying maps. From these maps, what can you determine about the relative rates of seafloor spreading along these two ridges? Explain.

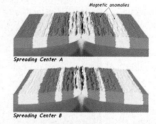

Magnetic anomalies

Spreading Center A

Spreading Center B

7. Australian marsupials (kangaroos, koala bears, etc.) have direct fossil links to marsupial opossums found in the Americas. Yet the modern marsupials in Australia are markedly different from their American relatives. How does the breakup of Pangaea help to explain these differences (see Figure 7.24)?
8. Density is a key component in the behavior of Earth materials and is especially important in understanding key aspects of plate tectonics. Describe three different ways that density and/or density differences play a role in plate tectonics.

Carefully selected art and photos aid understanding, add realism, and heighten student interest.

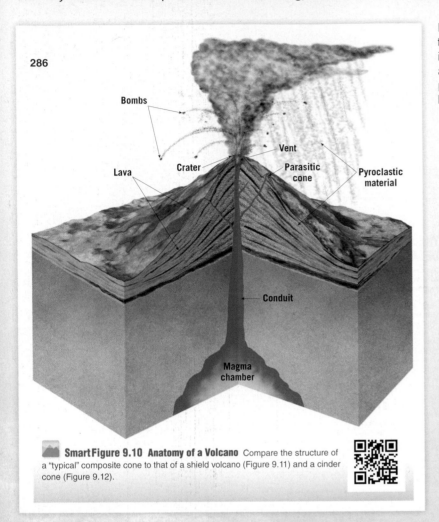

SmartFigure 9.10 Anatomy of a Volcano Compare the structure of a "typical" composite cone to that of a shield volcano (Figure 9.11) and a cinder cone (Figure 9.12).

NEW! SmartFigures bring key chapter illustrations to life! Found throughout the book, SmartFigures are sophisticated, annotated illustrations that are also narrated videos. The SmartFigure videos are accessible on mobile devices via scannable Quick Response (QR) codes printed in the text and through the Study Area in MasteringGeology. See the Preface for more detailed information on SmartFigures.

Callan Bentley, SmartFigure author, is an assistant professor of geology at Northern Virginia Community College (NOVA) in Annandale, Virginia. Trained as a structural geologist, Callan teaches introductory level geology at NOVA, including field-based and hybrid courses. Callan writes a popular geology blog called *Mountain Beltway*, contributes cartoons, travel articles, and book reviews to *EARTH Magazine*, and is a leader in the two-year college geoscience community.

Mobile Field Trips

Scattered through this new edition of Earth Science are thirteen video field trips. On each trip, you will accompany geologist-pilot-photographer Michael Collier in the air and on the ground to see and learn about landscapes that relate to discussions in the chapter. These extraordinary field trips are accessed in the same way as SmartFigures. You will scan a QR code that accompanies a figure in the chapter—usually one of Michael's outstanding photos.

Mobile Field Trip 9.25 Sill Exposed in Utah's Sinbad Country
The dark, essentially horizontal bands are sills of basaltic composition that intruded horizontal layers of sedimentary rock. (Photo by Michael Collier)

As you turn the pages of this book, you will see dozens of extraordinary photographs by Michael Collier. Most are aerial shots taken from his nearly 60-year-old Cessna 180. Michael is an award-winning geologist, author, and photographer. Michael's photographs are the next best thing to being there. We were fortunate to have had Michael's assistance on Earth Science, Fourteenth edition.

NEW! GEOgraphics use contemporary, compelling visual representations to illustrate complex concepts, enhancing students' ability to synthesize and recall information.

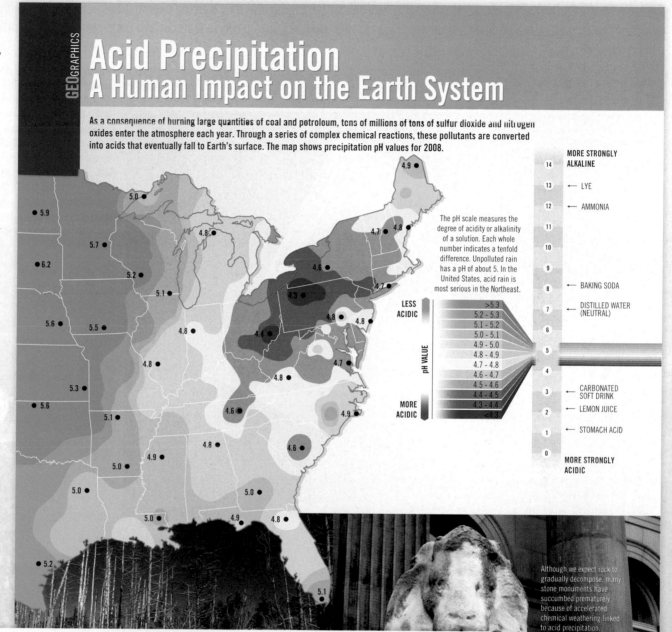

GEOGRAPHICS

Acid Precipitation
A Human Impact on the Earth System

As a consequence of burning large quantities of coal and petroleum, tens of millions of tons of sulfur dioxide and nitrogen oxides enter the atmosphere each year. Through a series of complex chemical reactions, these pollutants are converted into acids that eventually fall to Earth's surface. The map shows precipitation pH values for 2008.

The pH scale measures the degree of acidity or alkalinity of a solution. Each whole number indicates a tenfold difference. Unpolluted rain has a pH of about 5. In the United States, acid rain is most serious in the Northeast.

MORE STRONGLY ALKALINE

- 14
- 13 ← LYE
- 12 ← AMMONIA
- 11
- 10
- 9
- 8 ← BAKING SODA
- 7 ← DISTILLED WATER (NEUTRAL)
- 6
- 5
- 4
- 3 ← CARBONATED SOFT DRINK
- 2 ← LEMON JUICE
- 1 ← STOMACH ACID
- 0

MORE STRONGLY ACIDIC

pH VALUE

LESS ACIDIC

| >5.3 |
| 5.2 – 5.3 |
| 5.1 – 5.2 |
| 5.0 – 5.1 |
| 4.9 – 5.0 |
| 4.8 – 4.9 |
| 4.7 – 4.8 |
| 4.6 – 4.7 |
| 4.5 – 4.6 |
| 4.4 – 4.5 |
| 4.3 – 4.4 |
| <4.3 |

MORE ACIDIC

Although we expect rock to gradually decompose, many stone monuments have succumbed prematurely because of accelerated chemical weathering linked to acid precipitation.

NEW! Eye on Earth features engage students in active learning, asking them to perform critical thinking and visual analysis tasks to evaluate data and make predictions.

EYE ON EARTH

This image was obtained during the 1991 eruption of Mount Pinatubo in the Philippines. This was the largest eruption to affect a densely populated area in recent times. Timely forecasts of the event by scientists were credited with saving at least 5000 lives. (Alberto Garcia/CORBIS)

QUESTION 1 What name is given to the ash- and pumice-laden cloud that is racing toward the photographer?

QUESTION 2 At what speeds can these fiery clouds move down steep mountain slopes?

— Jeep

MasteringGeology™ www.masteringgeology.com

Available for the Earth science course, MasteringGeology delivers engaging, dynamic learning opportunities—focused on course objectives and responsive to each student's progress—that are proven to help students absorb course material and understand difficult Earth science concepts.

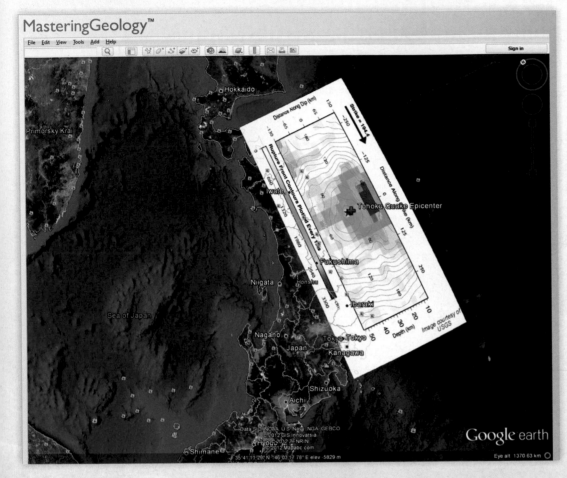

Encounter Activities provide rich, interactive explorations of Earth science concepts using the dynamic features of Google Earth™ to visualize and explore Earth's varied physical landscapes. Dynamic assessment includes questions related to core Earth science concepts. All explorations include corresponding Google Earth KMZ media files, and questions include hints and specific wrong-answer feedback to help coach students toward mastery of the concepts.

NEW! Inquiry-based interactive simulations, developed to allow students to manipulate Earth processes, assist students in mastering the most difficult Earth science processes as identified by today's instructors.

NEW! GigaPan® Activities take advantage of the GigaPan high-resolution panoramic picture technology developed by Carnegie Mellon University in conjunction with NASA. Photos and accompanying questions correlate with concepts in the student book.

Geoscience Animations and Activities illuminate difficult-to-visualize topics from across the physical geosciences. MasteringGeology allows instructors to easily assign the animations and corresponding assessment questions, all of which include hints and specific wrong-answer feedback.

Motion at Plate Boundaries

| Divergent Boundaries | Convergent Boundaries | Transform Boundaries | **Plate Motion and Tectonics** |

SHOW TEXT

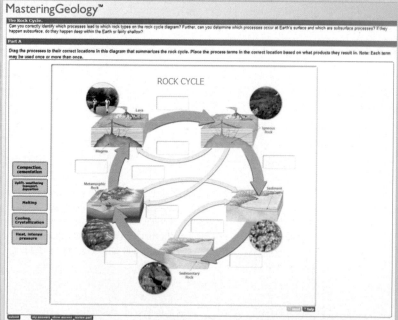

PEARSON

GEODe Tutorials provide an interactive visual walkthrough of core content through animations, videos, illustrations, photographs, and narration. Activities include assessment questions to test those concepts with hints and specific wrong-answer feedback.

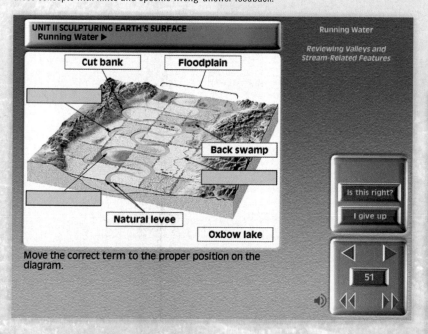

Give It Some Thought questions and problems relate back to each chapter's learning objectives and challenge learners by involving them in activities that require higher-order thinking skills such as synthesis, analysis, and application.

MasteringGeology™

The Rock Cycle

Can you correctly identify which processes lead to which rock types on the rock cycle diagram? Further, can you determine which processes occur at Earth's surface and which are subsurface processes? If they happen subsurface, do they happen deep within the Earth or fairly shallow?

Part A

Drag the processes to their correct locations in this diagram that summarizes the rock cycle. Place the process terms in the correct location based on what products they result in. Note: Each term may be used once or more than once.

ROCK CYCLE

- Compaction, cementation
- Uplift, weathering, transport, deposition
- Melting
- Cooling, Crystallization
- Heat, intense pressure

MasteringGeology™

www.masteringgeology.com

Quickly monitor and display student results

With the Mastering gradebook and diagnostics, instructors will be better informed about students' progress than ever before. Mastering captures the step-by-step work of every student—including wrong answers submitted, hints requested, and time taken at every step of every problem—all providing unique insight into the most common misconceptions of the class.

The **Gradebook** records all scores for automatically graded assignments. Shades of red highlight struggling students and challenging assignments.

Diagnostics provide unique insight into class and student performance. With a single click, charts summarize the most difficult items, vulnerable students, grade distribution, and score improvement over the duration of the course.

With a single click, **Individual Student Performance Data** provides at-a-glance statistics into each individual student's performance, including time spent on the item, number of hints opened, and number of wrong and correct answers submitted.

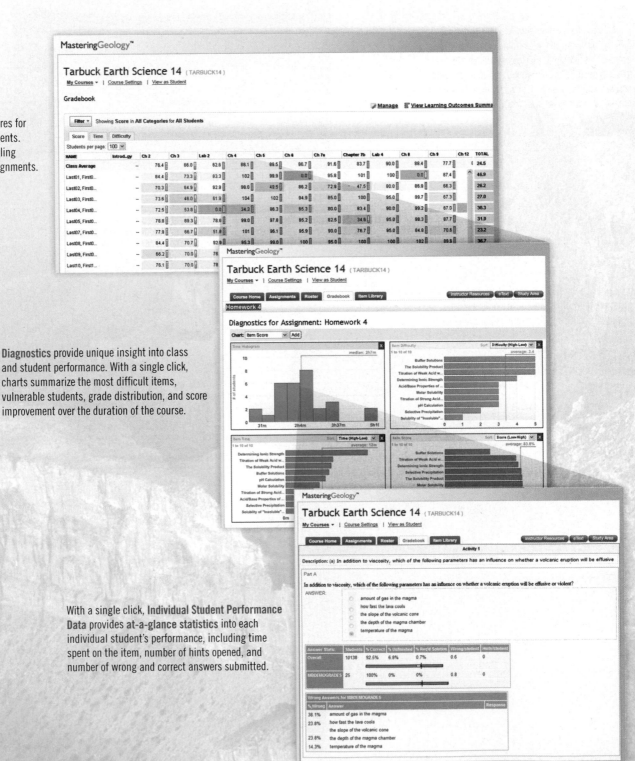

Easily measure student performance against learning outcomes

Learning Outcomes

MasteringGeology provides quick and easy access to information on student performance against learning outcomes and makes it easy for instructors to share those results.

Instructors can:

- Quickly add learning outcomes or use publisher-provided ones to track student performance and report it to administration.
- View class and individual student performance against specific learning outcomes.
- Effortlessly export results to a spreadsheet and further customize and/or share with chairs, deans, administrators, and/or accreditation boards.

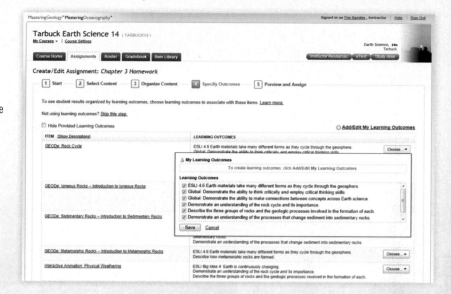

Easy to Customize

Instructors can customize publisher-provided problems or quickly add their own. MasteringGeology makes it easy for instructors to edit any questions or answers, import their own questions, and quickly add images, links, and files to further enhance the student experience.

Instructors can upload their own video and audio files from their hard drives to share with students, as well as record video from their computer's webcam directly into MasteringGeology—no plug-ins required. Students can download video and audio files to their local computer or launch them in Mastering to view the content.

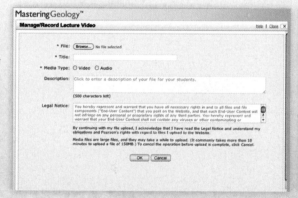 

Pearson eText gives students access to *Earth Science*, Fourteenth Edition whenever and wherever they can access the Internet. The eText pages look exactly like the printed text, and include powerful interactive and customization functions. Users can create notes, highlight text in different colors, create bookmarks, zoom, click hyperlinked words and phrases to view definitions, and view as a single page or as two pages. Pearson eText links students to associated media files, enabling them to view an animation as they read the text, and offers a full-text search and the ability to save and export notes. The Pearson eText also includes embedded URLs in the chapter text with active links to the Internet.

NEW! The Pearson eText app is a great companion to Pearson's eText browser-based book reader. It allows existing subscribers who view their Pearson eText titles on a Mac or PC to additionally access their titles in a bookshelf on their iPad™ or Android™ device either online or via download.

1

Introduction to Earth Science

1.1 List and describe the sciences that collectively make up Earth science. Discuss the scales of space and time in Earth science.

1.2 Discuss the nature of scientific inquiry and distinguish between a hypothesis and a theory.

1.3 Outline the stages in the formation of our solar system.

1.4 List and describe Earth's four major spheres.

1.5 Label a diagram that shows Earth's internal structure. Briefly explain why the geosphere can be described as being mobile.

1.6 List and describe the major features of the continents and ocean basins.

1.7 Define *system* and explain why Earth is considered to be a system.

An afternoon rainstorm near Muddy Creek in southern Utah.
(Photo by Michael Collier)

The spectacular eruption of a volcano, the magnificent scenery of a rocky coast, and the destruction created by a hurricane are all subjects for an Earth scientist. The study of Earth science deals with many fascinating and practical questions about our environment. What forces produce mountains? Why is our daily weather variable? Is climate really changing? How old is Earth, and how is our planet related to the other planets in the solar system? What causes ocean tides? What was the Ice Age like? Will there be another? Can a successful well be located at a particular site?

The subject of this text is *Earth science*. To understand Earth is not an easy task because our planet is not a static and unchanging mass. Rather, it is a dynamic body with many interacting parts and a long and complex history.

1.1 | WHAT IS EARTH SCIENCE? List and describe the sciences that collectively make up Earth science. Discuss the scales of space and time in Earth science.

Mobile Field Trip 1.1 Internal and External Processes The processes that operate beneath and upon Earth's surface are an important focus of physical geology. (Volcano photo by Lucas Jackson/ Reuters; glacier photo by Michael Collier)

Earth science is the name for all the sciences that collectively seek to understand Earth and its neighbors in space. It includes geology, oceanography, meteorology, and astronomy. Understanding Earth science is challenging because our planet is a dynamic body with many interacting parts and a complex history. Throughout its long existence, Earth has been changing. In fact, it is changing as you read this page and will continue to do so into the foreseeable future. Sometimes the changes are rapid and violent, as when severe storms, landslides, and volcanic eruptions occur. Conversely, many changes take place so gradually that they go unnoticed during a lifetime. Scales of size and space also vary greatly among the phenomena studied in Earth science.

Earth science is often perceived as science that is performed in the out of doors, and rightly so. A great deal of an Earth scientist's study is based on observations and experiments conducted in the field. But Earth science is also conducted in the laboratory, where, for example, the study of various Earth materials provides insights into many basic processes, and the creation of complex computer models allows for the simulation of our planet's complicated climate system. Frequently, Earth scientists require an understanding and application of knowledge and principles from physics, chemistry, and biology. Geology, oceanography, meteorology, and astronomy are sciences that seek to expand our knowledge of the natural world and our place in it.

Geology

In this book, Units 1–4 focus on the science of **geology**, a word that literally means "study of Earth." Geology is traditionally divided into two broad areas: physical and historical.

Physical geology examines the materials composing Earth and seeks to understand the many processes that operate beneath and upon its surface (**FIGURE 1.1**). Earth is a dynamic, ever-changing planet. Internal forces create earthquakes, build mountains, and produce volcanic structures. At the surface, external processes break rock apart and sculpt a broad array of landforms. The erosional

Internal processes are those that occur beneath Earth's surface. Sometimes they lead to the formation of major features at the surface.

External processes, such as landslides, rivers, and glaciers, erode and sculpt surface features. The Colorado River played a major role in creating the Grand Canyon.

effects of water, wind, and ice result in a great diversity of landscapes. Because rocks and minerals form in response to Earth's internal and external processes, their interpretation is basic to an understanding of our planet.

In contrast to physical geology, the aim of *historical geology* is to understand the origin of Earth and the development of the planet through its 4.6-billion-year history. It strives to establish an orderly chronological arrangement of the multitude of physical and biological changes that have occurred in the geologic past. The study of physical geology logically precedes the study of Earth history because we must first understand how Earth works before we attempt to unravel its past.

Oceanography

Earth is often called the "water planet" or the "blue planet." Such terms relate to the fact that more than 70 percent of Earth's surface is covered by the global ocean. If we are to understand Earth, we must learn about its oceans. Unit 5, *The Global Ocean*, is devoted to **oceanography**. Oceanography is actually not a separate and distinct science. Rather, it involves the application of all sciences in a comprehensive and interrelated study of the oceans in all their aspects and relationships. Oceanography integrates chemistry, physics, geology, and biology. It includes the study of the composition and movements of seawater, as well as coastal processes, seafloor topography, and marine life.

Meteorology

The continents and oceans are surrounded by an atmosphere. Unit 6, *Earth's Dynamic Atmosphere*, examines the mixture of gases that is held to the planet by gravity and thins rapidly with altitude. Acted on by the combined effects of Earth's motions and energy from the Sun, and influenced by Earth's land and sea surface, the formless and invisible atmosphere reacts by producing an infinite variety of weather, which in turn creates the basic pattern of global climates. **Meteorology** is the study of the atmosphere and the processes that produce weather and climate. Like oceanography, meteorology involves the application of other sciences in an integrated study of the thin layer of air that surrounds Earth.

Astronomy

Unit 7, *Earth's Place in the Universe*, demonstrates that an understanding of Earth requires that we relate our planet to the larger universe. Because Earth is related to all the other objects in space, the science of **astronomy**— the study of the universe—is very useful in probing the origins of our own environment. Because we are so closely acquainted with the planet on which we live, it is easy to forget that Earth is just a tiny object in a vast universe. Indeed, Earth is subject to the same physical laws that govern the many other objects populating the great expanses of space. Thus, to understand explanations of our planet's origin, it is useful to learn something about the other members of our solar system. Moreover, it is helpful to view the solar system as a part of the great assemblage of stars that comprise our galaxy, which is but one of many galaxies.

Earth Science Is Environmental Science

Earth science is an environmental science that explores many important relationships between people and the natural environment. Many of the problems and issues addressed by Earth science are of practical value to people.

Natural Hazards Natural hazards are a part of living on Earth. Every day they adversely affect literally millions of people worldwide and are responsible for staggering damages. Among the hazardous Earth processes studied by Earth scientists are volcanoes, floods, tsunami, earthquakes, landslides, and hurricanes. Of course, these hazards are *natural* processes. They become hazards only when people try to live where these processes occur.

For most of history, most people lived in rural areas. According to the United Nations, that changed in 2008, and today more people live in cities than in rural areas. This global trend toward urbanization concentrates millions of people into megacities, many of which are vulnerable to natural hazards (**FIGURE 1.2**). Coastal sites are becoming more vulnerable because development often destroys

FIGURE 1.2 Hurricane Sandy A portion of the New Jersey shoreline shortly after this huge storm struck in late October 2012. The storm was especially destructive because it struck a region with a high population density and extensive development. Shifting shoreline sands and the desire of people to occupy these areas are often in conflict. (Photo by AP Photo/Mike Groll)

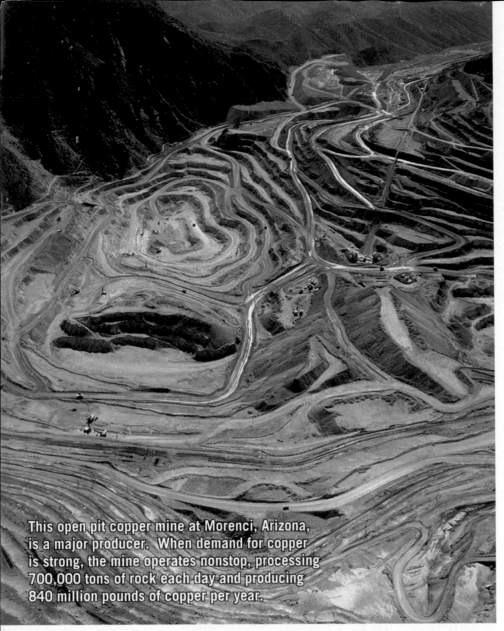

This open pit copper mine at Morenci, Arizona, is a major producer. When demand for copper is strong, the mine operates nonstop, processing 700,000 tons of rock each day and producing 840 million pounds of copper per year.

natural defenses such as wetlands and sand dunes. In addition, there is a growing threat associated with human influences on the Earth system such as sea level rise that is linked to global climate change.[1] Other megacities are exposed to seismic (earthquake) and volcanic hazards where inappropriate land use and poor construction practices, coupled with rapid population growth, are increasing vulnerability.

Resources Resources represent another important focus that is of great practical value to people. They include water and soil, a great variety of metallic and nonmetallic minerals, and energy (**FIGURE 1.3**). Together they form the very foundation of modern civilization. Earth science deals with the formation and occurrence of these vital resources and also with maintaining supplies and with the environmental impact of their extraction and use.

People Influence Earth Processes Not only do Earth processes have an impact on people, but we humans can dramatically influence Earth processes as well. Human activities alter the composition of the atmosphere that trigger air pollution episodes and cause global climate change (**FIGURE 1.4**). River flooding is natural, but the magnitude and frequency of flooding can be changed significantly by human activities such as clearing forests, building cities, and constructing dams. Unfortunately, natural systems do not always adjust to artificial changes in ways that we can anticipate. Thus, an alteration to the environment that was intended to benefit society often has the opposite effect.

At various places throughout this book, you will have opportunities to examine different aspects of our relationship with the physical environment. It will be rare to find

FIGURE 1.3 Copper Mine Resources represent an important link between people and Earth science. (Photo by Michael Collier)

[1]The idea of the Earth system is explored later in the chapter. Global climate change and its effects are a focus of Chapter 20.

FIGURE 1.4 Urban Air Pollution A severe air pollution episode at Beijing, China, on March 18, 2008. Fuel combustion by factories, power plants, and motor vehicles provided a high proportion of the pollutants. Meteorological factors determine whether pollutants remain trapped in the city or are dispersed. (Photo by AP Photo/Ng Han)

Earth science involves investigations of phenomena that range in size from the atomic level to those that involve large portions of the universe.

FIGURE 1.5 From Atoms to Galaxies Earth science studies phenomena on many different scales.

a chapter that does not address some aspect of natural hazards, environmental issues, or resources. Significant parts of some chapters provide the basic knowledge and principles needed to understand environmental problems.

Scales of Space and Time in Earth Science

When we study Earth, we must contend with a broad array of space and time scales (**FIGURE 1.5**). Some phenomena are relatively easy for us to imagine, such as the size and duration of an afternoon thunderstorm or the dimensions of a sand dune. Other phenomena are so vast or so small that they are difficult to imagine. The number of stars and distances in our galaxy (and beyond!) or the internal arrangement of atoms in a mineral crystal are examples of such phenomena.

Some of the events we study occur in fractions of a second. Lightning is an example. Other processes extend over spans of tens or hundreds of millions of years. For example, the lofty Himalaya Mountains began forming nearly 50 million years ago, and they continue to develop today.

The concept of **geologic time**, the span of time since the formation of Earth, is new to many nonscientists. People are accustomed to dealing with increments of time that are measured in hours, days, weeks, and years. Our history books often examine events over spans of centuries, but even a century is difficult to appreciate fully. For most of us, someone or something that is 90 years old is *very old*, and a 1000-year-old artifact is *ancient*.

Those who study Earth science must routinely deal with vast time periods—millions or billions (thousands of millions) of years. When viewed in the context of Earth's 4.6-billion-year history, an event that occurred 100 million years ago may be characterized as "recent" by a geologist, and a rock sample that has been dated at 10 million years may be called "young."

An appreciation for the magnitude of geologic time is important in the study of our planet because many processes are so gradual that vast spans of time are needed before significant changes occur. How long is 4.6 billion years? If you were to begin counting at the rate of one number per second and continued 24 hours a day, seven

SmartFigure 1.6
Magnitude of Geologic Time

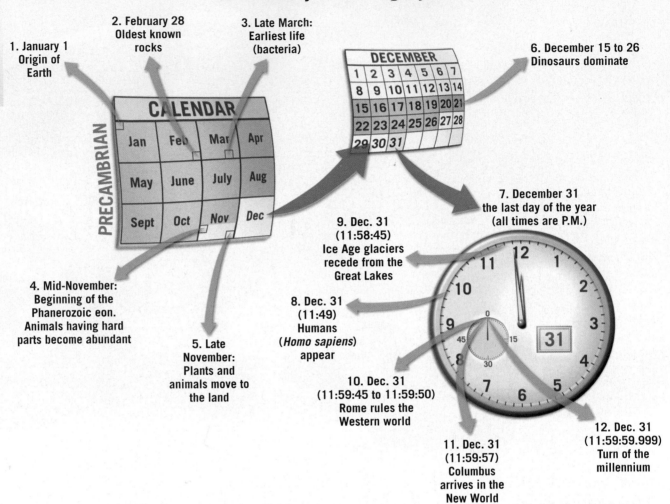

What if we compress the 4.6 billion years of Earth history into a single year?

1. January 1 Origin of Earth

2. February 28 Oldest known rocks

3. Late March: Earliest life (bacteria)

6. December 15 to 26 Dinosaurs dominate

PRECAMBRIAN

CALENDAR

Jan | Feb | Mar | Apr

May | June | July | Aug

Sept | Oct | Nov | Dec

DECEMBER
1 2 3 4 5 6 7
8 9 10 11 12 13 14
15 16 17 18 19 20 21
22 23 24 25 26 27 28
29 30 31

7. December 31 the last day of the year (all times are P.M.)

9. Dec. 31 (11:58:45) Ice Age glaciers recede from the Great Lakes

4. Mid-November: Beginning of the Phanerozoic eon. Animals having hard parts become abundant

8. Dec. 31 (11:49) Humans (*Homo sapiens*) appear

5. Late November: Plants and animals move to the land

10. Dec. 31 (11:59:45 to 11:59:50) Rome rules the Western world

11. Dec. 31 (11:59:57) Columbus arrives in the New World

12. Dec. 31 (11:59:59.999) Turn of the millennium

days a week and never stopped, it would take about two lifetimes (150 years) to reach 4.6 billion!

The preceding analogy is just one of many that have been conceived in an attempt to convey the magnitude of geologic time. Although helpful, all of them, no matter how clever, only begin to help us comprehend the vast expanse of Earth history. **FIGURE 1.6** provides another interesting way of viewing the age of Earth.

Over the past 200 years or so, Earth scientists have developed the *geologic time scale* of Earth history. It divides the 4.6-billion-year history of Earth into many different units and provides a meaningful time frame within which the events of the geologic past are arranged (see Figure 11.24, page 364). The principles used to develop the geologic time scale are examined in some detail in Chapter 11.

1.1 CONCEPT CHECKS

1   List and briefly describe the sciences that collectively make up Earth science.

2   Name the two broad subdivisions of geology and distinguish between them.

3   List at least four different natural hazards.

4   Aside from natural hazards, describe another important connection between people and Earth science.

5   List two examples of size/space scales in Earth science that are at opposite ends of the spectrum.

6   How old is Earth?

7   If you compress geologic time into a single year, how much time has elapsed since Columbus arrived in the New World?

1.2 | THE NATURE OF SCIENTIFIC INQUIRY Discuss the nature of scientific inquiry and distinguish between a hypothesis and a theory.

As members of a modern society, we are constantly reminded of the benefits derived from science. But what exactly is the nature of scientific inquiry? Developing an understanding of how science is done and how scientists work is another important theme that appears throughout this book. You will explore the difficulties in gathering data and some of the ingenious methods that have been developed to overcome these difficulties. You will also see many examples of how hypotheses are formulated and tested, as well as learn about the evolution and development of some major scientific theories.

All science is based on the assumption that the natural world behaves in a consistent and predictable manner that is comprehensible through careful, systematic study. The overall goal of science is to discover the underlying patterns in nature and then to use this knowledge to make predictions about what should or should not be expected, given certain facts or circumstances. For example, by understanding the processes that produce certain cloud types, meteorologists are often able to predict the approximate time and place of their formation.

The development of new scientific knowledge involves some basic logical processes that are universally accepted. To determine what is occurring in the natural world, scientists collect scientific facts through observation and measurement (**FIGURE 1.7**). The types of facts or data that are collected generally seek to answer a well-defined question about the natural world. Because some error is inevitable, the accuracy of a particular measurement or observation is always open to question. Nevertheless, these data are essential to science and serve as the springboard for the development of scientific hypotheses and theories.

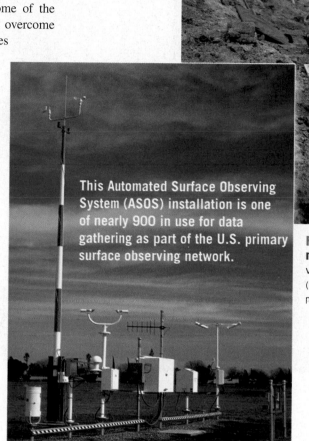

This paleontologist is collecting fossils in Antarctica. Later, a detailed analysis will occur in the lab.

This Automated Surface Observing System (ASOS) installation is one of nearly 900 in use for data gathering as part of the U.S. primary surface observing network.

FIGURE 1.7 Observation and Measurement Gathering data and making careful observations are basic parts of scientific inquiry. (Instrument photo by Bobbé Christopherson; paleontologist photo by British Antarctic Survey/Science Source)

Hypothesis

Once facts have been gathered and principles have been formulated to describe a natural phenomenon, investigators try to explain how or why things happened in the manner observed. They often do this by constructing a tentative (or untested) explanation, which is called a scientific **hypothesis**. It is best if an investigator can formulate more than one hypothesis to explain a given set of observations. If an individual scientist is unable to devise multiple hypotheses, others in the scientific community will almost always develop alternative explanations. A spirited debate frequently ensues. As a result, extensive research is conducted by proponents of opposing hypotheses, and the results are made available to the wider scientific community in scientific journals.

Before a hypothesis can become an accepted part of scientific knowledge, it must pass objective testing and analysis. If a hypothesis cannot be tested, it is not scientifically useful, no matter how interesting it might seem. The verification process requires that *predictions* be made based on the hypothesis being considered and that the predictions be tested by comparing them against objective observations of nature. Put another way, hypotheses must fit observations other than those used to formulate them in the first place. Hypotheses that fail rigorous testing are ultimately discarded. The history of science is littered with discarded hypotheses. One of the best known is the Earth-centered model of the universe—a proposal that was supported by the apparent daily motion of the Sun, Moon, and stars around Earth. As the mathematician Jacob Bronowski so ably stated, "Science is a great many things, but in the end they all return to this: Science is the acceptance of what works and the rejection of what does not."

World Population Passes 7 BILLION

Complicating all environmental issues is rapid world population growth and everyone's aspiration to a better standard of living. There is a ballooning demand for resources and a growing pressure for people to live in environments having significant geologic hazards.

NEW YORK, USA
19,430,000

MEXICO CITY, MEXICO
19,460,000

This composite satellite image of Earth's city lights helps us appreciate the intensity of human occupation in many parts of the world. In the year 1800, only about 3 percent of the world's people were urban. Today about 51 percent are classified as urban.

WORLD'S 10 LARGEST METRO AREAS IN 2010
MILLIONS OF CITIZENS

SAO PAULO, BRAZIL
20,260,000

Theory

When a hypothesis has survived extensive scrutiny and when competing hypotheses have been eliminated, it may be elevated to the status of a scientific **theory**. In everyday language, we might say, "That's only a theory." But a scientific theory is a well-tested and widely accepted view that the scientific community agrees best explains certain observable facts.

Some theories that are extensively documented and extremely well supported are comprehensive in scope. For example, the theory of plate tectonics provides the framework for understanding the origin of mountains, earthquakes, and volcanic activity. In addition, plate tectonics explains the evolution of the continents and the ocean basins through time—ideas that are explored in some detail in Chapters 7 through 10.

Scientific Methods

The process just described, in which researchers gather facts through observations and formulate scientific hypotheses and theories, is called the *scientific method*. Contrary to popular belief, the scientific method is not a standard recipe that scientists apply in a routine manner to unravel the secrets of our natural world. Rather, it is an endeavor that involves creativity and insight. Rutherford and Ahlgren put it this way: "Inventing hypotheses or theories to imagine how the world works and then figuring out how they can be put to the test of reality is as creative as writing poetry, composing music, or designing skyscrapers."[2]

There is not a fixed path that scientists can always follow unerringly to scientific knowledge. However, many scientific investigations involve the following processes:

- A question is raised about the natural world.
- Scientific data that relate to the question are collected.
- Questions that relate to the data are posed, and one or more working hypotheses are developed that may answer these questions.
- Observations and experiments are developed to test the hypotheses.

---

[2]F. James Rutherford and Andrew Ahlgren, *Science for All Americans* (New York: Oxford University Press, 1990), p. 7.

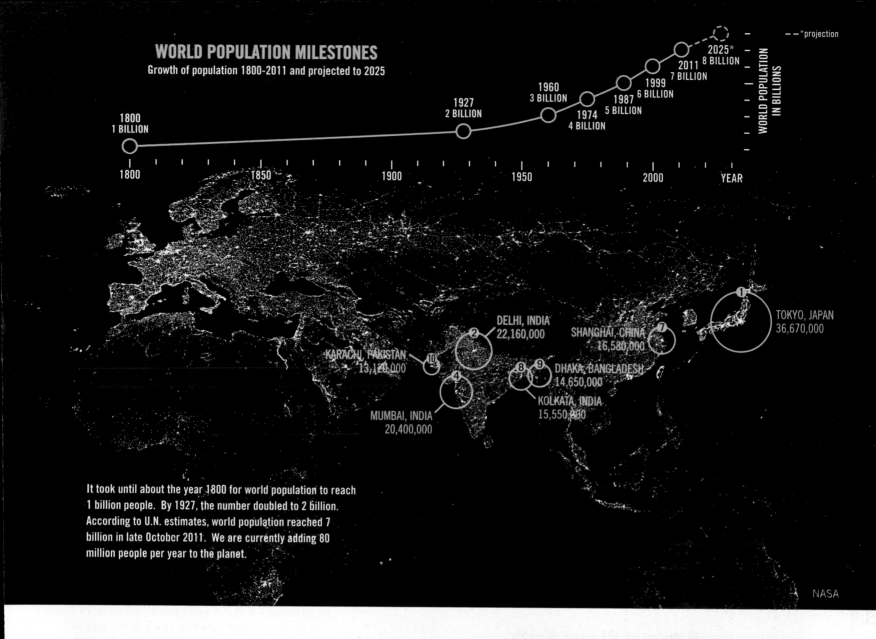

WORLD POPULATION MILESTONES
Growth of population 1800-2011 and projected to 2025

- - *projection

2025*
8 BILLION

2011
7 BILLION

1999
6 BILLION

1987
5 BILLION

1960
3 BILLION

1974
4 BILLION

1927
2 BILLION

1800
1 BILLION

WORLD POPULATION IN BILLIONS

1800   1850   1900   1950   2000   YEAR

DELHI, INDIA
22,160,000

KARACHI, PAKISTAN
13,120,000

SHANGHAI, CHINA
16,580,000

TOKYO, JAPAN
36,670,000

DHAKA, BANGLADESH
14,650,000

KOLKATA, INDIA
15,550,000

MUMBAI, INDIA
20,400,000

It took until about the year 1800 for world population to reach
1 billion people. By 1927, the number doubled to 2 billion.
According to U.N. estimates, world population reached 7
billion in late October 2011. We are currently adding 80
million people per year to the planet.

NASA

EYE ON
EARTH

This image shows rainfall data for December 7–13, 2004, in Malaysia. More than 800 millimeters (32 inches) of rain fell along the east coast of the peninsula (darkest red area). The extraordinary rains caused extensive flooding. The data for this image are from NASA's *Tropical Rainfall Measuring Mission* (*TRMM*). This is just one of hundreds of satellites that provide scientists with all kinds of data about our planet.

QUESTION 1 *Gathering data is a basic part of scientific inquiry. Suggest some advantages that satellites provide scientists as a way of gaining information about Earth.*

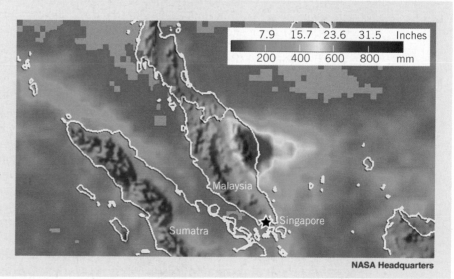

7.9	15.7	23.6	31.5	Inches
200	400	600	800	mm

Malaysia

Singapore

Sumatra

NASA Headquarters

11

- The hypotheses are accepted, modified, or rejected, based on extensive testing.
- Data and results are shared with the scientific community for critical examination and further testing.

Some scientific discoveries may result from purely theoretical ideas, which stand up to extensive examination. Some researchers use high-speed computers to simulate what is happening in the "real" world. These models are useful when dealing with natural processes that occur on very long time scales or that take place in extreme or inaccessible locations. Still other scientific advancements are made when a totally unexpected happening occurs during an experiment. These serendipitous discoveries are more than pure luck, for as Louis Pasteur said, "In the field of observation, chance favors only the prepared mind."

Scientific knowledge is acquired through several avenues, so it might be best to describe the nature of scientific inquiry as the *methods of science* rather than as the *scientific method*. In addition, it should always be remembered that even the most compelling scientific theories are still simplified explanations of the natural world.

In this book, you will discover the results of centuries of scientific work. You will see the end product of millions of observations, thousands of hypotheses, and hundreds of theories. We have distilled all of this to give you a "briefing" on Earth science.

But realize that our knowledge of Earth is changing daily, as thousands of scientists worldwide make satellite observations, analyze drill cores from the seafloor, measure earthquakes, develop computer models to predict climate, examine the genetic codes of organisms, and discover new facts about our planet's long history. This new knowledge often updates hypotheses and theories. Expect to see many new discoveries and changes in scientific thinking in your lifetime.

1.2 CONCEPT CHECKS

1 How is a scientific hypothesis different from a scientific theory?

2 Summarize the basic steps followed in many scientific investigations.

1.3 | EARLY EVOLUTION OF EARTH

Outline the stages in the formation of our solar system.

This section describes the most widely accepted views on the origin of our solar system. The theory summarized here represents the most consistent set of ideas available to explain what we know about our solar system today.

Origin of Planet Earth

Our story begins about 13.7 billion years ago, with the *Big Bang*, an incomprehensibly large explosion that sent all matter of the universe flying outward at incredible speeds. In time, the debris from this explosion, which was almost entirely hydrogen and helium, began to cool and condense into the first stars and galaxies. It was in one of these galaxies, the Milky Way, that our solar system and planet Earth took form.

Earth is one of eight planets that, along with more than 160 moons and numerous smaller bodies, revolve around the Sun. The orderly nature of our solar system leads most researchers to conclude that Earth and the other planets formed at essentially the same time and from the same primordial material as the Sun. The **nebular theory** proposes that the bodies of our solar system evolved from an enormous rotating cloud called the *solar nebula* (**FIGURE 1.8**). Besides the hydrogen and helium atoms generated during the Big Bang, the solar nebula consisted of microscopic dust grains and the ejected matter of long-dead stars. (Nuclear fusion in stars converts

hydrogen and helium into the other elements found in the universe.)

Nearly 5 billion years ago, this huge cloud of gases and minute grains of heavier elements began to slowly contract due to the gravitational interactions among its particles (see Figure 1.8). Some external influence, such as a shock wave traveling from a catastrophic explosion (*supernova*), may have triggered the collapse. As this slowly spiraling nebula contracted, it rotated faster and faster, much as spinning ice skaters do when they draw their arms toward their bodies. Eventually the inward pull of gravity came into balance with the outward force caused by the rotational motion of the nebula (see Figure 1.8). By this time, the once-vast cloud had assumed a flat disk shape with a large concentration of material at its center called the *protosun* (pre-Sun). Astronomers are fairly confident that the nebular cloud formed a disk because similar structures have been detected around other stars.

During the collapse, gravitational energy was converted to thermal energy (heat), causing the temperature of the inner portion of the nebula to dramatically rise. At these high temperatures, the dust grains broke up into molecules and extremely energetic atomic particles. However, at distances beyond the orbit of Mars, the temperatures probably remained quite low. At $-200°C$ ($-328°F$), the tiny particles in the outer portion of the nebula were likely covered with a thick layer of ices made of frozen water, carbon dioxide, ammonia, and methane. (Some of this material still resides

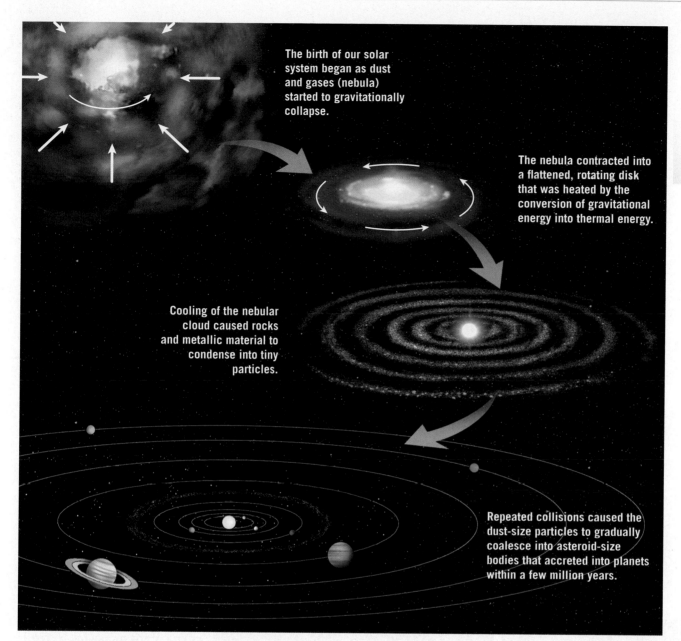

The birth of our solar system began as dust and gases (nebula) started to gravitationally collapse.

The nebula contracted into a flattened, rotating disk that was heated by the conversion of gravitational energy into thermal energy.

Cooling of the nebular cloud caused rocks and metallic material to condense into tiny particles.

Repeated collisions caused the dust-size particles to gradually coalesce into asteroid-size bodies that accreted into planets within a few million years.

SmartFigure 1.8 Nebular Theory Formation of the solar system according to the nebular theory.

in the outermost reaches of the solar system, in a region called the *Oort cloud*.) The disk-shaped cloud also contained appreciable amounts of the lighter gases hydrogen and helium.

The Inner Planets Form

The formation of the Sun marked the end of the period of contraction and thus the end of gravitational heating. Temperatures in the region where the inner planets now reside began to decline. The decrease in temperature caused substances with high melting points to condense into tiny particles that began to coalesce (that is, join together). Materials such as iron and nickel and the elements of which the rock-forming minerals are composed—silicon, calcium, sodium, and so forth—formed metallic and rocky clumps that orbited

the Sun (see Figure 1.8). Repeated collisions caused these masses to coalesce into larger asteroid-size bodies, called *planetesimals*, which in a few tens of millions of years accreted into the four inner planets we call Mercury, Venus, Earth, and Mars (**FIGURE 1.9**). Not all of these clumps of matter were incorporated into the planetesimals. Those rocky and metallic pieces that remained in orbit are called asteroids and become *meteorites* if they impact Earth's surface.

As more and more material was swept up by these growing planetary bodies, the high-velocity impact of nebular debris caused their temperatures to rise. Because of their relatively high temperatures and weak gravitational fields, the inner planets were unable to accumulate much of the lighter components of the nebular cloud. The lightest of these, hydrogen and helium, were eventually whisked from the inner solar system by the solar wind.

FIGURE 1.9 A Remnant Planetesimal This image of Asteroid 21 Lutetia was obtained by special cameras aboard the *Rosetta* spacecraft on July 10, 2010. Spacecraft instruments showed that Lutetia is a primitive body (planetesimal) left over from when the solar system formed. (Image courtesy of ESA)

The Outer Planets Develop

At the same time that the inner planets were forming, the larger, outer planets (Jupiter, Saturn, Uranus, and Neptune), along with their extensive satellite systems, were also developing. Because of low temperatures far from the Sun, the material from which these planets formed contained a high percentage of ices—water, carbon dioxide, ammonia, and methane—as well as rocky and metallic debris. The accumulation of ices accounts in part for the large size and low density of the outer planets. The two most massive planets, Jupiter and Saturn, had a surface gravity sufficient to attract and hold large quantities of even the lightest elements—hydrogen and helium.

1.3 CONCEPT CHECKS

1 Name and briefly outline the theory that describes the formation of our solar system.

2 List the inner planets and the outer planets. Describe basic differences in size and composition.

1.4 | EARTH'S SPHERES

List and describe Earth's four major spheres.

The images in **FIGURE 1.10** are considered to be classics because they let humanity see Earth differently than ever before. These early views profoundly altered our conceptualizations of Earth and remain powerful images decades after they were first viewed. Seen from space, Earth is breathtaking in its beauty and startling in its solitude. The photos remind us that our home is, after all, a planet—small, self-contained, and in some ways even fragile.

View called "Earthrise" that greeted *Apollo 8* astronauts as their spacecraft emerged from behind the Moon in December 1968. This classic image let people see Earth differently than ever before.

This image taken from *Apollo 17* in December 1972 is perhaps the first to be called "The Blue Marble." The dark blue ocean and swirling cloud patterns remind us of the importance of the oceans and atmosphere.

FIGURE 1.10 Two Classic Views of Earth from Space. (Photo on left courtesy of NASA Headquarters; photo on right courtesy of NASA/Johnson Space Center)

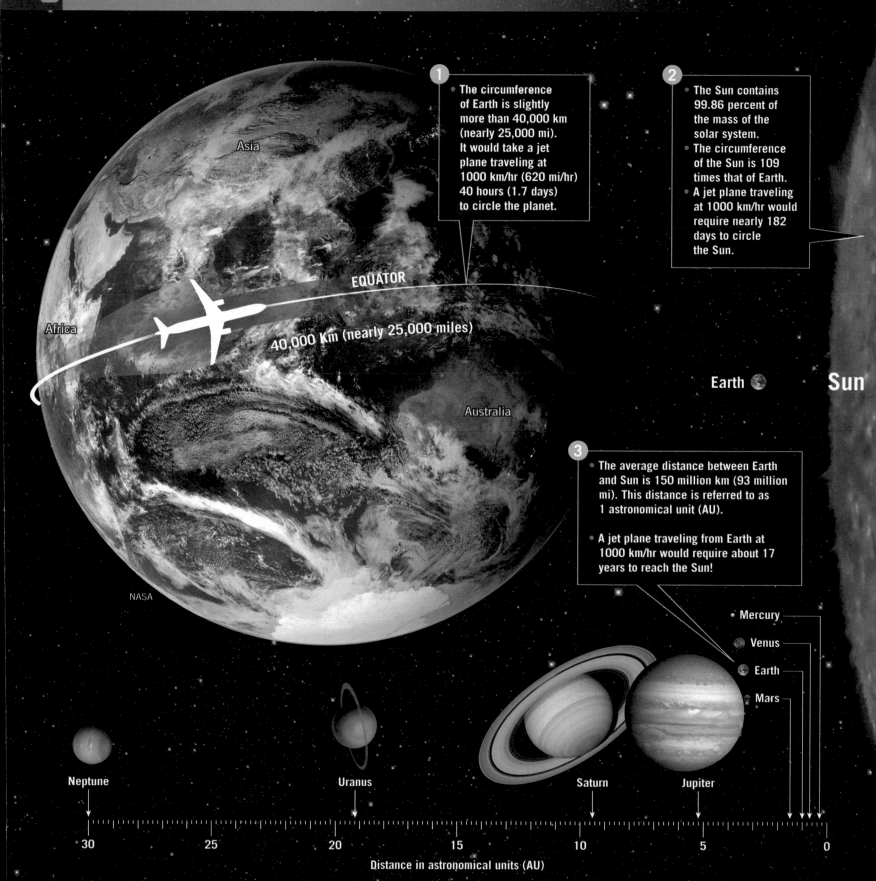

GEOGRAPHICS

Solar System: Size and Scale

The Sun is the center of a revolving system trillions of miles across, consisting of 8 planets, their satellites, and numerous dwarf planets, asteroids, comets, and meteoroids.

1
- The circumference of Earth is slightly more than 40,000 km (nearly 25,000 mi). It would take a jet plane traveling at 1000 km/hr (620 mi/hr) 40 hours (1.7 days) to circle the planet.

2
- The Sun contains 99.86 percent of the mass of the solar system.
- The circumference of the Sun is 109 times that of Earth.
- A jet plane traveling at 1000 km/hr would require nearly 182 days to circle the Sun.

Asia

Africa

EQUATOR

40,000 Km (nearly 25,000 miles)

Australia

NASA

Earth ◉

Sun

3
- The average distance between Earth and Sun is 150 million km (93 million mi). This distance is referred to as 1 astronomical unit (AU).

- A jet plane traveling from Earth at 1000 km/hr would require about 17 years to reach the Sun!

Mercury

Venus

Earth

Mars

Neptune

Uranus

Saturn

Jupiter

30   25   20   15   10   5   0

Distance in astronomical units (AU)

FIGURE 1.11 Interactions among Earth's Spheres The shoreline is considered an interface—a common boundary where different parts of a system interact. In this scene, ocean waves (hydrosphere) that were created by the force of moving air (atmosphere) break against a rocky shore (geosphere). (Photo by Michael Collier)

Bill Anders, the *Apollo 8* astronaut who took the "Earthrise" photo, expressed it this way: "We came all this way to explore the Moon, and the most important thing is that we discovered the Earth."

As we look closely at our planet from space, it becomes apparent that Earth is much more than rock and soil. In fact, the most conspicuous features in Figure 1.10 are not continents but swirling clouds suspended above the surface of the vast global ocean. These features emphasize the importance of water on our planet.

The closer view of Earth from space shown in Figure 1.10 helps us appreciate why the physical environment is traditionally divided into three major spheres: the water portion of our planet, the hydrosphere; Earth's gaseous envelope, the atmosphere; and, of course, the solid Earth, or geosphere.

It should be emphasized that our environment is highly integrated and is not dominated by rock, water, or air alone. It is instead characterized by continuous interactions as air comes in contact with rock, rock with water, and water with air. Moreover, the biosphere, the totality of life-forms on our planet, extends into each of the three physical realms and is an equally integral part of the planet. Thus, Earth can be thought of as consisting of four major spheres: the hydrosphere, atmosphere, geosphere, and biosphere.

The interactions among the spheres of Earth's environment are incalculable. **FIGURE 1.11** provides an easy-to-visualize example. The shoreline is an obvious meeting place for rock, water, and air. In this scene, ocean waves that were created by the drag of air moving across the water are breaking against the rocky shore. The force of the water can be powerful, and the erosional work that is accomplished can be great.

Hydrosphere

Earth is sometimes called the *blue planet* or, as we saw in Figure 1.10, "The Blue Marble." Water more than anything else makes Earth unique. The **hydrosphere** is a dynamic mass of water that is continually on the move, evaporating from the oceans to the atmosphere, precipitating to the land, and running back to the ocean again. The global ocean is certainly the most prominent feature of the hydrosphere, blanketing nearly 71 percent of Earth's surface to an average depth of about 3800 meters (12,500 feet). It accounts for about 97 percent of Earth's water (**FIGURE 1.12**). However, the hydrosphere also includes the freshwater found underground and in streams, lakes, and glaciers. Moreover, water is an important component of all living things.

Although these latter sources constitute just a tiny fraction of the total, they are much more important than their meager percentages indicate. In addition to providing the freshwater that is so vital to life on land, streams, glaciers, and groundwater are responsible for sculpturing and creating many of our planet's varied landforms.

Atmosphere

Earth is surrounded by a life-giving gaseous envelope called the **atmosphere** (**FIGURE 1.13**). When we watch a high-flying jet plane cross the sky, it seems that the atmosphere

FIGURE 1.12 The Water Planet Distribution of water in the hydrosphere.

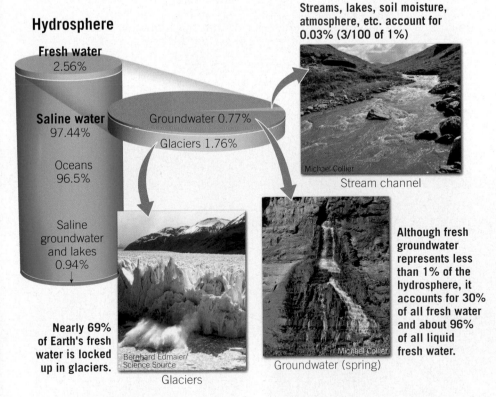

Hydrosphere

Fresh water 2.56%

Saline water 97.44%

Oceans 96.5%

Saline groundwater and lakes 0.94%

Groundwater 0.77%

Glaciers 1.76%

Streams, lakes, soil moisture, atmosphere, etc. account for 0.03% (3/100 of 1%)

Michael Collier

Stream channel

Although fresh groundwater represents less than 1% of the hydrosphere, it accounts for 30% of all fresh water and about 96% of all liquid fresh water.

Nearly 69% of Earth's fresh water is locked up in glaciers.

Bernhard Edmaier/ Science Source

Glaciers

Michael Collier

Groundwater (spring)

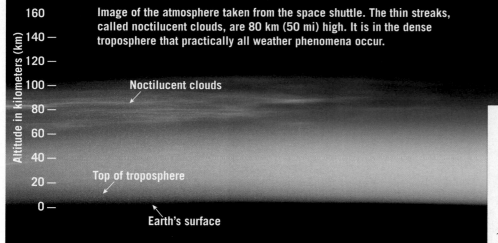

Image of the atmosphere taken from the space shuttle. The thin streaks, called noctilucent clouds, are 80 km (50 mi) high. It is in the dense troposphere that practically all weather phenomena occur.

Altitude in kilometers (km)

160
140
120
100
80
60
40
20
0

Noctilucent clouds

Top of troposphere

Earth's surface

NASA

FIGURE 1.13 A Shallow Layer The atmosphere is an integral part of the planet. (NASA)

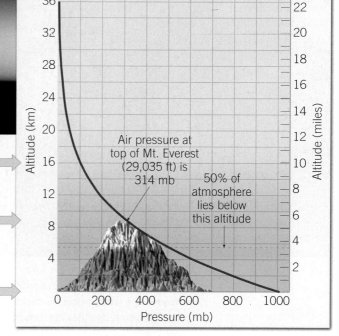

90% of the atmosphere is below 16 km (10 mi)

The air pressure atop Mt. Everest is about one-third that at sea level.

Average sea-level pressure is slightly more than 1000 millibars (about 14.7 lbs/sq. in)

Air pressure at top of Mt. Everest (29,035 ft) is 314 mb

50% of atmosphere lies below this altitude

Altitude (km)
Altitude (miles)
Pressure (mb)

extends upward for a great distance. However, when compared to the thickness (radius) of the solid Earth (about 6400 kilometers [4000 miles]), the atmosphere is a very shallow layer. Despite its modest dimensions, this thin blanket of air is nevertheless an integral part of the planet. It not only provides the air that we breathe but also protects us from the Sun's dangerous ultraviolet radiation. The energy exchanges that continually occur between the atmosphere and Earth's surface and between the atmosphere and space produce the effects we call *weather* and *climate*. Climate has a strong influence on the nature and intensity of Earth's surface processes. When climate changes, these processes respond.

If, like the Moon, Earth had no atmosphere, our planet would be lifeless because many of the processes and interactions that make the surface such a dynamic place could not operate. Without weathering and erosion, the face of our planet might more closely resemble the lunar surface, which has not changed appreciably in nearly 3 billion years.

Biosphere

The **biosphere** includes all life on Earth (**FIGURE 1.14**). Ocean life is concentrated in the sunlit surface waters of the sea. Most life on land is also concentrated near the surface, with tree roots and burrowing animals reaching a few meters underground and flying insects and birds reaching a kilometer or so above the surface. A surprising variety of life-forms are also adapted to extreme environments. For example, on the ocean floor, where pressures are extreme and no light penetrates, there are places where vents spew hot,

EYE ON EARTH

This jet is cruising at an altitude of 10 kilometers (6.2 miles).

QUESTION 1 *Refer to the graph in Figure 1.13. What is the approximate air pressure at the altitude where the jet is flying?*

QUESTION 2 *About what percentage of the atmosphere is below the jet (assuming that the pressure at the surface is 1000 millibars)?*

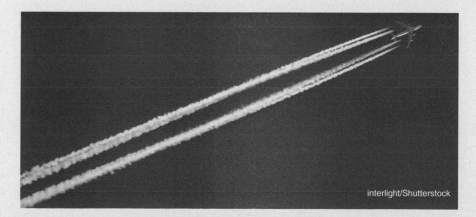

interlight/Shutterstock

FIGURE 1.14
The Biosphere The bio-
sphere, one of Earth's four
spheres, includes all life.
(Coral reef photo by Darryl Leniuk/
AGE Fotostock; rain forest photo by
AGE Fotostock/SuperStock)

The ocean contains a significant portion of Earth's biosphere. Modern coral reefs are unique and complex examples and are home to about 25% of all marine species. Because of this diversity, they are sometimes referred to as the ocean equivalent of a rain forest.

Tropical rain forests are characterized by hundreds of different species per square kilometer.

Geosphere

Lying beneath the atmosphere and the ocean is the solid Earth, or **geosphere**. The geosphere extends from the surface to the center of the planet, a depth of 6400 kilometers [4000 miles], making it by far the largest of Earth's four spheres. Much of our study of the solid Earth focuses on the more accessible surface features. Fortunately, many of these features represent the outward expressions of the dynamic behavior of Earth's interior. By examining the most prominent surface features and their global extent, we can obtain clues to the dynamic processes that have shaped our planet. The next section of this chapter takes a first look at the structure of Earth's interior and at the major surface features of the geosphere.

Soil, the thin veneer of material at Earth's surface that supports the growth of plants, may be thought of as part of all four spheres. The solid portion is a mixture of weathered rock debris (geosphere) and organic matter from decayed plant and animal life (biosphere). The decomposed and disintegrated rock debris is the product of weathering processes that require air (atmosphere) and water (hydrosphere). Air and water also occupy the open spaces between the solid particles.

mineral-rich fluids that support communities of exotic life-forms. On land, some bacteria thrive in rocks as deep as 4 kilometers (2.5 miles) and in boiling hot springs. Moreover, air currents can carry microorganisms many kilometers into the atmosphere. But even when we consider these extremes, life still must be thought of as being confined to a narrow band very near Earth's surface.

Plants and animals depend on the physical environment for the basics of life. However, organisms do more than just respond to their physical environment. Through countless interactions, life-forms help maintain and alter their physical environment. Without life, the makeup and nature of the geosphere, hydrosphere, and atmosphere would be very different than they are.

1.4 CONCEPT CHECKS

1 List Earth's four spheres.
2 Compare the height of the atmosphere to the thickness of the geosphere.
3 How much of Earth's surface do oceans cover? How much of the planet's total water supply do oceans represent?
4 To which sphere does soil belong?

1.5 | A CLOSER LOOK AT THE GEOSPHERE Label a diagram that shows Earth's internal structure. Briefly explain why the geosphere can be described as being mobile.

In this section and the next, we make a preliminary examination of the solid Earth. You will become more familiar with the internal and external "anatomy" of our planet and begin to understand that the geosphere is truly dynamic. The diagrams should help a great deal as you begin to develop a mental image of the geosphere's internal structure and major surface features, so study the figures carefully. We begin with a look at Earth's interior—its structure and mobility. Then we conduct a brief survey of the surface of the solid Earth. Although portions of the surface, such as mountains and river valleys, are familiar to most of us,

areas that are out of sight on the floor of the ocean are not so familiar.

Earth's Internal Structure

Early in Earth's history, when the planet was very hot, the sorting of material by compositional (density) differences resulted in the formation of three layers—the crust, mantle, and core. In addition to these compositionally distinct layers, Earth is also divided into layers based on *physical properties*. The physical properties that define these zones include whether the layer is solid or liquid and how weak or strong it is. Knowledge of both types of layers is essential to an understanding of our planet. **FIGURE 1.15** summarizes the two types of layers that characterize Earth's interior.

Earth's Crust   The **crust**, Earth's relatively thin, rocky outer skin, is of two different types—*continental crust* and *oceanic crust*. Both share the word *crust*, but the similarity ends there. The oceanic crust is roughly 7 kilometers (5 miles) thick and composed of the dark igneous rock *basalt*. By contrast, the continental crust averages about 35 kilometers (22 miles) thick but may exceed 70 kilometers (40 miles) in some mountainous regions, such as the Rockies and Himalayas. Unlike the oceanic crust, which has a relatively homogeneous chemical composition, the continental crust consists of many rock types. Although the upper crust has an

average composition of a *granitic rock* called *granodiorite*, it varies considerably from place to place.

Continental rocks have an average density of about 2.7 g/cm^3, and some have been discovered that are more than 4 billion years old. The rocks of the oceanic crust are younger (180 million years or less) and denser (about 3.0 g/cm^3) than continental rocks.[3]

Earth's Mantle   More than 82 percent of Earth's volume is contained in the **mantle**, a solid, rocky shell that extends to a depth of nearly 2900 kilometers (1800 miles). The boundary between the crust and mantle is the site of a marked change in chemical composition. The dominant rock

---

[3] Liquid water has a density of 1 g/cm^3; therefore, the density of basalt is three times that of water.

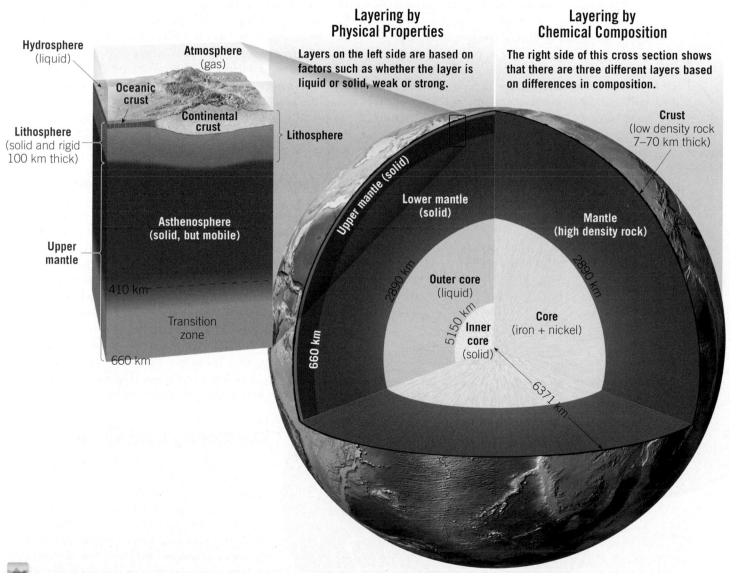

Layering by Physical Properties

Layers on the left side are based on factors such as whether the layer is liquid or solid, weak or strong.

Layering by Chemical Composition

The right side of this cross section shows that there are three different layers based on differences in composition.

Hydrosphere (liquid)

Atmosphere (gas)

Oceanic crust

Continental crust

Lithosphere (solid and rigid 100 km thick)

Lithosphere

Upper mantle

Asthenosphere (solid, but mobile)

410 km

Transition zone

660 km

Upper mantle (solid)

Lower mantle (solid)

660 km

2890 km

Outer core (liquid)

5150 km

Inner core (solid)

Core (iron + nickel)

6371 km

Crust (low density rock 7–70 km thick)

Mantle (high density rock)

2890 km

SmartFigure 1.15 Earth's Layers Structure of Earth's interior.

FIGURE 1.16 The Supercontinent of Pangaea Earth as it looked about 200 million years ago. At this time, the modern continents that we are familiar with were joined to form a supercontinent that we call Pangaea ("all land").

type in the uppermost mantle is *peridotite*, which is richer in the metals magnesium and iron than the minerals found in either the continental or oceanic crust.

The upper mantle extends from the crust–mantle boundary to a depth of about 660 kilometers (410 miles). The upper mantle can be divided into two different parts. The top portion of the upper mantle is part of the stiff *lithosphere*, and beneath that is the weaker *asthenosphere*.

The **lithosphere** ("sphere of rock") consists of the entire crust and uppermost mantle and forms Earth's relatively cool, rigid outer shell. Averaging about 100 kilometers (60 miles) in thickness, the lithosphere is more than 250 kilometers (150 miles) thick below the oldest portions of the continents (see Figure 1.15). Beneath this stiff layer to a depth of about 350 kilometers (220 miles) lies a soft, comparatively weak layer known as the **asthenosphere** ("weak sphere"). The top portion of the asthenosphere has a temperature/pressure regime

FIGURE 1.17 Some of Earth's Major Lithospheric Plates

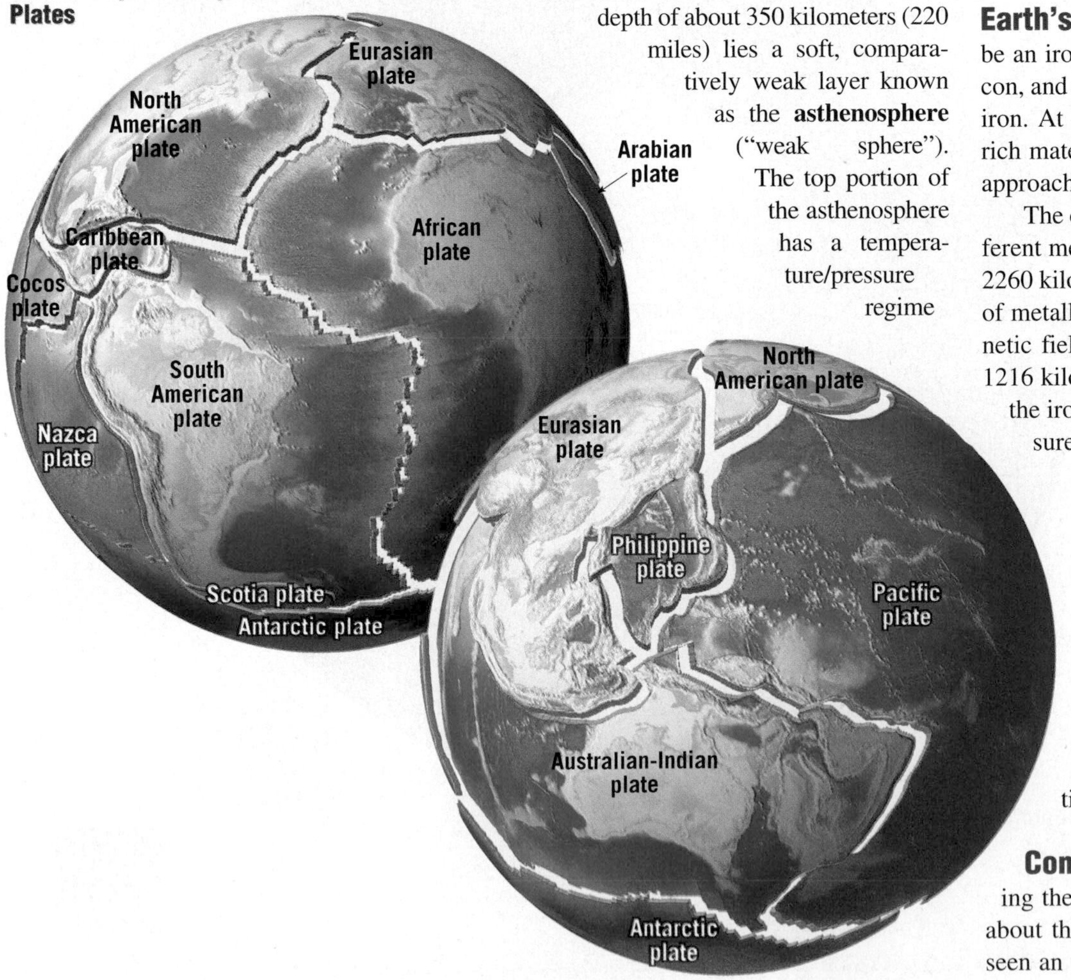

that results in a small amount of melting. Within this very weak zone, the lithosphere is mechanically detached from the layer below. The result is that the lithosphere is able to move independently of the asthenosphere, a fact we consider in more detail in Chapter 7.

It is important to emphasize that the strength of various Earth materials is a function of both their composition and the temperature and pressure of their environment. You should not get the idea that the entire lithosphere behaves like a brittle solid similar to rocks found on the surface. Rather, the rocks of the lithosphere get progressively hotter and weaker (more easily deformed) with increasing depth. At the depth of the uppermost asthenosphere, the rocks are close enough to their melting temperature (some melting may actually occur) that they are very easily deformed. Thus, the uppermost asthenosphere is weak because it is near its melting point, just as hot wax is weaker than cold wax.

From a depth of 660 kilometers (410 miles) to the top of the core, at a depth of 2900 kilometers (1800 miles), is the **lower mantle**. Because of an increase in pressure (caused by the weight of the rock above), the mantle gradually strengthens with depth. Despite their strength, however, the rocks within the lower mantle are very hot and capable of very gradual flow.

Earth's Core   The composition of the **core** is thought to be an iron–nickel alloy with minor amounts of oxygen, silicon, and sulfur—elements that readily form compounds with iron. At the extreme pressure found in the core, this iron-rich material has an average density of nearly 11 g/cm^3 and approaches 14 times the density of water at Earth's center.

The core is divided into two regions that exhibit very different mechanical strengths. The **outer core** is a *liquid layer* 2260 kilometers (about 1400 miles) thick. It is the movement of metallic iron within this zone that generates Earth's magnetic field. The **inner core** is a sphere that has a radius of 1216 kilometers (754 miles). Despite its higher temperature, the iron in the inner core is *solid* due to the immense pressures that exist in the center of the planet.

The Mobile Geosphere

Earth is a dynamic planet! If we could go back in time a few hundred million years, we would find the face of our planet dramatically different from what we see today. There would be no Mount St. Helens, Rocky Mountains, or Gulf of Mexico. Moreover, we would find continents having different sizes and shapes and located in different positions than today's landmasses (**FIGURE 1.16**).

Continental Drift and Plate Tectonics   During the past several decades, a great deal has been learned about the workings of our dynamic planet. This period has seen an unequaled revolution in our understanding of Earth.

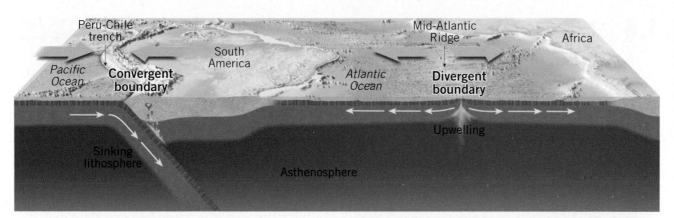

FIGURE 1.18 **Convergent and Divergent Bounda-ries** Convergent boundaries occur where two lithospheric plates move together, as along the western margin of South America in this diagram. Divergent boundaries occur where adjacent plates move away from one another. The Mid-Atlantic Ridge is such a boundary.

The revolution began in the early part of the twentieth century with the radical proposal of *continental drift*—the idea that the continents move about the face of the planet. This proposal contradicted the established view that the continents and ocean basins are permanent and stationary features on the face of Earth. For that reason, the notion of drifting continents was received with great skepticism and even ridicule. More than 50 years passed before enough data were gathered to transform this controversial hypothesis into a sound theory that wove together the basic processes known to operate on Earth. The theory that finally emerged, called **plate tectonics**, provided geologists with the first comprehensive model of Earth's internal workings.

According to the theory of plate tectonics, Earth's rigid outer shell (the *lithosphere*) is broken into numerous slabs called **lithospheric plates**, which are in continual motion. More than a dozen plates exist (**FIGURE 1.17**). The largest is the Pacific plate, covering much of the Pacific Ocean basin. Notice that several of the large lithospheric plates include an entire continent plus a large area of the seafloor. Note also that none of the plates are defined entirely by the margins of a continent.

Plate Motion Driven by the unequal distribution of heat within our planet, lithospheric plates move relative to each other at a very slow but continuous rate that averages about 5 centimeters (2 inches) per year—about as fast as your fingernails grow. Because plates move as coherent units relative to all other plates, they interact along their margins. Where two plates move together, called a *convergent boundary*, one of the plates plunges beneath the other and descends into the mantle (**FIGURE 1.18**). The lithospheric plates that sink into the mantle are those that are capped with relatively dense oceanic crust.

Any portion of a plate that is capped by continental crust is too buoyant to be carried into the mantle. As a result, when two plates carrying continental crust converge, a collision of the two continental margins occurs. The result is the formation of a major mountain belt, as exemplified by the Himalayas.

Divergent boundaries are located where plates pull apart (see Figure 1.18). Here the fractures created as the plates separate are filled with molten rock that wells up from the mantle. This hot material slowly cools to form solid rock, producing new slivers of seafloor. This process

FIGURE 1.19 **Transform Fault Boundary** California's San Andreas Fault is an example of a transform fault boundary where lithospheric plates slide past one another. (Photo by Michael Collier)

occurs along oceanic ridges where, over spans of millions of years, hundreds of thousands of square kilometers of new seafloor have been generated (see Figure 1.18). Thus, while new seafloor is constantly being added at the oceanic ridges, equal amounts are returned to the mantle along boundaries where two plates converge.

At other sites, plates do not push together or pull apart. Instead, they slide past one another, so that seafloor is neither created nor destroyed. These zones are called *transform fault boundaries*. California's San Andreas Fault is a well-known example (**FIGURE 1.19**).

1.5 CONCEPT CHECKS

1 List and briefly describe Earth's compositional layers.

2 Contrast the lithosphere and the asthenosphere.

3 What are lithospheric plates? List the three types of boundaries that separate plates.

1.6 | THE FACE OF EARTH

List and describe the major features of the continents and ocean basins.

The two principal divisions of Earth's surface are the continents and the ocean basins (**FIGURE 1.20**). A significant difference between these two areas is their relative levels. The elevation difference between the continents and ocean basins is primarily a result of differences in their respective densities and thicknesses:

• The **continents** are remarkably flat features that have the appearance of plateaus protruding above sea level. With an average elevation of about 0.8 kilometer (0.5 mile), continents lie relatively close to sea level, except for limited areas of mountainous terrain. Recall that the continents average about 35 kilometers (22 miles) in

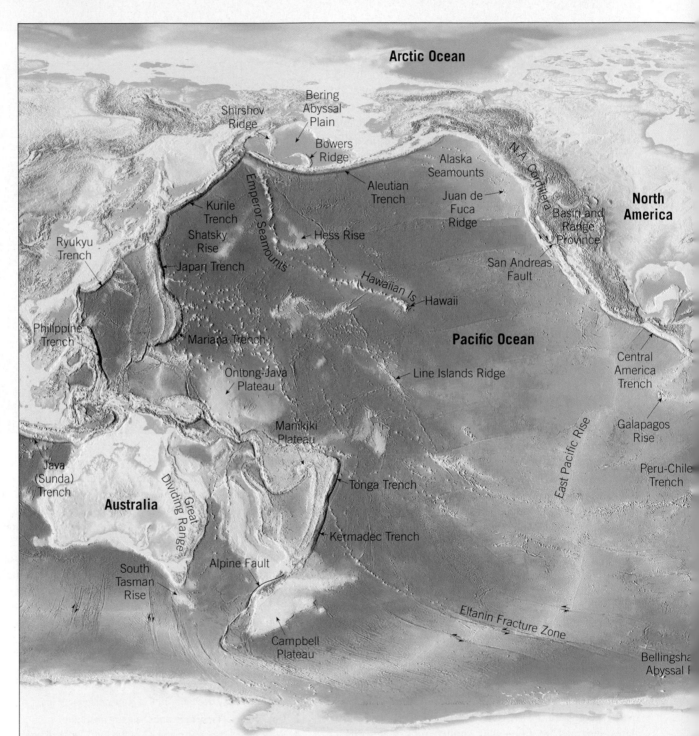

FIGURE 1.20 The Face of Earth Major surface features of the geosphere.

thickness and are composed of granitic rocks that have a density of about 2.7 g/cm³.

- The average depth of the **ocean basin** is about 3.8 kilometers (2.4 miles) below sea level, or about 4.5 kilometers (2.8 miles) lower than the average elevation of the continents. The basaltic rocks that comprise the oceanic crust average only 7 kilometers (5 miles) thick and have an average density of about 3.0 g/cm³.

Thus, the thicker and less dense continental crust is more buoyant than the oceanic crust. As a result, continental crust floats on top of the deformable rocks of the mantle at a higher level than oceanic crust for the same reason that a large, empty (less dense) cargo ship rides higher than a small, loaded (more dense) one.

Major Features of the Continents

The largest features of the continents can be grouped into two distinct categories: extensive, flat, stable areas that have been eroded nearly to sea level, and uplifted regions of deformed rocks that make up present-day mountain belts. Notice in

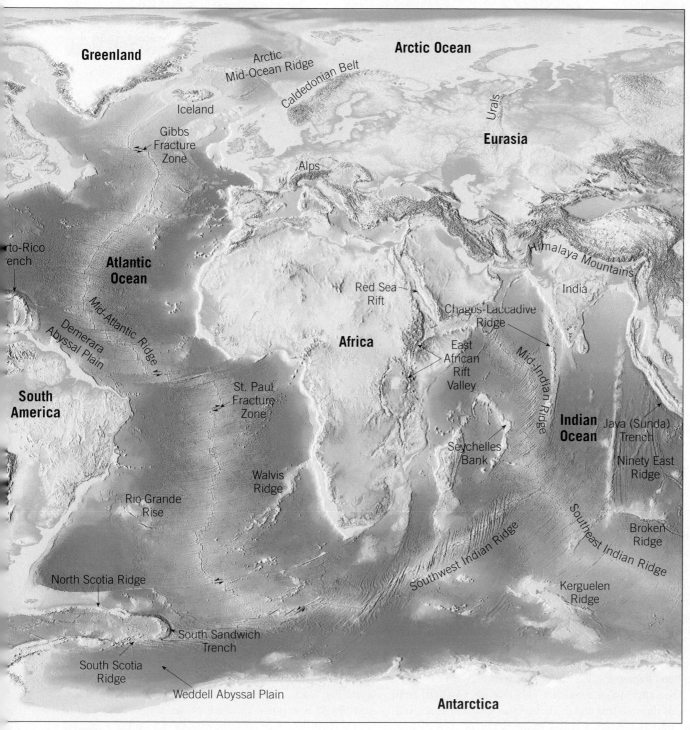

FIGURE 1.20 (Continued)

The Canadian shield is an expansive region of ancient Precambrian rocks, some more than 4 billion years old. It was recently scoured by Ice Age glaciers.

The Appalachians are old mountains. Mountain building began about 480 million years ago and continued for more than 200 million years. Erosion has lowered these once lofty peaks.

The rugged Himalayas are the highest mountains on Earth and are geologically young. They began forming about 50 million years ago and uplift continues today.

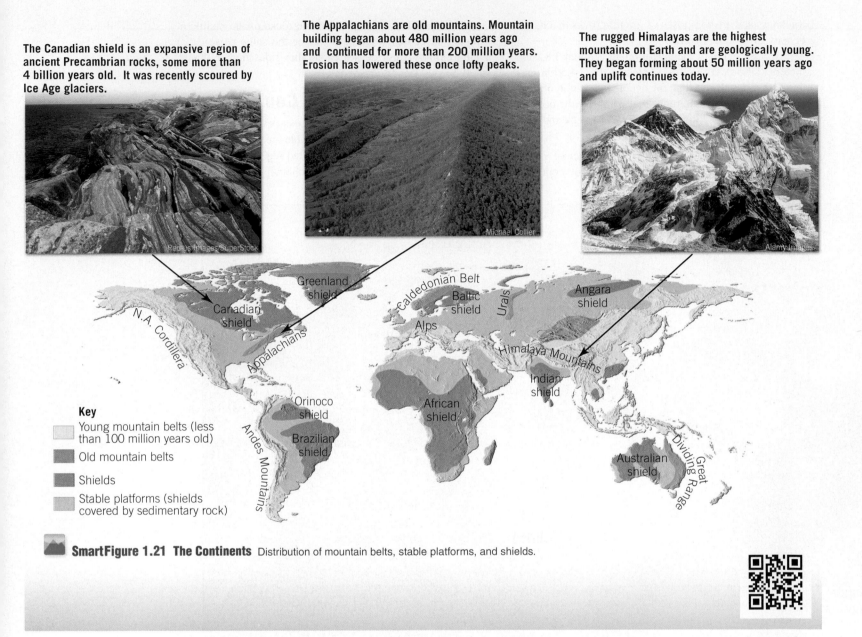

Key
- Young mountain belts (less than 100 million years old)
- Old mountain belts
- Shields
- Stable platforms (shields covered by sedimentary rock)

SmartFigure 1.21 The Continents Distribution of mountain belts, stable platforms, and shields.

FIGURE 1.21 that young mountain belts tend to be long, narrow features at the margins of continents and that the flat, stable areas are typically located in the interior of continents.

Mountain Belts   The most prominent topographic features of the continents are linear **mountain belts**. Although the distribution of mountains appears to be random, this is not the case. When the youngest mountains are considered (those less than 100 million years old), we find that they are located principally in two major zones. The circum-Pacific belt (the region surrounding the Pacific Ocean) includes the mountains of the western Americas and continues into the western Pacific in the form of volcanic islands such as the Aleutians, Japan, and the Philippines (see Figure 1.20).

The other major mountain belt extends eastward from the Alps through Iran and the Himalayas and then dips southward into Indonesia. Careful examination of mountainous terrains reveals that most are places where thick sequences of rocks have been squeezed and highly deformed, as if placed in a gigantic vise. Older mountains are also found on the continents. Examples include the Appalachians in the eastern United States and the Urals in Russia. Their once lofty peaks are now worn low, as a result of millions of years of erosion.

The Stable Interior   Unlike the young mountain belts, which have formed within the past 100 million years, the interiors of the continents have been relatively stable (undisturbed) for the past 600 million years or even longer. Typically, these regions were involved in mountain-building episodes much earlier in Earth's history.

Within the stable interiors are areas known as **shields**, which are expansive, flat regions composed of deformed crystalline rock. Notice in Figure 1.21 that the Canadian Shield is exposed in much of the northeastern part of North America. Age determinations for various shields have shown that they are truly ancient regions. All contain

Precambrian-age rocks that are over 1 billion years old, with some samples approaching 4 billion years in age. These oldest-known rocks exhibit evidence of enormous forces that have folded and faulted them and altered them with great heat and pressure. Thus, we conclude that these rocks were once part of an ancient mountain system that has since been eroded away to produce these expansive, flat regions.

Other flat areas of the stable interior exist in which highly deformed rocks, like those found in the shields, are covered by a relatively thin veneer of sedimentary rocks. These areas are called **stable platforms**. The sedimentary rocks in stable platforms are nearly horizontal except where they have been warped to form large basins or domes. In North America a major portion of the stable platform is located between the Canadian Shield and the Rocky Mountains (Figure 1.21).

Major Features of the Ocean Basins

If all water were drained from the ocean basins, a great variety of features would be seen, including linear chains of volcanoes, deep canyons, extensive plateaus, and large expanses of monotonously flat plains. In fact, the scenery would be nearly as diverse as that on the continents (see Figure 1.20).

During the past 70 years, oceanographers using modern depth-sounding equipment have gradually mapped significant portions of the ocean floor. From these studies they have defined three major regions: *continental margins*, *deep-ocean basins*, and *oceanic (mid-ocean) ridges*.

Continental Margins
The **continental margin** is the portion of the seafloor adjacent to major landmasses. It may include the *continental shelf*, the *continental slope*, and the *continental rise*.

Although land and sea meet at the shoreline, this is not the boundary between the continents and the ocean basins. Rather, along most coasts a gently sloping platform of material, called the **continental shelf**, extends seaward from the shore. Because it is underlain by continental crust, it is considered a flooded extension of the continents. A glance at Figure 1.20 shows that the width of the continental shelf is variable. For example, it is broad along the East and Gulf coasts of the United States but relatively narrow along the Pacific margin of the continent.

The boundary between the continents and the deep-ocean basins lies along the **continental slope**, which is a relatively steep dropoff that extends from the outer edge of the continental shelf to the floor of the deep ocean (see Figure 1.20). Using this as the dividing line, we find that about 60 percent of Earth's surface is represented by ocean basins and the remaining 40 percent by continents.

In regions where trenches do not exist, the steep continental slope merges into a more gradual incline known as the **continental rise**. The continental rise consists of a thick accumulation of sediments that moved downslope from the continental shelf to the deep-ocean floor.

Deep-Ocean Basins
Between the continental margins and oceanic ridges lie the **deep-ocean basins**. Parts of these regions consist of incredibly flat features called **abyssal plains**. The ocean floor also contains extremely deep depressions that are occasionally more than 11,000 meters (36,000 feet) deep. Although these **deep-ocean trenches** are relatively narrow and represent only a small fraction of the ocean floor, they are nevertheless very significant features. Some trenches are located adjacent to young mountains that flank the continents. For example, in Figure 1.20 the Peru–Chile trench off the west coast of South America parallels the Andes Mountains. Other trenches parallel linear island chains called *volcanic island arcs*.

Dotting the ocean floor are submerged volcanic structures called **seamounts**, which sometimes form long, narrow chains. Volcanic activity has also produced several large *lava plateaus*, such as the Ontong Java Plateau located northeast of New Guinea. In addition, some submerged plateaus are composed of continental-type crust. Examples include the Campbell Plateau southeast of New Zealand and the Seychelles Bank northeast of Madagascar.

EYE ON EARTH

This photo shows the picturesque coastal bluffs and rocky shoreline along a portion of the California coast south of San Simeon State Park.

QUESTION 1 *This area, like other shorelines, is described as an interface. What does that mean?*

QUESTION 2 *Does the shoreline represent the boundary between the continent and ocean basin? Explain.*

Michael Collier

Oceanic Ridges The most prominent feature on the ocean floor is the **oceanic ridge**, or **mid-ocean ridge**. As shown in Figure 1.20, the Mid-Atlantic Ridge and the East Pacific Rise are parts of this system. This broad elevated feature forms a continuous belt that winds for more than 70,000 kilometers (43,000 miles) around the globe in a manner similar to the seam of a baseball. Rather than consisting of highly deformed rock, such as most of the mountains on the continents, the oceanic ridge system consists of layer upon layer of igneous rock that has been fractured and uplifted.

Understanding the topographic features that comprise the face of Earth is critical to our understanding of the mechanisms that have shaped our planet. What is the significance of the enormous ridge system that extends through all the world's oceans? What is the connection, if any, between young, active mountain belts and deep-ocean trenches? What forces crumple rocks to produce majestic mountain ranges? These are questions that are addressed in some of the coming chapters, as we investigate the dynamic processes that shaped our planet in the geologic past and will continue to shape it in the future.

1.6 CONCEPT CHECKS

1 Contrast continents and ocean basins.

2 Describe the general distribution of Earth's youngest mountains.

3 What is the difference between shields and stable platforms?

4 What are the three major regions of the ocean floor and some features associated with each region?

1.7 | EARTH AS A SYSTEM

Define *system* and explain why Earth is considered to be a system.

Anyone who studies Earth soon learns that our planet is a dynamic body with many separate but interacting parts, or *spheres*. The hydrosphere, atmosphere, biosphere, and geosphere and all of their components can be studied separately. However, the parts are not isolated. Each is related in some way to the others, and together they produce a complex and continuously interacting whole that we call the *Earth system*.

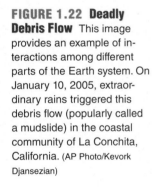

FIGURE 1.22 Deadly Debris Flow This image provides an example of interactions among different parts of the Earth system. On January 10, 2005, extraordinary rains triggered this debris flow (popularly called a mudslide) in the coastal community of La Conchita, California. (AP Photo/Kevork Djansezian)

Earth System Science

A simple example of the interactions among different parts of the Earth system occurs every winter, as moisture evaporates from the Pacific Ocean and subsequently falls as rain in the hills and mountains of southern California, triggering destructive debris flows (**FIGURE 1.22**). The processes that move water from the hydrosphere to the atmosphere and then to the solid Earth have a profound impact on the plants and animals (including humans) that inhabit the affected regions.

Scientists have recognized that in order to more fully understand our planet, they must learn how its individual components (land, water, air, and life-forms) are interconnected. This endeavor, called **Earth system science**, aims to study Earth as a *system* composed of numerous interacting parts, or *subsystems*. Rather than looking through the limited lens of only one of the traditional sciences—geology, atmospheric science, chemistry, biology, and so on—Earth system science attempts to integrate the knowledge of several academic fields. Using an interdisciplinary approach, those engaged in Earth system science attempt to achieve the level of understanding necessary to comprehend and solve many of our global environmental problems.

A **system** is a group of interacting, or interdependent, parts that form a complex whole. Most of us hear and use the term *system* frequently. We may service our car's cooling *system*, make use of the city's transportation *system*, and be a participant in the political *system*. A news report might inform us of an approaching weather *system*. Further, we know that Earth is just a small part of a larger system known as the *solar system*, which in turn is a subsystem of an even larger system called the Milky Way Galaxy.

The Earth System

The Earth system has a nearly endless array of subsystems in which matter is recycled over and over. One familiar loop, or subsystem, is the *hydrologic cycle.* It represents the unending circulation of Earth's water among the hydrosphere, atmosphere, biosphere, and geosphere. Water enters the atmosphere through evaporation from Earth's surface and transpiration from plants. Water vapor condenses in the atmosphere to form clouds, which in turn produce precipitation that falls back to Earth's surface. Some of the rain that falls onto the land sinks in and then is taken up by plants or becomes groundwater, and some flows across the surface toward the ocean.

Viewed over long time spans, the rocks of the geosphere are constantly forming, changing, and re-forming. The loop that involves the processes by which one rock changes to another is called the *rock cycle* and is discussed at some length in Chapter 3. The cycles of the Earth system are not independent of one another. To the contrary, there are many places where the cycles come in contact and interact.

The Parts Are Linked

The parts of the Earth system are linked so that a change in one part can produce changes in any or all of the other parts. For example, when a volcano erupts, lava from Earth's interior may flow out at the surface and block a nearby valley. This new obstruction influences the region's drainage system by creating a lake or causing streams to change course. The large quantities of volcanic ash and gases that can be emitted during an eruption might be blown high into the atmosphere and influence the amount of solar energy that can reach Earth's surface. The result could be a drop in air temperatures over the entire hemisphere.

Where the surface is covered by lava flows or a thick layer of volcanic ash, existing soils are buried. This causes the soil-forming processes to begin anew to transform the new surface material into soil (**FIGURE 1.23**). The soil that eventually forms will reflect the interactions among many parts of the Earth system—the volcanic parent material, the climate, and the impact of biological activity. Of course, there would also be significant changes in the biosphere. Some organisms and their habitats would be eliminated by the lava and ash, and new settings for life, such as a lake formed by a lava dam, would be created. The potential climate change could also impact sensitive life-forms.

Time and Space Scales

The Earth system is characterized by processes that vary on spatial scales from fractions of millimeters to thousands of kilometers. Time scales for Earth's processes range from milliseconds to billions of years. As we learn about Earth, it becomes increasingly clear that despite significant separations in distance or time, many processes are connected, and a change in one component can influence the entire system.

Energy for the Earth System

The Earth system is powered by energy from two sources. The Sun drives external processes that occur in the atmosphere, in the hydrosphere, and at Earth's surface. Weather and climate, ocean circulation,

FIGURE 1.23 Change Is a Constant When Mount St. Helens erupted in May 1980, the area shown here was buried by a volcanic mudflow. Now plants are reestablished and new soil is forming. (Photo by Terry Donnelly/Alamy Images)

and erosional processes are driven by energy from the Sun. Earth's interior is the second source of energy. Heat remaining from when our planet formed and heat that is continuously generated by radioactive decay power the internal processes that produce volcanoes, earthquakes, and mountains.

People and the Earth System

Humans are *part of* the Earth system, a system in which the living and nonliving components are entwined and interconnected. Therefore, our actions produce changes in all the other parts. When we burn gasoline and coal, dispose of our wastes, and clear the land, we cause other parts of the system to respond, often in unforeseen ways. Throughout this book, you will learn about many of Earth's subsystems, including the hydrologic system, the tectonic (mountain-building) system, the rock cycle, and the climate system. Remember that these components *and we humans* are all part of the complex interacting whole we call the Earth system.

The organization of this text involves traditional groupings of chapters that focus on closely related topics. Nevertheless, the theme of *Earth as a system* keeps recurring through *all* major units of *Earth Science.* It is a thread that weaves through the chapters and helps tie them together. At the end of each chapter is a section titled "Examining the Earth System." The questions and problems found there will help you develop an awareness and appreciation for some of the Earth system's important interrelationships.

1.7 CONCEPT CHECKS

1 What is a system? List three examples of systems.

2 What are the two sources of energy for the Earth system?

3 Predict how a change in the hydrologic cycle, such as increased rainfall in an area, might influence the biosphere and geosphere in that area.

Introduction to Earth Science

1.1 WHAT IS EARTH SCIENCE?

List and describe the sciences that collectively make up Earth science. Discuss the scales of space and time in Earth science.

KEY TERMS: Earth science, geology, oceanography, meteorology, astronomy, geologic time

- Earth science includes geology, oceanography, meteorology, and astronomy.
- There are two broad subdivisions of geology. Physical geology studies Earth materials and the internal and external processes that create and shape Earth's landscape. Historical geology examines Earth history.
- The other Earth sciences seek to understand the oceans, the atmosphere's weather and climate, and Earth's place in the universe.
- Important relationships between people and the environment include the quest for resources, the impact of people on the natural environment, and the effects of natural hazards.
- Earth science must deal with processes and phenomena that vary from the subatomic scale of matter to the nearly infinite scale of the universe. The time scales of phenomena studied in Earth science range from tiny fractions of a second to many billions of years.
- Geologic time, the span of time since the formation of Earth, is about 4.6 billion years.

1.2 THE NATURE OF SCIENTIFIC INQUIRY

Discuss the nature of scientific inquiry and distinguish between a hypothesis and a theory.

KEY TERMS: hypothesis, theory

- Scientists make careful observations, construct tentative explanations for those observations (hypotheses), and then test those hypotheses with field investigations and laboratory work. In science, a theory is a well-tested and widely accepted explanation that the scientific community agrees best fits certain observable facts.
- As failed hypotheses are discarded, scientific knowledge moves closer to a correct understanding, but we can never be fully confident that we know all the answers. Scientists must always be open to new information that forces change in our model of the world.

1.3 EARLY EVOLUTION OF EARTH

Outline the stages in the formation of our solar system.

KEY TERM: nebular theory

- The nebular theory describes the formation of the solar system. The planets and Sun began forming about 5 billion years ago from a large cloud of dust and gases.
- As the cloud contracted, it began to rotate and assume a disk shape. Material that was gravitationally pulled toward the center became the protosun. Within the rotating disk, small centers, called planetesimals, swept up more and more of the cloud's debris.
- Because of their high temperatures and weak gravitational fields, the inner planets were unable to accumulate and retain many of the lighter components. Because of the very cold temperatures existing far from the Sun, the large outer planets consist of huge amounts of lighter materials. These gaseous substances account for the comparatively large sizes and low densities of the outer planets.

Q Earth is about 4.6 billion years old. If all the planets in our solar system formed at about the same time, how old would you expect Mars to be? Jupiter? The Sun?

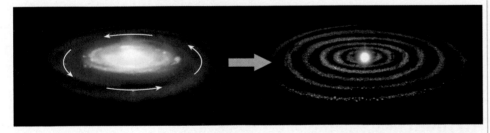

1.4 EARTH'S SPHERES

List and describe Earth's four major spheres.

KEY TERMS: hydrosphere, atmosphere, biosphere, geosphere

- Earth's physical environment is traditionally divided into three major parts: the solid Earth, called the geosphere; the water portion of our planet, called the hydrosphere; and Earth's gaseous envelope, called the atmosphere.
- A fourth Earth sphere is the biosphere, the totality of life on Earth. It is concentrated in a relatively thin zone that extends a few kilometers into the hydrosphere and geosphere and a few kilometers up into the atmosphere.
- Of all the water on Earth, more than 96 percent is in the oceans, which cover nearly 71 percent of the planet's surface.

Q Is glacial ice part of the geosphere, or does it belong to the hydrosphere? Explain your answer.

Michael Collier

1.5 A CLOSER LOOK AT THE GEOSPHERE

Label a diagram that shows Earth's internal structure. Briefly explain why the geosphere can be described as being mobile.

KEY TERMS: crust, mantle, lithosphere, asthenosphere, lower mantle, core, outer core, inner core, plate tectonics, lithospheric plate

■ Compositionally, the solid Earth has three layers: core, mantle, and crust. The core is most dense, and the crust is least dense.

■ Earth's interior can also be divided into layers based on physical properties. The crust and upper mantle make a two-part layer called the lithosphere, which is broken into the plates of plate tectonics. Beneath that is the "weak" asthenosphere. The lower mantle is stronger than the asthenosphere and overlies the molten outer core. This liquid is made of the same iron–nickel alloy as the inner core, but the extremely high pressure of Earth's center compacts the inner core into a solid form.

Q **The diagram represents Earth's layered structure. Does it show layering based on physical properties or layering based on composition? Identify the lettered layers.**

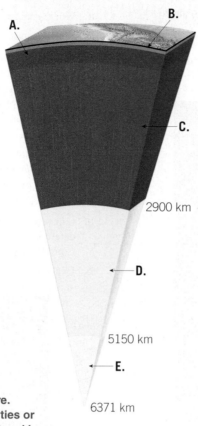

A.
B.
C.
2900 km
D.
5150 km
E.
6371 km

1.6 THE FACE OF EARTH

List and describe the major features of the continents and ocean basins.

KEY TERMS: continent, ocean basin, mountain belt, shield, stable platform, continental margin, continental shelf, continental slope, continental rise, deep-ocean basin, abyssal plain, deep-ocean trench, seamount, oceanic ridge (mid-ocean ridge)

■ Two principal divisions of Earth's surface are the continents and ocean basins. A significant difference is their relative levels. The elevation differences between continents and ocean basins is primarily the result of differences in their respective densities and thicknesses.

■ The largest features of the continents can be divided into two categories: mountain belts and the stable interior. The ocean floor is divided into three major topographic units: continental margin, deep-ocean basin, and oceanic (mid-ocean) ridge.

Q **Put these features of the ocean floor in order from shallowest to deepest: continental slope, deep-ocean trench, continental shelf, abyssal plain, continental rise.**

1.7 EARTH AS A SYSTEM

Define *system* and explain why Earth is considered to be a system.

KEY TERMS: Earth system science, system

■ Although each of Earth's four spheres can be studied separately, they are all related in a complex and continuously interacting whole that is called the Earth system.

■ Earth system science uses an interdisciplinary approach to integrate the knowledge of several academic fields in the study of our planet and its global environmental problems.

■ The two sources of energy that power the Earth system are (1) the Sun, which drives the external processes that occur in the atmosphere, hydrosphere, and at Earth's surface, and (2) heat from Earth's interior that powers the internal processes that produce volcanoes, earthquakes, and mountains.

Q **Give a specific example of how humans are affected by the Earth system and another example of how humans affect the Earth system.**

GIVE IT SOME **THOUGHT**

1. After entering a dark room, you turn on a wall switch, but the light does not come on. Suggest at least three hypotheses that might explain this observation. How would you determine which one of your hypotheses (if any) is correct?

2. Each of the following statements may either be a hypothesis (H), a theory (T), or an observation (O). Use one of these letters to identify each statement. Briefly explain each choice.

  a. A scientist proposes that a recently discovered large ring-shaped structure on the Canadian Shield is the remains of an ancient meteorite crater.

  b. The Redwall Formation in the Grand Canyon is composed primarily of limestone.

  c. The outer part of Earth consists of several large plates that move and interact with each other.

  d. Since 1885, the terminus of Canada's Athabasca Glacier has receded 1.5 kilometers.

3. Making accurate measurements and observations is a basic part of scientific inquiry. The accompanying radar image, showing the distribution and intensity of precipitation associated with a storm, provides one example. Identify another image in this chapter that illustrates a way in which scientific data are gathered. Suggest an advantage that might be associated with the example you select.

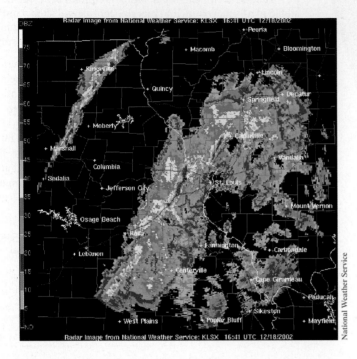

National Weather Service

4. The length of recorded history for humankind is about 5000 years. Clearly, most people view this span as being very long. How does it compare to the length of geologic time? Calculate the percentage or fraction of geologic time that is represented by recorded history. To make calculations easier, round the age of Earth to the nearest billion.

5. Refer to the graph in Figure 1.13 to answer the following questions.

   a. If you were to climb to the top of Mount Everest, how many breaths of air would you have to take at that altitude to equal one breath at sea level?

   b. If you are flying in a commercial jet at an altitude of 12 kilometers (about 39,000 feet), about what percentage of the atmosphere's mass is below you?

6. Examine Figure 1.12 to answer these questions.

   a. Where is most of Earth's freshwater stored?

   b. Where is most of Earth's liquid freshwater found?

7. Jupiter, the largest planet in our solar system, is 5.2 astronomical units (AU) from the Sun. How long would it take to go from Earth to Jupiter if you traveled as fast as a jet (1000 kilometers/hour)? Do the same calculation for Neptune, which is 30 AU from the Sun. Referring to the GEOgraphics feature on page 15 will be helpful.

8. These rock layers consist of materials such as sand, mud, and gravel that, over a span of millions of years, were deposited by rivers, waves, wind, and glaciers. Each layer was buried by subsequent deposits and eventually compacted and cemented into solid rock. Later, the region was uplifted, and erosion exposed the layers seen here.

   a. Can you establish a relative time scale for these rocks? That is, can you determine which one of the layers shown here is likely oldest and which is probably youngest?

   b. Explain the logic you used.

Michael Collier

EXAMINING THE **EARTH SYSTEM**

1. This scene is in British Columbia's Mount Robson Provincial Park. The park is named for the highest peak in the Canadian Rockies.

   a. List as many examples as possible of features associated with each of Earth's four spheres.

   b. Which, if any, of these features was created by internal processes? Describe the role of external processes in this scene.

2. Humans are a part of the Earth system. List at least three examples of how you, in particular, influence one or more of Earth's major spheres.

Michael Wheatley/AGE Fotostock

3. The accompanying photo provides an example of interactions among different parts of the Earth system. It is a view of a debris flow (popularly called a mudslide) that was triggered by extraordinary rains. Which of Earth's four spheres were involved in this natural disaster, which buried a small town on the Philippine island of Leyte? Describe how each contributed to or was influenced by the event.

Pat Roque/AP Photo

4. Examine the accompanying concept map that links the four spheres of the Earth system. All of the spheres are linked by arrows that represent processes by which the spheres interact and influence each other. For each arrow list at least one process.

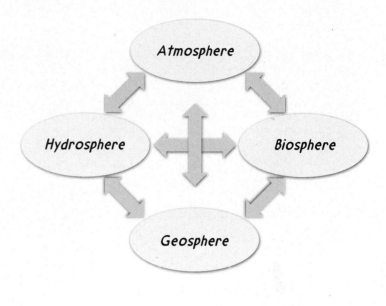

MasteringGeology™

Looking for additional review and test prep materials? Visit the Self Study area in **www.masteringgeology.com** to find practice quizzes, study tools, and multimedia that will aid in your understanding of this chapter's content. In **MasteringGeology™** you will find:

- GEODe: Earth Science: An interactive visual walkthrough of key concepts
- Geoscience Animation Library: More than 100 animations illuminating many difficult-to-understand Earth science concepts

- In The News RSS Feeds: Current Earth science events and news articles are pulled into the site with assessment
- Pearson eText
- Optional Self Study Quizzes
- Web Links
- Glossary
- Flashcards

2

Matter and Minerals

FOCUS ON CONCEPTS

Each statement represents the primary **LEARNING OBJECTIVE** for the corresponding major heading within the chapter. After you complete the chapter, you should be able to:

2.1 List the main characteristics that an Earth material must possess to be considered a mineral and describe each.

2.2 Compare and contrast the three primary particles contained in atoms.

2.3 Distinguish among ionic bonds, covalent bonds, and metallic bonds.

2.4 List and describe the properties that are used in mineral identification.

2.5 List the common silicate and nonsilicate minerals and describe the characteristics of each group.

2.6 Discuss Earth's natural resources in terms of renewability. Differentiate between mineral resources and ore deposits.

The Cave of Crystals, Chihuahua, Mexico, contains giant gypsum crystals, some of the largest natural crystals ever found.
(Photo by Carsten Peter/Speleoresearch & Films/National Geographic Stock/Getty Images)

Earth's crust and oceans are home to a wide variety of useful and essential minerals. Most people are familiar with the common uses of many basic metals, including aluminum in beverage cans, copper in electrical wiring, and gold and silver in jewelry. But some people are not aware that pencil "lead" contains the greasy-feeling mineral graphite and that bath powders and many cosmetics contain the mineral talc. Moreover, many do not know that dentists use drill bits impregnated with diamonds to drill through tooth enamel or that the common mineral quartz is the source of silicon for computer chips. In fact, practically every manufactured product contains materials obtained from minerals.

In addition to rocks and minerals having economic uses, all the processes that geologists study are in some way dependent on the properties of these basic Earth materials. Events such as volcanic eruptions, mountain building, weathering and erosion, and even earthquakes involve rocks and minerals. Consequently, a basic knowledge of Earth materials is essential to understanding all geologic phenomena.

2.1 | MINERALS: BUILDING BLOCKS OF ROCK List the main characteristics that an Earth material must possess to be considered a mineral and describe each.

We begin our discussion of Earth materials with an overview of **mineralogy** (*mineral* = mineral, *ology* = the study of) because minerals are the building blocks of rocks. In addition, humans have used minerals for both practical and decorative purposes for thousands of years (**FIGURE 2.1**). Flint and chert were the first minerals to be mined; they were fashioned into weapons and cutting tools. Egyptians began mining gold, silver, and copper as early as 3700 B.C. By 2200 B.C., humans had discovered how to combine copper with tin to make bronze, a strong, hard alloy. Later, a process was developed to extract iron from minerals such as hematite—a discovery that marked the decline of the Bronze Age. During the Middle Ages, mining of a variety of minerals became common, and the impetus for the formal study of minerals was in place.

The term *mineral* is used in several different ways. For example, those concerned with health and fitness extol the benefits of vitamins and minerals. The mining industry typically uses the word to refer to anything taken out of the ground, such as coal, iron ore, or sand and gravel. The guessing game known as *Twenty Questions* usually begins with the question "Is it animal, vegetable, or mineral?" What criteria do geologists use to determine whether something is a mineral?

Defining a Mineral

Geologists define **mineral** as *any naturally occurring inorganic solid that possesses an orderly crystalline structure and a definite chemical composition that allows for some variation*. Thus, Earth materials that are classified as minerals exhibit the following characteristics:

1. **Naturally occurring.** Minerals form through natural geologic processes. Synthetic materials—meaning those produced in a laboratory or by human intervention—are not considered minerals.

2. **Generally inorganic.** Inorganic crystalline solids, such as ordinary table salt (halite), that are found naturally in the ground are considered minerals. (Organic compounds, on the other hand, are generally not. Sugar, a crystalline solid like salt but that comes from sugarcane or sugar beets, is a common example of such an organic compound.) Many marine animals secrete inorganic compounds, such as calcium carbonate (calcite), in the form of shells and coral reefs. If these materials are buried and become part of the rock record, geologists consider them minerals.

3. **Solid substance.** Only solid crystalline substances are considered minerals. Ice (frozen water) fits this criterion and is considered a mineral, whereas liquid water and water vapor do not. The exception is mercury,

FIGURE 2.1 Quartz Crystals Well-developed quartz crystals found near Hot Springs, Arkansas. (Photo by BOL/TH FOTO/AGE fotostock)

which is found in its liquid form in nature.

4. **Orderly crystalline structure.** Minerals are crystalline substances, which means their atoms (ions) are arranged in an orderly, repetitive manner (**FIGURE 2.2**). This orderly packing of atoms is reflected in regularly shaped objects called *crystals*. Some naturally occurring solids, such as volcanic glass (obsidian), lack a repetitive atomic structure and are not considered minerals.

5. **Definite chemical composition that allows for some variation.** Minerals are chemical compounds having compositions that can be expressed by a chemical formula. For example, the common mineral quartz has the formula SiO_2, which indicates that quartz consists of silicon (Si) and oxygen (O) atoms, in a ratio of one-to-two. This proportion of silicon to oxygen is true for any sample of pure quartz, regardless of its origin. However, the compositions of some minerals vary *within specific, well-defined limits*. This occurs because certain elements can substitute for others of similar size without changing the mineral's internal structure.

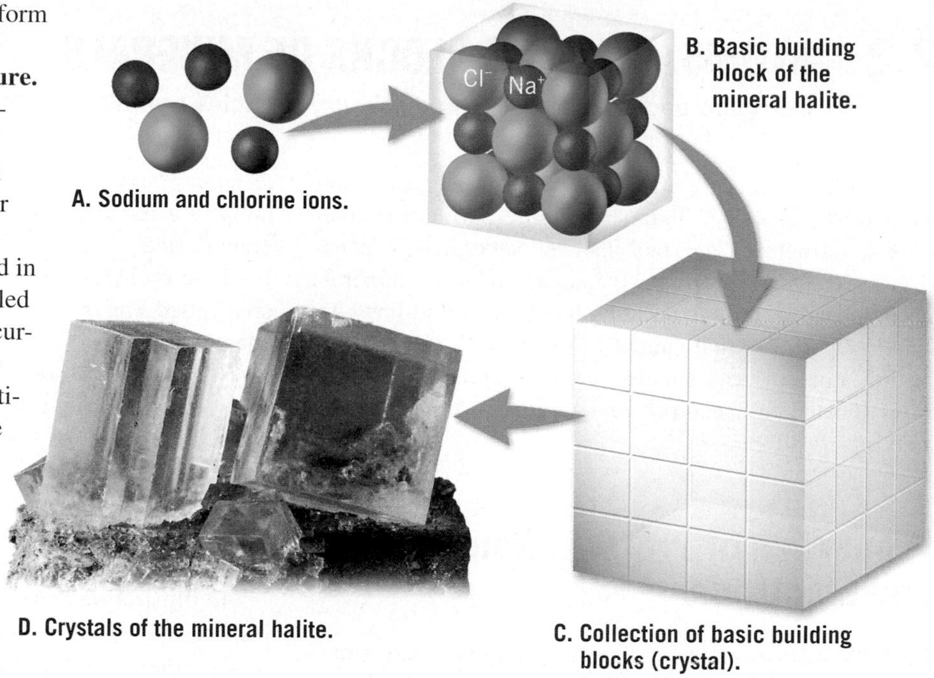

A. Sodium and chlorine ions.

B. Basic building block of the mineral halite.

C. Collection of basic building blocks (crystal).

D. Crystals of the mineral halite.

FIGURE 2.2 Arrangement of Sodium and Chloride Ions in the Mineral Halite The arrangement of atoms (ions) into basic building blocks that have a cubic shape results in regularly shaped cubic crystals. (Photo by Dennis Tasa)

What Is a Rock?

In contrast to minerals, rocks are more loosely defined. Simply, a **rock** is any solid mass of mineral, or mineral-like, matter that occurs naturally as part of our planet. Most rocks, like the sample of granite shown in **FIGURE 2.3**, occur as aggregates of several different minerals. The term *aggregate* implies that the minerals are joined in such a way that their individual properties are retained. Note that the different minerals that make up granite can be easily identified. However, some rocks are composed almost entirely of one mineral. A common example is the sedimentary rock *limestone*, which is an impure mass of the mineral calcite.

In addition, some rocks are composed of nonmineral matter. These include the volcanic rocks *obsidian* and *pumice*, which are noncrystalline glassy substances, and *coal*, which consists of solid organic debris.

Although this chapter deals primarily with the nature of minerals, keep in mind that most rocks are aggregates of minerals. Because the properties of rocks are determined largely by the chemical composition and crystalline structure of the minerals contained within them, we will first consider these Earth materials.

Granite (Rock)

SmartFigure 2.3 Most Rocks Are Aggregates of Minerals Shown here is a hand sample of the igneous rock granite and three of its major constituent minerals. (Photos by E. J. Tarbuck)

Quartz (Mineral)

Hornblende (Mineral)

Feldspar (Mineral)

2.1 CONCEPT CHECKS

1 List five characteristics an Earth material must have in order to be considered a mineral.

2 Based on the definition of *mineral*, which of the following materials are not classified as minerals and why: gold, water, synthetic diamonds, ice, and wood?

3 Define the term *rock*. How do rocks differ from minerals?

2.2 | ATOMS: BUILDING BLOCKS OF MINERALS

Compare and contrast the three primary particles contained in atoms.

When minerals are carefully examined, even under optical microscopes, the innumerable tiny particles of their internal structures are not visible. Nevertheless, scientists have discovered that all matter, including minerals, is composed of minute building blocks called **atoms**—the smallest particles that cannot be chemically split. Atoms, in turn, contain even smaller particles—*protons* and *neutrons* located in a central **nucleus** that is surrounded by *electrons* (**FIGURE 2.4**).

Properties of Protons, Neutrons, and Electrons

Protons and **neutrons** are very dense particles with almost identical masses. By contrast, **electrons** have a negligible mass, about 1/2000 that of a proton. To illustrate this difference, assume that on a scale where a proton has the mass of a baseball, an electron has the mass of a single grain of rice.

Both protons and electrons share a fundamental property called *electrical charge*. Protons have an electrical charge of +1, and electrons have a charge of −1. Neutrons, as the name suggests, have no charge. The charges of protons and electrons are equal in magnitude but opposite in polarity, so when these two particles are paired, the charges cancel each other out. Since matter typically contains equal numbers of positively charged protons and negatively charged electrons, most substances are electrically neutral.

In illustrations, electrons are sometimes shown orbiting the nucleus in a manner that resembles the planets of our solar system orbiting the Sun (see Figure 2.4A). However, electrons do not actually behave this way. A more realistic depiction shows electrons as a cloud of negative charges surrounding the nucleus (see Figure 2.4B). Studies of the arrangements of electrons show that they move about the nucleus in regions called *principal shells*, each with an associated energy level. In addition, each shell can hold a specific number of electrons, with the outermost shell generally containing **valence electrons** that interact with other atoms to form chemical bonds.

Most of the atoms in the universe (except hydrogen and helium) were created inside massive stars by nuclear fusion and released into interstellar space during hot, fiery supernova explosions. As this ejected material cooled, the newly formed nuclei attracted electrons to complete their atomic structure. At the temperatures found at Earth's surface, all free atoms (those not bonded to other atoms) have a full complement of electrons—one for each proton in the nucleus.

FIGURE 2.4 Two Models of an Atom A. Simplified view of an atom consisting of a central nucleus composed of protons and neutrons encircled by high-speed electrons. **B.** This model of an atom shows spherically shaped electron clouds (shells) surrounding a central nucleus. The nucleus contains virtually all of the mass of the atom. The remainder of the atom is the space occupied by negatively charged electrons. (The relative sizes of the nuclei shown are greatly exaggerated.)

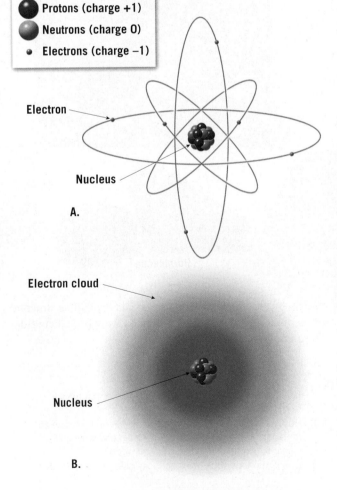

Protons (charge +1)
Neutrons (charge 0)
Electrons (charge −1)

Electron

Nucleus

A.

Electron cloud

Nucleus

B.

Elements: Defined by Their Number of Protons

The simplest atoms have only 1 proton in their nuclei, whereas others have more than 100. The number of protons in the nucleus of an atom, called the **atomic number**, determines its chemical nature. All atoms with the same number of protons have the same chemical and physical properties. Together, a group of the same kind of atoms is called an **element**. There are about 90 naturally occurring elements, and several more have been synthesized in the laboratory. You are probably familiar with the names of many elements, including carbon, nitrogen, and oxygen. All carbon atoms have six protons, whereas all nitrogen

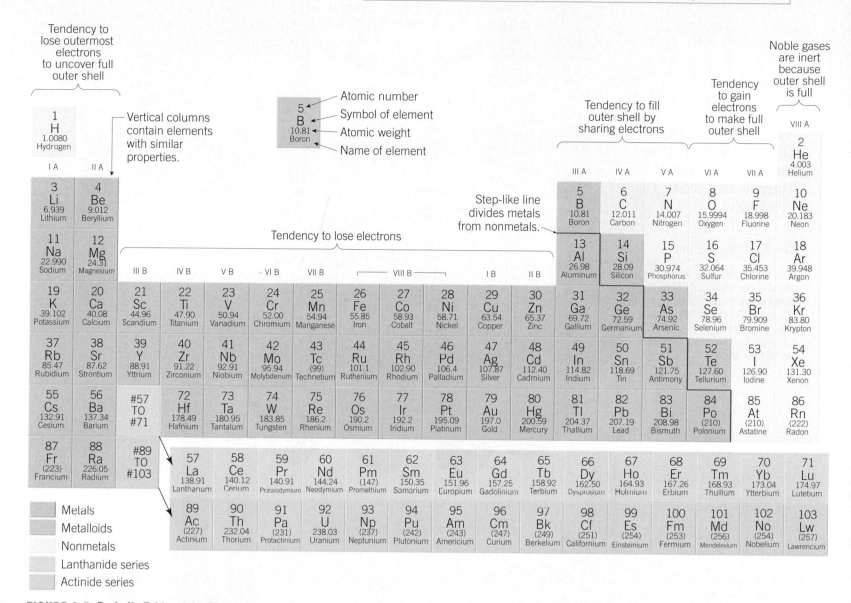

FIGURE 2.5 Periodic Table of the Elements

atoms have seven protons, and all oxygen atoms have eight protons.

Elements are organized so that those with similar properties line up in columns, referred to as groups. This arrangement, called the **periodic table**, is shown in **FIGURE 2.5**. Each element has been assigned a one- or two-letter symbol. The atomic numbers and masses for each element are also shown on the periodic table.

Atoms of the naturally occurring elements are the basic building blocks of Earth's minerals. Most elements join with atoms of other elements to form **chemical compounds**. Therefore, most minerals are chemical compounds composed of atoms of two or more elements. These include the minerals quartz (SiO_2), halite (NaCl), and calcite ($CaCO_3$). However, a few minerals, such as native copper, diamonds, sulfur, and gold, are made entirely of atoms of only one element (**FIGURE 2.6**).

A. Gold on quartz

B. Sulfur

C. Copper

FIGURE 2.6 **These Are among the Few Minerals That Are Composed of a Single Element** (Photos by Dennis Tasa)

2.2 CONCEPT CHECKS

1 List the three main particles of an atom and explain how they differ from one another.

2 Make a simple sketch of an atom and label its three main particles.

3 What is the significance of valence electrons?

Gold

Gold has been treasured since long before recorded history for its beauty. Even today, its most common use is in jewelry.

УКРАЇНА
1000г
ЗОЛОТО
999,9

Shutterstock

How valuable is gold?

In early 2013, the value of one troy ounce of gold was about US$1,677. Based on that value, a 1000-gram (32-ounce) bar of gold, like the one shown, was worth $53,664. In 1970, the price of gold was less than $40 per troy ounce!

$53,664

Where is the world's gold produced?

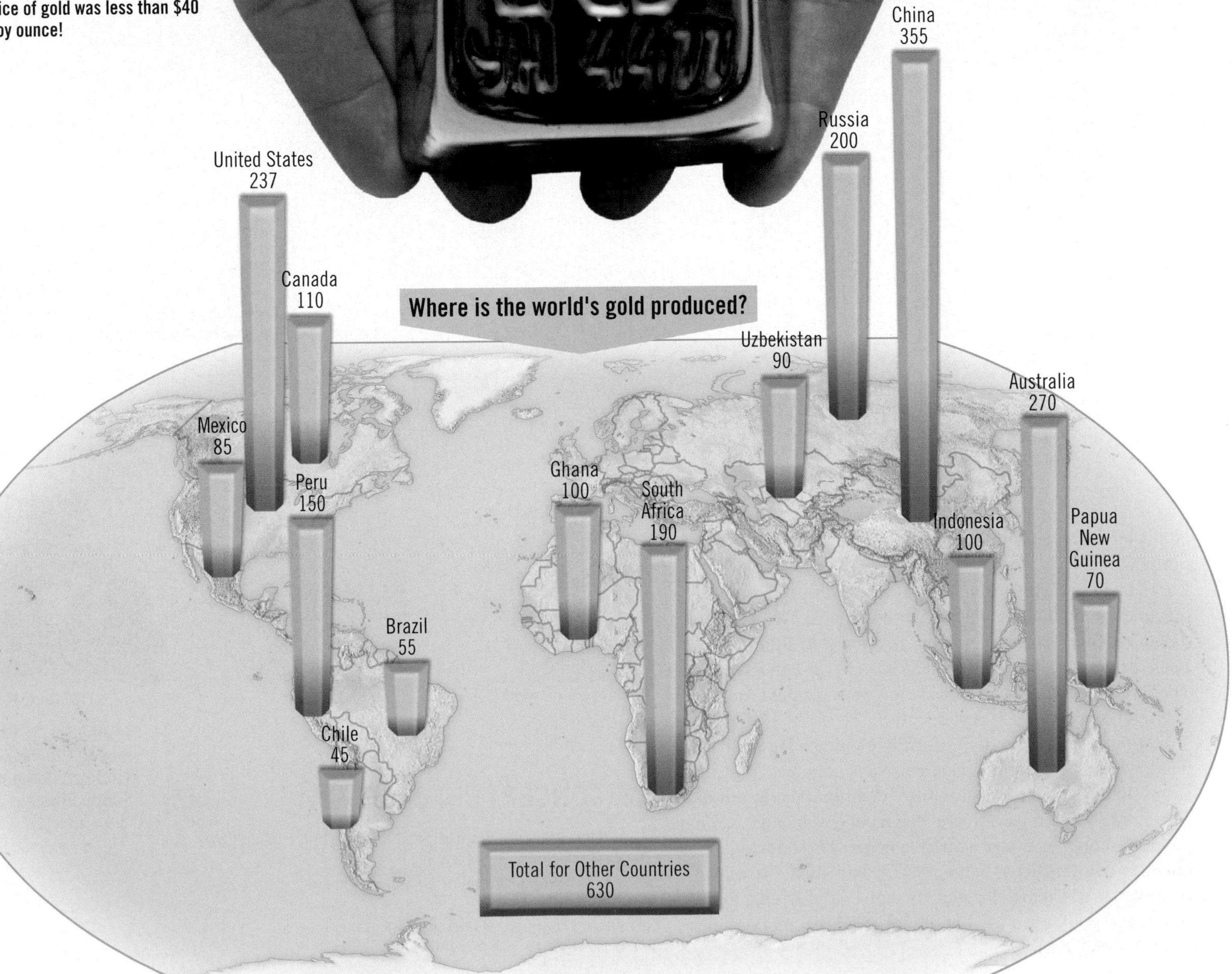

China 355

Russia 200

United States 237

Canada 110

Uzbekistan 90

Australia 270

Mexico 85

Peru 150

Ghana 100

South Africa 190

Indonesia 100

Papua New Guinea 70

Brazil 55

Chile 45

Total for Other Countries 630

(Data from USGS)

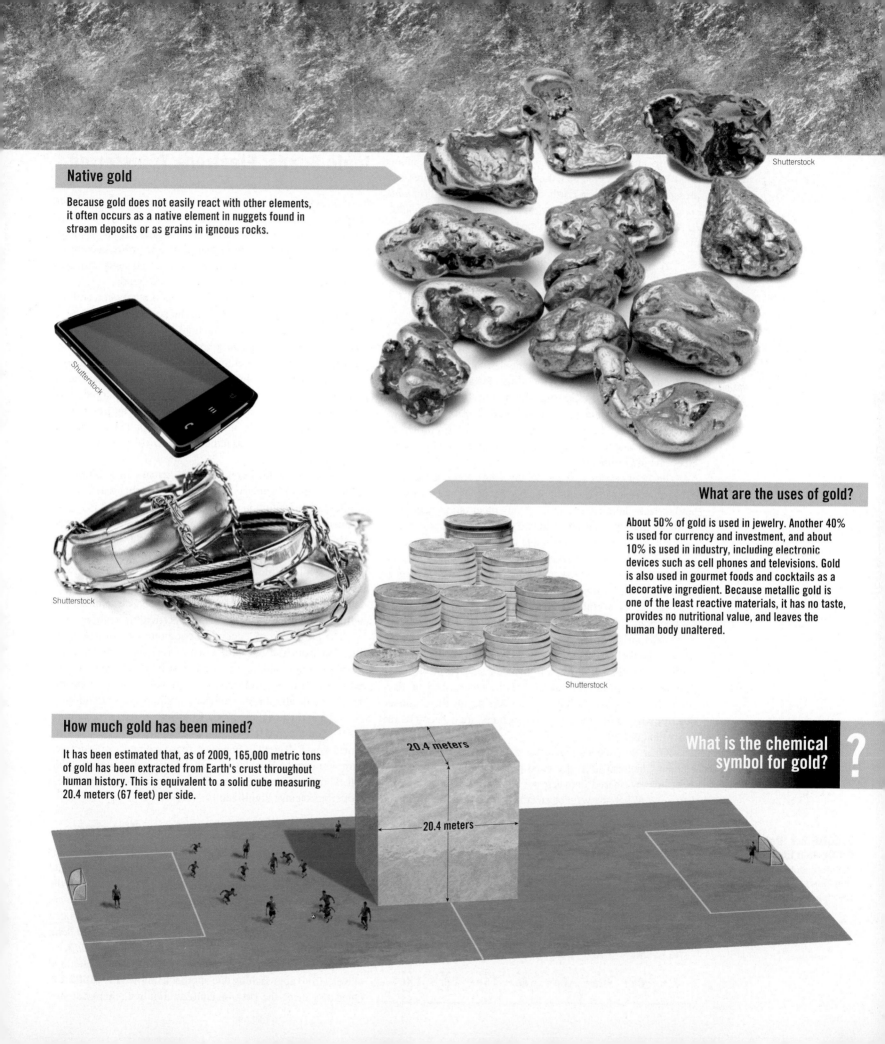

Native gold

Because gold does not easily react with other elements, it often occurs as a native element in nuggets found in stream deposits or as grains in igneous rocks.

Shutterstock

Shutterstock

Shutterstock

What are the uses of gold?

About 50% of gold is used in jewelry. Another 40% is used for currency and investment, and about 10% is used in industry, including electronic devices such as cell phones and televisions. Gold is also used in gourmet foods and cocktails as a decorative ingredient. Because metallic gold is one of the least reactive materials, it has no taste, provides no nutritional value, and leaves the human body unaltered.

Shutterstock

How much gold has been mined?

It has been estimated that, as of 2009, 165,000 metric tons of gold has been extracted from Earth's crust throughout human history. This is equivalent to a solid cube measuring 20.4 meters (67 feet) per side.

20.4 meters

20.4 meters

What is the chemical symbol for gold? ?

2.3 | WHY ATOMS BOND

Distinguish among ionic bonds, covalent bonds, and metallic bonds.

Except in a group of elements known as the noble gases, atoms bond to one another under the conditions (temperatures and pressures) that occur on Earth. Some atoms bond to form *ionic compounds*, some form *molecules*, and still others form *metallic substances*. Why does this happen? Experiments show that electrical forces hold atoms together and bond them to each other. These electrical attractions lower the total energy of the bonded atoms, which, in turn, generally makes them more stable. Consequently, atoms that are bonded in compounds tend to be more stable than atoms that are free (not bonded).

The Octet Rule and Chemical Bonds

As was noted earlier, valence (outer shell) electrons are generally involved in chemical bonding. **FIGURE 2.7** shows a shorthand way of representing the number of valence electrons for selected elements in each group. Notice that the elements in Group I have one valence electron, those in Group II have two valence electrons, and so on, up to eight valence electrons in Group VIII.

The noble gases (except helium) have very stable electron arrangements with eight valence electrons and, therefore, tend to lack chemical reactivity. Many other atoms gain, lose, or share electrons during chemical reactions, ending up with electron arrangements of the noble gases. This observation led to a chemical guideline known as the **octet rule**: *Atoms tend to gain, lose, or share electrons until they are surrounded by eight valence electrons.* Although there are exceptions to the octet rule, it is a useful rule of thumb for understanding chemical bonding.

When an atom's outer shell does not contain eight electrons, it is likely to chemically bond to other atoms to fill its shell. A **chemical bond** is a transfer or sharing of electrons that allows each atom to attain a full valence shell of electrons. Some atoms do this by transferring all their valence electrons to other atoms so that an inner shell becomes the full valence shell.

When the valence electrons are transferred between the elements to form ions, the bond is an *ionic bond*. When the electrons are shared between the atoms, the bond is a *covalent bond*. When the valence electrons are shared among all the atoms in a substance, the bonding is *metallic*.

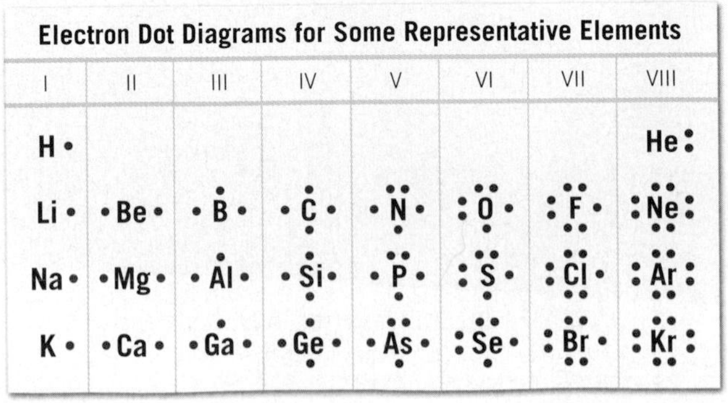

Ionic Bonds: Electrons Transferred

Perhaps the easiest type of bond to visualize is the *ionic bond*, in which one atom gives up one or more of its valence electrons to another atom to form **ions**—*positively and negatively charged atoms*. The atom that loses electrons becomes a positive ion, and the atom that gains electrons becomes a negative ion. Oppositely charged ions are strongly attracted to one another and join to form ionic compounds.

Consider the ionic bonding that occurs between sodium (Na) and chlorine (Cl) to produce the solid ionic compound sodium chloride—the mineral halite (common table salt). Notice in **FIGURE 2.8A** that a sodium atom gives up its single valence electron to chlorine and, as a result, becomes a positively charged sodium ion. Chlorine, on the other hand, gains one electron and becomes a negatively charged chloride ion. We know that ions having unlike charges attract. Thus, an **ionic bond** is an attraction of oppositely charged ions to one another, producing an electrically neutral ionic compound.

FIGURE 2.8B illustrates the arrangement of sodium and chlorine ions in ordinary table salt. Notice that salt consists of alternating sodium and chlorine ions, positioned in such a manner that each positive ion is attracted to and surrounded on all sides by negative ions and vice versa. This arrangement maximizes the attraction between ions with opposite charges while minimizing the repulsion between ions with identical charges. Thus, ionic compounds consist of an orderly arrangement of oppositely charged ions assembled in a definite ratio that provides overall electrical neutrality.

The properties of a chemical compound are dramatically different from the properties of the various elements comprising it. For example, sodium is a soft silvery metal that is extremely reactive and poisonous. If you were to consume even a small amount of elemental sodium, you would need immediate medical attention. Chlorine, a green poisonous gas, is so toxic that it was used as a chemical weapon during World War I. Together, however, these elements produce sodium chloride, a harmless flavor enhancer that we call table salt. Thus, when elements combine to form compounds, their properties change significantly.

Covalent Bonds: Electron Sharing

Sometimes the forces that hold atoms together cannot be understood on the basis of the attraction of oppositely charged ions. One example is the hydrogen molecule (H_2), in which the two hydrogen atoms are held together tightly, and no ions are present. The strong attractive force that holds two hydrogen atoms together results from a **covalent bond**, *a chemical bond formed by the sharing of a pair of electrons between atoms*.

Imagine two hydrogen atoms (each with one proton and one electron) approaching one another as shown in **FIGURE 2.9**. Once they meet, the electron configuration will change so that

FIGURE 2.7 Dot Diagrams for Certain Elements Each dot represents a valence electron found in the outermost principal shell.

Electron Dot Diagrams for Some Representative Elements							
I	II	III	IV	V	VI	VII	VIII
H •							He :
Li •	• Be •	• Ḃ •	• Ċ •	• N̈ •	: Ö •	: F̈ •	: N̈e :
Na •	• Mg •	• A̋l •	• Si •	• P̈ •	: S̈ •	: C̈l •	: Ȧr :
K •	• Ca •	• Ga •	• Ge •	• Äs •	: Se •	: Br •	: K̈r :

A. The transfer of an electron from a sodium (Na) to a chlorine (Cl) atom leads to the formation of a Na⁺ ion and a Cl⁻ ion.

B. The arrangement of Na⁺ and Cl⁻ in the solid ionic compound sodium chloride (NaCl), table salt.

FIGURE 2.8 Formation of the Ionic Compound Sodium Chloride

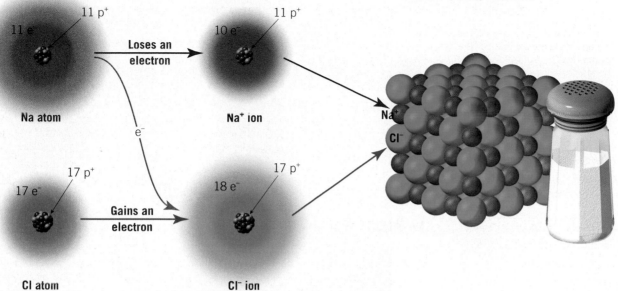

both electrons will primarily occupy the space between the atoms. In other words, the two electrons are shared by both hydrogen atoms and are attracted simultaneously by the positive charge of the proton in the nucleus of each atom. Although hydrogen atoms do not form ions, the force that holds these atoms together arises from the attraction of oppositely charged particles—positively charged protons in the nuclei and negatively charged electrons that surround these nuclei.

Two hydrogen atoms combine to form a hydrogen molecule, held together by the attraction of oppositely charged particles—positively charged protons in each nucleus and negatively charged electrons that surround these nuclei.

$$H \cdot + H \cdot \longrightarrow H : H$$

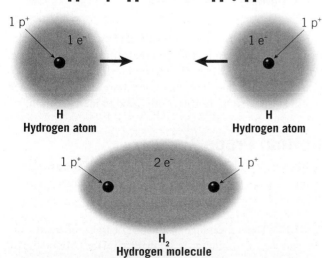

FIGURE 2.9 Covalent Bonding of Two Hydrogen Atoms (H) to Form a Hydrogen Molecule (H₂) When hydrogen atoms bond, the negatively charged electrons are shared by both hydrogen atoms and attracted simultaneously by the positive charge of the proton in the nucleus of each atom.

Metallic Bonds: Electrons Free to Move

A few minerals, such as native gold, silver, and copper, are made entirely of metal atoms that are packed tightly together in an orderly way. The bonding that holds these atoms together is the result of each atom contributing its valence electrons to a common pool of electrons that are free to move throughout the entire metallic structure. The contribution of one or more valence electrons leaves an array of positive ions immersed in a "sea" of valence electrons, as shown in **FIGURE 2.10**.

The attraction between the "sea" of negatively charged electrons and the positive ions produces the **metallic bonds** that give metals their unique properties. Metals are good conductors

A. The central core of each metallic atom, which has an overall positive charge, consists of the nucleus and inner electrons.

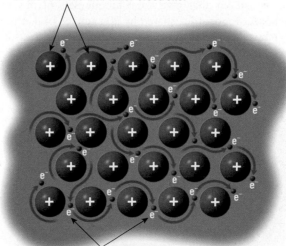

B. A "sea" of negatively charged outer electrons, that are free to move throughout the structure, surrounds the metallic atoms.

FIGURE 2.10 Metallic Bonds Metallic bonding is a result of each atom contributing its valence electrons to a common pool of electrons that are free to move throughout the entire metallic structure. The attraction between the "sea" of negatively charged electrons and the positive ions produces the metallic bonds that give metals their unique properties.

EYE ON EARTH

The accompanying image is of the world's largest open-pit gold mine, located near Kalgoorlie, Australia. Known as the Super Pit, it originally consisted of a number of small underground mines that were consolidated into a single open-pit mine. Each year, about 28 tons of gold are extracted from the 15 million tons of rock that are shattered by blasting and then transported to the surface.

QUESTION 1 *Calculate the average percentage of gold extracted from the rock removed each year from the Super Pit.*

QUESTION 2 *What is one environmental advantage that underground mining has over open-pit mining?*

QUESTION 3 *If you had been employed at this mine, what change in working conditions would you have seen as it evolved from an underground mine to an open-pit mine?*

McPHOTO/AGE Fotostock

of electricity because the valence electrons are free to move from one atom to another. Metals are also *malleable*, which means they can be hammered into thin sheets, and *ductile*, which means they can be drawn into thin wires. By contrast, ionic and covalent solids tend to be brittle and fracture when stress is applied. Consider the difference between dropping a metal frying pan and a ceramic plate onto a concrete floor.

2.3 CONCEPT CHECKS

1 What is the difference between an atom and an ion?
2 What occurs in an atom to produce a positive ion? A negative ion?
3 Briefly distinguish among ionic, covalent, and metallic bonding and describe the role that electrons play in each.

2.4 | PROPERTIES OF A MINERAL

List and describe the properties that are used in mineral identification.

Minerals have definite crystalline structures and chemical compositions that give them unique sets of physical and chemical properties shared by all samples of that mineral. For example, all specimens of halite have the same hardness, have the same density, and break in a similar manner. Because a mineral's internal structure and chemical composition are difficult to determine without the aid of sophisticated tests and equipment, the more easily recognized physical properties are frequently used in identification.

Optical Properties

Of the many optical properties of minerals, their luster, their ability to transmit light, their color, and their streak are most frequently used for mineral identification.

Luster   The appearance or quality of light reflected from the surface of a mineral is known as **luster**. Minerals that have the appearance of metals, regardless of color, are said to have a *metallic luster* (**FIGURE 2.11**). Some metallic minerals, such as native copper and galena, develop a dull coating, or tarnish, when exposed to the atmosphere. Because they

FIGURE 2.11 Metallic versus Submetallic Luster
(Photo courtesy of E. J. Tarbuck)

The freshly broken sample of galena (right) displays a metallic luster, while the sample on the left is tarnished and has a submetallic luster.

FIGURE 2.13 Streak

A. Fluorite　　　　**B. Quartz**

 SmartFigure 2.12 Color Variations in Minerals Some minerals, such as fluorite and quartz, occur in a variety of colors. (Photo A by Dennis Tasa; photo B by E. J. Tarbuck)

are not as shiny as samples with freshly broken surfaces, these samples are often said to exhibit a *submetallic luster*.

Most minerals have a *nonmetallic luster* and are described using adjectives such as *vitreous* or *glassy*. Other nonmetallic minerals are described as having a *dull*, or *earthy*, *luster* (a dull appearance like soil) or a *pearly luster* (such as a pearl or the inside of a clamshell). Still others exhibit a *silky luster* (like satin cloth) or a *greasy luster* (as though coated in oil).

Ability to Transmit Light Another optical property used in the identification of minerals is the ability to transmit light. When no light is transmitted, the mineral is described as *opaque*; when light, but not an image, is transmitted through a mineral sample, the mineral is said to be *translucent*. When both light and an image are visible through the sample, the mineral is described as *transparent*.

Color Although **color** is generally the most conspicuous characteristic of any mineral, it is considered a diagnostic property of only a few minerals. Slight impurities in the common mineral quartz, for example, give it a variety of tints, including pink, purple, yellow, white, gray, and even black (**FIGURE 2.12**). Other minerals, such as tourmaline, also exhibit a variety of hues, with multiple colors sometimes occurring in the same sample. Thus, the use of color as a means of identification is often ambiguous or even misleading.

Streak The color of a mineral in powdered form, called **streak**, is often useful in identification. A mineral's streak is obtained by rubbing the mineral across a *streak plate* (a piece of unglazed porcelain) and observing the color of the mark it leaves (**FIGURE 2.13**). Although the color of a mineral may vary from sample to sample, its streak is usually consistent in color.

Streak can also help distinguish between minerals with metallic luster and those with nonmetallic luster. Metallic minerals generally have a dense, dark streak, whereas minerals with nonmetallic luster typically have a light-colored streak.

It should be noted that not all minerals produce a streak when rubbed across a streak plate. For example, the mineral

Although the color of a mineral is not always helpful in identification, the streak, which is the color of the powdered mineral, can be very useful.

quartz is harder than a porcelain streak plate. Therefore, no streak is observed for quartz.

Crystal Shape, or Habit

Mineralogists use the term **crystal shape**, or **habit**, to refer to the common or characteristic shape of individual crystals or aggregates of crystals. Some minerals tend to grow equally in all three dimensions, whereas others tend to be elongated in one direction or flattened if growth in one dimension is suppressed. A few minerals have crystals that exhibit regular polygons that are helpful in their identification. For example, magnetite crystals sometimes occur as octahedrons, garnets often form dodecahedrons, and halite and fluorite crystals tend to grow as cubes or near cubes. While minerals tend to have one common crystal shape, a few have two or more characteristic crystal shapes, such as the pyrite samples shown in **FIGURE 2.14**.

In addition, some mineral samples consist of numerous intergrown crystals that exhibit characteristic shapes that

FIGURE 2.14 Common Crystal Shapes of Pyrite

Although most minerals exhibit only one common crystal shape, some, such as pyrite, have two or more characteristic habits.

Dennis Tasa

SmartFigure 2.15 Common Crystal Habits **A.** Thin, rounded crystals that break into fibers. **B.** Elongated crystals that are flattened in one direction. **C.** Minerals that have stripes or bands of different color or texture. **D.** Groups of crystals that are shaped like cubes.

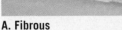

A. Fibrous
E.J. Tarbuck

B. Bladed
Dennis Tasa

C. Banded
Dennis Tasa

D. Cubic crystals
Dennis Tasa

are useful for identification. Terms commonly used to describe these and other crystal habits include *equant* (equidimensional), *bladed*, *fibrous*, *tabular*, *prismatic*, *platy*, *blocky*, *banded*, *granular*, and *botryoidal*. Some of these habits are pictured in **FIGURE 2.15**.

Mineral Strength

How easily minerals break or deform under stress is determined by the type and strength of the chemical bonds that hold the crystals together. Mineralogists use terms including *tenacity*, *hardness*, *cleavage*, and *fracture* to describe mineral strength and how minerals break when stress is applied.

Hardness One of the most useful diagnostic properties is **hardness**, a measure of the resistance of a mineral to abrasion or scratching. This property can be determined by rubbing a mineral of unknown hardness against one of known hardness or vice versa. A numerical value of hardness can be obtained by using the **Mohs scale** of hardness, which consists of 10 minerals arranged in order from 1 (softest) to 10 (hardest), as shown in **FIGURE 2.16A**. It should be noted that the Mohs scale is a relative ranking, and it does not imply that mineral number 2, gypsum, is twice as hard as mineral 1, talc. In fact, gypsum is only slightly harder than talc, as **FIGURE 2.16B** indicates.

In the laboratory, other common objects can be used to determine the hardness of a mineral. These include a human fingernail, which has a hardness of about 2.5, a copper penny (3.5), and a piece of glass (5.5). The mineral gypsum, which has a hardness of 2, can be easily scratched with a fingernail. On the other hand, the mineral calcite, which has a hardness of 3, will scratch a fingernail but will not scratch glass. Quartz, one of the hardest common minerals, will easily scratch glass. Diamonds, hardest of all, scratch anything, including other diamonds.

Cleavage In the crystal structure of many minerals, some atomic bonds are weaker than others. It is along these weak bonds that minerals tend to break when they are stressed. **Cleavage** (*kleiben* = carve) is the tendency of a mineral to break (cleave)

SmartFigure 2.16 Hardness Scales **A.** The Mohs scale of hardness, showing the hardness of some common objects. **B.** Relationship between the Mohs relative hardness scale and an absolute hardness scale.

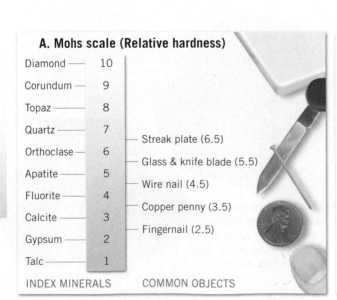
A. Mohs scale (Relative hardness)

INDEX MINERALS		COMMON OBJECTS
Diamond	10	
Corundum	9	
Topaz	8	
Quartz	7	
Orthoclase	6	Streak plate (6.5)
Apatite	5	Glass & knife blade (5.5)
Fluorite	4	Wire nail (4.5)
Calcite	3	Copper penny (3.5)
Gypsum	2	Fingernail (2.5)
Talc	1	

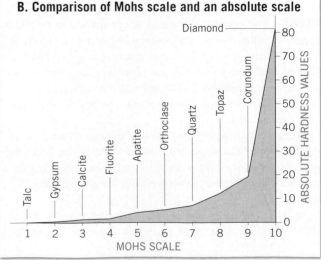
B. Comparison of Mohs scale and an absolute scale

(graph with MOHS SCALE on x-axis 1–10 and ABSOLUTE HARDNESS VALUES on y-axis 0–80, labeled with Talc, Gypsum, Calcite, Fluorite, Apatite, Orthoclase, Quartz, Topaz, Corundum, Diamond)

FIGURE 2.17 Micas Exhibit Perfect Cleavage The thin sheets shown here exhibit one plane of cleavage. (Photo by Chip Clark/ Fundamental Photographs, NYC)

along planes of weak bonding. Not all minerals have cleavage, but those that do can be identified by the relatively smooth, flat surfaces that are produced when the mineral is broken.

The simplest type of cleavage is exhibited by the micas (**FIGURE 2.17**). Because these minerals have very weak bonds in one direction, they cleave to form thin, flat sheets. Some minerals have excellent cleavage in one, two, three, or more directions,

whereas others exhibit fair or poor cleavage, and still others have no cleavage at all. When minerals break evenly in more than one direction, cleavage is described by *the number of cleavage directions and the angle(s) at which they meet* (**FIGURE 2.18**).

Each cleavage surface that has a different orientation is counted as a different direction of cleavage. For example, some minerals cleave to form six-sided cubes. Because cubes are defined by three different sets of parallel planes that intersect at 90-degree angles, cleavage is described as *three directions of cleavage that meet at 90 degrees*.

Do not confuse cleavage with crystal shape. When a mineral exhibits cleavage, it will break into pieces that all have the same geometry. By contrast, the smooth-sided quartz crystals shown in Figure 2.1 do not have cleavage. If broken, they fracture into shapes that do not resemble one another or the original crystals.

Fracture Minerals that have chemical bonds that are equally, or nearly equally, strong in all directions exhibit a property called **fracture**. When minerals fracture, most produce uneven surfaces and are described as exhibiting *irregular fracture*. However, some minerals, such as quartz, break into smooth, curved surfaces resembling broken glass. Such breaks are called *conchoidal fractures* (**FIGURE 2.19**). Still other minerals exhibit fractures that produce splinters or fibers that are referred to as *splintery fracture* and *fibrous fracture*, respectively.

Tenacity The term **tenacity** describes a mineral's resistance to breaking, bending, cutting, or other forms of

A. Cleavage in one direction.
Example: Muscovite

B. Cleavage in two directions at 90° angles.
Example: Feldspar

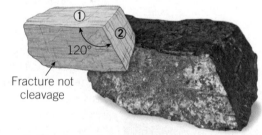

C. Cleavage in two directions not at 90° angles.
Example: Hornblende

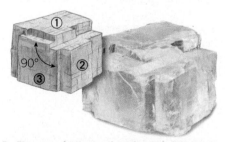

D. Cleavage in three directions at 90° angles.
Example: Halite

E. Cleavage in three directions not at 90° angles.
Example: Calcite

F. Cleavage in four directions.
Example: Fluorite

 SmartFigure 2.18 Cleavage Directions Exhibited by Minerals (Photos by E. J. Tarbuck and Dennis Tasa)

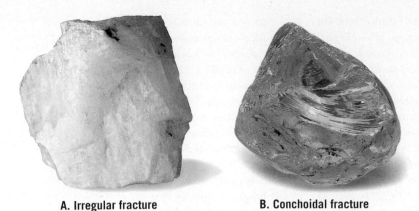

A. Irregular fracture   B. Conchoidal fracture

FIGURE 2.19 Irregular versus Conchoidal Fracture (Photos by Dennis Tasa and E. J. Tarbuck)

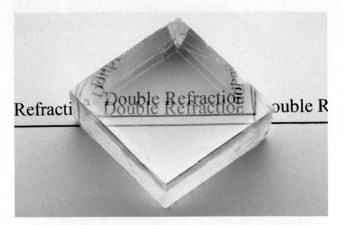

FIGURE 2.20 Double Refraction This sample of calcite exhibits double refraction. (Photo by Chip Clark/Fundamental Photographs, NYC)

deformation. Nonmetallic minerals, such as quartz and halite, tend to be *brittle* and fracture or exhibit cleavage when struck. By contrast, native metals such as copper and gold are *malleable*, and can be hammered into different shapes. In addition, minerals, including gypsum and talc, that can be cut into thin shavings are described as *sectile*. Still others, notably the micas, are *elastic* and will bend and snap back to their original shape after stress is released.

Density and Specific Gravity

Density, an important property of matter, is defined as mass per unit volume. Mineralogists often use a related measure called **specific gravity** to describe the density of minerals. Specific gravity is a number representing the ratio of a mineral's weight to the weight of an equal volume of water.

Most common minerals have a specific gravity between 2 and 3. For example, quartz has a specific gravity of 2.65. By contrast, some metallic minerals, such as pyrite, native copper, and magnetite, are more than twice as dense and thus have more than twice the specific gravity of quartz. Galena, an ore of lead, has a specific gravity of roughly 7.5, whereas the specific gravity of 24-karat gold is approximately 20.

With a little practice, you can estimate the specific gravity of a mineral by hefting it in your hand. Does this mineral feel about as "heavy" as similar sized rocks you have handled? If the answer is "yes," the specific gravity of the sample will likely be between 2.5 and 3.

Other Properties of Minerals

FIGURE 2.21 Calcite Reacting with a Weak Acid (Photo by Chip Clark/Fundamental Photographs, NYC)

In addition to the properties discussed thus far, some minerals can be recognized by other distinctive properties. For example, halite is ordinary salt, so it can be quickly identified

through taste. Talc and graphite both have distinctive feels: Talc feels soapy, and graphite feels greasy. Further, the streaks of many sulfur-bearing minerals smell like rotten eggs. A few minerals, such as magnetite (see Figure 2.31F), have a high iron content and can be picked up with a magnet, while some varieties (such as lodestone) are natural magnets and will pick up small iron-based objects such as pins and paper clips.

Moreover, some minerals exhibit special optical properties. For example, when a transparent piece of calcite is placed over printed text, the letters appear twice. This optical property is known as *double refraction* (**FIGURE 2.20**).

One very simple chemical test involves placing a drop of dilute hydrochloric acid from a dropper bottle onto a freshly broken mineral surface. Using this technique, certain minerals, called carbonates, will effervesce (fizz) as carbon dioxide gas is released (**FIGURE 2.21**). This test is especially useful in identifying the common carbonate mineral calcite.

2.4 CONCEPT CHECKS

1. Define *luster*.
2. Why is color not always a useful property in mineral identification? Give an example of a mineral that supports your answer.
3. What differentiates cleavage from fracture?
4. What do we mean when we refer to a mineral's *tenacity*? List three terms that describe tenacity.
5. What simple chemical test is useful in the identification of the mineral calcite?

2.5 | MINERAL GROUPS List the common silicate and nonsilicate minerals and describe the characteristics of each group.

More than 4000 minerals have been named, and several new ones are identified each year. Fortunately for students who are beginning to study minerals, no more than a few dozen are abundant! Collectively, these few make up most of the

rocks of Earth's crust and, as such, they are often referred to as the **rock-forming minerals**.

Although less abundant, many other minerals are used extensively in the manufacture of products and are called

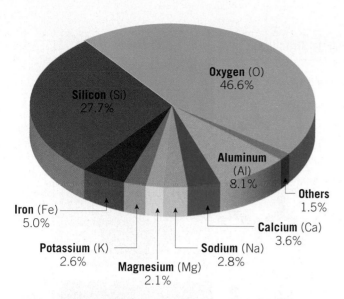

FIGURE 2.22 The Eight Most Abundant Elements in the Continental Crust

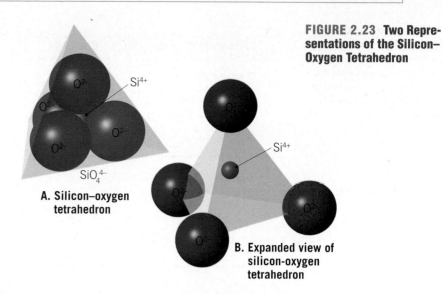

FIGURE 2.23 Two Representations of the Silicon–Oxygen Tetrahedron

economic minerals. However, rock-forming minerals and economic minerals are not mutually exclusive groups. When found in large deposits, some rock-forming minerals are economically significant. One example is the mineral calcite, which is the primary component of the sedimentary rock limestone and has many uses, including being used in the production of cement.

It is worth noting that *only eight elements* make up the vast majority of the rock-forming minerals and represent more than 98 percent (by weight) of the continental crust (**FIGURE 2.22**). These elements, in order of abundance from most to least, are oxygen (O), silicon (Si), aluminum (Al), iron (Fe), calcium (Ca), sodium (Na), potassium (K), and magnesium (Mg). As shown in Figure 2.22, silicon and oxygen are by far the most common elements in Earth's crust. Furthermore, these two elements readily combine to form the basic "building block" for the most common mineral group, the **silicates**. More than 800 silicate minerals are known, and they account for more than 90 percent of Earth's crust.

Because other mineral groups are far less abundant in Earth's crust than the silicates, they are often grouped together under the heading **nonsilicates**. Although not as common as silicates, some nonsilicate minerals are very important economically. They provide us with iron and aluminum to build automobiles, gypsum for plaster and drywall for home construction, and copper wire that carries electricity and connects us to the Internet. Some common nonsilicate mineral groups include the carbonates, sulfates, and halides. In addition to their economic importance, these mineral groups include members that are major constituents in sediments and sedimentary rocks.

Silicate Minerals

Each of the silicate minerals contains oxygen and silicon atoms. Except for a few silicate minerals such as quartz, most silicate minerals also contain one or more additional elements in their crystalline structure. These elements give rise to the great variety of silicate minerals and their varied properties.

All silicates have the same fundamental building block, the **silicon–oxygen tetrahedron** (*tetra* = four, *hedra* = a base). This structure consists of four oxygen atoms surrounding a much smaller silicon atom, as shown in **FIGURE 2.23**. In some minerals, the tetrahedra are joined into chains, sheets, or three-dimensional networks by sharing oxygen atoms (**FIGURE 2.24**). These larger silicate structures are then connected to one another by other elements. The primary elements that join silicate structures are iron (Fe), magnesium (Mg), potassium (K), sodium (Na), and calcium (Ca).

Major groups of silicate minerals and common examples are given in Figure 2.24. The *feldspars* are by far the most plentiful group, comprising over 50 percent of Earth's crust. *Quartz*, the second-most-abundant mineral in the continental crust, is the only common mineral made completely of silicon and oxygen.

Notice in Figure 2.24 that each mineral *group* has a particular silicate *structure*. A relationship exists between this internal structure of a mineral and the *cleavage* it exhibits. Because

EYE ON EARTH

Glass bottles, like most other manufactured products, contain substances obtained from minerals extracted from Earth's crust and oceans. The primary ingredient in commercially produced glass bottles is the mineral quartz. Glass also contains lesser amounts of the mineral calcite. (Photo by Chris Brignell/Shutterstock)

QUESTION 1 *In what mineral group does quartz belong?*

QUESTION 2 *Glass beer bottles are usually clear, green, or brown. Based on what you know about how the mineral quartz is colored, what do glass manufacturers do to get green and brown bottles?*

QUESTION 3 *Why do you suppose some brewers start using brown bottles rather than the green bottles that were popular until the 1930s?*

FIGURE 2.24 Common Silicate Minerals Note that the complexity of the silicate structure increases from the top of the chart to the bottom. (Photos by Dennis Tasa and E. J. Tarbuck)

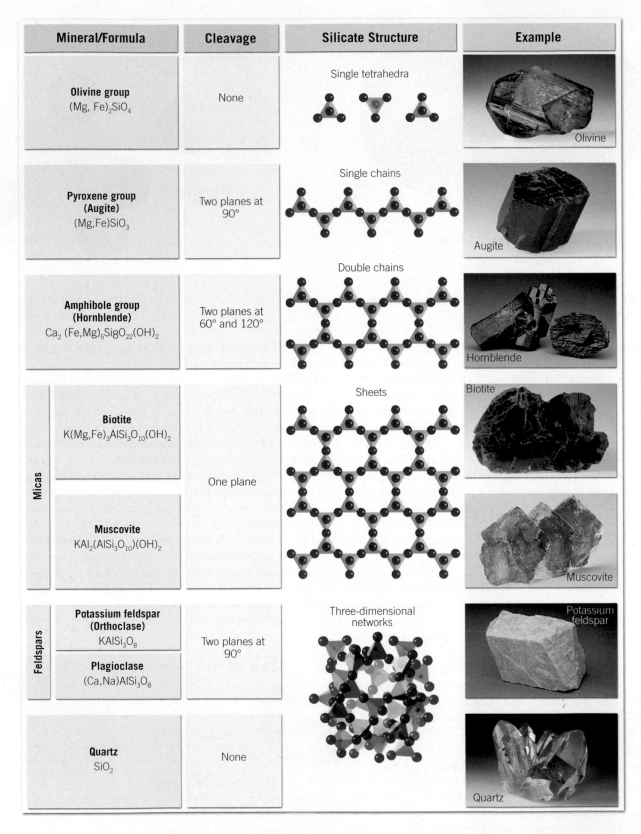

Mineral/Formula	Cleavage	Silicate Structure	Example
Olivine group $(Mg, Fe)_2SiO_4$	None	Single tetrahedra	Olivine
Pyroxene group (Augite) $(Mg,Fe)SiO_3$	Two planes at 90°	Single chains	Augite
Amphibole group (Hornblende) $Ca_2(Fe,Mg)_5Si8O_{22}(OH)_2$	Two planes at 60° and 120°	Double chains	Hornblende
Micas **Biotite** $K(Mg,Fe)_3AlSi_3O_{10}(OH)_2$	One plane	Sheets	Biotite
Micas **Muscovite** $KAl_2(AlSi_3O_{10})(OH)_2$	One plane	Sheets	Muscovite
Feldspars **Potassium feldspar (Orthoclase)** $KAlSi_3O_8$	Two planes at 90°	Three-dimensional networks	Potassium feldspar
Feldspars **Plagioclase** $(Ca,Na)AlSi_3O_8$	Two planes at 90°	Three-dimensional networks	
Quartz SiO_2	None		Quartz

the silicon–oxygen bonds are strong, silicate minerals tend to cleave between the silicon–oxygen structures rather than across them. For example, the micas have a sheet structure and thus tend to cleave into flat plates (see muscovite in Figure 2.18A). Quartz, which has equally strong silicon–oxygen bonds in all directions, has no cleavage but fractures instead.

How do silicate minerals form? Most of them crystallize from molten rock as it cools. This cooling can occur at or near Earth's surface (low temperature and pressure) or at great depths (high temperature and pressure). The *environment during crystallization* and the *chemical composition of the molten rock* mainly determine which minerals are produced.

For example, the silicate mineral olivine crystallizes at high temperatures (about 1200°C [2200°F]), whereas quartz crystallizes at much lower temperatures (about 700°C [1300°F]).

In addition, some silicate minerals form at Earth's surface from the weathered products of other silicate minerals. Clay minerals are an example. Still other silicate minerals are formed under the extreme pressures associated with mountain building. Each silicate mineral, therefore, has a structure and a chemical composition that *indicate the conditions under which it formed*. Thus, by carefully examining the mineral makeup of rocks, geologists can often determine the circumstances under which the rocks formed.

We will now examine some of the most common silicate minerals, which are divided into two major groups based on their chemical makeup.

Common Light Silicate Minerals   The most common light silicate minerals are the feldspars, quartz, muscovite, and the clay minerals. Generally light in color and having a specific gravity of about 2.7, **light silicate minerals** contain varying amounts of aluminum, potassium, calcium, and sodium.

The most abundant mineral group, the *feldspars*, are found in many igneous, sedimentary, and metamorphic rocks (**FIGURE 2.25**). One group of feldspar minerals contains potassium ions in its crystalline structure and is referred to

A. Smoky quartz   **B. Rose quartz**

C. Milky quartz   **D. Jasper**

FIGURE 2.26 Quartz Is One of the Most Common Minerals and Has Many Varieties A. Smoky quartz is commonly found in coarse-grained igneous rocks. **B.** Rose quartz owes its color to small amounts of titanium. **C.** Milky quartz often occurs in veins that occasionally contain gold. **D.** Jasper is a variety of quartz composed of minute crystals. (Photos by Dennis Tasa and E. J. Tarbuck)

as *potassium feldspar*. The other group, called *plagioclase feldspar*, contains calcium and/or sodium ions (see Figure 2.25). All feldspar minerals have two directions of cleavage that meet at 90-degree angles and are relatively hard (6 on the Mohs scale). The only reliable way to physically differentiate the feldspars is to look for striations that are present on some cleavage surfaces of plagioclase feldspar (see Figure 2.25) but do not appear in potassium feldspar.

Quartz is a major constituent of many igneous, sedimentary, and metamorphic rocks. Found in a wide variety of colors (caused by impurities), quartz is quite hard (7 on the Mohs scale) and exhibits conchoidal fracture when broken (**FIGURE 2.26**). Pure quartz is clear, and if allowed to grow without interference, it will develop hexagonal crystals with pyramid-shaped ends (see Figure 2.1).

Another light silicate mineral, *muscovite*, is an abundant member of the mica family and has excellent cleavage in one direction. Muscovite is relatively soft (2.5 to 3 on the Mohs scale).

Clay minerals are light silicates that typically form as products of chemical weathering of igneous rocks. They make up much of the surface material we call soil, and nearly half of the volume of sedimentary rocks is composed of clay minerals. Kaolinite is a common clay mineral formed from the weathering of feldspar (**FIGURE 2.27**).

Potassium Feldspar

A. Potassium feldspar crystal (orthoclase)   **B. Potassium feldspar showing cleavage (orthoclase)**

Plagioclase Feldspar

C. Sodium-rich plagioclase feldspar (albite)   **D. Plagioclase feldspar showing striations (labradorite)**

FIGURE 2.25 Feldspar Minerals   A. Characteristic crystal form of potassium feldspar. **B.** Like this sample, most salmon-colored feldspar belongs to the potassium feldspar subgroup. **C.** Most sodium-rich plagioclase feldspar is light colored and has a porcelain luster. **D.** Calcium-rich plagioclase feldspar tends to be gray, blue-gray, or black in color. Labradorite, the variety shown here, exhibits striations on one of its crystal faces. (Photos by Dennis Tasa and E. J. Tarbuck)

Dennis Tasa

FIGURE 2.27 Kaolinite Kaolinite is a common clay mineral formed by weathering of feldspar minerals.

Olivine-rich peridotite (variety dunite)

FIGURE 2.28 Olivine Commonly black to olive green in color, olivine has a glassy luster and a granular appearance. Olivine is commonly found in the igneous rock basalt.

Common Dark Silicate Minerals The **dark silicate minerals** contain iron and magnesium in their crystalline structures and include the pyroxenes, amphiboles, olivine, biotite, and garnet. Iron gives the dark silicates their color and contributes to their high specific gravity, which is between 3.2 and 3.6, significantly greater than the specific gravity of the light silicate minerals.

Olivine is an important group of dark silicate minerals that are major constituents of dark-colored igneous rocks. Abundant in Earth's upper mantle, olivine is black to olive green in color, has a glassy luster, and often forms small crystals which gives it a granular appearance (**FIGURE 2.28**).

The *pyroxenes* are a group of dark silicate minerals that are important components of dark-colored igneous rocks. The most common member, *augite*, is a black, opaque mineral with two directions of cleavage that meet at nearly 90-degree angles (**FIGURE 2.29A**).

The *amphibole group*, the most common of which is *hornblende*, is usually dark green to black in color (**FIGURE 2.29B**). Except for its cleavage angles, which are about 60 degrees and 120 degrees, hornblende is very similar in appearance to augite. Found in igneous rocks, hornblende makes up the dark portion of otherwise light-colored rocks.

Biotite is the dark, iron-rich member of the mica family. Like other micas, biotite possesses a sheet structure that gives it excellent cleavage in one direction. Biotite's shiny appearance helps distinguish it from other dark silicate minerals. Like hornblende, biotite is a common constituent of most light-colored igneous rocks, including granite.

Another dark silicate is *garnet* (**FIGURE 2.30**). Much like olivine, garnet has a glassy luster, lacks cleavage, and exhibits conchoidal fracture. Although the colors of garnet are varied, the mineral is most often brown to deep red, and in transparent form, it is used as a gemstone.

FIGURE 2.29 Augite and Hornblende These dark-colored silicate minerals are common constituents of a variety of igneous rocks. (Photos by E. J. Tarbuck)

A. Augite

B. Hornblende

Important Nonsilicate Minerals

Nonsilicate minerals are typically divided into groups, based on the negatively charged ion or complex ion that the members have in common. For example, the *oxides* contain the negative oxygen ions (O^{-2}), which are bonded to one or more kinds of positive ions. Thus, within each mineral group, the basic structure and type of bonding is similar. As a result, the minerals in each group have similar physical properties that are useful in mineral identification.

Although the nonsilicates make up only about 8 percent of Earth's crust, some minerals, such as gypsum, calcite, and halite, occur as constituents in sedimentary rocks in significant amounts. Furthermore, many others are important economically. **TABLE 2.1** lists some of the nonsilicate mineral groups and a few examples of each. A brief

FIGURE 2.30 Well-formed Garnet Crystal Garnets come in a variety of colors and are commonly found in mica-rich metamorphic rocks. (Photo by E. J. Tarbuck)

TABLE 2.1 Common Nonsilicate Mineral Groups

Mineral Groups (key ion[s] or element[s])	Mineral Name	Chemical Formula	Economic Use
Carbonates (CO_3^{22})	Calcite	$CaCO_3$	Portland cement, lime
	Dolomite	$CaMg(CO_3)_2$	Portland cement, lime
Halides (Cl^{12}, F^{12}, Br^{12})	Halite	$NaCl$	Common salt
	Fluorite	CaF_2	Used in steelmaking
	Sylvite	KCl	Fertilizer
Oxides (O^{22})	Hematite	Fe_2O_3	Ore of iron, pigment
	Magnetite	Fe_3O_4	Ore of iron
	Corundum	Al_2O_3	Gemstone, abrasive
	Ice	H_2O	Solid form of water
Sulfides (S^{22})	Galena	PbS	Ore of lead
	Sphalerite	ZnS	Ore of zinc
	Pyrite	FeS_2	Sulfuric acid production
	Chalcopyrite	$CuFeS_2$	Ore of copper
	Cinnabar	HgS	Ore of mercury
Sulfates (SO_4^{22})	Gypsum	$CaSO_4 \cdot 2H_2O$	Plaster
	Anhydrite	$CaSO_4$	Plaster
	Barite	$BaSO_4$	Drilling mud
Native elements (single elements)	Gold	Au	Trade, jewelry
	Copper	Cu	Electrical conductor
	Diamond	C	Gemstone, abrasive
	Sulfur	S	Sulfa drugs, chemicals
	Graphite	C	Pencil lead, dry lubricant
	Silver	Ag	Jewelry, photography
	Platinum	Pt	Catalyst

A. Calcite

B. Dolomite

C. Halite

D. Gypsum

E. Hematite

F. Magnetite

G. Galena

H. Chalcopyrite

I. Fluorite

FIGURE 2.31 Important Nonsilicate Minerals (Photos by Dennis Tasa and E. J. Tarbuck)

discussion of a few of the most common nonsilicate minerals follows.

Some of the most common nonsilicate minerals belong to one of three classes of minerals: the carbonates (CO_3^{-2}), the sulfates (SO_4^{-2}), and the halides (Cl^{-1}, F^{-1}, Br^{-1}). The carbonate minerals are much simpler structurally than the silicates. These minerals are *calcite*, $CaCO_3$ (calcium carbonate), and *dolomite*, $CaMg(CO_3)_2$ (calcium/magnesium carbonate) (**FIGURE 2.31A, B**). Calcite and dolomite are usually found together as the primary constituents in the sedimentary rocks limestone and dolostone. When calcite is the dominant mineral, the rock is called *limestone*, whereas *dolostone* results from a predominance of dolomite. Limestone has many uses, including as road aggregate, as building stone, and as the main ingredient in Portland cement.

Two other nonsilicate minerals frequently found in sedimentary rocks are *halite* and *gypsum* (**FIGURE 2.31C, D**). Both minerals are commonly found in thick layers that are the last vestiges of ancient seas that have long since evaporated (**FIGURE 2.32**). Like limestone, both halite and gypsum are important nonmetallic resources. Halite is the mineral name for common table salt (NaCl). Gypsum ($CaSO_4 \cdot 2H_2O$), which is calcium sulfate with water bound into the

FIGURE 2.32 Thick Bed of Halite Exposed in an Underground Mine Halite (salt) mine in Grand Saline, Texas. Note the person for scale. (Photo by Tom Bochsler/ Pearson Education)

structure, is the mineral of which plaster and other similar building materials are composed.

Most nonsilicate mineral classes contain members that are prized for their economic value. This includes the oxides, whose members *hematite* and *magnetite* are important ores of iron (see **FIGURE 2.31E, F**). Also significant are the sulfides, which are basically compounds of sulfur (S) and one or more metals. Examples of important sulfide minerals include galena (lead), sphalerite (zinc), and chalcopyrite (copper). In addition, native elements—including gold, silver, and carbon (diamonds)—plus a host of other nonsilicate minerals—fluorite (flux in making steel), corundum (gemstone, abrasive), and uraninite (a uranium source)—are important economically.

2.6 | NATURAL RESOURCES Discuss Earth's natural resources in terms of renewability. Differentiate between mineral resources and ore deposits.

Earth's crust and oceans are the source of a wide variety of useful and valuable materials. From the first use of clay to make pottery nearly 10,000 years ago, the use of Earth materials has expanded and contributed to societies becoming more complex. Materials we extract from Earth are the basis of modern civilization. Mineral and energy resources from Earth's crust are the raw materials from which we make all the products we use.

Natural resources are typically grouped into broad categories according to their ability to be regenerated (renewable or nonrenewable) or their origin, or type. Here we will consider mineral resources. However, other indispensable natural resources exist, including air, water, and solar energy.

Renewable Versus Nonrenewable Resources

Resources classified as **renewable** can be replenished over relatively short time spans. Common examples are corn used for food and for making ethanol, natural fibers such as cotton for clothing, and forest products for lumber and paper. Energy from flowing water, wind, and the Sun are also considered renewable (**FIGURE 2.33**).

By contrast, many other basic resources are classified as **nonrenewable**. Important metals such as iron, aluminum, and copper fall into this category, as do our most important fuels: oil, natural gas, and coal. Although these and other resources form continuously, the processes that create them are so slow that significant deposits take millions of years to accumulate. Thus, for all practical purposes, Earth contains fixed quantities of these substances. The present supplies will be depleted as they are mined or pumped from the ground. Although some nonrenewable resources, such as the aluminum we use for containers, can be recycled, others, such as the oil used for fuel, cannot.

Mineral Resources

Today, practically every manufactured product contains materials obtained from minerals. Table 2.1 lists some of the most economically important mineral groups. **Mineral resources** are those occurrences of useful minerals that are formed in such quantities that eventual extraction is reasonably certain. Mineral resources include deposits of metallic minerals that can be presently extracted profitably, as well as known deposits that are not yet economically or technologically recoverable. Materials used for such purposes as building stone, road aggregate, abrasives, ceramics, and fertilizers are not usually called mineral resources; rather, they are classified as industrial rocks and minerals.

An **ore deposit** is a naturally occurring concentration of one or more metallic minerals that can be extracted economically. In common usage, the term *ore* is also applied to some nonmetallic minerals such as fluorite and sulfur. Recall that more than

FIGURE 2.33 Solar Energy Is Renewable These solar collectors focus sunlight onto collection pipes filled with a fluid. The heat is used to make steam that drives turbines used to generate electricity. (Photo by Jim West/Alamy)

98 percent of Earth's crust is composed of only eight elements, and except for oxygen and silicon, all other elements make up a relatively small fraction of common crustal rocks (see Figure 2.22). Indeed, the natural concentrations of many elements are exceedingly small. A deposit containing the average concentration of an element such as gold has no economic value because the cost of extracting it greatly exceeds the value of the gold that could be recovered.

In order to have economic value, an ore deposit must be highly concentrated. For example, copper makes up about 0.0135 percent of the crust. For a deposit to be considered a copper ore, it must contain a concentration of copper that is about 100 times this amount, or 2.35 percent. Aluminum, on the other hand, represents 8.13 percent of the crust and can be extracted profitably when it is found in concentrations about four times that amount.

It is important to understand that due to economic or technological changes, a deposit may either become profitable to extract or lose its profitability. If the demand for a metal increases and its value rises sufficiently, the status of a previously unprofitable deposit can be upgraded from a mineral to an ore. Technological advances that allow a resource to be extracted more efficiently and, thus, more profitably than before may also trigger a change of status.

Conversely, changing economic factors can turn what was once a profitable ore deposit into an unprofitable mineral deposit. This situation was illustrated at the copper mining operation located at Bingham Canyon, Utah, one of the largest open-pit mines on Earth (**FIGURE 2.34**). Mining was halted there in 1985 because outmoded equipment had driven the cost of extracting the copper beyond the current selling price. The mine owners responded by replacing an antiquated 1000-car railroad with more modern conveyor belts and dump trucks for efficiently transporting the ore and waste. The advanced equipment accounted for a cost reduction of nearly 30 percent, ultimately returning the copper mine operation to profitability. Today Bingham Canyon Mine produces about 25 percent of the refined copper in the United States.

Over the years, geologists have been keenly interested in learning how natural processes produce localized concentrations of essential minerals. One well-established fact is that occurrences of valuable mineral resources are closely related to the rock cycle. That is, the mechanisms that generate igneous, sedimentary, and metamorphic rocks, including the processes of weathering and erosion, play major roles in producing concentrated accumulations of useful elements.

Moreover, with the development of the theory of plate tectonics, geologists have added another tool for understanding the processes by which one rock is transformed into another. As these rock-forming processes are examined in the following chapters, we consider their role in producing some of our important mineral resources.

2.6 CONCEPT CHECKS

1. List three examples of renewable resources and three examples of nonrenewable resources.
2. Compare and contrast a *mineral resource* and an *ore deposit*.
3. Explain how a mineral deposit that previously could not be mined profitably might be upgraded to an ore deposit.

FIGURE 2.34 Aerial View of the Bingham Canyon Copper Mine Near Salt Lake City, Utah Although the amount of copper in the rock is less than 0.5 percent, the huge volume of material removed and processed each day (over 250,000 tons) yields enough metal to be profitable. In addition to copper, this mine produces various amounts of gold, silver, and molybdenum. (Photo by Michael Collier)

Gemstones

Precious stones have been prized since antiquity. Although most gemstones are varieties of a particular mineral, misinformation abounds regarding gems and their mineral makeup.

Important Gemstones

Gemstones are classified in one of two categories: precious or semiprecious. Precious gems are rare and generally have hardnesses that exceed 9 on the Mohs scale. Therefore, they are more valuable and thus more expensive than semiprecious gems.

GEM	MINERAL NAME		PRIZED HUES
PRECIOUS			
Diamond	Diamond	●●●	Colorless, pinks, blues
Emerald	Beryl	●	Greens
Ruby	Corundum	●	Reds
Sapphire	Corundum	●	Blues
Opal	Opal	●	Brilliant hues
SEMIPRECIOUS			
Alexandrite	Chrysoberyl		Variable
Amethyst	Quartz	●	Purples
Cat's-eye	Chrysoberyl	●	Yellows
Chalcedony	Quartz (agate)	●	Banded
Citrine	Quartz	●	Yellows
Garnet	Garnet	●●	Red, greens
Jade	Jadeite or nephrite	●	Greens
Moonstone	Feldspar	●	Transparent blues
Peridot	Olivine	●	Olive greens
Smoky quartz	Quartz	●	Browns
Spinel	Spinel	●	Reds
Topaz	Topaz	●●	Purples, reds
Tourmaline	Tourmaline	●●	Reds, blue-greens
Turquoise	Turquoise	●	Blues
Zircon	Zircon	●	Reds

Reuters

The Famous Hope Diamond

The deep-blue Hope Diamond is a 45.52-carat gem that is thought to have been cut from a much larger 115-carat stone discovered in India in the mid-1600s. The original 115-carrat stone was cut into a smaller gem that became part of the crown jewels of France and was in the possession of King Louis XVI and Marie Antoinette before they attempted to escape France. Stolen during the French Revolution in 1792, the gem is thought to have been recut to its present size and shape. In the 1800s, it became part of the collection of Henry Hope (hence its name) and is on display at the Smithsonian in Washington, DC.

What Constitutes a Gemstone?

When found in their natural state, most gemstones are dull and would be passed over by most people as "just another rock." Gems must be cut and polished by experienced professionals before their true beauty is displayed. Cutting and polishing is accomplished using abrasive material, most often tiny fragments of diamonds that are embedded in a metal disk.

Naming Gemstones

Most precious stones are given names that differ from their parent mineral. For example, sapphire is one of two gems that are varieties of the same mineral, corundum. Trace elements can produce vivid sapphires of nearly every color. Tiny amounts of titanium and iron in corundum produce the most prized blue sapphires. When the mineral corundum contains a sufficient quantity of chromium, it exhibits a brilliant red color. This variety of corundum is called ruby.

Shutterstock

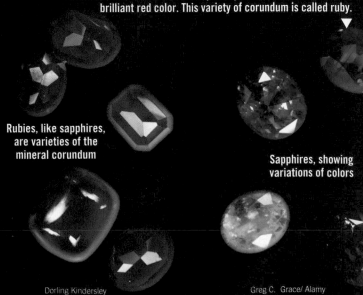

Rubies, like sapphires, are varieties of the mineral corundum

Sapphires, showing variations of colors

? Why are diamonds used as an abrasive material to cut and polish gemstones?

Charles D. Winters/ Photo Researchers, Inc.

Dorling Kindersley

Greg C. Grace/ Alamy

2.1 MINERALS: BUILDING BLOCKS OF ROCKS

List the main characteristics that an Earth material must possess to be considered a mineral and describe each.

KEY TERMS: mineralogy, mineral, rock

- In Earth science, the word *mineral* refers to naturally occurring inorganic solids that possess an orderly crystalline structure and a characteristic chemical composition. The study of minerals is mineralogy.
- Minerals are the building blocks of rocks. Rocks are naturally occurring masses of minerals or mineral-like matter, such as natural glass or organic material.

2.2 ATOMS: BUILDING BLOCKS OF MINERALS

Compare and contrast the three primary particles contained in atoms.

KEY TERMS: atom, nucleus, proton, neutron, electron, valence electron, atomic number, element, periodic table, chemical compound

- Minerals are composed of atoms of one or more elements. The atoms of any element consist of the same three basic ingredients: protons, neutrons, and electrons.
- The number of protons in an atom is its atomic number. For example, an oxygen atom has eight protons, so its atomic number is eight. Protons and neutrons are approximately the same size and mass, but while protons are positively charged, neutrons have no charge.
- Electrons are much smaller than both protons and neutrons, and they weigh about 2000 times less. Each electron has a negative charge, equal in magnitude to the positive charge of a proton. Electrons swarm around an atom's nucleus at a distance, in several distinctive energy levels called principal shells. The electrons in the outermost principal shell, called valence electrons, are important when one atom bonds with other atoms to form chemical compounds.
- Elements with similar numbers of valence electrons tend to behave in similar ways. The periodic table displays these similarities in its graphical arrangement of the elements.

Q Use the periodic table to identify these geologically important elements by their number of protons: (A) 14, (B) 6, (C) 13, (D) 17, and (E) 26.

2.3 WHY ATOMS BOND

Distinguish among ionic bonds, covalent bonds, and metallic bonds.

KEY TERMS: octet rule, chemical bond, ion, ionic bond, covalent bond, metallic bond

- When atoms attract to other atoms, they can form chemical bonds, which generally involve the transfer or sharing of valence electrons. For most atoms, the most stable arrangement is to have eight electrons in the outermost principal shell. This idea is called the octet rule.

- Ionic bonds involve atoms of one element giving up electrons to atoms of another element, forming positively and negatively charged atoms called ions. Positively charged ions bond with negative ions to form ionic bonds.
- Covalent bonds involve the sharing of electrons between two adjacent atoms. The forces that holds these atoms together arises from the attraction of oppositely charged particles—protons in the nuclei and the electrons shared by the atoms.
- In metallic bonds, the sharing is more extensive: Electrons can freely move from one atom to the next throughout the entire mass.

Q Which of the situations in the diagram shows ionic bonding? What are its distinguishing characteristics?

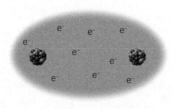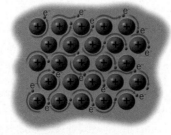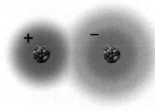

2.4 PHYSICAL PROPERTIES OF MINERALS

List and describe the properties that are used in mineral identification.

KEY TERMS: luster, color, streak, crystal shape (habit), hardness, Mohs scale, cleavage, fracture, tenacity, density, specific gravity

- The composition and internal crystalline structure of a mineral give it specific physical properties. These properties can be used to tell minerals apart from each other and make minerals useful for specific human tasks.

- Luster is a mineral's ability to reflect light. The terms *transparent*, *translucent*, and *opaque* are used to describe the degree to which a mineral can transmit light. Color is an unreliable characteristic for identification, as slight impurities can "stain" minerals into misleading colors. A trustworthy characteristic is streak, the color of the powder generated by scraping a mineral against a porcelain streak plate.
- The shape that a crystal assumes as it grows is often useful for identification.
- Variations in the strength of chemical bonds give minerals properties such as tenacity (whether a mineral breaks in a brittle fashion or bends when stressed) and hardness (resistance to being scratched). Cleavage, the preferential breakage of a mineral along planes of weakly bonded atoms, is often useful in identifying minerals.
- The amount of matter packed into a given volume determines a mineral's density. To compare the density of minerals, mineralogists find it simplest to use a related property, specific gravity, which is the ratio between a mineral's density and the density of water.
- Other properties are diagnostic for certain minerals but rare in most others. A particular smell, taste, or feel; reaction to hydrochloric acid; magnetism; and double refraction are examples.

Quartz   Dennis Tasa   Calcite   Dennis Tasa

Q Research the minerals *quartz* and *calcite*. List three physical properties that may be used to distinguish one from the other.

2.5 MINERAL GROUPS

List the common silicate and nonsilicate minerals and describe the characteristics of each group.

KEY TERMS: rock-forming mineral, economic mineral, silicate, nonsilicate, silicon–oxygen tetrahedron, light silicate mineral, dark silicate mineral

- Silicate minerals have a basic building block in common: a small pyramid-shaped structure that consists of one silicon atom surrounded by four oxygen atoms. Because this structure has four sides, it is called the silicon–oxygen tetrahedron. Neighboring tetrahedra can share some of their oxygen atoms, causing them to develop long chains or sheet structures.
- Silicate minerals make up the most common mineral class on Earth. They are subdivided into minerals that contain iron and/or magnesium (dark silicates) and those that do not (light silicates). The light silicate minerals are generally light in color and have relatively low specific gravities. Feldspar, quartz, muscovite, and clay minerals are examples. The dark silicate minerals are generally dark in color and relatively dense. Olivine, pyroxene, amphibole, biotite, and garnet are examples.
- Nonsilicate minerals include oxides, which contain oxygen ions that bond to other elements (usually metals); carbonates, which have CO_3 as a critical part of their crystal structure; sulfates, which have SO_4 as their basic building block; and halides, which contain a nonmetal ion such as chlorine, bromine, or fluorine that bonds to a metal ion such as sodium or calcium.

2.6 NATURAL RESOURCES

Discuss Earth's natural resources in terms of renewability. Differentiate between mineral resources and ore deposits.

KEY TERMS: renewable, nonrenewable, mineral resource, ore deposit

- Resources are classified as renewable when they can be replenished over short time spans and nonrenewable when they can't.
- Ore deposits are naturally occurring concentrations of one or more metallic minerals that can be extracted economically using current technology. A mineral resource can be upgraded to an ore deposit if the price of the commodity increases sufficiently or when the cost of extraction decreases.

GIVE IT SOME **THOUGHT**

1. Using the geologic definition of *mineral* as your guide, determine which of the items in this list are minerals and which are not. If something in this list is not a mineral, explain.

   a. Gold nugget   **d.** Cubic zirconia   **g.** Glacial ice

   b. Seawater   **e.** Obsidian   **h.** Amber

   c. Quartz   **f.** Ruby

Refer to the periodic table of the elements (see Figure 2.5) to help you answer Questions 2 and 3.

2. Assume that the number of protons in a neutral atom is 92 and its mass number is 238.

   a. What is the name of the element?

   b. How many electrons does it have?

   c. How many neutrons does it have?

3. Which of the following elements is more likely to form chemical bonds: xenon (Xe) or sodium (Na)? Explain why.

4. Referring to the accompanying photos of five minerals, determine which of these specimens exhibit a metallic luster and which have a nonmetallic luster. (Photos by Dennis Tasa and E. J. Tarbuck)

A.   B.   C.

D.   E.

5. Gold has a specific gravity of almost 20. A 5-gallon bucket of water weighs 40 pounds. How much would a 5-gallon bucket of gold weigh?

6. Examine the accompanying photo of a mineral that has several smooth, flat surfaces that resulted when the specimen was broken.

   a. How many flat surfaces are present on this specimen?

   b. How many different directions of cleavage does this specimen have?

   c. Do the cleavage directions meet at 90-degree angles?

Dennis Tasa

Cleaved sample

7. Each of the following statements describes a silicate mineral or mineral group. In each case, provide the appropriate name.

   a. The most common member of the amphibole group

   b. The most common light-colored member of the mica family

   c. The only common silicate mineral made entirely of silicon and oxygen

   d. A silicate mineral with a name that is based on its color

   e. A silicate mineral that is characterized by striations

   f. A silicate mineral that originates as a product of chemical weathering

8. What mineral property is illustrated in the accompanying photo?

Dennis Tasa

9. Do an Internet search to determine what minerals are extracted from the ground during the manufacture of the following products.

   a. Stainless steel utensils

   b. Cat litter

   c. Tums brand antacid tablets

   d. Lithium batteries

   e. Aluminum beverage cans

10. Most states have designated a state mineral, rock, or gemstone to promote interest in the state's natural resources. Describe your state mineral, rock, or gemstone and explain why it was selected. If your state does not have a state mineral, rock, or gemstone, complete the exercise by selecting one from a state adjacent to yours.

EXAMINING THE **EARTH SYSTEM**

1. Perhaps one of the most significant interrelationships between humans and the Earth system involves the extraction, refinement, and distribution of the planet's mineral wealth. To help you understand these associations, begin by thoroughly researching a mineral commodity that is mined in your local region or state. (You might find useful the information at the U.S. Geological Survey [USGS] Website: **http://minerals.er.usgs.gov/minerals/pubs/state/.**) What products are made from this mineral? Do you use any of these products? Describe the mining and refining of the mineral and the local impact these processes have on each of Earth's spheres (atmosphere, hydrosphere, geosphere, and biosphere). Are any of the effects negative? If so, what, if anything, is being done to end or minimize the damage?

2. Referring to the mineral you described in Question 1, in your opinion does the environmental impact of extracting this mineral outweigh the benefits derived from its products?

MasteringGeology™

Looking for additional review and test prep materials? Visit the Self Study area in **www.masteringgeology.com** to find practice quizzes, study tools, and multimedia that will aid in your understanding of this chapter's content. In **MasteringGeology™** you will find:

- GEODe: Earth Science: An interactive visual walkthrough of key concepts
- Geoscience Animation Library: More than 100 animations illuminating many difficult-to-understand Earth science concepts

- In The News RSS Feeds: Current Earth science events and news articles are pulled into the site with assessment
- Pearson eText
- Optional Self Study Quizzes
- Web Links
- Glossary
- Flashcards

3

Rocks: Materials of the Solid Earth

Each statement represents the primary **LEARNING OBJECTIVE** for the corresponding major heading within the chapter. After you complete the chapter, you should be able to:

3.1 Sketch, label, and explain the rock cycle.

3.2 Describe the two criteria used to classify igneous rocks and explain how cooling influences the crystal size of minerals.

3.3 List and describe the different categories of sedimentary rocks and discuss the processes that change sediment into sedimentary rock.

3.4 Define *metamorphism*, explain how metamorphic rocks form, and describe the agents of metamorphism.

3.5 Distinguish between metallic and nonmetallic mineral resources and list at least two examples of each. Compare and contrast the three traditional fossil fuels.

Sedimentary strata in Canyonlands National Park, Utah with the La Sal. (SeBuKi/Alamy)

Why study rocks? You have already learned that some rocks and minerals have great economic value. In addition, all Earth processes depend in some way on the properties of these basic Earth materials. Events such as volcanic eruptions, mountain building, weathering, erosion, and even earthquakes involve rocks and minerals. Consequently, a basic knowledge of Earth materials is essential to understanding most geologic phenomena.

Every rock contains clues about the environment in which it formed. For example, some rocks are composed entirely of small shell fragments. This tells Earth scientists that the rock likely originated in a shallow marine environment. Other rocks contain clues which indicate that they formed from a volcanic eruption or deep in the Earth during mountain building. Thus, rocks contain a wealth of information about events that have occurred over Earth's long history.

3.1 | EARTH AS A SYSTEM: THE ROCK CYCLE

Sketch, label, and explain the rock cycle.

Earth as a system is illustrated most vividly when we examine the rock cycle. The **rock cycle** allows us to see many of the interactions among the numerous components and processes of the Earth system (**FIGURE 3.1**). It helps us understand the origins of igneous, sedimentary, and metamorphic rocks and how they are connected. In addition, the rock cycle demonstrates that any rock type, under the right circumstances, can be transformed into any other type.

The Basic Cycle

We begin our discussion of the rock cycle with molten rock, called *magma*, which forms by melting that occurs primarily within Earth's crust and upper mantle (see Figure 3.1). Once formed, a magma body often rises toward the surface because it is less dense than the surrounding rock. Occasionally, magma reaches Earth's surface, where it erupts as *lava.* Eventually, molten rock cools and solidifies, a process called *crystallization* or *solidification.* Molten rock may solidify either beneath the surface or, following a volcanic eruption, at the surface. In either situation, the resulting rocks are called *igneous rocks.*

If igneous rocks are exposed at the surface, they undergo *weathering*, in which the daily influences of the atmosphere slowly disintegrate and decompose rocks. The loose materials that result are often moved downslope by gravity and then picked up and transported by one or more erosional agents—running water, glaciers, wind, or waves. Eventually, these particles and dissolved substances, called *sediment*, are deposited. Although most sediment ultimately comes to rest in the ocean, other sites of deposition include river floodplains, desert basins, lakes, inland seas, and sand dunes.

Next, the sediments undergo *lithification*, a term meaning "conversion into rock." Sediment is usually lithified into *sedimentary rock* when compacted by the weight of overlying materials or when cemented as percolating groundwater fills the pores with mineral matter.

If the resulting sedimentary rock becomes deeply buried or is involved in the dynamics of mountain building, it will be subjected to great pressures and intense heat. The sedimentary rock may react to the changing environment by turning into the third rock type, *metamorphic rock.* If metamorphic rock is subjected to still higher temperatures, it may melt, creating magma, and the cycle begins again.

Although rocks may appear to be stable, unchanging masses, the rock cycle shows that they are not. The changes, however, take time—sometimes millions or even billions of years. In addition, the rock cycle operates continuously around the globe, but in different stages, depending on the location. Today, new magma is forming under the island of Hawaii, whereas the rocks that comprise the Colorado Rockies are slowly being worn down by weathering and erosion. Some of this weathered debris will eventually be carried to the Gulf of Mexico, where it will add to the already substantial mass of sediment that has accumulated there.

Alternative Paths

Rocks do not necessarily go through the cycle in the order just described. Other paths are also possible. For example, rather than being exposed to weathering and erosion at Earth's surface, igneous rocks may remain deeply buried (see Figure 3.1). Eventually, these masses may be subjected to the strong compressional forces and high temperatures associated with mountain building. When this occurs, they are transformed directly into metamorphic rocks.

Metamorphic and sedimentary rocks, as well as sediment, do not always remain buried. Rather, overlying layers may be eroded away, exposing the once-buried rock. When this happens, the material is attacked by weathering processes and turned into new raw materials for sedimentary rocks.

In a similar manner, igneous rocks that formed at depth can be uplifted, weathered, and turned into sedimentary rocks. Alternatively, igneous rocks may remain at depth, where the high temperatures and forces associated with mountain building may metamorphose or even melt them. Over time, rocks may be transformed into any other rock type, or even into a different form of the original type. Rocks may take many paths through the rock cycle.

ROCK CYCLE

Viewed over long time spans, rocks are constantly forming, changing, and reforming.

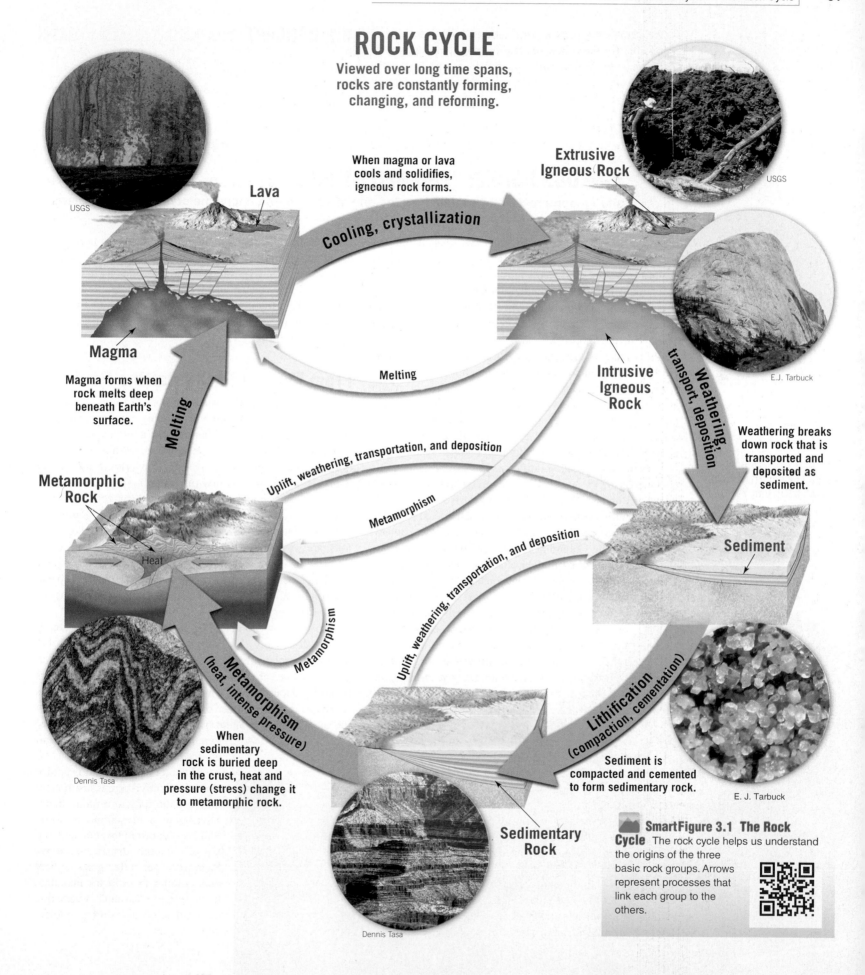

USGS

Lava

When magma or lava cools and solidifies, igneous rock forms.

Extrusive Igneous Rock

USGS

Cooling, crystallization

Magma

Magma forms when rock melts deep beneath Earth's surface.

Melting

Melting

Intrusive Igneous Rock

E.J. Tarbuck

Weathering, transport, deposition

Weathering breaks down rock that is transported and deposited as sediment.

Metamorphic Rock

Heat

Uplift, weathering, transportation, and deposition

Metamorphism

Uplift, weathering, transportation, and deposition

Sediment

Metamorphism

Dennis Tasa

Metamorphism (heat, intense pressure)

When sedimentary rock is buried deep in the crust, heat and pressure (stress) change it to metamorphic rock.

Lithification (compaction, cementation)

Sediment is compacted and cemented to form sedimentary rock.

E. J. Tarbuck

Sedimentary Rock

Dennis Tasa

SmartFigure 3.1 The Rock Cycle The rock cycle helps us understand the origins of the three basic rock groups. Arrows represent processes that link each group to the others.

What drives the rock cycle? Earth's internal heat is responsible for the processes that form igneous and metamorphic rocks. Weathering and the transport of weathered material are external processes, powered by energy from the Sun. External processes produce sedimentary rocks.

3.1 CONCEPT CHECKS

1   Sketch and label the rock cycle. Make sure your sketch includes alternative paths.

2   Use the rock cycle to explain the statement "One rock is the raw material for another."

3.2 | IGNEOUS ROCKS: "FORMED BY FIRE" Describe the two criteria used to classify igneous rocks and explain how the rate of cooling influences the crystal size of minerals.

In the discussion of the rock cycle, we pointed out that **igneous rocks** form as *magma* or *lava* cools and crystallizes. **Magma** is molten rock that is most often generated by melting of rocks in Earth's mantle, although melting of crustal rock generates some magma. Once formed, a magma body buoyantly rises toward the surface because it is less dense than the surrounding rocks.

When magma reaches the surface, it is called **lava** (**FIGURE 3.2**). Sometimes, lava is emitted as fountains produced when escaping gases propel molten rock skyward. On other occasions, magma is explosively ejected from vents, producing a spectacular eruption such as the 1980 eruption of Mount St. Helens. However, most eruptions are not violent; rather, volcanoes most often emit quiet outpourings of lava.

When molten rock solidifies *at the surface*, the resulting igneous rocks are classified as **extrusive**, or **volcanic** (after the Roman fire god, Vulcan). Extrusive igneous rocks are abundant in western portions of the Americas, including the volcanic cones of the Cascade Range and the extensive lava flows of the Columbia Plateau. In addition, many oceanic islands, including the Hawaiian islands, are composed almost entirely of volcanic igneous rocks.

As we will see later, most magma loses its mobility before reaching Earth's surface and eventually crystallizes deep below the surface. Igneous rocks that *form at depth* are termed **intrusive**, or **plutonic** (after Pluto, the god of the underworld in classical mythology).

FIGURE 3.2 Fluid Basaltic Lava Emitted from Hawaii's Kilauea Volcano Kilauea, on the Big Island, is one of the most active volcanoes on Earth. (Photo courtesy of USGS)

Intrusive igneous rocks remain at depth unless portions of the crust are uplifted and the overlying rocks are stripped away by erosion. Exposures of intrusive igneous rocks occur in many places, including Mount Washington, New Hampshire; Stone Mountain, Georgia; Mount Rushmore in the Black Hills of South Dakota; and Yosemite National Park, California (**FIGURE 3.3**).

From Magma to Crystalline Rock

Magma is molten rock (*melt*) composed of ions of the elements found in silicate minerals, mainly silicon and oxygen that move about freely. Magma also contains gases, particularly water vapor, that are confined within the magma body by the weight (pressure) of the overlying rocks, and it may contain some solids (mineral crystals). As magma cools, the once-mobile ions begin to arrange themselves into orderly patterns—a process called *crystallization*. As cooling continues, numerous small crystals develop, and ions are systematically added to these centers of crystal growth. When the crystals grow large enough for their edges to meet, their growth ceases because of lack of space. Eventually, all the liquid is transformed into a solid mass of interlocking crystals.

The rate of cooling strongly influences crystal size. If magma cools very slowly, ions can migrate over great distances. Consequently, *slow cooling results in the formation of fewer, larger crystals*. On the other hand, if cooling occurs rapidly, the ions lose their motion and quickly combine. This results in a large number of tiny crystals all competing for the available ions. Therefore, *rapid cooling results in the formation of a solid mass of small intergrown crystals*.

If the molten material is cooled almost instantly, there is insufficient time for the ions to arrange themselves into a crystalline network. Solids produced in this manner consist of randomly distributed atoms. Such rocks are called *glass* and are quite similar to ordinary manufactured glass. "Instant" quenching sometimes occurs during violent

FIGURE 3.3 Mount Rushmore National Memorial, Black Hills of South Dakota This memorial is carved from the intrusive igneous rock granite. This massive igneous body cooled very slowly at depth and has since been uplifted, with the overlying rocks stripped away by erosion. (Photo by Barbara A. Harvey/Shutterstock)

volcanic eruptions that produce tiny shards of glass called *volcanic ash*.

In addition to the rate of cooling, the composition of a magma and the amount of dissolved gases influence crystallization. Because magmas differ in each of these aspects, the physical appearance and mineral composition of igneous rocks vary widely.

Igneous Compositions

Igneous rocks are composed mainly of silicate minerals. Chemical analysis shows that silicon and oxygen—usually expressed as the silica (SiO_2) content of a magma—are by far the most abundant constituents of igneous rocks. These two elements, plus ions of aluminum (Al), calcium (Ca), sodium (Na), potassium (K), magnesium (Mg), and iron (Fe), make up roughly 98 percent by weight of most magmas. In addition, magma contains small amounts of many other elements, including titanium and manganese, and trace amounts of much rarer elements, such as gold, silver, and uranium.

As magma cools and solidifies, these elements combine to form two major groups of silicate minerals. The *dark silicates* are rich in iron and/or magnesium and are comparatively low in silica (SiO_2). *Olivine*, *pyroxene*, *amphibole*, and *biotite mica* are the common dark silicate minerals of Earth's crust. By contrast, the *light silicates* contain greater amounts of potassium, sodium, and calcium and are richer in silica than dark silicates. Light silicates include *quartz*, *muscovite mica*, and the most abundant mineral group, the *feldspars*. Feldspars make up at least 40 percent of most igneous rocks.

Thus, in addition to feldspar, igneous rocks contain some combination of the other light and/or dark silicates listed earlier.

Granitic (Felsic) Versus Basaltic (Mafic) Compositions
Despite their great compositional diversity, igneous rocks (and the magmas from which they form) can be divided into broad groups according to their proportions of light and dark minerals (**FIGURE 3.4**). Near one end of the continuum are rocks composed almost entirely of light-colored silicates—quartz and potassium feldspar. Igneous rocks in which these are the dominant minerals have a **granitic composition**. Geologists refer to granitic rocks as being **felsic**, a term derived from *feld*spar and *si*lica (quartz). In addition to quartz and feldspar, most granitic rocks contain about 10 percent dark silicate minerals, usually biotite mica and amphibole. Granitic rocks are major constituents of the continental crust.

Rocks that contain at least 45 percent dark silicate minerals and calcium-rich plagioclase feldspar (but no quartz) are said to have a **basaltic composition**. Basaltic rocks contain a high percentage of dark silicate minerals, so geologists refer to them as **mafic** (from *ma*gnesium and *fer*rum, the Latin word for iron). Because of their iron content, mafic rocks are typically darker and denser than granitic rocks. Basaltic rocks make up the ocean floor as well as many of the volcanic islands located within the ocean basins.

Other Compositional Groups
As you can see in Figure 3.4, rocks with a composition between granitic and basaltic rocks are said to have an **andesitic composition**, or

SmartFigure 3.4 Composition of Common Igneous Rocks (After Dietrich, Daily, and Larsen)

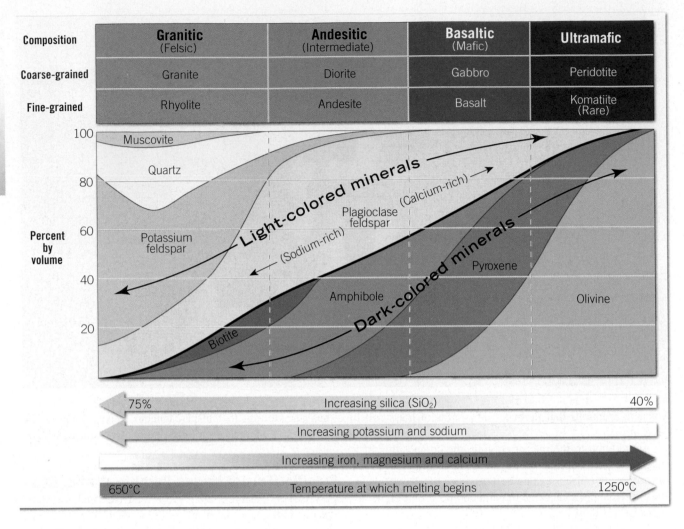

Composition	**Granitic** (Felsic)	**Andesitic** (Intermediate)	**Basaltic** (Mafic)	**Ultramafic**
Coarse-grained	Granite	Diorite	Gabbro	Peridotite
Fine-grained	Rhyolite	Andesite	Basalt	Komatiite (Rare)

Increasing silica (SiO₂) — 75% ← → 40%

Increasing potassium and sodium ←

Increasing iron, magnesium and calcium →

Temperature at which melting begins — 650°C ← → 1250°C

intermediate composition, after the common volcanic rock *andesite*. Intermediate rocks contain at least 25 percent dark silicate minerals, mainly amphibole, pyroxene, and biotite mica, with the other dominant mineral being plagioclase feldspar. This important category of igneous rocks is associated with volcanic activity that is typically confined to the seaward margins of the continents and on volcanic island arcs such as the Aleutian chain.

Another important igneous rock, *peridotite*, contains mostly olivine and pyroxene and thus falls on the opposite side of the compositional spectrum from granitic rocks (see Figure 3.4). Because peridotite is composed almost entirely of ferromagnesian minerals, its chemical composition is referred to as **ultramafic**. Although ultramafic rocks are rare at Earth's surface, peridotite is the main constituent of the upper mantle.

Silica Content as an Indicator of Composition

An important aspect of the chemical composition of igneous rocks is silica (SiO_2) content. Typically, the silica content of crustal rocks ranges from a low of about 40 percent in ultramafic rocks to a high of more than 70 percent in felsic rocks (see Figure 3.4). The percentage of silica in igneous rocks varies in a systematic manner that parallels the abundance of other elements. For example, rocks that are relatively low in silica contain large amounts of iron, magnesium, and calcium. By contrast, rocks high in silica contain very little iron, magnesium, or calcium but are enriched with sodium and potassium. Consequently, the chemical makeup of an igneous rock can be inferred directly from its silica content.

Further, the amount of silica present in magma strongly influences the magma's behavior. Granitic magma, which has a high silica content, is quite viscous ("thick") and may erupt at temperatures as low as 650°C (1200°F). On the other hand, basaltic magmas are low in silica and are generally more fluid. Basaltic magmas also erupt at higher temperatures than granitic magmas—usually at temperatures between 1050° and 1250°C (1920° and 2280°F).

What Can Igneous Textures Tell Us?

Geologists describe the overall appearance of a rock, based on the *size*, *shape*, and *arrangement* of its mineral grains, as its **texture**. Texture is an important property because it allows geologists to make inferences about a rock's origin, based on

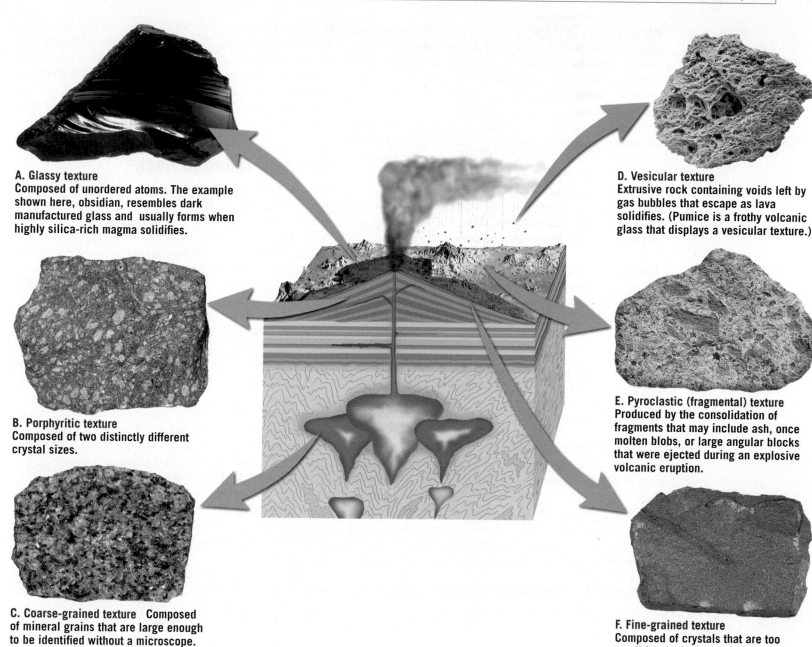

A. Glassy texture
Composed of unordered atoms. The example shown here, obsidian, resembles dark manufactured glass and usually forms when highly silica-rich magma solidifies.

B. Porphyritic texture
Composed of two distinctly different crystal sizes.

C. Coarse-grained texture Composed of mineral grains that are large enough to be identified without a microscope.

D. Vesicular texture
Extrusive rock containing voids left by gas bubbles that escape as lava solidifies. (Pumice is a frothy volcanic glass that displays a vesicular texture.)

E. Pyroclastic (fragmental) texture
Produced by the consolidation of fragments that may include ash, once molten blobs, or large angular blocks that were ejected during an explosive volcanic eruption.

F. Fine-grained texture
Composed of crystals that are too small for the individual minerals to be identified without a microscope.

careful observations of crystal size and other characteristics (**FIGURE 3.5**). Rapid cooling produces small crystals, whereas very slow cooling produces much larger crystals. As you might expect, the rate of cooling is slow in magma chambers that lie deep within the crust, whereas a thin layer of lava extruded upon Earth's surface may chill to form solid rock in a matter of hours. Small molten blobs ejected from a volcano during a violent eruption can solidify in mid-air.

Fine-Grained Texture Igneous rocks that form at Earth's surface or as small intrusive masses within the upper crust, where cooling is relatively rapid, exhibit a **fine-grained texture** (see Figure 3.5F). By definition, the crystals that make up fine-grained igneous rocks are so small that individual minerals can be distinguished only with the aid of a polarizing microscope or other sophisticated techniques. Therefore, we commonly characterize fine-grained rocks as being light, intermediate, or dark in color.

Coarse-Grained Texture When large masses of magma slowly crystallize at great depth, they form igneous rocks that exhibit a **coarse-grained texture**. Coarse-grained rocks consist of a mass of intergrown crystals that are roughly equal in size and large enough so that the individual minerals can be identified without the aid of a microscope (see Figure 3.5C). Geologists often use a small magnifying lens to aid in identifying minerals in coarse-grained igneous rocks.

SmartFigure 3.5
Igneous Rock Textures
(Photos courtesy of E. J. Tarbuck and Dennis Tasa)

Porphyritic Texture

A large mass of magma may require thousands or even millions of years to solidify. Because different minerals crystallize under different environmental conditions (temperatures and pressure), it is possible for crystals of one mineral to become quite large before others even begin to form. If molten rock containing some large crystals moves to a different environment—for example, by erupting at the surface—the remaining liquid portion of the lava cools more quickly. The resulting rock, which has large crystals embedded in a matrix of smaller crystals, is said to have a **porphyritic texture** (see Figure 3.5B). The large crystals in porphyritic rocks are referred to as **phenocrysts** (*pheno* = show, *cryst* = crystal), whereas the matrix of smaller crystals is called **groundmass**.

Vesicular Texture

Common features of many extrusive rocks are the voids left by gas bubbles that escape as lava solidifies. These nearly spherical openings are called *vesicles*, and the rocks that contain them are said to have a **vesicular texture**. Rocks that exhibit a vesicular texture often form in the upper zone of a lava flow, where cooling occurs rapidly enough to preserve the openings produced by the expanding gas bubbles (**FIGURE 3.6**). Another common vesicular rock, called *pumice*, forms when silica-rich lava is ejected during an explosive eruption (see Figure 3.5D).

Glassy Texture

During some volcanic eruptions, molten rock is ejected into the atmosphere, where it is quenched and cools quickly to become a solid. Rapid cooling of this type may generate rocks having a **glassy texture** (see Figure 3.5A). Glass results when unordered ions are "frozen in place" before they are able to unite into an orderly crystalline structure. *Obsidian*, a common type of natural glass, is similar in appearance to dark chunks of manufactured glass.

Pyroclastic (Fragmental) Texture

Another group of igneous rocks is formed from the consolidation of individual rock fragments ejected during explosive volcanic eruptions. The ejected particles might be very fine ash, molten blobs, or large angular blocks torn from the walls of a vent during an eruption. Igneous rocks composed of these rock fragments are said to have a **pyroclastic texture**, or **fragmental texture** (see Figure 3.5E). A common type of pyroclastic rock, called *welded tuff*, is composed of fine fragments of glass that remained hot enough to eventually fuse together.

Common Igneous Rocks

Igneous rocks are classified by their texture and mineral composition. The texture of an igneous rock is mainly a result of its cooling history, whereas its mineral composition is largely a result of the chemical makeup of the parent magma (**FIGURE 3.7**). Because igneous rocks are classified on the basis of both mineral composition and texture, some rocks having similar mineral constituents but exhibiting different textures are given different names.

Granitic (Felsic) Rocks

Granite is a coarse-grained igneous rock that forms where large masses of magma slowly solidify at depth. During episodes of mountain building, granite and related crystalline rocks may be uplifted, with the processes of weathering and erosion stripping away the overlying crust. Areas where large quantities of granite are exposed at the surface include Pikes Peak in the Rockies, Mount Rushmore in the Black Hills, Stone Mountain in Georgia, and Yosemite National Park in the Sierra Nevada (**FIGURE 3.8**).

Granite is perhaps the best-known igneous rock, in part because of its natural beauty, which is enhanced when polished, and partly because of its abundance. Slabs of polished granite are commonly used for tombstones, monuments, and countertops.

Rhyolite is the extrusive equivalent of granite (same chemical composition but different texture) and, likewise, is composed essentially of light-colored silicates (see Figure 3.7). This fact accounts for its color, which is usually buff to pink or light gray. Rhyolite is fine grained and frequently contains glass fragments and voids, indicating rapid cooling in a surface environment. In contrast to granite, which is widely distributed as large intrusive masses, rhyolite deposits are less common and generally less voluminous. Yellowstone Park is one well-known exception where extensive lava flows and thick ash deposits of rhyolitic composition are found.

Obsidian is a common type of natural glass that is similar in appearance to a dark chunk of manufactured glass. Although dark in color, obsidian usually has a higher silica content and a chemical composition similar to that of light-colored igneous rocks.

FIGURE 3.6 Vesicular Texture Vesicles form as gas bubbles escape near the top of a lava flow.

E. J. Tarbuck

Scoria, a volcanic rock with a vesicular texture.

IGNEOUS ROCK CLASSIFICATION CHART

MINERAL COMPOSITION			
Granitic (Felsic)	**Andesitic** (Intermediate)	**Basaltic** (Mafic)	**Ultramafic**

	Granitic (Felsic)	Andesitic (Intermediate)	Basaltic (Mafic)	Ultramafic
Dominant Minerals	Quartz Potassium feldspar	Amphibole Plagioclase feldspar	Pyroxene Plagioclase feldspar	Olivine Pyroxene
Accessory Minerals	Plagioclase feldspar Amphibole Muscovite Biotite	Pyroxene Biotite	Amphibole Olivine	Plagioclase feldspar

TEXTURE

	Granitic (Felsic)	Andesitic (Intermediate)	Basaltic (Mafic)	Ultramafic
Coarse-grained	Granite	Diorite	Gabbro	Peridotite
Fine-grained	Rhyolite	Andesite	Basalt	Komatiite (rare)
Porphyritic (two distinct grain sizes)	Granite porphyry	Andesite porphyry	Basalt porphyry	Uncommon
Glassy	Obsidian	Less common	Less common	Uncommon
Vesicular (contains voids)	Pumice (also glassy)		Scoria	Uncommon
Pyroclastic (fragmental)	Most fragments < 4mm — Tuff or welded tuff		Most fragments > 4mm — Volcanic breccia	Uncommon

Rock Color (based on % of dark minerals)	0% to 25%	25% to 45%	45% to 85%	85% to 100%

SmartFigure 3.7 Classification of Igneous Rocks, Based on Their Mineral Composition and Texture Coarse-grained rocks are plutonic, solidifying deep underground. Fine-grained rocks are volcanic, or solidify as shallow, thin plutons. Ultramafic rocks are dark, dense rocks, composed almost entirely of minerals containing iron and magnesium. Although relatively rare on Earth's surface, these rocks are major constituents of the upper mantle. (Photos by E. J. Tarbuck and Dennis Tasa)

Corey Rich/
Getty Images

Henrik Lehneren/
Glow Images

Granite

FIGURE 3.8 Granitic Rock Exposed in California's Yosemite National Park This rock formed from magma that crystallized deep beneath the surface. (Granite inset photo by E. J. Tarbuck)

FIGURE 3.9 Obsidian, a Natural Glass Native Americans used obsidian to make arrowheads and cutting tools. (Photo by Mark Thiessen/Getty Images)

Obsidian's dark color results from small amounts of metallic ions in an otherwise clear, glassy substance. Because of its excellent conchoidal fracture and ability to hold a sharp, hard edge, obsidian was a prized material from which Native Americans chipped arrowheads and cutting tools (**FIGURE 3.9**).

Another silica-rich volcanic rock that exhibits a glassy and also vesicular texture is *pumice*. Often found with obsidian, pumice forms when large amounts of gas escape from molten rock to generate a gray, frothy mass (**FIGURE 3.10**). In some samples, the vesicles are quite noticeable, whereas in others, the pumice resembles fine shards of intertwined glass. Because of the large volume of air-filled voids, many samples of pumice float in water (see Figure 3.10).

Andesitic (Intermediate) Rocks *Andesite* is a medium-gray, fine-grained rock, typically of volcanic origin. Its name comes from South America's Andes Mountains, where numerous volcanoes are composed of this rock type. In addition, the volcanoes of the Cascade Range and many of the volcanic structures occupying the continental margins that surround the Pacific Ocean have an andesitic composition. Andesite commonly exhibits a porphyritic texture (see Figure 3.7) consisting of phenocrysts that are often light, rectangular crystals of plagioclase feldspar or black, elongated amphibole crystals.

Diorite, the intrusive equivalent of andesite, is a coarse-grained rock that resembles gray granite. However, it can be distinguished from granite because it contains little or no

visible quartz crystals and has a higher percentage of dark silicate minerals.

Basaltic (Mafic) Rocks *Basalt*, the most common extrusive igneous rock, is a very dark green to black, fine-grained volcanic rock composed primarily of pyroxene, olivine, and plagioclase feldspar. Many volcanic islands, such as the Hawaiian islands and Iceland, are composed mainly of basalt (**FIGURE 3.11**). Furthermore, the upper layers of the oceanic crust consist of basalt. In the United States, large portions of central Oregon and Washington were the sites of extensive basaltic outpourings.

← 2 cm →

FIGURE 3.10 Pumice, a Vesicular Glassy Rock Pumice is very lightweight because it contains numerous vesicles. (Photo by E. J. Tarbuck; inset photo by Chip Clark)

FIGURE 3.11 Basaltic Lava Flowing from Kilauea Volcano, Hawaii (Photo by David Reggie/Getty Images)

The coarse-grained, intrusive equivalent of basalt is *gabbro* (see Figure 3.7). Although gabbro is not commonly exposed at the surface, it makes up a significant percentage of the oceanic crust.

How Different Igneous Rocks Form

Because a large variety of igneous rocks exist, it is logical to assume that an equally large variety of magmas also exist. However, geologists have observed that a single volcano may extrude lavas exhibiting quite different compositions. Data of this type led them to examine the possibility that magma might change (evolve) and thus become the parent to a variety of igneous rocks. To explore this idea, a pioneering investigation into the crystallization of magma was carried out by N. L. Bowen in the first quarter of the twentieth century.

Bowen's Reaction Series In a laboratory setting, Bowen demonstrated that magma, with its diverse chemistry, crystallizes over a temperature range of at least 200°C, unlike simple compounds (such as water), which solidify at specific temperatures. As magma cools, certain minerals crystallize first at relatively high temperatures. At successively lower temperatures, other minerals begin to crystallize. This arrangement of minerals, shown in **FIGURE 3.12**, became known as **Bowen's reaction series**.

Bowen discovered that the first mineral to crystallize from a body of magma is *olivine*. Further cooling results in the formation of *pyroxene*, as well as *plagioclase feldspar*. At intermediate temperatures, the minerals *amphibole* and *biotite* begin to crystallize.

During the last stage of crystallization, after most of the magma has solidified, the minerals *muscovite* and *potassium feldspar* may form (see Figure 3.12). Finally, *quartz* crystallizes from any remaining liquid. Olivine and quartz are seldom found in the same igneous rock because quartz crystallizes at much lower temperatures than olivine.

Analysis of igneous rocks provides evidence that this crystallization model approximates what can happen in nature. In particular, we find that minerals that form in the same general range on Bowen's reaction series are found together in the same igneous rocks. For example, notice in Figure 3.12 that the minerals quartz, potassium feldspar, and muscovite, located in the same region of Bowen's

EYE ON EARTH

These two types of hardened lava, called Pele's hair and Pele's tears, are commonly generated during lava fountaining that occurs at Kilauea Volcano. They are named after Pele, the Hawaiian goddess of fire and volcanoes. Pele's hair consists of goldish strands that are created when tiny blobs of lava are stretched by strong winds. When lava droplets cool quickly, they form Pele's tears, which are sometimes connected to strands of Pele's hair.

QUESTION 1 *What is the texture of Pele's hair and Pele's tears?*

QUESTION 2 *What is the igneous rock name for Pele's tears?*

1. Pele's hair   Marli Miller

2. Pele's tears   USGS

FIGURE 3.12 Bowen's Reaction Series This diagram shows the sequence in which minerals crystallize from a magma. Compare this figure to the mineral composition of the rock groups in Figure 3.7. Note that each rock group consists of minerals that crystallize in the same temperature range.

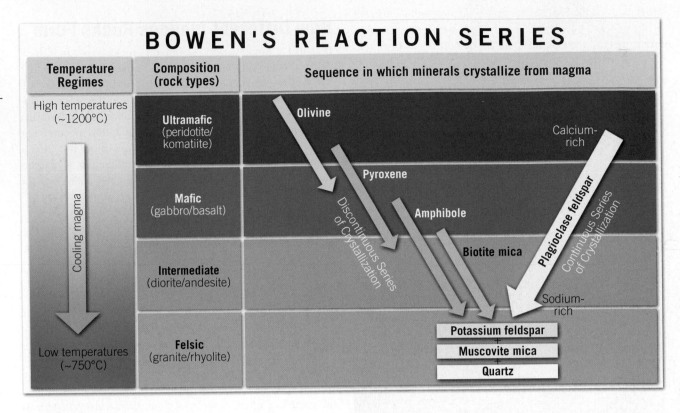

BOWEN'S REACTION SERIES

Temperature Regimes	Composition (rock types)	Sequence in which minerals crystallize from magma
High temperatures (~1200°C)	**Ultramafic** (peridotite/komatiite)	Olivine · Calcium-rich
	Mafic (gabbro/basalt)	Pyroxene · Amphibole · Discontinuous Series of Crystallization · Plagioclase feldspar · Continuous Series of Crystallization
	Intermediate (diorite/andesite)	Biotite mica · Sodium-rich
Low temperatures (~750°C)	**Felsic** (granite/rhyolite)	**Potassium feldspar + Muscovite mica + Quartz**

Cooling magma

diagram, are typically found together as major constituents of the igneous rock *granite*.

Magmatic Differentiation

Bowen demonstrated that different minerals crystallize from magma according to a predictable pattern. But how do Bowen's findings account for the great diversity of igneous rocks? During the crystallization process, the composition of magma continually changes. This occurs because as crystals form, they selectively remove certain elements from the magma,

which leaves the remaining liquid portion (melt) depleted in these elements. Occasionally, separation of the solid and liquid components of magma occurs during crystallization, which creates different sets of minerals and, thus, different types of igneous rocks. One such scenario, called **crystal settling**, occurs when the earlier formed minerals are more dense (heavier) than the liquid portion and sink toward the bottom of the magma chamber, as shown in **FIGURE 3.13**. When the remaining molten material solidifies—either in place or in another location, if it migrates

FIGURE 3.13 Magmatic Differentiation and Crystal Settling Illustration of how a magma evolves as the earliest-formed minerals (those richer in iron, magnesium, and calcium) crystallize and settle to the bottom of the magma chamber, leaving the remaining melt richer in sodium, potassium, and silica (SiO_2).

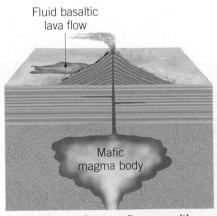

A. A magma having a mafic composition erupts fluid basaltic lavas.

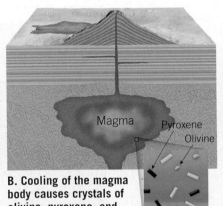

B. Cooling of the magma body causes crystals of olivine, pyroxene, and calcium-rich plagioclase to form and settle out, or crystallize along the magma body's cool margins.

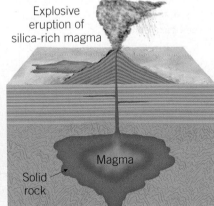

C. The remaining melt will be enriched with silica, and should a subsequent eruption occur, the rocks generated will be more silica-rich and closer to the felsic end of the compositional range than the initial magma.

into fractures in the surrounding rocks—it will form a rock with a chemical composition much different from the parent magma (see Figure 3.13). The formation of one or more secondary magmas from a single parent magma is called **magmatic differentiation**.

At any stage in the evolution of a magma, the solid and liquid components can separate into two chemically distinct units. Furthermore, magmatic differentiation within the secondary magma can generate other chemically distinct masses of molten rock. Consequently, magmatic differentiation and separation of the solid and liquid components at various stages of crystallization can produce several chemically diverse magmas and, ultimately, a variety of igneous rocks.

3.2 CONCEPT CHECKS

1 What is magma? How does magma differ from lava?
2 In what basic settings do intrusive and extrusive igneous rocks originate?
3 How does the rate of cooling influence crystal size? What other factors influence the texture of igneous rocks?
4 What does a porphyritic texture indicate about the history of an igneous rock?
5 List and distinguish among the four basic compositional groups of igneous rocks.
6 How are granite and rhyolite different? In what way are they similar?
7 What is magmatic differentiation? How might this process lead to the formation of several different igneous rocks from a single magma?

3.3 | SEDIMENTARY ROCKS: COMPACTED AND CEMENTED SEDIMENT

List and describe the different categories of sedimentary rocks and discuss the processes that change sediment into sedimentary rock.

Recall the rock cycle, which shows the origin of **sedimentary rocks**. Weathering begins the process. Next, gravity and erosional agents (running water, wind, waves, and glacial ice) remove the products of weathering and carry them to a new location, where they are deposited. Usually, the particles are broken down further during this transport phase. Following deposition, this **sediment** may become lithified, or "turned to rock." Commonly, *compaction* and *cementation* transform the sediment into solid sedimentary rock.

The word *sedimentary* indicates the nature of these rocks, for it is derived from the Latin *sedimentum*, which means "settling," a reference to a solid material settling out of a fluid. Most sediment is deposited in this fashion. Weathered debris is constantly being swept from bedrock and carried away by water, ice, or wind. Eventually, the material is deposited in lakes, river valleys, seas, and countless other places. The particles in a desert sand dune, the mud on the floor of a swamp, the gravels in a streambed, and even household dust are examples of sediment produced by this never-ending process.

The weathering of bedrock and the transport and deposition of the weathering products are continuous. Therefore, sediment is found almost everywhere. As piles of sediment accumulate, the materials near the bottom are compacted by the weight of the overlying layers. Over long periods, these sediments are cemented together by mineral matter deposited from water in the spaces between particles. This forms solid sedimentary rock.

Geologists estimate that sedimentary rocks account for only about 5 percent (by volume) of Earth's outer 16 kilometers (10 miles). However, the importance of this group of rocks is far greater than this percentage implies. If you sampled the rocks exposed at Earth's surface, you would find that the great majority are sedimentary (**FIGURE 3.14**).

 Mobile Field Trip 3.14 Sedimentary Rocks Exposed in Capitol Reef National Park, Utah
About 75 percent of all rock exposures on continents are sedimentary rocks.
(Photo by Michael Collier)

Indeed, about 75 percent of all rock outcrops on the continents are sedimentary. Therefore, we can think of sedimentary rocks as comprising a relatively thin and somewhat discontinuous layer in the uppermost portion of the crust. This makes sense because sediment accumulates at the surface.

It is from sedimentary rocks that geologists reconstruct many details of Earth's history. Because sediments are deposited in a variety of different settings at the surface, the rock layers that they eventually form hold many clues to past surface environments. They may also exhibit characteristics that allow geologists to decipher information about the method and distance of sediment transport. Furthermore, sedimentary rocks contain fossils, which are vital evidence in the study of the geologic past.

Finally, many sedimentary rocks are important economically. Coal, which is burned to provide a significant portion of U.S. electrical energy, is classified as a sedimentary rock. Other major energy resources (such as petroleum and natural gas) occur in pores within sedimentary rocks. Other sedimentary rocks are major sources of iron, aluminum, manganese, and fertilizer, plus numerous materials essential to the construction industry.

Classifying Sedimentary Rocks

Materials that accumulate as sediment have two principal sources. First, sediments may originate as solid particles from weathered rocks, such as the igneous rocks described earlier. These particles are called *detritus*, and the sedimentary rocks they form are called **detrital sedimentary rocks** (**FIGURE 3.15**).

Detrital Sedimentary Rocks Though a wide variety of minerals and rock fragments may be found in detrital rocks, clay minerals and quartz dominate. As you learned earlier, clay minerals are the most abundant product of the chemical weathering of silicate minerals, especially the feldspars. Quartz, on the other hand, is abundant because it is extremely durable and very resistant to chemical weathering. Thus, when igneous rocks such as granite are weathered, individual quartz grains are set free.

Geologists use particle size to distinguish among detrital sedimentary rocks. Figure 3.15 presents the four size categories for particles making up detrital rocks. When gravel-size particles predominate, the rock is called *conglomerate* if the sediment is rounded and *breccia* if the pieces are angular (see Figure 3.15). Angular fragments indicate that the particles were not transported very far from their source prior to deposition and so have not had corners and rough edges abraded. *Sandstone* is the name given to rocks when sand-size grains prevail. *Shale*, the most common sedimentary rock, is made of very fine-grained sediment and composed mainly of clay minerals (see Figure 3.15). *Siltstone*, another rather fine-grained rock, is composed of clay-sized sediment intermixed with slightly larger silt-sized grains.

Using particle size is not only a convenient method of dividing detrital rocks; the sizes of the component grains

FIGURE 3.15 Detrital Sedimentary Rocks (Photos by E. J. Tarbuck and Dennis Tasa)

Detrital Sedimentary Rocks

ClasticTexture (particle size)	Sediment Name	Rock Name
Coarse (over 2 mm)	Gravel (Rounded particles)	Conglomerate 
	Gravel (Angular particles)	Breccia 
Medium (1/16 to 2 mm)	Sand	Sandstone 
		Arkose* 
Fine (1/16 to 1/256 mm)	Silt	Siltstone 
Very fine (less than 1/256 mm)	Clay	Shale or Mudstone 

*If abundant feldspar is present the rock is called arkose.

also provide useful information about the environment in which the sediment was deposited. Currents of water or air sort the particles by size. The stronger the current, the larger the particle size carried. Gravels, for example, are moved by swiftly flowing rivers, rockslides, and glaciers. Less energy is required to transport sand; thus, it is common in wind-blown dunes, river deposits, and beaches. Because silts and clays settle very slowly, accumulations of these materials are generally associated with the quiet waters of a lake, lagoon, swamp, or marine environment.

Although detrital sedimentary rocks are classified by particle size, in certain cases, the mineral composition is also part of naming a rock. For example, most sand-stones are predominantly quartz rich, and they are often referred to as quartz sandstone. In addition, rocks consist-ing of detrital sediments are rarely composed of grains of just one size. Consequently, a rock containing quantities of both sand and silt can be correctly classified as sandy siltstone or silty sandstone, depending on which particle size dominates.

Chemical and Biochemical Sedimentary Rocks

In contrast to detrital rocks, which form from the solid products of weathering, **chemical sedimentary rocks** and **biochemical sedimentary rocks** are derived from material (ions) that is carried in solution to lakes and seas (**FIGURE 3.16**). This material does not remain dissolved in water indefinitely. Under certain conditions, it precipi-tates (settles out) to form *chemical sediments* as a result of physical processes. An example of chemical sediments resulting from physical processes is the salt left behind as a body of saltwater evaporates.

Precipitation may also occur indirectly through life processes of water-dwelling organisms that form materi-als called *biochemical sediments*. Many water-dwelling animals and plants extract dissolved mineral matter to form shells and other hard parts. After the organisms die, their skeletons may accumulate on the floor of a lake or an ocean.

Limestone, an abundant sedimentary rock, is composed chiefly of the mineral calcite ($CaCO_3$). Nearly 90 per-cent of limestone is formed from biochemical sediments secreted by marine organisms, and the remaining amount consists of chemical sediments that precipitated directly from seawater.

One easily identified biochemical limestone is *coquina*, a coarse rock composed of loosely cemented shells and shell fragments (**FIGURE 3.17**). Another less obvious but familiar example is *chalk*, a soft, porous rock made up almost entirely of the hard parts of microscopic organisms that are no larger than the head of a pin. Among the most famous chalk depos-its are the White Chalk Cliffs exposed along the southeast coast of England (**FIGURE 3.18**).

Inorganic limestone forms when chemical changes or high water temperatures increase the concentration of cal-cium carbonate to the point that it precipitates. *Travertine*, the type of limestone that decorates caverns, is one example.

Chemical, Biochemical, and Organic Sedimentary Rocks

Composition		Texture	Rock Name
Calcite, $CaCO_3$		Nonclastic: Fine to coarse crystalline	Crystalline Limestone 
		Nonclastic: Microcrystalline calcite	Microcrystalline Limestone 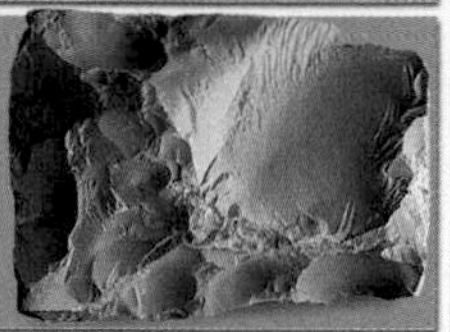
		Nonclastic: Fine to coarse crystalline	Travertine 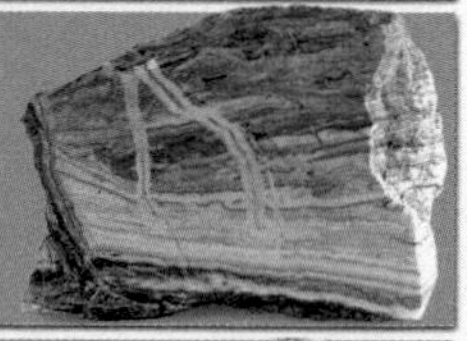
	Biochemical Limestone	Clastic: Visible shells and shell fragments loosely cemented	Coquina 
		Clastic: Various size shells and shell fragments cemented with calcite cement	Fossiliferous Limestone 
		Clastic: Microscopic shells and clay	Chalk 
Quartz, SiO_2		Nonclastic: Very fine crystalline	Chert (light colored) 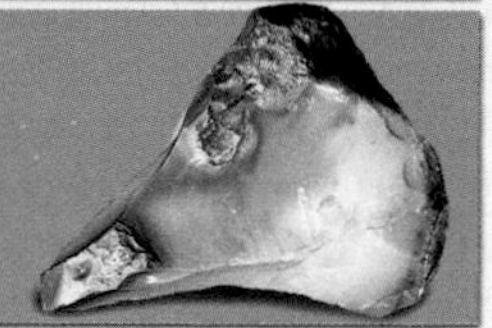
Gypsum $CaSO_4 \bullet 2H_2O$		Nonclastic: Fine to coarse crystalline	Rock Gypsum 
Halite, NaCl		Nonclastic: Fine to coarse crystalline	Rock Salt 
Altered plant fragments (organic)		Nonclastic: Fine-grained organic matter	Bituminous Coal 

FIGURE 3.16 Chemical, Biochemical, and Organic Sedimentary Rocks. (Photos by E. J. Tarbuck and Dennis Tasa)

← 5 cm →

FIGURE 3.17 Coquina
This variety of limestone consists of shell fragments; therefore, it has a biochemical origin. (Rock photo by E. J. Tarbuck; beach photo by Donald R. Frazier Photolibrary, Inc./Alamy)

Groundwater is the source of travertine that is deposited in caves. As water drops reach the air in a cavern, some of the carbon dioxide dissolved in the water escapes, causing calcium carbonate to precipitate.

Dissolved silica (SiO_2) precipitates to form varieties of microcrystalline quartz (**FIGURE 3.19**). Sedimentary rocks composed of microcrystalline quartz include chert (light color), flint (dark), jasper (red), and agate (banded). These chemical sedimentary rocks may have either an inorganic or biochemical origin, but the mode of origin is usually difficult to determine.

Very often, evaporation causes minerals to precipitate from water. Such minerals include halite, the chief component of *rock salt*, and gypsum, the main ingredient of *rock gypsum*. Both materials have significant commercial importance. Halite is familiar to everyone as the common salt used in cooking and seasoning foods. Of course, it has many other uses and has been considered important enough that people have sought, traded, and fought over it for much of human history. Gypsum is the basic ingredient of plaster of Paris. This material is used most extensively in the construction industry for "drywall" and plaster.

In the geologic past, many areas that are now dry land were covered by shallow arms of the sea that had only narrow connections to the open ocean. Under these conditions, water continually moved into the bay to replace

FIGURE 3.18 The White Chalk Cliffs This prominent deposit underlies large portions of southern England as well as parts of northern France. (White Chalk Cliffs photo by BL Images Ltd/Alamy; Coccolithophores photo by Science Photo Library/Alamy Images)

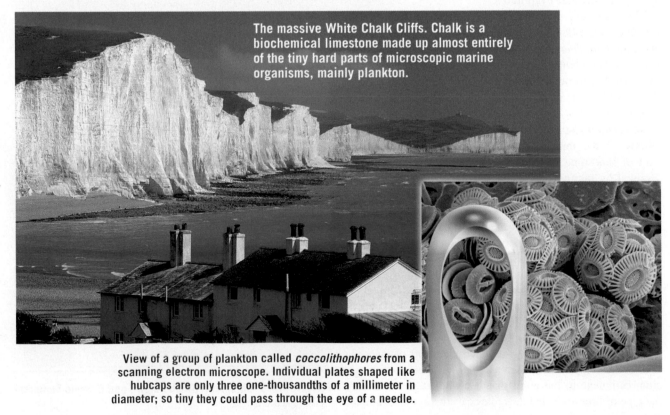

The massive White Chalk Cliffs. Chalk is a biochemical limestone made up almost entirely of the tiny hard parts of microscopic marine organisms, mainly plankton.

View of a group of plankton called *coccolithophores* from a scanning electron microscope. Individual plates shaped like hubcaps are only three one-thousandths of a millimeter in diameter; so tiny they could pass through the eye of a needle.

Agate

Flint

Jasper

Arrowhead

Petrified wood

FIGURE 3.19 Varieties of Chert Chert is the name applied to a number of dense, hard chemical sedimentary rocks made of microcrystalline quartz. (Agate photo by The Natural History Museum/Alamy Images; petrified wood photo by gracious_tiger/Shutterstock; flint and jasper photos by E. J. Tarbuck; arrowhead photo by Daniel Sambraus/Science Source)

water lost by evaporation. Eventually, the waters of the bay became saturated, and salt deposition began. Today, these arms of the sea are gone, and the remaining deposits are called **evaporite deposits**.

On a smaller scale, evaporite deposits can be seen in such places as Death Valley, California. Here, following rains or periods of snowmelt in the mountains, streams flow from surrounding mountains into an enclosed basin. As the water evaporates, dissolved materials left behind as a white crust on the ground form *salt flats* (**FIGURE 3.20**).

Coal—An Organic Sedimentary Rock

Coal, in contrast to sedimentary rocks that are rich in calcite or silica, consists mostly of **organic matter**. Close examination of a piece of coal under a microscope or magnifying glass often reveals plant structures such as leaves, bark, and wood that have been chemically altered but remain identifiable. This supports the conclusion that coal is the end product of the burial of large amounts of plant material over extended periods (**FIGURE 3.21**).

The initial stage in coal formation is the accumulation of large quantities of plant remains. However, special conditions are required for such accumulations because dead plants normally decompose when exposed to the atmosphere. An ideal environment that allows for the accumulation of plant material is a swamp. Because stagnant swamp water is oxygen deficient, complete decay (oxidation) of the

plant material is not possible. At various times during Earth history, such environments have been common. Coal undergoes successive stages of formation. With each successive stage, higher temperatures and pressures drive off impurities and volatiles, as shown in Figure 3.21.

Lignite and bituminous coals are sedimentary rocks, but anthracite is a metamorphic rock. Anthracite forms when sedimentary layers are subjected to the folding and deformation associated with mountain building.

Lithification of Sediment

Lithification refers to the processes by which sediments are transformed into solid sedimentary rocks. One of the most common processes is **compaction** (**FIGURE 3.22**). As

This extensive evaporite deposit is a 30,000-acre expanse of hard white salt that in places is nearly 2 meters thick.

SmartFigure 3.20 Bonneville Salt Flats This well-known Utah site was once a large salt lake. (Photo by Jupiterimages/Glow Images)

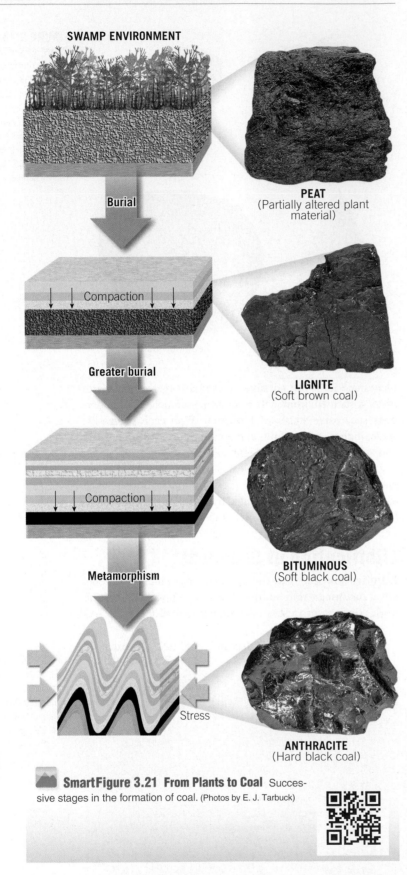

SWAMP ENVIRONMENT

Burial

PEAT
(Partially altered plant material)

Compaction

Greater burial

LIGNITE
(Soft brown coal)

Compaction

Metamorphism

BITUMINOUS
(Soft black coal)

Stress

ANTHRACITE
(Hard black coal)

SmartFigure 3.21 From Plants to Coal Successive stages in the formation of coal. (Photos by E. J. Tarbuck)

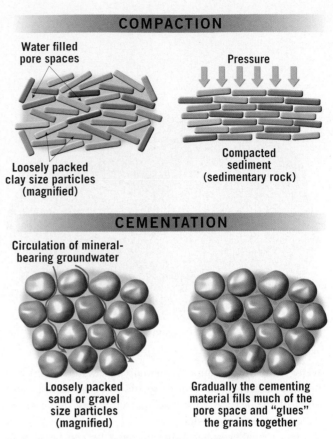

COMPACTION

Water filled pore spaces

Pressure

Loosely packed clay size particles (magnified)

Compacted sediment (sedimentary rock)

CEMENTATION

Circulation of mineral-bearing groundwater

Loosely packed sand or gravel size particles (magnified)

Gradually the cementing material fills much of the pore space and "glues" the grains together

FIGURE 3.22 Compaction and Cementation

several thousand meters of material, the volume of the clay may be reduced as much as 40 percent. Compaction is most effective in converting very fine-grained sediments, such as clay-size particles, into sedimentary rocks.

Because sand and coarse sediments (gravel) are not easily compressed, they are generally transformed into sedimentary rocks by the process of **cementation** (see Figure 3.22). The cementing materials are carried in a water-rich solution that percolates through the pore spaces between particles. Over time, the cement precipitates onto the sediment grains, fills the open spaces, and acts like a "glue" to join the particles together. Calcite, silica, and iron oxide are the most common cements. Identification of the cementing material is simple. Calcite cement will effervesce (fizz) with dilute hydrochloric acid. Silica is the hardest cement and thus produces the hardest sedimentary rocks. When a sedimentary rock has an orange or red color, this usually means iron oxide is present.

Features of Sedimentary Rocks

Sedimentary rocks are particularly important in the study of Earth history. These rocks form at Earth's surface, and as layer upon layer of sediment accumulates, each records the nature of the environment at the time the sediment was deposited. These layers, called **strata**, or **beds**, are the *single most characteristic feature of sedimentary rocks* (see Figure 3.14).

sediments accumulate through time, the weight of overlying material compresses the deeper sediments. As the grains are pressed closer and closer, pore space is greatly reduced. For example, when clays are buried beneath

FIGURE 3.23 Sedimentary Environments A. Ripple marks preserved in sedimentary rocks may indicate a beach or stream channel environment. (Photo by Tim Graham/Alamy Images) **B.** Mud cracks form when wet mud or clay dries and shrinks, perhaps signifying a tidal flat or desert basin. (Photo by Marli Miller)

The thickness of beds ranges from microscopically thin to tens of meters thick. Separating the strata are *bedding planes*, flat surfaces along which rocks tend to separate or break. Generally, each bedding plane marks the end of one episode of sedimentation and the beginning of another.

Sedimentary rocks provide geologists with evidence for deciphering past environments. A conglomerate, for example, indicates a high-energy environment, such as a rushing stream, where only the coarse materials can settle out. By contrast, black shale and coal are associated with a low-energy, organic-rich environment, such as a swamp or lagoon. Other features found in some sedimentary rocks also give clues to past environments (**FIGURE 3.23**).

Fossils, the traces or remains of prehistoric life, are perhaps the most important inclusions found in some sedimentary rock (**FIGURE 3.24**). Knowing the nature of the life-forms that existed at a particular time may help answer many questions about the environment. Was it land or ocean, lake or swamp? Was the climate hot or cold, rainy or dry? Was the ocean water shallow or deep, turbid or clear? Furthermore, fossils are important time indicators and play a key role in matching up rocks from different places that are the same age. Fossils are important tools used in interpreting the geologic past and will be examined in some detail in Chapter 11.

FIGURE 3.24 Fossils—Clues to the Past Fossils, the remains or traces of prehistoric life, are primarily associated with sediments and sedimentary rocks. A large variety of trilobites are associated with the Paleozoic era. (Photo by Russell Shively/Shutterstock)

3.3 CONCEPT CHECKS

1 Why are sedimentary rocks important?

2 What minerals are most abundant in detrital sedimentary rocks? In which rocks do these sediments predominate?

3 Distinguish between conglomerate and breccia.

4 What are the two categories of chemical sedimentary rock? Give an example of a rock that belongs to each category.

5 How do evaporites form? Give an example.

6 Describe the two processes by which sediments are transformed into sedimentary rocks. Which is the most effective process in the lithification of sand- and gravel-sized sediments?

7 List three common cements. How might each be identified?

8 What is the most characteristic feature of sedimentary rocks?

3.4 | METAMORPHIC ROCKS: NEW ROCK FROM OLD Define *metamorphism*, explain how metamorphic rocks form, and describe the agents of metamorphism.

Recall from the discussion of the rock cycle that metamorphism is the transformation of one rock type into another. **Metamorphic rocks** are produced from preexisting igneous, sedimentary, or even other metamorphic rocks (**FIGURE 3.25**). Thus, every metamorphic rock has a *parent rock*—the rock from which it was formed.

Metamorphism, which means "to change form," is a process that leads to changes in the mineralogy, texture (for example, grain size), and sometimes chemical composition of rocks. Metamorphism takes place when preexisting rock is subjected to a physical or chemical environment that is significantly different from that in which it initially formed. In response to these new conditions, the rock gradually changes until a state of equilibrium with the new environment is reached. Most metamorphic changes occur at the elevated temperatures and pressures that exist in the zone beginning a few kilometers below Earth's surface and extending into the mantle.

Metamorphism often progresses incrementally, from slight changes (*low-grade metamorphism*) to substantial changes (*high-grade metamorphism*) (**FIGURE 3.26**). For example, under low-grade metamorphism, the common sedimentary rock *shale* becomes the more compact metamorphic rock called *slate* (see Figure 3.26A). Hand samples of these rocks are sometimes difficult to distinguish, illustrating that the transition from sedimentary to metamorphic rock is often gradual, and the changes can be subtle.

In more extreme environments, metamorphism causes a transformation so complete that the identity of the parent rock cannot be determined. In high-grade metamorphism, such features as bedding planes, fossils, and vesicles that may have existed in the parent rock are obliterated. Furthermore, when rocks deep in the crust (where temperatures are high) are subjected to directed pressure, the entire mass may deform, producing large-scale structures such as folds (see Figure 3.26B).

In the most extreme metamorphic environments, the temperatures approach those at which rocks melt. However, *during metamorphism, the rock must remain essentially solid*, for if complete melting occurs, we have entered the realm of igneous activity.

Most metamorphism occurs in one of two settings:

1. When rock is intruded by magma, **contact metamorphism**, or **thermal metamorphism**, may take place. In such a situation, change is caused by the rise in temperature within the rock surrounding the mass of molten material.
2. During mountain building, great quantities of rock are subjected to pressures and high temperatures associated with large-scale deformation called **regional metamorphism**.

Extensive areas of metamorphic rocks are exposed on every continent. Metamorphic rocks are an important component of many mountain belts, where they make up a large portion of a mountain's crystalline core. Even the stable continental interiors, which are generally covered by sedimentary rocks, are underlain by metamorphic basement rocks. In each of these settings, the metamorphic rocks are usually highly deformed and intruded by igneous masses. Consequently, significant parts of Earth's continental crust are composed of metamorphic and associated igneous rocks.

Mobile Field Trip FIGURE 3.25 Metamorphic Rocks in the Adirondacks, New York.
(Photo by Michael Collier)

What Drives Metamorphism?

The agents of metamorphism include *heat*, *confining pressure*, *differential stress*, and *chemically active fluids*. During metamorphism, rocks are often subjected to all four metamorphic agents simultaneously. However, the degree of metamorphism and the contribution of each agent vary greatly from one environment to another.

Heat as a Metamorphic Agent

Thermal energy (*heat*) is the most important factor driving metamorphism. It triggers chemical reactions that result in the recrystallization of existing minerals and the formation of new minerals. Thermal energy for metamorphism comes mainly from two sources. Rocks experience a rise in temperature when they are intruded by magma rising from below. This is called *contact*, or *thermal*, *metamorphism*. In this situation, the adjacent host rock is "baked" by the emplaced magma.

By contrast, rocks that formed at Earth's surface will experience a gradual increase in temperature and pressure as they are taken to greater depths. In the upper crust, this increase in temperature averages about 25°C per kilometer. When buried to a depth of about 8 kilometers (5 miles), where temperatures are between 150° and 200°C, clay minerals tend to become unstable and begin to recrystallize into other minerals, such as chlorite and muscovite, that are stable in this environment. (Chlorite is a mica-like mineral formed by the metamorphism of iron- and magnesium-rich silicates.) However, many silicate minerals, particularly those found in crystalline igneous rocks—quartz and feldspar, for example—remain stable at these temperatures. Thus, metamorphic changes in these minerals require much higher temperatures in order to recrystallize.

Confining Pressure and Differential Stress as Metamorphic Agents

Pressure, like temperature, increases with depth as the thickness of the overlying rock increases. Buried rocks are subjected to *confining pressure*—similar to water pressure in that the forces are equally applied in all directions (**FIGURE 3.27A**). The deeper you go in the ocean, the greater the confining pressure. The same is true for buried rock. Confining pressure causes the spaces between mineral grains to close, producing a more compact rock that has greater density. Further, at great depths,

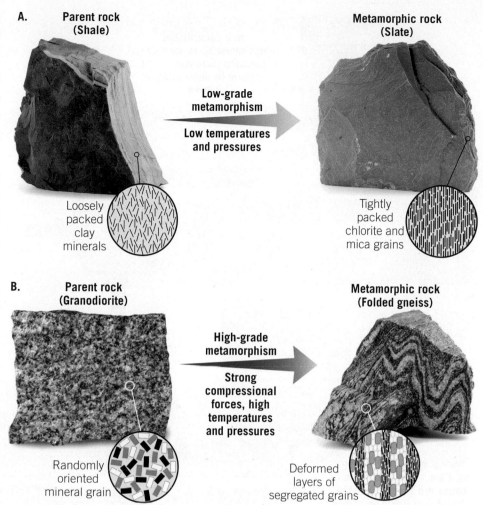

A. **Parent rock (Shale)** → Low-grade metamorphism / Low temperatures and pressures → **Metamorphic rock (Slate)**

Loosely packed clay minerals

Tightly packed chlorite and mica grains

B. **Parent rock (Granodiorite)** → High-grade metamorphism / Strong compressional forces, high temperatures and pressures → **Metamorphic rock (Folded gneiss)**

Randomly oriented mineral grain

Deformed layers of segregated grains

FIGURE 3.26 Metamorphic Grade A. Low-grade metamorphism illustrated by the transformation of the common sedimentary rock shale to the more compact metamorphic rock slate. **B.** High-grade metamorphic environments obliterate the existing texture and often change the mineralogy of the parent rock. High-grade metamorphism occurs at temperatures that approach those at which rocks melt. (Photos by Dennis Tasa)

EYE ON EARTH

This rock outcrop, located in Joshua Tree National Park, consists of dark-colored metamorphic rocks that overlie light-colored igneous rocks.

QUESTION 1 *Name the type of metamorphism—contact (thermal), or regional metamorphism—that likely produced these metamorphic rocks.*

QUESTION 2 *Write a brief statement that describes the geologic history of this area, based on what you observe in this image.*

E.J. Tarbuck

**SmartFigure 3.27
Confining Pressure and
Differential Stress**

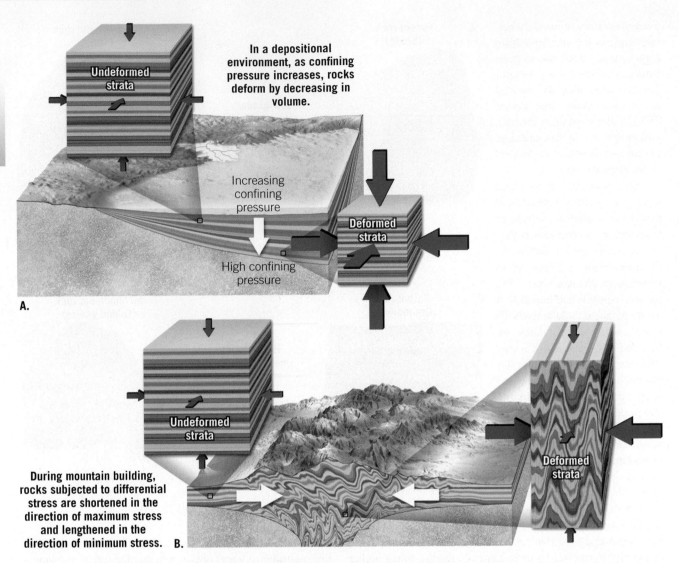

In a depositional environment, as confining pressure increases, rocks deform by decreasing in volume.

Undeformed strata

Increasing confining pressure

High confining pressure

Deformed strata

A.

Undeformed strata

During mountain building, rocks subjected to differential stress are shortened in the direction of maximum stress and lengthened in the direction of minimum stress. B.

Deformed strata

confining pressure may cause minerals to recrystallize into new minerals that display more compact crystalline forms.

During episodes of mountain building, large rock bodies become highly crumpled and metamorphosed (**FIGURE 3.27B**). The forces that generate mountains are unequal in different directions and are called *differential stress*. Unlike confining pressure, which "squeezes" rock equally in all directions, differential stresses are greater in one direction than in others. As shown in Figure 3.27B, rocks subjected to differential stress are shortened in the direction of greatest stress, and they are elongated, or lengthened, in the direction perpendicular to that stress. The deformation caused by differential stresses plays a major role in developing metamorphic textures.

In surface environments where temperatures are relatively low, rocks are *brittle* and tend to fracture when subjected to differential stress. (Think of a heavy boot crushing a piece of fine crystal.) Continued deformation grinds and pulverizes the mineral grains into small fragments. By contrast, in high-temperature, high-pressure environments deep in Earth's crust, rocks are *ductile* and tend to flow rather than break. (Think of a heavy boot crushing a soda can.) When rocks exhibit ductile behavior, their mineral grains tend to flatten and elongate when subjected to differential stress.

This accounts for their ability to deform by flowing (rather than fracturing) to generate intricate folds.

Chemically Active Fluids Ion-rich fluids composed mainly of water and other volatiles (materials that readily change to gases at surface conditions) are believed to play an important role in some types of metamorphism. Fluids that surround mineral grains act as catalysts to promote recrystallization by enhancing ion migration. In progressively hotter environments, these ion-rich fluids become correspondingly more reactive. Chemically active fluids can produce two types of metamorphism, explained below. The first type changes the arrangement and shape of mineral grains within a rock; the second type changes the rock's chemical composition.

When two mineral grains are squeezed together, the parts of their crystalline structures that touch are the most highly stressed. Atoms at these sites are readily dissolved by the hot fluids and move to the voids between individual grains. Thus, hot fluids aid in the recrystallization of mineral grains by dissolving material from regions of high stress and then precipitating (depositing) this material in areas of low stress. As a result, *minerals tend to recrystallize and grow longer in a direction perpendicular to compressional stresses.*

When hot fluids circulate freely through rocks, ionic exchange may occur between adjacent rock layers, or ions may migrate great distances before they are finally deposited. The latter situation is particularly common when we consider hot fluids that escape during the crystallization of an intrusive mass of magma. If the rocks surrounding the magma differ markedly in composition from the invading fluids, there may be a substantial exchange of ions between the fluids and host rocks. When this occurs, the overall composition of the surrounding rock changes.

Metamorphic Textures

The degree of metamorphism is reflected in a rock's *texture* and *mineralogy*. (Recall that the term *texture* is used to describe the size, shape, and arrangement of grains within a rock.) When rocks are subjected to low-grade metamorphism, they become more compact and thus denser. A common example is the metamorphic rock slate, which forms when shale is subjected to temperatures and pressures only slightly greater than those associated with the compaction that lithifies sediment. In this case, differential stress causes the microscopic clay minerals in shale to align into the more compact arrangement found in slate.

Under more extreme conditions, stress causes certain minerals to recrystallize. In general, recrystallization encourages the growth of larger crystals. Consequently, many metamorphic rocks consist of visible crystals, much like coarse-grained igneous rocks.

Foliation The term **foliation** refers to any nearly flat arrangement of mineral grains or structural features within a rock. Although foliation may occur in some sedimentary and even a few types of igneous rocks, it is a fundamental characteristic of regionally metamorphosed rocks—that is, rock units that have been strongly deformed, mainly during folding. In metamorphic environments, foliation is ultimately driven by compressional stresses that shorten rock units, causing mineral grains in preexisting rocks to develop parallel, or nearly parallel, alignments (**FIGURE 3.28**). Examples of foliation include the *parallel alignment of platy (flat and disk-like) minerals* such as the micas; the *parallel alignment of flattened pebbles*; *compositional banding*, in which dark and light minerals separate generating a layered appearance; and *rock cleavage*, in which rocks can be easily split into tabular slabs.

Nonfoliated Textures Not all metamorphic rocks exhibit a foliated texture. Those that do not are referred to as **nonfoliated** and typically develop in environments where deformation is minimal and the parent rocks are composed of minerals that have a relatively simple chemical composition, such as quartz or calcite. For example, when a fine-grained limestone (made of calcite) is metamorphosed by the intrusion of a hot magma body (contact metamorphism), the small calcite grains recrystallize and form larger interlocking crystals. The resulting rock, *marble*, exhibits large,

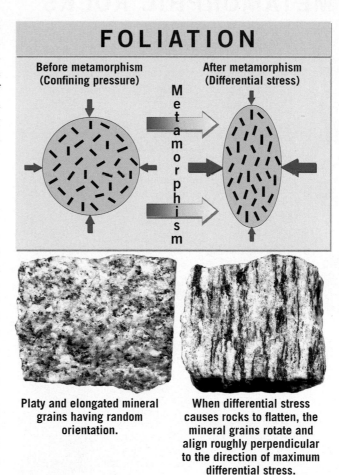

FIGURE 3.28 Rotation of Platy and Elongated Mineral Grains to Produce a Foliated Texture Under the pressures of metamorphism, some mineral grains become reoriented and aligned at right angles to the stress. The resulting orientation of mineral grains gives the rock a foliated (layered) texture. If the coarse-grained igneous rock (granite) on the left underwent intense metamorphism, it could end up closely resembling the metamorphic rock on the right (gneiss). (Photos by E. J. Tarbuck)

FOLIATION

Before metamorphism (Confining pressure)

After metamorphism (Differential stress)

Metamorphism

Platy and elongated mineral grains having random orientation.

When differential stress causes rocks to flatten, the mineral grains rotate and align roughly perpendicular to the direction of maximum differential stress.

equidimensional grains that are randomly oriented, similar to those in a coarse-grained igneous rock.

Common Metamorphic Rocks

A chart depicting the common rocks produced by metamorphic processes is found in **FIGURE 3.29**, and a description of each follows.

Foliated Rocks *Slate* is a very fine-grained foliated rock composed of minute mica flakes that are too small to be visible (see Figure 3.29). A noteworthy characteristic of slate is its excellent rock cleavage, or tendency to break into flat slabs. This property has made slate a useful rock for roof and floor tile, as well as billiard tables (**FIGURE 3.30**). Slate is usually generated by the low-grade metamorphism of shale. Less frequently, it is produced when volcanic ash is metamorphosed. Slate's color is variable. Black slate contains organic material, red slate gets its color from iron oxide, and green slate is usually composed of chlorite, a greenish micalike mineral.

Phyllite represents a degree of metamorphism between slate and schist. Its constituent platy minerals, mainly muscovite and chlorite, are larger than those in slate but not large enough to be readily identifiable with the unaided eye. Although phyllite appears similar to slate, it can be easily distinguished from slate by its glossy sheen and wavy surface (see Figure 3.29).

COMMON METAMORPHIC ROCKS

Metamorphic Rock	Texture	Comments	Parent Rock
Slate 	Foliated	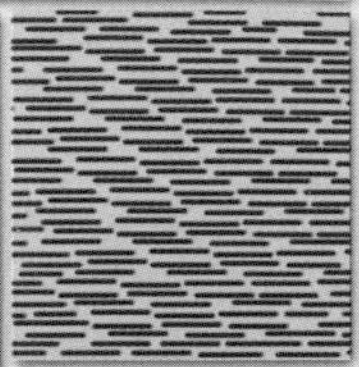 Composed of tiny chlorite and mica flakes, breaks in flat slabs called slaty cleavage, smooth dull surfaces	Shale, mudstone, or siltstone
Phyllite 	Foliated	 Fine-grained, glossy sheen, breaks along wavy surfaces	Shale, mudstone, or siltstone
Schist 	Foliated	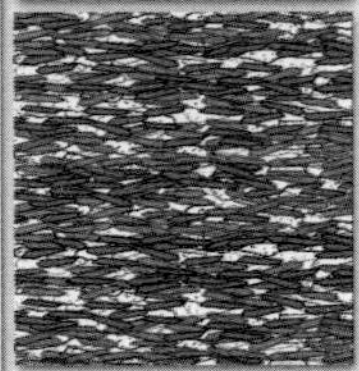 Medium- to coarse-grained, scaly foliation, micas dominate	Shale, mudstone, or siltstone
Gneiss 	Foliated	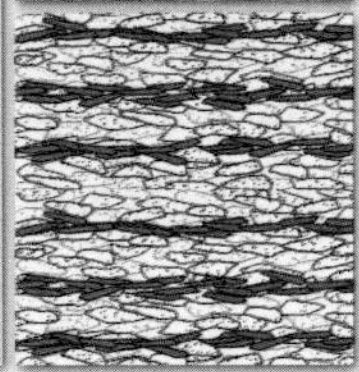 Coarse-grained, compositional banding due to segregation of light and dark colored minerals	Shale, granite, or volcanic rocks
Marble 	Nonfoliated	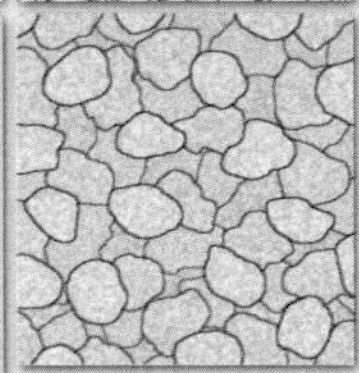 Medium- to coarse-grained, relatively soft (3 on the Mohs scale), interlocking calcite or dolomite grains	Limestone, dolostone
Quartzite 	Nonfoliated	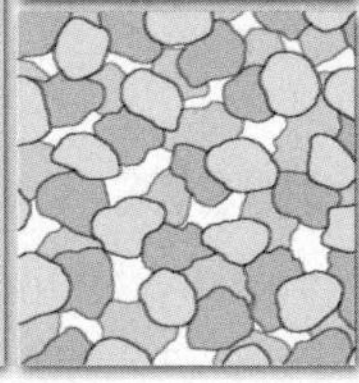 Medium- to coarse-grained, very hard, massive, fused quartz grains	Quartz sandstone

FIGURE 3.29 Classification of Common Metamorphic Rocks (Photos by E. J. Tarbuck)

FIGURE 3.30 Slate Exhibits Rock Cleavage Because slate breaks into flat slabs, it has many uses. The larger image shows a quarry near Alta, Norway. (Photo by Fred Bruemmer/Getty Images) In the inset photo, slate is used as a roof on a house in Switzerland. (Photo by E. J. Tarbuck)

Schists are moderately to strongly foliated rocks formed by regional metamorphism (see Figure 3.29). They are platy and can be readily split into thin flakes or slabs. Many schists originate from shale parent rock. The term *schist* describes the *texture* of a rock regardless of composition. For example, schists composed primarily of muscovite and biotite are called *mica schists*.

Gneiss (pronounced "nice") is the term applied to banded metamorphic rocks in which elongated and granular (as opposed to platy) minerals predominate (see Figure 3.29). The most common minerals in gneisses are quartz and feldspar, with lesser amounts of muscovite, biotite, and hornblende. Gneisses exhibit strong segregation of light and dark silicates, giving them a characteristic banded texture. While still deep below the surface where temperatures and pressures are great, banded gneisses can be deformed into intricate folds.

Nonfoliated Rocks *Marble* is a coarse, crystalline rock whose parent rock is limestone. Marble is composed of large interlocking calcite crystals, which form from the recrystallization of smaller grains in the parent rock. Because of its color and relative softness (hardness of only 3 on the Mohs scale), marble is a popular building stone. White marble is particularly prized as a stone from which to carve monuments and statues, such as the Lincoln Memorial in Washington, DC, and the Taj Mahal in India. The parent rocks from which various marbles form contain impurities that color the stone. Thus, marble can be pink, gray, green, or even black.

Quartzite is a very hard metamorphic rock most often formed from quartz sandstone. Under moderate- to high-grade metamorphism, the quartz grains in sandstone fuse. Pure quartzite is white, but iron oxide may produce reddish or pinkish stains, and dark minerals may impart a gray color.

3.4 CONCEPT CHECKS

1. Metamorphism means "change form." Describe how a rock may change during metamorphism.
2. Briefly describe what is meant by the statement "every metamorphic rock has a parent rock."
3. List the four agents of metamorphism and describe the role of each.
4. Distinguish between regional and contact metamorphism.
5. What feature easily distinguishes schist and gneiss from quartzite and marble?
6. In what ways do metamorphic rocks differ from the igneous and sedimentary rocks from which they formed?

3.5 | RESOURCES FROM ROCKS AND MINERALS

Distinguish between metallic and nonmetallic mineral resources and list at least two examples of each. Compare and contrast the three traditional fossil fuels.

The outer layer of Earth, which we call the crust, is only as thick when compared to the remainder of the Earth as a peach skin is to a peach, yet it is of supreme importance to us. We depend on it for fossil fuels and as a source of such diverse minerals as the iron for automobiles, salt to flavor food, and gold for world trade. In fact, on occasion, the availability or absence of certain Earth materials has altered the course of history. As the world population grows and the material requirements of modern society increase, the need to locate additional supplies of useful minerals becomes more challenging.

Most of the energy and mineral resources used by humans are nonrenewable. **FIGURE 3.31** shows the annual per capita consumption of several important mineral and energy resources. These data reflect an individually prorated share of the materials required by industry to support the needs of our

modern society—including a vast array of homes, cars, electronics, cosmetics, packaging, and other products and services.

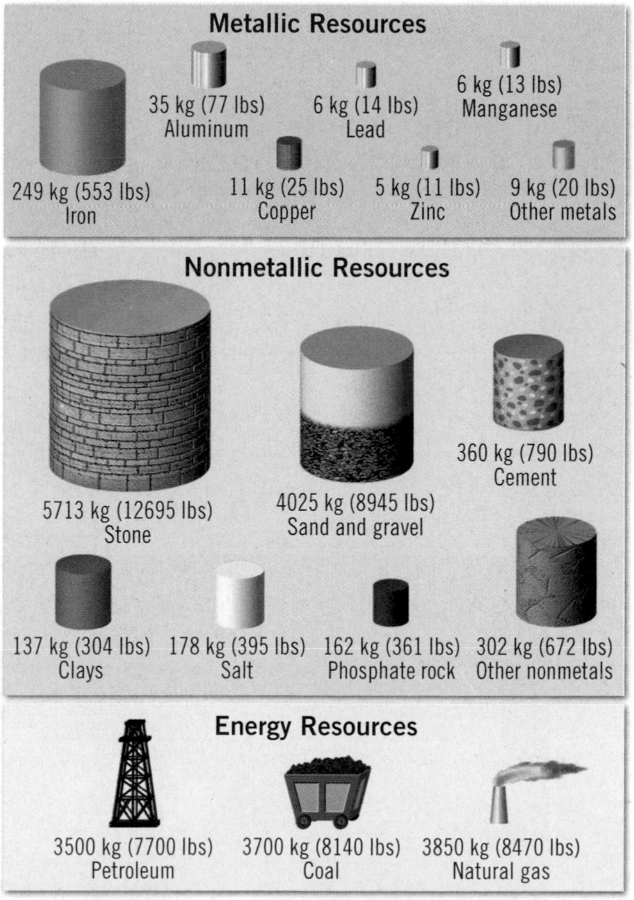

Metallic Mineral Resources

Some of the most important accumulations of metals, including gold, silver, copper, platinum, and nickel, are produced by igneous and metamorphic processes that concentrate desirable materials to the extent that extraction is economically feasible (**TABLE 3.1**). Igneous processes that generate some metal deposits are straightforward. For example, as a large magma body cools, minerals that crystallize early and are heavy tend to settle to the lower portion of the magma chamber. This type of *magmatic differentiation* is particularly important in large

FIGURE 3.31 How Much Do Each of Us Use? The annual per capita consumption of metallic and nonmetallic resources for the United States is about 11,000 kilograms (12 tons). About 97 percent of the materials used are nonmetallic. The per capita use of oil, coal, and natural gas exceeds 11,000 kilograms. (U.S. Geological Survey)

TABLE 3.1 Occurrence of Metallic Minerals

Metal	Principal Ores	Geologic Occurrences
Aluminum	Bauxite	Residual product of weathering
Chromium	Chromite	Magmatic segregation
Copper	Chalcopyrite	Hydrothermal deposits; contact metamorphism; enrichment by weathering processes
	Bornite	
	Chalcocite	
Gold	Native gold	Hydrothermal deposits; placers
Iron	Hematite	Banded sedimentary formations; magmatic segregation
	Magnetite	
	Limonite	
Lead	Galena	Hydrothermal deposits
Magnesium	Magnesite	Hydrothermal deposits
	Dolomite	
Manganese	Pyrolusite	Residual product of weathering
Mercury	Cinnabar	Hydrothermal deposits
Molybdenum	Molybdenite	Hydrothermal deposits
Nickel	Pentlandite	Magmatic segregation
Platinum	Native platinum	Magmatic segregation; placers
Silver	Native silver	Hydrothermal deposits; enrichment by weathering processes
	Argentite	
Tin	Cassiterite	Hydrothermal deposits; placers
Titanium	Ilmenite	Magmatic segregation; placers
	Rutile	
Tungsten	Wolframite	Pegmatites; contact metamorphic deposits; placers
	Scheelite	
Uranium	Uraninite (pitchblende)	Pegmatites; sedimentary deposits
Zinc	Sphalerite	Hydrothermal deposits

Marble

Marble is crystalline metamorphic rock whose parent rock was limestone or dolostone. Pure marble is white; however, most marble contains impurities such as iron oxide, chlorite, and organic debris which renders the marble pink, green, gray, or sometimes black.

Pure white marbles are prized for sculpture

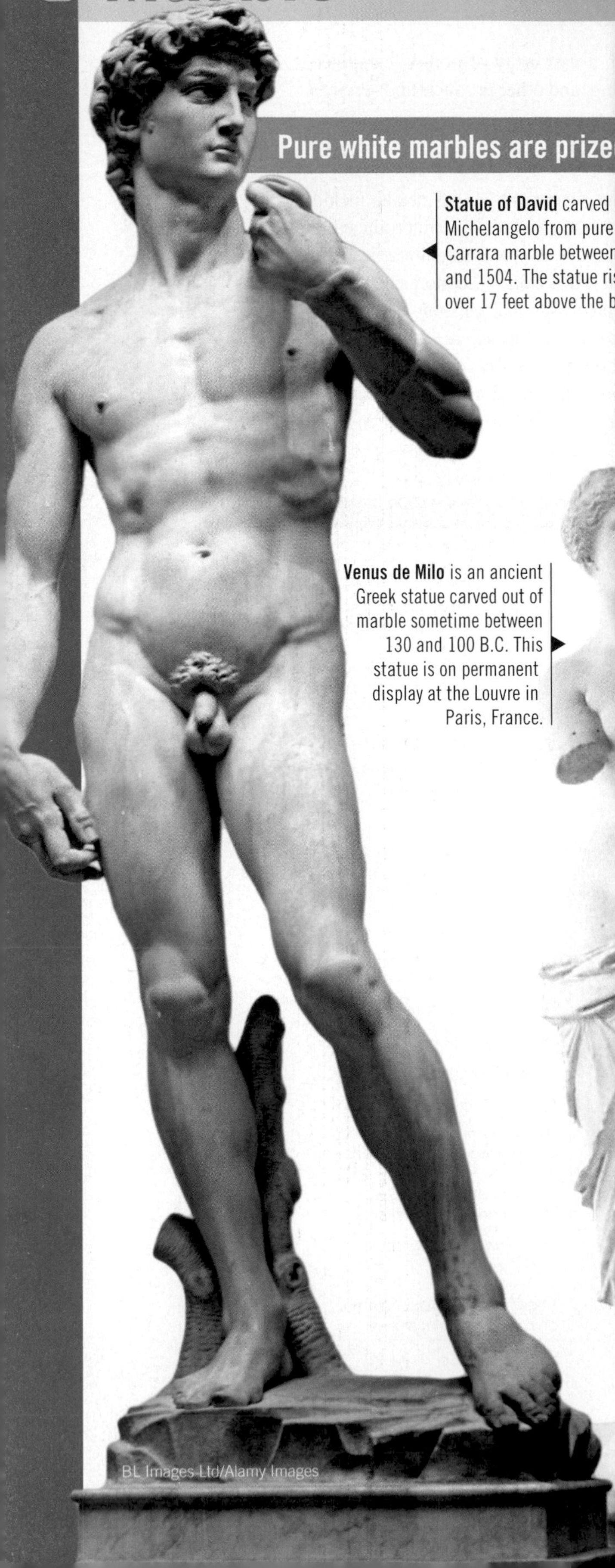

Statue of David carved by Michelangelo from pure Carrara marble between 1501 and 1504. The statue rises over 17 feet above the base.

Carrara Marble quarries located in the mountains near Carrara, Italy, have produced some of the finest marbles for thousands of years.

Venus de Milo is an ancient Greek statue carved out of marble sometime between 130 and 100 B.C. This statue is on permanent display at the Louvre in Paris, France.

These marble quarries located near Carrara, Italy, are so large they can be seen from space.

Fotografiche/Shutterstock

BL Images Ltd/Alamy Images

Ray Roberts/Alamy Images

NASA

Marble, because of its workability, is a widely used building stone

Taj Mahal Constructed primarily of marble, the Taj Mahal is considered one of the world's most spectacular buildings. Built by Mughal Emperor Shah Jahin in memory of his third wife, the structure is located in northcentral India. Construction began around 1632 and was completed in 1653.

Steve Vider/Superstock

The white exterior of the Lincoln Memorial in Washington, DC, is constructed mainly of marble that was quarried near the town of Marble, Colorado. Inside, pink Tennessee "marble" was used for the floors, Alabama marble for the ceilings, and Georgia marble for Lincoln's statue.

Marble is also widely used for floor tile and countertops. The marble in these floor tiles contains impurities that impart a variety of color and also show evidence of deformation.

Vilo Vad/Shutterstock

Orhan Cam/Shutterstock

FIGURE 3.32 Pegmatites This pegmatite in the Black Hills of South Dakota was mined for its large crystals of spodumene, a source of lithium which is used in the manufacture of batteries for cell phones and computers. Arrows are pointing to impressions left by crystals. Note the person in the upper center of the photo for scale. (Photo by James G. Kirchner)

basaltic magmas where chromite (ore of chromium), magnetite, and platinum are occasionally generated. Layers of chromite, along with other heavy minerals, are mined from such deposits in the Bushveld Complex in South Africa, which contains over 70 percent of the world's known reserves of platinum.

Igneous processes are also important in generating other types of mineral deposits. For example, as a granitic magma cools and crystallizes, the residual melt becomes enriched in rare elements and heavy metals, including gold and silver. Furthermore, because water and other volatile substances do not crystallize along with the bulk of the magma body, these fluids make up a high percentage of the melt during the final phase of solidification. Crystallization in a fluid-rich environment enhances the migration of ions and results in the formation of crystals several centimeters, or even a few meters, in length. The resulting rocks, called **pegmatites**, are composed of these unusually large crystals (**FIGURE 3.32**).

Most pegmatites are granitic in composition and consist of large crystals of quartz, feldspar, and muscovite. Feldspar is used in the production of ceramics, and muscovite is used for electrical insulation and glitter. In addition to these common silicates, some pegmatites contain semiprecious gems such as beryl, topaz, and tourmaline. Moreover, minerals containing the elements lithium, gold, silver, uranium, and the rare earths are sometimes found. Most pegmatites are located within large igneous masses or as dikes or veins that cut into the host rock surrounding a magma chamber (**FIGURE 3.33**).

Among the most important ore deposits are those generated from hydrothermal (*hydra* = water, *therm* = heat) solutions. Included in this group are the gold deposits of the Homestake Mine in South Dakota; the lead, zinc, and silver ores near Coeur d'Alene, Idaho; the silver deposits of the Comstock Lode in Nevada; and the copper ores of the Keweenaw Peninsula in Michigan.

The majority of hydrothermal deposits originate from hot, metal-rich fluids that are associated with cooling magma bodies. During solidification, liquids plus various metallic ions accumulate near the top of the magma chamber. Because these hot fluids are very mobile, they can migrate great distances through the surrounding host rock before they are eventually deposited. Some of this fluid moves along fractures or bedding planes, where it cools and precipitates the metallic ions to produce **vein deposits** (see Figure 3.33). Many of the most productive deposits of gold, silver, and mercury occur as hydrothermal vein deposits.

Another important type of accumulation generated by hydrothermal activity is called a **disseminated deposit**. Rather than being concentrated in narrow veins and dikes, these ores are distributed as minute masses throughout the entire rock mass (see Figure 3.33). Much of the world's copper is extracted from disseminated deposits, including the huge Bingham Canyon copper mine in Utah (see Figure 2.34). Because these accumulations contain only 0.4 to 0.8 percent copper, between 125 and 250 metric tons of ore must be mined for every ton of metal recovered. The environmental impact of these large excavations, including the problems of waste disposal, is significant.

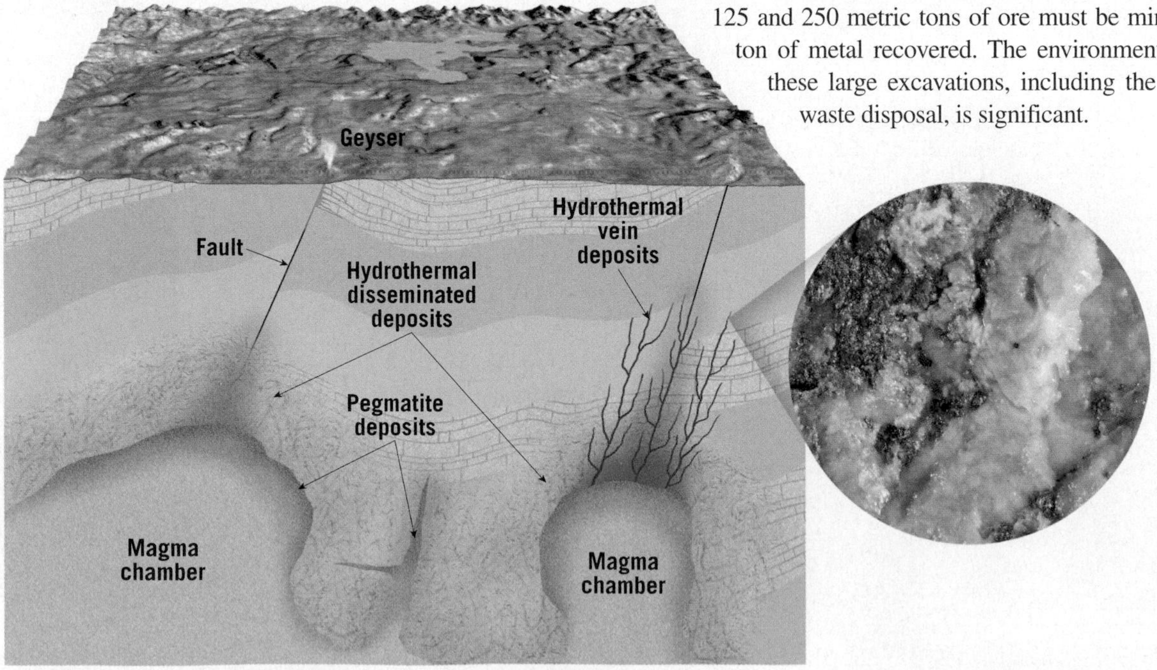

Figure 3.33 Relationship Between An Igneous Body and Associated Pegmatites and Hydrothermal Mineral Deposits (Photo by Greenshoots Communications/Alamy)

Geyser

Fault

Hydrothermal disseminated deposits

Hydrothermal vein deposits

Pegmatite deposits

Magma chamber

Magma chamber

TABLE 3.2   Occurrences and Uses of Nonmetallic Minerals

Mineral	Uses	Geologic Occurrences
Apatite	Phosphorus fertilizers	Sedimentary deposits
Asbestos	Incombustible fibers	Metamorphic alteration (chrysotile)
Calcite	Aggregate; steelmaking; soil conditioning; chemicals; cement; building stone	Sedimentary deposits
Clay minerals	Ceramics; china	Residual product of weathering (kaolinite)
Corundum	Gemstones; abrasives	Metamorphic deposits
Diamond	Gemstones; abrasives	Kimberlite pipes; placers
Fluorite	Steelmaking; aluminum refining; glass; chemicals	Hydrothermal deposits
Garnet	Abrasives; gemstones	Metamorphic deposits
Graphite	Pencil lead; lubricant; refractories	Metamorphic deposits
Gypsum	Plaster of Paris	Evaporite deposits
Halite	Table salt; chemicals; ice control	Evaporite deposits; salt domes
Muscovite	Insulator in electrical applications	Pegmatites
Quartz	Primary ingredient in glass	Igneous intrusions; sedimentary deposits
Sulfur	Chemicals; fertilizer manufacture	Sedimentary deposits; hydrothermal deposits
Sylvite	Potassium fertilizers	Evaporite deposits
Talc	Powder used in paints, cosmetics, etc.	Metamorphic deposits

Nonmetallic Mineral Resources

Mineral resources not used as fuels or processed for the metals they contain are referred to as *nonmetallic mineral resources.* These materials are extracted and processed either to make use of the nonmetallic elements they contain or for the physical and chemical properties they possess (**TABLE 3.2**). Nonmetallic mineral resources are commonly divided into two broad groups: *building materials* and *industrial minerals.* Some substances have many different uses and are found in both categories. Limestone, perhaps the most versatile and widely used rock of all, is perhaps the best example. As a building material, it is used as crushed rock, as building stone, and in the making of cement. Moreover, as an industrial mineral, it is an ingredient in the manufacture of steel and is used in agriculture to neutralize acidic soils.

Besides aggregate (sand, gravel, and crushed rock) and cut stone, the other important building materials include gypsum for plaster and wallboard, clay for tile and bricks, and cement, which is made from limestone and shale. Cement and aggregate are used to make concrete, a material that is essential to construction of roads, bridges, and all large buildings.

Many and various nonmetallic resources are classified as industrial minerals. People often do not realize the importance of industrial minerals because they see only the products that result from their use and not the minerals themselves. That is, many nonmetallics are used up in the process of creating other products. For example, fluorite and limestone are part of the steelmaking process; corundum and garnet are used as abrasives to make machinery parts; and sylvite, a potassium-rich mineral, is used in the production of fertilizers.

Energy Resources: Fossil Fuels

Coal, petroleum, and natural gas are the primary fuels of our modern industrial economy. Currently, about 82 percent of the energy consumed in the United States comes from these basic fossil fuels. Although major shortages of oil and gas will not occur for many years, known global reserves are declining. Unless large new petroleum sources are discovered, a greater share of our future needs will be required from alternative energy sources, such as nuclear, geothermal, solar, wind, tidal, and hydroelectric power. Two fossil-fuel alternatives, oil sands and oil shale, are also promising sources of liquid fuels.

Coal   In addition to oil and natural gas, coal is commonly called a **fossil fuel**. This designation is appropriate because when coal is burned, energy from the Sun that was stored by plants many millions of years ago is being used, hence the actual burning of "fossils."

In the United States, coal fields are widespread and contain supplies that should last hundreds of years (**FIGURE 3.34**). Although coal is plentiful, its recovery and utilization present a number of challenges. Surface mining can turn the countryside into a scarred wasteland if careful and costly reclamation is not carried out to restore the land. Although underground mining does not scar the landscape to the same degree, it presents significant risks to human health and safety.

Air pollution is another significant problem associated with the burning of coal. Much coal contains significant quantities of sulfur. Despite efforts to remove sulfur before the coal is burned, some remains; when coal is burned, the sulfur is converted into noxious sulfur dioxide gas. Through

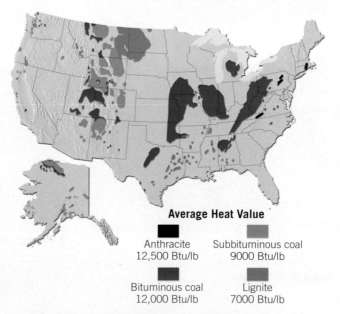

Average Heat Value

Anthracite
12,500 Btu/lb

Subbituminous coal
9000 Btu/lb

Bituminous coal
12,000 Btu/lb

Lignite
7000 Btu/lb

FIGURE 3.34 Coal Fields of the United States Most of the coal mined is subbituminous and bituminous. Wyoming and West Virginia are the leading producers. (Data from the U.S. Geological Survey)

a series of complex chemical reactions in the atmosphere, sulfur dioxide is converted to sulfuric acid, which eventually falls to Earth's surface in rain or snow. This acid precipitation, known as acid rain, can have adverse ecological effects over widespread areas.

As is the case when other fossil fuels are burned, the combustion of coal produces carbon dioxide. This major greenhouse gas plays a significant role in the heating of our atmosphere. Chapter 20 examines this issue in more detail.

Oil and Natural Gas   Together, petroleum and natural gas provided more than 60 percent of the energy consumed in the United States in 2012. Like coal, petroleum and natural gas are biological products derived from the remains of organisms. However, the environments in which they form are very different, as are the organisms. Coal is formed mostly from plant material that accumulated in swampy environments on land, as shown in Figure 3.21.

Oil and natural gas are generated from the remains of marine plants and animals, mainly microscopic plankton, which died in ancient seas millions of years ago. This organic material was deposited in sedimentary basins together with mud, sand, and other sediments that protected it from oxidation. With increased temperature and burial, chemical reactions gradually transform this organic matter into the liquid and gaseous hydrocarbons we call petroleum and natural gas.

Unlike the organic matter from which they formed, the newly created petroleum and natural gas are mobile. These fluids are gradually squeezed from the compacting, mud-rich layers where they originate, called the **source rock**, into adjacent permeable beds such as sandstone, where openings between sediment grains are larger. Because this occurs in a marine environment, the rock layers containing the oil and gas are saturated with water. Oil and gas, being less dense than water, migrate upward through the water-filled pore spaces of the enclosing rocks.

A geologic environment that allows for economically significant amounts of oil and gas to accumulate underground is termed an **oil trap**. Several geologic structures may act as oil traps and all share have two basic features: a porous, permeable **reservoir rock** that will yield petroleum and natural gas in sufficient quantities to make drilling worthwhile and a **cap rock**, such as shale, that is virtually impermeable to oil and gas. The cap rock halts the upwardly mobile oil and gas and keeps the oil and gas from escaping at the surface. **FIGURE 3.35** illustrates some common oil and natural-gas traps, which are described in the following list:

■ **Anticline**—One of the simplest traps is an *anticline*, an uparched series of sedimentary strata (see Figure 3.35A). As the strata are bent, the rising oil and gas collect at the apex (top) of the fold. Because of its

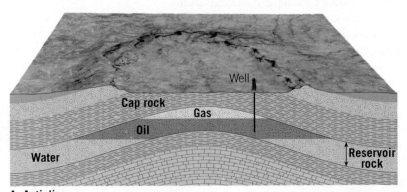

A. Anticline

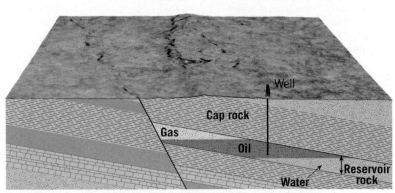

B. Fault trap

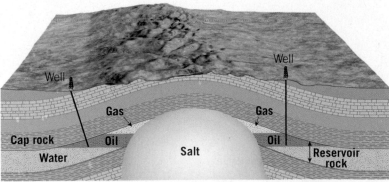

C. Salt dome

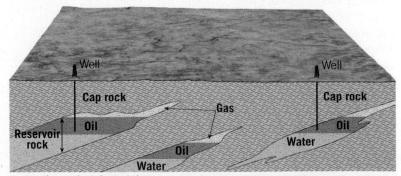

D. Stratigraphic (pinchout) trap

  **SmartFigure 3.35 Common Oil Traps**

lower density, the natural gas collects above the oil. Both rest upon the denser water that saturates the reservoir rock.

- **Fault trap**—This type of trap forms when strata are displaced in such a manner as to bring a dipping reservoir rock into position opposite an impermeable bed, as shown in Figure 3.35B. In this case, the upward migration of the oil and gas is halted where it encounters the fault.

- **Salt dome**—In the Gulf coastal plain region of the United States, important accumulations of oil occur in association with *salt domes*. Such areas have thick accumulations of sedimentary strata, including layers of rock salt. Salt occurring at great depths has been forced to rise in columns by the pressure of overlying beds. These rising salt columns gradually deform the overlying strata. Because oil and gas migrate to the highest level possible, they accumulate in the upturned sandstone beds adjacent to the salt column (see Figure 3.35C).

- **Stratigraphic (pinchout) trap**—Yet another important geologic circumstance that may lead to significant accumulations of oil and gas is termed a *stratigraphic trap*. These oil-bearing structures result primarily from the original pattern of sedimentation rather than structural deformation. The stratigraphic trap illustrated in Figure 3.35D exists because a sloping bed of sandstone thins to the point of disappearance.

When drilling punctures the lid created by the cap rock, the oil and natural gas, which are under pressure, migrate from the pore spaces of the reservoir rock to the drill pipe. On rare occasions, when fluid pressure is great, it may force oil up the drill hole to the surface, causing a "gusher" at the surface. Usually, however, a pump is required to extract the oil.

Hydraulic Fracturing In some shale deposits, there are significant reserves of natural gas and petroleum that cannot naturally leave because of the rock's low permeability. The practice of **hydraulic fracturing** (often called "fracking") shatters the shale, opening cracks through which these fluids can flow into wells and then be brought to Earth's surface. The fracturing of the shale is initiated by pumping liquids into the rock at very high pressures. The pressurized liquid is mostly water but also includes sand and other chemicals that aid in the fracturing process. Some of these chemicals may be toxic, and there are concerns about fracking fluids leaking into aquifers that supply communities with freshwater. The injection fluid also includes sand, so when fractures open in the shale, the sand grains can keep the fractures propped open and permit the gas to continue to flow. Because of concerns about potential groundwater contamination and induced seismicity, hydraulic fracturing remains a controversial practice. Its environmental effects are a focus of continuing research.

3.5 CONCEPT CHECKS

1 List two general types of hydrothermal deposits.
2 Nonmetallic resources are commonly divided into two broad groups. List the two groups and give some examples of materials that belong to each.
3 Why are coal, oil, and natural gas called *fossil fuels*?
4 What is an oil trap? Sketch two examples.
5 What do all oil traps have in common?

EYE ON EARTH

The image on the left shows an active landfill where tons of trash and garbage are dumped every day. Eventually this site will be reclaimed to resemble the area shown on the right and become a source of energy. (Left photo by NHPA/SuperStock; Right photo by Jim West/Alamy)

QUESTION 1 *Explain how an area filled with trash and waste could become a source of energy. What form of energy will it be? How might it be used?*

QUESTION 2 *Will this energy be considered renewable or nonrenewable? Explain.*

3.1 EARTH AS A SYSTEM: THE ROCK CYCLE

Sketch, label, and explain the rock cycle.

KEY TERM: rock cycle

Rock cycle

■ The rock cycle is a good model for thinking about the transformation of one rock to another due to Earth processes. All igneous rocks are made from molten rock. All sedimentary rocks are made from weathered products of other rocks. All metamorphic rocks are the products of preexisting rocks that are heated or squeezed at high temperatures and/or pressures. Given the right conditions, any kind of rock can be transformed into any other kind of rock.

Q Name the processes that are represented by each of the letters in this rock cycle diagram.

3.2 IGNEOUS ROCKS: "FORMED BY FIRE"

Describe the two criteria used to classify igneous rocks and explain how the rate of cooling influences the crystal size of minerals.

KEY TERMS: igneous rock, magma, lava, extrusive (volcanic), intrusive (plutonic), granitic (felsic) composition, basaltic (mafic) composition, andesitic (intermediate) composition, ultramafic, texture, fine-grained texture, coarse-grained texture, porphyritic texture, phenocrysts, groundmass, vesicular texture, glassy texture, pyroclastic (fragmental) texture, Bowen's reaction series, crystal settling, magmatic differentiation

■ Completely or partly molten rock is called magma if it is below Earth's surface and lava if it has erupted onto the surface. It consists of a liquid melt that contains gases (volatiles) such as water vapor, and it may contain solids (mineral crystals).

■ Magmas that cool at depth produce intrusive igneous rocks, whereas those that erupt onto Earth's surface produce extrusive igneous rocks.

■ To geologists, texture is a description of the size, shape, and arrangement of mineral grains in a rock. A careful observation of the texture of igneous rocks can lead to insights about the conditions under which they formed. Lava on or near the surface cools rapidly, resulting in a

large number of very small crystals that gives it a fine-grained texture. When magma cools at depth, the surrounding rock insulates it, and heat is lost more slowly. This allows sufficient time for the magma's ions to be organized into larger crystals, resulting in a rock with a coarse-grained texture. If crystals begin to form at depth and then the magma moves to a shallow depth or erupts at the surface, it will have a two-stage cooling history. The result is a rock with a porphyritic texture.

■ Pioneering experimentation by N. L. Bowen revealed that in a cooling magma, minerals crystallize in a specific order. The dark-colored silicate minerals, such as olivine, crystallize first at the highest temperatures (1250°C [2300°F]), whereas the light silicates, such as quartz, crystallize last at the lowest temperatures (650°C [1200°F]). Separation of minerals by mechanisms such as crystal settling result in igneous rocks having a wide variety of chemical compositions.

■ Igneous rocks are divided into broad compositional groups based on the percentage of dark and light silicate minerals they contain. Granitic (or felsic) rocks (such as granite and rhyolite) are composed mostly of the light-colored silicate minerals potassium feldspar and quartz. Rocks of andesitic (or intermediate) composition (rocks such as andesite) contain plagioclase feldspar and amphibole. Basaltic (or mafic) rocks (such as basalt) contain abundant pyroxene and calcium-rich plagioclase feldspar.

3.3 SEDIMENTARY ROCKS: COMPACTED AND CEMENTED SEDIMENT

List and describe the different categories of sedimentary rocks and discuss the processes that change sediment into sedimentary rock.

KEY TERMS: sedimentary rock, sediment, detrital sedimentary rock, chemical sedimentary rock, biochemical sedimentary rock, evaporite deposit, organic matter, lithification, compaction, cementation, strata (beds), fossil

■ Although igneous and metamorphic rocks make up most of Earth's crust by volume, sediment and sedimentary rocks are concentrated near the surface.

■ Detrital sedimentary rocks are made of solid particles, mostly quartz grains and microscopic clay minerals. Common detrital sedimentary rocks include shale (the most abundant sedimentary rock), sandstone, and conglomerate.

■ Chemical and biochemical sedimentary rocks are derived from mineral matter (ions) that is carried in solution to lakes and seas. Under certain conditions, ions in solution precipitate (settle out) to form chemical sediments as a result of physical processes, such as evaporation. Precipitation may also occur indirectly through life processes of water-dwelling organisms that form materials called biochemical sediments. Many water-dwelling animals and plants extract dissolved mineral matter to form shells and other hard parts. After the organisms die, their skeletons may accumulate on the floor of a lake or an ocean.

- Limestone, an abundant sedimentary rock, is composed chiefly of the mineral calcite ($CaCO_3$). Rock gypsum and rock salt are chemical rocks that form as water evaporates.
- Coal forms from the burial of large amounts of plant matter in low-oxygen depositional environments such as swamps and bogs.
- The transformation of sediment into sedimentary rock is lithification. The two main processes that contribute to lithification are compaction (a reduction in pore space by packing grains more tightly together) and cementation (a reduction in pore space by adding new mineral material that acts as a "glue" to bind the grains to each other).

Q **This photo was taken in the Grand Canyon, Arizona. What name(s) do geologists use for the characteristic features found in the sedimentary rocks shown in this image?**

Dennis Tasa

3.4 METAMORPHIC ROCKS: NEW ROCK FROM OLD

Define *metamorphism*, explain how metamorphic rocks form, and describe the agents of metamorphism.

KEY TERMS: metamorphic rock, metamorphism, contact (thermal) metamorphism, regional metamorphism, foliation, nonfoliated

- When rocks are subjected to elevated temperatures and pressures, they can change form, producing metamorphic rocks. Every metamorphic rock has a parent rock—the rock it used to be prior to metamorphism. When the minerals in parent rocks are subjected to heat and pressure, new minerals can form.
- Heat, confining pressure, differential stress, and chemically active fluids are four agents that drive metamorphic reactions. Any one alone may trigger metamorphism, or all four may exert influence simultaneously.
- Confining pressure results from burial. The pressure is of the same magnitude in every direction, like the pressure that a swimmer experiences when diving to the bottom of a swimming pool. An increase in confining pressure causes rocks to compact into more dense configurations.
- Differential stress results from tectonic forces. The pressure is greater in some directions and less in other directions. Rocks subjected to differential stress under ductile conditions deep in the crust tend to shorten in the direction of greatest stress and elongate in the direction(s) of least stress, producing flattened or stretched grains. If the same differential stress is applied to a rock in the shallow crust, it may deform in a brittle fashion instead, breaking into smaller pieces.

- A common kind of texture is foliation, the planar arrangement of mineral grains. Common foliated metamorphic rocks include slate, phyllite, schist, and gneiss. The order listed here is the order of increasing metamorphic grade: Slate is least metamorphosed, and gneiss is the most metamorphosed.
- Common nonfoliated metamorphic rocks include quartzite and marble, recrystallized rocks that form from quartz sandstone and limestone.

Q **Examine the photograph. Determine whether this rock is foliated or nonfoliated and then determine whether it formed under confining pressure or differential stress. Which of the pairs of arrows shows the direction of maximum stress?**

Dennis Tasa

3.5 RESOURCES FROM ROCKS AND MINERALS

Distinguish between metallic and nonmetallic mineral resources and list some examples of each. Compare and contrast the three traditional fossil fuels.

KEY TERMS: pegmatite, vein deposit, disseminated deposit, fossil fuel, source rock, oil trap, reservoir rock, cap rock, hydraulic fracturing

- Igneous processes concentrate some economically important elements through both magmatic differentiation and the emplacement of pegmatites. Magmas may also release hydrothermal (hot-water) solutions that penetrate surrounding rock, carrying dissolved metals in them. The metal ores may be precipitated as fracture-filling deposits (veins) or may "soak" into surrounding strata, producing countless tiny deposits disseminated throughout the host rock.
- Earth materials that are not used as fuels or processed for the metals they contain are referred to as nonmetallic resources. The two broad groups of nonmetallic resources are building materials (such as gypsum, used for plaster) and industrial minerals (including sylvite, a potassium-rich mineral used to make fertilizers).
- Coal, oil, and natural gas are all fossil fuels. In each, the energy of ancient sunlight, captured by photosynthesis, is stored in the hydrocarbons of plants or other living things buried by sediments. Although coal is abundant, coal mining can be dangerous and environmentally harmful, and burning coal generates several kinds of pollution.
- Oil and natural gas are formed from the heated remains of tiny marine plants and animals. Both oil and natural gas leave their source rock (typically shale) and migrate to an oil trap made up of more porous rocks, called reservoir rocks, which are covered by an impermeable cap rock.

Q **As you can see here, the Sinclair Oil Company's logo speaks directly to the "fossil" nature of the fuel it sells. However, is it likely that any *dinosaur* carbon has ended up in Sinclair's oil? Explain.**

Vespasian/Alamy

GIVE IT SOME **THOUGHT**

1. Refer to Figure 3.1. How does the rock cycle diagram—in particular, the labeled arrows—support the fact that sedimentary rocks are the most abundant rock type on Earth's surface?

2. Would you expect all the crystals in an intrusive igneous rock to be the same size? Explain why or why not.

3. Apply your understanding of igneous rock textures to describe the cooling history of each of the igneous rocks pictured on the right.

4. Is it possible for two igneous rocks to have the same mineral composition but be different rocks? Support your answer with an example.

5. Use your understanding of magmatic differentiation to explain how magmas of different composition can be generated in a cooling magma chamber.

6. Dust collecting on furniture is an everyday example of a sedimentary process. Provide another example of a sedimentary process that might be observed in or around where you live.

7. Describe two reasons sedimentary rocks are more likely to contain fossils than igneous rocks.

8. If you hiked to a mountain peak and found limestone at the top, what would that indicate about the likely geologic history of the rock there?

9. The accompanying photos each illustrate either a typical igneous, sedimentary, or metamorphic rock body. Which do you think is a metamorphic rock? Explain why you ruled out the other rock bodies. (Photos by E. J. Tarbuck)

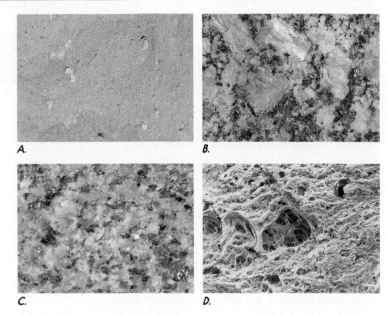
A.

B.

C.

D.

A.

B.

C.

10. Examine the accompanying photos, which show the geology of the Grand Canyon. Notice that most of the canyon consists of layers of sedimentary rocks, but if you were to hike down into the Inner Gorge, you would encounter the Vishnu Schist, which is metamorphic rock.

    a. What process might have been responsible for the formation of the Vishnu Schist? How does this process differ from the processes that formed the sedimentary rocks that are atop the Vishnu Schist?

    b. What does the Vishnu Schist tell you about the history of the Grand Canyon prior to the formation of the canyon itself?

    c. Why is the Vishnu Schist visible at Earth's surface?

    d. Is it likely that rocks similar to the Vishnu Schist exist elsewhere but are not exposed at Earth's surface? Explain.

A. Inner Gorge of the Grand Canyon

B. Close up of Vishnu Schist
(dark color)

EXAMINING THE **EARTH SYSTEM**

Coquina

E.J. Tarbuck

1. The sedimentary rock coquina, shown at right, formed in response to interactions among two or more of Earth's spheres. List the spheres associated with the formation of this rock and write a short explanation for each of your choices.

2. Of the two main sources of energy that drive the rock cycle—Earth's internal heat and solar energy—which is primarily responsible for each of the three groups of rocks found on and within Earth? Explain your reasoning.

3. Every year about 20,000 pounds of stone, sand, and gravel are mined for each person in the United States.
   a. Calculate how many pounds of stone, sand, and gravel will be needed for an individual during an 80-year lifespan.
   b. If 1 cubic yard of rocks weighs roughly 1700 pounds, calculate (in cubic yards) how large a hole must be dug to supply an individual with 80 years' worth of stone, sand, and gravel?
   c. A typical pickup truck can carry about a half cubic yard of rock. How many pickup truck loads would be necessary during the 80-year span?

MasteringGeology™

Looking for additional review and test prep materials? Visit the Self Study area in **www.masteringgeology.com** to find practice quizzes, study tools, and multimedia that will aid in your understanding of this chapter's content. In **MasteringGeology**™ you will find:

- GEODe: Earth Science: An interactive visual walkthrough of key concepts
- Geoscience Animation Library: More than 100 animations illuminating many difficult-to-understand Earth science concepts

- In The News RSS Feeds: Current Earth science events and news articles are pulled into the site with assessment
- Pearson eText
- Optional Self Study Quizzes
- Web Links
- Glossary
- Flashcards

4 Weathering, Soil, and Mass Wasting

In March 2012 a portion of England's White Cliffs of Dover, weakened by weathering and battered by waves, succumbed to the pull of gravity. Events such as this, popularly called landslides, are spectacular examples of mass wasting.

(Photo by AP Photo/Rex Features)

E arth's surface is constantly changing. Rock is disintegrated and decomposed, moved to lower elevations by gravity, and carried away by water, wind, or ice. In this manner, Earth's physical landscape is sculpted. This chapter focuses on the first two steps of this never-ending process: weathering and mass wasting. What causes solid rock to crumble, and why do the type and rate of weathering vary from place to place? What mechanisms act to move weathered debris downslope? Soil, an important product of the weathering process and a vital resource, is also examined.

4.1 | EARTH'S EXTERNAL PROCESSES
List three types of external processes and discuss the role each plays in the rock cycle.

Weathering, mass wasting, and erosion are called **external processes** because they occur at or near Earth's surface and are powered by energy from the Sun. External processes are a basic part of the rock cycle because they are responsible for transforming solid rock into sediment.

To the casual observer, the face of Earth may appear to be without change, unaffected by time. In fact, 200 years ago most people believed that mountains, lakes, and deserts were permanent features of an Earth that was thought to be no more than a few thousand years old. Today we know that Earth is 4.6 billion years old and that mountains eventually succumb to weathering and erosion, lakes fill with sediment

or are drained by streams, and deserts come and go with changes in climate.

Earth is a dynamic body. Some parts of Earth's surface are gradually elevated by mountain building and volcanic activity. These **internal processes** derive their energy from Earth's interior. Meanwhile, opposing external processes are continually breaking rock apart and moving the debris to lower elevations (**FIGURE 4.1**). The latter processes include:

1. **Weathering**—The physical breakdown (disintegration) and chemical alteration (decomposition) of rocks at or near Earth's surface

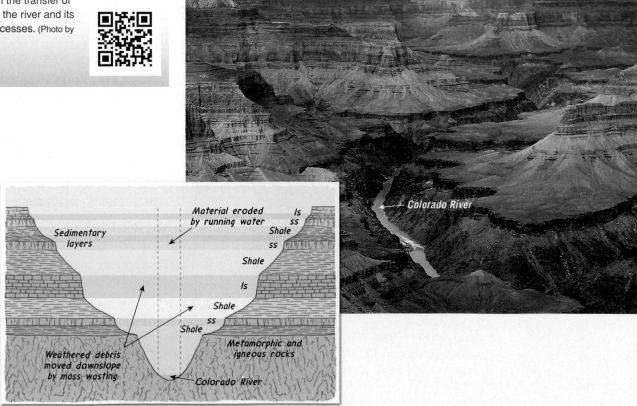

SmartFigure 4.1 Excavating the Grand Canyon The walls of the canyon extend far from the channel of the Colorado River. This results primarily from the transfer of weathered debris downslope to the river and its tributaries by mass-wasting processes. (Photo by Bryan Brazil/Shutterstock)

Geologist's Sketch

2. **Mass wasting**—The transfer of rock and soil downslope under the influence of gravity
3. **Erosion**—The physical removal of material by mobile agents such as water, wind, or ice

We first turn our attention to the process of weathering and the products generated by this activity. However, weathering cannot be easily separated from the other two processes because, as weathering breaks rocks apart, it facilitates the movement of rock debris by mass wasting and erosion. Conversely, the transport of material by mass wasting and erosion further disintegrates and decomposes the rock.

4.1 CONCEPT CHECKS

1 List examples of Earth's external and internal processes.

2 From where do these processes derive their energy?

4.2 | WEATHERING

Define *weathering* and distinguish between the two main categories of weathering.

Weathering goes on all around us, but it seems like such a slow and subtle process that it is easy to underestimate its importance. Yet, it is worth remembering that weathering is a basic part of the rock cycle and thus a key process in the Earth system. Weathering is also important to humans—even to those of us who are not studying geology. For example, many of the life-sustaining minerals and elements found in soil, and ultimately in the food we eat, were freed from solid rock by weathering processes. As **FIGURE 4.2** and many other images in this book illustrate, weathering also contributes to the formation of some of Earth's most spectacular scenery. Of course, these same processes are also responsible for causing the deterioration of many of the structures we build.

All materials are susceptible to weathering. Consider, for example, the fabricated product concrete, which closely resembles the sedimentary rock called conglomerate. A newly poured concrete sidewalk has a smooth, unweathered look. However, not many years later, the same sidewalk will appear chipped, cracked, and rough, with pebbles exposed at the surface. If a tree is nearby, its roots may grow under the concrete, heaving and buckling

FIGURE 4.2 Arches National Park Mechanical and chemical weathering contributed greatly to the formation of the arches and other rock formations in Utah's Arches National Park. (Photo by Whit Richardson/Aurora Open/SuperStock)

Some Everyday Examples of Weathering

Weathering processes are responsible for the deterioration of objects and materials at Earth's surface. Day to day changes are not obvious, but over time the effects are significant.

Weathering gradually deteriorated the paint protecting this wood siding.

Even the most "solid" stone structure that people erect gradually succumbs to weathering processes.

Michelle Milano/Dreamstime

William Britten/iStockphoto

Plant roots can break concrete and rock.

Potholes are especially common in places that experience wintry conditions. The freeze-thaw cycle coupled with water and road salt contribute to pavement destruction.

The chemical weathering process called oxidation is responsible for the rust on this car.

EuroStyle Graphics/Alamy

Christian Delbert/Dreamstime

Ondrejparaj/Dreamstime

it. The same natural processes that eventually break apart a concrete sidewalk also act to disintegrate rocks, regardless of their type or strength.

Weathering occurs when rock is mechanically fragmented (disintegrated) and/or chemically altered (decomposed). **Mechanical weathering** is accomplished by physical forces that break rock into smaller and smaller pieces without changing the rock's mineral composition. **Chemical weathering** involves a chemical transformation of rock into one or more new compounds. These two concepts can be illustrated with a large log. The log disintegrates when it is split into smaller and smaller pieces by, whereas decomposition occurs when the log is set afire and burned.

In the following sections, we discuss the various modes of mechanical and chemical weathering. Although we consider these two categories separately, keep in mind that mechanical and chemical weathering processes usually work simultaneously in nature and reinforce each other.

Mechanical Weathering

When a rock undergoes mechanical weathering, it is broken into smaller and smaller pieces, each retaining the characteristics of the original material. The end result is many small pieces from a single large one. **FIGURE 4.3** shows that breaking a rock into smaller pieces increases the surface area available for chemical attack. Hence, by breaking rocks into smaller pieces, mechanical weathering increases the amount of surface area available for chemical weathering.

In nature, four important physical processes lead to the fragmentation of rock: frost wedging, salt crystal growth,

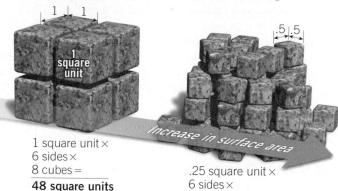

As mechanical weathering breaks rock into smaller pieces, more surface area is exposed to chemical weathering.

4 square units × 6 sides × 1 cube =
24 square units

1 square unit × 6 sides × 8 cubes =
48 square units

.25 square unit × 6 sides × 64 cubes =
96 square units

Increase in surface area

SmartFigure 4.3 Mechanical Weathering Increases Surface Area Mechanical weathering adds to the effectiveness of chemical weathering because chemical weathering can occur only on exposed surfaces.

expansion resulting from unloading (sheeting), and biological activity. In addition, although the work of erosional agents such as waves, wind, glacial ice, and running water is usually considered separately from mechanical weathering, it is nevertheless related. As these mobile agents move rock debris, particles continue to be broken and abraded.

Frost Wedging   If you leave a glass bottle of water in the freezer a bit too long, you will find the bottle fractured, as in **FIGURE 4.4**. The bottle breaks because liquid water has the unique property of expanding about 9 percent upon freezing. This is also the reason that poorly insulated or exposed water pipes rupture during frigid weather. You might also expect this same process to fracture rocks in nature. This is, in fact, the basis for the traditional explanation of **frost wedging**. After water works its way into the cracks in rock, the freezing water enlarges the cracks, and angular fragments are eventually produced (**FIGURE 4.5**).

For many years, the conventional wisdom was that most frost wedging occurred in the manner just described.

SmartFigure 4.5 Ice Breaks Rock In mountainous areas, frost wedging creates angular rock fragments that accumulate to form talus slopes. (Photo by Marli Miller)

FIGURE 4.4 Ice Breaks a Bottle The bottle broke because water expands about 9 percent when it freezes. (Photo by Martyn F. Chillmaid/Science Source)

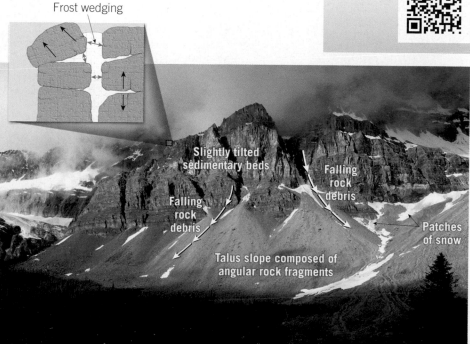

Frost wedging

Slightly tilted sedimentary beds

Falling rock debris

Falling rock debris

Patches of snow

Talus slope composed of angular rock fragments

Recently, however, research has shown that frost wedging can also occur in a different way.[1] It has long been known that when moist soils freeze, they expand, or *frost heave*, due to the growth of ice lenses. These masses of ice grow larger because they are supplied with water migrating from unfrozen areas as thin liquid films. As more water accumulates and freezes, the soil is heaved upward. A similar process occurs within the cracks and pore spaces of rocks. Lenses of ice grow larger as they attract liquid water from surrounding pores. The growth of these ice masses gradually weakens the rock, causing it to fracture.

Salt Crystal Growth

Another expansive force that can split rocks is created by the growth of salt crystals. Rocky shorelines and arid regions are common settings for this process. It begins when sea spray from breaking waves or salty groundwater penetrates crevices and pore spaces in rock. As this water evaporates, salt crystals form. As these crystals gradually grow larger, they weaken the rock by pushing apart the surrounding grains or enlarging tiny cracks.

This same process can also contribute to crumbling roadways where salt is spread to melt snow and ice in winter. The salt dissolves in water and seeps into cracks that quite likely originated from frost action. When the water evaporates, the growth of salt crystals further breaks the pavement.

Sheeting

When large masses of igneous rock, particularly those composed of granite, are exposed by erosion,

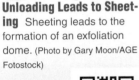

SmartFigure 4.6 Unloading Leads to Sheeting Sheeting leads to the formation of an exfoliation dome. (Photo by Gary Moon/AGE Fotostock)

[1]Bernard Hallet, "Why Do Freezing Rocks Break?" *Science* 314 (November 2006): 1092–1099.

concentric slabs begin to break loose. The process generating these onion-like layers is called **sheeting**. It takes place, at least in part, due to the great reduction in pressure that occurs when the overlying rock is eroded away in a process called *unloading*. **FIGURE 4.6** illustrates what happens: As the overburden is removed, the outer parts of the granitic mass expand more than the rock below and separate from the rock body. Continued weathering eventually causes the slabs to separate and spall off, creating **exfoliation domes** (*ex* = off, *folium* = leaf). Excellent examples of exfoliation domes include Stone Mountain in Georgia and Half Dome in Yosemite National Park.

Other examples illustrate how rock behaves when the confining pressure is greatly reduced. These examples support identifying the process of unloading as the cause of sheeting. In deep mines, large rock slabs have been known to explode off the walls of newly cut tunnels. In quarries, fractures occur parallel to the floor when large blocks of rock are removed.

Although many fractures are created by expansion, others are produced by contraction as igneous materials cool (see Figure 9.27, page 303), and still others are produced by tectonic forces during mountain building. Fractures produced by these activities often form a definite pattern and are called *joints* (see Figure 4.8, page 103, and Figure 10.21, page 328). Joints are important rock structures that allow water to penetrate deeply and start the process of weathering long before the rock is exposed.

Biological Activity

The activities of organisms, including plants, burrowing animals, and humans can cause weathering. Plant roots in search of minerals and water grow into fractures, and as the roots grow, they wedge the rock

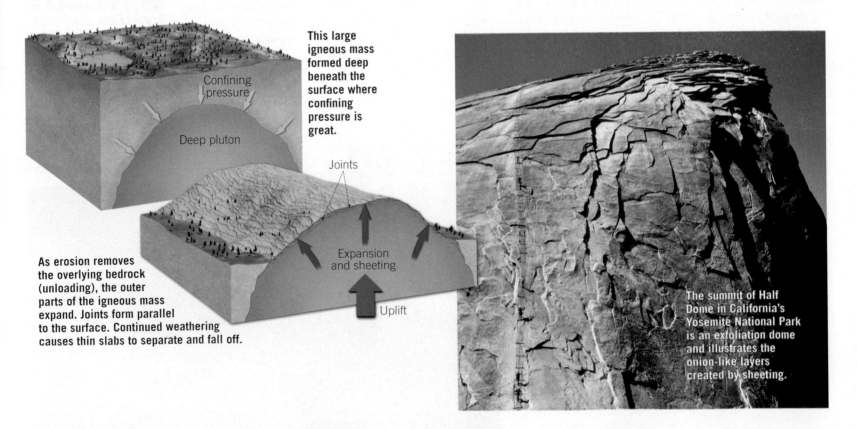

This large igneous mass formed deep beneath the surface where confining pressure is great.

Confining pressure

Deep pluton

Joints

Expansion and sheeting

Uplift

As erosion removes the overlying bedrock (unloading), the outer parts of the igneous mass expand. Joints form parallel to the surface. Continued weathering causes thin slabs to separate and fall off.

The summit of Half Dome in California's Yosemite National Park is an exfoliation dome and illustrates the onion-like layers created by sheeting.

apart (**FIGURE 4.7**). Burrowing animals further break down the rock by moving fresh material to the surface, where physical and chemical processes can more effectively attack it. Decaying organisms also produce acids, which contribute to chemical weathering. Where rock has been blasted in search of minerals or for road construction, the impact of humans is particularly noticeable.

Organisms play many roles in chemical weathering. For example, plant roots, fungi, and lichens that occupy fractures or that may encrust a rock produce acids that promote decomposition. Moreover, some bacteria are capable of extracting compounds from minerals and using the energy from a compound's chemical bonds to supply their life needs. These primitive "mineral eating" life-forms can live at depths as great as a few kilometers.

Chemical Weathering

In the preceding discussion of mechanical weathering, you learned that breaking rock into smaller pieces aids chemical weathering by increasing the surface area available for chemical attack. It should also be pointed out that chemical weathering contributes to mechanical weathering. It does so by weakening the outer portions of some rocks, which, in turn, makes them more susceptible to being broken by mechanical weathering processes.

Chemical weathering involves the complex processes that alter the internal structures of minerals by removing and/or adding elements. During this transformation, the original rock decomposes into substances that are stable in the surface environment. Consequently, the products of chemical weathering will remain essentially unchanged as long as they remain in an environment similar to the one in which they formed.

Water and Carbonic Acid   Water is by far the most important agent of chemical weathering. Although pure water is nonreactive, a small amount of dissolved material is generally all that is needed to activate it. Oxygen dissolved in water will *oxidize* some materials. For example, when an iron nail is found in moist soil, it will have a coating of rust (iron oxide), and if the time of exposure has been long, the nail will be so weak that it can be broken as easily as a

FIGURE 4.7 Plants Can Break Rock Root wedging near Boulder, Colorado. (Photo by Kristin Piljay)

Plant roots can extend into joints and grow in diameter and length. This process enlarges fractures and breaks rock.

toothpick. When rocks containing iron-rich minerals oxidize, a yellow to reddish-brown rust will appear on the surface.

Carbon dioxide (CO_2) dissolved in water (H_2O) forms **carbonic acid** (H_2CO_3)—the same weak acid produced when soft drinks are carbonated. Rain dissolves some carbon

EYE ON EARTH

This is a close-up view of a massive granite feature in the Sierra Nevada of California. (Photo by Marli Miller)

QUESTION 1 *Relatively thin slabs of granite are separating from this rock mass. Describe the process that caused this to occur.*

QUESTION 2 *What term is applied to the process that you described in Question 1? What term describes the domelike feature that results from this process?*

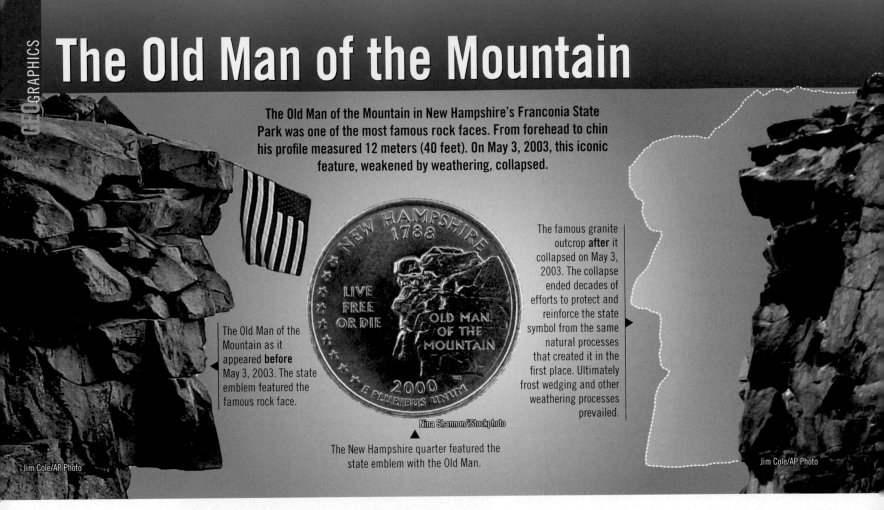

The Old Man of the Mountain

The Old Man of the Mountain in New Hampshire's Franconia State Park was one of the most famous rock faces. From forehead to chin his profile measured 12 meters (40 feet). On May 3, 2003, this iconic feature, weakened by weathering, collapsed.

The Old Man of the Mountain as it appeared **before** May 3, 2003. The state emblem featured the famous rock face.

The famous granite outcrop **after** it collapsed on May 3, 2003. The collapse ended decades of efforts to protect and reinforce the state symbol from the same natural processes that created it in the first place. Ultimately frost wedging and other weathering processes prevailed.

Nina Shannon/iStockphoto

The New Hampshire quarter featured the state emblem with the Old Man.

Jim Cole/AP Photo

Jim Cole/AP Photo

dioxide as it falls through the atmosphere, and additional amounts released by decaying organic matter are acquired as the water percolates through the soil. Carbonic acid ionizes to form the very reactive hydrogen ion (H^+) and the bicarbonate ion (HCO_3).

Acids such as carbonic acid readily decompose many rocks and produce certain products that are water soluble. For example, the mineral calcite ($CaCO_3$), which composes the common building stones marble and limestone, is easily attacked by even a weakly acidic solution. In nature, over spans of thousands of years, large quantities of limestone are dissolved and carried away by groundwater. This activity is largely responsible for the formation of limestone caverns.

How Granite Weathers To illustrate how a rock rich in silicate minerals chemically weathers when attacked by carbonic acid, we will consider the weathering of granite, an abundant continental rock. Recall that granite consists mainly of quartz and potassium feldspar. The weathering of the potassium feldspar component of granite takes place as follows:

$$2\ KAlSi_3O_8 + 2(H^+ + HCO_3) + H_2O$$
potassium feldspar   carbonic acid   water

$$Al_2Si_2O_5(OH)_4 + 2K^+ + 2\ HCO_3 + SiO_2$$
clay mineral   potassium   bicarbonite   silica
ion   ion

in solution

In this reaction, the hydrogen ions (H^+) attack and replace potassium ions (K^+) in the feldspar structure, thereby disrupting the crystalline network. Once removed, the potassium is available as a nutrient for plants or becomes the soluble salt potassium bicarbonate ($KHCO_3$), which may be incorporated into other minerals or carried to the ocean in dissolved form by groundwater and streams.

The most abundant products of the chemical breakdown of feldspar are residual clay minerals. Clay minerals are the end product of weathering and are very stable under surface conditions. Consequently, clay minerals make up a high percentage of the inorganic material in soils. Moreover, the most abundant sedimentary rock, shale, contains a high proportion of clay minerals.

In addition to the formation of clay minerals during this reaction, some silica is removed from the feldspar structure and is carried away by groundwater. This dissolved silica will eventually precipitate to produce nodules of chert or flint, fill in the pore spaces between sediment grains, or be carried to the ocean, where microscopic animals will remove it to build hard silica shells.

To summarize, the weathering of potassium feldspar generates a residual clay mineral, a soluble salt (potassium bicarbonate), and some silica that enters into solution.

Quartz, the other main component of granite, is *very resistant* to chemical weathering; it remains substantially unaltered when attacked by weakly acidic solutions. As a result, when granite weathers, the feldspar crystals dull and slowly turn to clay, releasing the once interlocked quartz grains, which still retain their fresh, glassy appearance. Although some quartz remains in the soil, much is transported

TABLE 4.1 Products of Weathering

Mineral	Residual Products	Material in Solution
Quartz	Quartz grains	Silica
Feldspars	Clay minerals	Silica, K^+, Na^+, Ca^{2+}
Amphibole	Clay minerals	Silica, Ca^{2+}, Mg^{2+}
	Iron oxides	
Olivine	Iron oxides	Silica, Mg^{2+}

to the sea or to other sites of deposition, where it becomes the main constituent of such features as sandy beaches and sand dunes. In time it may become lithified to form the sedimentary rock *sandstone*.

Weathering of Silicate Minerals **TABLE 4.1** lists the weathered products of some of the most common silicate minerals. Remember that silicate minerals make up most of Earth's crust and that these minerals are composed essentially of only eight elements. When chemically weathered, silicate minerals yield sodium, calcium, potassium, and magnesium ions, which form soluble products that may be removed by groundwater. The element iron combines with oxygen, producing relatively insoluble iron oxides, which give soil a reddish-brown or yellowish color. Under most conditions, the three remaining elements—aluminum, silicon, and oxygen—join with water to produce residual clay minerals. However, even the

highly insoluble clay minerals are very slowly removed by subsurface water.

Spheroidal Weathering Many rock outcrops have a rounded appearance. This occurs because chemical weathering works inward from exposed surfaces. **FIGURE 4.8** illustrates how angular masses of jointed rock change through time. The process is aptly called **spheroidal weathering**. Because weathering attacks edges from two sides and corners from three sides, these areas wear down faster than does a single flat surface. Gradually, sharp edges and corners become smooth and rounded. Eventually an angular block may evolve into a nearly spherical boulder. Once this occurs, the boulder's shape does not change, but the spherical mass continues to get smaller.

4.2 CONCEPT CHECKS

1 When a rock is mechanically weathered, how does its surface area change? How does this influence chemical weathering?
2 Explain how water can cause mechanical weathering.
3 How does an exfoliation dome form?
4 How does biological activity contribute to weathering?
5 How is carbonic acid formed in nature? What products result when carbonic acid reacts with potassium feldspar?
6 Explain how angular masses of rock often become spherical boulders.

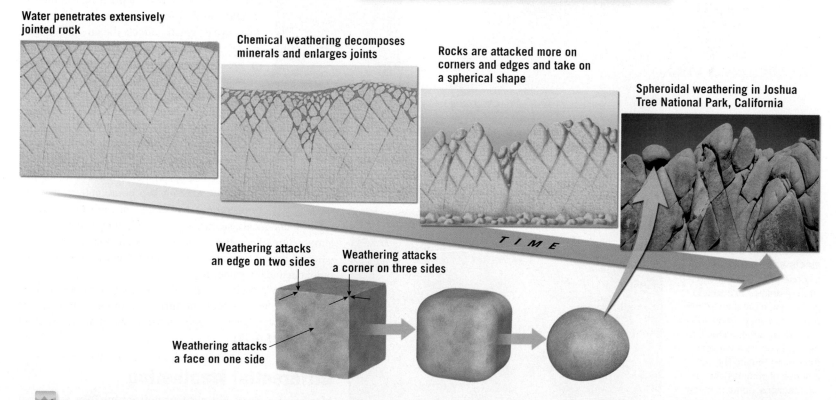

SmartFigure 4.8 The Formation of Rounded Boulders Spheroidal weathering of extensively jointed rock in California's Joshua Tree National Park. (Photo by E. J. Tarbuck)

4.3 | RATES OF WEATHERING

Summarize the factors that influence the type and rate of rock weathering.

Several factors influence the type and rate of rock weathering. We have already seen how mechanical weathering affects the rate of weathering. When rocks are broken into smaller pieces, the amount of surface area exposed to chemical weathering is increased. Here we examine two other important factors: rock characteristics and climate.

Rock Characteristics

Rock characteristics encompass all the chemical traits of rocks, including mineral composition and solubility. In addition, physical features such as joints (cracks) can be important because they influence the ability of water to penetrate rock.

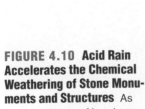
SmartFigure 4.9 Rock Type Influences Weathering An examination of headstones in the same cemetery shows that the rate of chemical weathering depends on the rock type. (Photos by E. J. Tarbuck)

This granite headstone was erected in 1868. The inscription is still fresh looking.

This monument is composed of marble, a calcite-rich metamorphic rock. The headstone was erected in 1872, four years after the granite stone. The inscription is nearly illegible.

The variations in weathering rates due to mineral constituents can be demonstrated by comparing old headstones made from different rock types. Headstones of granite, which are composed of silicate minerals, are relatively resistant to chemical weathering. We can see this by examining the inscriptions on the headstones shown in **FIGURE 4.9**. On the other hand, the marble headstone shows signs of extensive chemical alteration over a relatively short period. Marble is composed of calcite (calcium carbonate), which readily dissolves even in a weakly acidic solution.

The silicates, the most abundant mineral group, weather in essentially the same sequence as their order of crystallization. By examining Bowen's reaction series (see Figure 3.12, page 70), you can see that olivine crystallizes first and is therefore the least resistant to chemical weathering, whereas quartz, which crystallizes last, is the most resistant.

Climate

Climatic factors, particularly temperature and moisture, are crucial to the rate of rock weathering. One important example from mechanical weathering is that the frequency of freeze–thaw cycles greatly affects the amount of frost wedging. Temperature and moisture also exert a strong influence on the rates of chemical weathering and on the kind and amount of vegetation present. Regions with lush vegetation generally have a thick mantle of soil rich in decayed organic matter from which chemically active fluids such as carbonic and humic acids are derived.

The optimal environment for chemical weathering is a combination of warm temperatures and abundant moisture. In polar regions chemical weathering is ineffective because frigid temperatures keep the available moisture locked up as ice, whereas in arid regions there is insufficient moisture to foster rapid chemical weathering.

Human activities can influence the composition of the atmosphere, which in turn can impact the rate of chemical weathering. One well-known example is acid rain (**FIGURE 4.10**).

Differential Weathering

Masses of rock do not weather uniformly. Take a moment to look at the photo of Shiprock, New Mexico, in Figure 9.23 (page 300). The durable igneous mass is more resistant to weathering than the surrounding rock layers and so protrudes high above the surface. A glance at Figure 4.2 shows an additional example of this phenomenon, called

FIGURE 4.10 Acid Rain Accelerates the Chemical Weathering of Stone Monuments and Structures As a consequence of burning large quantities of coal and petroleum, more than 20 million tons of sulfur and nitrogen oxides are released into the atmosphere each year in the United States. Through a series of complex chemical reactions, some of these pollutants are converted into acids that then fall to Earth's surface as rain or snow. This decomposing building facade is in Leipzig, Germany. (Photo by Doug Plummer/Science Source)

Mobile Field Trip 4.11 Monuments to Weathering

An example of differential weathering in New Mexico's Bisti Badlands. When weathering accentuates differences in rocks, spectacular landforms are sometimes created.

(Photo by Michael Collier)

differential weathering. The results vary in scale from the rough, uneven surface of the marble headstone in Figure 4.10 to the boldly sculpted exposures of bedrock in New Mexico's Bisti Badlands (**FIGURE 4.11**).

Many factors influence the rate of rock weathering. Among the most important are variations in the composition of the rock. More resistant rock protrudes as ridges or pinnacles, or as steeper cliffs on a canyon wall (see Figure 11.2, page 350). The number and spacing of joints can also be a significant factor (see Figure 4.9 and Figure 10.21, page 328). Differential weathering and subsequent erosion are responsible for creating many unusual and sometimes spectacular rock formations and landforms.

4.3 CONCEPT CHECKS

1 Why did the headstones in Figure 4.9 weather so differently?

2 How does climate influence weathering?

4.4 | SOIL

Define *soil* and explain why soil is referred to as an interface.

Weathering is a key process in the formation of soil. Along with air and water, soil is one of our most indispensable resources. Also, like air and water, soil is often taken for granted. The following quote helps put this vital layer in perspective:

> Science, in recent years, has focused more and more on the Earth as a planet, one that for all we know is unique—where a thin blanket of air, a thinner film of water, and the thinnest veneer of soil combine to support a web of life of wondrous diversity in continuous change.[2]

[2]Jack Eddy, "A Fragile Seam of Dark Blue Light," in *Proceedings of the Global Change Research Forum*. U.S. Geological Survey Circular 1086, 1993, p. 15.

Soil has accurately been called "the bridge between life and the inanimate world." All life—the entire biosphere—owes its existence to a dozen or so elements that must ultimately come from Earth's crust. After weathering and other processes create soil, plants carry out the intermediary role of assimilating the necessary elements and making them available to animals, including humans.

An Interface in the Earth System

When Earth is viewed as a system, soil is referred to as an *interface*—a common boundary where different parts of a system interact. This is an appropriate designation because

FIGURE 4.12 What Is Soil? The graph depicts the composition (by volume) of a soil in good condition for plant growth. Although percentages vary, each soil is composed of mineral and organic matter, water, and air. (Photo by i love images/gardening/Alamy Images)

25% air

45% mineral matter

25% water

5% organic matter

soil forms where the geosphere, the atmosphere, the hydrosphere, and the biosphere meet. Soil develops in response to complex environmental interactions among different parts of the Earth system. Over time, soil gradually evolves to a state of equilibrium, or balance, with the environment. Soil is dynamic and sensitive to almost every aspect of its surroundings. Thus, when environmental changes occur—in climate, vegetative cover, or animal (including human) activity—soil responds. Any such change produces a gradual alteration of soil characteristics until a new balance is reached. Although thinly distributed over the land surface, soil functions as a fundamental interface, providing an excellent example of the integration among many parts of the Earth system.

What Is Soil?

With few exceptions, Earth's land surface is covered by **regolith** (*rhegos* = blanket, *lithos* = stone), a layer of rock and mineral fragments produced by weathering. Some would call this material soil, but soil is more than an accumulation of weathered debris. **Soil** is a combination of mineral and organic matter, water, and air—the portion of the regolith that supports the growth of plants. Although the proportions of the major components in soil vary, the same four components are always present to some extent (**FIGURE 4.12**). About one-half of the total volume of a good-quality surface soil is a mixture of disintegrated and decomposed rock (mineral matter) and *humus*, the decayed remains of animal and plant life (organic matter). The remaining half consists of pore spaces among the solid particles where air and water circulate.

Although the mineral portion of the soil is usually much greater than the organic portion, humus is an essential component. In addition to being an important source of plant nutrients, humus enhances the soil's ability to retain water. Because plants require air and water to live and grow, the portion of the soil consisting of pore spaces that allow for the circulation of these fluids is as vital as the solid soil constituents.

Soil water is far from "pure" water; instead, it is a complex solution containing many soluble nutrients. Soil water not only provides the necessary moisture for the chemical reactions that sustain life; it also supplies plants with nutrients in a form they can use. The pore spaces not filled with water contain air. This air is the source of necessary oxygen and carbon dioxide for most microorganisms and plants that live in the soil.

Soil Texture and Structure

Most soils are far from uniform and contain particles of different sizes. **Soil texture** refers to the proportions of different particle sizes. Texture is a very basic soil property because it strongly influences the soil's ability to retain and transmit water and air, both of which are essential to plant growth. Sandy soils may drain too rapidly and dry out quickly. At the opposite extreme, the pore spaces of clay-rich soils may be so small that they inhibit drainage, and long-lasting puddles result. Moreover, when the clay and silt content is very high, plant roots may have difficulty penetrating the soil.

Because soils rarely consist of particles of only one size, *textural categories* have been established, based on the varying proportions of clay, silt, and sand. The standard system of classes used by the U.S. Department of Agriculture is shown in **FIGURE 4.13**. For example, point *A* on this triangular diagram (left center) represents a soil composed of 10 percent silt, 40 percent clay, and 50 percent sand. Such a soil is called a *sandy clay*. The soils called *loam*, which occupy the central portion of the diagram, are those in which no single particle size predominates over the other two. Loam soils are best suited to support plant life because they generally have better moisture characteristics and nutrient storage ability than do soils composed predominantly of clay or coarse sand.

Soil particles are seldom completely independent of one another. Rather, they usually form clumps called *peds* that give soils a particular structure. Four basic soil structures are recognized: platy, prismatic, blocky, and spheroidal. Soil structure is important because it influences the

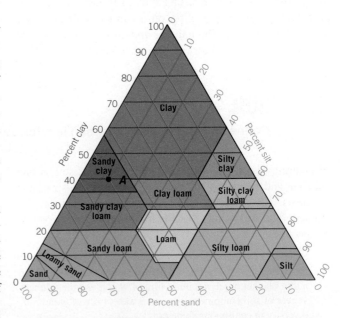

FIGURE 4.13 Soil-texture Diagram The texture of any soil can be represented by a point on this diagram. Soil texture is one of the factors used to estimate agricultural potential and engineering characteristics. (U.S. Department of Agriculture)

ease of a soil's cultivation as well as the susceptibility of a soil to erosion. In addition, soil structure affects the porosity and permeability of soil (that is, the ease with which water can penetrate). This in turn influences the movement of nutrients to plant roots. Prismatic and blocky peds usually allow for moderate water infiltration, whereas platy and spheroidal structures are characterized by slower infiltration rates.

4.4 CONCEPT CHECKS

1 Why is soil considered an interface in the Earth system?

2 How is regolith different from soil?

3 Why is texture an important soil property?

4 Using the soil texture diagram in Figure 4.14, name the soil that consists of 60 percent sand, 30 percent silt, and 10 percent clay.

4.5 | CONTROLS OF SOIL FORMATION

List and briefly discuss five controls of soil formation.

Soil is the product of the complex interplay of several factors. The most important of these are parent material, time, climate, plants and animals, and topography. Although all these factors are interdependent, their roles are examined separately.

learned that similar soils often develop from different parent materials and that dissimilar soils can develop from the same parent material. Such discoveries reinforce the importance of the other soil-forming factors.

Parent Material

The source of the weathered mineral matter from which soils develop is called the **parent material**, and it is a major factor influencing a newly forming soil. Gradually it undergoes physical and chemical changes, as the processes of soil formation progress. Parent material may be the underlying bedrock, or it can be a layer of unconsolidated deposits, as in a stream valley. When the parent material is bedrock, the soils are termed *residual soils*. By contrast, soils developed on loose sediment are called *transported soils* (**FIGURE 4.14**). Note that transported soils form *in place* on parent materials that have been carried from elsewhere and deposited by gravity, water, wind, or ice.

The nature of the parent material influences soils in two ways. First, the type of parent material affects the rate of weathering and thus the rate of soil formation. (Consider the weathering rates of granite versus limestone.) Also, because sediments are already partly weathered and provide more surface area for chemical weathering, soil development on such material usually progresses more rapidly. Second, the chemical makeup of the parent material affects the soil's fertility. This influences the character of the natural vegetation the soil can support.

At one time parent material was thought to be the primary factor causing differences among soils. However, soil scientists came to understand that other factors, especially climate, are more important. In fact, they

Time

Time is an important component of *every* geological process, and soil formation is no exception. The nature of soil is strongly influenced by the length of time that processes have been operating. If weathering has been going on for a comparatively short time, the parent material strongly influences the characteristics of the soil. As weathering processes continue, the influence of parent material on soil is overshadowed by the other soil-forming factors, especially climate. The amount of time required for various soils to evolve cannot be specified because the soil-forming processes act at varying rates under different circumstances. However, as a rule, the longer a soil has been forming, the thicker it becomes and the less it resembles the parent material.

No soil development because of very steep slope

Transported soil developed on unconsolidated stream deposits

Residual soil is developed on bedrock

Thicker soil develops on flat terrain

Bedrock

Thinner soil on steep slope because of erosion

Unconsolidated deposits

FIGURE 4.14 Slopes and Soil Development The parent material for residual soils is the underlying bedrock. Transported soils form on unconsolidated deposits. Also note that as slopes become steeper, soil becomes thinner. (Left and center photos by E. J. Tarbuck; right photo by Lucarelli Temistocle/Shutterstock)

Climate

Climate is the most influential control of soil formation. Just as temperature and precipitation are the climatic elements that influence people the most, so too are they the elements that exert the strongest impact on soil formation. Variations in temperature and precipitation determine whether chemical or mechanical weathering predominates. They also greatly influence the rate and depth of weathering. For instance, a hot, wet climate may produce a thick layer of chemically weathered soil in the same amount of time that a cold, dry climate produces a thin mantle of mechanically weathered debris. Also, the amount of precipitation influences the degree to which various materials are removed (leached) from the soil, thereby affecting soil fertility. Finally, climatic conditions are important factors controlling the type of plant and animal life present.

Plants and Animals

The biosphere plays a vital role in soil formation. The types and abundance of organisms present have a strong influence on the physical and chemical properties of a soil. In fact, for well-developed soils in many regions, the significance of natural vegetation is frequently implied in the description used by soil scientists. Phrases such as *prairie soil*, *forest soil*, and *tundra soil* are common (**FIGURE 4.15**).

Plants and animals furnish organic matter to the soil. Certain bog soils are composed almost entirely of organic matter, whereas desert soils may contain only a tiny percentage. Although the quantity of organic matter varies substantially among soils, almost no soils completely lack organic matter.

The primary source of organic matter is plants, although animals and microorganisms also contribute. When organic matter decomposes, important nutrients are supplied to plants, as well as to animals and microorganisms living in the soil. Consequently, soil fertility depends in part on the amount of organic matter present. Furthermore, the decay of plant and animal remains causes the formation of various organic acids. These complex acids hasten the weathering process. Organic matter also has a high water-holding ability and thus aids water retention in a soil.

Microorganisms play an active role in the decay of plant and animal remains. The end product is *humus*, a material that no longer resembles the plants and animals from which it is formed. In addition, certain microorganisms aid soil fertility because they have the ability to convert atmospheric nitrogen gas into soil nitrogen compounds.

Earthworms and other burrowing animals mix the mineral and organic portions of a soil. Earthworms, for example, feed on organic matter and thoroughly mix soils in which they live, often moving and enriching many tons per acre each year. Burrows and holes also aid the passage of water and air through the soil.

Topography

The lay of the land can vary greatly over short distances. Such variations in topography can lead to the development of a variety of localized soil types. Many of the differences exist because the length and steepness of slopes have a significant impact on the amount of erosion and the water content of soil.

On steep slopes, soils are often poorly developed. In such situations, little water can soak in, and as a result, soil moisture may be insufficient for vigorous plant growth. Furthermore, because of accelerated erosion on steep slopes, the soils are thin or nonexistent (see Figure 4.14).

In contrast, waterlogged soils in poorly drained bottomlands have a much different character. Such soils are usually thick and dark. The dark color results from the large quantity of organic matter that accumulates because saturated conditions retard the decay of vegetation. The optimum terrain for soil development is a flat-to-undulating upland surface. This terrain experiences good drainage, minimum erosion, and sufficient infiltration of water into the soil.

Slope orientation, the direction a slope is facing, also is significant. In the midlatitudes of the Northern Hemisphere, a south-facing slope receives a great deal more sunlight than does a north-facing slope. In fact, a steep north-facing slope may receive no direct sunlight at all. The difference in the amount of solar radiation received causes substantial differences in soil temperature and moisture, which in turn influences the nature of the vegetation and the character of the soil.

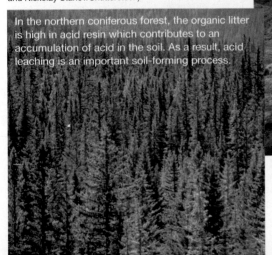

FIGURE 4.15 Plants Influence Soil The nature of the vegetation in an area can have a significant influence on soil formation. (Photos by Bill Brooks/Alamy Images; Nickolay Stanev/Shutterstock; and Nickolay Stanev/Shutterstock)

Meager desert rainfall means reduced rates of weathering and relatively meager vegetation. Desert soils are typically thin and lack much organic matter.

In the northern coniferous forest, the organic litter is high in acid resin which contributes to an accumulation of acid in the soil. As a result, acid leaching is an important soil-forming process.

Soils that develop in well-drained prairie regions typically have a humus-rich surface horizon that is rich in calcium and magnesium. Fertility is usually excellent.

Although we have dealt separately with each of the soil-forming factors, remember that *all of them work together* to form soil. No single factor is responsible for a soil being as it is. Rather, the combined influence of parent material, time, climate, plants and animals, and topography determines a soil's character.

4.6 | THE SOIL PROFILE

Sketch, label, and describe an idealized soil profile.

Because soil-forming processes operate from the surface downward, variations in composition, texture, structure, and color gradually evolve at varying depths. These vertical differences, which usually become more pronounced as time passes, divide the soil into zones or layers known as **soil horizons**. If you were to dig a trench in soil, you would see that its walls are layered. Such a vertical section through all the soil horizons constitutes the **soil profile**.

FIGURE 4.16 presents an idealized view of a well-developed soil profile in which five horizons are identified. From the surface downward, they are designated as *O*, *A*, *E*, *B*, and *C*. These five horizons are common to soils in temperate regions:

- The *O* soil horizon consists largely of organic material. This is in contrast to the layers beneath it, which consist mainly of mineral matter. The upper portion of the *O* horizon is primarily plant litter, such as loose leaves and other organic debris that is still recognizable. By contrast, the lower portion of the *O* horizon is made up of partly decomposed organic matter (humus) in which plant structures can no longer be identified. In addition to plants, the *O* horizon is teeming with microscopic life, including bacteria, fungi, algae, and insects. All these organisms contribute oxygen, carbon dioxide, and organic acids to the developing soil.
- The *A* horizon is largely mineral matter, yet biological activity is high and humus is generally present—up to 30 percent in some instances. Together the *O* and *A* horizons make up what is commonly called the *topsoil*.
- The *E* horizon is a light-colored layer that contains little organic material. As water percolates downward through this zone, finer particles are carried away. This washing out of fine soil components is termed **eluviation** (*elu* = get away from, *via* = a way). Water percolating downward also dissolves soluble inorganic soil components and carries them to deeper zones. This depletion of soluble materials from the upper soil is termed **leaching**.
- The *B* horizon, or *subsoil*, is where much of the material removed from the *E* horizon by eluviation is deposited. Thus, the *B* horizon is often referred to as the *zone of accumulation*. The accumulation of the fine clay particles enhances water retention in the subsoil. In extreme cases, clay accumulation can form a very compact and impermeable layer called *hardpan*.

- The *O*, *A*, *E*, and *B* horizons together constitute the **solum**, or "true soil." It is in the solum that the soil-forming processes are active and that living roots and other plant and animal life are largely confined.
- The *C horizon* is characterized by partially altered parent material. Whereas the *O*, *A*, *E*, and *B* horizons bear little resemblance to the parent material, it is easily identifiable in the *C* horizon. Although this material is undergoing changes that will eventually transform it into soil, it has not yet crossed the threshold that separates regolith from soil.

The characteristics and extent of development can vary greatly among soils in different environments (**FIGURE 4.17**). Thus, different localities exhibit soil profiles that can

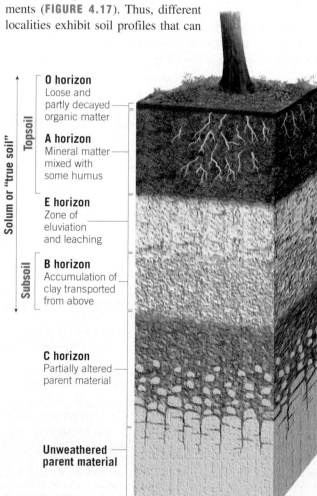

FIGURE 4.16 Soil Horizons Idealized soil profile from a humid climate in the middle latitudes.

Solum or "true soil"

Topsoil

Subsoil

O horizon Loose and partly decayed organic matter

A horizon Mineral matter mixed with some humus

E horizon Zone of eluviation and leaching

B horizon Accumulation of clay transported from above

C horizon Partially altered parent material

Unweathered parent material

FIGURE 4.17 Contrasting Soil Profiles Soil characteristics and development vary greatly in different environments. (Left photo by USDA; right photo courtesy of E. J. Tarbuck)

Horizons are indistinct in this soil in Puerto Rico, giving it a relatively uniform appearance.

This profile shows a soil in southeastern South Dakota with well-developed horizons.

contrast greatly with one another. The boundaries between soil horizons may be sharp, or the horizons may blend gradually from one to another. Consequently, a well-developed soil profile indicates that environmental conditions have been relatively stable over an extended time span and that the soil is *mature*. By contrast, some soils lack horizons altogether. Such soils are called *immature* because soil building has been going on for only a short time. Immature soils are also characteristic of steep slopes, where erosion continually strips away the soil, preventing full development.

4.6 CONCEPT CHECKS

1 Sketch and label the main soil horizons in a well-developed soil profile.

2 Describe the following features or processes: eluviation, leaching, zone of accumulation, and hardpan.

4.7 | CLASSIFYING SOILS

Explain the need for classifying soils.

There are many variations from place to place and from time to time among the factors that control soil formation. These differences lead to a bewildering variety of soil types. To cope with such variety, it is essential to devise some means of classifying the vast array of data to be studied. One way to do this is to establish groups consisting of items that have

TABLE 4.2 Basic Soil Orders

Soil Order	Description	Percentage*
Alfisols	Moderately weathered soils that form under boreal forests or broadleaf deciduous forests, rich in iron and aluminum. Clay particles accumulate in a subsurface layer in response to leaching in moist environments. Fertile, productive soils, because they are neither too wet nor too dry.	9.65
Andisols	Young soils in which the parent material is volcanic ash and cinders, deposited by recent volcanic activity.	0.7
Aridosols	Soils that develop in dry places; insufficient water to remove soluble minerals; may have an accumulation of calcium carbonate, gypsum, or salt in subsoil; low organic content.	12.02
Entisols	Young soils having limited development and exhibit properties of the parent material. Productivity ranges from very high for some that form on recent river deposits to very low for those that form on shifting sand or rocky slopes.	16.16
Gelisols	Young soils with little profile development that occur in regions with permafrost. Low temperatures and frozen conditions for much of the year slow soil-forming processes.	8.61
Histosols	Organic soils with little or no climatic implications. Can be found in any climate where organic debris can accumulate to form a bog soil. Dark, partially decomposed organic material commonly referred to as *peat*.	1.71
Inceptisols	Weakly developed young soils in which the beginning (inception) of profile development is evident. Most common in humid climates, they exist from the Arctic to the tropics. Native vegetation is most often forest.	9.81
Mollisols	Dark, soft soils that have developed under grass vegetation, generally found in prairie areas. Humus-rich surface horizon that is rich in calcium and magnesium. Soil fertility is excellent. Also found in hardwood forests with significant earthworm activity. Climatic range is boreal or alpine to tropical. Dry seasons are normal.	6.89
Oxisols	Soils that occur on old land surfaces unless parent materials were strongly weathered before they were deposited. Generally found in the tropics and subtropical regions. Rich in iron and aluminum oxides, Oxisols are heavily leached, hence are poor soils for agricultural activity.	7.5
Spodosols	Soils found only in humid regions on sandy material. Common in northern coniferous forests and cool humid forests. Beneath the dark upper horizon of weathered organic material lies a light-colored horizon of leached material, the distinctive property of this soil.	2.56
Ultisols	Soils that represent the products of long periods of weathering. Water percolating through the soil concentrates clay particles in the lower horizons (argillic horizons). Restricted to humid climates in the temperate regions and the tropics, where the growing season is long. Abundant water and a long frost-free period contribute to extensive leaching, hence poorer soil quality.	8.45
Vertisols	Soils containing large amounts of clay, which shrink upon drying and swell with the addition of water. Found in subhumid to arid climates, provided that adequate supplies of water are available to saturate the soil after periods of drought. Soil expansion and contraction exert stresses on human structures.	2.24

* Percentages refer to the world's ice-free surface.

FIGURE 4.18 Global Soil Regions Worldwide distribution of the Soil Taxonomy's 12 soil orders. Points A and B are references for a *Give It Some Thought* item at the end of the chapter. (Natural Resources Conservation Service/ USDA)

certain important characteristics in common. Bringing order to large quantities of information not only aids comprehension and understanding but also facilitates analysis and explanation.

In the United States, soil scientists use a system for classifying soils known as the **Soil Taxonomy**. It emphasizes the physical and chemical properties of the soil profile and is organized on the basis of observable soil characteristics. There are six hierarchical categories of classification, ranging from *order*, the broadest category, to *series*, the most specific category. The system recognizes 12 soil orders and more than 19,000 soil series.

The names of the classification units are combinations of syllables, most of which are derived from Latin or Greek. The names are descriptive. For example, soils of the order Aridosols (*aridus* = dry, *solum* = soil) are characteristically dry soils in arid regions. Soils in the order Inceptisols (*inceptum* = beginning, *solum* = soil) are soils with only the beginning or inception of profile development.

TABLE 4.2 provides brief descriptions of the 12 basic soil orders. **FIGURE 4.18** shows the complex worldwide distribution pattern of the Soil Taxonomy's 12 soil orders. Like many other classification systems, the Soil Taxonomy is not suitable for every purpose. It is especially useful for agricultural and related land-use purposes, but it is not a useful system for engineers who are preparing evaluations of potential construction sites.

4.7 CONCEPT CHECKS

1 Why are soils classified?

2 Refer to Figure 4.19 and identify three particularly extensive soil orders that occur in the contiguous 48 United States. Describe two soil orders that occur in Alaska.

EYE ON EARTH

This thick red soil is exposed in one of the states in the United States and is either from the Gelisols, Molisols, or Oxisols order. (Photo by Sandra A. Dunlap/Shutterstock)

QUESTION 1 *Refer to the descriptions in Table 4.2 and determine the likely soil order shown in this image. Explain your choice.*

QUESTION 2 *Which one of these states is the most likely location of the soil—Alaska, Illinois, or Hawaii?*

4.8 | SOIL EROSION: LOSING A VITAL RESOURCE

Discuss the detrimental impact of human activities on soil and some ways that soil erosion is controlled.

Many people do not realize that soil erosion is a serious environmental problem. Perhaps this is the case because a substantial amount of soil seems to remain even where soil erosion is serious. Nevertheless, although the loss of fertile topsoil may not be obvious to the untrained eye, it is a significant and growing problem as human activities expand and disturb more and more of Earth's surface.

Soil erosion is a natural process; it is part of the constant recycling of Earth materials that we call the *rock cycle.* Once soil forms, erosional forces, especially water and wind, move soil components from one place to another.

Water Erosion Raindrops strike the land with surprising force (**FIGURE 4.19**). Each drop acts like a tiny bomb, blasting movable soil particles out of their positions in the soil mass. Water flowing across the surface carries away the dislodged soil particles. Because the soil is moved by thin sheets of water, this process is termed *sheet erosion.*

After this thin, unconfined sheet flows for a relatively short distance, threads of current typically develop, and tiny channels called *rills* begin to form. Still deeper cuts in the soil, known as *gullies*, are created as rills enlarge (**FIGURE 4.20**). When normal farm cultivation cannot eliminate the channels, the rills have grown large enough to be called gullies. Although most dislodged soil particles move only a short distance during each rainfall, substantial quantities eventually leave the fields and make their way downslope to a stream. Once in the stream channel, these soil particles, which can now be called *sediment*, are transported downstream and are eventually deposited.

Wind Erosion It is estimated that flowing water is responsible for about two-thirds of the soil erosion in the United States. Much of the remainder is caused by wind. When dry conditions prevail, strong winds can remove large quantities of soil from unprotected fields.

Rates of Erosion Soil erosion is the ultimate fate of practically all soils. In the past, erosion occurred at slower rates than it does today because more of the land surface was covered and protected by trees, shrubs, grasses, and other plants. However, human

FIGURE 4.19 Raindrop Impact Soil dislodged by raindrop impact is more easily moved by sheet erosion. (Photo courtesy U.S. Department of the Navy/Soil Conservation Service/USDA)

Raindrops may strike the surface at velocities approaching 35 km per hour. When a drop strikes an exposed surface, soil particles may splash as high as one meter and land more than a meter away from the point of raindrop impact.

FIGURE 4.20 Soil Erosion Sheetflow, rills, and gullies on unprotected soils. (Photos by Lynn Betts /NRCS and D. P. Burnside/Science Source)

Severe sheet and rill erosion on an Iowa farm following heavy rains.

Gully erosion on unprotected soil on a Wisconsin farm. Just one millimeter of soil from a single acre amounts to about 5 tons.

Crops planted in strips along contours of hillside

Corn

Grass planted along drainage

Hay

Corn and hay have been planted in strips that follow the contours of the hillside. This pattern reduces soil loss because it slows the rate of water runoff.

FIGURE 4.21 Soil Conservation Planting crops to decrease water erosion on a farm in northeastern Iowa. (Photo courtesy of USDA/NRCS/Natural Resources Conservation Service)

activities such as farming, logging, and construction, which remove or disrupt the natural vegetation, have greatly accelerated the rate of soil erosion. Without the stabilizing effect of plants, the soil is more easily swept away by the wind or carried downslope by sheet wash.

Natural rates of soil erosion vary greatly from one place to another and depend on soil characteristics as well as factors such as climate, slope, and type of vegetation. Over a broad area, erosion caused by surface runoff may be estimated by determining the sediment loads of the streams that drain the region. When studies of this kind were made on a global scale, they indicated that prior to the appearance of humans, sediment transport by rivers to the ocean amounted to just over 9 billion metric tons per year. In contrast, the amount of material currently transported to the sea by rivers is about 24 billion metric tons per year, or more than 2.5 times the earlier rate.

Controlling Soil Erosion

At present, it is estimated that topsoil is eroding faster than it forms on more than one-third of the world's croplands. The results are lower productivity, poorer crop quality, reduced agricultural income, and an ominous future. On every continent, unnecessary soil loss is occurring because appropriate conservation measures are not being used. Although it is a recognized fact that soil erosion can never be completely eliminated, soil conservation programs can substantially reduce the loss of this basic resource.

Steepness of slope is an important factor in soil erosion. The steeper the slope, the faster water runs off and the greater the erosion. It is best to leave steep slopes undisturbed. When such

slopes are farmed, terraces can be constructed. These nearly flat, steplike surfaces slow runoff and thus decrease soil loss while allowing more water to soak into the ground.

Soil erosion by water also occurs on gentle slopes. **FIGURE 4.21** illustrates one conservation method of planting crops parallel to the contours of the slope. This pattern reduces soil loss by slowing runoff. Strips of grass or cover crops such as hay slow runoff even more, promoting water infiltration and trapping sediment.

Creating grassed waterways is another common practice that helps to reduce soil erosion (**FIGURE 4.22**). Natural drainageways are shaped to form smooth, shallow channels and then planted with grass. The grass prevents the formation of gullies and traps soil washed from cropland. Frequently, crop residues are also left on fields. This debris protects the surface

The grassed waterway prevents the formation of gullies and traps soil washed from cropland.

FIGURE 4.22 Reducing Erosion by Water Grassed waterway on a Pennsylvania farm. (Photo courtesy Bob Nichols/NRCS)

FIGURE 4.23
**Reducing Wind
Erosion** Windbreaks
protecting wheat
fields in North Dakota.
(Photo courtesy Erwin C.
Cole/USDA)

These flat expanses are susceptible to wind erosion, especially when the fields are bare. The rows of trees slow and deflect the wind, which decreases the loss of top soil.

from both water and wind erosion. To protect fields from excessive wind erosion, rows of trees and shrubs are planted to act as windbreaks that slow the wind and deflect it upward (**FIGURE 4.23**).

4.8 CONCEPT CHECKS

1. Link these phenomena in a description of soil erosion: sheet erosion, gullies, rain drop impact, rills, stream.
2. How have human activities affected the rate of soil erosion?
3. Briefly describe three methods of controlling soil erosion.

4.9 | MASS WASTING: THE WORK OF GRAVITY

Discuss the role that mass wasting plays in the development of landforms.

Earth's surface is never perfectly flat but instead consists of a great variety of slopes. Some are steep and precipitous; others are moderate or gentle. Some are long and gradual; others are short and abrupt. Some slopes are mantled with soil and covered by vegetation; others consist of barren rock and rubble. Although most slopes appear to be stable and unchanging, they are not static features because the force of gravity causes material to move. At one extreme, the movement may be gradual and practically imperceptible. At the other extreme, it may consist of a roaring debris flow or a thundering rock avalanche. Landslides are a worldwide natural hazard. When these natural processes lead to loss of life and property, they become natural disasters.

Landslides as Geologic Hazards

The chapter-opening photo is a spectacular example of a phenomenon that many people would call a landslide. For most of us, the word *landslide* implies a sudden event in which large quantities of rock and soil plunge down steep slopes. When people and communities are in the way, a natural disaster may result. Landslides constitute major geologic hazards that each year in the United States cause billions of dollars in damages and the loss of 25 to 50 lives. As you will see, many landslides occur in connection with other major natural disasters, including earthquakes, volcanic eruptions, wildfires, and severe storms.

If you were to view several images of events called landslides, you would likely notice that the term seems to refer to several different things—from mudflows to rock avalanches. This variety reflects the fact that although many people, including geologists, frequently use the word *landslide*, the term has no specific definition in geology. Rather, it is a popular, nontechnical word used to describe any or all relatively rapid forms of mass wasting.

Landslides are spectacular examples of a basic geologic process called **mass wasting**. Mass wasting refers to the downslope movement of rock, regolith, and soil under the direct influence of gravity. It is distinct from the erosional processes that are examined in subsequent chapters because mass wasting does not require a transporting medium such as water, wind, or glacial ice.

The Role of Mass Wasting in Landform Development

In the evolution of most landforms, mass wasting is the step that follows weathering. By itself, weathering does not produce significant landforms. Rather, landforms develop as the products of weathering are removed from the places where they originate. Once weathering weakens rock and breaks it apart, mass wasting transfers the debris downslope, where a stream, acting as a conveyor belt, usually carries it away (see Figure 4.1). Although there may be many intermediate stops along the way, the sediment is eventually transported to its ultimate destination: the sea.

The combined effects of mass wasting and running water produce stream valleys, which are the most common and conspicuous of Earth's landforms. If streams alone were responsible for creating the valleys in which they flow, the valleys would be very narrow features. However, the fact that most stream valleys are much wider than they are deep is a strong indication of the significance of mass-wasting processes in supplying material to streams. The walls of a canyon extend far from the stream because of the transfer of weathered debris downslope to the stream and its tributaries by mass-wasting processes. In this manner, streams and mass wasting combine to modify and sculpt the surface. Of course, glaciers, groundwater, waves, and wind are also important agents in shaping landforms and developing landscapes.

Slopes Change Through Time

It is clear that if mass wasting is to occur, there must be slopes that rock, soil, and regolith can move down. It is Earth's

Landslides as Natural Disasters

Landslides don't just occur in remote mountains and canyons. People frequently live where rapid mass wasting events occur. Here are a few examples.

Even in areas with steep slopes, catastrophic landslides are relatively rare, thus, people living in susceptible areas often do not appreciate the risks of living where they do. This deadly event, triggered by an earthquake in January 2001, buried 300 homes in Santa Tecla, El Salvador. Many unsuspecting residents perished.

Ed Harp/USGS

Thevenot Laurent/PHOTOPQR/LE PROGRES/Newscom

On March 1, 2012, this boulder broke loose from a steep mountain slope in the French Alps, crushing a car and then crashing into this house. There was no warning and no obvious trigger for this event.

Weathered rock, precipitous slopes, and the shock of an earthquake combined to produce many rockfalls and rock avalanches in and around Christchurch, New Zealand on February 22, 2011.

Neil Sands/AFP/Getty Images

Marty Melville/AFP/Getty Images

In January 2011, following a period of torrential rains, this thick slurry of mud, appropriately called a mudflow, buried these cars in Nova Friburgo, Brazil. Heavy rains are an important trigger of mass wasting processes.

Ed Harp/USGS

www.ghapneus.com.br

mountain-building and volcanic processes that produce these slopes through sporadic changes in the elevations of landmasses and the ocean floor. If dynamic internal processes did not continually produce regions having higher elevations, the system that moves debris to lower elevations would gradually slow and eventually cease.

Most rapid and spectacular mass-wasting events occur in areas of rugged, geologically young mountains. Newly formed mountains are rapidly eroded by rivers and glaciers into regions characterized by steep and unstable slopes. It is in such settings that massive destructive landslides occur. As mountain building subsides, mass wasting and erosional processes lower the land. Through time, steep and rugged mountain slopes give way to gentler, more subdued terrain. Thus, as a landscape ages, massive and rapid mass-wasting processes give way to smaller, less dramatic downslope movements that are often imperceptibly slow.

4.9 CONCEPT CHECKS

1 Discuss the meaning of the term *landslide*.

2 What is the controlling force of mass wasting?

3 In what environment are rapid mass-wasting processes most likely to occur?

4 Sketch or describe how mass wasting contributes to the development of a valley.

4.10 | CONTROLS AND TRIGGERS OF MASS WASTING

Summarize the factors that control and trigger mass-wasting processes.

Gravity is the controlling force of mass wasting, but several factors play important roles in overcoming inertia and creating downslope movements. Long before a landslide occurs, various processes work to weaken slope material, gradually making it more and more susceptible to the pull of gravity. During this span, the slope remains stable but gets closer and closer to being unstable. Eventually, the strength of the slope is weakened to the point that something causes it to cross the threshold from stability to instability. Such an event that initiates downslope movement is called a *trigger*. Remember that a trigger is not the sole cause of a mass-wasting event but just the last of many causes. Among the common factors that trigger mass-wasting processes are saturation of material with water, oversteepening of slopes, removal of anchoring vegetation, and ground vibrations from earthquakes.

The Role of Water

Mass wasting is sometimes triggered when heavy rains or periods of snow melting saturate surface materials. The water does not transport the material. Rather, it allows gravity to more easily set the material in motion. This was the case in January 2005, when a massive debris flow (popularly called a mudslide) swept through La Conchita, California, a small coastal community northwest of Los Angeles (**FIGURE 4.24**).

FIGURE 4.24 Heavy Rains Trigger Debris Flow On January 10, 2005, a massive debris flow swept through La Conchita, California, a small coastal town situated on a narrow coastal strip between the shoreline and a steep bluff. The event occurred after a span of near-record amounts of rainfall. (Photos by AP Photo/Kevork Djansezian)

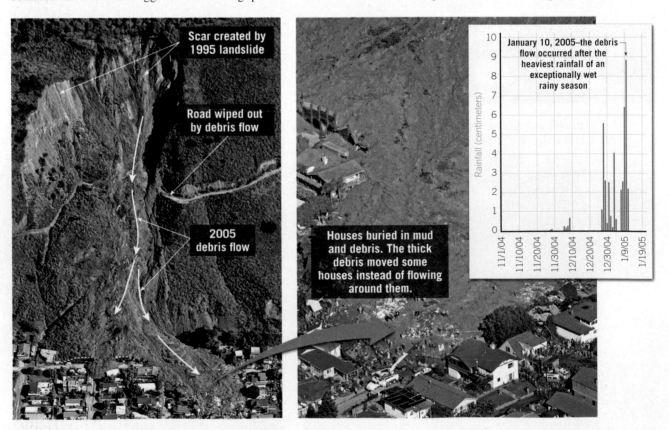

Scar created by 1995 landslide

Road wiped out by debris flow

2005 debris flow

Houses buried in mud and debris. The thick debris moved some houses instead of flowing around them.

January 10, 2005—the debris flow occurred after the heaviest rainfall of an exceptionally wet rainy season

When the pores in sediment become filled with water, the cohesion among particles is destroyed, and they can slide past one another with relative ease. For example, when sand is slightly moist, it sticks together quite well. However, if enough water is added to fill the openings between the grains, the sand will ooze out in all directions (**FIGURE 4.25**). Thus, saturation reduces the internal resistance of materials, which are then easily set in motion by the force of gravity. When clay is wetted, it becomes very slick—another example of the "lubricating" effect of water. Water also adds considerable weight to a mass of material. The added weight in itself may be enough to cause the material to slide or flow downslope.

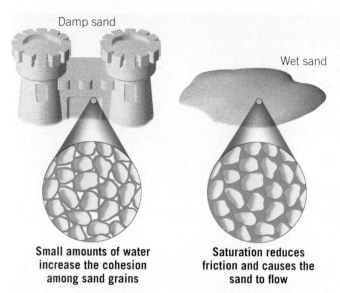

Dry sand

Dry sand grains are bound mainly by friction with one another

Damp sand

Wet sand

Small amounts of water increase the cohesion among sand grains

Saturation reduces friction and causes the sand to flow

FIGURE 4.25 Saturation Reduces Friction When water saturates soil, friction among particles is reduced, allowing soil to move downslope.

Oversteepened Slopes

Oversteepening of slopes is another trigger of many mass movements. This takes place in many situations in nature. A stream undercutting a valley wall and waves pounding against the base of a cliff are two familiar examples. Furthermore, through their activities, people often create oversteepened and unstable slopes that become prime sites for mass wasting (**FIGURE 4.26**).

Loose, granular (sand-size or coarser) particles assume a stable slope called the **angle of repose** (*reposen* = to be at rest). This is the steepest angle at which material remains stable (**FIGURE 4.27**). Depending on the size and shape of the particles, the angle varies from 25 to 40 degrees. The larger, more angular particles maintain the steepest slopes. If the angle is increased, the rock debris will adjust by moving downslope.

Oversteepening is important not only because it triggers movements of unconsolidated granular materials, but also because it produces unstable slopes and mass movements in cohesive soils, regolith, and bedrock. The response will not be immediate, as with loose, granular material, but sooner or later, one or more mass-wasting processes will eliminate the oversteepening and restore stability to the slope.

Removal of Vegetation

Plants protect against erosion and contribute to the stability of slopes because their root systems bind soil and regolith together. In addition, plants shield the soil surface from the erosional effects of raindrop impact (see Figure 4.20). Where plants are lacking, mass wasting is enhanced, especially if slopes are steep and water is plentiful. When anchoring vegetation is removed by forest fires or by people (for timber, farming, or development), surface materials frequently move downslope.

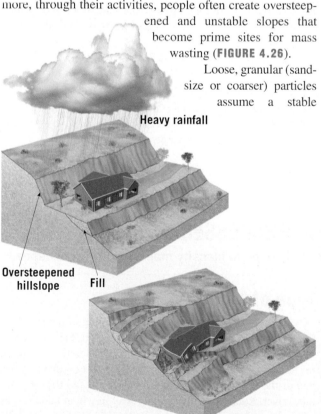

Heavy rainfall

Oversteepened hillslope **Fill**

FIGURE 4.26 Unstable Slopes Natural processes such as stream and wave erosion can oversteepen slopes. Changing the slope to accommodate a new house or road can also lead to instability and a destructive mass-wasting event.

Angle of repose

G. Leavens/Science Source

FIGURE 4.27 Angle of Repose The angle of repose is the steepest angle at which an accumulation of granular particles remains stable. Larger, more angular particles maintain the steepest slopes.

FIGURE 4.28 Wildfires Contribute to Mass Wasting During the summer, wildfires are common occurrences in many parts of the western United States. Millions of acres are burned each year. The loss of anchoring vegetation sets the stage for accelerated mass wasting. (Photo by Raymond Gehman/National Geographic Stock)

FIGURE 4.29 Earthquakes as Triggers A major earthquake in China's mountainous Sichuan Province in May 2008 triggered hundreds of landslides. The landslide shown here took 51 lives. (Photo by Lynn Highland/USGS)

In July 1994 a severe wildfire swept Storm King Mountain, west of Glenwood Springs, Colorado, denuding the slopes of vegetation. Two months later, heavy rains resulted in numerous debris flows, one of which blocked Interstate 70 and threatened to dam the Colorado River. A 5-kilometer (3-mile) length of the highway was inundated with tons of rock, mud, and burned trees. The closure of Interstate 70 imposed costly delays on this major highway. Events such as this are relatively common in the American West, where slopes are often steep and summer wildfires may affect millions of acres each year (**FIGURE 4.28**).

Earthquakes as Triggers

Conditions favoring mass wasting can exist in an area for a long time without movement occurring. An additional factor is sometimes necessary to trigger the movement. Among the most important and dramatic triggers are earthquakes.

An earthquake and its aftershocks can dislodge enormous volumes of rock and unconsolidated material. The mass-wasting event shown in **FIGURE 4.29** was triggered by an earthquake. In many areas that are jolted by earthquakes, it is not ground vibrations directly but landslides and ground subsidence triggered by the vibrations that cause the greatest damage.

4.10 CONCEPT CHECKS

1 How does water affect mass-wasting processes?
2 Describe the significance of the angle of repose.
3 How might a forest fire influence mass wasting?
4 Link earthquakes to landslides.

4.11 | CLASSIFYING MASS-WASTING PROCESSES

List and explain the criteria that are commonly used to classify mass-wasting processes.

Geologists include several processes under the umbrella of mass wasting. Four of them are illustrated in **FIGURE 4.30**. Generally, each process is defined by the type of material involved, the kind of motion, and the velocity of the movement.

If soil and regolith dominate, terms such as *debris*, *mud*, and *earth* are used. In contrast, when a mass of bedrock breaks loose and moves downslope, the term *rock* may be part of the description. Generally, the kind of motion is described as either a fall, a slide, or a flow.

Type of Motion

When the movement involves the free falling of detached individual pieces of any size, it is termed a **fall**. Falls are common on slopes that are too steep for loose material to remain on the surface. Many falls result when freeze–thaw cycles and/or the action of plant roots loosen rock and then gravity takes over. Rockfall is the primary way in which talus slopes are built and maintained (see Figure 4.5). Sometimes falls may trigger other forms of downslope movement.

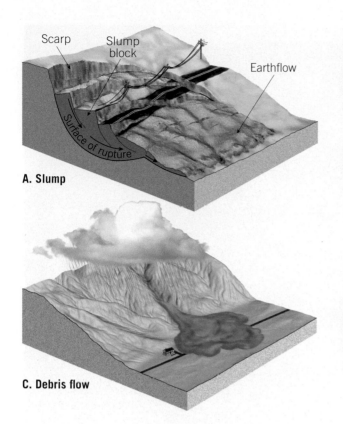

A. Slump

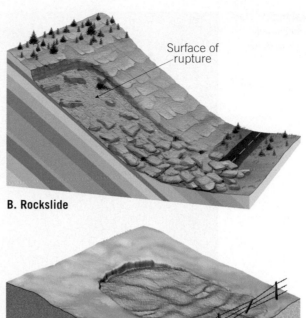

B. Rockslide

C. Debris flow

D. Earthflow

FIGURE 4.30 Four Relatively Rapid Forms of Mass Wasting Because materials in slumps (**A**) and rockslides (**B**) move along well-defined surfaces, they are said to move by sliding. By contrast, when materials move downslope as a viscous fluid, the movement is described as a flow. Debris flow (**C**) and earthflow (**D**) advance downslope in this manner.

Many mass-wasting processes are **slides**, which occur whenever material remains fairly coherent and moves along a well-defined surface. Sometimes the surface is a joint, a fault, or a bedding plane that is roughly parallel to the slope. However, in the movement called *slump*, the descending material moves en masse along a curved surface of rupture.

A third type of movement common to mass wasting is termed **flow**. Flow occurs when material moves downslope as a viscous fluid. Most flows are saturated with water and typically move as lobes or tongues.

Rate of Movement

When mass-wasting events make the news, a large quantity of material has in all likelihood moved rapidly downslope and has had a disastrous effect on people and property. Indeed, during events called **rock avalanches**, rock and debris can hurtle downslope at speeds exceeding 200 kilometers (125 miles) per hour. Researchers now understand that rock avalanches, such as the one that produced the scene in **FIGURE 4.31**, must literally "float on air" as they move downslope. That is,

EYE ON EARTH

The mountainous terrain in this photo is in California's Sierra Nevada. Notice the accumulation of angular rock fragments at the base of the sheer rock wall. (Photo by Taylor S. Kennedy/National Geographic RF/glow Images)

QUESTION 1 *What term is applied to a pile of rock debris such as this?*

QUESTION 2 *Describe the probable combination of weathering and mass-wasting processes that produced this feature.*

QUESTION 3 *Did the feature probably form rapidly in hours or days, or gradually over many years? Explain.*

FIGURE 4.31 Blackhawk Rock Avalanche This prehistoric event is considered one of the largest-known landslides in North America. (Photo by Michael Collier)

San Bernardino Mountains

8 kilometers

Between 9 and 30 meters thick

The mass of rocky debris raced downslope on a cushion of compressed air.

high velocities result when air becomes trapped and compressed beneath the falling mass of debris, allowing it to move as a buoyant, flexible sheet across the surface.

Most mass movements, however, do not move with the speed of a rock avalanche. In fact, a great deal of mass wasting is imperceptibly slow. One process that we will examine later, termed *creep*, results in particle movements that are usually measured in millimeters or centimeters per year. Thus, as you can see, rates of movement can be spectacularly sudden or exceptionally gradual. Although various types of mass wasting are often classified as either rapid or slow, such a distinction is highly subjective because a wide range of rates exists between the two extremes. Even the velocity of a single process at a particular site can vary considerably from one time to another.

4.11 CONCEPT CHECKS

1 What terms describe the way material moves during mass wasting?

2 Why can rock avalanches move at such great speeds?

4.12 | RAPID FORMS OF MASS WASTING

Distinguish among slump, rockslide, debris flow, and earthflow.

The common rapid mass-wasting processes discussed in this section include slump, rockslide, debris flow, and earthflow. They are most common where slopes are steep. The speed of movement varies from barely perceptible to very rapid.

Slump

Slump refers to the downward sliding of a mass of rock or unconsolidated material moving as a unit along a *curved* surface (see Figure 4.30A). Usually the slumped material does not travel spectacularly fast or very far. This is a common form of mass wasting, especially in thick accumulations of cohesive materials such as clay. As the movement occurs, a crescent-shaped scarp (cliff) is created at the head, and the block's upper surface is sometimes tilted backward.

Slump commonly occurs because a slope has been oversteepened. The material on the upper portion of a slope is held in place by the material at the bottom of the slope. As this anchoring material at the base is removed, the material above is made unstable and reacts to the pull of gravity. A common example is a valley wall that becomes oversteepened by a meandering river. Another is a coastal area that has been undercut by wave activity at its base.

Rockslide

Rockslides occur when blocks of bedrock break loose and slide down a slope (see Figure 4.30B). If the material is mostly soil and regolith, the term *debris slide* is used instead. Such events are among the fastest and most destructive mass movements. Usually rockslides take place in a geologic setting where the rock strata are inclined or where joints and fractures exist parallel to the slope. When such a rock unit is undercut at the base of the slope, it loses support, and the rock eventually gives way.

Sometimes an earthquake is the trigger. On other occasions, a rockslide is triggered when rain or melting snow lubricates the underlying surface to the point that friction is no longer sufficient to hold the rock unit in place. As a result, rockslides tend to be most common during the spring, when heavy rains and melting snow are most prevalent. The massive Gros Ventre slide shown in **FIGURE 4.32** is a classic example.

Debris Flow

Debris flow is a relatively rapid type of mass wasting that involves a flow of soil and regolith containing a large amount of water (see Figure 4.25 and Figure 4.30C). Debris flows are sometimes called **mudflows** when the material is primarily fine grained. Although they can occur in many different climate settings, they tend to occur most frequently in semiarid mountainous regions. Debris flows called *lahars* are also common on the steep slopes of some volcanoes. Because of their fluid properties, debris flows frequently follow canyons and stream channels. In populated areas, debris flows can pose a significant hazard to life and property.

Debris Flows in Semiarid Regions

When a cloudburst or rapidly melting mountain snows create a sudden flood in a semiarid region, large quantities of soil and regolith are washed into nearby stream channels because there is usually little vegetation to anchor the surface material. The end product is a flowing tongue of well-mixed mud, soil, rock, and water. Its consistency may range from that of wet concrete to a soupy mixture not much thicker than muddy water. The rate of flow therefore depends not only on the slope but also on the water content. When dense, debris flows

The side of the mountain gave way when the tilted sandstone bed, that had been cut through by the river, could no longer maintain its position atop the saturated bed of clay.

Gros Ventre landslide debris

Former land surface

Scar

Lake

Clay bed

Rupture surface

Sandstone

Limestone

0   0.5   1
Kilometer

Even though the Gros Ventre rockslide occurred in 1925, the scar left on the side of Sheep Mountain is still a prominent feature.

Smart Figure 4.32 Gros Ventre Rockslide This massive slide occurred on June 23, 1925, just east of the small town of Kelly, Wyoming. (Photo by Michael Collier)

are capable of carrying or pushing large boulders, trees, and even houses with relative ease.

Debris flows pose a serious hazard to development in dry mountainous areas such as southern California. The construction of homes on canyon hillsides and the removal of anchoring vegetation by brush fires and other means have increased the frequency of these destructive events.

Lahars   Debris flows composed mostly of volcanic materials on the flanks of volcanoes are called **lahars**. The word originated in Indonesia, a volcanic region that has experienced many of these often-destructive events. Historically, lahars have been some of the deadliest volcano hazards. They can occur either during an eruption or when a volcano is quiet. They take place when highly unstable layers of ash and debris become saturated with water and flow down steep volcanic slopes, generally following existing stream channels. Heavy rainfalls often trigger these flows. Others are triggered when large volumes of ice and snow are suddenly melted by heat flowing to the surface from within the volcano or by the hot gases and near-molten debris emitted during a violent eruption.

In November 1985 lahars were produced when Nevado del Ruiz, a 5300-meter (17,400-foot) volcano in the Andes Mountains of Colombia, erupted. The eruption melted much of the snow and ice that capped the uppermost 600 meters (2000 feet) of the peak, producing torrents of hot, thick mud, ash, and debris. The lahars moved outward from the volcano,

FIGURE 4.33 Earth-flow This small tongue-shaped earthflow occurred on a newly formed slope along a recently constructed highway in central Illinois. It formed in clay-rich material following a period of heavy rain. Notice the small slump at the head of the earthflow. (Photo by E. J. Tarbuck)

FIGURE 4.33 Earth-flow This small tongue-shaped earthflow occurred on a newly formed slope along a recently constructed highway in central Illinois. It formed in clay-rich material following a period of heavy rain. Notice the small slump at the head of the earthflow. (Photo by E. J. Tarbuck)

following the valleys of three rain-swollen rivers that radiate from the peak. The flow that moved down the valley of the Lagunilla River was the most destructive, devastating the town of Armero, 48 kilometers (30 miles) from the mountain. Most of the more than 25,000 deaths caused by the event occurred in this once-thriving agricultural community.

Earthflow

We have seen that debris flows are frequently confined to channels in semiarid regions. In contrast, **earthflows** most often form on hillsides in humid areas during times of heavy precipitation or snowmelt (see Figure 4.30D). When water saturates the soil and regolith on a hillside, the material may break away, leaving a scar on the slope and forming a tongue- or teardrop-shaped mass that flows downslope (**FIGURE 4.33**). The materials most commonly involved are rich in clay and silt and contain only small proportions of sand and coarser particles. Earthflows range in size from bodies a few meters long, a few meters wide, and less than 1 meter deep to masses more than 1 kilometer long, several

hundred meters wide, and more than 10 meters deep.

Because earthflows are quite viscous, they generally move at slower rates than the more fluid debris flows described in the preceding section. They move slowly and persistently for periods ranging from days to years. Depending on slope steepness and the material's consistency, velocities range from less than 1 millimeter per day up to several meters per day. Movement is typically faster during wet periods. In addition to occurring as isolated hillside phenomena, earthflows commonly take place in association with large slumps. In this situation, they may be seen as tonguelike flows at the base of the slump.

A special type of earthflow, known as *liquefaction*, sometimes occurs in association with earthquakes. Porous clay- to sand-size sediments that are saturated with water are most vulnerable. When shaken suddenly, the grains lose cohesion, and the ground flows. Liquefaction can cause buildings to sink or tip on their sides and underground storage tanks and sewer lines to float upward. Damage can be substantial. You will learn more about this in Chapter 8.

4.12 CONCEPT CHECKS

1 Without looking at Figure 4.31A, sketch and label a simple cross section (side view) of a slump.

2 Both slumps and rockslides move by sliding. How do these processes differ from one another?

3 What factors contributed to the massive rockslide at Gros Ventre, Wyoming?

4 How is a lahar different from a debris flow that might occur in southern California?

5 Contrast earthflows and debris flows.

4.13 | SLOW FORMS OF MASS WASTING
Contrast creep and solifluction.

Movements such as rockslides, rock avalanches, and lahars are certainly the most spectacular and catastrophic forms of mass wasting. These dangerous events deserve intensive study to enable more effective prediction, timely warnings, and better controls to save lives. However, because of their large size and spectacular nature, they give us a false impression of their importance as a mass-wasting process. Indeed, sudden movements transport less material than does the slow, subtle action of creep. Whereas rapid types of mass wasting

are characteristic of mountains and steep hillsides, creep takes place on both steep and gentle slopes and is thus much more widespread.

Creep

Creep is a type of mass wasting that involves the gradual downhill movement of soil and regolith. One factor that contributes to creep is the alternate expansion and contraction

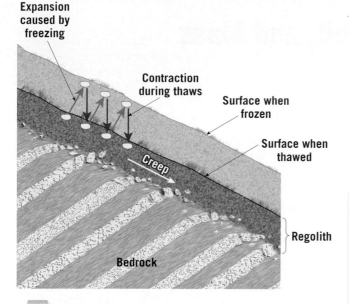

Expansion caused by freezing

Contraction during thaws

Surface when frozen

Surface when thawed

Creep

Regolith

Bedrock

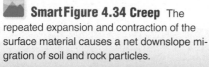 **SmartFigure 4.34 Creep** The repeated expansion and contraction of the surface material causes a net downslope migration of soil and rock particles.

FIGURE 4.35 Solifluction Lobes Near the Arctic Circle in Alaska Solifluction occurs in permafrost regions when the active layer thaws in summer. (Photo by James E. Patterson Collection)

Active layer (thawed in summer)

Frozen layer (permafrost)

Geologist's Sketch

of surface material caused by freezing and thawing or wetting and drying. As shown in **FIGURE 4.34**, freezing or wetting lifts particles at right angles to the slope, and thawing or drying allows the particles to fall back to a slightly lower level. Each cycle therefore moves the material a short distance downhill.

Creep is aided by anything that disturbs the soil. For example, raindrop impact and disturbance by plant roots and burrowing animals may contribute. Creep is also promoted when the ground becomes saturated with water. Following a heavy rain or snowmelt, a waterlogged soil may lose its internal cohesion, allowing gravity to pull the material downslope. Because creep is imperceptibly slow, the process cannot be observed in action. However, the effects of creep can be observed. Creep causes fences and utility poles to tilt and retaining walls to be displaced.

Solifluction

When soil is saturated with water, the soggy mass may flow downslope at a rate of a few millimeters or a few centimeters per day or per year. Such a process is called **solifluction** (which means "soil flow"). It is a type of mass wasting that is common wherever water cannot escape from the saturated surface layer by infiltrating to deeper levels. A dense clay hardpan in soil or an impermeable bedrock layer can promote solifluction.

Solifluction is common in regions underlain by **permafrost**. Permafrost refers to the permanently frozen ground that occurs in Earth's harsh tundra and ice-cap climates. It occurs in a zone above the permafrost called the *active layer*, which thaws in summer and refreezes in winter. During summer, water is unable to percolate into the impervious permafrost layer below. As a result, the active layer becomes saturated and slowly flows. The process can occur on slopes as gentle as 2–3 degrees. Where there is a well-developed mat of vegetation, a solifluction sheet may move in a series of well-defined lobes or overriding folds (**FIGURE 4.35**).

4.13 CONCEPT CHECKS

1 Describe the basic mechanisms that contribute to creep.

2 During what season does solifluction occur? Explain why it occurs only during that time of year.

4.1 EARTH'S EXTERNAL PROCESSES

List three types of external processes and discuss the role each plays in the rock cycle.

KEY TERMS: external process, internal process, weathering, mass wasting, erosion

- Weathering, mass wasting, and erosion are responsible for creating, transporting, and depositing sediment. They are called external processes because they occur at or near Earth's surface and are powered by energy from the Sun.
- Internal processes lead to volcanic activity and mountain building and derive their energy from Earth's interior.

4.2 WEATHERING

Define *weathering* and distinguish between the two main categories of weathering.

KEY TERMS mechanical weathering, chemical weathering, frost wedging, sheeting, exfoliation dome, carbonic acid, spheroidal weathering

- Mechanical weathering is the physical breaking up of rock into smaller pieces. Rocks can be broken into smaller fragments by frost wedging, salt crystal growth, unloading, and biological activity.
- Chemical weathering alters a rock's chemistry, changing it into different substances.
- Water is by far the most important agent of chemical weathering. Oxygen in water can oxidize some materials, while carbon dioxide (CO_2) dissolved in water forms carbonic acid. The chemical weathering of silicate minerals produces soluble products containing sodium, calcium, potassium, and magnesium; insoluble iron oxides; and clay minerals.

Michael Collier

Q How are the two main categories of weathering represented in this image, which shows human-made objects?

4.3 RATES OF WEATHERING

Summarize the factors that influence the type and rate of rock weathering.

KEY TERM: differential weathering

- The rate at which rock weathers is influenced by particle size; when rocks are broken into smaller pieces, the rate of weathering increases.
- The mineral composition of a rock affects the rate of weathering. For example, calcite readily dissolves in mildly acidic solutions. Among the silicate minerals, those that form first from magma are less resistant to chemical weathering than those that form later.
- Rock weathers most rapidly in an environment with lots of heat to drive reactions and water to facilitate those reactions. Consequently, rocks decompose relatively quickly in hot, wet climates and slowly in cold, dry conditions. Masses of rock do not weather uniformly. Differential weathering refers to the variation in the rate and degree of weathering caused by factors such as rock type, climate, and degree of jointing.

4.4 SOIL

Define *soil* and explain why soil is referred to as an interface.

KEY TERMS: regolith, soil, soil texture

- Soils are vital combinations of organic and non-organic components found at the interface where the geosphere, atmosphere, hydrosphere, and biosphere meet. This dynamic zone is the overlap between different parts of the Earth system. It includes the regolith's rocky debris, mixed with humus, water, and air.
- Soil texture refers to the proportions of different particle sizes (clay, silt, and sand) found in soil.

Q Label the four components of a soil on this pie chart.

4.5 CONTROLS OF SOIL FORMATION

List and briefly discuss five controls of soil formation.

KEY TERM: parent material

- Residual soils form in place due to the weathering of bedrock, whereas transported soils develop on unconsolidated sediment.
- Forming soil takes time. Soils that have developed for a longer period of time will have different characteristics than young soils. In addition, some minerals break down more readily than others. Soils produced from the weathering of different parent rocks will be produced at different rates.
- Soils formed in different climates will be different in part due to temperature and moisture differences and also due to the organisms that live in those different environments. Those organisms can add organic matter or chemical compounds to the developing soil or can help mix the soil through their growth and movement.
- The steepness of the slope on which a soil is forming is a key variable, with shallow slopes retaining their soils and steeper slopes shedding them to accumulate elsewhere.

4.6 THE SOIL PROFILE

Sketch, label, and describe an idealized soil profile.

KEY TERMS: soil horizon, soil profile, eluviation, leaching, solum

- Despite the great diversity of soils around the world, there are some broad patterns to the vertical anatomy of soil layers. Organic material, called humus, is added at the top (*O* horizon), mainly from plant sources. There, it mixes with mineral matter (*A* horizon). At the bottom, bedrock breaks down and contributes mineral matter (*C* horizon). In between, some materials are leached out or eluviated from higher levels (*E* horizon) and transported to lower levels (*B* horizon), where they may form an impermeable layer called hardpan.

4.7 CLASSIFYING SOILS

Explain the need for classifying soils.

KEY TERM: Soil Taxonomy

- The need to bring order to huge quantities of data motivated the establishment of a classification scheme for the world's soils. This Soil Taxonomy features 12 broad orders.

Q **Which soil order would likely contain a higher proportion of humus: Inceptisols or Histosols? Which soil order would be more likely found in Brazil: Gelisols or Oxisols?**

4.8 SOIL EROSION: LOSING A VITAL RESOURCE

Discuss the detrimental impact of human activities on soil and some ways that soil erosion is controlled.

- Soil erosion is a natural process; it is part of the constant recycling of Earth materials that we call the rock cycle.
- Because of human activities, soil erosion rates have increased over the past several hundred years. Natural soil production rates are constant, so there is a net loss of soil at a time when a record-breaking number of people live on the planet. Using windbreaks, terracing the land, installing grassed waterways, and plowing the land along horizontal contour lines are all practices that have been shown to reduce soil erosion.

Q **Why was this row of evergreens planted on this Indiana farm?** (Photo by Erwin C. Cole/USDA/NRCS/Natural Resources Conservation Science)

4.9 MASS WASTING: THE WORK OF GRAVITY

Discuss the role that mass wasting plays in the development of landforms.

KEY TERM: mass wasting

- After weathering breaks rock apart, gravity moves the debris downslope, in a process called mass wasting. Sometimes this occurs rapidly as a landslide, and at other times the movement is slower. Landslides are a significant geologic hazard, taking many lives and destroying property every year.
- Mass wasting serves an important role in landscape development. It widens stream-cut valleys and helps tear down mountains thrust up by internal processes.

4.10 CONTROLS AND TRIGGERS OF MASS WASTING

Summarize the factors that control and trigger mass-wasting processes.

KEY TERM angle of repose

- An event that initiates a mass-wasting process is referred to as a trigger. The addition of water, oversteepening of the slope, removal of vegetation, and shaking due to an earthquake are four important examples. Not all landslides are triggered by one of these four processes, but many are.
- Water added to a slope can expand the pore space between grains, causing them to lose their cohesion. Water also adds a significant amount of mass to a wetted slope.
- Granular materials can pile up to a certain angle of slope, but granular piles steeper than that critical angle will spontaneously collapse outward to form a gentler slope. For most geologic materials, this angle of repose varies between 25 and 40 degrees from horizontal.
- The roots of plants (especially plants with deep roots) act as a three-dimensional "net" that holds soil and regolith particles in place. The removal of vegetation such as by wildfires or various human activities that clear the land on steep slopes can set the stage for significant mass wasting.
- Earthquakes are significant triggers that deliver an energetic jolt to slopes poised on the brink of failure.

4.11 CLASSIFYING MASS-WASTING PROCESSES

List and explain the criteria that are commonly used to classify mass-wasting processes.

KEY TERMS fall, slide, flow, rock avalanche

- There are a variety of Earth materials and a variety of rates of mass-wasting movement. The type of material and the nature of the motion are used to classified based on the different types of mass wasting.
- Rockfalls occur when pieces of bedrock detach and fall freely through the air, slamming into the ground below with tremendous force. Repeated rockfalls generate a talus slope, the characteristic "apron" of angular rock debris that accumulates below mountain cliffs. Slides occur when discrete blocks of rock or unconsolidated material slip downslope on a fairly flat or curved surface. Unlike in a fall, the material in a slide does not drop through the air. Flows occur when individual grains or particles move randomly in a slurry, a viscous mixture of water-saturated materials.
- Rock avalanches move incredibly rapidly over surprising distances (tens of kilometers in a few minutes) because they ride along a layer of compressed air, in the same way that a hovercraft slides over the land with minimal friction. Other forms of mass wasting are much slower.

Q What type of motion is illustrated by this cautionary highway sign?

4.12 RAPID FORMS OF MASS WASTING

Distinguish among slump, rockslide, debris flow, and earthflow.

KEY TERMS slump, rockslide, debris flow, mudflow, lahar, earthflow

- A slump is a distinctive form of mass wasting in which coherent blocks of material move downhill along a spoon-shaped slip surface. They are frequently triggered by oversteepening, which may be caused by stream erosion of a valley wall.
- Rockslides are rapid forms of mass wasting in which blocks of rock slide downhill along a relatively flat surface. Often this is a preexisting structure such as a joint or a bedding plane. Situations where these surfaces dip into a valley at an angle are especially risky.
- Debris flows occur when unconsolidated soil or regolith becomes saturated with water and moves downhill in a slurry, picking up other objects (such as trees, houses, or boulders) along the way. Varieties of debris flow include mudflows, which are dominated by small particle sizes, and lahars, which involve volcanic material.
- Earthflow is characterized by a similar loss of coherence between grains in unconsolidated material, but it is much slower than a debris flow. Typically, sites of earthflow show an uphill scarp and a lobe of viscous soil on the downhill side.

Q This sketch shows the result of a mass-wasting event along the California coast. What rapid form of mass wasting occurred here? Speculate on what might have triggered the event.

4.13 SLOW FORMS OF MASS WASTING

Contrast creep and solifluction.

KEY TERMS creep, solifluction, permafrost

- Creep is a widespread and important form of mass wasting that is very slow. It occurs when freezing (or wetting) causes soil particles to be pushed out away from the slope, only to drop down to a lower position following thawing (or drying). In contrast, solifluction is the gradual flow of a saturated surface layer that is underlain by an impermeable zone. In arctic regions, the impermeable zone is permafrost.

GIVE IT SOME **THOUGHT**

1. Describe how plants promote mechanical and chemical weathering but inhibit erosion.

2. Granite and basalt are exposed at Earth's surface in a hot, wet region. Will mechanical weathering or chemical weathering predominate? Which rock will weather more rapidly? Why?

3. The accompanying photo shows Shiprock, a well-known landmark in the northwestern corner of New Mexico. It is a mass of igneous rock that represents the "plumbing" of a now-vanished volcano. Extending toward the upper left is a related wall-like igneous structure known as a dike. The igneous features are surrounded by sedimentary rocks. Explain why these once deeply buried igneous features now stand high above the surrounding terrain. What term from Section 4.3, "Rates of Weathering," applies to this situation?

Michael Collier

4. The accompanying photo shows a footprint on the Moon left by an *Apollo* astronaut in material popularly called *lunar soil*. Does this material satisfy the definition we use for soil on Earth? Explain why or why not. You may want to refer to Figure 4.12.

NASA

5. What might cause different soils to develop from the same kind of parent material or similar soils to form from different parent materials?

6. Using the map of global soil regions in Figure 4.18, identify the main soil order in the region adjacent to South America's Amazon River (point A on the map) and the predominant soil order in the American Southwest (point B). Briefly contrast these soils. Do they have anything in common? Referring to Table 4.2 might be helpful.

7. This soil sample is from a farm in the Midwest. From which horizon was the sample most likely taken— *A*, *E*, *B*, or *C*? Explain.

Lynn Betts/NRCS

8. The concept of external and internal processes was introduced at the beginning of the chapter. In the accompanying photo, external processes are clearly active. Describe the role that mass wasting is playing. Identify one feature that is a result of mass wasting.

Michael Collier

9. Describe at least one situation in which an internal process might cause or contribute to a mass-wasting process.

10. Mass wasting is influenced by many processes associated with all four spheres of the Earth system. Select three items from the list below. For each, outline a series of events that relate the item to various spheres and to a mass-wasting process. Here is an example which assumes that "frost wedging" is an item on the list: *Frost wedging involves rock (geosphere) being broken when water (hydrosphere) freezes. Freeze–thaw cycles (atmosphere) promote frost wedging. When frost wedging loosens a rock on a cliff, the fragment tumbles to the base of the cliff. This event, called a rockfall, is an example of mass wasting.* Now you give it a try. Use your imagination.

   a. Wildfire

   b. Spring thaw/melting snow

   c. Highway road cut

   d. Crashing waves

   e. Cavern formation (see Figure 5.40)

EXAMINING THE **EARTH SYSTEM**

1. Because of the burning of fossil fuels such as coal and petroleum, the level of carbon dioxide (CO_2) in the atmosphere has been increasing for more than 150 years. Should this increase tend to accelerate or slow down the rate of chemical weathering of Earth's surface rocks? Explain how you arrived at your conclusion.

2. Discuss the interaction of the atmosphere, geosphere, biosphere, and hydrosphere in the formation of soil.

3. The aerial view at right shows landslide debris atop Buckskin Glacier in Denali National Park in the rugged Alaska Range. Where Buckskin Glacier ends, its meltwater feeds a river that flows into Cook Inlet, just west of Anchorage. Cook Inlet is an arm of the North Pacific. Many processes have been responsible for creating this scene. Prepare a brief outline or summary to explain the formation and evolution of this landscape. Include internal and external processes and be sure to mention which spheres of the Earth system were involved.

Michael Collier

4. During summer, wildfires are common occurrences in many parts of the western United States. Millions of acres are burned each year. The wildfire shown here occurred near Santa Clarita, California, in July 2004. Earth is a system in which various parts of its four major spheres interact in uncountable ways. Relate this idea to the situation pictured here.

   a. What atmospheric conditions might have preceded and thus set the stage for and/or contributed to this wildfire?
   b. What might have ignited the blaze? Suggest a natural possibility and a human possibility.
   c. Discuss how wildfires like the one shown here might influence future mass-wasting events in the area.

5. Heavy rains in late July 2010 triggered the mass wasting that occurred in this mountain valley near Durango, Colorado. Heavy equipment is clearing away material that blocked railroad tracks and significantly narrowed the adjacent stream channel.

   a. What type of mass wasting likely occurred here? Explain your choice.
   b. Most of us are familiar with the phrase "One thing leads to another." It certainly applies to the Earth system. Suppose the material from the Durango mass-wasting event had completely filled the stream. What other natural hazard might have developed?

Soaring Tree Adventures

Joshua Gates Weisburg/EPA Newscom

MasteringGeology™

Looking for additional review and test prep materials? Visit the Self Study area in **www.masteringgeology.com** to find practice quizzes, study tools, and multimedia that will aid in your understanding of this chapter's content. In **MasteringGeology™** you will find:

- GEODe: Earth Science: An interactive visual walkthrough of key concepts
- Geoscience Animation Library: More than 100 animations illuminating many difficult-to-understand Earth science concepts

- In The News RSS Feeds: Current Earth science events and news articles are pulled into the site with assessment
- Pearson eText
- Optional Self Study Quizzes
- Web Links
- Glossary
- Flashcards

5

Running Water and Groundwater

Turbulent flow in the Little River in northeastern Alabama.

(Photo by Michael Collier)

ater is continually on the move, from the ocean to the land and back again, in an endless cycle. This chapter deals with the part of the hydrologic cycle that returns water to the sea. Some water travels quickly via a rushing stream, and some moves more slowly below the surface. When viewed as part of the Earth system, streams and groundwater represent basic links in the constant cycling of the planet's water. In this chapter, we examine the factors that influence the distribution and movement of water, as well as look at how water sculpts the landscape. To a great extent, the Grand Canyon, Niagara Falls, Old Faithful, and Mammoth Cave all owe their existence to the action of water on its way to the sea.

5.1 | EARTH AS A SYSTEM: THE HYDROLOGIC CYCLE List the hydrosphere's major reservoirs and describe the different paths that water takes through the hydrologic cycle.

Water is constantly moving among Earth's different spheres—the *hydrosphere*, the *atmosphere*, the *geosphere*, and the *biosphere*. This unending circulation of water is called the **hydrologic cycle**. Earth is the only planet in the solar system that has a global ocean and a hydrologic cycle.

Earth's Water

Water is almost everywhere on Earth—in the oceans, glaciers, rivers, lakes, air, soil, and living tissue. All these "reservoirs" constitute Earth's hydrosphere. In all, the water content of the hydrosphere is an estimated 1.36 billion cubic kilometers (326 million cubic miles). The vast bulk of it, about 96.5 percent, is stored in the global ocean. Ice sheets and glaciers account for an additional 1.76 percent, leaving just slightly more than 2 percent to be divided among lakes, streams, groundwater, and the atmosphere (**FIGURE 5.1**). Although the percentage of Earth's total water found in each of the latter sources is just a small fraction of the total inventory, the absolute quantities are great.

Water's Paths

The hydrologic cycle is a gigantic, worldwide system powered by energy from the Sun, in which the atmosphere provides a vital link between the oceans and continents (**FIGURE 5.2**). **Evaporation**, the process by which liquid water changes into water vapor (gas), is how water enters the atmosphere from the ocean and, to a much lesser extent, from the continents. Winds transport this moisture-laden air, often great distances. Complex processes of cloud formation eventually result in precipitation. The precipitation that falls into the ocean has completed its cycle and is ready to begin another. The water that falls on the continents, however, must make its way back to the ocean.

What happens to precipitation once it has fallen on land? A portion of the water soaks into the ground (called **infiltration**), slowly moving downward, then moving laterally, and finally seeping into lakes, streams, or directly into the ocean. When the rate of rainfall exceeds Earth's ability to absorb it, the surplus water flows over the surface into lakes and streams, a process called **runoff**. Much of the water that infiltrates or runs off eventually returns to the atmosphere because of evaporation from the soil, lakes, and streams. Also, some of the water that soaks into the ground is absorbed by plants, which then release it into the atmosphere. This process is called **transpiration**. Because both evaporation and transpiration involve the transfer of water from the surface directly to the atmosphere, they are often considered together as the combined process of **evapotranspiration**.

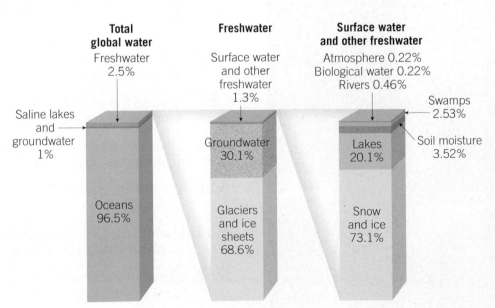

FIGURE 5.1 Distribution of Earth's Water

Total global water
- Freshwater 2.5%
- Saline lakes and groundwater 1%
- Oceans 96.5%

Freshwater
- Surface water and other freshwater 1.3%
- Groundwater 30.1%
- Glaciers and ice sheets 68.6%

Surface water and other freshwater
- Atmosphere 0.22%
- Biological water 0.22%
- Rivers 0.46%
- Swamps 2.53%
- Soil moisture 3.52%
- Lakes 20.1%
- Snow and ice 73.1%

Storage in Glaciers

When precipitation falls in very cold places—at high elevations or high latitudes—the water may not immediately soak in, run off, or evaporate. Instead, it may become part of a snowfield or a glacier. In this way, glaciers store large quantities of water on land. If present-day glaciers were to melt and release all their water, sea level would rise by several dozen meters. Such a rise would submerge many heavily populated coastal areas. As you will see in Chapter 6, over the past 2 million years, huge ice sheets have formed and melted on several occasions, each time affecting the balance of the hydrologic cycle.

Water Balance

Figure 5.2 also shows Earth's overall *water balance*, or the volume that passes through each part of the cycle annually. The amount of water vapor in the air at any one time is just a tiny fraction of Earth's total water supply. But the *absolute* quantities that are cycled through the atmosphere over a 1-year period are immense—some 380,000 cubic kilometers (91,000 cubic miles)—enough to cover Earth's entire surface to a depth of about 1 meter (39 inches).

It is important to know that the hydrologic cycle is *balanced*. Because the total amount of water vapor in the atmosphere remains about the same, the average annual precipitation worldwide must be equal to the quantity of water evaporated. However, for all the continents taken together, precipitation exceeds evaporation. Conversely, over the oceans, evaporation exceeds precipitation. Because the level of the world ocean is not dropping, the system must be in balance. In Figure 5.2, the 36,000 cubic kilometers (8600 cubic miles) of water that annually makes its way from the land to the ocean causes enormous erosion. In fact, this immense volume of moving water is *the single most important agent sculpting Earth's land surface.*

In the rest of this chapter, we will observe the work of water running over the surface, including floods, erosion, and the formation of valleys. Then we will look underground

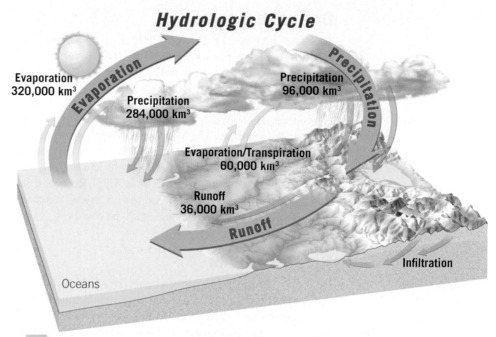

Hydrologic Cycle

Evaporation 320,000 km³

Precipitation 284,000 km³

Precipitation 96,000 km³

Evaporation/Transpiration 60,000 km³

Runoff 36,000 km³

Runoff

Infiltration

Oceans

SmartFigure 5.2 The Hydrologic Cycle The primary movement of water through the cycle is shown by the large arrows. A number refers to the annual amount of water taking a particular path.

at the slow labors of groundwater as it forms springs and caverns and provides drinking water on its long migration to the sea.

5.1 CONCEPT CHECKS

1 Describe or sketch the movement of water through the hydrologic cycle. Once precipitation has fallen on land, what paths might it take?
2 What is meant by the term *evapotranspiration*?
3 Over the oceans, evaporation exceeds precipitation, yet sea level does not drop. Explain why.

5.2 | RUNNING WATER Describe the nature of drainage basins and river systems. Sketch four basic drainage patterns.

Much of the precipitation that falls on land either enters the soil (infiltration) or remains at the surface, moving downslope as runoff. The amount of water that runs off rather than soaking into the ground depends on several factors: (1) the intensity and duration of rainfall, (2) the amount of water already in the soil, (3) the nature of the surface material, (4) the slope of the land, and (5) the extent and type of vegetation. When the surface material is highly impermeable, or when it becomes saturated, runoff

is the dominant process. Runoff is also high in urban areas because large areas are covered by impermeable buildings, roads, and parking lots.

Runoff initially flows in broad, thin sheets. This unconfined flow eventually develops threads of current that form tiny channels called rills. Rills meet to form gullies, which join to form streams. At first streams are small, but as one intersects another, larger and larger ones form. Eventually rivers develop that carry water from a broad region.

Drainage Basins

The land area that contributes water to a river system is called a **drainage basin** (**FIGURE 5.3**). The drainage basin of one stream is separated from the drainage basin of another by an imaginary line called a **divide**. Divides range in scale from a ridge separating two small gullies on a hillside to a *continental divide*, which splits whole continents into enormous drainage basins. The Mississippi River has the largest drainage basin in North America (**FIGURE 5.4**). Extending between the Rocky Mountains in the west and the Appalachian Mountains in the east, the Mississippi River and its tributaries collect water from more than 3.2 million square kilometers (1.2 million square miles) of the continent.

River Systems

River systems involve not only a network of stream channels but the entire drainage basin. Based on the dominant processes operating within them, river systems can be divided into three zones: *sediment production*—where erosion dominates, *sediment transport*, and *sediment deposition* (**FIGURE 5.5**). It is important to recognize that sediment is being eroded, transported, and deposited along the entire length of a stream, regardless of which process is dominant within each zone.

Sediment Production   The zone of *sediment production*, where most of the sediment is derived, is located in the headwater region of the river system. Much of the sediment carried by streams begins as bedrock that is subsequently broken down by weathering and then transported downslope by mass wasting and overland flow. Bank erosion can also contribute significant amounts of sediment. In addition,

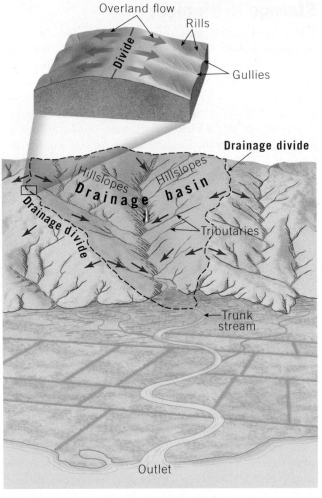

FIGURE 5.3 Drainage Basin and Divide A drainage basin is the area drained by a stream and its tributaries. Boundaries between basins are called divides.

 SmartFigure 5.4 Mississippi River Drainage Basin The drainage basin of the Mississippi River, North America's largest, covers about 3 million square kilometers (almost 1.2 million square miles) and consists of many smaller drainage basins. The drainage basin of the Yellowstone River is one of many that contribute water to the Missouri River, which, in turn, is one of many that make up the drainage basin of the Mississippi River.

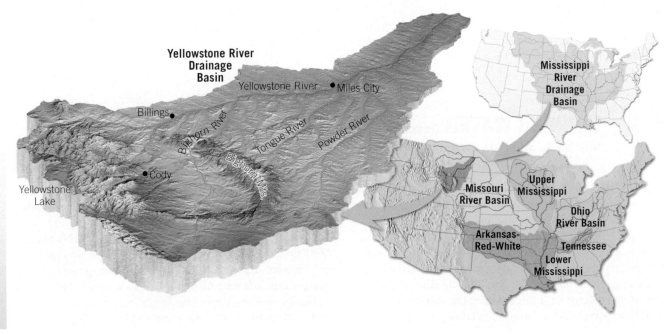

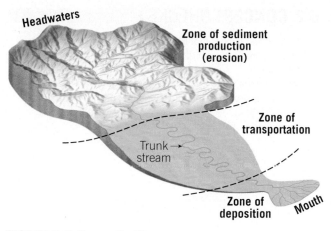

FIGURE 5.5 Zones of a River Each of the three zones is based on the dominant process that is operating in that part of the river system.

scouring of the channel bed deepens the channel and adds to the stream's sediment load.

Sediment Transport   Sediment acquired by a stream is transported through the channel network along sections referred to as *trunk streams*. When trunk streams are in balance, the amount of sediment eroded from their banks equals the amount deposited elsewhere in the channel. Although trunk streams rework their channels over time, they are not a source of sediment, nor do they accumulate or store it.

Sediment Deposition   When a river reaches the ocean, or another large body of water, it slows, and the energy to

transport sediment is greatly reduced. Most of the sediments either accumulate at the mouth of the river to form a delta, are reconfigured by wave action to form a variety of coastal features, or are moved far offshore by ocean currents. Because coarse sediments tend to be deposited upstream, it is primarily the fine sediments (clay, silt, and fine sand) that eventually reach the ocean. Taken together, erosion, transportation, and deposition are the processes by which rivers move Earth's surface materials and sculpt landscapes.

Drainage Patterns

Drainage systems are networks of streams that together form distinctive patterns. The nature of a drainage pattern can vary greatly from one type of terrain to another, primarily in response to the kinds of rock on which the streams developed and/or the structural pattern of faults and folds. **FIGURE 5.6** illustrates four drainage patterns.

The most commonly encountered drainage pattern is the **dendritic pattern**. This pattern of irregularly branching tributary streams resembles the branching pattern of a deciduous tree. In fact, the word *dendritic* means "treelike." The dendritic pattern forms where the underlying material is relatively uniform. Because the surface material is essentially uniform in its resistance to erosion, it does not control the pattern of streamflow. Rather, the pattern is determined chiefly by the direction of slope of the land.

When streams diverge from a central area like spokes from the hub of a wheel, the pattern is said to be **radial**. This pattern typically develops on isolated volcanic cones and domal uplifts.

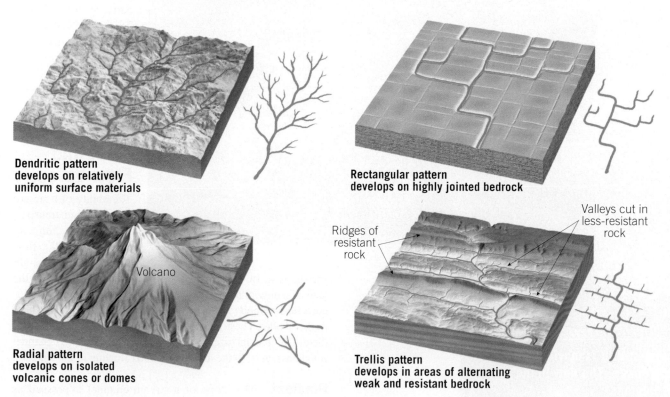

Dendritic pattern develops on relatively uniform surface materials

Rectangular pattern develops on highly jointed bedrock

Radial pattern develops on isolated volcanic cones or domes

Volcano

Trellis pattern develops in areas of alternating weak and resistant bedrock

Ridges of resistant rock

Valleys cut in less-resistant rock

FIGURE 5.6 Drainage Patterns Networks of streams form a variety of patterns.

A **rectangular pattern** exhibits many right-angle bends. This pattern develops when the bedrock is crisscrossed by a series of joints and/or faults. Because these structures are eroded more easily than unbroken rock, their geometric pattern guides the directions of valleys.

A **trellis pattern** is a rectangular drainage pattern in which tributary streams are nearly parallel to one another and have the appearance of a garden trellis. This pattern forms in areas underlain by alternating bands of resistant and less-resistant rock.

5.2 CONCEPT CHECKS

1 List several factors that influence infiltration.
2 Draw a simple sketch of a drainage basin and divide and label each.
3 What are the three main parts (zones) of a river system?
4 Prepare a sketch of the four drainage patterns discussed in this section.

5.3 STREAMFLOW

Discuss streamflow and the factors that cause it to change.

Water may flow in one of two ways, either as **laminar flow** or **turbulent flow**. In slow-moving streams, the flow is often laminar, which means that the water moves in roughly straight-line paths that parallel the stream channel. However, streamflow is usually turbulent, with the water moving in an erratic fashion that can be characterized as a swirling motion. Strong, turbulent flow may be seen in whirlpools and eddies, as well as rolling whitewater rapids (**FIGURE 5.7**). Even streams that appear smooth on the surface often exhibit turbulent flow near the bottom and sides of the channel. Turbulence contributes to a stream's ability to erode its channel because it acts to lift sediment from the streambed.

An important factor influencing stream turbulence is the water's flow velocity. As the velocity of a stream increases, the flow becomes more turbulent. Flow velocities can vary significantly from place to place along a stream, as well as over time, in response to variations in the amount and intensity of precipitation. If you have ever waded into a stream, you may have noticed that the strength of the current increased as you moved into deeper parts of the channel. This is related to the fact that frictional resistance is greatest near the banks and bed of the stream channel.

FIGURE 5.7 Laminar and Turbulent Flow Most often streamflow is turbulent.
(Photos by Michael Collier)

A.

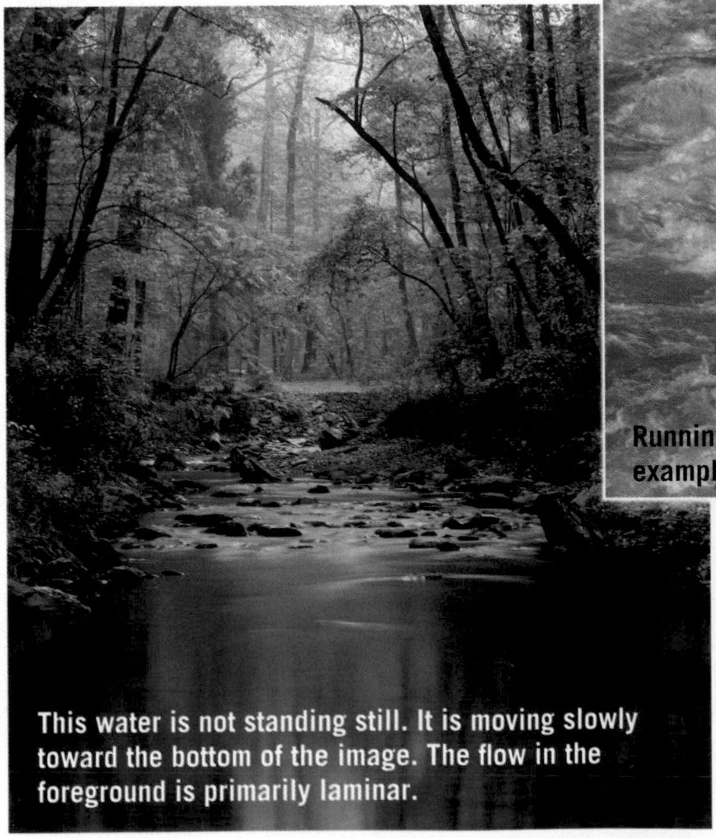
This water is not standing still. It is moving slowly toward the bottom of the image. The flow in the foreground is primarily laminar.

B.

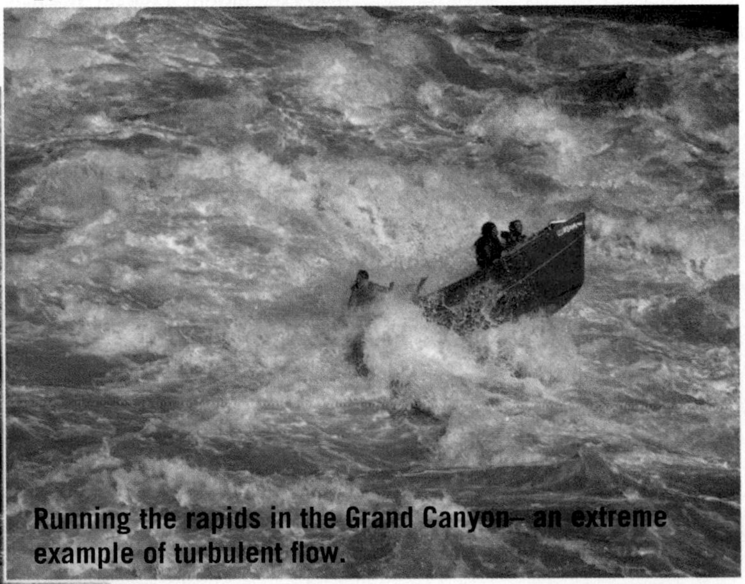
Running the rapids in the Grand Canyon—an extreme example of turbulent flow.

Factors Affecting Flow Velocity

The ability of a stream to erode and transport material is directly related to its flow velocity. Even slight variations in flow rate can lead to significant changes in the load of sediment that water can transport. Several factors influence flow velocity and, therefore, control a stream's potential to do "work." These factors include (1) channel slope or gradient, (2) channel size and cross-sectional shape, (3) channel roughness, and (4) the amount of water flowing in the channel.

Gradient The slope of a stream channel expressed as the vertical drop of a stream over a specified distance is the **gradient**. Portions of the lower Mississippi River have very low gradients of 10 centimeters per kilometer or less. By contrast, some mountain stream channels decrease in elevation at a rate of more than 40 meters per kilometer, a gradient 400 times steeper than the lower Mississippi. Gradient varies not only among different streams but also over a particular stream's length. The steeper the gradient, the more energy available for streamflow. If two streams were identical in every respect except gradient, the stream with the higher gradient would have the greater velocity.

Channel Shape, Size, and Roughness A stream's channel is a conduit that guides the flow of water, but the water encounters friction as it flows. The shape, size, and roughness of the channel affect the amount of friction. Larger channels have more efficient flow because a smaller proportion of water is in contact with the channel. A smooth channel promotes a more uniform flow, whereas an irregular channel filled with boulders creates enough turbulence to slow the stream significantly.

Discharge Streams vary in size from small headwater creeks less than 1 meter wide to large rivers as wide as several kilometers. The size of a stream channel is largely determined by the amount of water supplied from the drainage basin. The measure most often used to compare the sizes of streams is **discharge**—the volume of water flowing past a certain point in a given unit of time. Discharge, usually measured in cubic meters per second or cubic feet per second, is determined by multiplying a stream's cross-sectional area by its velocity.

The largest river in North America, the Mississippi, discharges an average of 17,300 cubic meters (611,000 cubic feet) per second. Although this is a huge quantity of water, it is dwarfed by the mighty Amazon in South America, the world's largest river. Fed by a vast rainy region that is nearly three-fourths the size of the conterminous United States, the Amazon discharges 12 times more water than the Mississippi.

The discharges of most rivers are far from constant. This is true because of variables such as rainfall and snowmelt. In areas with seasonal variations in precipitation, streamflow will tend to be highest during the wet season or during spring snowmelt, and it will be lowest during the dry season or during periods when high temperatures increase water losses through evaporation. However, not all channels maintain a continuous flow of water. Streams that exhibit flow only during wet periods are referred to as *intermittent streams*. In arid climates, many streams carry water only occasionally, after a heavy rainstorm, and are called *ephemeral streams*.

Changes from Upstream to Downstream

One useful way of studying a stream is to examine its **longitudinal profile**. Such a profile is simply a cross-sectional view of a stream from its source area (called the *head* or *headwaters*) to its *mouth*, the point downstream where it empties into

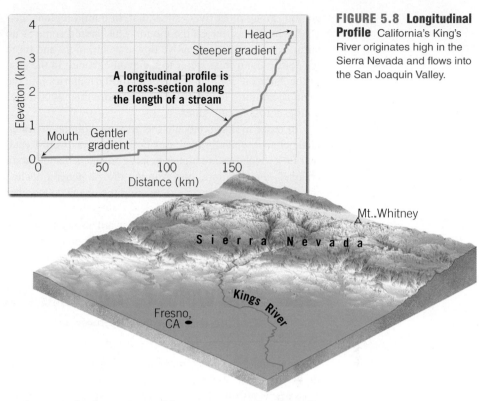

FIGURE 5.8 Longitudinal Profile California's King's River originates high in the Sierra Nevada and flows into the San Joaquin Valley.

another water body—a river, a lake, or an ocean. As shown in **FIGURE 5.8**, the most obvious feature of a typical longitudinal profile is its concave shape—a result of the decrease in slope that occurs from the headwaters to the mouth. In addition, local irregularities exist in the profiles of most streams; the flatter sections may be associated with lakes or reservoirs, and the steeper sections are sites of rapids or waterfalls.

The change in slope observed on most stream profiles is usually accompanied by an increase in discharge and channel size, as well as a reduction in sediment particle size (**FIGURE 5.9**). For example, data from successive gaging

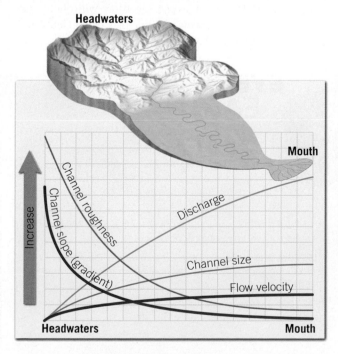

SmartFigure 5.9 Channel Changes from Head to Mouth Although the gradient decreases toward the mouth, increases in discharge and channel size and decreases in roughness more than offset the decrease in slope. Consequently, flow velocity usually increases toward the mouth.

stations along most rivers show that, in humid regions, discharge increases toward the mouth. This should come as no surprise because, as we move downstream, more and more tributaries contribute water to the main channel. In the case of the Amazon, for example, about 1000 tributaries join the main river along its 6500-kilometer (more than 4000-mile) course across South America.

In order to accommodate the growing volume of water, channel size typically increases downstream as well. Recall that flow velocities are higher in large channels than in small channels. Furthermore, observations show a general decline in sediment size downstream, making the channel smoother and more efficient.

Although the gradient decreases toward a stream's mouth, the flow velocity generally increases. This fact contradicts our intuitive assumptions of swift, narrow headwater streams and wide, placid rivers flowing across more subtle topography. Increases in channel size and discharge, and decreases in channel roughness that occur downstream, compensate for the decrease in slope—thereby making the stream more efficient.

5.3 CONCEPT CHECKS

1   Contrast laminar flow and turbulent flow.
2   Summarize the factors that influence flow velocity.
3   What is a longitudinal profile?
4   What typically happens to channel width, channel depth, flow velocity, and discharge between the head and mouth of a stream? Briefly explain why these changes occur.

5.4 | THE WORK OF RUNNING WATER

Outline the ways in which streams erode, transport, and deposit sediment.

Streams are Earth's most important erosional agents. Not only do they have the ability to downcut and widen their channels, but streams also have the capacity to transport the enormous quantities of sediment that are delivered to the stream by sheetflow, mass wasting, and groundwater. Eventually much of this material is deposited to create a variety of landforms.

Stream Erosion

A stream's ability to accumulate and transport soil and weathered rock is aided by the work of raindrops, which knock sediment particles loose (see Figure 4.19, page 112). When the ground is saturated, rainwater cannot infiltrate, so it flows downslope, transporting some of the material it has dislodged. On barren slopes the sheetflow will often erode small channels, or *rills*, which in time may evolve into larger *gullies* (see Figure 4.20 on page 112).

Once flow is confined in a channel, the erosional power of a stream is related to its slope and discharge. When the flow of water is sufficiently strong, it can dislodge particles from the channel and lift them into the moving water. In this manner, the force of running water swiftly erodes poorly consolidated materials on the bed and sides of a stream channel. On occasion, the banks of the channel may be undercut, dumping even more loose debris into the water to be carried downstream.

In addition to eroding unconsolidated materials, the hydraulic force of streamflow can also cut a channel into solid bedrock. A stream's ability to erode bedrock is greatly enhanced by the particles it carries. These particles can be any size, from large boulders in very fast-flowing waters to sand and gravel-size particles in somewhat slower flow. Just as the particles of grit on sandpaper can wear away a piece of wood, so too can the sand and gravel carried by a stream abrade a bedrock channel. Moreover, pebbles caught

EYE ON EARTH

The White River in Arkansas is a tributary of the Mississippi River. As you can see in this aerial image, this part of the river has many twists and turns called meanders. There is more about meanders later in the chapter.

QUESTION 1 *The color of the White River is brown. What part of the stream's load gives it this color?*

QUESTION 2 *If a channel were created across the narrow neck of land shown by the arrow, how would the river's gradient change?*

QUESTION 3 *How would the flow velocity be affected by the formation of such a channel?*

Michael Collier

What Are the Largest Rivers?

When rivers are ranked, the criterion most often used is the amount of water the river delivers to the ocean expressed in cubic feet (ft³) or cubic meters (m³) per second.

The flow of the Amazon accounts for about 15 percent of all the fresh water that flows into the oceans. Much of its huge drainage basin is tropical rainforest that receives 80 inches or more of rainfall each year.

Aris Michicl/AGE Fotostock

WORLD'S 10 LARGEST RIVERS

#6 YENISEI RIVER
Drainage basin:
1,000,000 square miles
Average discharge:
614,000 cubic feet per second

#9 LENA RIVER
Drainage basin:
936,000 square miles
Average discharge:
547,000 cubic feet per second

#8 MISSISSIPPI RIVER
Drainage basin:
1,150,000 square miles
Average discharge:
593,000 cubic feet per second

#7 ORINOCO RIVER
Drainage basin:
340,000 square miles
Average discharge:
600,000 cubic feet per second

#5 GANGES RIVER
Drainage basin:
409,000 square miles
Average discharge:
660,000 cubic feet per second

#3 YANGTZE RIVER
Drainage basin:
750,000 square miles
Average discharge:
770,000 cubic feet per second

#1 AMAZON RIVER
Drainage basin:
2,231,000 square miles
Average discharge:
7,500,000 cubic feet per second

#2 CONGO RIVER
Drainage basin:
1,550,000 square miles
Average discharge:
1,400,000 cubic feet per second

#4 BRAHMAPUTRA RIVER
Drainage basin:
361,000 square miles
Average discharge:
700,000 cubic feet per second

#10 PARANA RIVER
Drainage basin:
890,000 square miles
Average discharge:
526,000 cubic feet per second

The drainage basin of the "Mighty Mississippi" covers about 40 percent of the lower 48 states and includes all or parts of 31 states and 2 Canadian provinces. The discharge of North America's largest river is just one-twelfth that of the Amazon.

The size of the drainage basin and the amount of rainfall it receives are the primary factors influencing discharge.

LARGEST U.S. RIVERS

COLUMBIA RIVER

MISSOURI RIVER

ST. LAWRENCE RIVER

OHIO RIVER

MISSISSIPPI RIVER

TENNESSEE RIVER

YUKON RIVER

RANK	RIVER	DRAINAGE BASIN (1000 mi²)	AVERAGE DISCHARGE AT MOUTH (1000 ft³/sec)
1	Mississippi	1,150	593
2	St. Lawrence	396	348
3	Ohio	203	281
4	Columbia	258	265
5	Yukon	328	225
6	Missouri	529	76
7	Tennessee	41	68

Guido Alberto Rossi/AGE Fotostock

Abrasion by long-lived whirlpools armed with sand and pebbles creates bowl-shaped bedrock depressions called potholes.

FIGURE 5.10 Potholes The rotational motion of swirling pebbles acts like a drill, creating potholes. (Photo by Elmari Joubert/Alamy)

in swirling eddies can act like "drills" and bore circular **potholes** into the channel floor (**FIGURE 5.10**).

Transportation of Sediment

All streams, regardless of size, transport some weathered rock material (**FIGURE 5.11**). Streams also sort the solid sediment they transport because finer, lighter material is carried more readily than larger, heavier particles. Streams transport their load of sediment in three ways: (1) in solution (**dissolved load**), (2) in suspension (**suspended load**), and (3) sliding or rolling along the bottom (**bed load**).

Dissolved Load　Most of the dissolved load is brought to a stream by groundwater and is dispersed throughout the flow. When water percolates through the ground, it acquires soluble soil compounds. Then it seeps through cracks and pores in bedrock, dissolving additional mineral matter. Eventually much of this mineral-rich water finds its way into streams.

The velocity of streamflow has essentially no effect on a stream's ability to carry its dissolved load; material in the solution goes wherever the stream goes. Precipitation of the dissolved mineral matter occurs when the chemistry of the water changes, when organisms create hard parts, or when the water enters an inland "sea," located in an arid climate where the rate of evaporation is high.

Suspended Load　Most streams carry the largest part of their load in *suspension*. Indeed, the muddy appearance created by suspended sediment is the most obvious portion of a

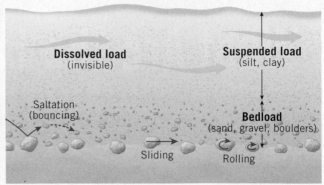

FIGURE 5.11 Transportation of Sediment Streams transport their load of sediment in three ways. The dissolved and suspended loads are carried in the general flow. The bed load includes coarse sand, gravel, and boulders that move by rolling, sliding, and saltation.

stream's load (**FIGURE 5.12**). Usually only fine particles consisting of silt and clay can be carried this way, but during a flood, larger particles are transported as well. During a flood, the total quantity of material carried in suspension increases dramatically, as people whose homes have been sites for the deposition of this material can attest.

The type and amount of material carried in suspension are controlled by two factors: the flow velocity and the settling velocity of each sediment grain. **Settling velocity** is defined as the speed at which a particle falls through a still fluid. The larger the particle, the more rapidly it settles toward the stream bed. In addition to size, the shape and

This river's muddy appearance is a result of suspended sediment.

FIGURE 5.12 Suspended Load An aerial view of the Colorado River in the Grand Canyon. Heavy rains washed sediment into the river. (Photo by Michael Collier)

specific gravity of particles also influence settling velocity. Flat grains sink through water more slowly than do spherical grains, and dense particles fall toward the bottom more rapidly than do less dense particles. The slower the settling velocity and the higher the flow velocity, the longer a sediment particle will stay in suspension, and the farther it will be carried downstream.

Bed Load A portion of a stream's load of solid material consists of sediment that is too large to be carried in suspension. These coarser particles move along the bottom (bed) of the stream and constitute the *bed load*. In terms of the erosional work accomplished by a downcutting stream, the grinding action of the bed load is of great importance.

The particles that make up the bed load move by rolling, sliding, and saltation. Sediment moving by **saltation** (*saltare* = to leap) appears to jump or skip along the stream bed. This occurs as particles are propelled upward by collisions or lifted by the current and then carried downstream a short distance until gravity pulls them back to the bed of the stream. Particles that are too large or heavy to move by saltation either roll or slide along the bottom, depending on their shapes.

Compared with the movement of suspended load, the movement of bed load through a stream network tends to be less rapid and more localized. A study conducted on a glacially fed river in Norway determined that suspended sediments took only a day to exit the drainage basin, while the bed load required several decades to travel the same distance. Depending on the discharge and slope of the channel, coarse gravels may only be moved during times of high flow, while boulders move only during exceptional floods. Once set in motion, large particles are usually carried short distances. Along some stretches of a stream, bed load cannot be carried at all until it is broken into smaller particles.

Competence and Capacity A stream's ability to carry solid particles is described using two criteria: *capacity* and *competence*. **Capacity** is the maximum load of solid particles a stream can transport per unit of time. The greater the discharge, the greater the stream's capacity for hauling sediment. Consequently, large rivers with high flow velocities have large capacities.

Competence is a measure of a stream's ability to transport particles based on size rather than quantity. Flow velocity is the key: Swift streams have greater competencies than slow streams, regardless of channel size. A stream's competence increases proportionally to the square of its velocity. Thus, if the velocity of a stream doubles, the impact force of the water increases four times; if the velocity triples, the force increases nine times, and so forth. Hence, large boulders that are often visible during low water and seem immovable can, in fact, be transported during exceptional floods because of the stream's increased competence.

By now it should be clear why the greatest erosion and transportation of sediment occur during floods. The increase in discharge results in greater capacity, and the increased velocity produces greater competency. Rising velocity makes the water more turbulent, and larger particles are set in motion. In just a few days, or perhaps a few hours, a stream at flood stage can erode and transport more sediment than it does during many months of normal flow.

Deposition of Sediment

Deposition occurs whenever a stream slows, causing a reduction in competence. Put another way, particles are deposited when flow velocity is less than the settling velocity; as a stream's flow velocity decreases, sediment begins to settle, largest particles first. In this manner, stream transport provides a mechanism by which solid particles of various sizes are separated. This process, called **sorting**, explains why particles of similar size are deposited together.

The general term for sediment deposited by streams is **alluvium**. Many different depositional features are composed of alluvium. Some occur within stream channels, some occur on the valley floor adjacent to a channel, and some are found at the mouth of a stream. We will consider the nature of these features later in the chapter.

5.4 CONCEPT CHECKS

1 List two ways in which streams erode their channels.
2 In what three ways does a stream transport its load? Which part of the load moves slowest?
3 What is the difference between capacity and competency?
4 What is settling velocity? What factors influence settling velocity?

5.5 | STREAM CHANNELS Contrast bedrock and alluvial stream channels. Distinguish between two types of alluvial channels.

A basic characteristic of streamflow that distinguishes it from sheetflow is that it is usually confined to a channel. A stream channel can be thought of as an open conduit that consists of the streambed and banks that act to confine the flow, except during floods.

Although somewhat oversimplified, we can divide stream channels into two types. *Bedrock channels* are those in which the streams are actively cutting into solid rock. In contrast, when the bed and banks are composed mainly of unconsolidated sediment or alluvium, the channel is called an *alluvial channel*.

Bedrock Channels

As the name suggests, bedrock channels are cut into the underlying strata and typically form in the headwaters of river systems where streams have steep slopes. The energetic flow tends to transport coarse particles that actively abrade the bedrock channel. Potholes are often visible evidence of the erosional forces at work.

Steep bedrock channels often develop a sequence of *steps* and *pools*. Steps are steep segments where bedrock is exposed. These steep areas contain rapids or, occasionally, waterfalls. Pools are relatively flat segments where alluvium tends to accumulate.

The channel pattern exhibited by streams cutting into bedrock is controlled by the underlying geologic structure. Even when flowing over rather uniform bedrock, streams tend to exhibit winding or irregular patterns rather than flowing in straight channels. Anyone who has gone white-water rafting has observed the steep, winding nature of a stream flowing in a bedrock channel.

Alluvial Channels

Many stream channels are composed of loosely consolidated sediment (alluvium) and therefore can undergo significant changes in shape because the sediments are continually being eroded, transported, and redeposited. The major factors affecting the shapes of these channels are the average size of the sediment being transported, the channel gradient, and the discharge.

Alluvial channel patterns reflect a stream's ability to transport its load at a uniform rate, while expending the least amount of energy. Thus, the size and type of sediment being carried help determine the nature of the stream channel. Two common types of alluvial channels are *meandering channels* and *braided channels*.

Meandering Streams Streams that transport much of their load in suspension generally move in sweeping bends called **meanders**. These streams flow in relatively deep, smooth channels and transport mainly mud (silt and clay), sand, and occasionally fine gravel. The lower Mississippi River exhibits a channel of this type.

Meandering channels evolve over time as individual meanders migrate across the floodplain. Most of the erosion is focused at the outside of the meander, where velocity and turbulence are greatest. In time, the outside bank is undermined, especially during periods of high water. Because the outside of a meander is a zone

 SmartFigure 5.13 Formation of Cut Banks and Point Bars By eroding its outer bank and depositing material on the inside of the bend, a stream is able to shift its channel.

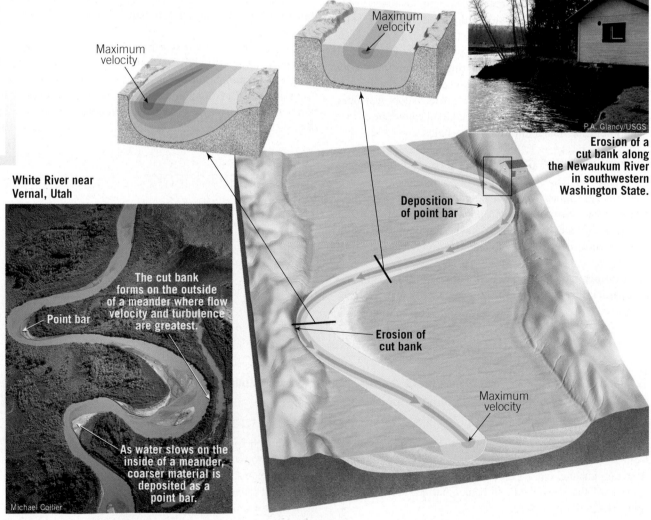

Maximum velocity

Maximum velocity

Maximum velocity

Erosion of a cut bank along the Newaukum River in southwestern Washington State.

P.A. Glancy/USGS

Deposition of point bar

Erosion of cut bank

White River near Vernal, Utah

The cut bank forms on the outside of a meander where flow velocity and turbulence are greatest.

Point bar

As water slows on the inside of a meander, coarser material is deposited as a point bar.

Michael Collier

of active erosion, it is often referred to as the **cut bank** (**FIGURE 5.13**). Debris acquired by the stream at the cut bank moves downstream, where the coarser material is generally deposited as **point bars** on the insides of meanders. In this manner, meanders migrate laterally by eroding the outside of the bends and depositing sediment on the inside without appreciably changing their shape.

In addition to migrating laterally, the bends in a channel also migrate down the valley. This occurs because erosion is more effective on the downstream (downslope) side of the meander. Sometimes the downstream migration of a meander is slowed when it reaches a more resistant bank material. This allows the next meander upstream to gradually erode the material between the two meanders, as shown in **FIGURE 5.14**. Eventually, the river may erode through the narrow neck of land, forming a new, shorter channel segment called a **cutoff**. Because of its shape, the abandoned bend is called an **oxbow lake**.

Braided Streams

Some streams consist of a complex network of converging and diverging channels that thread their way among numerous islands or gravel bars (**FIGURE 5.15**). Because these channels have an interwoven appearance, these streams are said to be **braided channels**. Braided channels form where a large proportion of the stream's load consists of coarse material (sand and gravel) and the stream has a highly variable discharge. Because the bank material is readily erodible, braided channels are wide and shallow.

One setting in which braided streams form is at the end of a glacier where there is a large seasonal variation in discharge. During the summer, large amounts of ice-eroded sediment are dumped into the meltwater streams flowing away from the glacier. However, when flow is sluggish, the stream is unable to move all the sediment and therefore deposits the coarsest material as bars that force the flow to split and follow several paths. Usually the laterally shifting channels completely rework most of the surface sediments each year, thereby transforming the entire streambed. In some braided streams, however, the bars have built up to form islands that are anchored by vegetation.

In summary, meandering channels develop where the load consists largely of fine-grained particles that are transported as suspended load in deep, relatively smooth channels. By contrast, wide, shallow braided channels develop where coarse-grained material is transported as bed load.

5.5 CONCEPT CHECKS

1 Are bedrock channels more likely to be found near the head or the mouth of a stream?
2 Describe or sketch the development of a meander, including how an oxbow lake forms.
3 Describe a situation that might cause a stream to become braided.

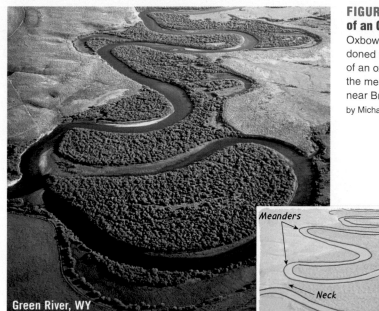

FIGURE 5.14 Formation of an Oxbow Lake Oxbow lakes occupy abandoned meanders. Aerial view of an oxbow lake created by the meandering Green River near Bronx, Wyoming. (Photo by Michael Collier)

Green River, WY

Geologist's Sketch

FIGURE 5.15 Braided Stream The Knik River is a classic braided stream with multiple channels separated by migrating gravel bars. The Knik is choked with sediment from four melting glaciers in the Chugach Mountains north of Anchorage, Alaska. (Photo by Michael Collier)

5.6 | SHAPING STREAM VALLEYS Contrast narrow V-shaped valleys, broad valleys with floodplains, and valleys that display incised meanders.

Streams, with the aid of weathering and mass wasting, shape the landscape through which they flow. As a result, streams continuously modify the valleys they occupy.

A **stream valley** consists of not only the channel but also the surrounding terrain that directly contributes water to the stream. Thus, it includes the valley bottom, which is the lower, flatter area that is partially or totally occupied by the stream channel, and the sloping valley walls that rise above the valley bottom on both sides. Most stream valleys are much broader at the top than their channels are wide at the bottom. This would not be the case if the only agent responsible for eroding valleys were the streams flowing through them. The sides of most valleys are shaped by a combination of weathering, overland flow, and mass wasting. In some arid regions, where weathering is slow and where rock is particularly resistant, narrow valleys (sometimes called *slot canyons*) that have nearly vertical walls are common.

Stream valleys can be divided into two general types—narrow V-shaped valleys and wide valleys with flat floors, with many gradations between.

Base Level and Stream Erosion

Streams cannot endlessly erode their channels deeper and deeper. There is a lower limit to how deep a stream can erode, and that limit is called **base level**. Although the idea is relatively straightforward, it is nevertheless a key concept in the study of stream activity. Base level is defined as the lowest elevation to which a stream can erode its channel. Essentially this is the level at which the mouth of a stream enters the ocean, a lake, or another stream. Base level accounts for the fact that most stream profiles have low gradients near their mouths, because the streams are approaching the elevation below which they cannot erode their beds.

Two general types of base level are recognized. Sea level is considered the *ultimate base level* because it is the lowest level to which stream erosion could lower the land. *Temporary*, or *local*, *base levels* include lakes, resistant layers of rock, and main streams that act as base levels for their tributaries. For example, when a stream enters a lake, its velocity quickly approaches zero, and its ability to erode ceases. Thus, the lake prevents the stream from eroding below its level at any point upstream from the lake. However, because the outlet of the lake can cut downward and drain the lake, the lake is only a temporary hindrance to the stream's ability to downcut its channel. In a similar manner, the layer of resistant rock at the lip of the waterfall in **FIGURE 5.16** acts as a temporary base level. Until the ledge of hard rock is eliminated, it will limit the amount of downcutting upstream.

Any change in base level will cause a corresponding readjustment of stream activities. When a dam is built along a stream, the reservoir that forms behind it raises the base level of the stream (**FIGURE 5.17**). Upstream from the dam the gradient is reduced, lowering the stream's velocity and, hence, its sediment-transporting ability. The stream, now having too little energy to transport its entire load, will deposit sediment. This builds up its channel. Deposition will be the dominant process until the stream's gradient increases sufficiently to transport its load.

FIGURE 5.16 Temporary Base Level This series of diagrams illustrates what would happen if a fault raised a layer of resistant rock across the path of a stream. **A**. Stream course with a smooth profile prior to faulting. **B**. After faulting, the resistant layer acts as a temporary base level and forms a waterfall. Because of the steep gradient, the stream's erosive energy is focused on the resistant bed. **C**. The waterfall evolves to rapids. **D**. Eventually the river reestablishes a smooth profile.

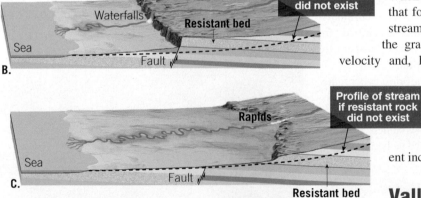

Valley Deepening

When a stream's gradient is steep and the channel is well above base level, downcutting is the dominant activity. Abrasion caused by bed load sliding and rolling along the bottom, and the hydraulic power of fast-moving water, slowly lower the streambed. The result is usually a V-shaped valley with steep sides. A classic example of a V-shaped valley is the section of Yellowstone River shown in **FIGURE 5.18**.

The most prominent features of a V-shaped valley are *rapids* and *waterfalls*. Both occur where the stream's gradient increases significantly, a situation usually caused by variations in the erodibility of the bedrock into which a stream channel is cutting. Resistant beds create rapids by acting as a temporary base level upstream while allowing downcutting to continue downstream. In time, erosion usually eliminates the resistant rock. Waterfalls are places where the stream makes an abrupt vertical drop.

Valley Widening

Once a stream has cut its channel closer to base level, downward erosion becomes less dominant. At this point, the stream's channel takes on a meandering pattern, and more of the stream's energy is directed from side to side. The result is a widening of the valley as the river cuts away first at one bank and then at the other (**FIGURE 5.19**). The continuous lateral erosion caused by shifting of the stream's meanders produces an increasingly broader, flat valley floor covered with alluvium. This feature, called a **floodplain**, is appropriately named because when a river overflows its banks during flood stage, it inundates the floodplain.

Over time the floodplain will widen to the point that the stream is actively eroding the valley walls only in a few places. In fact, in large rivers such as the lower Mississippi River Valley, the distance from one valley wall to another can exceed 160 kilometers (100 miles).

Changing Base Level and Incised Meanders

We usually expect a stream with a highly meandering course to be on a floodplain in a wide valley. However, certain rivers exhibit meandering channels that flow in steep, narrow valleys. Such meanders are called **incised** (*incisum* = to cut into) **meanders** (**FIGURE 5.20**). How do such features form?

Originally, the meanders probably developed on the floodplain of a stream that was relatively near base level. Then, a change in base level caused the stream to begin downcutting. One of two events could have occurred: Either base level

FIGURE 5.17 Building a Dam The base level upstream from the reservoir is raised, which reduces the stream's flow velocity and leads to deposition and a reduced gradient.

dropped or the land on which the river was flowing was uplifted.

An example of the first circumstance happened during the Ice Age, when large quantities of water were withdrawn from the ocean and locked up in glaciers on land.

FIGURE 5.18 Yellowstone River The V-shaped valley, rapids, and waterfalls indicate that the river is vigorously downcutting. (Photo by Charles A. Blakeslee/AGE Fotostock)

Geologist's Sketch

FIGURE 5.19 Development of an Erosional Floodplain
Continuous side-to-side erosion by shifting meanders gradually produces a broad, flat valley floor. Alluvium deposited during floods covers the valley floor.

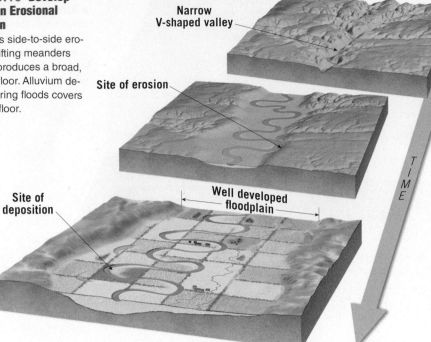

The result was that sea level (the ultimate base level) dropped, causing rivers flowing into the ocean to begin to downcut. Of course, this activity ceased at the close of the Ice Age, when ice sheets melted and sea level rose.

Regional uplift of the land, the second cause of incised meanders, is exemplified by the Colorado Plateau in the southwestern United States. Here, as the plateau was gradually uplifted, numerous meandering rivers adjusted to being higher above base level by downcutting.

5.6 CONCEPT CHECKS

1 Define *base level* and distinguish between ultimate base level and temporary base level.
2 Explain why V-shaped valleys often contain rapids and waterfalls.
3 Describe or sketch how an erosional floodplain develops.
4 Relate the formation of incised meanders to changes in base level.

SmartFigure 5.20 Incised Meanders Aerial view of incised meanders of the Colorado River on the Colorado Plateau. (Photo by Michael Collier)

Before uplift of the Colorado Plateau, the river was meandering on a floodplain.

During uplift of the plateau, the meanders downcut because of the steepening gradient.

5.7 | DEPOSITIONAL LANDFORMS

Discuss the formation of deltas, natural levees, and alluvial fans.

Recall that streams continually pick up sediment in one part of their channel and deposit it downstream. These small-scale channel deposits are most often composed of sand and gravel and are commonly referred to as **bars**. Such features, however, are only temporary because the material will be picked up again and eventually carried to the ocean. In addition to sand and gravel bars, streams also create other depositional features that have a somewhat longer life span. These include *deltas*, *natural levees*, and *alluvial fans*.

deltas are characterized by these shifting channels that act in an opposite way to that of tributaries.

Rather than carry water into the main channel, distributaries carry water away from the main channel. After numerous shifts of the channel, a delta may grow into a roughly triangular shape like the Greek letter delta (Δ), for which it is named. Note, however, that many deltas do not exhibit the idealized shape. Differences in the configurations of shorelines and variations in the nature and strength of wave activity result in many shapes. Many large rivers have deltas

Deltas

A **delta** forms where a sediment-charged stream enters the relatively still waters of a lake, an inland sea, or the ocean (**FIGURE 5.21**). As the stream's forward motion slows, sediments are deposited by the dying current. As the delta grows outward, the stream's gradient continually lessens. This circumstance eventually causes the channel to become choked with sediment deposited from the slowing water. As a consequence, the river seeks a shorter, higher-gradient route to base level, as illustrated in Figure 5.21. This illustration shows the main channel dividing into several smaller ones, called **distributaries**. Most

Topset beds are deposited atop the foreset beds during floods.

Distributaries

Foreset beds consist of coarse particles that drop soon after entering the water body. As the delta grows, these beds cover the bottomset beds.

Bottomset beds consist of fine silt and clay particles that settled beyond the mouth of the river.

As the stream extends its channel, the gradient is reduced. During flood stage some of the flow is diverted to a shorter, higher-gradient route forming a new distributary.

FIGURE 5.21 Formation of a Simple Delta Structure and growth of a simple delta that forms in relatively quiet waters.

EYE ON EARTH

This satellite image shows the delta of the Yukon River. The river originates in northern British Columbia. It flows through the Yukon Territory and across the tundra of Alaska before entering the Bering Sea, a distance of nearly 3200 kilometers (about 2000 miles).

QUESTION 1 *Explain why the river breaks into numerous channels as it crosses the delta.*

QUESTION 2 *What term is applied to the channels that radiate across the delta?*

QUESTION 3 *Notice the cloud of sediment in the water surrounding the delta. Are these sediments more likely sand and gravel or silt and clay? Explain.*

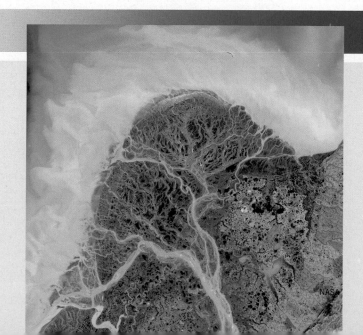
NASA

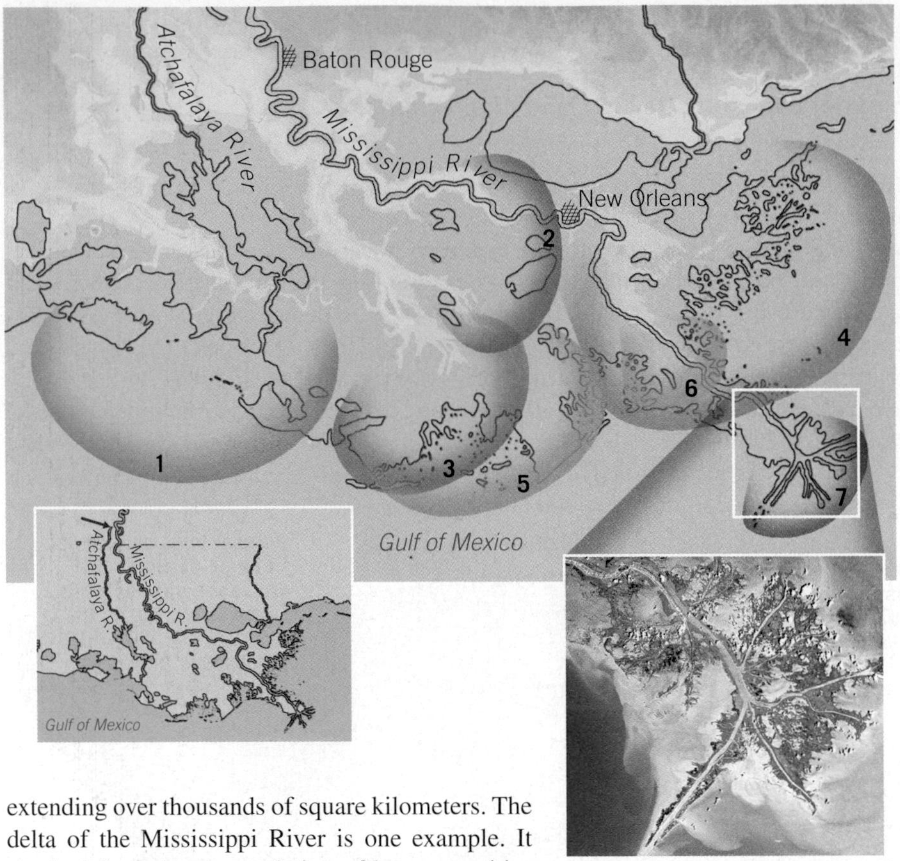

FIGURE 5.22 Growth of the Mississippi River Delta During the past 6000 years, the river has built a series of seven coalescing subdeltas. The numbers indicate the order in which the subdeltas were deposited. The present bird-foot delta (number 7) represents the activity of the past 500 years. The left inset shows the point where the Mississippi may sometime break through (arrow) and the shorter path it would take to the Gulf of Mexico. (Image courtesy of JPL/Cal Tech/NASA)

extending over thousands of square kilometers. The delta of the Mississippi River is one example. It resulted from the accumulation of huge quantities of sediment derived from the vast region drained by the river and its tributaries (see Figure 5.4). Today, New Orleans rests where there was ocean less than 5000 years ago. **FIGURE 5.22** shows that portion of the Mississippi delta that has been built over the past 6000 years. As you can see, the delta is actually a series of seven coalescing subdeltas. Each

formed when the river left its existing channel in favor of a shorter, more direct path to the Gulf of Mexico. The individual subdeltas interfinger and partially cover one another, producing a very complex structure. The present subdelta, called a *bird-foot* delta because of the configuration of its distributaries, has been built by the Mississippi in the past 500 years.

Natural Levees

Some rivers occupy valleys with broad floodplains and build **natural levees** that parallel their channels on both banks (**FIGURE 5.23**). Natural levees are built by successive floods over many years. When a stream overflows its banks, its velocity immediately diminishes, leaving coarse sediment deposited in strips bordering the channel. As the water spreads out over the valley, a lesser amount of fine sediment is deposited over the valley floor. This uneven distribution of material produces the very gentle slope of the natural levee.

The natural levees of the lower Mississippi rise 6 meters (20 feet) above the floodplain. The area behind the levee is characteristically poorly drained for the obvious reason that water cannot flow up the levee and into the river. Marshes called **back swamps** result. A tributary stream that cannot

FIGURE 5.23 Formation of a Natural Levee These gently sloping structures that parallel a river channel are created by repeated floods. Because the ground next to the channel is higher than the adjacent floodplain, back swamps and yazoo tributaries may develop.

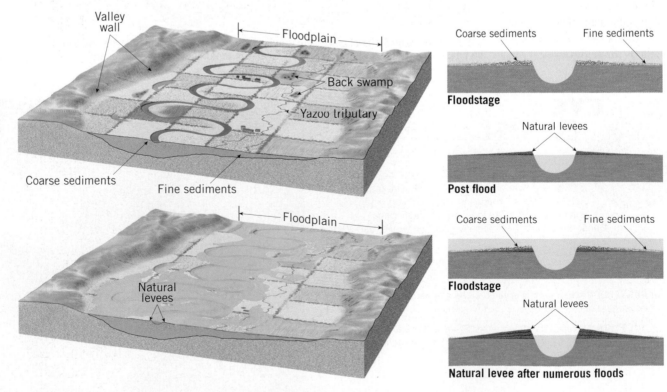

enter a river because levees block the way often has to flow parallel to the river until it can breach the levee. Such streams are called **yazoo tributaries**, after the Yazoo River, which parallels the Mississippi for more than 300 kilometers (about 190 miles).

Alluvial Fans

Alluvial fans typically develop where a high-gradient stream leaves a narrow valley in mountainous terrain and comes out suddenly onto a broad, flat plain or valley floor (see Figure 6.33, page 196). Alluvial fans form in response to the abrupt drop in gradient combined with the change from a narrow channel of a mountain stream to less confined channels at the base of the mountains. The sudden drop in velocity causes the stream to dump its load of sediment quickly in a distinctive cone- or fan-shaped accumulation. As illustrated by Figure 6.33, the surface of the fan slopes outward in a broad arc from an apex at the mouth of the steep valley. Usually, coarser material is dropped near the apex of the fan, while finer material is carried toward the base of the deposit.

5.7 CONCEPT CHECKS

1 What feature may form where a stream enters the relatively still waters of a lake, an inland sea, or the ocean?
2 What are distributaries, and why do they form?
3 Briefly describe the formation of a natural levee. How is this feature related to back swamps and yazoo tributaries?
4 How does an alluvial fan differ from a delta?

5.8 | FLOODS AND FLOOD CONTROL

Discuss the causes of floods and some common flood control measures.

A **flood** occurs when the flow of a stream becomes so great that it exceeds the capacity of its channel and overflows its banks. Among the most deadly and most destructive of all geologic hazards, floods are, nevertheless, simply part of the *natural* behavior of streams.

Causes of Floods

Rivers flood because of the weather. Rapid melting of snow in the spring and/or major storms that bring heavy rains over a large region cause most floods. Exceptional rains caused the devastating floods in the upper Mississippi River Valley during the summer of 1993 (**FIGURE 5.24**).

Unlike the extensive *regional floods* just mentioned, *flash floods* are more limited in extent. Flash floods occur with little warning and can be deadly because they produce a rapid rise in water levels and can have a devastating flow velocity. Several factors influence flash flooding. Among them are rainfall intensity and duration, topography, and surface conditions. Mountainous areas are susceptible because steep slopes can quickly funnel runoff into narrow canyons. Urban areas are susceptible to flash floods because a high percentage of the surface area is composed of impervious surfaces such as roofs, streets, and parking lots, where runoff is very rapid.

Human interference with a stream system can worsen or even cause floods. A prime example is the failure of a dam or an artificial levee. These structures are built for flood protection. They are designed to contain floods of a certain magnitude. If a larger flood occurs, the dam or levee is overtopped. If the dam or levee fails or is washed out, the water behind it is released and becomes a flash flood. The bursting of a dam in 1889 on the Little Conemaugh River caused the devastating Johnstown, Pennsylvania, flood that took some 3000 lives. A second dam failure occurred there in 1977, causing 77 fatalities.

Flood Control

Several strategies have been devised to eliminate or lessen the catastrophic effects of floods. Engineering efforts include the construction of artificial levees, the building of flood-control dams, and river channelization.

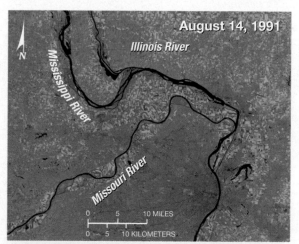
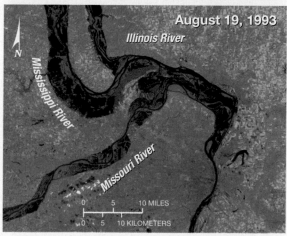

FIGURE 5.24 The Great Flood of 1993 These satellite images show the St. Louis area where the Mississippi, Illinois, and Missouri Rivers join. The upper image shows the rivers during a relatively normal period in August 1991. The lower image shows the peak of the 1993 flood. In all, nearly 14 million acres were inundated, displacing at least 50,000 people. (Courtesy of NASA)

Flash floods

Flash floods are local floods of great volume and short duration. The rapidly rising surge of water usually occurs with little advance warning and can destroy roads, bridges, homes, and other substantial structures.

USGS

Michael Collier

The power of a flash flood is illustrated by the Big Thompson River flood of July 31, 1976, in Colorado. During a four-hour span more than 30 centimeters (12 inches) of rain fell on portions of the river's small drainage basin. This amounted to nearly three-quarters of the average yearly total. The flash flood in the narrow canyon lasted only a few hours, but cost 139 people their lives.

Urban development increases runoff. As a result, peak discharge and flood frequency increase. A recent study indicated that the area of impervious surfaces in the 48 contiguous United States is roughly equal to the area of the state of Ohio (44,000 mi^2).

Most people do not appreciate the power of moving water. Many automobiles will float and be swept away in a strong current that is only 2 feet deep. More than half of all U.S. flash-flood fatalities are auto related!

AP Photo/Sue Ogrocki

Average Annual Storm-Related Deaths in the United States

In most years floods are responsible for the greatest number of storm-related deaths. The average number of hurricane deaths was dramatically affected by Hurricane Katrina in 2005 (more than 1000). For all other years on this graph, hurricane fatalities numbered fewer than 20.

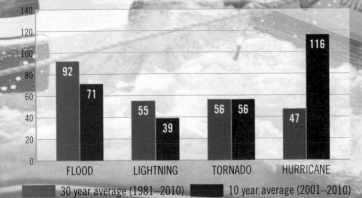

	FLOOD	LIGHTNING	TORNADO	HURRICANE
30 year average (1981–2010)	92	55	56	47
10 year average (2001–2010)	71	39	56	116

Effect of Urban Development on Flooding

Streamflow in Mercer Creek, an urban stream in western Washington, increases more quickly, reaches a higher peak discharge, and has a larger volume during a one-day storm on February 1, 2000, than streamflow in Newaukum Creek, a nearby rural stream. Streamflow during the following week, however, was greater in Newaukum Creek.

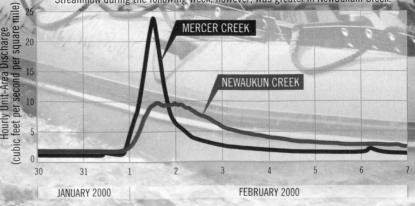

MERCER CREEK

NEWAUKUN CREEK

Hourly Unit-Area Discharge (cubic feet per second per square mile)

JANUARY 2000   FEBRUARY 2000

Artificial Levees   *Artificial levees* are earthen mounds built on the banks of a river to increase the volume of water the channel can hold. Levees, used since ancient times, are the most commonly used stream-containment structures.

Artificial levees are usually easy to distinguish from natural levees because their slopes are much steeper. In some locations, especially urban areas, concrete floodwalls are sometimes constructed that serve the same purpose as artificial levees.

Many artificial levees were not built to withstand periods of extreme flooding. For example, levee failures were numerous in the Midwest during the summer of 1993, when the upper Mississippi and many of its tributaries experienced record floods (**FIGURE 5.25**). During that same event, floodwalls at St. Louis, Missouri, created a bottleneck for the river that led to increased flooding upstream of the city.

Flood-Control Dams   *Flood-control dams* are built to store floodwater and then let it out slowly. This lowers the flood crest by spreading it out over a longer time span. Since the 1920s, thousands of dams have been built on nearly every major river in the United States. Many dams have significant non-flood-related functions, such as providing water for irrigated agriculture and for hydroelectric power generation. Many reservoirs are also major regional recreational facilities.

Although dams may reduce flooding and provide other benefits, building these structures also has significant costs and consequences. For example, reservoirs created by dams may cover fertile farmland, useful forests, historic sites, and scenic valleys. Of course, dams trap sediment. Therefore, deltas and floodplains downstream erode because they are no longer replenished with silt during floods. Large dams can also cause significant ecological damage to river environments that took thousands of years to establish.

Building a dam is not a permanent solution to flooding. Sedimentation behind a dam causes the volume of its reservoir to gradually diminish, reducing the effectiveness of this flood-control measure.

Channelization   *Channelization* involves altering a stream channel in order to speed the flow of water to prevent it from reaching flood height. This may simply involve clearing a channel of obstructions or dredging a channel to make it wider and deeper.

Another alteration involves straightening a channel by creating *artificial cutoffs*. The idea is that by shortening the stream, the gradient and hence the flow velocity are both increased. By increasing velocity, the larger discharge associated with flooding can be dispersed more rapidly.

Since the early 1930s, the U.S. Army Corps of Engineers has created many artificial cutoffs on the Mississippi for the purpose of increasing the efficiency of the channel and reducing the threat of flooding. In all, the river has been shortened more than 240 kilometers (150 miles). These efforts have been somewhat successful in reducing the height of the river in flood stage. However, channel shortening led to higher gradients and accelerated erosion of riverbank material, both of which necessitated further intervention. Following the creation of artificial cutoffs, massive riverbank protection to reduce erosion was installed along several stretches of the lower Mississippi.

A Nonstructural Approach   All of the flood-control measures described so far have involved structural solutions aimed at "controlling" a river. These solutions are expensive

River

Break in levee

 Mobile Field Trip 5.25 Broken Levee   Water from the swollen Mississippi River rushes through a break in an artificial levee in Monroe County, Illinois. During the record-breaking 1993 Midwest floods, many artificial levees could not withstand the force of the floodwaters. Sections of many weakened structures were overtopped or collapsed. (Photo by AP Photo/ James A. Finley)

and often give people residing on the floodplain a false sense of security.

Today, many scientists and engineers advocate a nonstructural approach to flood control. They suggest that an alternative to artificial levees, dams, and channelization is sound floodplain management. By identifying high-risk areas, appropriate zoning regulations can be implemented to minimize development and promote more appropriate land use.

5.9 | GROUNDWATER: WATER BENEATH THE SURFACE

Discuss the importance of groundwater and describe its distribution and movement.

Groundwater is one of our most important and widely available resources. Yet people's perceptions of the subsurface environment from which it comes are often unclear and incorrect. The reason is that groundwater is hidden from view except in caves and mines, and the impressions people gain from these subsurface openings are often misleading. Observations on the land surface give an impression that Earth is "solid." This view is not changed very much when we enter a cave and see water flowing in a channel that appears to have been cut into solid rock.

Because of such observations, many people believe that groundwater occurs only in underground "rivers." But actual rivers underground are extremely rare. In reality, most of the subsurface environment is not "solid" at all. Rather, it includes countless tiny *pore spaces* between grains of soil and sediment plus narrow joints and fractures in bedrock. Together, these spaces add up to an immense volume. Groundwater collects and moves in these tiny openings.

The Importance of Groundwater

Of the entire hydrosphere, or all of Earth's water, only a small percentage occurs underground. Nevertheless, this small percentage, stored in the rocks and sediments beneath Earth's surface, is a vast quantity. When the oceans are excluded and only sources of freshwater are considered, the significance of groundwater becomes more apparent.

Take a look back at Figure 5.1, which shows the distribution of freshwater in the hydrosphere. Clearly, the largest volume of freshwater is in the form of glacial ice. Second in rank is groundwater, with slightly more than 30 percent of the total. However, when ice is excluded and just liquid water is considered, nearly 96 percent is groundwater. Without question, *groundwater represents the largest reservoir of freshwater that is readily available to humans.* Its value in terms of economics and human well-being is incalculable.

Worldwide, wells and springs provide water for cities, crops, livestock, and industry. In the United States, groundwater is the source of about 40 percent of the water used for all purposes (except hydroelectric power generation and power plant cooling). Groundwater is the drinking water for more than 50 percent of the population, it is 40 percent of the water for irrigation, and it provides more than 25 percent of industry's needs. In some areas, however, overuse of this basic resource has caused serious problems, including streamflow depletion, land subsidence, and increased pumping costs. In addition, groundwater contamination resulting from human activities is a real and growing threat in many places.

Groundwater's Geologic Roles

Geologically, groundwater is important as an erosional agent. The dissolving action of groundwater slowly removes rock, allowing surface depressions known as sinkholes to form and creating subterranean caverns (**FIGURE 5.26**). Groundwater is also an equalizer of streamflow. Much of the water that flows in rivers is not direct runoff from rain and snowmelt. Rather, a large percentage of precipitation soaks in and then moves slowly underground to stream channels. Groundwater is thus a form of storage that sustains streams during periods when rain does not fall. When we see water flowing in a river during a dry period, it is water from rain that fell at some earlier time and was stored underground.

Distribution of Groundwater

When rain falls, some of the water runs off, some returns to the atmosphere through evaporation and transpiration, and the remainder soaks into the ground. This last path is the primary source of practically all groundwater. The amount of water that takes each of these paths, however, varies greatly from time to time and place to place. Influential factors include the steepness of the slope, the nature of the surface material, the intensity of the rainfall, and the type and amount of vegetation. Heavy rains falling on steep slopes underlain by impervious materials will obviously result in a high percentage of the water running off. Conversely, if rain falls steadily and gently on more gradual slopes composed of materials that are more easily penetrated by water, a much larger percentage of the water soaks into the ground.

Underground Zones Some of the water that soaks in does not travel far because it is held by molecular attraction as a surface film on soil particles. This near-surface zone is called the *belt of soil moisture*. It is crisscrossed by roots, voids left by decayed roots, and animal and worm burrows that enhance the infiltration of rainwater into the soil. Soil water is used by plants for life functions and transpiration. Some of this water also evaporates directly back into the atmosphere.

Water that is not held as soil moisture penetrates downward until it reaches a zone where all the open spaces in sediment and rock are completely filled with water. This is the **zone of saturation**. Water within it is called **groundwater**. The upper limit of this zone is known as the **water table**. The area above the water table where the soil, sediment, and rock are not saturated is called the **unsaturated zone** (**FIGURE 5.27**). Although a considerable amount of water can be present in the unsaturated zone, this water cannot be pumped by wells because it clings too tightly to rock and soil particles. By contrast, below the water table, the water pressure is great enough to allow water to enter

A.

B.

FIGURE 5.26 Caverns and Sinkholes A. A view of the interior of New Mexico's Carlsbad Caverns. The dissolving action of acidic groundwater created the caverns. Later, groundwater deposited the limestone decorations. (Photo by Clint Farlinger/Alamy)
B. Groundwater was responsible for creating these depressions, called sinkholes, west of Timaru on New Zealand's South Island. The white dots in this photo are grazing sheep. (Photo by David Wall/Alamy)

EYE ON EARTH

Marshes and swamps are characterized by saturated, poorly drained soils. This wetland is located southwest of Fort McMurray, Alberta, Canada. (Photo by Michael Collier)

QUESTION 1 *Describe the position of the water table in this image.*

QUESTION 2 *Suggest two situations that might cause this marsh to disappear. Include one natural cause and one cause related to human activities.*

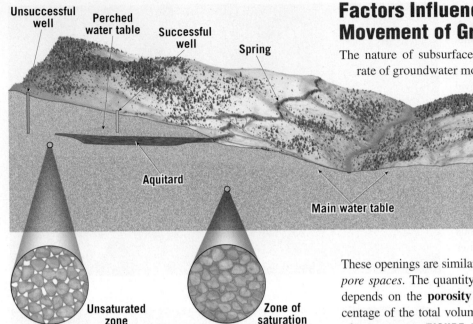

Unsuccessful well
Perched water table
Successful well
Spring
Aquitard
Main water table
Unsaturated zone
Zone of saturation

Factors Influencing the Storage and Movement of Groundwater

The nature of subsurface materials strongly influences the rate of groundwater movement and the amount of groundwater that can be stored. Two factors are especially important—porosity and permeability.

Porosity Water soaks into the ground because bedrock, sediment, and soil contain countless voids or openings. These openings are similar to those of a sponge and are called *pore spaces*. The quantity of groundwater that can be stored depends on the **porosity** of the material, which is the percentage of the total volume of rock or sediment that consists of pore spaces (**FIGURE 5.28**). Voids most often are spaces between sedimentary particles, but also common are joints, faults, cavities formed by the dissolving of soluble rock such as limestone, and vesicles (voids left by gases escaping from lava).

Variations in porosity can be great. Sediment is commonly quite porous, and open spaces may occupy 10 percent to 50 percent of the sediment's total volume. Pore space depends on the size and shape of the grains, how they are packed together, the degree of sorting, and, in sedimentary rocks, the amount of cementing material. Most igneous and metamorphic rocks, as well as some sedimentary rocks, are composed of tightly interlocking crystals, so the voids between grains may be negligible. In these rocks, fractures must provide the voids.

Permeability Porosity alone cannot measure a material's capacity to yield groundwater. Rock or sediment may be very porous and still prohibit water from moving through it. The **permeability** of a material indicates its ability to *transmit* a fluid. Groundwater moves by twisting and turning through interconnected small openings. The smaller the pore spaces, the slower the groundwater moves. If the spaces between particles are too small, water cannot move at all. For example, clay's ability to store water can be great, due to its high porosity, but its pore spaces are so small that water is unable to move through it. Thus, we say that clay is *impermeable*.

Aquitards and Aquifers

Impermeable layers such as clay that hinder or prevent water movement are termed **aquitards** (*aqua* = water, *tard*

wells, thus permitting groundwater to be withdrawn for use. We will examine wells more closely later in the chapter.

Water Table The water table is rarely level, as we might expect a table to be. Instead, its shape is usually a subdued replica of the surface, reaching its highest elevations beneath hills and decreasing in height toward valleys (see Figure 5.27). The water table of a wetland (swamp) is right at the surface. Lakes and streams generally occupy areas low enough that the water table is above the land surface.

Several factors contribute to the irregular surface of the water table. One important influence is the fact that groundwater moves very slowly. Because of this, water tends to "pile up" beneath high areas between stream valleys. If rainfall were to cease completely, these water "hills" would slowly subside and gradually approach the level of the adjacent valleys. However, new supplies of rainwater are usually added often enough to prevent this. Nevertheless, in times of extended drought, the water table may drop enough to dry up shallow wells. Other causes for the uneven water table are variations in rainfall and permeability of Earth materials from place to place.

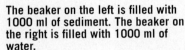

The beaker on the left is filled with 1000 ml of sediment. The beaker on the right is filled with 1000 ml of water.

Before

The sediment-filled beaker now contains 500 ml of water. Pore spaces (porosity) must represent 50 percent of the volume of the sediment.

After

= slow). In contrast, larger particles, such as sand or gravel, have larger pore spaces. Therefore, water moves with relative ease. Permeable rock strata or sediments that transmit groundwater freely are called **aquifers** ("water carriers"). Aquifers are important because they are the water-bearing layers sought after by well drillers.

Groundwater Movement

The movement of most groundwater is exceedingly slow, from pore to pore. A typical rate is a few centimeters per day. The energy that makes the water move is provided by the force of gravity. In response to gravity, water moves from areas where the water table is high to zones where the water table is lower. This means that water usually gravitates toward a stream channel, lake, or spring. Although some water takes the most direct path down the slope of the water table, much of the water follows long, curving paths toward the zone of discharge.

FIGURE 5.29 shows how water percolates into a stream from all possible directions. Some paths clearly turn upward, apparently against the force of gravity, and enter through the bottom of the channel. This is easily explained: The deeper you go into the zone of saturation, the greater the water pressure. Thus, the looping curves followed by water in the saturated zone may be thought of as a compromise between the downward pull of gravity and the tendency of water to move toward areas of reduced pressure.

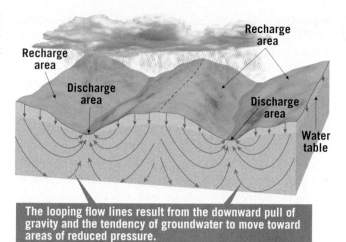

FIGURE 5.29 Groundwater Movement Arrows show paths of groundwater movement through uniformly permeable material.

The looping flow lines result from the downward pull of gravity and the tendency of groundwater to move toward areas of reduced pressure.

5.9 CONCEPT CHECKS

1 About what percentage of freshwater is groundwater? How does this change if glacial ice is excluded?
2 What are two geologic roles for groundwater?
3 When it rains, what factors influence the amount of water that soaks in?
4 Define *groundwater* and relate it to the water table.
5 Distinguish between porosity and permeability. Contrast aquifer and aquitard.
6 What factors cause water to follow the paths shown in Figure 5.29?

5.10 | SPRINGS, WELLS, AND ARTESIAN SYSTEMS

Compare and contrast springs, wells, and artesian systems.

A great deal of groundwater eventually makes its way to the surface. Sometimes this occurs as a naturally flowing spring or as a spectacularly erupting geyser. Much of the groundwater that people use is brought to the surface by pumping it from a well. To understand these phenomena, it is necessary to understand Earth's sometimes complex underground "plumbing."

Springs

Springs have aroused the curiosity and wonder of people for thousands of years. The fact that springs were (and to some people still are) rather mysterious phenomena is not difficult to understand, for here water is flowing freely from the ground in all kinds of weather, in seemingly inexhaustible supply but with no obvious source. Today, we know that the source of springs is water from the zone of saturation and that the ultimate source of this water is precipitation.

Whenever the water table intersects the ground surface, a natural flow of groundwater results, which we call a **spring** (FIGURE 5.30). Many springs form when an aquitard blocks the downward movement of groundwater and forces it to move laterally. When the permeable bed (aquifer) outcrops in a valley, one or more springs result.

Another situation that can produce a spring is illustrated in Figure 5.27. Here an aquitard is situated above the main water table. As water percolates downward, a portion

FIGURE 5.30 Thunder Spring Water gushes from a bedrock wall in the Grand Canyon. (Photo by Michael Collier)

accumulates above the aquitard to create a localized zone of saturation and a **perched water table**. Springs, however, are not confined to places where a perched water table creates a flow at the surface. Many geological situations lead to the formation of springs because subsurface conditions vary greatly from place to place.

Hot Springs

There is no universally accepted definition of **hot spring**. One frequently used definition is that the water in a hot spring is 6° to 9°C (10° to 15°F) warmer than the average annual air temperature for the locality where it occurs. In the United States alone, there are well over 1000 such springs.

Temperatures in deep mines and oil wells usually rise with increasing depth, an average of about 2°C per 100 meters (1°F per 100 feet), a figure known as the *geothermal gradient*. Therefore, when groundwater circulates at great depths, it becomes heated. If the hot water rises rapidly to the surface, it may emerge as a hot spring. The water of some hot springs in the eastern United States is heated in this manner. The springs at Hot Springs National Park in Arkansas are one example. Water temperatures of these springs average about 60°C (140°F).

The great majority (more than 95 percent) of the hot springs (and geysers) in the United States are found in the West. The reason for this distribution is that the sources of heat for most hot springs are magma bodies and hot igneous rocks, and it is in the West that igneous activity has occurred most recently. The hot springs and geysers of the Yellowstone region are well-known examples.

Geysers

Intermittent fountains in which columns of hot water and steam are ejected with great force, often rising 30 to 60 meters (100 to 200 feet) into the air, are called **geysers**. After the jet of water ceases, a column of steam rushes out, often with a thunderous roar. Perhaps the most famous geyser in the world is Old Faithful in Yellowstone National Park (**FIGURE 5.31**). The great abundance, diversity, and spectacular nature of Yellowstone's geysers and other thermal features undoubtedly was the primary reason for its becoming the first national park in the United States. Geysers are also found in other parts of the world, notably New Zealand and Iceland. In fact, the Icelandic word *geysa*, meaning "to gush," gives us the name *geyser*.

Geysers occur where extensive underground chambers exist within hot igneous rocks. As relatively cool groundwater enters the chambers, it is heated by the surrounding rock. At the bottom of the chamber, the water is under great pressure because of the weight of the overlying water. This great pressure prevents the water from boiling at the normal surface temperature of 100°C (212°F). For example, at the bottom of a 300-meter (1000-foot) water-filled chamber, water must attain a temperature of nearly 230°C (450°F) before it will boil. The heating causes the water to expand, and as a result, some of the water is forced out at the surface. This loss of water reduces the pressure on the remaining water in the chamber, which lowers the boiling point. As a result, a portion of the water deep within the chamber quickly turns to an expanding mass of steam, which causes the geyser to erupt. Following the eruption, cool groundwater again seeps into the chamber, and the cycle begins anew.

Wells

The most common method for removing groundwater is to use a **well**, a hole bored into the zone of saturation. Wells serve as small reservoirs into which groundwater migrates and from which it can be pumped to the surface.

The use of wells dates back many centuries and continues to be an important method of obtaining water. According to the National Groundwater Association, there are about 16 million water wells for various purposes in the United States. Private household wells constitute the largest share—more than 13 million. About 500,000 new residential wells are drilled each year.

By far the single greatest use of well water in the United States is irrigation for agriculture. More than 65 percent of the groundwater used each year is for this purpose. Industrial uses rank a distant second, followed by the amount used by homes in cities and rural areas.

The water table level may fluctuate considerably during the course of a

FIGURE 5.31 Old Faithful This geyser in Wyoming's Yellowstone National Park is one of the most famous in the world. Contrary to popular legend, it does not erupt every hour on the hour. Time spans between eruptions vary from about 65 minutes to more than 90 minutes and have generally increased over the years, thanks to changes in the geyser's plumbing. (Photo by Jeff Vanuga/Corbis RF)

year, dropping during dry seasons and rising following periods of precipitation. Therefore, to ensure a continuous supply of water, a well must penetrate below the water table. Whenever a substantial amount of water is withdrawn from a well, the water table around the well is lowered. This effect, termed **drawdown**, decreases with increasing distance from the well. The result is a depression in the water table, roughly conical in shape, known as a **cone of depression** (**FIGURE 5.32**). For most small domestic wells, the cone of depression is negligible. However, when wells are used for irrigation or for industrial purposes, the withdrawal of water can be great enough to create a very wide and steep cone of depression that may substantially lower the water table in an area and cause nearby shallow wells to become dry. Figure 5.32 illustrates this situation.

Artesian Systems

In most wells, water cannot rise on its own. If water is first encountered at 30 meters (100 feet) depth, it remains at that level, fluctuating perhaps 1 or 2 meters with seasonal wet and dry periods. However, in some wells, water rises, sometimes overflowing at the surface.

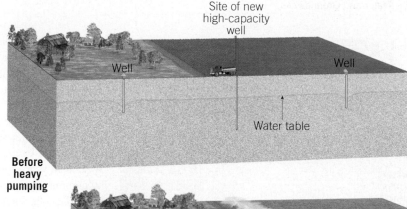
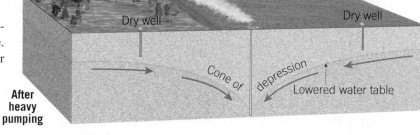

Artesian system refers to a situation in which groundwater rises in a well above the level where it was initially encountered. For such a situation to occur, two conditions must exist (**FIGURE 5.33**): (1) Water must be

SmartFigure 5.32 Cone of Depression For most small domestic wells, the cone of depression is negligible. When wells are heavily pumped, the cone of depression can be large and may lower the water table such that nearby shallower wells may be left dry.

SmartFigure 5.33 Artesian Systems These groundwater systems occur where an inclined aquifer is surrounded by impermeable beds (aquitards). Such aquifers are called *confined aquifers*. The photo shows a flowing artesian well. (Photo by James E. Patterson)

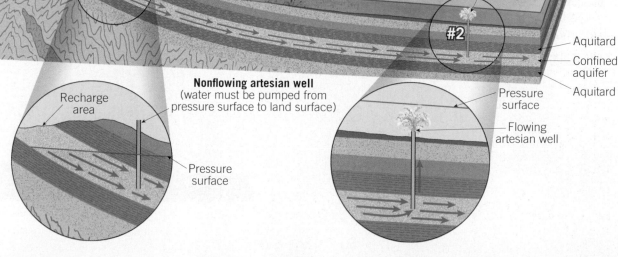

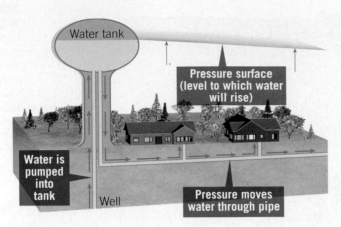

FIGURE 5.34 City Water Systems City water systems can be considered to be artificial artesian systems.

Water tank

Pressure surface (level to which water will rise)

Water is pumped into tank

Well

Pressure moves water through pipe

confined to an aquifer that is inclined so that one end is exposed at the surface, where it can receive water; and (2) aquitards both above and below the aquifer must be present to prevent the water from escaping. Such an aquifer is called a **confined aquifer**. When such a layer is tapped, the pressure created by the weight of the water above will force the water to rise. If there were no friction, the water in the well would rise to the level of the water at the top of the aquifer. However, friction reduces the height of this pressure surface. The greater the distance from the recharge area (the area where water enters the inclined aquifer), the greater the friction and the smaller the rise of water.

In Figure 5.33, Well 1 is a *nonflowing artesian well* because at this location, the pressure surface is below ground level. When the pressure surface is above the ground and a well is drilled into the aquifer, a *flowing artesian well* is created (Well 2 in Figure 5.33). Not all artesian systems are

wells. *Artesian springs* also exist. In such situations, groundwater may reach the surface by rising along a natural fracture such as a fault rather than through an artificially produced hole. In deserts, artesian springs are sometimes responsible for creating oases.

Artesian systems act as "natural pipelines," transmitting water from remote areas of recharge great distances to the points of discharge. In this manner, water that fell in central Wisconsin years ago is now taken from the ground and used by communities many kilometers to the south, in Illinois. In South Dakota, such a system brings water from the western Black Hills eastward across the state.

On a different scale, city water systems may be considered examples of artificial artesian systems (**FIGURE 5.34**). A water tower, into which water is pumped, may be considered the area of recharge, the pipes the confined aquifer, and the faucets in homes the flowing artesian wells.

5.10 CONCEPT CHECKS

1 Describe the circumstances that created the spring in Figure 5.27.

2 What is the source of heat for most hot springs and geysers?

3 Describe what occurs to cause a geyser to erupt.

4 Relate drawdown to cone of depression.

5 In Figure 5.27, two wells are at the same level. Why is one successful and the other not?

6 Sketch a simple cross section of an artesian system with a flowing artesian well. Label aquitards, the aquifer, and the pressure surface.

EYE ON EARTH

In 1900, when this well was drilled near Woonsocket in eastern South Dakota, a "gusher" of water resulted. The stream of water from a 3-inch pipe reached a height of nearly 30 meters (100 feet). Thousands of additional wells now tap the same aquifer. (Photo by N.H. Darton/USGS)

QUESTION 1 *Describe or sketch the subsurface geologic situation that was responsible for this fountain of water.*

QUESTION 2 *What term is applied to a well such as this?*

QUESTION 3 *Today wells that tap this aquifer do not flow freely at the surface but must be pumped. Suggest a likely reason.*

5.11 | ENVIRONMENTAL PROBLEMS OF GROUNDWATER

List and discuss three important environmental problems associated with groundwater.

Like many of our other valuable natural resources, groundwater is being exploited at an increasing rate. In some areas, overuse threatens the groundwater supply. In other places, groundwater withdrawal has caused the ground and everything resting on it to sink. Still other localities are concerned with the possible contamination of their groundwater supply.

Treating Groundwater as a Nonrenewable Resource

For many, groundwater appears to be an endlessly renewable resource, for it is continually replenished by rainfall and melting snow. But in some regions, groundwater has been and continues to be treated as a *nonrenewable* resource. Where this occurs, the amount of water available to recharge the aquifer is significantly less than the amount being withdrawn.

The High Plains, a relatively dry region that extends from South Dakota to western Texas, is one example of an extensive agricultural economy that is largely dependent on irrigation using groundwater (**FIGURE 5.35**). Underlying about 111 million acres (450,000 square kilometers [174,000 square miles]) in parts of eight states, the High Plains aquifer is one of the largest and most agriculturally significant aquifers in the United States. It accounts for about 30 percent of all groundwater withdrawn for irrigation in the country. In the southern part of this region, which includes the Texas panhandle, the natural recharge of the aquifer is very slow, and the problem of declining groundwater levels is acute. In fact, in years of average or below-average precipitation, recharge is negligible because all or nearly all of the meager rainfall is returned to the atmosphere by evaporation and transpiration.

Therefore, where intense irrigation has been practiced for an extended period, depletion of groundwater can be severe. Declines in the water table at rates as great as 1 meter (3 feet) per year have led to an overall drop of between 15 and 60 meters (50 and 200 feet) in some areas. Under these circumstances, it can be said that the groundwater is literally being "mined." Even if pumping were to cease immediately, it would take thousands of years for the groundwater to be fully replenished.

Groundwater depletion has been a concern in the High Plains and other areas of the West for many years, but it is worth pointing out that the problem is not confined to that part of the country. Increased demands on groundwater resources have overstressed aquifers in many areas, not just in arid and semiarid regions.

Land Subsidence Caused by Groundwater Withdrawal

As you will see later in this chapter, surface subsidence can result from natural processes related to groundwater. However, the ground may also sink when water is pumped from wells faster than natural recharge processes can replace it. This effect is particularly pronounced in areas underlain by thick layers of loose sediments. As water is withdrawn, the water pressure drops, and the weight of the overburden is transferred to the sediment. The greater pressure packs

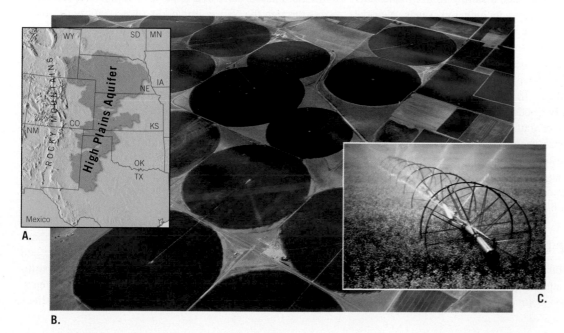

FIGURE 5.35 Mining Groundwater A. The High Plains aquifer is one of the largest aquifers in the United States. **B.** In parts of the High Plains aquifer, water is pumped from the ground faster than it is replenished. In such instances, groundwater is being treated as a nonrenewable resource. This aerial view shows circular crop fields irrigated by center-pivot irrigating systems in semiarid eastern Colorado. (Photo by James L. Amos/CORBIS). **C.** Groundwater provides more than 54 billion gallons per day in support of agriculture in the United States. (Photo by Michael Collier)

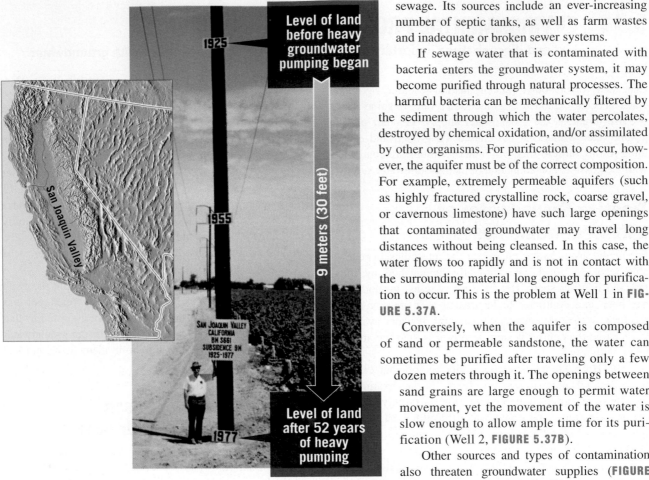

FIGURE 5.36 That Sinking Feeling! The San Joaquin Valley, an important agricultural area, relies heavily on irrigation. Between 1925 and 1975, this part of the valley subsided almost 9 meters (30 feet) due to the withdrawal of groundwater and the resulting compaction of sediments. (Photo courtesy of U.S. Geological Survey, Denver)

the sediment grains more tightly together, and the ground subsides.

Many areas can be used to illustrate such land subsidence. A classic example in the United States occurred in the San Joaquin Valley of California (**FIGURE 5.36**). Other well-known cases of land subsidence resulting from groundwater pumping in the United States include Las Vegas, Nevada; New Orleans and Baton Rouge, Louisiana; portions of southern Arizona; and the Houston–Galveston area of Texas. In the low-lying coastal area between Houston and Galveston, land subsidence ranges from 1.5 to 3 meters (5 to 10 feet). The result is that about 78 square kilometers (30 square miles) are permanently flooded.

Outside the United States, one of the most spectacular examples of subsidence occurred in Mexico City, a portion of which is built on a former lake bed. In the first half of the twentieth century, thousands of wells were sunk into the water-saturated sediments beneath the city. As water was withdrawn, portions of the city subsided by 6 meters (20 feet) or more.

Groundwater Contamination

The pollution of groundwater is a serious matter, particularly in areas where aquifers provide a large part of the water supply. One common source of groundwater pollution is

sewage. Its sources include an ever-increasing number of septic tanks, as well as farm wastes and inadequate or broken sewer systems.

If sewage water that is contaminated with bacteria enters the groundwater system, it may become purified through natural processes. The harmful bacteria can be mechanically filtered by the sediment through which the water percolates, destroyed by chemical oxidation, and/or assimilated by other organisms. For purification to occur, however, the aquifer must be of the correct composition. For example, extremely permeable aquifers (such as highly fractured crystalline rock, coarse gravel, or cavernous limestone) have such large openings that contaminated groundwater may travel long distances without being cleansed. In this case, the water flows too rapidly and is not in contact with the surrounding material long enough for purification to occur. This is the problem at Well 1 in **FIGURE 5.37A**.

Conversely, when the aquifer is composed of sand or permeable sandstone, the water can sometimes be purified after traveling only a few dozen meters through it. The openings between sand grains are large enough to permit water movement, yet the movement of the water is slow enough to allow ample time for its purification (Well 2, **FIGURE 5.37B**).

Other sources and types of contamination also threaten groundwater supplies (**FIGURE 5.38**). These include widely used substances such as highway salt, fertilizers that are spread across the land surface, and pesticides. In addition, a wide array of chemicals and industrial materials may leak from pipelines, storage tanks, landfills, and holding ponds. Some of these pollutants are classified as *hazardous*, meaning that they are either flammable, corrosive, explosive, or toxic. As rainwater oozes through the refuse, it may dissolve a variety of potential contaminants. If the leached material reaches the water table, it will mix with the groundwater and contaminate the supply.

Because groundwater movement is usually slow, polluted water might go undetected for a long time. In fact, contamination is sometimes discovered only after drinking water has been affected and people become ill. By this time, the volume of polluted water might be very large, and even if the source of contamination is removed immediately, the problem is not solved. Although the sources of groundwater contamination are numerous, there are relatively few solutions.

Once the source of the problem has been identified and eliminated, the most common practice is simply to abandon the water supply and allow the pollutants to be flushed away gradually. This is the least costly and easiest solution, but the aquifer must remain unused for many years. To accelerate this process, polluted water is sometimes

Although the contaminated water has traveled more than 100 meters before reaching Well 1, the water moves too rapidly through the cavernous limestone to be purified.

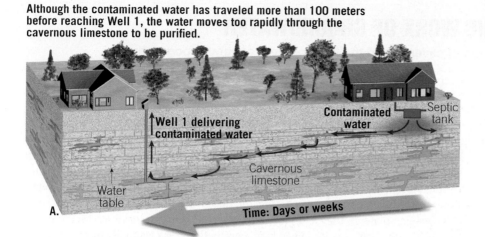

FIGURE 5.37 Comparing Two Aquifers In this example, the limestone aquifer allowed the contamination to reach a well but the sandstone aquifer did not.

As the discharge from the septic tank percolates through the permeable sandstone, it moves more slowly and is purified in a relatively short distance.

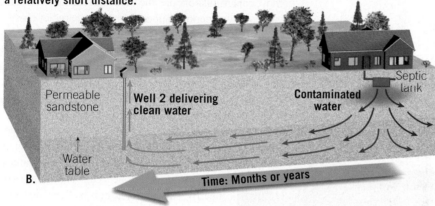

FIGURE 5.38 Potential Sources of Groundwater Contamination Sometimes leaking gasoline storage tanks and materials leached from landfills contaminate aquifers. (Storage tank photo by Earth Gallery Environment/ Alamy; landfill photo by Picsfive/ Shutterstock)

pumped out and treated. Following removal of the tainted water, the aquifer is allowed to recharge naturally or, in some cases, the treated water or other freshwater is pumped back in. This process is costly and time-consuming, and it may be risky because there is no way to be certain that all the contamination has been removed. Clearly, the most effective solution to groundwater contamination is prevention.

5.11 CONCEPT CHECKS

1 Describe the problem associated with pumping groundwater for irrigation in parts of the High Plains.

2 Explain why ground may subside after groundwater is pumped to the surface.

3 Which aquifer would be most effective in purifying polluted groundwater: coarse gravel, sand, or cavernous limestone?

5.12 | THE GEOLOGIC WORK OF GROUNDWATER

Explain the formation of caverns and the development of karst topography.

Groundwater dissolves rock. This fact is key to understanding how caverns and sinkholes form. Because soluble rocks, especially limestone, underlie millions of square kilometers of Earth's surface, it is here that groundwater carries on its important role as an erosional agent. Limestone is nearly insoluble in pure water but is quite easily dissolved by water containing small quantities of carbonic acid. Most natural water contains this weak acid because rainwater readily dissolves carbon dioxide from the air and from decaying plants. Therefore, when groundwater comes in contact with limestone, the carbonic acid reacts with calcite in the rocks to form calcium bicarbonate, a soluble material that is then carried away in solution.

Caverns

The most spectacular results of groundwater's erosional handiwork are limestone **caverns**. In the United States alone, about 17,000 caves have been discovered. Although most are relatively small, some have spectacular dimensions. Carlsbad Caverns in southeastern New Mexico and Mammoth Cave in Kentucky are famous examples. One chamber in Carlsbad Caverns has an area equivalent to 14 football fields and enough height to accommodate the U.S. Capitol Building. At Mammoth Cave, the total length of interconnected caverns extends for more than 540 kilometers (340 miles).

Most caverns are created at or below the water table, in the zone of saturation. Here acidic groundwater follows lines of weakness in the rock, such as joints and bedding planes. As time passes, the dissolving process slowly creates cavities and gradually enlarges them into caverns. Material that is dissolved by the groundwater is eventually discharged into streams and carried to the ocean.

Certainly the features that arouse the greatest curiosity for most cavern visitors are the stone formations that give some caverns a wonderland appearance. These are not erosional features, like the caverns in which they reside, but depositional features. They are created by the seemingly endless dripping of water over great spans of time. The calcium carbonate that is left behind produces the limestone we call *travertine*. These cave deposits, however, are also commonly called *dripstone*, an obvious reference to their mode of origin.

Although the formation of caverns takes place in the zone of saturation, the deposition of dripstone is not possible until the caverns are above the water table in the unsaturated zone. This commonly occurs as nearby streams cut their valleys deeper, lowering the water table as the elevation of the rivers drops. As soon as the chamber is filled with air, the conditions are right for the decoration phase of cavern building to begin.

Of the various dripstone features found in caverns, perhaps the most familiar are **stalactites**. These icicle-like pendants hang from the ceiling of a cavern and form where water seeps through cracks above. When water reaches air in the cave, some of the dissolved carbon dioxide escapes from the drop, and calcite begins to precipitate. Deposition occurs as a ring around the edge of the water drop. As drop after drop follows, each leaves an infinitesimal trace of calcite behind, and a hollow limestone tube is created. Water then moves through the tube, remains suspended momentarily at the end, contributes a tiny ring of calcite, and falls to the cavern floor. The stalactite just described is appropriately called a *soda straw* (**FIGURE 5.39A**). Often the hollow tube of the soda straw becomes plugged, or its supply of water increases. In either case, the water is forced to flow and deposit along the outside of the tube. As deposition continues, the stalactite takes on the more common conical shape.

Formations that develop on the floor of a cavern and reach upward toward the ceiling are

FIGURE 5.39 Cave Decorations A. Close-up of a delicate live soda-straw stalactite in Chinn Springs Cave, Independence County, Arkansas. (Photo by Dante Fenolio/ Science Source) **B.** Stalagmites and stalactites in Carlsbad Caverns National Park, New Mexico. (Photo by Fritz Poelking/ Glow Images)

A.

B.

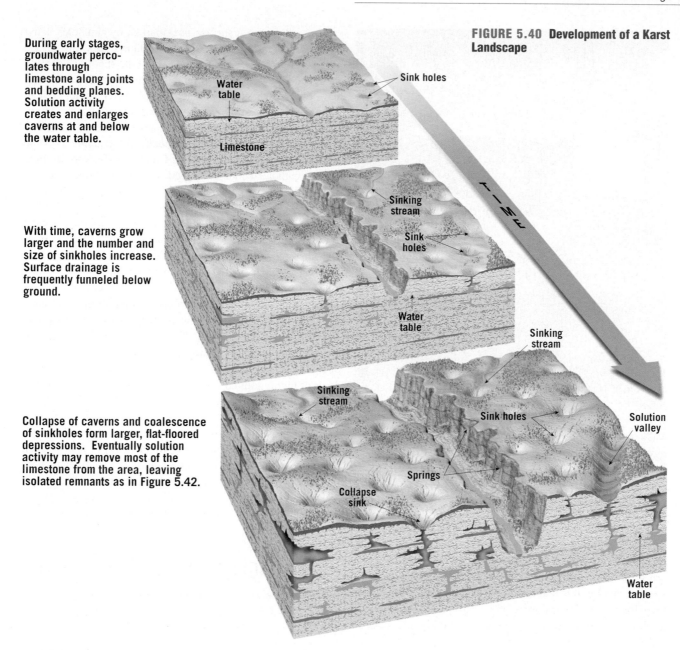

During early stages, groundwater percolates through limestone along joints and bedding planes. Solution activity creates and enlarges caverns at and below the water table.

With time, caverns grow larger and the number and size of sinkholes increase. Surface drainage is frequently funneled below ground.

Collapse of caverns and coalescence of sinkholes form larger, flat-floored depressions. Eventually solution activity may remove most of the limestone from the area, leaving isolated remnants as in Figure 5.42.

FIGURE 5.40 Development of a Karst Landscape

called **stalagmites** (**FIGURE 5.39B**). The water supplying the calcite for stalagmite growth falls from the ceiling and splatters over the surface. As a result, stalagmites do not have a central tube and are usually more massive in appearance and more rounded on their upper ends than stalactites. Given enough time, a downward-growing stalactite and an upward-growing stalagmite may join to form a *column*.

Karst Topography

Many areas of the world have landscapes that, to a large extent, have been shaped by the dissolving power of groundwater. Such areas are said to exhibit **karst topography**, named for the *Krs* region in the border area between Slovenia and Italy, where such topography is strikingly developed. In the United States, karst landscapes occur in many areas that are underlain by limestone, including portions of Kentucky, Tennessee, Alabama, southern Indiana, and central and

northern Florida (**FIGURE 5.40**). Generally, arid and semi-arid areas do not develop karst topography because there is insufficient groundwater. When karst features exist in such regions, they are likely to be remnants of a time when rainier conditions prevailed.

Sinkholes Karst areas typically have irregular terrain punctuated with many depressions called **sinkholes** or, simply, **sinks** (see Figure 5.26). In the limestone areas of Florida, Kentucky, and southern Indiana, literally tens of thousands of these depressions vary in depth from just 1 to 2 meters (3 to 6 feet) to a maximum of more than 50 meters (165 feet).

Sinkholes commonly form in one of two ways. Some develop gradually over many years, without any physical disturbance to the rock. In these situations, the limestone immediately below the soil is dissolved by downward-seeping rainwater that is freshly charged with carbon dioxide. These

FIGURE 5.41 Sinkholes Can Be Geologic Hazards This sinkhole formed suddenly in the backyard of a home in Lake City, Florida. Sinkholes such as this one form when the roof of a cavern collapses. (AP Photo/*The Florida Times-Union*, Jon M. Fletcher)

In addition to a surface pockmarked by sinkholes, karst regions characteristically show a striking lack of surface drainage (streams). Following a rainfall, runoff is quickly funneled below ground, through sinks. It then flows through caverns until it finally reaches the water table. Where streams do exist at the surface, their paths are usually short. The names of such streams often give a clue to their fate. The Mammoth Cave area of Kentucky, for example, is home to Sinking Creek, Little Sinking Creek, and Sinking Branch. Some sinkholes become plugged with clay and debris, creating small lakes or ponds.

Tower Karst Landscapes Some regions of karst development exhibit landscapes that look very different from the sinkhole-studded terrain depicted in Figure 5.40. One striking example is an extensive region in southern China that is described as exhibiting *tower karst*. As **FIGURE 5.42** shows, the term *tower* is appropriate because the landscape consists of a maze of isolated steep-sided hills that rise abruptly from the ground. Each is riddled with interconnected caves and passageways. This type of karst topography forms in wet tropical and subtropical regions having thick beds of highly jointed limestone. In such settings, groundwater dissolves large volumes of limestone, leaving only these residual towers. Karst development occurs more rapidly in tropical climates due to the abundant rainfall and greater availability of carbon dioxide from the decay of lush tropical vegetation. The extra carbon dioxide in the soil means there is more carbonic acid for dissolving limestone. Other tropical areas of advanced karst development include portions of Puerto Rico, western Cuba, and northern Vietnam.

depressions are usually not deep and are characterized by relatively gentle slopes. By contrast, sinkholes can also form suddenly and without warning when the roof of a cavern collapses under its own weight. Typically, the depressions created in this manner are steep sided and deep. When they form in populous areas, they may represent a serious geologic hazard. Such a situation is clearly the case in **FIGURE 5.41**.

5.12 CONCEPT CHECKS

1 How does groundwater create caverns?
2 How do stalactites and stalagmites form?
3 Describe two ways in which sinkholes form.

FIGURE 5.42 Tower Karst Landscape in China One of the best-known and most distinctive regions of tower karst development is along the Li River in the Guilin District of southeastern China. (Photo by Philippe Michel/AGE Fotostock)

5 CONCEPTS IN REVIEW

Running Water and Groundwater

5.1 EARTH AS A SYSTEM: THE HYDROLOGIC CYCLE

List the hydrosphere's major reservoirs and describe the different paths that water takes through the hydrologic cycle.

KEY TERMS: hydrologic cycle, evaporation, infiltration, runoff, transpiration, evapotranspiration

- Water moves through the hydrosphere's many reservoirs by evaporating, condensing into clouds, and falling as precipitation. Once it reaches the ground, rain can either soak in, evaporate, be returned to the atmosphere by plant transpiration, or run off. Running water is the most important agent sculpting Earth's varied landscapes.

5.2 RUNNING WATER

Describe the nature of drainage basins and river systems. Sketch four basic drainage patterns.

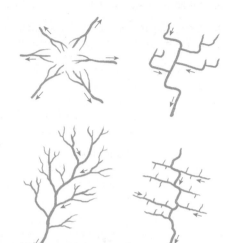

KEY TERMS: drainage basin, divide, dendritic pattern, radial pattern, rectangular pattern, trellis pattern

- The land area that contributes water to a stream is its drainage basin. Drainage basins are separated by imaginary lines called divides.
- As a generalization, river systems tend to erode at the upstream end, transport sediment through the middle section, and deposit sediment at the downstream end.

Q Identify each of the drainage patterns depicted in the accompanying sketch.

5.3 STREAMFLOW

Discuss streamflow and the factors that cause it to change.

KEY TERMS: laminar flow, turbulent flow, gradient, discharge, longitudinal profile

- The flow of water in a stream may be laminar or turbulent. A stream's flow velocity is influenced by channel gradient; size, shape, and roughness of the channel; and discharge.
- A cross-sectional view of a stream from head to mouth is a longitudinal profile. Usually the gradient and roughness of the stream channel decrease going downstream, whereas the size of the channel, stream discharge, and flow velocity increase in the downstream direction.

Q Sketch a typical longitudinal profile. Where does most erosion happen? Where is sediment transport the dominant process?

5.4 THE WORK OF RUNNING WATER

Outline the ways in which streams erode, transport, and deposit sediment.

KEY TERMS: pothole, dissolved load, suspended load, bed load, settling velocity, saltation, capacity, competence, sorting, alluvium

- Streams erode when turbulent water lifts loose particles from the streambed. The focused "drilling" of the stream armed with swirling particles also creates potholes in solid rock.
- Streams transport their load of sediment in solution, in suspension, and along the bottom (bed) of the channel. A stream's ability to transport solid particles is described using two criteria: Capacity refers to how much sediment a stream is transporting, and competence refers to the particle sizes the stream is capable of moving.
- Streams deposit sediment when velocity slows and competence is reduced. This results in sorting, the process by which like-size particles are deposited together.

5.5 STREAM CHANNELS

Contrast bedrock and alluvial stream channels. Distinguish between two types of alluvial channels.

KEY TERMS: meander, cut bank, point bar, cutoff, oxbow lake, braided channel

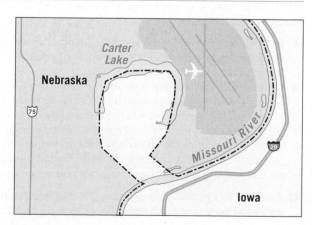

- Bedrock channels are cut into solid rock and are most common in headwaters areas where gradients are steep. Rapids and waterfalls are common features.
- Alluvial channels are dominated by streamflow through alluvium previously deposited by the stream. A floodplain usually covers the valley floor, with the river meandering or moving through braided channels.
- Meanders change shape through erosion at the cut bank (the outer edge of the meander) and deposition of sediment on point bars (the inside of a meander). A meander may become cut off and form an oxbow lake.

Q The town of Carter Lake is the *only* place in the state of Iowa that is west of the Missouri River. Examine the map and hypothesize how this unusual situation could have occurred.

5.6 SHAPING STREAM VALLEYS

Contrast narrow V-shaped valleys, broad valleys with floodplains, and valleys that display incised meanders.

KEY TERMS: stream valley, base level, floodplain, incised meander

- A stream valley includes the channel itself, the adjacent floodplain, and the relatively steep valley walls. Streams erode downward until they approach base level, the lowest point to which a stream can erode its channel. A river flowing toward the ocean (the ultimate base level) may encounter several local base levels along its route. These could be lakes or resistant rock layers that retard downcutting by the stream.
- A stream valley is widened through the meandering action of the stream, which erodes the valley walls and widens the floodplain. If base level were to drop or if the land were uplifted, a meandering stream might start downcutting and develop incised meanders.

Michael Collier

Q Meanders are associated with a river that is eroding from side to side, whereas narrow canyons are associated with rivers that are vigorously downcutting. The river in this image is confined to a narrow canyon but is also meandering. Explain.

5.7 DEPOSITIONAL LANDFORMS

Discuss the formation of deltas, natural levees, and alluvial fans.

KEY TERMS: bar, delta, distributary, natural levee, back swamp, yazoo tributary, alluvial fan

- A delta may form where a river deposits sediment in another water body at its mouth. The partitioning of streamflow into multiple distributaries spreads sediment in different directions.
- Natural levees result from sediment deposited along the margins of a stream channel by many flooding events. Because the levees slope gently away from the channel, the adjacent floodplain is poorly drained, resulting in back swamps and yazoo tributaries flowing parallel to the main river.
- Alluvial fans are fan-shaped deposits of alluvium that form where steep mountain fronts drop down into adjacent valleys.

Q At first glance, deltas and alluvial fans may look quite similar. How are they similar, and what sets them apart?

5.8 FLOODS AND FLOOD CONTROL

Discuss the causes of floods and some common flood control measures.

KEY TERM: flood

- Floods are triggered by heavy rains and/or snowmelt. Sometimes human interference can worsen or even cause floods. Flood control measures include the building of artificial levees and dams. Channelization may involve creating artificial cutoffs. Many scientists and engineers advocate a nonstructural approach to flood control that involves more appropriate land use.

Q Artificial levees are constructed to protect property from floods. Sometimes artificial levees in rural areas are intentionally opened up to protect a city from experiencing flooding. How does this work?

5.9 GROUNDWATER: WATER BENEATH THE SURFACE

Discuss the importance of groundwater and describe its distribution and movement.

KEY TERMS: zone of saturation, groundwater, water table, unsaturated zone, porosity, permeability, aquitard, aquifer

- Groundwater represents the largest reservoir of freshwater that is readily available to humans. Geologically, groundwater is an equalizer of streamflow, and the dissolving action of groundwater produces caverns and sinkholes.
- Groundwater is water that occupies the pore spaces in sediment and rock in a zone beneath the surface called the zone of saturation. The upper limit of this zone is called the water table. The zone above the water table where the material is not saturated is called the unsaturated zone.
- The quantity of water that can be stored in the open spaces in rock or sediment is termed porosity. Permeability, the ability of a material to transmit a fluid through interconnected pore spaces, is a key factor affecting the movement of groundwater. Aquifers are permeable materials that transmit groundwater freely, whereas aquitards are impermeable materials.

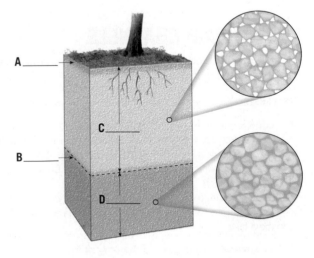

Q Examine this profile view showing the distribution of water in relatively uniform unconsolidated sediments and label the various portions of the groundwater complex.

5.10 SPRINGS, WELLS, AND ARTESIAN SYSTEMS

Compare and contrast springs, wells, and artesian systems.

KEY TERMS: spring, perched water table, hot spring, geyser, well, drawdown, cone of depression, artesian system, confined aquifer

- Springs occur where the water table intersects the land surface and a natural flow of groundwater results. When groundwater circulates deep below the surface, it may become heated and emerge at the surface as a hot spring. Geysers occur when groundwater is heated in underground chambers and expands, with some water quickly changing to steam and causing the geyser to erupt. The source of heat for most hot springs and geysers is hot igneous rock.
- Wells, which are openings bored into the zone of saturation, withdraw groundwater and may create roughly conical depressions in the water table known as cones of depression.
- Artesian wells tap into inclined aquifers bounded above and below by aquitards. For a system to qualify as artesian, the water in the well must be under sufficient pressure for the water to rise above the top of confined aquifer. Artesian wells may be flowing or nonflowing, depending on whether the pressure surface is above or below the ground surface.

5.11 ENVIRONMENTAL PROBLEMS OF GROUNDWATER

List and discuss three important environmental problems associated with groundwater.

- Groundwater can be "mined" by being extracted at a rate that is greater than the rate of replenishment. When groundwater is treated as a nonrenewable resource, as it is in parts of the High Plains aquifer, the water table drops, in some cases by more than 45 meters (150 feet).
- The extraction of groundwater can cause pore space to decrease in volume and the grains of loose Earth materials to pack more closely together. This overall compaction of sediment volume results in the subsidence of the land surface.
- Contamination of groundwater with sewage, highway salt, fertilizer, or industrial chemicals is another issue of critical concern. Once groundwater is contaminated, the problem is very difficult to solve, requiring expensive remediation or simply abandonment of the aquifer.

5.12 THE GEOLOGIC WORK OF GROUNDWATER

Explain the formation of caverns and the development of karst topography.

KEY TERMS: cavern, stalactite, stalagmite, karst topography, sinkhole (sink)

- Groundwater dissolves rock, in particular limestone, leaving behind void spaces in the rock. Caverns form at the zone of saturation, but later dropping of the water table may leave them open and dry—and available for people to explore.
- Dripstone is rock deposited by dripping of water containing dissolved calcium carbonate inside caverns. Features made of dripstone include stalactites, stalagmites, and columns.
- Karst topography develops in limestone regions and exhibits irregular terrain punctuated with many depressions called sinkholes. Some sinkholes form when the cavern roofs collapse.

Q Identify the three cavern deposits labeled in this photograph.

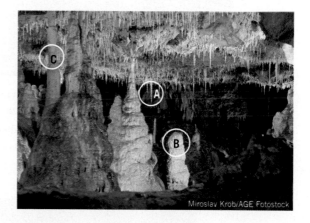
Miroslav Krob/AGE Fotostock

GIVE IT SOME **THOUGHT**

1. A river system consists of three zones, based on the dominant process operating in each part of the river system. On the accompanying illustration, match each process with one of the three zones:

   a. Sediment production (erosion)

   b. Sediment deposition

   c. Sediment transport

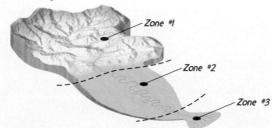

2. What factors influence how much of this rain will soak into the ground compared to how much will run off?

elwynn/Shutterstock

3. If you collect a jar of water from a stream, what part of its load will settle to the bottom of the jar? What portion will remain in the water indefinitely? What part of the stream's load would probably not be represented in your sample?

4. This satellite image shows portions of the Ohio and Wabash Rivers in May 2011. What is the base level for the Wabash River? What is base level for the Ohio River? (*Hint:* Referring to Figure 5.4 will help.) Are either of the base levels you just noted considered *ultimate base level*? Explain.

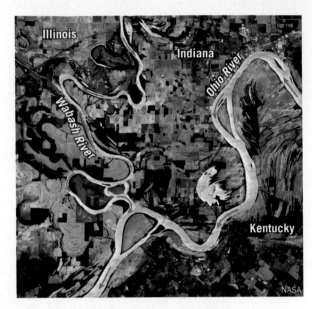

NASA

5. The Middle Fork of the Salmon River flows for about 175 kilometers (110 miles) through a rugged wilderness area in central Idaho.

    a. Is the river flowing in an alluvial channel or a bedrock channel? Explain.

    b. What process is dominant here: valley deepening or valley widening?

    c. Is the area shown in this image more likely near the mouth or the head of the river?

Michael Collier

6. Which one of the three basic rock types (igneous, sedimentary, or metamorphic) is most likely to be a good aquifer? Why?

7. What is the likely difference between an intermittent stream (one that flows off and on) and a stream that flows all the time, even during extended dry periods?

8. The cemetery in this photo is located in New Orleans, Louisiana. As in other cemeteries in the area, all of the burial plots are above ground. Based on what you have learned in this chapter, suggest a reason for this rather unusual practice.

Caitlin Mirra/Shutterstock

9. During a trip to the grocery store, your friend wants to buy some bottled water. Some brands promote the fact that their product is artesian. Other brands boast that their water comes from a spring. Your friend asks, "Is artesian water or spring water necessarily better than water from other sources?" How would you answer?

10. Imagine that you are an environmental scientist who has been hired to solve a groundwater contamination problem. Several homeowners have noticed that their well water has a funny smell and taste. Some think the contamination is coming from a landfill, but others think it might be a nearby cattle feedlot or chemical plant. Your first step is to gather data from wells in the area and prepare the map of the water table shown here.

    a. Based on your map, can any of the three potential sources of contamination be eliminated? If so, explain.

    b. What other steps would you take to determine the source of the contamination?

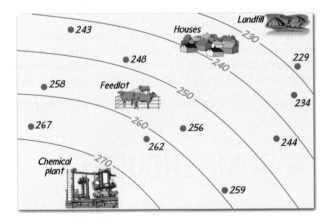

11. This black-and-white photo from the 1930s shows Franklin Roosevelt enjoying the hot springs at the presidential retreat at Warm Springs, Georgia. The temperature of these hot springs is always near 32°C (90°F). This area has no history of recent volcanic activity. What is the likely reason these springs are so warm?

New York Daily News/Getty Images

EXAMINING THE **EARTH SYSTEM**

1. This photo shows the famous Thousand Springs along the Snake River in southern Idaho. The springs are natural outlets for groundwater. Is there water in this image that can't be seen? If so, where might it be? Prepare a story that speculates on the answers to the following questions. Suggest several possibilities for both questions:

David R. Frazier/Alamy

    a. What journeys did the water in this image take to get to this place?

    b. What paths might the water follow when it makes its way back to the ocean?

2. Building a dam is one method of regulating the flow of a river to control flooding. Dams and their reservoirs may also provide recreational opportunities and water for irrigation and hydroelectric power generation. This image, from near Page, Arizona, shows Glen Canyon Dam on the Colorado River upstream from the Grand Canyon and a portion of Lake Powell, the reservoir it created.

    a. How did the behavior of the stream likely change upstream from Lake Powell?

    b. How might the behavior of the Colorado River downstream from the dam have been affected?

    c. Given enough time, how might the reservoir change?

    d. Speculate on the possible environmental impacts of building a dam such as this one.

Michael Collier

3. Imagine a water molecule that is part of a groundwater system in an area of gently rolling hills in the eastern United States. Describe some possible paths the molecule might take through the hydrologic cycle if:

    a. It were pumped from the ground to irrigate a farm field.

    b. There was a long period of heavy rainfall.

    c. The water table in the vicinity of the molecule developed a steep cone of depression due to heavy pumping from a nearby well.

4. Combine your understanding of the hydrologic cycle with your imagination and include possible short-term and long-term destinations and information about how the molecule gets to these places via evaporation, transpiration, condensation, precipitation, infiltration, and runoff. Remember to consider possible interactions with streams, lakes, groundwater, the ocean, and the atmosphere.

5. A glance at the map shows that the drainage basin of the Republican River occupies portions of Colorado, Nebraska, and Kansas. A significant part of the basin is considered semiarid. In 1943, the three states made a legal agreement regarding sharing the river's water. In 1998, Kansas went to court to force farmers in southern Nebraska to substantially reduce the amount of groundwater they used for irrigation. Nebraska officials claimed that the farmers were not taking water from the Republican River and thus were not violating the 1943 agreement. The court ruled in favor of Kansas.

    a. Explain why the court ruled that groundwater in southern Nebraska should be considered part of the Republican River system.

    b. How might heavy irrigation in a drainage basin influence the flow of a river?

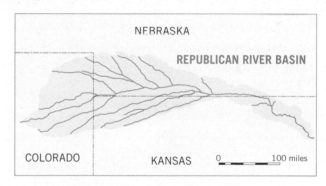

6. Over the oceans, evaporation exceeds precipitation, yet sea level does not drop. Why?

MasteringGeology™

6

Glaciers, Deserts, and Wind

California's Death Valley, a desert in the rain shadow of the Sierra Nevada, exhibits many classic features associated with a mountainous desert landscape. (Photo by Michael Collier)

Like the running water and groundwater that were the focus of Chapter 5, glaciers and wind are significant erosional processes. They are responsible for creating many different landforms and are part of an important link in the rock cycle, in which the products of weathering are transported and deposited as sediment.

Climate has a strong influence on the nature and intensity of Earth's external processes. This fact is dramatically illustrated in this chapter. The existence and extent of glaciers are largely controlled by Earth's changing climate. Another excellent example of the strong link between climate and geology is seen when we examine the development of arid landscapes.

Today, glaciers cover nearly 10 percent of Earth's land surface; however, in the recent geologic past, ice sheets were three times more extensive, covering vast areas with ice thousands of meters thick. Many regions still bear the mark of these glaciers. The first part of this chapter examines glaciers and the erosional and depositional features they create. The second part explores dry lands and the geologic work of wind. Because desert and near-desert conditions prevail over an area as large as that affected by the massive glaciers of the Ice Age, the nature of such landscapes is indeed worth investigating.

6.1 | GLACIERS AND THE EARTH SYSTEM Explain the role of glaciers in the hydrologic and rock cycles and describe the different types of glaciers, their characteristics, and their present-day distribution.

A **glacier** is a thick ice mass that forms over hundreds or thousands of years. It originates on land from the accumulation, compaction, and recrystallization of snow. A glacier appears to be motionless, but it is not; glaciers move very slowly. Although glaciers are found in many parts of the world, most are located in remote areas, either near Earth's poles or in high mountains.

Many present-day landscapes were modified by the widespread glaciers of the most recent Ice Age and still strongly reflect the handiwork of ice. The basic features of such diverse places as the Alps, Cape Cod, and Yosemite Valley were fashioned by now vanished masses of glacial ice. Moreover, Long Island, the Great Lakes, and the fiords of Norway and Alaska all owe their existence to glaciers.

Glaciers: A Part of Two Basic Cycles

Glaciers are part of two important cycles in the Earth system—the hydrologic cycle and the rock cycle. In Chapter 5 you learned that the water of the hydrosphere is constantly cycled through the atmosphere, biosphere, and geosphere. Over and over again, water evaporates from the oceans into the atmosphere, precipitates on the land, and flows in rivers and underground back to the sea. However, when precipitation falls at high elevations or high latitudes, the water may not immediately make its way toward the sea. Instead, it may become part of a glacier. Although the ice will eventually melt, allowing the water to continue its path to the sea, water can be stored as glacial ice for many tens, hundreds, or even thousands of years. During the time that the water is part of a glacier, the moving mass of ice can do enormous amounts of work. Like running water, groundwater, wind, and waves, glaciers are dynamic erosional agents that accumulate, transport, and deposit sediment. This activity is a basic part of the rock cycle (**FIGURE 6.1**).

Valley (Alpine) Glaciers

Literally thousands of relatively small glaciers exist in lofty mountain areas, where they usually follow valleys originally occupied by streams. Unlike the rivers that previously flowed in these valleys, the glaciers advance slowly, perhaps only a few centimeters each day. Because of their setting, these moving ice masses are termed **valley glaciers**, or **alpine glaciers** (see Figure 6.1). Each glacier is a stream of ice, bounded by precipitous rock walls, that flows downvalley from a snow accumulation center near its head. Like rivers, valley glaciers can be long or short, wide or narrow, single or with branching tributaries. Generally, alpine glaciers are longer than they are wide; some extend for just a fraction of a kilometer, whereas others go on for many dozens of kilometers. The west branch of the Hubbard Glacier, for example, runs through 112 kilometers (nearly 70 miles) of mountainous terrain in Alaska and the Yukon Territory.

Ice Sheets

Ice sheets exist on a much larger scale than valley glaciers. These enormous masses flow out in all directions from one or more snow-accumulation centers and completely obscure all but the highest areas of underlying terrain. The low total annual solar energy reaching the poles makes these regions eligible for such great ice accumulations. Presently each of Earth's polar regions supports an ice sheet—Greenland in the Northern Hemisphere and Antarctica in the Southern Hemisphere (**FIGURE 6.2**).

Ice Age Ice Sheets About 18,000 years ago, glacial ice not only covered Greenland and Antarctica but also covered large portions of North America, Europe, and

FIGURE 6.1
Kennicott Glacier
This 43-kilometer-long (27-mile-long) valley glacier is sculpting the mountains in Alaska's Wrangell–St. Elias National Park. According to the U.S. Geological Survey, it is one of 616 officially named glaciers in Alaska. There are thousands of unnamed ones as well. The dark stripes of sediment are called medial moraines. The transport and deposition of sediment make glaciers a basic part of the rock cycle. (Photo by Michael Collier)

Siberia. This date in Earth history is appropriately known as the *Last Glacial Maximum*. The term implies that there were other glacial maximums, and this is indeed the case. Throughout the Quaternary period, which began about 2.6 million years ago and extends to the present, ice sheets formed, advanced over broad areas, and then melted away. These alternating glacial and interglacial periods have occurred over and over again.

Greenland and Antarctica Some people mistakenly think that the North Pole is covered by glacial ice, but this is not the case. The ice that covers the Arctic Ocean is **sea ice**—frozen seawater. Sea ice floats because ice is less dense than liquid water. Although sea ice never completely disappears from the Arctic, the area covered expands and contracts with the seasons. The thickness of sea ice ranges from a few centimeters for new ice to 4 meters for sea ice that has survived for years. By contrast, glaciers can be hundreds or thousands of meters thick.

Glaciers form on land, and in the Northern Hemisphere, Greenland supports an ice sheet. Greenland extends between about 60° and 80° north latitude. This largest island on Earth is covered by an imposing ice sheet that occupies 1.7 million square kilometers (more than 660,000 square miles), or about 80 percent of the island. Averaging nearly 1500 meters (5000 feet) thick, the ice extends 3000 meters (10,000 feet) above the island's bedrock floor in some places.

In the Southern Hemisphere, the huge Antarctic Ice Sheet attains a maximum thickness of about 4300 meters (14,000 feet) and covers an area of more than 13.9 million square kilometers (5.4 million square miles)—nearly

Greenland's ice sheet occupies 1.7 million square kilometers (663,000 square miles), about 80 percent of the island.

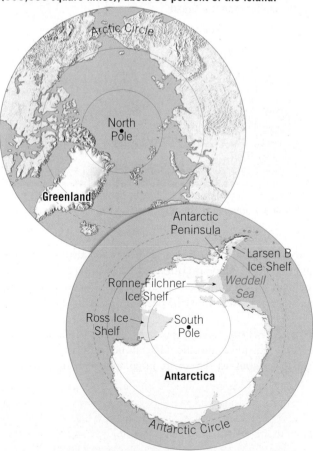

The area of the Antarctic Ice Sheet is almost 14 million square kilometers (5,460,000 square miles). Ice shelves occupy an additional 1.4 million square kilometers (546,000 square miles).

FIGURE 6.2 Ice Sheets
The only present-day ice sheets are those covering Greenland and Antarctica. Their combined areas represent nearly 10 percent of Earth's land area.

173

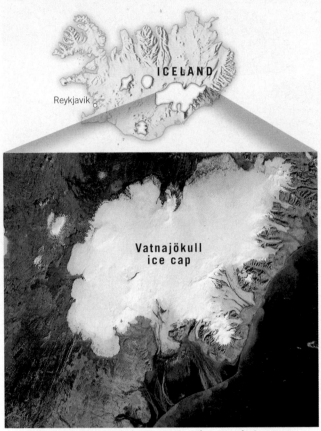

A. Ice caps completely bury the underlying terrain but are much smaller than ice sheets.

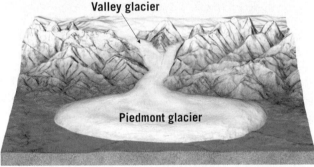

B. When a valley glacier is no longer confined, it spreads out to become a piedmont glacier.

floating ice that extend seaward from the coast but remain attached to the land along one or more sides. The shelves are thickest on their landward sides, and they become thinner seaward. They are sustained by ice from the adjacent ice sheet as well as being nourished by snowfall and the freezing of seawater to their bases. Antarctica's ice shelves extend over approximately 1.4 million square kilometers (0.6 million square miles). The Ross and Ronne-Filchner Ice Shelves are the largest; the Ross Ice Shelf covers an area approximately the size of Texas (see Figure 6.2).

Other Types of Glaciers

In addition to valley glaciers and ice sheets, other types of glaciers are also recognized. Covering some uplands and plateaus are masses of glacial ice called **ice caps**. Like ice sheets, ice caps completely bury the underlying landscape, but they are much smaller than the continental-scale features. Ice caps occur in many places, including Iceland and several of the large islands in the Arctic Ocean (**FIGURE 6.3A**).

Another type, known as **piedmont glaciers**, occupy broad lowlands at the bases of steep mountains and form when one or more valley glaciers emerge from the confining walls of mountain valleys (**FIGURE 6.3B**). The advancing ice spreads out to form a broad sheet. The size of individual piedmont glaciers varies greatly. Among the largest is the broad Malaspina Glacier along the coast of southern Alaska. It covers more than 5000 square kilometers (2000 square miles) of the flat coastal plain at the foot of the lofty St. Elias range.

Often ice caps and ice sheets feed **outlet glaciers**. These tongues of ice flow down valleys, extending outward from the margins of these larger ice masses. The tongues are essentially valley glaciers that are avenues for ice movement from an ice cap or ice sheet through mountainous terrain to the sea. Where they encounter the ocean, some outlet glaciers spread out as floating ice shelves. Often large numbers of icebergs are produced.

the entire continent. Because of the proportions of these huge features, they are often called *continental ice sheets*. Indeed, the combined areas of present-day continental ice sheets represent almost 10 percent of Earth's land area.

Ice Shelves Along portions of the Antarctic coast, glacial ice flows into the adjacent ocean, creating features called **ice shelves**. These are large, relatively flat masses of

6.1 CONCEPT CHECKS

1 Where are glaciers found on Earth today, and what percentage of Earth's land surface do they cover?

2 Describe how glaciers fit into the hydrologic cycle. What role do they play in the rock cycle?

3 List and briefly distinguish among four types of glaciers.

4 What is the difference between an ice sheet, sea ice, and an ice shelf?

6.2 | HOW GLACIERS MOVE Describe how glaciers move, the rates at which they move, and the significance of the glacial budget.

The way in which ice moves is complex, and there are two basic types of ice movement. The first of these, *plastic flow*, involves movement *within* the ice. Ice behaves as a brittle solid until the pressure on it is equivalent to the weight of about 50 meters (165 feet) of ice. Once that load is surpassed, ice behaves as a plastic material, and flow begins. Such flow occurs because of the molecular structure of ice. Glacial ice consists of layers of molecules stacked one upon the other. The bonds between layers are weaker than those within each layer. There-

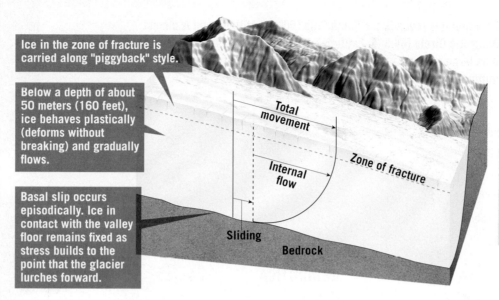

Ice in the zone of fracture is carried along "piggyback" style.

Below a depth of about 50 meters (160 feet), ice behaves plastically (deforms without breaking) and gradually flows.

Basal slip occurs episodically. Ice in contact with the valley floor remains fixed as stress builds to the point that the glacier lurches forward.

Total movement

Zone of fracture

Internal flow

Sliding

Bedrock

SmartFigure 6.4 Movement of a Glacier This vertical cross section through a glacier shows that movement is divided into two components. Also notice that the rate of movement is slowest at the base of the glacier, where frictional drag is greatest.

fore, when a stress exceeds the strength of the bonds between the layers, the layers remain intact and slide over one another.

A second and often equally important mechanism of glacial movement occurs when an entire ice mass slips along the ground. The lowest portions of most glaciers probably move by this sliding process. **FIGURE 6.4** illustrates the effects of the two basic types of glacial motion. This vertical profile through a glacier also shows that not all the ice flows forward at the same rate. Frictional drag with the bedrock floor causes the lower portions of the glacier to move more slowly.

The uppermost 50 meters (165 feet) of a glacier is appropriately referred to as the **zone of fracture**. Because there is not enough overlying ice to cause plastic flow, this upper part of the glacier consists of brittle ice. Consequently, the ice in this zone is carried along piggyback-style by the ice below. When the glacier moves over irregular terrain, the zone of fracture is subjected to tension, resulting in cracks called **crevasses** (**FIGURE 6.5**). These gaping cracks, which often make travel across glaciers dangerous, may extend to depths of 50 meters (165 feet). Below this depth, plastic flow seals them off.

Observing and Measuring Movement

Unlike the movement of water in streams, the movement of glacial ice is not obvious. If we could watch a valley glacier move, we would see that as with the water in a river, the ice does not all move downstream at the same rate. Flow is greatest in the center of the glacier because of the drag created by the walls and floor of the valley.

Early in the nineteenth century, the first experiments involving the movement of glaciers were designed and carried out in the Alps. Markers were placed in a straight line across an alpine glacier. The position of the line was marked on the valley

walls so that if the ice moved, the change in position could be detected. Periodically the positions of the markers were noted, showing the movement just described. Although most glaciers move too slowly for direct visual detection, the experiments succeeded in demonstrating that movement nevertheless occurs. The experiment illustrated in **FIGURE 6.6A** was carried out at Switzerland's Rhône Glacier later in the nineteenth century. It not only traced the movement of markers within the ice but also mapped the position of the glacier's terminus.

FIGURE 6.5 Crevasses As a glacier moves, internal stresses cause large cracks to develop in the brittle upper portion of the glacier, called the zone of fracture. Crevasses can extend to depths of 50 meters (165 feet) and can make travel across glaciers dangerous. (Photo by Wave/Glow Images)

Antarctica Fact File

Earth's southernmost continent surrounds the South Pole (90° S. Latitude) and is almost entirely south of the Antarctic Circle (66.5° S. Latitude). This icy landmass is the fifth largest continent and is twice as large as Australia. It also has the distinction of being the coldest, driest, and windiest continent and also has the highest average elevation.

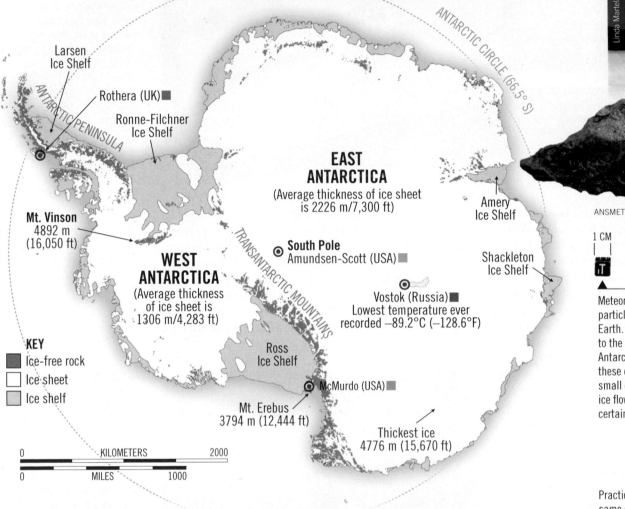

Larsen Ice Shelf

ANTARCTIC CIRCLE (66.5° S)

ANTARCTIC PENINSULA

Rothera (UK)

Ronne-Filchner Ice Shelf

EAST ANTARCTICA
(Average thickness of ice sheet is 2226 m/7,300 ft)

Amery Ice Shelf

Mt. Vinson
4892 m (16,050 ft)

TRANSANTARCTIC MOUNTAINS

⊙ South Pole
Amundsen-Scott (USA)

WEST ANTARCTICA
(Average thickness of ice sheet is 1306 m/4,283 ft)

Shackleton Ice Shelf

Vostok (Russia)
Lowest temperature ever recorded −89.2°C (−128.6°F)

KEY
- ▨ Ice-free rock
- ☐ Ice sheet
- ▨ Ice shelf

Ross Ice Shelf

McMurdo (USA)

Mt. Erebus
3794 m (12,444 ft)

Thickest ice
4776 m (15,670 ft)

KILOMETERS
0 ─────────── 2000

MILES
0 ─────────── 1000

Antarctica is 1.4 times larger than the United States and about 58 times bigger than the United Kingdom. The continent is almost completely ice covered. The ice-free area amounts to only 44,890 square kilometers (17,330 square miles) or just 0.32 percent (32/100 of 1 percent) of the continent.

Linda Martel/ANSMET

ANSMET

1 CM

Meteorites are rocky or metallic particles from space that have fallen on Earth. These ancient fragments provide clues to the origin and history of our solar system. Antarctica is an especially good place to collect these dark masses from space because even small ones are relatively easy to spot. In addition, ice flow patterns tend to concentrate them in certain areas.

Practically all of the continent belongs to the same climate classification, aptly termed ice cap climate, in which the average temperature of the warmest month is 0° C (32° F) or below. Take a look at the graph and you will see that some areas are much colder than others.

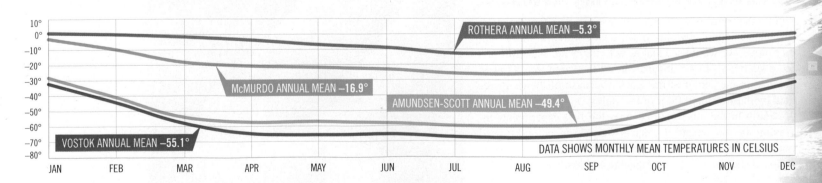

ROTHERA ANNUAL MEAN −5.3°

McMURDO ANNUAL MEAN −16.9°

AMUNDSEN-SCOTT ANNUAL MEAN −49.4°

VOSTOK ANNUAL MEAN −55.1°

DATA SHOWS MONTHLY MEAN TEMPERATURES IN CELSIUS

JAN   FEB   MAR   APR   MAY   JUN   JUL   AUG   SEP   OCT   NOV   DEC

John Goodge/NSF

Rick Price/Getty Images

What if the ice melted? The discharge at the mouth of the Mississippi River is 17,300 cubic meters (593,000 cubic feet) per second. If Antarctica's ice sheets melted at a suitable rate, they could maintain the flow of the Mississippi River for more than 50,000 years! If all of the continent's ice were to melt, sea level would rise by an estimated 56 meters (more than 180 feet). Antarctica's ice represents about 65 percent of Earth's entire supply of freshwater.

The Transantarctic Mountains are a 3300-kilometer- (2600-mile-) long range that separates the West Antarctic Ice Sheet and the East Antarctic Ice Sheet. Vinson Massif is the highest peak at 4892 meters (16,050 feet). Most of the mountains are buried beneath the continent's huge ice sheets.

About half of the continent's coastal areas are characterized by ice shelves. The Ross Ice Shelf is about the size of France, whereas the Ronne-Filchner Ice Shelf has an area similar to that of Spain.

McMurdo Station is the main U.S. scientific research station. It is the largest installation on the continent, capable of supporting more than 1200 residents. The total population at all research stations is about 4000 in summer and 1000 in winter. There are no permanent (indigenous) human residents on the continent.

ZUMA Wire Service/Alamy

MCMURDO STATION ANTARCTICA

Dan Leeth/Alamy

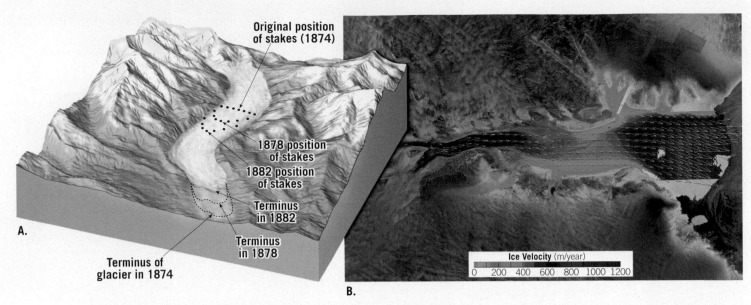

FIGURE 6.6 Measuring the Movement of a Glacier **A.** Ice movement and changes in the terminus of Rhône Glacier, Switzerland. In this classic study of a valley glacier, the movement of stakes clearly shows that glacial ice moves and that movement along the sides of the glacier is slower than movement in the center. Also notice that even though the ice front was retreating, the ice within the glacier was advancing. **B.** This satellite image provides detailed information about the movement of Antarctica's Lambert Glacier. Ice velocities are determined from pairs of images obtained 24 days apart using radar data. (NASA)

For many years, time lapse photography has allowed us to observe glacial movement. Images are taken from the same vantage point on a regular basis (for example, once per day) over an extended span and then played back like a movie. More recently, satellites let us track the movement of glaciers and observe glacial behavior (**FIGURE 6.6B**). This is especially useful because the remoteness and extreme weather associated with many glacial areas limits on-site study.

How rapidly does glacial ice move? Average rates vary considerably from one glacier to another. Some glaciers move so slowly that trees and other vegetation may become well established in the debris that accumulates on the glacier's surface. Others advance up to several meters each day. Recent satellite radar imaging provided insights into movements within the Antarctic ice sheet. Portions of some outlet glaciers move at rates greater than 800 meters (2600 feet) per year; on the other hand, ice in some interior regions creeps along at less than 2 meters (6.5 feet) per year.

Movement of some glaciers is characterized by occasional periods of extremely rapid advance called *surges*, followed by periods during which movement is much slower.

Budget of a Glacier: Accumulation Versus Wastage

Snow is the raw material from which glacial ice originates. Therefore, glaciers form in areas where more snow falls in winter than can melt during the summer. Glaciers are constantly gaining and losing ice.

Glacial Zones Snow accumulation and ice formation occur in the **zone of accumulation** (**FIGURE 6.7**). Its outer limits are defined by the snowline. The elevation of the snowline varies greatly, from sea level in polar regions to altitudes approaching 5000 meters (16,000 feet) near the equator. Above the snowline, in the zone of accumulation, the addition of snow thickens the glacier and promotes movement. Below the snowline is the **zone of wastage**. Here there is a net loss to the glacier as all the snow from the previous winter melts, as does some of the glacial ice.

In addition to melting, glaciers also waste as large pieces of ice break off the front of a glacier, in a process

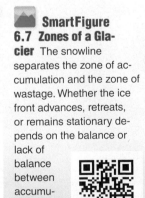

SmartFigure 6.7 Zones of a Glacier The snowline separates the zone of accumulation and the zone of wastage. Whether the ice front advances, retreats, or remains stationary depends on the balance or lack of balance between accumulation and wastage.

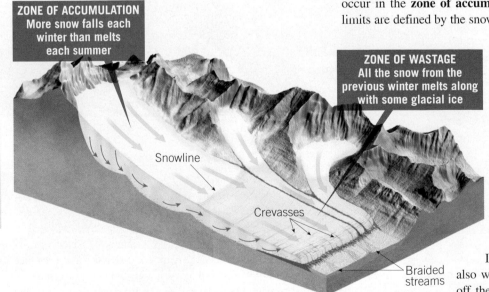

ZONE OF ACCUMULATION
More snow falls each winter than melts each summer

ZONE OF WASTAGE
All the snow from the previous winter melts along with some glacial ice

Snowline

Crevasses

Braided streams

Only about 20 percent or less of an iceberg protrudes above the waterline.

Geologist's Sketch

FIGURE 6.8 Icebergs Icebergs form when large masses of ice break off from the front of a glacier after it reaches a water body, in a process known as calving. (Photo by Radius Images/Photolibrary)

called **calving**. Calving creates **icebergs** in places where the glacier has reached the sea or a lake (**FIGURE 6.8**). Because icebergs are just slightly less dense than seawater, they float very low in the water, with more than 80 percent of their mass submerged. Along the margins of Antarctica's ice shelves, calving is the primary means by which the ice shelves lose mass. The relatively flat icebergs produced here can be several kilometers across and 600 meters (2000 feet) thick. By comparison, thousands of irregularly shaped icebergs are produced by outlet glaciers flowing from the margins of the Greenland Ice Sheet. Many drift southward and find their way into the North Atlantic, where they can be hazardous to navigation.

Glacial Budget Whether the margin of a glacier is advancing, retreating, or remaining stationary depends on the *budget* of the glacier. The **glacial budget** is the balance or lack of balance between accumulation at the upper end of a glacier and loss at the lower end. If ice accumulation exceeds wastage, the glacial front advances until the two factors balance. At this point, the terminus of the glacier becomes stationary.

If a warming trend increases wastage and/or if a drop in snowfall decreases accumulation, the ice front will retreat. As the terminus of the glacier retreats, the extent of the zone of wastage diminishes. Therefore, in time a new balance will be reached between accumulation and wastage, and the ice front will again become stationary.

Whether the margin of a glacier is advancing, retreating, or stationary, the ice within the glacier continues to flow forward. In the case of a receding glacier, the ice still flows

forward, but not rapidly enough to offset wastage. This point is illustrated in Figure 6.6A. As the line of stakes within the Rhône Glacier continued to move downvalley, the terminus of the glacier slowly retreated upvalley.

Glaciers in Retreat Because glaciers are sensitive to changes in temperature and precipitation, they provide clues about changes in climate. With few exceptions, valley glaciers around the world have been retreating at unprecedented rates over the past century. The photos in **FIGURE 6.9** provide an example. Many valley glaciers have disappeared altogether. For example, 150 years ago, there were 147 glaciers in Montana's Glacier National Park. Today, only 37 remain, and they may vanish by 2030.

Riggs Glacier

Muir Glacier

1941

Riggs Glacier

2004

FIGURE 6.9 Retreating Glaciers Two images taken 63 years apart from the same vantage point in Alaska's Glacier Bay National Park. Muir Glacier, which is prominent in the 1941 photo, has retreated out of the field of view in the 2004 image. Also, Riggs Glacier (upper right) has thinned and retreated significantly. More glacier images can be seen in Chapter 16, in the GEOgraphic "Earth's Shrinking Glaciers." (Photos courtesy of National Snow and Ice Data Center)

6.2 CONCEPT CHECKS

1 Describe two components of glacial movement.
2 How rapidly does glacial ice move? Provide some examples.
3 What are crevasses, and where do they form?
4 Relate glacial budget to two zones of a glacier.
5 Under what circumstances will the front of a glacier advance? Retreat? Remain stationary?

6.3 | GLACIAL EROSION

Discuss the processes of glacial erosion and the major features created by these processes.

Glaciers erode tremendous volumes of rock. For anyone who has observed the terminus of an alpine glacier, the evidence of its erosive force is clear. You can witness firsthand the release of rock fragments of various sizes from the ice as it melts (**FIGURE 6.10**). All signs lead to the conclusion that the ice has scraped, scoured, and torn rock debris from the floor and walls of the valley and carried it downvalley. In addition, in mountainous regions, mass-wasting processes also make substantial

contributions to the sediment load of a glacier. The photo that accompanies *Examining the Earth System* Problem 3 in Chapter 4 (page 128) provides an excellent example.

Once a glacier acquires rock debris, that debris cannot settle out as does the load carried by a stream or by the wind. Consequently, glaciers can carry huge blocks that no other erosional agent could possibly budge. Although today's glaciers are of limited importance as erosional agents, many landscapes that

FIGURE 6.10 Evidence of Glacial Erosion As the terminus of this glacier in Alaska wastes away, it deposits large quantities of sediment. The close-up view shows that the rock debris dropped by the melting ice is a jumbled mixture of different-size sediments. (Photos by Michael Collier)

loose. In this manner, sediment of all sizes becomes part of the glacier's load.

The second major erosional process is **abrasion**. As the ice and its load of rock fragments slide over bedrock, they function like sandpaper to smooth and polish the surface below. The pulverized rock produced by the glacial gristmill is appropriately called **rock flour**. So much rock flour may be produced that meltwater streams leaving a glacier often have the grayish appearance of skim milk—visible evidence of the grinding power of the ice.

When the ice at the bottom of a glacier contains large rock fragments, long scratches and grooves called **glacial striations** may be gouged into the bedrock (**FIGURE 6.11A**). These linear scratches on the bedrock surface provide clues to the direction of glacial movement. By mapping the striations over large areas, glacial flow patterns can often be reconstructed.

Not all abrasive action produces striations. The rock surface over which the glacier moves may also become highly polished by the ice and its load of finer particles. The broad expanses of smoothly polished granite in California's Yosemite National Park provide an excellent example (**FIGURE 6.11B**).

As is the case with other agents of erosion, the rate of glacial erosion is highly variable. This differential erosion by ice is largely controlled by four factors: (1) rate of glacial movement; (2) thickness of the ice; (3) shape, abundance, and hardness of the rock fragments contained in the ice at the base of the glacier; and (4) erodibility of the surface beneath the glacier. Variations in any or all of these factors from time to time and/or from place to place mean that the features, effects, and degree of landscape modification in glaciated regions can vary greatly.

were modified by the widespread glaciers of the recent Ice Age still reflect to a high degree the work of ice.

How Glaciers Erode

Glaciers erode land primarily in two ways: plucking and abrasion. First, as a glacier flows over a fractured bedrock surface, it loosens and lifts blocks of rock and incorporates them into the ice. This process, known as **plucking**, occurs when meltwater penetrates the cracks and joints along the rock floor of the glacier and freezes. When water freezes, it expands, exerting tremendous leverage that pries the rock

Mobile Field Trip 6.11 Glacial Abrasion

Moving glacial ice, armed with sediment, acts like sandpaper, scratching and polishing rock. (Photos by Michael Collier)

Glacial abrasion created the scratches and grooves in this bedrock.

A.

Glacially polished granite in California's Yosemite National Park.

B.

Landforms Created by Glacial Erosion

Although the erosional accomplishments of ice sheets can be tremendous, landforms carved by these huge ice masses usually do not inspire the same awe as do the erosional features created by valley glaciers. In regions where the erosional effects of ice

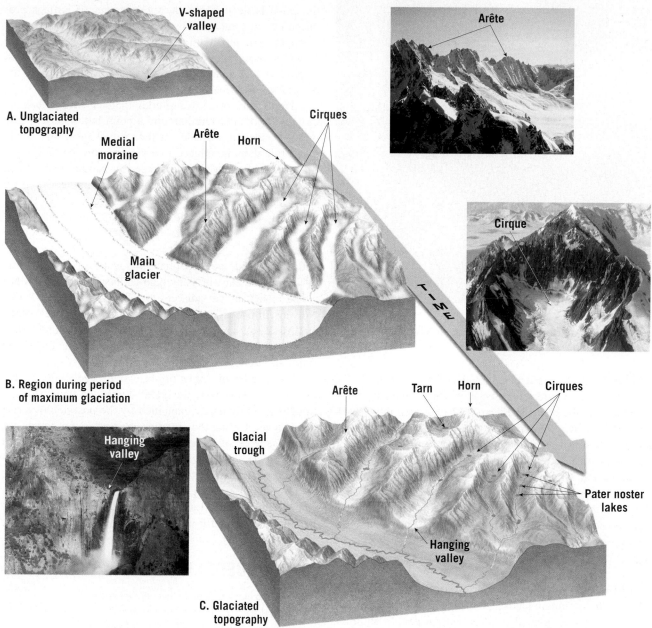

A. Unglaciated topography

Medial moraine

Arête

Horn

Cirques

Main glacier

B. Region during period of maximum glaciation

V-shaped valley

TIME

Hanging valley

Arête

Glacial trough

Arête   Tarn   Horn   Cirques

Pater noster lakes

Hanging valley

C. Glaciated topography

Arête

Cirque

SmartFigure 6.12 Erosional Landforms Created by Alpine Glaciers The unglaciated landscape in part A is modified by valley glaciers in part B. After the ice recedes in part C, the terrain looks very different than before glaciation. (Arête photo by James E. Patterson Collection; cirque photo by Marli Miller; hanging valley photo by John Warden/ Superstock)

sheets are significant, glacially scoured surfaces and subdued terrain are the rule. By contrast, in mountainous areas, erosion by valley glaciers produces many truly spectacular features. Much of the rugged mountain scenery so celebrated for its majestic beauty was produced by erosion due to valley glaciers.

Take a moment to study **FIGURE 6.12**, which shows a mountain setting before, during, and after glaciation. You will refer to this figure often in the following discussion.

Glaciated Valleys   A hike up a glaciated valley reveals a number of striking ice-created features. The valley itself is often a dramatic sight. Unlike streams, which create their own valleys, glaciers take the path of least resistance, following the paths of existing stream valleys. Prior to glaciation, mountain valleys are characteristically narrow and V-shaped because streams are well above base level and are therefore downcutting. However, during glaciation, these narrow valleys undergo a transformation

as the glacier widens and deepens them, creating a U-shaped **glacial trough** (see Figure 6.12C). In addition to producing a broader and deeper valley, the glacier also straightens the valley. As ice flows around sharp curves, its great erosional force removes the spurs of land that extend into the valley.

The amount of glacial erosion that takes place in different valleys in a mountainous area varies. Prior to glaciation, the mouths of tributary streams join the main valley (or *trunk* valley) at the elevation of the stream in that valley. During glaciation, the amount of ice flowing through the main valley can be much greater than the amount advancing down each tributary. Consequently, the valley containing the main glacier (or *trunk* glacier) is eroded deeper than the smaller valleys that feed it. Thus, after the ice has receded, the valleys of tributary glaciers are left standing above the main glacial trough and are termed **hanging valleys** (see Figure 6.12C). Rivers flowing through hanging valleys can produce

FIGURE 6.13 The Matterhorn Horns are sharp, pyramid-like peaks that were shaped by alpine glaciers. The Matterhorn, in the Swiss Alps, is a famous example. (Photo by Andy Selinger/AGE Fotostock)

spectacular waterfalls, such as those in Yosemite National Park, California.

Cirques   At the head of a glacial valley is a characteristic and often imposing feature associated with an alpine glacier— a **cirque**. As the photo in Figure 6.12 illustrates, these bowl-shaped depressions have precipitous walls on three sides but are open on the downvalley side. The cirque is the focal point of the glacier's growth because it is the area of snow accumulation and ice formation. Cirques begin as irregularities in the mountainside that are subsequently enlarged by frost wedging and plucking along the sides and bottom of the glacier. The glacier in turn acts as a conveyor belt that carries away the debris. After the glacier has melted away, the cirque basin is sometimes occupied by a small lake called a *tarn* (see Figure 6.12C).

Arêtes and Horns   The Alps, Northern Rockies, and many other mountain landscapes sculpted by valley glaciers reveal more than glacial troughs and cirques. In addition, sinuous, sharp-edged ridges called **arêtes** and sharp, pyramid-like peaks termed **horns** project above the surroundings (see Figure 6.12C). Both features can originate from

the same basic process: the enlargement of cirques produced by plucking and frost action. Several cirques around a single high mountain create the spires of rock called *horns*. As the cirques enlarge and converge, an isolated horn is produced. A famous example is the Matterhorn in the Swiss Alps (**FIGURE 6.13**).

Arêtes can form in a similar manner except that the cirques are not clustered around a point but rather exist on opposite sides of a divide. As the cirques grow, the divide separating them is reduced to a very narrow, knifelike partition. An arête can also be created in another way. When two glaciers occupy parallel valleys, an arête can form when the land separating the moving tongues of ice is progressively narrowed as the glaciers scour and widen their valleys.

Fiords   **Fiords** are deep, often spectacular, steep-sided inlets of the sea that exist in many high-latitude areas of the world where mountains are adjacent to the ocean (**FIGURE 6.14**). Norway, British Columbia, Greenland, New Zealand, Chile, and Alaska all have coastlines characterized by fiords. They are glacial troughs that became submerged as the ice left the valley and sea level rose following the Ice Age.

The depths of some fiords can exceed 1000 meters (3300 feet). However, the great depths of these flooded troughs are only partly explained by the post–Ice Age rise in sea level. Unlike the situation governing the downward erosional work of rivers, sea level does not act as a base level for glaciers. As a consequence, glaciers are capable of eroding their beds far below the surface of the sea. For example, a valley glacier 300 meters (1000 feet) thick can carve its valley floor more than 250 meters (800 feet) below sea level before downward erosion ceases and the ice begins to float.

6.3 CONCEPT CHECKS

1 How do glaciers acquire their load of sediment?
2 How does a glaciated mountain valley differ in appearance from a mountain valley that was not glaciated?
3 Describe the features created by glacial erosion that you might see in an area where valley glaciers recently existed.

FIGURE 6.14 Fiords The coast of Norway is known for its many fiords. Frequently these ice-sculpted inlets of the sea are hundreds of meters deep. (Satellite images courtesy of NASA Earth Observing System; photo by Wolfgang Meier/ Getty Images)

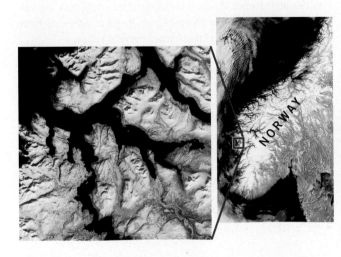

6.4 | GLACIAL DEPOSITS Distinguish between the two basic types of glacial deposits and briefly describe the features associated with each type.

A glacier picks up and transports a huge load of debris as it slowly advances across the land. Ultimately these materials are deposited when the ice melts. In regions where glacial sediment is deposited, it can play a significant role in forming the physical landscape. For example, in many areas once covered by the ice sheets of the recent Ice Age, the bedrock is rarely exposed because glacial deposits that are dozens or even hundreds of meters thick completely mantle the terrain. The general effect of these deposits is to reduce the local relief and thus level the topography. Indeed, rural country scenes familiar to many of us—rocky pastures in New England, wheat fields in the Dakotas, rolling farmland in the Midwest—result directly from glacial deposition.

Types of Glacial Drift

Long before the theory of an extensive Ice Age was proposed, much of the soil and rock debris covering portions of Europe was recognized as coming from elsewhere. At the time, these foreign materials were believed to have been "drifted" into their present positions by floating ice during an ancient flood. As a consequence, the term *drift* was applied to this sediment. Although rooted in a concept that was not correct, this term was so well established by the time the true glacial origin of the debris became widely recognized that it remained in the glacial vocabulary. Today, **glacial drift** is an all-embracing term for sediments of glacial origin, no matter how, where, or in what form they were deposited.

Glacial drift is divided into two distinct types: (1) materials deposited directly by the glacier, which are known as *till*, and (2) sediments laid down by glacial meltwater, called *stratified drift*. **Till** is deposited as glacial ice melts and drops its load of rock debris. Unlike moving water and wind, ice cannot sort the sediment it carries; therefore, deposits of till are characteristically unsorted mixtures of many particle

FIGURE 6.15 Glacial Till Unlike sediment deposited by running water and wind, material deposited directly by a glacier is not sorted. (Photos by E. J. Tarbuck)

Glacial till is an unsorted mixture of many different sediment sizes.

A close examination of glacial till often reveals cobbles that have been scratched as they were dragged along by the ice.

sizes (**FIGURE 6.15**). **Stratified drift** is sorted according to the size and weight of the fragments. Because ice is not capable of such sorting activity, these sediments are not deposited

EYE ON EARTH

This photo shows an iceberg floating in the ocean near the coast of Greenland.

QUESTION 1 *How do icebergs form? What term applies to this process?*

QUESTION 2 *Using the knowledge you have gained about these features, explain the common phrase "It's only the tip of the iceberg."*

QUESTION 3 *Is an iceberg the same as sea ice? Explain.*

QUESTION 4 *If this iceberg were to melt, how would sea level be affected?*

(Photo by Andrzej Gibasiewicz/Shutterstock)

FIGURE 6.16 Glacial Erratics Two large glacially transported boulders in Denali National Park, Alaska. (Photo by Michael Collier)

Glacial erratic

Glacial erratic

directly by the glacier. Rather, they reflect the sorting action of glacial meltwater.

Some deposits of stratified drift are made by streams issuing directly from the glacier. Other stratified deposits involve sediment that was originally laid down as till and later picked up, transported, and redeposited by meltwater beyond the margin of the ice. Accumulations of stratified drift often consist largely of sand and gravel because the meltwater is not capable of moving larger material and because the finer rock flour remains suspended and is commonly carried far from the glacier. An indication that stratified drift consists primarily of sand and gravel can be seen in many areas where these deposits are actively mined as aggregate for road work and other construction projects.

Boulders found in the till or lying free on the surface are called **glacial erratics** if they are different from the bedrock below (**FIGURE 6.16**). Of course, this means that they must have been derived from a source outside the area where they are found. Although the locality of origin for most erratics is unknown, the origin of some can be determined. Therefore, by studying glacial erratics as well as the mineral composition of till, geologists can sometimes trace the path of a lobe of ice. In portions of New England as well as other areas, erratics can be seen dotting pastures and farm fields. In some places, these rocks are cleared from fields and piled to make fences and walls.

Moraines, Outwash Plains, and Kettles

Perhaps the most widespread features created by glacial deposition are *moraines*, which are simply layers or ridges of till. Several types of moraines are identified; some are common only to mountain valleys, and others are associated with areas affected by either ice sheets or valley glaciers. Lateral and medial moraines fall in the first category, whereas end moraines and ground moraines are in the second.

Lateral and Medial Moraines The sides of a valley glacier accumulate large quantities of debris from the valley walls. When the glacier wastes away, these materials are left

FIGURE 6.17 Formation of an End Moraine This end moraine is forming at the terminus of Exit Glacier in Kenai Fjords National Park. (Photo by Michael Collier)

When accumulation equals wastage, the terminus of a glacier is stationary and an end moraine forms.

as ridges, called **lateral moraines**, along the sides of the valley. **Medial moraines** are formed when two valley glaciers coalesce to form a single ice stream. The till that was once carried along the edges of each glacier joins to form a single dark stripe of debris within the newly enlarged glacier. Creation of these dark stripes within the ice stream is obvious proof that glacial ice moves because the medial moraine could not form if the ice did not flow downvalley. It is common to see several medial moraines within a large alpine glacier; a streak forms whenever a tributary glacier joins the main valley. Kennicott Glacier in Figure 6.1 provides an excellent example.

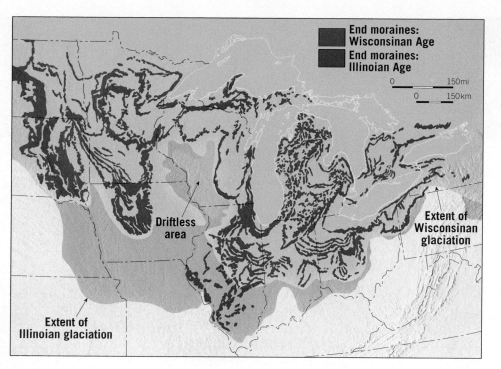

FIGURE 6.18 **End Moraines of the Great Lakes Region** End moraines deposited during the most recent stage of glaciation, called the Wisconsinan, are the most prominent.

End Moraines and Ground Moraines
An **end moraine** is a ridge of till that forms at the terminus of a glacier and is characteristic of ice sheets and valley glaciers alike. These relatively common landforms are deposited when a state of equilibrium is attained between wastage and ice accumulation. That is, the end moraine forms when the ice is melting near the end of the glacier at a rate equal to the forward advance of the glacier from its region of nourishment (**FIGURE 6.17**). Although the terminus of the glacier is stationary, the ice continues to flow forward, delivering a continuous supply of sediment in the same manner a conveyor belt delivers goods to the end of a production line. As the ice melts, the till is dropped, and the end moraine grows. The longer the ice front remains stable, the larger the ridge of till becomes.

Eventually, wastage exceeds nourishment. At this point, the front of the glacier begins to recede in the direction from which it originally advanced. However, as the ice front retreats, the conveyor-belt action of the glacier continues to provide fresh supplies of sediment to the terminus. In this manner, a large quantity of till is deposited as the ice melts away, creating a rock-strewn, undulating surface. This gently rolling layer of till deposited as the ice front recedes is termed **ground moraine**. Ground moraine has a leveling effect, filling in low spots and clogging old stream channels, often leading to a derangement of the existing drainage system. In areas where this layer of till is still relatively fresh, such as the northern Great Lakes region, poorly drained swampy lands are quite common.

Periodically, a glacier will retreat to a point where wastage and nourishment once again balance. When this happens, the ice front stabilizes and a new end moraine forms.

The pattern of end moraine formation and ground moraine deposition may be repeated many times before the glacier has completely vanished. Such a pattern is illustrated in **FIGURE 6.18**. The very first end moraine to form marks the farthest advance of the glacier and is called the *terminal end moraine*. Those end moraines that form as the ice front occasionally stabilizes during retreat are termed *recessional end moraines*. Terminal and recessional moraines are essentially alike; the only difference between them is their relative positions.

End moraines deposited by the most recent major stage of Ice Age glaciation are prominent features in many parts of the Midwest and Northeast. In Wisconsin, the wooded, hilly terrain of the Kettle Moraine near Milwaukee is a particularly picturesque example. A well-known example in the Northeast is Long Island. This linear strip of glacial sediment that extends northeastward from New York City is part of an end moraine complex that stretches from eastern Pennsylvania to Cape Cod, Massachusetts (**FIGURE 6.19**).

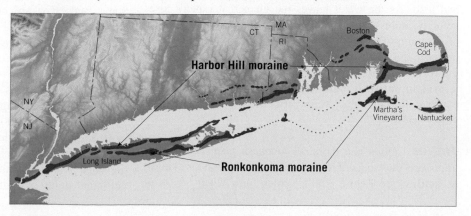

FIGURE 6.19 **Two Significant End Moraines in the Northeast** The Ronkonkoma moraine, which was deposited about 20,000 years ago, extends through central Long Island, Martha's Vineyard, and Nantucket. The Harbor Hill moraine formed about 14,000 years ago and extends along the north shore of Long Island, through southern Rhode Island and Cape Cod.

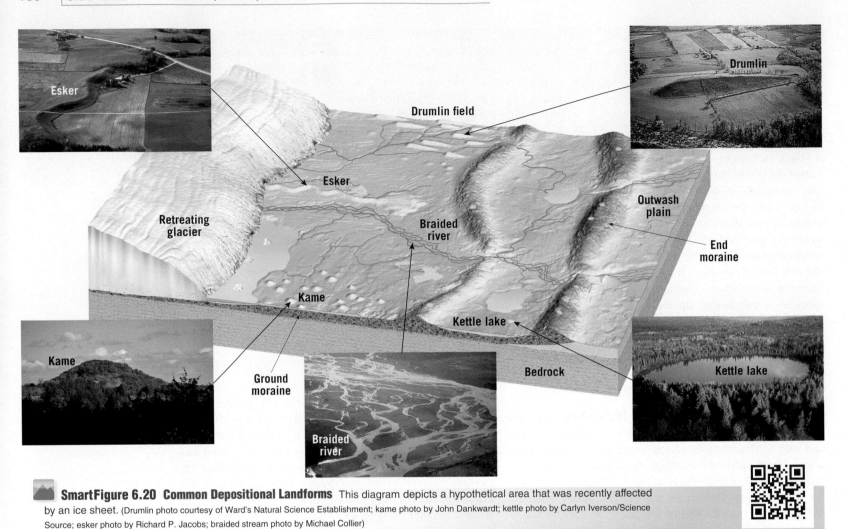

SmartFigure 6.20 Common Depositional Landforms This diagram depicts a hypothetical area that was recently affected by an ice sheet. (Drumlin photo courtesy of Ward's Natural Science Establishment; kame photo by John Dankwardt; kettle photo by Carlyn Iverson/Science Source; esker photo by Richard P. Jacobs; braided stream photo by Michael Collier)

FIGURE 6.20 represents a hypothetical area during and following glaciation. It shows end moraines as well as the depositional features that are discussed in the sections that follow. This figure depicts landscape features similar to what you might encounter if you were traveling in the upper Midwest or New England. As you read about other glacial deposits, you will refer to this figure.

Outwash Plains and Valley Trains

At the same time that an end moraine is forming, meltwater emerges from the ice in rapidly moving streams. Often these streams are choked with suspended material and carry a substantial bed load. As the water leaves the glacier, it rapidly loses velocity, and much of its bed load is dropped. In this way, a broad, ramplike accumulation of stratified drift is built adjacent to the downstream edge of most end moraines. When the feature is formed in association with an ice sheet, it is termed an **outwash plain** (see Figure 6.20), and when it is confined to a mountain valley, it is usually referred to as a **valley train**.

Kettles

Often end moraines, outwash plains, and valley trains are pockmarked with basins, or depressions, known as **kettles** (see Figure 6.20). Kettles form when blocks of stagnant ice become buried in drift and eventually melt, leaving pits in the glacial sediment. Most kettles do not exceed 2 kilometers in diameter, and the typical depth of most kettles is less than 10 meters (33 feet). Water often fills the depression and forms a pond or lake. One well-known example is Walden Pond near Concord, Massachusetts. It is here that Henry David Thoreau lived alone for 2 years in the 1840s and about which he wrote *Walden*, his classic of American literature.

Drumlins, Eskers, and Kames

Moraines are not the only landforms deposited by glaciers. Some landscapes are characterized by numerous elongate parallel hills made of till. Other areas exhibit conical hills and relatively narrow winding ridges composed mainly of stratified drift.

Drumlins   Drumlins are streamlined asymmetrical hills composed of till (see Figure 6.20). They range in height from 15 to 60 meters (50–200 feet) and average 0.4 to 0.8 kilometer (0.25–0.50 mile) in length. The steep side of the hill faces the direction from which the ice advanced, whereas the gentler slope points in the direction the ice moved. Drumlins are not found singly but rather occur in clusters, called *drumlin* *fields*. One such cluster, east of Rochester, New York, is estimated to contain about 10,000 drumlins. Their streamlined shape indicates that they were molded in the zone of flow within an active glacier. It is thought that drumlins originate when glaciers advance over previously deposited drift and reshape the material.

Eskers and Kames   In some areas that were once occupied by glaciers, sinuous ridges composed largely of sand and gravel might be found. These ridges, called **eskers**, are deposits made by streams flowing in tunnels beneath the ice, near the terminus of a glacier (see Figure 6.20). They may be several meters high and extend for many kilometers. In some areas they are mined for sand and gravel, and for this reason, eskers are disappearing in some localities.

Kames are steep-sided hills that, like eskers, are composed of sand and gravel (see Figure 6.20). Kames originate when glacial meltwater washes sediment into openings and depressions in the stagnant wasting terminus of a glacier. When the ice eventually melts away, the stratified drift is left behind as mounds or hills.

6.4 CONCEPT CHECKS

1 What is the difference between till and stratified drift?
2 Distinguish among terminal end moraine, recessional end moraine, and ground moraine. Relate these moraines to the budget of a glacier.
3 Describe the formation of a medial moraine.
4 List four depositional features other than moraines.

6.5 | OTHER EFFECTS OF ICE AGE GLACIERS Describe and explain several important effects of Ice Age glaciers other than erosional and depositional landforms.

In addition to the massive erosional and depositional work carried out by Ice Age glaciers, the ice sheets had other effects, sometimes profound, on the landscape. For example, as the ice advanced and retreated, animals and plants were forced to migrate. This led to stresses that some organisms could not tolerate. Hence, a number of plants and animals became extinct.

Other effects of Ice Age glaciers that are described in this section involve adjustments in Earth's crust due to the addition and removal of ice and sea-level changes associated with the formation and melting of ice sheets. The advance and retreat of ice sheets also led to significant changes in the routes taken by rivers. In some regions, glaciers acted as dams that created large lakes. When these ice dams failed, the effects on the landscape were profound. In areas that today are deserts, lakes of another type, called pluvial lakes, formed.

Changing Rivers

Many present-day stream courses bear little resemblance to their preglacial routes. The Missouri River once flowed northward toward Hudson Bay in Canada. The Mississippi River followed a path through central Illinois, and the head of the Ohio River reached only as far as Indiana (**FIGURE 6.21**). A comparison of the two parts of Figure 6.21 shows that the Great Lakes were created by glacial erosion during the Ice Age. Prior to the Quaternary period, the basins occupied by these huge lakes were lowlands with rivers that ran eastward to the Gulf of St. Lawrence.

Crustal Subsidence and Rebound

In areas that were centers of ice accumulation, such as Scandinavia and northern Canada, the land has been slowly rising

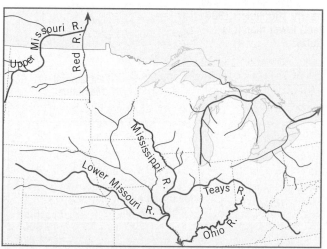

FIGURE 6.21 Changing Rivers The advance and retreat of ice sheets caused major changes in the routes followed by rivers in the central United States.

This map shows the Great Lakes and the familiar present-day pattern of rivers. Quaternary ice sheets played a major role in creating this pattern.

Reconstruction of drainage systems prior to the Ice Age. The pattern was very different from today, and the Great Lakes did not exist.

FIGURE 6.22 Glacial Lake Agassiz This lake was an immense feature—bigger than all of the present-day Great Lakes combined. Modern-day remnants of this proglacial water body are still major landscape features.

for the past several thousand years. The land had down-warped under the tremendous weight of 3-kilometer-thick (almost 2-mile-thick) ice sheets. Since the removal of this immense load, the crust has been adjusting by gradually rebounding upward.* Uplift of nearly 300 meters (1000 feet) has occurred in the Hudson Bay region.

Proglacial Lakes Created by Ice Dams

Ice sheets and alpine glaciers can act as dams to create lakes by trapping glacial meltwater and blocking the flow of rivers. Some of these lakes are relatively small and short-lived.

*For a more complete discussion of this concept, termed *isostatic adjustment*, see the section "Principle of Isostasy" in Chapter 10, page 340.

FIGURE 6.23 Changing Sea Level As ice sheets form and then melt away, sea level falls and rises, causing the shoreline to shift.

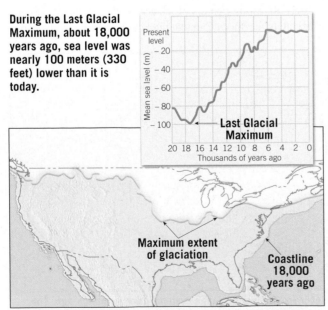

During the Last Glacial Maximum, about 18,000 years ago, sea level was nearly 100 meters (330 feet) lower than it is today.

During the Last Glacial Maximum, the shoreline extended out onto the present day continental shelf.

Others can be large and exist for hundreds or thousands of years.

FIGURE 6.22 is a map of Lake Agassiz—the largest lake to form during the Ice Age in North America. With the retreat of the ice sheet came enormous volumes of meltwater. The Great Plains generally slope upward to the west. As the terminus of the ice sheet receded northeastward, meltwater was trapped between the ice on one side and the sloping land on the other, causing Lake Agassiz to deepen and spread across the landscape. It came into existence about 12,000 years ago and lasted for about 4500 years. Such water bodies are termed **proglacial lakes**, referring to their position just beyond the outer limits of a glacier or ice sheet. Research shows that the shifting of glaciers and the failure of ice dams can cause the rapid release of huge volumes of water. Such events occurred during the history of Lake Agassiz.

Sea-Level Changes

A far-reaching effect of the Ice Age was the world-wide change in sea level that accompanied each advance and retreat of the ice sheets. The snow that nourishes glaciers ultimately comes from moisture evaporated from the oceans. Therefore, when the ice sheets increased in size, sea level fell, and the shoreline moved seaward (**FIGURE 6.23**). Estimates suggest that sea level was as much as 100 meters (330 feet) lower than it is today. Consequently, the Atlantic coast of the

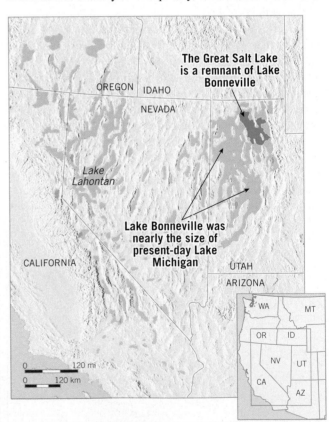

FIGURE 6.24 Pluvial Lakes During the Ice Age, the Basin and Range region experienced a wetter climate than it has today. Many basins were turned into large lakes.

United States was located more than 100 kilometers (60 miles) to the east of New York City. Moreover, France and Britain were joined where the English Channel is today. Alaska and Siberia were connected across the Bering Strait, and Southeast Asia was tied by dry land to the islands of Indonesia.

Pluvial Lakes

The formation and growth of ice sheets was an obvious response to significant changes in climate. But the existence of the glaciers themselves triggered climatic changes in the regions beyond their margins. In arid and semiarid areas on all continents, temperatures were lower, which meant evaporation rates were also lower. At the same time, precipitation was moderate. This cooler, wetter climate resulted in the formation of many lakes called **pluvial lakes** (from the Latin term *pluvia*, meaning

"rain"). In North America, pluvial lakes were concentrated in the vast Basin and Range region of Nevada and Utah (**FIGURE 6.24**). Although most pluvial lakes are now gone, a few remnants remain, the largest being Utah's Great Salt Lake.

6.5 CONCEPT CHECKS

1 List four effects of Ice Age glaciers, aside from the formation of major erosional and depositional features.

2 Compare the two parts of Figure 6.21 and identify three major changes to the flow of rivers in the central United States during the Ice Age.

3 Examine Figure 6.23 and determine how much sea level has changed since the Last Glacial Maximum.

4 Contrast proglacial lakes and pluvial lakes.

6.6 | EXTENT OF ICE AGE GLACIATION

Discuss the extent of glaciation and climate variability during the Quaternary Ice Age.

During the Ice Age, ice sheets and alpine glaciers were far more extensive than they are today. There was a time when the most popular explanation for what we now know to be glacial deposits was that the material had been drifted in by means of icebergs or perhaps simply swept across the landscape by a catastrophic flood. However, during the nineteenth century, field investigations by many scientists provided convincing proof that an extensive Ice Age was responsible for these deposits and for many other features.

By the beginning of the twentieth century, geologists had largely determined the extent of Ice Age glaciation. Further, they had discovered that many glaciated regions did not have just one layer of drift but several. Close examination of these older deposits showed well-developed zones of chemical weathering and soil formation, as well as the remains of plants that require warm temperatures. The evidence was clear: There had not been just one glacial advance but several, each separated by extended periods when climates were as warm as or warmer than at present. The Ice Age was not simply a time when the ice advanced over the land, lingered for a while, and then receded. Rather, it was a complex period characterized by a number of advances and withdrawals of glacial ice.

The glacial record on land is punctuated by many erosional gaps. This makes it difficult to reconstruct the episodes of the Ice Age. But sediment on the ocean floor provides an uninterrupted record of climate cycles for this period. Studies of these seafloor sediments show that glacial/interglacial cycles have occurred about every 100,000 years. About 20 such cycles of cooling and warming were identified for the span we call the Ice Age.

During the Ice Age, ice left its imprint on almost 30 percent of Earth's land area, including about 10 million square kilometers (nearly 4 million square miles) of North America, 5 million square kilometers (2 million square miles) of Europe, and 4 million square kilometers (1.6 million square miles) of Siberia (**FIGURE 6.25**). The amount of glacial ice in the Northern Hemisphere was roughly twice that in the Southern Hemisphere. The primary reason is that the Southern Hemisphere has little land in the middle latitudes, and, therefore, the southern polar ice could not spread far beyond the margins of Antarctica. By contrast, North America and Eurasia provided great expanses of land for the spread of ice sheets.

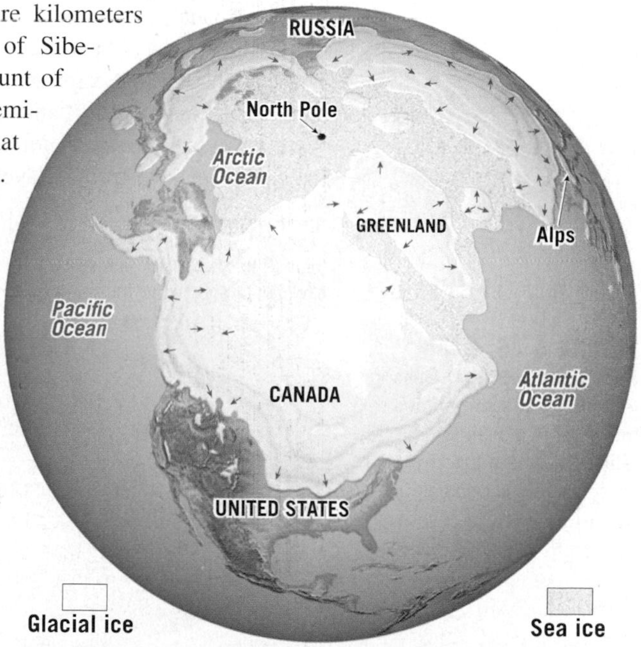

We now know that the Ice Age began between 2 million and 3 million years ago. This means that most of the major glacial episodes occurred during a division of the geologic time scale called the **Quaternary period**. Although the Quaternary is commonly used as a synonym for the Ice Age, this period does not encompass it all. The Antarctic Ice Sheet, for example, formed at least 30 million years ago.

FIGURE 6.25 **Where Was the Ice?** This map shows the maximum extent of ice sheets in the Northern Hemisphere during the Ice Age.

6.6 CONCEPT CHECKS

1 About what percentage of Earth's land surface was affected by glaciers during the Quaternary period?

2 Where were ice sheets more extensive during the Ice Age: the Northern Hemisphere or the Southern Hemisphere? Why?

6.7 | CAUSES OF ICE AGES
Summarize some of the current ideas about the causes of ice ages.

A great deal is known about glaciers and glaciation. Much has been learned about glacier formation and movement, the extent of glaciers past and present, and the features created by glaciers, both erosional and depositional. However, the causes of glacial ages are not completely understood.

Although widespread glaciation has been rare in Earth's history, the Quaternary Ice Age is not the only glacial period for which a record exists. Earlier glaciations are indicated by rock layers called *tillite*, a sedimentary rock formed when glacial till becomes lithified. Such strata usually contain striated rock fragments, and some lie atop grooved and polished rock surfaces or are associated with sandstones and conglomerates that show features indicating they were deposited as stratified drift. Two Precambrian glacial episodes have been identified in the geologic record, the first approximately 2 billion years ago and the second about 600 million years ago. Furthermore, a well-documented record of an earlier glacial age is found in late Paleozoic rocks that are about 250 million years old and that exist on several landmasses.[†]

Any theory that attempts to explain the causes of ice ages must successfully answer two basic questions: (1) *What causes the onset of glacial conditions?* For continental ice sheets to have formed, average temperatures must have

[†]The terms *Precambrian* and *Paleozoic* refer to time spans on the geologic time scale of Earth history. For more on the geologic time scale, see Chapter 11 and Figure 11.24, page 364.

FIGURE 6.26 A Late Paleozoic Ice Age Shifting tectonic plates sometimes move landmasses to high latitudes, where the formation of ice sheets is possible.

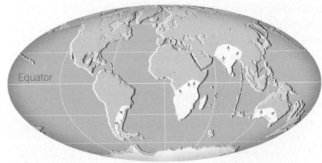

The supercontinent Pangaea showing the area covered by glacial ice near the end of the Paleozoic era.

The continents as they appear today. The white areas indicate where evidence of the late Paleozoic ice sheets exists.

been somewhat lower than at present and perhaps substantially lower than throughout much of geologic time. Thus, a successful theory would have to account for the cooling that finally leads to glacial conditions. (2) *What caused the alternation of glacial and interglacial stages that have been documented for the Quaternary period?* Whereas the first question deals with long-term trends in temperature on a scale of millions of years, the second question relates to much shorter-term changes.

Although the literature of science contains many hypotheses related to the possible causes of glacial periods, we discuss only a few major ideas to summarize current thought.

Plate Tectonics

Probably the most attractive proposal for explaining the fact that extensive glaciations have occurred only a few times in the geologic past comes from the theory of plate tectonics.[‡] Because glaciers can form only on land, we know that landmasses must exist somewhere in the higher latitudes before an ice age can commence. Many scientists suggest that ice ages have occurred only when Earth's shifting crustal plates have carried the continents from tropical latitudes to more poleward positions.

Glacial features in present-day Africa, Australia, South America, and India indicate that these regions, which are now tropical or subtropical, experienced an ice age near the end of the Paleozoic era, about 250 million years ago. However, there is no evidence that ice sheets existed during this same period in what are today the higher latitudes of North America and Eurasia. For many years, this puzzled scientists. Was the climate in these relatively tropical latitudes once like it is today in Greenland and Antarctica? Why did glaciers not form in North America and Eurasia? Until the plate tectonics theory was formulated, there had been no reasonable explanation.

Today, scientists realize that the areas containing these ancient glacial features were joined together as a single supercontinent (Pangaea) located at latitudes far to the south of their present positions. Later, this landmass broke apart, and its pieces, each moving on a different plate, migrated toward their present locations (**FIGURE 6.26**). It is now understood that during the geologic past, plate movements accounted for many dramatic climatic changes as landmasses shifted in relation to one another and moved to different latitudinal positions. Shifting landmasses also caused changes in oceanic circulation, altering the transport of heat and moisture and consequently the climate as well. Because the rate of plate movement is very slow—a few centimeters per year—appreciable changes in the positions of the continents occur only over great spans of geologic time. Thus, climate changes brought about by shifting

[‡]A complete discussion of plate tectonics is presented in Chapter 7.

plates are extremely gradual and occur on a scale of millions of years.

Variations in Earth's Orbit

Because climate changes brought about by moving plates are extremely gradual, the plate tectonics theory cannot be used to explain the alternation between glacial and interglacial climates that occurred during the Quaternary period. Therefore, we must look to some other triggering mechanism that may cause climate change on a scale of thousands rather than millions of years. Today, many scientists strongly suspect that the climatic oscillations that characterized the Quaternary may be linked to changes in Earth's orbit. This hypothesis was first developed and strongly advocated by Serbian scientist Milutin Milankovitch and is based on the premise that variations in incoming solar radiation are a principal factor in controlling Earth's climate.

Milankovitch formulated a comprehensive mathematical model based on the following elements (FIGURE 6.27):

- Variations in the shape (*eccentricity*) of Earth's orbit about the Sun
- Changes in *obliquity*—that is, changes in the angle that the axis makes with the plane of Earth's orbit
- The wobbling of Earth's axis, called *precession*

Using these factors, Milankovitch calculated variations in the receipt of solar energy and the corresponding surface temperature of Earth back into time, in an attempt to correlate these changes with the climate fluctuations of the Quaternary. In explaining climate changes that result from these three variables, note that they cause little or no variation in the *total* solar energy reaching the ground. Instead, their impact is felt because they change the degree of contrast between the seasons. Somewhat milder winters in the middle to high latitudes means greater snowfall totals, whereas cooler summers bring a reduction in snowmelt.

Among the studies that added considerable credibility to this astronomical hypothesis is one in which deep-sea sediments containing certain climatically sensitive microorganisms were analyzed to establish a chronology of temperature changes going back nearly 500,000 years.[§] This time scale of climatic change was then compared to astronomical calculations of eccentricity, obliquity, and precession to determine whether a correlation did indeed exist. Although the study was very involved and mathematically complex, the conclusions were straightforward. The researchers found that major variations in climate over the past several hundred thousand years were closely associated with changes in the geometry of Earth's orbit; that is, cycles of climate change were shown to correspond closely with the periods of obliquity, precession, and orbital eccentricity. More specifically, the researchers stated: "It is concluded that changes in the earth's orbital geometry are

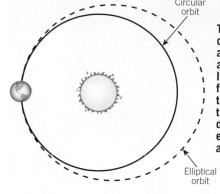

The shape of Earth's orbit changes during a cycle that spans about 100,000 years. It gradually changes from nearly circular to more elliptical and then back again. This diagram greatly exaggerates the amount of change.

SmartFigure 6.27 Orbital Variations Periodic variations in Earth's orbit are linked to alternating glacial and interglacial conditions during the Ice Age.

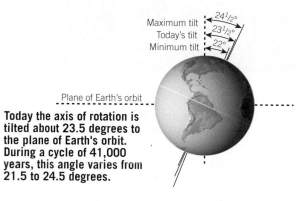

Today the axis of rotation is tilted about 23.5 degrees to the plane of Earth's orbit. During a cycle of 41,000 years, this angle varies from 21.5 to 24.5 degrees.

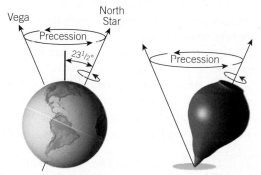

Earth's axis wobbles like a spinning top. Consequently, the axis points to different spots in the sky during a cycle of about 26,000 years.

the fundamental cause of the succession of Quaternary ice ages."[**]

Let us briefly summarize the ideas that were just described. The theory of plate tectonics provides an explanation for the widely spaced and nonperiodic onset of glacial conditions at various times in the geologic past, whereas the astronomical model proposed by Milankovitch and supported by the work of J. D. Hays and his colleagues furnishes an explanation for the alternating glacial and interglacial episodes of the Pleistocene.

Other Factors

Variations in Earth's orbit correlate closely with the timing of glacial–interglacial cycles. However, the variations in solar

---

[§]J. D. Hays, John Imbrie, and N. J. Shackelton, "Variations in the Earth's Orbit: Pacemaker of the Ice Ages," *Science* 194 (1976): 1121–1132.

[**]J. D. Hays et al., p. 1131.

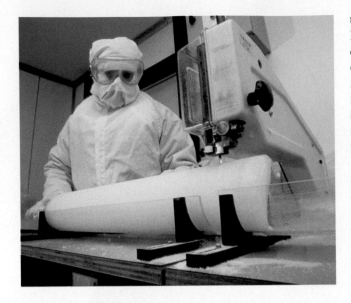

FIGURE 6.28 Ice Cores Contain Clues to Shifts in Climate This scientist is slicing an ice core from Antarctica for analysis. He is wearing protective clothing and a mask to minimize contamination of the sample. Chemical analyses of ice cores can provide important data about past climates. (Photo by British Antarctic Survey/Science Source)

energy reaching Earth's surface caused by these orbital changes do not adequately explain the magnitude of the temperature changes that occurred during the most recent ice age. Other factors must also have contributed. One factor involves variations in the chemical composition of the atmosphere. Other influences involve changes in the reflectivity of Earth's surface and in ocean circulation. Let's take a brief look at these factors.

Chemical analyses of air bubbles that become trapped in glacial ice at the time of ice formation indicate that the Ice Age atmosphere contained less of the gases carbon dioxide and methane than the post–Ice Age atmosphere (**FIGURE 6.28**). Carbon dioxide and methane are important "greenhouse" gases, which means they trap radiation emitted by Earth and contribute to the heating of the atmosphere.[††] When the amount of carbon dioxide and methane in the atmosphere increases, global temperatures rise, and when

──────────

[††]For more on this idea, see the section "Heating the Atmosphere: The Greenhouse Effect" in Chapter 16 and the section "Carbon Dioxide, Trace Gases, and Global Warming" in Chapter 20.

there is a reduction in these gases, as occurred during the Ice Age, temperatures fall. Therefore, reductions in the concentrations of greenhouse gases help explain the magnitude of the temperature drop that occurred during glacial times. Although scientists know that concentrations of carbon dioxide and methane dropped, they do not know what caused the drop. As often occurs in science, observations gathered during one investigation yield information and raise questions that require further analysis and explanation.

Whenever Earth enters an ice age, extensive areas of land that were once ice free are covered with ice and snow. In addition, a colder climate causes the area covered by sea ice (frozen surface sea water) to expand as well. Ice and snow reflect a large portion of incoming solar energy back to space. Thus, energy that would have warmed Earth's surface and the air above is lost, and global cooling is reinforced.

Yet another factor that influences climate during glacial times relates to ocean currents, which, as you will learn in Chapter 15, are a complex matter. Research has shown that ocean circulation changes during ice ages. For example, studies suggest that the warm current that transports large amounts of heat from the tropics toward higher latitudes in the North Atlantic was significantly weaker during the Ice Age. This would lead to a colder climate in Europe, amplifying the cooling that can be attributed to orbital variations.

In conclusion, it should be noted that our understanding of the causes of glacial ages is not complete. The ideas that were just discussed do not represent all the possible explanations. Additional factors may be, and probably are, involved.

6.7 CONCEPT CHECKS

1 How does the theory of plate tectonics help us understand the causes of ice ages?

2 Does the theory of plate tectonics explain alternating glacial–interglacial climates during the Quaternary period? Why or why not?

6.8 | DESERTS Describe the general distribution and extent of Earth's dry lands and the role that water plays in modifying desert landscapes.

The dry regions of the world encompass about 42 million square kilometers (about 16 million square miles)—a surprising 30 percent of Earth's land surface. No other climate group covers so large a land area. The word *desert* literally means "deserted," or "unoccupied." For many dry regions, this is a very appropriate description. Yet where water is available in deserts, plants and animals thrive. Nevertheless, the world's dry regions are among the least familiar land areas on Earth outside the polar realm.

Desert landscapes frequently appear stark. Their profiles are not softened by a carpet of soil and abundant plant life. Instead, barren rocky outcrops with steep, angular slopes are common. At some places the rocks are tinted orange and red. At others they are gray and brown and streaked with black. For many visitors desert scenery exhibits a striking beauty; to others the terrain seems bleak. No matter which feeling is elicited, it is clear that deserts are very different from the more humid places where most people live.

As you will see, arid regions are not dominated by a single geologic process. Rather, the effects of mountain-building forces, running water, and wind are all apparent. Because these processes combine in different ways from place to place, the appearance of desert landscapes varies a great deal as well (**FIGURE 6.29**).

FIGURE 6.29 Sonoran Desert near Tucson, Arizona Some of these giant saguaro cacti are more than 100 years old. The appearance of desert landscapes varies a great deal from place to place. (Photo by Michael Collier)

Distribution and Causes of Dry Lands

We all recognize that deserts are dry places, but just what is meant by the word *dry*? That is, how much rain defines the boundary between humid and dry regions?

Sometimes, it is arbitrarily defined by a single rainfall figure, such as 25 centimeters (10 inches) per year of precipitation. However, the concept of dryness is relative; it refers to *any situation in which a water deficiency exists*. Climatologists define **dry climate** as a climate in which

yearly precipitation is less than the potential loss of water by evaporation.

Within these water-deficient regions, two climatic types are commonly recognized: **desert**, or arid, and **steppe**, or semiarid. The two categories have many features in common; their differences are primarily a matter of degree. The steppe is a marginal and more humid variant of the desert and represents a transition zone that surrounds the desert and separates it from bordering humid climates. Maps showing the distribution of desert and steppe regions reveal that dry lands are concentrated in the subtropics and in the middle latitudes (**FIGURE 6.30**).

Deserts in places such as Africa, Arabia, and Australia primarily result from the prevailing global distribution of air pressure and winds. Coinciding with dry regions in the lower latitudes are zones of high air pressure known as the *subtropical highs*. These pressure systems are characterized by subsiding air currents. When air sinks, it is compressed and warmed. Such conditions are just the opposite of what is needed to produce clouds and precipitation. Consequently, these regions are known for their clear skies, sunshine, and dryness. There is more about pressure systems and the distribution of precipitation in Chapter 18.

Middle-latitude deserts and steppes exist principally because they are sheltered in the deep interiors of large landmasses. They are far removed from the ocean, which is the ultimate source of moisture for cloud formation and precipitation. In addition, the presence of high mountains across the paths of prevailing winds further acts to separate these areas from water-bearing maritime air masses. In North America, the Coast Ranges, Sierra Nevada, and Cascades are the foremost mountain barriers to moisture from the Pacific (see

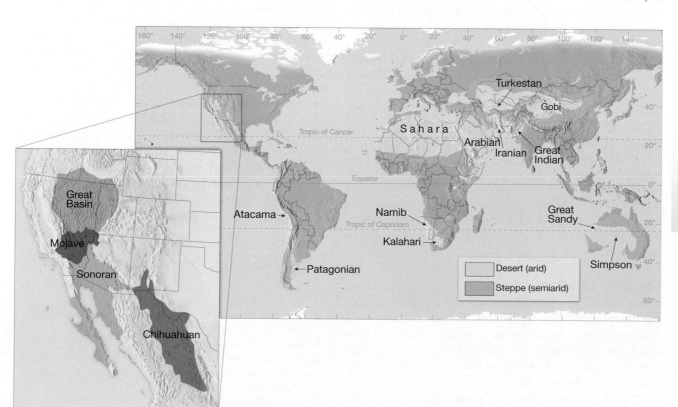

SmartFigure 6.30 Dry Climates Arid and semiarid climates cover about 30 percent of Earth's land surface. The dry region of the American West is commonly divided into four deserts, two of which extend into Mexico.

An ephemeral stream shortly after a heavy shower. Although such floods are short-lived, they cause large amounts of erosion.

Most of the time desert stream channels are dry.

A familiar sign in desert areas. Roads dip into washes which can rapidly fill with water following a heavy rain.

POTENTIAL FLASH FLOOD AREAS

NEXT 21 MILES

FIGURE 6.31 Ephemeral Stream This example is near Arches National Park in southern Utah. (Photos by E. J. Tarbuck; flash flood sign by Demetrio Carrasco/DK Images)

Figure 17.11, page 526). In their rain shadow lies the dry and expansive Basin and Range region of the American West.

Middle-latitude deserts provide an example of how mountain-building processes affect climate. Without mountains, wetter climates would prevail where dry regions exist today.

Geologic Processes in Arid Climates

The angular rock exposures, the sheer canyon walls, and the rocky and pebble- or sand-covered surfaces of deserts contrast sharply with the rounded hills and curving slopes of more humid places. To a visitor from a humid region, a desert landscape may seem to have been shaped by forces different from those operating in wetter areas. However, although the contrasts might be striking, they do not reflect different processes. They merely disclose the differing effects of the same processes that operate under contrasting climate conditions.

Weathering   In humid regions, relatively well-developed soils support an almost continuous cover of vegetation. Here the slopes and rock edges are rounded. Such a landscape reflects the strong influence of chemical weathering in a humid climate. By contrast, much of the weathered debris in deserts consists of unaltered rock and mineral fragments—the result of mechanical weathering processes. In dry lands rock weathering of any type is greatly reduced because of the lack of moisture and the scarcity of organic acids from decaying plants. Chemical weathering, however, is not completely lacking in deserts. Over long spans of time, clays and thin soils do form, and many iron-bearing silicate minerals oxidize, producing the rust-colored stain that tints some desert landscapes.

The Role of Water   Permanent streams are normal in humid regions, but almost all desert streams are dry most of the time. Deserts have **ephemeral streams**, which means they carry water only in response to specific episodes of rainfall. A typical ephemeral stream might flow only a few days or perhaps just a few hours during the year. In some years the channel may carry no water at all.

This fact is obvious even to a casual observer who, while traveling in a dry region, notices the number of bridges with no

streams beneath them or the number of dips in the road where dry channels cross. However, when the rare heavy showers do occur, so much rain falls in such a short time that all of it cannot soak in. Because the vegetative cover is sparse, runoff is largely unhindered and consequently rapid, often creating flash floods along valley floors (**FIGURE 6.31**). Such floods, however, are quite unlike floods in humid regions. Whereas a flood on a river such as the Mississippi may take many days to reach its crest and then subside, a desert flood arrives suddenly and subsides quickly. Because much of the surface material is not anchored by vegetation, the amount of erosional work that occurs during a single short-lived rain event is impressive. The muddy torrent in Figure 6.31 illustrates this fact.

Throughout the world, a number of names are used for ephemeral streams. Two of the most common in the dry western United States are *wash* and *arroyo*. In other parts of the world, a dry desert stream may be called a *wadi* (Arabia and North Africa), a *donga* (South America), or a *nullah* (India).

Humid regions are notable for their integrated drainage systems. But in arid regions, streams usually lack an extensive system of tributaries. In fact, a basic characteristic of desert streams is that they are small and die out before reaching the sea. Because the water table is usually far below the surface, few desert streams can draw upon it as streams do in humid regions. Without a steady supply of water, the combination of evaporation and infiltration soon depletes the stream.

The few permanent streams that do cross arid regions, such as the Colorado and Nile Rivers, originate *outside* the desert, often in well-watered mountains. In these situations, the water supply must be great to compensate for the losses occurring as the stream crosses the desert. For example, after the Nile leaves the lakes and mountains of central Africa that are its source, it traverses almost 3000 kilometers (nearly 1900 miles) of the Sahara *without a single tributary*. By contrast, in humid regions, the discharge of a river usually increases in the downstream direction because tributaries and groundwater contribute additional water along the way. It should be emphasized that *running water, although it occurs infrequently in deserts, nevertheless does most of the erosional work in deserts*. This is contrary to a common belief that wind is the most important erosional agent sculpting desert landscapes. Although wind erosion is indeed more significant in dry areas than elsewhere, most desert landforms are nevertheless carved by running water. As you will see shortly, the main role of wind is in the transportation and deposition of sediment, which creates and shapes the ridges and mounds we call *dunes*.

6.8 CONCEPT CHECKS

1   Define *dry climate*. How extensive are the desert and steppe regions of Earth?

2   How does the rate of rock weathering in dry climates compare to the rate in humid regions?

3   What is an ephemeral stream?

4   When a permanent stream such as the Nile River crosses a desert, does discharge increase or decrease? How does this compare to a river in a humid area?

5   What is the most important agent of erosion in deserts?

6.9 | BASIN AND RANGE: THE EVOLUTION OF A MOUNTAINOUS DESERT LANDSCAPE Discuss the stages of landscape evolution in the Basin and Range region of the western United States.

Dry regions typically lack permanent streams and often have **interior drainage**. This means they have a discontinuous pattern of ephemeral streams that do not flow out of the desert to the ocean. In the United States, the dry Basin and Range region provides an excellent example. The region includes southern Oregon, all of Nevada, western Utah, southeastern California, southern Arizona, and southern New Mexico. The name *Basin and Range* is an apt description for this almost 800,000-square-kilometer (312,000-square-mile) region, which is characterized by more than 200 relatively small mountain ranges that rise 900 to 1500 meters (3000–5000 feet) above the basins that separate them. The origin of these fault-block mountains is examined in Chapter 10. In this section, we look at how surface processes change the landscape.

In the Basin and Range region, as in other regions like it around the world, most erosion occurs without reference to the ocean (ultimate base level) because the interior drainage never reaches the sea. Even where permanent streams flow to the ocean, few tributaries exist, and thus only a narrow strip of land adjacent to the stream has sea level as its ultimate level of land reduction.

The block models in **FIGURE 6.32** depict how the landscape has evolved in the Basin and Range region. During and following uplift of the mountains, running water carves the elevated mass and deposits large quantities of debris in adjacent basins. In this early stage, relief (the difference in elevation between high and low points in an area) is greatest. As erosion lowers the mountains and sediment fills the basins, elevation differences diminish.

When the occasional torrents of water produced by sporadic rains or periods of snowmelt high in the mountains move down the mountain canyons, they are heavily loaded with sediment. Emerging from the confines of the canyon, the runoff spreads over the gentler slopes at the base of the mountains and quickly loses velocity. Consequently, most of the sediment load is dumped within a short distance. The result is a cone of debris known as an **alluvial fan** at the mouth of a canyon. Over the years, a fan enlarges, eventually coalescing with fans from adjacent canyons to produce an apron of sediment, called a **bajada**, along the mountain front.

On the rare occasions of abundant rainfall, streams may flow across the alluvial fans to the center of the basin, converting the basin floor into a shallow **playa lake**. Playa lakes last only a few days or weeks, before evaporation and infiltration remove the water. The dry, flat lake bed that remains is termed a *playa*. Playas occasionally become encrusted with salts (*salt flats*) that are left behind when the water in which they were dissolved evaporates. **FIGURE 6.33** includes a satellite view (larger image) and an aerial view of a portion of California's Death

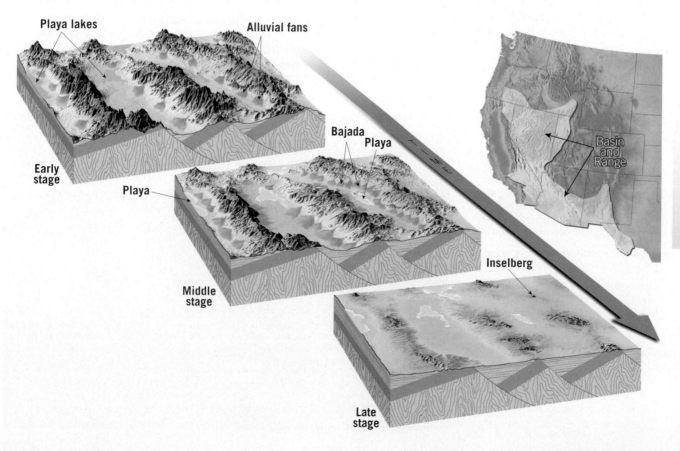

Playa lakes   Alluvial fans

Early stage

Playa

Bajada   Playa

Middle stage

Inselberg

Late stage

SmartFigure 6.32 Landscape Evolution in the Basin and Range Region As erosion of the mountains and deposition in the basins continue, relief diminishes. (*Relief* refers to the difference in elevation between the high and low spots in an area.)

FIGURE 6.33 Death Valley: A Classic Basin and Range Landscape Shortly before the satellite image (left) was taken in February 2005, heavy rains led to the formation of a small playa lake—the pool of greenish water on the basin floor. By May 2005, the lake had reverted to a salt-covered playa. (NASA) The small photo is a closer view of one of Death Valley's many alluvial fans. (Photo by Michael Collier)

Geologist's Sketch

Valley, a classic Basin and Range landscape. Many of the features just described are prominent, including a bajada (left side of valley), alluvial fans, a playa lake, and extensive salt flats.

With the ongoing erosion of the mountain mass and the accompanying sedimentation, the local relief continues to diminish. Eventually nearly the entire mountain mass is gone. Thus, by the late stages of erosion, the mountain areas are reduced to a few large bedrock knobs (called *inselbergs*) projecting above the sediment-filled basin.

Each of the stages of landscape evolution in an arid climate depicted in Figure 6.32 can be observed in the Basin and Range region. Recently, uplifted mountains in an early stage of erosion were found in southern Oregon and northern Nevada. Death Valley, California, and southern Nevada fit into the more advanced middle stage, whereas the late stage, with its inselbergs, can be seen in southern Arizona.

6.9 CONCEPT CHECKS

1 What is meant by *interior drainage*?
2 Describe the features and characteristics associated with each stage in the evolution of a mountainous desert.
3 Where in the United States can each stage of desert landscape evolution be observed?

EYE ON EARTH

This satellite image shows a small portion of the Zagros Mountains in dry southern Iran. Streams in this region flow only occasionally. The green tones on the image identify productive agricultural areas.

QUESTION 1 *Identify the large feature labeled with a question mark.*

QUESTION 2 *Explain how the feature named in Question 1 formed.*

QUESTION 3 *What term is used to describe streams like the ones that occur in this region?*

QUESTION 4 *Speculate on the likely source of water for the agricultural areas in this image.*

6.10 | WIND EROSION

Describe the ways that wind transports sediment and the features created by wind erosion.

Moving air, like moving water, is turbulent and able to pick up loose debris and transport it to other locations. Just as in a stream, the velocity of wind increases with height above the surface. Also like a stream, wind transports fine particles in suspension, while heavier ones are carried as bed load. However, the transport of sediment by wind differs from that of running water in two significant ways. First, wind's lower density compared to water renders it less capable of picking up and transporting coarse materials. Second, because wind is not confined to channels, it can spread sediment over large areas, as well as high into the atmosphere.

Compared to running water and glaciers, wind is a relatively insignificant erosional agent. Recall that even in deserts, most erosion is performed by running water, not by the wind. It is also important to point out that wind erosion is more effective in arid lands than in humid areas because in humid places, moisture binds particles together, and vegetation anchors the soil. For wind to be an effective erosional force, dryness and scanty vegetation are essential. When such circumstances exist, wind may pick up, transport, and deposit great quantities of fine sediment. During the 1930s, parts of the Great Plains experienced vast dust storms. The plowing under of the natural vegetative cover for farming, followed by severe drought, exposed the land to wind erosion and led to the area being labeled the Dust Bowl. The *GEOgraphic* "The 1930s Dust Bowl: An Environmental Disaster" in Chapter 18 takes a closer look at this event.

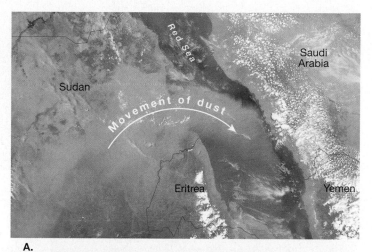

A.

Saltating sand grains.

B.

FIGURE 6.34 How Wind Transports Sediment A. This satellite image shows thick plumes of dust from the Sahara Desert blowing across the Red Sea on June 30, 2009. Such dust storms are common in arid North Africa. In fact, this region is the largest dust source in the world. Satellites are an excellent tool for studying the transport of dust on a global scale. They show that dust storms can cover huge areas and that dust can be transported great distances. (NASA) **B.** The bed load carried by wind consists of sand grains, many of which move by bouncing along the surface. Sand never travels far from the surface, even when winds are very strong. (Photo by Bernd Zoller/Getty Images)

Deflation, Blowouts, and Desert Pavement

One way that wind erodes is by **deflation** (*de* = out, *flat* = blow)—the lifting and removal of loose material. Because the competence (ability to transport different-sized particles) of moving air is low, it can suspend only fine sediment, such as clay and silt (**FIGURE 6.34A**). Larger grains of sand are rolled or skipped along the surface (a process called *saltation*) and comprise the bed load (**FIGURE 6.34B**). Particles larger than sand are usually not transported by wind. The effects of deflation are sometimes difficult to notice because the entire surface is being lowered at the same time, but the impact of this process can be significant.

Blowouts The most noticeable results of deflation in some places are shallow depressions called **blowouts** (**FIGURE 6.35**). In the Great Plains region, from Texas north to Montana, thousands of blowouts can be seen. They range from small dimples less than 1 meter (3 feet) deep and 3 meters (10 feet) wide to depressions that are over 45 meters (50 feet) deep and several kilometers across.

Desert Pavement In portions of many deserts, the surface is characterized by a layer of coarse pebbles and cobbles that are too large to be moved by the wind. This stony veneer, called **desert pavement**, may form as deflation lowers the surface by removing sand and silt from poorly sorted materials. As **FIGURE 6.36A** illustrates, the concentration of larger particles at the surface gradually increases as the finer particles are blown away. Eventually, a continuous cover of coarse particles remains.

FIGURE 6.35 Blowouts
Deflation is especially effective in creating these depressions when the land is dry and largely unprotected by anchoring vegetation. (Photo courtesy of USDA/Natural Resources Conservation Service)

The man is pointing to where the ground surface was when the grasses began to grow. Wind erosion lowered the land surface to the level of his feet.

Clumps of anchored soil

Unanchored soil

Sand dune

1.2 meters

FIGURE 6.36 Formation of Desert Pavement A. This model shows an area with poorly sorted surface deposits. Over time, deflation lowers the surface, and coarse particles become concentrated. **B.** In this model, the surface is initially covered with cobbles and pebbles. Over time, windblown dust accumulates at the surface and gradually sifts downward. Infiltrating rainwater aids the process.

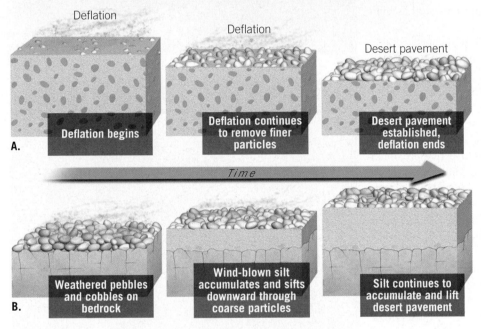

Deflation

Deflation

Desert pavement

A. Deflation begins

Deflation continues to remove finer particles

Desert pavement established, deflation ends

Time

Weathered pebbles and cobbles on bedrock

Wind-blown silt accumulates and sifts downward through coarse particles

Silt continues to accumulate and lift desert pavement

B.

Studies have shown that the process depicted in Figure 6.36A is not an adequate explanation for all environments in which desert pavement exists. As a result, an alternate explanation was formulated and is illustrated in **FIGURE 6.36B**. This hypothesis suggests that pavement develops on a surface that initially consists of coarse pebbles. Over time, protruding cobbles trap fine wind-blown grains that settle and sift downward through the spaces between the larger surface stones. The process is aided by infiltrating rainwater.

Once desert pavement becomes established, a process that might take hundreds of years, the surface is effectively protected from further deflation if left undisturbed. However, as the layer is only one or two stones thick, the passage of vehicles or animals can dislodge the pavement and expose the fine-grained material below. If this happens, the surface is no longer protected from deflation.

Wind Abrasion

Like glaciers, streams, and waves, wind erodes in part by *abrasion*. In dry regions as well as along some beaches, windblown sand cuts and polishes exposed rock surfaces. Abrasion is often credited for accomplishments beyond its actual capabilities. Such features as balanced rocks that stand high atop narrow pedestals and intricate detailing on tall pinnacles are *not* the results of wind abrasion. Sand seldom travels more than 1 meter above the surface, so the wind's sandblasting effect is obviously limited in vertical extent. However, in areas prone to such activity, telephone poles have actually been cut through near their bases. For this reason, collars may be fitted on the poles to protect them from being "sawed" down.

6.10 CONCEPT CHECKS

1 Why is wind erosion relatively more effective in arid regions than in humid areas?

2 What are blowouts? What term describes the process that creates these features?

3 Briefly describe two hypotheses used to explain the formation of desert pavement.

6.11 | WIND DEPOSITS Explain how loess deposits differ from deposits of sand. Discuss the movement of dunes and distinguish among different dune types.

Although wind is relatively unimportant in producing *erosional* landforms, significant *depositional* landforms are created by wind in some regions. Accumulations of windblown sediment are particularly conspicuous in the world's dry lands and along many sandy coasts. Wind deposits are of two distinctive types: (1) extensive blankets of silt, called *loess*, which once were carried in suspension, and (2) mounds and ridges of sand from the wind's bed load, which we call *dunes*.

This vertical bluff near the Mississippi River in southern Illinois is about 3 meters (10 feet) high.

In parts of China, loess has sufficient structural strength to permit excavation of cave-like dwellings.

FIGURE 6.37 Loess In some regions, the surface is mantled with deposits of windblown silt. (Illinois photo by James E. Patterson Collection; China photo by Ashley Cooper/Alamy)

Loess

In some parts of the world, the surface topography is mantled with deposits of windblown silt termed **loess**. Over thousands of years, dust storms deposited this material. When loess is breached by streams or road cuts, it tends to maintain vertical cliffs and lacks any visible layers, as you can see in **FIGURE 6.37**.

The distribution of loess worldwide indicates that there are two primary sources for this sediment: deserts and glacial deposits of stratified drift. The thickest and most extensive deposits of loess on Earth occur in western and northern China. They were blown there from the extensive desert basins of central Asia. Accumulations of 30 meters (100 feet) are not uncommon, and thicknesses of more than 100 meters (325 feet) have been measured. It is this fine, buff-colored sediment that gives China's Yellow River (Hwang Ho) its name.

In the United States, deposits of loess are significant in many areas, including South Dakota, Nebraska, Iowa, Missouri, and Illinois, as well as portions of the Columbia Plateau in the Pacific Northwest. Unlike the deposits in China, the loess in the United States, as well as in Europe, is an indirect product of glaciation. Its source is deposits of stratified drift. During the retreat of the ice sheets, many river valleys were choked with sediment deposited by meltwater. Strong westerly winds sweeping across the barren floodplains picked up the finer sediment and dropped it as a blanket on areas adjacent to the valleys.

Sand Dunes

Like running water, wind deposits its load of sediment when its velocity falls and the energy available for transport diminishes. Thus, sand begins to accumulate wherever an obstruction across the path of the wind slows its movement. Unlike deposits of loess, which form blanketlike layers over broad areas, winds commonly deposit sand in mounds or ridges called **dunes**.

As moving air encounters an object, such as a clump of vegetation or a rock, the wind sweeps around and over it, leaving a shadow of more slowly moving air behind the obstacle as well as a smaller zone of quieter air just in front of the obstacle. Some of the saltating sand grains moving with the wind come to rest in these wind shadows. As the accumulation of sand continues, it forms an increasingly efficient wind barrier to trap even more sand. If there is a sufficient supply of sand and the wind blows steadily long enough, the mound of sand grows into a dune.

Many dunes have an asymmetrical profile, with the leeward (sheltered) slope being steep and the windward slope being more gently inclined. The dunes in **FIGURE 6.38** are a good example. Sand moves up

Wind

Strong winds move sand up the relatively gentle windward slope.

As sand accumulates at the dune crest, the slope steepens and some of the sand slides down the steep *slip face*.

 Mobile Field Trip 6.38 White Sands National Monument

The dunes at this landmark in southeastern New Mexico are composed of gypsum. The dunes slowly migrate with the wind. (Photos by Michael Collier)

Smart Figure
6.39 Cross Bedding As sand is deposited on the slip face, layers form that are inclined in the direction the wind is blowing. With time, complex patterns develop in response to changes in wind direction. (Photo by Dennis Tasa)

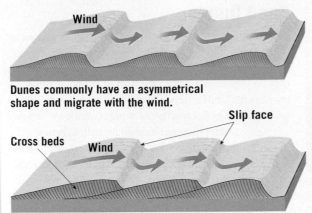

Dunes commonly have an asymmetrical shape and migrate with the wind.

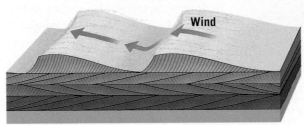

Sand grains deposited on the slip face at the angle of repose create the cross-bedding of dunes.

When dunes are buried and become part of the sedimentary rock record, the cross bedding is preserved.

Cross beds are an obvious characteristic of the Navajo Sandstone in Zion National Park, Utah.

the gentler slope on the windward side by saltation. Just beyond the crest of the dune, where wind velocity is reduced, the sand accumulates. As more sand collects, the slope steepens, and eventually some of it slides or slumps under the pull of gravity. You can see this in the bottom image of **FIGURE 6.39**. In this way, the leeward slope of the dune, called the **slip face**, maintains an angle of about 34 degrees. Continued sand accumulation, coupled with periodic slides down the slip face, results in the slow migration of the dune in the direction of air movement.

As sand is deposited on the slip face, it forms layers inclined in the direction the wind is blowing. These sloping layers are referred to as **cross bedding** (see Figure 6.39). When the dunes are eventually buried under other layers of sediment and become part of the sedimentary rock record, their asymmetrical shape is destroyed, but the cross beds remain as a testimony to their origin. Nowhere is cross bedding more prominent than in the sandstone walls of Zion Canyon in Utah, shown on the right side of **FIGURE 6.40**.

Types of Sand Dunes

Dunes are not just random heaps of windblown sediment. Rather, they are accumulations that usually assume patterns that are surprisingly consistent (see Figure 6.40). A broad assortment of dune forms exist, generally simplified to a few major types for discussion. Of course, gradations exist among different forms as well as irregularly shaped dunes that do not fit easily into any category. Several factors influence the form and size that dunes ultimately assume. These include wind direction and velocity, availability of sand, and the amount of vegetation present. Six basic dune types are shown in Figure 6.40, with arrows indicating wind directions.

Barchan Dunes Solitary sand dunes shaped like crescents and with their tips pointing downwind are called **barchan dunes** (see Figure 6.40A). These dunes form where supplies of sand are limited and the surface is relatively flat, hard, and lacking vegetation. They migrate slowly with the wind at a rate of up to 15 meters (50 feet) annually. Their size is usually modest, with the largest barchans dunes reaching heights of about 30 meters (100 feet) and the maximum spread between their horns approaching 300 meters (nearly 1000 feet). When the wind direction is nearly constant, the crescent form of these dunes is nearly symmetrical. However, when the wind direction is not perfectly fixed, one tip becomes larger than the other.

Transverse Dunes In regions where the prevailing winds are steady, sand is plentiful, and vegetation is sparse or absent, the dunes form a series of long ridges that are separated by troughs and oriented at right angles to the prevailing wind. Because of this orientation, they are termed **transverse dunes** (see Figure 6.40B). Typically, many coastal dunes are of this type. In addition, transverse dunes are common in many arid regions where the extensive surface of wavy sand is sometimes called a *sand sea*. In some parts of the Sahara and Arabian Deserts, transverse dunes reach heights of 200 meters (650 feet), are 1 to 3 kilometers (0.5 to 2 miles) across, and can extend for distances of 100 kilometers (60 miles) or more.

Barchanoid Dunes There is a relatively common dune form that is intermediate between isolated barchans and extensive waves of transverse dunes. Such dunes, called **barchanoid dunes**, form scalloped rows of sand oriented at right angles to the wind (see Figure 6.40C). The rows

A. Barchan

B. Transverse

C. Barchanoid

D. Longitudinal

E. Parabolic

F. Star

Wind

SmartFigure 6.40 Types of Sand Dunes Factors that influence the form and size of dunes include wind direction and velocity, the availability of sand, and the amount of vegetation.

resemble a series of barchans that have been positioned side by side. Visitors exploring the gypsum dunes at White Sands National Monument in New Mexico will recognize this form (see Figure 6.38).

Longitudinal Dunes Longitudinal dunes are long ridges of sand that form more or less parallel to the prevailing wind and where sand supplies are moderate (see Figure 6.40D). Apparently the prevailing wind direction varies somewhat but remains in the same quadrant of the compass. Although the smaller types are only 3 or 4 meters high and several tens of meters long, in some large deserts, longitudinal dunes can reach great size. For example, in portions of North Africa, Arabia, and central Australia, these dunes may approach a height of 100 meters and extend for distances of more than 100 kilometers (62 miles).

EYE ON EARTH

This satellite image shows a large plume of wind-blown sediment covering large portions of Iran, Afghanistan, and Pakistan in March 2012. The airborne material is thick enough to completely hide the area beneath it. On either side of the plume, skies are mostly clear.

QUESTION 1 What term is applied to the erosional process that was responsible for producing this plume?

QUESTION 2 Is the wind-transported material in the image more likely bed load or suspended load?

QUESTION 3 People sometimes refer to events like the one pictured here as "sandstorms." Is that an appropriate description? Why or why not?

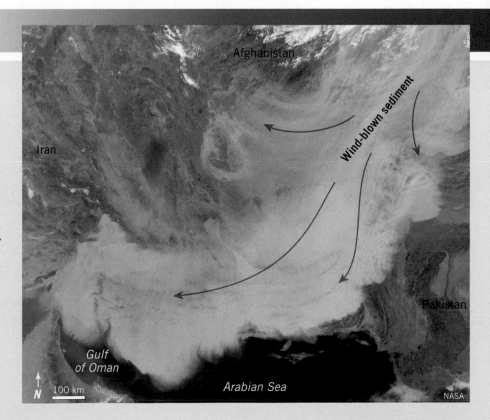

Parabolic Dunes Unlike the other dunes that have been described thus far, **parabolic dunes** form where vegetation partially covers the sand. The shape of these dunes resembles the shape of barchans except that their tips point into the wind rather than downwind (see Figure 6.40E). Parabolic dunes often form along coasts where there are strong onshore winds and abundant sand. If the sand's sparse vegetative cover is disturbed at some spot, deflation creates a blowout. Sand is then transported out of the depression and deposited as a curved rim that grows higher as deflation enlarges the blowout.

Star Dunes Confined largely to parts of the Sahara and Arabian Deserts, **star dunes** are isolated hills of sand that exhibit a complex form (see Figure 6.40F). Their name is derived from the fact that the bases of these dunes resemble multipointed stars. Usually three or four sharp-crested ridges diverge from a central high point that in some cases may approach a height of 90 meters (300 feet). As their form suggests, star dunes develop where wind directions are variable.

6.11 CONCEPT CHECKS

1 Contrast loess and sand dunes in terms of composition and how they form.
2 How are some loess deposits related to glaciers?
3 Describe how sand dunes migrate.
4 What is cross bedding?
5 List and briefly distinguish among basic dune types.

6 CONCEPTS IN REVIEW | Glaciers, Deserts, and Wind

6.1 GLACIERS AND THE EARTH SYSTEM

Explain the role of glaciers in the hydrologic and rock cycles and describe the different types of glaciers, their characteristics, and their present-day distribution

KEY TERMS: glacier, valley (alpine) glacier, ice sheet, sea ice, ice shelf, ice cap, piedmont glacier, outlet glacier

- A glacier is a thick mass of ice that originates on land from the compaction and recrystallization of snow, and it shows evidence of past or present flow. Glaciers are part of both the hydrologic cycle and the rock cycle. They store and release freshwater, and they transport and deposit large quantities of sediment.
- Valley glaciers flow down mountain valleys, whereas ice sheets are very large masses, such as those that cover Greenland and Antarctica. During the Last Glacial Maximum, around 18,000 years ago, Earth was in an Ice Age that covered large areas with glacial ice.
- When valley glaciers exit confining mountains, they spread out into broad lobes called piedmont glaciers. Similarly, ice shelves form when glaciers flow into the ocean, producing a layer of floating ice.
- Ice caps are like small ice sheets. Both ice sheets and ice caps may be drained by outlet glaciers, which often resemble valley glaciers.

•North pole

Greenland

NASA

Q This satellite image shows ice in the high latitudes of the Northern Hemisphere. What term is applied to the ice at the North Pole? What term best describes Greenland's ice? Are both considered glaciers? Explain.

6.2 HOW GLACIERS MOVE

Describe how glaciers move, the rates at which they move, and the significance of the glacial budget.

KEY TERMS: zone of fracture, crevasse, zone of accumulation, zone of wastage, calving, iceberg, glacial budget

- Glaciers move in part by flowing under pressure. On the surface of a glacier, ice is brittle. Below about 50 meters (165 feet), pressure is great, and ice behaves like a plastic material and flows. Another type of movement consists of the bottom of the glacier sliding along the ground.
- Fast glaciers may move 800 meters (2600 feet) per year, while slow glaciers may move only 2 meters (6.5 feet) per year. Some glaciers experience periodic surges of rapid movement.
- When a glacier has a positive budget, the terminus will advance. This occurs when the glacier gains more snow in its upper zone of accumulation that it loses at its downstream zone of wastage. If wastage exceeds the input of new ice, the glacier's terminus will retreat.

Q If a glacier gains 0.5 cubic kilometer of ice annually and loses 0.75 cubic kilometer through wastage, will its terminus advance or retreat? Describe the movement of ice within the glacier. What figure in this section illustrates your answer?

6.3 GLACIAL EROSION

Discuss the processes of glacial erosion and the major features created by these processes.

KEY TERMS: plucking, abrasion, rock flour, glacial striations, glacial trough, hanging valley, cirque, arête, horn, fiord

- Glaciers acquire sediment through plucking from the bedrock beneath the glacier, by abrasion of the bedrock using sediment already in the ice, and when mass-wasting processes drop debris on top of the glacier. Grinding of the bedrock produces grooves and scratches called glacial striations.
- Erosional features produced by valley glaciers include glacial troughs, hanging valleys, cirques, arêtes, horns, and fiords.

Q **Examine the illustration of a mountainous landscape after glaciation. Identify the landforms that resulted from glacial erosion.**

6.4 GLACIAL DEPOSITS

Distinguish between the two basic types of glacial deposits and briefly describe the features associated with each type.

KEY TERMS: glacial drift, till, stratified drift, glacial erratic, lateral moraine, medial moraine, end moraine, ground moraine, outwash plain, valley train, kettle, drumlin, esker, kame

- Any sediment of glacial origin is called drift. The two distinct types of glacial drift are till, which is unsorted material deposited directly by the ice, and stratified drift, which is sediment sorted and deposited by meltwater from a glacier.
- The most widespread features created by glacial deposition are layers or ridges of till, called moraines. Associated with valley glaciers are lateral moraines, formed along the sides of the valley, and medial moraines, formed between two valley glaciers that have merged. End moraines, which mark the former position of the front of a glacier, and ground moraines, undulating layers of till deposited as the ice front retreats, are common to both valley glaciers and ice sheets.

Q **Examine the illustration of depositional features left behind by a retreating ice sheet. Identify the features and indicate which landforms are composed of till and which are composed of stratified drift.**

6.5 OTHER EFFECTS OF ICE AGE GLACIERS

Describe and explain several important effects of Ice Age glaciers other than erosional and depositional landforms.

KEY TERMS: proglacial lake, pluvial lake

- In addition to erosional and depositional features, other effects of Ice Age glaciers include the forced migration of organisms and adjustments of the crust by rebounding upward after removal of the immense load of ice.
- Advance and retreat of ice sheets caused significant changes to the paths followed by rivers. Proglacial lakes formed when glaciers acted as dams to create lakes by trapping glacial meltwater or blocking rivers. In response to the cooler and wetter glacial climate, pluvial lakes formed in areas such as present-day Nevada.
- Ice sheets are nourished by water that ultimately comes from the ocean, so when ice sheets grow, sea level falls, and when they melt, sea level rises.

6.6 EXTENT OF ICE AGE GLACIATION

Discuss the extent of glaciation and climate variability during the Quaternary Ice Age.

KEY TERM: Quaternary period

- The Ice Age that began between 2 and 3 million years ago was a complex period characterized by numerous advances and withdrawals of glacial ice. Most of the major glacial episodes occurred during a span on the geologic time scale called the Quaternary period. Evidence for the occurrence of several glacial advances during the Ice Age is the existence of multiple layers of drift on land and an uninterrupted record of climate cycles preserved in seafloor sediments.

6.7 CAUSES OF ICE AGES

Summarize some of the current ideas about the causes of ice ages.

- While rare, glacial episodes have occurred in Earth history prior to the recent glaciations we call the Ice Age. Lithified till, called tillite, is a major line of evidence for these ancient ice ages. There are several reasons that glacial ice might accumulate globally, including the position of the continents, which is driven by plate tectonics. Antarctica's position over the South Pole is doubtless a key reason for its massive ice sheets, for instance.
- The Quaternary period is marked by not only glacial advances but also intervening episodes of glacial retreat. One way to explain these oscillations is through variations in Earth's orbit, which lead to seasonal variations in the distribution of solar radiation. The orbit's shape varies (eccentricity), the tilt of the planet's rotational axis varies (obliquity), and the axis slowly "wobbles" over time (precession). These three effects, each of which occur on different time scales, collectively do a good job of accounting for alternating colder and warmer periods during the Quaternary.
- Additional factors that may be important for initiating or ending glaciations include rising or falling levels of greenhouse gases, changes in the reflectivity of Earth's surface, and variations in the ocean currents that redistribute heat energy from warmer to colder regions.

Q About 250 million years ago, parts of India, Africa, and Australia were covered by ice sheets, while Greenland, Siberia, and Canada were ice free. Explain why this was the case.

6.8 DESERTS

Describe the general distribution and extent of Earth's dry lands and the role that water plays in modifying desert landscapes.

KEY TERMS: dry climate, desert, steppe, ephemeral stream

- Dry climates cover about 30 percent of Earth's land area. These regions have yearly precipitation totals that are less than the potential loss of water through evaporation. Deserts are drier than steppes, but both climate types are considered water deficient.
- Dry regions in the lower latitudes coincide with zones of subsiding air and high air pressure known as subtropical highs. Middle-latitude deserts exist because of their positions in the deep interiors of large continents far removed from oceans. Mountains also act to shield these regions from humid marine air masses.
- Practically all desert streams are dry most of the time and are said to be ephemeral. Nevertheless, running water is responsible for most of the erosional work in a desert. Although wind erosion is more significant in dry areas than elsewhere, the main role of wind in a desert is transportation and deposition of sediment.

6.9 BASIN AND RANGE: THE EVOLUTION OF A MOUNTAINOUS DESERT LANDSCAPE

Discuss the stages of landscape evolution in the Basin and Range region of the western United States.

KEY TERMS: interior drainage, alluvial fan, bajada, playa lake

- The Basin and Range region of the western United States is characterized by interior drainage with streams eroding uplifted mountain blocks and depositing sediment in interior basins. Alluvial fans, bajadas, playas, playa lakes, salt flats, and insel-bergs are features often associated with these landscapes.

Q Identify the lettered features in this photo. How did they form?

Michael Collier

6.10 WIND EROSION

Describe the ways that wind transports sediment and the features created by wind erosion.

KEY TERMS: deflation, blowout, desert pavement

- For wind erosion to be effective, dryness and scant vegetation are essential. Deflation, the lifting and removal of loose material, often produces shallow depressions called blowouts.
- Abrasion, the sandblasting effect of wind, is often given too much credit for producing desert features. However, abrasion does cut and polish rock near the surface.

Q What term is applied to the layer of coarse pebbles covering this desert surface? How might it have formed?

Lens cap for scale.

Bobbé Christopherson

6.11 WIND DEPOSITS

Explain how loess deposits differ from deposits of sand. Discuss the movement of dunes and distinguish among different dune types.

KEY TERMS: loess, dune, slip face, cross bedding, barchan dunes, transverse dunes, barchanoid dunes, longitudinal dunes, parabolic dunes, star dunes

- Wind deposits are of two distinct types: extensive blankets of silt, called loess, carried by wind in suspension, and mounds and ridges of sand, called dunes, which are formed from sediment that is carried as part of the wind's bed load.
- Loess is windblown silt, deposited over large areas sometimes in thick blankets. Most loess is derived from either deserts or areas that have recently been glaciated. Winds blow across stratified drift and pick up silt-size grains, carrying them in suspension to the site of deposition.
- Dunes accumulate due to the difference in wind energy on the upwind and downwind sides of some obstacle. Wind moves sand up the more gently sloping upwind side, across the crest of the dune, where it settles out in the calmer air on the downwind side of the dune, the steeply sloped slip face. When the sand on the slip face is piled up past the angle of repose, it will spontaneously collapse in small "avalanches" of sand. Over time, this will cause a dune to slowly move in the direction of the prevailing wind, with the slip face "leading the way." Inside the dune, the buried slip faces may be preserved as cross beds.
- There are six major kinds of dunes. Their shapes result from the pattern of prevailing winds, the amount of available sand, and the presence of vegetation.

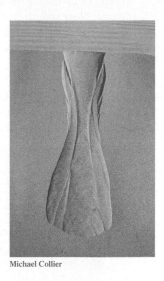

Michael Collier

Q **This close-up photo shows a small portion of one side of a barchan dune. What term is applied to this side of the dune? Is the prevailing wind direction "coming out" of the photo or "going into" the photo? Explain. Why did some of the sand break away and slide?**

GIVE IT SOME **THOUGHT**

1. The accompanying diagram shows the results of a classic experiment used to determine how glacial ice moves in a mountain valley. The experiment was carried out over an 8-year span. Refer to this diagram and answer the following:

   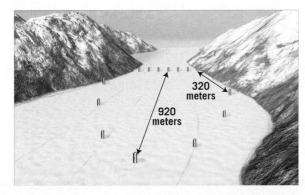

   320 meters

   920 meters

   a. What was the average yearly rate of ice advance in the center of the glacier?
   b. About how fast was the center of the glacier advancing *per day*?
   c. What was the average rate at which ice advanced along the sides of the glacier?
   d. Why was the rate at the center different than the rate along the sides?

2. Studies have shown that during the Ice Age, the margins of some ice sheets advanced southward from the Hudson Bay region at rates ranging from about 50 to 320 meters per year.

   a. Determine the maximum amount of time required for an ice sheet to move from the southern end of Hudson Bay to the south shore of present-day Lake Erie, a distance of 1600 kilometers.
   b. Calculate the minimum number of years required for an ice sheet to move this distance.

3. The accompanying image shows the top of a valley glacier in which the ice is fractured.

   a. What term is applied to fractures such as these?
   b. In what vertical zone do these breaks occur?
   c. Do the fractures likely extend to the base of the glacier? Explain.

   Glow Images

4. If the budget of a valley glacier were balanced for an extended span of time, what feature would you expect to find at the terminus of the glacier? Now assume that the glacier's budget changes so that wastage exceeds accumulation. How would the terminus of the glacier change? Describe the deposit you would expect to form under these conditions.

5. Assume that you and a nongeologist friend are visiting Alaska's Hubbard Glacier, shown here. After studying the glacier for quite a long time, your friend asks, "Do these things really move?" How would you convince your companion that this glacier does indeed move, using evidence that is clearly visible in this image?

Michael Collier

6. If Earth were to experience another Ice Age, one hemisphere would have substantially more expansive ice sheets than the other. Would it be the Northern Hemisphere or the Southern Hemisphere? What is the reason for the large disparity?

7. Is either of the following statements true? Are they both true? Explain.
    a. Wind does its most effective erosional work in dry places.
    b. Wind is the most important agent of erosion in deserts.

8. This is an aerial view of the Preston Mesa dunes in northern Arizona.
    a. Which one of the basic dune types is shown here?
    b. Sketch a simple profile (side view) of one of these dunes. Add an arrow to show the prevailing wind direction and label the dune's slip face.
    c. These dunes gradually migrate across the surface. Describe this process.

Michael Collier

9. Bryce Canyon National Park, shown in the accompanying photo, is in dry southern Utah. It is carved into the eastern edge of the Paunsaugunt Plateau. Erosion has sculpted the colorful limestone into bizarre shapes, including spires called "hoodoos." As you and a companion (who has not studied geology) are viewing the scenery in Bryce Canyon, your friend says, "It's amazing how wind has created this incredible scenery!" Now that you have studied arid landscapes, how would you respond to your companion's statement?

ozoptimest/Shutterstock

10. Compare the sediment deposited by a stream, the wind, and a glacier. Which deposit should have the most uniform grain size? Which one would exhibit the poorest sorting? Explain your choices.

EXAMINING THE **EARTH SYSTEM**

1. Assume that you are teaching an introductory Earth science class and that you have just assigned Chapter 6 of this text. A student in the class asks why glaciers, deserts, and wind are treated in the same chapter. Formulate a response that connects these topics.

2. This image is a close-up of Surprise Glacier in Alaska's Prince William Sound. Prepare two brief descriptions that relate to Surprise Glacier (and to all other glaciers as well). One should describe how Surprise Glacier fits into the hydrologic cycle, and the second description should explain how Surprise Glacier fits into the rock cycle. Is Surprise Glacier a part of the hydrosphere? Is it a part of the geosphere? Some scientists think that ice should be a separate sphere of the Earth system, called the *cryosphere*. Does such an idea have merit? Explain.

Michael Collier

3. These two images show an ephemeral stream in Niger, a country in North Africa's Sahara Desert. What is the local term for an ephemeral stream in this part of the world? Prepare a brief story that would explain the contrast between the two images. Try to include all four of Earth's spheres in your story.

NASA Earth Observing System

4. Wind erosion occurs at the interface of the atmosphere, geosphere, and biosphere and is influenced by the hydrosphere and human activity. With this in mind, describe how human activity contributed to the Dust Bowl, the period of intense wind erosion in the Great Plains

in the 1930s. You might find it helpful to research Dust Bowl on the Internet using a search engine such as Google (**www.google.com**) or Yahoo (**www.yahoo.com**). You may also find the Wind Erosion Research Unit site to be informative: **www.weru.ksu.edu**.

USDA/NRCS/Natural Resources Conversation Service

MasteringGeology™

Looking for additional review and test prep materials? Visit the Self Study area in **www.masteringgeology.com** to find practice quizzes, study tools, and multimedia that will aid in your understanding of this chapter's content. In **MasteringGeology**™ you will find:

- GEODe: Earth Science: An interactive visual walkthrough of key concepts
- Geoscience Animation Library: More than 100 animations illuminating many difficult-to-understand Earth science concepts

- In The News RSS Feeds: Current Earth science events and news articles are pulled into the site with assessment
- Pearson eText
- Optional Self Study Quizzes
- Web Links
- Glossary
- Flashcards

UNIT THREE

FORCES WITHIN

7 Plate Tectonics: A Scientific Revolution Unfolds

Climber ascending Chang Zheng Peak near Mount Everest.
(Photo by Stock Connection/SuperStock)

Plate tectonics is the first theory to provide a comprehensive view of the processes that produced Earth's major surface features, including the continents and ocean basins. Within the framework of this theory, geologists have found explanations for the basic causes and distribution of earthquakes, volcanoes, and mountain belts. Further, we are now better able to explain the distribution of plants and animals in the geologic past, as well as the distribution of economically significant mineral deposits.

7.1 | FROM CONTINENTAL DRIFT TO PLATE TECTONICS

Discuss the view that most geologists held prior to the 1960s regarding the geographic positions of the ocean basins and continents.

Prior to the late 1960s, most geologists held the view that the ocean basins and continents had fixed geographic positions and were of great antiquity. Researchers came to realize that Earth's continents are not static; instead, they gradually migrate across the globe. Because of these movements, blocks of continental material collide, deforming the intervening crust, thereby creating Earth's great mountain chains (**FIGURE 7.1**). Furthermore, landmasses occasionally split apart. As continental blocks separate, a new ocean basin emerges between them. Meanwhile, other portions of the seafloor plunge into the mantle. In short, a dramatically different model of Earth's tectonic processes emerged. Tectonic processes are processes that deform Earth's crust to create major structural features, such as mountains, continents, and ocean basins.

This profound reversal in scientific thought has been appropriately described as a *scientific revolution*. The revolution began early in the twentieth century as a relatively straightforward proposal called *continental drift*. For more than 50 years, the scientific establishment categorically rejected the idea that continents are capable of movement. Continental drift was particularly distasteful to North American geologists, perhaps because much of the supporting evidence had been gathered from the continents of Africa, South America, and Australia, with which most North American geologists were unfamiliar.

FIGURE 7.1 Rock Pinnacle near Mount Blanc The Alps were created by the collision of the African and Eurasian plates. (Photo by Bildagentur Walhaeus/AGE Fotostock)

Following World War II, modern instruments replaced rock hammers as the tools of choice for many researchers. Armed with more advanced tools, geologists and a new breed of researchers, including *geophysicists* and *geochemists*, made several surprising discoveries that began to rekindle interest in the drift hypothesis. By 1968 these developments had led to the unfolding of a far more encompassing explanation known as the *theory of plate tectonics*.

In this chapter, we will examine the events that led to this dramatic reversal of scientific opinion. We will also briefly trace the development of the *continental drift hypothesis*, examine why it was initially rejected, and consider the evidence that finally led to the acceptance of its direct descendant—the theory of plate tectonics.

7.1 CONCEPT CHECKS

1 Briefly describe the view held by most geologists regarding the ocean basins and continents prior to the 1960s.

2 What group of geologists were the least receptive to the continental drift hypothesis? Explain.

7.2 | CONTINENTAL DRIFT: AN IDEA BEFORE ITS TIME

List and explain the evidence Wegener presented to support his continental drift hypothesis.

The idea that continents, particularly South America and Africa, fit together like pieces of a jigsaw puzzle came about during the 1600s, as better world maps became available. However, little significance was given to this notion until 1915, when Alfred Wegener (1880–1930), a German meteorologist and geophysicist, wrote *The Origin of Continents and Oceans*. This book set forth the basic outline of Wegener's hypothesis, called **continental drift**, which dared to challenge the long-held assumption that the continents and ocean basins had fixed geographic positions.

Wegener suggested that a single **supercontinent** consisting of all Earth's landmasses once existed.* He named this giant landmass **Pangaea** (pronounced "Pan-jcc-ah," meaning "all lands") (**FIGURE 7.2**). Wegener further hypothesized that about 200 million years ago, during the early part of the Mesozoic era, this supercontinent began to fragment into smaller landmasses. These continental blocks then "drifted" to their present positions over a span of millions of years.

Wegener and others who advocated the continental drift hypothesis collected substantial evidence to support their point of view. The fit of South America and Africa and the geographic distribution of fossils and ancient climates all seemed to buttress the idea that these now separate landmasses were once joined. Let us examine some of this evidence.

Evidence: The Continental Jigsaw Puzzle

Like a few others before him, Wegener suspected that the continents might once have been joined when he noticed the remarkable similarity between the coastlines on opposite sides of the Atlantic Ocean. However, other Earth scientists immediately challenged Wegener's use of present-day shorelines to fit these continents together. These opponents correctly argued that shorelines are continually modified by wave erosion and depositional processes. Even if continental displacement had taken place, a good fit today would be unlikely. Because Wegener's original jigsaw fit of the continents was crude, it is assumed that he was aware of this problem (see Figure 7.2).

Scientists later determined that a much better approximation of the outer boundary of a continent is the seaward edge of its continental shelf, which lies submerged a few hundred meters below sea level. In the early 1960s, Sir Edward Bullard and two associates constructed a map that pieced together the edges of the continental shelves of South America and Africa at a depth of about 900 meters (3000 feet) (**FIGURE 7.3**). The remarkable fit that was obtained was more precise than even these researchers had expected.

Modern reconstruction of Pangaea

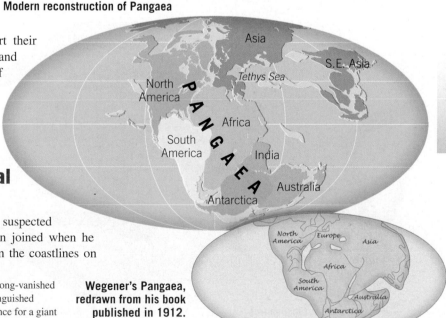

SmartFigure 7.2 Reconstructions of Pangaea This is as it is thought to have appeared 200 million years ago.

Wegener's Pangaea, redrawn from his book published in 1912.

---

*Wegener was not the first person to conceive of a long-vanished supercontinent. Edward Suess (1831–1914), a distinguished nineteenth-century geologist, pieced together evidence for a giant landmass consisting of the continents of South America, Africa, India, and Australia.

FIGURE 7.3 Two of the Puzzle Pieces Best fit of South America and Africa along the continental slope at a depth of 500 fathoms (about 900 meters [3000 feet]).

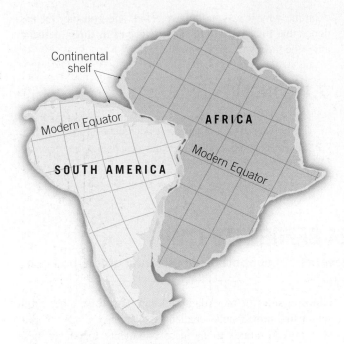

those of Africa and Australia, during the Mesozoic era, organisms on widely separated continents should have been distinctly different.

Mesosaurus To add credibility to his argument, Wegener documented cases of several fossil organisms found on different landmasses, despite the unlikely possibility that their living forms could have crossed the vast ocean presently separating them (**FIGURE 7.4**). A classic example is *Mesosaurus*, a small aquatic freshwater reptile whose fossil remains are limited to black shales of the Permian period (about 260 million years ago) in eastern South America and southwestern Africa. If *Mesosaurus* had been able to make the long journey across the South Atlantic, its remains would likely be more widely distributed. As this is not the case, Wegener asserted that South America and Africa must have been joined during that period of Earth history.

How did opponents of continental drift explain the existence of identical fossil organisms in places separated by thousands of kilometers of open ocean? Rafting, transoceanic land bridges (isthmian links), and island stepping stones were the most widely invoked explanations for these migrations (**FIGURE 7.5**). We know, for example, that during the Ice Age that ended about 8000 years ago, the lowering of sea level allowed mammals (including humans) to cross the narrow Bering Strait that separates Russia and Alaska. Was it possible that land bridges once connected Africa and South America but later subsided below sea level? Modern maps of the seafloor substantiate Wegener's contention that if land bridges of this magnitude once existed, their remnants would still lie below sea level.

Evidence: Fossils Matching Across the Seas

Although the seed for Wegener's hypothesis came from the remarkable similarities of the continental margins on opposite sides of the Atlantic, it was when he learned that identical fossil organisms had been discovered in rocks from both South America and Africa that his pursuit of continental drift became more focused. Through a review of the literature, Wegener learned that most paleontologists (scientists who study the fossilized remains of ancient organisms) were in agreement that some type of land connection was needed to explain the existence of similar Mesozoic age life-forms on widely separated landmasses. Just as modern life-forms native to North America are quite different from

Glossopteris Wegener also cited the distribution of the fossil "seed fern" *Glossopteris* as evidence for the existence of Pangaea (see Figure 7.4). This plant, identified by its tongue-shaped leaves and seeds that were too large to be carried by the wind, was known to be widely dispersed among Africa, Australia, India, and South America. Later, fossil remains of *Glossopteris* were also discovered in Antarctica.* Wegener also learned that these seed ferns and associated flora grew only in cool climates—similar to central Alaska. Therefore, he concluded that when these landmasses were joined, they were located much closer to the South Pole.

FIGURE 7.4 Fossil Evidence Supporting Continental Drift Fossils of identical organisms have been discovered in rocks of similar age in Australia, Africa, South America, Antarctica, and India—continents that are currently widely separated by ocean barriers. Wegener accounted for these occurrences by placing these continents in their pre-drift locations.

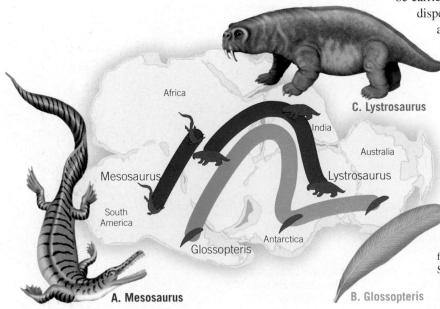

*In 1912 Captain Robert Scott and two companions froze to death lying beside 35 pounds (16 kilograms) of rock on their return from a failed attempt to be the first to reach the South Pole. These samples, collected on the moraines of Beardmore Glacier, contained fossil remains of *Glossopteris*.

FIGURE 7.5 **How Do Land Animals Cross Vast Oceans?** These sketches illustrate various explanations for the occurrence of similar species on landmasses that are presently separated by vast oceans. (Reprinted with permission of John Holden)

in Brazil that closely resembled similarly aged rocks in Africa.

Similar evidence can be found in mountain belts that terminate at one coastline and reappear on landmasses across the ocean. For instance, the mountain belt that includes the Appalachians trends northeastward through the eastern United States and disappears off the coast of Newfoundland (**FIGURE 7.6A**). Mountains of comparable age and structure are found in the British Isles, western Africa, and Scandinavia. When these landmasses are positioned as they were about 200 million years ago, as shown in **FIGURE 7.6B**, the mountain chains form a nearly continuous belt.

Evidence: Rock Types and Geologic Features

Anyone who has worked a jigsaw puzzle knows that its successful completion requires that you fit the pieces together while maintaining the continuity of the picture. The "picture" that must match in the "continental drift puzzle" is one of rock types and geologic features such as mountain belts. If the continents were together, the rocks found in a particular region on one continent should closely match in age and type those found in adjacent positions on the once adjoining continent. Wegener found evidence of 2.2-billion-year-old igneous rocks

Wegener described how the similarities in geologic features on both sides of the Atlantic linked these landmasses when he said, "It is just as if we were to refit the torn pieces of a newspaper by matching their edges and then check whether the lines of print run smoothly across. If they do, there is nothing left but to conclude that the pieces were in fact joined in this way."*

*Alfred Wegener, *The Origin of Continents and Oceans*, translated from the 4th revised German ed. of 1929 by J. Birman (London: Methuen, 1966).

FIGURE 7.6 **Matching Mountain Ranges Across the North Atlantic**

FIGURE 7.7 Paleoclimatic Evidence for Continental Drift **A.** About 300 million years ago, ice sheets covered extensive areas of the Southern Hemisphere and India. Arrows show the direction of ice movement that can be inferred from the pattern of glacial striations and grooves found in the bedrock. **B.** The continents restored to their pre-drift positions accounts for tropical coal swamps that existed in areas presently located in temperate climates.

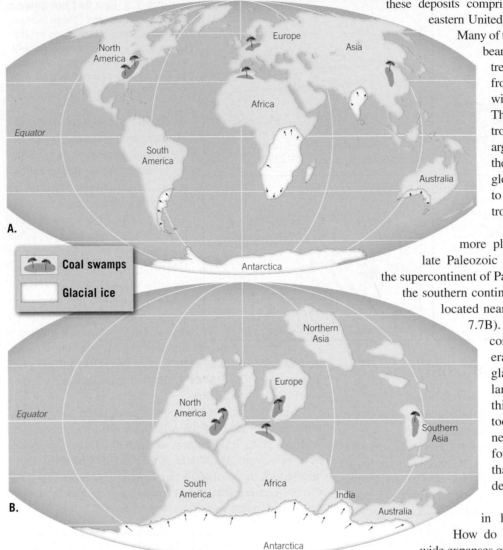

Evidence: Ancient Climates

Because Alfred Wegener was a student of world climates, he suspected that paleoclimatic (*paleo* = ancient, *climatic* = climate) data might also support the idea of mobile continents. His assertion was bolstered when he learned that evidence for a glacial period that dated to the late Paleozoic had been discovered in southern Africa, South America, Australia, and India. This meant that about 300 million years ago, vast ice sheets covered extensive portions of the Southern Hemisphere as well as India (**FIGURE 7.7A**). Much of the land area that contains evidence of this period of Paleozoic glaciation presently lies within 30° of the equator, in subtropical or tropical climates.

How could extensive ice sheets form near the equator? One proposal suggested that our planet experienced a period of extreme global cooling. Wegener rejected this explanation because during the same span of geologic time, large tropical swamps existed in several locations in the Northern Hemisphere. The lush vegetation in these swamps was eventually buried and converted to coal (**FIGURE 7.7B**). Today

these deposits comprise major coal fields in the eastern United States and Northern Europe. Many of the fossils found in these coal-bearing rocks were produced by tree ferns that possessed large fronds—a feature consistent with warm, moist climates.** The existence of these large tropical swamps, Wegener argued, was inconsistent with the proposals that extreme global cooling caused glaciers to form in what are currently tropical areas.

Wegener suggested that a more plausible explanation for the late Paleozoic glaciation was provided by the supercontinent of Pangaea. In this configuration, the southern continents are joined together and located near the South Pole (see Figure 7.7B). This would account for the conditions necessary to generate extensive expanses of glacial ice over much of these landmasses. At the same time, this geography would place today's northern continents nearer the equator and account for the tropical swamps that generated the vast coal deposits.

How does a glacier develop in hot, arid central Australia? How do land animals migrate across wide expanses of the ocean? As compelling as this evidence may have been, 50 years passed before most of the scientific community accepted the concept of continental drift and the logical conclusions to which it led.

**It is important to note that coal can form in a variety of climates, provided that large quantities of plant life are buried.

7.2 CONCEPT CHECKS

1. What was the first line of evidence that led early investigators to suspect that the continents were once connected?

2 Explain why the discovery of the fossil remains of *Mesosaurus* in both South America and Africa, but nowhere else, supports the continental drift hypothesis.

3 Early in the twentieth century, what was the prevailing view of how land animals migrated across vast expanses of open ocean?

4 How did Wegener account for the existence of glaciers in the southern landmasses at a time when areas in North America, Europe, and Asia supported lush tropical swamps?

7.3 | THE GREAT DEBATE

Discuss the two main objections to the continental drift hypothesis.

Wegener's proposal did not attract much open criticism until 1924, when his book was translated into English, French, Spanish, and Russian. From that point until his death in 1930, the drift hypothesis encountered a great deal of hostile criticism. The respected American geologist R. T. Chamberlain stated, "Wegener's hypothesis in general is of the foot-loose type, in that it takes considerable liberty with our globe, and is less bound by restrictions or tied down by awkward, ugly facts than most of its rival theories."

Rejection of the Drift Hypothesis

One of the main objections to Wegener's hypothesis stemmed from his inability to identify a credible mechanism for continental drift. Wegener proposed that gravitational forces of the Moon and Sun that produce Earth's tides were also capable of gradually moving the continents across the globe. However, the prominent physicist Harold Jeffreys correctly countered that tidal forces strong enough to move Earth's continents would have resulted in halting our planet's rotation, which, of course, has not happened.

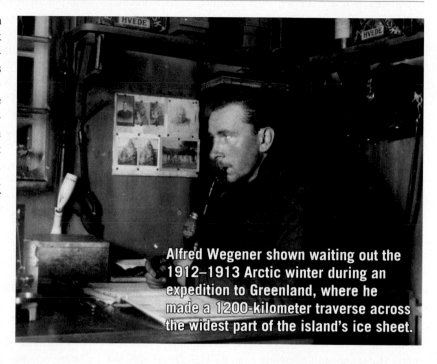

Alfred Wegener shown waiting out the 1912–1913 Arctic winter during an expedition to Greenland, where he made a 1200-kilometer traverse across the widest part of the island's ice sheet.

FIGURE 7.8 Alfred Wegener During an Expedition to Greenland (Photo courtesy of Archive of Alfred Wegener Institute for Polar and Marine Research)

Wegener also incorrectly suggested that the larger and sturdier continents broke through thinner oceanic crust, much as ice breakers cut through ice. However, no evidence existed to suggest that the ocean floor was weak enough to permit passage of the continents without the continents being appreciably deformed in the process.

FIGURE 7.9 Plate Movement Causes Destructive Earthquakes Photo of Pisco, Peru, following a powerful earthquake on August 16, 2007. (Sergio Urday/epa/Corbis)

In 1930 Wegener made his fourth and final trip to the Greenland Ice Sheet (**FIGURE 7.8**). Although the primary focus of this expedition was to study this great ice cap and its climate, Wegener continued to test his continental drift hypothesis. While returning from Eismitte, an experimental station located in the center of Greenland, Wegener perished along with his Greenland companion. His intriguing idea, however, did not die.

Why was Wegener unable to overturn the established scientific views of his day? Foremost was the fact that, although the central theme of Wegener's drift hypothesis was correct, it contained some incorrect details. For example, continents do not break through the ocean floor, and tidal energy is much too weak to cause continents to be displaced. Moreover, in order for any comprehensive scientific theory to gain wide acceptance, it must withstand critical testing from all areas of science. Despite Wegener's great contribution to our understanding of Earth, not *all* of the evidence supported the continental drift hypothesis as he had proposed it.

Although many of Wegener's contemporaries opposed his views, even to the point of open ridicule, some considered his ideas plausible. For those geologists who continued the search, the exciting concept of continents adrift held their interest. Others viewed continental drift as a solution to previously unexplainable observations such as the cause of earthquakes (**FIGURE 7.9**). Nevertheless, most of the scientific community, particularly in North America, either categorically rejected continental drift or treated it with considerable skepticism.

7.3 CONCEPT CHECKS

1 What two aspects of Wegener's continental drift hypothesis were objectionable to most Earth scientists?

7.4 | THE THEORY OF PLATE TECTONICS

List the major differences between Earth's lithosphere and asthenosphere and explain the importance of each in the plate tectonics theory.

Following World War II, oceanographers equipped with new marine tools and ample funding from the U.S. Office of Naval Research embarked on an unprecedented period of oceanographic exploration. Over the next two decades, a much better picture of large expanses of the seafloor slowly and painstakingly began to emerge. From this work came the discovery of a global **oceanic ridge system** that winds through all the major oceans in a manner similar to the seams on a baseball.

In other parts of the ocean, more new discoveries were being made. Studies conducted in the western Pacific demonstrated that earthquakes were occurring at great depths beneath deep-ocean trenches. Of equal importance was the fact that dredging of the seafloor did not bring up any oceanic crust that was older than 180 million years. Further, sediment accumulations in the deep-ocean basins were found to be thin, not the thousands of meters that were predicted. By 1968 these developments, among others, had led to the unfolding of a far more encompassing theory than continental drift, known as the **theory of plate tectonics** (*tekto* = to build).

SmartFigure 7.10 Rigid Lithosphere Overlies the Weak Asthenosphere

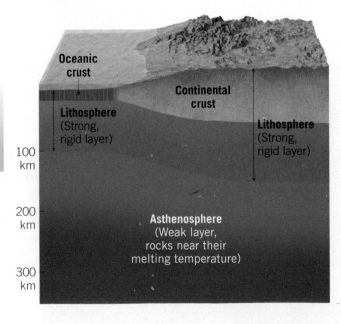

Oceanic crust

Continental crust

Lithosphere (Strong, rigid layer)

Lithosphere (Strong, rigid layer)

100 km

200 km

Asthenosphere (Weak layer, rocks near their melting temperature)

300 km

Rigid Lithosphere Overlies Weak Asthenosphere

According to the plate tectonics model, the crust and the uppermost, and therefore coolest, part of the mantle constitute Earth's strong outer layer, known as the **lithosphere** (*lithos* = stone, *sphere* = ball). The lithosphere varies in both thickness and density, depending on whether it is oceanic lithosphere or continental lithosphere (**FIGURE 7.10**). Oceanic lithosphere is about 100 kilometers (60 miles) thick in the deep-ocean basins but is considerably thinner along the crest of the oceanic ridge system—a topic we will consider later. By contrast, continental lithosphere averages about 150 kilometers (90 miles) thick but may extend to depths of 200 kilometers (125 miles) or more beneath the stable interiors of the continents. Further, the composition of both the oceanic and continental crusts affects their respective densities. Oceanic crust is composed of rocks that have a mafic (basaltic)

composition, and therefore oceanic lithosphere has a greater density than continental lithosphere. Continental crust is
composed largely of less dense felsic (granitic) rocks, making continental lithosphere less dense
than its oceanic counterpart.

The **asthenosphere** (*asthe-
nos* = weak, *sphere* = ball) is
a hotter, weaker region in the
mantle that lies below the lithosphere (see Figure 7.10). The
temperatures and pressures in
the upper asthenosphere (100 to
200 kilometers in depth) are such
that rocks at this depth are very near
their melting temperatures and, hence,
respond to forces by *flowing*, similarly to the
way a thick liquid would flow. By contrast, the
relatively cool and rigid lithosphere tends to respond to forces
acting on it by *bending or breaking but not flowing*. Because of
these differences, Earth's rigid outer shell is effectively detached
from the asthenosphere, which allows these layers to move
independently.

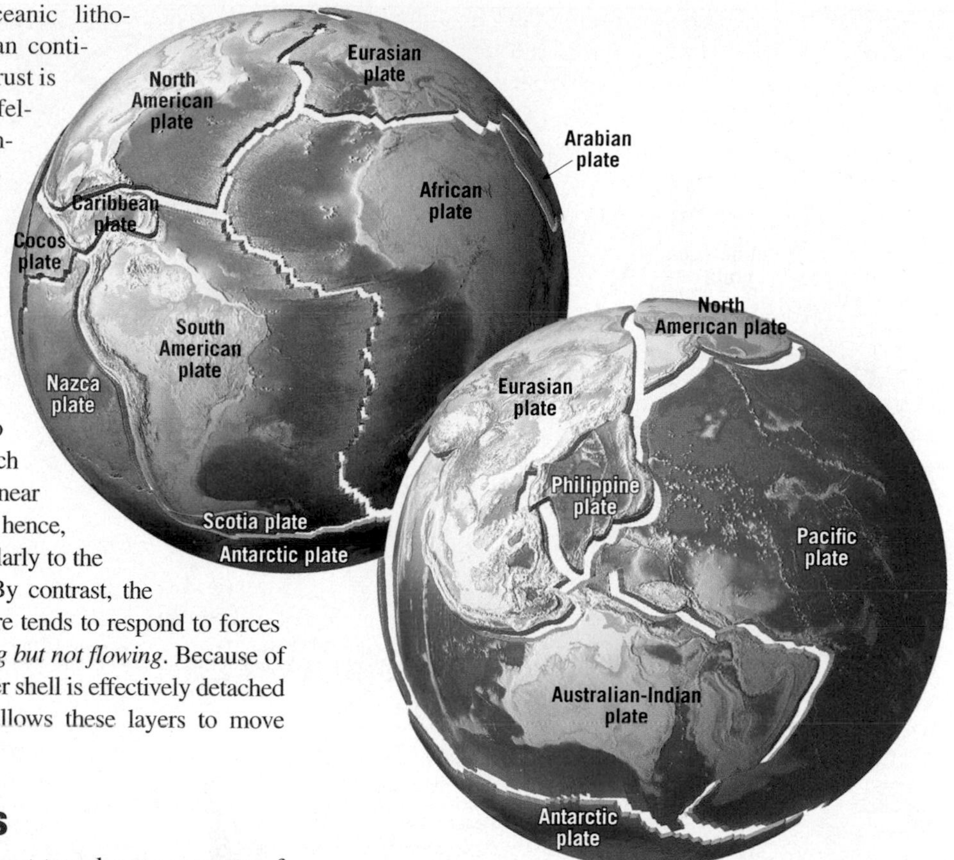

FIGURE 7.11 Earth's
Major Lithospheric Plates

Earth's Major Plates

The lithosphere is broken into about two dozen segments of
irregular size and shape called **lithospheric plates**, or simply **plates**, that are in constant motion with respect to one
another (**FIGURE 7.11**). Seven major lithospheric plates are
recognized and account for 94 percent of Earth's surface area:
the *North American*, *South American*, *Pacific*, *African*, *Eurasian*, *Australian-Indian*, and *Antarctic plates*. The largest is
the Pacific plate, which encompasses a significant portion of
the Pacific basin. Each of the six other large plates includes
an entire continent plus a significant amount of ocean floor.
Notice in **FIGURE 7.12** that the South American plate encompasses almost all of South America and about one-half of the
floor of the South Atlantic. This is a major departure from
Wegener's continental drift hypothesis, which proposed that
the continents move through the ocean floor, not with it. Note
also that none of the plates are defined entirely by the margins
of a single continent.

Intermediate-sized plates include the *Caribbean*, *Nazca*,
Philippine, *Arabian*, *Cocos*, *Scotia*, and *Juan de Fuca plates*.
These plates, with the exception of the Arabian plate, are
composed mostly of oceanic lithosphere. In addition, several
smaller plates (*microplates*) have been identified but are not
shown in Figure 7.12.

Plate Boundaries

One of the main tenets of the plate tectonics theory is that
plates move as somewhat rigid units relative to all other
plates. As plates move, the distance between two locations

on different plates, such as New York and London, gradually changes, whereas the distance between sites on
the same plate—New York and Denver, for example—
remains relatively constant. However, parts of some plates
are comparatively "soft," such as southern China, which is
literally being squeezed as the Indian subcontinent rams
into Asia proper.

Because plates are in constant motion relative to each
other, most major interactions among them (and, therefore, most deformation) occur along their *boundaries*.
In fact, plate boundaries were first established by plotting the locations of earthquakes and volcanoes. Plates
are bounded by three distinct types of boundaries, which
are differentiated by the type of movement they exhibit.
These boundaries are depicted in Figure 7.12 and are
briefly described here:

1. Divergent plate boundaries (*constructive margins*)—
   where two plates move apart, resulting in upwelling
   of hot material from the mantle to create new seafloor (**FIGURE 7.12A**).
2. Convergent plate boundaries (*destructive margins*)—
   where two plates move together, resulting in oceanic
   lithosphere descending beneath an overriding plate,
   eventually to be reabsorbed into the mantle or possibly
   in the collision of two continental blocks to create a
   mountain belt (**FIGURE 7.12B**).

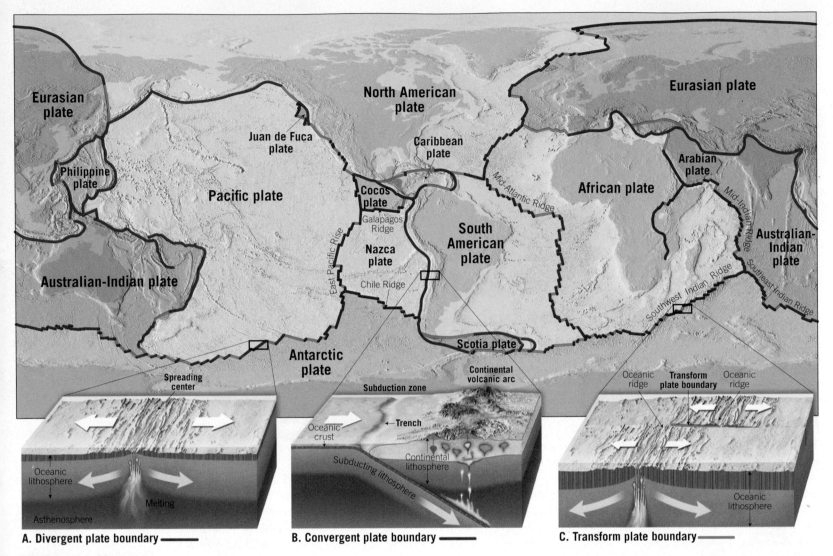

FIGURE 7.12 Divergent, Convergent, and Transform Plate Boundaries

3. Transform plate boundaries (*conservative margins*)—where two plates grind past each other without producing or destroying lithosphere (**FIGURE 7.12C**).

Divergent and convergent plate boundaries each account for about 40 percent of all plate boundaries. Transform faults account for the remaining 20 percent. In the following sections we will summarize the nature of the three types of plate boundaries.

7.4 CONCEPT CHECKS

1 What major ocean floor feature did oceanographers discover after World War II?

2 Compare and contrast the lithosphere and the asthenosphere.

3 List the seven largest lithospheric plates.

4 List the three types of plate boundaries and describe the relative motion at each of them.

7.5 | DIVERGENT PLATE BOUNDARIES AND SEAFLOOR SPREADING

Sketch and describe the movement along a divergent plate boundary that results in the formation of new oceanic lithosphere.

Most **divergent plate boundaries** (*di* = apart, *vergere* = to move) are located along the crests of oceanic ridges and can be thought of as *constructive plate margins* because this is where new ocean floor is generated (**FIGURE 7.13**). Here, two adjacent plates move away from each other, producing long, narrow fractures in the ocean crust. As a result, hot rock from the mantle below migrates upward and fills the voids left as the crust is being ripped apart. This molten material gradually cools, producing new slivers of seafloor. In a slow and unending manner, adjacent plates spread apart, and new oceanic lithosphere forms between them. For this reason, divergent plate boundaries are also referred to as **spreading centers**.

Oceanic Ridges and Seafloor Spreading

The majority of, but not all, divergent plate boundaries are associated with *oceanic ridges*: elevated areas of the seafloor characterized by high heat flow and volcanism. The global ridge system is the longest topographic feature on Earth's surface, exceeding 70,000 kilometers (43,000 miles) in length. As shown in Figure 7.12, various segments of the global ridge system have been named, including the Mid-Atlantic Ridge, East Pacific Rise, and Mid-Indian Ridge.

Representing 20 percent of Earth's surface, the oceanic ridge system winds through all major ocean basins like the seams on a baseball. Although the crest of the oceanic ridge is commonly 2 to 3 kilometers higher than the adjacent ocean basins, the term *ridge* may be misleading because it implies "narrow" when, in fact, ridges vary in width from 1000 kilometers (600 miles) to more than 4000 kilometers (2500 miles). Further, along the crest of some ridge segments is a deep canyonlike structure called a **rift valley** (FIGURE 7.14). This structure is evidence that tensional forces are actively pulling the ocean crust apart at the ridge crest.

The mechanism that operates along the oceanic ridge system to create new seafloor is appropriately called **seafloor spreading**. Typical rates of spreading average around 5 centimeters (2 inches) per year, roughly the same rate at which human fingernails grow. Comparatively slow spreading rates of 2 centimeters per year are found along the Mid-Atlantic Ridge, whereas spreading rates exceeding 15 centimeters (6 inches) per year have been measured along sections of the East Pacific Rise. Although these rates of seafloor production are slow on a human time scale, they are nevertheless rapid enough to have generated all of Earth's ocean basins within the past 200 million years.

The primary reason for the elevated position of the oceanic ridge is that newly created oceanic lithosphere is hot, which means it is less dense than cooler rocks found away from the ridge axis. (Geologists use the term *axis* to refer to a line that follows the general trend of the ridge

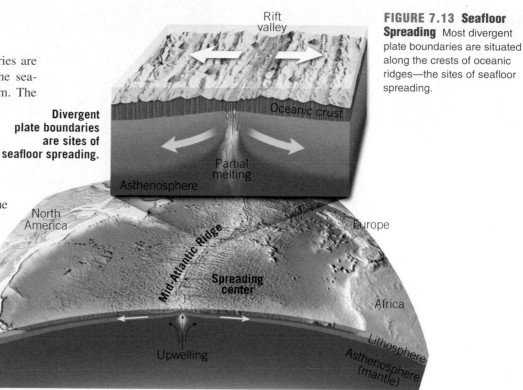

Divergent plate boundaries are sites of seafloor spreading.

FIGURE 7.13 **Seafloor Spreading** Most divergent plate boundaries are situated along the crests of oceanic ridges—the sites of seafloor spreading.

FIGURE 7.14 **Rift Valley** Thingvellir National Park, Iceland, is located on the western margin of a rift valley that is roughly 30 kilometers (20 miles) wide in this region. This rift valley is connected to a similar feature that extends along the crest of the Mid-Atlantic Ridge. The cliff in the left half of the image approximates the eastern edge of the North American plate. (Photo by Ragnar ThSigurdsson/Alamy)

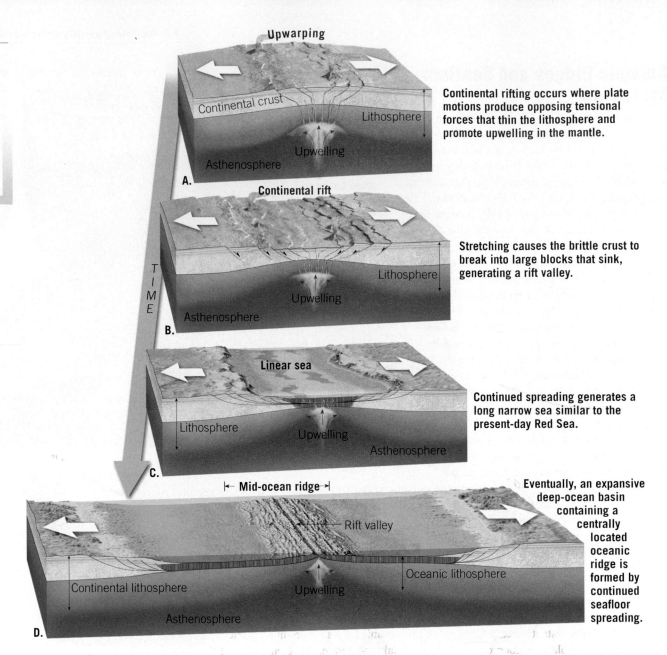

Upwarping

Continental crust

Lithosphere

Upwelling

Asthenosphere

A.

Continental rifting occurs where plate motions produce opposing tensional forces that thin the lithosphere and promote upwelling in the mantle.

Continental rift

Lithosphere

Upwelling

Asthenosphere

B.

Stretching causes the brittle crust to break into large blocks that sink, generating a rift valley.

Linear sea

Lithosphere

Upwelling

Asthenosphere

C.

Continued spreading generates a long narrow sea similar to the present-day Red Sea.

|← **Mid-ocean ridge** →|

Rift valley

Continental lithosphere

Oceanic lithosphere

Upwelling

Asthenosphere

D.

Eventually, an expansive deep-ocean basin containing a centrally located oceanic ridge is formed by continued seafloor spreading.

T I M E

crest.) As soon as new lithosphere forms, it is slowly yet continually displaced away from the zone of upwelling. Thus, it begins to cool and contract, thereby increasing in density. This thermal contraction accounts for the increase in ocean depths away from the ridge crest. It takes about 80 million years for the temperature of oceanic lithosphere to stabilize and contraction to cease. By this time, rock that was once part of the elevated oceanic ridge system is located in the deep-ocean basin, where it may be buried by substantial accumulations of sediment.

In addition, as the plate moves away from the ridge, cooling of the underlying asthenosphere causes it to become increasingly more rigid. Thus, oceanic lithosphere is generated by cooling of the asthenosphere from the top down. Stated another way, the thickness of oceanic lithosphere is age dependent. The older (cooler) it is, the greater its thickness. Oceanic lithosphere that exceeds 80 million years in age is about 100 kilometers (60 miles) thick—approximately its maximum thickness.

Continental Rifting

Divergent boundaries can develop within a continent, in which case the landmass may split into two or more smaller segments separated by an ocean basin. Continental rifting begins when plate motions produce opposing (tensional) forces that pull and stretch the lithosphere. Because the lower lithosphere is warm and weak it deforms without breaking. Stretching, in turn, thins the lithosphere, which promotes mantle upwelling and broad upwarping of the overlying lithosphere (**FIGURE 7.15A**). During this process the outermost crustal rocks, which are cool and brittle, break into large blocks. As the tectonic forces continue to pull apart the crust, the broken crustal fragments sink, generating an elongated depression called a **continental rift**, which eventually widens to form a narrow sea (**FIGURE 7.15B, C**) and then a new ocean basin (**FIGURE 7.15D**).

A modern example of an active continental rift is the East African Rift (**FIGURE 7.16**). Whether this rift will eventually

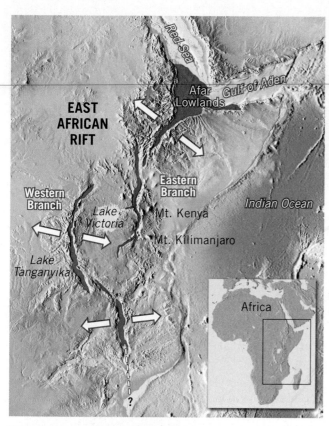

FIGURE 7.16 East African Rift Valley

result in the breakup of Africa is a topic of continued research. Nevertheless, the East African Rift is an excellent model of the initial stage in the breakup of a continent. Here, tensional forces have stretched and thinned the lithosphere, allowing molten rock to ascend from the mantle. Evidence for recent volcanic activity includes several large volcanic mountains, including Mount Kilimanjaro and Mount Kenya, the tallest peaks in Africa. Research suggests that if rifting continues, the rift valley will lengthen and deepen (see Figure 7.15C). At some point, the rift valley will become a narrow sea with an outlet to the ocean. The Red Sea, which formed when the Arabian Peninsula split from Africa, is a modern example of such a feature and provides us with a view of how the Atlantic Ocean may have looked in its infancy (see Figure 7.15D).

7.5 CONCEPT CHECKS

1 Sketch or describe how two plates move in relation to each other along divergent plate boundaries.

2 What is the average rate of seafloor spreading in modern oceans?

3 List four facts that characterize the oceanic ridge system.

4 Briefly describe the process of continental rifting. Where is it occurring today?

7.6 | CONVERGENT PLATE BOUNDARIES AND SUBDUCTION

Compare and contrast the three types of convergent plate boundaries and name a location where each type can be found.

New lithosphere is constantly being produced at the oceanic ridges. However, our planet is not growing larger; its total surface area remains constant. A balance is maintained because older, denser portions of oceanic lithosphere descend into the mantle at a rate equal to seafloor production. This activity occurs along **convergent plate boundaries**, where two plates move toward each other and the leading edge of one is bent downward, as it slides beneath the other (**FIGURE 7.17**).

Convergent boundaries are also called **subduction zones** because they are sites where lithosphere is descending (being subducted) into the mantle. Subduction occurs because the density of the descending lithospheric plate is greater than the density of the underlying asthenosphere. In general, old oceanic lithosphere is about 2 percent more dense than the underlying asthenosphere, which causes it to subduct. Continental lithosphere, in contrast, is less dense and resists subduction. As a consequence, only oceanic lithosphere will subduct to great depths.

Deep-ocean trenches are the surface manifestations produced as oceanic lithosphere descends into the mantle (see Figure 1.20). These large linear depressions are remarkably long and deep. The Peru–Chile trench along the west

coast of South America is more than 4500 kilometers (3000 miles) long, and its base is as much as 8 kilometers (5 miles) below sea level. The trenches in the western Pacific, including the Mariana and Tonga trenches, tend to be even deeper than those of the eastern Pacific.

Slabs of oceanic lithosphere descend into the mantle at angles that vary from a few degrees to nearly vertical (90 degrees). The angle at which an oceanic plate subducts depends largely on its age and therefore its density. For example, when seafloor spreading occurs near a subduction zone, as is the case along the coast of Chile, the subducting lithosphere is young and buoyant, which results in a low angle of descent. As the two plates converge, the overriding plate scrapes over the top of the subducting plate below—a type of forced subduction. Consequently, the region around the Peru–Chile trench experiences great earthquakes, including the 2010 Chilean earthquake—one of the 10 largest on record.

As oceanic lithosphere ages (gets farther from the spreading center), it gradually cools, which causes it to thicken and increase in density. In parts of the western Pacific, some oceanic lithosphere is 180 million years old—the thickest and densest in today's oceans. The very dense slabs in this region typically

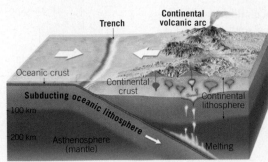

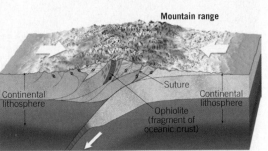

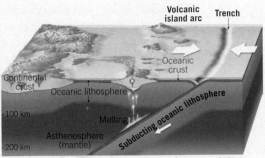

A. Convergent plate boundary where oceanic lithosphere is subducting beneath continental lithosphere.

B. Convergent plate boundary involving two slabs of oceanic lithosphere.

C. Continental collisions occur along convergent plate boundaries when both plates are capped with continental crust.

 SmartFigure 7.17 Three Types of Convergent Plate Boundaries

FIGURE 7.18 Oceanic–continental Convergent Plate Boundary Mount Hood, Oregon, is one of more than a dozen large composite volcanoes in the Cascade Range.

Wallace Garrison/Getty Images

Mt. Hood, Oregon

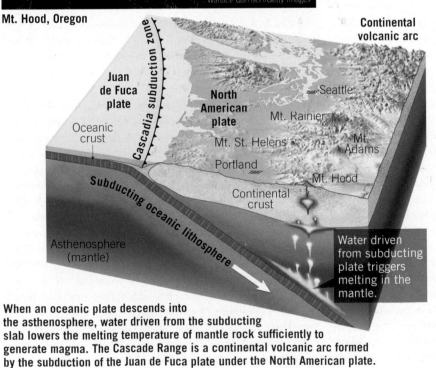

When an oceanic plate descends into the asthenosphere, water driven from the subducting slab lowers the melting temperature of mantle rock sufficiently to generate magma. The Cascade Range is a continental volcanic arc formed by the subduction of the Juan de Fuca plate under the North American plate.

plunge into the mantle at angles approaching 90 degrees. This largely explains the fact that most trenches in the western Pacific are deeper than trenches in the eastern Pacific.

Although all convergent zones have the same basic characteristics, they may vary considerably, based on the type of crustal material involved and the tectonic setting. Convergent boundaries can form between *two oceanic plates*, *one oceanic plate and one continental plate*, or *two continental plates*.

Oceanic–Continental Convergence

When the leading edge of a plate capped with continental crust converges with a slab of oceanic lithosphere, the buoyant continental block remains "floating," while the denser oceanic slab sinks into the mantle (see Figure 7.17A). When a descending oceanic slab reaches a depth of about 100 kilometers (60 miles), melting is triggered within the wedge of hot asthenosphere that lies above it. But how does the subduction of a cool slab of oceanic lithosphere cause mantle rock to melt? The answer lies in the fact that water contained in the descending plates acts the way salt does to melt ice. That is, "wet" rock in a high-pressure environment melts at substantially lower temperatures than does "dry" rock of the same composition.

Sediments and oceanic crust contain large amounts of water, which is carried to great depths by a subducting plate. As the plate plunges downward, heat and pressure drive water from the voids in the rock. At a depth of roughly 100 kilometers (60 miles), the wedge of mantle rock is sufficiently hot that the introduction of water from the slab below leads to some melting. This process, called **partial melting**, is thought to generate some molten material, which is mixed with unmelted mantle rock. Being less dense than the surrounding mantle, this hot mobile material gradually rises toward the surface. Depending on the environment, these mantle-derived masses of molten rock may ascend through the crust and give rise to a volcanic eruption. However, much of this material never reaches the surface; rather, it solidifies at depth—a process that thickens the crust.

The volcanoes of the towering Andes are the product of molten rock generated by the subduction of the Nazca plate beneath the South American continent (see Figure 7.12). Mountain systems, such as the Andes, which are produced in part by volcanic activity associated with the subduction of oceanic lithosphere, are called **continental volcanic arcs**. The Cascade Range in Washington, Oregon, and California is another mountain system that consists of several well-known volcanic mountains, including Mount Rainier, Mount Shasta, Mount St. Helens, and Mount Hood (**FIGURE 7.18**). This active volcanic arc also extends into Canada, where it includes Mount Garibaldi, Mount Silverthrone, and others.

Oceanic–Oceanic Convergence

An *oceanic–oceanic convergent boundary* has many features in common with oceanic–continental plate margins. Where two oceanic slabs converge, one descends beneath the other, initiating volcanic activity by the same mechanism that operates at all subduction zones (see Figure 7.12). Water squeezed from the subducting slab of oceanic lithosphere triggers melting in the hot wedge of mantle rock above. In this setting, volcanoes grow up from the ocean floor rather than on a continental platform. When subduction is sustained, it will eventually build a chain of volcanic structures large enough to emerge as islands. The newly formed land consisting of an arc-shaped chain of volcanic islands is called a **volcanic island arc**, or simply an **island arc** (**FIGURE 7.19**).

The Aleutian, Mariana, and Tonga islands are examples of relatively young volcanic island arcs. Island arcs are generally located 100 to 300 kilometers (60 to 200 miles) from a deep-ocean trench. Located adjacent to the island arcs just mentioned

FIGURE 7.19 Volcanoes of the Aleutian Islands

USGS

Gareloi
Kanaga
Great Sitkin
Cleveland
Shishaldin
Pavlof
Aniakchak
Katmai
Augustine
Redoubt

Active volcanoes of the Aleutian chain, Alaska

are the Aleutian trench, the Mariana trench, and the Tonga trench.

Most volcanic island arcs are located in the western Pacific. Only two are located in the Atlantic—the Lesser Antilles arc, on the eastern margin of the Caribbean Sea, and the Sandwich Islands, located off the tip of South America. The Lesser Antilles are a product of the subduction of the Atlantic seafloor beneath the Caribbean plate. Located within this volcanic arc are the Virgin Islands of the United States and Britain as well as the island of Martinique, where Mount Pelée erupted in 1902, destroying the town of St. Pierre and killing an estimated 28,000 people. This chain of islands also includes Montserrat, where there has been recent volcanic activity. More on these volcanic events is found in Chapter 9.

Island arcs are typically simple structures made of numerous volcanic cones underlain by oceanic crust that is generally less than 20 kilometers (12 miles) thick. By contrast, some island arcs are more complex and are underlain by highly deformed crust that may reach 35 kilometers (22 miles) in thickness. Examples include Japan, Indonesia, and the Alaskan Peninsula. These island arcs are built on material generated by earlier episodes of subduction or on small slivers of continental crust that have rafted away from the mainland.

FIGURE 7.20 The Collision of India and Eurasia Formed the Himalayas

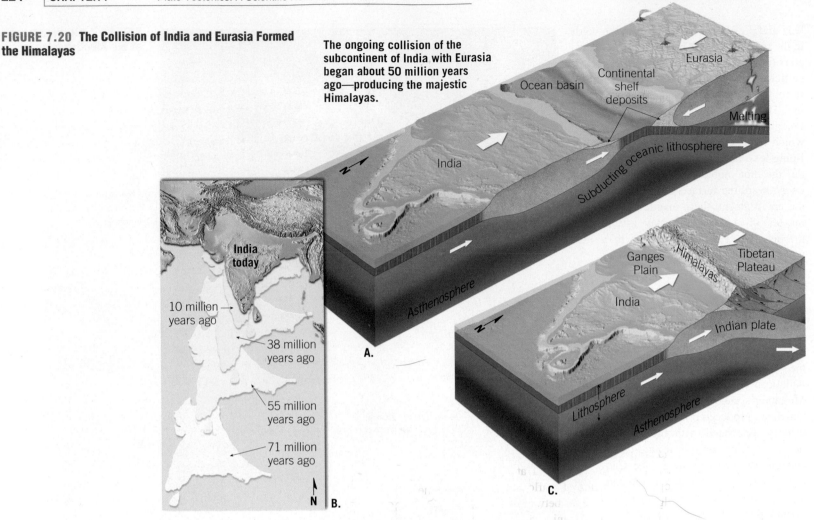

The ongoing collision of the subcontinent of India with Eurasia began about 50 million years ago—producing the majestic Himalayas.

Continental–Continental Convergence

The third type of convergent boundary results when one landmass moves toward the margin of another because of subduction of the intervening seafloor (**FIGURE 7.20A**). Whereas oceanic lithosphere tends to be dense and sink into the mantle, the buoyancy of continental material inhibits it from being subducted. Consequently, a collision between two converging continental fragments ensues (**FIGURE 7.20B**). This event folds and deforms the accumulation of sediments and sedimentary rocks along the continental margins as if they had been placed in a gigantic vise. The result is the formation of a new mountain belt composed of deformed sedimentary and metamorphic rocks that often contain slivers of oceanic crust.

Such a collision began about 50 million years ago, when the subcontinent of India "rammed" into Asia, producing the Himalayas—the most spectacular mountain range on Earth (see Figure 7.20B). During this collision, the continental crust buckled and fractured and was generally shortened horizontally and thickened vertically. In addition to the Himalayas, several other major mountain systems, including the Alps, Appalachians, and Urals, formed as continental fragments collided. This topic will be considered further in Chapter 10.

7.6 CONCEPT CHECKS

1 Explain why the rate of lithosphere production roughly balances with the rate of lithosphere destruction.

2 Compare a continental volcanic arc and a volcanic island arc.

3 Describe the process that leads to the formation of deep-ocean trenches.

4 Why does oceanic lithosphere subduct, while continental lithosphere does not?

5 Briefly describe how mountain belts such as the Himalayas form.

7.7 | TRANSFORM PLATE BOUNDARIES Describe the relative motion along a transform plate boundary and locate several examples on a plate boundary map.

Along a **transform plate boundary**, also called a **transform fault**, plates slide horizontally past one another without producing or destroying lithosphere. The nature of transform faults was discovered in 1965 by Canadian geologist J. Tuzo Wilson, who proposed that these large faults connect two spreading centers (divergent boundaries) or, less commonly, two trenches (convergent boundaries). Most transform faults are found on the ocean floor, where they offset segments of the oceanic ridge system, producing a steplike plate margin (**FIGURE 7.21A**). Notice that the zigzag shape of the Mid-Atlantic Ridge in Figure 7.12 roughly reflects the shape of the original rifting that caused the breakup of the supercontinent of Pangaea. (Compare the shapes of the continental margins of the landmasses on both sides of the Atlantic with the shape of the Mid-Atlantic Ridge.)

Typically, transform faults are part of prominent linear breaks in the seafloor known as **fracture zones**, which include both active transform faults and their inactive extensions into the plate interior (**FIGURE 7.21B**). Active transform faults lie *only between* the two offset ridge segments and are generally defined by weak, shallow earthquakes. Here seafloor produced at one ridge axis moves in the opposite direction of seafloor produced at an opposing ridge segment. Thus, between the ridge segments, these adjacent slabs of oceanic crust are grinding past each other along a transform fault. Beyond the ridge crests are inactive zones, where the fractures are preserved as linear topographic

depressions. The trend of these fracture zones roughly parallels the direction of plate motion at the time of their formation. Thus, these structures are useful in mapping the direction of plate motion in the geologic past.

In another role, transform faults provide the means by which the oceanic crust created at ridge crests can be transported to a site of destruction—the deep-ocean trenches. **FIGURE 7.22** illustrates this situation. Notice that the Juan de Fuca plate moves in a southeasterly direction, eventually being subducted under the west coast of the United States. The southern end of this plate is bounded by a transform fault called the Mendocino Fault. This transform boundary connects the Juan de Fuca Ridge to the Cascadia subduction zone. Therefore, it facilitates the movement of the crustal material created at the Juan de Fuca Ridge to its destination beneath the North American continent.

SmartFigure 7.21 Transform Plate Boundaries

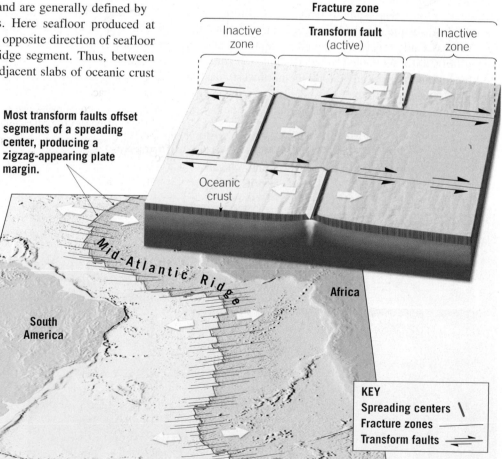

B. Fracture zones are long, narrow scar-like features in the seafloor that are roughly perpendicular to the offset ridge segments. They include both the active transform fault and its "fossilized" trace.

Fracture zone

Inactive zone | **Transform fault** (active) | Inactive zone

Oceanic crust

Most transform faults offset segments of a spreading center, producing a zigzag-appearing plate margin.

Mid-Atlantic Ridge

Africa

South America

KEY
Spreading centers \
Fracture zones ———
Transform faults ⇄

A. The Mid-Atlantic Ridge, with its zigzag pattern, roughly reflects the shape of the rifting zone that resulted in the breakup of Pangaea.

FIGURE 7.22 Transform Faults Facilitate Plate Motion Seafloor spreading generated along the Juan de Fuca Ridge moves southeastward, past the Pacific plate. Eventually it subducts beneath the North American plate. Thus, this transform fault connects a spreading center (divergent boundary) to a subduction zone (convergent boundary). Also shown is the San Andreas Fault, a transform fault that connects two spreading centers: the Juan de Fuca Ridge and a spreading center located in the Gulf of California.

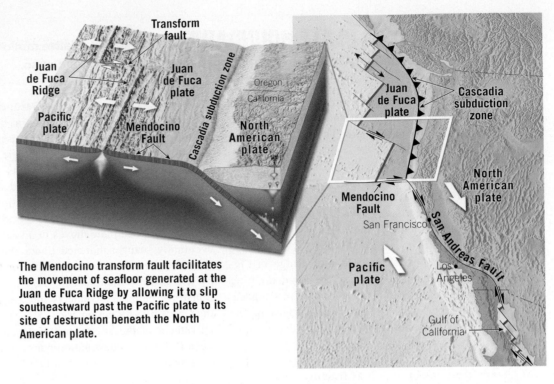

The Mendocino transform fault facilitates the movement of seafloor generated at the Juan de Fuca Ridge by allowing it to slip southeastward past the Pacific plate to its site of destruction beneath the North American plate.

Like the Mendocino Fault, most other transform fault boundaries are located within the ocean basins; however, a few cut through continental crust. Two examples are the earthquake-prone San Andreas Fault of California and New Zealand's Alpine Fault. Notice in Figure 7.22 that the San Andreas Fault connects a spreading center located in the Gulf of California to the Cascadia subduction zone and the Mendocino Fault located along the northwest coast of the United States. Along the San Andreas Fault, the Pacific plate is moving toward the northwest, past the North American plate (**FIGURE 7.23**). If this movement continues, the part of California west of the fault zone, including the Baja Peninsula of Mexico, will become an island off the west coast of the United States and Canada—with the potential to eventually reach Alaska. However, a more immediate concern is the earthquake activity triggered by movements along this fault system.

7.7 CONCEPT CHECKS

1 Sketch or describe how two plates move in relation to each other along a transform plate boundary.

2 Differentiate between transform faults and the two other types of plate boundaries.

Mobile Field Trip 7.23 Movement along the San Andreas Fault This aerial view shows the offset in the dry channel of Wallace Creek near Taft, California.

Geologist's Sketch

7.8 | HOW DO PLATES AND PLATE BOUNDARIES CHANGE?

Explain why plates such as the African and Antarctic plates are getting larger, while the Pacific plate is getting smaller.

Although the total surface area of Earth does not change, the size and shape of individual plates are constantly changing. For example, the African and Antarctic plates, which are mainly bounded by divergent boundaries—sites of sea-floor production—are continually growing in size as new lithosphere is added to their margins. By contrast, the Pacific plate is being consumed into the mantle along its northern and western flanks faster that it is growing along the East Pacific Rise and thus is diminishing in size.

Another result of plate motion is that boundaries also migrate. For example, the position of the Peru–Chile trench, which is the result of the Nazca plate being bent downward as it descends beneath the South American plate, has changed over time (see Figure 7.12). Because of the westward drift of the South American plate relative to the Nazca plate, the position of the Peru–Chile trench has migrated in a westerly direction as well.

Plate boundaries can also be created or destroyed in response to changes in the forces acting on the lithosphere. Recall that the Red Sea is the site of a relatively new spreading center that came into existence less than 20 million years ago, when the Arabian Peninsula began to split from Africa. At other locations, plates carrying continental crust are presently moving toward one another. Eventually these continental fragments may collide and be sutured together. This could occur, for example, in the South Pacific, where Australia is moving northward toward southern Asia. If Australia continues its northward migration, the boundary separating it from Asia will disappear as these plates become one. The breakup of Pangaea is a classic example of how plate boundaries change through geologic time.

The Breakup of Pangaea

Wegener used evidence from fossils, rock types, and ancient climates to create a jigsaw-puzzle fit of the continents, thereby creating his supercontinent of Pangaea. In a similar manner, but employing modern tools not available to Wegener, geologists have re-created the steps in the breakup of this supercontinent, an event that began about 180 million years ago. From this work, the dates when individual crustal fragments separated from one another and their relative motions have been well established (**FIGURE 7.24**).

An important consequence of the breakup of Pangaea was the creation of a "new" ocean basin: the Atlantic. As you can see in Figure 7.24, splitting of the supercontinent did not occur simultaneously along the margins of the Atlantic. The first split developed between North America and Africa. Here, the continental crust was highly fractured, providing pathways for huge quantities of fluid lavas to reach the surface. Today, these lavas are represented by weathered igneous rocks found along the eastern seaboard of the United States—primarily buried beneath the sedimentary rocks that form the continental shelf. Radiometric dating of these

EYE ON **EARTH**

Baja California is separated from mainland Mexico by a long narrow sea called the Gulf of California (also known to local residents as the Sea of Cortez). The Gulf of California contains many islands that were created by volcanic activity.

QUESTION 1 *What type of plate boundary is responsible for opening the Gulf of California?*

QUESTION 2 *What major U.S. river originates in the Colorado Rockies and created a large delta at the northern end of the Gulf of California?*

QUESTION 3 *If the material carried by the river in Question 2 had not been deposited, the Gulf of California would extend northward to include the inland sea shown in this satellite image. What is the name of this inland sea?*

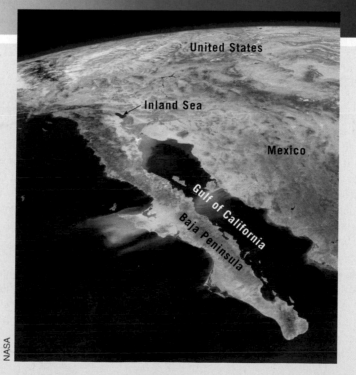

United States

Inland Sea

Mexico

Gulf of California

Baja Peninsula

NASA

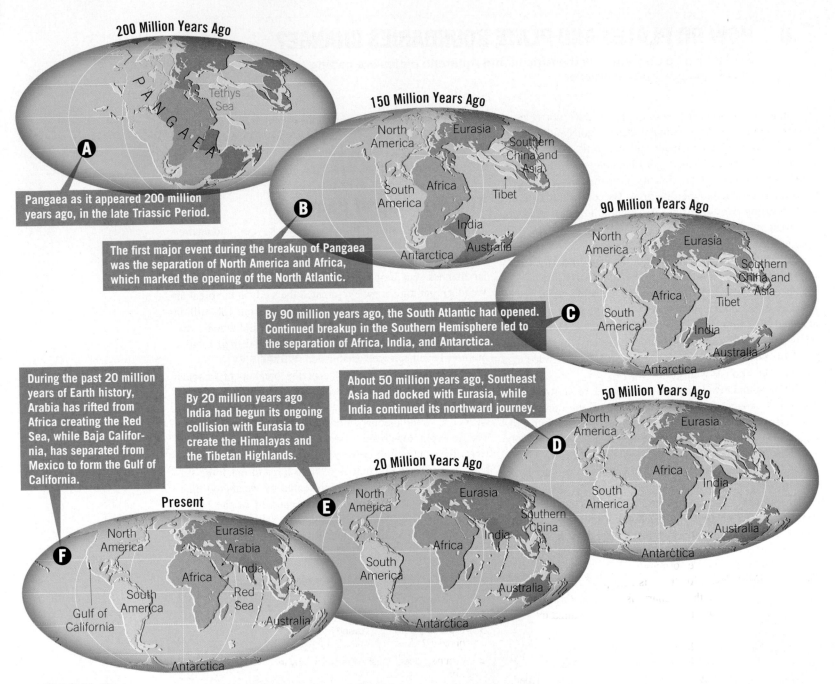

200 Million Years Ago

Tethys Sea

PANGAEA

Ⓐ

Pangaea as it appeared 200 million years ago, in the late Triassic Period.

150 Million Years Ago

North America
Eurasia
Southern China and Asia
South America
Africa
Tibet
India
Antarctica
Australia

Ⓑ

The first major event during the breakup of Pangaea was the separation of North America and Africa, which marked the opening of the North Atlantic.

90 Million Years Ago

North America
Eurasia
Southern China and Asia
Africa
Tibet
South America
India
Australia
Antarctica

Ⓒ

By 90 million years ago, the South Atlantic had opened. Continued breakup in the Southern Hemisphere led to the separation of Africa, India, and Antarctica.

During the past 20 million years of Earth history, Arabia has rifted from Africa creating the Red Sea, while Baja California, has separated from Mexico to form the Gulf of California.

By 20 million years ago India had begun its ongoing collision with Eurasia to create the Himalayas and the Tibetan Highlands.

About 50 million years ago, Southeast Asia had docked with Eurasia, while India continued its northward journey.

50 Million Years Ago

North America
Eurasia
Africa
India
South America
Australia
Antarctica

Ⓓ

20 Million Years Ago

North America
Eurasia
Southern China
Africa
India
South America
Australia
Antarctica

Ⓔ

Present

North America
Eurasia
Arabia
India
Africa
Red Sea
South America
Gulf of California
Australia
Antarctica

Ⓕ

FIGURE 7.24 The Breakup of Pangaea

solidified lavas indicates that rifting began between 180 million and 165 million years ago. This time span represents the "birth date" for this section of the North Atlantic.

By 130 million years ago, the South Atlantic began to open near the tip of what is now South Africa. As this zone of rifting migrated northward, it gradually opened the South Atlantic (Figure 7.24B,C). Continued breakup of the southern landmass led to the separation of Africa and Antarctica and sent India on a northward journey. By the early Cenozoic, about 50 million years ago, Australia had separated from Antarctica, and the South Atlantic had emerged as a full-fledged ocean (**FIGURE 7.24D**).

A modern map (**FIGURE 7.24F**) shows that India eventually collided with Asia, an event that began about 50 million years ago and created the Himalayas as well as the Tibetan Highlands. About the same time, the separation of Greenland

from Eurasia completed the breakup of the northern landmass. During the past 20 million years or so of Earth's history, Arabia rifted from Africa to form the Red Sea, and Baja California separated from Mexico to form the Gulf of California (**FIGURE 7.24E**). Meanwhile, the Panama Arc joined North America and South America to produce our globe's familiar modern appearance.

Plate Tectonics in the Future

Geologists have extrapolated present-day plate movements into the future. **FIGURE 7.25** illustrates where Earth's landmasses may be 50 million years from now if present plate movements persist during this time span.

In North America we see that the Baja Peninsula and the portion of southern California that lies west of the San

Andreas Fault will have slid past the North American plate. If this northward migration takes place, Los Angeles and San Francisco will pass each other in about 10 million years, and in about 60 million years the Baja Peninsula will begin to collide with the Aleutian islands.

If Africa continues on a northward path, it will continue to collide with Eurasia. The result will be the closing of the Mediterranean, the last remnant of a once vast ocean called Tethys Ocean, and the initiation of another major mountain-building episode (see Figure 7.25). In other parts of the world, Australia will be astride the equator and, along with New Guinea, will be on a collision course with Asia. Meanwhile, North and South America will begin to separate, while the Atlantic and Indian Oceans will continue to grow, at the expense of the Pacific Ocean.

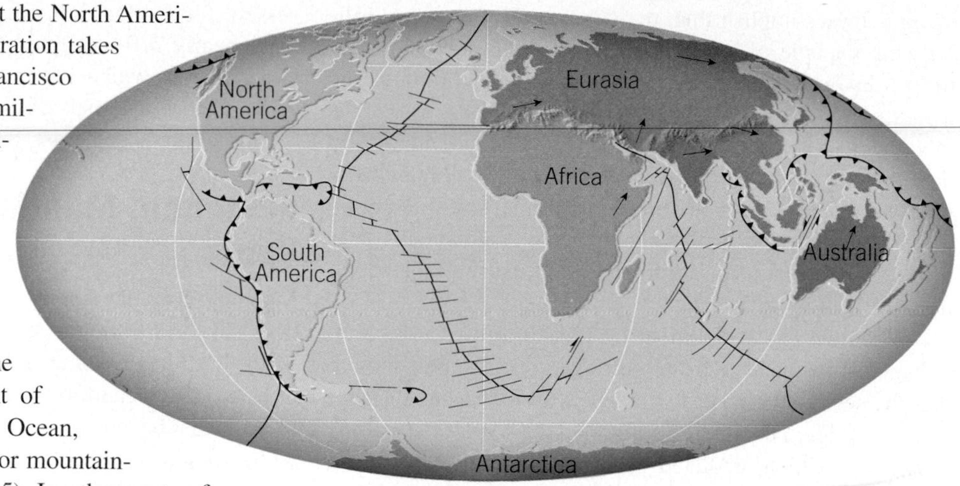

A few geologists have even speculated on the nature of the globe 250 million years in the future. As shown in **FIGURE 7.26**, the next supercontinent may form as a result of subduction of the floor of the Atlantic Ocean, resulting in the collision of the Americas with the Eurasian–African landmass. Support for the possible closing of the Atlantic comes from a similar event when an ocean that predates the Atlantic closed during the formation of Pangaea. During the next 250 million years, Australia is also projected to collide with Southeast Asia. If this scenario is accurate, the dispersal of Pangaea will end when the continents reorganize into the next supercontinent.

Such projections, although interesting, must be viewed with considerable skepticism because many assumptions must be correct for these events to unfold as just described. Nevertheless, changes in the shapes and positions of continents that are equally profound will undoubtedly occur for many hundreds of millions of years to come. Only after

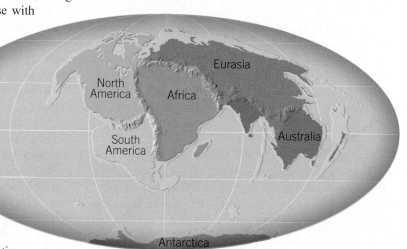

FIGURE 7.25 The World as it May Look 50 million Years from Now This reconstruction is highly idealized and based on the assumption that the processes that caused the breakup of Pangaea will continue to operate. (Modified after Robert S. Dietz, John C. Holden, C. Scotese, and others)

FIGURE 7.26 Earth as it May Appear 250 million Years from Now.

much more of Earth's internal heat has been lost will the engine that drives plate motions cease.

7.8 CONCEPT CHECKS

1 What two plates are growing in size? Which plate is shrinking?

2 What new ocean basin was created by the breakup of Pangaea?

3 Briefly describe some major changes to the globe when we extrapolate present-day plate movements 50 million years into the future.

7.9 | TESTING THE PLATE TECTONICS MODEL

List and explain the evidence used to support the plate tectonics theory.

Some of the evidence supporting continental drift has already been presented. With the development of the theory of plate tectonics, researchers began testing this new model of how Earth works. Although new supporting data were obtained, it has often been new interpretations of already existing data that have swayed the tide of opinion.

Evidence: Ocean Drilling

Some of the most convincing evidence for seafloor spreading came from the Deep Sea Drilling Project, which operated from 1968 until 1983. One of the early goals of the project was to gather samples of the ocean floor in order to establish its age. To accomplish this, the *Glomar Challenger*, a drilling ship capable of working in water thousands of meters deep, was built. Hundreds of holes were drilled through the layers of sediments that blanket the ocean crust, as well as into the basaltic rocks below. Rather than radiometrically date the crustal rocks, researchers used the fossil remains of microorganisms found in the sediments resting directly on the crust to date the seafloor at each site. Radiometric dates of the ocean crust itself are unreliable because of the alteration of basalt by seawater.

When the oldest sediment from each drill site was plotted against its distance from the ridge crest, the plot showed that the sediments increased in age with increasing distance from the ridge. This finding supported the seafloor-spreading hypothesis, which predicted that the youngest oceanic crust would be found at the ridge crest, the site of seafloor production, and the oldest oceanic crust would be located adjacent to the continents.

The thickness of ocean-floor sediments provided additional verification of seafloor spreading. Drill cores from the *Glomar Challenger* revealed that sediments are almost entirely absent on the ridge crest and that sediment thickness increases with increasing distance from the ridge (**FIGURE 7.27**). This pattern of sediment distribution should be expected if the seafloor-spreading hypothesis is correct.

The data collected by the Deep Sea Drilling Project also reinforced the idea that the ocean basins are geologically young because no seafloor with an age in excess of 180 million years was found. By comparison, most continental crust exceeds several hundred million years in age, and some has been located that exceeds 4 billion years in age.

In October 2003, the *JOIDES Resolution* became part of a new program, the Integrated Ocean Drilling Program (IODP). This new international effort uses multiple vessels for exploration, including the massive 210-meter-long (nearly 770-foot-long) *Chikyu* (meaning "planet Earth" in Japanese), which began operations in 2007 (see Figure 7.27). One of the goals of the IODP is to recover a complete section of the ocean crust, from top to bottom.

Evidence: Mantle Plumes and Hot Spots

Mapping volcanic islands and seamounts (submarine volcanoes) in the Pacific Ocean revealed several linear chains of volcanic structures. One of the most-studied chains consists of at least 129 volcanoes that extend from the Hawaiian islands to Midway Island and continue northwestward toward the Aleutian trench (**FIGURE 7.28**). Radiometric dating of this linear structure, called the *Hawaiian Island–Emperor Seamount chain*, showed that the volcanoes increase in age with increasing distance from the Big Island of Hawaii. The youngest volcanic island in the chain (Hawaii) rose from the ocean floor less than 1 million years ago, whereas Midway Island is 27 million years old, and Detroit Seamount, near the Aleutian trench, is about 80 million years old (see Figure 7.28).

Most researchers agree that a cylindrically shaped upwelling of hot rock, called a **mantle plume**, is located beneath the island of Hawaii. As the hot, rocky plume ascends through the mantle, the confining pressure drops, which triggers partial melting. (This process, called *decompression melting*, is discussed in Chapter 9.) The surface manifestation of this activity is a **hot spot**, an area of volcanism, high heat flow, and crustal uplifting that is a few hundred kilometers across. As the Pacific plate moved over a hot spot, a chain of volcanic structures known as a **hot-spot track** was built. As shown in Figure 7.28, the age of each volcano indicates how much time has elapsed since it was situated over the mantle plume.

A closer look at the five largest Hawaiian islands reveals a similar pattern of ages, from the volcanically active island of Hawaii to the inactive volcanoes that make up the oldest island, Kauai (see Figure 7.28). Five million years ago, when Kauai was positioned over the hot spot, it was the *only* modern Hawaiian island in existence. Visible evidence of the age of Kauai can be seen by examining its extinct volcanoes, which have been eroded into jagged peaks and vast canyons. By contrast, the relatively young island of Hawaii exhibits

Core samples show that the thickness of sediments increases with distance from the ridge crest.

Older ← Age of seafloor → Older
Younger

Drilling ship collects core samples of seafloor sediments and basaltic crust

Ocean crust (basalt)

FIGURE 7.27 Deep-sea Drilling Data collected through deep-sea drilling have shown that the ocean floor is indeed youngest at the ridge axis.

Chikyu is a state-of-the-art drilling ship designed to drill up to 7,000 meters (more than 4 miles) below the seafloor.

AP Photo/Itsuo Inouye

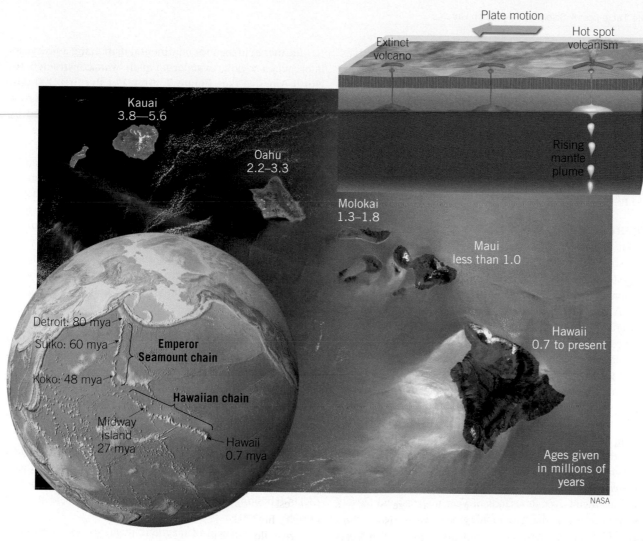

Plate motion

Extinct volcano

Hot spot volcanism

Rising mantle plume

Kauai
3.8—5.6

Oahu
2.2–3.3

Molokai
1.3–1.8

Maui
less than 1.0

Detroit: 80 mya

Suiko: 60 mya

Emperor Seamount chain

Koko: 48 mya

Hawaiian chain

Midway Island
27 mya

Hawaii
0.7 mya

Hawaii
0.7 to present

Ages given in millions of years

NASA

FIGURE 7.28 Hot Spots and Hot-Spot Tracks Radiometric dating of the Hawaiian islands shows that volcanic activity increases in age moving away from the Big Island of Hawaii.

many fresh lava flows, and one of its five major volcanoes, Kilauea, remains active today.

Evidence: Paleomagnetism

Anyone who has used a compass to find direction knows that Earth's magnetic field has north and south magnetic poles. Today these magnetic poles roughly align with the geographic poles. (The geographic poles are located where Earth's rotational axis intersects the surface.) Earth's magnetic field is similar to that produced by a simple bar magnet. Invisible lines of force pass through the planet and extend from one magnetic pole to the other (**FIGURE 7.29**). A compass needle, itself a small magnet free to rotate on an axis, becomes aligned with the magnetic lines of force and points to the magnetic poles.

Earth's magnetic field is less obvious than the pull of gravity because humans cannot feel it. Movement of a compass needle, however, confirms its presence. In addition, some naturally occurring minerals are magnetic and are influenced by Earth's magnetic field. One of the most common is the iron-rich mineral *magnetite*, which is abundant in lava flows of basaltic composition.*

*Some sediments and sedimentary rocks also contain enough iron-bearing mineral grains to acquire a measurable amount of magnetization.

Basaltic lavas erupt at the surface at temperatures greater than 1000°C (1800°F), exceeding a threshold temperature for magnetism known as the **Curie point** (about 585°C [1085° F]). Consequently, the magnetite grains in molten lava are nonmagnetic. However, as the lava cools, these

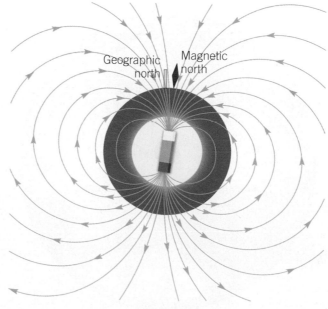

Geographic north

Magnetic north

FIGURE 7.29 Earth's Magnetic Field Earth's magnetic field consists of lines of force much like those a giant bar magnet would produce if placed at the center of Earth.

iron-rich grains become magnetized and align themselves in the direction of the existing magnetic lines of force. Once the minerals solidify, the magnetism they possess usually remains "frozen" in this position. Thus, they act like a compass needle because they "point" toward the position of the magnetic poles at the time of their formation. Rocks that formed thousands or millions of years ago and contain a "record" of the direction of the magnetic poles at the time of their formation are said to possess **paleomagnetism**, or **fossil magnetism**.

Apparent Polar Wandering

A study of paleomagnetism in ancient lava flows throughout Europe led to an interesting discovery. The magnetic alignment of iron-rich minerals in lava flows of different ages indicated that the position of the paleomagnetic poles had changed through time. A plot of the location of the magnetic north pole, as measured from Europe, revealed that during the past 500 million years, the pole had gradually wandered from a location near Hawaii northeastward to its present location over the Arctic Ocean (**FIGURE 7.30**). This was strong evidence that either the magnetic north pole had migrated, an idea known as *polar wandering*, or that the poles had remained in place and the continents had drifted beneath them—in other words, Europe had drifted in relation to the magnetic north pole.

Although the magnetic poles are known to move in a somewhat erratic path, studies of paleomagnetism from numerous locations show that the positions of the magnetic poles, averaged over thousands of years, correspond closely to the positions of the geographic poles. Therefore, a more acceptable explanation for the apparent polar wandering paths was provided by Wegener's hypothesis: If the magnetic poles remain stationary, their *apparent movement* is produced by continental drift.

Further evidence for continental drift came a few years later, when a polar-wandering path was constructed for North America (see Figure 7.30A). For the first 300 million years or so, the paths for North America and Europe were found to be similar in direction—but separated by about 5000 kilometers (3000 miles). Then, during the middle of the Mesozoic era (180 million years ago), they began to converge on the present North Pole. The explanation for these curves is that North America and Europe were joined until the Mesozoic, when the Atlantic began to open. From this time forward, these continents continuously moved apart. When North America and Europe are moved back to their pre-drift positions, as shown in Figure 7.30B, these apparent wandering paths coincide. This is evidence that North America and Europe were once joined and moved relative to the poles as part of the same continent.

Magnetic Reversals and Seafloor Spreading

Another discovery came when geophysicists learned that over periods of hundreds of thousands of years, Earth's magnetic field periodically reverses polarity. During a **magnetic reversal**, the magnetic north pole becomes the magnetic south pole and vice versa. Lava solidifying during a period of reverse polarity will be magnetized with the polarity opposite that of volcanic rocks being formed today. When rocks exhibit the same magnetism as the present magnetic field, they are said to possess **normal polarity**, whereas rocks exhibiting the opposite magnetism are said to have **reverse polarity**.

Once the concept of magnetic reversals was confirmed, researchers set out to establish a time scale for these occurrences. The task was to measure the magnetic polarity of hundreds of lava flows and use radiometric dating techniques

FIGURE 7.30 Apparent Polar Wandering Paths
A. The more westerly path determined from North American data is thought to have been caused by the westward drift of North America by about 24 degrees from Eurasia. **B.** The positions of the wandering paths when the landmasses are reassembled in their pre-drift locations.

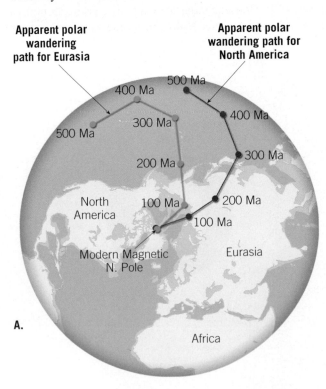

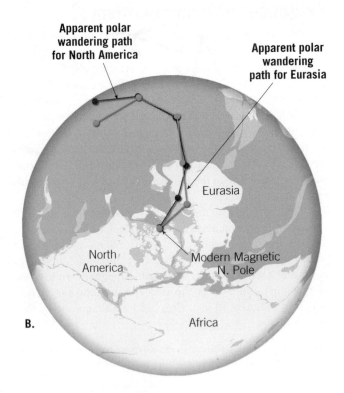

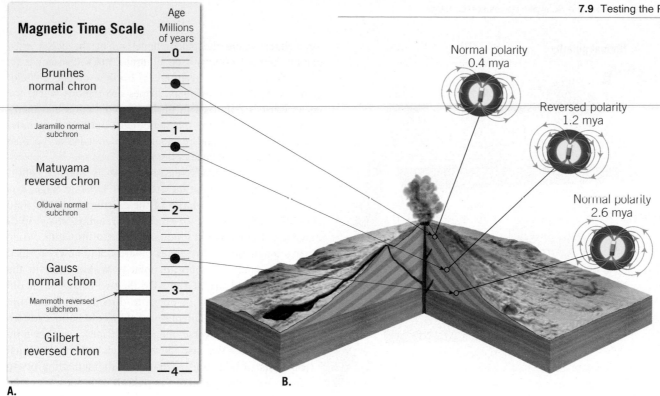

Magnetic Time Scale

Age
Millions
of years

Brunhes
normal chron

Jaramillo normal
subchron

Matuyama
reversed chron

Olduvai normal
subchron

Gauss
normal chron

Mammoth reversed
subchron

Gilbert
reversed chron

A.

B.

Normal polarity
0.4 mya

Reversed polarity
1.2 mya

Normal polarity
2.6 mya

SmartFigure 7.31 Time Scale of Magnetic Reversals A. Time scale of Earth's magnetic reversals for the past 4 million years. **B.** This time scale was developed by establishing the magnetic polarity for lava flows of known age. (Data from Allen Cox and G. B. Dalrymple)

to establish the age of each flow. **FIGURE 7.31** shows the **magnetic time scale** established using this technique for the past few million years. The major divisions of the magnetic time scale are called *chrons* and last roughly 1 million years. As more measurements became available, researchers realized that several short-lived reversals (less than 200,000 years long) often occurred during a single chron.

Meanwhile, oceanographers had begun to do magnetic surveys of the ocean floor in conjunction with their efforts to construct detailed maps of seafloor topography. These magnetic surveys were accomplished by towing very sensitive instruments, called **magnetometers**, behind research vessels (**FIGURE 7.32A**). The goal of these geophysical surveys was to map variations in the strength of Earth's magnetic field

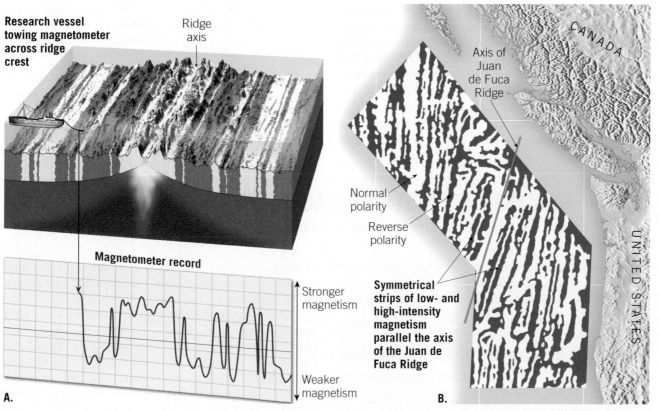

Research vessel towing magnetometer across ridge crest

Ridge axis

Magnetometer record

Stronger magnetism

Weaker magnetism

A.

CANADA

Axis of Juan de Fuca Ridge

Normal polarity

Reverse polarity

Symmetrical strips of low- and high-intensity magnetism parallel the axis of the Juan de Fuca Ridge

UNITED STATES

B.

FIGURE 7.32 Ocean Floor as a Magnetic Recorder A. Magnetic intensities are recorded when a magnetometer is towed across a segment of the ocean floor. **B.** Notice the symmetrical strips of low- and high-intensity magnetism that parallel the axis of the Juan de Fuca Ridge. Vine and Matthews suggested that the strips of high-intensity magnetism occur where normally magnetized oceanic rocks enhanced the existing magnetic field. Conversely, the low-intensity strips are regions where the crust is polarized in the reverse direction, which weakens the existing magnetic field.

FIGURE 7.33 Magnetic Reversals and Seafloor Spreading When new basaltic rocks form at mid-ocean ridges, they magnetize according to Earth's existing magnetic field. Hence, oceanic crust provides a permanent record of each reversal of our planet's magnetic field over the past 200 million years.

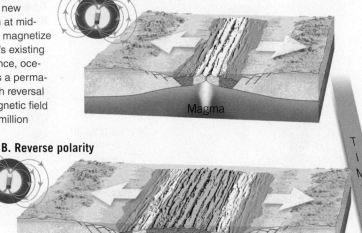

A. Normal polarity

B. Reverse polarity

C. Normal polarity

are regions where the paleomagnetism of the ocean crust exhibits normal polarity (see Figure 7.30A). Consequently, these rocks *enhance* (reinforce) Earth's magnetic field. Conversely, the low-intensity stripes are regions where the ocean crust is polarized in the reverse direction and therefore *weaken* the existing magnetic field. But how do parallel stripes of normally and reversely magnetized rock become distributed across the ocean floor?

Vine and Matthews reasoned that as magma solidifies along narrow rifts at the crest of an oceanic ridge, it is magnetized with the polarity of the existing magnetic field (**FIGURE 7.33**). Because of seafloor spreading, this strip of magnetized crust would gradually increase in width. When Earth's magnetic field reverses polarity, any newly formed seafloor having the opposite polarity would form in the middle of the old strip. Gradually, the two halves of the old strip would be carried in opposite directions, away from the ridge crest. Subsequent reversals would build a pattern of normal and reverse magnetic stripes, as shown in Figure 7.33. Because new rock is added in equal amounts to both trailing edges of the spreading ocean floor, we should expect the pattern of stripes (size and polarity) found on one side of an oceanic ridge to be a mirror image of those on the other side. In fact, a survey across the Mid-Atlantic Ridge just south of Iceland reveals a pattern of magnetic stripes exhibiting a remarkable degree of symmetry in relation to the ridge axis.

that arise from differences in the magnetic properties of the underlying crustal rocks.

The first comprehensive study of this type was carried out off the Pacific coast of North America and had an unexpected outcome. Researchers discovered alternating stripes of high- and low-intensity magnetism, as shown in **FIGURE 7.32B**. This relatively simple pattern of magnetic variation defied explanation until 1963, when Fred Vine and D. H. Matthews demonstrated that the high- and low-intensity stripes supported the concept of seafloor spreading. Vine and Matthews suggested that the stripes of high-intensity magnetism

7.9 CONCEPT CHECKS

1 What is the age of the oldest sediments recovered using deep-ocean drilling? How do the ages of these sediments compare to the ages of the oldest continental rocks?

2 Assuming that hot spots remain fixed, in what direction was the Pacific plate moving while the Hawaiian islands were forming? When Suiko Seamount was forming?

3 How do sediment cores from the ocean floor support the concept of seafloor spreading?

4 Describe how Fred Vine and D. H. Matthews related the seafloor-spreading hypothesis to magnetic reversals.

7.10 | HOW IS PLATE MOTION MEASURED

Describe two methods researchers use to measure relative plate motion.

A number of methods are used to establish the direction and rate of plate motion. Some of these techniques not only confirm that lithospheric plates move but allow us to trace those movements back in geologic time.

Geologic Evidence for Plate Motion

Using ocean-drilling ships, researchers have obtained radiometric dates for hundreds of locations on the ocean floor. By

knowing the age of a sample and its distance from the ridge axis where it was generated, an average rate of plate motion can be calculated.

Scientists have used these data, combined with paleomagnetism stored in hardened lavas on the ocean floor, to create maps that show the age of the ocean floor. The map in **FIGURE 7.34** shows that the reddish-orange colored bands range in age from the present to about 30 million years ago. The width of the bands indicates how much crust formed

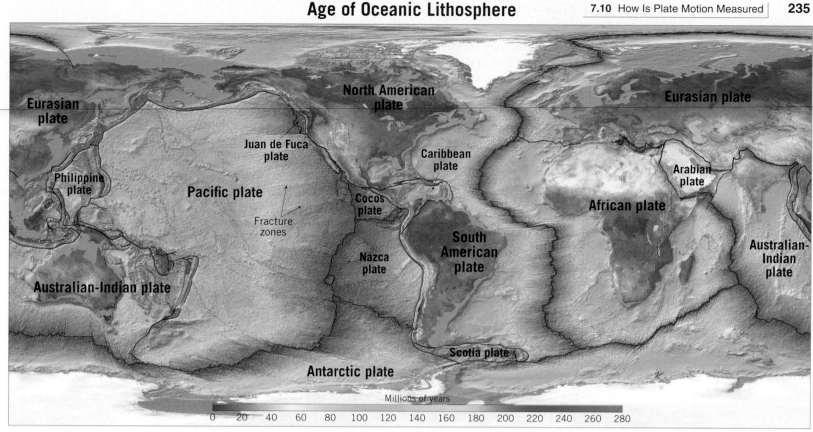

Millions of years

0 20 40 60 80 100 120 140 160 180 200 220 240 260 280

FIGURE 7.34 Age of the Ocean Floor

during that time period. For example, the reddish-orange band along the East Pacific Rise is more than three times wider than the same-color band along the Mid-Atlantic Ridge. Therefore, the rate of seafloor spreading has been approximately three times faster in the Pacific basin than in the Atlantic.

Maps of this type also provide clues to the current direction of plate movement. Notice the offsets in the ridges;

these are transform faults that connect the spreading centers. Recall that transform faults are aligned parallel to the direction of spreading. Therefore, when measured carefully, transform faults yield the direction of plate movement.

One way geologists can establish the direction of plate motion in the past is by examining the long fracture zones that extend for hundreds or even thousands of kilometers from ridge crests. Fracture zones are inactive extensions

Directions and rates of plate motions measured in centimeters per year

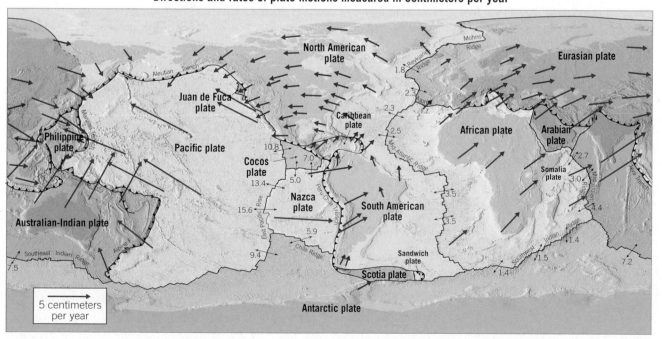

FIGURE 7.35 Rates of Plate Motion The red arrows show plate motion at selected locations based on GPS data. The small black arrows and labels show seafloor spreading velocities based mainly on paleomagnetic data. (Seafloor data from DeMets and others; GPS data from Jet Propulsion Laboratory)

of transform faults and therefore preserve a record of past directions of plate motion. Unfortunately, most of the ocean floor is less than 180 million years old, so to look deeper into the past, researchers must rely on paleomagnetic evidence provided by continental rocks.

Measuring Plate Motion from Space

You are likely familiar with the Global Positioning System (GPS), which is part of the navigation system used to locate one's position in order to provide directions to some other location. The GPS employs satellites that send radio signals that are intercepted by GPS receivers located at Earth's surface. The exact position of a site is determined by simultaneously establishing the distance from the receiver to four or more satellites. Researchers use specifically designed equipment to locate the position of a point on Earth to within a few millimeters (about the diameter of a small pea). To establish plate motions, GPS data are collected at numerous sites repeatedly over a number of years.

Data obtained from these and other techniques are shown in **FIGURE 7.35**. Calculations show that Hawaii is moving in a northwesterly direction and approaching Japan at 8.3 centimeters per year. A location in Maryland is retreating from a location in England at a speed of 1.7 centimeters per year—a value that is close to the 2.0-centimeters-per-year spreading rate established from paleomagnetic evidence obtained for the North Atlantic. Techniques involving GPS devices have also been useful in establishing small-scale crustal movements such as those that occur along faults in regions known to be tectonically active (for example, the San Andreas Fault).

7.10 CONCEPT CHECKS

1. What do transform faults that connect spreading centers indicate about plate motion?

2. Refer to Figure 7.35 and determine which three plates appear to exhibit the highest rates of motion.

7.11 | WHAT DRIVES PLATE MOTIONS? Summarize what is meant by plate–mantle convection and explain two of the primary driving forces of plate motion.

Researchers are in general agreement that some type of convection—where hot mantle rocks rise and cold, dense oceanic lithosphere sinks—is the ultimate driver of plate tectonics. Many of the details of this convective flow, however, remain topics of debate in the scientific community.

Forces That Drive Plate Motion

From geophysical evidence, we have learned that although the mantle consists almost entirely of solid rock, it is hot and weak enough to exhibit a slow, fluid-like convective flow. The simplest type of **convection** is analogous to heating a pot of water on a stove (**FIGURE 7.36**). Heating the base of a pot causes the water to become less dense (more buoyant), causing it to rise in relatively thin sheets or blobs that spread out at the surface. As the surface

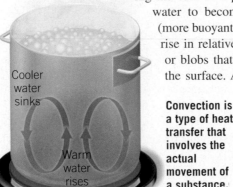

FIGURE 7.36 Convection in a Cooking Pot As a stove warms the water in the bottom of a cooking pot, the heated water expands, becomes less dense (more buoyant), and rises. Simultaneously, the cooler, denser water near the top sinks.

Cooler water sinks

Warm water rises

Convection is a type of heat transfer that involves the actual movement of a substance.

layer cools, its density increases, and the cooler water sinks back to the bottom of the pot, where it is reheated until it achieves enough buoyancy to rise again. Mantle convection is similar to, but considerably more complex than, the model just described.

There is general agreement that subduction of cold, dense slabs of oceanic lithosphere is a major driving force of plate motion (**FIGURE 7.37**). This phenomenon, called **slab pull**, occurs because cold slabs of oceanic lithosphere are more dense than the underlying warm asthenosphere and hence "sink like a rock" that is pulled down into the mantle by gravity.

Another important driving force is **ridge push** (see Figure 7.37). This gravity-driven mechanism results from the elevated position of the oceanic ridge, which causes slabs of lithosphere to "slide" down the flanks of the ridge. Ridge push appears to contribute far less to plate motions than slab pull. The primary evidence for this is that the fastest-moving plates—the Pacific, Nazca, and Cocos plates—have extensive subduction zones along their margins. By contrast, the spreading rate in the North Atlantic basin, which is nearly devoid of subduction zones, is one of the lowest, at about 2.5 centimeters (1 inch) per year.

Although the subduction of cold, dense lithospheric plates appears to be the dominant force acting on plates, other factors are at work as well. Flow in the mantle, perhaps best described as "mantle drag," is also thought to affect plate motion (see Figure 7.37). When flow in the asthenosphere is

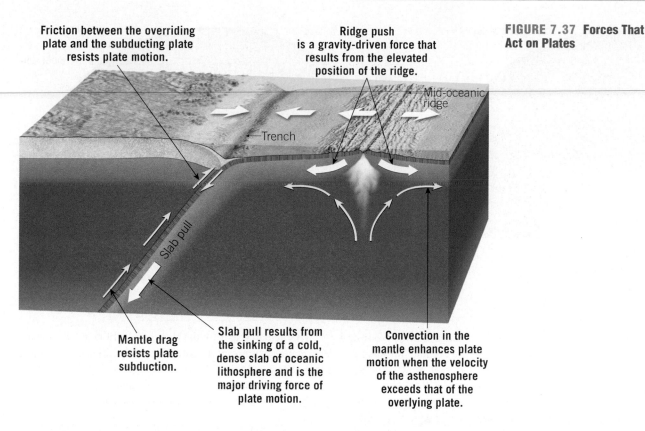

Friction between the overriding plate and the subducting plate resists plate motion.

Ridge push is a gravity-driven force that results from the elevated position of the ridge.

FIGURE 7.37 Forces That Act on Plates

Trench

Mid-oceanic ridge

Slab pull

Mantle drag resists plate subduction.

Slab pull results from the sinking of a cold, dense slab of oceanic lithosphere and is the major driving force of plate motion.

Convection in the mantle enhances plate motion when the velocity of the asthenosphere exceeds that of the overlying plate.

moving at a velocity that exceeds that of the plate, mantle drag enhances plate motion. However, if the asthenosphere is moving more slowly than the plate, or if it is moving in the opposite direction, this force tends to resist plate motion. Another type of resistance to plate motion occurs along some subduction zones, where friction between the overriding plate and the descending slab generates significant earthquake activity.

Models of Plate–Mantle Convection

Although convection in the mantle has yet to be fully understood, researchers generally agree on the following:

- Convective flow in the rocky 2900-kilometer- (1800-mile-) thick mantle—in which warm, buoyant rock rises and cooler, denser material sinks—is the underlying driving force for plate movement.
- Mantle convection and plate tectonics are part of the same system. Subducting oceanic plates drive the cold downward-moving portion of convective flow, while shallow upwelling of hot rock along the oceanic ridge and buoyant mantle plumes are the upward-flowing arms of the convective mechanism.
- Convective flow in the mantle is a major mechanism for transporting heat away from Earth's interior to the surface, where it is eventually radiated into space.

What is not known with certainty is the exact structure of this convective flow. Several models have been proposed for plate–mantle convection, and we will look at two of them.

Whole-Mantle Convection Most researchers favor some type of *whole-mantle convection* model, also called the *plume model*, in which cold oceanic lithosphere sinks to great depths and stirs the entire mantle (**FIGURE 7.38A**). The whole-mantle model suggests that the ultimate burial ground for subducting slabs is the core–mantle boundary. This downward flow is balanced by buoyantly rising mantle plumes that transport hot material toward the surface (see Figure 7.38A).

Two kinds of plumes have been proposed—narrow tube-like plumes and giant upwellings. The long, narrow plumes that extend from the core–mantle boundary are thought to produce hot-spot volcanism of the type associated with the Hawaiian islands, Iceland, and Yellowstone. Areas of large mega-plumes, as shown in Figure 7.38A, are thought to occur beneath the Pacific basin and southern Africa. The latter structure is thought to account for the fact that southern Africa has an elevation that is much higher than would be predicted for a stable continental landmass. Heat for both types of plumes is thought to arise mainly from the core, while the deep mantle provides a source for chemically distinct magmas.

Layer Cake Model Some researchers argue that the mantle resembles a "layer cake" divided at a depth of perhaps 660 kilometers (410 miles) but no deeper than 1000 kilometers (620 miles). As shown in **FIGURE 7.38B**, this layered model has two zones of convection—a thin, dynamic layer in the upper mantle and a thicker, sluggish one located below. As

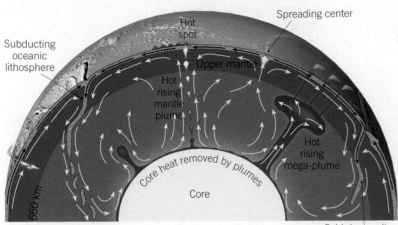

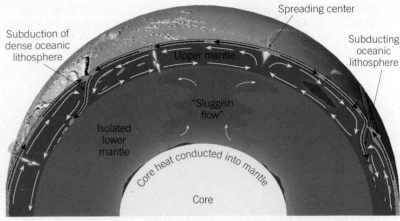

A. In the "whole-mantle model," sinking slabs of cold oceanic lithosphere are the downward limbs of convection cells, while rising mantle plumes carry hot material from the core-mantle boundary toward the surface.

B. The "layer cake model" has two largely disconnected convective layers. A dynamic upper layer driven by descending slabs of cold oceanic lithosphere and a sluggish lower layer that carries heat upward without appreciably mixing with the layer above.

FIGURE 7.38 Models of Mantle Convection

with the whole-mantle model, the downward convective flow is driven by the subduction of cold, dense oceanic lithosphere. However, rather than reach the lower mantle, these subducting slabs penetrate to depths of no more than 1000 kilometers (620 miles). Notice in Figure 7.38B that the upper layer in the layer cake model is littered with recycled oceanic lithosphere of various ages. Melting of these fragments is thought to be the source of magma for some of the volcanism that occurs away from plate boundaries, as in the volcanoes of Hawaii.

In contrast to the active upper mantle, the lower mantle is sluggish and does not provide material to support volcanism at the surface. However, very slow convection within this layer likely carries heat upward, but very little mixing between these two layers is thought to occur.

The actual shape of the convective flow in the mantle is a matter of current scientific investigation and is highly debated by geologists. Perhaps a hypothesis that combines features from the layer cake model and the whole-mantle convection model will emerge. McPHOTO/AGE Fotostock

7.11 CONCEPT CHECKS

1 Describe slab pull and ridge push. Which of these forces appears to contribute more to plate motion?

2 What role are mantle plumes thought to play in the convective flow in the mantle?

3 Briefly describe the two models proposed for mantle–plate convection.

EYE ON EARTH

In December 2011 a new volcanic island formed near the southern end of the Red Sea. People fishing in the area witnessed lava fountains reaching up to 30 meters (100 feet). This volcanic activity occurred off the west coast of Yemen, along the Red Sea Rift, among a collection of small islands in the Zubair Group. (NASA)

QUESTION 1 *What type of plate boundary produced this new volcanic island?*

QUESTION 2 *What two plates border the Red Sea Rift?*

QUESTION 3 *Are these two plates moving toward or away from each other?*

7 CONCEPTS IN REVIEW | Plate Tectonics: A Scientific Revolution Unfolds

7.1 FROM CONTINENTAL DRIFT TO PLATE TECTONICS

Discuss the view that most geologists held prior to the 1960s regarding the geographic positions of the ocean basins and continents.

- Fifty years ago, most geologists thought that ocean basins were very old and that continents were fixed in place. Those ideas were discarded with a scientific revolution that revitalized geology: the theory of plate tectonics. Supported by multiple kinds of evidence, plate tectonics is the foundation of modern Earth science.

7.2 CONTINENTAL DRIFT: AN IDEA BEFORE ITS TIME

List and explain the evidence Wegener presented to support his continental drift hypothesis.

KEY TERMS continental drift, supercontinent, Pangaea

- German meteorologist Alfred Wegener formulated the idea of continental drift in 1917. He suggested that Earth's continents are not fixed in place but have moved slowly over geologic time.
- Wegener reconstructed a super-continent called Pangaea that existed about 200 million years ago, during the late Paleozoic and early Mesozoic.
- Wegener's evidence that Pangaea existed but later broke into pieces that drifted apart included (1) the shape of the continents, (2) continental fossil organisms that matched across oceans, (3) matching rock types and modern mountain belts, and (4) sedimentary rocks that recorded ancient climates, including glaciers on the southern portion of Pangaea.

Q Why did Wegener choose organisms such as *Glossopteris* and *Mesosaurus* as evidence for continental drift, as opposed to other fossil organisms such as sharks or jellyfish?

0   10 cm
0   5 inches

John Cancalosi/AGE Fotostock

7.3 THE GREAT DEBATE

Discuss the two main objections to the continental drift hypothesis.

- Wegener's hypothesis suffered from two flaws: It proposed tidal forces as the mechanism for the motion of continents, and it implied that the continents would have plowed their way through weaker oceanic crust, like a boat cutting through a thin layer of sea ice. Geologists rejected the idea of continental drift when Wegener proposed it, and it wasn't resurrected for another 50 years.

Q Today, we know that the early twentieth-century scientists who rejected the idea of continental drift were wrong. Were they therefore bad scientists? Why or why not?

7.4 THE THEORY OF PLATE TECTONICS

List the major differences between Earth's lithosphere and asthenosphere and explain the importance of each in the plate tectonics theory.

KEY TERMS ocean ridge system, theory of plate tectonics, lithosphere, asthenosphere, lithospheric plate (plate)

- Research conducted during World War II led to new insights that helped revive Wegener's hypothesis of continental drift. Exploration of the seafloor revealed previously unknown features, including an extremely long mid-ocean ridge system. Sampling of the oceanic crust revealed that it was quite young relative to the continents.
- The lithosphere is Earth's outermost rocky layer that is broken into plates. It is relatively stiff and deforms by breaking or bending. Beneath the lithosphere is the asthenosphere, a relatively weak layer that deforms by flowing. The lithosphere consists both of crust (either oceanic or continental) and underlying upper mantle.
- There are seven large plates, another seven intermediate-size plates, and numerous relatively small microplates. Plates meet along boundaries that may either be divergent (moving apart from each other), convergent (moving toward each other), or transform (moving laterally past each other).

7.5 DIVERGENT PLATE BOUNDARIES AND SEAFLOOR SPREADING

Sketch and describe the movement along a divergent plate boundary that results in the formation of new oceanic lithosphere.

KEY TERMS divergent plate boundary (spreading center), rift valley, seafloor spreading, continental rift

- Seafloor spreading leads to the generation of new oceanic lithosphere at mid-ocean ridge systems. As two plates move apart from one another, tensional forces open cracks in the plate, allowing magma to well up and generate new slivers of seafloor. This process generates new oceanic lithosphere at a rate of 2 to 15 centimeters (1 to 6 inches) each year.
- As it ages, oceanic lithosphere cools and becomes denser. As it is transported away from the mid-ocean ridge, it therefore subsides. At the same time, new material is added to its underside, causing the plate to grow thicker.
- Divergent boundaries are not limited to the seafloor. Continents can break apart, too, starting with a continental rift (as in modern-day eastern Africa) and eventually leading to a new ocean basin opening between the two sides of the rift.

7.6 CONVERGENT PLATE BOUNDARIES AND SUBDUCTION

Compare and contrast the three types of convergent plate boundaries and name a location where each type can be found.

KEY TERMS convergent plate boundary, subduction zone, deep-ocean trench, partial melting, continental volcanic arc, volcanic island arc (island arc)

- When plates move toward one another, oceanic lithosphere is subducted into the mantle, where it is recycled. Subduction manifests itself on the ocean floor as a deep linear trench. The subducting slab of oceanic lithosphere can descend at a variety of angles, from nearly horizontal to nearly vertical.
- Aided by the presence of water, the subducted oceanic lithosphere triggers melting in the mantle, which produces magma. The magma is less dense than the surrounding rock and will rise. It may cool at depth, thickening the crust, or it may make it all the way to Earth's surface, where it erupts as a volcano.
- A line of volcanoes that erupt through continental crust is termed a continental volcanic arc, while a line of volcanoes that erupt through an overriding plate of oceanic lithosphere is a volcanic island arc.
- Continental crust resists subduction due to its relatively low density, and so when an intervening ocean basin is completely destroyed through subduction, the continents on either side collide, generating a new mountain range.

Q Sketch a typical continental volcanic arc and label the key parts. Then repeat the drawing with an overriding plate made of oceanic lithosphere.

7.7 TRANSFORM PLATE BOUNDARIES

Describe the relative motion along a transform plate boundary and locate several examples on a plate boundary map.

KEY TERMS transform plate boundary (transform fault), fracture zone

- At a transform boundary, lithospheric plates slide horizontally past one another. No new lithosphere is generated, and no old lithosphere is consumed. Shallow earthquakes signal the movement of these slabs of rock as they grind past their neighbors.
- The San Andreas Fault in California is an example of a transform boundary in continental crust, while the fracture zones between segments of the Mid-Atlantic Ridge are examples of transform faults in oceanic crust.

Q On the accompanying tectonic map of the Caribbean, find the Enriquillo Fault. (The location of the 2010 Haiti earthquake is shown as a yellow star.) What type a fault is the Enriquillo Fault? Are there any other faults in the area that show the same type of motion?

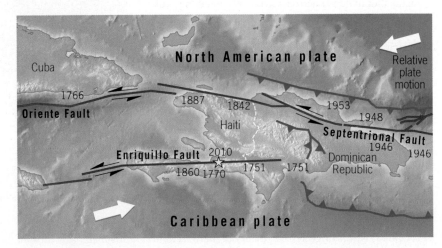

7.8 HOW DO PLATES AND PLATE BOUNDARIES CHANGE?

Explain why plates such as the African and Antarctic plates are getting larger, while the Pacific plate is getting smaller.

- Although the total surface area of Earth does not change, the shape and size of individual plates are constantly changing as a result of subduction and seafloor spreading. Plate boundaries can also be created or destroyed in response to changes in the forces acting on the lithosphere.
- The breakup of Pangaea and the collision of India with Eurasia are two examples of how plates change through geologic time.

7.9 TESTING THE PLATE TECTONICS MODEL

List and explain the evidence used to support the plate tectonics theory.

KEY TERMS mantle plume, hot spot, hot-spot track, Curie point, paleomagnetism (fossil magnetism), magnetic reversal, normal polarity, reverse polarity, magnetic time scale, magnetometer

- Multiple lines of evidence have verified the plate tectonics model. For instance, the Deep Sea Drilling Project found that the age of the seafloor increases with distance from a mid-ocean ridge. The thickness of sediment atop this seafloor is also proportional to distance from the ridge: Older lithosphere has had more time to accumulate sediment.
- Overall, oceanic lithosphere is quite young, with none older than about 180 million years.
- A hot spot is an area of volcanic activity where a mantle plume reaches the surface. Volcanic rocks generated by hot-spot volcanism provide evidence of both the direction and rate of plate movement over time.
- Magnetic minerals such as magnetite align themselves with Earth's magnetic field as rock forms. These fossil magnets act as recorders of the ancient orientation of Earth's magnetic field. This is useful to geologists in two ways: (1) It allows a given stack of rock layers to be interpreted in terms of changing position relative to the magnetic poles through time, and (2) reversals in the orientation of the magnetic field are preserved as "stripes" of normal and reversed polarity in the oceanic crust. Magnetometers reveal this signature of seafloor spreading as a symmetrical pattern of magnetic stripes parallel to the axis of the mid-ocean ridge.

7.10 HOW IS PLATE MOTION MEASURED?

Describe two methods researchers use to measure relative plate motion.

- Data collected from the ocean floor have established the direction and rate of motion of lithospheric plates. Transform faults point in the direction the plate is moving. Sediments with diagnostic fossils and radiometric dates of igneous rocks both help to calibrate the rate of motion.
- GPS satellites can be used to accurately measure the motion of special receivers to within a few millimeters. These "real-time" data back up the inferences made from seafloor observations. Plates move at about the same rate your fingernails grow: an average of about 5 centimeters (2 inches) per year.

7.11 WHAT DRIVES PLATE MOTIONS?

Summarize what is meant by plate–mantle convection and explain two of the primary driving forces of plate motion.

KEY TERMS convection, slab pull, ridge push

- Some kind of convection (upward movement of less dense material and downward movement of more dense material) appears to drive the motion of plates.
- Slabs of oceanic lithosphere sink at subduction zones because a subducted slab is denser than the underlying asthenosphere. In this process, called slab pull, Earth's gravity tugs at the slab, drawing the rest of the plate toward the subduction zone. As lithosphere descends the mid-ocean ridge, it exerts a small additional force, the outward-directed ridge push. Frictional drag exerted on the underside of plates by flowing mantle is another force that acts on plates and influences their direction and rate of movement.
- The exact patterns of mantle convection are not fully understood. Convection may occur throughout the entire mantle, as suggested by the whole-mantle model. Or it may occur in two layers within the mantle—the active upper mantle and the sluggish lower mantle—as proposed in the layer cake model.

Q Compare and contrast mantle convection with the operation of a lava lamp.

Photo by Steve Bower/Shutterstock

GIVE IT SOME **THOUGHT**

1. After referring to the section in the Introduction titled "The Nature of Scientific Inquiry," answer the following:

   a. What observation led Alfred Wegener to develop his continental drift hypothesis?

   b. Why did the majority of the scientific community reject the continental drift hypothesis?

   c. Do you think Wegener followed the basic principles of scientific inquiry? Support your answer.

2. Referring to the accompanying diagrams that illustrate the three types of convergent plate boundaries, complete the following:

   a. Identify each type of convergent boundary.

   b. On what type of crust do volcanic island arcs develop?

   c. Why are volcanoes largely absent where two continental blocks collide?

   d. Describe two ways that oceanic–oceanic convergent boundaries are different from oceanic–continental boundaries. How are they similar?

A.   B.   C.

3. Some predict that California will sink into the ocean. Is this idea consistent with the theory of plate tectonics? Explain.

4. Refer to the accompanying hypothetical plate map to answer the following questions:

   a. How many portions of plates are shown?

   b. Are continents A, B, and C moving toward or away from each other? How did you determine your answer?

   c. Explain why active volcanoes are more likely to be found on continents A and B than on continent C.

   d. Provide at least one scenario in which volcanic activity might be triggered on continent C.

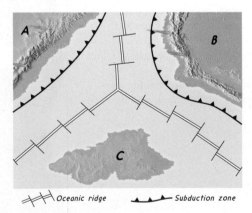

━┿┿━ Oceanic ridge   ▲▲▲▲ Subduction zone

5. Volcanoes, such as the Hawaiian chain, that form over mantle plumes are some of the largest shield volcanoes on Earth. However, several shield volcanoes on Mars are gigantic compared to those on Earth. What does this difference tell us about the role of plate motion in shaping the Martian surface?

6. Imagine that you are studying seafloor spreading along two different oceanic ridges. Using data from a magnetometer, you produced the two accompanying maps. From these maps, what can you determine about the relative rates of seafloor spreading along these two ridges? Explain.

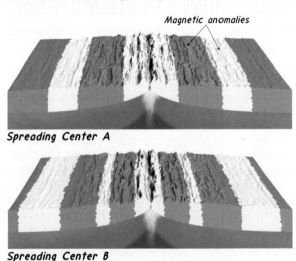

Magnetic anomalies

Spreading Center A

Spreading Center B

7. Australian marsupials (kangaroos, koala bears, etc.) have direct fossil links to marsupial opossums found in the Americas. Yet the modern marsupials in Australia are markedly different from their American relatives. How does the breakup of Pangaea help to explain these differences (see Figure 7.24)?

8. Density is a key component in the behavior of Earth materials and is especially important in understanding key aspects of plate tectonics. Describe three different ways that density and/or density differences play a role in plate tectonics.

9. Refer to the accompanying map and the pairs of cities below to complete the following:
(Boston, Denver), (London, Boston), (Honolulu, Beijing)

   a. List the pair of cities that is moving apart as a result of plate motion.
   b. List the pair of cities that is moving closer as a result of plate motion.
   c. List the pair of cities that is presently not moving relative to each other.

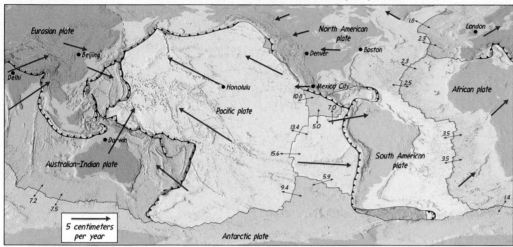

Plate motion measured in centimeters per year

EXAMINING THE **EARTH SYSTEM**

1. As an integral subsystem of the Earth system, plate tectonics has played a major role in determining the events that have taken place on Earth since its formation about 4.6 billion years ago. Briefly comment on the general effect that the changing positions of the continents and the redistribution of land and water over Earth's surface have had on the atmosphere, hydrosphere, and biosphere through time. You may want to review plate tectonics by investigating the topic on the Internet using a search engine or visit the U.S. Geological Survey's (USGS's) This Dynamic Earth: The Story of Plate Tectonics Website, at **http://pubs.usgs.gov/gip/dynamic/ dynamic.html.**

2. Assume that plate tectonics did not cause the breakup of the supercontinent Pangaea. Use the accompanying map of Pangaea to describe how the climate (atmosphere), vegetation and animal life (biosphere), and geologic features (geosphere) would be different from the conditions that exist today in the areas currently occupied by the cities of Miami, Florida; Chicago, Illinois; New York, New York; and your college campus location.

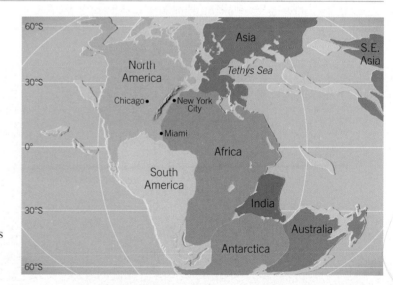

MasteringGeology™

Looking for additional review and test prep materials? Visit the Self Study area in **www.masteringgeology.com** to find practice quizzes, study tools, and multimedia that will aid in your understanding of this chapter's content. In **MasteringGeology™** you will find:

- GEODe: Earth Science: An interactive visual walkthrough of key concepts
- Geoscience Animation Library: More than 100 animations illuminating many difficult-to-understand Earth science concepts

- In The News RSS Feeds: Current Earth science events and news articles are pulled into the site with assessment
- Pearson eText
- Optional Self Study Quizzes
- Web Links
- Glossary
- Flashcards

8
Earthquakes and Earth's Interior

Tsunami striking the coast of Japan on March 11, 2011.
(Photo by Sadatsugu Tomizawa/AFP/Getty Images)

On January 12, 2010, an estimated 316,000 people lost their lives when a magnitude 7.0 earthquake struck the small Caribbean nation of Haiti, the poorest country in the Western Hemisphere. In addition to the staggering death toll, there were more than 300,000 injuries, and more than 280,000 houses were destroyed or damaged.

The quake originated only 25 kilometers (15 miles) from the country's densely populated capital city of Port-au-Prince. It occurred along a San Andreas–like fault at a depth of just 10 kilometers (6 miles). Because of the quake's shallow depth,

ground shaking was exceptional for an event of this magnitude. Other factors that contributed to the Port-au-Prince disaster included the city's geologic setting and the nature of its buildings. The city is built on sediment, which is quite susceptible to ground shaking during an earthquake. More importantly, inadequate or nonexistent building codes meant that buildings collapsed far more readily than they should have. At least 52 aftershocks, measuring magnitude 4.5 or greater, jolted the area and added to the trauma survivors experienced for days after the original quake.

8.1 | WHAT IS AN EARTHQUAKE?

Sketch and describe the mechanism that generates most earthquakes.

An **earthquake** is ground shaking caused by the sudden and rapid movement of one block of rock slipping past another along fractures in Earth's crust called **faults**. Most faults are locked, except for brief, abrupt movements when sudden slippage produces an earthquake (**FIGURE 8.1**). Faults are locked because the confining pressure exerted by the overlying crust is enormous, causing these fractures in the crust to be "squeezed shut."

Earthquakes tend to occur along preexisting faults where internal stresses have caused the crustal rocks to rupture or break into two or more units. The location where slippage begins is called the **hypocenter**, or **focus**. Earthquake waves radiate from this spot outward into the surrounding rock. The point on Earth's surface directly above the hypocenter is called the **epicenter** (**FIGURE 8.2**).

Large earthquakes release huge amounts of stored-up energy as **seismic waves**—a form of energy that travels through the lithosphere and Earth's interior. The energy carried by these waves causes the material that transmits them to shake. Seismic waves are analogous to waves produced when a stone is dropped into a calm pond. Just as the impact of the stone creates a pattern of circular waves, an earthquake generates waves that radiate outward in all directions from the hypocenter. Although seismic energy dissipates rapidly as it moves away from the quake's hypocenter, sensitive instruments can detect earthquakes even when they occur on the opposite side of Earth.

Thousands of earthquakes occur around the world every day. Fortunately, most are small and cannot be detected by people. Of these, only about 15 strong earthquakes

FIGURE 8.1 Presidential Palace Damaged During the 2010 Haiti Earthquake
(Photo by Luis Acosta/Getty Images)

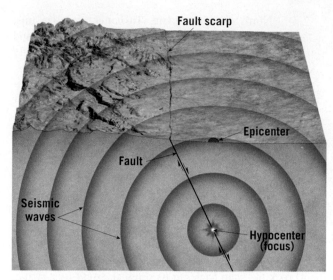

FIGURE 8.2 Earthquake Hypocenter and Epicenter The *hypocenter* is the zone at depth where the initial displacement occurs. The *epicenter* is the surface location directly above the hypocenter.

(magnitude 7 or greater) are recorded each year, many of them occurring in remote regions. Occasionally, a large earthquake is triggered near a major population center. Such events are among the most destructive natural forces on Earth. The shaking of the ground, coupled with the liquefaction of soils, wreaks havoc on buildings, roadways, and other structures. In addition, when a quake occurs in a populated area, power and gas lines are often ruptured, causing

numerous fires. In the famous 1906 San Francisco earthquake, much of the damage was caused by fires that became uncontrollable when broken water mains left firefighters with only trickles of water (**FIGURE 8.3**).

Discovering the Causes of Earthquakes

The energy released by volcanic eruptions, massive landslides, and meteorite impacts can generate earthquakelike waves, but these events are usually weak. What mechanism produces a destructive earthquake? As you have learned, Earth is not a static planet. We know that large sections of Earth's crust have been thrust upward because fossils of marine organisms have been discovered thousands of meters above sea level. Other regions, such as California's Death Valley, exhibit evidence of extensive subsidence. In addition to these vertical displacements, offsets in fences, roads, and other structures indicate that horizontal movements between blocks of Earth's crust are also common (**FIGURE 8.4**).

The actual mechanism of earthquake generation eluded geologists until H. F. Reid of Johns Hopkins University conducted a landmark study following the 1906 San Francisco earthquake. This earthquake was accompanied by horizontal surface displacements of several meters along the northern portion of the San Andreas Fault. Field studies determined that during this single earthquake, the Pacific plate lurched as much as 9.7 meters (32 feet) northward, past the adjacent

San Francisco in flames following the 1906 quake. Broken water mains left firefighters without water.

FIGURE 8.3 Earthquakes Can Trigger Fires
(Reproduced from the collection of the Library of Congress; inset photo by Hal Garb/AFP/Getty Images)

Fire triggered when a gas line ruptured during the Northridge earthquake in southern California in 1994.

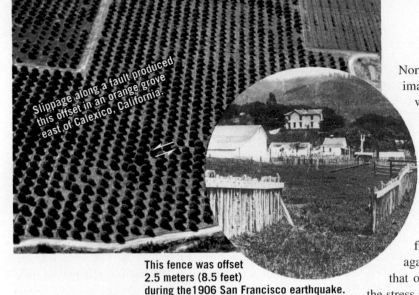

Slippage along a fault produced this offset in an orange grove east of Calexico, California.

This fence was offset 2.5 meters (8.5 feet) during the 1906 San Francisco earthquake.

North American plate. To better visualize this, imagine standing on one side of the fault and watching a person on the other side suddenly slide horizontally 32 feet to your right.

What Reid concluded from his investigations is illustrated in **FIGURE 8.5**. Over tens to hundreds of years, differential stress slowly bends the crustal rocks on both sides of a fault. This is much like a person bending a limber wooden stick, as shown in Figure 8.5. Frictional resistance keeps the fault from rupturing and slipping. (Friction acts against slippage and is enhanced by irregularities that occur along the fault surface.) At some point, the stress along the fault overcomes the frictional resistance, and slip initiates. Slippage allows the deformed (bent) rock to "snap back" to its original, stress-free, shape; a series of earthquake waves radiate as it slides (see Figure 8.5C,D). Reid termed the "springing back" **elastic rebound** because the rock behaves elastically, much like a stretched rubber band does when it is released.

SmartFigure 8.5 Elastic Rebound

Deformation of rocks   **Deformation of a limber stick**

A. Original position of rocks on opposite sides of a fault.

B. The movement of tectonic plates causes the rocks to bend and store elastic energy.

C. Once the strength of the rocks is exceeded, slippage along the fault produces an earthquake.

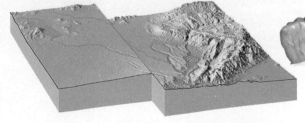

D. The rocks return to their original shape, but in a new location.

Tens to hundreds of years

Seconds to a few minutes

Time

Aftershocks and Foreshocks

Strong earthquakes are followed by numerous earthquakes of lesser magnitude, called **aftershocks**, which are the result of crust along the fault surface adjusting to the displacement caused by the main shock. Aftershocks gradually diminish in frequency and intensity over a period of several months following an earthquake. In a little more than a month following the 2010 Haiti earthquake, the U.S. Geological Survey detected nearly 60 aftershocks with magnitudes of 4.5 or greater. The two largest aftershocks had magnitudes of 6.0 and 5.9, both large enough to inflict damage. Hundreds of minor tremors were also felt. Although aftershocks are weaker than the main earthquake, they often trigger the destruction of already weakened structures. For example, in northwestern Armenia in 1988, where many people lived in large apartment buildings constructed of brick and concrete slabs, a moderate earthquake of magnitude 6.9 weakened many structures, and a strong aftershock of magnitude 5.8 completed the demolition.

In contrast to aftershocks, small earthquakes called **foreshocks** often, but not always, precede major earthquakes by days or, in some cases, several years. Monitoring of foreshocks to predict forthcoming earthquakes has been attempted with only limited success.

Faults and Large Earthquakes

The slippage that occurs along faults can be explained by the plate tectonics theory, which states that large slabs of Earth's lithosphere are continually grinding past one another. These mobile plates interact with neighboring plates, straining and deforming the rocks at their edges. Faults associated with plate boundaries are the source of most large earthquakes.

Transform Fault Boundaries Faults in which the dominant displacement is horizontal and parallel to the *strike* (direction) of the fault surface are called **strike-slip faults**. Large strike-slip faults that slice through Earth's lithosphere and accommodate motion between two tectonic plates are called **transform faults**. For example, the San Andreas Fault is a large strike-slip fault that separates the North American plate and the Pacific plate. Most transform faults, including the San Andreas Fault, are not perfectly straight or continuous; instead, they consist of numerous branches and smaller fractures that display kinks and offsets (**FIGURE 8.6**). In addition, geologists have learned that displacement along transform faults occurs in discrete segments that often behave differently from one another. Some sections of the San Andreas Fault exhibit slow, gradual displacement, known as **fault creep**, and produce little seismic shaking. Other segments slip at relatively closely spaced intervals, producing numerous small to moderate earthquakes. Still other segments remain locked and store elastic energy for a few hundred years before they break loose; ruptures on these segments usually result in major earthquakes.

Earthquakes that occur along locked segments of the San Andreas Fault tend to be repetitive: As soon as one is

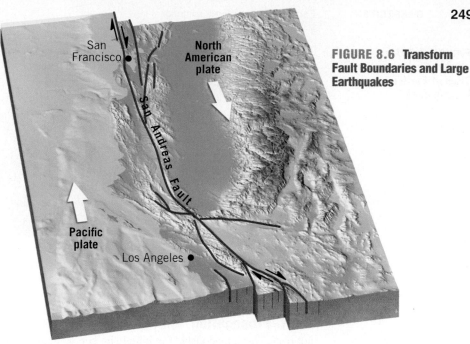

FIGURE 8.6 Transform Fault Boundaries and Large Earthquakes

over, strain immediately begins accumulating due to the continuous motion of the plates. Decades or centuries later, the fault fails again.

Faults Associated with Convergent Plate Boundaries Strong earthquakes also occur along large faults associated with convergent plate boundaries. Compressional forces associated with continental collisions that result in mountain building generate many *thrust faults*. Displacement along a thrust fault results in rock above the fault being forced (or *thrust*) over rock below the fault surface, causing an earthquake (**FIGURE 8.7**).

In addition, the plate boundary between a subducting slab of oceanic lithosphere and the overlying plate form a fault referred to as a **megathrust fault** (see Figure 8.7). Because these large thrust faults lie partially beneath the ocean floor, movement along these faults may displace the overlying seawater, generating destructive tsunami. Megathrust faults have produced the majority of Earth's most powerful and destructive earthquakes, including the 2011 Japan quake (M 9.0), the 2004

EYE ON EARTH

The Calaveras Fault, a branch of the San Andreas Fault system, cuts directly through the town of Hollister, California. Rather than being "locked," this section of the fault is slowly slipping, producing noticeable offsets and damage to curbs, sidewalks, roads, and buildings. The concrete wall and sidewalk shown here were straight when they were originally constructed.

QUESTION 1 What term is used to describe the phenomenon observed in Hollister along the Calaveras Fault?

QUESTION 2 Are faults that exhibit this type of slippage considered likely to generate a major earthquake? Explain.

FIGURE 8.7 Megathrust Faults are the Sites of Earth's Largest Earthquakes

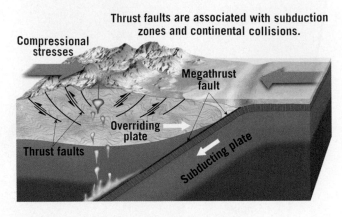

Thrust faults are associated with subduction zones and continental collisions.

Compressional stresses

Megathrust fault

Overriding plate

Thrust faults

Subducting plate

Indian Ocean (Sumatra) quake (M 9.1), the 1964 Alaska quake (M 9.2), and the largest earthquake yet recorded, the 1960 Chile quake (M 9.5).

Fault Propagation   Slippage along large faults does not occur instantaneously. The initial slip begins at the hypocenter and propagates (travels) along the fault surface, at 2 to 4 kilometers per second—faster than a rifle shot. Slippage on one section of the fault adds strain to the adjacent segment, which may also slip. As this zone of slippage advances, it can slow down, speed up, or even jump to a nearby fault segment. The propagation of the rupture zone along a 300-kilometer-(200-mile-) long fault, for example, takes about 1.5 minutes, as compared to about 30 seconds for a 100-kilometer- (60-mile-) long fault. Earthquake waves are generated at every point along the fault as that portion of the fault begins to slip.

8.1 CONCEPT CHECKS

1 What is an earthquake? Under what circumstances do most large earthquakes occur?

2 How are faults, hypocenters, and epicenters related?

3 Who was the first person to explain the mechanism by which most earthquakes are generated?

4 Explain what is meant by *elastic rebound*.

5 What is the approximate duration of an earthquake that occurs along a 300-kilometer-long fault?

6 Defend or rebut this statement: Faults that do not experience fault creep may be considered safe.

7 What type of faults tend to produce the most destructive earthquakes?

8.2 | SEISMOLOGY: THE STUDY OF EARTHQUAKE WAVES Compare and contrast the types of seismic waves and describe the principle of the seismograph.

The study of earthquake waves, **seismology**, dates back to attempts made in China almost 2000 years ago to determine the direction from which these waves originated. The earliest known instrument, invented by Zhang Heng, was a large hollow jar that contained a weight suspended from the top (**FIGURE 8.8**). The suspended weight (similar to a clock pendulum) was connected to the jaws of several large dragon figurines that encircled the container. The jaws of each dragon held a metal ball. When earthquake waves reached the instrument, the relative motion between the suspended mass and the jar would dislodge some of the metal balls into the waiting mouths of frogs directly below.

Ball dropping

FIGURE 8.8 Ancient Chinese Seismograph During an Earth tremor, the dragons located in the direction of the main vibrations would drop a ball into the mouth of a frog below. (Photo by James E. Patterson Collection)

Instruments That Record Earthquakes

In principle, modern **seismographs**, or **seismometers**, are similar to the instruments used in ancient China. A seismograph has a weight freely suspended from a support

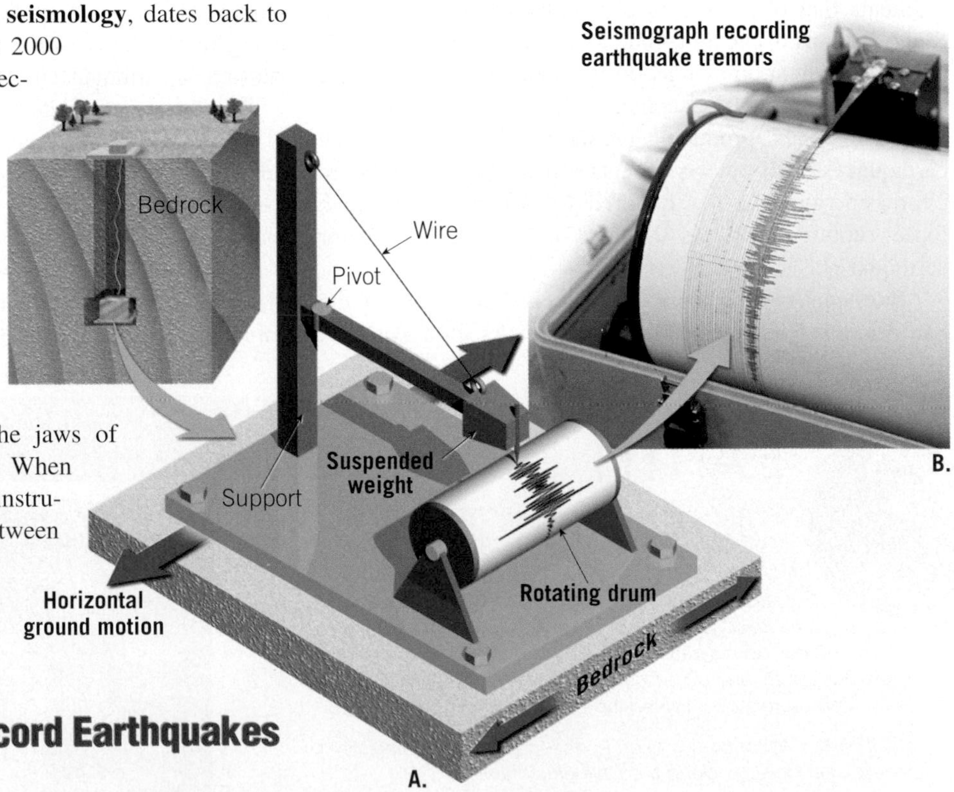

Seismograph recording earthquake tremors

Bedrock

Wire

Pivot

Suspended weight

Support

Horizontal ground motion

Rotating drum

Bedrock

A.

B.

FIGURE 8.9 Principle of the Seismograph The inertia of the suspended weight tends to keep it motionless, while the recording drum, which is anchored to bedrock, vibrates in response to seismic waves. The stationary weight provides a reference point from which to measure the amount of displacement occurring as a seismic wave passes through the ground. (Photo courtesy of Zephyr/Science Source)

that is securely attached to bedrock (**FIGURE 8.9**). When vibrations from an earthquake reach the instrument, the **inertia** of the weight keeps it relatively stationary, while Earth and the support move. Inertia can be simply described by this statement: Objects at rest tend to stay at rest, and objects in motion tend to remain in motion, unless acted upon by an outside force. You probably have experienced inertia when you have tried to stop your automobile quickly and your body has continued to move forward.

To detect very weak earthquakes or a great earthquake that has occurred in another part of the world, most seismographs are designed to amplify ground motion. In earthquake-prone areas, the instruments used are designed to withstand the violent shaking that can occur near the quake's epicenter.

SmartFigure 8.10 Body Waves (P and S waves) versus Surface Waves P and S waves travel through Earth's interior, while surface waves travel in the layer directly below the surface. P waves are the first to arrive at a seismic station, followed by S waves, and then surface waves.

Seismic Waves

The records obtained from seismographs, called **seismograms**, provide useful information about the nature of seismic waves. Seismograms reveal that two main types of seismic waves are generated by the slippage of a rock mass. One of these wave types, called **surface waves**, travels in the rock layers just below Earth's surface (**FIGURE 8.10**). The other wave types travel through Earth's interior and are called **body waves**.

Body Waves
Body waves are further divided into two types—called **primary waves**, or **P waves**, and **secondary waves**, or **S waves**—and are identified by their mode of travel through intervening materials. P waves are "push/pull" waves; they momentarily push (compress) and pull (stretch) rocks in the direction the wave is traveling (**FIGURE 8.11A**). This wave motion is similar to that generated by human vocal cords as they move air back and forth to create sound. Solids, liquids, and gases resist stresses that change their volume when compressed and, therefore, elastically spring back once the stress is removed. Therefore, P waves can travel through all these materials.

By contrast, S waves "shake" the particles at right angles to their direction of travel. This can be illustrated by fastening one end of a rope and shaking the other end, as shown in **FIGURE 8.11B**. Unlike P waves, which temporarily change the *volume* of intervening material by alternately squeezing and stretching it, S waves change the *shape* of the material that transmits them. Because fluids (gases and liquids) do not resist stresses that cause changes in shape—meaning fluids do not return to their original shape once the stress is removed—liquids and gases will not transmit S waves.

Surface Waves
There are two types of surface waves. One type causes Earth's surface and anything resting on it

A. As illustrated by a toy Slinky, P waves alternately compress and expand the material through which they pass.

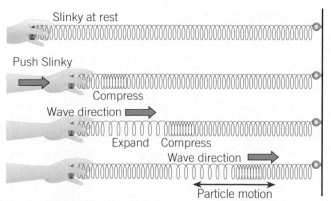

B. S waves cause material to oscillate at right angles to the direction of wave motion.

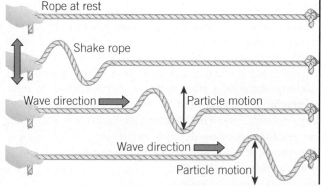

FIGURE 8.11 The Characteristic Motion of P Waves and S Waves During a strong earthquake, ground shaking consists of a combination of various kinds of seismic waves.

FIGURE 8.12 Two Types of Surface Waves

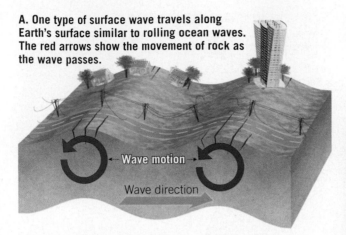

A. One type of surface wave travels along Earth's surface similar to rolling ocean waves. The red arrows show the movement of rock as the wave passes.

Wave motion

Wave direction

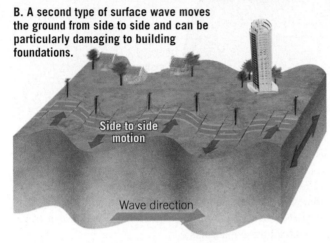

B. A second type of surface wave moves the ground from side to side and can be particularly damaging to building foundations.

Side to side motion

Wave direction

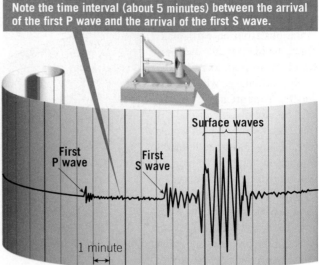

Note the time interval (about 5 minutes) between the arrival of the first P wave and the arrival of the first S wave.

Surface waves

First P wave

First S wave

1 minute

(Earlier)   T I M E   (Later)

FIGURE 8.13 Typical Seismogram

to move, much as ocean swells toss a ship (**FIGURE 8.12A**). The second type of surface wave causes Earth's materials to move side to side. This motion is particularly damaging to the foundations of structures (**FIGURE 8.12B**).

Body Waves Versus Surface Waves By examining the seismogram shown in **FIGURE 8.13**, you can see that the major difference among seismic waves is their speed of travel. P waves are the first to arrive at a recording station, then S waves, and finally surface waves. Generally, in any

solid Earth material, P waves travel about 1.7 times faster than S waves, and S waves are roughly 10 percent faster than surface waves.

In addition to velocity differences, notice in Figure 8.13 that the height, or *amplitude*, of these wave types also varies. S waves have slightly greater amplitudes than P waves, and surface waves exhibit even greater amplitudes. Surface waves also retain their maximum amplitude longer than P and S waves. As a result, surface waves tend to cause greater ground shaking and, hence, greater property damage, than either P or S waves.

8.2 CONCEPT CHECKS

1 Describe the principle of the seismograph.

2 List the major differences between P, S, and surface waves.

3 Which type of seismic waves tend to cause the greatest destruction to buildings?

8.3 | DETERMINING THE SIZE OF EARTHQUAKES

Distinguish between intensity scales and magnitude scales.

Seismologists use a variety of methods to determine two fundamentally different measures that describe the size of an earthquake: *intensity* and *magnitude*. The first of these to be used was **intensity**—a measure of the amount of ground shaking at a particular location, based on observed property damage. Later, with the development of seismographs, it became possible to measure ground motion using instruments. This quantitative measurement, called **magnitude**, relies on data gleaned from seismic records to estimate the amount of energy released at an earthquake's source.

Intensity Scales

Until the mid-1800s, historical records provided the only accounts of the severity of earthquake shaking and destruction. Perhaps the first attempt to scientifically describe the aftermath of an earthquake came following the great Italian earthquake of 1857. By systematically mapping effects of the earthquake, a measure of the intensity of ground shaking was established. The map generated by this study used lines to connect places of equal damage and hence equal ground shaking. Using this technique, zones of intensity were identified, with the zone of highest intensity located near the center of maximum ground shaking and often (but not always) the earthquake epicenter.

In 1902, Giuseppe Mercalli developed a more reliable intensity scale, which in a modified form is still used today. The **Modified Mercalli Intensity scale**, shown in **TABLE 8.1**, was developed using California buildings as its standard. For example, on the 12-point Mercalli Intensity scale, when some well-built wood structures and most masonry buildings are destroyed by an earthquake, the affected area is assigned a Roman numeral X (10). **FIGURE 8.14** shows the zone of destruction for the 1989 Loma Prieta earthquake, where the intensity of ground shaking was based on the Modified Mercalli Intensity scale.

Magnitude Scales

To more accurately compare earthquakes around the globe, scientists searched for a way to describe the energy released by earthquakes that did not rely on factors, such as building practices, that vary considerably from one part of the world to another. As a result, several magnitude scales were developed.

Richter Magnitude   In 1935 Charles Richter of the California Institute of Technology developed the first magnitude scale to use seismic records. As shown in **FIGURE 8.15** (top), the **Richter scale** is calculated by measuring the amplitude of the largest seismic wave (usually an S, or surface, wave)

TABLE 8.1   Modified Mercalli Intensity Scale

I	Not felt except by a very few under especially favorable circumstances.
II	Felt only by a few persons at rest, especially on upper floors of buildings.
III	Felt quite noticeably indoors, especially on upper floors of buildings, but many people do not recognize it as an earthquake.
IV	During the day, felt indoors by many, outdoors by few. Sensation like heavy truck striking building.
V	Felt by nearly everyone, many awakened. Disturbances of trees, poles, and other tall objects sometimes noticed.
VI	Felt by all; many frightened and run outdoors. Some heavy furniture moved; few instances of fallen plaster or damaged chimneys. Damage slight.
VII	Everybody runs outdoors. Damage negligible in buildings of good design and construction; slight to moderate in well-built ordinary structures; considerable in poorly built or badly designed structures.
VIII	Damage slight in specially designed structures; considerable in ordinary substantial buildings with partial collapse; great in poorly built structures. (Fall of chimneys, factory stacks, columns, monuments, walls.)
IX	Damage considerable in specially designed structures. Buildings shifted off foundations. Ground cracked conspicuously.
X	Some well-built wooden structures destroyed. Most masonry and frame structures destroyed. Ground badly cracked.
XI	Few, if any, (masonry) structures remain standing. Bridges destroyed. Broad fissures in ground.
XII	Damage total. Waves seen on ground surfaces. Objects thrown upward into air.

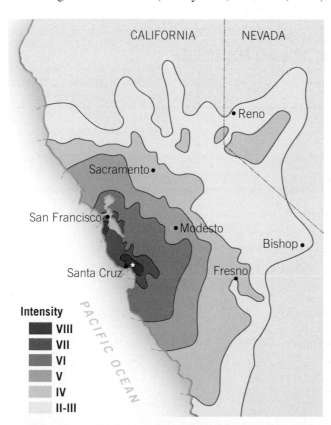

FIGURE 8.14 Seismic Intensity Map, Loma Prieta, 1989 Intensity levels are based on the Modified Mercalli Intensity scale, which uses Roman numerals to represent the intensity categories. The zone of maximum intensity during this event roughly corresponded to the epicenter, but this is not always the case.

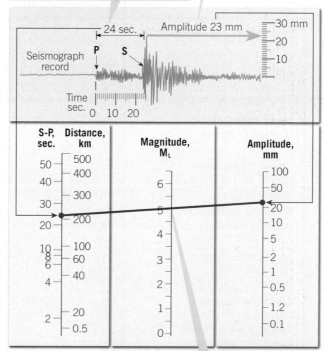

1. Measure the height (amplitude) of the largest wave on the seismogram (23 mm) and plot it on the amplitude scale (right).

2. Determine the distance to the earthquake using the time interval separating the arrival of the first P wave and the arrival of the first S wave (24 seconds) and plot it on the distance scale (left).

3. Draw a line connecting the two plots and read the Richter magnitude (M$_L$ 5) from the magnitude scale (center).

FIGURE 8.15 Determining the Richter Magnitude of an Earthquake

FIGURE 8.16 Magnitude versus Ground Motion and Energy Released An earthquake that is one magnitude stronger than another (M 6 versus M 5) produces seismic waves that have a maximum amplitude 10 times greater, and they release about 32 times more energy than the waves of the weaker quake.

Magnitude vs. Ground Motion and Energy

Magnitude Change	Ground Motion Change (amplitude)	Energy Change (approximate)
4.0	10,000 times	1,000,000 times
3.0	1000 times	32,000 times
2.0	100 times	1000 times
1.0	10.0 times	32 times
0.5	3.2 times	5.5 times
0.1	1.3 times	1.4 times

recorded on a seismogram. Because seismic waves weaken as the distance between the hypocenter and the seismograph increases, Richter developed a method that accounts for the decrease in wave amplitude with increasing distance. Theoretically, as long as equivalent instruments are used, monitoring stations at different locations will obtain the same Richter magnitude for each recorded earthquake. In practice, however, different recording stations often obtain slightly different magnitudes for the same earthquake—a consequence of the variations in rock types through which the waves travel.

Earthquakes vary enormously in strength, and great earthquakes produce wave amplitudes that are thousands of times larger than those generated by weak tremors. To accommodate this wide variation, Richter used a *logarithmic scale* to express magnitude, in which a *10-fold* increase in wave amplitude corresponds to an increase of 1 on the magnitude scale. Thus, the

intensity of ground shaking for a magnitude 5 earthquake is 10 times greater than that produced by an earthquake having a Richter magnitude of 4 (**FIGURE 8.16**).

In addition, each unit of increase in Richter magnitude equates to roughly a *32-fold increase in the energy released*. Thus, an earthquake with a magnitude of 6.5 releases 32 times more energy than one with a magnitude of 5.5 and roughly 1000 times (32 × 32) more energy than a magnitude 4.5 quake. A major earthquake with a magnitude of 8.5 releases millions of times more energy than the smallest earthquakes felt by humans (**FIGURE 8.17**).

The convenience of describing the size of an earthquake by a single number that can be calculated quickly from seismograms makes the Richter scale a powerful tool. Seismologists have since modified Richter's work and developed other Richter-like magnitude scales.

Despite its usefulness, the Richter scale is not adequate for describing very large earthquakes. For example, the 1906 San Francisco earthquake and the 1964 Alaska earthquake have roughly the same Richter magnitudes. However, based on the relative size of the affected areas and the associated tectonic changes, the Alaska earthquake released considerably more energy than the San Francisco quake. Thus, the Richter scale is considered *saturated* for major earthquakes because it cannot distinguish among them. Despite this shortcoming, Richter-like scales are still used because they can be calculated quickly.

FIGURE 8.17 Annual Occurrence of Earthquakes with Various Magnitudes

Frequency and Energy Released by Earthquakes of Different Magnitudes

Magnitude (Mw)	Average Per Year	Description	Examples	Energy Release (equivalent kilograms of explosive)
9	<1	**Largest recorded earthquakes–** destruction over vast area massive loss of life possible	Chile, 1960 (M 9.5); Alaska, 1964 (M 9.0); Japan, 2011 (M 9.0)	56,000,000,000,000
8	1	**Great earthquakes–** severe economic impact large loss of life	Sumatra, 2006 (M 8.6); Mexico City, 1980 (M 8.1)	1,800,000,000,000
7	15	**Major earthquakes–** damage ($ billions) loss of life	New Madrid, Missouri 1812 (M 7.7); Turkey, 1999 (M 7.6); Charleston, South Carolina, 1886 (M 7.3)	56,000,000,000
6	134	**Strong earthquakes–** can be destructive in populated areas	Kobe, Japan, 1995 (M 6.9); Loma Prieta, California, 1989 (M 6.9); Northridge, California, 1994 (M 6.7)	1,800,000,000
5	1319	**Moderate earthquakes–** property damage to poorly constructed buildings	Mineral, Virginia, 2011 (M 5.8); Northern New York, 1994 (M 5.8); East of Oklahoma City, Oklahoma, 2011 (M5.6)	56,000,000
4	13,000	**Light earthquakes–** noticeable shaking of items indoors, some property damage	Western Minnesota, 1975 (M 4.6); Arkansas, 2011 (M 4.7)	1,800,000
3	130,000	**Minor earthquakes–** felt by humans, very light property damage, if any	New Jersey, 2009 (M3.0) Maine, 2006 (M3.8)	56,000
2	1,300,000	**Very minor earthquakes–** felt by humans, no property damage		1,800
	Unknown	**Very minor earthquakes–** generally not felt by humans, but may be recorded		56

Finding the Epicenter of an Earthquake

The difference in the velocities of P and S waves provides a method for locating the epicenter of an earthquake. Since P waves travel faster than S waves, the further the epicenter is from the recording instrument, the greater the difference in the arrival times of the first P wave compared to the first S wave.

THREE SEISMOGRAMS

Seismogram A – New York, NY

1 minute

FIRST P WAVE

FIRST S WAVE

Seismogram B – Nome, Alaska

FIRST P WAVE

FIRST S WAVE

Seismogram C – Mexico City, Mexico

FIRST P WAVE

FIRST S WAVE

STEP 1 ①

Using the seismogram on the left for a seismic recording station in New York, determine the time difference between the arrival of the first P wave and the arrival of the first S wave. In this example, the P-S time interval is 5 minutes.

STEP 2

Find the place on the travel-time graph where the vertical separation between the P and S curves is equal to the P-S interval determined in Step 1.

STEP 3

From this position, draw a vertical line that extends to the bottom of the graph and read the distance to the epicenter. The distance from our seismograph in New York to the earthquake epicenter is 2300 miles.

TRAVEL-TIME GRAPH

Distance from epicenter in kilometers

S-WAVE CURVE ②

P-WAVE CURVE

Time in minutes since earthquake

③

Distance from epicenter in miles

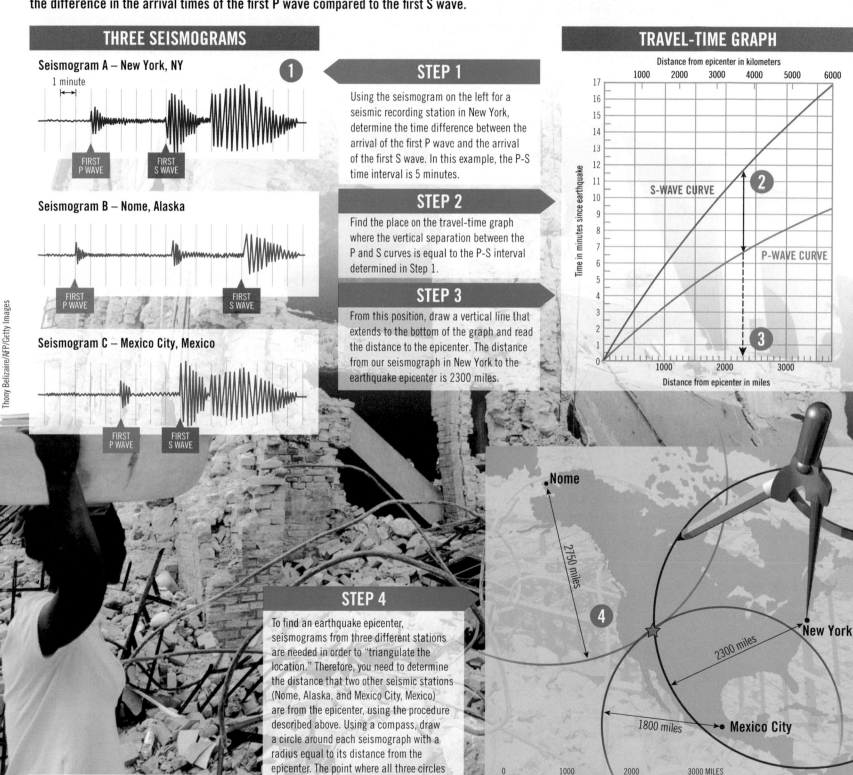

STEP 4 ④

To find an earthquake epicenter, seismograms from three different stations are needed in order to "triangulate the location." Therefore, you need to determine the distance that two other seismic stations (Nome, Alaska, and Mexico City, Mexico) are from the epicenter, using the procedure described above. Using a compass, draw a circle around each seismograph with a radius equal to its distance from the epicenter. The point where all three circles intersect is the approximate location of the earthquake epicenter.

Nome

2750 miles

New York

2300 miles

1800 miles

Mexico City

0 1000 2000 3000 MILES

0 1000 2000 3000 4000 5000 KILOMETERS

Building destroyed by 2010 earthquake, Port-au-Prince, Haiti

Thony Belizaire/AFP/Getty Images

Moment Magnitude For measuring medium and large earthquakes, seismologists have come to favor a newer scale, called **moment magnitude** (M_W), which measures the total energy released during an earthquake. Moment magnitude is calculated by determining the average amount of slip on the fault, the area of the fault surface that slipped, and the strength of the faulted rock.

Moment magnitude can also be calculated by modeling data obtained from seismograms. The results are converted to a magnitude number, as in other magnitude scales. In addition, as in the Richter scale, each unit increase in moment magnitude equates to roughly a 32-fold increase in the energy released.

Because moment magnitude estimates the total energy released, it is better for measuring very large earthquakes. Seismologists have recalculated the magnitudes of older, strong earthquakes using the moment magnitude scale. For example, the 1964 Alaska earthquake, which was originally given a Richter magnitude of 8.3, has since been recalculated using the moment magnitude scale, resulting in an upgrade to 9.2. Conversely, the 1906 San Francisco earthquake that was given a Richter magnitude of 8.3, was downgraded to a M_W 7.9. The strongest earthquake on record is the 1960 Chilean subduction zone earthquake, with a moment magnitude of 9.5.

8.3 CONCEPT CHECKS

1 What does the Modified Mercalli Intensity scale tell us about an earthquake?

2 What information is used to establish the lower numbers on the Mercalli scale?

3 How much more energy does a magnitude 7.0 earthquake release than a 6.0 earthquake?

4 Why is the moment magnitude scale favored over the Richter scale?

8.4 | EARTHQUAKE DESTRUCTION List and describe the major destructive forces that can be triggered by earthquake vibrations.

The most violent earthquake ever recorded in North America—the Alaska earthquake—occurred at 5:36 P.M. on March 27, 1964. Felt over most of the state, the earthquake had a moment magnitude (M_W) of 9.2 and lasted 3 to 4 minutes. This event left 128 people dead and thousands homeless, and it badly disrupted the economy of the state. Within 24 hours of the initial shock, 28 aftershocks were recorded, 10 of them exceeding a magnitude of 6. The location of the epicenter and the towns hardest hit by the quake are shown in **FIGURE 8.18**.

Many factors determine the degree of destruction that accompanies an earthquake. The most obvious is *the magnitude of the earthquake and its proximity to a populated area.* During an earthquake, the region within 20 to 50 kilometers (12 to 30 miles) of the epicenter tends to experience roughly the same degree of ground shaking, and beyond that limit, vibrations usually diminish rapidly. Earthquakes that occur in the stable continental interior, such as the New Madrid earthquake of 1811–1812, are generally felt more over a much larger area than those in earthquake-prone areas such as California.

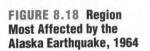

FIGURE 8.18 Region Most Affected by the Alaska Earthquake, 1964

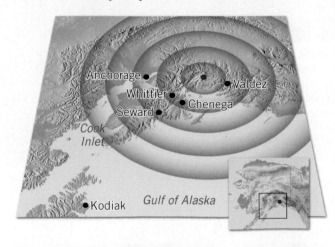

Destruction from Seismic Vibrations

The 1964 Alaska earthquake provided geologists with insights into the role of ground shaking as a destructive force. As the energy released by an earthquake travels along Earth's surface, it causes the ground to vibrate in a complex manner by involving up-and-down as well as side-to-side motion. The amount of damage to human-made structures attributable to the vibrations depends on several factors, including (1) *the intensity* and (2) *the duration of the vibrations*, (3) *the nature of the material on which structures rest*, and (4) *the nature of building materials and construction practices of the region*.

All the multi-story structures in Anchorage were damaged by the vibrations. The more flexible wood-frame residential buildings fared best. A striking example of how construction variations affect earthquake damage is shown in **FIGURE 8.19**. You can see that the steel-frame building on the left withstood the vibrations, whereas the poorly designed JCPenney building was badly damaged. Engineers have learned that buildings built of blocks and bricks that are not reinforced with steel rods are the most serious safety threats in earthquakes. Unfortunately, most of the structures in the developing world are constructed of unreinforced concrete slabs and bricks made of dried mud—a primary reason the death toll in poor countries such as Haiti is usually higher than for earthquakes of similar size in the United States.

The 1964 Alaska earthquake damaged most large structures in Anchorage, even though they were built according to the earthquake provisions of the Uniform Building Code. Perhaps some of that destruction can be attributed to the unusually long duration of the earthquake. Most quakes involve tremors that last less than a minute. For example, the 1994 Northridge earthquake was felt for about 40 seconds, and the

FIGURE 8.19 Comparing Damage to Structures The poorly designed five-story JCPenney building in Anchorage, Alaska, sustained extensive damage. The steel-frame adjacent building incurred very little structural damage. (Courtesy of NOAA)

strong vibrations of the 1989 Loma Prieta earthquake lasted less than 15 seconds. But the Alaska quake reverberated for 3 to 4 minutes.

Amplification of Seismic Waves

Although the region near the epicenter experiences about the same intensity of ground shaking, destruction may vary considerably in this area. These differences are usually attributable to the nature of the ground on which the structures are built. Soft sediments, for example, generally amplify the vibrations more than solid bedrock. Thus, the buildings in Anchorage that were situated on unconsolidated sediments experienced heavy structural damage (**FIGURE 8.20**). In contrast, most of the town of Whittier, although much nearer the epicenter, rested on a firm foundation of solid

bedrock and suffered much less damage from seismic vibrations.

Liquefaction The intense shaking of an earthquake can cause loosely packed water-logged materials, such as sandy stream deposits or fill, to be transformed into a substance that acts like fluids. The phenomenon of transforming a somewhat stable soil into mobile material capable of rising toward Earth's surface is known as **liquefaction**. When liquefaction occurs, the ground may not be capable of supporting buildings, and underground storage tanks and sewer lines may literally float toward the surface (**FIGURE 8.21**).

During the 1989 Loma Prieta earthquake, in San Francisco's Marina District, foundations failed, and geysers of sand and water shot from the ground, evidence that liquefaction had occurred (**FIGURE 8.22**).

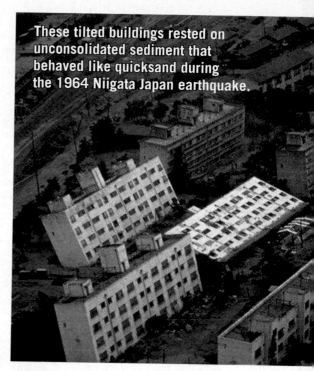

These tilted buildings rested on unconsolidated sediment that behaved like quicksand during the 1964 Niigata Japan earthquake.

FIGURE 8.21 Effects of Liquefaction on Buildings (Photo courtesy of USGS)

Sand volcanoes

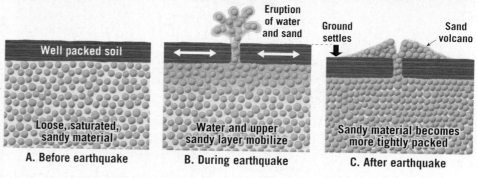

Well packed soil — Loose, saturated, sandy material — **A. Before earthquake**

Eruption of water and sand — Water and upper sandy layer mobilize — **B. During earthquake**

Ground settles — Sand volcano — Sandy material becomes more tightly packed — **C. After earthquake**

FIGURE 8.22 Liquefaction These "sand volcanoes," produced by the Christchurch New Zealand earthquake of 2011, formed when "geysers" of sand and water shot from the ground, an indication that liquefaction occurred. (Photo by Thorsten Blackwood/AFP Getty Images/Newscom)

FIGURE 8.20 Ground Failure Caused This Street in Anchorage, Alaska, to Collapse (Photo by USGS)

Downtown Anchorage following the 1964 Alaskan earthquake.

A. Vibrations from the Alaskan earthquake caused cracks to appear near the edge of the Turnagain Heights bluff.

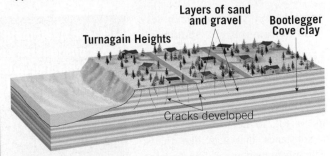

B. Blocks of land began to slide toward the sea on a weak layer called the Bootlegger Cove clay and in less than 5 minutes, as much as 200 meters of the Turnagain Heights bluff area had been destroyed.

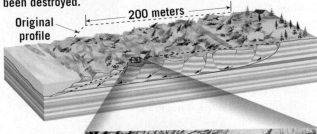

C. Photo of a small area of destruction caused by the Turnagain Heights slide.

Liquefaction also contributed to the damage inflicted on San Francisco's water system during the 1906 earthquake. During the 2011 Japan earthquake, liquefaction caused entire buildings to sink several feet.

Landslides and Ground Subsidence

The greatest earthquake-related damage to structures is often caused by landslides and ground subsidence triggered by earthquake vibrations. This was the case during the 1964 Alaska earthquake in Valdez and Seward, where the violent shaking caused coastal sediments to slump, carrying both waterfronts away. In Valdez, 31 people died when a dock slid into the sea. Because of the threat of recurrence, the entire town of Valdez was relocated to more stable ground about 7 kilometers (4 miles) away.

Much of the damage in Anchorage was attributed to landslides. Homes were destroyed in Turnagain Heights when a layer of clay lost its strength and over 200 acres of land slid toward the ocean (**FIGURE 8.23**). A portion of this spectacular landslide was left in its natural condition, as a reminder of this destructive event. The site was appropriately named "Earthquake Park." Downtown Anchorage was also disrupted as sections of the main business district dropped by as much as 3 meters (10 feet).

Fire

More than 100 years ago, San Francisco was the economic center of the western United States, largely because of gold and silver mining. Then, at dawn on April 18, 1906, a violent earthquake struck, triggering an enormous firestorm (see Figure 8.3). Much of the city was reduced to ashes and ruins. It is estimated that 3000 people died and more than half of the city's 400,000 residents were left homeless.

The historic San Francisco earthquake reminds us of the formidable threat of fire, which started in that quake when gas and electrical lines were severed. The initial ground shaking broke the city's water lines into hundreds of disconnected pieces, which made controlling the fires virtually impossible. The fires, which raged out of control for 3 days, were finally contained when buildings were dynamited along a wide boulevard to provide a fire break, similar to the strategy used in fighting forest fires.

While few deaths were attributed to the San Francisco fires, other earthquake-initiated fires have been more destructive and have claimed many more lives. For example, the 1923 earthquake in Japan triggered an estimated 250 fires, devastating the city of Yokohama and destroying more than half the homes in Tokyo. More than 100,000 deaths were attributed to the fires, which were driven by unusually high winds.

EYE ON EARTH

Water-saturated sandy soil provides students an opportunity to experience the phenomenon of liquefaction. Liquefaction may occur when ground shaking causes a layer of saturated material to lose strength and act like a fluid.

QUESTION 1 Describe what you think would happen to a structure built on sandy soil that suddenly experienced liquefaction during an earthquake.

QUESTION 2 How might a nearly empty underground storage tank be affected by liquefaction of the surrounding soil?

Marli Miller

Tsunami speed: 800 km/hr at water depth of 5000 meters

Tsunami speed: 340 km/hr at water depth of 900 meters

Tsunami speed: 50 km/hr near shore

Overriding plate

Displacement on megathrust fault

Subducting plate

Hypocenter

A.

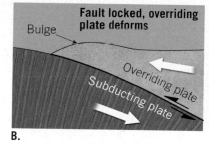

Bulge

Fault locked, overriding plate deforms

Overriding plate

Subducting plate

B.

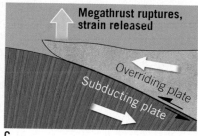

Megathrust ruptures, strain released

Overriding plate

Subducting plate

C.

FIGURE 8.24 Tsunami Generated by Displacement of the Ocean Floor
The speed of a wave correlates with ocean depth. Waves moving in deep water advance at speeds in excess of 800 kilometers (500 miles) per hour. Speed gradually slows to 50 kilometers (30 miles) per hour at depths of 20 meters (65 feet). As waves slow in shallow water, they grow in height until they rush onto shore with tremendous force. The size and spacing of these swells are not to scale.

What Is a Tsunami?

Major undersea earthquakes occasionally set in motion a series of large ocean waves that are known by the Japanese name **tsunami** (which means "harbor wave"). Most tsunami are generated by displacement along a megathrust fault that suddenly lifts a large slab of seafloor (**FIGURE 8.24**). Once generated, a tsunami resembles a series of ripples formed when a pebble is dropped into a pond. In contrast to ripples, tsunami advance across the ocean at amazing speeds, about 800 kilometers (500 miles) per hour—equivalent to the speed of a commercial airliner. Despite this striking characteristic, a tsunami in the open ocean can pass undetected because its height (amplitude) is usually less than 1 meter, and the distance separating wave crests ranges from 100 to 700 kilometers. However, upon entering shallow coastal waters, these destructive waves "feel bottom" and slow, causing the water to pile up (see Figure 8.24). A few exceptional tsunami have exceeded 30 meters (100 feet) in height. As the crest of a tsunami approaches the shore, it appears as a rapid rise in sea level with a turbulent and chaotic surface; it does not resemble a breaking wave (**FIGURE 8.25**).

The first warning of an approaching tsunami is often the rapid withdrawal of water from beaches, the result of the trough of the first large wave preceding the crest. Some inhabitants of the Pacific basin have learned to heed this warning and quickly move to higher ground. Approximately 5 to 30 minutes after the retreat of water, a surge capable of extending several kilometers inland occurs. In a successive fashion, each surge is followed by a rapid oceanward retreat

of the sea. Therefore, people experiencing a tsunami should not return to the shore once the first surge of water retreats.

Tsunami Damage from the 2004 Indonesian Earthquake
A massive undersea earthquake of M_W 9.1 occurred near the island of Sumatra on December 26, 2004, sending waves of water racing across the Indian Ocean and Bay of Bengal (see Figure 8.25). It was one of the deadliest natural disasters of any kind in modern times, claiming more than 230,000 lives. As water surged several kilometers inland, cars and trucks were flung around like toys in a bathtub, and fishing boats were rammed into homes. In some locations, the backwash of water dragged bodies and huge amounts of debris out to sea.

The destruction was indiscriminate, destroying luxury resorts as well as poor fishing hamlets along the Indian Ocean. Damages were reported as far away as the coast of

FIGURE 8.25 Tsunami Generated Off the Coast of Sumatra, 2004 (AFP/Getty Images)

Somalia in Africa, 4100 kilometers (2500 miles) west of the earthquake epicenter.

Japan Tsunami

Because of Japan's location along the circum-Pacific belt and its expansive coastline, it is especially vulnerable to tsunami destruction. The most powerful earthquake to strike Japan in the age of modern seismology was the 2011 Tōhoku earthquake (M_W 9.0). This historic earthquake and devastating tsunami resulted in at least 15,861 deaths, more than 3000 people missing, and 6107 injured. Nearly 400,000 buildings, 56 bridges, and 26 railways were destroyed or damaged.

The majority of human casualties and damage were caused by a Pacific-wide tsunami that reached a maximum height of about 40 meters (130 feet) and traveled inland 10 kilometers (6 miles) in the region of Sendai, Japan (**FIGURE 8.26**). The chapter-opening photo (pages 244–245) shows this dramatic event. In addition, meltdowns occurred at three nuclear reactors in Japan's Fukushima Daiichi Nuclear Complex. Across the Pacific in California, Oregon, Peru, and Chile, some loss of life occurred, and several houses, boats, and docks were destroyed. The tsunami was generated when a slab of seafloor located 60 kilometers (37 miles) off the east coast of Japan was suddenly "thrust up" an estimated 5 to 8 meters (16 to 26 feet).

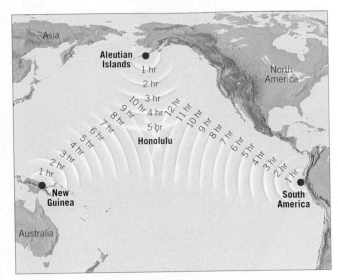

FIGURE 8.27 Tsunami Travel Times Travel times to Honolulu, Hawaii, from selected locations throughout the Pacific. (Data from NOAA)

Tsunami Warning System

In 1946, a large tsunami struck the Hawaiian islands without warning. A wave more than 15 meters (50 feet) high left several coastal villages in shambles. This destruction motivated the U.S. Coast and Geodetic Survey to establish a tsunami warning system for coastal areas of the Pacific that today includes 26 countries. Seismic observatories throughout the region report large earthquakes to the Tsunami Warning Center in Honolulu. Scientists at the center use deep-sea buoys equipped with pressure sensors to detect energy released by an earthquake. In addition, tidal gauges measure the rise and fall in sea level that accompany tsunami, and warnings are issued within an hour. Although tsunami travel very rapidly, there is sufficient time to warn all except those in the areas nearest the epicenter. For example, a tsunami generated near the Aleutian islands would take 5 hours to reach Hawaii, and one generated near the coast of Chile would travel 15 hours before reaching the shores of Hawaii (**FIGURE 8.27**).

8.4 CONCEPT CHECKS

1 List four factors that affect the amount of destruction that seismic vibrations cause to human-made structures.

2 In addition to the destruction created directly by seismic vibrations, list three other types of destruction associated with earthquakes.

3 What is a tsunami? How are tsunami generated?

4 List at least three reasons an earthquake with a magnitude of 7.0 might result in more death and destruction than a quake with a magnitude of 8.0.

FIGURE 8.26 2011 Japan Tsunami A massive tsunami hit the coastal area of northeastern Japan, March 11, 2011, following a massive 9.0 magnitude earthquake. (Photo by REUTERS/KYODO)

8.5 | EARTHQUAKE BELTS AND PLATE BOUNDARIES

Locate Earth's major earthquake belts on a world map and label the regions associated with the largest earthquakes.

About 95 percent of the energy released by earthquakes originates in the few relatively narrow zones, shown in **FIGURE 8.28**. The zone of greatest seismic activity, called the **circum-Pacific belt**, encompasses the coastal regions of Chile, Central America, Indonesia, Japan, and Alaska, including the Aleutian islands (see Figure 8.28). Most earthquakes in the circum-Pacific belt occur along convergent plate boundaries where one plate slides at a low angle beneath another. The contacts between the subducting and overlying plates are *megathrust faults*, along which Earth's largest earthquakes are generated (**FIGURE 8.29**).

There are more than 40,000 kilometers (25,000 miles) of subduction boundaries in the circum-Pacific belt where displacement is dominated by thrust faulting. Ruptures occasionally occur along segments that are nearly 1000 kilometers (600 miles) in length, generating catastrophic *megathrust earthquakes* with magnitudes of (M_W) 8 or greater.

Another major concentration of strong seismic activity, referred to as the *Alpine–Himalayan belt*, runs through the mountainous regions that flank the Mediterranean Sea and extends past the Himalayan Mountains (see Figure 8.28). Tectonic activity in this region is mainly attributed to collisions of the African plate with Eurasia and of the Indian plate with Southeast Asia. These plate interactions created many thrust and strike-slip faults that remain active. In addition, numerous faults located a considerable distance from these plate boundaries have been reactivated as India continues its northward advance into Asia. For example, slippage on a complex fault system in 2008 in the Sichuan Province of China killed at least 70,000 people and left 1.5 million others homeless. The cause was the Indian subcontinent continually shoving the Tibetan Plateau eastward against the rocks of the Sichuan basin.

Figure 8.28 shows another continuous earthquake belt that extends for thousands of kilometers through the world's oceans. This zone coincides with the oceanic ridge system, which is an area of frequent but weak seismic activity. As tensional forces pull the plates apart during seafloor spreading, displacement along normal faults generates most of the earthquakes in this zone. The remaining seismic activity is associated with slippage along transform faults located between ridge segments.

Transform faults and smaller strike-slip faults also run through continental crust, where they may generate large earthquakes that tend to occur on a cyclical basis. Examples

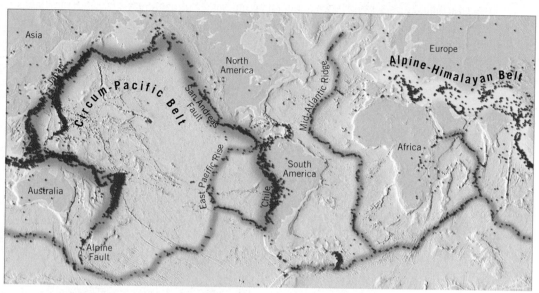

FIGURE 8.28 Global Earthquake Belts Distribution of nearly 15,000 earthquakes with magnitudes equal to or greater than 5 for a 10-year period. (Data from USGS)

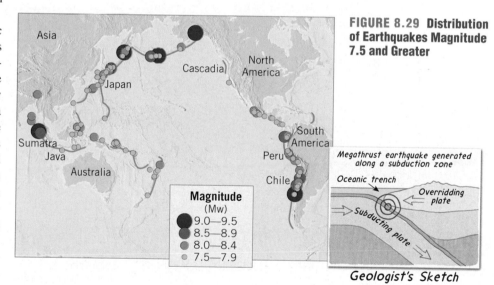

FIGURE 8.29 Distribution of Earthquakes Magnitude 7.5 and Greater

Magnitude (Mw)
- 9.0—9.5
- 8.5—8.9
- 8.0—8.4
- 7.5—7.9

Geologist's Sketch

include California's San Andreas Fault, New Zealand's Alpine Fault, and Turkey's North Anatolian Fault, which produced a deadly earthquake in 1999.

8.5 CONCEPT CHECKS

1 Where does the greatest amount of seismic activity occur?

2 What type of plate boundary is associated with Earth's largest earthquakes?

3 Name another major concentration of strong earthquake activity.

Historic Earthquakes East of the Rockies

Large earthquakes are relatively uncommon in the middle of continents, far from the places where plates collide or grind past one another, or where one plate slides beneath another. Nevertheless, several damaging earthquakes have occurred in the central and eastern United States since colonial times.

Compare the damage to the brick building in Charleston (below) with that to the wood-frame house barely visible to the building's left. **?**

Residence damaged during the 1886 Charleston, South Carolina, earthquake. (USGS)

Historic Earthquakes 1755–2011

	LOCATION	DATE	INTENSITY	MAGNITUDE*	COMMENTS
1	East of Oklahoma City	2011	VII	5.6	Fourteen homes destroyed
2	Mineral, Virginia	2011	VII	5.8	Felt by many due to its proximity to large population centers
3	Southeastern Illinois	2008	VII	5.4	Occurred along the Wabash Valley Seismic Zone
4	Northeast Kentucky	1980	VII	5.2	Largest earthquake ever recorded in Kentucky
5	Merriman, Nebraska	1964	VII	5.1	Largest earthquake ever recorded in Nebraska
6	Northern New York	1944	VIII	5.8	Left several structures unsafe for occupancy
7	Ossipee Lake, New Hampshire	1947	VII	5.5	Two earthquakes occurred four days apart
8	Western Ohio	1937	VIII	5.4	Extensive damage to chimneys and plaster walls
9	Valentine, Texas	1931	VIII	5.8	Brick buildings were severely damaged
10	Giles County, Virginia	1897	VIII	5.9	Changed the flow of natural springs
11	Charleston, Missouri	1895	VIII	6.6	Structural damage and liquefaction reported
12	Charleston, South Carolina	1886	X	7.3	Caused 60 deaths, destroyed many buildings
13	New Madrid, Missouri	1811-1812	X	7.7	Three strong earthquakes occurred in remote areas
14	Cape Ann, Massachusetts	1755	VIII	?	Chimneys leveled and brick buildings damaged in Boston

Source: U.S. Geological Survey
*Intensity and magnitudes have been estimated for many of these events.

New Madrid Seismic Zone

Three large earthquakes that were centered near the Mississippi River Valley in southeastern Missouri occurred on December 16, 1811, January 23, 1812, and February 7, 1812. Having estimated magnitudes of 7.7, 7.5, and 7.7, these earthquakes and numerous smaller aftershocks destroyed the town of New Madrid, Missouri. These quakes also triggered massive landslides, inflicted damage in the surrounding states of Illinois, Indiana, Kentucky, Arkansas, and Tennessee and were felt over the entire Midwest because of the rigid bedrock that underlies that region. Chimneys collapsed as far away as Cincinnati, Ohio. Destruction of property from the New Madrid earthquakes was minimal, primarily because the Midwest was sparsely populated in the early 1800s. Memphis, Tennessee, located near the epicenter, had not yet been established, and St. Louis, Missouri, was a small frontier town. Today, these metropolitan areas each have populations that exceed one million residents.

Eric Foltz/Getty Images

Reelfoot Lake, Tennessee, located about 14 miles south of New Madrid, formed in an area of subsidence that was produced during the February 7, 1812, event. Subsidence that exceeded 6 meters (20 feet) in some places occurred on the down-dropped side of the Reelfoot Fault.

The largest historical earthquake in the eastern United States occurred in Charleston, South Carolina, in 1886. This one-minute event resulted in 60 deaths, numerous injuries, and great economic loss within 200 kilometers (120 miles) of Charleston. Minutes after the quake, strong vibrations shook the upper floors of buildings in Chicago, Illinois, and St. Louis, Missouri, causing people to rush outdoors. In Charleston, more than one hundred buildings were destroyed, and 90 percent of the remaining structures were damaged.

(USGS)

8.6 | CAN EARTHQUAKES BE PREDICTED? Compare and contrast the goals of short-range earthquake predictions and long-range forecasts.

The vibrations that shook the San Francisco area in 1989 caused 63 deaths, heavily damaged the Marina District, and caused the collapse of a double-decked section of I-880 in Oakland, California (**FIGURE 8.30**). This level of destruction was the result of an earthquake of moderate intensity (M_W 6.9). Seismologists warn that other earthquakes of comparable or greater strength can be expected along the San Andreas system, which cuts a nearly 1300-kilometer (800-mile) path through the western one-third of the state. An obvious question is: Can these earthquakes be predicted?

Short-Range Predictions

The goal of short-range earthquake prediction is to provide a warning of the location and magnitude of a large earthquake within a narrow time frame (**TABLE 8.2**). Substantial efforts to achieve this objective have been attempted in Japan, the United States, China, and Russia—countries where earthquake risks are high. This research has concentrated on monitoring possible *precursors*—events or changes that precede a forthcoming earthquake and thus may provide warning. In California, for example, seismologists monitor changes in ground elevation and variations in strain levels near active faults. Other researchers measure changes in groundwater levels, while still others try to predict earthquakes based on an increase in the frequency of foreshocks that precede some, but not all, earthquakes.

Japanese and Chinese scientists have been known to monitor anomalous animal behavior. A few days before the May 12, 2008, earthquake in China's Sichuan Province, the streets of a village near the fault were filled with toads migrating from the mountains. Was this a warning? Perhaps. Walter Mooney, a USGS seismologist, put it best: "Everyone hopes that animals can tell us something we don't know . . . but animal behavior is way too unreliable." Although precursors may exist, we have yet to determine effective ways to interpret and utilize the information.

One claim of a successful short-range prediction, based on an increase in foreshocks, was made by the Chinese government after the February 4, 1975, earthquake in Liaoning Province. According to reports, very few people were killed—even though more than 1 million lived near the epicenter—because the earthquake was "predicted," and the residents were evacuated. Some Western seismologists have questioned this claim and suggest instead that an intense swarm of foreshocks, which began 24 hours before the main earthquake, may have caused many people to evacuate of their own accord.

One year after the Liaoning earthquake, an estimated 240,000 people perished in the Tangshan, China, earthquake, which was *not* predicted. There were no foreshocks. Predictions can also lead to false alarms. In a province near Hong Kong, people reportedly evacuated their dwellings for over a month, but no earthquake followed.

In order for a short-range prediction scheme to be generally accepted, it must be both accurate and reliable. Thus, *it must have a small range of uncertainty in regard to location and timing, and it must produce few failures or false alarms.* Can you imagine the debate that would precede an order to evacuate a large city in the United States, such as Los Angeles or San Francisco? The cost of evacuating millions of people, arranging for living accommodations, and providing for their lost work time and wages would be staggering.

Currently, no reliable method exists for making short-range earthquake predictions. In fact, leading seismologists in the past 100 years have generally concluded that short-range earthquake prediction is not feasible.

Long-Range Forecasts

In contrast to short-range predictions, which aim to predict earthquakes within a time frame of hours or days, long-range forecasts are estimates of how likely it is for an earthquake of a certain magnitude to occur on a time scale of 30 to 100 years or more. These forecasts give statistical estimates of the expected intensity of ground motion for a given area over a specified time frame. Although long-range forecasts are not as informative as we might like, these data are useful for providing important guides for building codes so that buildings, dams, and roadways are constructed to withstand expected levels of ground shaking.

Most long-range forecasting strategies are based on evidence that many large faults break in a cyclical manner, producing similar quakes at roughly similar intervals. In other words, as soon as a section of a fault ruptures, the continuing motions of Earth's plates begin to build strain in the rocks again until they fail once more. Seismologists have therefore studied historical records of earthquakes to see if there are any discernible patterns so that they can establish the probability of recurrence.

FIGURE 8.30 Collapse of the Double-decked Section of I-880 This section of a double-decked highway, known as the Cypress Viaduct, collapsed during the 1989 Loma Prieta earthquake. (Photo by Paul Sakuma/AP Photo)

TABLE 8.2 Some Notable Earthquakes

Year	Location	Deaths (est.)	Magnitude*	Comments
856	Iran	200,000		
893	Iran	150,000		
1138	Syria	230,000		
1268	Asia Minor	60,000		
1290	China	100,000		
1556	Shensi, China	830,000		Possibly the greatest natural disaster
1667	Caucasia	80,000		
1727	Iran	77,000		
1755	Lisbon, Portugal	70,000		Tsunami damage extensive
1783	Italy	50,000		
1908	Messina, Italy	120,000		
1920	China	200,000	7.5	Landslide buried a village
1923	Tokyo, Japan	143,000	7.9	Fire caused extensive destruction
1948	Turkmenistan	110,000	7.3	Almost all brick buildings near epicenter collapsed
1960	Southern Chile	5700	9.5	The largest-magnitude earthquake ever recorded
1964	Alaska	131	9.2	Greatest-magnitude North American earthquake
1970	Peru	70,000	7.9	Great rockslide
1976	Tangshan, China	242,000	7.5	Estimates for the death toll are as high as 655,000
1985	Mexico City	9500	8.1	Major damage occurred 400 km from epicenter
1988	Armenia	25,000	6.9	Poor construction practices resulted in many deaths
1990	Iran	50,000	7.4	Landslides and poor construction practices led to great damage
1993	Latur, India	10,000	6.4	Located in stable continental interior
1995	Kobe, Japan	5472	6.9	Damages estimated to exceed $100 billion
1999	Izmit, Turkey	17,127	7.4	Nearly 44,000 injured and more than 250,000 displaced
2001	Gujarat, India	20,000	7.9	Millions homeless
2003	Bam, Iran	31,000	6.6	Ancient city with poor construction
2004	Indian Ocean (Sumatra)	230,000	9.1	Devastating tsunami damage
2005	Pakistan/Kashmir	86,000	7.6	Many landslides; 4 million homeless
2008	Sichuan, China	87,000	7.9	Millions homeless, some towns will not be rebuilt
2010	Port-au-Prince, Haiti	316,000	7.0	More than 300,000 injured and 1.3 million homeless
2011	Japan	20,000	9.0	Majority of the casualties due to a tsunami

* Widely differing magnitudes have been estimated for some of these earthquakes. When available, moment magnitudes are used.

Source: U.S. Geological Survey.

Seismic Gaps

Seismologists began to plot the distribution of rupture zones associated with great earthquakes around the globe. The maps revealed that individual rupture zones tend to occur adjacent to one another, without appreciable overlap, thereby tracing out a plate boundary. Because plates are moving at known velocities, the rate at which strain builds can also be estimated.

When these researchers studied historical records, they discovered that some seismic zones had not produced a large earthquake in more than a century or, in some locations, for several centuries. These quiet zones, called **seismic gaps**, are believed to be zones that are storing strain that will be released during a future earthquake. **FIGURE 8.31** shows a patch (seismic gap) of the megathrust fault that lies offshore of Padang, a low-lying city of 800,000 people off the coast of Sumatra that has not ruptured since 1797. Scientists are particularly concerned about this seismic gap because, in 2004, a rupture of an adjacent segment of this megathrust fault that lies to the north generated a tsunami that claimed 230,000 lives.

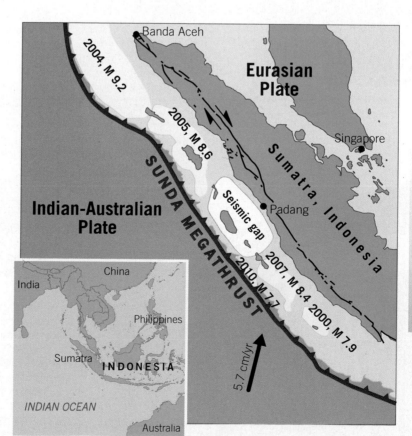

SmartFigure 8.31 Seismic Gaps: Tools for Forecasting Earthquakes Seismic gaps are "quiet zones" thought to be inactive zones that are storing elastic strain that will eventually produce major earthquakes. This seismic gap occurs along a patch of the megathrust fault where oceanic lithosphere is being subducted beneath Sumatra, near Padang, a low-lying coastal city with a population of 800,000 people.

Seismic Risks on the San Andreas Fault System

California's San Andreas Fault runs diagonally from southeast to northwest for nearly 1300 kilometers (800 miles) through much of the western part of the state. For years researchers have been trying to predict the location of the next "Big One"—an earthquake with a magnitude of 8 or greater—along this fault system.

CALIFORNIA

The 1906 San Francisco earthquake caused displacement on the 477-kilometer-long northernmost section of the fault. This event, which had an estimated magnitude of 7.8, likely relieved much of the strain that had been building during the previous 200 years or so.

1906 epicenter ★
San Francisco ●

Located just south of the 1906 rupture is a section of the San Andreas Fault that exhibits fault creep. When plates gradually slide past each other, less strain accumulates than when the fault is locked, diminishing the possibilities of an eventual large quake.

1857 epicenter ★

The 1906 San Francisco earthquake was the most devastating in California's history. The quake and resulting fires caused an estimated 3000 deaths and extensive damaged buildings throughout the city. The business district was devastated largely because it was built on land made by filling in a cove. (USG

This 300-kilometer-long section of the San Andreas Fault System produced the Fort Tejon earthquake of 1857 that had an estimated magnitude of 7.9. Because a portion of the fault has likely accumulated considerable strain since the Fort Tejon quake, the U.S. Geological Survey gives it a 60 percent probability of producing a major earthquake in the next 30 years.

● **Los Angeles**

The next major quake on the San Andreas may well be on its southernmost 200 kilometers—an area that has not produced a large event in about 300 years.

Adam Teitelbaum/AFP/Getty Images

On October 17, 1989, millions of television viewers around the world were settling in to watch the third game of the World Series. Instead, they saw their TVs go to black as tremors hit San Francisco's Candlestick Park, where power and communication lines were severed in what came to be known as the Loma Prieta earthquake. Although the quake was centered in a remote section of the Santa Cruz Mountains—100 miles to the south—major damage occurred in the Marina District of San Francisco.

The U.S. Geological Survey concluded that between 2003 and 2032 there is a 62 percent probability of at least one magnitude 6.7 or greater earthquake striking somewhere in the San Francisco Bay area. They also predict that this quake would be roughly comparable to the 1989 Loma Prieta event (MW 6.9) and capable of causing significant damage.

This is the probability for one or more magnitude 6.7 or greater earthquakes between 2003 to 2032. This result incorporates 14% odds of quakes occurring on faults not shown on this map.

% Probability of magnitude 6.7 or greater quake before 2032 on the indicated faults

Increasing probability along fault segments

In mid-January 1994, less than five years after the Loma Prieta event, the Northridge earthquake struck an area slightly north of Los Angeles. Although not the fabled "Big One," this moderate 6.7-magnitude earthquake claimed the lives of 57 people. Nearly 300 schools were severely damaged, and one dozen major roadways buckled. Among these were two of California's major arteries—sections of the Santa Monica Freeway and the Golden State Freeway (I-5) where an overpass collapsed completely and blocked the highway.

Rogers Creek fault

Napa

Concord-Green Valley fault

27%

4%

Walnut Creek

Greenville fault

Mt. Diablo thrust fault

3%

San Andreas fault

Hayward fault

San Francisco

Oakland

San Francisco Bay

21%

3%

Calaveras fault

Palo Alto

San Jose

10%

11%

San Gregorio fault

EXTENT OF RUPTURE IN THE LOMA PRIETA QUAKE

PACIFIC OCEAN

Monterey Bay

Monterey

5 NORTH 14
Autos

FIGURE 8.32 Paleoseismology: The Study of Prehistoric Earthquakes These studies are often conducted by digging a trench across a fault zone and then looking for evidence of ancient displacements, such as offset sedimentary strata. This simplified diagram shows that vertical displacement occurred on this fault three different times, with each event producing an earthquake. Based on the size of the vertical displacement, these ancient earthquakes had estimated magnitudes of 6.8 and 7.4. (Photo courtesy of USGS)

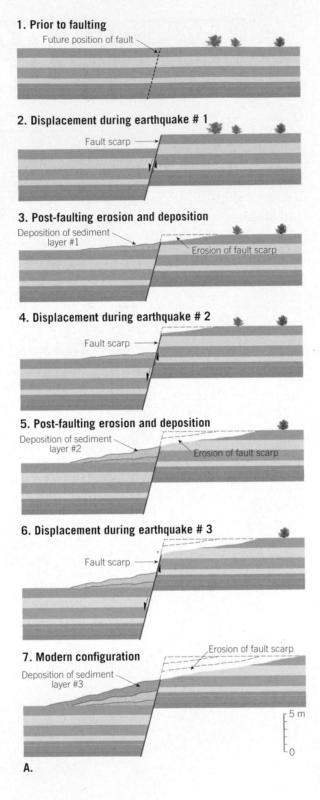

1. **Prior to faulting**
   Future position of fault

2. **Displacement during earthquake # 1**
   Fault scarp

3. **Post-faulting erosion and deposition**
   Deposition of sediment layer #1
   Erosion of fault scarp

4. **Displacement during earthquake # 2**
   Fault scarp

5. **Post-faulting erosion and deposition**
   Deposition of sediment layer #2
   Erosion of fault scarp

6. **Displacement during earthquake # 3**
   Fault scarp

7. **Modern configuration**
   Erosion of fault scarp
   Deposition of sediment layer #3

   5 m
   0

A.

B. The events depicted in the accompanying diagram were deciphered by digging a trench (shown here) across the fault zone and studying the displaced sedimentary beds.

(**FIGURE 8.32**). A large vertical offset of the layers of sediments indicates a large earthquake. Sometimes buried plant debris can be carbon dated, allowing for the timing of recurrence to be established.

One investigation that used this method focused on a segment of the San Andreas Fault that lies north and east of Los Angeles. At this site, the drainage of Pallet Creek has been repeatedly disturbed by successive ruptures along the fault zone. Trenches excavated across the creek bed have exposed sediments that have been displaced by several large earthquakes over a span of 1500 years. From these data, it was determined that strong earthquakes occur an average of once every 135 years. The last major event, called the Fort Tejon earthquake, occurred on this segment of the San Andreas Fault in 1857, roughly 150 years ago. Because earthquakes occur on a cyclical basis, a major event in southern California may be imminent.

Using other paleoseismology techniques, researchers determined that several powerful earthquakes (magnitude 8 or larger) have repeatedly struck the coastal Pacific Northwest over the past several thousand years. The most recent event, which occurred about 300 years ago, generated a destructive tsunami. As a result of these findings, public officials have taken steps to strengthen some of the region's existing buildings, dams, bridges, and water systems. Even the private sector responded. The U.S. Bancorp building in Portland, Oregon, was strengthened at a cost of $8 million.

Paleoseismology Another method of long-term forecasting involves *paleoseismology* (*paleo* = ancient, *seismos* = shake, *ology* = study of), the study of the timing, location, and size of prehistoric earthquakes. Paleoseismology studies are often conducted by digging a trench across a suspected fault zone and then looking for evidence of ancient faulting, such as offset sedimentary strata or mud volcanoes

8.6 CONCEPT CHECKS

1 Are accurate, short-range earthquake predictions currently possible using modern seismic instruments? Explain.

2 What is the value of long-range earthquake forecasts?

8.7 | EARTH'S INTERIOR Explain how Earth acquired its layered structure and briefly describe how seismic waves are used to probe Earth's interior.

If you could slice Earth in half, the first thing you would notice is that it has distinct layers. The heaviest materials (metals) would be in the center. Lighter solids (rocks) would be in the middle, and liquids and gases would be on top. Within Earth we know these layers as the iron core, the rocky mantle and crust, the liquid ocean, and the gaseous atmosphere. More than 95 percent of the variations in composition and temperature in Earth are due to layering. However, this is not the end of the story. If it were, Earth would be a dead, lifeless cinder floating in space.

There are also variations in composition and temperature with depth, which indicate that the interior of our planet is very dynamic. The rocks of the mantle and crust are in constant motion, not only moving about through plate tectonics but also continuously recycling between the surface and the deep interior. Furthermore, it is from Earth's deep interior that the water and air of our oceans and atmosphere are replenished, allowing life to exist at the surface.

Formation of Earth's Layered Structure

As material accumulated to form Earth (and for a short period afterward), the high-velocity impact of nebular debris and the decay of radioactive elements caused the temperature of our planet to steadily increase. During this time of intense heating, Earth became hot enough that iron and nickel began to melt. Melting produced liquid blobs of heavy metal that sank toward the center of the planet. This process occurred rapidly on the scale of geologic time and produced Earth's dense iron-rich core.

The early period of heating resulted in another process of chemical differentiation, whereby melting formed buoyant masses of molten rock that rose toward the surface and solidified to produce a primitive crust. These rocky materials were rich in oxygen and "oxygen-seeking" elements, particularly silicon and aluminum, along with lesser amounts of calcium, sodium, potassium, iron, and magnesium. In addition, some heavy metals such as gold, lead, and uranium, which have low melting points or were highly soluble in the ascending molten masses, were scavenged from Earth's interior and concentrated in the developing crust. This early period of chemical segregation established the three basic divisions of Earth's interior: (1) the iron-rich *core*, (2) the thin *primitive crust*, and (3) Earth's largest layer, called the *mantle*, which is located between the core and crust.

Probing Earth's Interior: "Seeing" Seismic Waves

Discovering the structure and properties of Earth's deep interior has not been easy. Light does not travel through rock, so we must find other ways to "see" into our planet. The best way to learn about Earth's interior is to dig or drill a hole and examine it directly. Unfortunately, this is possible only at shallow depths. The deepest a drilling rig has ever penetrated is only 12.3 kilometers (8 miles), which is about 1/500 of the way to Earth's center! Even this was an extraordinary accomplishment because temperature and pressure increase rapidly with depth.

Many earthquakes are large enough that their seismic waves travel all the way through Earth and can be recorded on the other side (**FIGURE 8.33**). This means that the seismic waves act like medical x-rays used to take images of a person's insides. About 100 to 200 earthquakes each year are large enough (about M_w 6) to be well recorded by seismographs all around the globe. These large earthquakes provide the means to "see" into our planet and have been the source of most of the data that have allowed us to figure out the nature of Earth's interior.

Interpreting the waves recorded on seismograms in order to identify Earth structures is challenging. Seismic waves do not travel along straight paths; instead, seismic waves are *reflected*, *refracted*, and *diffracted* as they pass through our planet. They reflect off boundaries between different layers, they refract (or bend) when passing from one layer to another layer, and they diffract around any obstacles they encounter (see Figure 8.33). These different wave behaviors have been used to identify the boundaries that exist within Earth.

One of the most noticeable behaviors of seismic waves is that they follow strongly curved paths (see Figure 8.33). This occurs because the velocity of seismic waves generally increases with depth. In addition, seismic waves travel faster when rock is stiffer or less compressible. These properties of stiffness and compressibility are then used to interpret the composition and temperature of the rock. For instance,

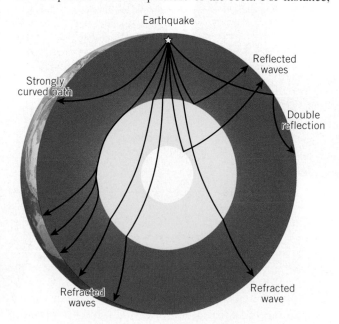

FIGURE 8.33 Possible Paths that Earthquake Waves Take

Earthquake

Reflected waves

Strongly curved path

Double reflection

Refracted waves

Refracted wave

when rock is hotter, it becomes less stiff (imagine heating up a frozen chocolate bar!), and waves travel more slowly. Waves also travel at different speeds through rocks of different compositions. Thus, the speed at which seismic waves travel can help determine both the kind of rock that is inside Earth and how hot it is.

8.8 | EARTH'S LAYERS

List and describe each of Earth's major layers.

Earth's three compositionally distinct layers—the crust, mantle, and core—can be further subdivided into zones based on physical properties. The physical properties used to define such regions include whether the layer is solid or liquid and how weak or strong it is. Knowledge of both types of layers is essential to our understanding of basic geologic processes, such as volcanism, earthquakes, and mountain building (**FIGURE 8.34**).

Crust

The **crust** is Earth's relatively thin, rocky outer skin, and there are two types: continental crust and oceanic crust. Both share the word *crust*, but the similarity ends there. The oceanic crust is roughly 7 kilometers (4 miles) thick and composed of the dark igneous rock *basalt*. By contrast, the continental crust averages 35 to 40 kilometers (22 to 25 miles) thick but may exceed 70 kilometers (40 miles) in some mountainous regions such as the Rockies and Himalayas. Unlike the oceanic crust, which has a relatively homogeneous chemical composition, the continental crust consists of many rock types. Although the upper crust has an average composition of a *granitic rock* called *granodiorite*, it varies considerably from place to place.

Continental rocks have an average density of about 2.7 grams per cubic centimeter, and some are 4 billion years old. The rocks of the oceanic crust are younger (180 million years or less) and denser (about 3.0 grams per cubic centimeter) than continental rocks.

FIGURE 8.34 Views of Earth's Layered Structure

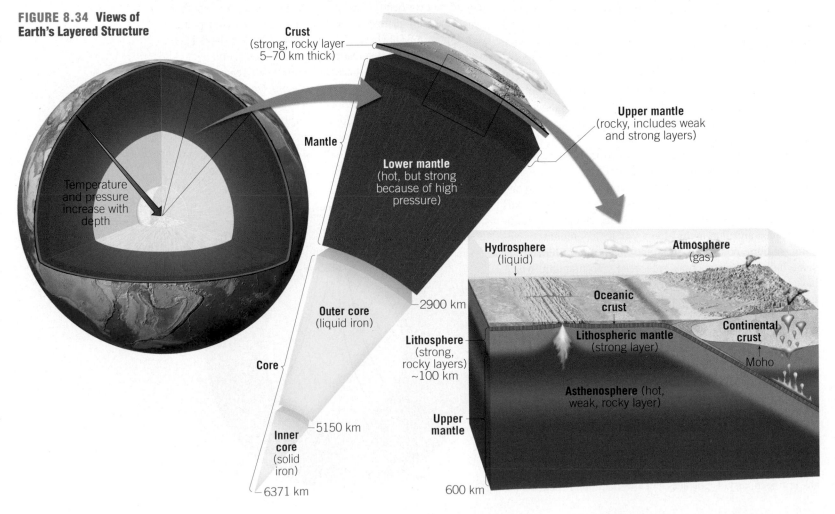

Crust
(strong, rocky layer
5–70 km thick)

Mantle

Temperature and pressure increase with depth

Lower mantle
(hot, but strong because of high pressure)

Upper mantle
(rocky, includes weak and strong layers)

Outer core
(liquid iron)

—2900 km

Core

Inner core
(solid iron)

—5150 km

—6371 km

Lithosphere
(strong, rocky layers)
~100 km

Upper mantle

600 km

Hydrosphere
(liquid)

Atmosphere
(gas)

Oceanic crust

Continental crust

Lithospheric mantle
(strong layer)

Moho

Asthenosphere (hot, weak, rocky layer)

Mantle

Beneath Earth's crust lies the mantle. More than 82 percent of Earth's volume is contained in the **mantle**, a solid, rocky shell that extends to a depth of about 2900 kilometers (1800 miles). The boundary between the crust and mantle represents a marked change in chemical composition. The dominant rock type in the uppermost mantle is *peridotite*, which is richer in the metals magnesium and iron than the minerals found in either the continental or oceanic crust.

The upper mantle extends from the crust–mantle boundary down to a depth of about 660 kilometers (410 miles). The upper mantle can be divided into two different parts. The top portion of the upper mantle is part of the stiff *lithosphere*, and beneath that is the weaker *asthenosphere*. The **lithosphere** ("sphere of rock") consists of the entire crust and uppermost mantle and forms Earth's relatively cool, rigid outer shell. Averaging about 100 kilometers (62 miles) thick, the lithosphere is more than 250 kilometers (155 miles) thick below the oldest portions of the continents (see Figure 8.34). Beneath this stiff layer to a depth of about 350 kilometers (217 miles) lies a soft, comparatively weak layer known as the **asthenosphere** ("weak sphere"). The top portion of the asthenosphere has a temperature/pressure regime that results in a small amount of melting. Within this very weak zone, the asthenosphere and lithosphere are mechanically detached from each other. The result is that the lithosphere is able to move independently of the asthenosphere.

It is important to emphasize that the strength of various Earth materials is a function of both their composition and the temperature and pressure of their environment. The entire lithosphere does *not* behave like a brittle solid similar to rocks found on the surface. Rather, the rocks of the lithosphere get progressively hotter and weaker (more easily deformed) with increasing depth. At the depth of the uppermost asthenosphere, the rocks are close enough to their melting temperature that they are very easily deformed, and some melting may actually occur. Thus, the uppermost asthenosphere is weak because it is near its melting point, just as hot wax is weaker than cold wax.

From 660 kilometers (410 miles) deep to the top of the core, at a depth of 2900 kilometers (1800 miles), is the lower mantle. Because of an increase in pressure (caused by the weight of the rock above), the mantle gradually strengthens with depth. Despite their strength, however, the rocks within the lower mantle are very hot and capable of very gradual flow.

Core

The composition of the **core** is thought to be an iron–nickel alloy, with minor amounts of oxygen, silicon, and sulfur—elements that readily form compounds with iron. At the extreme pressure found in the core, this iron-rich material has an average density of more than 10 grams per cubic centimeter and is about 13 grams per cubic centimeter at Earth's center. The core is divided into two regions that exhibit very different mechanical strengths. The **outer core** is a liquid layer 2270 kilometers (1410 miles) thick. It is the movement of metallic iron within this zone that generates Earth's magnetic field. The **inner core** is a sphere with a radius of 1216 kilometers (754 miles). Despite its higher temperature, the iron in the inner core is solid due to the immense pressures that exist in the center of the planet.

8.8 CONCEPT CHECKS

1 How do continental crust and oceanic crust differ?
2 Contrast the physical makeup of the asthenosphere and the lithosphere.
3 How are Earth's inner and outer cores different? How are they similar?

8 CONCEPTS IN REVIEW | Earthquakes and Earth's Interior

8.1 WHAT IS AN EARTHQUAKE?

Sketch and describe the mechanism that generates most earthquakes.

KEY TERMS: earthquake, fault, hypocenter (focus), epicenter, seismic wave, elastic rebound, aftershock, foreshock, strike-slip fault, transform fault, fault creep, megathrust fault

- Earthquakes are caused by the sudden movement of blocks of rock on opposite sides of faults. The spot where the rock begins to slip is the hypocenter (or focus). Seismic waves radiate from this spot outward into the surrounding rock. The point on Earth's surface directly above the hypocenter is the epicenter.
- Elastic rebound explains why most earthquakes happen: Rock is deformed by movement of Earth's crust. However, frictional resistance keeps the fault locked in place, and the rock bends elastically. Strain builds up until it is greater than the resistance, and the blocks of rock suddenly slip, releasing the pent-up energy in the form of seismic waves. As elastic rebound occurs, the blocks of rock on either side of the fault return to their original shapes, but they are now in new positions.
- Foreshocks are smaller earthquakes that precede larger earthquakes. Aftershocks are smaller earthquakes that happen after large earthquakes, as the crust readjusts to the new, post-earthquake conditions.
- Faults associated with plate boundaries are the source of most large earthquakes.
- The San Andreas Fault in California is an example of a transform fault boundary capable of generating destructive earthquakes.
- Subduction zones are marked by megathrust faults, large faults that are responsible for the largest earthquakes in recorded history. Megathrust faults are also capable of generating tsunami.

Q Label the blanks on the diagram to show the relationship between earthquakes and faults.

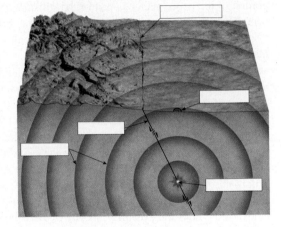

8.2 SEISMOLOGY: THE STUDY OF EARTHQUAKE WAVES

Compare and contrast the types of seismic waves and describe the principle of the seismograph.

KEY TERMS: seismology, seismograph (seismometer), inertia, seismogram, surface waves, body waves, P waves (primary waves), S waves (secondary waves)

- Seismology is the study of seismic waves. An instrument called a seismograph measures these waves, using the principle of inertia. While the body of the machine moves with the waves, the inertia of a suspended weight keeps a pen (or an electronic sensor) stationary and records the relative difference between the two. The resulting record of the waves is called a seismogram.
- Seismograms reveal that there are two main categories of earthquake waves: body waves (P waves and S waves) that are capable of moving through

Earth's interior and surface waves, which travel only along the upper layers of the crust. P waves are the fastest, S waves are intermediate in speed, and surface waves are the slowest. However, surface waves tend to have the greatest amplitude, while S waves are intermediate, and P waves have the least amplitude. Large-amplitude waves produce the most shaking, so surface waves are responsible for most damage during earthquakes.

- P waves and S waves exhibit different kinds of motion. P waves show compressional motion: They change the volume of the rock as they pass through it. In contrast, S waves impart a shaking motion to rock as they pass through it. This changes the shape of the rock but doesn't alter its volume. As a result, P waves can travel through liquids, but S waves cannot. Liquids can be compressed but not sheared.

Q How could you physically demonstrate the difference between P waves and S waves to a friend who hasn't taken a geology course? (*Caution:* Don't hurt your friend!)

8.3 DETERMINING THE SIZE OF EARTHQUAKES

Distinguish between intensity scales and magnitude scales.

KEY TERMS: intensity, magnitude, Modified Mercalli Intensity scale, Richter scale, moment magnitude

- Intensity and magnitude are different measures of earthquake strength. Intensity measures the amount of ground shaking a place experiences due to an earthquake, and magnitude is an estimate of the actual amount of energy released during the earthquake.
- The Modified Mercalli Intensity scale is a tool for measuring an earthquake's intensity at different locations. The scale is based on verifiable physical evidence to quantify intensity on a 12-point scale. However, it's hard to measure

intensity if there are no buildings or people present at a particular location.

- The Richter scale was the first attempt at measuring an earthquake's magnitude. The scale takes into account both the maximum amplitude of the seismic waves measured at a given seismograph and the distance of that seismograph from the earthquake. The Richter scale is logarithmic, meaning that each numerical unit up on the scale represents seismic wave amplitudes that are 10 times greater than those of the next number down. Seismic waves with larger amplitudes possess more energy, so each numerical Richter unit represents about 32 times as much energy release as the next number below it.
- Because the Richter scale does not effectively differentiate between very large earthquakes, the moment magnitude scale was devised. The moment magnitude scale measures the total energy released from an earthquake by considering the strength of the faulted rock, the amount of slip, and the area of the fault that slipped. Moment magnitude is the modern standard for measuring the size of earthquakes.

Q On the Richter scale diagram, first determine the M_L for an earthquake at 400 kilometers distance, with a maximum amplitude of 0.5 millimeters. Second, for this same earthquake (same M_L), determine the amplitude of the biggest waves for a seismograph only 40 kilometers from the hypocenter.

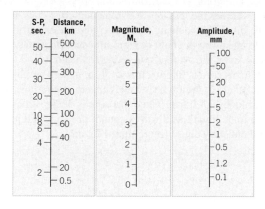

8.4 EARTHQUAKE DESTRUCTION

List and describe the major destructive forces that can be triggered by earthquake vibrations.

KEY TERMS: liquefaction, tsunami

- Several factors influence how much destruction results from an earthquake. The earthquake's magnitude and distance are important, but we must also consider (1) the intensity of the shaking, (2) how long shaking persists, (3) what sort of ground underlies buildings, and (4) building construction standards. Buildings constructed of unreinforced masonry (bricks and blocks) are more likely than other types of buildings to collapse in a quake.
- In general, bedrock-grounded buildings fare better in an earthquake, as loose sediments amplify seismic shaking. A particular hazard occurs with water-logged sediments or soil: When shaken at certain frequencies, the material will flow like a liquid. This phenomenon, called liquefaction, sometimes produces "sand volcanoes." Buildings may also sink into the ground, or underground tanks or sewer lines may float up to the surface when the Earth's subsurface experiences liquefaction.
- Earthquakes may also trigger landslides or ground subsidence, and they may break gas lines, which can initiate devastating fires.
- Tsunami are large ocean waves that form when water is displaced, usually by a megathrust fault rupturing on the seafloor. Crossing the ocean at the speed of a commercial airplane, a tsunami is hardly noticeable in deep water. However, upon arrival in shallower coastal waters, the tsunami slows down and piles up, sometimes producing a wall of water more than 30 meters (100 feet) in height. A tsunami does not look like a curling breaker wave. Instead, it resembles a rapid rise in sea level and may be choked with sediment and other debris. Tsunami warning systems have been established in several ocean basins, including the Pacific.

Q Of the secondary earthquake hazards discussed in this section, which is (are) of the greatest concern in the region where you live? Why?

8.5 EARTHQUAKE BELTS AND PLATE BOUNDARIES

Locate Earth's major earthquake belts on a world map and label the regions associated with the largest earthquakes.

KEY TERM: circum-Pacific belt

- Most earthquake energy is released in the circum-Pacific belt, the ring of megathrust faults rimming the Pacific Ocean. Another earthquake belt is the Alpine–Himalayan belt, which runs along the collisional zone between Eurasia and the Indian–Australian and African plates.
- Earth's divergent oceanic ridge system produces another belt of earthquake activity where seafloor spreading generates many frequent quakes of small magnitude. Transform faults in the continental crust, including the San Andreas Fault, can produce large earthquakes.

Q From memory, outline Earth's major earthquake belts on this map. Describe the tectonic stresses responsible for producing each belt.

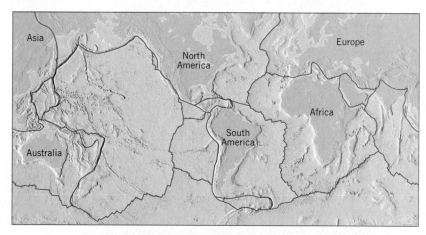

8.6 CAN EARTHQUAKES BE PREDICTED?

Compare and contrast the goals of short-range earthquake predictions to long-range forecasts.

KEY TERM: seismic gap

- Successful earthquake prediction has been an elusive goal of seismology for many years. Shorter-range predictions (for hours or days) are based on precursor events such as changes in ground elevation or variations in strain levels near a fault. Unfortunately, they haven't been found to be consistently reliable.
- Long-range forecasts (for time scales of 30 to 100 years) are statistical estimates of how likely it is for an earthquake of a given magnitude to

occur. Long-range forecasts are useful because they can be used to guide building codes and infrastructure development.

- Scientists have identified seismic gaps, portions of faults that have been storing strain for a long time without releasing it. Knowing the location of seismic gaps can help in the preparation of long-range forecasts. Another tool used in making long-range forecasts is paleoseismology, the study of ancient earthquakes. Because earthquakes occur on a cyclical basis, determining how frequently they have occurred in the past can suggest when they are most likely to occur again.

Q If you were considering moving to a city located in a seismic gap, how would you determine whether it was safe or foolhardy? Which factors would be most influential in your decision?

8.7 EARTH'S INTERIOR

Explain how Earth acquired its layered structure and briefly describe how seismic waves are used to probe Earth's interior.

- The layered internal structure of Earth developed due to gravitational sorting of Earth materials early in the history of the planet. The densest material settled to form the center, while the least dense material rose to form the surface.
- Seismic waves allow geoscientists to "look" into Earth's interior, which would otherwise be invisible to scientific investigation. Like the x-rays used to image human bodies, seismic waves generated by large earthquakes reveal details about Earth's layered structure.

8.8 EARTH'S LAYERS

List and describe each of Earth's major layers.

KEY TERMS: crust, mantle, lithosphere, asthenosphere, core, outer core, inner core

- Earth has two distinct kinds of crust: oceanic and continental. Oceanic crust is thinner, denser, and younger than continental crust. Oceanic crust also readily subducts, whereas the less dense continental crust does not.
- Earth's mantle may be divided by density into upper and lower portions. The uppermost mantle makes up the bulk of rigid lithospheric plates, while a relatively weak layer, the asthenosphere, lies beneath it.
- The composition of Earth's core is likely a mix of iron, nickel, and lighter elements. Iron and nickel are common heavy elements in meteorites, the leftover "building blocks" of Earth. The outer core is dense (around 10 g/cm^3). It is known to be liquid, as S waves cannot pass through it. The inner core is solid and very dense (more than 13 g/cm^3).

GIVE IT SOME **THOUGHT**

1. Draw a sketch that illustrates the concept of elastic rebound. Develop an analogy other than a rubber band to illustrate this concept.

2. The accompanying map shows the locations of many of the largest earthquakes in the world since 1900. Refer to the map of Earth's plate boundaries in Figure 7.12 (page 218) and determine which type of plate boundary is most often associated with these destructive events.

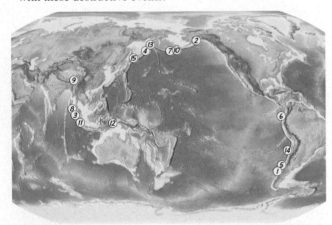

3. Use the seismogram located in the upper right column to answer the following questions:

    a. Which of the three types of seismic waves reached the seismograph first?

    b. What is the time interval between the arrival of the first P wave and the arrival of the first S wave?

    c. Use your answer from Question b and the travel-time graph on page 255 to determine the distance from the seismic station to the earthquake.

    d. Which of the three types of seismic waves had the highest amplitude when they reached the seismic station?

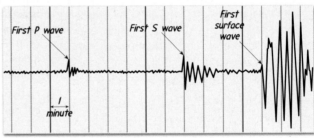

4. You go for a jog on a beach and choose to run near the water, where the sand is well packed and solid under your feet. With each step, you notice that your footprint quickly fills with water but not water coming in from the ocean. What is this water's source? For what earthquake-related hazard is this phenomenon a good analogy?

5. Make a sketch that illustrates why a tsunami often causes a rapid withdrawal of water from beaches before the first surge.

6. Why is it possible to issue a tsunami warning but not a warning for an impending earthquake? Describe a scenario in which a tsunami warning would be of little value.

7. Using the accompanying map of the San Andreas Fault, answer the following questions:

    a. Which of the four segments (1–4) of the San Andreas Fault do you think is experiencing fault creep?

    b. Paleoseismology studies have found that the section of the San Andreas Fault that failed during the Fort Tejon quake (segment 3) produces a major earthquake every 135 years, on average. Based on this information, how would you rate the chances of a major earthquake occurring along this section in the next 30 years? Explain.

    c. Do you think San Francisco or Los Angeles has the greater risk of experiencing a major earthquake in the near future? Defend your selection.

8. The accompanying image shows a double-decked section of Interstate 880 (the Nimitz Freeway) that collapsed during the 1989 Loma Prieta earthquake and caused 42 deaths. About 1.4 kilometers (0.9 mile) of this freeway section, commonly called the Cypress Viaduct, collapsed, while a similar section survived the vibration. Both sections were subsequently demolished and rebuilt as a single-level structure, at a cost of $1.2 billion. Examine the map and seismograms from an aftershock that shows the intensity of shaking observed at three nearby locations to answer the following questions:

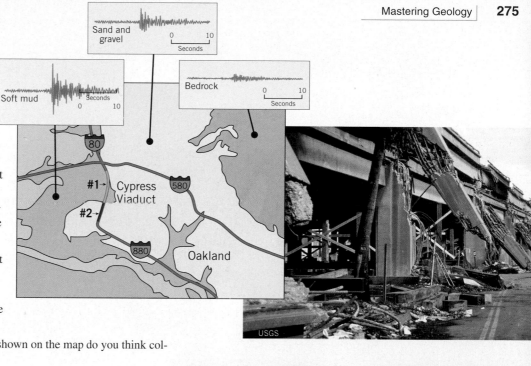

   a. What type of ground material experienced the least amount of shaking during the aftershock?

   b. What type of ground materials experienced the greatest amount of ground shaking during the same event?

   c. Which of the two sections of the Cypress Viaduct shown on the map do you think collapsed? Explain.

9. Strike-slip faults, like the San Andreas Fault, are not perfectly straight but bend gradually back and forth. In some locations, the bends are oriented such that blocks on opposite sides of the fault pull away from each other, as shown in the accompanying sketch. As a result, the ground between the bends sags, forming a depression or basin. These depressions often fill with water.

   a. What name is given to the depression in the accompanying photo?

   b. Describe what would happen if these two blocks began moving in opposite directions.

10. Using the Internet, compare and contrast the 2010 Haiti earthquake with the 2011 Japan earthquake. Include magnitude, type of plate boundary, and extent of destruction. Explain why the Japan earthquake produced a tsunami, while the Haiti quake did not.

11. Describe the two different ways that Earth's layers are defined.

12. Based on the properties of Earth's layers and the mode of travel of body waves, predict the location in Earth's interior where waves should (a) travel fastest and (b) travel slowest. Is there an exception for these generalities? Explain your answers.

Michael Collier

Before offset   After offset

Extension and subsidence

EXAMINING THE **EARTH SYSTEM**

What potentially disastrous phenomenon often occurs when the energy of an earthquake is transferred from the solid earth to the hydrosphere (ocean) at their interface on the floor of the ocean? When the energy from this event is expended along a coast, how might coastal lands and the biosphere be altered?

MasteringGeology™

Looking for additional review and test prep materials? Visit the Self Study area in **www.masteringgeology.com** to find practice quizzes, study tools, and multimedia that will aid in your understanding of this chapter's content. In **MasteringGeology™** you will find:

- GEODe: Earth Science: An interactive visual walkthrough of key concepts
- Geoscience Animation Library: More than 100 animations illuminating many difficult-to-understand Earth science concepts

- In The News RSS Feeds: Current Earth science events and news articles are pulled into the site with assessment
- Pearson eText
- Optional Self Study Quizzes
- Web Links
- Glossary
- Flashcards

9

Volcanoes and Other Igneous Activity

Fluid basaltic lava erupting from Mount Etna, Italy. (Photo by Westend61 GmbH/Alamy)

The significance of igneous activity may not be obvious at first glance. However, because volcanoes extrude molten rock that formed at great depth, they provide our only means of directly observing processes that occur many kilometers below Earth's surface. Furthermore, the atmosphere and oceans have evolved from gases emitted during volcanic eruptions. Either of these facts is reason enough for igneous activity to warrant our attention.

9.1 | MOUNT ST. HELENS VERSUS KILAUEA

Compare and contrast the 1980 eruption of Mount St. Helens with the eruption of Kilauea, which began in 1983 and continues today.

On Sunday, May 18, 1980, the largest volcanic eruption to occur in North America in historic times transformed a picturesque volcano into a decapitated remnant (**FIGURE 9.1**). On that date in southwestern Washington State, Mount St. Helens erupted with tremendous force. The blast blew out the entire north flank of the volcano, leaving a gaping hole. In one brief moment, a prominent volcano whose summit had been more than 2900 meters (9500 feet) above sea level was lowered by more than 400 meters (1350 feet).

The event devastated a wide swath of timber-rich land on the north side of the mountain (**FIGURE 9.2**). Trees within a 400-square-kilometer (160-square-mile) area lay intertwined and flattened, stripped of their branches and appearing from the air like toothpicks strewn about. The accompanying mudflows carried ash, trees, and water-saturated rock debris 29 kilometers (18 miles) down the Toutle River. The eruption claimed 59 lives; some died from the intense heat and the suffocating cloud of ash and gases, others from the impact of the blast, and still others from being entrapped in mudflows.

The eruption ejected nearly 1 cubic kilometer of ash and rock debris. Following the devastating explosion, Mount St. Helens continued to emit great quantities of hot gases and ash. The force of the blast was so strong that some ash was propelled more than 18 kilometers (over 11 miles) into the stratosphere. During the next few days, this very fine-grained material was carried around Earth by strong upper-air winds. Measurable deposits were reported in Oklahoma and Minnesota, and crop damage occurred as far away as central Montana. Meanwhile, ash fallout in the immediate vicinity exceeded 2 meters (6 feet) in depth. The

FIGURE 9.1 Before-and-After Photographs Show the Transformation of Mount St. Helens The May 18, 1980, eruption of Mount St. Helens occurred in southwestern Washington.

Spirit Lake

USGS

The blast blew out the entire north flank of Mount St. Helen's, leaving a gaping hole. In a brief moment, a prominent volcano was lowered by 1350 feet.

1350 feet

Spirit Lake

USGS

air over Yakima, Washington (130 kilometers [80 miles] to the east), was so filled with ash that residents experienced midnight-like darkness at noon.

Not all volcanic eruptions are as violent as the 1980 Mount St. Helens event. Some volcanoes, such as Hawaii's Kilauea volcano, generate relatively quiet outpourings of fluid lavas. These "gentle" eruptions are not without some fiery displays; occasionally fountains of incandescent lava spray hundreds of meters into the air. Nevertheless, during Kilauea's most recent active phase, which began in 1983, more than 180 homes and a national park visitor center have been destroyed.

Testimony to the "quiet nature" of Kilauea's eruptions is the fact that the Hawaiian Volcanoes Observatory has operated on its summit since 1912, despite the fact that Kilauea has had more than 50 eruptive phases since record keeping began in 1823.

9.1 CONCEPT CHECKS

1 Briefly compare the May 18, 1980, eruption of Mount St. Helens to a typical eruption of Hawaii's Kilauea volcano.

FIGURE 9.2 Douglas Fir Trees Snapped Off or Uprooted by the Lateral Blast of Mount St. Helens (Large photo by USGS Cascades Volcano Observatory/Lyn Topinka/AP Photo; inset photo by John M. Burnley/Science Source)

9.2 | THE NATURE OF VOLCANIC ERUPTIONS

Explain why some volcanic eruptions are explosive and others are quiescent.

Volcanic activity is commonly perceived as a process that produces a picturesque, cone-shaped structure that periodically erupts in a violent manner. However, many eruptions are not explosive. What determines the manner in which volcanoes erupt?

Factors Affecting Viscosity

The source material for volcanic eruptions is **magma**, molten rock that usually contains some crystals and varying amounts of dissolved gas. Erupted magma is called **lava**. The primary factors that affect the behavior of magma and lava are its *temperature* and *composition* and, to a lesser extent, the amount of *dissolved gases* it contains. To varying degrees, these factors determine a magma's mobility, or **viscosity** (*viscos* = sticky). The more viscous the material, the greater its resistance to flow. For example, syrup is more viscous and, thus, more resistant to flow, than water.

Temperature   The effect of temperature on viscosity is easily seen. Just as heating syrup makes it more fluid (less viscous), temperature also strongly influences the mobility of lava. As lava cools and begins to congeal, its viscosity increases, and eventually the flow halts.

Composition   Another significant factor influencing volcanic behavior is the chemical composition of the magma. Recall that a major difference among various igneous rocks is their silica (SiO_2) content (**TABLE 9.1**). Magmas

TABLE 9.1   Different Compositions of Magmas Cause Properties to Vary

Composition	Silica Content	Gas Content	Eruptive Temperatures	Viscosity	Tendency to Form Pyroclastics	Volcanic Landform
Basaltic (mafic)	Least (~50%)	Least (1–2%)	1000–1250°C	Least	Least	Shield volcanoes, basalt plateaus, cinder cones
Andesitic (intermediate)	Intermediate (~60%)	Intermediate (3–4%)	800–1050°C	Intermediate	Intermediate	Composite cones
Rhyolitic (felsic)	Most (~70%)	Most (4–6%)	650–900°C	Greatest	Greatest	Pyroclastic flows, lava domes

that produce mafic rocks such as basalt contain about 50 percent silica, whereas magmas that produce felsic rocks (granite and its extrusive equivalent, rhyolite) contain more than 70 percent silica. Intermediate rock types—andesite and diorite—contain about 60 percent silica.

A magma's viscosity is directly related to its silica content: *The more silica in magma, the greater its viscosity.* Silica impedes the flow of magma because silicate structures start to link together into long chains early in the crystallization process. Consequently, felsic (rhyolitic) lavas are very viscous and tend to form comparatively short, thick flows. By contrast, mafic (basaltic) lavas, which contain less silica, are relatively fluid and have been known to travel 150 kilometers (90 miles) or more before solidifying.

Dissolved Gases

The gaseous components in magma (mainly dissolved water and carbon dioxide), called **volatiles**, also affect the mobility of magma. Other factors being equal, water dissolved in magma tends to increase fluidity because it reduces formation of long silicate chains by breaking silicon–oxygen bonds. It follows, therefore, that the loss of gases renders magma (lava) more viscous. Gases also give magmas their explosive character.

Quiescent Versus Explosive Eruptions

Most magma is generated by partial melting of the rock in the upper mantle and has a basaltic composition, a topic we will consider later in the chapter. The newly formed magma, which is less dense than the surrounding rock, slowly rises toward the surface. In some settings, high-temperature basaltic magmas reach Earth's surface, where they produce highly fluid lavas. This most commonly occurs on the ocean floor, in association with seafloor spreading. In continental settings, however, the density of crustal rocks is less than that of the ascending material, causing the magma to pond at the crust–mantle boundary. Heat from the hot magma is often sufficient to partially melt the overlying crustal rocks, generating a less dense, silica-rich magma, which continues the journey toward Earth's surface.

Quiescent Hawaiian-Type Eruptions Eruptions that involve very fluid basaltic lavas, such as the eruptions of Kilauea on Hawaii's Big Island, are often triggered by the arrival of a new batch of molten rock into a near-surface magma chamber. Such an event can usually be detected because the summit of the volcano begins to inflate and rise months or even years before an eruption. The injection of a fresh supply of hot molten rock heats and remobilizes the semi-liquid magma chamber. In addition, swelling of the magma chamber fractures the rock above, allowing the fluid magma to move upward along the newly formed openings, often generating outpourings of lava for weeks, months, or possibly years. The eruption of Kilauea that began in 1983 has been ongoing for more than 30 years.

Triggering Explosive Eruptions All magmas contain some water vapor and other gases that are kept in solution by the immense pressure of the overlying rock. As magma rises (or the rocks confining the magma fail), the pressure reduces, and the dissolved gases begin to separate from the melt, forming tiny bubbles. This is analogous to opening a can of soda and allowing the carbon dioxide bubbles to escape.

When fluid basaltic magmas erupt, the pressurized gases readily escape. At temperatures that often exceed 1100°C (2000°F), these gases can quickly expand to occupy hundreds of times their original volumes. Occasionally, these expanding gases propel incandescent lava hundreds of meters into the air, producing lava fountains (**FIGURE 9.3**). Although spectacular, these fountains are usually harmless and generally not associated with major explosive events that cause great loss of life and property.

At the other extreme, highly viscous magmas expel particles of fragmented lava and gases at nearly supersonic speeds, creating buoyant plumes called **eruption columns**. Eruption columns can rise perhaps 40 kilometers (25 miles) into the atmosphere (**FIGURE 9.4**). Because silica-rich magmas are sticky (viscous), a significant portion of the gaseous material remains dissolved until the magma nears Earth's surface, at which time tiny bubbles begin to form and grow. When the pressure of the expanding magma body exceeds the strength of the overlying rock, fracturing occurs. As magma moves up the fractures, a further drop in confining pressure causes more gas bubbles to form and grow. This chain reaction may generate an explosive event in which magma is literally blown into fragments (ash and pumice) that are carried to great heights

FIGURE 9.3 Lava Fountain Produced by Gases Escaping Fluid Basaltic Lava Lava erupting from Mount Etna, Italy. (Photo by D. Szczepanski terras/AGE Fotostock)

Gases readily escape hot fluid basaltic flows, producing lava fountains. Although often spectacular, these features generally do not cause great loss of life or property.

by the hot gases. (As exemplified by the 1980 eruption of Mount St. Helens, the collapse of a volcano's flank can result in a reduction in the pressure on the magma below, triggering an explosive eruption.)

When magma in the uppermost portion of the magma chamber is forcefully ejected by the escaping gases, the confining pressure on the molten rock directly below drops suddenly. Thus, rather than a single "bang," volcanic eruptions are really a series of explosions. Following explosive eruptions, degassed lava may slowly ooze out of the vent to form short rhyolitic flows or dome-shaped lava bodies that grow over the vent.

FIGURE 9.4 Eruption Column Generated by Viscous, Silica-Rich Magma Steam and ash eruption column from Mount Augustine, Cook Inlet, Alaska. (Photo by Steve Kaufman/Getty Images)

Eruptions of highly viscous lavas may produce explosive clouds of hot ash and gases called eruption columns.

9.2 CONCEPT CHECKS

1 Define *viscosity* and list three factors that influence the viscosity of magma.

2 Explain how the viscosity of magma influences the explosiveness of a volcano.

3 List these three magmas in order from *most* silica rich to *least* silica rich, based on their compositions: mafic (basaltic), felsic (rhyolitic), intermediate (andesitic).

4 The eruption of what type of magma may produce an eruption column?

5 Why is a volcano that is fed by highly viscous magma likely to be a greater threat to life and property than a volcano supplied with very fluid magma?

9.3 | MATERIALS EXTRUDED DURING AN ERUPTION

List and describe the three categories of materials extruded during volcanic eruptions.

Volcanoes extrude lava, large volumes of gas, and pyroclastic materials (broken rock, lava "bombs," fine ash, and dust). In this section, we examine each of these materials.

Lava Flows

The vast majority of lava on Earth—more than 90 percent of the total volume—is estimated to be basaltic in composition. Andesites and other lavas of intermediate composition account for most of the rest, while rhyolitic (felsic) flows make up as little as 1 percent of the total.

Hot basaltic lavas, which are usually very fluid, generally flow in thin, broad sheets or streamlike ribbons. On the Big Island of Hawaii, these lavas have been clocked at 30 kilometers (19 miles) per hour down steep slopes. However, flow rates of 10 to 300 meters (30 to 1000 feet) per hour are more common. By contrast, the movement of silica-rich, rhyolitic lava may be too slow to perceive. Furthermore, most

A. Active aa flow overriding an older pahoehoe flow.

Aa flow

Pahoehoe flow

B. Pahoehoe flow displaying the characteristic ropy appearance.

FIGURE 9.5 Lava Flows A. A typical slow-moving, basaltic, aa flow. **B.** A typical fluid pahoehoe (ropy) lava. Both of these lava flows erupted from a rift on the flank of Hawaii's Kilauea volcano. (Photos courtesy of USGS)

281

A. Lava tubes are cave-like tunnels that once served as conduits carrying lava from an active vent to the flow's leading edge.

Valentine Cave, a lava tube at Lava Beds National Monument, California.

FIGURE 9.6 Lava Tubes A lava flow may develop a solid upper crust while the molten lava below continues to advance in a conduit called a *lava tube.* Some lava tubes exhibit extraordinary dimensions. One such structure, Kazumura Cave, located on the southeastern slope of Hawaii's Mauna Loa Volcano, extends more than 60 kilometers (40 miles). (Photo A. by Dave Bunnell and Photo B. Courtesy of USGS)

rhyolitic lavas seldom travel more than a few kilometers from their vents. As you might expect, andesitic lavas, which are intermediate in composition, exhibit characteristics that are between the extremes.

B. Skylights develop where the roofs of lava tubes collapse and reveal the hot lava flowing through the tube.

Aa and Pahoehoe Flows

Fluid basaltic magmas tend to generate two types of lava flows known by Hawaiian names. One of these, called **aa** (pronounced "ah-ah") **flows**, have surfaces of rough, jagged blocks with dangerously sharp edges and spiny projections (**FIGURE 9.5A**). Crossing a hardened aa flow can be a trying and miserable experience. The second type, **pahoehoe** (pronounced "pah-hoy-hoy") **flows**, exhibit smooth surfaces that sometimes resemble twisted braids of ropes (**FIGURE 9.5B**). Pahoehoe means "on which one can walk."

Although both lava types can erupt from the same volcano, pahoehoe lavas are hotter and more fluid than aa flows. In addition, pahoehoe lavas can change into aa lava flows, although the reverse (aa to pahoehoe) does not occur.

Cooling that occurs as the flow moves away from the vent is one factor that facilitates the change from pahoehoe to aa. The reduction in temperature increases viscosity and promotes bubble formation. Escaping gas bubbles produce numerous voids (vesicles) and sharp spines in the surface of the congealing lava. As the molten interior advances, the outer crust is broken, transforming a relatively smooth surface of a pahoehoe flow into an aa flow made up of an advancing mass of rough, sharp, broken lava blocks.

Pahoehoe flows often develop cave-like tunnels called **lava tubes** that were previously conduits for carrying lava from an active vent to the flow's leading edge (**FIGURE 9.6**). Lava tubes form in the interior of a lava flow where the temperature remains high long after the surface cools and hardens. Because they serve as insulated pathways that facilitate the advance of lava great distances from its source, lava tubes are important features of fluid lava flows.

FIGURE 9.7 Pillow Lava This image shows pillow lava that formed off the coast of Hawaii.

Pillow lavas form on the ocean floor and have elongated shapes, resembling toothpaste coming out of a tube.

Pillow Lavas

Recall that much of Earth's volcanic output occurs along oceanic ridges (divergent plate boundaries). When outpourings of lava occur on the ocean floor, the flow's outer skin quickly freezes to form obsidian. However, the interior lava is able to move forward by breaking through the hardened surface. This process occurs over and over, as molten basalt is extruded—like toothpaste from a tightly squeezed tube. The result is a lava flow composed of numerous tube-like structures called **pillow lavas**, stacked one atop the other (**FIGURE 9.7**). Pillow lavas are useful in the reconstruction of geologic history because their presence indicates that the lava flow formed below the surface of a water body.

Gases

Magmas contain varying amounts of dissolved gases (*volatiles*) held in the molten rock by confining pressure, just as carbon dioxide is held in cans and bottles of soft drinks. As with soft drinks, as soon as the pressure is reduced, the gases begin to escape. Obtaining gas samples from an erupting volcano is difficult and dangerous, so geologists usually must estimate the amount of gas originally contained in the magma.

The gaseous portion of most magmas makes up 1 to 6 percent of the total weight, with most of this in the form of water vapor. Although the percentage may be small, the actual quantity of emitted gas can exceed thousands of tons per day. Occasionally, eruptions emit colossal amounts of volcanic gases that rise high into the atmosphere, where they may reside for several years. Some of these eruptions may have an impact on Earth's climate, a topic we consider later in this chapter.

The composition of volcanic gases is important because these gases contribute significantly to our planet's atmosphere. Analyses of samples taken during Hawaiian eruptions indicate that the gas component is about 70 percent water vapor, 15 percent carbon dioxide, 5 percent nitrogen, and 5 percent sulfur dioxide, with lesser amounts of chlorine, hydrogen, and argon. (The relative proportion of each gas varies significantly from one volcanic region to another.) Sulfur compounds are easily recognized by their pungent odor. Volcanoes are also natural sources of air pollution; some emit large quantities of sulfur dioxide, which readily combines with atmospheric gases to form sulfuric acid and other sulfate compounds.

Pyroclastic Materials

When volcanoes erupt energetically, they eject pulverized rock, lava, and glass fragments from the vent. The particles produced are called **pyroclastic materials** (*pyro* = fire, *clast* = fragment) and are also referred to as *tephra*. These fragments range in size from very fine dust and sand-sized volcanic ash (less than 2 millimeters) to pieces that weigh several tons (**FIGURE 9.8**).

Pyroclastic Materials (Tephra)		
Particle name	**Particle size**	**Image**
Volcanic Ash*	Less than 2mm (0.08 inch)	
Lapilli (Cinders)	Between 2mm and 64 mm (0.08–2.5 inches)	
Volcanic Bombs	More than 64 mm (2.5 inches)	
Volcanic Blocks		

FIGURE 9.8 Types of Pyroclastic Materials Pyroclastic materials are also commonly referred to as tephra.

*The term volcanic dust is used for fine volcanic ash less than 0.063 mm (0.0025 inch).

A. Scoria is a vesicular rock commonly having a basaltic or andesitic composition. Pea-to-basketball size scoria fragments make up a large portion of most cinder cones (also called *scoria cones*).

B. Pumice is a low density vesicular rock that forms during explosive eruptions of viscous magma having an andesitic to rhyolitic composition.

FIGURE 9.9 Common Vesicular Rocks Scoria and pumice are volcanic rocks that exhibit a vesicular texture. Vesicles are small holes left by escaping gas bubbles. (Photos by E. J. Tarbuck)

Somewhat larger pyroclasts that range in size from small beads to walnuts are known as *lapilli* ("little stones"). These ejecta are commonly called *cinders* (2–64 millimeters). Particles larger than 64 millimeters (2.5 inches) in diameter are called *blocks* when they are made of hardened lava and *bombs* when they are ejected as incandescent lava (see Figure 9.8). Because bombs are semimolten upon ejection, they often take on a streamlined shape as they hurtle through the air. Because of their size, bombs and blocks usually fall near the vent; however, they are occasionally propelled great distances. For instance, bombs 6 meters (20 feet) long and weighing about 200 tons were blown 600 meters (2000 feet) from the vent during an eruption of the Japanese volcano Asama.

So far we have distinguished various pyroclastic materials based largely on the sizes of the fragments. Some materials are also identified by their texture and composition. In particular, **scoria** is the name applied to vesicular ejecta that is a product of basaltic magma (**FIGURE 9.9A**). These black to reddish-brown fragments are generally found in the size range of lapilli and resemble cinders and clinkers produced by furnaces used to smelt iron. When magmas with intermediate (andesitic) or felsic (rhyolitic) compositions erupt explosively, they emit ash and the vesicular rock **pumice** (**FIGURE 9.9B**). Pumice is usually lighter in color and less dense than scoria, and many pumice fragments have so many vesicles that they are light enough to float.

Ash and *dust* particles are produced when gas-rich viscous magma erupts explosively (see Figure 9.7). As magma moves up in the vent, the gases rapidly expand, generating a melt that resembles the froth that flows from a bottle of champagne. As the hot gases expand explosively, the froth is blown into very fine glassy fragments. When the hot ash falls, the glassy shards often fuse to form a rock called *welded tuff*. Sheets of this material, as well as ash deposits that later consolidate, cover vast portions of the western United States.

9.3 CONCEPT CHECKS

1 Describe pahoehoe and aa lava flows.

2 How do lava tubes form?

3 List the main gases released during a volcanic eruption. What role do gases play in eruptions?

4 How do volcanic bombs differ from blocks of pyroclastic debris?

5 What is scoria? How is scoria different from pumice?

9.4 | ANATOMY OF A VOLCANO
Label a diagram that illustrates the basic features of a typical volcanic cone.

A popular image of a volcano is a solitary, graceful, snow-capped cone, such as Mount Hood in Oregon or Japan's Fujiyama. These picturesque conical mountains are produced by volcanic activity that occurred intermittently over thousands, or even hundreds of thousands, of years. However, many volcanoes do not fit this image. Cinder cones are quite small and form during a single eruptive phase that lasts a few days to a few years. Alaska's Valley of Ten Thousand Smokes is a flat-topped deposit consisting of 15 cubic kilometers (more than 3.5 cubic miles) of ash, more than 20 times the volume of the 1980 Mount St. Helens eruption. It formed in less than 60 hours and blanketed a river valley to a depth of 200 meters (600 feet).

Volcanic landforms come in a wide variety of shapes and sizes, and each volcano has a unique eruptive history.

Nevertheless, volcanologists have been able to classify volcanic landforms and determine their eruptive patterns. In this section, we will consider the general anatomy of an idealized volcanic cone. We will follow this discussion by exploring the three major types of volcanic cones—shield volcanoes, cinder cones, and composite volcanoes—as well as their associated hazards.

Volcanic activity frequently begins when a fissure (crack) develops in Earth's crust, as magma moves forcefully toward the surface. As the gas-rich magma moves up through a fissure, its path is usually localized into a circular **conduit** that terminates at a surface opening called a **vent** (**FIGURE 9.10**). The cone-shaped structure we call a **volcanic cone** is often created by successive eruptions of lava,

Comparison of Three Types of Volcanic Cones

These images and drawings compare the relative sizes and shapes of three different volcanoes. Note the size comparison between a large shield volcano and a large composite volcano.

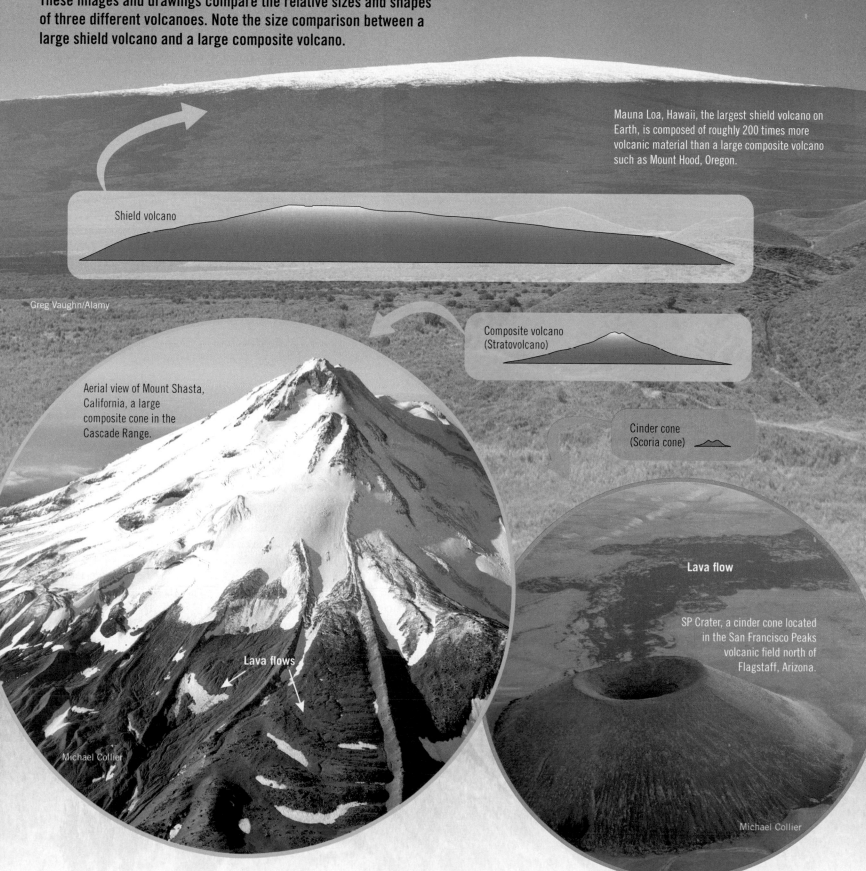

Mauna Loa, Hawaii, the largest shield volcano on Earth, is composed of roughly 200 times more volcanic material than a large composite volcano such as Mount Hood, Oregon.

Shield volcano

Greg Vaughn/Alamy

Composite volcano (Stratovolcano)

Cinder cone (Scoria cone)

Aerial view of Mount Shasta, California, a large composite cone in the Cascade Range.

Lava flows

Michael Collier

Lava flow

SP Crater, a cinder cone located in the San Francisco Peaks volcanic field north of Flagstaff, Arizona.

Michael Collier

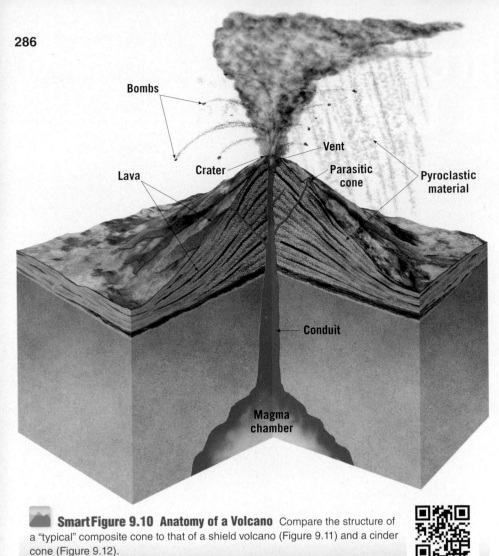

Bombs

Lava

Crater

Vent

Parasitic cone

Pyroclastic material

Conduit

Magma chamber

SmartFigure 9.10 Anatomy of a Volcano Compare the structure of a "typical" composite cone to that of a shield volcano (Figure 9.11) and a cinder cone (Figure 9.12).

pyroclastic material, or frequently a combination of both, often separated by long periods of inactivity.

Located at the summit of most volcanic cones is a somewhat funnel-shaped depression, called a **crater** (*crater* = bowl). Volcanoes built primarily of pyroclastic materials typically have craters that form by gradual accumulation of volcanic debris on the surrounding rim. Other craters form during explosive eruptions as the rapidly ejected particles erode the crater walls. Craters also form when the summit area of a volcano collapses following an eruption. Some volcanoes have very large circular depressions called *calderas* that have diameters greater than 1 kilometer (0.6 mile) and in some cases can exceed 50 kilometers (30 miles). We consider the formation of various types of calderas later in this chapter.

During early stages of growth, most volcanic discharges come from a central summit vent. As a volcano matures, material also tends to be emitted from fissures that develop along the flanks or at the base of the volcano. Continued activity from a flank eruption may produce one or more small **parasitic cones** (*parasitus* = one who eats at the table of another). Italy's Mount Etna, for example, has more than 200 secondary vents, some of which have built parasitic cones. Many of these vents, however, emit only gases and are appropriately called **fumaroles** (*fumus* = smoke).

9.4 CONCEPT CHECKS

1 How is a crater different from a caldera?

2 Distinguish among conduit, vent, and crater.

3 What is a parasitic cone, and where does it form?

4 What is emitted from a fumarole?

9.5 | SHIELD VOLCANOES

Summarize the characteristics of shield volcanoes and provide one example.

Shield volcanoes are produced by the accumulation of fluid basaltic lavas and exhibit the shape of a broad, slightly domed structure that resembles a warrior's shield (**FIGURE 9.11**). Most shield volcanoes begin on the ocean floor as seamounts, a few of which grow large enough to form volcanic islands. In fact, with the exception of the volcanic islands that form above subduction zones, most other small oceanic islands are either a single shield volcano or, more often, the coalescence of two or more shields built on massive amounts of pillow lavas. Examples include the Canary islands, the Hawaiian islands, the Galapagos, and Easter Island. In addition, some shield volcanoes form on continental crust. Included in this group are Nyamuragira, Africa's most active volcano, and Newberry volcano, located in Oregon.

Mauna Loa: Earth's Largest Shield Volcano

Extensive study of the Hawaiian islands revealed that they are constructed of a myriad of thin basaltic lava flows averaging a few meters thick intermixed with relatively minor amounts of pyroclastic ejected material. Mauna Loa is one of five overlapping shield volcanoes that comprise the Big Island of Hawaii (see Figure 9.11). From its base on the floor of the Pacific Ocean to its summit, Mauna Loa is over 9 kilometers (6 miles) high, exceeding the elevation of Mount Everest. The volume of material composing Mauna Loa is roughly 200 times greater than that of a large composite cone such as Mount Rainier.

The flanks of Mauna Loa have gentle slopes of only a few degrees. This low angle results because very hot, fluid lava travelled "fast and far" from the vent. In addition, most of the lava (perhaps 80 percent) flowed through a well-developed system of lava tubes. Another feature common to many active shield volcanoes is a large, steep-walled caldera that occupies the summit (see Figure 9.11). A caldera usually forms when the roof of solid rock above a magma chamber collapses. This usually occurs as the magma reservoir empties following a large eruption, or as magma migrates to the flank of a volcano to feed a fissure eruption.

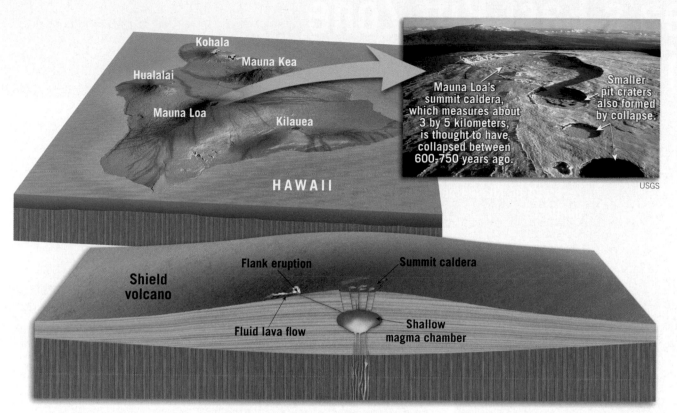

FIGURE 9.11 Mauna Loa: Earth's Largest Volcano Mauna Loa is one of five shield volcanoes that collectively make up the Big Island of Hawaii. Shield volcanoes are built primarily of fluid basaltic lava flows and contain only a small percentage of pyroclastic materials.

In the final stage of growth, shield volcanoes erupt more sporadically, and pyroclastic ejections are more common. Further, lavas increase in viscosity, resulting in thicker, shorter flows. These eruptions tend to steepen the slope of the summit area, which often becomes capped with clusters of cinder cones. This may explain why Mauna Kea, which is a more mature volcano that has not erupted in historic times, has a steeper summit than Mauna Loa, which erupted as recently as 1984. Astronomers are so certain that Mauna Kea is "over the hill" that they built an elaborate astronomical observatory on its summit, housing some of the most advanced and expensive telescopes.

Kilauea, Hawaii: Eruption of a Shield Volcano

Kilauea, the most active and intensely studied shield volcano in the world, is located on the island of Hawaii, in the shadow of Mauna Loa. More than 50 eruptions have been witnessed here since record keeping began in 1823. Several months before each eruptive phase, Kilauea inflates as magma gradually migrates upward and accumulates in a central reservoir located a few kilometers below the summit. For up to 24 hours before an eruption, swarms of small earthquakes warn of the impending activity.

Most of the recent activity on Kilauea has occurred along the flanks of the volcano, in a region called the East Rift Zone. The longest and largest rift eruption ever recorded on Kilauea began in 1983 and continues to this day, with no signs of abating (see the GEOgraphics on pages 288–289).

The first discharge began along a 6-kilometer (4 mile) fissure where a 100-meter- (300-foot-) high "curtain of fire" formed as red-hot lava was ejected skyward. When the activity became localized, a cinder and spatter cone, given the Hawaiian name *Puu Oo*, was built. Over the next 3 years, the general eruptive pattern consisted of short periods (hours to days) when fountains of gas-rich lava sprayed skyward. Each event was followed by nearly a month of inactivity.

By the summer of 1986, a new vent opened 3 kilometers (nearly 2 miles) downrift. Here smooth-surfaced pahoehoe lava formed a lava lake. Occasionally the lake overflowed, but more often lava escaped through tunnels to feed flows that moved down the southeastern flank of the volcano toward the sea. These flows destroyed nearly 180 rural homes, covered a major roadway, and eventually reached the sea. Lava has been intermittently pouring into the ocean ever since, adding new land to the island of Hawaii.

9.5 CONCEPT CHECKS

1 Describe the composition and viscosity of the lava associated with shield volcanoes.

2 Are pyroclastic materials a significant component of shield volcanoes?

3 Where do most shield volcanoes form—on the ocean floor or on the continents?

4 Relate lava tubes to the extent of lava flows associated with shield volcanoes.

5 Where are the best-known shield volcanoes in the United States? Name some examples in other parts of the world.

Kilauea's East Rift Zone Eruption

Kilauea, one of the most active volcanoes in the world, is located on the island of Hawaii in the shadow of Mauna Loa. Most of the recent activity on Kilauea has occurred along the flanks of the volcano in a region called the East Rift Zone. The longest and largest eruption ever recorded on Kilauea began in 1983, and continues with no signs of abating.

Greg Vaughn/Alamy

1 Kilauea's most recent eruptive phase began along a 6-kilometer (4 mile) fissure where a 100-meter (300-foot) high "curtain of fire" formed as red-hot basaltic lava was ejected skyward.

2 The activity became localized at a single vent and a series of 44 short-lived episodes of lava fountaining built a cinder and spatter cone—given the Hawaiian name *Puu Oo*.

3 One of many fluid pahoehoe flows that have moved down the flanks of Kilauea since 1983.

David Reggie/ Getty Images

4 By the summer of 1986 a new vent opened along the rift. Pahoehoe lava flowing from this vent cut off the coastal highway and destroyed more than 180 structures including the National Park Visitor Center.

Michael Collier

USGS

1983 to Present

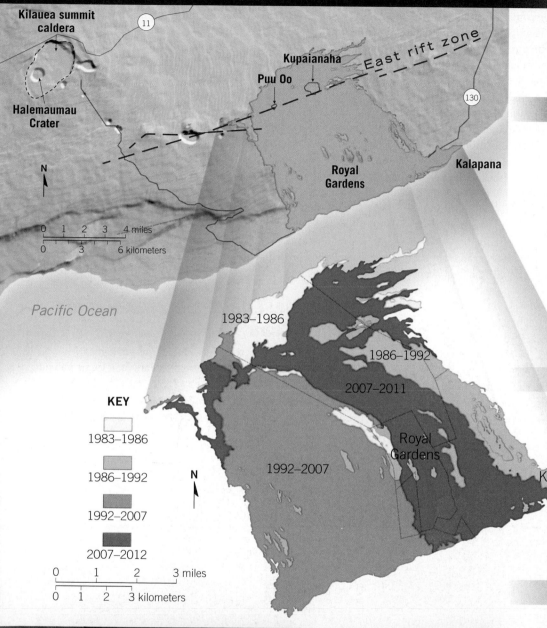

Kilauea summit caldera

11

Kupaianaha

East rift zone

Puu Oo

130

Halemaumau Crater

Royal Gardens

Kalapana

N

| 0 | 1 | 2 | 3 | 4 miles |

| 0 | 3 | 6 kilometers |

Pacific Ocean

1983–1986

1986–1992

2007–2011

KEY

1983–1986

1986–1992

1992–2007

2007–2012

N

1992–2007

Royal Gardens

Kalapana

| 0 | 1 | 2 | 3 miles |

| 0 | 1 | 2 | 3 kilometers |

Other vents have since opened around Puu Oo and lava continues to flow toward the sea

5

USGS

ERUPTION SUMMARY

January 3, 1983

Fissure eruption began along East Rift Zone in an area southeast of Kilauea's summit caldera.

1983-1986

Lava fountains built a cinder-and-spatter cone called Puu Oo and produced mainly aa flows that destroyed 16 homes in the sparsely populated Royal Gardens subdivision.

1986-1992

Eruption moved eastward and built a small shield volcano called Kupaianaha which contained an extensive lava lake. Lava flows buried most of the homes in the village of Kalapana under 15-25 meters (50-80 feet) of lava.

1992-2007

Activity returned to the flanks of Puu Oo. Lava tubes carried lava nearly continuously to the ocean.

2007-2012

Several new vents open, including a small vent within Kilauea's summit caldera.

9.6 | CINDER CONES

Describe the formation, size, and composition of cinder cones.

As the name suggests, **cinder cones** (also called **scoria cones**) are built from ejected lava fragments that begin to harden in flight to produce the vesicular rock *scoria* (**FIGURE 9.12**). These pyroclastic fragments range in size from fine ash to bombs that may exceed 1 meter (3 feet) in diameter. However, most of the volume of a cinder cone consists of pea- to walnut-sized fragments that are markedly vesicular and have a black to reddish-brown color. In addition, this pyroclastic material tends to have basaltic composition.

Although cinder cones are composed mostly of loose scoria fragments, some produce extensive lava fields. These lava flows generally form in the final stages of the volcano's life span, when the magma body has lost most of its gas content. Because cinder cones are composed of loose fragments rather than solid rock, the lava usually flows out from the unconsolidated base of the cone rather than from the crater.

Cinder cones have very simple, distinct shapes determined by the slope that loose pyroclastic material maintains as it comes to rest (see Figure 9.12). Cinder cones have a high angle of repose (the steepest angle at which material remains stable) and are therefore steep-sided, with slopes between 30 and 40 degrees. In addition, cinder cones have quite large, deep craters in relation to the overall size of the structure. Although relatively symmetrical, some cinder cones are elongated and higher on the side that was downwind during the final eruptive phase.

Most cinder cones are produced by a single, short-lived eruptive event. In one study, half of all cinder cones examined were constructed in less than 1 month, and 95 percent were formed in less than 1 year. Once the event ceases, the magma in the "plumbing" connecting the vent to the magma source solidifies, and the volcano usually does not erupt again. (One exception is Cerro Negro, a cinder cone in Nicaragua, which has erupted more than 20 times since it formed in 1850.) As a consequence of this typically short life span, cinder cones are small, usually between 30 meters (100 feet) and 300 meters (1000 feet). A few rare examples exceed 700 meters (2100 feet) in height.

Cinder cones number in the thousands around the globe. Some occur in groups such as the volcanic field near Flagstaff, Arizona, which consists of about 600 cones. Others are parasitic cones that are found on the flanks or within the calderas of larger volcanic structures.

Parícutin: Life of a Garden-Variety Cinder Cone

One of the very few volcanoes studied by geologists from its very beginning is the cinder cone called Parícutin, located about 320 kilometers (200 miles) west of Mexico City. In 1943 its eruptive phase began, in a cornfield owned by Dionisio Pulido, who witnessed the event as he prepared the field for planting.

Mobile Field Trip 9.12 Cinder Cone Cinder cones are built from ejected lava fragments (mostly cinders and bombs) and are relatively small—usually less than 300 meters (1000 feet) in height.

Lava flow

Crater

Pyroclastic material

Central vent filled with rock fragments

SP Crater is a classic cinder cone located north of Flagstaff, Arizona.

Michael Collier

Parícutin, a cinder cone located in Mexico, erupted for nine years.

Lava flow

Photos by Michael Collier

An aa flow emanating from the base of the cone buried much of the village of San Juan Parangaricutiro, leaving only remnants of the village's church.

FIGURE 9.13 Parícutin, a Well-Known Cinder Cone The village of San Juan Parangaricutiro was engulfed by aa lava from Parícutin. Only the church towers remain.

For 2 weeks prior to the first eruption, numerous Earth tremors caused apprehension in the nearby village of Parícutin. Then, on February 20, sulfurous gases began billowing from a small depression that had been in the cornfield for as long as local residents could remember. During the night, hot, glowing rock fragments were ejected from the vent, producing a spectacular fireworks display. Explosive discharges continued, throwing hot fragments and ash occasionally as high as 6000 meters (20,000 feet) into the air. Larger fragments fell near the crater, some remaining incandescent as they rolled down the slope. These built an aesthetically pleasing cone, while finer ash fell over a much larger area, burning and eventually covering the village of Parícutin. In the first day the cone grew to 40 meters (130 feet), and by the fifth day it was more than 100 meters (330 feet) high.

The first lava flow came from a fissure that opened just north of the cone, but after a few months, flows began to emerge from the base of the cone. In June 1944, a clinkery aa flow 10 meters (30 feet) thick moved over much of the village of San Juan Parangaricutiro, leaving only the church steeple exposed (**FIGURE 9.13**). After 9 years of intermittent pyroclastic explosions and nearly continuous discharge of lava from vents at its base, the activity ceased almost as quickly as it had begun. Today, Parícutin is just one of the scores of cinder cones dotting the landscape in this region of Mexico. Like the others, it will not erupt again.

9.6 CONCEPT CHECKS

1 Describe the composition of cinder cones.

2 How do the size and steepness of slopes of a cinder cone compare with those of a shield volcano?

3 Over what time span does a typical cinder cone form?

9.7 | COMPOSITE VOLCANOES

Explain the formation, distribution, and characteristics of composite volcanoes.

Earth's most picturesque yet potentially dangerous volcanoes are **composite volcanoes**, also known as **stratovolcanoes**. Most are located in a relatively narrow zone that rims the Pacific Ocean, appropriately called the *Ring of Fire* (see Figure 9.32). This active zone consists in part of a chain of continental volcanoes that are distributed along the west coast of the Americas, including the large cones of the Andes in South America and the Cascade Range of the western United States and Canada.

Classic composite cones are large, nearly symmetrical structures consisting of alternating layers of explosively erupted cinders and ash interbedded with lava flows. A few composite cones, notably Italy's Etna and Stromboli, display very persistent eruption activity, and molten lava has been observed in their summit craters for decades. Stromboli is so well known for eruptions that eject incandescent blobs of lava that it has been referred to as the "Lighthouse of the Mediterranean."

Eruption of Mount Vesuvius, AD 79

One well-documented volcanic eruption of historic proportions was the AD 79 eruption of the Italian volcano we now call Mount Vesuvius. For centuries prior to this eruption, Vesuvius had been relatively quiescent and its slopes were adorned with vineyards. However, the tranquility abruptly ended, and in less than 24 hours the entire city of Pompeii and a few thousand of its residents were entombed beneath a layer of volcanic ash and pumice that fell like rain.

ITALY
CAMPANIA

Naples

Bay of Naples

Herculaneum

▲ Vesuvius

Pompeii

© Google Earth

Mount Vesuvius has had more than two dozen explosi[ve] eruptions since AD 79, the most recent occurring in 194[4]. Today, roughly 3 million people inhabit the area arou[nd] Mount Vesuvius, making it potentially one of the m[ost] dangerous volcanoes in the world.

Olivier Goujon/Robert Harding

Ruins of Pompeii Nearly 17 centuries after the eruption, excavation of Pompeii gave archeologists a superb picture of Roman life. Most of the roofs in Pompeii collapsed under the weight of volcanic debris that fell at the rate of 5 to 6 inches per hour. Herculaneum, located closer to the modern city of Naples, was buried under 75 feet of ash and pumice.

Leonard von Matt/Photo Researchers, Inc.

Victims of the AD 79 eruption of Mount Vesuvius The remains of human victims were quickly buried by falling ash and subsequent rainfall caused the ash to harden. Over the centuries, the remains decomposed which created cavities discovered by nineteenth-century excavators. Casts were then produced by pouring plaster into the voids.

José Fuste Rega/Corbis/Glow Ima[ges]

Mount Vesuvius today This image of Nap[les] Italy, with Vesuvius in the background, should promp[t] to consider how volcanic crises might be managed in [the] future.

FIGURE 9.14 Fujiyama, a Classic Composite Volcano Japan's Fujiyama exhibits the classic form of a composite cone—steep summit and gently sloping flanks. (Photo by Koji Nakano/Sebun Photo/Getty Images)

Mount Etna, on the other hand, has erupted, on average, once every 2 years since 1979 (see the chapter-opening photo).

Just as shield volcanoes owe their shape to fluid basaltic lavas, composite cones reflect the viscous nature of the material from which they are made. In general, composite cones are the product of silica-rich magma having an andesitic composition. However, many composite cones also emit various amounts of fluid basaltic lava and, occasionally, pyroclastic material having felsic (rhyolitic) composition. The silica-rich magmas typical of composite cones generate thick, viscous lavas that travel less than a few kilometers. In addition, composite cones are noted for generating explosive eruptions that eject huge quantities of pyroclastic material.

A conical shape with a steep summit area and gradually sloping flanks is typical of most large composite cones. This classic profile, which adorns calendars and postcards, is partially a consequence of the way viscous lavas and pyroclastic ejected material contribute to the growth of the cone. Coarse fragments ejected from the summit crater tend to accumulate near their source and contribute to the steep slopes of the summit area. Finer ejected materials, on the other hand, are deposited as a thin layer over a large area that acts to flatten the flank of the cone. In addition, during the early stages of growth, lavas tend to be more abundant and flow greater distances from the vent than later in the volcano's history; this contributes to the cone's broad base. As the volcano matures, the shorter flows that come from the central vent serve to armor and strengthen the summit area. Consequently, steep slopes exceeding 40 degrees are possible. Two of the most perfect cones—Mount Mayon in the Philippines and Fujiyama in Japan—exhibit the classic form we expect of composite cones, with steep summits and gently sloping flanks (**FIGURE 9.14**).

Despite the symmetrical forms of many composite cones, most have complex histories. Many composite volcanoes have secondary vents that have produced cinder cones or even much larger volcanic structures on their flanks. Huge mounds of volcanic debris surrounding these structures provide evidence that large sections of these volcanoes slid downslope as massive landslides. Some develop amphitheater-shaped depressions at their summits as a result of explosive lateral eruptions—as occurred during the 1980 eruption of Mount St. Helens. Often, so much rebuilding has occurred since these eruptions that no trace of the amphitheater-shaped scars remain. Others, such as Crater Lake, have been truncated by the collapse of their summit (see Figure 9.20).

9.7 CONCEPT CHECKS

1 What zone on Earth has the greatest concentration of composite volcanoes?

2 Describe the materials that compose composite volcanoes.

3 How does the composition and viscosity of lava flows differ between composite volcanoes and shield volcanoes?

9.8 | VOLCANIC HAZARDS

Discuss the major geologic hazards associated with volcanoes.

Roughly 1500 of Earth's known volcanoes have erupted at least once, and some several times, in the past 10,000 years. Based on historical records and studies of active volcanoes, 70 volcanic eruptions can be expected each year and 1 large-volume eruption every decade. These large-volume eruptions account for the vast majority of human fatalities. For example, the 1902 eruption of Mount Pelée killed 28,000 people, wiping out the entire population of the nearby city of St. Pierre.

Today, an estimated 500 million people from Japan and Indonesia to Italy and Oregon live near active volcanoes. They face a number of volcanic hazards, such as pyroclastic flows, lahars, lava flows, falling volcanic ash, and volcanic gases.

Pyroclastic flow racing down the flanks of Mount St. Helens, August, 1980. Pyroclastic flows composed of ash and pumice typically move at speeds of 100 kilometers (60 miles) per hour and have temperatures that may exceed 400° C (800° F).

USGS

Geologists examining pumice blocks deposited by a pyroclastic flow that occurred on May 18, 1980. Mount St. Helens is in the background shrouded in steam.

USGS

FIGURE 9.15 Pyroclastic Flows, One of the Most Destructive Volcanic Forces Pyroclastic flows are composed of hot ash and pumice and/or blocky lava fragments that race down the slope of volcanoes.

cloud of hot expanding gases containing fine ash particles and a ground-hugging portion that is often composed of pumice and other vesicular pyroclastic material.

Driven by Gravity Pyroclastic flows are propelled by the force of gravity and tend to move in a manner similar to snow avalanches. They are mobilized by expanding volcanic gases released from the lava fragments and by the expansion of heated air that is overtaken and trapped in the moving front. These gases reduce friction between ash and pumice fragments, which gravity propels downslope in a nearly frictionless environment. This is the reason some pyroclastic flow deposits are found many miles from their source.

Occasionally, powerful hot blasts that carry small amounts of ash separate from the main body of a pyroclastic flow. These low-density clouds, called *surges*, can be deadly but seldom have sufficient force to destroy buildings in their paths. Nevertheless, in 1991, a hot ash cloud from Japan's Unzen volcano engulfed and burned hundreds of homes and moved cars as much as 80 meters (250 feet).

Pyroclastic Flow: A Deadly Force of Nature

One of the most destructive forces of nature is **pyroclastic flows**, which consist of hot gases infused with incandescent ash and large lava fragments. Also referred to as **nuée ardentes** (*glowing avalanches*), these fiery flows are capable of racing down steep volcanic slopes at speeds that can exceed 100 kilometers (60 miles) per hour (**FIGURE 9.15**). Pyroclastic flows are composed of two parts: a low-density

Pyroclastic flows may originate in a variety of volcanic settings. Some occur when a powerful eruption blasts pyroclastic material out of the side of a volcano. More frequently, however, pyroclastic flows are generated by the collapse of tall eruption columns during an explosive event. When gravity eventually overcomes the initial upward thrust provided by the escaping gases, the ejected materials begin to fall, sending massive amounts of incandescent blocks, ash, and pumice cascading downslope.

The Destruction of St. Pierre In 1902 an infamous pyroclastic flow and associated surge from Mount Pelée, a small volcano on the Caribbean island of Martinique, destroyed the port town of St. Pierre. Although the main pyroclastic flow was largely confined to the valley of Riviere Blanche, a low-density fiery surge spread south of the river and quickly engulfed the entire city. The destruction happened in moments and was so devastating that nearly all of St. Pierre's 28,000 inhabitants were killed. Only 1 person on the outskirts of town—a prisoner protected in a dungeon—and a few people on ships in the harbor were spared (**FIGURE 9.16**).

Within days of this calamitous eruption, scientists arrived on the scene. Although St. Pierre was mantled by only a thin layer of volcanic debris, the scientists discovered that masonry walls nearly 1 meter (3 feet) thick had been knocked over like

St. Pierre before the 1902 eruption.

St. Pierre following the eruption of Mount Pelée.

FIGURE 9.16 Destruction of St. Pierre The photo on the left shows St. Pierre as it appeared shortly after the eruption of Mount Pelée in 1902. (Reproduced from the collection of the Library of Congress) The before image in the upper right shows many vessels anchored offshore, as was the case on the day of the eruption. (Photo by UPPA/Photoshot)

dominoes, large trees had been uprooted, and cannons had been torn from their mounts.

Lahars: Mudflows on Active and Inactive Cones

In addition to violent eruptions, large composite cones may generate a type of fluid mudflow referred to by its Indonesian name, **lahar**. These destructive flows occur when volcanic ash and debris becomes saturated with water and rapidly moves down steep volcanic slopes, generally following stream valleys. Some lahars are triggered when magma nears the surface of a glacial clad volcano, causing large volumes of ice and snow to melt. Others are generated when heavy rains saturate weathered volcanic deposits. Thus, lahars may occur even when a volcano is *not* erupting.

When Mount St. Helens erupted in 1980, several lahars were generated. These flows and accompanying floodwaters raced down nearby river valleys at speeds exceeding 30 kilometers (20 miles) per hour. These raging rivers of mud destroyed or severely damaged nearly all the homes and bridges along their paths (**FIGURE 9.17**). Fortunately, the area was not densely populated.

In 1985, deadly lahars were produced during a small eruption of Nevado del Ruiz, a 5300-meter (17,400-foot) volcano in the Andes Mountains of Colombia. Hot pyroclastic material melted ice and snow that capped the mountain (*nevado* means "snow" in Spanish) and sent torrents of ash and debris down three major river valleys that flank the volcano. Reaching speeds of 100 kilometers (60 miles) per hour, these mudflows tragically claimed 25,000 lives.

Many consider Mount Rainier, Washington, to be America's most dangerous volcano because, like Nevado del Ruiz, it has a thick, year-round mantle of snow and glacial ice. Adding to the risk is the fact that more than 100,000 people live in the valleys around Rainier, and many homes are built on deposits left by lahars that flowed down the volcano hundreds or thousands of years ago. A future eruption, or perhaps just a period of heavier-than-average rainfall, may produce lahars that could be similarly destructive.

Other Volcanic Hazards

Volcanoes can be hazardous to human health and property in other various ways. Ash and other pyroclastic material can collapse the roofs of buildings or can be drawn into the lungs of humans and other animals or into aircraft engines (**FIGURE 9.18**). Volcanoes also emit gases, most notably sulfur dioxide, which affects air quality and, when mixed with rainwater, may destroy vegetation and reduce the quality of groundwater. Despite the known risks, millions of people live in close proximity to active volcanoes.

Volcano-Related Tsunami One hazard associated with volcanoes is their ability to generate tsunami. Although they are usually caused by strong earthquakes, these destructive sea waves sometimes result from powerful volcanic explosions or the sudden collapse of flanks of volcanoes into the ocean. In 1883, a volcanic eruption on the Indonesian island Krakatoa literally tore the island apart and generated a tsunami that claimed an estimated 36,000 lives.

Volcanic Ash and Aviation During the past 15 years, at least 80 commercial jets have been damaged by inadvertently flying into clouds of volcanic ash. For example, in 1989 a Boeing 747 carrying more than 300 passengers encountered an ash cloud from Alaska's Redoubt Volcano. All four of the engines stalled when they became clogged with ash. Fortunately, the pilots were able to restart the engines and safely land the aircraft in Anchorage.

More recently, in 2010, the eruption of Iceland's Eyjafjallajökull volcano sent ash high into the atmosphere. The thick plume of ash drifted over Europe, causing airlines to cancel thousands of flights, leaving hundreds of thousands of travelers stranded. Several weeks passed before air travel resumed its normal schedule.

Volcanic Gases and Respiratory Health One of the most destructive volcanic events, called the Laki eruptions, began along a large fissure in southern Iceland in 1783. An estimated 14 cubic kilometers (3.4 cubic miles) of fluid basaltic lavas, in addition to 130 million tons of sulfur dioxide and other poisonous gases, were released. When sulfur dioxide is inhaled, it reacts with moisture in the lungs to produce sulfuric acid. This eruption's sulfur dioxide release caused the death of more than 50 percent of Iceland's livestock. The ensuing famine killed 25 percent of the island's human population. In addition, this huge eruption endangered people and property across Europe. Crop failure occurred in parts of Western Europe, and thousands of residents perished from lung-related diseases. A recent report estimates that a similar eruption today would cause more than 140,000 cardiopulmonary fatalities in Europe alone.

FIGURE 9.17 Lahars, Mudflows That Originate on Volcanic Slopes This lahar raced down the Muddy River, located southeast of Mount St. Helens, following the May 18, 1980, eruption. Notice the former height of this fluid mudflow, as recorded by the mudflow line on the tree trunks. Note the person (circled) for scale. (Photo by Lyn Topinka/USGS)

Volcanic ash can destroy plant life and be drawn into the lungs of humans and other animals.

Ash and other pyroclastic materials can collapse roofs, or completely cover buildings.

Lava flows can destroy homes, roads, and other structures in their paths.

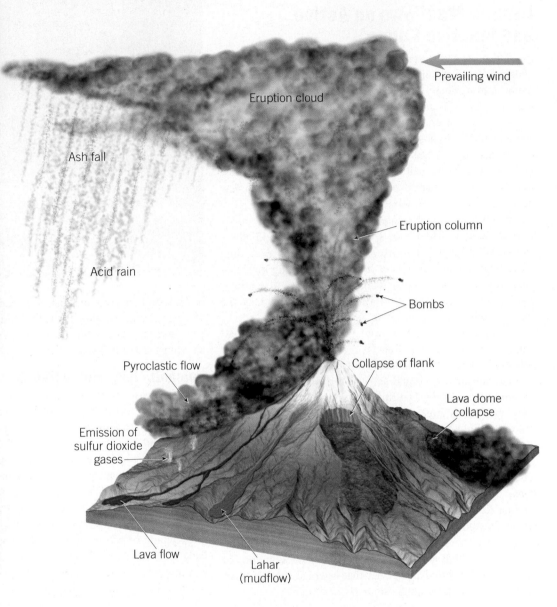

FIGURE 9.18 Volcanic Hazards In addition to generating pyroclastic flows and lahars, volcanoes can be hazardous to human health and property in many other ways.

Effects of Volcanic Ash and Gases on Weather and Climate

Volcanic eruptions can eject dust-sized particles of volcanic ash and sulfur dioxide gas high into the atmosphere. The ash particles reflect solar energy back toward space, producing temporary atmospheric cooling. In the case of the 1783 eruption in Iceland, the atmospheric circulation around the globe appears to have been affected. Drought conditions prevailed in the Nile River valley and in India, while the winter of 1784 brought the longest period of below-zero temperatures in New England's history.

Other eruptions that have produced significant effects on climate worldwide include the eruption of Indonesia's Mount Tambora in 1815, which produced the "year without a summer" (1816), and the eruption of El Chichón in Mexico in 1982. The eruption of El Chichón, although small, emitted an unusually large quantity of sulfur dioxide that reacted with water vapor in the atmosphere to produce a dense cloud of tiny sulfuric-acid droplets. These particles, called **aerosols**, take several years to settle out. Like fine ash, these aerosols lower the mean temperature of the atmosphere by reflecting solar radiation back to space.

9.8 CONCEPT CHECKS

1 Describe pyroclastic flows and explain why they are capable of traveling great distances.

2 What is a lahar?

3 List at least three volcanic hazards other than pyroclastic flows and lahars.

9.9 | OTHER VOLCANIC LANDFORMS

List and describe volcanic landforms other than volcanic cones.

The most widely recognized volcanic structures are the cone-shaped edifices of composite volcanoes that dot Earth's surface. However, other distinctive and important landforms are also produced by volcanic activity on Earth.

Calderas

Calderas (*caldaria* = cooking pot) are large steep-sided depressions with diameters that exceed 1 kilometer (0.6 mile) and have a somewhat circular form. (Those that are less than 1 kilometer across are called *collapse pits*, or *craters.*) Most calderas are formed by one of the following processes: (1) the collapse of the summit of a large composite volcano following an explosive eruption of silica-rich pumice and ash fragments (*Crater Lake–type calderas*); (2) the collapse of the top of a shield volcano caused by subterranean drainage from a central magma chamber (*Hawaiian-type calderas*); and (3) the collapse of a large area, caused by the discharge of colossal volumes of silica-rich pumice and ash along ring fractures (*Yellowstone-type calderas*).

Crater Lake–Type Calderas

Crater Lake, Oregon, is situated in a caldera approximately 10 kilometers (6 miles) wide and 1175 meters (more than 3800 feet) deep. This caldera formed about 7000 years ago, when a composite cone, named Mount Mazama, violently extruded 50 to 70 cubic kilometers (12 to 17 cubic miles) of pyroclastic material (**FIGURE 9.19**). With the loss of support, 1500 meters (nearly 1 mile) of the summit of this once-prominent cone collapsed,

producing a caldera that eventually filled with rainwater. Later, volcanic activity built a small cinder cone in the caldera. Today this cone, called Wizard Island, provides a mute reminder of past activity (see Figure 9.19).

Hawaiian-Type Calderas

Unlike Crater Lake–type calderas, many calderas form gradually because of the loss of lava from a shallow magma chamber underlying the volcano's

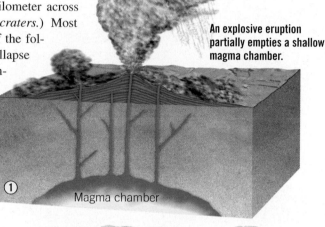

① An explosive eruption partially empties a shallow magma chamber.

Magma chamber

② Summit of volcano collapses, enhancing the eruption.

Newly formed caldera fills with rain and groundwater.

③

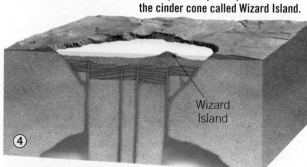

Subsequent eruptions produce the cinder cone called Wizard Island.

④

Wizard Island

Crater Lake

Wizard Island

Michael Collier

FIGURE 9.19 Formation of Crater Lake–Type Calderas About 7000 years ago, a violent eruption partly emptied the magma chamber of former Mount Mazama, causing its summit to collapse. Rainfall and groundwater contributed to form Crater Lake, the deepest lake in the United States—594 meters (1949 feet) deep—and the ninth deepest in the world. (Based on *The Ancient Volcanoes of Oregon*, H. Williams, University of Oregon Press, 1953, p. 47, Figure 8.)

SmartFigure 9.20 Super-Eruptions at Yellowstone The top map shows Yellowstone National Park. Eruptions of the Yellowstone caldera were responsible for the layers of ash shown in the bottom map. These eruptions were separated by relatively regular intervals of about 700,000 years. The largest of these eruptions was 10,000 times larger the 1980 eruption of Mount St. Helens.

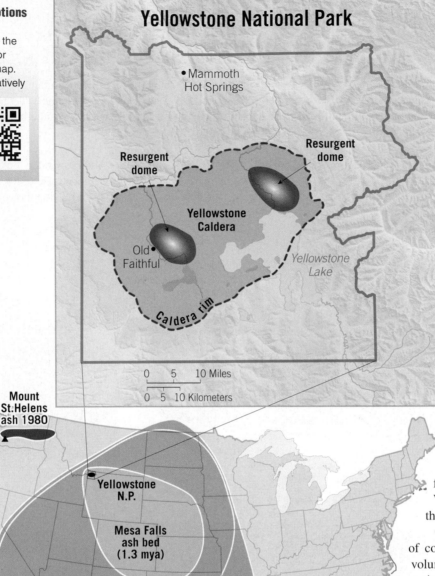

approximately 1000 cubic kilometers (240 cubic miles) of pyroclastic material erupted. This catastrophic eruption sent showers of ash as far as the Gulf of Mexico and resulted in the formation of a caldera 70 kilometers (43 miles) across (**FIGURE 9.20**). Vestiges of this event are the many hot springs and geysers in the Yellowstone region.

Based on the extraordinary volume of erupted material, researchers have determined that the magma chambers associated with Yellowstone-type calderas must be similarly monstrous. As more and more magma accumulates, the pressure within the magma chamber begins to exceed the pressure exerted by the weight of the overlying rocks. An eruption occurs when the gas-rich magma raises the overlying crust enough to create vertical fractures that extend to the surface. Magma surges upward along these cracks, forming a ring-shaped eruption. With a loss of support, the roof of the magma chamber collapses.

Caldera-forming eruptions are of colossal proportions, ejecting huge volumes of pyroclastic materials, mainly in the form of ash and pumice fragments. Typically, these materials form pyroclastic flows that sweep across the landscape, destroying most living things in their paths. Upon coming to rest, the hot fragments of ash and pumice fuse together, forming a welded tuff that closely resembles a solidified lava flow. Despite the immense size of these calderas, the eruptions that cause them are brief, lasting hours to perhaps a few days.

Large calderas tend to exhibit a complex eruptive history. In the Yellowstone region, for example, three caldera-forming episodes are known to have occurred over the past 2.1 million years. The most recent event (630,000 years ago) was followed by episodic outpourings of degassed rhyolitic and basaltic lavas. Geologic evidence suggests that a magma reservoir still exists beneath Yellowstone; thus, another caldera-forming eruption is likely but not necessarily imminent.

A distinctive characteristic of large caldera-forming eruptions is a slow upheaval of the floor of the caldera that

summit. For example, Hawaii's active shield volcanoes, Mauna Loa and Kilauea, both have large calderas at their summits. Kilauea's measures 3.3 by 4.4 kilometers (about 2 by 3 miles) and is 150 meters (500 feet) deep. The walls of this caldera are almost vertical, and as a result, it looks like a vast, nearly flat-bottomed pit. Kilauea's caldera formed by gradual subsidence as magma slowly drained laterally from the underlying magma chamber, leaving the summit unsupported.

Yellowstone-Type Calderas Historic and destructive eruptions such as Mount St. Helens and Vesuvius pale in comparison to what happened 630,000 years ago in the region now occupied by Yellowstone National Park, when

produces a central elevated region called a *resurgent dome* (see Figure 9.20). Unlike calderas associated with shield volcanoes or composite cones, Yellowstone-type depressions are so large and poorly defined that many remained undetected until high-quality aerial and satellite images became available. Other examples of large calderas located in the United States are California's Long Valley Caldera; LaGarita Caldera, located in the San Juan Mountains of southern Colorado; and the Valles Caldera, west of Los Alamos, New Mexico. These and similar calderas elsewhere around the globe are among the largest volcanic structures on Earth, hence the name "supervolcanoes." Volcanologists have compared their destructive force with that of the impact of a small asteroid. Fortunately, no caldera-forming eruption has occurred in historic times.

Fissure Eruptions and Basalt Plateaus

The greatest volume of volcanic material is extruded from fractures in Earth's crust called **fissures** (*fissura* = to split). Rather than build cones, **fissure eruptions** usually emit fluid basaltic lavas that blanket wide areas (**FIGURE 9.21**). Recall that the recent volcanic activity on Kilauea began along a series of fissures in the East Rift Zone.

In some locations, extraordinary amounts of lava have been extruded along fissures in a relatively short time, geologically speaking. These voluminous accumulations are commonly referred to as **basalt plateaus** because most have a basaltic composition and tend to be rather flat and broad. The Columbia Plateau located in the northwestern United States, which consists of the Columbia River basalts, is the product of this type of activity (**FIGURE 9.22**). Numerous fissure eruptions have buried the landscape, creating a lava plateau nearly 1500 meters (1 mile) thick. Some of the lava remained molten long enough to flow 150 kilometers

FIGURE 9.21 Basaltic Fissure Eruptions Lava fountaining from a fissure and formation of fluid lava flows called *flood basalts*. The lower photo shows flood basalt flows near Idaho Falls.

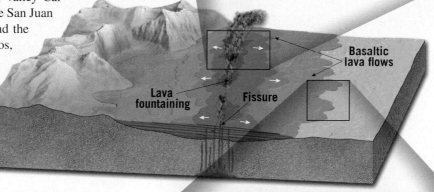

Lava fountaining — Fissure — Basaltic lava flows

Basaltic lava flows

John S. Shelton

(90 miles) from its source. The term **flood basalts** appropriately describe these extrusions.

Massive accumulations of basaltic lava, similar to those of the Columbia Plateau, occur elsewhere in the world.

EYE ON EARTH

This image was obtained during the 1991 eruption of Mount Pinatubo in the Philippines. This was the largest eruption to affect a densely populated area in recent times. Timely forecasts of the event by scientists were credited with saving at least 5000 lives. (Alberto Garcia/CORBIS)

QUESTION 1 *What name is given to the ash- and pumice-laden cloud that is racing toward the photographer?*

QUESTION 2 *At what speeds can these fiery clouds move down steep mountain slopes?*

Jeep

FIGURE 9.22 Columbia River Basalts **A.** The Columbia River basalts cover an area of nearly 164,000 square kilometers (63,000 square miles), commonly called the Columbia Plateau. Activity here began about 17 million years ago, as lava poured out of large fissures, eventually producing a basalt plateau with an average thickness of more than 1 kilometer (0.6 mile).
B. Columbia River basalt flows exposed in the Palouse River Canyon in southwestern Washington State. (Photo by Williamborg)

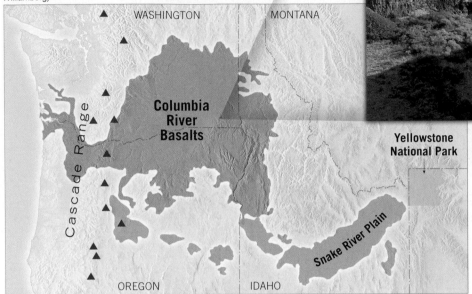

The Palouse River in Washington State has cut a canyon about 300 meters (1000 feet) deep into the flood basalts of the Columbia Plateau.

B.

KEY

■ Columbia River Basalts

■ Other basaltic rocks

▲ Large Cascade volcanoes

A.

One of the largest is the Deccan Traps, a thick sequence of flat-lying basalt flows covering nearly 500,000 square kilometers (195,000 square miles) of west-central India. When the Deccan Traps formed about 66 million years ago, nearly 2 million cubic kilometers (480,000 cubic miles) of lava were extruded over a period of approximately 1 million years. Several other huge deposits of flood basalts, including the Ontong Java Plateau, are found on the floor of the ocean.

Volcanic Necks and Pipes

Most volcanic eruptions are fed lava through short conduits that connect shallow magma chambers to vents located at the surface. When a volcano becomes inactive, congealed magma often becomes preserved in the feeding conduit of the volcano as a crudely cylindrical mass. However, all volcanoes will succumb to forces of weathering and erosion. As erosion progresses, the rock occupying the volcanic conduit, which is highly resistant to weathering, may remain standing

FIGURE 9.23 Volcanic Neck Shiprock, New Mexico, is a volcanic neck that stands over 420 meters (1380 feet) high. It consists of igneous rock that crystallized in the vent of a volcano that has long since been eroded.

Original volcanic structure

Conduit filled with fractured rock and magma

Mancos shale

Weathering and erosion have removed the volcano and sedimentary rock

Shiprock (Volcanic neck)

3000 ft.

Present day land surface

Mancos shale

Geologist's Sketch

1700 ft

Shiprock, New Mexico is a volcanic neck composed of igneous rock which solidified in the conduit of a volcano.

Dennis Tasa

above the surrounding terrain long after the cone has been worn away. Shiprock, New Mexico, is a widely recognized and spectacular example of these structures geologists call **volcanic necks** (or **plugs**) (**FIGURE 9.23**). More than 510 meters (1700 feet) high, Shiprock is higher than most skyscrapers and one of many such landforms that protrude conspicuously from the red desert landscapes of the American Southwest.

One rare type of conduit, called a **pipe**, carries magma that originated in the mantle at depths that may exceed 150 kilometers (93 miles). The gas-laden magmas that migrate through pipes travel rapidly enough to undergo minimal alteration during their ascent.

The best-known volcanic pipes are the diamond-bearing kimberlite pipes of South Africa. The rocks filling these pipes originated at great depths, where pressure is sufficiently high to generate diamonds and other high-pressure minerals. The process of transporting essentially unaltered magma (along with diamond inclusions) through 150 kilometers (93 miles) of solid rock is exceptional—and accounts for the scarcity of natural diamonds. Geologists consider pipes to be "windows" into Earth that allow us to view rocks that normally form only at great depths.

9.9 CONCEPT CHECKS

1 Describe the formation of Crater Lake. Compare it to the calderas found on shield volcanoes such as Kilauea.

2 Pyroclastic flows are associated with what volcanic structure that is not a cinder cone?

3 How do the eruptions that created the Columbia Plateau differ from eruptions that create large composite volcanoes?

4 Contrast the composition of a typical lava dome and a typical fissure eruption.

5 What type of volcanic structure is Shiprock, New Mexico, and how did it form?

9.10 | INTRUSIVE IGNEOUS ACTIVITY Compare and contrast these intrusive igneous structures: dikes, sills, batholiths, stocks, and laccoliths.

Although volcanic eruptions can be violent and spectacular events, most magma is emplaced and crystallizes at depth, without fanfare. Therefore, understanding the igneous processes that occur deep underground is as important to geologists as the study of volcanic events.

Nature of Intrusive Bodies

When magma rises through the crust, it forcefully displaces preexisting crustal rocks referred to as *host*, or *country*, *rock*. The structures that result from the emplacement of magma into preexisting rocks are called **intrusions** or **plutons**. Because all intrusions form far below Earth's surface, they are studied primarily after uplifting and erosion have exposed them. The challenge lies in reconstructing the events that generated these structures millions of years ago and in vastly different conditions deep underground.

Intrusions are known to occur in a great variety of sizes and shapes. Some of the most common types are illustrated in **FIGURE 9.24**. Notice that some plutons have a tabular (tablet) shape, whereas others are best described as massive (blob shaped). Also, observe that some of these bodies cut across existing structures, such as sedimentary strata, whereas others form when magma is injected between sedimentary layers. Because of these differences, intrusive igneous bodies are generally classified according to their shape as either **tabular** (*like a tablet*) or **massive** and by their orientation with respect to the host rock. Igneous bodies are said to be **discordant** (*discordare* = to disagree) if they cut across existing structures and **concordant** (*concordare* = to agree) if they inject parallel to features such as sedimentary strata.

EYE ON EARTH

The Palisades form impressive cliffs along the western side of the Hudson River for more than 80 kilometers (50 miles). This igneous feature, which is visible from Manhattan, formed when magma was injected between layers of sandstone and shale.

QUESTION 1 What type of igneous feature constitutes the Palisades?

QUESTION 2 Is this feature tabular or massive?

June Marie Sobrito/Shutterstock

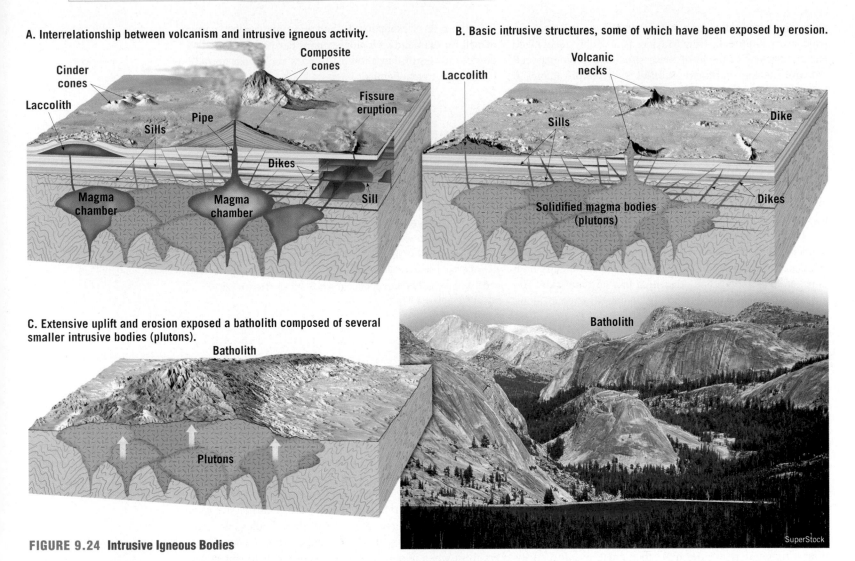

A. Interrelationship between volcanism and intrusive igneous activity.

Cinder cones

Composite cones

Laccolith

Pipe

Sills

Fissure eruption

Dikes

Magma chamber

Magma chamber

Sill

B. Basic intrusive structures, some of which have been exposed by erosion.

Laccolith

Volcanic necks

Dike

Sills

Solidified magma bodies (plutons)

Dikes

C. Extensive uplift and erosion exposed a batholith composed of several smaller intrusive bodies (plutons).

Batholith

Batholith

Plutons

SuperStock

FIGURE 9.24 Intrusive Igneous Bodies

Sill

Mobile Field Trip 9.25 Sill Exposed in Utah's Sinbad Country
The dark, essentially horizontal bands are sills of basaltic composition that intruded horizontal layers of sedimentary rock. (Photo by Michael Collier)

Tabular Intrusive Bodies: Dikes and Sills

Tabular intrusive bodies are produced when magma is forcibly injected into a fracture or zone of weakness, such as a bedding surface (see Figure 9.24). **Dikes** are discordant bodies that cut across bedding surfaces or other structures in the country rock. By contrast, **sills** are nearly horizontal, concordant bodies that form when magma exploits weaknesses between sedimentary beds or other structures (**FIGURE 9.25**). In general, dikes serve as tabular conduits that transport magma, whereas sills tend to accumulate magma and increase in thickness.

Dikes and sills are typically shallow features, occurring where the country rocks are sufficiently brittle to fracture. They can range in thickness from less than 1 millimeter to more than 1 kilometer (0.6 mile).

While dikes and sills can occur as solitary bodies, dikes tend to form in roughly parallel groups called *dike swarms*. These multiple structures reflect the tendency for fractures to form in sets when tensional forces stretch brittle country rock. Dikes can also occur radiating from an eroded volcanic neck, like spokes on a wheel. In these situations, the active

FIGURE 9.26 Dike Exposed in the Spanish Peaks, Colorado This elongated dike is composed of igneous rock that is more resistant to weathering than the surrounding material.

ascent of magma generated fissures in the volcanic cone out of which lava flowed. Dikes frequently weather more slowly than the surrounding rock. Consequently, when exposed by erosion, dikes tend to have a wall-like appearance, as shown in **FIGURE 9.26**.

Because dikes and sills are relatively uniform in thickness and can extend for many kilometers, they are assumed to be the product of very fluid and, therefore, mobile magmas. One of the largest and most studied of all sills in the United States is the Palisades Sill. Exposed for 80 kilometers (50 miles) along the western bank of the Hudson River in southeastern New York and northeastern New Jersey, this sill is about 300 meters (1000 feet) thick. Because it is resistant to erosion, the Palisades Sill forms an imposing cliff that can be easily seen from the opposite side of the Hudson.

In many respects, sills closely resemble buried lava flows. Both are tabular and can have a wide aerial extent, and both may exhibit columnar jointing. **Columnar jointing** occurs when igneous rocks cool and develop shrinkage fractures that produce elongated, pillar-like columns that most often have six sides (**FIGURE 9.27**). Further, because sills generally form in near-surface environments and may be only a few meters thick, the emplaced magma often cools quickly enough to generate a fine-grained texture. (Recall that most intrusive igneous bodies have a coarse-grained texture.)

Massive Intrusive Bodies: Batholiths, Stocks, and Laccoliths

By far the largest intrusive igneous bodies are **batholiths** (*bathos* = depth, *lithos* = stone). Batholiths occur as mammoth linear structures several hundreds of kilometers long and up to 100 kilometers wide (see Figure 9.24C). The Sierra Nevada batholith, for example, is a continuous granitic structure that forms much of the Sierra Nevada, in California. An even larger batholith extends for over 1800 kilometers (1100 miles) along the Coast Mountains of western Canada and into southern Alaska. Although batholiths can cover a large area, recent gravitational studies indicate that most are less than 10 kilometers (6 miles) thick. Some are even thinner; the Coastal batholith of Peru, for example, is essentially a flat slab with an average thickness of only 2 to 3 kilometers (1 to 2 miles).

Columnar jointing

Rapid cooling from the outside causes shrinkage cracks

Columnar jointing tends to produce 6-sided columns

Geologist's Sketch

FIGURE 9.27 Columnar Jointing Giant's Causeway in Northern Ireland is an excellent example of columnar jointing. (Photo by John Lawrence/Getty Images)

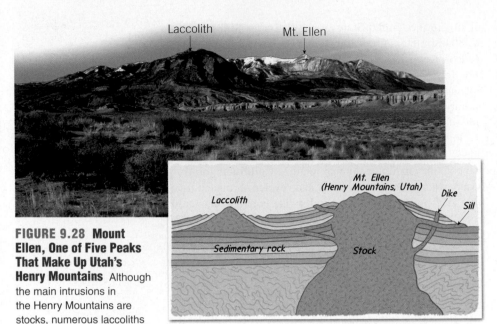

Laccolith   Mt. Ellen

FIGURE 9.28 Mount Ellen, One of Five Peaks That Make Up Utah's Henry Mountains Although the main intrusions in the Henry Mountains are stocks, numerous laccoliths formed as offshoots of these structures. (Photo by Michael DeFreitas North America/Alamy)

Geologist's Sketch

Batholiths are typically composed of felsic and intermediate rock types and are often referred to as "granite batholiths." Large granite batholiths consist of hundreds of plutons that intimately crowd against or penetrate one another. These bulbous masses were emplaced over spans of millions of years. The intrusive activity that created the Sierra Nevada batholith, for example, occurred nearly continuously over a 130-million-year period that ended about 80 million years ago.

By definition, a plutonic body must have a surface exposure greater than 100 square kilometers (40 square miles)

in order to be considered a batholith. Smaller plutons are termed **stocks**. However, many stocks appear to be portions of much larger intrusive bodies that would be classified as batholiths if they were fully exposed.

Laccoliths   A nineteenth-century study by G. K. Gilbert of the U.S. Geological Survey in the Henry Mountains of Utah produced the first clear evidence that igneous intrusions can lift the sedimentary strata they penetrate. Gilbert gave the name **laccoliths** to the igneous intrusions he observed, which he envisioned as igneous rock forcibly injected between sedimentary strata so as to arch the beds above while leaving those below relatively flat. It is now known that the five major peaks of the Henry Mountains are not laccoliths but stocks. However, these central magma bodies are the source material for branching offshoots that are true laccoliths, as Gilbert defined them (**FIGURE 9.28**).

Numerous other granitic laccoliths have since been identified in Utah. The largest is a part of the Pine Valley Mountains, located north of St. George, Utah. Others are found in the La Sal Mountains near Arches National Park and in the Abajo Mountains directly to the south.

9.10 CONCEPT CHECKS

1   What is meant by the term *country rock*?

2   Describe *dike* and *sill*, using the appropriate terms from the following list: massive, discordant, tabular, concordant.

3   Compare and contrast batholiths, stocks, and laccoliths in terms of size and shape.

9.11 | PARTIAL MELTING AND THE ORIGIN OF MAGMA
Summarize the major processes that generate magma from solid rock.

Evidence from the study of earthquake waves has shown that *Earth's crust and mantle are composed primarily of solid, not molten, rock.* Although the outer core is liquid, this iron-rich material is very dense and remains deep within Earth. So where does magma come from?

Partial Melting

Recall that igneous rocks are composed of a mixture of minerals. Since these minerals have different melting points, igneous rocks tend to melt over a temperature range of at least 200°C (360°F). As rock begins to melt, the minerals with the lowest melting temperatures are the first to melt. If melting continues, minerals with higher melting points begin to melt, and the composition of the melt steadily approaches the overall composition of the rock from which it was derived.

Most often, however, melting is not complete. The incomplete melting of rocks is known as **partial melting**, a process that produces most magma. Partial melting can be likened to a chocolate chip cookie containing nuts that is left out in the Sun. The chocolate chips represent the minerals

with the lowest melting points because they will begin to melt before the other ingredients. Unlike with the chocolate chip cookie, when rock partially melts, the molten material separates from the solid components. Also, because the molten material is less dense that the remaining solids, when enough of the melt collects, it rises toward Earth's surface.

Generating Magma from Solid Rock

Workers in underground mines know that temperatures increase as they descend deeper below Earth's surface. Although the rate of temperature change varies considerably from place to place, it *averages* about 25°C per kilometer in the *upper* crust. This increase in temperature with depth is known as the **geothermal gradient**. As shown in **FIGURE 9.29**, when a typical geothermal gradient is compared to the melting point curve for the mantle rock peridotite, the temperature at which peridotite melts is higher than the geothermal gradient. Thus, under normal conditions, the mantle is solid. However, tectonic processes exist that trigger melting by reducing the melting point (temperature) of mantle rock.

Decrease in Pressure: Decompression Melting

If temperature were the only factor that determines whether rock melts, our planet would be a molten ball covered with a thin, solid outer shell. This is not the case because pressure, which also increases with depth, influences the melting temperatures of rocks.

Melting, which is accompanied by an increase in volume, *occurs at progressively higher temperatures with increased depth*. This is the result of the steady increase in confining pressure exerted by the weight of overlying rocks. Conversely, reducing confining pressure lowers a rock's melting temperature. When confining pressure drops sufficiently, **decompression melting** is triggered.

Decompression melting occurs where hot, solid mantle rock ascends in zones of convective upwelling, thereby moving into regions of lower pressure. This process is responsible for generating magma along divergent plate boundaries (oceanic ridges) where plates are rifting apart (**FIGURE 9.30**). Below the ridge crest, hot mantle rock rises and melts, replacing the material that shifted horizontally away from the ridge axis. Decompression melting also occurs when ascending mantle plumes reach the uppermost mantle.

Addition of Water

Another important factor affecting the melting temperature of rock is its water content. Water and other volatiles act as salt does to melt ice. Water causes rock to melt at lower temperatures, just as putting rock salt on an icy sidewalk induces melting.

The introduction of water to generate magma occurs mainly at convergent plate boundaries where cool slabs of oceanic lithosphere descend into the mantle (**FIGURE 9.31**). As an oceanic plate sinks, heat and pressure drive water from the subducting oceanic crust and overlying sediments. These fluids migrate into the wedge of hot mantle that lies directly above. At a depth of about 100 kilometers (60 miles), the addition of water lowers the melting temperature of mantle rock sufficiently to trigger partial melting. Partial melting of the mantle rock peridotite generates hot basaltic magmas with temperatures that may exceed 1250°C (2300°F).

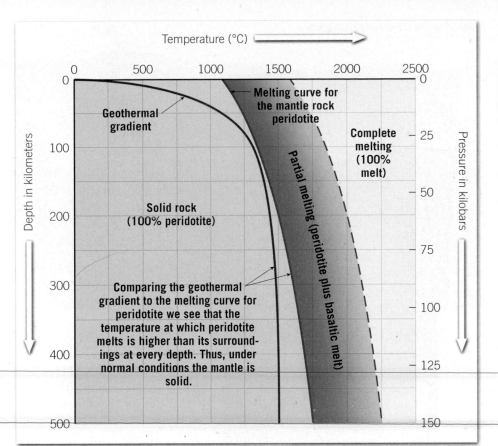

Figure 9.29 labels:
Temperature (°C)
Depth in kilometers
Pressure in kilobars
Geothermal gradient
Melting curve for the mantle rock peridotite
Complete melting (100% melt)
Solid rock (100% peridotite)
Partial melting (peridotite plus basaltic melt)
Comparing the geothermal gradient to the melting curve for peridotite we see that the temperature at which peridotite melts is higher than its surroundings at every depth. Thus, under normal conditions the mantle is solid.

FIGURE 9.29 Diagram Illustrating Why the Mantle Is Mainly Solid This diagram shows the geothermal gradient (increase in temperature with depth) for the crust and upper mantle. Also plotted is the melting point curve for the mantle rock peridotite. When we compare the change in temperature with depth (geothermal gradient) to the melting point of peridotite, the temperature at which peridotite melts is higher than the geothermal gradient at every depth. Thus, under normal conditions, the mantle is solid.

Temperature Increase: Melting Crustal Rocks

When enough mantle-derived basaltic magma forms, it buoyantly rises toward the surface. In continental settings, basaltic magma often "ponds" beneath crustal rocks, which have a lower density and are already near their melting temperature.

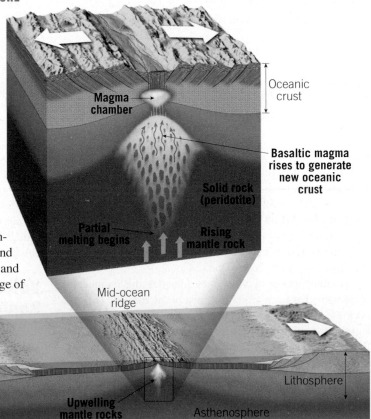

Figure 9.30 labels:
Magma chamber
Oceanic crust
Basaltic magma rises to generate new oceanic crust
Solid rock (peridotite)
Partial melting begins
Rising mantle rock
Mid-ocean ridge
Upwelling mantle rocks
Lithosphere
Asthenosphere

FIGURE 9.30 Decompression Melting As hot mantle rock ascends, it continually moves into zones of lower and lower pressure. This drop in confining pressure initiates *decompression melting* in the upper mantle. Decompression melting occurs without an outside source of energy (heat).

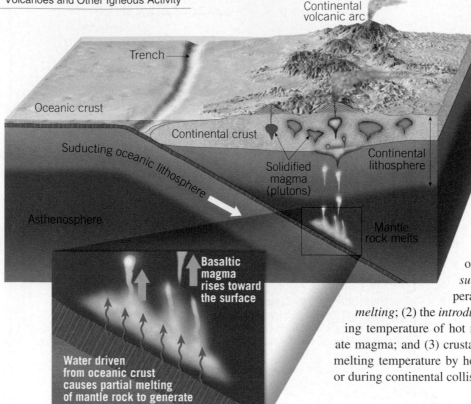

As an oceanic plate descends into the mantle, water and other volatiles are driven from the subducting crustal rocks into the wedge of mantle above. At a depth of about 100 kilometers (60 miles), mantle rock is hot enough that the addition of water can cause melting.

such events, the crust is greatly thickened, and some crustal rocks are buried to depths where the temperatures are elevated sufficiently to cause partial melting. The felsic (granitic) magmas produced in this manner usually solidify before reaching the surface, so volcanism in not typically associated with these collision-type mountain belts.

In summary, magma can be generated three ways: (1) in zones of upwelling, *a decrease in pressure* (without an increase in temperature) can result in *decompression melting*; (2) the *introduction of water* can lower the melting temperature of hot mantle rock sufficiently to generate magma; and (3) crustal rocks can be heated above their melting temperature by hot mantle-derived basaltic magma or during continental collisions.

The hot basaltic magma may heat the overlying crustal rocks sufficiently to generate a secondary, silica-rich magma. If these low-density silica-rich magmas reach the surface, they tend to produce explosive eruptions that we associate with convergent plate boundaries.

Crustal rocks can also melt during continental collisions that result in the formation of a large mountain belt. During

9.11 CONCEPT CHECKS

1 What is the geothermal gradient? Describe how the geothermal gradient compares with the melting temperatures of the mantle rock peridotite at various depths.

2 Describe the process of decompression melting.

3 What role do water and other volatiles play in the formation of magma?

9.12 | PLATE TECTONICS AND VOLCANIC ACTIVITY

Relate the distribution of volcanic activity to plate tectonics.

Geologists have known for decades that the global distribution of most of Earth's volcanoes is not random. Most active volcanoes on land are located along the margins of the ocean basins—notably within the circum-Pacific belt known as the **Ring of Fire** (**FIGURE 9.32**). These volcanoes consist mainly of composite cones that emit volatile-rich magma having an andesitic (intermediate) composition and occasionally produce awe-inspiring eruptions.

Another group of volcanoes includes the innumerable seamounts that form along the crest of the mid-ocean ridges. At these depths (1–3 kilometers below sea level), the pressures are so intense that the gases emitted quickly dissolve in the seawater and never reach the surface.

There are some volcanic structures, however, that appear to be somewhat randomly distributed around the globe. These volcanic structures comprise most of the islands of the deep-ocean basins, including the Hawaiian islands, the Galapagos islands, and Easter Island.

Prior to the development of the theory of plate tectonics, geologists did not have an acceptable explanation for the distribution of Earth's volcanoes. Recall that most magma originates in rocks from Earth's upper mantle that is essentially solid, *not molten*. The basic connection between plate tectonics and volcanism is that *plate motions provide the mechanisms by which mantle rocks undergo partial melting to generate magma.*

Volcanism at Convergent Plate Boundaries

Recall that along certain convergent plate boundaries, two plates move toward each other, and a slab of oceanic lithosphere descends into the mantle, generating a deep-ocean trench. As the slab sinks deeper into the mantle, the increase in temperature and pressure drives water and carbon dioxide from the oceanic crust. These mobile fluids migrate upward and reduce the melting point of hot mantle rock sufficiently to

FIGURE 9.32 Ring of Fire Most of Earth's major volcanoes are located in a zone around the Pacific called the Ring of Fire. Another large group of active volcanoes sits mainly undiscovered along the mid-ocean ridge system.

trigger some melting (**FIGURE 9.33A**). Recall that the partial melting of mantle rock (peridotite) generates magma having a basaltic composition. After a sufficient quantity of the rock has melted, blobs of buoyant magma slowly migrate upward.

Volcanism at a convergent plate margin results in the development of a slightly curved chain of volcanoes called a *volcanic arc*. These volcanic chains develop roughly parallel to the associated trench—at distances of 200 to 300 kilometers (100 to 200 miles). Volcanic arcs can be constructed on oceanic as well as continental lithosphere. Those that develop within the ocean and grow large enough for their tops to rise above the surface are labeled *archipelagos* in most atlases. Geologists prefer the more descriptive term **volcanic island arcs**, or simply **island arcs** (see Figure 9.33A). Several young volcanic island arcs border the western Pacific basin, including the Aleutians, the Tongas, and the Marianas.

Volcanism associated with convergent plate boundaries may also develop where slabs of oceanic lithosphere are subducted under continental lithosphere to produce a **continental volcanic arc** (**FIGURE 9.33E**). The mechanisms that generate these mantle-derived magmas are essentially the same as those that create volcanic island arcs. The most significant difference is that continental crust is much thicker and is composed of rocks having higher silica content than oceanic crust. Hence, a mantle-derived magma changes chemically as it rises by assimilating silica-rich crustal rocks. At the same time, extensive magmatic differentiation occurs. (Recall that magmatic differentiation is the formation of secondary magmas from parent magma.) Stated another way, the magma generated in the mantle may change from a fluid basaltic magma to a silica-rich andesitic or rhyolitic magma as it moves up through the continental crust.

The Ring of Fire, a belt of explosive volcanoes, surrounds the Pacific basin. The Ring of Fire exists because oceanic lithosphere is being subducted beneath most of the landmasses that surround the Pacific Ocean. The volcanoes of the Cascade Range in the northwestern United States, including Mount Hood, Mount Rainier, and Mount Shasta, are examples of volcanoes generated at a convergent plate boundary (**FIGURE 9.34**).

Volcanism at Divergent Plate Boundaries

The greatest volume of magma (perhaps 60 percent of Earth's total yearly output) is produced along the oceanic ridge system in association with seafloor spreading (**FIGURE 9.33B**). Below the ridge axis where lithospheric plates are continually being pulled apart, the solid yet mobile mantle responds by ascending to fill the rift. Recall that as hot rock rises, it experiences a decrease in confining pressure and undergoes melting without the addition of heat. Recall that this process is called *decompression melting*.

Partial melting of mantle rock at spreading centers produces basaltic magma. Because this newly formed magma is less dense than the mantle rock from which it was derived, it rises and collects in reservoirs located just below the ridge crest. This activity continuously adds new basaltic rock to plate margins, temporarily welding them together, only to have them break again as spreading continues. Along some ridge segments, outpourings of pillow lavas build numerous volcanic structures, the largest of which is Iceland.

Although most spreading centers are located along the axis of an oceanic ridge, some are not. In particular, the East African Rift is a site where continental lithosphere is being pulled apart (**FIGURE 9.33F**). Vast outpourings of fluid basaltic lavas as well as several active composite volcanoes are found in this region of the globe.

Intraplate Volcanism

We know why igneous activity is initiated along plate boundaries, but why do eruptions occur in the interiors of plates? Hawaii's Kilauea, one of the world's most active volcanoes, is situated thousands of kilometers from the nearest plate boundary, in the middle of the vast Pacific plate

A. *Convergent Plate Volcanism* When an oceanic plate subducts, melting in the mantle produces magma that gives rise to a volcanic island arc on the overlying oceanic crust.

Cleveland Volcano, Aleutian Islands (USGS)

C. *Intraplate Volcanism* When an oceanic plate moves over a hot spot, a chain of volcanic structures such as the Hawaiian Islands is created.

Kilauea, Hawaii (USGS)

E. *Convergent Plate Volcanism* When oceanic lithosphere descends beneath a continent, magma generated in the mantle rises to form a continental volcanic arc.

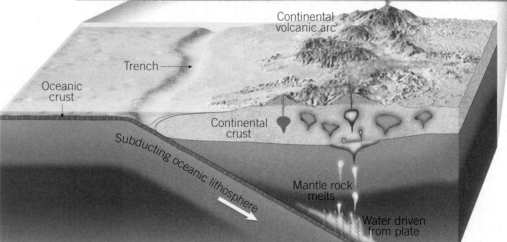

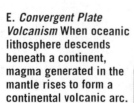

SmartFigure 9.33 Earth's Zones of Volcanism

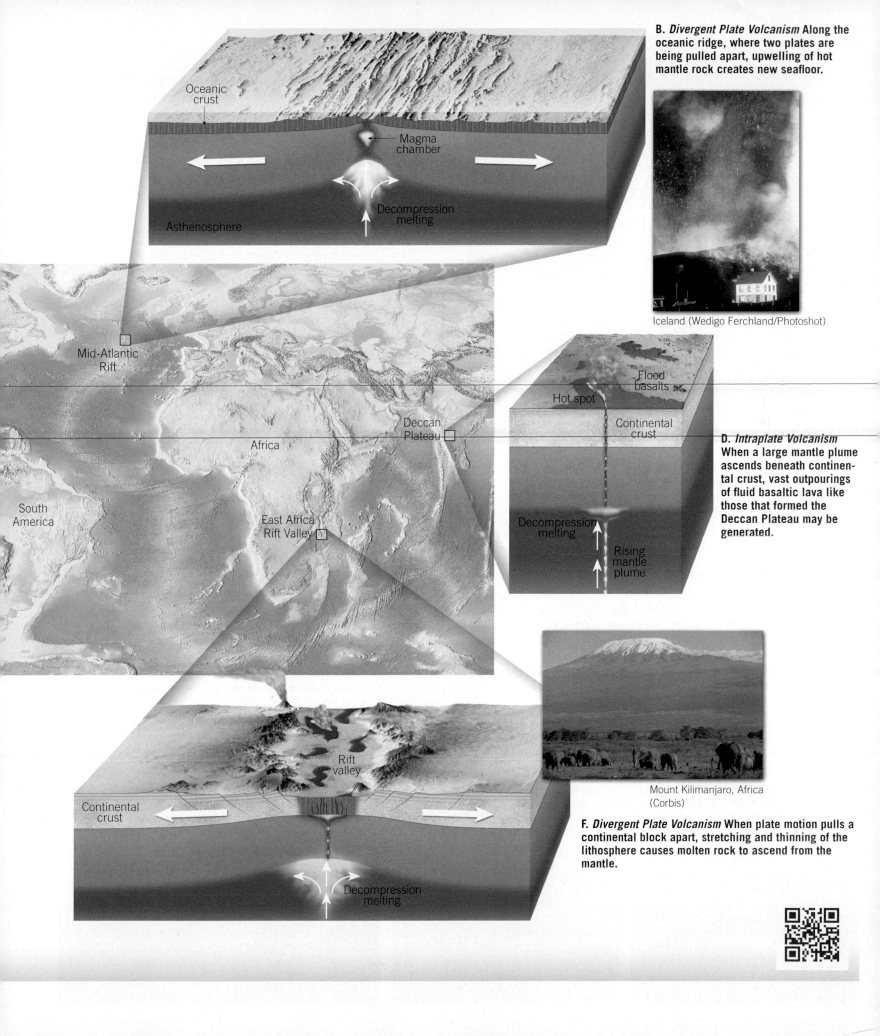

B. *Divergent Plate Volcanism* Along the oceanic ridge, where two plates are being pulled apart, upwelling of hot mantle rock creates new seafloor.

Oceanic crust

Magma chamber

Asthenosphere

Decompression melting

Iceland (Wedigo Ferchland/Photoshot)

Mid-Atlantic Rift

Africa

South America

Deccan Plateau

East Africa Rift Valley

Hot spot

Flood basalts

Continental crust

D. *Intraplate Volcanism* When a large mantle plume ascends beneath continental crust, vast outpourings of fluid basaltic lava like those that formed the Deccan Plateau may be generated.

Decompression melting

Rising mantle plume

Mount Kilimanjaro, Africa (Corbis)

Rift valley

Continental crust

Decompression melting

F. *Divergent Plate Volcanism* When plate motion pulls a continental block apart, stretching and thinning of the lithosphere causes molten rock to ascend from the mantle.

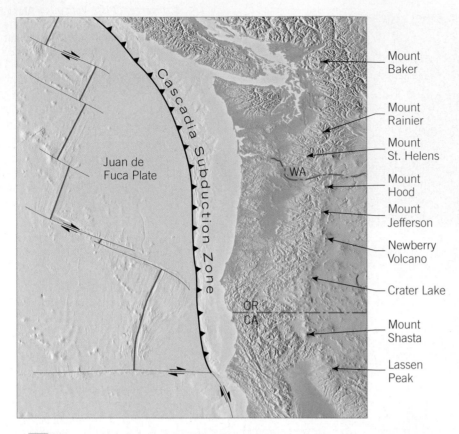

SmartFigure 9.34 Subduction of the Juan de Fuca Plate Produced the Cascade Volcanoes Major volcanic structures of the Cascade Range formed along a convergent plate boundary (Cascadia subduction zone) off the Pacific coast of North America.

a **mantle plume** ascends toward the surface (**FIGURE 9.35A**).[*] Although the depth at which mantle plumes originate is a topic of debate, some appear to form deep within Earth, at the core–mantle boundary. These plumes of solid yet mobile rock rise toward the surface in a manner similar to the blobs that form within a lava lamp. (These are the trendy lamps that contain two immiscible liquids in a glass container. As the base of the lamp is heated, the denser liquid at the bottom becomes buoyant and forms blobs that rise to the top.) Like the blobs in a lava lamp, a mantle plume has a bulbous head that draws out a narrow stalk beneath it as it rises. Once the plume head nears the top of the mantle, decompression melting generates basaltic magma that triggers volcanism at the surface.

Large mantle plumes, dubbed *superplumes*, are thought to be responsible for the vast outpourings of basaltic lava that created the large basalt provinces (**FIGURE 9.35A**). When the head of a superplume reaches the base of the lithosphere, melting progresses rapidly. This causes the burst of volcanism that emits voluminous outpourings of lava to form a huge basalt province in a matter of a million or so years (**FIGURE 9.35B**). Due to the extreme nature of the eruptions required to produce the large basalt provinces, some researchers believe the eruptions were responsible for the extinction of many of Earth's life-forms.

The comparatively short initial eruptive phase is often followed by millions of years of less voluminous activity, as the plume tail slowly rises to the surface. Extending away from some large flood basalt provinces is a chain of volcanic structures, similar to the Hawaiian chain (**FIGURE 9.35C**).

9.12 CONCEPT CHECKS

1 Are volcanoes in the Ring of Fire generally described as quiescent or explosive? Name an example that supports your answer.

2 How is magma generated along convergent plate boundaries?

3 Volcanism at divergent plate boundaries is most often associated with which rock type? What causes rocks to melt in these settings?

4 What is the source of magma for most intraplate volcanism?

5 At which of the three types of plate boundaries is the greatest quantity of magma generated?

(**FIGURE 9.33C**). Other sites of **intraplate volcanism** (meaning "within the plate") include large outpourings of fluid basaltic lavas like those that compose the Columbia River basalts, the Siberian Traps in Russia, India's Deccan Plateau, and several large oceanic plateaus, including the Ontong Java Plateau located in the western Pacific. These massive structures are estimated to be 10 to 40 kilometers (6 to 25 miles) thick. It is thought that most intraplate volcanism occurs where a mass of hotter-than-normal mantle material called

FIGURE 9.35 Hot Spots and Mantle Plumes This model of hot-spot volcanism explains the formation of oceanic plateaus, large basalt provinces on land, and chains of volcanic islands such as the Hawaiian chain.

[*]Recent research on the nature of mantle plumes has caused some geologists to question their role, if any, in the formation of at least some large basalt provinces.

A. A rising mantle plume with a large bulbous head is thought to generate Earth's large basalt provinces.

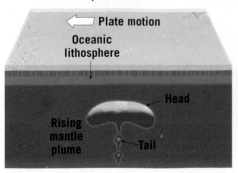

B. Rapid decompression melting of the plume head produces extensive outpourings of flood basalts over a relatively short time span.

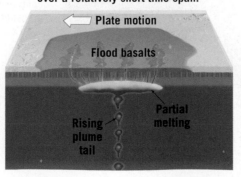

C. Because of plate movement, volcanic activity from the rising tail of the plume generates a linear chain of smaller volcanic structures.

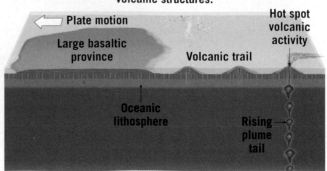

9.1 MOUNT ST. HELENS VERSUS KILAUEA

Compare and contrast the 1980 eruption of Mount St. Helens with the eruption of Kilauea, which began in 1983 and continues today.

- Volcanic eruptions cover a broad spectrum from explosive eruptions, like Mount St. Helens in 1980, to the quiescent eruptions of Kilauea.

Q Although Kilauea usually erupts in a gentle manner, what risks might you encounter if you lived nearby?

USGS

9.2 THE NATURE OF VOLCANIC ERUPTIONS

Explain why some volcanic eruptions are explosive and others are quiescent.

KEY TERMS: magma, lava, viscosity, volatiles, eruption column

- One important characteristic that differentiates various lavas is their viscosity (resistance to flow). In general, the higher silica content of a lava, the more viscous it is. The lower the silica content, the runnier the lava. Another factor that influences viscosity is temperature. Hot lavas are more fluid, while cool lavas are more viscous.
- High-silica, low-temperature lavas are the most viscous and allow the greatest amount of pressure to build up before they "let go" in an eruption. In contrast, lavas that are hot and low in silica are the most fluid. Because basaltic lavas are less viscous, they produce relatively gentle eruptions, while volcanoes that erupt felsic lavas (rhyolite and andesite) tend to be more explosive.

9.3 MATERIALS EXTRUDED DURING AN ERUPTION

List and describe the three categories of materials extruded during volcanic eruptions.

KEY TERMS: aa flow, pahoehoe flow, lava tube, pillow lava, pyroclastic materials, scoria, pumice

- Volcanoes bring hot molten lava, gases, and solid rocky chunks to Earth's surface.
- Because of their low viscosity, basaltic lava flows can extend great distances from a volcano, where they travel over the surface as pahoehoe or aa flows. Sometimes the surface of the flow congeals, but lava continues to flow below in tunnels called lava tubes.
- The gases most commonly emitted by volcanoes are water vapor and carbon dioxide. Upon reaching the surface, these gases rapidly expand, resulting in explosive eruptions that generate an eruptive column and produce a mass of lava fragments called pyroclastic materials.
- Pyroclastic materials come in several sizes. From smallest to largest, they are ash, lapilli, and blocks or bombs, depending on whether the material left the volcano as solid fragments or as liquid blobs.
- If bubbles of gas in lava don't pop before the lava solidifies, they are preserved as voids called vesicles. Especially frothy silica-rich lava can cool to make pumice, which may float in water. Basaltic (mafic) lava with lots of bubbles cools to make scoria.

9.4 ANATOMY OF A VOLCANO

Label a diagram that illustrates the basic features of a typical volcanic cone.

KEY TERMS: conduit, vent, volcanic cone, crater, parasitic cone, fumarole

- Volcanoes are varied in form but share a few common features. Most are roughly conical piles of extruded material that collect around a central vent. The vent is usually within a summit crater or caldera. On the flanks of the volcano, there may be smaller vents marked by small parasitic cones, or fumaroles, spots where gas is expelled.

Q Label the diagram with the following terms: conduit, vent, bombs, lava, parasitic cone, pyroclastic material

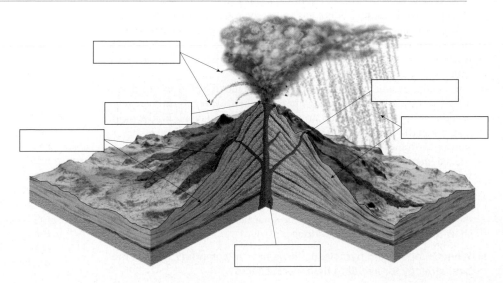

9.5 SHIELD VOLCANOES

Summarize the characteristics of shield volcanoes and provide one example.

KEY TERM: shield volcano

- Shield volcanoes consist of many successive layers of low-viscosity basaltic lava and lack significant amounts of pyroclastic debris. Lava tubes help transport lava far from the main vent, resulting in very gentle, shield-like profiles.
- Most large shield volcanoes are associated with hot-spot volcanism. The volcanoes Kilauea, Mauna Loa, and Mauna Kea in Hawaii are classic examples of the low, wide form.

9.6 CINDER CONES

Describe the formation, size, and composition of cinder cones.

KEY TERM: cinder cone (scoria cone)

- Cinder cones are steep-sided structures composed mainly of pyroclastic debris, typically having a basaltic composition. Lava flows sometimes emerge from the base of a cinder cone but typically do not flow out of the crater.
- Cinder cones are small relative to the other major kinds of volcanoes, reflecting the fact that they form quickly, in single eruptive events. Because they are unconsolidated, cinder cones easily succumb to weathering and erosion.

9.7 COMPOSITE VOLCANOES

Explain the formation, distribution, and characteristics of composite volcanoes.

KEY TERM: composite volcano (stratovolcano)

- Composite volcanoes are called "composite" because they consist of both pyroclastic material and lava flows. They typically erupt silica-rich lavas that cool to produce andesite or rhyolite. They are much larger than cinder cones and form from multiple eruptions over a million years or longer.

- Because the andesitic or rhyolitic lava erupted from composite volcanoes is more viscous than basaltic lava, it accumulates at a steeper angle than does the lava from shield volcanoes. Over time, a composite volcano's combination of lava and cinders produces towering volcanoes with a classic symmetrical shape.
- Mount Rainier and the other volcanoes of the Cascade Range in the northwestern United States are good examples of composite volcanoes, as are the other volcanoes of the Pacific Ocean's Ring of Fire.

Q **If your family had to live next to a volcano, would you rather it be a shield, cinder cone, or composite volcano? Explain.**

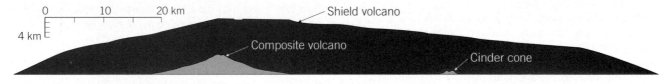

9.8 VOLCANIC HAZARDS

Discuss the major geologic hazards associated with volcanoes.

KEY TERMS: pyroclastic flow (nuée ardente), lahar, aerosol

- The greatest volcanic hazard to human life is pyroclastic flow, or nuée ardente. This dense mix of hot gas and pyroclastic fragments races downhill at great speed and incinerates everything in its path. A pyroclastic flow can travel many miles from its source volcano. Because pyroclastic flows are hot, their deposits frequently "weld" together into a solid rock called welded tuff.
- Lahars are volcanic mudflows. These rapidly moving slurries of ash and debris suspended in water can occur even when a volcano isn't actively erupting. They tend to follow stream valleys and can result in loss of life and/or significant damage to structures in their path.
- Volcanic ash in the atmosphere can be a risk to air travel when it is sucked into airplane engines. Volcanoes at sea level can generate tsunami when they erupt or their flanks collapse into the ocean. In addition, volcanoes that spew large amounts of gas such as sulfur dioxide can cause human respiratory problems. If volcanic gases reach the stratosphere, they screen out a portion of incoming solar radiation and can trigger short-term cooling at Earth's surface.

Q **What do lahars and pyroclastic flows have in common? What is the best strategy for avoiding their worst effects?**

9.9 OTHER VOLCANIC LANDFORMS

List and describe volcanic landforms other than volcanic cones.

KEY TERMS: caldera, fissure, fissure eruption, basalt plateau, flood basalt, volcanic neck (plug), pipe

- Calderas are among the largest volcanic structures. They form when the stiff, cold rock above a magma chamber cannot be supported and collapses to create a broad, bowl-like depression. On shield volcanoes, calderas form slowly as lava is drained from the magma chamber beneath the volcano. On composite volcanoes, a caldera collapse often follows an explosive eruption that can result in significant loss of life and destruction of property.
- Fissure eruptions produce massive floods of low-viscosity, silica-poor lava from large cracks in the crust. Layer upon layer of these flood basalts may build up to significant thicknesses, as in the Columbia Plateau or the Deccan Traps. The defining feature of a flood basalt is the broad area it covers.
- An example of a volcanic neck is preserved at Shiprock, New Mexico. The lava in the "throat" of this ancient volcano crystallized to form a "plug" of solid rock, and it weathers more slowly than the conical volcano in which it formed. Now, after the mound of pyroclastic debris has been eroded away, the resistant neck is a distinctive landform.

9.10 INTRUSIVE IGNEOUS ACTIVITY

Compare and contrast these intrusive igneous structures: dikes, sills, batholiths, stocks, and laccoliths.

KEY TERMS: intrusion, pluton, tabular, massive, discordant, concordant, dike, sill, columnar jointing, batholith, stock, laccolith

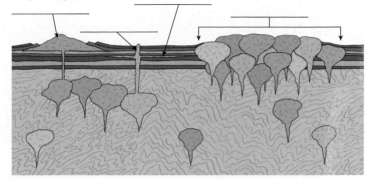

- When magma intrudes other rocks, it may cool and crystallize before reaching the surface, producing intrusions called plutons. Plutons come in many shapes. They may cut across the host rocks without regard for preexisting structures, or the magma may flow along weak zones in the host rock, such as between the horizontal layers of sedimentary bedding.

- Tabular intrusions may be concordant (sills) or discordant (dikes). Massive plutons may be small (stocks) or very large (batholiths). Blister-like intrusions also exist (laccoliths). As solid igneous rock cools, its volume decreases. Contraction can produce columnar jointing, a distinctive pattern consisting mainly of six-sided fractures.

- **Q** Label the accompanying diagram using the following terms: batholith, laccolith, sill, and dike.

9.11 PARTIAL MELTING AND THE ORIGIN OF MAGMA

Summarize the major processes that generate magma from solid rock.

KEY TERMS: partial melting, geothermal gradient, decompression melting

- Solid rock may melt under three geologic circumstances: when heat is added to the rock, raising its temperature; when already hot rock experiences lower pressures (decompression, as occurs at mid-ocean ridges); and when water is added to hot rock that is near its melting point (as occurs at subduction zones).

- **Q** Different processes produce magma in different tectonic settings. Consider situations A, B, and C in the accompanying diagram and describe the processes that would be most likely to trigger melting in each.

9.12 PLATE TECTONICS AND VOLCANIC ACTIVITY

Relate the distribution of volcanic activity to plate tectonics.

KEY TERMS: Ring of Fire, volcanic island arc (island arc), continental volcanic arc, intraplate volcanism, mantle plume

- Volcanoes occur at both convergent and divergent plate boundaries, as well as in intraplate settings.

- Convergent plate boundaries that involve the subduction of oceanic crust are the sites where the explosive volcanoes of the Pacific "Ring of Fire"

are most prevalent. Here, release of water from the subducted plate triggers melting in the overlying mantle. The resulting magma interacts with the lower crust of the overlying plate during its ascent and results in the formation of a volcanic arc at the surface.

- At divergent boundaries, decompression melting is the dominant generator of magma. As warm rock rises, it can begin to melt without the addition of heat. The result will be a rift valley if the overlying crust is continental or a mid-ocean ridge if it is oceanic.

- In an intraplate setting, the source of magma is a mantle plume: a column of warm, rising solid mantle rock that begins to melt in the uppermost mantle.

GIVE IT SOME **THOUGHT**

1. Match each of these volcanic regions with one of the three zones of volcanism (convergent plate boundaries, divergent plate boundaries, or intraplate volcanism):

   a. Crater Lake

   b. Hawaii's Kilauea

   c. Mount St. Helens

   d. East African Rift

   e. Yellowstone

   f. Vesuvius

   g. Deccan Plateau

   h. Mount Etna

2. Examine the accompanying photo and complete the following:

   a. What type of volcano is this? What features helped you make a decision?

b. What is the eruptive style of such volcanoes? Describe the likely composition and viscosity of the magma.

c. Which one of the three zones of volcanism is the likely setting for this volcano?

d. Name a city that is vulnerable to the effects of a volcano of this type.

USGS

3. Divergent boundaries, such as the Mid-Atlantic Ridge, are characterized by outpourings of basaltic lava. Answer the following questions about divergent boundaries and their associated lavas:

a. What is the source of these lavas?

b. What causes the source rocks to melt?

c. Describe a divergent boundary that would be associated with lava other than basalt. Why did you choose it, and what type of lava would you expect to erupt there?

4. Explain why volcanic activity occurs in places other than plate boundaries.

5. For each of the accompanying four sketches, identify the geologic setting (zone of volcanism). Which of these settings will most likely generate explosive eruptions? Which will produce outpouring of fluid basaltic lavas?

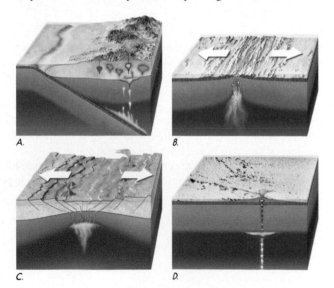

A.   B.

C.   D.

6. The following image shows the Buddhist monastery Taung Kalat, located in central Myanmar (Burma). The monastery sits high on a sheer-sided rock made mainly of magmas that solidified in the conduit of an ancient volcano. The volcano has since been worn away.

a. Based on this information, what volcanic structure do you think is shown in the photo?

b. Would this volcanic structure most likely have been associated with a composite volcano or a cinder cone? Explain how you arrived at your answer.

Dreamstime

7. Imagine that you are a geologist charged with the task of choosing three sites where state-of-the-art volcano monitoring systems will be deployed. The sites can be anywhere in the world, but the budget and number of experts you can employ to oversee the operations are limited. What criteria would you use to select these sites? List some potential choices and your reasons for considering them.

8. The accompanying image shows an igneous feature (dark color) located in southeastern Utah that was intruded into horizontal sedimentary strata.

a. What name is given to this intrusive igneous feature?

b. The light bands above and below the dark igneous body are metamorphic rock. Identify this type of metamorphism and briefly explain how it alters rock. (*Hint:* Refer to Chapter 2, if necessary.)

Michael Collier

9. Explain why an eruption of Mount Rainier would be considerably more destructive than the similar eruption of Mount St. Helens that occurred in 1980.

10. During a field trip with your geology class, you visit an exposure of rock layers similar to the one sketched on the following page. A fellow student suggests that the layer of basalt is a sill. You disagree. Why do you think the other student is incorrect? What is a more likely explanation for the basalt layer?

Shale

Vesicles

Basalt

Sandstone

Shale

Limestone

11. Each of the following descriptions indicates how an intrusive feature appears when exposed at Earth's surface by erosion. Name the feature.

a. A dome-shaped mountainous structure flanked by upturned layers of sedimentary rocks

b. A vertical wall-like feature a few meters wide and hundreds of meters long

c. A huge expanse of granitic rock forming a mountainous terrain tens of kilometers wide

d. A relatively thin layer of basalt sandwiched between layers of sedimentary rocks exposed on the side of a canyon

12. Mount Whitney, the highest summit (4421 meters [14,500 feet]) in the contiguous United States, is located in the Sierra Nevada batholith. Based on its location, is Mount Whitney likely composed of granitic, andesitic, or basaltic rocks?

Mount Whitney →

Don Smith/Getty Images

EXAMINING THE **EARTH SYSTEM**

1. Speculate about some of the possible consequences that a great and prolonged increase in explosive volcanic activity might have on each of Earth's four spheres.

Jon A. Helgason/AGE Fotostock

2. The accompanying image is of Mount Unzen, Japan, a composite volcano that produced several fiery pyroclastic flows between 1900 and 1995. Despite the potential for devastating destruction, humans continue to live on or near active volcanoes such as this. Based on this image, suggest a reason the villagers continue to inhabit this area. Identify the path of the most recent destructive pyroclastic flows and any protective measure that may have been employed to contain them.

Michael S. Yamashita/Corbis

MasteringGeology™

Looking for additional review and test prep materials? Visit the Self Study area in **www.masteringgeology.com** to find practice quizzes, study tools, and multimedia that will aid in your understanding of this chapter's content. In **MasteringGeology™** you will find:

■ GEODe: Earth Science: An interactive visual walkthrough of key concepts

■ Geoscience Animation Library: More than 100 animations illuminating many difficult-to-understand Earth science concepts

■ In The News RSS Feeds: Current Earth science events and news articles are pulled into the site with assessment

■ Pearson eText

■ Optional Self Study Quizzes

■ Web Links

■ Glossary

■ Flashcards

10

Crustal Deformation and Mountain Building

FOCUS ON CONCEPTS

Each statement represents the primary **LEARNING OBJECTIVE** for the corresponding major heading within the chapter. After you complete the chapter, you should be able to:

10.1 Describe the three types of differential stress. Differentiate stress from strain. Compare and contrast brittle and ductile deformation.

10.2 List and describe five types of folds.

10.3 Sketch and briefly describe the relative motion of rock bodies located on opposite sides of normal, reverse, and thrust faults as well as both types of strike-slip faults.

10.4 Locate and name Earth's major mountain belts on a world map.

10.5 Sketch a cross section of an Andean-type mountain belt and describe how its major features are generated.

10.6 Summarize the stages in the development of an Alpine-type mountain belt such as the Appalachians.

10.7 Explain the principle of isostasy and how it contributes to the elevated topography of young mountain belts like the Himalayas.

Sunrise over the Canadian Rockies, Banff National Park, Alberta, Canada. (Photo by Josh McCulloch/All Canada Photos/Corbis)

Mountains provide some of the most spectacular scenery on our planet. Poets, painters, and songwriters have captured their splendor. Geologists understand that at some time, all continental regions were mountainous masses and that continents grow by the addition of mountains to their flanks. As geologists unravel the secrets of mountain formation, they gain a deeper understanding of the evolution of Earth's continents. If continents do indeed grow by adding mountains to their flanks, how do geologists explain the existence of mountains that are located in the interior of landmasses? In order to answer this and related questions, this chapter pieces together the sequence of events that generate these lofty structures.

10.1 | CRUSTAL DEFORMATION Describe the three types of differential stress. Differentiate stress from strain. Compare and contrast brittle and ductile deformation.

Earth is a dynamic planet. Shifting lithospheric plates gradually change the face of our planet by moving continents across the globe. The results of this tectonic activity are perhaps most strikingly apparent in Earth's major mountain belts. Rocks containing fossils of marine organisms are found thousands of meters above sea level, and massive rock units are bent, contorted, overturned, and sometimes riddled with fractures.

In the Canadian Rockies, for example, rock strata have been thrust hundreds of kilometers over the top of other layers. On a smaller scale, crustal movements of a few meters occur along faults during major earthquakes. Even in the stable interiors of the continents, rocks reveal a history of deformation which shows that these areas were once sites of former mountain belts.

What Causes Rocks to Deform?

Every body of rock, no matter how strong, has a point at which it will deform by bending or breaking. **Deformation** (*de* = out, *forma* = form) is a general term that refers to the changes in the shape or position of a rock body in response to differential stress. Most crustal deformation occurs along plate boundaries. Plate motions and the interactions along plate margins generate the tectonic forces that cause rock to deform.

The basic geologic features that form as a result of the forces generated by the interactions of tectonic plates are called **rock structures**, or **geologic structures** (**FIGURE 10.1**). Rock structures include *folds* (wave-like undulations), *faults* (fractures along which one rock body slides past another), and *joints* (cracks).

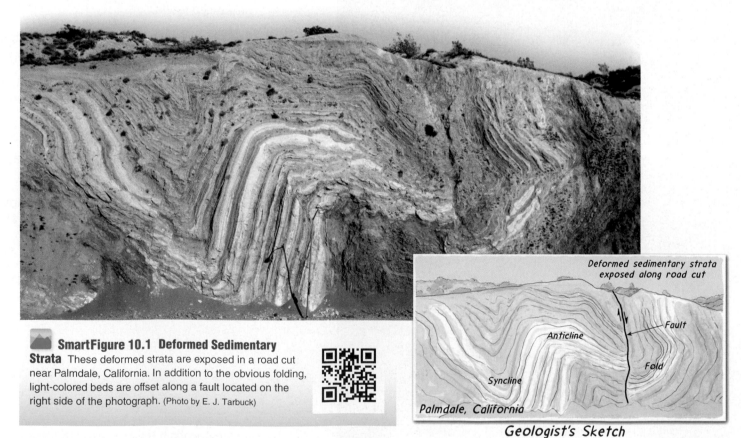

SmartFigure 10.1 Deformed Sedimentary Strata These deformed strata are exposed in a road cut near Palmdale, California. In addition to the obvious folding, light-colored beds are offset along a fault located on the right side of the photograph. (Photo by E. J. Tarbuck)

Deformed sedimentary strata exposed along road cut

Anticline

Fault

Fold

Syncline

Palmdale, California

Geologist's Sketch

Stress: The Force That Deforms Rocks

From everyday experience, you know that if a door is stuck, you must expend energy, called *force*, to open it. Geologists use the term **stress** to describe the forces that deform rocks. Whenever the stresses acting on a rock body exceed its strength, the rock will deform—usually by one or more of the following processes: folding, flowing, fracturing, or faulting.

As we discussed in Chapter 3, when stress is applied uniformly in all directions, it is called **confining pressure**. By contrast, when stress is applied unequally in different directions, it is termed **differential stress**. We will consider three types of differential stress:

1. **Compressional stress.** Differential stress that squeezes a rock mass as if placed in a vise is known as **compressional stress** (*com* = together, *premere* = to press) (**FIGURE 10.2B**). Compressional stresses are most often associated with convergent plate boundaries. When plates collide, Earth's crust is generally shortened horizontally and thickened vertically. Over millions of years, this deformation produces mountainous terrains.

2. **Tensional stress.** Differential stress that pulls apart or elongates rock bodies is known as **tensional stress** (*tendere* = to stretch) (**FIGURE 10.2C**). Along divergent plate boundaries where plates are moving apart, tensional stresses stretch and lengthen rock bodies. For example, in the Basin and Range Province in the western United States, tensional forces have fractured and stretched the crust to as much as twice its original width.

3. **Shear stress.** Differential stress can also cause rock to **shear**, which involves the movement of one part of a rock body past another (**FIGURE 10.2D**). Shear is similar to the slippage that occurs between individual playing cards when the top of the deck is moved relative to the bottom (**FIGURE 10.3**). Small-scale deformation of rocks by shear stresses occurs along closely spaced parallel surfaces of weakness, such as foliation surfaces and microscopic fractures, where slippage changes the shape of rocks. By contrast, at transform fault boundaries, such as the San Andreas Fault, shear stress causes large segments of Earth's crust to slip horizontally past one another.

Strain: A Change in Shape Caused by Stress

When flat-lying sedimentary layers are uplifted and tilted, their orientations change, but their shapes are often retained. Differential stresses can also change the shape of a rock body, referred to as **strain**. Like the circle shown in Figure 10.3, *strained bodies lose their original configuration during deformation*. In short, *stress* is the force that acts to deform rock bodies, while *strain* is the resulting deformation (distortion), or change in the shape of the rock body.

Types of Deformation

When rocks are subjected to stresses that exceed their strength, they deform, usually by bending or breaking. It is easy to visualize how rocks break because we normally think of rocks as being brittle. But how can large masses of rock be *bent* into intricate folds without fracturing in the process? To answer this question, geologists performed laboratory experiments in which they subjected rocks to differential stress under conditions experienced at various depths within Earth's crust. Geologists determined that although different rocks deform under somewhat different conditions and rates, rocks experience three types of deformation: *elastic*, *brittle*, and *ductile*.

Elastic Deformation

When stress is applied *gradually*, rocks initially respond by deforming elastically. Changes that result from **elastic deformation** are recoverable; that is, like a rubber band, the rock will snap back to nearly its original size and shape when the stress is removed. During elastic deformation, the chemical bonds of the minerals within a rock are stretched but do not break. When the stress is removed, the bonds snap back to their original length. As you saw in Chapter 8 the energy for most earthquakes comes from stored elastic energy that is released as rock snaps back to its original shape.

Brittle Deformation

When the elastic limit (strength) of a rock is surpassed, the rock either bends or breaks. Rocks that break into smaller pieces exhibit **brittle deformation** (*bryttian* = to shatter). From our everyday experience, we know that glass objects, wooden pencils, china plates, and even our bones exhibit brittle failure when their strength is surpassed. Brittle deformation occurs when stress breaks the chemical bonds that hold a material together.

Cube of rock

A. Undeformed rock

Compression

B. Compressional stress (shortening)

Tension

C. Tensional stress (stretching)

Shear

D. Shear stress (sliding and tearing)

FIGURE 10.2 Three Types of Stresses: Compressional, Tensional, and Shear

By sliding the top of the deck relative to the bottom, we can illustrate the type of shearing that commonly occurs along closely spaced planes of weakness in rocks.

Shear stress causes the circle in this deck of cards to become an ellipse, which can be used to measure the amount and type of strain.

FIGURE 10.3 Shearing and the Resulting Deformation (Strain) An ordinary deck of playing cards with a circle embossed on its side illustrates shearing and the resulting strain.

FIGURE 10.4 A Coin Deformed by a Passing Train (Photo by Anthony Pleva/Alamy)

Ductile Deformation

Ductile deformation is a type of solid-state flow that produces a change in the shape of an object without fracturing. Ordinary objects that display ductile behavior include modeling clay, beeswax, taffy, and some metals. For example, a coin placed on a railroad track will be flattened (without breaking) by the forces applied by a passing train (**FIGURE 10.4**).

Factors That Affect Rock Strength

The major factors that influence the strength of a rock and how it will deform include temperature, confining pressure, rock type, and time.

The Role of Temperature

The effect of temperature on the strength of a material can be easily demonstrated with a piece of glass tubing commonly found in a chemistry lab. If the tubing is dropped on a hard surface, it will shatter. However, if the tubing is heated over a Bunsen burner, it can be easily bent into a variety of shapes. Rocks respond to heat similarly. Where temperatures are high (deep in Earth's crust), rocks tend to soften and become more malleable, so they deform by folding or flowing (ductile deformation). Likewise, where temperatures are low (at or near the surface), rocks tend to behave like brittle solids and fracture.

The Role of Confining Pressure

Recall that pressure, like temperature, increases with depth as the thickness of the overlying rock increases. Buried rocks are subjected to confining pressure, which is much like water pressure, where the forces are applied equally in all directions. Confining pressure "squeezes" the materials in Earth's crust, which makes it stronger and thus harder to break. Therefore, rocks that are deeply buried are "held together" by the immense pressure and tend to bend rather than fracture.

The Influence of Rock Type

In addition to being influenced by the physical environment, deformation of rock is greatly influenced by its mineral composition and texture. For example, igneous and some metamorphic rocks (quartzite, for example) are composed of minerals that have strong internal chemical bonds. These strong, brittle rocks tend fail by fracturing when subjected to stresses that exceed their strength.

By contrast, sedimentary rocks that are weakly cemented or metamorphic rocks that contain zones of weakness, such as foliation, are more susceptible to ductile deformation. Weak rocks that are most likely to behave in a ductile manner (bend or flow) when subjected to differential stress include rock salt, shale, limestone, and schist. In fact, rock salt is so weak that large masses of it often rise through overlying beds of sedimentary rocks, much as hot magma rises toward Earth's surface. Perhaps the weakest naturally occurring solid to exhibit ductile flow is glacial ice.

Some outcrops consist of a sequence of interbedded weak and strong rock layers that have been moderately deformed by folding—for example, shale and well-cemented sandstone beds. In these settings, the strong sandstone layers are often highly fractured, while the weak shale beds form broad undulating folds. The formation of diverse structures (one brittle and one ductile) within the same rock mass can be illustrated by cooling a MilkyWay or similar chocolate-over-caramel candy bar in a refrigerator. When the cool candy bar is slowly bent, it exhibits brittle deformation in the chocolate and ductile deformation in the caramel.

Time as a Factor

One key factor that researchers are unable to duplicate in the laboratory is how rocks respond to small stresses applied gradually over long spans of *geologic time*. However, insights into the effects of time on deformation are provided in everyday settings. For example, marble benches have been known to sag under their own weight over a time span of 100 years or so, while wooden bookshelves may bend within a few months after being loaded with books.

In general, when tectonic forces are applied slowly over long time spans, rocks tend to display ductile behavior and deform by bending or flowing. However, the same rocks may shatter if force is applied suddenly. An analogous situation occurs when you take a taffy bar and slowly move its two ends together. The taffy will deform by folding. However, if you hit the taffy swiftly against the edge of a table, it will break into two or more pieces, exhibiting brittle failure.

10.1 CONCEPT CHECKS

1. What is rock deformation? How might a rock body change during deformation?
2. List the three types of differential stress and briefly describe the changes they impart to rock bodies.
3. What type of plate boundary is most commonly associated with compressional stress?
4. How is strain different from stress?
5. Describe elastic deformation.
6. How is brittle deformation different from ductile deformation?
7. List and describe the four factors that affect rock strength.

10.2 FOLDS: ROCK STRUCTURES FORMED BY DUCTILE DEFORMATION List and describe five types of folds.

Along convergent plate boundaries, flat-lying sedimentary strata, tabular intrusions, and volcanic rocks are often bent into a series of wavelike undulations called **folds**. Folds in sedimentary strata are much like those that would form if you were to hold the ends of a sheet of paper on a flat surface and then push them together. In nature, folds come in a wide variety of sizes and configurations. Some folds are broad flexures in which strata hundreds of meters thick have been slightly warped. Others are very tight microscopic structures found in metamorphic rocks. Size differences notwithstanding, most folds result from *compressional stresses that result in a shortening and thickening of the crust.*

To aid our understanding of folds and folding, it is important to become familiar with the terminology used to name the parts of a fold. Folds are geologic structures consisting of stacks of originally horizontal surfaces, such as sedimentary strata, that have been bent as a result of permanent deformation. Each layer is bent around an imaginary axis called a *hinge line*, or simply a *hinge* (**FIGURE 10.5**).

Folds are also described by their *axial plane*, which is a surface that connects all the hinge lines of the folded strata. In simple folds, the axial planc is vertical and divides the fold into two roughly symmetrical *limbs*. However, the axial plane often leans to one side so that one limb is steeper and shorter than the other.

Anticlines and Synclines

The two most common types of folds are anticlines and synclines (**FIGURE 10.6**). **Anticlines** usually arise by upfolding, or arching, of sedimentary layers and are sometimes spectacularly displayed along highways that have been cut through

deformed strata.[1] Typically found in association with anticlines are downfolds, or troughs, called **synclines**. Notice in Figure 10.6 that the limb of an anticline is also a limb of the adjacent syncline.

[1] By strict definition, an anticline is a structure in which the oldest strata are found in the center. A syncline is a structure in which the youngest strata are found in the center.

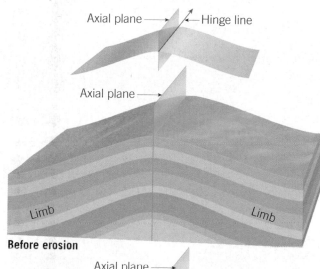

FIGURE 10.5 Features Associated with Symmetrical Folds The axial plane divides a fold as symmetrically as possible, while the hinge line traces the points of maximum curvature of any layer.

EYE ON EARTH

This image features a large geologic structure that outcrops in Death Valley National Park, California.

QUESTION 1 *What name would you give to this geologic structure?*

QUESTION 2 *Based on this image, would you describe this fold as symmetrical or asymmetrical?*

QUESTION 3 *Do these rock units mainly display ductile deformation or brittle deformation?*

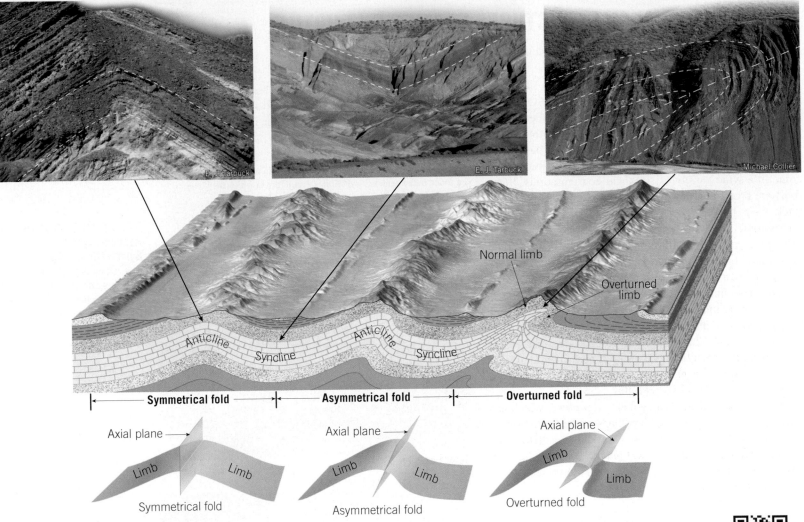

SmartFigure 10.6 Common Types of Folds The upfolded, or arched, structures are *anticlines*. The downfolds, or troughs, are *synclines*. Notice that the limb of an anticline is also the limb of the adjacent syncline.

Depending on their orientation, these basic folds are described as *symmetrical* when the limbs are mirror images of each other or *asymmetrical* when they are not. An asymmetrical fold is said to be *overturned* if one or both limbs are tilted beyond the vertical (see Figure 10.6). An overturned fold can also "lie on its side" so the axial plane is horizontal. These *recumbent* folds are common in highly deformed mountainous regions such as the Alps.

Folds can also by tilted by tectonic forces that cause their hinge lines to slope downward. Folds of this type are said to *plunge* because the hinge lines of the fold penetrate Earth's surface (**FIGURE 10.7A**). Sheep Mountain, Wyoming, is an example of a plunging anticline. **FIGURE 10.7B** shows the pattern produced when erosion removes the upper layers of a plunging fold and exposes its interior. Note that the outcrop pattern of Sheep Mountain, as with all other anticlines, points in the direction it is plunging (**FIGURE 10.7C**). The opposite is true for a syncline.

A good example of topography that results when erosional forces attack folded sedimentary strata is found in the Valley and Ridge Province of the Appalachians (see Figure 10.31). It is important to realize that ridges are not necessarily associated with anticlines, nor are valleys a characteristic of synclines. Rather, ridges and valleys result because of differential weathering and erosion. For example, in the Valley and Ridge Province, resistant sandstone beds remain as imposing ridges separated by valleys cut into more easily eroded shale or limestone beds.

Domes and Basins

Broad upwarps in basement rock may deform the overlying cover of sedimentary strata and generate large folds. When this upwarping produces a circular or slightly elongated structure, the feature is called a **dome** (**FIGURE 10.8A**). The Black Hills of western South Dakota is a large structural dome generated by upwarping. Here erosion has stripped away the highest portions of the overlying sedimentary beds, exposing older igneous and metamorphic rocks in the center (**FIGURE 10.9**).

Plunge

A.

B.

Oldest
rock

Youngest
rock

Direction
of
plunge

Youngest
rock

C. Sheep Mountain Anticline

**SmartFigure 10.7
Sheep Mountain,
Wyoming** Eroded
plunging anticlines, like
Sheep Mountain, have
an outcrop pattern that
"points" in the direction of
plunge.
(Photo by
Michael
Collier)

Structural domes can also be formed by the intrusion of magma (laccoliths), as shown in Figure 9.24. In addition, the upward migration of buried salt deposits can produce salt domes like those surrounding the Gulf of Mexico. Salt domes are economically important rock structures because when salt migrates upward, the surrounding oil-bearing sedimentary strata deforms to form oil reservoirs (see Figure 3.35C).

Downwarped structures having gently sloping beds similar to a saucer or bowl are termed **basins** (**FIGURE 10.8B**). Several large basins exist in the United States (**FIGURE 10.10**). The basins of Michigan and Illinois are thought to have resulted from large accumulations of sediment, whose weight caused the crust to subside (see Section 10.7). Some structural basins are the result of giant meteorite impacts.

Because large basins contain sedimentary beds sloping at low angles, they are usually identified by the age of the rocks composing them. The youngest rocks are found near the center, and the oldest rocks are at the flanks. This is the opposite order of a dome, such as the Black Hills, where the oldest rocks form the core.

Monoclines

Although we have separated our discussions of folds and faults, folds can be uniquely coupled with faults. Examples of this close association are broad, regional features

called *monoclines*. Particularly prominent features of the Colorado Plateau, **monoclines** (*mono* = one, *kleinen* = incline) are large, step-like folds in otherwise horizontal sedimentary strata (**FIGURE 10.11**). These folds appear to have resulted from the reactivation of ancient, steep-dipping reverse faults located in basement rocks beneath

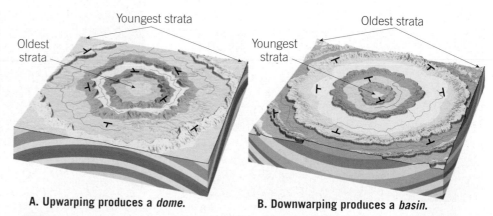

Youngest strata

Oldest
strata

Oldest strata

Youngest
strata

A. Upwarping produces a *dome*.

B. Downwarping produces a *basin*.

SmartFigure 10.8 Domes Versus Basins Gentle upwarping and downwarping of crustal rocks produce **A**. domes and **B**. basins. Erosion of these structures results in an outcrop pattern that is roughly circular or slightly elongate.

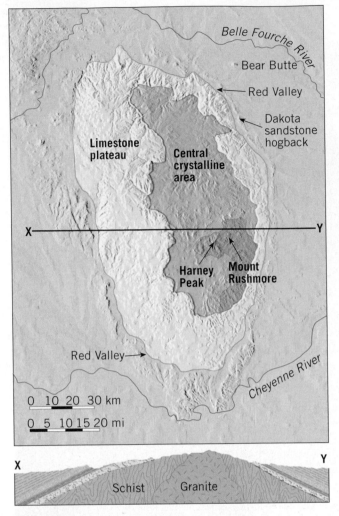

FIGURE 10.9 The Black Hills of South Dakota, a Large Structural Dome The central core of the Black Hills is composed of resistant Precambrian-age igneous and metamorphic rocks. The surrounding rocks are mainly younger limestones and sandstones.

the plateau. As large blocks of basement rock were displaced upward, the comparatively ductile sedimentary strata above responded by draping over the fault like clothes hanging over a bench. Displacement along these reactivated faults often exceeds 1 kilometer (0.6 miles) (see Figure 10.11).

Examples of monoclines found on the Colorado Plateau include the East Kaibab Monocline, the Raplee Anticline, the Waterpocket Fold, and the San Rafael Swell. The inclined strata shown in Figure 10.11 once extended over the sedimentary layers now exposed at the surface—evidence that a tremendous volume of rock has been eroded from this area.

10.2 CONCEPT CHECKS

1 Distinguish between anticlines and synclines, between domes and basins, and between anticlines and domes.

2 Draw a cross-sectional view of a symmetrical anticline. Include a line to represent the axial plane and label both limbs.

3 The Black Hills of South Dakota is a good example of what type of geologic structure?

4 Where are the youngest rocks in an eroded basin found: near the center or near the flanks?

5 Describe the formation of a monocline.

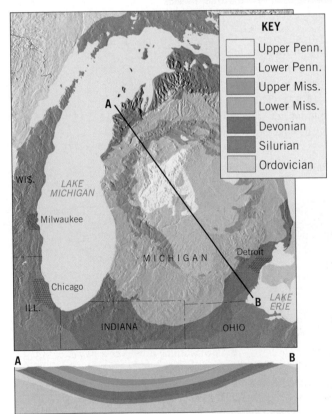

FIGURE 10.10 The Bedrock Geology of the Michigan Basin The youngest rocks are centrally located, whereas the oldest beds flank this structure.

KEY

- Upper Penn.
- Lower Penn.
- Upper Miss.
- Lower Miss.
- Devonian
- Silurian
- Ordovician

Michael Collier

FIGURE 10.11 The East Kaibab Monocline, Arizona This monocline consists of bent sedimentary beds that were deformed by faulting in the bedrock below. The thrust fault shown is called a *blind thrust* because it does not reach the surface.

10.3 | FAULTS AND JOINTS: ROCK STRUCTURES FORMED BY BRITTLE DEFORMATION
Sketch and briefly describe the relative motion of rock bodies located on opposite sides of normal, reverse, and thrust faults as well as both types of strike-slip faults.

Faults form where brittle deformation leads to fracturing and displacement of Earth's crust. Occasionally, small faults can be recognized in road cuts where sedimentary beds have been offset a few meters, as shown in **FIGURE 10.12**. Faults of this scale usually occur as single discrete breaks. By contrast, large faults, like the San Andreas Fault in California, have displacements of hundreds of kilometers and consist of many interconnecting fault surfaces. These structures, described as *fault zones*, can be several kilometers wide and are often easier to identify from aerial photographs than at ground level. Sudden movements along faults cause most earthquakes. However, the vast majority of faults are remnants of past deformation and are inactive.

Dip-Slip Faults

Faults in which movement is primarily parallel to the *inclination* (also called *dip*) of the fault surface are called **dip-slip faults**. Geologists identify the rock surface immediately above the fault as the **hanging wall block** and the rock surface below as the **footwall block** (**FIGURE 10.13**). These names were first used by prospectors and miners who excavated metallic ore deposits such as gold that had precipitated from hydrothermal solutions along inactive fault zones. The miners would walk on the rocks below the mineralized fault zone (the *footwall block*) and hang their lanterns on the rocks above (the *hanging wall block*).

Vertical displacements along dip-slip faults tend to produce long, low cliffs called **fault scarps** (*scarpe* = a slope). Fault scarps, such as the one shown in **FIGURE 10.14**, are produced by rapid vertical slips that generate earthquakes.

Normal Faults Dip-slip faults are classified as **normal faults** when the hanging wall block moves down relative to the footwall block (**FIGURE 10.15**). Normal faults are found in a variety of sizes. Some are small, having displacements of only a meter or so, like the one shown in the road cut in Figure 10.12. Others, however, extend for tens of kilometers and may sinuously trace the boundary of a mountain front. Most large normal faults have relatively steep dips that tend to flatten out with depth. Normal faults are associated with tensional stresses that pull rock units apart, thereby lengthening the crust. This "pulling apart" can be accomplished either by uplifting that causes the surface to stretch and break or by horizontal forces that have opposing orientations.

In the western United States, large normal faults are associated with structures called **fault-block mountains**. Excellent examples of fault-block mountains are found in the Basin and Range Province, a region that encompasses Nevada and portions of the surrounding states (**FIGURE 10.16**). Here the crust has been elongated and broken to create more than 200 relatively small mountain ranges. Averaging about 80 kilometers (50 miles) in length, the ranges rise 900 to 1500 meters (3000 to 5000 feet) above the adjacent down-faulted basins.

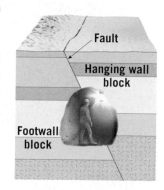

FIGURE 10.13 Hanging Wall Block and Footwall Block The rock immediately above a fault surface is the *hanging wall block*, and the one below is called the *footwall block*. These names came from miners who excavated ore deposits that formed along fault zones. The miners hung their lanterns on the rocks above the fault trace (hanging wall block) and walked on the rocks below the fault trace (footwall block).

FIGURE 10.12 Faults Are Fractures Where Slip Has Occurred (Photo by E. J. Tarbuck)

FIGURE 10.14 Fault Scarp This fault scarp was created during the Alaskan earthquake of 1964. (Photo courtesy of the USGS)

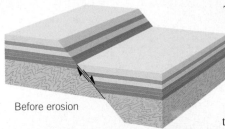

Before erosion

TENSIONAL STRESS

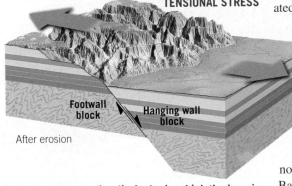

Footwall block

Hanging wall block

After erosion

Normal faults are dip-slip faults in which the hanging wall block moves down relative to the footwall block.

SmartFigure 10.15
Normal Dip-Slip Fault The upper diagram illustrates the relative displacement that occurs between the blocks on either side of a fault. The lower diagram show how erosion may alter the upfaulted blocks.

Mobile Field Trip 10.16 Normal Faulting in the Basin and Range Province Here, tensional stresses have elongated and fractured the crust into numerous blocks. Movement along these faults has tilted the blocks, producing parallel mountain ranges called *fault-block mountains*. The down-faulted blocks (*grabens*) form basins, whereas the upfaulted blocks (*horsts*) erode to form rugged mountainous topography. In addition, numerous tilted blocks (*half-grabens*) form both basins and mountains. (Photo by Michael Collier)

Graben

Horst

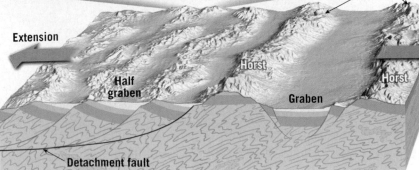

Fault-block mountains

Extension

Horst

Half graben

Horst

Graben

Detachment fault

The topography of the Basin and Range Province evolved in association with a system of normal faults trending roughly north–south. Movements along these faults produced alternating uplifted fault blocks called **horsts** (*horst* = hill) and down-dropped blocks called **grabens** (*graben* = ditch). Horsts form the ranges and are the source of sediments that have accumulated in the basins created by the grabens. As Figure 10.16 illustrates, structures called **half-grabens**, which are fault blocks that have been tilted, also contribute to the alternating topographic highs and lows in the Basin and Range Province.

Also notice in Figure 10.16 that the slopes of the large normal faults associated with the Basin and Range Province decrease with depth and eventually join to form a nearly horizontal fault called a **detachment fault**. These faults represent a major boundary between the rocks below, which exhibit ductile deformation, and the rocks above, which exhibit mainly brittle deformation.

Reverse and Thrust Faults Dip-slip faults in which the hanging wall block moves up relative to the footwall block are called **reverse faults** (**FIGURE 10.17A**). **Thrust faults** are a type of reverse faults having dips less than 45 degrees. Reverse and thrust faults result from compressional stresses that produce horizontal shortening of the crust.

Most high-angle reverse faults are small and accommodate local displacements in regions dominated by other types of faulting. Thrust faults, on the other hand, exist at all scales, with some large thrust faults having displacements ranging from tens to hundreds of kilometers. Movement along a thrust fault can cause the hanging wall block to be thrust nearly horizontally over the footwall block, as shown in **FIGURE 10.17B**.

Thrust faulting is most pronounced along convergent plate boundaries. Compressional forces associated with colliding plates generally create folds as well as thrust faults that thicken and shorten the crust to produce mountainous topography (**FIGURE 10.18**). Examples of mountainous belts produced by this type of compressional tectonics include the Alps, Northern Rockies, Himalayas, and Appalachians.

Strike-Slip Faults

A fault in which the dominant displacement is horizontal and parallel to the *trend* (direction) of the fault surface is called a **strike-slip fault** (**FIGURE 10.19**). The earliest scientific records of strike-slip faulting were made following surface ruptures that produced large earthquakes. One of the most noteworthy of these was the great San Francisco earthquake of 1906. During this strong earthquake, structures such as fences and roads that were built across the San Andreas Fault were displaced as much as 4.7 meters (15 feet). Because movement along the San Andreas Fault causes the crustal block on the opposite side of the fault to move to the right as you face the fault, it is called a *right-lateral* strike-slip fault.

The Great Glen Fault in Scotland, which exhibits the opposite sense of displacement, is a well-known example of a *left-lateral* strike-slip fault. The total displacement along the Great Glen Fault is estimated to exceed 100 kilometers (60 miles). Also associated with this fault trace are numerous lakes, including Loch Ness, home of the legendary monster.

Some strike-slip faults slice through Earth's crust and accommodate motion between two tectonic plates. These major strike-slip faults are called **transform faults** (*trans* = across, *forma* = form). Numerous transform faults cut the oceanic lithosphere and link spreading oceanic ridges. Others accommodate displacement between continental blocks that slip horizontally past each other. Some of the best-known transform faults include California's San Andreas Fault, New Zealand's Alpine Fault, the

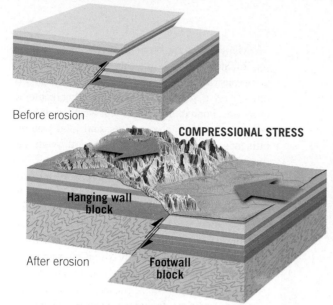

Before erosion

COMPRESSIONAL STRESS

Hanging wall block

After erosion

Footwall block

A. Reverse faults are dip-slip faults in which the hanging wall block moves up relative to the footwall block.

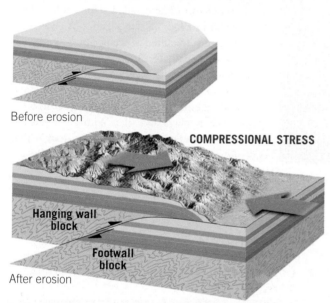

Before erosion

COMPRESSIONAL STRESS

Hanging wall block

Footwall block

After erosion

B. Thrust faults are low angle reverse faults in which the overlying blocks are thrust nearly horizontally over the underlying blocks.

FIGURE 10.17 Reverse and Thrust Faults Reverse and thrust faults are generated by compressional stresses that force one block of rock over another.

Middle East's Dead Sea Fault, and Turkey's North Anatolian Fault (see Figure 10.19). Large transform faults, like these, accommodate relative displacements of up to several hundred kilometers.

Rather than being a single fracture along which movement takes place, most continental transform faults consist of a zone of roughly parallel fractures. The zone may be up to several kilometers wide. The most recent movement, however, is often along a strand only a few meters wide, which may offset features such as stream channels. Crushed and broken rocks produced during faulting are more easily eroded, so linear valleys or sag ponds often mark the locations of large strike-slip faults.

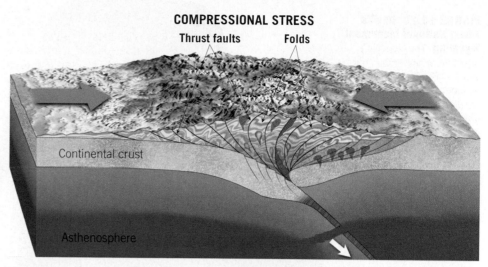

COMPRESSIONAL STRESS

Thrust faults **Folds**

Continental crust

Asthenosphere

FIGURE 10.18 Compressional Stresses Produce Folds and Thrust Faults At convergent plate boundaries, compressional stresses thicken and shorten the crust, resulting in mountainous topography.

Joints

Among the most common rock structures are fractures called joints. Unlike faults, **joints** are fractures along which no appreciable displacement has occurred. Although some joints have a random orientation, most occur in roughly parallel groups.

We have already considered two types of joints. In Chapter 9 we learned that *columnar joints* form when igneous rocks cool and develop shrinkage fractures that produce elongated, pillarlike columns (**FIGURE 10.20**). In addition, recall from Chapter 4 that sheeting produces a pattern of gently

A. The dominant displacement along a strike-slip fault is horizontal and parallel to the trend (direction) of the fault surface.

FIGURE 10.19 Strike-Slip Faulting
A. The block diagram illustrates the features associated with large strike-slip faults. Notice how the stream channels have been offset by fault movement. **B.** Aerial view of the San Andreas Fault. (Photo by David Parker/Science Source)

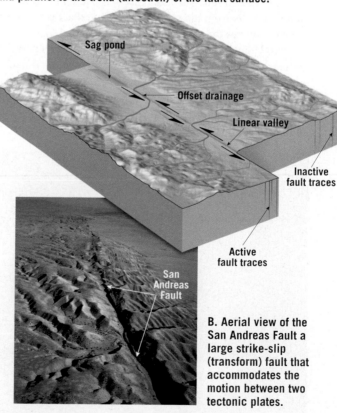

Sag pond

Offset drainage

Linear valley

Inactive fault traces

Active fault traces

San Andreas Fault

B. Aerial view of the San Andreas Fault a large strike-slip (transform) fault that accommodates the motion between two tectonic plates.

FIGURE 10.20 Devil's Tower National Monument, Wyoming This imposing structure exhibits columnar joints. Columnar joints form as igneous rocks cool and develop shrinkage fractures that produce five- to seven-sided elongated, pillarlike structures.

Established in September 1906, Devil's Tower became our nation's first national monument. This nearly vertical monolith rises more than 380 meters (1265 feet) above the surrounding grasslands and pine forests of eastern Wyoming.

Michael Collier

Many rocks are broken by two or even three sets of intersecting joints that slice the rock into numerous regularly shaped blocks. These joint sets often exert a strong influence on other geologic processes. For example, chemical weathering tends to be concentrated along joints, and joint patterns influence how groundwater moves through the crust (**FIGURE 10.21**).

Joints may also be significant from an economic standpoint. Some of the world's largest and most important mineral deposits are located along joint systems. Hydrothermal solutions (mineralized fluids) can migrate into fractured host rocks and precipitate economically significant amounts of copper, silver, gold, zinc, lead, and uranium.

Highly jointed rocks present a risk to construction projects, including bridges, highways, and dams. On June 5, 1976, 14 lives were lost and nearly $1 billion in property damage occurred when the Teton Dam in Idaho failed. This earthen dam, constructed of erodible clays and silts, was situated on highly fractured volcanic rocks. Although attempts were made to fill the voids in the jointed rock, water gradually penetrated the subsurface fractures and undermined the dam's foundation. Eventually the moving water cut a tunnel into the easily erodible clays and silts. Within minutes the dam failed, sending a 20-meter- (65-foot-) high wall of water down the Teton and Snake Rivers.

curved joints that develop more or less parallel to the surface of large exposed igneous bodies such as batholiths. Here the jointing results from the gradual expansion that occurs when erosion removes the overlying load (see Figure 4.6).

Most joints are produced when rocks in the outermost crust are deformed, causing the rock to fail by brittle fracture. Extensive joint patterns often develop in response to relatively subtle and often barely perceptible regional upwarping and downwarping of the crust. In many cases, the cause of jointing at a particular locale is not particularly obvious.

10.3 CONCEPT CHECKS

1 Contrast the movements that occur along normal and reverse faults. What type of stress is indicated by each fault?

2 What type of faults are associated with fault-block mountains?

3 How are reverse faults different from thrust faults? In what way are they similar?

4 Describe the relative movement along a strike-slip fault.

5 How are joints different from faults?

FIGURE 10.21 Nearly Parallel Joints in the Navajo Sandstone, Arches National Park, Utah Weathering and erosion along the joints have produced a topography known in the park as "fins." (Photo by Michael Collier)

10.4 | MOUNTAIN BUILDING

Locate and name Earth's major mountain belts on a world map.

Mountain building has occurred in the recent geologic past at several locations around the world. Young mountain belts include the American Cordillera (*cordillera* means "spine" or "backbone"), which runs along the western margin of the Americas from Cape Horn at the tip of South America to Alaska and includes the Andes; the Alpine–Himalaya chain, which extends along the margin of the Mediterranean, through Iran to northern India and into Indochina; and the mountainous terrains of the western Pacific, which include volcanic island arcs that comprise Japan, the Philippines, and Sumatra. Most of these young mountain belts have come into existence within the past 100 million years (**FIGURE 10.22**). Some, including the Himalayas, began their growth as recently as 50 million years ago.

In addition to these young mountain belts, there are several chains of Paleozoic-age mountains on Earth. Although these older mountain belts are deeply eroded and topographically less prominent, they exhibit the same structural features found in younger mountains. The Appalachians in the eastern United States and the Urals in Russia are classic examples of this group of older, well-worn mountain belts.

The term for the processes that collectively produce a mountain belt is **orogenesis** (*oros* = mountain, *genesis* = to come into being). Most major mountain belts display striking visual evidence of great compressional forces that have shortened the crust horizontally while thickening it vertically. These **compressional mountains** contain large quantities of preexisting sedimentary and crystalline rocks that have been faulted and contorted into a series of folds (see Figure 10.18).

Although folding and thrust faulting are often the most conspicuous signs of orogenesis, varying degrees of metamorphism and igneous activity are always present.

How do mountain belts form? Since the time of the ancient Greeks, this question has intrigued philosophers and scientists. One early proposal suggested that mountains are simply wrinkles in Earth's crust, produced as the planet cooled from its original semimolten state. According to this idea, Earth contracted and shrank as it lost heat, which caused the crust to deform in a manner similar to how an orange peel wrinkles as the fruit dries out. However, neither this nor any other early hypothesis withstood scientific scrutiny.

The theory of plate tectonics provides a model for orogenesis with excellent explanatory power and accounts for the origin of virtually all the present mountain belts and most of the ancient ones. According to this model, the tectonic processes that generate Earth's major mountainous terrains occurs along convergent plate boundaries, where oceanic lithosphere subducts into the mantle. We will first look at the nature of convergent plate boundaries and then examine how the process of subduction has driven mountain building around the globe.

10.4 CONCEPT CHECKS

1 Define *orogenesis*.

2 In the plate tectonics model, which type of plate boundary is most directly associated with Earth's major mountain belts?

FIGURE 10.22 Earth's Major Mountain Belts Notice the east–west trend of major mountain belts in Eurasia in contrast to the north–south trend of the North and South American Cordileras.

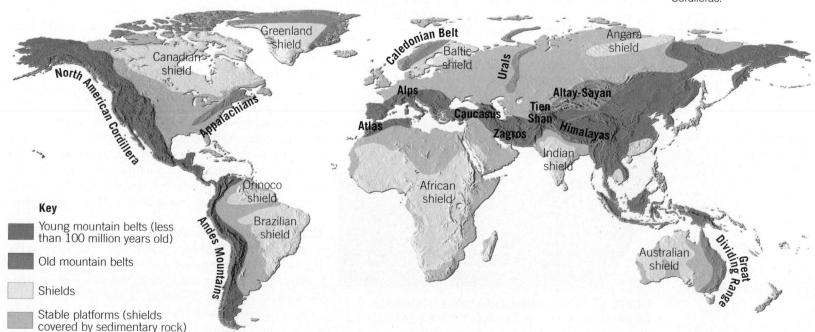

Key

- Young mountain belts (less than 100 million years old)
- Old mountain belts
- Shields
- Stable platforms (shields covered by sedimentary rock)

10.5 | SUBDUCTION AND MOUNTAIN BUILDING Sketch a cross section of an
Andean-type mountain belt and describe how its major features are generated.

The subduction of oceanic lithosphere is the driving force in mountain building (orogenesis). Where oceanic lithosphere subducts beneath an oceanic plate, a *volcanic island arc* and related tectonic features develop. Subduction beneath a continental block, on the other hand, results in the formation of a *continental volcanic arc* and mountainous topography along the margin of a continent. In addition, volcanic island arcs and other crustal fragments "drift" across the ocean basin until they reach a subduction zone, where they collide and become welded to another crustal fragment or a larger continental block. If subduction continues long enough, it can ultimately result in the collision of two or more continents.

Island Arc–Type Mountain Building

Island arcs result from the steady subduction of oceanic lithosphere, which may last for 200 million years or more. Periodic volcanic activity, the emplacement of igneous plutons at depth, and the accumulation of sediment that is scraped from the subducting plate gradually increase the volume of crustal material capping the upper plate (**FIGURE 10.23**). Some large volcanic island arcs, such as Japan, owe their size to having been built on fragments of continental crust that have rifted from a large landmass or the joining of multiple island arcs over time.

The continued growth of a volcanic island arc can result in the formation of mountainous topography consisting of nearly parallel belts of igneous and metamorphic rocks. This activity, however, is viewed as just one phase in the development of Earth's major mountain belts. As you will see later, some volcanic arcs are carried by subducting plates to the margin of large continental blocks, where they become involved in large-scale mountain-building episodes.

Continuous subduction of oceanic lithosphere results in the development of thick units of continental-like crust on the overlying plate.

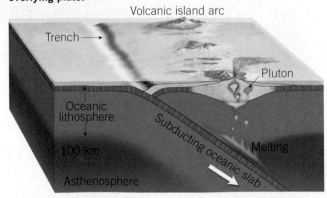

FIGURE 10.23 Development of a Volcanic Island Arc A volcanic island arc forms where one slab of oceanic lithosphere is subducted beneath another slab of the same material.

Andean-Type Mountain Building

Andean-type mountain building is characterized by subduction beneath a continent rather than oceanic lithosphere. Subduction along these active continental margins is associated with long-lasting magmatic activity that builds continental volcanic arcs. The result is crustal thickening, with the crust reaching thicknesses of more than 70 kilometers (45 miles).

The first stage in the development of Andean-type mountain belts, named after the Andes Mountains of South America, occurs along *passive continental margins*. The East coast of the United States provides a modern example of a passive continental margin where sedimentation has produced a thick platform of shallow-water sandstones, limestones, and shales (**FIGURE 10.24A**). At some point, the forces that drive plate motions change, and a subduction zone develops along the margin of the continent. In order for a new subduction zone to form, oceanic lithosphere must be old and dense enough to create a downward force capable of shearing the lithosphere. Alternatively, strong compressional forces tear off the oceanic lithosphere along a continental margin, thereby initiating subduction.

Building Volcanic Arcs Recall that as oceanic lithosphere descends into the mantle, increasing temperatures and pressures drive volatiles (mostly water and carbon dioxide) from the crustal rocks. These mobile fluids migrate upward into the wedge-shaped region of mantle between the subducting slab and upper plate. At a depth of about 100 kilometers (60 miles), these water-rich fluids sufficiently reduce the melting point of hot mantle rock to trigger some melting (**FIGURE 10.24B**). Partial melting of the ultramafic rock peridotite generates *primary magmas*, with mafic basaltic (mafic) compositions. Because these newly formed basaltic magmas are less dense than the rocks from which they originated, they will buoyantly rise. Upon reaching the base of the low-density materials of the continental crust, they typically collect, or pond.

Continued ascent through the thick continental crust is generally achieved through magmatic differentiation, in which heavy ferromagnesian minerals crystallize and settle out of the magma, leaving the remaining melt enriched in silica and other "light" components (see Chapter 3). Hence, through magmatic differentiation, a comparatively dense basaltic magma can generate low-density, buoyant *secondary magma* composed of intermediate and/or granitic (felsic) composition.

Emplacement of Batholiths Because of its low density and great thickness, continental crust significantly impedes the ascent of molten rock. Consequently, a high percentage of the magma that intrudes the crust never reaches the surface; instead, it crystallizes at depth to form massive igneous plutons called *batholiths*. The result of this activity is thickening of Earth's crust.

A. Passive continental margin with an extensive platform of sediments and sedimentary rocks.

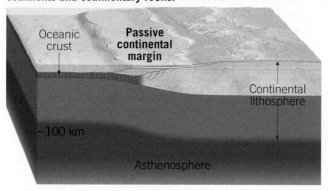

B. Plate convergence generates a subduction zone, and partial melting produces a continental volcanic arc. Compressional forces and igneous activity further deform and thicken the crust, elevating the mountain belt.

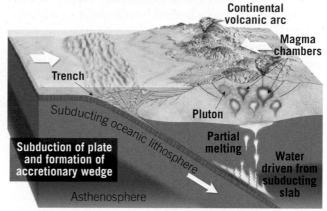

FIGURE 10.24 Andean-Type Mountain Building

C. Subduction ends and is followed by a period of uplift.

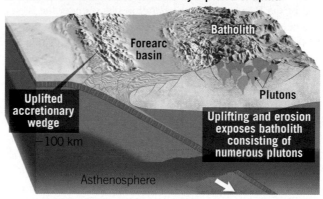

Eventually, uplifting and erosion exhume the batholiths that consist of numerous interconnected plutons. The American Cordillera contains several large batholiths, including the Sierra Nevada of California, the Coast Range Batholith of western Canada, and several large igneous bodies in the Andes. Most batholiths consist of intrusive igneous rocks that range in composition from granite to diorite.

Development of an Accretionary Wedge

During the development of volcanic arcs, unconsolidated sediments that are carried on the subducting plate, as well as fragments of oceanic crust, may be scraped off and plastered against the edge of the overriding plate. The resulting chaotic accumulation of deformed and thrust-faulted sediments and scraps of ocean crust is called an **accretionary wedge** (see Figure 10.24B). The processes that deform these sediments are comparable to a wedge of soil being scraped and pushed in front of an advancing bulldozer.

Some of the sediments that comprise an accretionary wedge are muds that accumulated on the ocean floor and were subsequently carried to the subduction zone by plate motion. Additional materials are derived from an adjacent continent or volcanic arc and consist of volcanic debris and products of weathering and erosion.

In regions where sediment is plentiful, prolonged subduction may thicken a developing accretionary wedge sufficiently that it protrudes above sea level. This has occurred along the southern end of the Puerto Rico trench, where the Orinoco River basin of Venezuela is a major source of sediments. The resulting wedge emerges to form the island of Barbados.

EYE ON EARTH

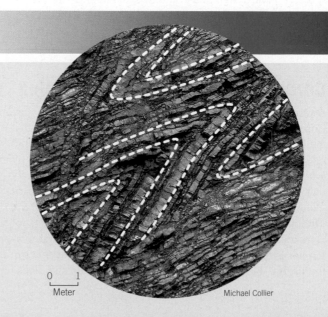

These interbedded layers of chert and shale have been strongly folded during the growth of an accretionary wedge. A recent period of uplift has exposed these deformed strata near the Marin Headlands, north of San Francisco, California.

QUESTION 1 *What is the nature of the stress that most likely generated these highly folded strata: compressional or tensional?*

QUESTION 2 *Along what type of plate boundaries do accretionary wedges form?*

QUESTION 3 *What type of plate boundary is found today in the San Francisco Bay area?*

Michael Collier

Forearc Basins As an accretionary wedge grows upward, it acts as a barrier to the movement of sediment from the volcanic arc to the trench. As a result, sediments begin to collect between the accretionary wedge and the volcanic arc. This region, which is composed of relatively undeformed layers of sediment and sedimentary rocks, is called a **forearc basin** (see Figure 10.24B,C). Subsidence and continued sedimentation in forearc basins can generate a sequence of nearly horizontal sedimentary strata that can attain thicknesses of several kilometers.

Sierra Nevada, Coast Ranges, and Great Valley

California's Sierra Nevada, Coast Ranges, and Great Valley are excellent examples of the tectonic structures that are typically generated along an Andean-type subduction zone. These structures were produced by the subduction of a portion of the Pacific basin (the Farallon plate) under the western margin of California (see Figure 10.24B). The Sierra Nevada Batholith is a remnant of the continental volcanic arc that was produced by many surges of magma during a time span that exceeded 100 million years. The Coast Ranges were built from the vast accumulation of sediments (accretionary wedge) that collected along the continental margin, or perhaps an island arc that lay offshore.

Beginning about 30 million years ago, subduction gradually ceased along much of the margin of North America, as the spreading center that produced the Farallon plate entered the California trench. The uplifting and erosion that followed removed most of the evidence of past volcanic activity and exposed a core of crystalline igneous and associated metamorphic rocks that make up the Sierra Nevada (see Figure 10.24C). The Coast Ranges were uplifted only recently, as evidenced by the young, unconsolidated sediments that currently blanket portions of these highlands.

California's Great Valley is a remnant of the forearc basin that formed between the Sierra Nevada and the accretionary prism and trench that lay offshore. Throughout much of its history, portions of the Great Valley lay below sea level. This sediment-laden basin contains thick marine deposits and debris eroded from the adjacent continental volcanic arc.

10.5 CONCEPT CHECKS

1 The formation of mountainous topography at a volcanic island arc is considered just one phase in the development of a major mountain belt. Explain.

2 Describe and give an example of a passive continental margin.

3 In what ways are the Sierra Nevada and the Andes similar?

4 What is an accretionary wedge? Briefly describe its formation.

5 What is a batholith? In what modern tectonic setting are batholiths being generated?

6 How are magmas that exhibit an intermediate-to-felsic composition thought to be generated from mantle-derived basaltic magmas?

10.6 | COLLISIONAL MOUNTAIN BELTS Summarize the stages in the development of an Alpine-type mountain belt such as the Appalachians.

Most major mountain belts are generated when one or more buoyant crustal fragments collide with a continental margin as a result of subduction. Whereas oceanic lithosphere, which is relatively dense, readily subducts, continental lithosphere contains significant amounts of low-density crustal rocks and is therefore too buoyant to undergo subduction. Consequently, the arrival of a crustal fragment at a trench results in a collision and usually ends further subduction.

Cordilleran-Type Mountain Building

A Cordilleran-type orogeny, named after the North American Cordillera, is associated with a Pacific-like ocean—in that unlike the Atlantic, the Pacific may never close. The rapid rate of seafloor spreading in the Pacific basin is balanced by a high rate of subduction. In this setting, it is highly likely that island arcs or small crustal fragments will be carried along until they collide with an active continental margin. The process of collision and accretion (joining together) of comparatively small crustal fragments to a continental margin has generated many of the mountainous regions that rim the Pacific. Geologists refer to these accreted crustal blocks as *terranes*. The term **terrane** is used to describe a crustal fragment that consists of a distinct and recognizable series of rock formations that has been transported by plate tectonic processes. By contrast, the term *terrain* is used when describing the shape of the surface topography, or "lay of the land."

The Nature of Terranes What is the nature of crustal fragments, and where did they originate? Research suggests that prior to their accretion to a continental block, some of these fragments may have been **microcontinents** similar to the modern-day island of Madagascar, located east of Africa in the Indian Ocean. Many others were island arcs similar to Japan, the Philippines, and the Aleutian Islands. Still others may have been submerged oceanic plateaus created by massive outpourings of basaltic lavas associated with mantle plumes (**FIGURE 10.25**). More than 100 of these relatively small crustal fragments, most of which are in the Pacific, are known to exist.

Accretion and Orogenesis As oceanic plates move, they carry embedded oceanic plateaus, volcanic island arcs, and microcontinents to an Andean-type subduction zone. When an oceanic plate contains small seamounts, these

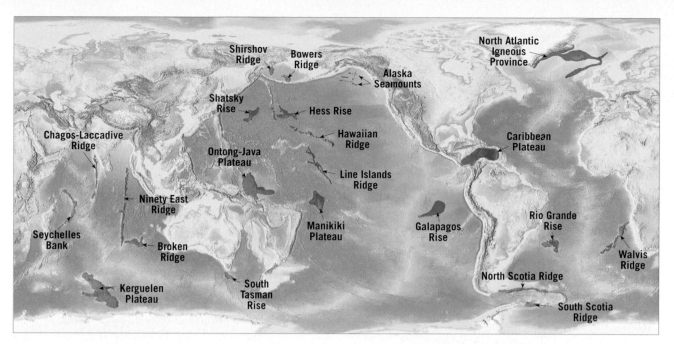

FIGURE 10.25
Distribution of Present-Day Oceanic Plateaus and Other Submerged Crustal Fragments (Data from Ben-Avraham and others)

structures are generally subducted along with the descending oceanic slab. However, large, thick units of oceanic crust, such as the Ontong Java Plateau, which is the size of Alaska, or an island arc composed of abundant "light" igneous rocks, render the oceanic lithosphere too buoyant to subduct. In these situations, a collision between the crustal fragment and the continental margin occurs.

The sequence of events that happen when small crustal fragments reach a Cordilleran-type margin is shown in **FIGURE 10.26**. Rather than subduct, the upper crustal layers of these thickened zones are "peeled" from the descending plate and thrust in relatively thin sheets on the adjacent continental block. Convergence does not generally end with the accretion of a crustal fragment. Rather, new subduction

A. A microcontinent and a volcanic island arc are being carried toward a subduction zone.

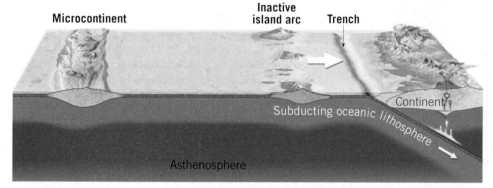

B. The volcanic island arc is sliced off the subducting plate and thrust onto the continent.

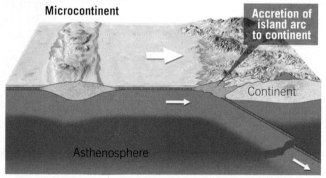

C. A new subduction zone forms seaward of the old subduction zone.

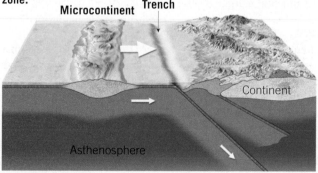

D. The accretion of the microcontinent to the continental margin shoves the remnant island arc further inland and grows the continental margin seaward.

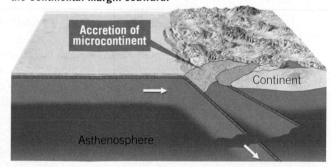

 SmartFigure 10.26 Collision and Accretion of Small Crustal Fragments to a Continental Margin

Paleomagnetic studies and fossil evidence indicate that some of these terranes originated thousands of kilometers to the south of their present locations. (After D. R. Hutchinson and others)

America. This activity resulted in the piecemeal addition of crustal fragments to the entire Pacific margin of the continent—from Mexico's Baja Peninsula to northern Alaska (see Figure 10.27). Geologists expect that many modern microcontinents will likewise be accreted to active continental margins surrounding the Pacific, producing new orogenic belts.

Alpine-Type Mountain Building: Continental Collisions

Alpine-type orogenies, named after the Alps that have been studied intensely for more than 200 years, occur where two continental masses collide. This type of orogeny may also involve the accretion of continental fragments or island arcs that occupied the ocean basin that once separated the two continental blocks. Mountain belts that were formed by the closure of major ocean basins include the Himalayas, Appalachians, Urals, and Alps. Continental collisions result in the development of mountains characterized by shortened and thickened crust, achieved through folding and large-scale thrust faulting.

The zone where two continents collide and are "welded" together is called a **suture**. This portion of a mountain belt often preserves slivers of oceanic lithosphere that were trapped between the colliding plates. As a result of their unique structure, these pieces of oceanic lithosphere help identify the collision boundary.

Next, we will take a closer look at two examples of collisional mountains—the Himalayas and the Appalachians. The Himalayas, Earth's youngest collisional mountains, are still rising. By contrast, the Appalachians are a much older mountain belt, in which active mountain building ceased about 250 million years ago.

The Himalayas

The mountain-building episode that created the Himalayas began roughly 50 million years ago, when India began to collide with Asia. Prior to the breakup of Pangaea, India was located between Africa and Antarctica in the Southern Hemisphere. As Pangaea fragmented, India moved rapidly, geologically speaking, a few thousand kilometers in a northward direction.

The subduction zone that facilitated India's northward migration was near the southern margin of Asia (**FIGURE 10.28A**). Continued subduction along Asia's margin created an Andean-type plate margin that contained a well-developed continental volcanic arc and an accretionary wedge. India's northern margin, on the other hand, was a passive continental margin consisting of a thick platform of shallow-water sediments and sedimentary rocks.

Geologists have determined that one, or perhaps more, small crustal fragments were positioned on the subducting plate somewhere between India and Asia. During the closing of the intervening ocean basin, a small crustal fragment, which now forms southern Tibet, reached the trench. This event was

zones typically form, and they can carry other island arcs or microcontinents toward a collision with the continental margin. Each collision displaces earlier accreted terranes further inland, adding to the zone of deformation as well as to the thickness and lateral extent of the continental margin.

The North American Cordillera

The correlation between mountain building and the accretion of crustal fragments arose primarily from studies conducted in the North American Cordillera (**FIGURE 10.27**). Researchers determined that some of the rocks in the orogenic belts of Alaska and British Columbia contained fossil and paleomagnetic evidence which indicated that these strata previously lay much closer to the equator.

It is now known that many of the terranes that make up the North American Cordillera were scattered throughout the eastern Pacific, like the island arcs and oceanic plateaus currently distributed in the western Pacific. During the breakup of Pangaea, the eastern portion of the Pacific basin (Farallon plate) began to subduct under the western margin of North

followed by the docking of India itself. The tectonic forces involved in the collision of India with Asia were immense and caused the more deformable materials located on the seaward edges of these landmasses to become highly folded and faulted (**FIGURE 10.28B**). The shortening and thickening of the crust elevated great quantities of crustal material, thereby generating the spectacular Himalaya Mountains. As a result, tropical marine limestones that formed along the continental shelf now lie at the summit of Mount Everest.

In addition to uplift, crustal shortening caused rocks at the "bottom of the pile" to become deeply buried—an environment where these rocks experienced elevated temperatures and pressures (see Figure 10.28B). Partial melting within the deepest and most-deformed region of the developing mountain belt produced magmas that intruded the overlying rocks. These environments generate the metamorphic and igneous cores of collisional mountains.

The formation of the Himalayas was followed by a period of uplift that raised the Tibetan Plateau. Seismic evidence suggests that a portion of the Indian subcontinent was thrust beneath Tibet—a distance of perhaps 400 kilometers (250 miles). If this occurred, the added crustal thickness would account for the lofty landscape of southern Tibet, which has an average elevation higher than Mount Whitney, the highest point in the contiguous United States.

The collision with Asia slowed but did not stop the northward migration of India, which has since penetrated at least 2000 kilometers (1200 miles) into the mainland of Asia. Crustal shortening accommodated some of this motion. Much of the remaining penetration into Asia caused lateral displacement of large blocks of the Asian crust by a mechanism described as *continental escape*. As shown in **FIGURE 10.29**, when India continued its northward trek, parts of Asia were "squeezed" eastward, out of the collision zone. These displaced crustal blocks include much of Southeast Asia (the region between India and China) and sections of China.

Why was the interior of Asia deformed to such a large extent, while India has remained essentially intact? The answer lies in the nature of these diverse crustal blocks. Much of India is a continental shield composed mainly of Precambrian rocks (see Figure 10.22). This thick, cold slab of crustal material has been intact for more than 2 billion years and is mechanically strong as a result. By contrast, Southeast Asia was assembled more recently, from the collision of several smaller crustal fragments. Consequently, it is still relatively "warm and weak" from recent periods of mountain building (see Figure 10.29).

A. Prior to the collision of India and Asia, India's northern margin consisted of a thick platform of continental shelf sediments, whereas Asia's was an active continental margin with a well developed accretionary wedge and volcanic arc.

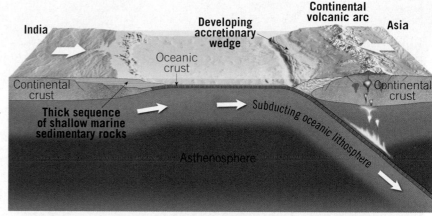

B. The continental collision folded and faulted the crustal rocks that lay along the margins of these continents to form the Himalayas. This event was followed by the gradual uplift of the Tibetan Plateau as the subcontinent of India was shoved under Asia.

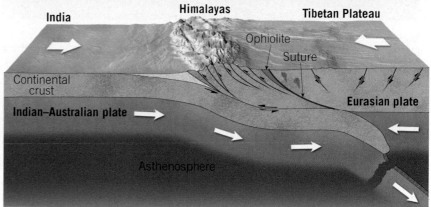

The Appalachians

The Appalachian Mountains provide great scenic beauty near the eastern margin of North America, from Alabama to New-foundland. In addition, mountain belts of similar origin that formed during the same period and were once contiguous are found in the British Isles, Scandinavia, northwestern Africa, and Greenland (see Figure 7.6). The orogeny that generated this extensive mountain system lasted a few hundred million years and was one of the stages in assembling the supercontinent of

FIGURE 10.28
Continental Collision: The Formation of the Himalayas These diagrams illustrate the collision of India with the Eurasian plate that produced the spectacular Himalayas.

Map view showing the southeastward displacement of China and the mainland of Southeast Asia as India plowed into Asia.

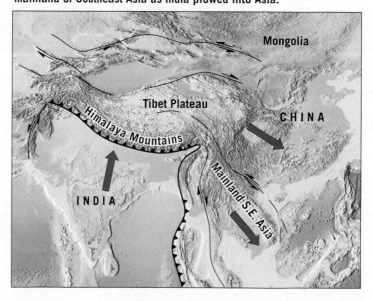

 SmartFigure 10.29
India's Continued Northward Migration Severely Deformed Much of China and Southeast Asia

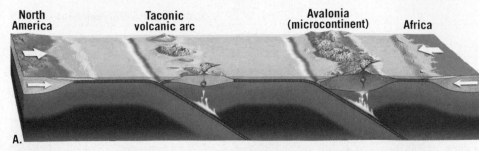

North America   **Taconic volcanic arc**   **Avalonia (microcontinent)**   **Africa**

A.

Closing of an Ocean Basin
About 600 million years ago, the precursor to the North Atlantic began to close. Located within this ocean basin was an active volcanic arc that lay off the coast of North America and a microcontinent situated closer to Africa.

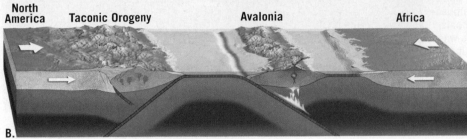

North America   **Taconic Orogeny**   **Avalonia**   **Africa**

B.

Taconic Orogeny Around 450 million years ago, the marginal sea between the volcanic island arc and North America closed. The collision, called the Taconic Orogeny, thrust the island arc over the eastern margin of North America.

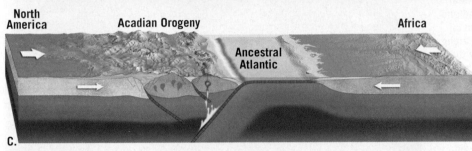

North America   **Acadian Orogeny**   **Ancestral Atlantic**   **Africa**

C.

Acadian Orogeny A second episode of mountain building, called the Acadian Orogeny, occurred about 350 million years ago and involved the collision of a microcontinent with North America.

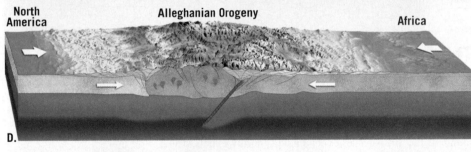

North America   **Alleghanian Orogeny**   **Africa**

D.

Alleghanian Orogeny The final event, the Alleghanian Orogeny, occurred between 250 and 300 million years ago, when Africa collided with North America. The result was the formation of the Appalachian Mountains, perhaps once as majestic as the Himalayas. The Appalachians lay in the interior of the newly assembled supercontinent of Pangaea.

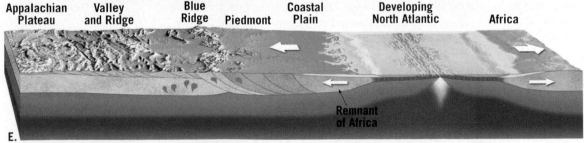

Appalachian Plateau   **Valley and Ridge**   **Blue Ridge**   **Piedmont**   **Coastal Plain**   **Developing North Atlantic**   **Africa**

Remnant of Africa

E.

Rifting of Pangaea About 180 million years ago, Pangaea began to break into smaller fragments, a process that ultimately created the modern Atlantic Ocean. Because this new zone of rifting occurred east of the suture that formed when Africa and North America collided, remnants of African crust remain "welded" to the North American plate.

SmartFigure 10.30 Formation of the Appalachian Mountains The Appalachians formed during the closing of a precursor to the Atlantic Ocean. This event involved three separate stages of mountain building that spanned more than 300 million years. (Based on Zve Ben-Avraham, Jack Oliver, Larry Brown, and Frederick Cook)

Pangaea. Detailed studies of the Appalachians indicate that the formation of this mountain belt was complex and resulted from three distinct episodes of mountain building.

Our simplified overview begins roughly 750 million years ago, with the breakup of a supercontinent called Rodinia that predates Pangaea. Similar to the breakup of Pangaea, this episode of continental rifting and seafloor spreading generated a new ocean between the rifted continental blocks. Located within this developing ocean basin was an active volcanic arc that lay off the coast of North America and a microcontinent situated closer to Africa (**FIGURE 10.30A**).

About 600 million years ago, for reasons geologists do not completely understand, plate motion changed dramatically, and this ancient ocean basin began to close. This led to three main orogenic events that culminated with the collision of North America and Africa.

Taconic Orogeny Around 450 million years ago, the marginal sea between the volcanic island arc and ancestral North America began to close. The collision that ensued, called the *Taconic Orogeny*, caused the volcanic arc along with ocean sediments that were located on the upper plate to be thrust over the larger continental block. The remnants of this volcanic arc and oceanic sediments are recognized today as the metamorphic rocks found across much of the western Appalachians, especially in New York (**FIGURE 10.30B**). In addition to the pervasive regional metamorphism, numerous magma bodies intruded the crustal rocks along the entire continental margin.

Acadian Orogeny A second episode of mountain building, called the *Acadian Orogeny*, occurred about 350 million years ago. The continued closing of this ancient ocean basin resulted in the collision of a microcontinent with North America (**FIGURE 10.30C**). This orogeny involved thrust faulting, metamorphism, and the intrusion of several large granite bodies. In addition, this event added substantially to the width of North American.

Alleghanian Orogeny The final orogeny, called the *Alleghanian Orogeny*, occurred between 250 and 300 million years ago, when Africa collided with North America. The result was the displacement of the material that was accreted earlier by as much as 250 kilometers (155 miles) toward the interior of North America. This event also displaced and further deformed the shelf sediments and sedimentary rocks that had once flanked the eastern margin of North America (**FIGURE 10.30D**). Today these folded and thrust-faulted sandstones, limestones, and shales make up the largely unmetamorphosed rocks of the Valley and Ridge Province (**FIGURE 10.31**). Outcrops of the folded and thrust-faulted structures that characterize collisional mountains are found as far inland as central Pennsylvania and West Virginia.

With the collision of Africa and North America, the Appalachians, perhaps as majestic as the Himalayas, lay in the interior of Pangaea. Then, about 180 million years ago, this newly formed supercontinent began to break into smaller fragments, a process that ultimately created the modern Atlantic Ocean. Because this new zone of rifting occurred east of the suture that formed when Africa and North America collided, remnants of Africa remain "welded" to the North American plate (**FIGURE 10.30E**). The crust underlying Florida is an example.

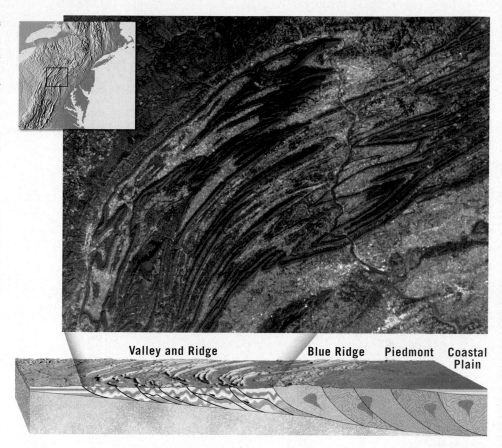

Valley and Ridge   Blue Ridge   Piedmont   Coastal Plain

Other mountain ranges that exhibit evidence of continental collisions include the Alps and the Urals. The Alps formed as Africa and at least two smaller crustal fragments collided with Europe during the closing of the Tethys Sea. Similarly, the Urals were uplifted during the assembly of Pangaea, when northern Europe and northern Asia collided, forming a major portion of Eurasia.

10.6 CONCEPT CHECKS

1 Differentiate between *terrane* and *terrain*.

2 During the formation of the Himalayas, the continental crust of Asia was deformed more than India proper. Why was this the case?

3 Where might magma be generated in a newly formed collisional mountain belt?

4 How does the plate tectonics theory help explain the existence of fossil marine life in rocks atop compressional mountains?

Mobile Field Trip 10.31 The Valley and Ridge Province This region of the Appalachian Mountains consists of folded and faulted sedimentary strata that were displaced landward along thrust faults as a result of the collision of Africa with North America. (NASA/GSFC/JPL, MISR Science Team)

10.7 | WHAT CAUSES EARTH'S VARIED TOPOGRAPHY?

Explain the principle of isostasy and how it contributes to the elevated topography of young mountain belts like the Himalayas.

The causes of Earth's varied topography are complex and cumulative. Geologists know that colliding plates provide the tectonic forces that thicken and elevate crustal rocks during mountain building. Simultaneously, weathering and erosion sculpt Earth's surface into a vast array of landforms, while up-and-down motions in the mantle can change the elevation of a region.

The Laramide Rockies

The portion of the Rocky Mountains that extends from southwestern Montana to New Mexico was produced during a period of deformation known as the Laramide Orogeny. This event, which created some of the most picturesque scenery in the United States, peaked about 60 million years ago.

NASA

Where are the Laramide Rockies?
Sometimes called the Central and Southern Rockies, this mountain belt lies to the east of the Colorado Plateau and includes the Bighorns of Wyoming, the Front Range of Colorado, the Uintas of Utah, and the Sangre de Cristo of southern Colorado and New Mexico.

Colorado Rockies near Steamboat Springs
photo by Michael Collier

What is the geologic history of the Laramide Rockies?
These mountain ranges formed when Precambrian age basement rocks were uplifted nearly vertically along reverse and thrust faults, upwarping the overlying layers of younger sedimentary rocks. Uplifting accelerated the processes of weathering and erosion, which removed much of the younger sedimentary cover from the highest portions of the uplifted blocks. Intrusion of igneous plutons and volcanism occurred simultaneously with this period of mountain building.

LARAMIDE UPLIFT

Basin — Crystalline core — Basin

Uplift

KEY
Paleogene sedimentary rocks
Mesozoic sedimentary rocks
Paleozoic sedimentary rocks
Precambrian sedimentary rocks
Precambrian crystalline rocks

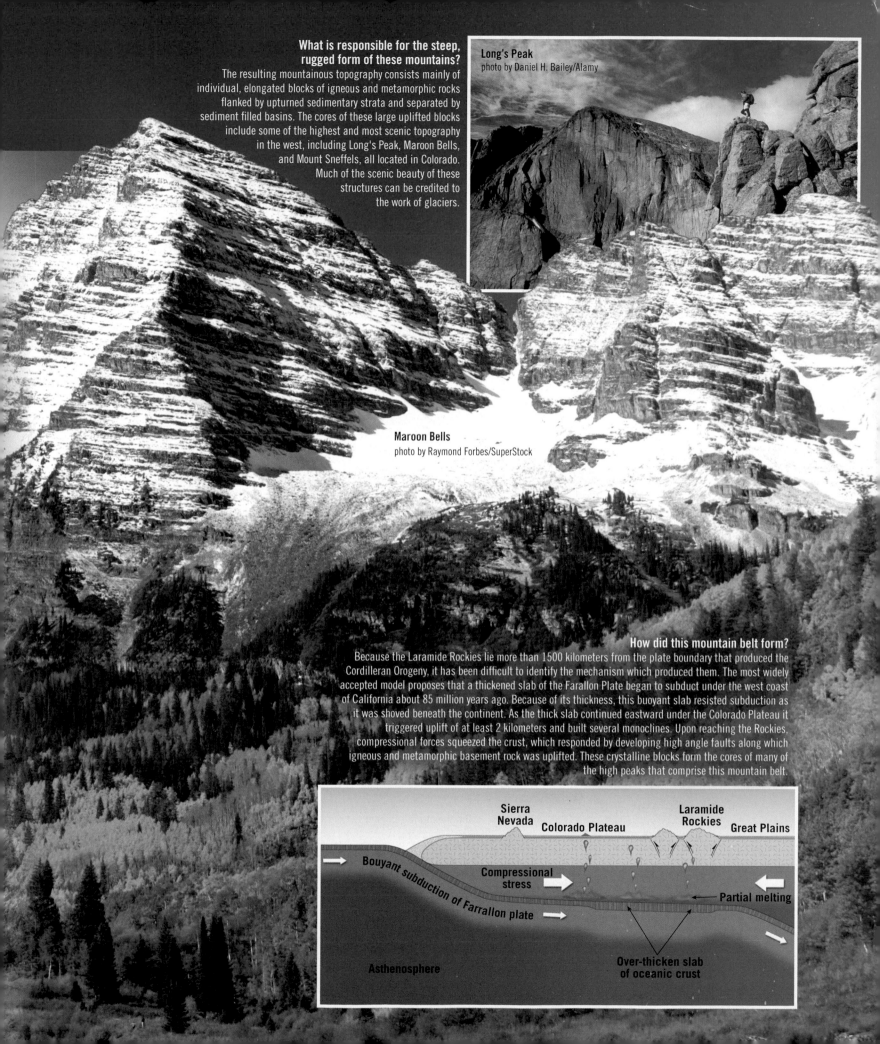

What is responsible for the steep, rugged form of these mountains?

The resulting mountainous topography consists mainly of individual, elongated blocks of igneous and metamorphic rocks flanked by upturned sedimentary strata and separated by sediment filled basins. The cores of these large uplifted blocks include some of the highest and most scenic topography in the west, including Long's Peak, Maroon Bells, and Mount Sneffels, all located in Colorado. Much of the scenic beauty of these structures can be credited to the work of glaciers.

Long's Peak
photo by Daniel H. Bailey/Alamy

Maroon Bells
photo by Raymond Forbes/SuperStock

How did this mountain belt form?

Because the Laramide Rockies lie more than 1500 kilometers from the plate boundary that produced the Cordilleran Orogeny, it has been difficult to identify the mechanism which produced them. The most widely accepted model proposes that a thickened slab of the Farallon Plate began to subduct under the west coast of California about 85 million years ago. Because of its thickness, this buoyant slab resisted subduction as it was shoved beneath the continent. As the thick slab continued eastward under the Colorado Plateau it triggered uplift of at least 2 kilometers and built several monoclines. Upon reaching the Rockies, compressional forces squeezed the crust, which responded by developing high angle faults along which igneous and metamorphic basement rock was uplifted. These crystalline blocks form the cores of many of the high peaks that comprise this mountain belt.

Sierra Nevada

Colorado Plateau

Laramide Rockies

Great Plains

Bouyant subduction of Farallon plate

Compressional stress

Partial melting

Over-thicken slab of oceanic crust

Asthenosphere

FIGURE 10.32 **The Principle of Isostasy** This drawing shows how wooden blocks of different thicknesses float in water. In a similar manner, thick sections of crustal material float higher than thinner crustal slabs.

FIGURE 10.32 **The Principle of Isostasy** This drawing shows how wooden blocks of different thicknesses float in water. In a similar manner, thick sections of crustal material float higher than thinner crustal slabs.

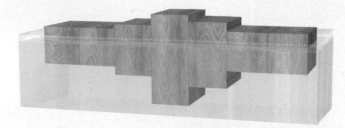

The Principle of Isostasy

During the 1840s, researchers discovered that Earth's low-density crust "floats" on top of the high-density, deformable rocks of the mantle. The concept of a floating crust in gravitational balance is called **isostasy**. The principle of isostasy helps us understand many large-scale variations on Earth's surface—from towering mountains to deep-ocean basins.

One way to explore the concept of isostasy is to envision a series of wooden blocks of different heights floating in water, as shown in **FIGURE 10.32**. Note that the thicker wooden blocks float higher than the thinner blocks. Similarly, compressional mountains stand high above the surrounding terrain because crustal thickening creates buoyant crustal "roots" that extend deep into the supporting material below. Thus, lofty mountains such as the Himalayas are much like the thicker wooden blocks shown in Figure 10.32.

How is isostasy related to elevation changes? Visualize what would happen if another small block of wood were placed atop one of the blocks in Figure 10.32. The combined block would sink until it reached a new isostatic (gravitational) balance. At this point, the top of the combined block would be higher than before, and the bottom would be lower. This process of establishing a new level of gravitational balance by loading or unloading is called **isostatic adjustment**.

Applying the concept of isostatic adjustment, we should expect that when weight is added to the crust, it will respond by subsiding and will rebound when weight is removed. (Visualize what happens when a ship's cargo is loaded or unloaded.) Evidence for crustal subsidence followed by crustal rebound is provided by Ice Age glaciers. When continental ice sheets occupied portions of North America during the Pleistocene epoch, the added weight of 3-kilometer- (2 mile-) thick masses of ice caused downwarping of Earth's crust by hundreds of meters. In the 8000 years since this last ice sheet melted, gradual uplift of as much as 330 meters (1000 feet) has occurred in Canada's Hudson Bay region, where the thickest ice had accumulated.

One of the consequences of isostatic adjustment is that, as erosion lowers a mountain range, the crust rises in response to the reduced load (**FIGURE 10.33**). The processes of uplift and erosion continue until the mountain block reaches "normal" crustal thickness. When this occurs, these once-elevated structures will be near sea level, and the once-deeply buried interior of the mountain will be exposed at the surface. In addition, as mountains are worn down, the eroded sediment is deposited on adjacent landscapes, causing these areas to subside (see Figure 10.33).

How High Is Too High?

Where compressional forces are great, such as those driving India into Asia, lofty mountains such as the Himalayas result. Is there

SmartFigure 10.33 The Effects of Isostatic Adjustment and Erosion on Mountainous Topography This sequence illustrates how the combined effects of erosion and isostatic adjustment result in a thinning of the crust in mountainous regions.

When compressional mountains are young they are composed of thick, low density crustal rocks that float on the denser asthenosphere below.

As erosion lowers the mountains, the crust rises in response to the reduced load in order to maintain isostatic balance.

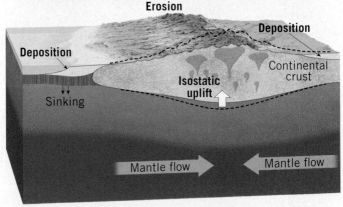

Erosion and uplift continue until the mountains reach "normal" crustal thickness.

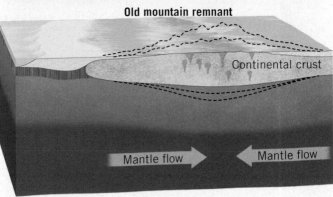

a limit on how high a mountain can rise? As mountaintops are elevated, gravity-driven processes such as erosion and mass wasting are accelerated, carving the deformed strata into rugged landscapes. Equally important, however, is the fact that gravity also acts on the rocks within these massive structures. The higher the mountain, the greater the downward force on rocks near the base. Eventually, the rocks deep within the developing mountain, which are relatively warm and weak, begin to flow laterally, as shown in **FIGURE 10.34**. This is analogous to what happens when a ladle of very thick pancake batter is poured onto a hot griddle. Similarly, mountains are altered by a process called **gravitational collapse**, which involves ductile spreading at depth and normal faulting and subsidence in the upper, brittle portion of Earth's crust.

Considering these factors, what keeps the Himalayas standing? Simply, the horizontal compressional forces that are driving India into Asia are greater than the vertical force of gravity. However, when India's northward trek ends, the downward pull of gravity, as well as weathering and erosion, will become the dominant forces acting on this mountainous region.

10.7 CONCEPT CHECKS

1 Define *isostasy*.

2 Give one example of evidence that supports the concept of crustal uplift.

3 What happens to a floating object when weight is added? Subtracted?

4 Briefly describe how the principle of isostatic adjustment applies to changes in the elevations of mountains.

5 Explain the process whereby mountainous regions experience gravitational collapse.

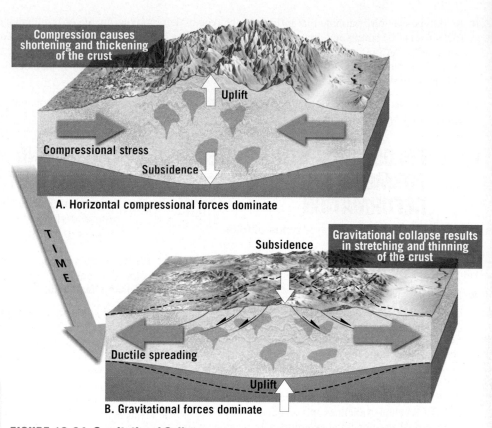

FIGURE 10.34 Gravitational Collapse Without compressional forces to support them, mountains gradually collapse under their own weight. Gravitational collapse involves normal faulting in the upper, brittle portion of the crust and ductile spreading in the warm, weak rocks at depth.

10 CONCEPTS IN REVIEW | Crustal Deformation and Mountain Building

10.1 CRUSTAL DEFORMATION

Describe the three types of differential stress. Differentiate stress from strain. Compare and contrast brittle and ductile deformation.

KEY TERMS: deformation, rock structure (geologic structure), stress, confining pressure, differential stress, compressional stress, tensional stress, shear, strain, elastic deformation, brittle deformation, ductile deformation

- Rock structures are generated when rocks are deformed by bending or breaking due to differential stress. Crustal deformation produces geologic structures that include folds, faults, and joints.

- Stress is the force that drives rock deformation. Stress may be the same magnitude in every direction, in which case we call it confining pressure. Alternatively, when the amount of stress coming from one direction is greater in magnitude than the stress coming from another direction, we

call it differential stress. There are three main types of differential stress: compressional, tensional, and shear stress.

- When the stresses are greater than the strength of a rock, the rock will deform, usually by folding or faulting.

- Elastic deformation is a temporary stretching of the chemical bonds in a rock. When the stress is released, the bonds snap back to their original lengths. When stress is greater than the strength of the bonds, the rock deforms in either a brittle or ductile fashion. Brittle deformation is the breaking of rocks into smaller pieces, whereas ductile deformation is flow in the manner of modeling clay or warm wax.

- Which type of deformation (ductile or brittle) occurs depends on four variables. One is temperature: If the rock is warm, it is more likely to flow than break. Second, if the rock is under high confining pressure, it is more likely to flow than break. Thus, conditions in the shallow crust are more likely to result in brittle deformation, while ductile deformation is more likely in the deeper crust.

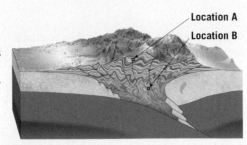

■ Rock type is a third variable that influences the type of deformation (ductile or brittle) that occurs. Igneous rocks tend to be strong and are more likely to deform in a brittle fashion, whereas many sedimentary layers will readily deform in a ductile fashion because they are weaker. The fourth variable is the amount of time over which the stress is applied. Sudden applications of stress are more likely to break rock, while slow applications of the same stress encourage ductile flow.

Q Examine this illustration of tectonic compression. Compare the deformation you would expect at locations A and B.

10.2 FOLDS: ROCK STRUCTURES FORMED BY DUCTILE DEFORMATION

List and describe five types of folds.

KEY TERMS: fold, anticline, syncline, dome, basin, monocline

■ Folds are wavelike undulations in layered rocks that develop through ductile deformation in rocks undergoing compressional stress.

■ Folds may be described in terms of their geometric configuration: If the limbs of a fold dip away from the hinge, the fold has an overall arch-like structure and is called an anticline. If the limbs of a fold dip upward, the fold has an overall trough-like structure and is called a syncline. Anticlines and synclines may be symmetrical, asymmetrical, overturned, or recumbent.

■ The shape of a fold does not necessarily correlate to the shape of the landscape above it. Rather, surface topography usually reflects patterns of differential weathering. A fold is said to plunge when its axis penetrates the ground at an angle. This results in a V-shaped outcrop pattern of the folded layers.

■ Domes and basins are large folds that produce roughly circular-shaped outcrop patterns. The overall shape of a dome or basin is like a saucer or a bowl, either right-side-up (basin) or inverted (dome).

■ Monoclines are large step-like folds in otherwise horizontal strata that result from subsurface faulting. Imagine a carpet draped over a staircase to envision how the strata can go from horizontal to tilted and back to horizontal again.

Q What do domes and anticlines have in common? What sets them apart from one another? Similarly, compare synclines and basins.

10.3 FAULTS AND JOINTS: STRUCTURES FORMED BY BRITTLE DEFORMATION

Sketch and briefly describe the relative motion of rock bodies located on opposite sides of normal, reverse, and thrust faults as well as both types of strike-slip faults.

KEY TERMS: fault, dip-slip fault, hanging wall block, footwall block, fault scarp, normal fault, fault-block mountain, horst, graben, half-graben, detachment fault, reverse fault, thrust fault, strike-slip fault, transform fault, joint

■ Faults and joints are fractures in rock that form through brittle deformation.

■ The offset on a fault may be determined by comparing the blocks of rock on either side of the fault surface. If the movement is in the direction of the fault's dip (or inclination), then the rock above the fault plane is the hanging wall block, and the rock below the fault is the footwall block. If the hanging wall moves *down* relative to the footwall, the fault is a normal fault. If the hanging wall moves *up* relative to the footwall, the fault is a reverse fault. Large normal faults with low dip angles are detachment faults. Large reverse faults with low dip angles are thrust faults.

■ Faults that intersect Earth's surface may produce a "step" in the land known as a fault scarp. Areas of tectonic extension, such as the Basin and Range Province, produce fault-block mountains—horsts separated by neighboring grabens or half-grabens.

■ Areas of tectonic compression, such as mountain belts, are dominated by reverse faults that shorten the crust horizontally while thickening it vertically.

■ Strike-slip faults have most of their movement in a horizontal direction along the trend of the fault trace. Transform faults are strike-slip faults that serve as tectonic boundaries between lithospheric plates.

■ Joints form in the shallow crust when rocks are stressed under brittle conditions. They facilitate groundwater movement and mineralization of economic resources, and they may result in human hazards.

East Africa

Q Examine the diagrams depicting the structure of East Africa and the Canadian Rockies. Characterize the type of faulting found at each location and infer the differential stresses and the type of plate boundary that were responsible.

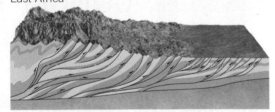

Canadian Rockies

10.4 MOUNTAIN BUILDING

Locate and name Earth's major mountain belts on a world map.

KEY TERMS: orogenesis, compressional mountain

■ Orogenesis is the making of mountains. Most orogenesis occurs along convergent plate boundaries, where compressional forces cause folding and faulting of the rock, thickening the crust vertically and shortening it horizontally. Some mountain belts are very old, while others are actively forming today.

Q Look at the South American plate on the map. Explain why the Andes Mountains are located on the western margin of South America rather than the eastern margin.

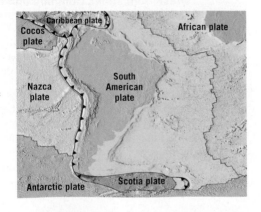

10.5 SUBDUCTION AND MOUNTAIN BUILDING

Sketch a cross section of an Andean-type mountain belt and describe how its major features are generated.

KEY TERMS: accretionary wedge, forearc basin

- Subduction leads to orogenesis. If a subducted plate is overridden by oceanic lithosphere, island arc–type mountain building results, with a thick accumulation of erupted volcanic rocks mixed with sediment scraped off the subducted plate. Andean-type mountain building occurs where subduction takes place beneath continental lithosphere.
- Release of water from a subducted slab triggers melting in the wedge of mantle above the slab. These primary magmas have a basaltic (mafic) composition. They rise to the base of the continental crust and pond. Fractional crystallization then separates ferromagnesian minerals from a melt that grows increasingly granitic (felsic) in composition.
- The evolved magma rises further and may either cool and crystallize below the surface to form a batholith or erupt at the surface and form a continental volcanic arc.
- Sediment scraped off the subducting plate builds up in an accretionary wedge. The thickness of the wedge varies in accordance with the amount of sediment available. Between the accretionary wedge and the volcanic arc is a relatively calm site of sedimentary deposition, the forearc basin.
- The geography of central California preserves an accretionary wedge (Coast Ranges), a forearc basin (Great Valley), and the roots of a continental volcanic arc (Sierra Nevada).

Q Over geologic time, what processes would have to take place in the Andes to make the landscape of western South America resemble the landscape of modern California?

10.6 COLLISIONAL MOUNTAIN BELTS

Summarize the stages in the development of an Alpine-type mountain belt such as the Appalachians.

KEY TERMS: terrane, microcontinent, suture

- Terranes are relatively small crustal fragments (microcontinents, volcanic island arcs, or oceanic plateaus). Terranes may be accreted to continents when subduction brings them to a trench, but they cannot subduct due to their relatively low density. The terranes are "peeled off" the subducted slab and thrust onto the leading edge of the continent.
- The Himalayas and Appalachians have similar origins in the collision of continents formerly separated by (now-subducted) ocean basins. The Appalachians are older and were caused by the collision of ancestral North America with ancestral Africa more than 250 million years ago. In contrast, the Himalayas are younger, having been formed by the collision of India and Eurasia starting around 50 million years ago. They are still rising today.

10.7 WHAT CAUSES EARTH'S VARIED TOPOGRAPHY?

Explain the principle of isostasy and how it contributes to the elevated topography of young mountain belts like the Himalayas.

KEY TERMS: isostasy, isostatic adjustment, gravitational collapse

- Isostasy is the principle that the elevation of the crust can change in accordance with gravitational balance. If additional weight is placed on the crust (a flood basalt, a thick layer of sediments, or an ice sheet), the crust will sink into the mantle until it comes to equilibrium with the extra load. If the new mass is removed, the crust will rebound upward.
- When mountains form and rock mass is pushed up to the height of a young compressional mountain belt such as the Himalayas, the rocks at the base of the mountains become warmer and weaker. As the overlying rock mass pushes down on them through gravity, these deeper rocks flow out of the way. This causes the mountain belt to collapse outward into a wider, lower-elevation region.

Q Based on the principle of isostasy, predict what would happen to the elevation of the highest peaks in a mountain range if rivers and glaciers eroded deep valleys through the mountains and removed large amounts of rock.

GIVE IT SOME **THOUGHT**

1. Is granite or mica schist more likely to fold or flow rather than fracture when subjected to differential stress? Explain.
2. Refer to the accompanying diagrams to answer the following:
   a. What type of dip-slip fault is shown in Diagram 1? Were the dominant forces during faulting tensional, compressional, or shear?
   b. What type of dip-slip fault is shown in Diagram 2? Were the dominant forces during faulting tensional, compressional, or shear?
   c. Match the correct pair of arrows in Diagram 3 to the faults in Diagrams 1 and 2.

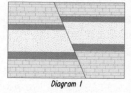
Diagram 1

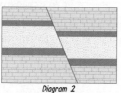
Diagram 2

Diagram 3

3. Refer to the accompanying photo to answer the following:
   a. The white line shows the approximate location of a fault that displaced these furrows created by a plow. What type of fault caused the offset shown?
   b. Is this a right-lateral or left-lateral fault? Explain.

USGS

4. With which of the three types of plate boundaries does normal faulting predominate? Thrust faulting? Strike-slip faulting?

5. Write a brief statement describing each of the accompanying photos, using terms from the following list: strike-slip, dip-slip, normal, reverse, right-lateral, left-lateral.

A.

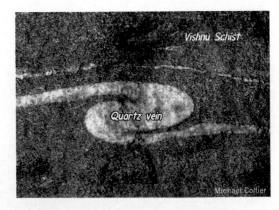

B.

6. The accompanying photo, taken near the bottom of the Grand Canyon, shows a quartz vein that has been deformed.

  a. What type of deformation is exhibited—ductile or brittle?

  b. Did this deformation most likely occur near Earth's surface or at great depth?

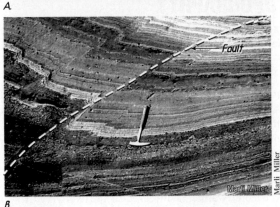

7. Refer to the accompanying photo to answer the following:

  a. Name the type of fold shown.

  b. Would you describe this fold as symmetrical or asymmetrical?

c. What name is given to the part of the fold labeled A?

d. Is point B located along the fold line, crest line, or axial plane of this particular fold?

8. Suppose that a sliver of oceanic crust were discovered in the interior of a continent. Would this refute the theory of plate tectonics? Explain.

9. Refer to the accompanying map, which shows the location of the Galapagos Rise and the Rio Grande Rise to answer the following questions:

  a. Compare the continental margin of the west coast of South America with the continental margin along the east coast.

  b. Based on your answer to the question above, is the Galapagos Rise or the Rio Grande Rise more likely to end up accreted to a continent? Explain your choice.

  c. In the distant future, how might a geologist determine that this accreted landmass is distinct from the continental crust to which it accreted?

10. The Ural Mountains exhibit a north–south orientation through Eurasia. How does the theory of plate tectonics explain the existence of this mountain belt in the interior of an expansive landmass?

11. Briefly describe the major differences between the evolution of the Appalachian Mountains and the North American Cordillera.

12. Which of the accompanying sketches best illustrates an Andean-type orogeny, a Cordilleran-type orogeny, and an Alpine-type orogeny?

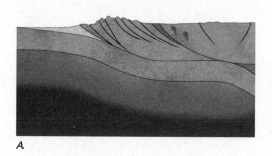

A.

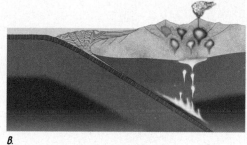

B.

C.

EXAMINING THE **EARTH SYSTEM**

1. A good example of the interaction among Earth's spheres is the influence of mountains on climate. Examine the accompanying temperature graph for the cities of Seattle and Spokane, Washington. Notice on the inset map that mountains (the Cascades) separate these two cities. The prevailing wind direction in the region is from west to east. (a) Contrast the summer and winter temperatures that occur at each city. Why are they different? *Hint:* Check out the sections "Land and Water" and "Geographic Position" in Chapter 16. The annual rainfall at Spokane (16.6 inches) is less than half that for Seattle (37.1 inches). Can you explain why?

2. The Cascades have had a profound effect on the amount and type of plant and animal life (biosphere) that inhabit the region around Spokane and Seattle. Provide several specific examples to support this statement.

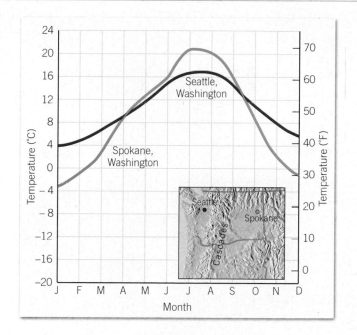

MasteringGeology™

UNIT FOUR | DECIPHERING EARTH'S HISTORY

11

Geologic Time

Rafting the Colorado River in Marble Canyon Shortly Before Entering the Grand Canyon Millions of years of Earth history are exposed in these walls of rock. (Photo by Michael Collier)

In the eighteenth century, James Hutton recognized the immensity of Earth history and the importance of time as a component in all geologic processes. In the nineteenth century, others effectively demonstrated that Earth had experienced many episodes of mountain building and erosion, which must have required great spans of geologic time. Although these pioneering scientists understood that Earth was very old, they had no way of knowing its true age. Was it tens of millions, hundreds of millions, or even billions of years old?

Because they didn't know exactly how old Earth was, they developed a geologic time scale that showed the sequence of events based on relative dating principles. What were these principles? What part did fossils play? With the discovery of radioactivity and the development of radiometric dating techniques, geologists can now assign fairly accurate dates to many of the events in Earth history. What is radioactivity? Why is it a good "clock" for dating the geologic past? This chapter answers these questions.

11.1 | A BRIEF HISTORY OF GEOLOGY

Explain the principle of uniformitarianism and discuss how it differs from catastrophism.

The hiker in **FIGURE 11.1** is perched on the rim of the Grand Canyon. Today we know that the strata beneath the hiker represent hundreds of millions of years of Earth history. We also have the knowledge and skills that are necessary to unravel the complex story contained in these rocks. Such an understanding is a relatively recent accomplishment.

The nature of our Earth—its materials and processes—has been a focus of study for centuries. However, the late 1700s is generally regarded as the beginning of modern geology. It was during this time that James Hutton published his important work *Theory of the Earth*. Prior to that time, a great many explanations about Earth history relied on supernatural events.

Catastrophism

In the mid-1600s, James Ussher, Anglican Archbishop of Armagh, Primate of All Ireland, published a work that had immediate and profound influence. A respected scholar of the Bible, Ussher constructed a chronology of human and Earth history in which he determined that Earth was only a few thousand years old, having been created in 4004 B.C. Ussher's treatise earned widespread acceptance among Europe's scientific and religious leaders, and his chronology was soon printed in the margins of the Bible itself.

During the 1600s and 1700s, the doctrine of **catastrophism** strongly influenced people's thinking about Earth. Briefly stated, catastrophists believed that Earth's varied landscapes had been fashioned primarily by great catastrophes. Features such as mountains and canyons, which today we know take great periods of time to form, were explained as having been produced by sudden and often worldwide disasters of unknowable causes that no longer operate. This philosophy was an attempt to fit the rate of Earth's processes to the prevailing ideas on Earth's age.

The Birth of Modern Geology

Modern geology began in the late 1700s, when James Hutton, a Scottish physician and gentleman farmer, published his *Theory of the Earth*. In this work, Hutton put forth a fundamental principle that is a pillar of geology today: **uniformitarianism**. It simply states that *the physical, chemical, and biological laws that operate today have also operated in the geologic past*. This means that the forces and processes that we observe presently shaping our planet have been at work for a very long time. Thus, to understand ancient rocks, we must first understand

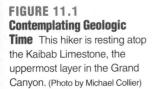

FIGURE 11.1
Contemplating Geologic Time This hiker is resting atop the Kaibab Limestone, the uppermost layer in the Grand Canyon. (Photo by Michael Collier)

present-day processes and their results. This idea is commonly expressed by saying "The present is the key to the past."

Prior to Hutton's *Theory of the Earth*, no one had effectively demonstrated that geologic processes occur over extremely long periods of time. However, Hutton persuasively argued that processes that appear to be weak and slow acting can, over long spans of time, produce effects that are just as great as those resulting from sudden catastrophic events. Unlike his predecessors, Hutton cited verifiable observations to support his ideas.

For example, when he argued that mountains are sculpted and ultimately destroyed by weathering and the work of running water and that their wastes are carried to the oceans by processes that can be observed, Hutton said, "We have a chain of facts which clearly demonstrates that the materials of the wasted mountains have traveled through the rivers"; and, further, "There is not one step in all this progress that is not to be actually perceived." He went on to summarize these thoughts by asking a question and immediately providing the answer: "What more can we require? Nothing but time."

Geology Today

The basic tenets of uniformitarianism are just as viable today as in Hutton's day. We realize more strongly than ever before that the present gives us insight into the past and that the physical, chemical, and biological laws that govern geologic processes remain unchanging through time. However, we also understand that the doctrine should not be taken too literally. To say that geologic processes in the past were the same as those occurring today is not to suggest that they always had the same relative importance or that they operated at precisely the same rate. Moreover, some important geologic processes

are not currently observable, but evidence that they occur is well established. For example, we know that Earth has experienced impacts from large meteorites even though we have no human witnesses. Such events altered Earth's crust, modified its climate, and strongly influenced life on the planet.

The acceptance of the concept of uniformitarianism, however, meant the acceptance of a very long history for Earth. Although Earth's processes vary in their intensity, they still take a long time to create or destroy major landscape features. For example, geologists have established that mountains once existed in portions of present-day Minnesota, Wisconsin, Michigan, and Manitoba. Today, the region consists of low hills and plains. Erosion (processes that wear away land) gradually destroyed those peaks. The rock record contains evidence that shows Earth has experienced *many cycles* of mountain building and erosion. Concerning the ever-changing nature of Earth through great expanses of geologic time, Hutton made a statement that was to become his most famous. In concluding his classic 1788 paper published in the *Transactions of the Royal Society of Edinburgh*, he stated, "The results, therefore, of our present enquiry is, that we find no vestige of a beginning—no prospect of an end."

It is important to remember that although many features of our physical landscape may seem to be unchanging over the decades we observe them, they are nevertheless changing—but on time scales of hundreds, thousands, or even many millions of years.

11.1 CONCEPT CHECKS

1 Contrast catastrophism and uniformitarianism.

2 How did each philosophy view the age of Earth?

11.2 CREATING A TIME SCALE: RELATIVE DATING PRINCIPLES

Distinguish between numerical dates and relative dates and apply relative dating principles to determine a time sequence of geologic events.

The Importance of a Time Scale

Like the pages in a long and complicated history book, rocks record the geologic events and changing life-forms of the past. The book, however, is not complete. Many pages, especially in the early chapters, are missing. Others are tattered, torn, or smudged. Yet enough of the book remains to allow much of the story to be deciphered.

Interpreting Earth history is a prime goal of the science of geology. Like a modern-day sleuth, a geologist must interpret the clues found preserved in the rocks. By studying rocks and the features they contain, geologists can unravel the complexities of the past.

Geologic events by themselves, however, have little meaning until they are put into a time perspective. Studying history, whether it is the Civil War or the age of dinosaurs, requires a calendar. Among geology's major contributions to human

knowledge are the *geologic time scale* and the discovery that Earth history is exceedingly long.

Numerical and Relative Dates

The geologists who developed the geologic time scale revolutionized the way people think about time and perceive our planet. They learned that Earth is much older than anyone had previously imagined and that its surface and interior have been changed over and over again by the same geologic processes that operate today.

Numerical Dates During the late 1800s and early 1900s, attempts were made to determine Earth's age. Although some of the methods appeared promising at the time, none of these early efforts proved to be reliable. What these scientists were seeking was a **numerical date**. Such

Dennis Tasa

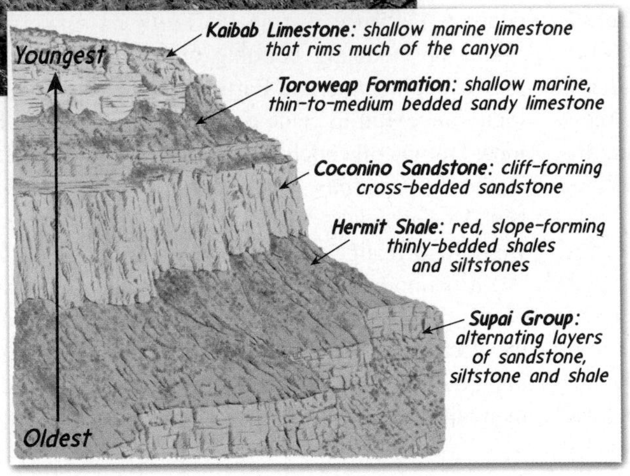

Geologist's Sketch

Youngest → **Kaibab Limestone:** *shallow marine limestone that rims much of the canyon*

Toroweap Formation: *shallow marine, thin-to-medium bedded sandy limestone*

Coconino Sandstone: *cliff-forming cross-bedded sandstone*

Hermit Shale: *red, slope-forming thinly-bedded shales and siltstones*

Supai Group: *alternating layers of sandstone, siltstone and shale*

Oldest

FIGURE 11.2 Superposition Based on the principle of superposition, among these layers in the upper portion of the Grand Canyon, the Supai Group is oldest, and the Kaibab Limestone is youngest.

dates cannot tell us how long ago something took place, only that it followed one event and preceded another. The relative dating techniques that were developed are valuable and still widely used. Numerical dating methods did not replace these techniques; rather, they supplemented them. To establish a relative time scale, a few basic principles or rules had to be discovered and applied. Although they may seem obvious to us today, they were major breakthroughs in thinking at the time, and their discovery was an important scientific achievement.

Principle of Superposition

Nicolas Steno, a Danish anatomist, geologist, and priest (1638–1686), is credited with being the first to recognize a sequence of historical events in an outcrop of sedimentary rock layers. Working in the mountains of western Italy, Steno applied a very simple rule that has come to be the most basic principle of relative dating—the **principle of superposition** (*super* = above; *positum* = to place). The principle simply states that in an undeformed sequence of sedimentary rocks, each bed is older than the one above and younger than the one below. Although it may seem obvious that a rock layer could not be deposited with nothing beneath it for support, it was not until 1669 that Steno clearly stated this principle.

This rule also applies to other surface-deposited materials, such as lava flows and beds of ash from volcanic eruptions. Applying superposition to the beds exposed in the upper portion of the Grand Canyon, we can easily place the layers in their proper order. Among those that are pictured in **FIGURE 11.2**, the sedimentary rocks in the Supai Group are the oldest, followed in order by the Hermit Shale, Coconino Sandstone, Toroweap Formation, and Kaibab Limestone.

dates specify the actual number of years that have passed since an event occurred. Today, our understanding of radiometric dating techniques allows us to accurately determine numerical dates for rocks that represent important events in Earth's distant past. We will study these techniques later in this chapter. Prior to the development of radiometric dating, geologists had no reliable method of numerical dating and had to rely solely on relative dating.

Relative Dates When we place rocks in their proper *sequence of formation*—which formed first, second, third, and so on—we are establishing **relative dates**. Such

Principle of Original Horizontality

Steno is also credited with recognizing the importance of another basic principle, called the **principle of original horizontality**. It states that layers of sediment are generally deposited in a horizontal position. Thus, if we observe rock layers that are flat, it means they have not been disturbed and still have their *original* horizontality. The layers in the Grand Canyon illustrate this in Figures 11.1 and 11.2. But if they are folded or inclined at a steep angle, they must have been moved into that position by crustal disturbances sometime *after* their deposition (**FIGURE 11.3**).

FIGURE 11.3 Original Horizontality Most layers of sediment are deposited in a nearly horizontal position. When we see strata that are folded or tilted, we can assume that they were moved into that position by crustal disturbances *after* their deposition. (Photo by Marco Simoni/Robert Harding World Imagery)

Principle of Lateral Continuity

The **principle of lateral continuity** refers to the fact that sedimentary beds originate as continuous layers that extend in all directions until they eventually grade into a different type of sediment or until they thin out at the edge of the basin of deposition (**FIGURE 11.4**). For example, when a river creates a canyon, we can assume that identical or similar strata on opposite sides once spanned the canyon. Although rock outcrops may be separated by a considerable distance, the principle of lateral continuity tells us those outcrops once formed a continuous layer. This principle allows geologists to relate rocks in isolated outcrops to one another. Combining the principles of lateral continuity and superposition lets us extend relative age relationships over broad areas. This process, called *correlation*, is examined later in the chapter.

Layer ends by thinning at margin of sedimentary basin

Layer ends by grading into a different kind of sediment

Lateral continuity allows us to infer that the layers were originally continuous across the canyon

FIGURE 11.4 Lateral Continuity Sediments are deposited over a large area in a continuous sheet. Sedimentary strata extend continuously in all directions until they thin out at the edge of a depositional basin or grade into a different type of sediment.

Principle of Cross-Cutting Relationships

FIGURE 11.5 shows a mass of rock that is offset by a fault, a fracture in rock along which displacement occurs. It is clear that the rocks must be older than the fault that broke them. The **principle of cross-cutting relationships** states that geologic features that cut across rocks must form *after* the rocks they cut through. Igneous intrusions provide another example. The dike shown in **FIGURE 11.6** is a tabular mass of igneous rock that cuts through the surrounding rocks. The magmatic heat from igneous intrusions often creates a narrow "baked" zone of contact metamorphism on the adjacent rock, also indicating that the intrusion occurred after the surrounding rocks were in place.

Fault

FIGURE 11.5 Cross-cutting Fault The rocks are older than the fault that displaced them. (Photo by Morley Read/Alamy Images)

EYE ON EARTH

This image shows West Cedar Mountain in southern Utah. The gray rocks are shale that originated as muddy river delta deposits. The sediments composing the orange sandstone were deposited by a river.

QUESTION 1 *Place the events related to the geologic history of this area in proper sequence. Explain your logic. Include the following: uplift, sandstone, erosion, and shale.*

QUESTION 2 *What term is applied to the type of dates you established for this site?*

Shale

Sandstone

Michael Collier

Dikes

FIGURE 11.6 Cross-cutting Dike An igneous intrusion is younger than the rocks that are intruded.
(Jonathan.s.kt)

Principle of Inclusions

Sometimes inclusions can aid in the relative dating process. **Inclusions** are fragments of one rock unit that have been enclosed within another. The principle of inclusions is logical and straightforward: The rock mass adjacent to the one containing the inclusions must have been there first in order to provide the rock fragments. Therefore, the rock mass that contains the inclusions is the younger of the two. For example, when magma intrudes into surrounding rock, blocks of the surrounding rock may become dislodged and incorporated into the magma. If these pieces do not melt, they remain as inclusions, known as *xenoliths*. In another example, when sediment is deposited atop a weathered mass of bedrock, pieces of the weathered rock become incorporated into the younger sedimentary layer (**FIGURE 11.7**).

SmartFigure 11.7 Inclusions The rock that contains the inclusions is younger than the inclusions.

These inclusions of igneous rock contained in the adjacent sedimentary layer indicate the sediments were deposited atop the weathered igneous mass and thus are younger.

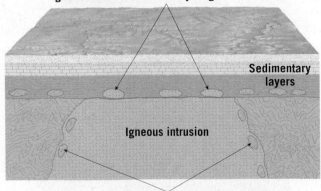

Sedimentary layers

Igneous intrusion

Xenoliths are inclusions in an igneous intrusion that form when pieces of surrounding rock are incorporated into magma.

Unconformities

When we observe layers of rock that have been deposited essentially without interruption, we call them **conformable**. Particular sites exhibit conformable beds representing certain spans of geologic time. However, no place on Earth has a complete set of conformable strata.

Throughout Earth history, the deposition of sediment has been interrupted over and over again. All such breaks in the rock record are termed unconformities. An **unconformity** represents a long period during which deposition ceased, erosion removed previously formed rocks, and then deposition resumed. In each case, uplift and erosion are followed by subsidence and renewed sedimentation. Unconformities are important features because they represent significant geologic events in Earth history. Moreover, their recognition helps us identify what intervals of time are not represented by strata and thus are missing from the geologic record.

There are three basic types of unconformities: angular unconformities, disconformities, and nonconformities.

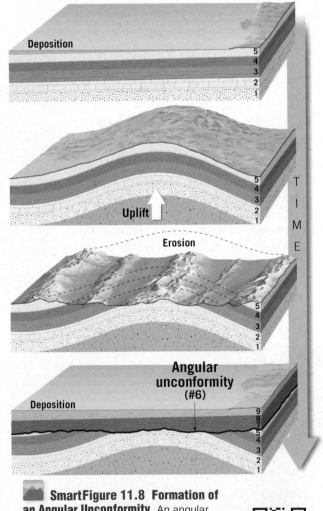

Deposition

Uplift

Erosion

Angular unconformity (#6)

Deposition

TIME

SmartFigure 11.8 Formation of an Angular Unconformity An angular unconformity represents an extended period during which deformation and erosion occurred.

FIGURE 11.9 Siccar Point, Scotland James Hutton studied this famous unconformity in the late 1700s. (Photo by Marli Miller)

Angular Unconformity

Perhaps the most easily recognized unconformity is an **angular unconformity**. It consists of tilted or folded sedimentary rocks that are overlain by younger, more flat-lying strata. An angular unconformity indicates that during the pause in deposition, a period of deformation (folding or tilting) and erosion occurred (**FIGURE 11.8**).

When James Hutton studied an angular unconformity in Scotland more than 200 years ago, it was clear to him that it represented a major episode of geologic activity (**FIGURE 11.9**). He and his colleagues also appreciated the immense time span implied by such relationships. When a companion later wrote of their visit to this site, he stated that "the mind seemed to grow giddy by looking so far into the abyss of time."

Disconformity

A **disconformity** is a gap in the rock record that represents a period during which erosion rather than deposition occurred. Imagine that a series of sedimentary layers are deposited in a shallow marine setting. Following this period of deposition, sea level falls or the land rises, exposing some the sedimentary layers. During this span, when the sedimentary beds are above sea level, no new sediment accumulates, and some of the existing layers are eroded away. Later sea level rises or the land subsides, submerging the landscape. Now the surface is again below sea level, and a new series of sedimentary beds is deposited. The boundary separating the two sets of beds is a disconformity—a span for which there is no rock record (**FIGURE 11.10**). Because

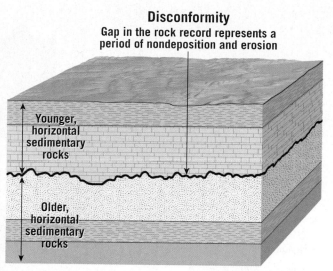

Disconformity
Gap in the rock record represents a period of nondeposition and erosion

FIGURE 11.10
Disconformity The layers on either side of this gap in the rock record are essentially parallel.

the layers above and below a disconformity are parallel, these features are sometimes difficult to identify unless you notice evidence of erosion such as a buried stream channel.

Nonconformity

The third basic type of unconformity is a **nonconformity**, in which younger sedimentary strata overlie older metamorphic or intrusive igneous rocks (**FIGURE 11.11**). Just as angular unconformities and some disconformities imply crustal movements, so too do nonconformities. Intrusive igneous masses and metamorphic rocks originate far below the surface. Thus, for a nonconformity to develop, there must be a period of uplift and the erosion of overlying rocks. Once exposed at the surface, the igneous or metamorphic rocks are subjected to weathering and erosion prior to subsidence and the renewal of sedimentation.

Unconformities in the Grand Canyon

The rocks exposed in the Grand Canyon of the Colorado River represent a tremendous span of geologic history. It is a wonderful place to take a trip through time. The canyon's colorful strata record a long history of sedimentation in a variety of

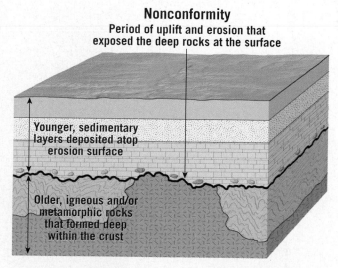

Nonconformity
Period of uplift and erosion that exposed the deep rocks at the surface

FIGURE 11.11
Nonconformity Younger sedimentary rocks rest atop older metamorphic or igneous rocks.

FIGURE 11.12 Cross Section of the Grand Canyon All three types of unconformities are present. (Center photo by Marli Miller; other photos by E. J. Tarbuck)

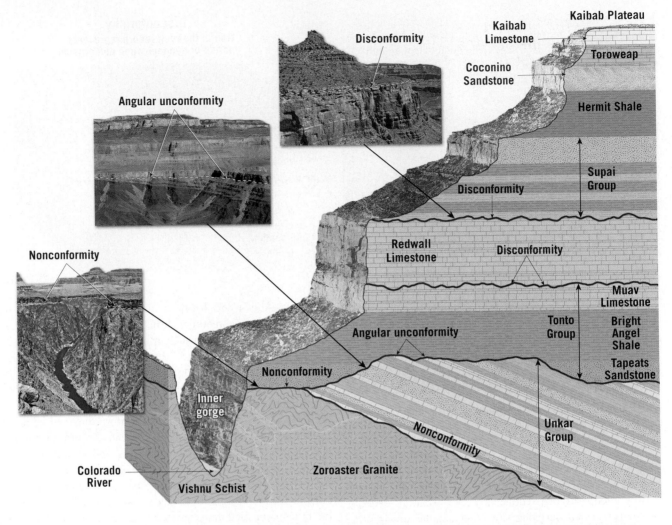

environments—advancing seas, rivers and deltas, tidal flats and sand dunes. But the record is not continuous. Unconformities represent vast amounts of time that have not been recorded in the canyon's layers. **FIGURE 11.12** is a geologic cross section of the Grand Canyon. All three types of unconformities can be seen in the canyon walls.

Applying Relative Dating Principles

By applying the principles of relative dating to the hypothetical geologic cross section shown in **FIGURE 11.13**, the rocks and the events in Earth history they represent can be placed in their proper sequence. The statements in the figure

EYE ON EARTH

This close-up shows pieces of diorite (darkest rock) in granite. The thin white line is a vein of quartz. Think of a vein as a tiny dike. (Photo by Marli Miller)

QUESTION 1 What term is applied to the pieces of diorite?

QUESTION 2 Place the quartz vein, diorite, and granite in order from oldest to youngest.

Working out the geologic history of a hypothetical region

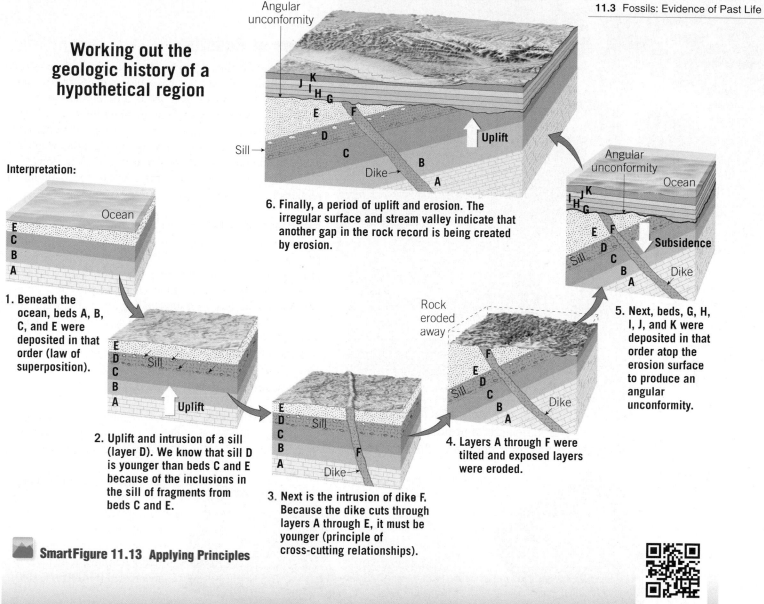

Interpretation:

1. Beneath the ocean, beds A, B, C, and E were deposited in that order (law of superposition).

2. Uplift and intrusion of a sill (layer D). We know that sill D is younger than beds C and E because of the inclusions in the sill of fragments from beds C and E.

3. Next is the intrusion of dike F. Because the dike cuts through layers A through E, it must be younger (principle of cross-cutting relationships).

4. Layers A through F were tilted and exposed layers were eroded.

5. Next, beds, G, H, I, J, and K were deposited in that order atop the erosion surface to produce an angular unconformity.

6. Finally, a period of uplift and erosion. The irregular surface and stream valley indicate that another gap in the rock record is being created by erosion.

SmartFigure 11.13 Applying Principles

summarize the logic used to interpret the cross section. In this example, we establish a relative time scale for the rocks and events in the area of the cross section. Remember, we do not know how many years of Earth history are represented, nor do we know how this area compares to any other.

11.2 CONCEPT CHECKS

1 Distinguish between numerical dates and relative dates.

2 Sketch and label four simple diagrams that illustrate each of the following: superposition, original horizontality, cross-cutting relationships, and inclusions.

3 What is the significance of an unconformity?

4 Distinguish among angular unconformity, disconformity, and nonconformity.

11.3 | FOSSILS: EVIDENCE OF PAST LIFE Define *fossil* and discuss the conditions that favor the preservation of organisms as fossils. List and describe various fossil types.

Fossils, the remains or traces of prehistoric life, are important inclusions in sediment and sedimentary rocks. They are basic and important tools for interpreting the geologic past. The scientific study of fossils is called **paleontology**. It is an interdisciplinary science that blends geology and biology in an attempt to understand all aspects of the succession of life over the vast expanse of geologic time. Knowing the nature of the life-forms that existed at a particular time helps researchers understand past environmental conditions. Further, fossils are important time indicators and play a key role in correlating rocks of similar ages that are from different places.

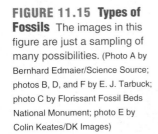

Excavating bones from pit 91. It is a site rich in unaltered Ice Age organisms. Scientists have been excavating here since 1915.

FIGURE 11.14 La Brea Tar Pits The fossils here are actual (unaltered) remains. (Excavation photo by AP Photo/ Reed Saxon; skeleton photo by Martin Shields/Alamy)

Skeleton of a mammoth, a prehistoric relative of modern elephant, from the La Brea tar pits.

FIGURE 11.15 Types of Fossils The images in this figure are just a sampling of many possibilities. (Photo A by Bernhard Edmaier/Science Source; photos B, D, and F by E. J. Tarbuck; photo C by Florissant Fossil Beds National Monument; photo E by Colin Keates/DK Images)

A. Petrified wood in Arizona's Petrified Forest National Park

B. This trilobite photo illustrates mold and cast

C. A fossil bee preserved as a thin carbon film

D. Impressions are common fossils and often show considerable detail

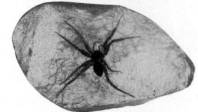

E. Spider in amber

F. Coprolite is fossil dung

Types of Fossils

Fossils are of many types. The remains of relatively recent organisms may not have been altered at all. Such objects as teeth, bones, and shells are common examples (FIGURE 11.14). Far less common are entire animals, flesh included, that have been preserved because of rather unusual circumstances. Remains of prehistoric elephants called mammoths that were frozen in the Arctic tundra of Siberia and Alaska are examples, as are the mummified remains of sloths preserved in a dry cave in Nevada.

Permineralization When mineral-rich groundwater permeates porous tissue such as bone or wood, minerals precipitate out of solution and fill pores and empty spaces, a process called *permineralization*. The formation of *petrified wood* involves permineralization with silica, often from a volcanic source such as a surrounding layer of volcanic ash. The wood is gradually transformed into chert, sometimes with colorful bands from impurities such as iron or carbon (FIGURE 11.15A). The word *petrified* literally means "turned into stone." Sometimes the microscopic details of the petrified structure are faithfully retained.

Molds and Casts Another common class of fossils is *molds* and *casts*. When a shell or another structure is buried in sediment and then dissolved by underground water, a *mold* is created. The mold faithfully reflects only the shape and surface marking of the organism; it does not reveal any information concerning its internal structure. If these hollow spaces are subsequently filled with mineral matter, *casts* are created (FIGURE 11.15B).

Carbonization and Impressions A type of fossilization called *carbonization* is particularly effective in preserving leaves and delicate animal forms. It occurs when fine sediment encases the remains of an organism. As time passes, pressure squeezes out the liquid and gaseous components and leaves behind a thin residue of carbon (FIGURE 11.15C). Black shale deposited as organic-rich mud in oxygen-poor environments often contains abundant carbonized remains. If the film of carbon is lost from a fossil preserved in fine-grained sediment, a replica of the surface, called an *impression*, may still show considerable detail (FIGURE 11.15D).

Amber Delicate organisms, such as insects, are difficult to preserve, and

How is paleontology different from archaeology?

People frequently confuse these two areas of study because a common perception of both paleontologists and archaeologists is of scientists carefully extracting important clues about the past from layers of rock or sediment. While it is true that scientists in both disciplines "dig" a lot, the focus of each is different.

Archaeology

Archaeologists help us learn about how our human ancestors met the challenges of life in the past.

Archaeologists focus on the material remains of past human life. These remains include both the objects used by people long ago, called *artifacts*, and the buildings and other structures associated with where people lived, called *sites*.

Paleontology

Paleontologists study fossils and are concerned with all life forms in the geologic past. These scientists are excavating the fossil remains of *Albertasaurus*, a carnivore similar to *Tyrannosaurus rex*.

Richard T. Nowitz/Science Source

Richard T. Nowitz/Science Source

Richard T. Nowitz/Science Source

consequently they are relatively rare in the fossil record. Not only must they be protected from decay but they must not be subjected to any pressure that would crush them. One way in which some insects have been preserved is in *amber*, the hardened resin of ancient trees. The spider in **FIGURE 11.15E** was preserved after being trapped in a drop of sticky resin. Resin sealed off the spider from the atmosphere and protected the remains from damage by water and air. As the resin hardened, a protective pressure-resistant case was formed.

Trace Fossils

In addition to the fossils already mentioned, there are numerous other types, many of them only traces of prehistoric life. Examples of such indirect evidence include:

- Tracks—animal footprints made in soft sediment that later turned into sedimentary rock.

- Burrows—tubes in sediment, wood, or rock made by an animal. These holes may later become filled with mineral matter and preserved. Some of the oldest-known fossils are believed to be worm burrows.

- Coprolites—fossil dung and stomach contents that can provide useful information pertaining to the size and food habits of organisms (**FIGURE 11.15F**).

- Gastroliths—highly polished stomach stones that were used in the grinding of food by some extinct reptiles.

Conditions Favoring Preservation

Only a tiny fraction of the organisms that lived during the geologic past have been preserved as fossils. The remains of an animal or a plant are normally destroyed. Under what circumstances are they preserved? Two special conditions appear to be necessary: rapid burial and the possession of hard parts.

When an organism perishes, its soft parts are usually quickly eaten by scavengers or decomposed by bacteria. Occasionally, however, the remains are buried by sediment. When this occurs, the remains are protected from the environment, where destructive processes operate. Rapid burial, therefore, is an important condition favoring preservation.

In addition, animals and plants have a much better chance of being preserved as part of the fossil record if they have hard parts. Although traces and imprints of soft-bodied animals such as jellyfish, worms, and insects exist, they are not common. Flesh usually decays so rapidly that preservation is exceedingly unlikely. Hard parts such as shells, bones, and teeth predominate in the record of past life.

Because preservation is contingent on special conditions, the record of life in the geologic past is biased. The fossil record of organisms with hard parts that lived in areas of sedimentation is quite abundant. However, we get only an occasional glimpse of the vast array of other life-forms that did not meet the special conditions favoring preservation.

11.3 CONCEPT CHECKS

1 Describe several ways that an animal or a plant can be preserved as a fossil.

2 List three examples of trace fossils.

3 What conditions favor the preservation of an organism as a fossil?

11.4 | CORRELATION OF ROCK LAYERS

Explain how rocks of similar age that are in different places can be matched up.

To develop a geologic time scale that is applicable to the entire Earth, rocks of similar age in different regions must be matched up. Such a task is referred to as **correlation**. Correlating the rocks from one place to another makes possible a more comprehensive view of the geologic history of a region. **FIGURE 11.16**, for example, shows the correlation of strata at three sites on the Colorado Plateau in southern Utah and northern Arizona. No single locale exhibits the entire sequence, but correlation reveals a more complete picture of the sedimentary rock record.

Correlation Within Limited Areas

Within a limited area, correlating rocks of one locality with those of another may be done simply by walking along the outcropping edges. However, this may not be possible when the rocks are mostly concealed by soil and vegetation. Correlation over short distances is often achieved by noting the position of a bed in a sequence of strata. Or a layer may be identified in another location if it is composed of distinctive or uncommon minerals.

Many geologic studies involve relatively small areas. Although they are important in their own right, their full value is realized only when they are correlated with other regions. Although the methods just described are sufficient to trace a rock formation over relatively short distances, they are not adequate for matching up rocks that are separated by great distances. When correlation between widely separated areas or between continents is the objective, geologists must rely on fossils.

Fossils and Correlation

The existence of fossils had been known for centuries, but it was not until the late 1700s and early 1800s that their significance as geologic tools was made evident. During

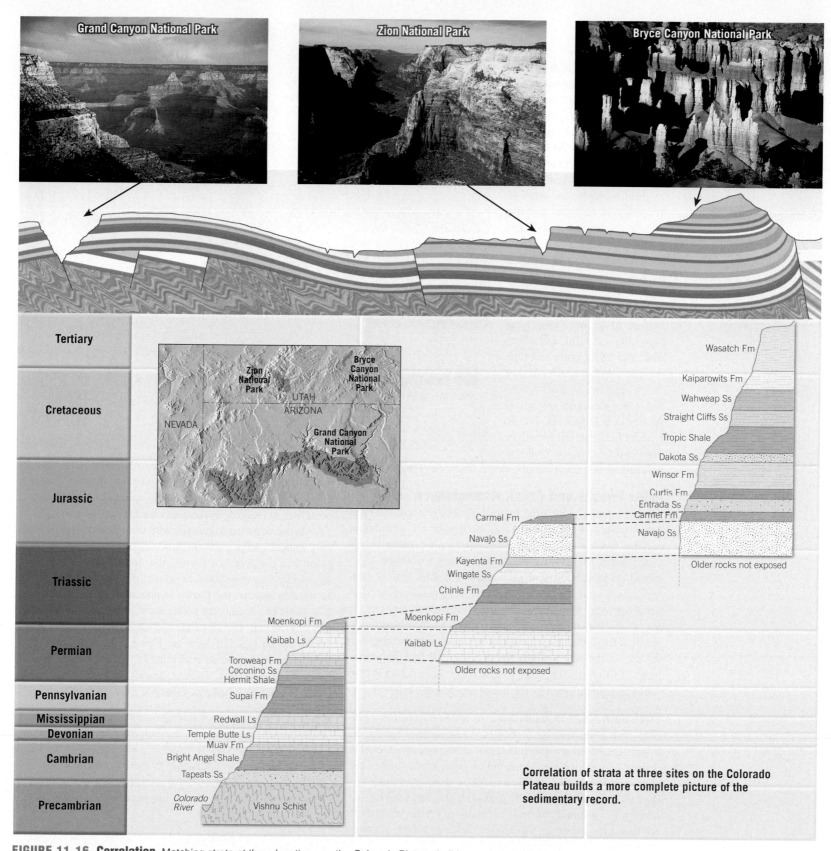

FIGURE 11.16 Correlation Matching strata at three locations on the Colorado Plateau builds a more complete picture of the sedimentary rocks in the region. (Photos by E. J. Tarbuck)

this period, an English engineer and canal builder, William Smith, discovered that each rock formation in the canals he worked on contained fossils unlike those in the beds either above or below. Further, he noted that sedimentary strata in widely separated areas could be identified—and correlated— by their distinctive fossil content.

Principle of Fossil Succession Based on Smith's classic observations and the findings of many geologists who followed, one of the most important and basic principles in historical geology was formulated: *Fossil organisms succeed one another in a definite and determinable order, and therefore any time period can be recognized by its fossil content.* This has

FIGURE 11.17 Index Fossils
Because microfossils are often very abundant, widespread, and quick to appear and become extinct, they constitute ideal index fossils. (Photo by Biophoto Associates/Science Source)

come to be known as the **principle of fossil succession**. In other words, when fossils are arranged according to their age, they do not present a random or haphazard picture. On the contrary, fossils document the evolution of life through time.

For example, an Age of Trilobites is recognized quite early in the fossil record. Then, in succession, paleontologists recognize an Age of Fishes, an Age of Coal Swamps, an Age of Reptiles, and an Age of Mammals. These "ages" pertain to groups that were especially plentiful and characteristic during particular time periods. Within each of the "ages" are many subdivisions based, for example, on certain species of trilobites and certain types of fish, reptiles, and so on. This same succession of dominant organisms, never out of order, is found on every continent.

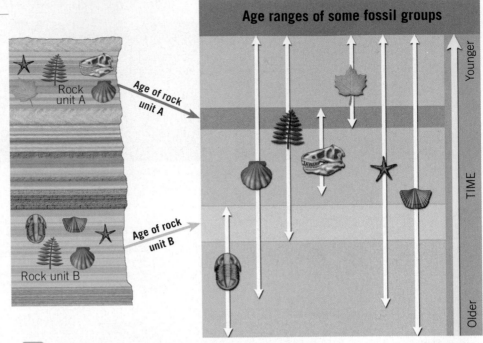

Age ranges of some fossil groups

SmartFigure 11.18 Fossil Assemblage Overlapping ranges of fossils help date rocks more exactly than using a single fossil.

Index Fossils and Fossil Assemblages When fossils were found to be time indicators, they became the most useful means of correlating rocks of similar age in different regions. Geologists pay particular attention to certain fossils called **index fossils** (**FIGURE 11.17**). These fossils are widespread geographically and are limited to a short span of geologic time, so their presence provides an important method of matching rocks of the same age. Rock formations, however, do not always contain a specific index fossil. In such situations a group of fossils, called a **fossil assemblage**, is used to establish the age of the bed. **FIGURE 11.18** illustrates how an assemblage of fossils may be used to date rocks more precisely than could be accomplished by the use of any one of the fossils.

Environmental Indicators In addition to being important, and often essential, tools for correlation, fossils are important environmental indicators. Although much can be deduced about past environments by studying the nature and characteristics of sedimentary rocks, a close examination of the fossils present can usually provide a great deal more information. For example, when the remains of certain clam shells are found in limestone, a geologist quite reasonably assumes

that the region was once covered by a shallow sea. Also, by using what we know of living organisms, we can conclude that fossil animals with thick shells, capable of withstanding pounding and surging waves, inhabited shorelines.

On the other hand, animals with thin, delicate shells probably indicate deep, calm offshore waters. Hence, by looking closely at the types of fossils, the approximate position of an ancient shoreline may be identified. Further, fossils can be used to indicate the former temperature of the water. Certain kinds of present-day corals must live in warm and shallow tropical seas like those around Florida and The Bahamas. When similar types of coral are found in ancient limestones, they indicate the marine environment that must have existed when they were alive. These examples illustrate how fossils can help unravel the complex story of Earth history.

11.4 CONCEPT CHECKS

1 What is the goal of correlation?
2 State the principle of fossil succession in your own words.
3 Contrast *index fossil* and *fossil assemblage*.
4 In addition to being important time indicators, how else are fossils useful to geologists?

11.5 | DATING WITH RADIOACTIVITY Discuss three types of radioactive decay and explain how radioactive isotopes are used to determine numerical dates.

In addition to establishing relative dates by using the principles described in the preceding sections, it is also possible to obtain reliable numerical dates for events in the geologic past. We know that Earth is about 4.6 billion years old and

that the dinosaurs became extinct about 65.5 million years ago. Dates that are expressed in millions and billions of years truly stretch our imagination because our personal calendars involve time measured in hours, weeks, and years.

Nevertheless, the vast expanse of geologic time is a reality, and radiometric dating allows us to measure it accurately. In this section, you will learn about radioactivity and its application in radiometric dating.

Reviewing Basic Atomic Structure

Recall from Chapter 2 that each atom has a *nucleus* that contains protons and neutrons and that the nucleus is orbited by electrons. An *electron* has a negative electrical charge, and a *proton* has a positive charge. A *neutron* is actually a proton and an electron combined, so it has no charge (it is neutral).

An element's *atomic number* (an element's identifying number) is the number of protons in the nucleus. Every element has a different number of protons in the nucleus and thus a different atomic number (hydrogen = 1, oxygen = 8, uranium = 92, etc.). Atoms of the same element always have the same number of protons, so the atomic number is constant.

Practically all (99.9 percent) of an atom's mass is found in the nucleus, indicating that electrons have practically no mass at all. By adding together the number of protons and neutrons in the nucleus, the *mass number* of the atom is determined. The number of neutrons in the nucleus can vary. These variants, called *isotopes*, have different mass numbers.

We can summarize with an example. Uranium's nucleus always has 92 protons, so its atomic number always is 92. But its neutron population varies, and uranium has three isotopes: uranium-234 (number of protons + neutrons = 234), uranium-235, and uranium-238. All three isotopes are found together in nature. They look the same and behave the same in chemical reactions.

Radioactivity

The forces that bind protons and neutrons together in the nucleus are usually strong. However, in some isotopes, the nuclei are unstable because the forces binding protons and neutrons together are not strong enough. As a result, the nuclei spontaneously break apart (decay) in a process called **radioactivity**.

Common Examples of Radioactive Decay What happens when unstable nuclei break apart? Three common types of radioactive decay are illustrated in **FIGURE 11.19** and are summarized as follows:

- Alpha emission—*Alpha particles* (α particles) may be emitted from the nucleus. An alpha particle consists of 2 protons and 2 neutrons. Consequently, the emission of an alpha particle means (a) the mass number of the isotope is reduced by 4, and (b) the atomic number is decreased by 2.

- Beta emission—When a *beta particle* (β particle), or an electron, is given off from a nucleus, the mass number remains unchanged because electrons have practically no mass. However, because the electron has come from a neutron (remember that a neutron is a combination of a proton and an electron), the nucleus contains 1 more proton than before. Therefore, the atomic number increases by 1.

- Electron capture—Sometimes an electron is captured by the nucleus. The electron combines with a proton and

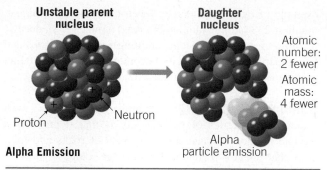

Alpha Emission

Atomic number: 2 fewer
Atomic mass: 4 fewer

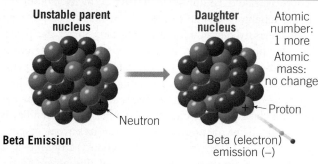

Beta Emission

Atomic number: 1 more
Atomic mass: no change

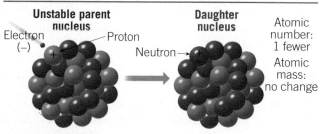

Electron Capture

Atomic number: 1 fewer
Atomic mass: no change

FIGURE 11.19 Common Types of Radioactive Decay Notice that in each example, the number of protons (atomic number) in the nucleus changes, thus producing a different element.

forms an additional neutron. As in the last example, the mass number remains unchanged. However, because the nucleus now contains 1 fewer proton, the atomic number decreases by 1.

An unstable (radioactive) isotope is referred to as the *parent*. The isotopes resulting from the decay of the parent are the *daughter products*. **FIGURE 11.20** provides an

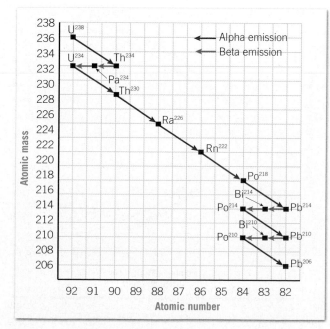

FIGURE 11.20 Decay of U-238 Uranium-238 is an example of a radioactive decay series. Before the stable end product (Pb-206) is reached, many different isotopes are produced as intermediate steps.

example of radioactive decay. When the radioactive parent, uranium-238 (atomic number 92, mass number 238), decays, it follows a number of steps, emitting 8 alpha particles and 6 beta particles before finally becoming the stable daughter product lead-206 (atomic number 82, mass number 206).

Radiometric Dating Certainly among the most important properties of radioactivity is that it provides a reliable method of calculating the ages of rocks and minerals that contain particular radioactive isotopes. The procedure is called **radiometric dating**. Why is radiometric dating reliable? The rates of decay for many isotopes have been precisely measured and do not vary under the physical conditions that exist in Earth's outer layers. Therefore, each radioactive isotope used for dating has been decaying at a fixed rate ever since the formation of the rocks in which it occurs, and the products of decay have been accumulating at a corresponding rate. For example, when uranium is incorporated into a mineral that crystallizes from magma, there is no lead (the stable daughter product) from previous decay. The radiometric "clock" starts at this point. As the uranium in this newly formed mineral disintegrates, atoms of the daughter product are trapped, and measurable amounts of lead eventually accumulate.

Half-Life

The time required for one-half of the nuclei in a sample to decay is called the **half-life** of the isotope. Half-life is a common way of expressing the rate of radioactive disintegration. **FIGURE 11.21** illustrates what occurs when a radioactive parent decays directly into its stable daughter product. When the quantities of parent and daughter are equal (ratio 1:1), we know that one half-life has transpired. When one-quarter of the original parent atoms remain and three-quarters have decayed to the daughter product, the parent/daughter ratio is 1:3, and we know that two half-lives

TABLE 11.1 Radioactive Isotopes Frequently Used in Radiometric Dating

Radioactive Parent	Stable Daughter Product	Currently Accepted Half-Life Values
Uranium-238	Lead-206	4.5 billion years
Uranium-235	Lead-207	704 million years
Thorium-232	Lead-208	14.1 billion years
Rubidium-87	Strontium-87	47.0 billion years
Potassium-40	Argon-40	1.3 billion years

have passed. After three half-lives, the ratio of parent atoms to daughter atoms is 1:7 (1 parent atom for every 7 daughter atoms).

If the half-life of a radioactive isotope is known and the parent/daughter ratio can be determined, the age of the sample can be calculated. For example, assume that the half-life of a hypothetical unstable isotope is 1 million years and the parent/daughter ratio in a sample is 1:15. Such a ratio indicates that four half-lives have passed and that the sample must be 4 million years old.

Using Various Isotopes

Notice that the *percentage* of radioactive atoms that decay during one half-life is always the same: 50 percent. However, the *actual number* of atoms that decay with the passing of each half-life continually decreases. As the percentage of radioactive parent atoms declines, the proportion of stable daughter atoms rises, with the increase in daughter atoms just matching the drop in parent atoms. This fact is the key to radiometric dating.

Of the many radioactive isotopes that exist in nature, five have proved particularly important in providing radiometric ages for ancient rocks (**TABLE 11.1**). Rubidium-87, uranium-238, and uranium-235 are used for dating rocks that are millions of years old, but potassium-40 is more versatile. Although the half-life of potassium-40 is 1.3 billion years, analytical techniques make possible the detection of tiny amounts of its stable daughter product, argon-40, in some rocks that are younger than 100,000 years. Another important reason for its frequent use is that potassium is abundant in many common minerals, particularly micas and feldspars.

It is important to realize that an accurate radiometric date can be obtained only if the mineral remained a closed system during the entire period since its formation. A correct date is not possible unless there was neither the addition nor loss of parent or daughter isotopes. This is not always the case. In fact, an important limitation of the potassium–argon method arises from the fact that argon is a gas, and it may leak from minerals, throwing off measurements.

A Complex Process Remember that although the basic principle of radiometric dating is simple, the actual procedure is quite complex. The analysis that determines the quantities of parent and daughter must be painstakingly precise. In addition, some radioactive materials do not decay

SmartFigure 11.21 Radioactive Decay Curve
Change is exponential. Half of the radioactive parent remains after one half-life. After a second half-life, one-quarter of the parent remains, and so forth.

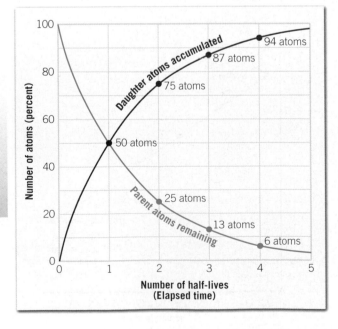

directly into the stable daughter product. As you saw in Figure 11.20, uranium-238 produces 13 intermediate unstable daughter products before the 14th and final daughter product, the stable isotope lead-206, is produced.

Earth's Oldest Rocks Radiometric dating methods have produced literally thousands of dates for events in Earth history. Rocks exceeding 3.5 billion years in age are found on all of the continents. Earth's oldest rocks (so far) may be as old as 4.28 billion years (b.y.). Discovered in northern Quebec, Canada, on the shores of Hudson Bay, these rocks may be remnants of Earth's earliest crust. Rocks from western Greenland have been dated at 3.7 to 3.8 b.y., and rocks nearly as old are found in the Minnesota River Valley and northern Michigan (3.5 to 3.7 b.y.), in southern Africa (3.4 to 3.5 b.y.), and in western Australia (3.4 to 3.6 b.y.). Tiny crystals of the mineral zircon with radiometric ages as old as 4.4 b.y. have been found in younger sedimentary rocks in western Australia. The source rocks for these tiny durable grains either no longer exist or have not yet been found.

Dating with Carbon-14

To date very recent events, carbon-14 is used. Carbon-14 is the radioactive isotope of carbon. The process is often called **radiocarbon dating**. Because the half-life of carbon-14 is only 5730 years, it can be used for dating events from the historic past as well as those from very recent geologic history. In some cases, carbon-14 can be used to date events as far back as 70,000 years.

Carbon-14 is continuously produced in the upper atmosphere as a result of cosmic-ray bombardment. Cosmic rays, which are high-energy particles, shatter the nuclei of gas atoms, releasing neutrons. Some of the neutrons are absorbed by nitrogen atoms (atomic number 7), causing their nuclei to emit a proton. As a result, the atomic number decreases by 1 (to 6), and a different element, carbon-14, is created (**FIGURE 11.22A**). This isotope of carbon quickly becomes incorporated into carbon dioxide, which circulates in the atmosphere and is absorbed by living matter. As a result, all organisms—including you—contain a small amount of carbon-14.

While an organism is alive, the decaying radiocarbon is continually replaced, and the proportions of carbon-14 and carbon-12 remain constant. Carbon-12 is the stable and most common isotope of carbon. However, when any plant or animal dies, the amount of carbon-14 gradually decreases as it decays to nitrogen-14 by beta emission (**FIGURE 11.22B**). By comparing the proportions of carbon-14 and carbon-12 in a sample, radiocarbon dates can be determined.

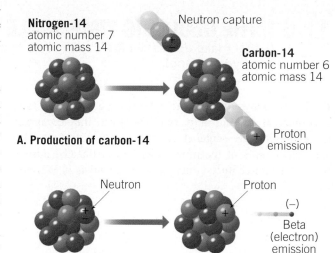

Nitrogen-14 atomic number 7 atomic mass 14

Neutron capture

Carbon-14 atomic number 6 atomic mass 14

Proton emission

A. Production of carbon-14

Neutron

Proton

(−) Beta (electron) emission

Carbon-14   **Nitrogen-14**

B. Decay of carbon-14

FIGURE 11.22 Carbon-14 Production and decay of radiocarbon. These sketches represent the nuclei of the respective atoms.

Although carbon-14 is useful in dating only the last small fraction of geologic time, it has become a valuable tool for anthropologists, archaeologists, and historians, as well as for geologists who study very recent Earth history (**FIGURE 11.23**). In fact, the development of radiocarbon dating was considered so important that the chemist who discovered this application, Willard F. Libby, received a Nobel Prize for the discovery.

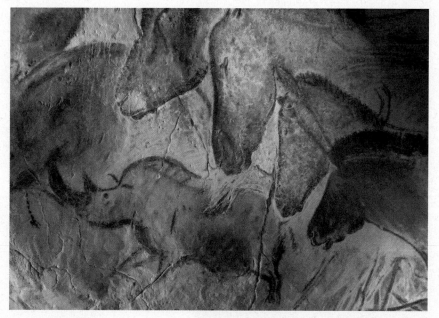

FIGURE 11.23 Cave Art Chauvet Cave in southern France, which was discovered in 1994, contains some of the earliest-known cave paintings. Radiocarbon dating indicates that most of the images were drawn between 30,000 and 32,000 years ago. (Photo by JAVIER TRUEBA/MSF/Science Source)

11.5 CONCEPT CHECKS

1. List three types of radioactive decay. For each type, describe how the atomic number and the atomic mass change.

2. Sketch a simple diagram that explains the idea of half-life.

3. Why is radiometric dating a reliable method for determining numerical dates?

4. For what time span does radiocarbon dating apply?

11.6 | THE GEOLOGIC TIME SCALE
Distinguish among the four basic time units that make up the geologic time scale and explain why the time scale is considered to be a dynamic tool.

Geologists have divided the whole of geologic history into units of varying magnitude. Together, they comprise the **geologic time scale** of Earth history (**FIGURE 11.24**). The major units of the time scale were delineated during the nineteenth century, principally by scientists in Western Europe and Great Britain. Because radiometric dating was unavailable at that time, the entire time scale was created using methods of relative dating. It was only in the twentieth century that radiometric dating permitted numerical dates to be added.

FIGURE 11.24 Geologic Time Scale Numbers on the time scale represent time in millions of years before the present. Numerical dates were added long after the time scale was established, using relative dating techniques.

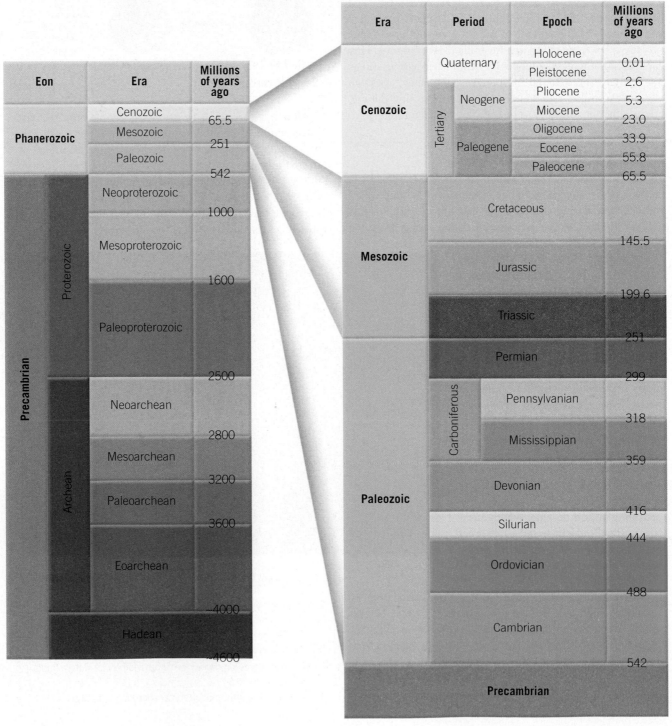

Structure of the Time Scale

The geologic time scale divides the 4.6-billion-year history of Earth into many different units and provides a meaningful time frame within which the events of the geologic past are arranged. As shown in Figure 11.24, **eons** represent the greatest expanses of time. The eon that began about 542 million years ago is the **Phanerozoic**, a term derived from Greek words meaning "visible life." It is an appropriate description because the rocks and deposits of the Phanerozoic eon contain abundant fossils that document major evolutionary trends.

Another glance at the time scale shows that eons are divided into **eras**. The three eras within the Phanerozoic are the **Paleozoic** ("ancient life"), the **Mesozoic** ("middle life"), and the **Cenozoic** ("recent life"). As the names imply, these eras are bounded by profound worldwide changes in life-forms.

Each era of the Phanerozoic eon is divided into units known as **periods**. The Paleozoic has seven, and the Mesozoic and Cenozoic each have three. Each of these periods is characterized by a somewhat less profound change in life-forms as compared with the eras.

Each of the periods is divided into still smaller units called **epochs**. As you can see in Figure 11.24, seven epochs have been named for the periods of the Cenozoic era. The epochs of other periods, however, are not usually referred to by specific names. Instead, the terms *early*, *middle*, and *late* are generally applied to the epochs of these earlier periods.

Precambrian Time

Notice that the detail of the geologic time scale does not begin until about 542 million years ago, the date for the beginning of the Cambrian period. The 4 billion years prior to the Cambrian is divided into two eons, the *Archean* and the *Proterozoic*, which are divided into four eras. It is also common for this vast expanse of time to simply be referred to as the **Precambrian**. Although it represents about 88 percent of Earth history, the Precambrian is not divided into nearly as many smaller time units as is the Phanerozoic eon.

Why is the huge expanse of Precambrian time not divided into numerous eras, periods, and epochs? The reason is that Precambrian history is not known in great enough detail. The

EYE ON **EARTH**

This image is from the bottom of the Grand Canyon, a zone called the Inner Gorge. The darker rock is the Vishnu Schist. The light-colored rock, called the Zoroaster Granite, is a series of dikes. Both rocks date from Precambrian time. Suppose you are on a raft trip through the Grand Canyon. Your companions are bright and curious but, unlike you, they are not trained in geology. Sitting around the campfire the night before you reached the site pictured here, you and your fellow rafters discussed the geologic time scale and the magnitude of geologic time.

QUESTION 1 *When you arrive at this site, someone asks why Precambrian history seems so sketchy and why this time span is not divided into nearly as many subdivisions as Phanerozoic time. Refer to this setting as you respond to these questions.*

QUESTION 2 *Which is older, the Vishnu Schist or the Zoroaster Granite? Explain.*

Phanerozoic

Precambrian

Vishnu schist

Vishnu schist

Zoroaster granite

Rafting the Colorado River

Michael Collier

quantity of information geologists have deciphered about Earth's past is somewhat analogous to the detail of human history. The farther back we go, the less we know. Certainly, more data and information exist about the past 10 years than for the first decade of the twentieth century; the events of the nineteenth century have been documented much better than the events of the first century A.D., and so on. So it is with Earth history. The more recent past has the freshest, least disturbed, and most observable record. The further back in time a geologist goes, the more fragmented the record and clues become.

Terminology and the Geologic Time Scale

Some terms are associated with the geologic time scale but are not "officially" recognized as being a part of it. The best-known and the most common example is *Precambrian*—the informal name for the eons that came before the current Phanerozoic eon. Although the term *Precambrian* has no formal status on the geologic time scale, it has been traditionally used as though it does.

Hadean is another informal term that is found on some versions of the geologic time scale and is used by some geologists. It refers to the earliest interval (eon) of Earth history—before the oldest-known rocks. When the term was coined in 1972, the age of Earth's oldest rocks was thought to be about 3.8 billion years. Today that number stands at slightly greater than 4 billion, and, of course, is subject to revision. The name *Hadean* derives from *Hades*, Greek for "underworld"—a reference to the "hellish" conditions that prevailed on Earth early in its history.

Effective communication in the geosciences requires that the geologic time scale consist of standardized divisions and dates. Who determines which names and dates on the geologic time scale are "official"? The organization that is largely responsible for maintaining and updating this important document is the International Commission on

---

*To view the current version of the ICS time scale, go to www .stratigraphy.org. *Stratigraphy* is the branch of geology that studies rock layers (strata) and layering (stratification), and thus its primary focus is sedimentary and layered volcanic rocks.

Stratigraphy (ICS), a committee of the International Union of Geological Sciences.* Advances in the geosciences require that the scale be periodically updated to include changes in unit names and boundary age estimates.

For example, the geologic time scale shown in Figure 11.24 was updated as recently as July 2009. After considerable dialogue among geologists who focus on very recent Earth history, the ICS changed the date for the start of the Quaternary period and the Pleistocene epoch from 1.8 million to 2.6 million years ago. Who knows? Perhaps by the time you read this, other changes will have been made.

If you were to examine a geologic time scale from just a few years ago, it is quite possible that you would see the Cenozoic era divided into the Tertiary and Quaternary periods. However, on more recent versions, the space formerly designated as Tertiary is divided into the Paleogene and Neogene periods. As our understanding of this time span has changed, so too has its designation on the geologic time scale. Today, the Tertiary period is considered as a "historic" name and is given no official status on the ICS version of the time scale. Nevertheless, many time scales still contain references to the Tertiary period, including Figure 11.24. One reason for this is that a great deal of past (and some current) geologic literature uses this name.

For those who study historical geology, it is important to realize that the geologic time scale is a dynamic tool that continues to be refined as our knowledge and understanding of Earth history evolves.

11.6 CONCEPT CHECKS

1. What are the four basic units that make up the geologic time scale? List the specific ones that apply to the present day.

2. Why is -*zoic* part of so many names on the geologic time scale?

3. What term applies to *all* geologic time prior to the Phanerozoic eon? Why is this span not divided into as many smaller units as the Phanerozoic eon?

4. To what does *Hadean* apply? Is it an "official" part of the geologic time scale?

11.7 | DETERMINING NUMERICAL DATES FOR SEDIMENTARY STRATA
Explain how reliable numerical dates are determined for layers of sedimentary rock.

Although reasonably accurate numerical dates have been worked out for the periods of the geologic time scale (see Figure 11.24), the task is not without difficulty. The primary problem in assigning numerical dates is the fact that not all rocks can be dated by using radiometric methods. For a radiometric date to be useful, all minerals in the rock must have formed at about the same time. For this reason, radioactive isotopes can be used to determine when minerals in an igneous rock crystallized and when pressure and heat created new minerals in a metamorphic rock.

However, samples of sedimentary rock can only rarely be dated directly by radiometric means. A sedimentary rock may include particles that contain radioactive isotopes, but the rock's age cannot be accurately determined because the grains that make up the rock are not the same age as the rock in which they occur. Rather, the sediments have been weathered from rocks of diverse ages.

Radiometric dates obtained from metamorphic rocks may also be difficult to interpret because the age of a particular mineral in a metamorphic rock does not necessarily represent the time when the rock initially formed. Instead, the date may indicate any one of a number of subsequent metamorphic phases.

If samples of sedimentary rocks rarely yield reliable radiometric ages, how can numerical dates be assigned to sedimentary layers? Usually a geologist must relate them to datable igneous masses, as in **FIGURE 11.25**. In this example, radiometric dating has determined the ages of the volcanic ash bed within the Morrison Formation and the dike cutting the Mancos Shale and Mesaverde Formation. The sedimentary beds below the ash are obviously older than the ash, and all the layers above the ash are younger (based on the principle of superposition). The dike is younger than the Mancos Shale and the Mesaverde Formation but older than the Wasatch Formation because the dike does not intrude the Paleogene rocks (based on the principle of cross-cutting relationships).

From this kind of evidence, geologists estimate that a part of the Morrison Formation was deposited about 160 million years ago, as indicated by the ash bed. Further, they conclude that the Wasatch Formation began after the intrusion of the dike, 66 million years ago. This is

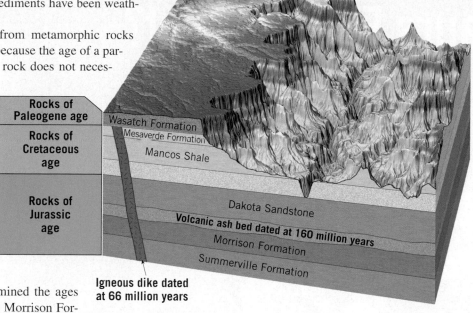

FIGURE 11.25 Dating Sedimentary Strata Numerical dates for sedimentary layers are usually determined by examining their relationship to igneous rocks.

Rocks of Paleogene age

Rocks of Cretaceous age

Rocks of Jurassic age

Wasatch Formation
Mesaverde Formation
Mancos Shale
Dakota Sandstone
Volcanic ash bed dated at 160 million years
Morrison Formation
Summerville Formation

Igneous dike dated at 66 million years

one example of literally thousands that illustrate how datable materials are used to "bracket" the various episodes in Earth history within specific time periods. It shows the necessity of combining laboratory dating methods with field observations of rocks.

11.7 CONCEPT CHECKS

1 Briefly explain why it is often difficult to assign a reliable numerical date to a sample of sedimentary rock.

2 How might a numerical date for a layer of sedimentary rock be determined?

EYE ON EARTH

This is a close-up view of the detrital sedimentary rock conglomerate. This rock contains radioactive isotopes that will yield numerical dates.

QUESTION 1 *Although radioactive isotopes are present, a reliable numerical date for this conglomerate cannot be accurately determined. Explain.*

QUESTION 2 *How might a numerical age range be established for the conglomerate layer?*

E.J. Tarbuck

Did Humans and Dinosaurs Ever Coexist?

Many old and not-so-old movies and TV shows show humans and dinosaurs living side-by-side. This is also true of many cartoons and comic strips. Although we know these stories are fiction, many people still have the impression that these two life forms coexisted.

Hanna-Barbera/Everett Collection

This well-known fossil of a human ancestor (*Australopithecus afarensis*) known as Lucy was discovered in Ethiopia and is 3.2 million years old. The oldest bones thus far assigned to the human genus *Homo*, are from the early Pleistocene epoch and are about 2.4 million years old.
(Andrew Holt/Alamy)

Tyrannosaurus rex dates from the late Cretaceous period. Dinosaurs became extinct more than 65 million years before humans evolved.
(Sabena Jane Black Bird/Alamy)

251 MYA	199.6 MYA	145.5 MYA	65.5 MYA	23 MYA	2.6 MY
	MESOZOIC			CENOZOIC	
TRIASSIC	JURASSIC	CRETACEOUS		TERTIARY	
				PALEOGENE	NEOGENE

QUATERNARY

Dinosaurs first appear in the Triassic period.

Dinosaurs did not become gigantic until the Jurassic period.

Dinosaurs became extinct at the close of the Cretaceous period 65.5 million years ago.

Humans (*Homo sapiens*) evolved in East Africa slightly more than 150,000 years ago.

11.1 A BRIEF HISTORY OF GEOLOGY

Explain the principle of uniformitarianism and discuss how it differs from catastrophism.

KEY TERMS catastrophism, uniformitarianism

- Early ideas about the nature of Earth were based on religious traditions and notions of great catastrophes.
- In the late 1700s, James Hutton emphasized that the same slow processes have acted over great spans of time and are responsible for Earth's rocks, mountains, and landforms. This similarity of processes over vast spans of time led to this principle being called uniformitarianism.

11.2 CREATING A TIME SCALE: RELATIVE DATING PRINCIPLES

Distinguish between numerical and relative dates and apply relative dating principles to determine a time sequence of geologic events.

KEY TERMS numerical date, relative date, principle of superposition, principle of original horizontality, principle of lateral continuity, principle of cross-cutting relationships, inclusion, conformable, unconformity, angular unconformity, disconformity, nonconformity

- The two types of dates that geologists use to interpret Earth history are (1) relative dates, which put events in their proper sequence of formation, and (2) numerical dates, which pinpoint the time in years when an event took place.
- Relative dates can be established using the principles of superposition, original horizontality, cross-cutting relationships, and inclusions. Unconformities, gaps in the geologic record, may be identified during the relative dating process.

Q The accompanying diagram is a cross section of a hypothetical area. Place the lettered features in the proper sequence, from oldest to youngest. Where in the sequence can you identify an unconformity? Which principles did you use to establish the sequence?

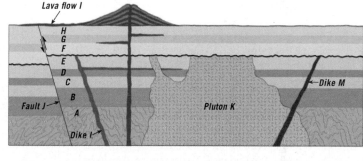

Oldest   Youngest

11.3 FOSSILS: EVIDENCE OF PAST LIFE

Define *fossil* and discuss the conditions that favor the preservation of organisms as fossils. List and describe various fossil types.

KEY TERMS fossil, paleontology

- Fossils are remains or traces of ancient life. Paleontology is the branch of science that studies fossils.
- Fossils can form through many processes. For an organism to be preserved as a fossil, it usually needs be buried rapidly. Also, an organism's hard parts are most likely to be preserved because soft tissue decomposes rapidly in most circumstances.

E.J. Tarbuck

Q What term is used to describe the type of fossil that is shown here? Briefly describe how it formed.

11.4 CORRELATION OF ROCK LAYERS

Explain how rocks of similar age that are in different places can be matched up.

KEY TERMS correlation, principle of fossil succession, index fossil, fossil assemblage

- Matching up exposures of rock that are the same age but are in different places is called correlation. By correlating rocks from around the world, geologists have developed the geologic time scale and obtained a fuller perspective on Earth history.
- Fossils can be used to correlate sedimentary rocks in widely separated places by using the rocks' distinctive fossil content and applying the principle of fossil succession. This principle states that fossil organisms succeed one another in a definite and determinable order, and, therefore, a time period can be recognized by examining its fossil content.
- Index fossils are particularly useful in correlation because they are widespread and associated with a relatively narrow time span. The overlapping ranges of fossils in an assemblage may be used to establish an age for a rock layer that contains multiple fossils.
- Fossils may be used to establish ancient environmental conditions that existed when sediment was deposited.

Q Which would make a more likely index fossil for modern times: a penguin or a pigeon? Why?

11.5 DATING WITH RADIOACTIVITY

Discuss three types of radioactive decay and explain how radioactive isotopes are used to determine numerical dates.

KEY TERMS   radioactivity, radiometric dating, half-life, radiocarbon dating

- Radioactivity is the spontaneous breaking apart (decay) of certain unstable atomic nuclei. Three common forms of radioactive decay are (1) emission of an alpha particle from the nucleus, (2) emission of a beta particle (electron) from the nucleus, and (3) capture of an electron by the nucleus.
- An unstable radioactive isotope, called a parent, will decay and form daughter products. The length of time for one-half of the nuclei of a radioactive isotope to decay is called the half-life of the isotope. If the half-life of an isotope is known, and the parent–daughter ratio can be measured, the age of the sample can be calculated.

Q Measurements of zircon crystals containing trace amounts of uranium from a specimen of granite yield parent/daughter ratios of 25 percent parent (uranium-235) and 75 percent daughter (lead-206). The half-life of uranium-235 is 704 million years. How old is the granite?

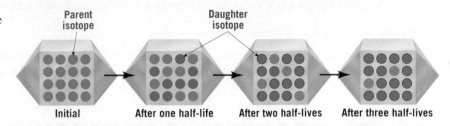

11.6 THE GEOLOGIC TIME SCALE

Distinguish among the four basic time units that make up the geologic time scale and explain why the time scale is considered to be a dynamic tool.

KEY TERMS   geologic time scale, eon, Phanerozoic eon, era, Paleozoic era, Mesozoic era, Cenozoic era, period, epoch, Precambrian

- Earth history is divided into units of time on the geologic time scale. Eons are divided into eras, which each contain multiple periods. Periods are divided into epochs.
- Precambrian time includes the Archean and Proterozoic eons. It is followed by the Phanerozoic eon, which is well documented by abundant fossil evidence, resulting in many subdivisions.
- The geologic time scale is a work in progress, continually being refined as new information becomes available.

Q Is the Mesozoic an example of an eon, an era, a period, or an epoch? What about the Jurassic?

11.7 DETERMINING NUMERICAL DATES FOR SEDIMENTARY STRATA

Explain how reliable numerical dates are determined for layers of sedimentary rock.

- Sedimentary strata are usually not directly datable using radiometric techniques because they consist of the material produced by the weathering of other rocks. A particle in a sedimentary rock comes from some older source rock. If you were to date the particle using isotopes, you would get the age of the source rock, not the age of the sedimentary deposit.
- One way geologists assign numerical dates to sedimentary rocks is to use relative dating principles to relate them to datable igneous masses, such as dikes and volcanic ash beds. A layer may be older than one igneous feature and younger than another.

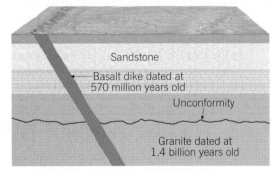

Q Express the numerical age of the sandstone layer in the diagram as accurately as possible.

GIVE IT SOME **THOUGHT**

1. The accompanying image shows the metamorphic rock gneiss, a basaltic dike, and a fault. Place these three features in their proper sequence (which came first, second, and third) and explain your logic.
2. A mass of granite is in contact with a layer of sandstone. Using a principle described in this chapter, explain how you might determine whether the sandstone was deposited on top of the granite or whether the magma that formed the granite was intruded after the sandstone was deposited.

Marli Miller

3. This scenic image is from Monument Valley in the northeastern corner of Arizona. The bedrock in this region consists of layers of sedimentary rocks. Although the prominent rock exposures ("monuments") in this photo are widely separated, we can infer that they represent a once continuous layer. Discuss the principle that allows us to make this inference.

Michael Collier

4. The accompanying photo shows two layers of sedimentary rock. The lower layer is shale from the late Mesozoic era. Note the old river channel that was carved into the shale after it was deposited. Above is a younger layer of boulder-rich breccia. Are these layers conformable? Explain why or why not. What term from relative dating applies to the line separating the two layers?

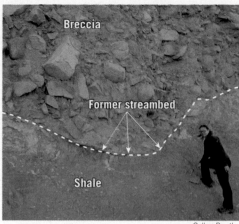

Breccia

Former streambed

Shale

Callan Bentley

5. If a radioactive isotope of thorium (atomic number 90, mass number 232) emits 6 alpha particles and 4 beta particles during the course of radioactive decay, what are the atomic number and mass number of the stable daughter product?

6. A hypothetical radioactive isotope has a half-life of 10,000 years. If the ratio of radioactive parent to stable daughter product is 1:3, how old is the rock that contains the radioactive material?

7. Solve the problems below that relate to the magnitude of Earth history. To make calculations easier, round Earth's age to 5 billion years.

   a. What percentage of geologic time is represented by recorded history? (Assume 5000 years for the length of recorded history.)

   b. Humanlike ancestors (hominids) have been around for roughly 5 million years. What percentage of geologic time is represented by these ancestors?

   c. The first abundant fossil evidence does not appear until the beginning of the Cambrian period, about 540 million years ago. What percentage of geologic time is represented by abundant fossil evidence?

8. These polished stones are called *gastroliths*. Explain how such objects can be considered fossils. What category of fossil are they? Name another example of a fossil in this category.

0 1 2
Centimeters

Francois Gohier/Photo Researchers, Inc.

9. A portion of a popular college text in historical geology includes 10 chapters (281 pages) in a unit titled "The Story of Earth." Two chapters (49 pages) are devoted to Precambrian time. By contrast, the last two chapters (67 pages) focus on the most recent 23 million years, with 25 of those pages devoted to the Holocene Epoch, which began 10,000 years ago.

   a. Compare the percentage of pages devoted to the Precambrian to the actual percentage of geologic time that this span represents.

   b. How does the number of pages about the Holocene compare to its actual percentage of geologic time?

   c. Suggest some reasons why the text seems to have such an unequal treatment of Earth history.

EXAMINING THE **EARTH SYSTEM**

1. The accompanying photo shows a large petrified log in Arizona's Petrified Forest National Park. Describe the transition of this tree from being part of the biosphere to being a component of the geosphere. How might the hydrosphere and/or atmosphere have played a role in the transition?

Buddy Mays/Alamy

2. The famous angular unconformity at Scotland's Siccar Point shown in this photo was originally studied by James Hutton in the late 1700s.

   a. Describe in a general way what occurred to produce this feature.

   b. Suggest ways in which all four spheres of the Earth system could have been involved in producing Siccar Point.

   c. The Earth system is powered by energy from two sources. How are both sources represented in the Siccar Point unconformity?

Marli Miller

3. This scene in Montana's Glacier National Park shows layers of Precambrian sedimentary rocks. The darker layer contained within the sedimentary layers is igneous. The narrow, light-colored areas adjacent to the igneous rock were created when molten material that formed the igneous rock baked the adjacent rock.

   a. Is the igneous layer more likely a lava flow that was laid down at the surface prior to the deposition of the layers above it or a sill that was intruded after all the sedimentary layers were deposited? Explain.

   b. Is it likely that the igneous layer will exhibit a vesicular texture? Explain.

   c. To which group (igneous, sedimentary, or metamorphic) does the light-colored rock belong? Relate your explanation to the rock cycle.

Marli Miller

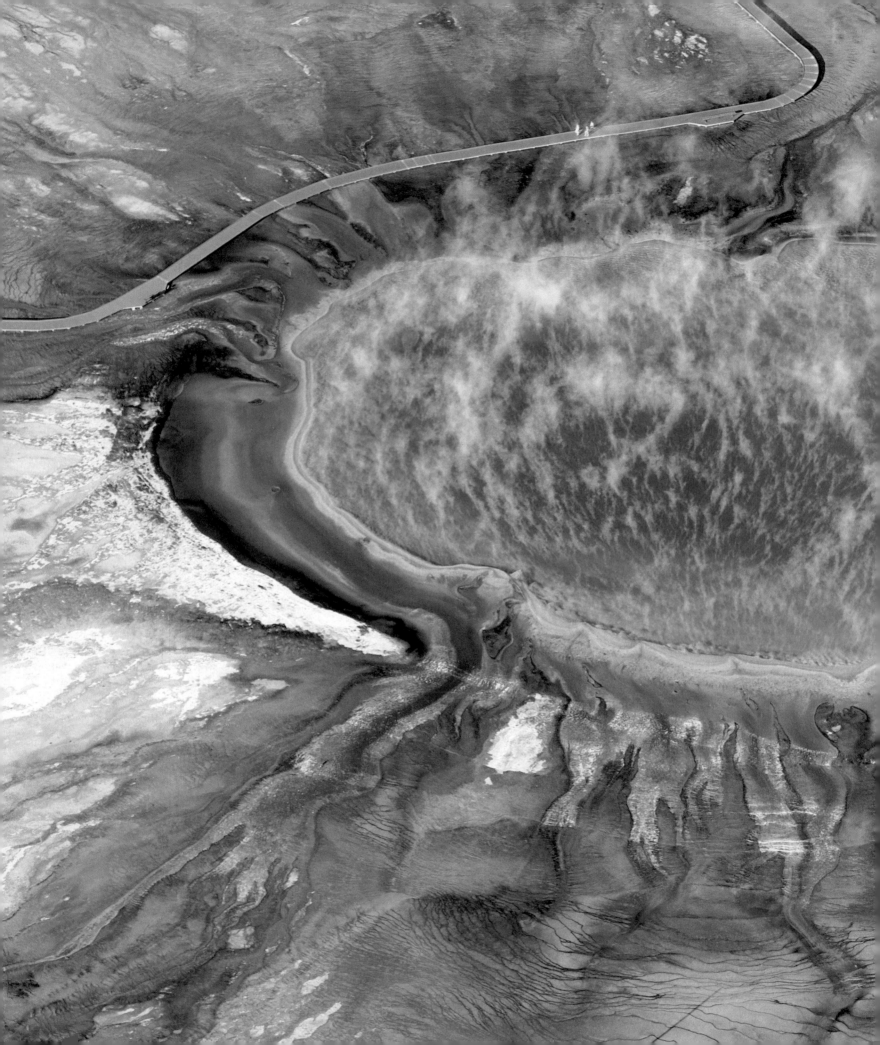

12

Earth's Evolution Through Geologic Time

Grand Prismatic Pool in Yellowstone National Park This hot-water pool gets its blue color from several species of heat-tolerant cyanobacteria. Microscopic fossils of organisms similar to modern cyanobacteria are among Earth's oldest fossils. (Photo by Michael Collier)

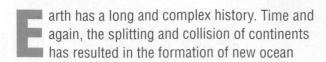

Earth has a long and complex history. Time and again, the splitting and collision of continents has resulted in the formation of new ocean basins and the creation of great mountain ranges. Furthermore, the nature of life on our planet has experienced dramatic changes through time.

12.1 | IS EARTH UNIQUE?

List the principal characteristics that make Earth unique among the planets.

There is only one place in the universe, as far as we know, that supports life—a modest-sized planet called Earth that orbits an average-sized star, the Sun. Life on Earth is ubiquitous; it is found in boiling mudpots and hot springs, in the deep abyss of the ocean, and even under the Antarctic Ice Sheet. Living space on our planet, however, is significantly limited when we consider the needs of individual organisms, particularly humans. The global ocean covers 71 percent of Earth's surface, but only a few hundred meters below the water's surface, pressures are so intense that humans cannot survive without an atmospheric diving suit. In addition, many continental areas are too steep, too high, or too cold for us to inhabit (**FIGURE 12.1**). Nevertheless, based on what we know about other bodies in the solar system—and the hundreds of planets recently discovered orbiting around other stars—Earth is still by far the most accommodating.

What fortuitous events produced a planet so hospitable to life? Earth was not always as we find it today. During its formative years, our planet became hot enough to support a magma ocean. It also survived a several-hundred-million-year period of extreme bombardment by asteroids, to which the heavily cratered surfaces of Mars and the Moon testify. The oxygen-rich atmosphere that makes higher life-forms possible developed relatively recently. Serendipitously, Earth seems to be the right planet, in the right location, at the right time.

The Right Planet

What are some of the characteristics that make Earth unique among the planets? Consider the following:

- If Earth were considerably larger (more massive), its force of gravity would be proportionately greater. Like the giant planets, Earth would have retained a thick, hostile atmosphere consisting of ammonia and methane, and possibly hydrogen and helium.

- If Earth were much smaller, oxygen, water vapor, and other volatiles would escape into space and be lost forever. Thus, like the Moon and Mercury, both of which lack atmospheres, Earth would be void of life.

- If Earth did not have a rigid lithosphere overlaying a weak asthenosphere, plate tectonics would not operate. The continental crust (Earth's "highlands") would not have formed without the recycling of plates. Consequently, the entire planet would likely be covered by an ocean a few kilometers deep. As author Bill Bryson so aptly stated, "There might be life in that lonesome ocean, but there certainly wouldn't be baseball."[1]

- Most surprisingly, perhaps, is the fact that if our planet did not have a molten metallic core, most of the life-forms on Earth would not exist. Fundamentally, without the flow of iron in the core, Earth could not support a magnetic field. It is the magnetic field that prevents lethal cosmic rays from showering Earth's surface and from stripping away our atmosphere.

FIGURE 12.1 Much of Earth's Surface Is Uninhabitable Climbers near the top of Mount Everest. At this altitude, the level of oxygen is only one-third the amount available at sea level. (Photo by STR/AFP/Getty Images)

[1]*A Short History of Nearly Everything* (Broadway Books, 2003).

The Right Location

One of the primary factors that determine whether a planet is suitable for higher life-forms is its location in the solar system. The following scenarios substantiate Earth's advantageous position:

- If Earth were about 10 percent closer to the Sun, our atmosphere would be more like that of Venus and consist mainly of the greenhouse gas carbon dioxide. As a result, Earth's surface temperature would be too hot to support higher life-forms.

- If Earth were about 10 percent farther from the Sun, the problem would be reversed—it would be too cold. The oceans would freeze over, and Earth's active water cycle would not exist. Without liquid water, all life would perish.

- Earth is near a star of modest size. Stars like the Sun have a life span of roughly 10 billion years. During most of this time, radiant energy is emitted at a fairly constant level. Giant stars, on the other hand, consume their nuclear fuel at very high rates and "burn out" in a few hundred million years. Therefore, Earth's proximity to a modest-sized star allowed enough time for the evolution of humans, who first appeared on this planet only a few million years ago.

The Right Time

The last, but certainly not the least, fortuitous factor for Earth is timing. The first organisms to inhabit Earth were extremely primitive and came into existence roughly 3.8 billion years ago. From that point in Earth's history, innumerable changes occurred—life-forms came and went, and there were many changes in the physical environment of our planet. Consider two of the many timely Earth-altering events:

- Earth's atmosphere has developed over time. Earth's primitive atmosphere is thought to have been composed mostly of methane, water vapor, ammonia, and carbon dioxide—but no free oxygen. Fortunately, microorganisms evolved that released oxygen into the atmosphere through the process of *photosynthesis*. About 2.5 billion years ago, an atmosphere with free oxygen came into existence. The result was the evolution of the ancestors of the vast array of organisms that occupy Earth today.

- About 65 million years ago, our planet was struck by an asteroid 10 kilometers (6 miles) in diameter. This impact likely caused a mass extinction during which nearly three-quarters of all plant and animal species were obliterated—including dinosaurs (**FIGURE 12.2**). Although this may not seem fortuitous, the extinction of dinosaurs opened new habitats for small mammals that survived the impact. These habitats, along with evolutionary forces, led to the development of many large mammals that occupy our modern world. Without this event, mammals may not have evolved beyond the small rodent-like creatures that live in burrows.

FIGURE 12.2 Paleontologist and Technicians Excavating a Dinosaur Known as *Titanosaurus* in Southern Argentina (Photo by BERNARDO GONZALEZ RIGA/EPA/Newscom)

As various observers have noted, Earth developed under "just right" conditions to support higher life-forms. Astronomers refer to this as the *Goldilocks scenario*. Like the classic "Goldilocks and the Three Bears" fable, Venus is too hot (Papa Bear's porridge), Mars is too cold (Mama Bear's porridge), but Earth is just right (Baby Bear's porridge).

Viewing Earth's History

The remainder of this chapter focuses on the origin and evolution of planet Earth—the one place in the universe we know fosters life. As you learned in Chapter 11, researchers utilize many tools to interpret clues about Earth's past. Using these tools, as well as clues contained in the rock record, scientists continue to unravel many complex events of the geologic past. The goal of this chapter is to provide a brief overview of the history of our planet and its life-forms—a journey that takes us back about 4.6 billion years, to the formation of Earth. Later, we will consider how our physical world assumed its present state and how Earth's inhabitants changed through time. We suggest that you reacquaint yourself with the *geologic time scale* presented in **FIGURE 12.3** and refer to it throughout the chapter.

12.1 CONCEPT CHECKS

1 In what way is Earth unique among the planets of our solar system?

2 Explain why Earth is just the right size.

3 Why is Earth's molten, metallic core important to humans living today?

4 Why is Earth's location in the solar system ideal for the development of higher life-forms?

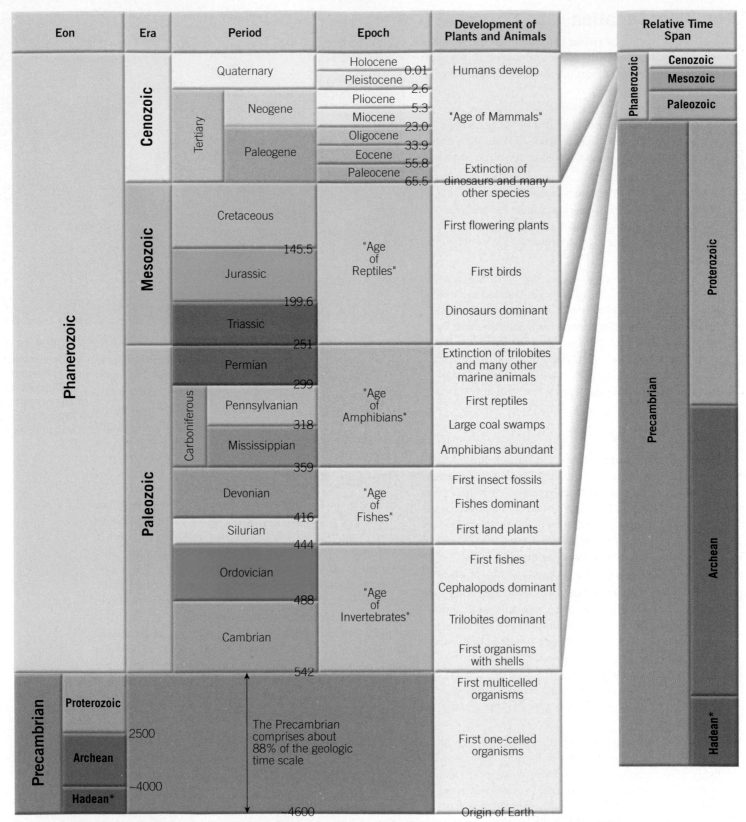

Eon	Era	Period		Epoch		Development of Plants and Animals
Phanerozoic	Cenozoic	Quaternary		Holocene	0.01	Humans develop
				Pleistocene	2.6	
		Tertiary	Neogene	Pliocene	5.3	"Age of Mammals"
				Miocene	23.0	
			Paleogene	Oligocene	33.9	
				Eocene	55.8	
				Paleocene	65.5	Extinction of dinosaurs and many other species
	Mesozoic	Cretaceous		"Age of Reptiles"		First flowering plants
					145.5	First birds
		Jurassic			199.6	Dinosaurs dominant
		Triassic			251	
	Paleozoic	Permian		"Age of Amphibians"		Extinction of trilobites and many other marine animals
					299	First reptiles
		Carboniferous	Pennsylvanian		318	Large coal swamps
			Mississippian		359	Amphibians abundant
		Devonian		"Age of Fishes"		First insect fossils
						Fishes dominant
		Silurian			416	First land plants
		Ordovician		"Age of Invertebrates"	444	First fishes
					488	Cephalopods dominant
		Cambrian				Trilobites dominant
					542	First organisms with shells
Precambrian	Proterozoic					First multicelled organisms
					2500	
	Archean			The Precambrian comprises about 88% of the geologic time scale		First one-celled organisms
					~4000	
	Hadean*				~4600	Origin of Earth

Relative Time Span

Phanerozoic: Cenozoic, Mesozoic, Paleozoic

Precambrian: Proterozoic, Archean, Hadean*

* Hadean is the informal name for the span that begins at Earth's formation and ends with Earth's earliest-known rocks.

FIGURE 12.3 The Geologic Time Scale Numbers represent time in millions of years before the present. The Precambrian accounts for about 88 percent of geologic time.

12.2 | BIRTH OF A PLANET Outline the major stages in the evolution of Earth, from the Big Bang to the formation of our planet's layered internal structure.

The universe had been evolving for several billion years before our solar system and Earth began to form. The universe began about 13.7 billion years ago, with the Big Bang, when all matter and space came into existence. Shortly thereafter, the two simplest elements, hydrogen and helium, formed. These basic elements were the ingredients for the first star systems. Several billion years later, our home galaxy, the Milky Way, came into existence. It was within a band of stars and nebular debris in an arm of this spiral galaxy that the Sun and planets took form nearly 4.6 billion years ago.

From the Big Bang to Heavy Elements

According to the Big Bang theory, the formation of our planet began about 13.7 billion years ago, with a cataclysmic explosion that created all matter and space (**FIGURE 12.4**). Initially, subatomic particles (protons, neutrons, and electrons) formed. Later, as this debris cooled, atoms of hydrogen and helium, the two lightest elements, began to form. Within a few hundred million years, clouds of these gases condensed and coalesced into stars that compose the galactic systems we now observe.

As these gases contracted to become the first stars, heating triggered the process of *nuclear fusion*. Within the interiors of stars, hydrogen nuclei convert to helium nuclei, releasing enormous amounts of radiant energy (heat, light, cosmic rays). Astronomers have determined that in stars more massive than our Sun, other thermonuclear reactions occur that generate all the elements on the periodic table up to number 26, iron. The heaviest elements (beyond number 26) are created only at extreme temperatures during the explosive death of a star eight or more times as massive as the Sun. During these cataclysmic **supernova** events, exploding stars produce all the elements heavier than iron and spew them into interstellar space. It is from such debris that our Sun and solar system formed. According to the Big Bang scenario, atoms in your body were produced billions of years ago, in the hot interior of now-defunct stars, and the gold in your jewelry was produced during a supernova explosion that occurred in some distant place.

From Planetesimals to Protoplanets

Recall that the solar system, including Earth, formed about 4.6 billion years ago, from the **solar nebula**, a large rotating cloud of interstellar dust and gas (see Figure 12.4E). As the solar nebula contracted, most of the matter collected in the center to create the hot *protosun*. The remaining materials formed a thick, flattened, rotating disk, within which matter gradually cooled and condensed into grains and clumps of icy, rocky, and metallic material. Repeated collisions resulted in most of the material eventually collecting into asteroid-sized objects called **planetesimals**.

The composition of planetesimals was largely determined by their proximity to the protosun. As you might expect, temperatures were highest in the inner solar system and decreased toward the outer edge of the disk. Therefore, between the present orbits of Mercury and Mars, the planetesimals were composed mainly of materials with high melting temperatures—metals and rocky substances. The planetesimals that formed beyond the orbit of Mars, where temperatures are low, contained high percentages of ices—water, carbon dioxide, ammonia, and methane—as well as smaller amounts of rocky and metallic debris.

Through repeated collisions and accretion (sticking together), these planetesimals grew into eight **protoplanets** and their moons (see Figure 12.4G). During this process, the same amount of matter was concentrated into fewer and fewer bodies, each having greater and greater masses.

At some point in Earth's evolution, a giant impact occurred between a Mars-sized object and a young, semi-molten Earth. This collision ejected huge amounts of debris into space, some of which coalesced to form the Moon (see Figure 12.4J,K,L).

Earth's Early Evolution

As material continued to accumulate, the high-velocity impact of interplanetary debris (planetesimals) and the decay of radioactive elements caused the temperature of our planet to steadily increase. This early period of heating resulted in a magma ocean, perhaps several hundred kilometers deep. Within the magma ocean, buoyant masses of molten rock rose toward the surface and eventually solidified to produce thin rafts of crustal rocks. Geologists call this early period of Earth's history the **Hadean**, which began with the formation of Earth about 4.6 billion years ago and ended roughly 3.8 billion years ago (**FIGURE 12.5**). The name *Hadean* is derived from the Greek word *Hades*, meaning "the underworld," referring to the "hellish" conditions on Earth at the time.

During this period of intense heating, Earth became hot enough that iron and nickel began to melt. Melting produced liquid blobs of heavy metal that sank under their own weight. This process occurred rapidly on the scale of geologic time and produced Earth's dense iron-rich core. The

A. Big Bang
13.7 Ga

B. Hydrogen and
helium atoms
created

C. Our galaxy forms
10 Ga

D. Heavy elements
synthesized by
supernova
explosions

E. Solar nebula
begins to contract
4.7 Ga

F. As material collects
to form the protosun
rotation flattens nebula

G. Accretion of planetesimals
to form Earth and the
other planets

Earth

H. Continual bombardment
and the decay of radioactive elements
produces magma ocean

I. Chemical differentiation
produces Earth's
layered structure

J. Mars-size
object impacts
young Earth
4.6 Ga

K. Debris orbits
Earth and
accretes

L. Formation of Earth–Moon system
4.5 Ga

M. Outgassing produces Earth's
primitive atmosphere and ocean

SmartFigure 12.4 Major Events That Led to the Formation of Early Earth

formation of a molten iron core was the first of many stages of chemical differentiation in which Earth converted from a homogeneous body, with roughly the same matter at all depths, to a layered planet with material sorted by density (see Figure 12.4I).

This period of chemical differentiation established the three major divisions of Earth's interior—the iron-rich *core*; the thin *primitive crust*; and Earth's thickest layer, the *mantle*, located between the core and the crust. In addition, the lightest materials—including water vapor, carbon dioxide, and other gases—escaped to form a primitive atmosphere and, shortly thereafter, the oceans.

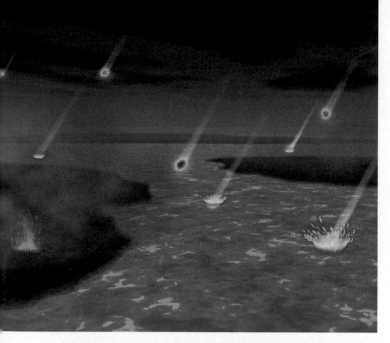

FIGURE 12.5 Artistic Depiction of Earth During the Hadean The Hadean is an unofficial eon of geologic time that occurred before the Archean. Its name refers to the "hellish" conditions on Earth. During the Hadean, Earth had a magma ocean and experienced intense bombardment by nebular debris.

12.2 CONCEPT CHECKS

1 What two elements made up most of the very early universe?

2 What is the name of the cataclysmic event in which an exploding star produces all the elements heavier than iron?

3 Briefly describe the formation of the planets from the solar nebula.

4 Describe the conditions on Earth during the Hadean.

12.3 | ORIGIN AND EVOLUTION OF THE ATMOSPHERE AND OCEANS

Describe how Earth's atmosphere and oceans have formed and evolved through time.

We can be thankful for our atmosphere; without it, there would be no greenhouse effect, and Earth would be nearly 60°F colder. Earth's water bodies would be frozen over, and the hydrologic cycle would be nonexistent.

The air we breathe is a stable mixture of 78 percent nitrogen, 21 percent oxygen, about 1 percent argon (an inert gas), and small amounts of gases such as carbon dioxide and water vapor. However, our planet's original atmosphere 4.6 billion years ago was substantially different.

Earth's Primitive Atmosphere

Early in Earth's formation, its atmosphere likely consisted of gases most common in the early solar system: hydrogen, helium, methane, ammonia, carbon dioxide, and water vapor. The lightest of these gases—hydrogen and helium—most likely escaped into space because Earth's gravity was too weak to hold them. The remaining gases—methane, ammonia, carbon dioxide, and water vapor—contain the basic ingredients of life: carbon, hydrogen, oxygen, and nitrogen. This early atmosphere was enhanced by a process called **outgassing**, through which gases trapped in the planet's interior are released. Outgassing from hundreds of active volcanoes still remains an important planetary function worldwide (**FIGURE 12.6**). However, early in Earth's history, when massive heating and fluidlike motion occurred in the mantle, the gas output would likely have been immense. These early eruptions probably released mainly water vapor, carbon dioxide, and sulfur dioxide, with minor amounts of other gases. Most importantly, free oxygen was not present in Earth's primitive atmosphere.

Oxygen in the Atmosphere

As Earth cooled, water vapor condensed to form clouds, and torrential rains began to fill low-lying areas, which became

FIGURE 12.6 Outgassing Produced Earth's First Enduring Atmosphere Outgassing continues today from hundreds of active volcanoes like this one in Iceland. (Photo by Lee Frost/Robert Harding)

the oceans. In those oceans, nearly 3.5 billion years ago, photosynthesizing bacteria began to release oxygen into the water. During *photosynthesis*, organisms use the Sun's energy to produce organic material (energetic molecules of sugar containing hydrogen and carbon) from carbon dioxide (CO_2) and water (H_2O). The first bacteria probably used hydrogen sulfide (H_2S) rather than water as the source of hydrogen. One of the earliest-known bacteria, *cyanobacteria* (once called blue-green algae), began to produce oxygen as a by-product of photosynthesis.

Initially, the newly released free oxygen was readily captured by chemical reactions with organic matter and dissolved iron in the ocean. It seems that large quantities of iron were released into the early ocean through submarine volcanism and associated hydrothermal vents. Iron has tremendous affinity for oxygen. When these two elements join, they become iron oxide (rust). As it accumulated on the seafloor, these early iron oxide deposits created alternating layers of iron-rich rocks and chert, called **banded iron formations**. Most banded iron deposits accumulated in the Precambrian, between 3.5 and 2 billion years ago, and represent the world's most important reservoir of iron ore.

As the number of oxygen-generating organisms increased, oxygen began to build in the atmosphere. Chemical analysis of rock suggests that oxygen first appeared in significant amounts in the atmosphere around 2.5 billion years ago, a phenomenon referred to as the **Great Oxygenation Event**. For the next billion years, oxygen levels in the atmosphere probably fluctuated but remained below 10 percent of current levels. Prior to the start of the Cambrian period 542 million years ago, which coincided with the evolution of complex life-forms, the level of free oxygen in the atmosphere began to climb. The availability of abundant oxygen in the atmosphere contributed to the proliferation of aerobic life-forms (oxygen-consuming organisms). One apparent spike in oxygen levels occurred during the Pennsylvanian period (300 million years ago), when oxygen made up about 35 percent of the atmosphere, compared to modern levels of only 21 percent. The large size of insects and amphibians during the Pennsylvanian period has been attributed to this abundance of oxygen.

Another positive benefit of the Great Oxygenation Event is that oxygen molecules (O_2) readily absorb ultraviolet radiation and form *ozone* (O_3), an oxygen molecule composed of three oxygen atoms. Ozone is concentrated between 10 and 50 kilometers (6 and 30 miles) above Earth's surface, in a layer called the *stratosphere*, where it absorbs much of the harmful ultraviolet radiation before it reaches the surface. For the first time, Earth's landmasses were protected from this form of solar radiation, which is particularly harmful to DNA—the stuff of which living organisms are made. Marine organisms had always been shielded from harmful ultraviolet radiation by seawater, but with the development of the atmosphere's protective ozone layer, the continents became more hospitable to plants and animals.

Evolution of the Oceans

When Earth cooled sufficiently to allow water vapor to condense, rainwater fell and collected in low-lying areas. By 4 billion years ago, it is estimated that as much as 90 percent of the current volume of seawater was contained in the developing ocean basins. Because volcanic eruptions released into the atmosphere large quantities of sulfur dioxide, which readily combines with water to form hydrochloric acid, the earliest rainwater was highly acidic. The level of acidity was even greater than the acid rain that damaged lakes and streams in eastern North America during the latter part of the twentieth century. Consequently, Earth's rocky surface weathered at an accelerated rate. The products released by chemical weathering included atoms and molecules of various substances—including sodium, calcium, potassium, and silica—that were carried

FIGURE 12.7 White Chalk Cliffs, England Similar deposits are also found in northern France. (Photo by Stuart Black/Robert Harding)

These prominent chalk cliffs are composed largely of tiny shells of marine organisms, such as foraminifera.

into the newly formed oceans. Some of these dissolved substances precipitated to become chemical sediment that mantled the ocean floor. Other substances formed soluble salts, which increased the salinity of seawater. Research suggests that the salinity of the oceans increased rapidly at first but has remained constant over the past 2 billion years.

Earth's oceans also serve as a repository for tremendous volumes of carbon dioxide, a major constituent in the primitive atmosphere. This is significant because carbon dioxide is a greenhouse gas that strongly influences the heating of the atmosphere. Venus, once thought to be very similar to Earth, has an atmosphere composed of 97 percent carbon dioxide that produced a "runaway" greenhouse effect. As a result, its surface temperature is 475°C (900°F).

Carbon dioxide is readily soluble in seawater, where it often joins other atoms or molecules to produce various chemical precipitates. The most common compound generated by this process is calcium carbonate ($CaCO_3$), which makes up limestone, the most abundant chemical sedimentary rock. About 542 million years ago, marine organisms began to extract calcium carbonate from seawater to make their shells and other hard parts. Included were trillions of tiny marine organisms, such as foraminifera, whose shells were deposited on the seafloor at the end of their life cycle. Today, some of these deposits can be observed in the chalk beds exposed along the White Cliffs of Dover, England, shown in **FIGURE 12.7**. By "locking up" carbon dioxide, these limestone deposits store this greenhouse gas so it cannot easily re-enter the atmosphere.

12.3 CONCEPT CHECKS

1 What is meant by *outgassing*, and what modern phenomenon serves that role today?

2 Identify the most abundant gases that were added to Earth's early atmosphere through the process of outgassing.

3 Why is the evolution of a type of bacteria that used photosynthesis to produce food important to most modern organisms?

4 Why was rainwater highly acidic early in Earth's history?

5 How does the ocean remove carbon dioxide from the atmosphere? What role do tiny marine organisms, such as foraminifera, play?

12.4 | PRECAMBRIAN HISTORY: THE FORMATION OF EARTH'S CONTINENTS Explain the formation of continental crust, how continental crust becomes assembled into continents, and the role that the supercontinent cycle has played in this process.

Earth's first 4 billion years are encompassed in the time span called the *Precambrian*. Representing nearly 90 percent of Earth's history, the Precambrian is divided into the *Archean eon* ("ancient age") and the *Proterozoic eon* ("early life"). Our knowledge of this ancient time is limited because much of the early rock record has been obscured by the very Earth processes you have been studying, especially plate tectonics, erosion, and deposition. Most Precambrian rocks lack fossils, which hinders correlation of rock units. In addition, rocks this old are often metamorphosed and deformed, extensively eroded, and frequently concealed by younger strata. Indeed, Precambrian history is written in scattered, speculative episodes, like a long book with many missing chapters.

Earth's First Continents

Geologists have discovered tiny crystals of the mineral zircon that formed 4.4 billion years ago—evidence that crustal rocks began to form early in Earth's history. What differentiates

EYE ON EARTH

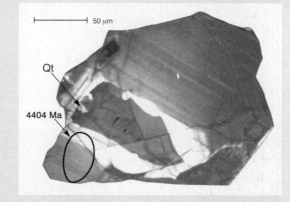

The oldest-known sample of Earth is a 4.4-billion-year-old zircon crystal found in a metaconglomerate in the Jack Hills area of western Australia. Zircon is a silicate mineral that commonly occurs in trace amounts in most granitic rocks. (Photo by John W. Valley/NSF)

QUESTION 1 What is the parent rock of a metaconglomerate?

QUESTION 2 Assuming that this zircon crystal originated as part of a granite intrusion, briefly describe its journey from the time of its formation until it was discovered in Jack Hill.

QUESTION 3 Is this zircon crystal younger or older than the metaconglomerate in which it was found? Explain.

FIGURE 12.8 Earth's Early Crust Was Continually Recycled (Photo by Mood Board/AGE Fotostock)

The crust covering this lava lake is continually being replaced with fresh lava from below, much like Earth's crust was recycled early in its history.

Making Continental Crust

Earth's first crust was probably ultramafic in composition, but because physical evidence no longer exists, we are not certain. The hot, turbulent mantle that most likely existed during the Archean eon recycled most of this material back into the mantle. In fact, it may have been continuously recycled, in much the same way that the "crust" that forms on a lava lake is repeatedly replaced with fresh lava from below (**FIGURE 12.8**).

The oldest preserved continental rocks occur as small, highly deformed terranes, which are incorporated within somewhat younger blocks of continental crust (**FIGURE 12.9**). One of the oldest of these is the 4-billion-year-old Acasta gneiss, located in Canada's Northwest Territories. Older crustal rocks may exist in the Nuvvuagittuq Greenstone Belt in Quebec, but the technique used to date them is controversial.

oceanic crust from continental crust? Recall that oceanic crust is a relatively dense (3.0 g/cm^3) homogeneous layer of basaltic rocks derived from partial melting of the rocky upper mantle. In addition, oceanic crust is thin, averaging only 7 kilometers (4.3 miles) thick. Continental crust, on the other hand, is composed of a variety of rock types, has an average thickness of nearly 40 kilometers (25 miles), and contains a large percentage of low-density (2.7 g/cm^3), silica-rich rocks such as granite.

The significance of the differences between continental crust and oceanic crust cannot be overstated in a review of Earth's geologic evolution. Oceanic crust, because it is relatively thin and dense, is found several kilometers below sea level—unless of course it has been pushed onto a landmass by tectonic forces. Continental crust, because of its great thickness and lower density, may extend well above sea level. Also, recall that oceanic crust of normal thickness readily subducts, whereas thick, buoyant blocks of continental crust resist being recycled into the mantle.

The formation of continental crust is a continuation of the gravitational segregation of Earth materials that began during the final stage of our planet's formation. Dense metallic material, mainly iron and nickel, sank to form Earth's core, leaving behind the less dense rocky material of which the mantle is composed. It is from Earth's rocky mantle that low-density, silica-rich minerals were gradually distilled to form continental crust. This process is analogous to making sour mash whiskeys. In the production of whiskeys, various grains, such as corn, are fermented to generate alcohol, with sour mash being the byproduct. This mixture is then heated or distilled, which drives off the lighter material (alcohol) and leaves behind the sour mash. In a similar manner, partial melting of mantle rocks generates low-density, silica-rich materials that buoyantly rise to the surface to form Earth's crust, leaving behind the dense mantle rocks (see Chapter 3). However, little is known about the details of the mechanisms that generated these silica-rich melts during the Archean.

Some geologists think that some type of plate-like motion operated early in Earth's history. In addition, hot-spot volcanism was likely active during this time. However, because the mantle was hotter in the Archean than it is today, both of these phenomena would have progressed at higher rates than their modern counterparts. Hot-spot volcanism is thought to

FIGURE 12.9 Earth's Oldest Preserved Continental Rocks Are About 4 Billion Years Old (Photo courtesy of James L. Amos/CORBIS)

These rocks at Isua, Greenland, some of the world's oldest, have been dated at 3.8 billion years.

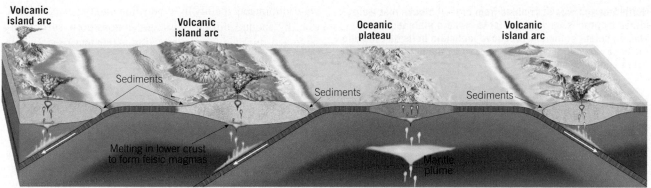

SmartFigure 12.10
Growth of Large Continental Masses Through the Collision and Accretion of Smaller Crustal Fragments

A. Scattered crustal fragments separated by ocean basins

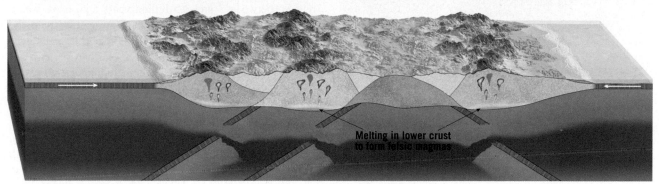

B. Collision of volcanic island arcs and oceanic plateau to form a larger crustal block

have created immense shield volcanoes as well as oceanic plateaus. Simultaneously, subduction of oceanic crust generated volcanic island arcs. Collectively, these relatively small crustal fragments represent the first phase in creating stable, continent-size landmasses.

From Continental Crust to Continents

The growth of larger continental masses was accomplished through collision and accretion of various types of crustal fragments, as illustrated in **FIGURE 12.10**. This type of collision tectonics deformed and metamorphosed sediments caught between converging crustal fragments, thereby shortening and thickening the developing crust. In the deepest regions of these collision zones, partial melting of the thickened crust generated silica-rich magmas that ascended and intruded the rocks above. The result was the formation of large crustal provinces that, in turn, accreted with others to form even larger crustal blocks called **cratons**.

The portion of a modern craton that is exposed at the surface is referred to as a **shield**. The assembly of a large craton involves the accretion of several crustal blocks that cause major mountain-building episodes similar to India's collision with Asia. **FIGURE 12.11** shows the extent of crustal material

FIGURE 12.11
Distribution of Crustal Material Remaining from the Archean and Proterozoic Eons

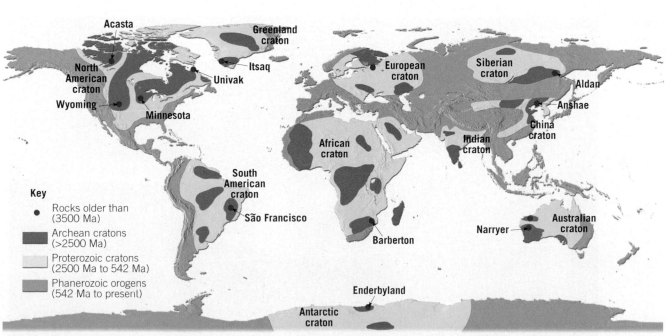

**SmartFigure 12.12
The Major Geologic
Provinces of North
America and Their Ages
in Billions of Years (Ga)**

North America was assembled from crustal blocks that were joined by processes very similar to modern plate tectonics. These ancient collisions produced mountain belts that include remnant volcanic island arcs, trapped by the colliding continental fragments.

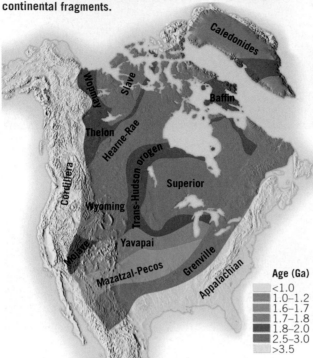

Age (Ga)
<1.0
1.0–1.2
1.6–1.7
1.7–1.8
1.8–2.0
2.5–3.0
>3.5

that was produced during the Archean and Proterozoic eons. This was accomplished by the collision and accretion of many thin, highly mobile terranes into nearly recognizable continental masses.

Although the Precambrian was a time when much of Earth's continental crust was generated, a substantial amount of crustal material was destroyed as well. Crust can be lost either by weathering and erosion or by direct reincorporation into the mantle through subduction. Evidence suggests that during much of the Archean, thin slabs of continental crust

were eliminated, mainly by subduction into the mantle. However, by about 3 billion years ago, cratons grew sufficiently large and thick to resist subduction. After that time, weathering and erosion became the primary processes of crustal destruction. By the close of the Precambrian, an estimated 85 percent of the modern continental crust had formed.

The Making of North America

North America provides an excellent example of the development of continental crust and its piecemeal assembly into a continent. Notice in **FIGURE 12.12** that very little continental crust older than 3.5 billion years remains. In the late Archean, between 3 and 2.5 billion years ago, there was a period of major continental growth. During this span, the accretion of numerous island arcs and other fragments generated several large crustal provinces. North America contains some of these crustal units, including the Superior and Hearne/Rae cratons shown in Figure 12.12. It remains unknown where these ancient continental blocks formed.

About 1.9 billion years ago, these crustal provinces collided to produce the Trans-Hudson mountain belt (see Figure 12.12). (Such mountain-building episodes were not restricted to North America because ancient deformed strata of similar age are also found on other continents.) This event built the North American craton, around which several large and numerous small crustal fragments were later added (**FIGURE 12.13**). One of these late arrivals is the Piedmont province of the Appalachians. In addition, several terranes were added to the western margin of North America during the Mesozoic and Cenozoic eras to generate the mountainous North American Cordillera.

Supercontinents of the Precambrian

At different times, parts of what is now North America have combined with other continental landmasses to form a supercontinent. **Supercontinents** are large landmasses that

FIGURE 12.13 Rocky Shoreline in the Superior Province The Superior province is part of the North American craton, around which several crustal fragments were later added. (Photo by Peter Van Rhijn/Getty Images)

contain all, or nearly all, the existing continents. Pangaea was the most recent, but certainly not the only, supercontinent to exist in the geologic past. The earliest well-documented supercontinent, *Rodinia*, formed during the Proterozoic eon, about 1.1 billion years ago. Although its reconstruction is still being researched, it is clear that Rodinia's configuration was quite different from Pangaea's (**FIGURE 12.14**). One obvious distinction is that North America was located near the center of this ancient landmass.

Between 800 and 600 million years ago, Rodinia gradually split apart. By the end of the Precambrian, many of the fragments reassembled, producing a large landmass in the Southern Hemisphere called *Gondwana*, comprised mainly of present-day South America, Africa, India, Australia, and Antarctica (**FIGURE 12.15**). Other continental fragments also developed—North America, Siberia, and Northern Europe. We consider the fate of these Precambrian landmasses later in the chapter.

Supercontinent Cycle

The idea that rifting and dispersal of one supercontinent is followed by a long period during which the fragments are gradually reassembled into a new supercontinent with a different configuration is called the **supercontinent cycle**. The assembly and dispersal of supercontinents had a profound impact on the evolution of Earth's continents. In addition, this phenomenon greatly influenced global climates and contributed to periodic episodes of rising and falling sea level.

Supercontinents, Mountain Building, and Climate

As continents move, the patterns of ocean currents and global winds change, which influences the global distribution of temperature and precipitation. One example of how a supercontinent's dispersal influenced climate is the formation of the Antarctic Ice Sheet. Although eastern Antarctica remained over the South Pole for more than 100 million years, it was not glaciated until about 25 million years ago. Prior to this period of glaciation, South America was connected to the Antarctic Peninsula. This arrangement of landmasses helped maintain a circulation pattern in which warm ocean currents reached the coast of Antarctica, as shown in **FIGURE 12.16A**. This is similar to the way in which the modern Gulf Stream keeps Iceland mostly ice free, despite its name. However, as South America separated from Antarctica, it moved northward, permitting ocean circulation to flow from west to east around the entire continent of Antarctica (**FIGURE 12.16B**). This cold current, called the West Wind Drift, effectively isolated the entire Antarctic coast from the warm,

poleward-directed currents in the southern oceans. As a result, most of the Antarctic landmass became covered with glacial ice.

Local and regional climates have also been impacted by large mountain systems created by the collision of large cratons. Because of their high elevations, mountains exhibit markedly lower average temperatures than surrounding lowlands. In addition, when air rises over these lofty structures, lifting "squeezes" moisture from the air, leaving the region downwind relatively dry. A modern analogy is the wet, heavily forested western slopes of the Sierra Nevada compared to the dry climate of the Great Basin Desert that lies directly to the east.

Supercontinents and Sea-Level Changes

Significant and numerous sea-level changes have been documented in geologic history, and many of them appear to be related to the assembly and dispersal of supercontinents. If sea level rises, shallow seas advance onto the continents. Evidence for periods when the seas advanced onto the continents include

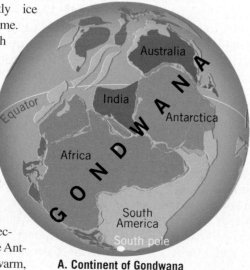

A. Continent of Gondwana

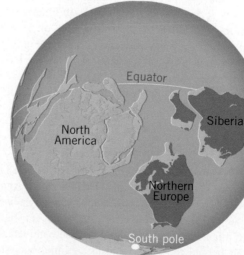

B. Continents not a part of Gondwana

50 million years ago warm ocean currents kept Antarctica nearly ice free.

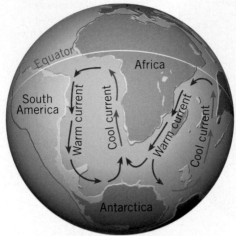

A. Not glaciated

As South America separated from Antarctica, the West Wind Drift developed. This newly formed ocean current effectively cut Antarctica off from warm currents and contributed to the formation of its vast ice sheets.

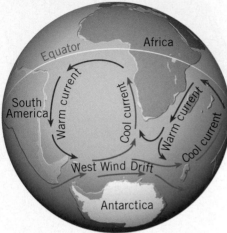

B. Glaciated

SmartFigure 12.16 Connection Between Ocean Circulation and the Climate in Antarctica

thick sequences of ancient marine sedimentary rocks that blanket large areas of modern landmasses—including much of the eastern two-thirds of the United States.

The supercontinent cycle and sea-level changes are directly related to rates of *seafloor spreading*. When the rate of spreading is rapid, as it is along the East Pacific Rise today, the production of warm oceanic crust is also high. Because warm oceanic crust is less dense (takes up more space) than cold crust, fast-spreading ridges occupy more volume in the ocean basins than do slow-spreading centers. (Think of getting into a tub filled with water.) As a result, when the rates of seafloor spreading increase, more seawater is displaced, which results in the sea level rising. This, in turn, causes shallow seas to advance onto the low-lying portions of the continents.

12.4 CONCEPT CHECKS

1 Briefly explain how low-density continental crust was produced from Earth's rocky mantle.

2 Describe how cratons came into being.

3 How can the movement of continents trigger climate change?

4 What is the supercontinent cycle? What supercontinent preceded Pangaea?

5 Explain how seafloor-spreading rates are related to sea-level changes.

12.5 | GEOLOGIC HISTORY OF THE PHANEROZOIC: THE FORMATION OF EARTH'S MODERN CONTINENTS List and discuss the major geologic events in the Paleozoic, Mesozoic, and Cenozoic eras.

The time span since the close of the Precambrian, called the *Phanerozoic eon*, encompasses 542 million years and is divided into three eras: *Paleozoic*, *Mesozoic*, and *Cenozoic*. The beginning of the Phanerozoic is marked by the appearance of the first life-forms with hard parts such as shells, scales, bones, or teeth—all of which greatly enhance the chances for an organism to be preserved in the fossil record. Thus, the study of Phanerozoic crustal history was aided by the availability of fossils, which improved our ability to date and correlate geologic events. Moreover, because every organism is associated with its own particular niche, the greatly improved fossil record provided invaluable information for deciphering ancient environments.

Paleozoic History

As the Paleozoic era opened, North America hosted no living things, neither plant nor animal. There were no Appalachian or Rocky Mountains; the continent was largely a barren lowland. Several times during the early Paleozoic, shallow seas moved inland and then receded from the continental interior and left behind the thick deposits of limestone, shale, and

clean sandstone that mark the shorelines of these previously midcontinent shallow seas.

Formation of Pangaea One of the major events of the Paleozoic was the formation of the supercontinent Pangaea, which began with a series of collisions that gradually joined North America, Europe, Siberia, and other smaller crustal fragments (**FIGURE 12.17**). These events eventually generated a large northern continent called *Laurasia*. This tropical landmass supported warm wet conditions that led to the formation of vast swamps that eventually converted to coal.

Simultaneously, the vast southern continent of *Gondwana* encompassed five continents—South America, Africa, Australia, Antarctica, and India—and perhaps portions of China. Evidence of extensive continental glaciation places this landmass near the South Pole. By the end of the Paleozoic, Gondwana had migrated northward and collided with Laurasia, culminating in the formation of the supercontinent *Pangaea*.

The accretion of Pangaea spans more than 300 million years and resulted in the formation of several mountain belts. The collision of Northern Europe (mainly Norway) with Greenland produced the Caledonian Mountains, whereas the joining of northern Asia (Siberia) and Europe created the Ural Mountains. Northern China is also thought to have accreted to Asia by the end of the Paleozoic, whereas southern China may not have become part of Asia until after Pangaea had begun to rift. (Recall that India did not begin to accrete to Asia until about 50 million years ago.)

Pangaea reached its maximum size about 250 million years ago, as Africa collided with North America (see Figure 12.17D). This event marked the final and most intense period of mountain building in the long history of the Appalachian Mountains (**FIGURE 12.18**).

Mesozoic History

Spanning about 186 million years, the Mesozoic era is divided into three periods: the *Triassic, Jurassic*, and *Cretaceous*. Major geologic events of the Mesozoic include the breakup of Pangaea and the evolution of our modern ocean basins.

Changes in Sea Levels   The Mesozoic era began with much of the world's continents above sea level. The exposed Triassic strata are primarily red sandstones and mudstones that lack marine fossils, features that indicate a terrestrial environment. (The red color in sandstone comes from the oxidation of iron.)

As the Jurassic period opened, the sea invaded western North America. Adjacent to this shallow sea, extensive continental sediments were deposited on what is now the Colorado Plateau. The most prominent is the Navajo Sandstone, a cross-bedded, quartz-rich layer that in some places approaches 300 meters (1000 feet) thick. These remnants of massive dunes indicate that an enormous desert occupied much of the American Southwest during early Jurassic times (**FIGURE 12.19**). Another well-known Jurassic deposit is the Morrison Formation—the world's richest storehouse of dinosaur fossils. Included are the fossilized bones of massive dinosaurs such as *Apatosaurus* (formerly *Brontosaurus*), *Brachiosaurus*, and *Stegosaurus*.

Coal Formation in Western North America   As the Jurassic period gave way to the Cretaceous, shallow seas again encroached upon much of western North America, as well as the Atlantic and Gulf coastal regions. This led to the formation of "coal swamps" similar to those of the Paleozoic era. Today, the Cretaceous coal deposits in the western United States and Canada are economically important. For example, on the Crow Native American reservation in Montana, there are nearly 20 billion tons of high-quality, Cretaceous-age coal.

The Breakup of Pangaea   Another major event of the Mesozoic era was the breakup of Pangaea. About 185 million years ago, a rift developed between what is now North America and western Africa, marking the birth of the Atlantic Ocean. As Pangaea gradually broke apart, the westward-moving North American plate began to override the Pacific basin. This tectonic event triggered a continuous wave of

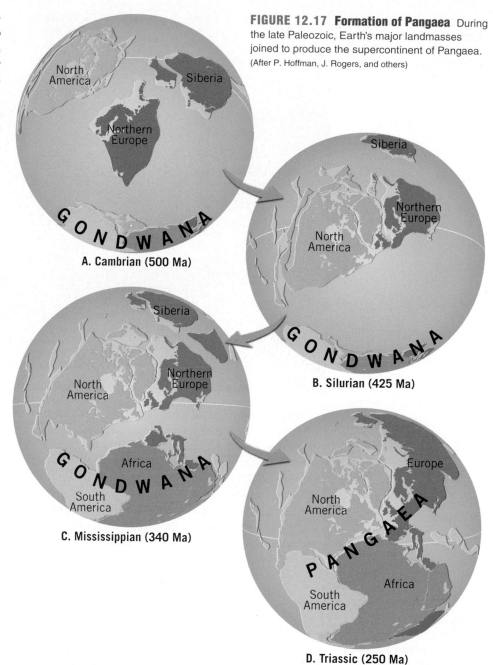

FIGURE 12.17 Formation of Pangaea During the late Paleozoic, Earth's major landmasses joined to produce the supercontinent of Pangaea. (After P. Hoffman, J. Rogers, and others)

A. Cambrian (500 Ma)

B. Silurian (425 Ma)

C. Mississippian (340 Ma)

D. Triassic (250 Ma)

deformation that moved inland along the entire western margin of North America.

Formation of the North American Cordillera   By Jurassic times, subduction of the Pacific basin under the North American plate began to produce the chaotic mixture of rocks that exist today in the Coast Ranges of California (see Figure 10.24). Further inland, igneous activity was widespread, and for more than 100 million years, volcanism was rampant as huge masses of magma rose within a few miles of Earth's surface. The remnants of this activity include the granitic plutons of the Sierra Nevada, as well as the Idaho batholith and British Columbia's Coast Range batholith.

The subduction of the Pacific basin under the western margin of North America also resulted in the piecemeal

FIGURE 12.18 **The Appalachian Mountains, a Product of a Continental Collision** (Photo by Jerry Whaley/AGE Fotostock)

addition of crustal fragments to the entire Pacific margin of the continent—from Mexico's Baja Peninsula to northern Alaska (see Figure 10.27). Each collision displaced earlier accreted terranes further inland, adding to the zone of deformation as well as to the thickness and lateral extent of the continental margin.

Compressional forces moved huge rock units in a shingle-like fashion toward the east. Across much of North America's western margin, older rocks were thrust eastward over younger strata, for distances exceeding 150 kilometers (90 miles). Ultimately, this activity was responsible for generating a vast portion of the North American Cordillera that extends from Wyoming to Alaska.

Toward the end of the Mesozoic, the southern portions of the Rocky Mountains developed. This mountain-building event, called the *Laramide Orogeny*, occurred when large blocks of deeply buried Precambrian rocks were lifted nearly vertically along steeply dipping faults, upwarping the overlying younger sedimentary strata. The mountain ranges produced by the Laramide Orogeny include Colorado's Front Range, the Sangre de Cristo of New Mexico and Colorado, and the Bighorns of Wyoming.

FIGURE 12.19 **Massive, Cross-Bedded Sandstone Cliffs in Zion National Park** These sandstone cliffs are the remnants of ancient sand dunes that were part of an enormous desert during the Jurassic period. (Photo by Michael Collier)

Cenozoic History

The Cenozoic era, or "era of recent life," encompasses the past 65.5 million years of Earth history. It was during this span that the physical landscapes and life-forms of our modern world came into existence. The Cenozoic era represents a considerably smaller fraction of geologic time than either the Paleozoic or the Mesozoic. Nevertheless, much more is known about this time span because the rock formations are more widespread and less disturbed than those of any preceding era.

Most of North America was above sea level during the Cenozoic era. However, the eastern and western margins of the continent experienced markedly contrasting events because of their different plate boundary relationships. The Atlantic and Gulf coastal regions, far removed from an active plate boundary, were tectonically stable. By contrast, western North America was the leading edge of the North American plate. As a result, plate interactions during the Cenozoic account for many events of mountain building, volcanism, and earthquakes.

Eastern North America The stable continental margin of eastern North America was the site of abundant marine sedimentation. The most extensive deposition surrounded the Gulf of Mexico, from the Yucatan Peninsula to Florida, where a massive buildup of sediment caused the crust to downwarp. In many instances, faulting created structures in which oil and natural gas accumulated. Today, these and other petroleum traps are the Gulf Coast's most economically important resource, evidenced by numerous offshore drilling platforms.

By early Cenozoic time, the Appalachians had eroded to create a low plain. Later, isostatic adjustments again raised the region and rejuvenated its rivers. Streams eroded with renewed vigor, gradually sculpting the surface into its present-day topography. Sediments from this erosion were deposited along the eastern continental margin, where they accumulated to a thickness of many kilometers. Today, portions of the strata deposited during the Cenozoic are exposed as the gently sloping Atlantic and Gulf coastal plains, where a large percentage of the eastern and southeastern U.S. population resides.

Western North America In the West, the Laramide Orogeny responsible for building the southern Rocky Mountains was coming to an end. As erosional forces lowered the mountains, the basins between uplifted ranges began to fill with sediment. East of the Rockies, a large wedge of sediment from the eroding mountains created the gently sloping Great Plains.

Beginning in the Miocene epoch, about 20 million years ago, a broad region from northern Nevada into Mexico experienced crustal extension that created more than 100 fault-block mountain ranges. Today, they rise abruptly above the adjacent basins, forming the Basin and Range Province (see Figure 10.16).

During the development of the Basin and Range Province, the entire western interior of the continent gradually uplifted. This event reelevated the Rockies and rejuvenated many of the West's major rivers. As the rivers became incised, many spectacular gorges were created, including the Grand Canyon of the Colorado River, the Grand Canyon of the Snake River, and the Black Canyon of the Gunnison River.

Volcanic activity was also common in the West during much of the Cenozoic. Beginning in the Miocene epoch, great volumes of fluid basaltic lava flowed from fissures in portions of present-day Washington, Oregon, and Idaho. These eruptions built the extensive (3.4-million-square-kilometer [1.3-million-square–mile]) Columbia Plateau. Immediately west of the Columbia Plateau, volcanic activity was different in character. Here, more viscous magmas with higher silica content erupted explosively, creating the Cascades, a chain of stratovolcanoes extending from northern California into Canada, some of which are still active (**FIGURE 12.20**).

As the Cenozoic was drawing to a close, the effects of mountain building, volcanic activity, isostatic adjustments, and extensive erosion and sedimentation created the physical landscape we know today. All that remained of Cenozoic time was the final 2.6-million-year episode called the Quaternary period. During this most recent, and ongoing, phase of Earth's history, humans evolved, and the action of glacial ice, wind, and running water added to our planet's long, complex geologic history.

FIGURE 12.20 Mount Shasta, California This volcano is one of several large composite cones that comprise the Cascade Range. (Photo by Michael Collier)

12.5 CONCEPT CHECKS

1 During which period of geologic history did the supercontinent Pangaea come into existence?

2 Where is most Cretaceous age coal found today in the United States?

3 During which period of geologic history did Pangaea begin to break apart?

4 Describe the climate during the early Jurassic period.

5 Compare and contrast eastern and western North America during the Cenozoic era.

12.6 | EARTH'S FIRST LIFE Describe some of the hypotheses on the origin of life and the characteristics of early prokaryotes, eukaryotes, and multicelled organisms.

The oldest fossils provide evidence that life on Earth was established at least 3.5 billion years ago. Microscopic fossils similar to modern cyanobacteria have been found in silica-rich chert deposits worldwide. Notable examples include southern Africa, where rocks date to more than 3.1 billion years, and the Lake Superior region of western Ontario and northern Minnesota, where the Gunflint Chert contains some fossils that are older than 2 billion years. Chemical traces of organic matter in rocks of greater age have led paleontologists to strongly suggest that life may have existed as early as 3.8 billion years ago.

Origin of Life

How did life begin? This question sparks considerable debate, and hypotheses abound. Requirements for life, assuming the presence of a hospitable environment, include the chemical raw materials that are found in essential molecules such as proteins. Proteins are made from organic compounds called *amino acids*. The first amino acids may have been synthesized from methane and ammonia, both of which were plentiful in Earth's primitive atmosphere. Some scientists suggest that these gases could have been easily reorganized into useful organic molecules by ultraviolet light. Others consider lightning to have been the impetus, as the well-known experiments conducted by Stanley Miller and Harold Urey attempted to demonstrate.

Still other researchers suggest that amino acids arrived "ready made," delivered by asteroids or comets that collided with a young Earth. A group of meteorites (debris from asteroids and comets that strike Earth) called *carbonaceous chrondrites*, known to contain amino acid–like organic compounds, led to a hypothesis which concludes that early life had an extraterrestrial beginning.

Yet another hypothesis proposes that the organic material needed for life came from the methane and hydrogen sulfide that spews from deep-sea hydrothermal vents (black smokers). It is also possible that life originated in hot springs similar to those found in Yellowstone National Park.

Earth's First Life: Prokaryotes

Regardless of where or how life originated, it is clear that the journey from "then" to "now" involved change. The first known organisms were simple single-cell bacteria called **prokaryotes**, which means their genetic material (DNA) is *not separated* from the rest of the cell by a nucleus. Because oxygen was largely absent from Earth's early atmosphere and oceans, the first organisms employed anaerobic (without oxygen) metabolism to extract energy from "food." Their food source was likely organic molecules in their surroundings, but that supply was very limited. Later, bacteria evolved that used solar energy to synthesize organic compounds (sugars). This event was an important turning point in biological evolution: For the first time, organisms had the capability of producing food for themselves as well as for other life-forms.

Recall that photosynthesis by ancient cyanobacteria, a type of prokaryote, contributed to the gradual rise in the level of oxygen, first in the ocean and later in the atmosphere. It was these early organisms, which began to inhabit Earth 3.5 billion years ago, that dramatically transformed our planet. Fossil evidence for the existence of these microscopic bacteria includes distinctively layered mats, called **stromatolites**, composed of slimy material secreted by these organisms, along with trapped sediments (**FIGURE 12.21A**). What is known about these ancient fossils comes mainly from the study of

FIGURE 12.21
Stromatolites Are Among the Most Common Precambrian Fossils A. Cross-section though fossil stromatolites deposited by cyanobacteria. (Photo by Sinclair Stammers/Science Source) **B.** Modern stromatolites exposed at low tide in western Australia. (Photo by Bill Bachman/Science Source)

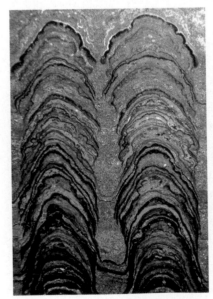

A.

B.

modern stromatolites like those found in Shark Bay, Australia (**FIGURE 12.21B**). Today's stromatolites look like stubby pillars built as microbes slowly move upward to avoid being buried by sediment that is continually being deposited on them.

Evolution of Eukaryotes
The oldest fossils of more advanced organisms, called **eukaryotes**, are about 2.1 billion years old. The first eukaryotes were microscopic, single-cell organisms, but unlike prokaryotes, eukaryotes contain nuclei. This distinctive cellular structure is what all multicellular organisms that now inhabit our planet—trees, birds, fish, reptiles, and humans—have in common.

During much of the Precambrian, life consisted exclusively of single-celled organisms. It wasn't until perhaps 1.2 billion years ago that multicelled eukaryotes evolved. Green algae, one of the first multicelled organisms, contained chloroplasts (used in photosynthesis)

FIGURE 12.22 Ediacaran Fossil The Ediacarans are a group of sea-dwelling animals that may have come into existence about 600 million years ago. These soft-bodied organisms were up to 1 meter in length and are the oldest animal fossils so far discovered. (Photo by Sinclair Stammers/Science Source)

and were the likely ancestors of modern plants. The first primitive marine animals did not appear until somewhat later, perhaps 600 million years ago (**FIGURE 12.22**).

Fossil evidence suggests that organic evolution progressed at an excruciatingly slow pace until the end of the Precambrian. At that time, Earth's continents were largely barren, and the oceans were populated by small organisms, many too small to be seen with the naked eye. Nevertheless, the stage was set for the evolution of larger and more complex plants and animals.

12.6 CONCEPT CHECKS

1 What group of organic compounds are essential for the formation of DNA and RNA and therefore necessary for life as we know it?

2 Why do some researchers think that a type of asteroid, called *carbonaceous chrondrites*, played an important role in the development of life on Earth?

3 What are stromatolites? What group of organisms is thought to have produced them?

4 Compare *prokaryotes* with *eukaryotes*. Within which of these two groups do all multicelled organisms belong?

12.7 | PALEOZOIC ERA: LIFE EXPLODES
Discuss the major developments in the history of life during the Paleozoic era.

The Cambrian period marks the beginning of the Paleozoic era, a time span that saw the emergence of a spectacular variety of new life-forms. All major invertebrate (animals lacking backbones) groups made their appearance, including jellyfish, sponges, worms, mollusks (clams and snails), and arthropods (insects and crabs). This huge expansion in biodiversity is often referred to as the **Cambrian explosion**.

the first truly large organisms on Earth, including one species that reached a length of nearly 10 meters (30 feet).

The early diversification of animals was driven, in part, by the emergence of predatory lifestyles. The larger mobile cephalopods preyed on trilobites that were typically smaller than a child's hand. The evolution of efficient movement was often associated with the development of greater sensory

Early Paleozoic Life-Forms

The Cambrian period was the golden age of *trilobites* (**FIGURE 12.23**). Trilobites developed a flexible exoskeleton of a protein called chitin (similar to a lobster shell), which permitted them to be mobile and search for food by burrowing through soft sediment. More than 600 genera of these mud-burrowing scavengers flourished worldwide.

The Ordovician marked the appearance of abundant cephalopods—mobile, highly developed mollusks that became the major predators of their time (**FIGURE 12.24**). Descendants of these cephalopods include the squid, octopus, and chambered nautilus that inhabit our modern oceans. Cephalopods were

FIGURE 12.23 Fossil of a Trilobite Trilobites dominated the early Paleozoic ocean, scavenging food from the bottom. (Photo by Ed Reschke/Getty Images)

Evolution of Life Through Geologic Time

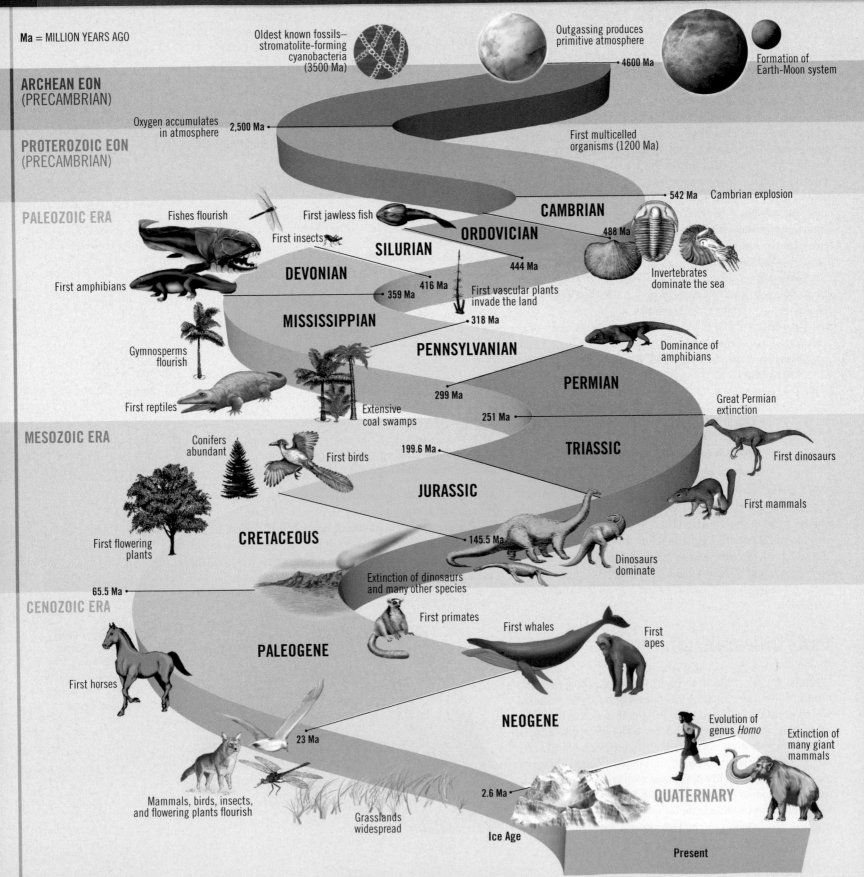

Ma = MILLION YEARS AGO

Oldest known fossils—stromatolite-forming cyanobacteria (3500 Ma)

Outgassing produces primitive atmosphere

4600 Ma

Formation of Earth-Moon system

ARCHEAN EON (PRECAMBRIAN)

Oxygen accumulates in atmosphere

2,500 Ma

First multicelled organisms (1200 Ma)

PROTEROZOIC EON (PRECAMBRIAN)

542 Ma Cambrian explosion

PALEOZOIC ERA

Fishes flourish

First jawless fish

CAMBRIAN

488 Ma

First insects

ORDOVICIAN

SILURIAN

444 Ma

Invertebrates dominate the sea

First amphibians

DEVONIAN

416 Ma

359 Ma

First vascular plants invade the land

MISSISSIPPIAN

318 Ma

Gymnosperms flourish

PENNSYLVANIAN

Dominance of amphibians

PERMIAN

299 Ma

First reptiles

Extensive coal swamps

251 Ma

Great Permian extinction

MESOZOIC ERA

Conifers abundant

First birds

199.6 Ma

TRIASSIC

First dinosaurs

JURASSIC

First mammals

CRETACEOUS

First flowering plants

145.5 Ma

Dinosaurs dominate

Extinction of dinosaurs and many other species

CENOZOIC ERA

65.5 Ma

First primates

First whales

First apes

PALEOGENE

First horses

NEOGENE

Evolution of genus *Homo*

Extinction of many giant mammals

23 Ma

2.6 Ma

QUATERNARY

Mammals, birds, insects, and flowering plants flourish

Grasslands widespread

Ice Age

Present

FIGURE 12.24 **Artistic Depiction of a Shallow Ordovician Sea** During the Ordovician period (488–444 million years ago), the shallow waters of an inland sea over central North America contained an abundance of marine invertebrates. Shown in this reconstruction are 1) corals, 2) trilobites, 3) snails, 4) brachiopods, and 5) straight-shelled cephalopods. (Field Museum/Getty Images)

capabilities and more complex nervous systems. These early animals developed sensory devices for detecting light, odor, and touch.

Approximately 400 million years ago, green algae that had adapted to survive at the water's edge gave rise to the first multicellular land plants. The primary difficulty in sustaining plant life on land was obtaining water and staying upright, despite gravity and winds. These earliest land plants were leafless, vertical spikes about the size of a human index finger (**FIGURE 12.25**). However, the fossil record indicates that by the beginning of the Mississippian period, there were forests with trees tens of meters tall (see Figure 12.25).

In the ocean, fish perfected an internal skeleton as a new form of support, and they were the first creatures to have jaws. Armor-plated fish that evolved during the Ordovician continued to adapt. Their armor plates thinned to lightweight scales that increased their speed and mobility. Other fish evolved during the Devonian, including primitive sharks with cartilage skeletons and bony fish—the groups in which many modern fish are classified. Fish, the first large vertebrates, proved to be faster swimmers than invertebrates and possessed more acute senses and larger brains. They became the dominant predators of the sea,

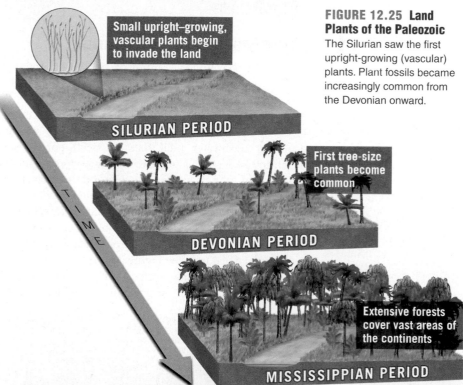

Small upright–growing, vascular plants begin to invade the land

SILURIAN PERIOD

TIME

First tree-size plants become common

DEVONIAN PERIOD

Extensive forests cover vast areas of the continents

MISSISSIPPIAN PERIOD

FIGURE 12.25 **Land Plants of the Paleozoic** The Silurian saw the first upright-growing (vascular) plants. Plant fossils became increasingly common from the Devonian onward.

EYE ON EARTH

The rocks shown here are Cambrian-age stromatolites of the Hoyt Limestone, exposed at Lester Park, near Saratoga Springs, New York. (Michael C. Rygel)

QUESTION 1 *Using Figure 12.3, determine approximately how many years ago these rocks were deposited.*

QUESTION 2 *What is the name of the group of organisms that likely produced these limestone deposits?*

QUESTION 3 *What was the environment like in this part of New York when these rocks were deposited?*

FIGURE 12.26 Comparison of the Anatomical Features of the Lobe-Finned Fish and Early Amphibians **A.** The fins on the lobe-finned fish contained the same basic elements (*h*, humerus, or upper arm; *r*, radius; and *u*, ulna, or lower arm) as those of the amphibians. **B.** This amphibian is shown with the standard five toes, but early amphibians had as many as eight toes. Eventually the amphibians evolved to have a standard toe count of five.

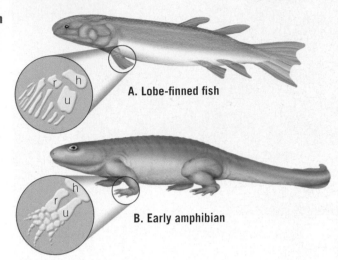

A. Lobe-finned fish

B. Early amphibian

which is why the Devonian period is often referred to as the "Age of the Fishes."

Vertebrates Move to Land

During the Devonian, a group of fish called the *lobe-finned fish* began to adapt to terrestrial environments (**FIGURE 12.26**). Lobe-finned fish had internal sacks that could be filled with air to supplement their "breathing" through gills. One group of lobe-finned fish probably occupied freshwater tidal flats or small ponds. Some began to use their fins to move from one pond to another in search of food or to evacuate deteriorating ponds. This favored the evolution of a group of animals able to stay out of water longer and move on land more efficiently. By the late Devonian, lobe-finned fish had evolved into air-breathing amphibians with strong legs yet retained a fishlike head and tail (see Figure 12.26).

Modern amphibians, such as frogs, toads, and salamanders, are small and occupy limited biological niches.

However, conditions during the late Paleozoic were ideal for these newcomers to land. Large tropical swamps that were teeming with large insects and millipedes extended across North America, Europe, and Siberia (**FIGURE 12.27**). With virtually no predatory risks, amphibians diversified rapidly. Some even took on lifestyles and forms similar to modern reptiles such as crocodiles.

Despite their success, amphibians were not fully adapted to life out of water. In fact, *amphibian* means "double life" because these animals need both the water from where they came and the land to which they moved. Amphibians are born in water, as exemplified by tadpoles, complete with gills and tails. These features disappear during the maturation process, resulting in air-breathing adults with legs.

Reptiles: The First True Terrestrial Vertebrates

Reptiles were the first true terrestrial animals with improved lungs for active lifestyles and "waterproof" skin that helped prevent the loss of body fluids. Most importantly, reptiles developed shell-covered eggs laid on land. The elimination of a water-dwelling stage (like the tadpole stage in frogs) was an important evolutionary step. Vertebrates advance from water-dwelling lobe-finned fish, to amphibians, to reptiles—the first true terrestrial vertebrates (**FIGURE 12.28**).

Of interest is the fact that the watery fluid within the reptilian egg closely resembles seawater in chemical composition. Because the reptile embryo develops in this watery environment, the shelled egg has been characterized as a "private aquarium" in which the embryos of these land vertebrates spend their water-dwelling stage of life. With this "sturdy egg," the remaining ties to the water were broken, and reptiles moved inland.

FIGURE 12.27 Artistic Depiction of a Pennsylvanian-Age Coal Swamp Shown are scale trees (left), seed ferns (lower left), and scouring rushes (right). Also note the large dragonfly. (Field Museum/Getty Images)

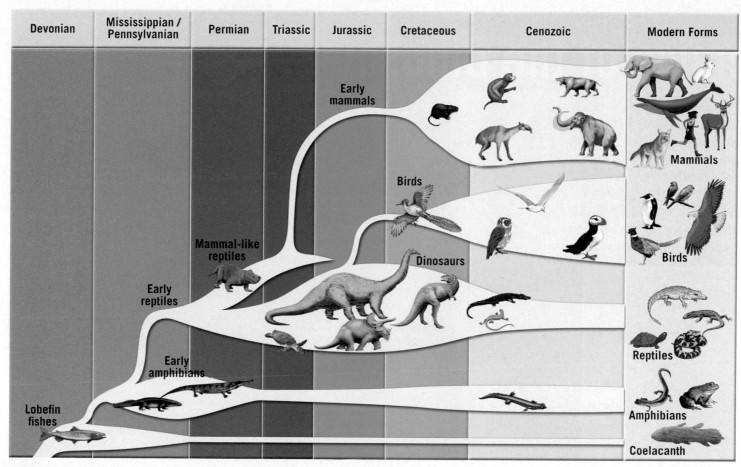

Devonian	Mississippian / Pennsylvanian	Permian	Triassic	Jurassic	Cretaceous	Cenozoic	Modern Forms

 SmartFigure 12.28 Relationships of Vertebrate Groups and Their Divergence from Lobefin Fish

The Great Permian Extinction

By the close of the Permian period, a **mass extinction** occurred in which a large number of Earth's species became extinct. During this mass extinction, 70 percent of all land-dwelling vertebrate species and perhaps 90 percent of all marine organisms were obliterated; it was the most significant of five mass extinctions that occurred over the past 500 million years. Each extinction wreaked havoc with the existing biosphere, wiping out large numbers of species. In each case, however, survivors entered new biological communities that were ultimately more diverse. Therefore, mass extinctions actually invigorated life on Earth, as the few hardy survivors eventually filled more environmental niches than those left behind by the victims.

Several mechanisms have been proposed to explain these ancient mass extinctions. Initially, paleontologists believed they were gradual events caused by a combination of climate change and biological forces, such as predation and competition. Other research groups have attempted to link certain mass extinctions to the explosive impact of a large asteroid striking Earth's surface.

The most widely held view is that the Permian mass extinction was driven mainly by volcanic activity because it coincided with a period of voluminous eruptions of flood basalts that blanketed about 1.6 million square kilometers (624,000 square miles), an area nearly the size of Alaska. This event, which lasted roughly 1 million years, occurred in northern Russia, in an area called the Siberian Traps. It was the largest volcanic eruption in the past 500 million years. The release of huge amounts of carbon dioxide likely generated a period of accelerated greenhouse warming, while the emissions of sulfur dioxide is credited with producing copious amounts of acid rain. These drastic changes in the environment likely put excessive stress on many of Earth's life-forms.

12.7 CONCEPT CHECKS

1 What is the Cambrian explosion?

2 What animal group was dominant in Cambrian seas?

3 What did plants have to overcome in order to move onto land?

4 What group of animals is thought to have left the ocean to become the first amphibians?

5 Why are amphibians not considered "true" land animals?

6 What major development allowed reptiles to move inland?

Demise of the Dinosaurs

The geologic time scale marks the transitions, or boundaries, of significant periods of geological and/or biological change. Of special interest to geoscientists is the boundary between the Mesozoic Era and the Cenozoic Era, about 65.5 million years ago, when roughly 70 percent of all plant and animal species died out in a mass extinction.

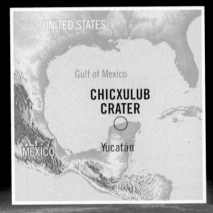

UNITED STATES

Gulf of Mexico

CHICXULUB CRATER

MEXICO

Yucatán

The most widely accepted hypothesis on the demise of the dinosaurs proposes that about 65.5 million years ago, Earth was struck by an asteroid—a relict from the formation of the solar system. The errant mass of rock was approximately 10 kilometers in diameter and struck Earth with 100 million megatons of force.

This asteroid struck a shallow tropical sea in the area of Mexico's Yucatan Peninsula and generated Chicxulub crater, which is about 180 kilometers in diameter and 10 kilometers deep. Much of the plant and animal life that survived the initial impact likely fell victim to firestorms triggered by atmospheric heating. Following the impact, the huge amount of debris ejected into the atmosphere greatly reduced the amount of sunlight reaching Earth's surface. This resulted in an extended period of global cooling that inhibited photosynthesis, which ultimately caused the demise of the dinosaurs.

Detlev van Ravenswaay/Science Source

Artistic recreation of Chicxulub crater

The transition from the Mesozoic to the Cenozoic era marks the end of the dinosaurs, a group that had dominated the landscape for more than 100 million years, and the beginning of the era when mammals became the most prominent land animals.

Pholri Inc./AGE Fotostock

What evidence supports a massive, catastrophic impact more than 65 million years ago? A thin layer of sediment has been discovered worldwide at Earth's physical boundary separating the Mesozoic and Cenozoic eras. This sediment contains a high level of iridium, an element that is rare in Earth's crust but is found in much higher proportions in stony meteorites. Scientists believe this layer contains debris from the impact.

The mass extinction of the dinosaurs opened habitats for the small mammals that survived the impact. These new habitats, along with evolutionary forces, led to the development of the large mammals that occupy our modern world.

Oleg Znamenskiy/Shutterstock

Mesozoic rocks

IRIDIUM LAYER

Wesley Aston/Shutterstock

VikOl/Shutterstock

Cenozoic rocks

Nationalparks

12.8 | MESOZOIC ERA: AGE OF THE DINOSAURS

Discuss the major developments in the history of life during the Mesozoic era.

As the Mesozoic era dawned, its life-forms were the survivors of the great Permian extinction. These organisms diversified in many ways to fill the biological voids created at the close of the Paleozoic. While life on land underwent a radical transformation with the rise of the dinosaurs, life in the sea also entered a dramatic phase of transformation that produced many of the animal groups that prevail in the oceans today, including modern groups of predatory fish, crustaceans, mollusks, and sand dollars.

Gymnosperms: The Dominant Mesozoic Trees

On land, conditions favored organisms that could adapt to drier climates. One such group of plants, **gymnosperms**, produced "naked" seeds that are exposed on modified leaves that usually form cones. The seeds are not enclosed in fruits as, for example, are apple seeds. Unlike the first plants to invade the land, seed-bearing gymnosperms did not depend on free-standing water for fertilization. Consequently, these plants were not restricted to a life near the water's edge.

The gymnosperms quickly became the dominant trees of the Mesozoic. Examples of this group include cycads that resembled large pineapple plants (**FIGURE 12.29**); ginkgo plants that had fan-shaped leaves, much like

FIGURE 12.30 Petrified Logs of Triassic Age, Arizona's Petrified Forest National Park (Photo by Bernd Siering/AGE Fotostock)

FIGURE 12.29 Cycads, a Type of Gymnosperm That Was Very Common in the Mesozoic These plants have palm-like leaves and large cones. (Photo by Jiri Loun/ Science Source)

their modern relatives; and the largest plants, the conifers, whose modern descendants include the pines, firs, and junipers. The best-known fossil occurrence of these ancient trees is in northern Arizona's Petrified Forest National Park. Here, huge petrified logs lie exposed at the surface, having been weathered from rocks of the Triassic Chinle Formation (**FIGURE 12.30**).

Reptiles: Dominating the Land, Sea, and Sky

Among the animals, reptiles readily adapted to the drier Mesozoic environment, thereby relegating amphibians to the swamps and wetlands, where most remain today. The first reptiles were small, but larger forms evolved rapidly, particularly the dinosaurs. One of the largest was *Apatosaurus*, which weighed more than 30 tons and measured over 25 meters (80 feet) from head to tail. Some of the largest dinosaurs were carnivorous (for example, *Tyrannosaurus*), whereas others were herbivorous (like ponderous *Apatosaurus*).

Some reptiles evolved specialized characteristics that enabled them to occupy drastically different environments. One group, the pterosaurs, became airborne. These "dragons of the sky" possessed huge membranous wings that allowed them rudimentary flight. How the largest pterosaurs (some had wing spans of 8 meters [26 feet] and weighed 90 kilograms [200 pounds]) took flight is still unknown. Another group, exemplified by

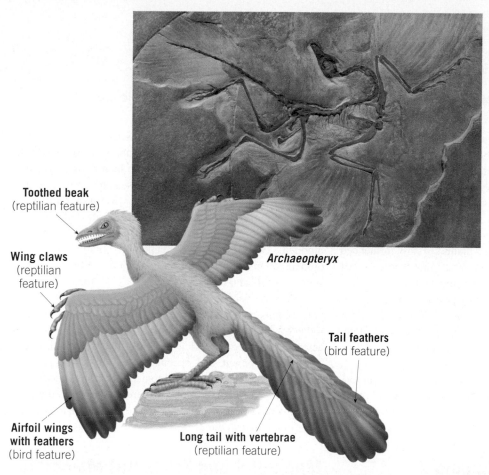

FIGURE 12.31 Flying Reptiles Like *Archaeopteryx* Are the Ancestors of Modern Birds Fossil evidence indicates that *Archaeopteryx* had feathers like modern birds but retained many characteristics of reptiles. The sketch shows an artist's reconstruction of *Archaeopteryx*. (Photo by Michael Collier)

Toothed beak
(reptilian feature)

Wing claws
(reptilian feature)

Archaeopteryx

Tail feathers
(bird feature)

Airfoil wings with feathers
(bird feature)

Long tail with vertebrae
(reptilian feature)

the fossil *Archaeopteryx*, led to more successful flyers—birds (**FIGURE 12.31**). These ancestors of modern birds had feathered wings but retained reptilian characteristics, such as sharp teeth, clawed digits in the wings, and a long tail with many vertebrae. A recent study concluded that *Archaeopteryx* were unable to use flapping flight. Rather, by running and leaping into the air, these bird-like reptiles escaped predators with glides and downstrokes. Other researchers disagree and see them as climbing animals that glided down to the ground, following the idea that birds evolved from tree-dwelling gliders. Whether birds took to the air from the ground *up* or from the trees *down* is a question scientists continue to debate.

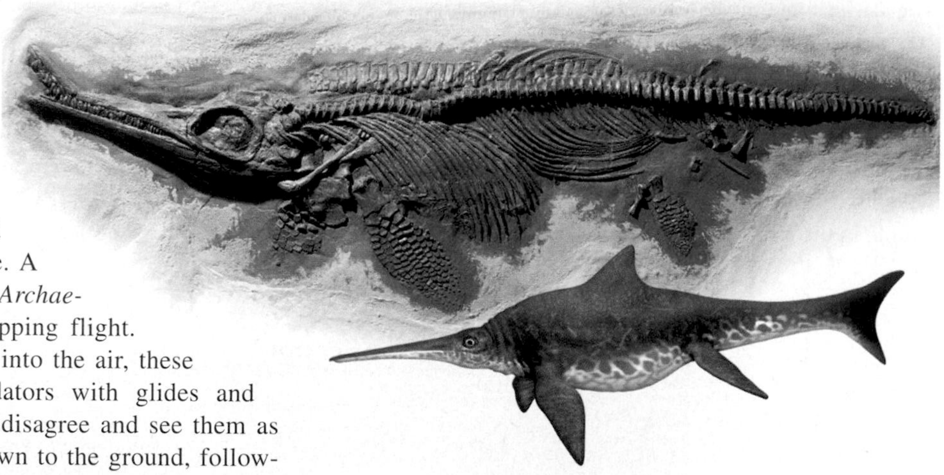

Other reptiles returned to the sea, including fish-eating *plesiosaurs* and *ichthyosaurs* (**FIGURE 12.32**). These reptiles became proficient swimmers but retained their reptilian teeth and breathed by means of lungs rather than gills.

For nearly 160 million years, dinosaurs reigned supreme. However, by the close of the Mesozoic, like many reptiles, they became extinct. Select reptile groups survived to recent times, including turtles, snakes, crocodiles, and lizards. The huge, land-dwelling dinosaurs, the marine plesiosaurs, and the flying pterosaurs are known only through the fossil record. What caused this great

FIGURE 12.32 During the Mesozoic, Some Reptiles Returned to the Sea Reptiles, including *Ichthyosaur*, became the dominant marine animals. (Photo by Chip Clark/Fundamental Photographs, NYC)

extinction? The GEOgraphics on page 396 provides the most plausible answer to this question.

12.8 CONCEPT CHECKS

1 What group of plants became the dominant trees during the Mesozoic era? Name a modern descendant of this group.

2 What group of reptiles led to the evolution of modern birds?

3 What was the dominant reptile group on land during the Mesozoic?

4 Name two reptiles that returned to life in the sea.

12.9 | CENOZOIC ERA: AGE OF MAMMALS

Discuss the major developments in the history of life during the Cenozoic era.

During the Cenozoic, mammals replaced reptiles as the dominant land animals. At nearly the same time, **angiosperms** (flowering plants with covered seeds) replaced gymnosperms as the dominant plants. The Cenozoic is often called the "Age of Mammals" but can also be considered the "Age of Flowering Plants" because, in the plant world, angiosperms enjoy a status similar to that of mammals in the animal world.

The development of flowering plants strongly influenced the evolution of both birds and mammals that feed on seeds and fruits, as well as many insect groups. During the middle of the Cenozoic, another type of angiosperm, grasses, developed and spread rapidly over the plains (**FIGURE 12.33**). This fostered the emergence of herbivorous (plant-eating) mammals, which, in turn, provided the evolutionary foundation for large predatory mammals.

During the Cenozoic, the ocean was teeming with modern fish such as tuna, swordfish, and barracuda. In addition, some mammals, including seals, whales, and walruses, took up life in the sea.

Mammals are distinct from reptiles in that they give birth to live young that suckle on milk and are warm blooded. This latter adaptation allowed mammals to lead more active lives and to occupy more diverse habitats than reptiles because they could survive in cold regions. (Modern reptiles are cold blooded and dormant during cold weather. However, recent studies suggest that dinosaurs may have been warm blooded.) Other mammalian adaptations included the development of insulating body hair and more efficient organs, such as hearts and lungs.

With the demise of the large Mesozoic reptiles, Cenozoic mammals diversified rapidly. The many forms that exist today evolved from small primitive mammals that were characterized by short legs; flat, five-toed feet; and small brains. Their development and specialization took four principal directions: increase in size, increase in brain capacity, specialization of teeth to better accommodate their diet, and specialization of limbs to be better equipped for a particular lifestyle or environment.

From Reptiles to Mammals

The earliest mammals coexisted with dinosaurs for nearly 100 million years but were small rodent-like creatures that gathered food at night, when dinosaurs were probably less active. Then, about 65 million years ago, fate intervened when a large asteroid collided with Earth and dealt a crashing blow to the reign of the dinosaurs. This transition, during which one dominant group is replaced by another, is clearly visible in the fossil record.

Marsupial and Placental Mammals

Two groups of mammals, the marsupials and the placentals, evolved and diversified during the Cenozoic. The groups differ principally in their modes of reproduction. Young marsupials are born live at a very early stage of development. At birth, the tiny and immature young enter the mother's pouch to suckle and complete their development. Today, marsupials are found primarily in Australia, where they underwent a separate evolutionary expansion, largely isolated from

FIGURE 12.33

Angiosperms Became the Dominant Plants During the Cenozoic This plant group, commonly known as flowering plants, consists of seed plants that have reproductive structures called flowers and fruits. **A.** The most diverse and widespread of modern plants, many angiosperms display easily recognizable flowers. **B.** Some angiosperms, including grasses, have very tiny flowers. The expansion of the grasslands during the Cenozoic era greatly increased the diversity of grazing mammals and the predators that feed on them. (Photo A by WDG Photo/Shutterstock; photo B by Torleif Svensson/Corbis)

FIGURE 12.34
Kangaroos, Examples of Marsupial Mammals After the breakup of Pangaea, the Australian marsupials evolved differently than their relatives in the Americas. (Photo by Martin Harvey/Getty Images)

placental mammals. Modern marsupials include kangaroos, opossums, and koalas (**FIGURE 12.34**).

Placental mammals (eutherians), conversely, develop within the mother's body for a much longer period, so birth occurs when the young are comparatively mature. Members of this group include wolves, elephants, bats, manatees, and monkeys. Most modern mammals, including humans, are placental.

Humans: Mammals with Large Brains and Bipedal Locomotion

Both fossil and genetic evidence suggest that around 7 or 8 million years ago in Africa, several populations of anthropoids (informally called apes) diverged. One line would eventually produce modern apes such as gorillas, orangutans, and chimpanzees, while the other population would produce several varieties of human ancestors. We have a good record of this evolution in fossils found in several sedimentary basins in Africa, in particular the rift valley system in East Africa.

The genus *Australopithecus*, which came into existence about 4.2 million years ago, showed skeletal characteristics that were intermediate between our apelike ancestors and modern humans. Over time, these human ancestors evolved features that suggest an upright posture and therefore a habit of walking around on two legs, rather than four. Evidence for this bipedal stride includes footprints preserved in 3.2-million-year-old ash deposits at Laetoli, Tanzania (**FIGURE 12.35**). This new way of moving around is correlated with our human ancestors leaving forest habitat in Africa and moving to open grasslands for hunting and gathering food.

The earliest fossils of our genus *Homo* include *Homo habilis*, nicknamed "handy man" because their remains were often found with sharp stone tools in sedimentary deposits

from 2.4 to 1.5 million years ago. *Homo habilis* had a shorter jaw and a larger brain than its ancestors. The development of a larger brain size is thought to be correlated with an increase in tool use.

During the next 1.3 million years of evolution, our ancestors developed substantially larger brains and long slender legs with hip joints adapted for long-distance walking. These species (including *Homo erectus*) ultimately gave rise to our species, *Homo sapiens*, as well as some extinct related species, including the Neanderthals (*Homo neanderthalis*). Despite having the same-sized brain as present-day humans and being able to fashion hunting tools from wood and stone, Neanderthals became extinct about 28,000 years ago. At one time, Neanderthals were considered a stage in the evolution of *Homo sapiens*, but that view has largely been abandoned.

Based on our current understanding, humans (*Homo sapiens*) originated in Africa about 200,000 years ago and began to spread around the globe. The oldest-known fossils of *Homo sapiens* outside Africa were found in the Middle East and date back to 115,000 years ago. Humans are also known to have coexisted with Neanderthals and other prehistoric populations, remains of which have been found in Siberia, China, and Indonesia. Further, there is mounting genetic evidence that our ancestors may have interbred with members of some of these groups. By 36,000 years ago, humans were producing spectacular cave paintings in Europe (**FIGURE 12.36**). About 11,500 years ago all prehistoric populations, except for modern humans (*Homo sapiens*), died out.

Large Mammals and Extinction

During the rapid mammal diversification of the Cenozoic era, some groups became very large. For example, by the Oligocene epoch, a hornless rhinoceros evolved that stood nearly 5 meters (16 feet) high. It is the largest land mammal known to have existed. As time passed, many other mammals evolved to larger forms—more, in fact, than now exist. Many of these large forms were common as recently as 11,000 years ago.

FIGURE 12.35 Footprints of *Australopithecus*, Human-Like Apes in Ash Deposits at Laetoli, Tanzania (Photo by John Reader/ Science Source)

FIGURE 12.36 Cave Painting of Animals by Early Humans (Photo courtesy of Sisse Brimberg/National Geographic Stock)

However, a wave of late Pleistocene extinctions rapidly eliminated these animals from the landscape.

North America experienced the extinction of mastodons and mammoths, both huge relatives of the modern elephant (**FIGURE 12.37**). In addition, saber-toothed cats, giant beavers, large ground sloths, horses, giant bison, and others died out. In Europe, late Pleistocene extinctions included woolly rhinos, large cave bears, and Irish elk. Scientists remain puzzled about the reasons for this recent wave of extinctions of large animals. Because these large animals survived several major glacial advances and interglacial periods, it is difficult to ascribe extinctions of these animals to climate change. Some scientists hypothesize that early humans hastened the decline of these mammals by selectively hunting large forms.

12.9 CONCEPT CHECKS

1 What animal group became the dominant land animals of the Cenozoic era?

2 Explain how the demise of the large Mesozoic reptiles impacted the development of mammals.

3 Where has most of the evidence for the early evolution of our ancestors been discovered?

4 What two characteristics best separate humans from other mammals?

5 Describe one hypothesis that explains the extinction of large mammals in the late Pleistocene.

FIGURE 12.37 Mammoths These relatives of modern elephants were among the large mammals that became extinct at the close of the Ice Age. (Image courtesy of INTERFOTO/Alamy)

12 CONCEPTS IN REVIEW | Earth's Evolution Through Geologic Time

12.1 IS EARTH UNIQUE?

List the principal characteristics that make Earth unique among the planets.

- As far as we know, Earth is unique among planets in the fact that it hosts life. The planet's size, composition, and location all contribute to conditions that support life (or at least our kind of life).

Q If you worked for NASA as an astrobiologist, what characteristics would you look for in a newly discovered planet as you considered whether it might host living organisms?

12.2 BIRTH OF A PLANET

Outline the major stages in the evolution of Earth, from the Big Bang to the formation of our planet's layered internal structure.

KEY TERMS: supernova, solar nebula, planetesimal, protoplanet, Hadean

- The universe is thought to have formed around 13.7 billion years ago, with the Big Bang, which generated space, time, energy, and matter. Early stars grew from the lightest elements, hydrogen and helium, and the process of nuclear fusion produced the other low mass elements. Some large stars exploded in supernovae, generating heavier atoms and spewing them into space.
- The story of Earth and the solar system began around 4.6 billion years ago, with the contraction of a solar nebula under the influence of gravity. Collisions between clumps of matter in this spinning disc resulted in the growth of planetesimals and then protoplanets. Over time, the matter of the solar nebula was concentrated into a smaller number of larger bodies: the Sun, the rocky inner planets, the gassy outer planets, moons, comets, and asteroids.
- Heat production within Earth during its formative years was much higher than it is today, thanks to the kinetic energy of impacting asteroids and planetesimals, as well as decay of short-lived radioactive isotopes. The high temperatures of our young Earth caused rock and iron to melt. This allowed iron to sink to form Earth's core and rocky material to rise to form the mantle and crust. This began the "hellish" first span of geologic time, the Hadean.

Q The photo shows Comet Shoemaker-Levy 9 impacting Jupiter in 1994. After this event, what happened to Jupiter's total mass? How was the number of objects in the solar system affected? Relate this example to the nebular theory and evolution of the solar system.

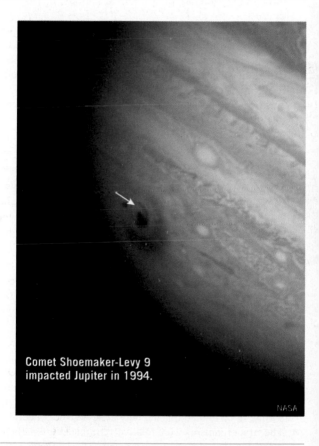

Comet Shoemaker-Levy 9 impacted Jupiter in 1994.

NASA

12.3 ORIGIN AND EVOLUTION OF THE ATMOSPHERE AND OCEANS

Describe how Earth's atmosphere and oceans have formed and evolved through time.

KEY TERMS: outgassing, banded iron formation, Great Oxygenation Event

- Earth's atmosphere is essential for life. It evolved as volcanic outgassing added mainly water vapor and carbon dioxide to the primordial atmosphere of gases common in the early solar system: methane and ammonia.

- Free oxygen began to accumulate partly through photosynthesis by cyanobacteria, which released oxygen as a waste product. Much of this early oxygen immediately reacted with iron dissolved in seawater and settled to the ocean floor as chemical sediments called banded iron formations. The Great Oxygenation Event of 2.5 billion years ago marks the first evidence of appreciable amounts of free oxygen in the atmosphere.
- Earth's oceans formed after the planet's surface had cooled. Soluble ions weathered from the crust were carried to the ocean, making it salty. The oceans also absorbed tremendous amounts of carbon dioxide from the atmosphere.

12.4 PRECAMBRIAN HISTORY: THE FORMATION OF EARTH'S CONTINENTS

Explain the formation of continental crust, how continental crust becomes assembled into continents, and the role that the supercontinent cycle has played in this process.

KEY TERMS: craton, shield, supercontinent, supercontinent cycle

- The Precambrian includes the Archean and Proterozoic eons. However, the geologic records of these eons are rather limited because the rock cycle operating over billions of years has destroyed much of the evidence.
- Continental crust was produced over time through the recycling of basaltic (mafic) crust in an early version of plate tectonics. Small crustal fragments formed and accreted to one another, amalgamating over time into large crustal provinces called cratons. Over time, North America and other continents grew through the accretion of new terranes around the edges of their central "nucleus" of crust.
- Early cratons not only merged but sometimes rifted apart, too. The supercontinent Rodinia formed around 1.1 billion years ago and then rifted apart, opening new ocean basins. In time, these also closed and formed a new supercontinent called Pangaea around 300 million years ago. Like Rodinia before it, Pangaea broke up as part of the ongoing supercontinent cycle.
- The formation of elevated oceanic ridges upon the breakup of a supercontinent displaced enough water that sea level rose, and shallow seas flooded low-lying portions of the continents. The breakup of continents also influenced the direction of ocean currents, with important consequences for climate.

Q Consult Figure 12.12 and use it to give a history of the North American continent's assembly over the past 3.5 billion years. Which pieces of the continent were added at what times?

12.5 GEOLOGIC HISTORY OF THE PHANEROZOIC: THE FORMATION OF EARTH'S MODERN CONTINENTS

List and discuss the major geologic events in the Paleozoic, Mesozoic, and Cenozoic eras.

- The Phanerozoic eon encompasses the most recent 542 million years of geologic time.
- In the Paleozoic era, North America experienced a series of collisions that resulted in the rise of the young Appalachian mountain belt and the assembly of Pangaea. High sea levels caused the ocean to cover vast areas of the continent and resulted in a thick sequence of sedimentary strata.
- In the Mesozoic, Pangaea broke up, and the Atlantic Ocean began to form. As the continent moved westward, the Cordillera began to rise due to subduction and the accretion of terranes along the west coast of North America. In the Southwest, vast deserts deposited thick layers of dune sand, while environments in the east were conducive to the formation and subsequent burial of coal swamps.
- In the Cenozoic era, a thick sequence of sediments was deposited along the Atlantic margin and the Gulf of Mexico. Meanwhile, western North America experienced an extraordinary episode of crustal extension; the Basin and Range Province resulted.

Q Contrast the tectonics of eastern and western North America during the Mesozoic era.

12.6 EARTH'S FIRST LIFE

Describe some of the hypotheses on the origin of life and the characteristics of early prokaryotes, eukaryotes, and multicelled organisms.

KEY TERMS: prokaryote, stromatolite, eukaryote

- Life began from nonlife. Whether the genetic code of life (DNA and RNA) or biologically active molecules (proteins) came first is a matter of some controversy. Amino acids are a necessary building block for proteins. They may have been assembled with energy from ultraviolet light or lightning, or in a hot spring, or on another planet, only to be delivered later to Earth on meteorites.
- The first organisms were relatively simple single-celled prokaryotes that thrived in low-oxygen environments. They formed around 3.8 billion years ago. The advent of photosynthesis allowed microbial mats to build up and form stromatolites.
- Eukaryotes have larger, more complex cells than prokaryotes. The oldest eukaryotic cells formed around 2.1 billion years ago. Eventually, some eukaryotic cells linked together and differentiated their structures and functions, producing the earliest multicellular organisms.

Q Are these stromatolites right-side up, or have they been flipped upside-down by folding of the crust? Explain.

Biophoto Associates/Science Source

12.7 PALEOZOIC ERA: LIFE EXPLODES

Discuss the major developments in the history of life during the Paleozoic era.

KEY TERMS: Cambrian explosion, mass extinction

- At the beginning of the Cambrian period, abundant fossil hard parts appear in sedimentary rocks. The source of these shells and other skeletal material were a profusion of new animals, including trilobites and cephalopods.
- Plants colonized the land around 400 million years ago and soon diversified into forests.
- In the Devonian, some lobe-finned fish began to spend time out of water and gradually evolved into the first amphibians. A subset of the amphibian population evolved waterproof skin and eggs and split off to become the reptile line.
- The Paleozoic era ended with the largest mass extinction in the geologic record. This deadly event may have been related to the eruption of the Siberian Traps flood basalts.

Q What advantages do reptiles have over amphibians? What advantages do amphibians have over fish? Suggest a reason that fish would still exist in a world that also contains reptiles.

12.8 MESOZOIC ERA: AGE OF THE DINOSAURS

Discuss the major developments in the history of life during the Mesozoic era.

KEY TERM: gymnosperm

- Plants diversified during the Mesozoic. The flora of that time was dominated by gymnosperms, the first plants with seeds that allowed them to migrate beyond the edge of water bodies.
- Reptiles diversified, too. The dinosaurs came to dominate the land, the pterosaurs dominated the air, and a suite of several different marine reptiles swam the seas. The first birds evolved during the Mesozoic, exemplified by *Archaeopteryx*, a transitional fossil.
- As with the Paleozoic, the Mesozoic ended with a mass extinction, probably due to a massive meteorite impact in what is now Chicxulub, Mexico.

Q Would a paleontologist be likely to announce that some dinosaurs ate apples? Explain.

12.9 CENOZOIC ERA: AGE OF MAMMALS

Discuss the major developments in the history of life during the Cenozoic era.

KEY TERM: angiosperm,

- After the giant Mesozoic reptiles were extinct, mammals were able to diversify on the land, in the air, and in the oceans. Mammals are warm blooded, have hair on their bodies, and nurse their young with milk. Marsupial mammals are born very young and then move to a pouch on the mother, while placental mammals spend a longer time *in utero* and are born in a relatively mature state compared to marsupials.
- Flowering plants, called angiosperms, diversified and spread around the world through the Cenozoic era.
- Humans evolved from primate ancestors in Africa over a period of 8 million years. They are distinguished from their ape ancestors by an upright, bipedal posture, large brains, and tool use. The oldest anatomically modern human fossils are 200,000 years old. Some of these humans migrated out of Africa and coexisted with Neanderthals and other related populations.

GIVE IT SOME **THOUGHT**

1. Refer to the geologic time scale in Figure 12.3. The Precambrian accounts for nearly 90 percent of geologic time. Why do you think it has fewer divisions than the rest of the time scale?
2. Referring to Figure 12.4, write a brief summary of the events that led to the formation of Earth.
3. Describe two ways in which the sudden appearance of oxygen in the atmosphere about 2.5 billion years ago influenced the development of modern life-forms.
4. The accompanying photograph shows layered iron-rich rocks called banded iron formations. What does the existence of these 2.5-billion-year-old rocks tell us about the evolution of Earth's atmosphere?

Blue Gum Pictures/Alamy

5. Five mass extinctions, in which 50 percent or more of Earth's marine species became extinct, are documented in the fossil record. Use the accompanying graph, which depicts the time and extent of each mass extinction, to answer the following:

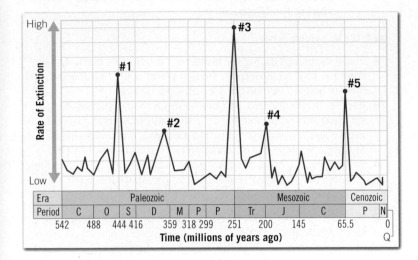

a. Which of the five mass extinctions was the *most extreme*? Identify this extinction by name and when it occurred.

b. What group of animals was most affected by the extinction referred to in Question a?

c. When did the *most recent* mass extinction occur?

d. During the most recent mass extinction, what prominent animal group was eliminated?

e. What animal group experienced a major period of diversification following the most recent mass extinction?

6. Currently, oceans cover about 71 percent of Earth's surface. This percentage was much higher early in Earth history. Explain.

7. Contrast the eastern and western margins of North America during the Cenozoic era in terms of their relationships to plate boundaries.

8. Suggest at least one reason plants moved onto land before large animals.

9. Some scientists have proposed that the environments around black smokers may be similar to the extreme conditions that existed early in Earth history. Therefore, these scientists look to the unusual life that exists around black smokers for clues about how earliest life may have survived. Compare and contrast the environment of a black smoker to the environment on Earth approximately 3 to 4 billion years ago. Do you think there are parallels between the two? If so, do you think black smokers are good examples of the environment that earliest life may have experienced? Explain.

10. About 250 million years ago, plate movement assembled all the previously separated landmasses together to form the supercontinent Pangaea. The formation of Pangaea resulted in deeper ocean basins, which caused a drop in sea level and caused shallow coastal areas to dry up. Thus, in addition to rearranging the geography of our planet, continental drift had a major impact on life on Earth. Use the accompanying diagram to answer the following:

a. Which of the following types of habitats would likely diminish in size during the formation of a supercontinent: deep-ocean habitats, wetlands, shallow marine environments, or terrestrial (land) habitats? Explain.

b. During the breakup of a supercontinent, what would happen to sea level? Would it remain the same, rise, or fall?

c. Explain how and why the development of an extensive oceanic ridge system that forms during the breakup of a supercontinent affects sea level.

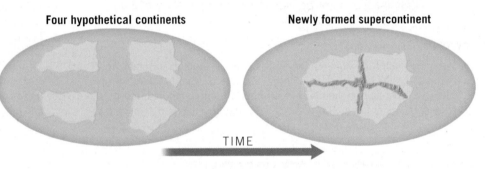

11. Suggest a geologic reason why the rift valley system of East Africa is so rich in human ancestor fossils.

EXAMINING THE **EARTH SYSTEM**

1. The Earth system has been responsible for both the conditions that favored the evolution of life on this planet and for the mass extinctions that have occurred throughout geologic time. Describe the role of the biosphere, hydrosphere, and solid Earth in forming the current level of atmospheric oxygen. How did Earth's outer-space environment interact with the atmosphere and biosphere to contribute to the great mass extinction that marked the end of the dinosaurs?

2. Most of the vast North American coal resources located from Pennsylvania to Illinois began forming during the Pennsylvanian and Mississippian periods of Earth history. (This time period is also referred to as the Carboniferous period.) Using Figure 12.27, a restoration of a Pennsylvania period coal swamp, describe the climatic and biological conditions associated with this unique environment. Next, examine the accompanying diagram that illustrates the geographic position of North America during the period of coal formation. Where, relative to the equator, was North America located during the time of coal formation? Why is it unlikely that a similar coal-forming environment will repeat itself in North America in the near future? (You may find it helpful to visit the University of California Time Machine Exhibit at www.ucmp.berkeley. edu/carboniferous/carboniferous.html.)

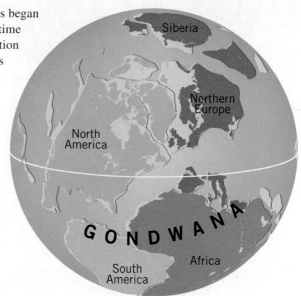

MasteringGeology™

Looking for additional review and test prep materials? Visit the Self Study area in **www.masteringgeology.com** to find practice quizzes, study tools, and multimedia that will aid in your understanding of this chapter's content. In **MasteringGeology**™ you will find:

- GEODe: Earth Science: An interactive visual walkthrough of key concepts
- Geoscience Animation Library: More than 100 animations illuminating many difficult-to-understand Earth science concepts

- In The News RSS Feeds: Current Earth science events and news articles are pulled into the site with assessment
- Pearson eText
- Optional Self Study Quizzes
- Web Links
- Glossary
- Flashcards

ちきゅう
CHIKYU

CHIKYU JAMSTEC

13 The Ocean Floor

The *Chikyu* (meaning "Earth" in Japanese) is one of the most advanced scientific drillings vessels. It is part of the Integrated Ocean Drilling Program (IODP). (Itsuo Inouye/AP Photo)

ow deep is the ocean? How much of Earth is covered by the global sea? What does the seafloor look like? Answers to these and other questions about the oceans and the basins they occupy are sometimes elusive and often difficult to determine. Suppose that all the water were drained from the ocean.

What would we see? Plains? Mountains? Canyons? Plateaus? Indeed, the ocean conceals all these features—and more. And what about the carpet of sediment that covers much of the seafloor? Where did it come from, and what can we learn by studying it? This chapter provides answers to these questions.

13.1 | THE VAST WORLD OCEAN Discuss the extent and distribution of oceans and continents on Earth. Identify Earth's four main ocean basins.

Oceans are a major part of our planet. In fact, Earth is frequently called the *water planet* or the *blue planet*. There is a good reason for these names: Nearly 71 percent of Earth's surface is covered by the global ocean (**FIGURE 13.1**). The focus of this chapter and Chapters 14 and 15 is oceanography. **Oceanography** is an interdisciplinary science that draws on the methods and knowledge of geology, chemistry, physics, and biology to study all aspects of the world ocean.

Geography of the Oceans

The area of Earth is about 510 million square kilometers (197 million square miles). Of this total, approximately 360 million square kilometers (140 million square miles), or 71 percent, is represented by oceans and marginal seas (meaning seas around the ocean's margin, like the Mediterranean Sea and Caribbean Sea). Continents and islands comprise the remaining 29 percent, or 150 million square kilometers (58 million square miles).

When we study a world map or globe, it is readily apparent that the continents and oceans are not evenly divided between the Northern and Southern Hemispheres (see Figure 13.1). When we compute the percentages of land and water in the Northern Hemisphere, we find that nearly 61 percent of the surface is water, and about 39 percent is land. In the Southern Hemisphere, on the other hand, almost 81 percent of the surface is water, and only 19 percent is land. It is no wonder then that the Northern Hemisphere is called the *land hemisphere* and the Southern Hemisphere the *water hemisphere*.

FIGURE 13.2A shows the distribution of land and water in the Northern and Southern Hemispheres. Between latitudes 45° north and 70° north, there is actually more land than water, whereas between 40° south and 65° south there is almost no land to interrupt the oceanic and atmospheric circulation.

The world ocean can be divided into four main ocean basins (**FIGURE 13.2B**):

FIGURE 13.1 North Versus South These views of Earth show the uneven distribution of land and water between the Northern and Southern Hemispheres. Almost 81 percent of the Southern Hemisphere is covered by the oceans—20 percent more than the Northern Hemisphere.

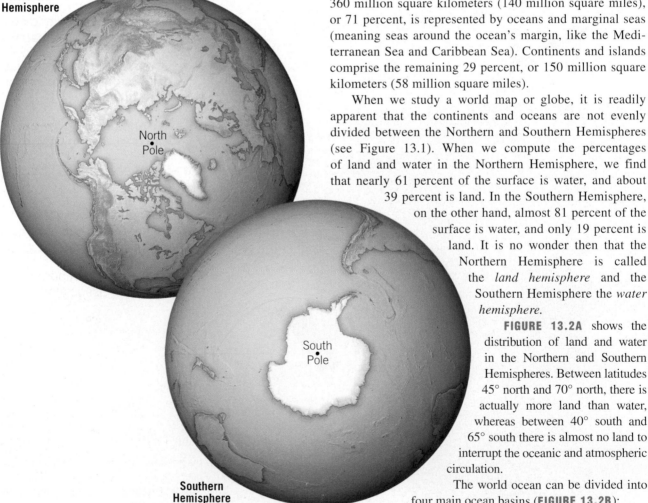

Northern Hemisphere

North Pole

South Pole

Southern Hemisphere

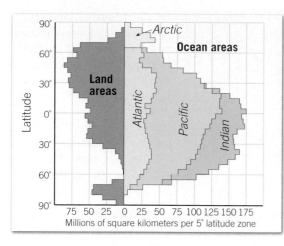

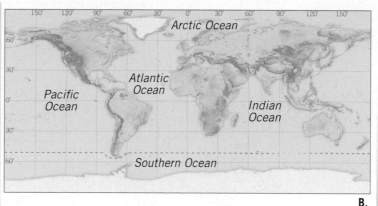

SmartFigure 13.2 Distribution of Land and Water A. The graph shows the amount of land and water in each 5° latitude belt. **B.** The world map provides a more familiar view.

1. The *Pacific Ocean*, which is the largest ocean and the largest single geographic feature on the planet, accounts for over half of the ocean surface area of Earth. In fact, the Pacific Ocean is so large that all the continents could fit into the space occupied by it—and have room left over! It is also the world's deepest ocean, with an average depth of 3940 meters (12,927 feet, or about 2.5 miles).
2. The *Atlantic Ocean* is about half the size of the Pacific Ocean and not quite as deep. It is a relatively narrow ocean compared to the Pacific and is bounded by almost parallel continental margins.
3. The *Indian Ocean* is slightly smaller than the Atlantic Ocean but has about the same average depth. Unlike the Pacific and Atlantic Oceans, it is largely a Southern Hemisphere water body.
4. The *Arctic Ocean* is about 7 percent the size of the Pacific Ocean and is only a little more than one-quarter as deep as the rest of the oceans.

Oceanographers also recognize an additional ocean near the continent of Antarctica in the Southern Hemisphere. Defined by the meeting of currents near Antarctica called the Antarctic Convergence, the *Southern Ocean*, or *Antarctic Ocean*, is actually those portions of the Pacific, Atlantic, and Indian Oceans south of about 50° south latitude.

Comparing the Oceans to the Continents

A major difference between continents and ocean basins is their relative levels. The average elevation of the continents above sea level is about 840 meters (2756 feet), whereas the average depth of the oceans is nearly four and a half times this amount—3729 meters (12,234 feet). The volume of ocean water is so large that if Earth's solid mass were perfectly smooth (level) and spherical, the oceans would cover Earth's entire surface to a uniform depth of more than 2000 meters (1.2 miles)!

13.1 CONCEPT CHECKS

1 How does the area of Earth's surface covered by the oceans compare with that of the continents?

2 Contrast the distribution of land and water in the Northern Hemisphere and the Southern Hemisphere.

3 Excluding the Southern Ocean, name the four main ocean basins. Contrast them in terms of area and depth.

4 How does the average depth of the oceans compare to the average elevation of the continents?

13.2 | AN EMERGING PICTURE OF THE OCEAN FLOOR

Define *bathymetry* and summarize the various techniques used to map the ocean floor.

If all water were removed from the ocean basins, a great variety of features would be seen, including broad volcanic peaks, deep trenches, extensive plains, linear mountain chains, and large plateaus. In fact, the scenery would be nearly as diverse as that on the continents.

Mapping the Seafloor

The complex nature of ocean-floor topography did not unfold until the historic 3½-year voyage of the HMS *Challenger* (**FIGURE 13.3**). From December 1872 to May 1876, the

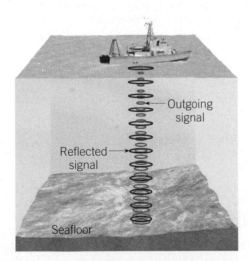

FIGURE 13.3 HMS *Challenger* The first systematic bathymetric measurements of the ocean were made aboard the HMS *Challenger*, which departed England in December 1872 and returned in May 1876. (Image courtesy of the Library of Congress)

Challenger expedition made the first comprehensive study of the global ocean ever attempted. During the 127,500-kilometer (79,200-mile) voyage, the ship and its crew of scientists traveled to every ocean except the Arctic. Throughout the voyage, they sampled a multitude of ocean properties, including water depth, which was accomplished

FIGURE 13.4 Echo Sounder An echo sounder determines water depth by measuring the time interval required for an acoustic wave to travel from a ship to the seafloor and back. The speed of sound in water is 1500 meters per second. Therefore, Depth = ½ (1500 m/sec × Echo travel time).

by laboriously lowering a long, weighted line overboard and then retrieving it. Using this laborious process, the *Challenger* made the first recording of the deepest-known point on the ocean floor in 1875. This spot, on the floor of the western Pacific, was later named the *Challenger Deep*.

Modern Bathymetric Techniques The measurement of ocean depths and the charting of the shape (topography) of the ocean floor is known as **bathymetry** (*bathos* = depth, *metry* = measurement). Today, sound energy is used to measure water depths. The basic approach employs **sonar**, an acronym for *so*und *na*vigation and *r*anging. The first devices that used sound to measure water depth, called **echo sounders**, were developed early in the twentieth century. Echo sounders work by transmitting a sound wave (called a *ping*) into the water in order to produce an echo when it bounces off any object, such as a large marine organism or the ocean floor (**FIGURE 13.4**). A sensitive receiver intercepts the echo reflected from the bottom, and a clock precisely measures the travel time to fractions of a second. By knowing the speed of sound waves in water—about 1500 meters (4900 feet) per second—and the time required for the energy pulse to reach the ocean floor and return, depth can be calculated. Depths determined from continuous monitoring of these echoes are plotted to create a profile of the ocean floor. By laboriously combining profiles from several adjacent traverses, a chart of a portion of the seafloor is produced.

Following World War II, the U.S. Navy developed *sidescan sonar* to look for explosive devices that had been deployed in shipping lanes (**FIGURE 13.5A**). These torpedo-shaped instruments can be towed behind a ship, where they send out a fan of sound extending to either side of the ship's track. By combining swaths of sidescan sonar data, researchers produced the first photograph-like images of the seafloor. Although sidescan sonar provides valuable views of the seafloor, it does not provide bathymetric (water depth) data.

This drawback was resolved in the 1990s, with the development of *high-resolution multibeam sonar* instruments (see Figure 13.5A). These systems use hull-mounted sound sources that send out a fan of sound and then record reflections from the seafloor through a set of narrowly focused receivers aimed at different angles. Rather than obtain the depth of a single point every few seconds, this technique makes it possible for a survey ship to map the features of the ocean floor along a strip

A. Sidescan sonar and multibeam sonar operating from the same research vessel.

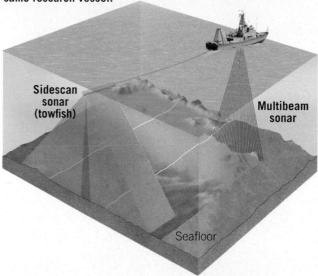

Sidescan sonar (towfish)

Multibeam sonar

Seafloor

B. Color-enhanced perspective map of the seafloor and coastal landforms in the Los Angeles area of California.

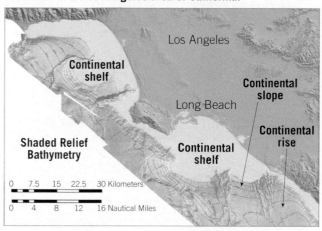

Los Angeles

Continental shelf

Long Beach

Continental slope

Continental rise

Shaded Relief Bathymetry

Continental shelf

0   7.5   15   22.5   30 Kilometers

0   4   8   12   16 Nautical Miles

FIGURE 13.5 Sidescan and Multibeam Sonar

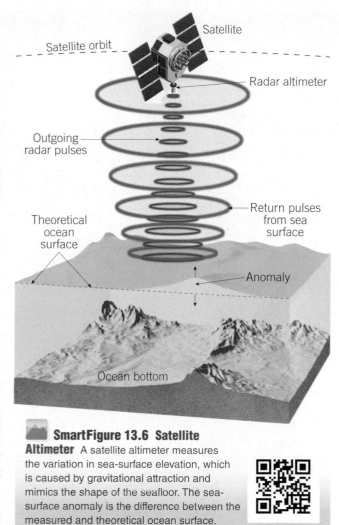

Satellite orbit

Satellite

Radar altimeter

Outgoing radar pulses

Return pulses from sea surface

Theoretical ocean surface

Anomaly

Ocean bottom

SmartFigure 13.6 Satellite Altimeter A satellite altimeter measures the variation in sea-surface elevation, which is caused by gravitational attraction and mimics the shape of the seafloor. The sea-surface anomaly is the difference between the measured and theoretical ocean surface.

tens of kilometers wide. These systems can collect bathymetric data of such high resolution that they can distinguish depths that differ by less than a meter (**FIGURE 13.5B**). When multibeam sonar is used to make a map of a section of seafloor, the ship travels through the area in a regularly spaced back-and-forth pattern known as "mowing the lawn."

Despite their greater efficiency and enhanced detail, research vessels equipped with multibeam sonar travel at a mere 10–20 kilometers (6 to 12 miles) per hour. It would take at least 100 vessels outfitted with this equipment hundreds of years to map the entire seafloor. This explains why only about 5 percent of the seafloor has been mapped in detail—and why large portions of the seafloor have not yet been mapped with sonar at all.

Mapping the Ocean Floor from Space Another technological breakthrough that has led to an enhanced understanding of the seafloor involves measuring the shape of the ocean surface from space. After compensating for waves, tides, currents, and atmospheric effects, it was discovered that the ocean surface is not perfectly "flat." Because massive seafloor features exert stronger-than-average gravitational attraction, they produce elevated areas on the ocean surface. Conversely, canyons and trenches create slight depressions.

Satellites equipped with *radar altimeters* are able to measure these subtle differences by bouncing microwaves off the sea surface (**FIGURE 13.6**). These devices can measure variations as small as a few centimeters. Such data have added greatly to the knowledge of ocean-floor topography. Combined with traditional sonar depth measurements, the data are used to produce detailed ocean-floor maps, such as the one shown in **FIGURE 13.7**.

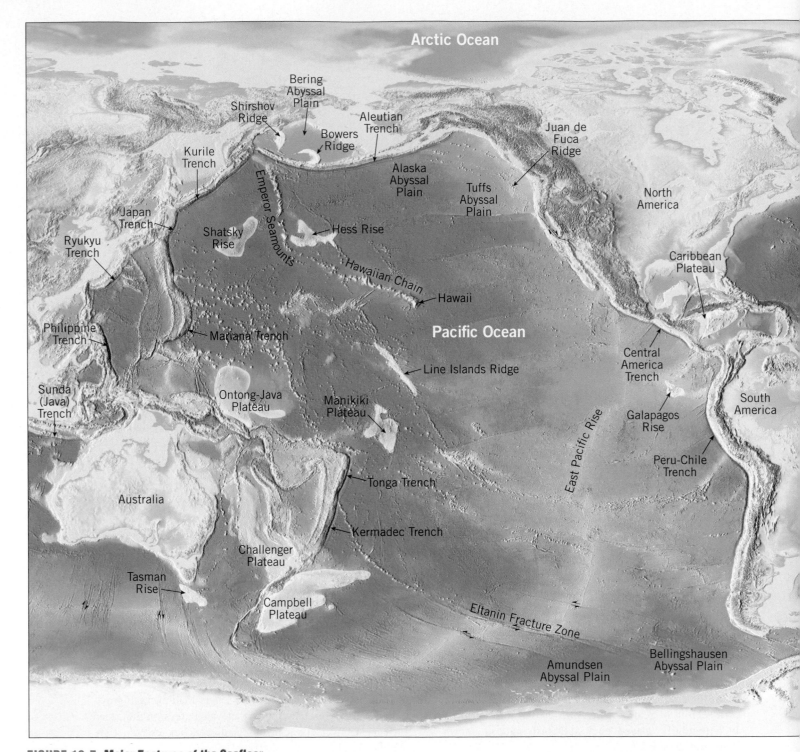

FIGURE 13.7 Major Features of the Seafloor

Provinces of the Ocean Floor

Oceanographers studying the topography of the ocean floor recognize three major units: *continental margins*, the *deep-ocean basin*, and the *oceanic (mid-ocean) ridge*.

The map in **FIGURE 13.8** outlines these provinces for the North Atlantic Ocean, and the profile at the bottom of the illustration shows the varied topography. Such profiles usually have their vertical dimension exaggerated many times—40 times in this case—to make topographic

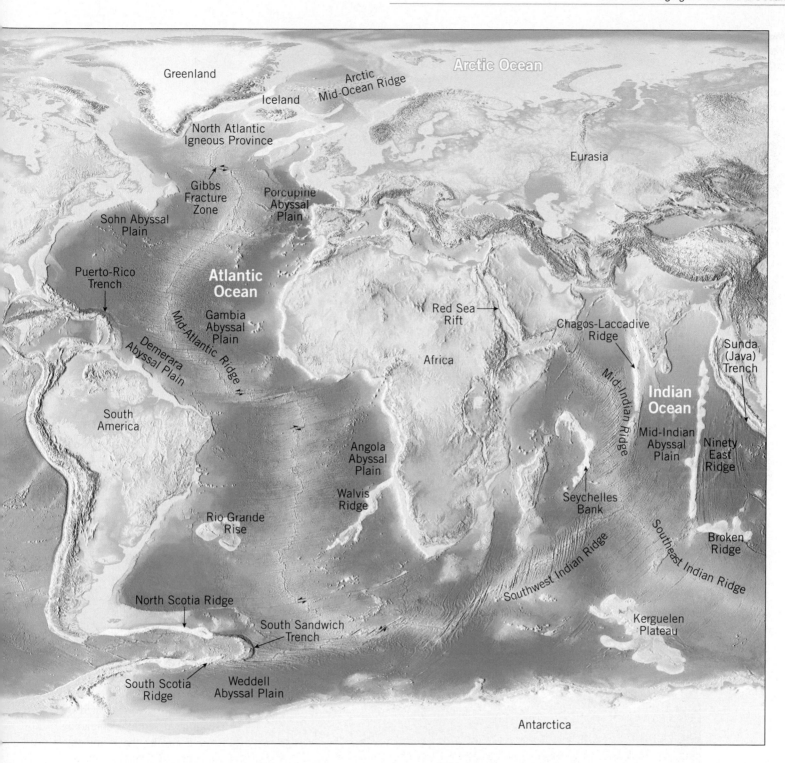

Greenland

Arctic Ocean

Arctic Mid-Ocean Ridge

Iceland

North Atlantic Igneous Province

Eurasia

Gibbs Fracture Zone

Porcupine Abyssal Plain

Sohn Abyssal Plain

Atlantic Ocean

Puerto-Rico Trench

Gambia Abyssal Plain

Mid-Atlantic Ridge

Red Sea Rift

Chagos-Laccadive Ridge

Sunda (Java) Trench

Demerara Abyssal Plain

Africa

Mid-Indian Ridge

Indian Ocean

South America

Angola Abyssal Plain

Mid-Indian Abyssal Plain

Ninety East Ridge

Walvis Ridge

Seychelles Bank

Rio Grande Rise

Broken Ridge

North Scotia Ridge

Southwest Indian Ridge

Southeast Indian Ridge

South Sandwich Trench

Kerguelen Plateau

South Scotia Ridge

Weddell Abyssal Plain

Antarctica

features more conspicuous. Vertical exaggeration, however, makes slopes shown in seafloor profiles appear to be *much* steeper than they actually are.

13.2 CONCEPT CHECKS

1 Define *bathymetry*.

2 Describe how satellites orbiting Earth can determine features on the seafloor without being able to directly observe them beneath several kilometers of seawater.

3 List the three major provinces of the ocean floor.

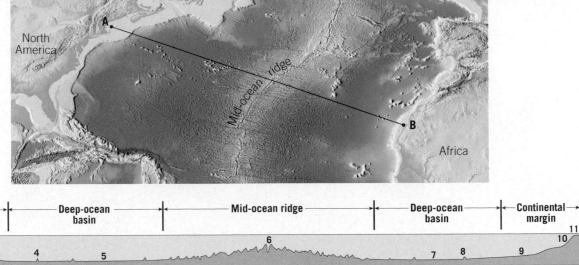

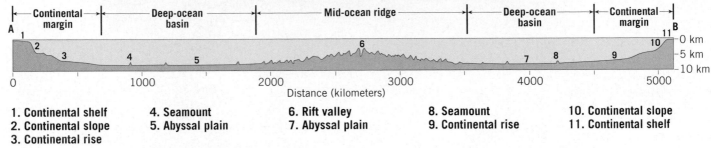

1. Continental shelf
2. Continental slope
3. Continental rise
4. Seamount
5. Abyssal plain
6. Rift valley
7. Abyssal plain
8. Seamount
9. Continental rise
10. Continental slope
11. Continental shelf

13.3 | CONTINENTAL MARGINS Compare a passive continental margin with an active continental margin and list the major features of each.

As the name implies, **continental margins** are the outer margins of the continents where continental crust transitions to oceanic crust. Two types of continental margin have been identified: *passive* and *active*. Nearly the entire Atlantic Ocean and a large portion of the Indian Ocean are surrounded by passive continental margins (see Figure 13.7). By contrast, most of the Pacific Ocean is bordered by active continental margins that are represented by subduction zones in **FIGURE 13.9**. Notice that many of those active subduction zones lie far beyond the margins of the continents.

Passive Continental Margins

Passive continental margins are geologically inactive regions located some distance from plate boundaries. As a result, they are not associated with strong earthquakes or volcanic activity. Passive continental margins develop when continental blocks rift apart and are separated by continued seafloor spreading. As a result, the continental blocks are firmly attached to the adjacent oceanic crust.

Most passive margins are relatively wide and are sites where large quantities of sediments are deposited. The features comprising passive continental margins include the continental shelf, the continental slope, and the continental rise (**FIGURE 13.10**).

FIGURE 13.9 Distribution of Earth's Subduction Zones Most of the active subduction zones surround the Pacific basin.

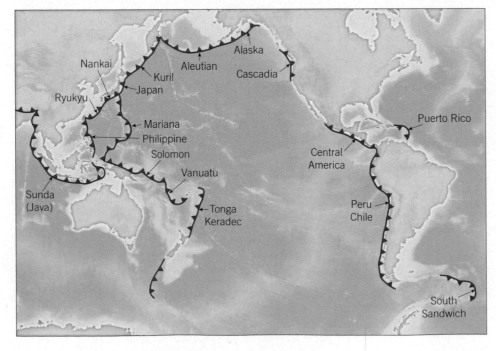

Continental Shelf

The **continental shelf** is a gently sloping, submerged surface that extends from the shoreline toward the deep-ocean basin. It consists mainly of continental crust capped with sedimentary rocks and sediments eroded from adjacent landmasses.

The width of the continental shelf varies greatly. The shelf is almost nonexistent along portions of some continents, and it extends seaward more than 1500 kilometers (930 miles) along others. The average inclination of the continental shelf is only about one-tenth of 1 degree, a slope so slight that it would appear to an observer to be a horizontal surface.

The continental shelf tends to be relatively featureless; however, some areas are mantled by extensive glacial deposits and thus are quite rugged. In addition, some continental shelves are dissected by large valleys that run from the coastline into deeper waters. Many of these *shelf valleys* are the seaward extensions of river valleys on the adjacent landmass. They were eroded during the last Ice Age (Quaternary period), when enormous quantities of water were stored in vast ice sheets on the continents, causing sea level to drop at least 100 meters (330 feet). Because of this sea-level drop, rivers extended their valleys, and land-dwelling plants and animals migrated to the newly exposed portions of the continents. Dredging off the coast of North America has retrieved the ancient remains of numerous land dwellers, including mammoths, mastodons, and horses, providing further evidence that portions of the continental shelves were once above sea level.

Although continental shelves represent only 7.5 percent of the total ocean area, they have economic and political significance because they contain important reservoirs of oil and natural gas, and they support important fishing grounds.

Continental Slope

Marking the seaward edge of the continental shelf is the **continental slope**, a relatively steep structure that marks the boundary between continental crust and oceanic crust. Although the inclination of the continental slope varies greatly from place to place, it averages about 5 degrees and in places exceeds 25 degrees.

Continental Rise

The continental slope merges into a more gradual incline known as the **continental rise** that may extend seaward for hundreds of kilometers. The continental rise consists of a thick accumulation of sediment that has moved down the continental slope and onto deep-ocean floor. Most of the sediments are delivered to the seafloor by *turbidity currents* that periodically flow down *submarine canyons*. (We will discuss these shortly.) When these muddy slurries emerge from the mouth of a canyon onto the relatively flat ocean floor, they deposit sediment that forms a **deep-sea fan**. As fans from adjacent submarine canyons grow, they merge to produce a continuous wedge of sediment at the base of the continental slope, forming the continental rise.

CONTINENTAL

Shelf **Slope** **Rise**

Submarine canyons

Shelf break

Deep-sea fan

Abyssal plain

Continental crust

Oceanic crust

FIGURE 13.10 Passive Continental Margin Note that the slopes shown for the continental shelf and continental slope are greatly exaggerated. The continental shelf has an average slope of one-tenth of 1 degree, whereas the continental slope has an average slope of about 5 degrees.

EYE ON EARTH

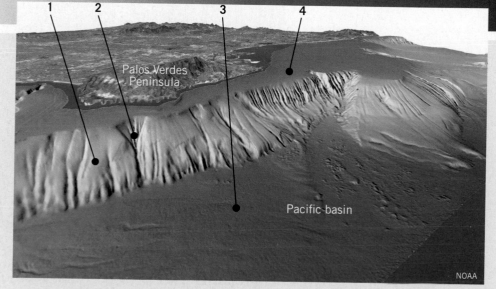

This image shows a perspective view of the continental margin, looking southwest toward the Palos Verdes Peninsula near Los Angeles.

QUESTION 1 *Match the features labeled 1 through 4 with the following terms: continental shelf, continental slope, continental rise, and submarine canyon.*

QUESTION 2 *Based on the features in this image, what type of continental margin is shown here?*

Palos Verdes Peninsula

Pacific basin

NOAA

FIGURE 13.11 Turbidity Currents and Submarine Canyons Turbidity currents are an important factor in the formation of submarine canyons. (Photo by Marli Miller)

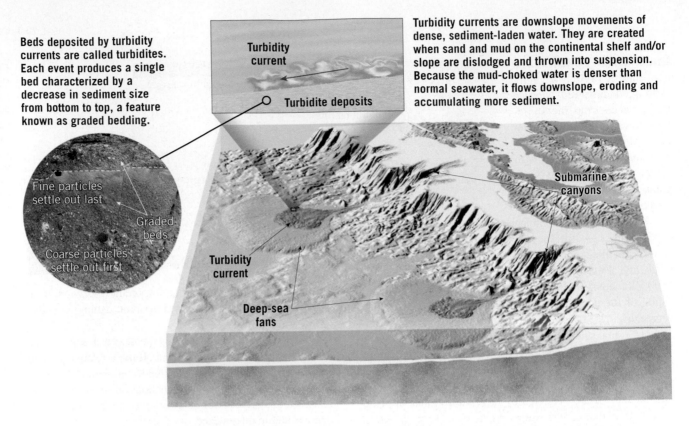

Beds deposited by turbidity currents are called turbidites. Each event produces a single bed characterized by a decrease in sediment size from bottom to top, a feature known as graded bedding.

Turbidity current

Turbidite deposits

Turbidity currents are downslope movements of dense, sediment-laden water. They are created when sand and mud on the continental shelf and/or slope are dislodged and thrown into suspension. Because the mud-choked water is denser than normal seawater, it flows downslope, eroding and accumulating more sediment.

Fine particles settle out last

Graded beds

Coarse particles settle out first

Submarine canyons

Turbidity current

Deep-sea fans

Submarine Canyons and Turbidity Currents

Deep, steep-sided valleys known as **submarine canyons** are cut into the continental slope and may extend across the entire continental rise to the deep-ocean basin (**FIGURE 13.11**). Although some of these canyons appear to be the seaward extensions of river valleys, many others do not line up in this manner. Furthermore, submarine canyons extend to depths far below the maximum lowering of sea level during the Ice Age, so we cannot attribute their formation to stream erosion.

These submarine canyons have probably been excavated by turbidity currents. **Turbidity currents** are downslope movements of dense, sediment-laden water. They are created when sand and mud on the continental shelf and slope are dislodged and thrown into suspension. Because the mud-choked water is denser than normal seawater, it moves downslope as a mass, eroding and accumulating more sediment as it goes. The erosional work repeatedly carried on by these muddy torrents is thought to be the major force in the excavation of most submarine canyons.

Turbidity currents usually originate along the continental slope and continue across the continental rise, still cutting channels. Eventually, they lose momentum and come to rest along the

SmartFigure 13.12 Active Continental Margins

Active continental margins are located along convergent plate boundaries where oceanic lithosphere is being subducted beneath the leading edge of a continent.

Continental volcanic arc

Trench

Oceanic crust

Continental crust

Continental lithosphere

- 100 km

Asthenosphere

Subducting oceanic lithosphere

Melting

- 200 km

Shallow trench

Accretionary wedge

Continental crust

Subducting oceanic lithosphere

A. Accretionary wedges develop along subduction zones where sediments from the ocean floor are scraped from the descending oceanic plate and pressed against the edge of the overriding plate.

Deep trench

Continental crust

Erosion

Subducting oceanic lithosphere

B. Subduction erosion occurs where sediment and rock scraped off the bottom of the overriding plate is carried into the mantle by the subducting plate.

floor of the deep-ocean basin. As these currents slow, suspended sediments begin to settle out. First, the coarser sand is dropped, followed by successively finer accumulations of silt and then clay. These deposits, called *turbidites*, display a decrease in sediment grain size from bottom to top, a feature known as *graded bedding*. Turbidity currents are an important mechanism of sediment transport in the ocean. By the action of turbidity currents, submarine canyons are excavated, and sediments are carried to the deep-ocean floor.

Active Continental Margins

Active continental margins are located along convergent plate boundaries where oceanic lithosphere is being subducted beneath the leading edge of a continent (**FIGURE 13.12**). Deep-ocean trenches are the major topographic expression at convergent plate boundaries. These deep, narrow furrows surround most of the Pacific Rim.

Along some subduction zones, sediments from the ocean floor and pieces of oceanic crust are scraped from the descending oceanic plate and plastered against the edge of the overriding plate. This chaotic accumulation of deformed sediment and scraps of oceanic crust is called an **accretionary wedge** (*ad* = toward, *crescere* = to grow). Prolonged

plate subduction can produce massive accumulations of sediment along active continental margins.

The opposite process, known as **subduction erosion**, characterizes many other active continental margins. Rather than sediment accumulating along the front of the overriding plate, sediment and rock are scraped off the bottom of the overriding plate and transported into the mantle by the subducting plate. Subduction erosion is particularly effective when the angle of descent is steep. Sharp bending of the subducting plate causes faulting in the ocean crust and a rough surface, as shown in Figure 13.12.

13.3 CONCEPT CHECKS

1 List the three major features of a passive continental margin. Which of these features is considered a flooded extension of the continent? Which one has the steepest slope?

2 Describe the differences between active and passive continental margins. Where is each type found?

3 Discuss the process that is responsible for creating most submarine canyons.

4 How are active continental margins related to plate tectonics?

5 Briefly explain how an accretionary wedge forms. What is meant by *subduction erosion*?

13.4 | FEATURES OF DEEP-OCEAN BASINS

List and describe the major features associated with deep-ocean basins.

Between the continental margin and the oceanic ridge lies the **deep-ocean basin** (see Figure 13.8). The size of this region—almost 30 percent of Earth's surface—is roughly comparable to the percentage of land that presently projects above sea level. This region includes *deep-ocean trenches*, which are extremely deep linear depressions in the ocean floor; remarkably flat areas known as *abyssal plains*; tall volcanic peaks called *seamounts* and *guyots*; and extensive areas of lava flows piled one atop the other called *oceanic plateaus*.

Deep-Ocean Trenches

Deep-ocean trenches are long, relatively narrow troughs that are the deepest parts of the ocean. Most trenches are located along the margins of the Pacific Ocean, where many exceed 10 kilometers (6 miles) in depth (see Figure 13.7). A portion of one—the Challenger Deep in the Mariana Trench—has been measured at a record 10,994 meters (36,069 feet) below sea level, making it the deepest-known part of the world ocean (**FIGURE 13.13**). Only two trenches are located in the Atlantic: the Puerto Rico Trench and the South Sandwich Trench.

Although deep-ocean trenches represent only a very small portion of the area of the ocean floor, they are nevertheless significant geologic features. Trenches are sites of plate convergence where slabs of oceanic lithosphere subduct and plunge back into the mantle. In addition to earthquakes being

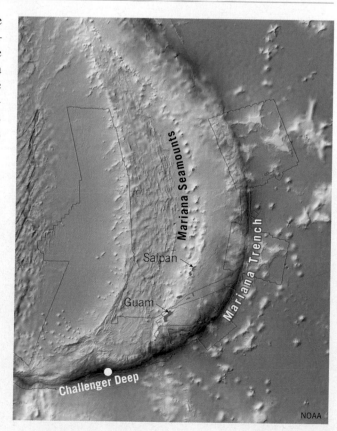

FIGURE 13.13 The Challenger Deep Located near the southern end of the Mariana Trench, the Challenger Deep is the deepest place in the global ocean, about 10,994 meters (36,069 feet) deep. Film director James Cameron (*Titanic* and *Avatar*) made news in March 2012 as the first person to take a deep-diving submersible to the bottom of the Challenger Deep in more than 50 years.

NOAA

Explaining Coral Atolls: Darwin's Hypothesis

Douglas Peebles Photography/Alamy

Coral atolls are ring-shaped structures that often extend from slightly above sea level to depths of several thousand meters. What causes atolls to form, and how do they attain such great thicknesses?

Reef-building corals only grow in warm, clear sunlit water. The depth of most active reef growth is limited to warm tropical water no more than about 45 meters (150 feet) deep. The strict environmental conditions required for coral growth create an interesting paradox: How can corals—which require warm, shallow, sunlit water no deeper than a few dozen meters—create massive structures such as coral atolls that extend to great depths?

Mark A. Wilson

Most corals secrete a hard external skeleton made of calcium carbonate. Colonies of these corals build large calcium carbonate structures, called reefs, where new colonies grow atop the strong skeletons of previous colonies. Sponges and algae may attach to the reef, further enlarging the structure.

2.0 cm

TENTACLE

POLYP

MOUTH

STOMACH

CALCIUM CARBONATE SUPPORT STRUCTURE

Corals: Tiny colonial animals
Corals are marine animals that typically live in compact colonies of many identical organisms called polyps.

NOAA

NOAA

Reinhard Dirsherl/AGE Fotostock

Fringing reef
David Hiser/Getty Images

Barrier reef
Chad Ehlers/Alamy

Atoll
imagebroker/Alamy Images

What Darwin observed Naturalist Charles Darwin was one of the first to formulate a hypothesis on the origin of these ringed-shaped atolls. Aboard the British ship HMS *Beagle* during its famous global circumnavigation, Darwin noticed a progression of stages in coral reef development from (1) a fringing reef along the margins of a volcano to (2) a barrier reef with a volcano in the middle to (3) an atoll, consisting of a continuous or broken ring of coral reef surrounding a central lagoon.

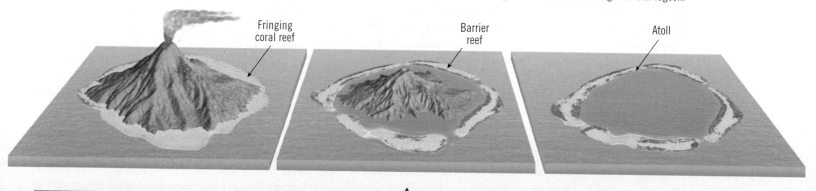

Fringing coral reef

Barrier reef

Atoll

Darwin's hypothesis Darwin's hypothesis asserts that, in addition to being lowered by erosional forces, many volcanic islands gradually sink. Darwin also suggested that corals respond to the gradual change in water depth caused by the subsiding volcano by building the reef complex upward. During Darwin's time, however, there was no plausible mechanism to account for how, or why so many volcanic islands sink.

Plate tectonics and coral atolls The plate tectonics theory provides the most current scientific explanation regarding how volcanic islands become extinct and sink to great depths over long periods of time. Some volcanic islands form over relatively stationary mantle plumes, causing the lithosphere to be buoyantly uplifted. Over spans of millions of years, these volcanic islands gradually sink as moving plates carry them away from the region of hot-spot volcanism because the oceanic lithosphere cools, becomes denser, and sinks.

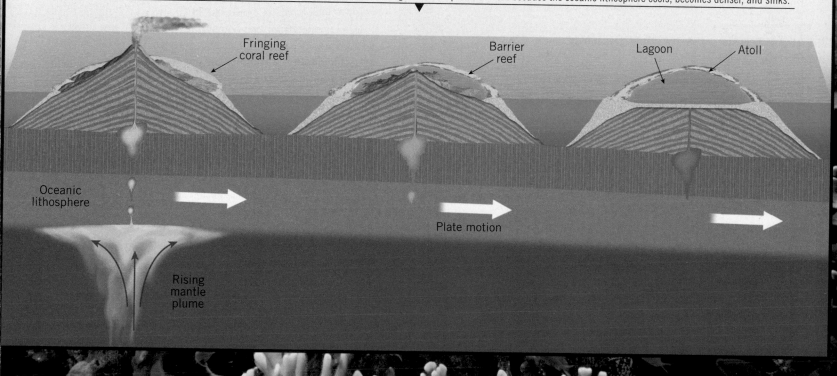

Fringing coral reef

Barrier reef

Lagoon   Atoll

Oceanic lithosphere

Plate motion

Rising mantle plume

created as one plate "scrapes" against another, volcanic activity is also triggered by plate subduction. As a result, some trenches run parallel to an arc-shaped row of active volcanoes called a **volcanic island arc**. In Figure 13.13, the Mariana Seamounts are such a feature. Furthermore, **continental volcanic arcs**, such as those making up portions of the Andes and Cascades, are located parallel to trenches that lie adjacent to continental margins. The volcanic activity associated with the trenches that surround the Pacific Ocean explains why the region is called the *Ring of Fire*.

Abyssal Plains

Abyssal (*a* = without, *byssus* = bottom) **plains** are deep, incredibly flat features; in fact, these regions are likely the most level places on Earth. The abyssal plain found off the coast of Argentina, for example, has less than 3 meters (10 feet) of relief over a distance exceeding 1300 kilometers (800 miles). The monotonous topography of abyssal plains is occasionally interrupted by the protruding summit of a partially buried volcanic peak (seamount).

Using *seismic reflection profilers*, instruments that generate signals designed to penetrate far below the ocean floor, researchers have determined that abyssal plains owe their relatively featureless topography to thick accumulations of sediment that have buried an otherwise rugged ocean floor (**FIGURE 13.14**). The nature of the sediment indicates that these plains consist primarily of three types of sediment: (1) fine sediments transported far out to sea by turbidity currents, (2) mineral matter that has precipitated out of seawater, and (3) shells and skeletons of microscopic marine organisms.

Abyssal plains occur in all oceans. However, the floor of the Atlantic has the most extensive abyssal plains because it has few trenches to act as traps for sediment carried down the continental slope.

FIGURE 13.14 Seismic Profile This seismic cross section and matching sketch across a portion of the Madeira Abyssal Plain in the eastern Atlantic show the irregular oceanic crust buried by sediments. (Image courtesy Woods Hole Oceanographic Institution, Charles Hollister, translator. Used with permission of the Woods Hole Oceanographic Institution.)

Seismic reflection profile

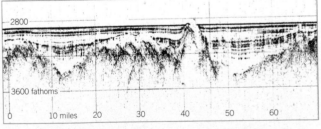

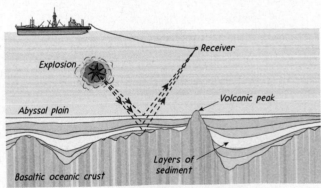

Geologist's Sketch

Volcanic Structures on the Ocean Floor

Dotting the seafloor are numerous volcanic structures of various sizes. Many occur as isolated features that resemble volcanoes on land. Others occur as long, narrow chains that stretch for thousands of kilometers, while still others are massive structures that cover an area the size of Texas.

Seamounts and Volcanic Islands   Submarine volcanoes, called **seamounts**, may rise hundreds of meters above the surrounding topography. It is estimated that more than 1 million seamounts exist. Some grow large enough to become oceanic islands, but most do not have a sufficiently long eruptive history to build a structure above sea level. Although seamounts are found on the floors of all the oceans, they are most common in the Pacific.

Some, like the Hawaiian Island–Emperor seamount chain, which stretches from the Hawaiian islands to the Aleutian trench, form over volcanic hot spots (see Figure 13.7 and Figure 7.28, page 231). Others are born near oceanic ridges.

If a volcano grows large enough before it is carried from its magma source by plate motion, the structure may emerge as a *volcanic island*. Examples of volcanic islands include Easter Island, Tahiti, Bora Bora, Galapagos, and the Canary islands.

Guyots   During their existence, inactive volcanic islands are gradually but inevitably lowered to near sea level by the forces of weathering and erosion. As a moving plate slowly carries inactive volcanic islands away from the elevated oceanic ridge or hot spot over which they formed, they gradually sink and disappear below the water surface. Submerged, flat-topped seamounts that formed in this manner are called **guyots**.*

Oceanic Plateaus   The ocean floor contains several massive **oceanic plateaus**, which resemble lava plateaus composed of flood basalts found on the continents. Oceanic plateaus, which can be more than 30 kilometers (20 miles) thick, are generated from vast outpourings of fluid basaltic lavas.

Some oceanic plateaus appear to have formed quickly in geologic terms. Examples include the Ontong Java Plateau, which formed in less than 3 million years, and the Kerguelen Plateau, which formed in 4.5 million years (see Figure 13.7). Like basalt plateaus on land, oceanic plateaus are thought to form when the bulbous head of a rising mantle plume melts and produces a vast outpouring of basalt.

---

*The term *guyot* comes from the name of Arnold Guyot, Princeton University's first geology professor. It is pronounced "GEE-oh," with a hard *g*, as in *give*.

13.4 CONCEPT CHECKS

1 Explain how deep-ocean trenches are related to plate boundaries.

2 Why are abyssal plains more extensive on the floor of the Atlantic than on the floor of the Pacific?

3 How does a flat-topped seamount, called a *guyot*, form?

4 What features on the ocean floor most resemble basalt plateaus on the continents?

13.5 | THE OCEANIC RIDGE

Summarize the basic characteristics of oceanic ridges.

Along well-developed divergent plate boundaries, the seafloor is elevated, forming a broad linear swell called the or **rise**, or **mid-ocean ridge**. Our knowledge of the oceanic ridge system comes from soundings of the ocean floor, core samples from deep-sea drilling, visual inspection using deep-diving submersibles (**FIGURE 13.15**), and firsthand inspection of slices of ocean floor that have been thrust onto dry land during continental collisions. At oceanic ridges we find extensive normal and strike-slip faulting, earthquakes, high heat flow, and volcanism.

FIGURE 13.15 The Deep-Diving Submersible *Alvin* This submersible is 7.6 meters (25 feet) long, weighs 16 tons, has a cruising speed of 1 knot, and can reach depths of 4000 meters (more than 13,000 feet). A pilot and two scientific observers are along during a normal 6- to 10-hour dive. (Photo by Rod Catanach KRT/Newscom)

Anatomy of the Oceanic Ridge

The oceanic ridge system winds through all major oceans in a manner similar to the seam on a baseball and is the longest topographic feature on Earth, at more than 70,000 kilometers (43,000 miles) in length (**FIGURE 13.16**). The crest of the ridge typically stands 2 to 3 kilometers above the adjacent deep-ocean basins and marks the plate boundary where new oceanic crust is created.

Notice in Figure 13.16 that large sections of the oceanic ridge system have been named based on their locations within the various ocean basins. Some ridges run through the middle of ocean basins, where they are appropriately called *mid-ocean* ridges. The Mid-Atlantic Ridge and the Mid-Indian Ridge are examples. By contrast, the East Pacific Rise is *not* a "mid-ocean" feature. Rather, as its name implies, it is located in the eastern Pacific, far from the center of the ocean.

The term *ridge* is somewhat misleading because these features are not narrow and steep as the term implies but have widths of 1000 to 4000 kilometers (600 to 2500 miles) and the appearance of broad, elongated swells that exhibit varying degrees of ruggedness. Furthermore, the ridge system is broken into segments that range from a few tens to hundreds of kilometers in length. Each segment is offset from the adjacent segment by a transform fault.

Oceanic ridges are as high as some mountains on the continents; but the similarities end there. Whereas most mountain ranges on land form when the compressional forces associated with continental collisions fold and metamorphose thick sequences of sedimentary rocks, oceanic ridges form where upwelling from the mantle generates new oceanic crust. Oceanic ridges consist of layers and piles of newly formed basaltic rocks that are buoyantly uplifted by the hot mantle rocks from which they formed.

Along the axes of some segments of the oceanic ridge system are deep, down-faulted structures that are called **rift valleys** because of their striking similarity to the continental rift valleys found in East Africa (**FIGURE 13.17**). Some rift valleys, including those along the rugged Mid-Atlantic Ridge, are typically 30 to 50 kilometers (20 to 30 miles) wide and have walls that tower 500 to 2500 meters (1640 to 8200 feet) above the valley floor. This makes them comparable to the deepest and widest part of Arizona's Grand Canyon.

Why Is the Oceanic Ridge Elevated?

The primary reason for the elevated position of the ridge system is that newly created oceanic lithosphere is hot and therefore less dense than cooler rocks of the deep-ocean basin. As the newly formed basaltic crust travels away from the ridge crest, it is cooled from above as seawater

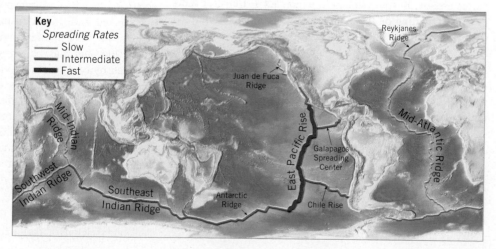

Key
Spreading Rates
— Slow
— Intermediate
— Fast

Reykjanes Ridge

Juan de Fuca Ridge

Mid-Indian Ridge

Southwest Indian Ridge

Southeast Indian Ridge

Antarctic Ridge

East Pacific Rise

Galapagos Spreading Center

Chile Rise

Mid-Atlantic Ridge

FIGURE 13.16 Distribution of the Oceanic Ridge System The map shows ridge segments that exhibit slow, intermediate, and fast spreading rates.

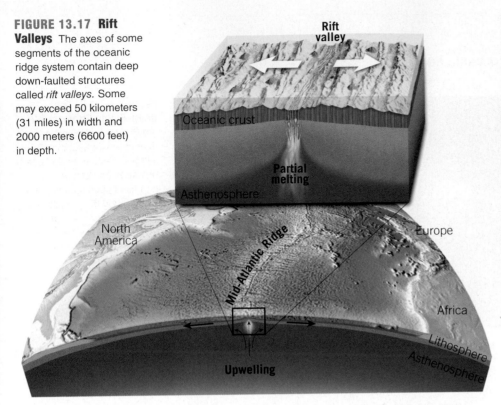

FIGURE 13.17 Rift Valleys The axes of some segments of the oceanic ridge system contain deep down-faulted structures called *rift valleys*. Some may exceed 50 kilometers (31 miles) in width and 2000 meters (6600 feet) in depth.

circulates through the pore spaces and fractures in the rock. In addition, it cools because it gets farther and farther from the zone of hot mantle upwelling. As a result, the lithosphere gradually cools, contracts, and becomes denser. This thermal contraction accounts for the greater ocean depths that occur away from the ridge. It takes almost 80 million years of cooling and contraction for rock that was once part of an elevated ocean ridge system to relocate to the deep-ocean basin.

As lithosphere is displaced away from the ridge crest, cooling also causes a gradual increase in lithospheric thickness. This happens because the boundary between the lithosphere and asthenosphere is a thermal (temperature) boundary. Recall that the lithosphere is Earth's cool, stiff outer layer, whereas the asthenosphere is a comparatively hot and weak layer. As material in the uppermost asthenosphere ages (cools), it becomes stiff and rigid. Thus, the upper portion of the asthenosphere is gradually converted to lithosphere simply by cooling. Oceanic lithosphere continues to thicken until it is about 80 to 100 kilometers (50 to 60 miles) thick. Thereafter, its thickness remains relatively unchanged until it is subducted.

13.5 CONCEPT CHECKS

1 Briefly describe oceanic ridges.

2 Although oceanic ridges can be as tall as some mountains on the continents, list some ways that oceanic ridges are different from mountains.

3 Where do rift valleys form along the oceanic ridge system?

4 What is the primary reason for the elevated position of the oceanic ridge system?

13.6 | SEAFLOOR SEDIMENTS Distinguish among three categories of seafloor sediment and explain why some of these sediments can be used to study climate change.

Except for steep areas of the continental slope and areas near the crest of the mid-ocean ridge, the ocean floor is covered with sediment. Part of this material has been deposited by turbidity currents, and the rest has slowly settled to the seafloor from above. The thickness of this carpet of debris varies greatly. In some trenches, which act as traps for sediments originating on the continental margin, accumulations may approach 10 kilometers (6 miles). In general, however, sediment accumulations are considerably less. In the Pacific Ocean, for example, uncompacted sediment measures about 600 meters (2000 feet) or less, whereas on the floor of the Atlantic, the thickness varies from 500 to 1000 meters (1600 to 3300 feet).

Types of Seafloor Sediments

Seafloor sediments can be classified according to their origin into three broad categories: **terrigenous** ("derived from land"), **biogenous** ("derived from organisms"), and **hydrogenous** ("derived from water"). Although each category is discussed separately, remember that all seafloor sediments are mixtures. No body of sediment comes entirely from a single source.

Terrigenous Sediment Terrigenous sediment consists primarily of mineral grains that were weathered from continental rocks and transported to the ocean. Larger particles (sand and gravel) usually settle rapidly near shore, whereas the very smallest particles take years to settle to the ocean floor and may be carried thousands of kilometers by ocean currents. As a result, virtually every area of the ocean receives some terrigenous sediment. The rate at which this sediment accumulates on the deep-ocean floor, though, is very slow. Forming a 1-centimeter (0.4-inch) abyssal clay layer, for example, requires as much as 50,000 years. Conversely, on the continental margins near the mouths of large rivers, terrigenous sediment accumulates rapidly and forms thick deposits.

Biogenous Sediment Biogenous sediment consists of shells and skeletons of marine animals and algae (**FIGURE 13.18**).

This debris is produced mostly by microscopic organisms living in the sunlit waters near the ocean surface. Once these organisms die, their hard *tests* (*testa* = shell) continually "rain" down and accumulate on the seafloor.

The most common biogenous sediment is *calcareous* ($CaCO_3$) *ooze*, which, as the name implies, has the consistency of thick mud. This sediment is produced from the tests of organisms that inhabit warm surface waters. When calcareous hard parts slowly sink through a cool layer of water, they begin to dissolve. This occurs because the deeper cold seawater is richer in carbon dioxide and is thus more acidic than warm water. In seawater deeper than about 4500 meters (15,000 feet), calcareous tests completely dissolve before they reach bottom. Consequently, calcareous ooze does not accumulate at these greater depths.

Other biogenous sediments include *siliceous* (SiO_2) *ooze* and phosphate-rich material. The former is composed primarily of tests of diatoms (single-celled algae) and radiolaria (single-celled animals), whereas the latter is derived from the bones, teeth, and scales of fish and other marine organisms.

Hydrogenous Sediment Hydrogenous sediment consists of minerals that crystallize directly from seawater through various chemical reactions. For example, some limestones are formed when calcium carbonate precipitates directly from the water; however, most limestone is composed of biogenous sediment.

Some of the most common types of hydrogenous sediment include the following:

- *Manganese nodules* are rounded, hard lumps of manganese, iron, and other metals that precipitate in concentric layers around a central object (such as a volcanic pebble or a grain of sand). The nodules can be up to 20 centimeters (8 inches) in diameter and are often littered across large areas of the deep seafloor (**FIGURE 13.19A**).
- *Calcium carbonates* form by precipitating directly from seawater in warm climates. If this material is buried and hardened, it forms limestone. Most limestone, however, is composed of biogenous sediment.
- *Metal sulfides* are usually precipitated as coatings on rocks near black smokers associated with the crest of a mid-ocean ridge (**FIGURE 13.19B**). These deposits contain iron, nickel, copper, zinc, silver, and other metals in varying proportions.
- *Evaporites* form where evaporation rates are high and there is restricted open-ocean circulation. As water evaporates from such areas, the remaining seawater becomes saturated with dissolved minerals,

FIGURE 13.18 Marine Microfossils: An Example of Biogenous Sediment These tiny, single-celled organisms are sensitive to even small fluctuations in temperature. Seafloor sediments containing these fossils are useful sources of data on climate change. (Photo by Mary Martin/Science Source)

which then begin to precipitate. Heavier than seawater, they sink to the bottom or form a characteristic white crust of evaporite minerals around the edges of these areas. Evaporites are collectively termed *salts*; some evaporite minerals taste salty, such as *halite* (common table salt, NaCl), and some do not, such as the calcium sulfate minerals *anhydrite* ($CaSO_4$) and *gypsum* ($CaSO_4 \cdot 2H_2O$).

Seafloor Sediment—A Storehouse of Climate Data

Reliable climate records go back only a couple hundred years, at best. How do scientists learn about climates and climate change prior to that time? They must reconstruct past climates from *indirect evidence*; that is, they must analyze phenomena that respond to and reflect changing atmospheric conditions. An interesting and important technique for analyzing Earth's climate history is the study of biogenous seafloor sediments.

We know that the parts of the Earth system are linked so that a change in one part can produce changes in any or all of the other parts. For example, changes in atmospheric and oceanic temperatures are reflected in the nature of life in the sea.

A.

B.

SmartFigure 13.19 Examples of Hydrogenous Sediment
A. Manganese nodules. (Photo by Charles A. Winter/Science Source) **B.** This black smoker is spewing hot, mineral-rich water. When the heated solutions meet cold seawater, metal sulfides precipitate and form mounds of minerals around these hydrothermal vents. (Photo by Fisheries and Oceans Canada/Uvic—Verena Tunnicliffe/Newscom)

FIGURE 13.20 Data from the Seafloor This scientist is holding a sediment core obtained by an oceanographic research vessel. Cores of sediment provide data that allow for a more complete understanding of past climates. Notice the "library" of sediment cores in the background. (Photo by INGO WAGNER/EPA/Newscom)

Most seafloor sediments contain the remains of organisms that once lived near the sea surface (the ocean–atmosphere interface). When such near-surface organisms die, their shells slowly settle to the floor of the ocean, where they become part of the sedimentary record. These seafloor sediments are useful recorders of worldwide climate change because the numbers and types of organisms living near the sea surface change with the climate.

Thus, in seeking to understand climate change as well as other environmental transformations, scientists are tapping the huge reservoir of data in seafloor sediments (**FIGURE 13.20**). The sediment cores gathered by drilling ships such as the one in the chapter-opening photo and other research vessels have provided invaluable data that have significantly expanded our knowledge and understanding of past climates.

One notable example of the importance of seafloor sediments to our understanding of climate change relates to unraveling the fluctuating atmospheric conditions of the Ice Age. The records of temperature changes contained in cores of sediment from the ocean floor have been critical to our present understanding of this recent span of Earth history.

13.6 CONCEPT CHECKS

1 Distinguish among the three basic types of seafloor sediments. Give an example of each.

2 Why are seafloor sediments useful in studying past climates?

13.7 | RESOURCES FROM THE SEAFLOOR Discuss some important resources and potential resources associated with the ocean floor.

The seafloor is rich in mineral and organic resources. Most, however, are not easily accessible, and recovery involves technological challenges and high cost. Nevertheless, certain resources have high value and thus make appealing exploration targets.

Energy Resources

Among the nonliving resources extracted from the oceans, more than 95 percent of the economic value comes from energy products. The main energy products are oil and natural gas, which are currently being extracted, and gas hydrates, which are not yet being utilized but have significant potential.

Oil and Natural Gas
The ancient remains of microscopic organisms, buried within marine sediments before they could completely decompose, are the source of today's deposits of oil and natural gas. The percentage of world oil produced from offshore regions has increased from trace amounts in the 1930s to more than 30 percent today. Most of this increase results from continuing technological advancements employed by offshore drilling platforms.

Major offshore reserves exist in the Persian Gulf, in the Gulf of Mexico, off the coast of southern California, in the North Sea, and in the East Indies. Additional reserves are located off the northern coast of Alaska and in the Canadian Arctic, Asian seas, Africa, and Brazil. Because the likelihood of finding major new reserves on land is small, future offshore exploration will continue to be important, especially in deeper waters of the continental margins. A major environmental concern about offshore petroleum exploration is the possibility of oil spills caused by inadvertent leaks or blowouts during the drilling process (**FIGURE 13.21**).

Gas Hydrates
Gas hydrates are unusually compact chemical structures made of water and natural gas. The most common type of natural gas is methane, which produces *methane hydrate*. Gas hydrates occur beneath permafrost areas on land and under the ocean floor at depths below 525 meters (1720 feet).

Most oceanic gas hydrates are created when bacteria break down organic matter trapped in seafloor sediments, producing methane gas with minor amounts of ethane and propane. These gases combine with water in deep-ocean sediments (where pressures are high and temperatures are low) in such a way that the gas is trapped inside a lattice-like cage of water molecules.

Vessels that have drilled into gas hydrates have retrieved cores of mud mixed with chunks or layers of gas hydrates (FIGURE 13.22) that fizzle and evaporate quickly when they are exposed to the relatively warm, low-pressure conditions at the ocean surface. Gas hydrates resemble chunks of ice but ignite when lit by a flame because methane and other flammable gases are released as gas hydrates vaporize.

Some estimates indicate that as much as 20 quadrillion cubic meters (700 quadrillion cubic feet) of methane are locked up in sediments containing gas hydrates. This is equivalent to about *twice* as much carbon as Earth's coal, oil, and conventional gas reserves combined, so gas hydrates have great potential. A major drawback in exploiting reserves of gas hydrates is that they rapidly decompose at surface temperatures and pressures. In the future, however, these vast seafloor reserves of energy may help power modern society.

Other Resources

Other major resources from the seafloor include sand and gravel, evaporative salts, and manganese nodules.

Sand and Gravel The offshore sand-and-gravel industry is second in economic value only to the petroleum industry. Sand and gravel, which includes rock fragments that are washed out to sea and shells of marine organisms, are mined by offshore barges using suction dredges. Sand and gravel are primarily used as an aggregate in concrete, as a fill material in grading projects, and on recreational beaches.

In some cases, materials of high economic value are associated with offshore sand and gravel deposits. Gem-quality diamonds, for example, are recovered from gravels on the continental shelf offshore of South Africa and Australia. Sediments rich in tin have been mined from some offshore areas of Southeast Asia. Platinum and gold have been found in deposits in gold-mining areas throughout the world, and some Florida beach sands are rich in titanium.

FIGURE 13.21 **Disaster in the Gulf of Mexico** On April 20, 2010, the mobile floating drilling platform named *Deepwater Horizon* exploded while working on a 10,680-meter (35,000-foot) well in approximately 1600 meters (5200 feet) of water, causing a devastating oil spill. (Photo by John Mosier/ZUMA Press/Newscom)

Evaporative Salts When seawater evaporates, the salts increase in concentration until they can no longer remain dissolved, so they precipitate out of solution and form salt deposits, which can then be harvested. The most economically important salt is *halite* (common table salt). Halite is widely used for seasoning, curing, and preserving foods. It is also used in water conditioners, in agriculture, in the clothing industry for dying fabric, and to de-ice roads. Since ancient times, the ocean has been an important source of salt for human consumption, and the sea remains a significant supplier (FIGURE 13.23).

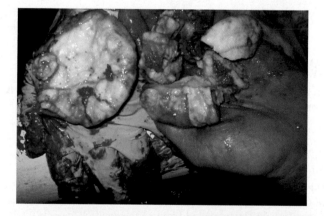

FIGURE 13.22 **Gas Hydrates** This sample was retrieved from the deep-ocean floor and shows white ice-like gas hydrate mixed with mud. Gas hydrates evaporate when exposed to surface conditions and release natural gas, which can be ignited. (Photo courtesy USGS National Center)

EYE ON EARTH

Gas hydrates are natural gas reservoirs in ice-like crystalline solids found in sediments on the deep-ocean floor and in Arctic permafrost areas deeper than 200 meters (660 feet). Gas hydrates contain methane molecules—the main ingredient of most natural gas—which is trapped within a lattice-like cage of water molecules. These deposits, called "ice that burns" are stable only at relatively high pressures and/or low temperatures. (Photo by Yonhap News/Newscom)

QUESTION 1 *At surface temperatures and pressures, in what state of matter is methane? With this in mind, what problems exist for the extraction of gas hydrate deposits?*

QUESTION 2 *Explain why gas hydrates are not found in sediments at shallow depths such as those found on the continental shelf.*

FIGURE 13.23 Solar Salt Seawater is one important source of common salt (sodium chloride). Saltwater is held in shallow ponds, where solar energy evaporates the water. The nearly pure salt deposits that eventually form are essentially artificial evaporate deposits. (Photo by Geof Kirb/ Alamy Images)

Manganese Nodules   Manganese nodules contain significant concentrations of manganese and iron, and they contain smaller concentrations of copper, nickel, and cobalt, all of which have a variety of economic uses. Cobalt, for example, is deemed "strategic" (essential to U.S. national security)

because it is required to produce dense, strong alloys with other metals and is used in high-speed cutting tools, powerful permanent magnets, and jet engine parts. Technologically, mining the deep-ocean floor for manganese nodules is possible but economically not profitable.

Nodules are widely distributed, but not all regions have the same potential for mining. Good locations have abundant nodules that contain the economically optimum mix of copper, nickel, and cobalt. Sites meeting these criteria, however, are relatively limited. In addition, there are political problems of establishing mining rights far from land and environmental concerns about disturbing large portions of the deep-ocean floor.

13.7 CONCEPT CHECKS

1   Which seafloor resource is presently most valuable?

2   What are gas hydrates? Will they likely be a significant energy source in the next 10 years?

3   What nonenergy seafloor resource is most valuable?

13 CONCEPTS IN REVIEW | The Ocean Floor

13.1 THE VAST WORLD OCEAN

Discuss the extent and distribution of oceans and continents on Earth. Identify Earth's four main ocean basins.

KEY TERM: oceanography

■ Oceanography is an interdisciplinary science that draws on the methods and knowledge of biology, chemistry, physics, and geology to study all aspects of the world ocean.

■ Earth's surface is dominated by oceans. Nearly 71 percent of the planet's surface area is oceans and marginal seas. In the Southern Hemisphere, about 81 percent of the surface is water.

■ Of the three major oceans—the Pacific, Atlantic, and Indian—the Pacific Ocean is the largest, contains slightly more than half of the water in the world ocean, and has the greatest average depth—3940 meters (12,927 feet).

Q Name the four principal oceans identified with letters on this world map.

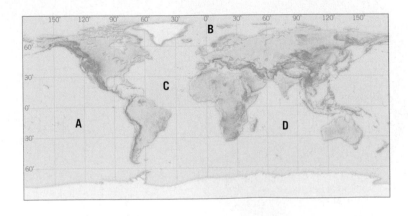

13.2 AN EMERGING PICTURE OF THE OCEAN FLOOR

Define *bathymetry* and summarize the various techniques used to map the ocean floor.

KEY TERMS: bathymetry, sonar, echo sounder

■ Seafloor mapping is done with sonar—shipboard instruments that emit pulses of sound that "echo" off the bottom. Satellites are also used to map the ocean floor. Their instruments measure slight variations in sea level that result from differences in the gravitational pull of features on the seafloor. Accurate maps of seafloor topography can be made using these data.

■ Mapping efforts have revealed three major areas of the ocean floor: continental margins, deep-ocean basins, and oceanic ridges.

Q Do the math: How many seconds would it take an echo sounder's ping to make the trip from a ship to the Challenger Deep (10,994 meters [36,069 feet]) and back? Recall that Depth = ½ (1500 m/sec × Echo travel time).

13.3 CONTINENTAL MARGINS

Compare a passive continental margin with an active continental margin and list the major features of each.

KEY TERMS: continental margin, passive continental margin, continental shelf, continental slope, continental rise, deep-sea fan, submarine canyon, turbidity current, active continental margin, accretionary wedge, subduction erosion

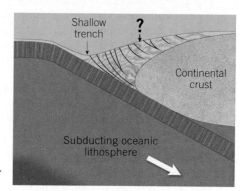

- Continental margins are transition zones between continental and oceanic crust. Active continental margins occur where a plate boundary and the edge of a continent coincide, usually on the leading edge of a plate. Passive continental margins are on the trailing edges of continents, far from plate boundaries.
- Heading offshore from the shoreline of a passive margin, a submarine traveler would first encounter the gently sloping continental shelf and then the steeper continental slope, which marks the end of the continental crust and the beginning of the oceanic crust. Beyond the continental slope is another gently sloping section, the continental rise: It is made of sediment transported by turbidity currents through submarine canyons and piled up in deep-sea fans atop the oceanic crust.
- Submarine canyons are deep, steep-sided valleys that originate on the continental slope and may extend to the deep-ocean basin. Many submarine canyons have been excavated by turbidity currents (downslope movements of dense, sediment-laden water).
- At an active continental margin, material may be added to the leading edge of a continent in the form of an accretionary wedge (common at shallow-angle subduction zones), or material may be scraped off the edge of a continent by subduction erosion (common at steeply dipping subduction zones).

Q What type of continental margin is depicted by this diagram? Be as specific as possible. Name the feature indicated by the question mark.

13.4 FEATURES OF DEEP-OCEAN BASINS

List and describe the major features associated with deep-ocean basins.

KEY TERMS: deep-ocean basin, deep-ocean trench, volcanic island arc, continental volcanic arc, abyssal plain, seamount, guyot, oceanic plateau

- The deep-ocean basin makes up about half of the ocean floor's area. Much of it is abyssal plain (deep, featureless sediment-draped crust). Subduction zones and deep-ocean trenches also occur in deep-ocean basins. Paralleling trenches are volcanic island arcs (if the subduction goes underneath oceanic lithosphere) or continental volcanic arcs (if the overriding plate has continental lithosphere on its leading edge).
- There are a variety of volcanic structures on the deep-ocean floor. Seamounts are submarine volcanoes; if they pierce the surface of the ocean, we call them volcanic islands. Guyots are old volcanic islands that have had their tops eroded off before they sink below sea level. Oceanic plateaus are unusually thick sections of oceanic crust formed by massive underwater eruptions of lava.

Q On this cross-sectional view of a deep-ocean basin, label the following features: a seamount, a guyot, a volcanic island, an oceanic plateau, and an abyssal plain.

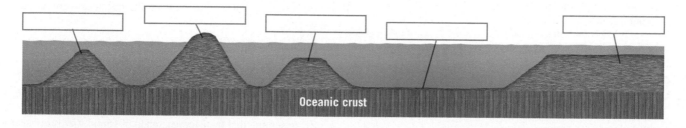

13.5 THE OCEANIC RIDGE

Summarize the basic characteristics of oceanic ridges.

KEY TERMS: oceanic ridge or rise (mid-ocean ridge), rift valley

- The oceanic ridge system is the longest topographic feature on Earth, wrapping around the world through all major ocean basins. It is a few kilometers tall, a few thousand kilometers wide, and a few tens of thousands of kilometers long. The summit is the place where new oceanic crust is generated, often marked by a rift valley.
- Oceanic ridges are elevated features because they are warm and therefore less dense than older, colder oceanic lithosphere. As oceanic crust moves away from the ridge crest, heat loss causes the oceanic crust to become denser and subside. After 80 million years, crust that was once part of an oceanic ridge is in the deep-ocean basin, far from the ridge.

13.6 SEAFLOOR SEDIMENTS

Distinguish among three categories of seafloor sediment and explain why some of these sediments can be used to study climate change.

KEY TERMS: terrigenous sediment, biogenous sediment, hydrogenous sediment

- There are three broad categories of seafloor sediments. Terrigenous sediment consists primarily of mineral grains that were weathered from continental rocks and transported to the ocean; biogenous sediment consists of shells and skeletons of marine animals and plants; and hydrogenous sediment includes minerals that crystallize directly from seawater through various chemical reactions.
- Seafloor sediments are helpful in studying worldwide climate change because they often contain the remains of organisms that once lived near the sea surface. The numbers and types of these organisms change as the climate changes, and their remains in seafloor sediments record these changes.

Q This satellite image from February 4, 2013, shows a plume of dust moving from the southern Arabian Peninsula over the Arabian Sea, the name given to a portion of the Indian Ocean. The wind will subside, and the dust will settle onto the water and slowly sink, eventually reaching the ocean floor. To which category of seafloor sediment will this material belong?

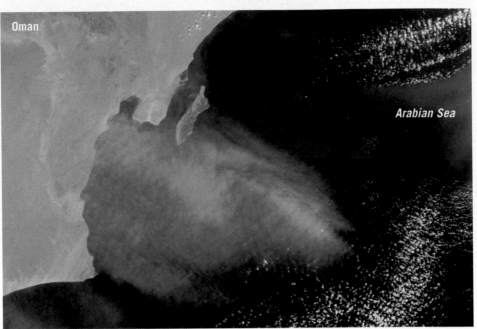

NASA

13.7 RESOURCES FROM THE SEAFLOOR

Discuss some important resources and potential resources associated with the ocean floor.

KEY TERM: gas hydrate

- Oil and natural gas represent more than 95 percent of the value of the nonliving resources extracted from the oceans.
- About 30 percent of world oil production is from offshore areas. Environmental concerns include leaks and blowouts.

- Gas hydrates are complex molecules made of natural gas and water. Some form under high pressure and low temperature in marine sediments. Gas hydrates also form in permafrost environments. When brought to Earth's surface, they rapidly decompose into methane and water.
- Significant nonenergy resources and potential resources include sand and gravel, salt, and manganese nodules. About 30 percent of the world's supply of salt is extracted from seawater.

Q Name two *potential* resources from the seafloor—one energy and one nonenergy.

GIVE IT SOME THOUGHT

1. Refer to the accompanying graph to answer the following questions:

   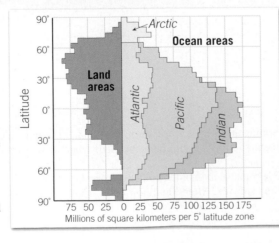

   a. Water dominates Earth's surface but not everywhere. In what Northern Hemisphere latitude belt is there more land than water?

   b. In what latitude belt is there no land at all?

2. Assuming that the average speed of sound waves in water is 1500 meters per second, determine the water depth if a signal sent out by an echo sounder on a research vessel requires 6 seconds to strike bottom and return to the recorder aboard the ship.

3. Refer to the accompanying map on the top of the next page, which shows the eastern seaboard of the United States to complete the following:

   a. Associate each of the following with a letter on the map: continental rise, continental shelf, continental slope, and shelf-break.

   b. How does the size of the continental shelf surrounding Florida compare to the size of the Florida peninsula?

   c. Why are there no deep-ocean trenches on this map?

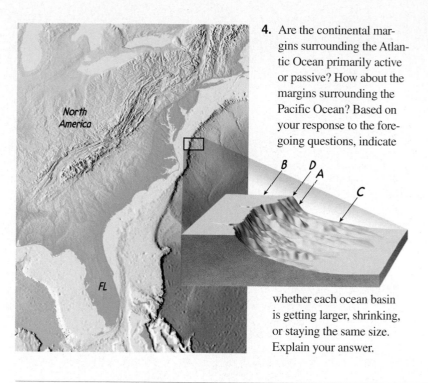

4. Are the continental margins surrounding the Atlantic Ocean primarily active or passive? How about the margins surrounding the Pacific Ocean? Based on your response to the foregoing questions, indicate whether each ocean basin is getting larger, shrinking, or staying the same size. Explain your answer.

5. Imagine that you and a passenger are in a deep-diving submersible such as the one shown here in the North Pacific Ocean near Alaska's Aleutian islands. During a typical 6- to 10-hour dive, you encounter a long, narrow depression on the ocean floor. Your passenger asks whether you think it is a submarine canyon, a rift valley, or a deep-ocean trench. How would you respond? Explain your choice.

Jeff Rotman/Alamy

6. Examine the accompanying sketch, showing three sediment layers on the ocean floor. What term is applied to such layers? What process was responsible for creating these layers? Are these layers more likely part of a deep-sea fan or an accretionary wedge?

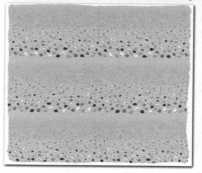

EXAMINING THE **EARTH SYSTEM**

1. Describe some of the material and energy exchanges that take place at the following interfaces:
   a. ocean surface and atmosphere
   b. ocean water and ocean floor
   c. ocean biosphere and ocean water

2. This image from a scanning electron microscope shows the shell (test) of a tiny organism called a foraminiferan that was collected from the ocean floor by a research vessel. When alive, the organism that created this shell lived in the shallow surface waters of the ocean. Relate this tiny shell to each of Earth's four major spheres. Why are microscopic fossils such as this one useful in the study of climate change?

Dee Breger/Science Source

3. Sediment on the seafloor often leaves clues about various conditions that existed during deposition. What do the following layers in a seafloor core indicate about the environment in which each layer was deposited?
   ▪ Layer 5 (top): A layer of fine clays
   ▪ Layer 4: Siliceous ooze
   ▪ Layer 3: Calcareous ooze
   ▪ Layer 2: Fragments of coral reef
   ▪ Layer 1 (bottom): Rocks of basaltic composition with some metal sulfide coatings

   Explain how one area of the seafloor could experience such varied conditions of deposition.

4. Reef-building corals are responsible for creating atolls—ring-shaped structures that extend from the surface of the ocean to depths of thousands of meters. These corals, however, can only live in warm, sunlit water no more than about 45 meters (150 feet) deep. This presents a paradox: How can corals, which require warm, sunlit water, create structures that extend to great depths? Explain the apparent contradiction.

Reinlhard Dirscheri/Getty Images

MasteringGeology™

14

Ocean Water and Ocean Life*

FOCUS ON CONCEPTS

Each statement represents the primary **LEARNING OBJECTIVE** for the corresponding major heading within the chapter. After you complete the chapter, you should be able to:

14.1 Define *salinity* and list the main elements that contribute to the ocean's salinity. Describe the sources of dissolved substances in seawater and causes of variations in salinity.

14.2 Discuss temperature, salinity, and density changes with depth in the open ocean.

14.3 Distinguish among plankton, nekton, and benthos. Summarize the factors used to divide the ocean into marine life zones.

14.4 Contrast ocean productivity in polar, midlatitude, and tropical settings.

14.5 Define *trophic level* and discuss the efficiency of energy transfer between different trophic levels.

Modern coral reefs are unique ecosystems of animals, plants, and their associated geologic framework. They are home to about 25 percent of all marine species. Because of this great diversity, they are sometimes referred to as the ocean equivalent of rain forests. (Photo by Image Marine Quest)

---

*This chapter was prepared with the assistance of Professor Alan P. Trujillo, Palomar College.

What is the difference between pure water and seawater? One of the most obvious differences is that seawater contains dissolved substances that give it a distinctly salty taste. These dissolved substances are not simply sodium chloride (table salt); they include various other salts, metals, and even dissolved gases. In fact, every known naturally occurring element is found dissolved in at least trace amounts in seawater. Unfortunately, the salt content of seawater makes it unsuitable for drinking or for irrigating most crops and causes it to be highly corrosive to many materials. Yet, many parts of the ocean are teeming with life that is superbly adapted to the marine environment.

There is an amazing variety of life in the ocean, from microscopic bacteria and algae to the largest organism alive today (the blue whale). Water is the major component of nearly every life-form on Earth, and our own body fluid chemistry is remarkably similar to the chemistry of seawater.

14.1 | COMPOSITION OF SEAWATER

Define *salinity* and list the main elements that contribute to the ocean's salinity. Describe the sources of dissolved substances in seawater and causes of variations in salinity.

Seawater consists of about 3.5 percent (by weight) dissolved mineral substances that are collectively termed "salts." Although the percentage of dissolved components may seem small, the actual quantity is huge because the ocean is so vast.

Salinity

Salinity (*salinus* = salt) is the total amount of solid material dissolved in water. More specifically, it is the ratio of the mass of dissolved substances to the mass of the water sample. Many common quantities are expressed in percent (%), which is really *parts per hundred*. Because the proportion of dissolved substances in seawater is such a small number, oceanographers typically express salinity in *parts per thousand* (‰). Thus, the average salinity of seawater is 3.5%, or 35‰.

FIGURE 14.1 shows the principal elements that contribute to the ocean's salinity. If one wanted to make artificial seawater, it could be approximated by following the recipe shown in **TABLE 14.1**. From this table, it is evident that most of the salt in seawater is sodium chloride—common table salt. Sodium chloride together with the next four most abundant salts comprise more than 99 percent of all dissolved substances in the sea. Although only eight elements make

TABLE 14.1   Recipe for Artificial Seawater

To Make Seawater, Combine:	Amount (grams)
Sodium chloride (NaCl)	23.48
Magnesium chloride (MgCl$_2$)	4.98
Sodium sulfate (Na$_2$SO$_4$)	3.92
Calcium chloride (CaCl$_2$)	1.10
Potassium chloride (KCl)	0.66
Sodium bicarbonate (NaHCO$_3$)	0.192
Potassium bromide (KBr)	0.096
Hydrogen borate (H$_3$BO$_3$)	0.026
Strontium chloride (SrCl$_2$)	0.024
Sodium fluoride (NaF)	0.003

Then Add: Pure water (H$_2$O) to form 1,000 grams of solution.

up these five most abundant salts, seawater contains all of Earth's other naturally occurring elements. Despite their presence in minute quantities, many of these elements are very important in maintaining the necessary chemical environment for life in the sea.

Sources of Sea Salts

What are the primary sources for the vast quantities of dissolved substances in the ocean? Chemical weathering of rocks on the continents is one important source. These dissolved materials are delivered to the oceans by streams at an estimated rate of more than 2.3 billion metric tons (2.5 billion short tons) annually.[1] The second major source of elements found in seawater is from Earth's interior. Through volcanic eruptions, large quantities of water vapor and other gases have been emitted during much of geologic time. This process, called *outgassing*, is the principal source of water

FIGURE 14.1 Composition of Seawater The diagram shows the relative proportions of water and dissolved components in a typical sample of seawater.

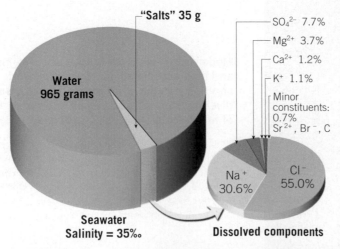

"Salts" 35 g

SO$_4^{2-}$ 7.7%
Mg^{2+} 3.7%
Ca^{2+} 1.2%
K$^+$ 1.1%
Minor constituents: 0.7%
Sr^{2+}, Br$^-$, C

Water 965 grams

Na$^+$ 30.6%

Cl$^-$ 55.0%

Seawater Salinity = 35‰

Dissolved components

---

[1] A metric ton equals 1,000 kilograms, or 2,205 pounds. Thus, it is larger than the "ton" that most Americans are familiar with (called a *short ton*). There are 2,000 pounds in a short ton.

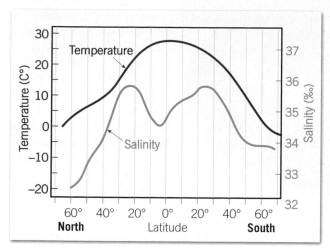

SmartFigure 14.2 **Variations in Surface Temperature and Salinity with Latitude** Average temperatures are highest near the equator and get colder toward the poles. Important factors influencing salinity are variations in rainfall and rates of evaporation. For example, in the dry subtropics in the vicinity of the Tropics of Cancer and Capricorn, high evaporation rates remove more water than is replaced by the meager rainfall, resulting in high surface salinities. In the wet equatorial region, abundant rainfall reduces surface salinities.

in the oceans. Certain elements—notably chlorine, bromine, sulfur, and boron—were outgassed along with water and exist in the ocean in much greater abundance than could be explained by weathering of continental rocks alone.

Although rivers and volcanic activity continually contribute dissolved substances to the oceans, the salinity of seawater is not increasing. In fact, evidence suggests that the composition of seawater has been relatively stable for millions of years. Why doesn't the sea get saltier? The answer is because material is being removed just as rapidly as it is added. For example, some dissolved components are withdrawn from seawater by organisms as they build hard parts. Other components are removed when they chemically precipitate out of the water as sediment. Still others are exchanged at the oceanic ridge by *hydrothermal* (*hydro* = water, *thermos* = hot) *activity*. The net effect is that the overall makeup of seawater has remained relatively constant through time.

Processes Affecting Seawater Salinity

Because the ocean is well mixed, the relative abundances of the major components in seawater are essentially constant, no matter where the ocean is sampled. Variations in salinity, therefore, primarily result from changes in the water content of the solution.

Various surface processes alter the amount of water in seawater, thereby affecting salinity. Processes that add large amounts of freshwater to seawater—and thereby decrease salinity—include the addition of rain and snow, the discharge of rivers, and the melting of icebergs and sea ice. Processes that remove large amounts of freshwater from seawater—and

thereby increase seawater salinity—include evaporation (in which liquid water becomes a gas and enters the atmosphere) and the formation of sea ice. High salinities, for example, are found where evaporation rates are high, as is the case in the dry subtropical regions (roughly between 25° and 35° north or south latitude). Conversely, where large amounts of precipitation dilute ocean waters, as in the midlatitudes (between 35° and 60° north or south latitude) and near the equator, lower salinities prevail. The graph in **FIGURE 14.2** shows this pattern.

Surface salinity in polar regions varies seasonally due to the formation and melting of sea ice. When seawater freezes in winter, sea salts do not become part of the ice. Therefore, the salinity of the remaining seawater increases. In summer when sea ice melts, the addition of the freshwater dilutes the solution, and salinity decreases (**FIGURE 14.3**).

The map in **FIGURE 14.4** was produced with data from the *Aquarius* satellite. It shows how ocean surface salinity varies worldwide. Notice how the overall pattern on the map matches the graph in Figure 14.2. Also notice the following:

- The Atlantic Ocean is saltiest, greater than 37‰ in some places. This is because compared to Earth's other oceans, the Atlantic has a greater inequality between the loss of freshwater via evaporation and the addition of freshwater from rainfall and river runoff.
- Notice how the huge discharge of freshwater from the Amazon River produces a zone of lower-salinity seawater in the western Atlantic off the northeast coast of South America.
- The impact of freshwater discharge from rivers is also illustrated when the salinity of the Bay of Bengal is compared to the salinity of the Arabian Sea. Freshwater from the Ganges River causes the surface salinity of the Bay of Bengal to be lower than the salinity of the Arabian Sea.

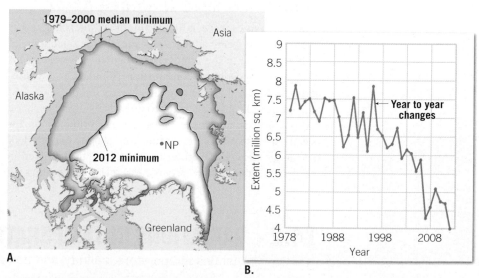

FIGURE 14.3 Tracking Sea Ice Changes A. Sea ice is frozen seawater. The formation and melting of sea ice influence the surface salinity of the ocean. In winter the Arctic Ocean is completely ice covered. In summer a portion of the ice melts. This map shows the extent of sea ice in early September 2012 compared to the average extent for the period 1979 to 2000. The extent in 2012 was the lowest on record. **B.** The graph clearly depicts the trend in the area covered by Arctic sea ice at the end of the summer melt period. The trend is very likely related to global climate change. (NASA)

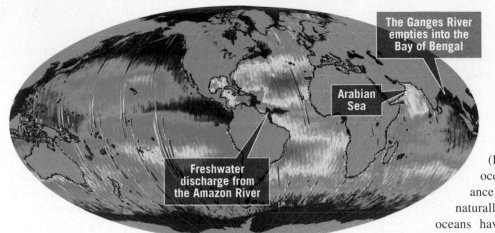

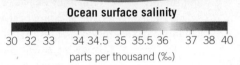

Ocean surface salinity

30 32 33   34 34.5 35 35.5 36   37 38 40

parts per thousand (‰)

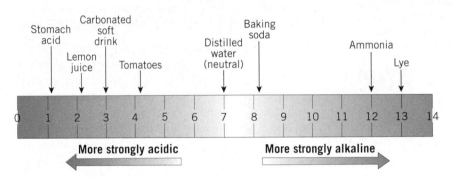

FIGURE 14.5 The pH Scale This is the common measure of the degree of acidity or alkalinity of a solution. The scale ranges from 0 to 14, with a value of 7 indicating a solution that is neutral. Values below 7 indicate greater acidity, whereas numbers above 7 indicate greater alkalinity. It is important to note that the pH scale is logarithmic; that is, each whole number increment indicates a tenfold difference. Thus, pH 4 is 10 times more acidic than pH 5 and 100 times (10 × 10) more acidic than pH 6.

seawater more acidic. The pH scale is shown and briefly described in **FIGURE 14.5**.

When carbon dioxide (CO_2) from the atmosphere dissolves in seawater (H_2O), it forms carbonic acid (H_2CO_3). This lowers the ocean's pH and changes the balance of certain chemicals found naturally in seawater. In fact, the oceans have already absorbed enough carbon dioxide for surface waters to have experienced a pH decrease of 0.1 since preindustrial times, with an additional pH decrease likely in the future. Moreover, if the current trend of human-induced carbon dioxide emissions continues, by the year 2100, the ocean will experience a pH decrease of at least 0.3, which represents a change in ocean chemistry that has not occurred for millions of years. This shift toward acidity and the changes in ocean chemistry that result make it more difficult for certain marine creatures to build hard parts out of calcium carbonate. The decline in pH thus threatens a variety of calcite-secreting organisms as diverse as microbes and corals, which concerns marine scientists because of the potential consequences for other sea life that depends on the health and availability of these organisms.

Recent Increase in Ocean Acidity

Human activities, especially burning fossil fuels and deforestation, have been adding ever-increasing quantities of carbon dioxide to the atmosphere for many decades. This change in atmospheric composition is linked in a cause-and-effect way to global climate change, a topic addressed at some length in Chapter 20. Rising levels of atmospheric carbon dioxide also have some serious implications for ocean chemistry and for marine life. About one-third of the human-generated carbon dioxide ends up dissolved in the oceans. This causes the ocean's pH to drop, making

14.1 CONCEPT CHECKS

1 What is salinity, and how is it usually expressed? What is the average salinity of the ocean?

2 What are the six most abundant elements dissolved in seawater? What is produced when the two most abundant elements combine?

3 What are the two primary sources for the elements that comprise the dissolved components in seawater?

4 List several factors that cause salinity to vary from place to place and from time to time.

5 Is the pH of the ocean increasing or decreasing? What is responsible for the changing pH?

14.2 | VARIATIONS IN TEMPERATURE AND DENSITY WITH DEPTH

Discuss temperature, salinity, and density changes with depth in the open ocean.

Temperature and density are basic ocean water properties that influence such things as deep-ocean circulation and the distribution and types of life-forms. A sampling of the open ocean from the surface to the seafloor shows

that these basic properties change with depth and that the changes are not the same everywhere. How and why temperature and density change with depth is the focus of this section.

Temperature Variations

If a thermometer were lowered from the surface of the ocean into deeper water, what temperature pattern would be found? Surface waters are warmed by the Sun, so they generally have higher temperatures than deeper waters. Surface water temperatures are also higher in the tropics and get colder toward the poles.

FIGURE 14.6 shows two graphs of temperature versus depth: one for high-latitude regions and one for low-latitude regions. The low-latitude curve begins at the surface with high temperature, but the temperature decreases rapidly with depth because of the inability of the Sun's rays to penetrate very far into the ocean. At a depth of about 1000 meters (3300 feet), the temperature remains just a few degrees above freezing and is relatively constant from this level down to the ocean floor. The layer of ocean water between about 300 meters (980 feet) and 1000 meters (3300 feet), where there is a rapid change of temperature with depth, is called the **thermocline** (*thermo* = heat, *cline* = slope). The thermocline is a very important zone in the ocean because it creates a vertical barrier to many types of marine life.

The high-latitude curve in Figure 14.6 displays a pattern quite different from the low-latitude curve. Surface-water temperatures in high latitudes are much cooler than in low latitudes, so the curve begins at the surface with low temperature. Deeper in the ocean, the temperature of the water is similar to that at the surface (just a few degrees above freezing), so the curve remains vertical, and there is no rapid change of temperature with depth. A thermocline is not present in high latitudes; instead, the water column is *isothermal* (*iso* = same, *thermal* = heat).

Some high-latitude waters can experience minor warming during the summer months. Thus, certain high-latitude regions experience an extremely weak seasonal thermocline. Midlatitude waters, on the other hand, experience a more dramatic seasonal thermocline and exhibit characteristics intermediate between high- and low-latitude regions.

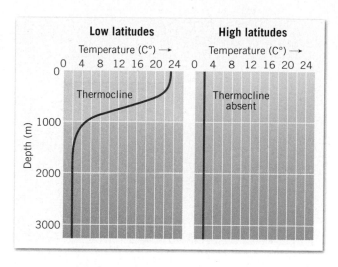

FIGURE 14.6 Variations in Ocean Water Temperature with Depth for Low- and High-Latitude Regions The layer of rapidly changing temperature called the *thermocline* is not present in the high latitudes.

Density Variations

Density is defined as mass per unit volume but can be thought of as a measure of *how heavy something is for its size.* For instance, an object that has low density is lightweight for its size—for example, a dry sponge, foam packing, or a surfboard. Conversely, an object that has high density is heavy for its size—for example, cement and many metals.

Density is an important property of ocean water because it determines the water's vertical position in the ocean. Furthermore, density differences cause large areas of ocean water to sink or float. For example, when high-density seawater is added to low-density freshwater, the denser seawater sinks below the freshwater.

Factors Affecting Seawater Density Seawater density is influenced by two main factors: *salinity* and *temperature.* An increase in salinity adds dissolved substances and results in an increase in seawater density (**FIGURE 14.7**). An increase in temperature, on the other hand, causes water to expand and results in a decrease in seawater density. Such a relationship, where one variable decreases as a result of another variable's increase, is known as an *inverse*

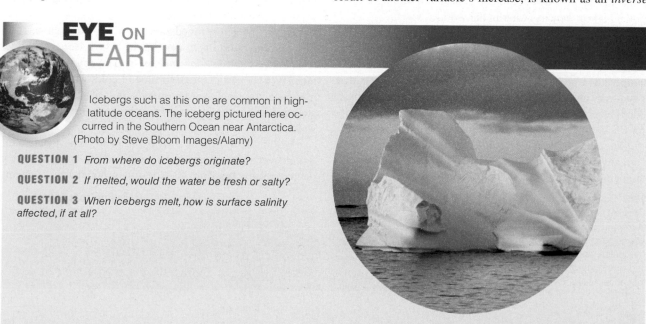

EYE ON EARTH

Icebergs such as this one are common in high-latitude oceans. The iceberg pictured here occurred in the Southern Ocean near Antarctica. (Photo by Steve Bloom Images/Alamy)

QUESTION 1 *From where do icebergs originate?*

QUESTION 2 *If melted, would the water be fresh or salty?*

QUESTION 3 *When icebergs melt, how is surface salinity affected, if at all?*

FIGURE 14.7 The Dead Sea This water body, which has a salinity of 330‰ (almost 10 times the average salinity of seawater), has high density. As a result, it also has high buoyancy that allows swimmers to float easily. (Photo by Peter Guttman/Corbis)

 SmartFigure 14.8 Variations in Ocean-Water Density with Depth for Low- and High-Latitude Regions The layer of rapidly changing density called the *pycnocline* is present in the low latitudes but absent in the high latitudes.

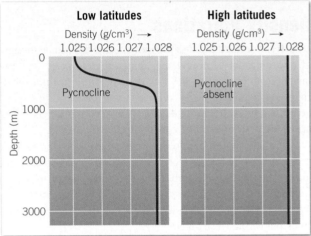

relationship, where one variable is *inversely proportional* with the other.

Temperature has the greatest influence on surface seawater density because variations in surface seawater temperature are greater than salinity variations. In fact, only in the extreme polar areas of the ocean, where temperatures are low and remain relatively constant, does salinity significantly affect density. Cold water that also has high salinity is some of the highest-density water in the world.

Density Variation with Depth By extensively sampling ocean waters, oceanographers have learned that temperature and salinity—and the water's resulting density—vary with depth. **FIGURE 14.8** shows two graphs of density versus depth: one for high-latitude regions and one for low-latitude regions.

The low-latitude curve in Figure 14.8 begins at the surface with low density (related to high surface-water temperatures). However, density increases rapidly with depth

because the water temperature is getting colder. At a depth of about 1000 meters (3300 feet), seawater density reaches a maximum value related to the water's low temperature. From this depth to the ocean floor, density remains constant and high. The layer of ocean water between about 300 meters (980 feet) and 1000 meters (3300 feet), where there is a rapid change of density with depth, is called the **pycnocline** (*pycno* = density, *cline* = slope). A pycnocline has a high gravitational stability and presents a significant barrier to mixing between low-density water above and high-density water below.

The high-latitude curve in Figure 14.8 is also related to the temperature curve for high latitudes shown in Figure 14.6. Figure 14.8 shows that in high latitudes, there is high-density (cold) water at the surface and high-density (cold) water below. Thus, the high-latitude density curve remains vertical, and there is no rapid change of density with depth. A pycnocline is not present in high latitudes; instead, the water column is *isopycnal* (*iso* = same, *pycno* = density).

Ocean Layering

The ocean, like Earth's interior, is layered according to density. Low-density water exists near the surface, and higher-density water occurs below. Except for some shallow inland seas with a high rate of evaporation, the highest-density water is found at the greatest ocean depths. Oceanographers generally recognize a three-layered structure in most parts of the open ocean: a shallow surface mixed zone, a transition zone, and a deep zone (**FIGURE 14.9**).

Because solar energy is received at the ocean surface, it is here that water temperatures are warmest. The mixing of these waters by waves as well as turbulence from currents and tides distributes heat gained at the surface through a shallow layer. Consequently, this *surface mixed zone* has nearly uniform temperatures. The thickness and temperature of this layer vary, depending on latitude and season. The zone usually extends to about 300 meters (980 feet) but may attain a thickness of 450 meters (1500 feet). The surface mixed zone accounts for only about 2 percent of ocean water.

Below the sun-warmed zone of mixing, the temperature falls abruptly with depth (see Figure 14.6). Here, a distinct layer called the *transition zone* exists between the warm surface layer above and the deep zone of cold water below. The transition zone includes a prominent thermocline and associated pycnocline and accounts for about 18 percent of ocean water.

Below the transition zone is the *deep zone*, where sunlight never reaches and water temperatures are just a few degrees above freezing. As a result, water density remains

FIGURE 14.9 The Ocean's Layers Oceanographers recognize three main layers in the ocean, based on water density, which varies with temperature and salinity. The warm *surface mixed layer* accounts for only 2 percent of ocean water; the *transition zone* includes the thermocline and the pycnocline and accounts for 18 percent of ocean water; and the *deep zone* contains cold, high-density water that accounts for 80 percent of ocean water.

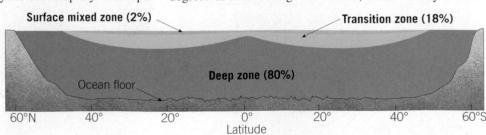

constant and high. Remarkably, the deep zone includes about 80 percent of ocean water, indicating the immense depth of the ocean. (The average depth of the ocean is more than 3700 meters [12,200 feet].)

In high latitudes, the three-layer structure does not exist because the water column is isothermal and isopycnal, which means there is no rapid change in temperature or density with depth. Consequently, good vertical mixing between surface and deep waters can occur in high-latitude regions. Here, cold, high-density water forms at the surface, sinks, and initiates deep-ocean currents, which are discussed in Chapter 15.

14.2 CONCEPT CHECKS

1 Contrast temperature variations with depth in the high and low latitudes. Why do high-latitude waters generally lack a thermocline?

2 What two factors influence seawater density? Which one has the greater influence on surface seawater density?

3 Contrast density variations with depth in the high and low latitudes. Why do high-latitude waters generally lack a pycnocline?

4 Describe the ocean's layered structure. Why does the three-layer structure not exist in high latitudes?

14.3 | THE DIVERSITY OF OCEAN LIFE Distinguish among plankton, nekton, and benthos. Summarize the factors used to divide the ocean into marine life zones.

A wide variety of organisms inhabit the marine environment. These organisms range in size from microscopic bacteria and algae to blue whales, which are as long as three buses lined up end to end. Marine biologists have identified more than 250,000 marine species, a number that is constantly increasing as new organisms are discovered and classified.

Most marine organisms live within the sunlit surface waters of the ocean. Strong sunlight supports **photosynthesis** (*photo* = light, *syn* = with, *thesis* = an arranging) by marine algae, which either directly or indirectly provide food for the vast majority of marine organisms. All marine algae live near the surface because they need sunlight; most marine animals also live near the surface because this is where food can be obtained. In shallow-water areas close to land, sunlight reaches all the way to the bottom, resulting in an abundance of marine life on the ocean floor.

There are advantages and disadvantages to living in the marine environment. One advantage is that there is an abundance of water available, which is necessary for supporting all types of life. One disadvantage is that maneuvering in water, which has high density and impedes movement, can be difficult. The success of marine organisms depends on their ability to avoid predators, find food, and cope with the physical challenges of their environment.

Classification of Marine Organisms

Marine organisms can be classified according to where they live (their habitat) and how they move (their mobility). Organisms that inhabit the water column can be classified as either *plankton* (floaters) or *nekton* (swimmers). All other organisms are *benthos* (bottom dwellers).

Plankton (Floaters) All organisms—algae, animals, and bacteria—that drift with ocean currents are **plankton** (*planktos* = wandering). Just because plankton drift does not mean they are unable to swim. Many plankton can swim but either move very weakly or move only vertically within the water column.

Among plankton, the algae (photosynthetic cells, most of which are microscopic) are called **phytoplankton**, and the animals are called **zooplankton** (*zoo* = animal, *planktos* = wandering). Representative members of each group are shown in **FIGURE 14.10**.

Phytoplankton:
(1) Coccolithophores; (2–4) Diatoms; (5–7) Dinoflagellates.

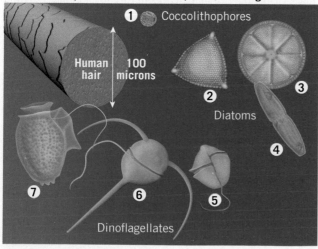

Zooplankton:
(1) Squid larva; (2) Copepod; (3) Snail larva; (4) Fish larva;
(5) Arrowworm; (6) Foraminifers; (7) Radiolarian.

FIGURE 14.10
Phytoplankton and Zooplankton (Floaters)
Schematic drawings of various phytoplankton and zooplankton. (After TRUJILLO, ALAN P.; THURMAN, HAROLD V., ESSENTIALS OF OCEANOGRAPHY, 11th Ed., ©2014, p.364. Reprinted and Electronically reproduced by permission of Pearson Education, Inc., Upper Saddle River, New Jersey.)

FIGURE 14.11 Nekton All animals capable of moving independently of ocean currents are nekton. **A.** Gray reef shark, Bikini Atoll. (Photo by Yann Hubert/Science Source); **B.** California market squid. (Photo by Tom McHugh/Science Source); **C.** School of grunts, Florida Keys. (Photo by Georgie Holland/AGE fotostock); **D.** Pygmy Killer Whales in shallow surface waters along the Kona Coast, Big Island, Hawaii. (Photo by Image Marine Quest)

A.

B.

C.

D.

Plankton are extremely abundant and very important within the marine environment. In fact, most of Earth's **biomass**—the mass of all living organisms—consists of plankton adrift in the oceans. Even though 98 percent of marine species are bottom dwelling, the vast majority of the ocean's biomass is planktonic.

Nekton (Swimmers)

All animals capable of moving independently of ocean currents, by swimming or other means of propulsion, are called **nekton** (*nektos* = swimming). They are capable not only of determining their position within the ocean but also, in many cases, of long migrations. Nekton include most adult fish and squid, marine mammals, and marine reptiles (**FIGURE 14.11**).

Although nekton move freely, they are unable to move throughout the breadth of the ocean. Gradual changes in temperature, salinity, density, and availability of nutrients effectively limit their lateral range. The deaths of large numbers of fish, for example, can be caused by temporary shifts of water masses in the ocean. High water pressure at depth normally limits the vertical range of nekton.

Fish may appear to exist everywhere in the oceans, but they are most abundant near continents and islands and in colder waters. Some fish, such as salmon, ascend freshwater rivers to spawn. Many eels do just the reverse, growing to maturity in freshwater and then descending streams to breed in the great depths of the ocean.

Benthos (Bottom Dwellers)

The term **benthos** (*benthos* = bottom) describes organisms living on or in the ocean bottom. *Epifauna* (*epi* = upon, *fauna* = animal) live on the surface of the seafloor, either attached to rocks or moving along the bottom. *Infauna* (*in* = inside, *fauna* = animal) live buried in the sand or mud. Some benthos, called *nektobenthos*, live on the bottom but also swim or crawl through the water above the ocean floor. Examples of benthos are shown in **FIGURE 14.12**.

The shallow coastal ocean floor contains a wide variety of physical conditions and nutrient levels, both of which have allowed a great number of species to evolve. Moving across the bottom from the shore into deeper water, the number of benthos species may remain relatively constant, but the biomass of benthos organisms decreases. In addition, shallow coastal areas are the only locations where large marine algae (often called "seaweeds") are found attached to the bottom. This is the case because these are the only areas of the seafloor that receive sufficient sunlight.

Throughout most of the deeper parts of the seafloor, animals live in perpetual darkness, where photosynthesis cannot occur. They must feed on each other, or on whatever nutrients fall from the productive surface waters above. The deep-sea bottom is an environment of coldness, stillness, and darkness. Under these conditions, life progresses slowly, and organisms that live in the deep sea usually are widely distributed because physical conditions vary little on the deep-ocean floor, even over great distances.

There is one exception to the situation just described. Instead of sparse life, some spots on the deep-ocean floor

A.

B.

C.

D.

SmartFigure 14.12
Benthos Organisms living on or in the ocean bottom are classified as benthos. **A.** Sea star. (Photo by David Hall/Science Source); **B.** Yellow tube sponge. (Photo by Andrew J. Martinez/Science Source); **C.** Green sea urchin. (Photo by Andrew J. Martinez/Science Source); **D.** Coral crab. (Photo by Images & Stories/Alamy)

exhibit abundant life-forms. These places, called *hydrothermal vents*, are found along portions of the oceanic ridge system. There is more about these unique features in the GEOgraphics "Deep-Sea Hydrothermal Vents," on page 442.

Marine Life Zones

The distribution of marine organisms is affected by the chemistry, physics, and geology of the oceans. Marine organisms are influenced by a variety of physical oceanographic factors. Some of these factors—such as availability of sunlight, distance from shore, and water depth—are used to divide the ocean into distinct marine life zones (**TABLE 14.2** and **FIGURE 14.13**).

Availability of Sunlight The upper part of the ocean into which sunlight penetrates is called the **photic** (*photos* = light) **zone**. The clarity of seawater is affected by many factors, such as the amount of plankton, suspended sediment, and decaying organic particles in the water. In addition, the amount of sunlight varies with atmospheric conditions, time of day, season of the year, and latitude.

The **euphotic** (*eu* = good, *photos* = light) **zone** is the portion of the photic zone near the surface where light is strong enough for photosynthesis to occur. In the open ocean, this zone can reach a depth of 100 meters (330 feet), but the zone is much shallower close to shore, where water clarity is typically reduced. In the euphotic zone, phytoplankton use sunlight to produce food molecules and become the basis of most oceanic food webs.

Different wavelengths of sunlight are absorbed as they pass through seawater. The longer-wavelength red and orange colors are absorbed first, the greens and yellows next, and the shorter-wavelength blues and violets penetrate the farthest. In fact, faint traces of blue and violet light can still be measured in extremely clear water at depths of 1,000 meters (3,300 feet).

Although photosynthesis cannot occur much below 100 meters (330 feet), there is enough light in the lower photic zone for marine animals to avoid predators, find food, recognize their species, and locate mates. Even deeper is the **aphotic** (*a* = without, *photos* = light) **zone**, where there is no sunlight.

TABLE 14.2 Marine Life Zones

Basis	Marine Life Zone	Subdivision	Characteristics
Available sunlight	Photic		Sunlit surface waters
		Euphotic	Has enough sunlight to support photosynthesis
	Aphotic		No sunlight; many organisms have bioluminescent capabilities
Distance from shore	Intertidal		Narrow strip of land between high and low tides; dynamic area
	Neritic		Above continental shelf; high biomass and diversity of species
	Oceanic		Open ocean beyond the continental shelf; low nutrient concentrations
Depth	Pelagic		All water above the ocean floor; organisms swim or float
	Benthic		Bottom of ocean; organisms attach to, burrow into, or crawl on seafloor
		Abyssal	Deep-sea bottom; dark, cold, high pressure; sparse life

Deep-Sea Hydrothermal Vents

Deep-sea hydrothermal vents are openings in the oceanic crust from which geothermally heated water rises. They are found mainly along the oceanic ridge system where tectonic plates rift apart, resulting in the production of new seafloor by upwelling magma.

When these hot, mineral-rich fluids reach the seafloor their temperatures can exceed 350 °C, but because of the extremely high pressures exerted by the water column above, they do not boil. When this hydrothermal fluid comes into contact with the much colder chemical-rich seawater, mineral matter rapidly precipitates to form shimmering smoke-like clouds called "*black smokers*." The particles that compose the black smokers eventually settle out of the seawater. These deposits may contain economically significant amounts of iron, copper, zinc, lead, and occasionally silver and gold.

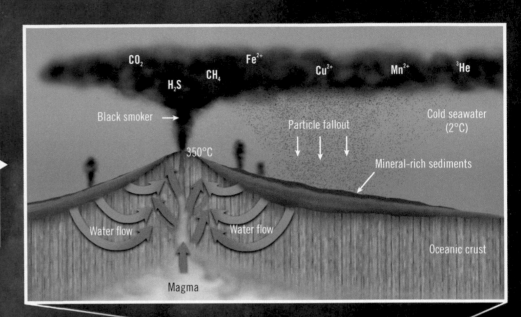

CO_2 Fe^{2+} Cu^{2+} Mn^{2+} ^3He

H_2S CH_4

Black smoker

Particle fallout

Cold seawater (2°C)

350°C

Mineral-rich sediments

Water flow   Water flow

Oceanic crust

Magma

At oceanic ridges, cold seawater circulates hundreds of meters down into the highly fractured basaltic crust, where it is heated by magmatic sources. Along the way, the hot water strips metals and other elements such as sulfur from surrounding rock. This heated fluid eventually becomes hot and buoyant enough to rise along conduits and fractures towards the surface.

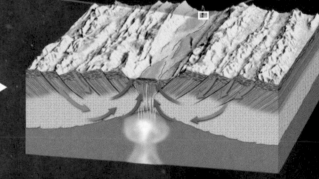

Some minerals immediately solidify and contribute to the formation of spectacular chimney-like structures, which can be as tall as 15-story buildings, and are appropriately given names like *Godzilla* and *Inferno*.

NOAA

Tubeworms

Fisheries and Oceans Canada
Uvic-Verena Tunnicliffe/Newscom

B. Murton/Southhampton Oceanography Center/
Science Source

Hydrothermal vents are also remarkable for the marine biology they support. In these environments, completely devoid of sunlight, microorganisms utilize mineral-rich hydrothermal fluid to perform chemosynthesis—the conversion of carbon atoms into organic compounds without sunlight for energy. The microbial communities, in turn, support larger, more complex animals such as fish, crabs, worms, mussels, clams, and perhaps the most unique, the tubeworm. With their white chitinous tubes and bright red plumes, these conspicuous creatures rely entirely on bacteria growing in their trophosome, an internal organ designed for harvesting bacteria. These symbiotic bacteria rely on the tubeworm to provide them with a suitable habitat and, in return, they use chemosynthesis to provide carbon-based building blocks (nutrients) to the tubeworms.

Where are hydrothermal vents found?

Most hydrothermal vents are found around the oceanic ridge system including some small spreading centers such as the Juan de Fuca Ridge and Galapagos Rift, as well as in the back arc basins that lie behind subduction zones.

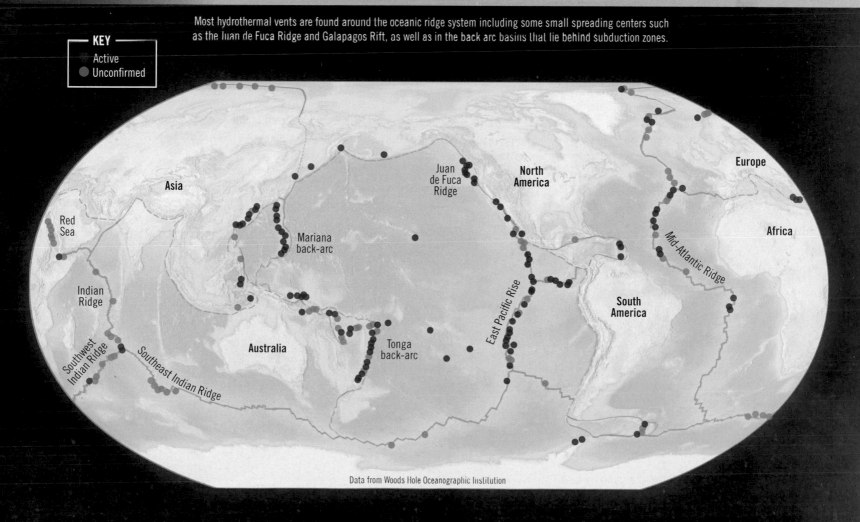

KEY
Active
Unconfirmed

Red Sea

Asia

Juan de Fuca Ridge

North America

Europe

Africa

Mariana back-arc

Mid-Atlantic Ridge

Indian Ridge

East Pacific Rise

South America

Southwest Indian Ridge

Southeast Indian Ridge

Australia

Tonga back-arc

Data from Woods Hole Oceanographic Institution

FIGURE 14.13 Marine Life Zones Criteria for determining life zones include availability of light, distance from shore, and water depth.

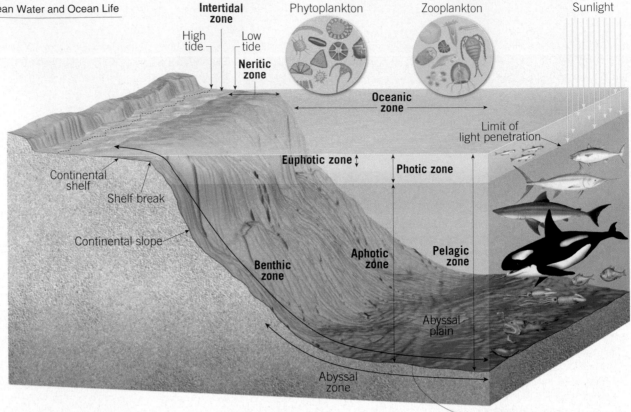

Distance from Shore

Marine life zones are also subdivided based on distance from shore. The area where the land and ocean meet and overlap is called the **intertidal zone**. This narrow strip of land between high and low tides is alternately covered and uncovered by seawater with each tidal change. Even though it appears to be a harsh place to live with crashing waves, periodic drying out, and rapid changes in temperature, salinity, and oxygen concentrations, many species live here that are superbly adapted to the dramatic environmental changes.

Seaward from the low-tide line is the **neritic** (*neritos* = of the coast) **zone**. This covers the gently sloping continental shelf out to the *shelf break* (see Figure 13.10, page 417). This zone can be very narrow or may extend hundreds of kilometers from shore. The neritic zone is often shallow enough for sunlight to reach all the way to the ocean floor, putting it entirely within the photic zone.

Although the neritic zone covers only about 5 percent of the world's oceans, it is rich in both biomass and number of species. Many organisms find the conditions here ideal because photosynthesis occurs readily, nutrients wash in from the land, and the bottom provides shelter and habitat. This zone is so rich, in fact, that it supports 90 percent of the world's commercial fisheries.

Beyond the continental shelf is the **oceanic zone**. The open ocean reaches great depths, and as a result, surface waters typically have lower nutrient concentrations because nutrients tend to sink out of the photic zone to the deep-ocean floor. This low nutrient concentration usually results in much smaller populations than the more productive neritic zone.

Water Depth

A third method of classifying marine habitats is based on water depth. Open ocean of *any* depth is called the **pelagic** (*pelagios* = of the sea) **zone**. Animals in this zone

EYE ON EARTH

These dolphins were photographed in the relatively shallow waters near the coast of southern California. (Photo by Danny Frank/AGE Fotostock)

QUESTION 1 *Which of these terms best fits the organisms shown here—plankton, nekton, or benthos? Explain.*

QUESTION 2 *Is the marine life zone shown in this photo intertidal, neritic, or oceanic?*

QUESTION 3 *Which term, benthic or pelagic, best fits the situation shown in the photo?*

swim or float freely. The photic part of the pelagic zone is home to phytoplankton, zooplankton, and nekton, such as tuna, sea turtles, and dolphins. The aphotic part has strange species like viperfish and giant squid that are adapted to life in deep water.

Benthos organisms such as giant kelp, sponges, crabs, sea anemones, sea stars, and marine worms that attach to, crawl upon, or burrow into the seafloor occupy parts of the **benthic** (*benthos* = bottom) **zone**. The benthic zone includes any sea-bottom surface, regardless of its distance from shore, and is mostly inhabited by benthos organisms.

The **abyssal** (*a* − without, *byssus* = bottom) **zone** is a subdivision of the benthic zone and includes the deep-ocean floor, such as *abyssal plains*. This zone is characterized by extremely high water pressure, consistently low temperature, no sunlight, and sparse life. Three food sources exist at abyssal depths: (1) tiny decaying particles steadily "raining" down from above, which provide food for filter-feeders, brittle stars, and burrowing worms; (2) large fragments or entire dead bodies falling at scattered sites, which supply meals for actively searching fish, such as the grenadier, tripodfish, and hagfish, which locate food by chemical sensing; and (3) the hydrothermal vents described in the GEOgraphics.

14.3 CONCEPT CHECKS

1 Describe the lifestyles of plankton, nekton, and benthos, and give examples of each. Which group comprises the largest biomass?

2 List three physical factors that are used to divide the ocean into marine life zones. How does each factor influence the abundance and distribution of marine life?

3 Why are there greater numbers and types of organisms in the neritic zone than in the oceanic zone?

14.4 | OCEAN PRODUCTIVITY

Contrast ocean productivity in polar, midlatitude, and tropical settings.

Why are some regions of the ocean teeming with life, while other areas seem barren? The answer is related to the amount of primary productivity in various parts of the ocean. **Primary productivity** is the amount of carbon fixed by organisms through the synthesis of organic matter using energy derived from solar radiation (*photosynthesis*) or chemical reactions (*chemosynthesis*). Although chemosynthesis supports hydrothermal vent biocommunities along the oceanic ridge, it is much less significant than photosynthesis in worldwide oceanic productivity.

Two factors influence a region's photosynthetic productivity: *availability of nutrients* (such as nitrates, phosphorus, iron, and silica) and the *amount of solar radiation* (sunlight). Thus, the most abundant marine life exists where there are ample nutrients and good sunlight. Oceanic productivity, however, varies dramatically because of the uneven distribution of nutrients throughout the photosynthetic zone and seasonal changes in the availability of solar energy.

A permanent *thermocline* (and resulting *pycnocline*) develops nearly everywhere in the oceans. This layer forms a barrier to vertical mixing and prevents the resupply of nutrients to sunlit surface waters. In the midlatitudes, a thermocline develops only during the summer season, and in polar regions a thermocline does not usually develop at all. The degree to which waters develop a thermocline profoundly affects the amount of productivity observed at different latitudes.

Productivity in Polar Oceans

Polar regions such as the Arctic Ocean's Barents Sea, which is off the northern coast of Europe, experience continuous darkness for about 3 months of winter and continuous illumination for about 3 months during summer. Productivity of phytoplankton—mostly single-celled algae called *diatoms*—peaks there during May (**FIGURE 14.14**), when the Sun rises high enough in the sky that there is deep penetration

of sunlight into the water. As soon as the diatoms develop, zooplankton—mostly small crustaceans called *copepods* and larger *krill*—begin feeding on them. The zooplankton biomass peaks in June and continues at a relatively high level until winter darkness begins in October.

Recall that temperature and density change very little with depth in polar regions (see Figures 14.6 and 14.8), so these waters are *isothermal*, and there is no barrier to mixing between surface waters and deeper, nutrient-rich waters. In the summer, however, melting ice creates a thin, low-salinity layer that does not readily mix with the deeper waters. This stratification is crucial to summer production because it helps prevent phytoplankton from being carried into deeper, darker waters. Instead, they are concentrated in the sunlit surface waters, where they reproduce continuously.

Because of the constant supply of nutrients rising from deeper waters below, high-latitude surface waters typically have high nutrient concentrations. The availability of solar energy, however, limits photosynthetic productivity in these areas.

Productivity in Tropical Oceans

You may be surprised to learn that productivity is low in tropical regions of the open ocean. Because the Sun is more directly overhead, light penetrates much deeper into tropical

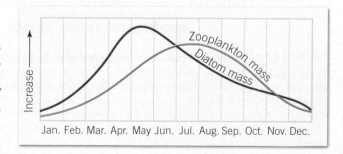

FIGURE 14.14 One Example of Productivity in Polar Oceans Illustrated by the Barents Sea A springtime increase in diatom mass is followed closely by an increase in zooplankton abundance.

FIGURE 14.15
Productivity in Tropical Oceans Although tropical regions receive adequate sunlight year-round, a permanent thermocline prevents the mixing of surface and deep water. As phytoplankton consume nutrients in the surface layer, productivity is limited because the thermocline prevents replenishment of nutrients from deeper water. Thus, productivity remains at a steady, low level.

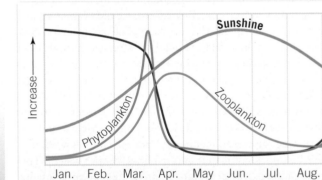

oceans than in temperate and polar waters, and solar energy is available year-round. However, productivity is low in tropical regions of the open ocean because a permanent thermocline produces a stratification of water masses that prevents mixing between surface waters and nutrient-rich deeper waters (**FIGURE 14.15**). In essence, the thermocline is a barrier that eliminates the supply of nutrients from deeper waters below. So, productivity in tropical regions is limited by the lack of nutrients (unlike in polar regions, where productivity is limited by the lack of sunlight). In fact, these areas have so few organisms that they are considered biological deserts.

Productivity in Midlatitude Oceans

Productivity is limited by available sunlight in polar regions and by nutrient supply in the tropics. In midlatitude regions, a combination of these two limiting factors controls productivity, as shown in **FIGURE 14.16** (which shows the pattern for the Northern Hemisphere; in the Southern Hemisphere, the seasons are reversed).

Winter Productivity is very low during winter, even though nutrient concentrations are highest at this time. The reason is that solar energy is limited because the length of daylight is short and the Sun angle is low. As a result, the depth at which photosynthesis can occur is so shallow that phytoplankton do not grow much.

Spring The Sun rises higher in the sky during spring, creating a greater depth at which photosynthesis can occur. A *spring bloom* of phytoplankton occurs because solar energy and nutrients are available, and a seasonal thermocline develops (due to increased solar heating) that traps algae in the euphotic zone (**FIGURE 14.17**). This creates a tremendous demand for nutrients in the euphotic zone, so the supply is quickly depleted, causing productivity to decrease sharply. Even though the days are lengthening and sunlight is increasing, productivity during the spring bloom is limited by the lack of nutrients.

Summer The Sun rises even higher in the summer, so surface waters continue to warm. A strong thermocline develops that prevents vertical mixing, so nutrients depleted from surface waters cannot be replaced by those from deeper waters. Throughout summer, the phytoplankton population remains relatively low (see Figure 14.16).

Fall Solar radiation diminishes in the fall as the Sun moves lower in the sky, so surface temperatures drop, and the summer thermocline breaks down. Nutrients return to the surface layer as increased wind strength mixes surface waters with deeper waters. These conditions create a *fall bloom* of phytoplankton, which is much less dramatic than the spring bloom (see Figure 14.16). The fall bloom is very short-lived because sunlight (not nutrient

**SmartFigure 14.16
Productivity in Temperate Oceans (Northern Hemisphere)** The graph shows the relationship among phytoplankton, zooplankton, amount of sunshine, and nutrient levels for surface waters.

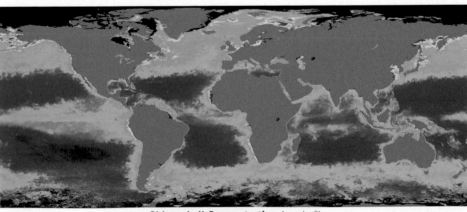

FIGURE 14.17 Chlorophyll Concentrations Measured by Satellite Instruments Bright greens, yellows, and reds indicate that the northern oceans were alive with plant life in the spring of 2006. High chlorophyll concentrations indicate high amounts of photosynthesis. Observations of global chlorophyll patterns tell scientists where ocean surface plants (phytoplankton) are growing, which is an indicator of where marine ecosystems are thriving. Such global maps also give scientists an idea of how much carbon the plants are soaking up, which is important in understanding the global carbon budget. (NASA)

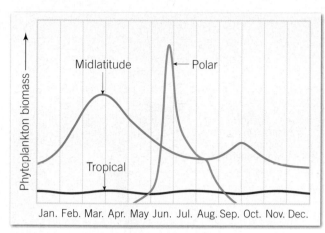

FIGURE 14.18 Comparing Productivity This comparison of tropical, midlatitude, and polar oceans in the Northern Hemisphere shows seasonal variations in phytoplankton biomass. The total area under each curve represents annual photosynthetic productivity.

supply, as in the spring bloom) becomes the limiting factor as winter approaches to repeat the seasonal cycle.

FIGURE 14.18 compares the seasonal variation in phytoplankton biomass of tropical, polar, and midlatitude regions. The total area under each curve represents photosynthetic productivity. The graph shows the dramatic peak in productivity in polar oceans during the summer; the steady, low rate of productivity year-round in the tropical oceans; and the seasonal productivity that occurs in midlatitude oceans. It also shows that the highest overall productivity occurs in the midlatitudes.

14.4 CONCEPT CHECKS

1 List two methods by which primary productivity is accomplished in the ocean. Which one is most significant? What two factors influence it?

2 Compare the biological productivity of polar, temperate, and tropical regions of the ocean.

14.5 | OCEANIC FEEDING RELATIONSHIPS Define *trophic level* and discuss the efficiency of energy transfer between different trophic levels.

Marine algae, plants, bacteria, and bacteria-like archaea are the main oceanic producers. As these producers make food (organic matter) available to the consuming animals of the ocean, it passes from one feeding population to the next. Only a small percentage of the energy taken in at any level is passed on to the next because energy is consumed and lost at each level. As a result, the producers' biomass in the ocean is many times greater than the mass of the top consumers, such as sharks or whales.

Trophic Levels

Chemical energy stored in the mass of the ocean's algae (the "grass of the sea") is transferred to the animal community mostly through feeding. Zooplankton are *herbivores* (*herba* = grass, *vora* = eat), so they eat diatoms and other microscopic marine algae. Larger herbivores feed on the larger algae and marine plants that grow attached to the ocean bottom near shore.

The herbivores (grazers) are then eaten by larger animals, the *carnivores* (*carni* = meat, *vora* = eat). They in turn are eaten by another population of larger carnivores, and so on. Each of these feeding stages is called a **trophic** (*tropho* = nourishment) **level**.

In the ocean, individual members of a feeding population are generally larger—but not too much larger—than the organisms they eat. There are conspicuous exceptions, however, such as the blue whale. Up to 30 meters (100 feet) long, it is possibly the largest animal that has ever existed on Earth, yet it feeds mostly on krill, which have a maximum length of only 6 centimeters (2.4 inches).

Transfer Efficiency

The transfer of energy between trophic levels is very inefficient. The efficiencies of different algal species vary, but the average is only about *2 percent*, which means that 2 percent of the light energy absorbed by algae is ultimately synthesized into food and made available to herbivores.

FIGURE 14.19 shows the passage of energy between trophic levels through an entire ecosystem, from the solar energy assimilated by phytoplankton through all trophic levels to the ultimate consumer—humans. Because energy is lost at each trophic level, it takes thousands of smaller marine organisms to produce a single fish that is so easily consumed during a meal!

Food Chains and Food Webs

A **food chain** is a sequence of organisms through which energy is transferred, starting with an organism that is the primary producer, then an herbivore, then one or more carnivores, and finally culminating with the "top carnivore," which is not usually preyed upon by any other organism.

Because energy transfer between trophic levels is inefficient, it is advantageous for fishers to choose a population that feeds as close to the primary producing population as possible. This increases the biomass available for food and the number of individuals available to be taken by the fishery. Newfoundland herring, for example, are an important fishery that usually represents the third trophic level in a food chain. They feed primarily on small copepods that feed, in turn, on diatoms (**FIGURE 14.20A**).

Feeding relationships are rarely as simple as that of the Newfoundland herring. More often, top carnivores in a food chain feed on a number of different animals, each of which has its own simple or complex feeding relationships. This constitutes a **food web**, as shown in **FIGURE 14.20B** for North Sea herring.

Animals that feed through a food web rather than a food chain are more likely to survive because they have

**SmartFigure 14.19
Ecosystem Energy Flow
and Efficiency** For every
500,000 units of radiant
energy input available to the
producers (phytoplankton),
only 1 unit of mass is added to
the fifth trophic level (humans).
Average phytoplankton
transfer efficiency is 2 percent
(98 percent loss), and all
other trophic levels average
10 percent efficiency (90
percent loss). The ultimate
effect of energy transfer
between trophic levels is that
the number of individuals and
the total biomass decrease
at successive trophic levels
because
the amount
of available
energy
decreases.

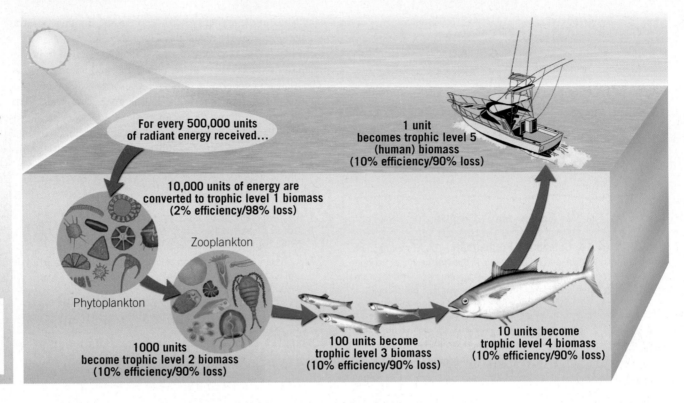

**FIGURE 14.20 Comparing
a Food Chain and a Food
Web A.** A food chain is the
passage of energy along
a single path, such as
from diatoms to copepods
to Newfoundland herring.
Feeding relationships are
rarely this simple. **B.** A food
web showing multiple paths
for food sources of the North
Sea herring

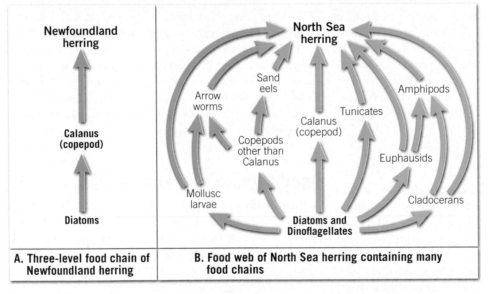

alternative foods to eat should one of their food sources
diminish in quantity or even disappear. Newfoundland herring, on the other hand, eat only copepods, so the disappearance of copepods would catastrophically affect their
population.

14.5 CONCEPT CHECKS

1 Discuss energy transfer between trophic levels.

2 Describe the advantage that a top carnivore gains by eating
from a food web rather than a food chain.

14 CONCEPTS IN REVIEW | Ocean Water and Ocean Life

14.1 COMPOSITION OF SEAWATER

Define *salinity* and list the main elements that contribute to the ocean's salinity. Describe the sources of dissolved substances in seawater and causes of variations in salinity.

KEY TERM: salinity

Michael Collier

- Salinity is the proportion of dissolved salts to pure water, usually expressed in parts per thousand (‰). The average salinity in the open ocean is about 35‰. The principal elements that contribute to the ocean's salinity are chlorine (55%) and sodium (31%). The primary sources for the elements in sea salt are chemical weathering of rocks on the continents and volcanic outgassing on the ocean floor.
- Variations in salinity are primarily caused by changing the water content of the seawater solution. Natural processes that add large amounts of freshwater to seawater and decrease salinity include precipitation, runoff from land, iceberg melting, and sea ice melting. Processes that remove large amounts of freshwater from seawater and increase salinity include the formation of sea ice and evaporation.

Q How are emissions from this coal-burning power plant influencing the acidity of seawater? Explain.

14.2 VARIATIONS IN TEMPERATURE AND DENSITY WITH DEPTH

Discuss temperature, salinity, and density changes with depth in the open ocean.

KEY TERMS: thermocline, density, pycnocline

- The ocean's surface temperature is related to the amount of solar energy received and varies as a function of latitude. Low-latitude regions have relatively warm surface water and distinctly colder water at depth, creating a thermocline, which is a layer of rapid temperature change. No thermocline exists in high-latitude regions because there is little temperature difference between the top and bottom of the water column. The water column is isothermal.
- Seawater density is mostly affected by water temperature but also by salinity. Cold, high-salinity water is densest. Low-latitude regions have distinctly denser (colder) water at depth than at the surface, creating a

pycnocline, which is a layer of rapidly changing density. No pycnocline exists in high-latitude regions because the water column is isopycnal.

- Most open-ocean regions exhibit a three-layered structure based on water density. The shallow surface mixed zone has warm and nearly uniform temperatures. The transition zone includes a prominent thermocline and associated pycnocline. The deep zone is continually dark and cold and accounts for 80 percent of the water in the ocean. In high latitudes, the three-layered structure does not exist.

Q Does this graph represent changes in ocean water temperature or ocean water density with depth? Does it more likely represent a location in the tropics or near the poles? Explain.

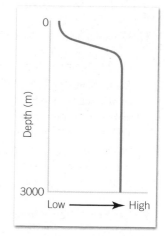

14.3 THE DIVERSITY OF OCEAN LIFE

Distinguish among plankton, nekton, and benthos. Summarize the factors used to divide the ocean into marine life zones.

KEY TERMS: photosynthesis, plankton, phytoplankton, zooplankton, biomass, nekton, benthos, photic zone, euphotic zone, aphotic zone, intertidal zone, neritic zone, oceanic zone, pelagic zone, benthic zone, abyssal zone

Arterra Picture Library/Alamy

- Marine organisms can be classified into one of three groups, based on habitat and mobility. Plankton are free-floating forms with little power of locomotion, nekton are swimmers, and benthos are bottom dwellers. Most of the ocean's biomass is planktonic.
- Three criteria are frequently used to establish marine life zones. Based on availability of sunlight, the ocean can be divided into the photic zone (which includes the euphotic zone) and the aphotic zone. Based on distance from shore, the ocean can be divided into the intertidal zone, the neritic zone, and the oceanic zone. Based on water depth, the ocean can be divided into the pelagic zone and the benthic zone (which includes the abyssal zone).

Q If you were a commercial fisher, in which of these marine life zones would you concentrate your efforts—Intertidal, neritic, or oceanic? Explain.

14.4 OCEANIC PRODUCTIVITY

Contrast ocean productivity in polar, midlatitude, and tropical settings.

KEY TERM: primary productivity

- Primary productivity is the amount of carbon fixed by organisms through the synthesis of organic matter using energy derived from solar radiation (photosynthesis) or chemical reactions (chemosynthesis). Chemosynthesis is much less significant than photosynthesis in worldwide oceanic productivity. Photosynthetic productivity in the ocean varies due to the availability of nutrients and amount of solar radiation.
- Oceanic photosynthetic productivity varies at different latitudes because of seasonal changes and the development of a thermocline. In polar oceans, the availability of solar radiation limits productivity even though nutrient levels are high. In tropical oceans, a strong thermocline exists year-round, so the lack of nutrients generally limits productivity. In midlatitude oceans, productivity peaks in the spring and fall and is limited by the lack of solar radiation in winter and by the lack of nutrients in summer.

14.5 OCEANIC FEEDING RELATIONSHIPS

Define *trophic level* and discuss the efficiency of energy transfer between different trophic levels.

KEY TERMS: trophic level, food chain, food web

- The Sun's energy is utilized by phytoplankton and converted to chemical energy, which is passed through different trophic levels. On average, only about 10 percent of the mass taken in at one trophic level is passed on to the next. As a result, the size of individuals increases but the number of individuals decreases with each trophic level of a food chain or a food web. Overall, the total biomass of populations decreases at successive trophic levels.

Q What trophic level is represented by the person in this photo?

Jupiterimages/Getty Images

GIVE IT SOME **THOUGHT**

1. The accompanying photo shows sea ice in the Beaufort Sea near Barrow, Alaska. How do seasonal changes in the amount of sea ice influence the salinity of the remaining surface water? Is water density greater before or after sea ice forms? Explain.

Michael Collier

2. Say that someone brings several water samples to your laboratory. His problem is that the labels are incomplete. He knows samples A and B are from the Atlantic Ocean and that one came from near the equator and the other from near the Tropic of Cancer. But he does not know which one is which. He has a similar problem with samples C and D. One is from the Red Sea and the other is from the Baltic Sea. Applying your knowledge of ocean salinity, how would you identify the location of each sample? How were you able to figure this out?

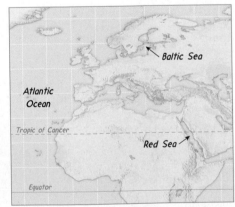

3. You are swimming in the open ocean near the equator. The thermocline in this location is about 1° C per 50 meters of depth. If the sea surface temperature is 24° C, how deep must you dive before you encounter a water temperature of 19°C?

4. The accompanying graph depicts variations in ocean water density and temperature with depth for a location near the equator. Which line represents temperature and which represents density? Explain.

5. After sampling a column of water from the surface to a depth of 3000 meters (nearly 10,000 feet), a colleague aboard an oceanographic research vessel tells you that the water column is *isopycnal*. What does this mean? What conditions create such a situation? What would have to happen in order to create a pycnocline?

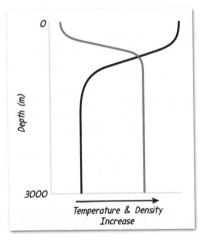

6. Tropical environments on land are well known for their abundant life; rain forests are an example. By contrast, biological productivity in tropical oceans is meager. Why is this the case?

7. The accompanying graph relates to the abundance of ocean life (productivity) in a polar region of the Northern Hemisphere. Which line represents phytoplankton, and

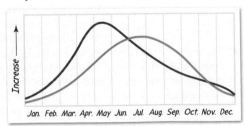

which represents zooplankton? How did you figure this out? Why are the curves so low from November through February?

8. Refer to Figure 14.19. What is the average efficiency of energy transfer between trophic levels? Use this efficiency to determine how much phytoplankton mass is required to add 1 gram of mass to a killer whale, which is a third-level carnivore.

9. How might the removal of the top carnivore affect a food web? How would the removal of the primary producer affect the food web? Which change would be more significant?

EXAMINING THE **EARTH SYSTEM**

1. Reef-building corals are tiny invertebrate colonial animals that live in warm, sunlit marine environments. They extract calcium carbonate from seawater and secrete an external skeleton. Although individuals are small, colonies are

Dirscherl Reinhard/AGE Fotostock

capable of creating massive reefs. Many other organisms also make the reef structure their home. Corals are a part of the biosphere that inhabit the hydrosphere. The solid calcium carbonate reefs they build may ultimately become the sedimentary rock limestone, a part of the geosphere. Can you relate coral reefs to the atmosphere? Can you come up with more than one connection? Explain.

2. A storm near the coast produces sediment-rich runoff that causes water in the euphotic zone to become cloudy. Describe how this would affect the plankton, nekton, and benthos.

Connecticut River

Long Island Sound

NASA

3. Hydrothermal vents, such as the one shown here, occur on the ocean floor along mid-ocean ridges. The GEOgraphics in this chapter on (page 442) examined these features, where very hot, mineral-rich water is emitted into the cold water of the deep-ocean environment.

   a. What category of sediment is created by the process described above— hydrogenous, biogenous, or terrigenous?

   b. How are hydrothermal vents related to the biosphere? How can there be life in these deep, cold, and dark environments? Explain.

NOAA/ Science Source

MasteringGeology™

Looking for additional review and test prep materials? Visit the Self Study area in **www.masteringgeology.com** to find practice quizzes, study tools, and multimedia that will aid in your understanding of this chapter's content. In **MasteringGeology™** you will find:

- GEODe: Earth Science: An interactive visual walkthrough of key concepts
- Geoscience Animation Library: More than 100 animations illuminating many difficult-to-understand Earth science concepts

- In The News RSS Feeds: Current Earth science events and news articles are pulled into the site with assessment
- Pearson eText
- Optional Self Study Quizzes
- Web Links
- Glossary
- Flashcards

15

The Dynamic Ocean

15.1 Discuss the factors that create and influence ocean currents and describe the affect ocean currents have on climate.

15.2 Explain the processes that produce coastal upwelling and the ocean's deep circulation.

15.3 Explain why the shoreline is considered a dynamic interface and identify the basic parts of the coastal zone.

15.4 List and discuss the factors that influence the height, length, and period of a wave and describe the motion of water within a wave.

15.5 Describe how waves erode and move sediment along the shore.

15.6 Describe the features typically created by wave erosion and those resulting from sediment deposited by longshore transport processes.

15.7 Summarize the ways in which people deal with shoreline erosion problems.

15.8 Contrast the erosion problems faced along different parts of America's coasts. Distinguish between emergent and submergent coasts.

15.9 Explain the cause of tides, their monthly cycles, and patterns. Describe the horizontal flow of water that accompanies the rise and fall of tides.

North Carolina's Outer Banks. Oregon Inlet Bridge connects Bodie Island (foreground) and Hatteras Island. These narrow wisps of sand are part of an extensive barrier island system. The beach and dunes on the left face the Atlantic Ocean. On the right are the quieter waters of Pamlico Sound. (Photo by Michael Collier)

The restless waters of the ocean are constantly in motion, powered by many different forces. Winds, for example, generate surface currents, which influence coastal climate and provide nutrients that affect the abundance of algae and other marine life in surface waters. Winds also produce waves that carry energy from storms to distant shores, where their impact erodes the land. In some regions, density differences create deep-ocean circulation, which is important for ocean mixing and recycling nutrients. In addition, the Moon and the Sun produce tides, which periodically raise and lower average sea level. This chapter examines these movements of ocean waters and their effect on coastal regions.

15.1 | THE OCEAN'S SURFACE CIRCULATION Discuss the factors that create and influence ocean currents and describe the affect ocean currents have on climate.

You may have heard of the Gulf Stream, an important surface current in the Atlantic Ocean that flows northward along the East coast of the United States (**FIGURE 15.1**). Surface currents like this one are set in motion by the wind. At the water surface, where the atmosphere and ocean meet, energy is passed from moving air to the water through friction. The drag exerted by winds blowing steadily across the ocean causes the surface layer of water to move. Thus, major horizontal movements of surface waters are closely related to the global pattern of prevailing winds.[1] As an example, the small map in **FIGURE 15.2** shows how the wind belts known as the *trade winds* and the *westerlies* create large, circular-moving loops of water in the Atlantic Ocean. The same wind belts influence the other oceans as well, so that a similar pattern of currents can also be seen in the Pacific and Indian Oceans. Essentially, the pattern of surface-ocean circulation closely matches the pattern of global winds but is also strongly influenced by the distribution of major landmasses and by the spinning of Earth on its axis.

The Pattern of Ocean Currents

Huge, circular-moving current systems dominate the surfaces of the oceans. These large whirls of water within an ocean basin are called **gyres** (*gyros* = circle). The large map in Figure 15.2 shows the world's five main gyres: the *North Pacific Gyre*, the *South Pacific Gyre*, the *North Atlantic Gyre*, the *South Atlantic Gyre*, and the *Indian Ocean Gyre* (which exists mostly within the Southern Hemisphere). The center of each gyre coincides with the subtropics at about 30° north or south latitude, so they are often called *subtropical gyres*.

Coriolis Effect As shown in Figure 15.2, subtropical gyres rotate clockwise in the Northern Hemisphere and counterclockwise in the Southern Hemisphere. Why do the gyres flow in different directions in the two hemispheres? Although wind is the force that generates surface currents, other factors also influence the movement of ocean waters. The most significant of these is the **Coriolis effect**. Because of Earth's rotation, currents are deflected to the *right* in the Northern Hemisphere and to the *left* in the Southern Hemisphere.[2] As a consequence, gyres flow in opposite directions in the two different hemispheres.

North Pacific Currents Four main currents generally exist within each gyre (see Figure 15.2). The North Pacific Gyre, for example, consists of the North Equatorial Current, the Kuroshio Current, the North Pacific Current, and the California Current. Tracking of floating objects that are released into the ocean intentionally or accidentally reveals that it takes about 6 years for the objects to go all the way around the loop.

North Atlantic Currents The North Atlantic Ocean has four main currents, too (see Figure 15.2). Beginning near the equator, the North Equatorial Current is deflected northward through the Caribbean, where it becomes the Gulf Stream. As the Gulf Stream moves along the East Coast of the United States, it is strengthened by the prevailing westerly winds and is deflected to the east (to the right) offshore of North Carolina into the North Atlantic. As it continues northeastward, it gradually widens and slows until it becomes a vast, slowly moving

FIGURE 15.1 The Gulf Stream In this satellite image off the East coast of the United States, orange and yellow represent higher water temperatures, and blue indicates cooler water temperatures. The current transports heat from the subtropics far into the North Atlantic. (NOAA)

Gulf Stream

---

[1]Details about the global pattern of winds appear in Chapter 18.
[2]The Coriolis effect is more fully explained in Chapter 18.

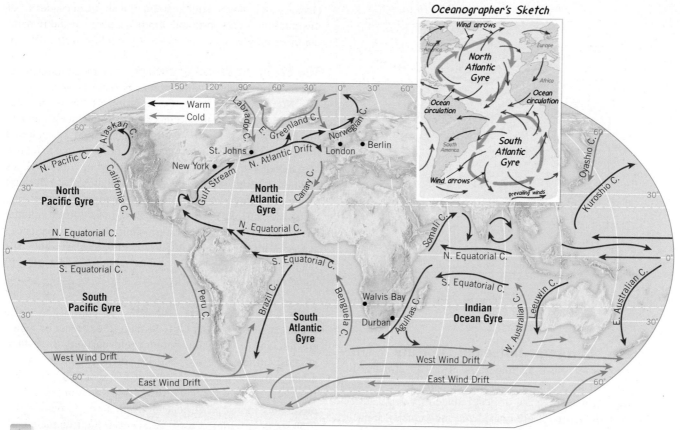

Oceanographer's Sketch

SmartFigure 15.2 Major Surface-Ocean Currents The ocean's surface circulation is organized into five major gyres. Poleward-moving currents are warm, and equatorward-moving currents are cold. Ocean currents play an important role in redistributing heat around the globe. Note that cities mentioned in the text discussion are shown on this map. In the smaller inset map, broad arrows show the idealized surface circulation for the Atlantic, and the thin arrows show prevailing winds. Winds provide the energy that drives the ocean's surface circulation. (NOAA)

current known as the North Atlantic Current, which, because of its sluggish nature, is also known as the North Atlantic Drift.

As the North Atlantic Current approaches Western Europe, it splits, part of it moving northward past Great Britain, Norway, and Iceland, carrying heat to these otherwise chilly areas. The other part is deflected southward as the cool Canary Current. As the Canary Current moves southward, it eventually merges into the North Equatorial Current, completing the gyre. Because the North Atlantic Ocean basin is about half the size of the North Pacific, it takes floating objects about 3 years to go completely around this gyre.

The circular motion of gyres leaves a large central area that has no well-defined currents. In the North Atlantic, this zone of calmer waters is known as the Sargasso Sea, named for the large quantities of *Sargassum*, a type of floating seaweed encountered there.

Southern Hemisphere Currents The ocean basins in the Southern Hemisphere exhibit a similar pattern of flow as the Northern Hemisphere basins, with surface currents that are influenced by wind belts, the position of continents, and the Coriolis effect. In the South Atlantic and South Pacific, for example, surface ocean circulation is very much the same as in their Northern Hemisphere counterparts except that the direction of flow is counterclockwise (see Figure 15.2).

Indian Ocean Currents The Indian Ocean exists mostly in the Southern Hemisphere, so it follows a surface circulation pattern similar to other Southern Hemisphere ocean basins (see Figure 15.2). The small portion of the Indian Ocean in the Northern Hemisphere, however, is influenced by the seasonal wind shifts known as the summer and winter *monsoons* (*mausim* = season). When the winds change direction, the surface currents also reverse direction. During the summer, winds blow from the Indian Ocean toward the Asian landmass. In the winter, the winds reverse and blow out from Asia over the Indian Ocean. You can see this reversal when you compare the January and July winds in Figure 18.18. When the winds change direction, the surface currents also reverse direction.

West Wind Drift The only current that completely encircles Earth is the *West Wind Drift* (see Figure 15.2). It flows around the ice-covered continent of Antarctica, where no large landmasses are in the way, so its cold surface waters circulate in a continuous loop. It moves in response to the Southern Hemisphere prevailing westerly winds, and portions of it split off into the adjoining southern ocean basins. Its strong flow also helps define the Southern Ocean or Antarctic Ocean, which is really the portions of the Pacific, Atlantic, and Indian Oceans south of about 50° south latitude.

FIGURE 15.3 The Chilling Effect of a Cold Current Monthly mean temperatures for Rio de Janeiro, Brazil, and Arica, Chile, both of which are coastal cities near sea level. Even though Arica is closer to the equator, its temperatures are cooler than Rio de Janeiro's. Arica is influenced by the cold Peru Current, whereas Rio de Janeiro is adjacent to the warm Brazil Current.

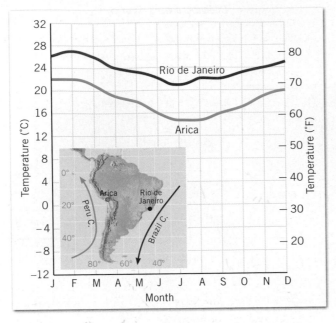

Ocean water movements account for about one-quarter of this total heat transport, and winds account for the remaining three-quarters.

The Effect of Warm Currents

The moderating effect of poleward-moving warm ocean currents is well known. The North Atlantic Drift, an extension of the warm Gulf Stream, keeps wintertime temperatures in Great Britain and much of Western Europe warmer than would be expected for their latitudes. London is farther north than St. John's, Newfoundland, yet is not nearly so frigid in winter. (Cities mentioned in this section are shown in Figure 15.2.) Because of the prevailing westerly winds, the moderating effects are carried far inland. For example, Berlin (52° north latitude) has a mean January temperature similar to that experienced at New York City, which lies 12° latitude farther south. The January mean at London (51° north latitude) is 4.5°C (8.1°F) higher than at New York City.

Cold Currents Chill the Air

In contrast to warm ocean currents like the Gulf Stream, the effects of which are felt most during the winter, cold currents exert their greatest influence in the tropics and during the summer months in the middle latitudes. For example, the cool Benguela Current off the western coast of southern Africa moderates the tropical heat along this coast. Walvis Bay (23° south latitude), a town adjacent to the Benguela Current, is 5°C (9°F) cooler in summer than Durban, which is 6° latitude farther poleward but on the eastern side of South Africa, away from the influence of the cold current. The east and west coasts of South America provide another example. **FIGURE 15.3** shows monthly mean temperatures for Rio de Janeiro, Brazil, which is influenced by the warm Brazil Current, and Arica, Chile, which is adjacent to the cold Peru Current. Closer to home, because of the cold California Current, summer temperatures in subtropical coastal southern California are lower by 6°C (10.8°F) or more compared to East coast stations.

Ocean Currents Influence Climate

Surface-ocean currents have an important effect on climate. It is known that for Earth as a whole, the gains in solar energy equal the losses to space of heat radiated from the surface. When most latitudes are considered individually, however, this is not the case. There is a net gain of energy in lower latitudes and a net loss at higher latitudes. Because the tropics are not becoming progressively warmer, nor are the polar regions becoming colder, there must be a large-scale transfer of heat from areas of excess to areas of deficit. This is indeed the case. *The transfer of heat by winds and ocean currents equalizes these latitudinal energy imbalances.*

FIGURE 15.4 Chile's Atacama Desert This is the driest desert on Earth. Average rainfall at the wettest locations is not more than 3 millimeters (0.12 inch) per year. Stretching nearly 1000 kilometers (600 miles), the Atacama is situated between the Pacific Ocean and the towering Andes Mountains. The cold Peru Current makes this slender zone cooler and drier than it would otherwise be. (Photo by Jaques Jangoux/Science Source)

Cold Currents Increase Aridity

In addition to influencing temperatures of adjacent land areas, cold currents have other climatic influences. For example, where tropical deserts exist along the west coasts of continents, cold ocean currents have a dramatic impact. The principal west coast deserts are the Atacama in Peru and Chile and the Namib in southwestern Africa (**FIGURE 15.4**). The aridity along these coasts is intensified because the lower atmosphere is chilled by cold offshore waters. When this occurs, the air becomes very stable and resists the upward movement necessary to create precipitation-producing clouds.

In addition, the presence of cold currents causes temperatures to approach and often reach the dew point, the temperature at which water vapor condenses. As a result, these areas are characterized by high relative humidities and much fog. Thus, not all subtropical deserts are hot with low humidities and clear skies. Rather, the presence of cold currents transforms some subtropical deserts into relatively cool, damp places that are often shrouded in fog.

15.2 | UPWELLING AND DEEP-OCEAN CIRCULATION

Explain the processes that produce coastal upwelling and the ocean's deep circulation.

The preceding discussion focused mainly on the horizontal movements of the ocean's surface waters. In this section, you will learn that the ocean also exhibits significant vertical movements and a slow-moving, multilayered deep-ocean circulation. Some vertical movements are related to wind-driven surface currents, whereas deep-ocean circulation is strongly influenced by density differences.

Coastal Upwelling

In addition to producing surface currents, winds can also cause *vertical* water movements. **Upwelling**, the rising of cold water from deeper layers to replace warmer surface water, is a common wind-induced vertical movement. One type of upwelling, called *coastal upwelling*, is most characteristic along the west coasts of continents, most notably along California, western South America, and West Africa.

Coastal upwelling occurs in these areas when winds blow toward the equator and parallel to the coast (**FIGURE 15.5**). Coastal winds combined with the Coriolis effect cause surface water to move away from shore. As the surface layer moves away from the coast, it is replaced by water that "upwells" from below the surface. This slow upward movement of water from depths of 50 to 300 meters (165 to 1000 feet) brings water that is cooler than the original surface water and results in lower surface-water temperatures near the shore.

For swimmers who are accustomed to the warm waters along the mid-Atlantic shore of the United States, a swim in the Pacific off the coast of central California can be a chilling surprise. In August, when temperatures in the Atlantic are 21°C (70°F) or higher, central California's surf is only about 15°C (60°F).

Upwelling brings greater concentrations of dissolved nutrients, such as nitrates and phosphates, to the ocean surface. These nutrient-enriched waters from below promote the growth of microscopic plankton, which in turn support extensive populations of fish and other marine organisms. Figure 15.5 includes a satellite image that shows high productivity due to coastal upwelling off the southwest coast of Africa.

Deep-Ocean Circulation

Deep-ocean circulation has a significant vertical component and accounts for the thorough mixing of deep-water masses. This component of ocean circulation is a response to density differences among water masses that cause

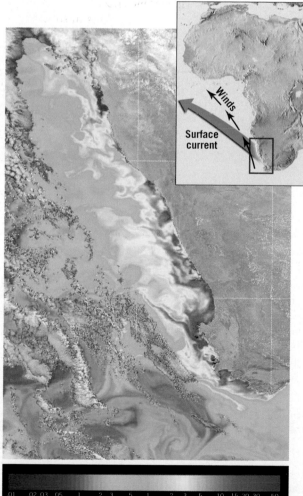

Chlorophyll a Concentration
mg/m³

.01 .02 .03 .05 .1   .2 .3   .5 1   2 3 5   10 15 20 30 50

SmartFigure 15.5 Coastal Upwelling
Coastal upwelling occurs along the west coasts of continents, where winds blow toward the equator and parallel to the coast. The Coriolis effect (deflection to the left in the Southern Hemisphere) causes surface water to move away from the shore, which brings cold, nutrient-rich water to the surface. This satellite image shows chlorophyll concentration along the southwest coast of Africa (February 21, 2001). An instrument aboard the satellite detects changes in seawater color caused by changing concentrations of chlorophyll. High chlorophyll concentrations indicate high amounts of photosynthesis, which is linked to the upwelling nutrients. Red indicates high concentrations, and blue indicates low concentrations. (Provided by from the SeaWIFS Project, NASA/GODDARD Space Flight Center and ORBIMAGE.)

FIGURE 15.6 Sea Ice Near Antarctica When seawater freezes, sea salts do not become part of the ice. Consequently, the surface salinity of the remaining seawater increases, which makes it denser and prone to sink. (Photo by John Higdon/AGE Fotostock)

Most water involved in deep-ocean currents (thermohaline circulation) begins in high latitudes at the surface. In these regions, where surface waters are cold, salinity increases when sea ice forms (**FIGURE 15.6**). When seawater freezes to form sea ice, salts do not become part of the ice. As a result, the salinity (and therefore the density) of the remaining seawater increases. When this surface water becomes dense enough, it sinks, initiating deep-ocean currents. Once this water sinks, it is removed from the physical processes that increased its density in the first place, and so its temperature and salinity remain largely unchanged for the duration of the time it spends in the deep ocean.

Near Antarctica, surface conditions create the highest-density water in the world. This cold saline brine slowly sinks to the seafloor, where it moves throughout the ocean basins in sluggish currents. After sinking from the surface of the ocean, deep waters will not reappear at the surface for an average of 500 to 2000 years.

A simplified model of ocean circulation is similar to a conveyor belt that travels from the Atlantic Ocean through the Indian and Pacific Oceans and back again (**FIGURE 15.7**). In this model, warm water in the ocean's upper layers flows poleward, converts to dense water, and returns toward the equator as cold deep water that eventually upwells to complete the circuit. As this "conveyor belt" moves around the globe, it influences global climate by converting warm water to cold and liberating heat to the atmosphere.

FIGURE 15.7 The Ocean Conveyor Belt Source areas for dense water masses exist in high-latitude regions where cold, high-salinity water sinks and flows into all the oceans. This water eventually ascends and completes the conveyor by returning to the source areas as warm surface currents.

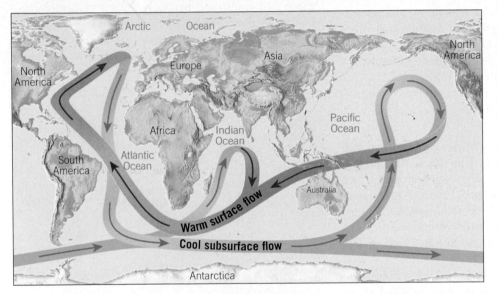

denser water to sink and slowly spread out beneath the surface. Because the density variations that cause deep-ocean circulation are caused by differences in temperature and salinity, deep-ocean circulation is also referred to as **thermohaline** (*thermo* = heat, *haline* = salt) **circulation**.

An increase in seawater density can be caused by either a decrease in temperature or an increase in salinity. Density changes due to salinity variations are important in very high latitudes, where water temperature remains low and relatively constant.

15.2 CONCEPT CHECKS

1 Describe the process of coastal upwelling. Why is an abundance of marine life associated with these areas?

2 Why is deep-ocean circulation referred to as *thermohaline circulation*?

3 Describe or make a simple sketch of the ocean's conveyor-belt circulation.

15.3 | THE SHORELINE: A DYNAMIC INTERFACE Explain why the shoreline is considered a dynamic interface and identify the basic parts of the coastal zone.

Shorelines are dynamic environments. Their topography, geologic makeup, and climate vary greatly from place to place. Continental and oceanic processes converge along coasts to create landscapes that frequently undergo rapid change. When it comes to the deposition of sediment, coasts are transition zones between marine and continental environments.

Nowhere is the restless nature of the ocean's water more noticeable than along the shore—the dynamic interface among air, land, and sea. An *interface* is a common boundary where different parts of a system interact. This is certainly an appropriate designation for the coastal zone. Here we can see the rhythmic rise and fall of tides and observe waves rolling in and breaking. Sometimes the waves are low and gentle. At other times they pound the shore with awesome fury.

The Coastal Zone

Although it may not be obvious, the shoreline is constantly being modified by waves. Crashing surf can erode the adjacent land. Wave activity also moves sediment toward and away from the shore, as well as along it. Such activity sometimes produces narrow sandbars that frequently change size and shape as storm waves come and go.

Present-Day Shorelines The nature of present-day shorelines is not just the result of the relentless attack of the land by the sea. The shore has a complex character that results from multiple geologic processes. For example, practically all coastal areas were affected by the worldwide rise in sea level that accompanied the melting of glaciers at the close of the Pleistocene epoch. As the sea encroached landward, the shoreline retreated, becoming superimposed upon existing landscapes that had resulted from such diverse processes as stream erosion, glaciation, volcanic activity, and the forces of mountain building.

Human Activity Today, the coastal zone is experiencing intensive human activity. Unfortunately, people often treat the shoreline as if it were a stable platform on which structures can safely be built. This attitude inevitably leads to conflicts between people and nature. In October 2012, this fact was tragically reinforced when the storm surge from Hurricane Sandy struck parts of New York City and the narrow barrier islands along the coast of New Jersey (**FIGURE 15.8**). Many coastal landforms, especially beaches and barrier islands, are relatively fragile, short-lived features that are often inappropriate sites for development.

Basic Features

In general conversation, several terms are used when referring to the boundary between land and sea. In the preceding paragraphs, the terms *shore, shoreline, coastal zone,* and *coast* were all used. Moreover, when many think of the land–sea interface, the word *beach* comes to mind. Let's take a moment to clarify these terms and introduce some other terminology used by those who study the land–sea boundary

FIGURE 15.8 Hurricane Sandy A portion of the New Jersey shoreline shortly after this huge storm struck in late October 2012. The extraordinary storm surge caused much of the damage pictured here. Many shoreline areas are intensively developed. Often the shifting shoreline sands and the desire of people to occupy these areas are in conflict. There is more about hurricanes and hurricane damage in Chapter 19. (Photo by REUTERS/Tim Larsen/Governor's Office/Handout)

zone. You will find it helpful to refer to **FIGURE 15.9**, which is an idealized profile of the coastal zone.

The **shoreline** is the line that marks the contact between land and sea. Each day, as tides rise and fall, the position of the shoreline migrates. Over longer time spans, the average position of the shoreline gradually shifts as sea level rises or falls.

The **shore** is the area that extends between the lowest tide level and the highest elevation on land that is affected by storm waves. By contrast, the **coast** extends inland from the shore as far as ocean-related features can be found. The **coastline** marks the coast's seaward edge, whereas the inland boundary is not always obvious or easy to determine.

As Figure 15.9 illustrates, the shore is divided into the *foreshore* and the *backshore*. The **foreshore** is the area exposed when the tide is out (low tide) and submerged when the tide is in (high tide). The **backshore** is landward of the high-tide shoreline. It is usually dry, being affected

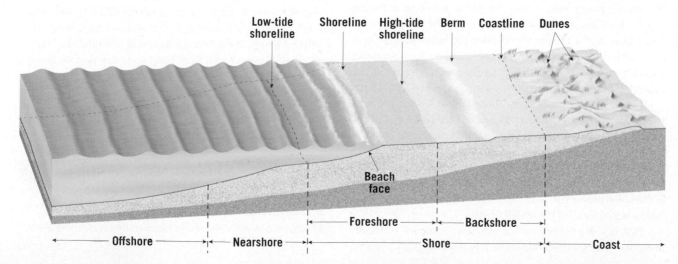

FIGURE 15.9 The Coastal Zone This transition zone between land and sea consists of several parts.

FIGURE 15.10 Beaches
A beach is an accumulation of sediment on the landward margin of an ocean or a lake and can be thought of as material in transit along the shore. Beaches are composed of whatever material is locally available. (Photo A by David R. Frazier Photolibrary, Inc./Alamy; photo B by E. J. Tarbuck)

This beach on Florida's Sanibel Island consists of shells and shell fragments.

A.

The black sands on this beach in Hawaii were derived from the weathering of nearby basaltic lava flows.

B.

by waves only during storms. Two other zones are commonly identified. The **nearshore zone** lies between the low-tide shoreline and the line where waves break at low tide. Seaward of the nearshore zone is the **offshore zone**.

Beaches

For many a beach is the sandy area where people lie in the sun and walk along the water's edge. Technically, a **beach** is an accumulation of sediment found along the landward margin of an ocean or a lake. Along straight coasts, beaches may extend for tens or hundreds of kilometers. Where coasts are irregular, beach formation may be confined to the relatively quiet waters of bays.

Beaches consist of one or more **berms**, which are relatively flat platforms often composed of sand that are adjacent to coastal dunes or cliffs and marked by a change in slope at the seaward edge. Another part of the beach is the **beach face**, which is the wet sloping surface that extends from the berm to the shoreline. Where beaches are sandy, sunbathers usually prefer the berm, whereas joggers prefer the wet, hard-packed sand of the beach face.

Beaches are composed of whatever material is locally abundant. The sediment for some beaches is derived from the

erosion of adjacent cliffs or nearby coastal mountains. Other beaches are built from sediment delivered to the coast by rivers.

Although the mineral makeup of many beaches is dominated by durable quartz grains, other minerals may be dominant. For example, in areas such as southern Florida, where there are no mountains or other sources of rock-forming minerals nearby, most beaches are composed of shell fragments and the remains of organisms that live in coastal waters (**FIGURE 15.10A**). Some beaches on volcanic islands in the open ocean are composed of weathered grains of the basaltic lava that comprise the islands or of coarse debris eroded from coral reefs that develop around islands in low latitudes (**FIGURE 15.10B**).

Regardless of the composition, the material that comprises the beach does not stay in one place. Instead, crashing waves are constantly moving it. Thus, beaches can be thought of as material in transit along the shore.

15.3 CONCEPT CHECKS

1 Why is the shoreline considered an *interface*?

2 Distinguish among shore, shoreline, coast, and coastline.

3 What is a *beach*? Distinguish between *beach face* and *berm*.

15.4 | OCEAN WAVES List and discuss the factors that influence the height, length, and period of a wave and describe the motion of water within a wave.

Ocean waves are energy traveling along the interface between ocean and atmosphere, often transferring energy from a storm far out at sea over distances of several thousand kilometers. That's why even on calm days, the ocean still has waves that travel across its surface. When observing waves, always remember that you are watching *energy* travel through a medium (water). If you make waves by tossing a pebble into a pond, splashing in a pool, or blowing across the surface of a cup of coffee, you are imparting *energy* to the water, and the waves you see are just the visible evidence of the energy passing through.

Wind-generated waves provide most of the energy that shapes and modifies shorelines. Where the land and sea meet, waves that may have traveled unimpeded for hundreds or thousands of kilometers suddenly encounter a barrier that will not allow them to advance farther and must absorb their energy. Stated another way, the shore is the location where a practically irresistible force confronts an almost immovable object. The conflict that results is never-ending and sometimes dramatic.

Wave Characteristics

Most ocean waves derive their energy and motion from the wind. When a breeze is less than 3 kilometers (2 miles) per hour, only wavelets appear. At greater wind speeds, more stable waves gradually form and advance with the wind.

Characteristics of ocean waves are illustrated in **FIGURE 15.11**, which shows a simple, nonbreaking waveform. The tops of the waves are the *crests*, which are separated by *troughs*. Halfway between the crests and troughs is the *still water level*, which is the level that the water would occupy if there were no waves. The vertical distance between trough and crest is called the **wave height**, and

the horizontal distance between successive crests or successive troughs is the **wavelength**. The time it takes one full wave—one wavelength—to pass a fixed position is the **wave period**.

The height, length, and period that are eventually achieved by a wave depend on three factors: (1) wind speed, (2) length of time the wind has blown, and (3) **fetch**, the distance that wind has traveled across open water. As the quantity of energy transferred from the wind to the water increases, both the height and steepness of the waves increase. Eventually, a critical point is reached where waves grow so tall that they topple over, forming ocean breakers called *whitecaps*.

For a particular wind speed, there is a maximum fetch and duration of wind beyond which waves will no longer increase in size. When the maximum fetch and duration are reached for a given wind velocity, the waves are said to be "fully developed." The reason that waves can grow no further is that they are losing as much energy through the breaking of whitecaps as they are receiving energy from the wind.

When the wind stops or changes direction, or the waves leave the storm area where they were created, they continue on without relation to local winds. The waves also undergo a gradual change to *swells*, which are lower in height and longer in length and may carry a storm's energy to distant shores. Because many independent wave systems exist at the same

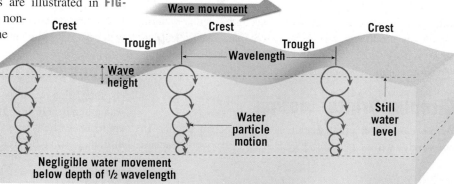

FIGURE 15.11 Wave Basics An idealized drawing of a nonbreaking wave, showing its basic parts and the movement of water with increasing depth.

EYE ON EARTH

This surfer is enjoying a ride on a large wave along the coast of Maui. (Photo by Ron Dahlquist/Getty Images)

QUESTION 1 *What was the source of energy that created this wave?*

QUESTION 2 *How was the wavelength changing just prior to the time when this photo was taken?*

QUESTION 3 *Why was the wavelength changing?*

QUESTION 4 *Many waves exhibit circular orbital motion. Is that true of the wave in this photo? Explain.*

SmartFigure 15.12 Passage of a Wave The movements of the toy boat show that the waveform advances, but the water does not advance appreciably from the original position. In this sequence, the wave moves from left to right as the boat (and the water in which it is floating) rotates in an imaginary circle.

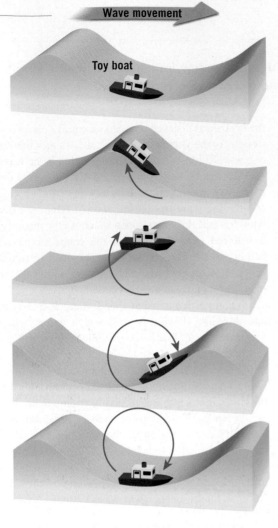

Wave movement

Toy boat

as they traveled through the Pacific Ocean basin. After more than 10,000 kilometers (more than 6,000 miles), the waves finally expended their energy a week later, along the shoreline of the Aleutian Islands of Alaska. The water itself doesn't travel the entire distance, but the waveform does. As the wave travels, the water passes the energy along by moving in a circle. This movement is called **circular orbital motion**.

Observation of an object floating in waves shows that it moves not only up and down but also slightly forward and backward with each successive wave. **FIGURE 15.12** shows that a floating object moves up and backward as the crest approaches, up and forward as the crest passes, down and forward after the crest, down and backward as the trough approaches, and rises and moves backward again as the next crest advances. When the movement of the floating toy boat shown in Figure 15.12 is traced as a wave passes, it can be seen that the boat moves in a circle and it returns to essentially the same place. Circular orbital motion allows a waveform (the wave's shape) to move forward *through the water* while the individual water particles that transmit the wave move around in a circle. Wind moving across a field of wheat causes a similar phenomenon: The wheat itself doesn't travel across the field, but the waves do.

The energy contributed by the wind to the water is transmitted not only along the surface of the sea but also downward. However, beneath the surface the circular motion rapidly diminishes until, at a depth equal to one-half the wavelength measured from still water level, the movement of water particles becomes negligible. This depth is known as the **wave base**. The dramatic decrease of wave energy with depth is shown by the rapidly diminishing diameters of water-particle orbits in Figure 15.11.

time, the sea surface acquires a complex and irregular pattern, sometimes producing very large waves. The sea waves that are seen from shore are usually a mixture of swells from faraway storms and waves created by local winds.

Circular Orbital Motion

Waves can travel great distances across ocean basins. In one study, waves generated near Antarctica were tracked

Waves in the Surf Zone

As long as a wave is in deep water, it is unaffected by water depth (**FIGURE 15.13**, left). However, when a wave approaches the shore, the water becomes shallower and influences wave behavior. The wave begins to "feel bottom"

FIGURE 15.13 Waves Approaching the Shore Waves touch bottom as they encounter water depths that are less than half a wavelength. As a result, the wave speed decreases, and the faster-moving waves farther from shore begin to catch up, which causes the distance between waves (the wavelength) to decrease. This causes an increase in wave height to the point where the waves finally pitch forward and break in the surf zone.

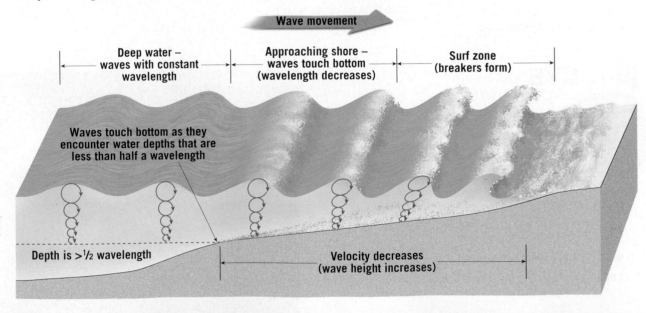

Wave movement

Deep water – waves with constant wavelength | Approaching shore – waves touch bottom (wavelength decreases) | Surf zone (breakers form)

Waves touch bottom as they encounter water depths that are less than half a wavelength

Depth is >½ wavelength

Velocity decreases (wave height increases)

at a water depth equal to its wave base. Such depths interfere with water movement at the base of the wave and slow its advance (see Figure 15.13, center).

As a wave advances toward the shore, the slightly faster waves farther out to sea catch up, decreasing the wavelength. As the speed and length of the wave diminish, the wave steadily grows higher. Finally, a critical point is reached when the wave is too steep to support itself, and the wave front collapses, or *breaks* (see Figure 15.13, right), causing water to advance up the shore.

The turbulent water created by breaking waves is called **surf**. On the landward margin of the surf zone, the turbulent sheet of water from collapsing breakers, called *swash*, moves up the slope of the beach. When the energy of the swash has been expended, the water flows back down the beach toward the surf zone as *backwash*.

15.4 CONCEPT CHECKS

1 List three factors that determine the height, length, and period of a wave.

2 Describe the motion of a floating object as a wave passes.

3 How do the speed, length, and height of a wave change as the wave moves into shallow water and breaks?

15.5 | THE WORK OF WAVES

Describe how waves erode and move sediment along the shore.

During calm weather, wave action is minimal. However, just as streams do most of their work during floods, waves accomplish most of their work during storms. The impact of high, storm-induced waves against the shore can be awesome in its violence (**FIGURE 15.14**).

Wave Erosion

Each breaking wave may hurl thousands of tons of water against the land, sometimes causing the ground to literally tremble. The pressures exerted by Atlantic waves in wintertime, for example, average nearly 10,000 kilograms per square meter (more than 2000 pounds per square foot). The force during storms is even greater. It is no wonder that cracks and crevices are quickly opened in cliffs, seawalls, breakwaters, and anything else that is subjected to these enormous shocks. Water is forced into every opening, causing air in the cracks to become highly compressed by the thrust of crashing waves. When the wave subsides, the air expands rapidly, dislodging rock fragments and enlarging and extending fractures.

In addition to the erosion caused by wave impact and pressure, **abrasion**—the sawing and grinding action of the water armed with rock fragments—is also important. In fact, abrasion is probably more intense in the surf zone than in any other environment. Smooth, rounded stones and pebbles along the shore are obvious reminders of the relentless grinding action of rock against rock in the surf zone (**FIGURE 15.15A**). Further, the waves use such fragments as "tools" as they cut horizontally into the land (**FIGURE 15.15B**).

Sand Movement on the Beach

Beaches are sometimes called "rivers of sand." The reason is that the energy from breaking waves often causes large quantities of sand to move along the beach face and in the surf zone roughly parallel to the shoreline. Wave energy also causes sand to move perpendicular to (toward and away from) the shoreline.

Movement Perpendicular to the Shoreline
If you stand ankle deep in water at the beach, you will see that swash and backwash move sand toward and away from the shoreline. Whether there is a net loss or addition of sand depends on the level of wave activity. When wave activity is relatively light (less energetic waves), much of the swash soaks into the beach, which reduces the backwash. Consequently, the swash dominates and causes a net movement of sand up the beach face toward the berm.

When high-energy waves prevail, the beach is saturated from previous waves, so much less of the swash soaks in. As a result, the berm erodes because backwash is strong and causes a net movement of sand down the beach face.

Along many beaches, light wave activity is the rule during the summer. Therefore, a wide sand berm gradually develops. During winter, when storms are frequent and more powerful, strong wave activity erodes and narrows the berm. A wide berm that may have taken months to build can be dramatically narrowed in just a few hours by the high-energy waves created by a strong winter storm.

Wave Refraction   The bending of waves, called **wave refraction**, plays an important part in shoreline processes (**FIGURE 15.16**). It affects the distribution of energy along the shore and thus strongly influences where and to what degree erosion, sediment transport, and deposition will take place.

Waves seldom approach the shore straight on. Rather, most waves move toward the shore at an angle. However,

FIGURE 15.14 Storm Waves When large waves break against the shore, the force of the water can be powerful, and the erosional work that is accomplished can be great. These storm waves are breaking along the coast of Wales. (The Photolibrary Wales/Alamy)

FIGURE 15.15
Abrasion—Sawing and Grinding Breaking waves armed with rock debris can do a great deal of erosional work. (Photo A by Michael Collier; photo B by Fletcher & Baylis/ Science Source)

A.

Smooth, rounded rocks along the shore are an obvious reminder that abrasion can be intense in the surf zone.

B.

This sandstone cliff at Gabriola Island, British Columbia, was undercut by wave action.

when they reach the shallow water of a smoothly sloping bottom, they are bent and tend to become parallel to the shore. Such bending occurs because the part of the wave nearest the shore reaches shallow water and slows first, whereas the end that is still in deep water continues forward at its full speed. The net result is a wave front that may approach nearly parallel to the shore, regardless of the original direction of the wave.

Because of refraction, wave impact is concentrated against the sides and ends of headlands that project into the water,

whereas wave attack is weakened in bays. This differential wave attack along irregular coastlines is illustrated in Figure 15.16. As the waves reach the shallow water in front of the headland sooner than they do in adjacent bays, they are bent more nearly parallel to the protruding land and strike it from all three sides. By contrast, refraction in the bays causes waves to diverge and expend less energy. In these zones of weakened wave activity, sediments can accumulate and form sandy beaches. Over a long period, erosion of the headlands and deposition in the bays will straighten an irregular shoreline.

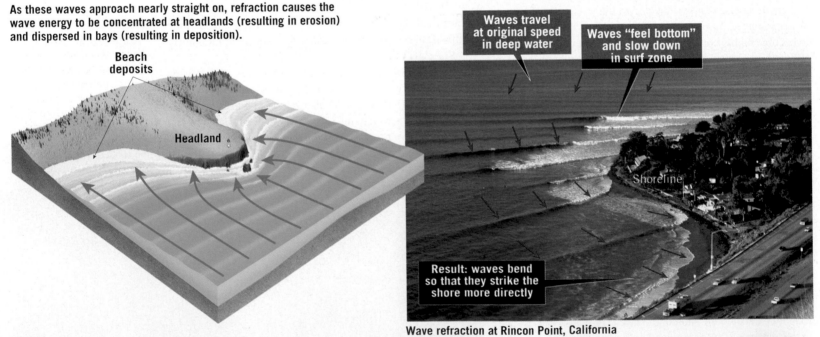

As these waves approach nearly straight on, refraction causes the wave energy to be concentrated at headlands (resulting in erosion) and dispersed in bays (resulting in deposition).

Beach deposits

Headland

Waves travel at original speed in deep water

Waves "feel bottom" and slow down in surf zone

Shoreline

Result: waves bend so that they strike the shore more directly

Wave refraction at Rincon Point, California

SmartFigure 15.16 Wave Refraction As waves first touch bottom in the shallows along an irregular coast, they are slowed, causing them to bend (refract) and align nearly parallel to the shoreline. (Photo by Rich Reid/Getty Images, Inc.)

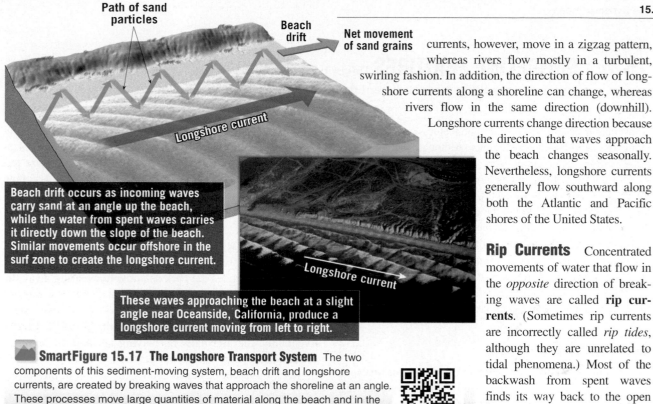

Path of sand particles

Beach drift

Net movement of sand grains

Beach drift occurs as incoming waves carry sand at an angle up the beach, while the water from spent waves carries it directly down the slope of the beach. Similar movements occur offshore in the surf zone to create the longshore current.

Longshore current

These waves approaching the beach at a slight angle near Oceanside, California, produce a longshore current moving from left to right.

SmartFigure 15.17 **The Longshore Transport System** The two components of this sediment-moving system, beach drift and longshore currents, are created by breaking waves that approach the shoreline at an angle. These processes move large quantities of material along the beach and in the surf zone. (Photo by University of Washington Libraries, Special Collections, John Shelton Collection, KC14461)

Longshore Transport Although waves are refracted, most still reach the shore at some angle, however slight. Consequently, the uprush of water from each breaking wave (the swash) is at an oblique angle to the shoreline. However, the backwash is straight down the slope of the beach. The effect of this pattern of water movement is to transport sediment in a zigzag pattern along the beach face (**FIGURE 15.17**). This movement is called **beach drift**, and it can transport sand and pebbles hundreds or even thousands of meters each day. However, a more typical rate is 5 to 10 meters (16 to 33 feet) per day.

Waves that approach the shore at an angle also produce currents within the surf zone that flow parallel to the shore and move substantially more sediment than beach drift. Because the water here is turbulent, these **longshore currents** easily move the fine suspended sand and roll larger sand and gravel along the bottom. When the sediment transported by longshore currents is added to the quantity moved by beach drift, the total amount can be very large. At Sandy Hook, New Jersey, for example, the quantity of sand transported along the shore over a 48-year period averaged almost 750,000 tons annually. For a 10-year period in Oxnard, California, more than 1.5 million tons of sediment moved along the shore each year.

Both rivers and coastal zones move water and sediment from one area (*upstream*) to another (*downstream*). This is why the beach has often been characterized as a "river of sand." Beach drift and longshore

currents, however, move in a zigzag pattern, whereas rivers flow mostly in a turbulent, swirling fashion. In addition, the direction of flow of longshore currents along a shoreline can change, whereas rivers flow in the same direction (downhill). Longshore currents change direction because the direction that waves approach the beach changes seasonally. Nevertheless, longshore currents generally flow southward along both the Atlantic and Pacific shores of the United States.

Rip Currents Concentrated movements of water that flow in the *opposite* direction of breaking waves are called **rip currents**. (Sometimes rip currents are incorrectly called *rip tides*, although they are unrelated to tidal phenomena.) Most of the backwash from spent waves finds its way back to the open ocean as an unconfined flow across the ocean bottom called *sheet flow*. However, sometimes a portion of the returning water moves seaward in the form of surface rip currents. These currents do not travel far beyond the surf zone before breaking up and can be recognized by the way they interfere with incoming waves or by the sediment that is often suspended within the rip current (**FIGURE 15.18**). They can be hazardous to swimmers, who, if caught in them, can be carried out away from shore. The best strategy for exiting a rip current is to swim *parallel* to the shore for a few tens of meters.

15.5 CONCEPT CHECKS

1 Describe two ways in which waves cause erosion.
2 Why do waves that are approaching the shoreline often bend?
3 What is the effect of wave refraction along an irregular coastline?
4 Describe the two processes that contribute to longshore transport.

Rip current extends outward from shore and interferes with incoming waves.

WARNING
DANGEROUS RIP CURRENTS
NO BOARD SURFING ZONE
SURF BOARDS, SURF MATS, SURF SKIS, BODY BOARDS, HAND BOARDS, KAYAKS ARE PROHIBITED

FIGURE 15.18 **Rip Current** These concentrated movements of water flow opposite the direction of breaking waves. (Photo by A.P. Trujillo/APT Photos)

15.6 | SHORELINE FEATURES Describe the features typically created by wave erosion and those resulting from sediment deposited by longshore transport processes.

A fascinating assortment of shoreline features can be observed along the world's coastal regions. Although the same processes cause change along every coast, not all coasts respond in the same way. Interactions among different processes and the relative importance of each process depend on local factors. The factors include (1) the proximity of a coast to sediment-laden rivers, (2) the degree of tectonic activity, (3) the topography and composition of the land, (4) prevailing winds and weather patterns, and (5) the configuration of the coastline and nearshore areas. Features that originate primarily because of erosion are called *erosional features*, whereas accumulations of sediment produce *depositional features*.

Erosional Features

Many coastal landforms owe their origin to erosional processes. Such erosional features are common along the rugged and irregular New England coast and along the steep shorelines of the West coast of the United States.

Wave-Cut Cliffs, Wave-Cut Platforms, and Marine Terraces
As the name implies, **wave-cut cliffs** originate in the cutting action of the surf against the base of coastal land. As erosion progresses, rocks overhanging the notch at the base of the cliff crumble into the surf, and the cliff retreats. A relatively flat, benchlike surface, called a **wave-cut platform**, is left behind by the receding cliff (**FIGURE 15.19**, left). The platform broadens as wave attack continues. Some debris produced by the breaking waves may remain along the water's edge as sediment on the beach, and

the remainder is transported farther seaward. If a wave-cut platform is uplifted above sea level by tectonic forces, it becomes a **marine terrace** (Figure 15.19, right). Marine terraces are easily recognized by their gentle seaward-sloping shape and are often desirable sites for coastal roads, buildings, or agriculture.

Sea Arches and Sea Stacks
Because of refraction, waves vigorously attack headlands that extend into the sea. The surf erodes the rock selectively, wearing away the softer or more highly fractured rock at the fastest rate. At first, sea caves may form. When caves on opposite sides of a headland unite, a **sea arch** results (**FIGURE 15.20**). Eventually, the arch falls in, leaving an isolated remnant, or **sea stack**, on the wave-cut platform (see Figure 15.20). In time, it too will be consumed by the action of the waves.

Depositional Features

Sediment eroded from the beach is transported along the shore and deposited in areas where wave energy is low. Such processes produce a variety of depositional features.

Spits, Bars, and Tombolos
Where beach drift and longshore currents are active, several features related to the movement of sediment along the shore may develop. A **spit** (*spit* = spine) is an elongated ridge of sand that projects from the land into the mouth of an adjacent bay. Often the end in the water hooks landward in response to the dominant direction of the longshore current. Both images in **FIGURE 15.21** show spits. The term **baymouth bar** is applied

FIGURE 15.19 Wave-Cut Platform and Marine Terrace This wave-cut platform is exposed at low tide along the California coast at Bolinas Point near San Francisco. A wave-cut platform was uplifted to create the marine terrace. (Photo by University of Washington Libraries, Special Collections, John Shelton Collection, KC5902)

FIGURE 15.20 Sea Arch and Sea Stack These features at the tip of Mexico's Baja Peninsula resulted from the vigorous wave attack of a headland. (Photo by Lew Robertson/Getty Images)

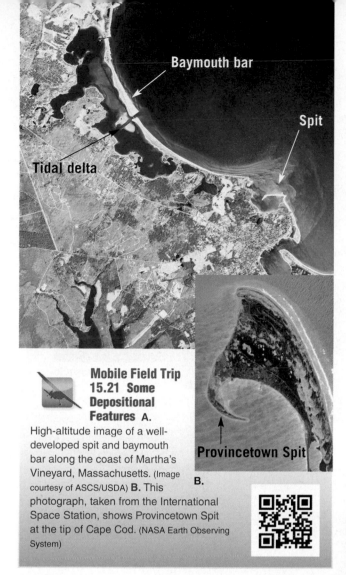

Mobile Field Trip 15.21 Some Depositional Features A. High-altitude image of a well-developed spit and baymouth bar along the coast of Martha's Vineyard, Massachusetts. (Image courtesy of ASCS/USDA) **B.** This photograph, taken from the International Space Station, shows Provincetown Spit at the tip of Cape Cod. (NASA Earth Observing System)

to a sandbar that completely crosses a bay, sealing it off from the open ocean (see Figure 15.21A). Such a feature tends to form across bays where currents are weak, allowing a spit to extend to the other side. A **tombolo** (*tombolo* = mound), a ridge of sand that connects an island to the mainland or to another island, forms in much the same manner as a spit.

Barrier Islands The Atlantic and Gulf Coastal Plains are relatively flat and slope gently seaward. The shore zone is characterized by **barrier islands**. These low ridges of sand parallel the coast at distances from 3 to 30 kilometers (2 to 19 miles) offshore. From Cape Cod, Massachusetts, to Padre Island, Texas, nearly 300 barrier islands rim the coast. The chapter-opening photo and **FIGURE 15.22** show examples from North Carolina.

Most barrier islands are 1 to 5 kilometers (0.6–3 miles) wide and between 15 and 30 kilometers (9–18 miles) long. The tallest features are sand dunes, which usually reach heights of 5 to 10 meters (16–33 feet); in a few areas, unvegetated dunes are more than 30 meters (100 feet) high. The lagoons separating these narrow islands from the shore are zones of relatively quiet water that allow small craft traveling between New York and northern Florida to avoid the rough waters of the North Atlantic.

Barrier islands probably formed in several ways. Some originated as spits that were subsequently severed from the mainland by wave erosion or by the general rise in sea level following the last episode of glaciation. Others were created

when turbulent waters in the line of breakers heaped up sand that had been scoured from the bottom. Finally, some barrier islands may be former sand dune ridges that originated along the shore during the last glacial period, when sea level was lower. As the ice sheets melted, sea level rose and flooded the area behind the beach–dune complex.

The Evolving Shore

A shoreline continually undergoes modification, regardless of its initial configuration. At first most coastlines are irregular, although the degree of and reason for the irregularity may differ considerably from place to place. Along a coastline that is characterized by varied geology, the pounding surf may initially increase its irregularity because the waves will erode the weaker rocks more easily than the stronger ones. However, if a shoreline remains stable, marine erosion and deposition will eventually produce a straighter, more regular coast.

FIGURE 15.23 illustrates the evolution of an initially irregular coast that remains relatively stable and shows many of the coastal features discussed in the previous section. As

FIGURE 15.22 Barrier Islands Nearly 300 barrier islands line the Gulf and Atlantic coasts. The islands along the coast of North Carolina are excellent examples. (Photo by Michael Collier)

FIGURE 15.23 The Evolving Shore These diagrams illustrate changes that can take place through time along an initially irregular coastline that remains tectonically stable. The diagrams also serve to illustrate many of the features described in the section on shoreline features. (Top and bottom photos by E. J. Tarbuck; middle photo by Michael Collier)

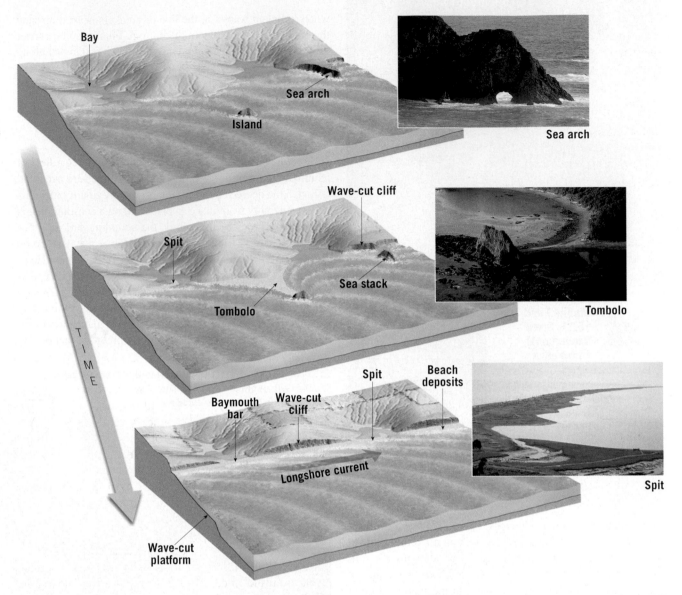

FIGURE 15.23 The Evolving Shore These diagrams illustrate changes that can take place through time along an initially irregular coastline that remains tectonically stable. The diagrams also serve to illustrate many of the features described in the section on shoreline features. (Top and bottom photos by E. J. Tarbuck; middle photo by Michael Collier)

headlands are eroded and erosional features such as wave-cut cliffs and wave-cut platforms are created, sediment is produced that is carried along the shore by beach drift and longshore currents. Some material is deposited in the bays, while other debris is formed into depositional features such as spits and baymouth bars. At the same time, rivers fill the bays with sediment. Ultimately, a smooth coast results.

15.6 CONCEPT CHECKS

1. How is a marine terrace related to a wave-cut platform?
2. Describe the formation of the features labeled in Figure 15.20 and Figure 15.21.
3. List three ways that barrier islands may form.

15.7 | STABILIZING THE SHORE

Summarize the ways in which people deal with shoreline erosion problems.

The coastal zone teems with human activity. Unfortunately, people often treat the shoreline as if it were a stable platform on which structures can be built safely. This approach jeopardizes both people and the shoreline because many coastal landforms are relatively fragile, short-lived features that are easily damaged by development. As anyone who has endured a tsunami or a strong coastal storm knows, the shoreline is not always a safe place to live. Figure 15.8 illustrates this point.

Compared with natural hazards, such as earthquakes, volcanic eruptions, and landslides, shoreline erosion appears to be a more continuous and predictable process that causes relatively modest damage to limited areas. In reality, the shoreline is one of Earth's most dynamic places that changes rapidly in response to natural forces. Storms, for example, are capable of eroding beaches and cliffs at rates that far exceed the long-term average. Such bursts of accelerated erosion not only have a significant impact on the natural evolution of a coast but can also have a profound impact on people who reside in the coastal zone. Erosion along the coast causes significant property damage. Huge sums are spent annually not only to repair damage but also in an attempt to prevent or control erosion. Already a problem at many sites, shoreline erosion is certain to become increasingly serious as extensive coastal development continues.

During the past 100 years, growing affluence and increasing demands for recreation have brought unprecedented development to many coastal areas. As both the number and the value of buildings have increased, so too have efforts to protect property from storm waves by stabilizing the shore. Also, controlling the natural migration of sand is an ongoing struggle in many coastal areas. Such interference can result in unwanted changes that are difficult and expensive to correct.

Hard Stabilization

Structures built to protect a coast from erosion or to prevent the movement of sand along a beach are known as **hard stabilization**. Hard stabilization can take many forms and often results in predictable yet unwanted outcomes. Hard stabilization includes jetties, groins, breakwaters, and seawalls.

Jetties   Since relatively early in America's history, a principal goal in coastal areas has been the development and maintenance of harbors. In many cases, this has involved the construction of jetty systems. **Jetties** are usually built in pairs and extend into the ocean at the entrances to rivers and harbors. With the flow of water confined to a narrow zone, the ebb and flow caused by the rise and fall of the tides keep the sand in motion and prevent deposition in the channel. However, as illustrated in **FIGURE 15.24**, a jetty may act as a dam against which the longshore current and beach drift deposit sand. At the same time, wave activity removes sand on the other side. Because the other side is not receiving any new sand, there is soon no beach at all.

Groins   To maintain or widen beaches that are losing sand, groins are sometimes constructed. A **groin** (*groin* = ground) is a barrier built at a right angle to the beach to trap sand that is moving parallel to the shore. Groins are usually constructed of large rocks but may also be composed of wood. These structures often do their job so effectively that the longshore current beyond the groin

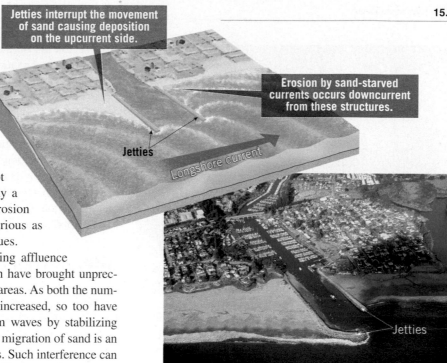

Jetties interrupt the movement of sand causing deposition on the upcurrent side.

Erosion by sand-starved currents occurs downcurrent from these structures.

Jetties

Longshore current

Jetties

FIGURE 15.24 Jetties These structures are built at entrances to rivers and harbors and are intended to prevent deposition in the navigation channel. The photo is an aerial view at Santa Cruz Harbor, California. (Photo by U.S. Army Corps of Engineers, Washington)

becomes sand-starved. As a result, the current erodes sand from the beach on the downstream side of the groin.

To offset this effect, property owners downstream from the structure may erect a groin on their property. In this manner, the number of groins multiplies, resulting in a *groin field* (**FIGURE 15.25**). An example of such proliferation is the shoreline of New Jersey, where hundreds of these structures have been built. Because it has been shown that groins often do not provide a satisfactory solution, they are no longer the preferred method of keeping beach erosion in check.

Breakwaters and Seawalls   Hard stabilization can also be built parallel to the shoreline. One such structure is a **breakwater**, the purpose of which is to protect boats from the force of large breaking waves by creating a quiet water zone near the shore. However, when this is done, the reduced wave activity along the shore behind the structure may allow sand to accumulate. If this happens, the boat anchorage will eventually fill with sand, while the

FIGURE 15.25 Groins These wall-like structures trap sand that is moving parallel to the shore. This series of groins is along the shoreline near Chichester, Sussex, England. (Photo by Sandy Stockwell/London Aerial Photo Library/Corbis)

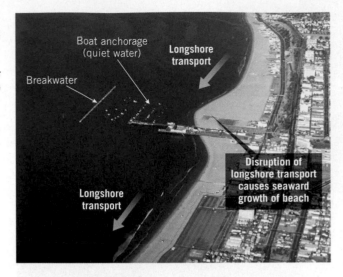

FIGURE 15.26
Breakwater Aerial view of a breakwater at Santa Monica, California. The structure appears as a line in the water behind which many boats are anchored. The construction of the breakwater disrupted longshore transport and caused the seaward growth of the beach. (Photo by University of Washington Libraries, Special Collections, John Shelton Collection, KC8275)

downstream beach erodes and retreats. At Santa Monica, California, where the building of a breakwater has created such a problem, the city uses a dredge to remove sand from the protected quiet water zone and deposit it farther downstream, where longshore currents continue to move the sand down the coast (**FIGURE 15.26**).

Another type of hard stabilization built parallel to the shore is a **seawall**, which is designed to armor the coast and defend property from the force of breaking waves. Waves expend much of their energy as they move across an open beach. Seawalls cut this process short by reflecting the force of unspent waves seaward. As a consequence, the beach to the seaward side of the seawall experiences significant erosion and may, in some instances, be eliminated entirely (**FIGURE 15.27**). Once the width of the beach is reduced, the seawall is subjected to even greater pounding by the waves. Eventually this battering

takes such a toll on the seawall that it will fail and a larger, more expensive structure must be built to take its place.

The wisdom of building temporary protective structures along shorelines is increasingly questioned. The opinions of many coastal scientists and engineers is that halting an eroding shoreline with protective structures benefits only a few and seriously degrades or destroys the natural beach and the value it holds for the majority. Protective structures divert the ocean's energy temporarily from private properties but usually refocus that energy on the adjacent beaches. Many structures interrupt the natural sand flow in coastal currents, robbing affected beaches of vital sand replacement.

Alternatives to Hard Stabilization

Armoring the coast with hard stabilization has several potential drawbacks, including the cost of the structure and the loss of sand on the beach. Alternatives to hard stabilization include beach nourishment and relocation.

Beach Nourishment   One approach to stabilizing shoreline sands without hard stabilization is **beach nourishment**. As the term implies, this practice involves adding large quantities of sand to the beach system (**FIGURE 15.28**). Extending beaches seaward makes buildings along the shoreline less vulnerable to destruction by storm waves and enhances recreational uses.

The process of beach nourishment is straightforward. Sand is pumped by dredges from offshore or trucked from inland locations. The "new" beach, however, will not be the same as the former beach. Because replenishment sand is from somewhere else, typically not another beach, it is new to the beach environment. The new sand is usually different

FIGURE 15.27 Seawall Seabright in northern New Jersey once had a broad, sandy beach. A seawall 5 to 6 meters (16 to 18 feet) high and 8 kilometers (5 miles) long was built to protect the town and the railroad that brought tourists to the beach. After the wall was built, the beach narrowed dramatically. (Photo by Rafael Macia/Science Source)

in size, shape, sorting, and composition. These differences pose problems in terms of erodibility and the kinds of life the new sand will support.

Beach nourishment is not a permanent solution to the problem of shrinking beaches. The same processes that removed the sand in the first place will eventually remove the replacement sand as well. Nevertheless, the number of nourishment projects has increased in recent years, and many beaches, especially along the Atlantic coast, have had their sand replenished many times. Virginia Beach, Virginia, has been nourished more than 50 times.

Beach nourishment is costly. For example, a modest project might involve 50,000 cubic yards of sand distributed across a half mile of shoreline. A good-sized dump truck holds about 10 cubic yards of sand. So this small project would require about 5000 dump-truck loads. Many projects extend for many miles. Nourishing beaches typically costs millions of dollars per mile.

Dredge

Offshore sand pouring onto the beach

FIGURE 15.28 Beach Nourishment If you visit a beach along the Atlantic coast, it is more and more likely that you will walk into the surf zone atop an artificial beach. In this image, an offshore dredge is pumping sand to a beach. (Photo by Michael Weber/Alamy)

Relocation Instead of building structures such as groins and seawalls to hold the beach in place or adding sand to replenish eroding beaches, another option is available. Many coastal scientists and planners are calling for a policy shift from defending and rebuilding beaches and coastal property in high-hazard areas to *relocating* storm-damaged buildings in those places and letting nature reclaim the beach. This approach is similar to an approach the federal government adopted for river floodplains following the devastating 1993 Mississippi River floods, in which vulnerable structures were abandoned and relocated on higher, safer ground.

Such proposals, of course, are controversial. People with significant nearshore investments want to rebuild and defend coastal developments from the erosional wrath of the sea. Others, however, argue that with sea level rising, the impact of coastal storms will get worse in the decades to come, and oft-damaged structures should be abandoned or relocated to improve personal safety and reduce costs. Such ideas will no doubt be the focus of much study and debate as states and communities evaluate and revise coastal land-use policies.

15.7 CONCEPT CHECKS

1 List three examples of *hard stabilization* and describe what each is intended to do. How does each affect sand distribution on a beach?

2 What are two alternatives to hard stabilization, and what potential problems are associated with each?

EYE ON EARTH

This structure along the eastern shore of Lake Michigan was built at the entrance to Port Shelton, Michigan.

QUESTION 1 *What is the purpose of the artificial structure pictured here?*

QUESTION 2 *What term is applied to structures such as this?*

QUESTION 3 *Suggest a reason there is a greater accumulation of sand on one side of the structure than the other.*

Michael Collier

15.8 | CONTRASTING AMERICA'S COASTS

Contrast the erosion problems faced along different parts of America's coasts. Distinguish between emergent and submergent coasts.

The shoreline along the Pacific coast of the United States is strikingly different from that of the Atlantic and Gulf coast regions. Some of the differences are related to plate tectonics. The West coast represents the leading edge of the North American plate; therefore, it experiences active uplift and deformation. By contrast, the East coast is a tectonically quiet region that is far from any active plate margin. Because of this basic geologic difference, the nature of shoreline erosion problems along America's opposite coasts is different.

Atlantic and Gulf Coasts

Much of the coastal development along the Atlantic and Gulf coasts has occurred on barrier islands. Typically, a barrier island consists of a wide beach that is backed by dunes and separated from the mainland by marshy lagoons. The broad expanses of sand and exposure to the ocean have made barrier islands exceedingly attractive sites for development. Unfortunately, development has grown more rapidly than has our understanding of barrier island dynamics.

Because barrier islands face the open ocean, they receive the full force of major storms that strike the coast. When a storm occurs, the barriers absorb the energy of the waves primarily through the movement of sand. **FIGURE 15.29**, which shows changes at Cape Hatteras National Seashore,

reinforces this point. This process and the dilemma that results have been described as follows:

> Waves may move sand from the beach to offshore areas or, conversely, into the dunes; they may erode the dunes, depositing sand onto the beach or carrying it out to sea; or they may carry sand from the beach and the dunes into the marshes behind the barrier, a process known as *overwash*. The common factor is movement. Just as a flexible reed may survive a wind that destroys an oak tree, so the barriers survive hurricanes and nor'easters not through unyielding strength but by giving before the storm.
>
> This picture changes when a barrier is developed for homes or a resort. Storm waves that previously rushed harmlessly through gaps between the dunes now encounter buildings and roadways. Moreover, since the dynamic nature of the barriers is readily perceived only during storms, homeowners tend to attribute damage to a particular storm, rather than to the basic mobility of coastal barriers. With their homes or investments at stake, local residents are more likely to seek to hold the sand in place and the waves at bay than to admit that development was improperly placed to begin with.[3]

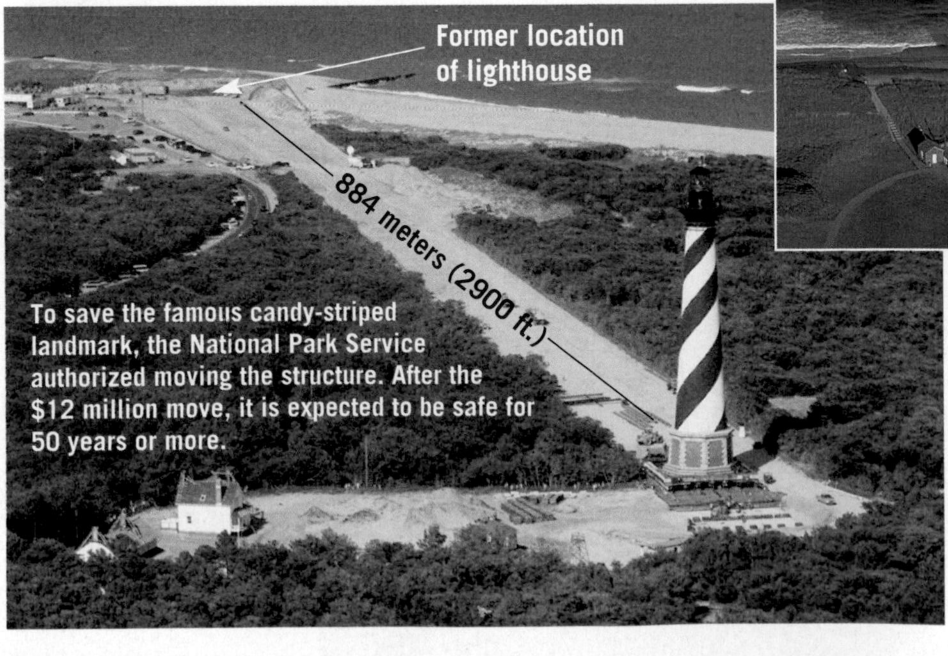

Pacific Coast

In contrast to the broad, gently sloping coastal plains of the Atlantic and Gulf coasts, much of the Pacific coast is characterized by relatively narrow beaches that are backed by steep cliffs and mountain ranges (**FIGURE 15.30**). Recall that America's western margin is a more rugged and tectonically active region than the eastern margin. Because uplift continues, the rise in sea level in the West is not so readily apparent. Nevertheless, like the shoreline erosion problems facing the East's barrier islands, West coast difficulties also stem largely from the alteration of natural systems by people.

A major problem facing the Pacific shoreline—particularly along southern California—is a significant narrowing of many beaches. The bulk of the sand on many of these beaches is supplied by rivers that transport it from the mountainous regions to the coast. Over the years, this natural flow of material to the coast has been interrupted by dams built for irrigation and flood control. The reservoirs effectively trap the sand that would otherwise nourish the beach environment. When the beaches were wider,

FIGURE 15.29 Relocating the Cape Hatteras Lighthouse Despite a number of efforts to protect this 21-story lighthouse, the nation's tallest, from being destroyed due to a receding shoreline, the structure finally had to be moved. (Photos by USGS National Center and AP Photo/Virginian–Pilot, DREW C. WILSON)

Various attempts to protect the lighthouse failed. They included building groins and beach nourishment. By 1999, when this photo was taken, the lighthouse was only 36 meters (120 ft.) from the water.

Former location of lighthouse

884 meters (2900 ft.)

To save the famous candy-striped landmark, the National Park Service authorized moving the structure. After the $12 million move, it is expected to be safe for 50 years or more.

---

[3]Frank Lowenstein, "Beaches or Bedrooms—The Choice as Sea Level Rises," Oceanus 28 (No. 3, Fall 1985): 22. Copyright © 1985 Woods Hole Oceanographic Institution. Reprinted with permission.

they protected the cliffs behind them from the force of storm waves. Now, however, the waves move across the narrowed beaches without losing much energy and cause more rapid erosion of the sea cliffs.

Although the retreat of the cliffs provides material to replace some of the sand impounded behind dams, it also endangers homes and roads built on the bluffs. In addition, development atop the cliffs aggravates the problem. Urbanization increases runoff, which, if not carefully controlled, can result in serious bluff erosion. Watering lawns and gardens adds significant quantities of water to the slope. This water percolates downward toward the base of the cliff, where it may emerge in small seeps. This action reduces the slope's stability and facilitates mass wasting.

Shoreline erosion along the Pacific coast varies considerably from one year to the next, largely because of the sporadic occurrence of storms. As a result, when the infrequent but serious episodes of erosion occur, the damage is often blamed on the unusual storms and not on coastal development or the sediment-trapping dams that may be great distances away. If, as predicted, sea level experiences a significant rise in the years to come because of global climate change, increased shoreline erosion and sea-cliff retreat should be expected along many parts of the Pacific coast.

Coastal Classification

The great variety of shorelines demonstrates their complexity. Indeed, to understand any particular coastal area, many factors must be considered, including rock types, size and direction of waves, frequency of storms, tidal range, and offshore topography. In addition, practically all coastal areas were affected by the worldwide rise in sea level that accompanied the melting of Ice Age glaciers following the Last Glacial Maximum. Finally, tectonic events that uplift or downdrop the land or change the volume of ocean basins must be taken into account. The large number of factors that influence coastal areas makes shoreline classification difficult.

FIGURE 15.30 **Pacific Coast** Wave refraction along the steep cliffs of the California coast south of Shelter Cove. (Photo by Michael Collier)

Many geologists classify coasts based on changes that have occurred with respect to sea level. This commonly used classification system divides coasts into two general categories: emergent and submergent. **Emergent coasts** develop either because an area experiences uplift or as a result of a drop in sea level. Conversely, **submergent coasts** are created when sea level rises or the land adjacent to the sea subsides.

EYE ON EARTH

This is an aerial view of a small portion of a coastal area in the United States.

QUESTION 1 *Is this an emergent coast or a submergent coast?*

QUESTION 2 *Provide an easily seen line of evidence to support your answer to Question 1.*

QUESTION 3 *Is the location more likely along the coast of North Carolina or California? Explain.*

Michael Collier

A Brief Tour of America's Coasts*

1 A small portion of the Cape Cod coast shows the entrance to Nauset Bay. Depending on the whims of recent storms and the strength of coastal currents, the opening into the bay may only be a few hundred feet wide. Tidal currents have created an underwater sandbar just inside the harbor.

10 Sea ice hugs Alaska's north slope near Barrow. The Arctic shore is locked in ice for much of the year.

7 This highway clings to the California Coast south of Big Sur. Uplift is occurring in this area near the boundary separating the Pacific and North American plates.

3 North Carolina's Outer Banks. Oregon Inlet Bridge connecting Bodie Island (foreground) and Hatteras Island. These narrow wisps of sand are part of an extensive barrier island system. The beach and dunes on the left face the Atlantic Ocean. On the right are the quieter waters of Pamlico Sound.

Prepared with the assistance of Michael Collier. All photos by Michael Collier.

Coasts are among Earth's most dynamic landscapes. Waves, tides, and currents continuously shape this interface between land and sea. Coasts may also exhibit the effects of mountain building, sea level changes, rivers, glaciers, and people. Here is a very small glimpse at the diversity and beauty of America's coasts.

5 The delta of the Mississippi River is a major feature in the Gulf of Mexico. This low-lying coastal zone is a maze of low soggy islands that are barely above sea level with a myriad of natural distributaries and artificial channels.

9 A small glacier flows down a steep mountain slope into the Cook Inlet southwest of Anchorage, Alaska. The total length of Alaska's irregular coastline is nearly 71,000 kilometers (44,000 miles).

6 Waves have aggressively attacked the coast north of La Push Harbor, Washington. In places, remnants of the cliff remain as sea stacks.

2 Spruce and fir cover Turtle Island, part of Maine's Acadia National Park. This region was sculpted by Ice Age glaciers, then flooded when sea level rose as the ice sheets melted.

4 Mobjack Bay near Gloucester, Virginia, is part of the much larger Chesapeake Bay. These bays, called estuaries, are actually river valleys that were drowned when sea level rose at the end of the Ice Age.

8 Cliffs and waterfalls along the north coast of Hawaii's Big Island. This portion of the island receives abundant rainfall. There are still active volcanoes elsewhere on the island.

11 Padre Island is a barrier island that protects the coast of southern Texas from storms. Winds off the Gulf of Mexico create dunes, which offer a temporary footing for vegetation that is occasionally stripped away by hurricanes.

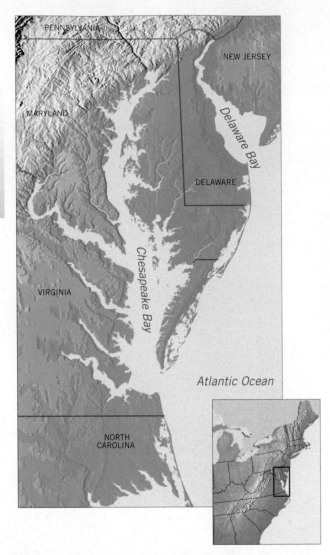

SmartFigure 15.31 East Coast Estuaries The lower portions of many river valleys were flooded by the rise in sea level that followed the end of the Quaternary Ice Age, creating large estuaries such as the Chesapeake and Delaware Bays.

seven episodes of uplift. The ever-persistent sea is now cutting a new platform at the base of the cliff. If uplift follows, it, too, will become an elevated marine terrace.

Other examples of emergent shores include regions that were once buried beneath great ice sheets. When glaciers were present, their weight depressed the crust, and when the ice melted, the crust began gradually to spring back. Consequently, prehistoric shoreline features today are found high above sea level. The Hudson Bay region of Canada is one such area; portions of it are still rising at a rate of more than 1 centimeter (0.4 inch) per year.

Submergent Coasts In contrast to the preceding examples, other coastal areas show definite signs of submergence. Shorelines that have been submerged in the relatively recent past are often highly irregular because the sea typically floods the lower reaches of river valleys flowing into the ocean. The ridges separating the valleys, however, remain above sea level and project into the sea as headlands. These drowned river mouths, which are called **estuaries**, characterize many coasts today. Along the Atlantic coastline, Chesapeake and Delaware Bays are examples of estuaries created by submergence (**FIGURE 15.31**). The picturesque coast of Maine, particularly in the vicinity of Acadia National Park, is another excellent example of an area that was flooded by the postglacial rise in sea level and transformed into a highly irregular submerged coastline.

Keep in mind that most coasts have a complicated geologic history. With respect to sea level, many coasts have at various times emerged and then submerged again. Each time, they retain some of the features created during the previous event.

15.8 CONCEPT CHECKS

1 Briefly describe what happens when storm waves strike an undeveloped barrier island.

2 How might building a dam on a river that flows to the sea affect a beach?

3 What is an observable feature that would lead you to classify a coastal area as emergent?

4 Are estuaries associated with submergent or emergent coasts? Explain.

Emergent Coasts In some areas, the coast is clearly emergent because rising land or falling water levels expose wave-cut cliffs and marine terraces above sea level. Excellent examples include portions of coastal California where uplift has occurred in the recent geologic past. The elevated marine terrace in Figure 15.19 illustrates this situation. In the case of the Palos Verdes Hills, south of Los Angeles, California, seven different terrace levels exist, indicating at least

---

15.9 | TIDES Explain the cause of tides, their monthly cycles, and patterns. Describe the horizontal flow of water that accompanies the rise and fall of tides.

Tides are daily changes in the elevation of the ocean surface. Their rhythmic rise and fall along coastlines have been known since antiquity. Other than waves, they are the easiest ocean movements to observe (**FIGURE 15.32**).

Although known for centuries, tides were not explained satisfactorily until Sir Isaac Newton applied the law of gravitation to them. Newton showed that there is a mutual attractive force between two bodies, as between Earth and the Moon. Because both the atmosphere and the ocean are fluids

and are free to move, both are deformed by this force. Hence, ocean tides result from the gravitational attraction exerted upon Earth by the Moon and, to a lesser extent, by the Sun.

Causes of Tides

To illustrate how tides are produced, consider an idealized case in which Earth is a rotating sphere covered to a uniform depth with water (**FIGURE 15.33**). Furthermore, ignore the effect of the Sun for now. It is easy to see how the Moon's gravitational force can cause the water to bulge on the side of Earth nearer the Moon. In addition, however, an equally large tidal bulge is produced on the side of Earth directly *opposite* the Moon.

Both tidal bulges are caused, as Newton discovered, by the pull of gravity. Gravity is inversely proportional to the square of the distance between two objects, meaning simply that it quickly weakens with distance. In this case, the two objects are the Moon and Earth. Because the force of gravity decreases with distance, the Moon's gravitational pull on Earth is slightly greater on the near side of Earth than on the far side. The result of this differential pulling is to stretch (elongate) the "solid" Earth very slightly. In contrast, the world ocean, which is mobile, is deformed quite dramatically by this effect, producing the two opposing tidal bulges.

Because the position of the Moon changes only moderately in a single day, the tidal bulges remain in place while Earth rotates "through" them. For this reason, if you stand on the seashore for 24 hours, Earth will rotate you through alternating areas of higher and lower water. As you are carried into each tidal bulge, the tide rises, and as you are carried into the intervening troughs between the tidal bulges, the tide falls. Therefore, most places on Earth experience two high tides and two low tides each day.

In addition, the tidal bulges migrate as the Moon revolves around Earth about every 29 days. As a result, the tides, like the time of moonrise, shift about 50 minutes later each day. In essence, the tidal bulges exist in fixed positions relative to the Moon, which slowly moves progressively eastward as it orbits Earth. After 29 days the cycle is complete, and a new one begins.

Many locations may show an inequality between the high tides during a given day. Depending on the Moon's position, the tidal bulges may be inclined to the equator, as in Figure 15.33. This figure illustrates that one high tide experienced by an observer in the Northern Hemisphere is considerably higher than the high tide half a day later. In contrast, a Southern Hemisphere observer would experience the opposite effect.

Monthly Tidal Cycle

The primary body that influences the tides is the Moon, which makes one complete revolution around Earth every 29 days. The Sun, however, also influences the tides. It is far larger than the Moon, but because it is much farther away, its effect is considerably less. In fact, the Sun's tide-generating effect is only about 46 percent that of the Moon's.

Near the times of new and full moons, the Sun and Moon are aligned, and their forces are added together

High tide

Low tide

Tidal flat

NEW BRUNSWICK Minas Basin

MAINE

Bay of Fundy NOVA SCOTIA ATLANTIC OCEAN

FIGURE 15.32 Bay of Fundy Tides High tide and low tide on Nova Scotia's Minas Basin in the Bay of Fundy. Tidal flats are exposed during low tide. (Photos courtesy of left: Ray Coleman/Science Source; right: Jeffrey Greenberg/Science Source)

(**FIGURE 15.34A**). The combined gravity of these two tide-producing bodies causes larger tidal bulges (higher high tides) and larger tidal troughs (lower low tides), producing a large tidal range. These are called the **spring** (*springen* = to rise up) **tides**, which have no connection with the spring season but occur twice a month, during the time when the Earth–Moon–Sun system is aligned. Conversely, at about the time of the first and third quarters of the Moon, the gravitational forces of the Moon and Sun act on Earth at right angles, and each partially offsets the influence of the other (**FIGURE 15.34B**). As a result, the daily tidal range is less. These are called **neap** (*nep* = scarcely or barely touching) **tides**, and they also occur twice each month. Each month, then, there are two spring tides and two neap tides, each about 1 week apart.

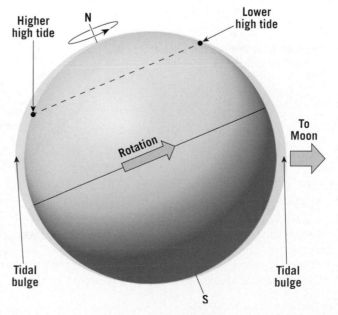

Higher high tide

N

Lower high tide

Rotation

To Moon

Tidal bulge

Tidal bulge

S

FIGURE 15.33 Idealized Tidal Bulges Caused by the Moon If Earth were covered to a uniform depth with water, there would be two tidal bulges: one on the side of Earth facing the Moon (right) and the other on the opposite side of Earth (left). Depending on the Moon's position, tidal bulges may be inclined relative to Earth's equator. In this situation, Earth's rotation causes an observer to experience two unequal high tides in a day.

FIGURE 15.34 Spring and Neap Tides Earth–Moon–Sun positions influence the tides.

A. Spring Tide When the Moon is in the full or new position, the tidal bulges created by the Sun and Moon are aligned and there is a large tidal range.

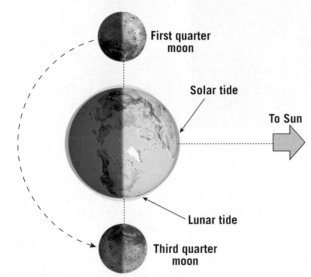

B. Neap Tide When the Moon is in the first-or third-quarter position, the tidal bulges produced by the Moon are at right angles to the bulges created by the Sun and the tidal range is smaller.

Tidal Patterns

The basic causes and types of tides have been explained. Keep in mind, however, that these theoretical considerations cannot be used to predict either the height or the time of actual tides at a particular place. Many factors—including the shape of the coastline, the configuration of ocean basins, the Coriolis effect, and water depth—greatly influence the tides. Consequently, tides at various locations respond differently to the tide-producing forces. Thus, the nature of the tide at any coastal location can be determined most accurately by actual observation. The predictions in tidal tables and tidal data on nautical charts are based on such observations.

Three main tidal patterns exist worldwide. A **diurnal** (*diurnal* = daily) **tidal pattern** is characterized by a single high tide and a single low tide each tidal day (**FIGURE 15.35**). Tides of this type occur along the northern shore of the Gulf of Mexico, among other locations. A **semidiurnal** (*semi* = twice, *diurnal* = daily) **tidal pattern** exhibits two high tides and two low tides each tidal day, with the two highs about the same height and the two lows about the same height (see Figure 15.35). This type of tidal pattern is common along the Atlantic coast of the United States. A **mixed tidal pattern** is similar to a semidiurnal pattern except that it is characterized by a large inequality in high water heights, low water heights, or both (see Figure 15.34). In this case, there are usually two high and two low tides each day, with high tides of different heights and low tides of different heights. Such tides are prevalent along the Pacific coast of the United States and in many other parts of the world.

Tidal Currents

Tidal current is the term used to describe the *horizontal* flow of water accompanying the rise and fall of the tides. These water movements induced by tidal forces can be important in some coastal areas. Tidal currents that advance into the coastal zone as the tide rises are called *flood currents*. As the tide falls, seaward-moving water generates *ebb currents*. Periods of little or no current, called *slack water*, separate flood and ebb. The areas affected by these alternating tidal currents are called **tidal flats** (see Figure 15.32). Depending on the nature of the coastal zone, tidal flats vary

SmartFigure 15.35 Tidal Patterns A diurnal tidal pattern (lower right) shows one high tide and one low tide each tidal day. A semidiurnal pattern (upper right) shows two high tides and two low tides of approximately equal heights during each tidal day. A mixed tidal pattern (left) shows two high tides and two low tides of unequal heights during each tidal day.

Two high tides and two low tides of unequal heights during each tidal day

MIXED TIDAL PATTERN

Two high tides and two low tides of approximately equal heights during each tidal day

SEMIDIURNAL TIDAL PATTERN

One high tide and one low tide each tidal day

DIURNAL TIDAL PATTERN

Diurnal Semidiurnal
Mixed 2.7 Spring tidal range (m)

from narrow strips seaward of the beach to zones that may extend for several kilometers.

Although tidal currents are not important in the open sea, they can be rapid in bays, river estuaries, straits, and other narrow places. Off the coast of Brittany in France, for example, tidal currents that accompany a high tide of 12 meters (40 feet) may attain a speed of 20 kilometers (12 miles) per hour. Tidal currents are not generally considered to be major agents of erosion and sediment transport, but notable exceptions occur where tides move through narrow inlets. Here they scour the narrow entrances to many harbors that would otherwise be blocked.

Sometimes deposits called **tidal deltas** are created by tidal currents (**FIGURE 15.36**). They may develop either as *flood deltas* landward of an inlet or as *ebb deltas* on the seaward side of an inlet. Because wave activity and longshore currents are reduced on the sheltered landward side, flood deltas are more common and are actually more prominent (see Figure 15.21A). They form after the tidal current moves rapidly through an inlet. As the current emerges into more open waters from the narrow passage, it slows and deposits its load of sediment.

Because this tidal delta has developed on the landward side of the inlet, it is called a *flood delta*.

Tidal flats

Barrier island

Lagoon

FIGURE 15.36 Tidal Deltas As a rapidly moving tidal current (flood current) moves through a barrier island's inlet into the quiet waters of the lagoon, the current slows and deposits sediment, creating a tidal delta. Because this tidal delta has developed on the landward side of the inlet, it is called a *flood delta*. Such a tidal delta is shown in Figure 15.21A.

15.9 CONCEPT CHECKS

1 Explain why an observer can experience two unequal high tides during one day.

2 Distinguish between *neap tides* and *spring tides*.

3 How is a mixed tidal pattern different from a semidiurnal tidal pattern?

4 Contrast *flood current* and *ebb current*.

15 CONCEPTS IN REVIEW | The Dynamic Ocean

15.1 THE OCEAN'S SURFACE CIRCULATION

Discuss the factors that create and influence ocean currents and describe the affect ocean currents have on climate.

KEY TERMS: gyre, Coriolis effect

- The ocean's surface currents follow the general pattern of the world's major wind belts. Surface currents are parts of huge, slowly moving loops of water called *gyres* that are centered in the subtropics of each ocean basin. The positions of the continents and the Coriolis effect also influence the movement of ocean water within gyres. Because of the Coriolis effect, subtropical gyres move clockwise in the Northern Hemisphere and counterclockwise in the Southern Hemisphere. Generally, four main currents comprise each subtropical gyre.
- Ocean currents can have a significant effect on climate. Poleward-moving *warm* ocean currents moderate winter temperatures in the middle latitudes. Cold currents exert their greatest influence during summer in middle latitudes and year-round in the tropics. In addition to cooler temperatures, cold currents are associated with greater fog frequency and drought.

Q Assume that arrow A represents prevailing winds in a Northern Hemisphere ocean. Which arrow, A, B, or C, best represents the surface-ocean current in this region? Explain.

15.2 UPWELLING AND DEEP-OCEAN CIRCULATION

Explain the processes that produce coastal upwelling and the ocean's deep circulation.

KEY TERMS: upwelling, thermohaline circulation

- Upwelling, the rising of colder water from deeper layers, is a wind-induced movement that brings cold, nutrient-rich water to the surface. Coastal upwelling is most characteristic along the west coasts of continents.
- In contrast to surface currents, deep-ocean circulation is governed by gravity and driven by density differences. The two factors that are most significant in creating a dense mass of water are temperature and salinity, so the movement of deep-ocean water is often termed thermohaline circulation. Most water involved in thermohaline circulation begins in high latitudes at the surface, when the salinity of the cold water increases as a result of sea ice formation. This dense water sinks, initiating deep-ocean currents.

15.3 THE SHORELINE: A DYNAMIC INTERFACE

Explain why the shoreline is considered a dynamic interface and identify the basic parts of the coastal zone.

KEY TERMS: shoreline, shore, coast, coastline, foreshore, backshore, nearshore zone, offshore zone, beach, berm, beach face

- The shore is the area extending between the lowest tide level and the highest elevation on land that is affected by storm waves. The coast extends inland from the shore as far as ocean-related features can be found. The shore is divided into the foreshore and backshore. Seaward of the foreshore are the nearshore and offshore zones.
- A beach is an accumulation of sediment found along the landward margin of the ocean or a lake. Among its parts are one or more berms and the beach face. Beaches are composed of whatever material is locally abundant and should be thought of as material in transit along the shore.

Q Assume that you are the photographer who took this photo and that the photo was taken at high tide. On which part of the shore are you standing?

Shutterstock

15.4 OCEAN WAVES

List and discuss the factors that influence the height, length, and period of a wave and describe the motion of water within a wave.

KEY TERMS: wave height, wavelength, wave period, fetch, circular orbital motion, wave base, surf

- Waves are moving energy, and most ocean waves are initiated by wind. The three factors that influence the height, wavelength, and period of a wave are (1) wind speed, (2) length of time the wind has blown, and (3) fetch, the distance that the wind has traveled across open water. Once waves leave a storm area, they are termed *swells*, which are symmetrical, longer-wavelength waves.
- As waves travel, water particles transmit energy by circular orbital motion, which extends to a depth equal to one-half the wavelength (the wave base). When a wave enters water that is shallower than the wave base, it slows down, which allows waves farther from shore to catch up. As a result, wavelength decreases and wave height increases. Eventually the wave breaks, creating turbulent surf in which water rushes toward the shore.

15.5 SHORELINE PROCESSES

Describe how waves erode and move sediment along the shore.

KEY TERMS: abrasion, wave refraction, beach drift, longshore current, rip current

- Wind-generated waves provide most of the energy that modifies shorelines. Each time a wave hits, it can impart tremendous force. The impact of waves, coupled with abrasion from the grinding action of rock particles, erodes material exposed along the shoreline.
- Wave refraction is a consequence of a wave encountering shallower water as it approaches shore. The shallowest part of the wave (closest to shore) slows the most, allowing the faster part (still in deeper water) to catch up. This modifies a wave's trajectory so that the wave front becomes almost parallel to the shore by the time it hits. Wave refraction concentrates impacting energy on headlands and dissipates that energy in bays, which become sites of sediment accumulation.
- Beach drift describes the movement of sediment in a zigzag pattern along a beach face. The swash of incoming waves pushes the sediment up the beach at an oblique angle, but the backwash transports it directly downhill. Net movement along the beach can be many meters per day. Longshore currents are a similar phenomenon in the surf zone, capable of transporting very large quantities of sediment parallel to a shoreline.

Q What process is causing wave energy to be concentrated on the headland? Predict how this area will appear in the future.

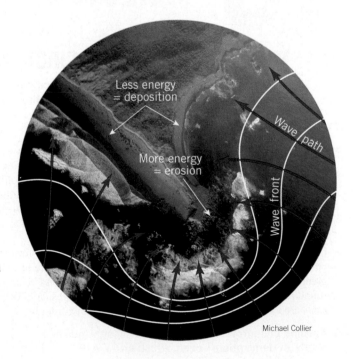

Michael Collier

15.6 SHORELINE FEATURES

Describe the features typically created by wave erosion and those resulting from sediment deposited by longshore transport processes.

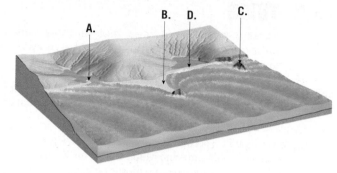

KEY TERMS: wave-cut cliff, wave-cut platform, marine terrace, sea arch, sea stack, spit, baymouth bar, tombolo, barrier island

■ Erosional features include wave-cut cliffs (which originate from the cutting action of the surf against the base of coastal land), wave-cut platforms (relatively flat, bench-like surfaces left behind by receding cliffs), and marine terraces (uplifted wave-cut platforms). Erosional features also include sea arches (formed when a headland is eroded and two sea caves from opposite sides unite) and sea stacks (formed when the roof of a sea arch collapses).

■ Some of the depositional features that form when sediment is moved by beach drift and longshore currents are spits (elongated ridges of sand that project from the land into the mouth of an adjacent bay), baymouth bars (sandbars that completely cross a bay), and tombolos (ridges of sand that connect an island to the mainland or to another island). Along the Atlantic and Gulf coastal plains, the coastal region is characterized by offshore barrier islands, which are low ridges of sand that parallel the coast.

Q Identify the lettered features in this diagram.

15.7 STABILIZING THE SHORE

Summarize the ways in which people deal with shoreline erosion problems.

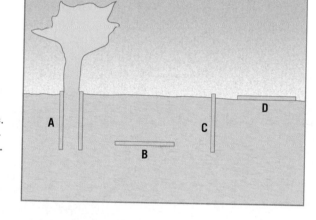

KEY TERMS: hard stabilization, jetty, groin, breakwater, seawall, beach nourishment

■ Hard stabilization is a term that refers to any structures built along the coastline to prevent movement of sand. Jetties project out from the coast, with the goal of keeping inlets open. Groins are also oriented perpendicular to the coast, but with the goal of slowing beach erosion by longshore currents. Breakwaters are parallel to the coast but located some distance offshore. Their goal is to blunt the force of incoming ocean waves, often to protect boats. Like breakwaters, seawalls are parallel to the coast, but they are built on the shoreline itself. Often the installation of hard stabilization results in increased erosion elsewhere.

■ Beach nourishment is an expensive alternative to hard stabilization. Sand is pumped onto a beach from some other area, temporarily replenishing the sediment supply. Another possibility is relocating buildings away from high-risk areas and leaving the beach to be shaped by natural processes.

Q Based on their position and orientation, identify the four kinds of hard stabilization illustrated in this diagram.

15.8 CONTRASTING AMERICA'S COASTS

Contrast the erosion problems faced along different parts of America's coasts. Distinguish between emergent and submergent coasts.

KEY TERMS: emergent coast, submergent coast, estuary

■ The Atlantic and Gulf coasts of the United States are markedly different from the Pacific coast. The Atlantic and Gulf coasts are lined in many places by barrier islands—dynamic expanses of sand that see a lot of change during storm events. Many of these low and narrow islands have also been prime sites for real estate development.

■ The Pacific coast's big issue is the thinning of beaches due to sediment starvation. Rivers that drain to the coast (bringing it sand) have been dammed, resulting in reservoirs that trap sand before it can make it to the coast. Thinner beaches offer less resistance to incoming waves, often leading to erosion of bluffs behind the beach.

■ Coasts may be classified by their changes relative to sea level. Emergent coasts are sites of either land uplift or sea-level fall. Marine terraces are features of emergent coasts. Submergent coasts are sites of land subsidence or sea-level rise. One characteristic of submergent coasts is drowned river valleys called estuaries.

Q What term is applied to the masses of rock protruding from the water in this photo? How did they form? Is the location more likely along the Gulf coast or the coast of California? Explain.

Michael Collier

15.9 TIDES

Explain the cause of tides, their monthly cycles, and patterns. Describe the horizontal flow of water that accompanies the rise and fall of tides.

KEY TERMS: tide, spring tide, neap tide, diurnal tidal pattern, semidiurnal tidal pattern, mixed tidal pattern, tidal current, tidal flat, tidal delta

- Tides are daily changes in ocean surface elevation. They are caused by gravitational pull on ocean water by the Moon and, to a lesser extent, the Sun. When the Sun, Earth, and Moon all line up about every 2 weeks (full moon or new moon), the tides are most exaggerated. When a quarter moon is in the sky, the Moon is pulling on Earth's water at a right angle relative to the Sun, and the daily tidal range is minimized as the two forces partially counteract one another.

- Tides are strongly influenced by local conditions, including the shape of the local coastline and the depth of the ocean basin. Tidal patterns may be diurnal (one high tide per day), semidiurnal (two high tides per day), or mixed (similar to semidiurnal but with significant inequality between high tides).

- A flood current is the landward movement of water during the shift between low tide and high tide. When high tide transitions to low tide again, the movement of water away from the land is an ebb current. Ebb currents may expose tidal flats to the air. If a tide passes through an inlet, the current may carry sediment that gets deposited as a tidal delta.

Q Would spring tides and neap tides occur on an Earth-like planet that had no moon? Explain.

GIVE IT SOME THOUGHT

1. In this chapter you learned that global winds are the force that drive surface ocean currents. A glance at the accompanying map, however, shows a surface current that does not exactly coincide with the prevailing wind. Provide an explanation.

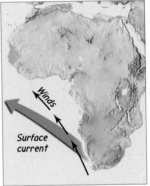

2. During a visit to the beach, you get in a small rubber raft and paddle out *beyond* the surf zone. Tiring, you stop and take a rest. Describe the movement of your raft during your rest. How does this movement differ, if at all, from what you would have experienced if you had stopped paddling while *in* the surf zone?

3. You and a friend set up an umbrella and chairs at a beach. Your friend then goes into the surf zone to play Frisbee with another person. Several minutes later your friend looks back toward the beach and is surprised to see that she is no longer near where the umbrella and chairs were set up. Although she is still in the surf zone, she is 30 or 40 yards away from where she started. How would you explain to your friend why she moved along the shore?

4. Examine the accompanying aerial photo that shows a portion of the New Jersey shoreline. What term is applied to the wall-like structures that extend into the water? What is their purpose? In what direction are beach drift and longshore currents moving sand? Is sand moving toward the top or toward the bottom of the photo?

5. A friend wants to purchase a vacation home on a barrier island. If consulted, what advice would you give your friend?

6. The force of gravity plays a critical role in creating ocean tides. The more massive an object, the stronger the pull of gravity. Explain why the Sun's influence is only half that of the Moon, even though the Sun is much more massive than the Moon.

7. This photo shows a portion of the Maine coast. The brown muddy area in the foreground is influenced by tidal currents. What term is applied to this muddy area? Name the type of tidal current this area will experience in the hours to come.

EXAMINING THE **EARTH SYSTEM**

1. Palm trees in Scotland? Yes, in the 1850s and 1860s, amateur gardeners planted palm trees on the western shore of Scotland. The latitude here is 57° north, about the same as the northern portion of Labrador across the Atlantic in Canada. Surprisingly, these exotic plants flourished. Suggest a possible explanation for how these palms can survive at such a high latitude.

South West Images Scotland / Alamy

2. If wastage (melting and calving) of the Greenland Ice Sheet were to dramatically increase, how would the salinity of the adjacent North Atlantic be affected? How might this influence thermohaline circulation?

Paul Souders/Corbis

3. In this chapter, the shoreline was described as a "dynamic interface." What is an interface? List and briefly describe some other interfaces in the Earth system. You need not confine yourself to examples from this chapter.

4. This dam in the San Gabriel Mountains near Los Angeles was built on a river that flows into the Pacific Ocean. What impact might this artificial structure have on coastal beaches?

Dam

Michael Collier

MasteringGeology™

Looking for additional review and test prep materials? Visit the Self Study area in **www.masteringgeology.com** to find practice quizzes, study tools, and multimedia that will aid in your understanding of this chapter's content. In **MasteringGeology™** you will find:

- GEODe: Earth Science: An interactive visual walkthrough of key concepts
- Geoscience Animation Library: More than 100 animations illuminating many difficult-to-understand Earth science concepts

- In The News RSS Feeds: Current Earth science events and news articles are pulled into the site with assessment
- Pearson eText
- Optional Self Study Quizzes
- Web Links
- Glossary
- Flashcards

16

The Atmosphere: Composition, Structure, and Temperature

BENGOA SOLAR

FOCUS ON CONCEPTS

Each statement represents the primary **LEARNING OBJECTIVE** for the corresponding major heading within the chapter. After you complete the chapter, you should be able to:

16.1 Distinguish between weather and climate and name the basic elements of weather and climate.

16.2 List the major gases composing Earth's atmosphere and identify the components that are most important to understanding weather and climate.

16.3 Interpret a graph that shows changes in air pressure from Earth's surface to the top of the atmosphere. Sketch and label a graph that shows atmospheric layers based on temperature.

16.4 Explain what causes the Sun angle and length of daylight to change during the year and describe how these changes produce the seasons.

16.5 Distinguish between heat and temperature. List and describe the three mechanisms of heat transfer.

16.6 Sketch and label a diagram that shows the paths taken by incoming solar radiation. Summarize the greenhouse effect.

16.7 Calculate five commonly used types of temperature data and interpret a map that depicts temperature data using isotherms.

16.8 Discuss the principal controls of temperature and use examples to describe their effects.

16.9 Interpret the patterns depicted on world maps of January and July temperatures.

This power plant in Andalucia, Spain, produces clean thermo-electric power from the Sun. Solar radiation provides practically all the energy that heats Earth's surface and atmosphere. (Photo by Kevin Foy/Alamy)

Earth's atmosphere is unique. No other planet in our solar system has an atmosphere with the exact mixture of gases or the heat and moisture conditions necessary to sustain life as we know it. The gases that make up Earth's atmosphere and the controls to which they are subject are vital to our existence. In this chapter we begin our examination of the ocean of air in which we all must live. We try to answer a number of basic questions: What is the composition of the atmosphere? Where does the atmosphere end, and where does outer space begin? What causes the seasons? How is air heated? What factors control temperature variations around the globe?

16.1 | FOCUS ON THE ATMOSPHERE Distinguish between weather and climate and name the basic elements of weather and climate.

Weather influences our everyday activities, our jobs, and our health and comfort. Many of us pay little attention to the weather unless we are inconvenienced by it or when it adds to our enjoyment outdoors. Nevertheless, there are few other aspects of our physical environment that affect our lives more than the phenomena we collectively call the weather.

Weather in the United States

The United States occupies an area that stretches from the tropics of Hawaii to beyond the Arctic Circle in Alaska. It has thousands of miles of coastline and extensive regions that are far from the influence of the ocean. Some landscapes are mountainous, and others are dominated by plains. It is a place where Pacific storms strike the West coast, and the East is sometimes influenced by events in the Atlantic and the Gulf of Mexico. Those in the center of the country commonly experience weather events triggered when frigid southward-bound Canadian air masses clash with northward-moving ones from the Gulf of Mexico.

Stories about weather are a routine part of the daily news. Articles and items about the effects of heat, cold, floods, drought, fog, snow, ice, and strong winds are commonplace. Of course, storms of all kinds are frequently front-page news (**FIGURE 16.1**). Beyond its direct impact on the lives of individuals, the weather has a strong effect on the world economy, influencing agriculture, energy use, water resources, transportation, and industry.

Weather clearly influences our lives a great deal. Yet, it is important to realize that people influence the atmosphere and its behavior as well (**FIGURE 16.2**). There are, and will continue to be, significant political and scientific decisions that must be made involving these impacts. Important examples are air pollution control and the effects of human activities on global climate and the atmosphere's protective ozone layer. So there is a need for increased awareness and understanding of our atmosphere and its behavior.

FIGURE 16.1 Memorable Weather Events Few aspects of our physical environment influence our daily lives more than the weather. The image on the right shows hundreds of cars stranded on Chicago's Lake Shore Drive during a blizzard on February 2, 2011. (Kiichiro Sato/AP Photo) The top photo shows the aftermath of the tornado that devastated Moore, Oklahoma, on May 20, 2013. (Photo by Julie Dermansky/Corbis)

Weather and Climate

Acted on by the combined effects of Earth's motions and energy from the Sun, our planet's formless and invisible envelope of air reacts by producing an infinite variety of weather, which in turn creates the basic pattern of global climates. Although not identical, weather and climate have much in common.

Weather is constantly changing, sometimes from hour to hour and at other times from day to day. It is a term that refers to the state of the atmosphere at a given time and place. Whereas changes in the weather are continuous and sometimes seemingly erratic, it is nevertheless possible to arrive at a generalization of these variations. Such a description of aggregate weather conditions is termed **climate**. It is based on observations that have been accumulated over many years. Climate is often defined simply as "average weather," but this is an inadequate definition. To more accurately portray the character of an area, variations and extremes must also be included, as well as the probabilities that such departures will take place. For example, farmers need to know the average rainfall during the growing season, and they also need to know the frequency of extremely wet and extremely dry years. Thus, climate is the sum of all statistical weather information that helps describe a place or region.

Suppose you were planning a vacation trip to an unfamiliar place. You would probably want to know what kind of weather to expect. Such information would help as you selected clothes to pack and could influence decisions regarding activities you might engage in during your stay. Unfortunately, weather forecasts that go beyond a few days are not very dependable. Therefore, you might ask someone who is familiar with the area about what kind of weather to expect. "Are thunderstorms common?" "Does it get cold at night?" "Are the afternoons sunny?" What you are seeking is information about the climate, the conditions that are typical for that place. Another useful source of such information is the great variety of climate tables, maps, and graphs that are available. For example, the graph in **FIGURE 16.3** shows average daily high and low temperatures for each month, as well as extremes for New York City.

Such information could no doubt help as you planned your trip. But it is important to realize that *climate data cannot predict the weather*. Although a place may usually

FIGURE 16.2 People Influence the Atmosphere Smoke bellows from a coal-fired electricity-generating plant in New Delhi, India, in June 2008. (Gurinder Osan/AP Photo)

EYE ON EARTH

This is a scene on a summer day in a portion of southern Utah called Sinbad Country. Hebes Mountain is in the center of the image. (Photo by Michael Collier)

QUESTION 1 *Write two brief statements about this place—one that clearly relates to weather and another that might be part of a description of climate.*

QUESTION 2 *Explain the reasoning associated with each statement.*

FIGURE 16.3 Graphs Can Display Climate Data This graph shows daily temperature data for New York City. In addition to the average daily maximum and minimum temperatures for each month, extremes are also shown. As this graph shows, there can be significant departures from the average.

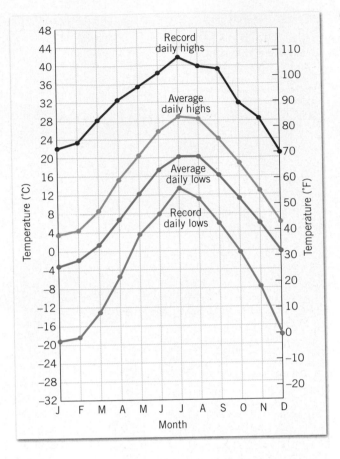

(climatically) be warm, sunny, and dry during the time of your planned vacation, you may actually experience cool, overcast, and rainy weather. There is a well-known saying that summarizes this idea: "Climate is what you expect, but weather is what you get."

The nature of weather and climate is expressed in terms of the same basic **elements**—quantities or properties that are measured regularly. The most important elements are (1) air temperature, (2) humidity, (3) type and amount of cloudiness, (4) type and amount of precipitation, (5) air pressure, and (6) the speed and direction of the wind. These elements are the major variables from which weather patterns and climate types are deciphered. Although you will study these elements separately at first, keep in mind that they are very much interrelated. A change in any one of the elements will often bring about changes in the others.

16.1 CONCEPT CHECKS

1 Distinguish between weather and climate.

2 Write two brief statements about your current location: one that relates to weather and one that relates to climate.

3 What is an *element*?

4 List the basic elements of weather and climate

16.2 | COMPOSITION OF THE ATMOSPHERE

List the major gases composing Earth's atmosphere and identify the components that are most important to understanding weather and climate.

Sometimes the term *air* is used as if it were a specific gas, but it is not. Rather, **air** is a *mixture* of many discrete gases, each with its own physical properties, in which varying quantities of tiny solid and liquid particles are suspended.

Major Components

The composition of air is not constant; it varies from time to time and from place to place. If the water vapor, dust, and other variable components were removed from the atmosphere, we would find that its makeup is very stable worldwide up to an altitude of about 80 kilometers (50 miles).

As you can see in **FIGURE 16.4**, two gases—nitrogen and oxygen—make up 99 percent of the volume of clean, dry air. Although these gases are the most plentiful components of air and are of great significance to life on Earth, they are of minor importance in affecting weather phenomena. The remaining 1 percent of dry air is mostly the inert gas argon (0.93 percent) plus tiny quantities of a number of other gases.

Carbon Dioxide (CO_2)

Carbon dioxide, although present in only minute amounts (0.0397 percent, or 397 parts per million [ppm]), is nevertheless an important constituent of air. Carbon dioxide is of great interest to meteorologists because it is an efficient absorber of energy emitted by Earth and thus influences the heating of the atmosphere. Although the proportion of carbon dioxide in

FIGURE 16.4 Composition of the Atmosphere This graph shows the proportional volume of gases composing dry air. Nitrogen and oxygen clearly dominate.

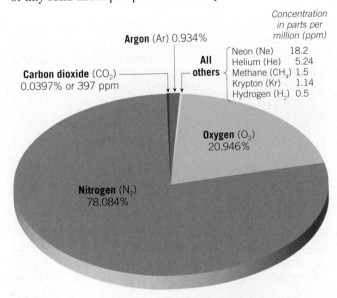

the atmosphere is relatively uniform, its percentage has been rising steadily for 200 years. **FIGURE 16.5** is a graph that shows the growth in atmospheric CO_2 since 1958. Much of this rise is attributed to the burning of ever-increasing quantities of fossil fuels, such as coal and oil. Some of this additional carbon dioxide is absorbed by the ocean or is used by plants, but more than 40 percent remains in the air. Estimates project that by sometime in the second half of the twenty-first century, CO_2 levels will be twice as high as the pre-industrial level.

Most atmospheric scientists agree that increased carbon dioxide concentrations have contributed to a warming of Earth's atmosphere over the past several decades and will continue to do so in the decades to come. The magnitude of such temperature changes is uncertain and depends partly on the quantities of CO_2 contributed by human activities in the years ahead. The role of carbon dioxide in the atmosphere and its possible effects on climate are examined in Chapter 20.

Variable Components

Air includes many gases and particles whose quantities vary significantly from time to time and place to place. Important examples include water vapor, dust particles, and ozone. Although usually present in small percentages, they can have significant effects on weather and climate.

Water Vapor You are probably familiar with the term *humidity* from watching weather reports on TV. Humidity is a reference to water vapor in the air. As you will learn in Chapter 17, there are several ways to express humidity. The amount of water vapor in the air varies considerably, from practically none at all up to about 4 percent by volume. Why is such a small fraction of the atmosphere so significant? The fact that water vapor is the source of all clouds and precipitation would be enough to explain its importance. However, water vapor has other roles. Like carbon dioxide, water vapor absorbs heat given off by Earth as well as some solar energy. It is therefore important when we examine the heating of the atmosphere.

When water changes from one state to another (see Figure 17.2, page 519), it absorbs or releases heat. This energy is termed *latent heat*, which means "hidden heat." As we shall see in later chapters, water vapor in the atmosphere transports this latent heat from one region to another, and it is the energy source that helps drive many storms.

Aerosols The movements of the atmosphere are sufficient to keep a large quantity of solid and liquid particles suspended within it. Although visible dust sometimes clouds the sky, these relatively large particles are too heavy to stay in the air very long. Many other particles are microscopic and remain suspended for considerable periods of time. They may originate from many sources, both natural and human made, and include sea salts from breaking waves, fine soil blown into the air, smoke and soot from fires, pollen and microorganisms lifted by the wind, ash and dust from volcanic eruptions, and more (**FIGURE 16.6A**). Collectively, these tiny solid and liquid particles are called **aerosols**.

From a meteorological standpoint, these tiny, often invisible particles can be significant. First, many act as surfaces on which water vapor can condense, an important function in the formation of clouds and fog. Second, aerosols can absorb, reflect, and scatter incoming solar radiation. Thus, when an air-pollution episode is occurring or when ash fills the sky following a volcanic eruption, the amount of sunlight reaching Earth's surface can be measurably reduced. Finally, aerosols contribute to an optical phenomenon we have all observed—the varied hues of red and orange at sunrise and sunset (**FIGURE 16.6B**).

Ozone Another important component of the atmosphere is **ozone**. It is a form of oxygen that combines three oxygen atoms into each molecule (O_3). Ozone is not the same as oxygen we breathe, which has two atoms per molecule (O_2). There is very little ozone in the atmosphere, and its distribution is not uniform. It is concentrated between 10 and 50 kilometers (6 and 31 miles) above the surface, in a layer called the *stratosphere*.

In this altitude range, oxygen molecules (O_2) are split into single atoms of oxygen (O) when they absorb ultraviolet radiation emitted by the Sun. Ozone is then created when a single atom of oxygen (O) and a molecule of oxygen (O_2) collide. This must happen in the presence of a third, neutral molecule that acts as a *catalyst* by allowing the reaction to take place without itself being consumed in the process. Ozone is concentrated in the 10- to 50-kilometer height

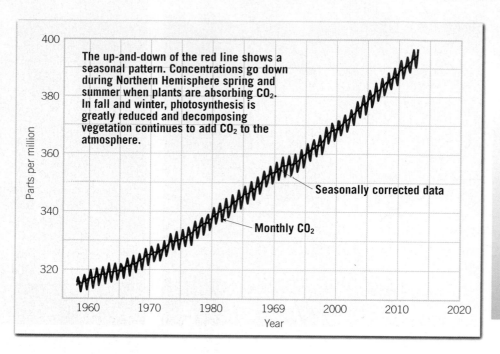

The up-and-down of the red line shows a seasonal pattern. Concentrations go down during Northern Hemisphere spring and summer when plants are absorbing CO_2. In fall and winter, photosynthesis is greatly reduced and decomposing vegetation continues to add CO_2 to the atmosphere.

Seasonally corrected data

Monthly CO_2

Parts per million

Year

SmartFigure 16.5 Monthly CO_2 Concentrations Atmospheric CO_2 has been measured at Mauna Loa Observatory, Hawaii, since 1958. There has been a consistent increase since monitoring began. This graphic portrayal is known as the *Keeling Curve*, in honor of the scientist who originated the measurements. (NOAA)

490

FIGURE 16.6 Aerosols
A. This satellite image from November 11, 2002, shows two examples of aerosols. First, a large dust storm is blowing across northeastern China, toward the Korean Peninsula. Second, a dense haze toward the south (bottom center) is human-generated air pollution. (NASA)
B. Dust in the air can cause sunsets to be especially colorful. (Photo by elwynn/Shutterstock)

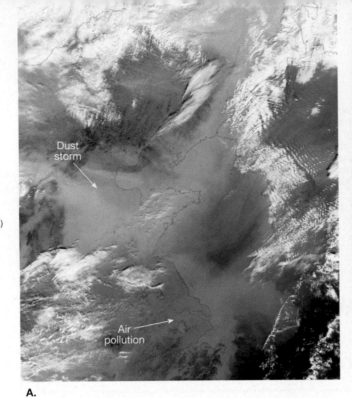

A.

B.

range because a crucial balance exists there: The ultraviolet radiation from the Sun is sufficient to produce single atoms of oxygen, and there are enough gas molecules to bring about the required collisions.

The presence of the ozone layer in our atmosphere is crucial to those of us who dwell on Earth. The reason is that ozone absorbs much of the potentially harmful ultraviolet (UV) radiation from the Sun. If ozone did not filter a great deal of the ultraviolet radiation, and if the Sun's UV rays reached the surface of Earth undiminished, our planet would be uninhabitable for most life as we know it. Thus, anything that reduces the amount of ozone in the atmosphere could affect the well-being of life on Earth. Just such a problem exists and is described in the next section.

Ozone Depletion: A Global Issue

Although stratospheric ozone is 10 to 50 kilometers (6 to 31 miles) above Earth's surface, it is vulnerable to human activities. Chemicals produced by people are breaking up ozone molecules in the stratosphere, weakening our shield against UV rays. This loss of ozone is a serious global-scale environmental problem. Measurements over the past three decades confirm that ozone depletion is occurring worldwide and is especially pronounced above Earth's poles. You can see this effect over the South Pole in **FIGURE 16.7**.

Over the past 60 years, people have unintentionally placed the ozone layer in jeopardy by polluting the

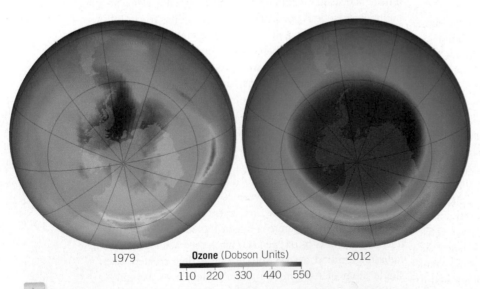

1979

Ozone (Dobson Units)

110   220   330   440   550

2012

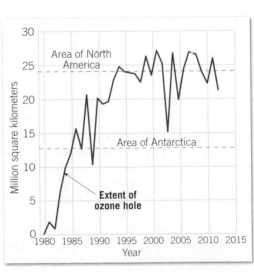

SmartFigure 16.7 Antarctic Ozone Hole The two satellite images show ozone distribution in the Southern Hemisphere on the days in September 1979 and 2012 when the ozone hole was largest. The dark blue shades over Antarctica correspond to the region with the sparsest ozone. The ozone hole is not technically a "hole" where no ozone is present but is actually a region of exceptionally depleted ozone in the stratosphere over the Antarctic that occurs in the spring. The small graph traces changes in the maximum size of the ozone hole, 1980 to 2012. (NASA)

Acid Precipitation
A Human Impact on the Earth System

As a consequence of burning large quantities of coal and petroleum, tens of millions of tons of sulfur dioxide and nitrogen oxides enter the atmosphere each year. Through a series of complex chemical reactions, these pollutants are converted into acids that eventually fall to Earth's surface. The map shows precipitation pH values for 2008.

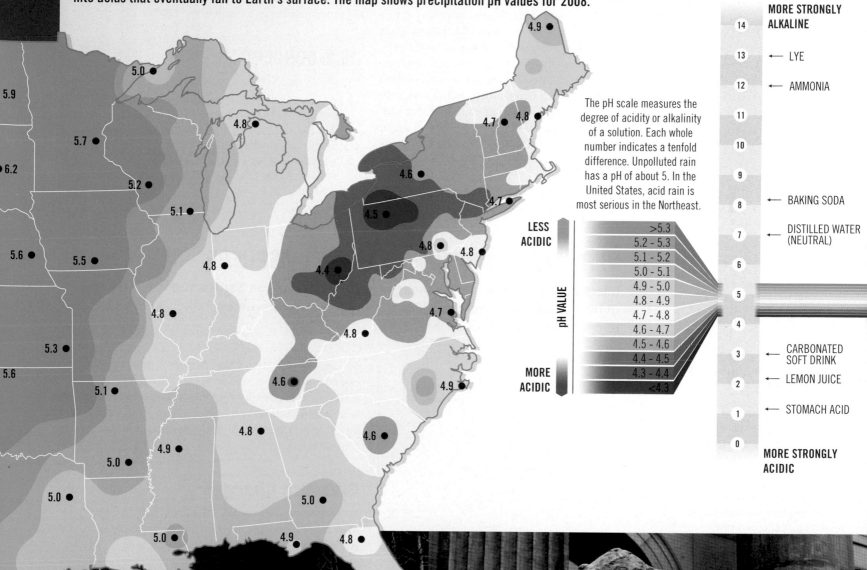

The pH scale measures the degree of acidity or alkalinity of a solution. Each whole number indicates a tenfold difference. Unpolluted rain has a pH of about 5. In the United States, acid rain is most serious in the Northeast.

MORE STRONGLY ALKALINE

14
13 ← LYE
12 ← AMMONIA
11
10
9 ← BAKING SODA
8
7 ← DISTILLED WATER (NEUTRAL)
6
5
4
3 ← CARBONATED SOFT DRINK
2 ← LEMON JUICE
1 ← STOMACH ACID
0

MORE STRONGLY ACIDIC

LESS ACIDIC

pH VALUE

MORE ACIDIC

pH VALUE
>5.3
5.2 – 5.3
5.1 – 5.2
5.0 – 5.1
4.9 – 5.0
4.8 – 4.9
4.7 – 4.8
4.6 – 4.7
4.5 – 4.6
4.4 – 4.5
4.3 – 4.4
<4.3

Damage to forests by acid precipitation is well documented in Europe and eastern North America. These trees in the Appalachian Mountains of North Carolina are one example. In addition, increased acidity in many lakes has had a detrimental impact on aquatic ecosystems.

Andre Jenny Stock Connection Worldwide/Newscom

Although we expect rock to gradually decompose, many stone monuments have succumbed prematurely because of accelerated chemical weathering linked to acid precipitation.

Ryan McGinnis/Alamy Images

atmosphere. The most significant of the offending chemicals are known as *chlorofluorocarbons* (CFCs for short). Over the decades, many uses were developed for CFCs: They were used as coolants for air-conditioning and refrigeration equipment, cleaning solvents for electronic components, and propellants for aerosol sprays, and they were used in the production of certain plastic foams.

Because CFCs are practically inert (that is, not chemically active) in the lower atmosphere, some of these gases gradually make their way to the ozone layer, where sunlight separates the chemicals into their constituent atoms. The chlorine atoms released this way break up some of the ozone molecules.

Because ozone filters out most of the UV radiation from the Sun, a decrease in its concentration permits more of these harmful wavelengths to reach Earth's surface. The most serious threat to human health is an increased risk of skin cancer. An increase in damaging UV radiation also can impair the human immune system as well as promote cataracts, a clouding of the eye lens that reduces vision and may cause blindness if not treated.

In response to this problem, an international agreement known as the *Montreal Protocol* was developed under the sponsorship of the United Nations to eliminate the production and use of CFCs. More than 180 nations eventually ratified the treaty. Although relatively strong action has been taken, CFC levels in the atmosphere will not drop rapidly. Once in the atmosphere, CFC molecules can take many years to reach the ozone layer, and once there, they can remain active for decades. This does not promise a near-term reprieve for the ozone layer. Between 2060 and 2075, the abundance of ozone-depleting gases is projected to fall to values that existed before the ozone hole began to form in the 1980s.

16.2 CONCEPT CHECKS

1 Is *air* a specific gas? Explain.

2 What are the two major components of clean, dry air? What proportion does each represent?

3 Why are water vapor and aerosols important constituents of Earth's atmosphere?

4 What is ozone? Why is ozone important to life on Earth? What are CFCs, and what is their connection to the ozone problem?

16.3 | VERTICAL STRUCTURE OF THE ATMOSPHERE

Interpret a graph that shows changes in air pressure from Earth's surface to the top of the atmosphere. Sketch and label a graph that shows atmospheric layers based on temperature.

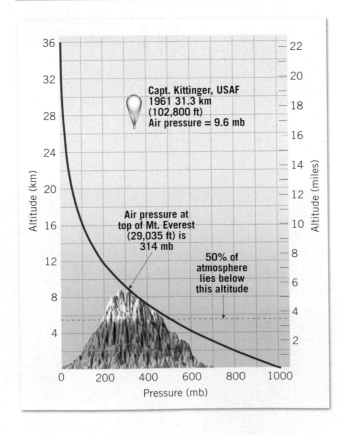

FIGURE 16.8 Air Pressure Changes with Altitude The rate of pressure decrease with an increase in altitude is not constant. Pressure decreases rapidly near Earth's surface and more gradually at greater heights. Put another way, the graph shows that the vast bulk of the gases making up the atmosphere is very near Earth's surface and that the gases gradually merge with the emptiness of space.

To say that the atmosphere begins at Earth's surface and extends upward is obvious. However, where does the atmosphere end, and where does outer space begin? There is no sharp boundary; the atmosphere rapidly thins as you travel away from Earth, until there are too few gas molecules to detect.

Pressure Changes

To understand the vertical extent of the atmosphere, let us examine the changes in atmospheric pressure with height. Atmospheric pressure is simply the weight of the air above. At sea level, the average pressure is slightly more than 1000 millibars (mb). This corresponds to a weight of slightly more than 1 kilogram per square centimeter (14.7 pounds per square inch). The pressure at higher altitudes is less (**FIGURE 16.8**).

One-half of the atmosphere lies below an altitude of 5.6 kilometers (3.5 miles). At about 16 kilometers (10 miles), 90 percent of the atmosphere has been traversed, and above 100 kilometers (62 miles), only 0.00003 percent of all the gases making up the atmosphere remains. Even so, traces of our atmosphere extend far beyond this altitude, gradually merging with the emptiness of space.

Temperature Changes

By the early twentieth century, much had been learned about the lower atmosphere. The upper atmosphere was partly known from indirect methods. Data from balloons and kites showed that near Earth's surface, air temperature drops with increasing height. This phenomenon is felt by anyone who has climbed a high mountain and is obvious in pictures of snowcapped mountaintops rising above snow-free lowlands (**FIGURE 16.9**). We divide the atmosphere vertically into four layers, on the basis of temperature (**FIGURE 16.10**).

FIGURE 16.9
Temperatures Drop with an Increase in Altitude in the Troposphere Snow-capped mountains and snow-free lowlands are a reminder that temperatures decrease as we go higher in the troposphere. (Photo by David Wall/Alamy)

Troposphere The lowermost layer in which we live, where temperature decreases with an increase in altitude, is the **troposphere**. The term literally means the region where air "turns over," a reference to the appreciable vertical mixing of air in this lowermost zone. The troposphere is the chief focus of meteorologists because it is in this layer that essentially all important weather phenomena occur.

The temperature decrease in the troposphere is called the **environmental lapse rate**. Although its average value is 6.5°C per kilometer (3.5°F per 1000 feet), a figure known as the *normal lapse rate*, its value is variable. To determine the actual environmental lapse rate as well as gather information about vertical changes in pressure, wind, and humidity, radiosondes are used. A **radiosonde** is an instrument package that is attached to a balloon and transmits data by radio as it ascends through the atmosphere (**FIGURE 16.11**).

The thickness of the troposphere is not the same everywhere; it varies with latitude and season. On average, the temperature drop continues to a height of about 12 kilometers (7.4 miles). The outer boundary of the troposphere is the *tropopause*.

Stratosphere Beyond the tropopause is the **stratosphere**. In the stratosphere, the temperature remains constant to a height of about 20 kilometers (12 miles) and then begins a gradual increase that continues until the *stratopause*, at a height of nearly 50 kilometers (30 miles) above

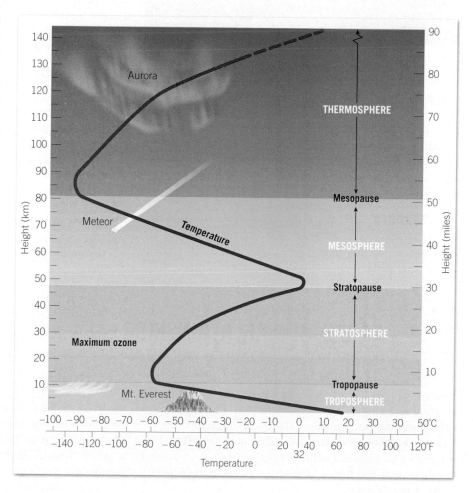

FIGURE 16.10 Thermal Structure of the Atmosphere Earth's atmosphere is traditionally divided into four layers, based on temperature.

FIGURE 16.11
Radiosonde A radiosonde is a lightweight package of instruments that is carried aloft by a small weather balloon. It transmits data on vertical changes in temperature, pressure, and humidity in the troposphere. The troposphere is where practically all weather phenomena occur, so it is very important to have frequent measurements. (Photo by David R. Frazier/Danita Delimont/Newscom)

Radar unit for tracking radiosonde

Weather balloon

Radiosonde (instrument package)

Earth's surface. Below the tropopause, atmospheric properties such as temperature and humidity are readily transferred by large-scale turbulence and mixing. Above the tropopause, in the stratosphere, they are not. Temperatures increase in the stratosphere because it is in this layer that the atmosphere's ozone is concentrated. Recall that ozone absorbs ultraviolet radiation from the Sun. As a consequence, the stratosphere is heated.

Mesosphere In the third layer, the **mesosphere**, temperatures again decrease with height until, at the *mesopause*—approximately 80 kilometers (50 miles) above the surface, the temperature approaches −90°C

(−130°F). The coldest temperatures anywhere in the atmosphere occur at the mesopause. Because accessibility is difficult, the mesosphere is one of the least explored regions of the atmosphere. It cannot be reached by the highest research balloons, nor is it accessible to the lowest orbiting satellites. Recent technological developments are just beginning to fill this knowledge gap.

Thermosphere The fourth layer extends outward from the mesopause and has no well-defined upper limit. It is the **thermosphere**, a layer that contains only a tiny fraction of the atmosphere's mass. In the extremely rarefied air of this outermost layer, temperatures again increase, due to the absorption of very short-wave, high-energy solar radiation by atoms of oxygen and nitrogen.

Temperatures rise to extremely high values of more than 1000°C (1800°F) in the thermosphere. But such temperatures are not comparable to those experienced near Earth's surface. Temperature is defined in terms of the average speed at which molecules move. Because the gases of the thermosphere are moving at very high speeds, the temperature is very high. But the gases are so sparse that, collectively, they possess only an insignificant quantity of heat. For this reason, the temperature of a satellite orbiting Earth in the thermosphere is determined chiefly by the amount of solar radiation it absorbs and not by the high temperature of the almost nonexistent surrounding air. If an astronaut inside were to expose his or her hand, the atmosphere would not feel hot.

16.3 CONCEPT CHECKS

1 Does air pressure increase or decrease with an increase in altitude? Is the rate of change constant or variable? Explain.

2 Is the outer edge of the atmosphere clearly defined? Explain.

3 The atmosphere is divided vertically into four layers, on the basis of temperature. List and describe these layers in order, from lowest to highest. In which layer does practically all our weather occur?

4 What is the environmental lapse rate, and how is it determined?

5 Why do temperatures increase in the stratosphere?

6 Why are temperatures in the thermosphere not strictly comparable to those experienced near Earth's surface?

16.4 | EARTH–SUN RELATIONSHIPS Explain what causes the Sun angle and length of daylight to change during the year and describe how these changes produce the seasons.

Nearly all of the energy that drives Earth's variable weather and climate comes from the Sun. Earth intercepts only a minute percentage of the energy given off by the Sun—less than 1 two-billionth. This may seem to be an insignificant amount, until we realize that it is several hundred thousand times the electrical-generating capacity of the United States.

Solar energy is not distributed evenly over Earth's land–sea surface. The amount of energy received varies with latitude, time of day, and season of the year. Contrasting images of polar bears on ice rafts and palm trees along a remote tropical beach serve to illustrate the extremes. It is the unequal heating of Earth that creates winds and drives the ocean's currents. These movements of air and water, in turn, transport heat from the tropics toward the poles, in an unending attempt to balance energy inequalities. The consequences of these processes are the phenomena we call weather.

If the Sun were "turned off," global winds and ocean currents would quickly cease. Yet as long as the Sun shines, the winds *will* blow and weather *will* persist. So to understand how the atmosphere's dynamic weather machine works, we must first know why different latitudes receive varying quantities of solar energy and why the amount of solar energy changes to produce the seasons. As you will see, the variations in solar heating are caused by the motions of Earth relative to the Sun and by variations in Earth's land–sea surface.

Earth's Motions

Earth has two principal motions—rotation and revolution. **Rotation** is the spinning of Earth about its axis. The axis is an imaginary line running through the poles. Our planet rotates once every 24 hours, producing the daily cycle of daylight and darkness. At any moment, half of Earth is experiencing daylight and the other half darkness. The line separating the dark half of Earth from the lighted half is called the **circle of illumination**.

Revolution refers to the movement of Earth in a slightly elliptical orbit around the Sun. The distance between Earth and Sun averages about 150 million kilometers (93 million miles). Because Earth's orbit is not perfectly circular, however, the distance varies during the course of a year. Each year, on about January 3, our planet is about 147.3 million kilometers (91.5 million miles) from the Sun, closer than at any other time—a position known as *perihelion*. About 6 months later, on July 4, Earth is about 152 million kilometers (94.5 million miles) from the Sun, farther away than at any other time—a position called *aphelion*. Although Earth is closest to the Sun

and receives up to 7 percent more energy in January than in July, this difference plays only a minor role in producing seasonal temperature variations, as evidenced by the fact that Earth is closest to the Sun during the Northern Hemisphere winter.

What Causes the Seasons?

If variations in the distance between the Sun and Earth are not responsible for seasonal temperature changes, what is? The gradual but significant change in the length of daylight certainly accounts for some of the difference we notice between summer and winter. Furthermore, a gradual change in the angle (altitude) of the Sun above the horizon is also a contributing factor (**FIGURE 16.12**). For example, someone living in Chicago, Illinois, experiences the noon Sun highest in the sky in late June. But as summer gives way to autumn, the noon Sun appears lower in the sky, and sunset occurs earlier each evening.

The seasonal variation in the angle of the Sun above the horizon affects the amount of energy received at Earth's surface in two ways. First, when the Sun is directly overhead (at a 90-degree angle), the solar rays are most concentrated and thus most intense (**FIGURE 16.13A**). The lower the angle, the more spread out and less intense is the solar radiation that reaches the surface (**FIGURE 16.13B,C**). To illustrate this principle, hold a flashlight at a right angle to a surface and then change the angle.

Second, but of lesser importance, the angle of the Sun determines the path solar rays take as they pass through the atmosphere (**FIGURE 16.14**). When the Sun is directly overhead, the rays strike the atmosphere at a 90-degree angle and travel the shortest possible route to the surface. This distance is referred to as *1 atmosphere*. However, rays entering at a 30-degree angle travel through twice this distance before reaching the surface, while rays at a 5-degree angle travel through a distance roughly equal to the thickness of 11

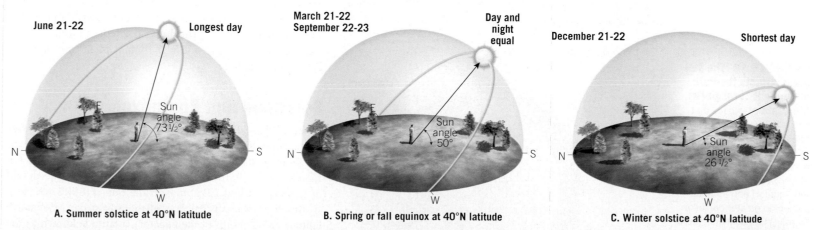

A. Summer solstice at 40°N latitude

B. Spring or fall equinox at 40°N latitude

C. Winter solstice at 40°N latitude

SmartFigure 16.12 The Changing Sun Angle Daily paths of the Sun for a place located at 40° north latitude for **A.** summer solstice, **B.** spring or fall equinox, and **C.** winter solstice. As we move from summer to winter, the angle of the noon Sun decreases from 73½ to 26½ degrees—a difference of 47 degrees. Notice also how the location of sunrise (east) and sunset (west) changes during a year.

FIGURE 16.13 The Sun Angle Influences the Intensity of Solar Radiation at Earth's Surface Changes in the Sun's angle cause variations in the amount of solar energy reaching Earth's surface. The nearer the angle to 90 degrees, the more intense the solar radiation.

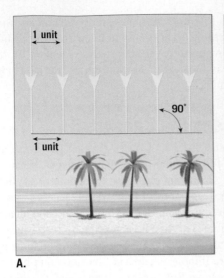
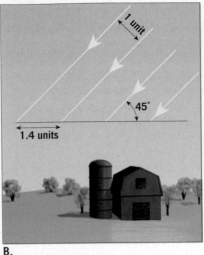
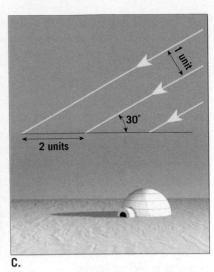

atmospheres. The longer the path, the greater the chance that sunlight will be dispersed by the atmosphere, which reduces the intensity at the surface. These conditions account for the fact that we cannot look directly at the midday Sun, but we can enjoy gazing at a sunset.

It is important to remember that Earth's shape is spherical. On any given day, the only places that will receive vertical (90-degree) rays from the Sun are located along one particular line of latitude. As we move either north or south of this location, the Sun's rays strike at ever-decreasing angles. The nearer a place is to the latitude receiving vertical rays of the Sun, the higher will be its noon Sun and the more concentrated will be the radiation it receives (see Figure 16.14).

Earth's Orientation

What causes fluctuations in Sun angle and length of daylight that occur during the course of a year? Variations occur because Earth's orientation to the Sun continually changes as it travels along its orbit. Earth's axis (the imaginary line through the poles around which Earth rotates) is not

FIGURE 16.14 The Sun Angle Affects the Path of Sunlight Through the Atmosphere Notice that rays striking Earth at a low angle (toward the poles) must travel through more of the atmosphere than rays striking at a high angle (around the equator) and are thus subject to greater depletion by reflection, scattering, and absorption.

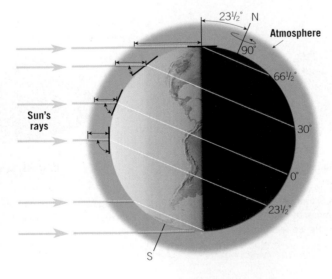

perpendicular to the plane of its orbit around the Sun. Instead, it is tilted 23½ degrees from the perpendicular, as shown in **FIGURE 16.15**. This is called the **inclination of the axis**. If the axis were not inclined, Earth would lack seasons. Because the axis remains pointed in the same direction (toward the North Star), the orientation of Earth's axis to the Sun's rays is constantly changing (see Figure 16.15).

For example, on one day in June each year, the axis is such that the Northern Hemisphere is "leaning" 23½ degrees *toward* the Sun (see Figure 16.15, left). Six months later, in December, when Earth has moved to the opposite side of its orbit, the Northern Hemisphere "leans" 23½ degrees *away* from the Sun (see Figure 16.15, right). On days between these extremes, Earth's axis is leaning at amounts less than 23½ degrees to the rays of the Sun. This change in orientation causes the spot where the Sun's rays are vertical to make an annual migration from 23½ degrees north of the equator to 23½ degrees south of the equator.

In turn, this migration causes the angle of the noon Sun to vary by up to 47° (23½° + 23½°) for many locations during the year. For example, a midlatitude city like New York (about 40° north latitude) has a maximum noon Sun angle of 73½° when the Sun's vertical rays reach their farthest northward location in June and a minimum noon Sun angle of 26½° 6 months later.

Solstices and Equinoxes

Historically, 4 days each year have been given special significance, based on the annual migration of the direct rays of the Sun and its importance to the yearly weather cycle. On June 21 or 22, Earth is in a position such that the north end of its axis is tilted 23½° *toward* the Sun (**FIGURE 16.16A**). At this time, the vertical rays of the Sun strike 23½° north latitude (23½° north of the equator), a latitude known as the **Tropic of Cancer**. For people in the Northern Hemisphere, June 21 or 22 is known as the **summer solstice**, the first "official" day of summer.

Six months later, on about December 21 or 22, Earth is in the opposite position, with the Sun's vertical rays

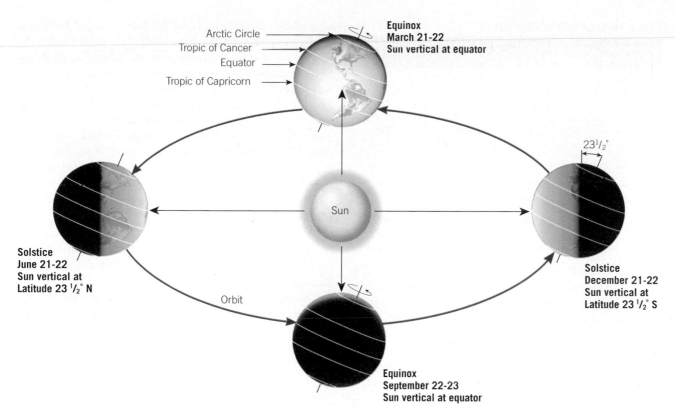

Arctic Circle
Tropic of Cancer
Equator
Tropic of Capricorn

Equinox
March 21-22
Sun vertical at equator

FIGURE 16.15 Earth–Sun Relationships

Sun

Orbit

Solstice
June 21-22
Sun vertical at
Latitude 23 ½° N

Solstice
December 21-22
Sun vertical at
Latitude 23 ½° S

23½°

Equinox
September 22-23
Sun vertical at equator

SmartFigure 16.16
Characteristics of the
Solstices and Equinoxes

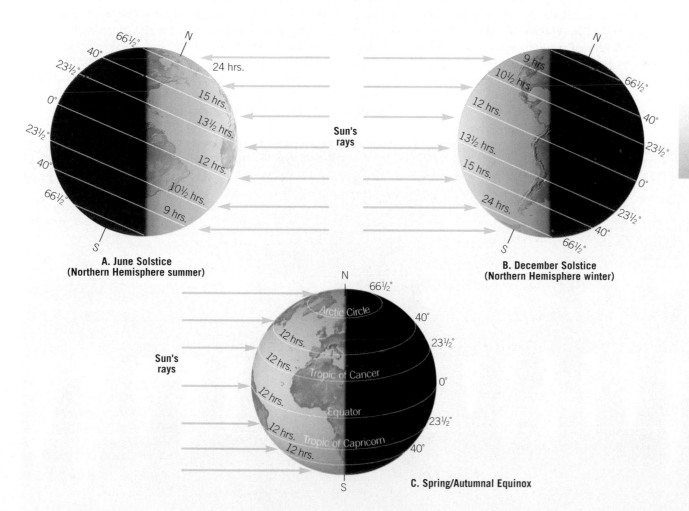

24 hrs.

Sun's
rays

15 hrs.

13½ hrs.

12 hrs.

10½ hrs.

9 hrs.

A. June Solstice
(Northern Hemisphere summer)

9 hrs.

10½ hrs.

12 hrs.

13½ hrs.

15 hrs.

24 hrs.

B. December Solstice
(Northern Hemisphere winter)

Arctic Circle

12 hrs.

12 hrs.

Tropic of Cancer

12 hrs.

Equator

12 hrs.

Tropic of Capricorn

12 hrs.

Sun's
rays

C. Spring/Autumnal Equinox

TABLE 16.1 Length of Daylight

Latitude (degrees)	Summer Solstice	Winter Solstice	Equinoxes
0	12 h	12 h	12 h
10	12 h 35 min	11 h 25 min	12 h
20	13 h 12 min	10 h 48 min	12 h
30	13 h 56 min	10 h 04 min	12 h
40	14 h 52 min	9 h 08 min	12 h
50	16 h 18 min	7 h 42 min	12 h
60	18 h 27 min	5 h 33 min	12 h
70	24 h (for 2 mo)	0 h 00 min	12 h
80	24 h (for 4 mo)	0 h 00 min	12 h
90	24 h (for 6 mo)	0 h 00 min	12 h

striking at 23½° south latitude (**FIGURE 16.16B**). This parallel is known as the **Tropic of Capricorn**. For those in the Northern Hemisphere, December 21 and 22 is the **winter solstice**. However, at the same time in the Southern Hemisphere, people are experiencing just the opposite—the summer solstice.

Midway between the solstices are the equinoxes. September 22 or 23 is the date of the **autumnal (fall) equinox** in the Northern Hemisphere, and March 21 or 22 is the date of the **spring equinox**. On these dates, the vertical rays of the Sun strike the equator (0° latitude) because Earth is in such a position in its orbit that the axis is tilted neither toward nor away from the Sun (**FIGURE 16.16C**).

The length of daylight versus darkness is also determined by Earth's position in orbit. The length of daylight on June 21, the summer solstice in the Northern Hemisphere, is greater than the length of night. This fact can be established from Figure 16.16A by comparing the fraction of a given latitude that is on the "day" side of the circle of illumination with the fraction on the "night" side. The opposite is true for the winter solstice, when the nights are longer than the days. Again for comparison, let us consider New York City, which has about 15 hours of daylight on June 21 and only about 9 hours on December 21. (You can see this in Figure 16.16 and **TABLE 16.1**.) Also note from Table 16.1 that the farther north of the equator you are on June 21, the longer the period of daylight. When you reach the Arctic Circle (66½° north latitude), the length of daylight is 24 hours. This is the land of the "midnight Sun," which does not set for about 6 months at the North Pole (**FIGURE 16.17**).

During an equinox (meaning "equal night"), the length of daylight is 12 hours *everywhere* on Earth, because the circle of illumination passes directly through the poles, dividing the latitudes in half (see Figure 16.16C).

As a review of the characteristics of the summer solstice for the Northern Hemisphere, examine Figure 16.16A and Table 16.1 and consider the following facts:

- The solstice occurs on June 21 or 22.
- The vertical rays of the Sun are striking the Tropic of Cancer (23½° north latitude).
- Locations in the Northern Hemisphere are experiencing their greatest length of daylight (opposite for the Southern Hemisphere).
- Locations north of the Tropic of Cancer are experiencing their highest noon Sun angles (opposite for places south of the Tropic of Capricorn).
- The farther you are north of the equator, the longer the period of daylight, until the Arctic Circle is reached, where daylight lasts for 24 hours (opposite for the Southern Hemisphere).

FIGURE 16.17 Midnight Sun Multiple exposures of the Sun, representative of midsummer in the high latitudes. This example shows the midnight Sun in Norway. (Photo by Martin Woike/ AGE Fotostock)

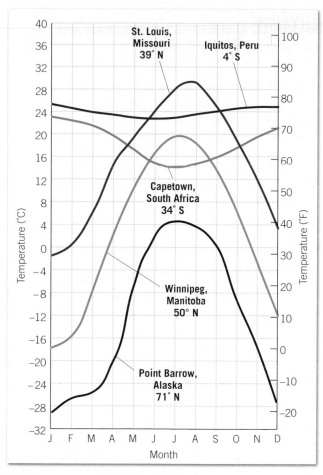

FIGURE 16.18 Monthly Temperatures for Cities at Different Latitudes Places located at higher latitudes experience larger temperature differences between summer and winter. Note that Cape Town, South Africa, experiences winter in June, July, and August.

The facts about the winter solstice are just the opposite. It should now be apparent why a midlatitude location is warmest in the summer, for it is then that days are longest and Sun's altitude is highest.

These seasonal changes, in turn, cause the month-to-month variations in temperature observed at most locations outside the tropics. **FIGURE 16.18** shows mean monthly temperatures for selected cities at different latitudes. Notice that the cities located at more poleward latitudes experience larger temperature differences from summer to winter than do cities located nearer the equator. Also notice that temperature minimums for Southern Hemisphere locations occur in July, whereas they occur in January for most places in the Northern Hemisphere.

All places at the same latitude have identical Sun angles and lengths of daylight. If the Earth–Sun relationships just described were the only controls of temperature, we would expect these places to have identical temperatures as well. Obviously, this is not the case. Other factors that influence temperature are discussed later in the chapter.

16.4 CONCEPT CHECKS

1. Do the annual variations in Earth–Sun distance adequately account for seasonal temperature changes? Explain.

2. Create a simple sketch to show why the intensity of solar radiation striking Earth's surface changes when the Sun angle changes.

3. Briefly explain the primary cause of the seasons.

4. What is the significance of the Tropic of Cancer and the Tropic of Capricorn?

5. After examining Table 16.1, write a general statement that relates the season, latitude, and the length of daylight.

EYE ON EARTH

This image shows the first sunrise of 2008 at Amundsen-Scott Station, a research facility at the South Pole. At the moment the Sun cleared the horizon, the weathered American flag was seen whipping in the wind above a sign marking the location of the geographic South Pole. (NASA)

QUESTION 1 *What was the approximate date that this photograph was taken?*

QUESTION 2 *How long after this photo was taken did the Sun set at the South Pole?*

QUESTION 3 *On what date would you expect Amundsen-Scott Station to experience its highest noon Sun angle? Explain.*

16.5 | ENERGY, HEAT, AND TEMPERATURE Distinguish between heat and temperature. List and describe the three mechanisms of heat transfer.

The universe is made up of a combination of matter and energy. The concept of matter is easy to grasp because it is the "stuff" we can see, smell, and touch. Energy, on the other hand, is abstract and therefore more difficult to describe. For our purposes, we define energy simply as *the capacity to do work*. We can think of work as being accomplished whenever matter is moved. You are likely familiar with some of the common forms of energy, such as thermal, chemical, nuclear, radiant (light), and gravitational energy. One type of energy is described as *kinetic energy*, which is energy of motion. Recall that matter is composed of atoms or molecules that are constantly in motion and therefore possesses kinetic energy.

Heat is a term that is commonly used synonymously with *thermal energy*. In this usage, heat is energy possessed by a material arising from the internal motions of its atoms or molecules. Whenever a substance is heated, its atoms move faster and faster, which leads to an increase in its heat content. **Temperature**, on the other hand, is related to the average kinetic energy of a material's atoms or molecules. Stated another way, the term *heat* generally refers to the quantity of energy present, whereas the word *temperature* refers to the intensity—that is, the degree of "hotness."

Heat and temperature are closely related concepts. Heat is the energy that flows because of temperature differences. In all situations, *heat is transferred from warmer to cooler objects*. Thus, if two objects of different temperature are in contact, the warmer object will become cooler and the cooler object will become warmer until they both reach the same temperature.

Three mechanisms of heat transfer are recognized: conduction, convection, and radiation. Although we present them separately, all three processes go on simultaneously in the atmosphere. In addition, these mechanisms operate to transfer heat between Earth's surface (both land and water) and the atmosphere.

SmartFigure 16.19
The Three Mechanisms of Heat Transfer

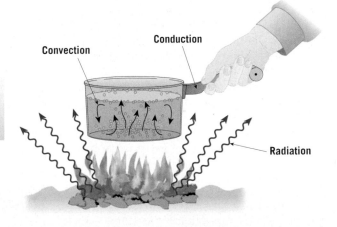

Mechanism of Heat Transfer: Conduction

Conduction is familiar to all of us. Anyone who has touched a metal spoon that was left in a hot pan has discovered that heat was conducted through the spoon. **Conduction** is *the transfer of heat through matter by molecular activity*. The energy of molecules is transferred through collisions from one molecule to another, with the heat flowing from the higher temperature to the lower temperature.

The ability of substances to conduct heat varies considerably. Metals are good conductors, as those of us who have touched hot metal have quickly learned (**FIGURE 16.19**). Air, conversely, is a very poor conductor of heat. Consequently,

EYE ON EARTH

This infrared (IR) image produced by the *GOES-14* satellite displays cold objects as bright white and hot objects as black. The hottest (blackest) features shown are land surfaces, and the coldest (whitest) features are the tops of towering storm clouds. Recall that we cannot see infrared (thermal) radiation, but we have developed instruments that are capable of extending our vision into the long-wavelength portion of the electromagnetic spectrum.

QUESTION 1 *Several areas of cloud development and potential storms are shown on this IR image. One is a well-developed tropical storm named Hurricane Bill. Can you locate this storm?*

QUESTION 2 *What is an advantage of IR images over visible images?*

(NASA)

conduction is important only between Earth's surface and the air directly in contact with the surface. As a means of heat transfer for the atmosphere as a whole, conduction is the least significant.

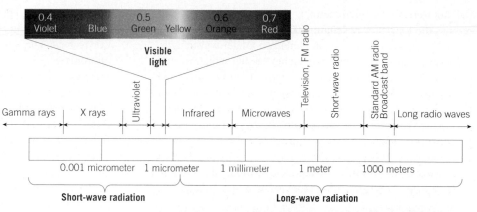

FIGURE 16.20 The Electromagnetic Spectrum This diagram illustrates the wavelengths and names of various types of radiation. Visible light consists of an array of colors we commonly call the "colors of the rainbow."

Mechanism of Heat Transfer: Convection

Much of the heat transport that occurs in the atmosphere occurs via convection. **Convection** is *the transfer of heat by mass movement or circulation within a substance*. It takes place in fluids (e.g., liquids like the ocean and gases like air) where the atoms and molecules are free to move about.

The pan of water in Figure 16.19 illustrates the nature of simple convective circulation. Radiation from the fire warms the bottom of the pan, which conducts heat to the water near the bottom of the container. As the water is heated, it expands and becomes less dense than the water above. Because of this new buoyancy, the warmer water rises. At the same time, cooler, denser water near the top of the pan sinks to the bottom, where it becomes heated. As long as the water is heated unequally—that is, from the bottom up—the water will continue to "turn over," producing a *convective circulation*. In a similar manner, most of the heat acquired in the lowest portion of the atmosphere by way of radiation and conduction is transferred by convective flow.

On a global scale, convection in the atmosphere creates a huge, worldwide air circulation. This is responsible for the redistribution of heat between hot equatorial regions and the frigid poles. This important process is discussed in detail in Chapter 18.

Mechanism of Heat Transfer: Radiation

The third mechanism of heat transfer is **radiation**. As shown in Figure 16.19, radiation travels out in all directions from its source. Unlike conduction and convection, which need a medium to travel through, radiant energy readily travels through the vacuum of space. Thus, radiation is the heat-transfer mechanism by which solar energy reaches our planet.

Solar Radiation From our everyday experience, we know that the Sun emits light and heat as well as the ultraviolet rays that cause suntan. Although these forms of energy comprise a major portion of the total energy that radiates from the Sun, they are only part of a large array of energy called radiation, or **electromagnetic radiation**. This array or spectrum of electromagnetic energy is shown in **FIGURE 16.20**. All radiation, whether x-rays, radio waves, or heat waves, travels through the vacuum of space at 300,000 kilometers (186,000 miles) per second and only slightly slower through our atmosphere.

Nineteenth-century physicists were so puzzled by the seemingly impossible phenomenon of energy traveling through the vacuum of space without a medium to transmit it that they assumed that a material, which they named *ether*, existed between the Sun and Earth. This medium was thought to transmit radiant energy in much the same way that air transmits sound waves. Of course, this was incorrect. We now know that, like gravity, radiation requires no material for transmission.

In some respects, the transmission of radiant energy parallels the motion of the gentle swells in the open ocean. Like ocean swells, electromagnetic waves come in various sizes. For our purpose, the most important characteristic is their *wavelength*, or the distance from one crest to the next. Radio waves have the longest wavelengths, ranging to tens of kilometers, whereas gamma waves are the shortest, being less than one-billionth of a centimeter long.

Visible light, as the name implies, is the only portion of the spectrum we can see. We often refer to visible light as "white" light because it appears "white" in color. However, it is easy to show that white light is really a mixture of colors, each corresponding to a specific wavelength. Using a prism, white light can be divided into the colors of the rainbow. Figure 16.20 shows that violet has the shortest wavelength—

0.4 micrometer (1 micrometer is 0.0001 centimeter)—and red has the longest wavelength—0.7 micrometer.

Located adjacent to red, and having a longer wavelength, is **infrared** radiation, which we cannot see but which we can detect as heat. The closest invisible waves to violet are called **ultraviolet (UV)** rays. They are responsible for the sunburn that can occur after intense exposure to the Sun. Although we divide radiant energy into groups based on our ability to perceive the different types, all forms of radiation are basically the same. When any form of radiant energy is absorbed by an object, the result is an increase in molecular motion, which causes a corresponding increase in temperature.

Laws of Radiation   To obtain a better understanding of how the Sun's radiant energy interacts with Earth's atmosphere and land–sea surface, it is helpful to have a general understanding of the basic laws governing radiation:

- *All objects, at whatever temperature, emit radiant energy.* Thus, not only hot objects like the Sun but also Earth, including its polar ice caps, continually emit energy.
- *Hotter objects radiate more total energy per unit area than do colder objects.* The Sun, which has a surface temperature of nearly 6000°C (10,000°F), emits about 160,000 times more energy per unit area than does Earth, which has an average surface temperature of about 15°C (59°F).
- *Hotter objects radiate more energy in the form of short-wavelength radiation than do cooler objects.* We can visualize this law by imagining a piece of metal that, when heated sufficiently (as occurs in a blacksmith shop), produces a white glow. As the metal cools, it emits more of its energy in longer wavelengths and glows a reddish color. Eventually, no

light is given off, but if you place your hand near the metal, you will detect the still-longer infrared radiation as heat. The Sun radiates maximum energy at 0.5 micrometer, which is in the visible range. The maximum radiation for Earth occurs at a wavelength of 10 micrometers, well within the infrared (heat) range. Because the maximum Earth radiation is roughly 20 times longer than the maximum solar radiation, Earth radiation is often called *long-wave radiation*, and solar radiation is called *short-wave radiation*.

- *Objects that are good absorbers of radiation are good emitters as well.* Earth's surface and the Sun are nearly perfect radiators because they absorb and radiate with nearly 100 percent efficiency for their respective temperatures. On the other hand, *gases are selective absorbers and radiators*. Thus, the atmosphere, which is nearly transparent to (does not absorb) certain wavelengths of radiation, is nearly opaque (a good absorber) to others. Our experience tells us that the atmosphere is transparent to visible light; hence, it readily reaches Earth's surface. This is not the case for the longer-wavelength radiation emitted by Earth.

16.5 CONCEPT CHECKS

1 Distinguish between heat and temperature.

2 Describe the three basic mechanisms of heat transfer. Which mechanism is *least* important as a means of heat transfer in the atmosphere?

3 In what part of the electromagnetic spectrum does the Sun radiate maximum energy? How does this compare to Earth?

4 Describe the relationship between the temperature of a radiating body and the wavelengths it emits.

16.6 | HEATING THE ATMOSPHERE   Sketch and label a diagram that shows the paths taken by incoming solar radiation. Summarize the greenhouse effect.

The goal of this section is to describe how energy from the Sun heats Earth's surface and atmosphere. It is important to know the paths taken by incoming solar radiation and the factors that cause the amount of solar radiation taking each path to vary.

What Happens to Incoming Solar Radiation?

When radiation strikes an object, three different results usually occur. First, some of the energy is *absorbed* by the object. Recall that when radiant energy is absorbed, it is converted to heat, which causes an increase in temperature. Second, substances such as water and air are transparent to certain wavelengths of radiation. Such materials simply *transmit* this energy. Radiation that is transmitted does not contribute energy to the object. Third, some radiation may "bounce off" the object without being absorbed or transmitted. *Reflection* and *scattering* are responsible for redirecting

incoming solar radiation. In summary, *radiation may be absorbed, transmitted, or redirected (reflected or scattered)*.

FIGURE 16.21 shows the fate of incoming solar radiation averaged for the entire globe. Notice that the atmosphere is quite transparent to incoming solar radiation. On average, about 50 percent of the solar energy that reaches the top of the atmosphere is absorbed at Earth's surface. Another 30 percent is reflected back to space by the atmosphere, clouds, and reflective surfaces. The remaining 20 percent is absorbed by clouds and the atmosphere's gases. What determines whether solar radiation will be transmitted to the surface, scattered, reflected outward, or absorbed by the atmosphere? As you will see, it depends greatly on the wavelength of the energy being transmitted, as well as on the nature of the intervening material.

Reflection and Scattering

Reflection is the process whereby light bounces back from an object at the same angle at which it encounters a surface

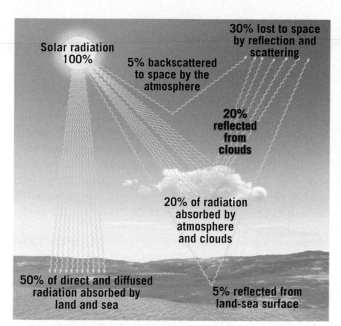

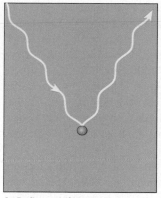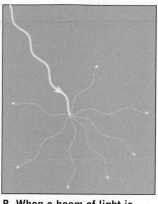

FIGURE 16.22
Reflection and
Scattering

A. Reflected light bounces back from a surface at the same angle at which it strikes that surface and with the same intensity.

B. When a beam of light is scattered, it results in a larger number of weaker rays, traveling in all directions. Usually more energy is scattered in the forward direction than is backscattered.

 SmartFigure 16.21 Paths Taken by Solar Radiation This diagram shows the average distribution of incoming solar radiation, by percentage. More solar radiation is absorbed by Earth's surface than by the atmosphere.

and with the same intensity (**FIGURE 16.22A**). By contrast, **scattering** produces a larger number of weaker rays that travel in different directions. Although scattering disperses light both forward and backward (*backscattering*), more energy is dispersed in the forward direction (**FIGURE 16.22B**).

Reflection and Earth's Albedo
Energy is returned to space from Earth in two ways: reflection and emission of radiant energy. The portion of solar energy that is reflected back to space leaves in the same short wavelengths in which it came to Earth. About 30 percent of the solar energy that reaches the outer atmosphere is reflected back to space. Included in this figure is the amount sent skyward by backscattering. This energy is lost to Earth and does not play a role in heating the atmosphere.

The fraction of the total radiation that is reflected by a surface is called its **albedo**. Thus, the albedo for Earth as a whole (the *planetary albedo*) is 30 percent. However, the albedo from place to place as well as from time to time in the same locale varies considerably, depending on the amount of cloud cover and particulate matter in the air, as well as on the angle of the Sun's rays and the nature of the surface. A lower Sun angle means that more atmosphere must be penetrated, thus making the "obstacle course" longer and the loss of solar radiation greater (see Figure 16.14). **FIGURE 16.23** shows the albedos for various surfaces. Note that the angle at which the Sun's rays strike a water surface greatly affects the albedo of that surface.

Scattering
Although incoming solar radiation travels in a straight line, small dust particles and gas molecules in the atmosphere scatter some of this energy in all directions. The result, called **diffused light**, explains how light reaches into the

area beneath a shade tree and how a room is lit in the absence of direct sunlight. Further, scattering accounts for the brightness and even the blue color of the daytime sky. In contrast, bodies such as the Moon and Mercury, which are without atmospheres, have dark skies and "pitch-black" shadows, even during daylight hours. Overall, about half of the solar radiation that is absorbed at Earth's surface arrives as diffused (scattered) light.

Absorption

As stated earlier, gases are **selective absorbers**, meaning that they absorb strongly in some wavelengths, moderately in others, and only slightly in still others. When a gas molecule absorbs radiation, the energy is transformed into internal molecular motion, which is detectable as a rise in temperature.

Nitrogen, the most abundant constituent in the atmosphere, is a poor absorber of all types of incoming solar

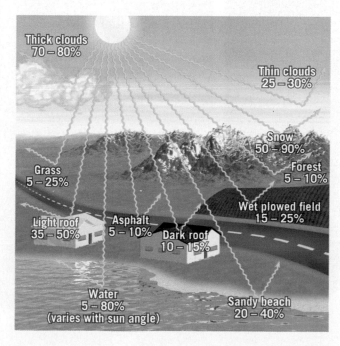

FIGURE 16.23 Albedo (Reflectivity) of Various Surfaces In general, light-colored surfaces tend to reflect more sunlight than dark-colored surfaces and thus have higher albedos.

radiation. Oxygen and ozone are efficient absorbers of ultraviolet radiation. Oxygen removes most of the shorter ultraviolet radiation high in the atmosphere, and ozone absorbs most of the remaining UV rays in the stratosphere. The absorption of UV radiation in the stratosphere accounts for the high temperatures experienced there. The only other significant absorber of incoming solar radiation is water vapor, which, along with oxygen and ozone, accounts for most of the solar radiation absorbed directly by the atmosphere.

For the atmosphere as a whole, none of the gases are effective absorbers of visible radiation. This explains why most visible radiation reaches Earth's surface and why we say that the atmosphere is *transparent* to incoming solar radiation. Thus, the atmosphere does not acquire the bulk of its energy directly from the Sun. Rather, it is heated chiefly by energy that is first absorbed by Earth's surface and then reradiated to the sky.

SmartFigure 16.24
The Greenhouse Effect
Earth's greenhouse effect is compared with two of our close solar system neighbors.

Heating the Atmosphere: The Greenhouse Effect

Approximately 50 percent of the solar energy that strikes the top of the atmosphere reaches Earth's surface and is absorbed. Most of this energy is then reradiated skyward. Because Earth has a much lower surface temperature than the Sun, the radiation that it emits has longer wavelengths than solar radiation.

The atmosphere as a whole is an efficient absorber of the longer wavelengths emitted by Earth (*terrestrial radiation*). Water vapor and carbon dioxide are the principal absorbing gases. Water vapor absorbs roughly five times more terrestrial radiation than do all the other gases combined and accounts for the warm temperatures found in the lower troposphere, where it is most highly concentrated. Because the atmosphere is quite transparent to shorter-wavelength solar radiation and more readily absorbs longer-wavelength radiation emitted by Earth, the atmosphere is heated from the ground up rather than vice versa. This explains the general drop in temperature with increasing altitude experienced in the troposphere. The farther from the "radiator," the colder it becomes.

When the gases in the atmosphere absorb terrestrial radiation, they warm; but they eventually radiate this energy away. Some energy travels skyward, where it may be reabsorbed by other gas molecules, a possibility that is less likely with increasing height because the concentration of water vapor decreases with altitude. The remainder travels Earthward and is again absorbed by Earth. For this reason, Earth's surface is continually being supplied with heat from the atmosphere as well as from the Sun. Without these absorptive gases in our atmosphere, Earth would not be a suitable habitat for humans and numerous other life-forms. This very important phenomenon has been termed the **greenhouse effect** because it was once thought that greenhouses were heated in a similar manner (**FIGURE 16.24**).

Airless bodies like the Moon All incoming solar radiation reaches the surface. Some is reflected back to space. The rest is absorbed by the surface and radiated directly back to space. As a result the lunar surface has a much lower average surface temperature than Earth.

Bodies with modest amounts of greenhouse gases like Earth The atmosphere absorbs some of the longwave radiation emitted by the surface. A portion of this energy is radiated back to the surface and is responsible for keeping Earth's surface 33°C (59°F) warmer than it would otherwise be.

Bodies with abundant greenhouse gases like Venus Venus experiences extraordinary greenhouse warming, which is estimated to raise its surface temperature by 523°C (941°F).

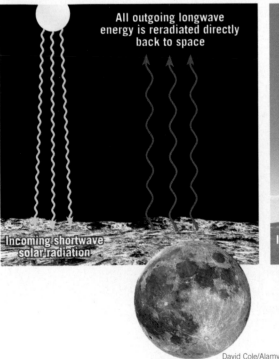
All outgoing longwave energy is reradiated directly back to space

Incoming shortwave solar radiation

David Cole/Alamy

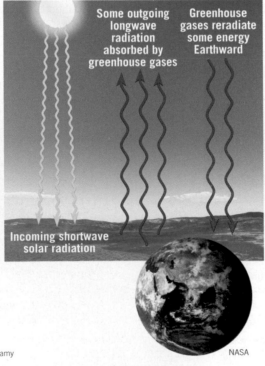
Some outgoing longwave radiation absorbed by greenhouse gases

Greenhouse gases reradiate some energy Earthward

Incoming shortwave solar radiation

NASA

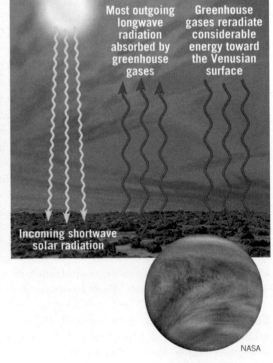
Most outgoing longwave radiation absorbed by greenhouse gases

Greenhouse gases reradiate considerable energy toward the Venusian surface

Incoming shortwave solar radiation

NASA

The gases of our atmosphere, especially water vapor and carbon dioxide, act very much like the glass in the greenhouse. They allow shorter-wavelength solar radiation to enter, where it is absorbed by the objects inside. These objects in turn radiate energy, but at longer wavelengths, to which glass is nearly opaque. The heat therefore is "trapped" in the greenhouse. However, a more important factor in keeping a greenhouse warm is the fact that the greenhouse itself prevents mixing of air inside with cooler air outside. Nevertheless, the term *greenhouse effect* is still used.

16.6 CONCEPT CHECKS

1 What three paths does incoming solar radiation take?

2 What factors cause albedo to vary from time to time and from place to place?

3 Explain why the atmosphere is heated chiefly by radiation emitted from Earth's surface rather than by direct solar radiation.

4 Prepare a sketch with labels that explains the greenhouse effect.

16.7 | FOR THE RECORD: AIR TEMPERATURE DATA

Calculate five commonly used types of temperature data and interpret a map that depicts temperature data using isotherms.

People probably notice changes in air temperature more often than they notice changes in any other element of weather. At a weather station, the temperature is monitored on a regular basis from instruments mounted in an instrument shelter (**FIGURE 16.25**). The shelter protects the instruments from direct sunlight and allows a free flow of air.

The daily maximum and minimum temperatures are the bases for much of the basic temperature data compiled by meteorologists:

- By adding the maximum and minimum temperatures and then dividing by two, the **daily mean temperature** is calculated.
- The **daily range** of temperature is computed by finding the difference between the maximum and minimum temperatures for a given day.
- The **monthly mean** is calculated by adding together the daily means for each day of the month and dividing by the number of days in the month.
- The **annual mean** is an average of the 12 monthly means.
- The **annual temperature range** is computed by finding the difference between the highest and lowest monthly means.

Mean temperatures are particularly useful for making comparisons, whether on a daily, monthly, or annual basis. It is quite common to hear a weather reporter state, "Last month was the hottest July on record," or "Today, Chicago was 10 degrees warmer than Miami." Temperature ranges are also useful statistics because they give an indication of extremes.

To examine the distribution of air temperatures over large areas, isotherms are commonly used. An **isotherm** is a line that connects points on a map that have the same temperature (*iso* = equal, *therm* = temperature). Therefore, all points through which an isotherm passes have identical temperatures for the time period indicated. Generally, isotherms representing 5° or 10° temperature differences are used, but any interval may be chosen. **FIGURE 16.26** illustrates how isotherms are drawn on a map. Notice that most

FIGURE 16.25 Measuring Temperature This modern shelter contains an electrical thermometer called a *thermistor.* A shelter protects instruments from direct sunlight and allows for the free flow of air. (Photo by Bobbé Christopherson)

isotherms do not pass directly through the observing stations because the station readings may not coincide with the values chosen for the isotherms. Only an occasional station temperature will be exactly the same as the value of the isotherm, so it is usually necessary to draw the lines by estimating the proper position between stations.

Maps with isotherms are valuable tools because they clearly make temperature distribution visible at a glance. Areas of low and high temperatures are easy to pick out. In addition, the amount of temperature change per unit of distance, called the **temperature gradient**, is easy to visualize. Closely spaced isotherms indicate a rapid rate of temperature change, whereas more widely spaced lines indicate a more gradual rate of change. You can see this in Figure 16.26. The

SmartFigure 16.26
Isotherms Map showing high temperatures for a spring day. Isotherms are lines that connect points of equal temperature. The temperatures on this map are in degrees Fahrenheit. Showing temperature distribution in this way makes patterns easier to see. Notice that most isotherms do not pass directly through the observing stations. It is usually necessary to draw isotherms by estimating their proper position between stations. On television and in many newspapers, temperature maps are in color. Rather than labeling isotherms, the area *between* isotherms is labeled. For example, the zone between the 60° and 70° isotherms is labeled "60s."

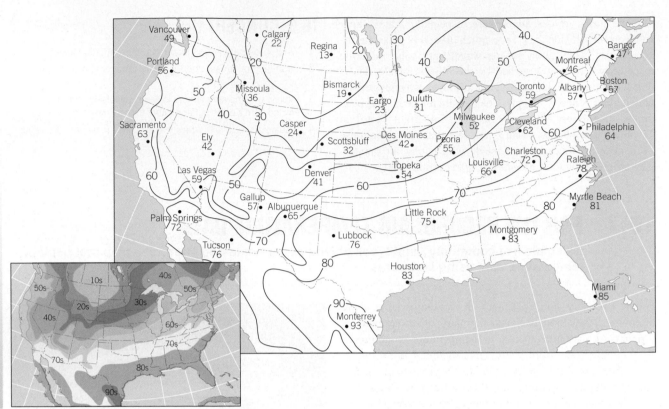

isotherms are closer in Colorado and Utah (steeper temperature gradient), whereas the isotherms are spread farther in Texas (gentler temperature gradient). Without isotherms, a map would be covered with numbers representing temperatures at dozens or hundreds of places, which would make patterns difficult to see.

16.7 CONCEPT CHECKS

1 How are the following temperature data calculated: daily mean, daily range, monthly mean, annual mean, and annual range?

2 What are isotherms, and what is their purpose?

16.8 | WHY TEMPERATURES VARY: THE CONTROLS OF TEMPERATURE

Discuss the principal controls of temperature and use examples to describe their effects.

A **temperature control** is any factor that causes temperature to vary from place to place and from time to time. Earlier in this chapter we examined the most important cause for temperature variations—differences in the receipt of solar radiation. Because variations in Sun angle and length of daylight depend on latitude, they are responsible for warm temperatures in the tropics and colder temperatures at more poleward locations. Of course, seasonal temperature changes at a given latitude occur as the Sun's vertical rays migrate toward and away from a place during the year.

But latitude is not the only control of temperature; if it were, we would expect all places along the same parallel of latitude to have identical temperatures. This is clearly not the case. For example, Eureka, California, and New York City are both coastal cities at about the same latitude, and both have an annual mean temperature of 11°C (52°F). However, New York City is 9°C (16°F) warmer than Eureka in July and 10°C (18°F) cooler in January. In another example, two cities in Ecuador—Quito and Guayaquil—are relatively close to each other, yet the annual mean temperatures

of these two cities differ by 12°C (21°F). To explain these situations and countless others, we must realize that factors other than latitude also exert a strong influence on temperature. In the next sections, we examine these other controls, which include differential heating of land and water, altitude, geographic position, cloud cover and albedo, and ocean currents.[1]

Land and Water

The heating of Earth's surface directly influences the heating of the air above it. Therefore, to understand variations in air temperature, we must understand the variations in heating properties of the different surfaces that Earth presents to the Sun—soil, water, trees, ice, and so on. Different land surfaces absorb varying amounts of incoming solar energy, which in turn cause variations in the temperature of the air above. The greatest contrast, however, is not between

---

[1]For a discussion of the effects of ocean currents on temperature, see Chapter 15.

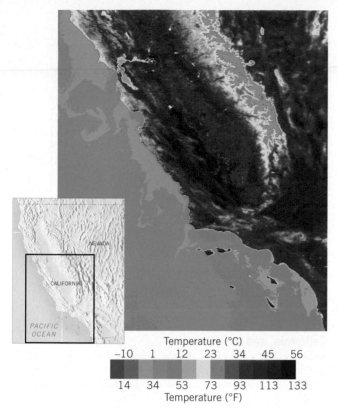

Temperature (°C)

| −10 | 1 | 12 | 23 | 34 | 45 | 56 |

| 14 | 34 | 53 | 73 | 93 | 113 | 133 |

Temperature (°F)

FIGURE 16.27 Differential Heating of Land and Water This satellite image from the afternoon of May 2, 2004, illustrates an important control of air temperature. Water-surface temperatures in the Pacific Ocean are much lower than land-surface temperatures in California and Nevada. The narrow band of cool temperatures in the center of the image is associated with mountains (the Sierra Nevada). The cooler water temperatures immediately offshore are associated with the California Current. (NASA)

different land surfaces but between land and water. **FIGURE 16.27** illustrates this idea nicely. This satellite image shows surface temperatures in portions of Nevada, California, and the adjacent Pacific Ocean on the afternoon of May 2, 2004, during a spring heat wave. Land-surface temperatures are clearly much higher than water-surface temperatures. The image shows the extreme high surface temperatures in southern California and Nevada in dark red.[2] Surface temperatures in the Pacific Ocean are much lower. The peaks of the Sierra Nevada, still capped with snow, form a cool blue line down the eastern side of California.

In side-by-side areas of land and water, such as those shown in Figure 16.27, *land heats more rapidly and to higher temperatures than water, and it cools more rapidly and to lower temperatures than water.* Variations in air temperatures, therefore, are much greater over land than over water.

Why do land and water heat and cool differently? Several factors are responsible:

- The **specific heat** (the amount of energy needed to raise the temperature of 1 gram of a substance 1°C) is far greater for water than for land. Thus, water requires a great deal more heat to raise its

temperature the same amount than does an equal quantity of land.
- Land surfaces are opaque, so heat is absorbed only at the surface. Water, being more transparent, allows heat to penetrate to a depth of many meters.
- The water that is heated often mixes with water below, thus distributing the heat through an even larger mass.
- Evaporation (a cooling process) from water bodies is greater than that from land surfaces.

All these factors collectively cause water to warm more slowly, store greater quantities of heat, and cool more slowly than land.

Monthly temperature data for two cities will demonstrate the moderating influence of a large water body and the extremes associated with land (**FIGURE 16.28**). Vancouver, British Columbia, is located along the windward Pacific coast, whereas Winnipeg, Manitoba, is in a continental position far from the influence of water. Both cities are at about the same latitude and thus experience similar Sun angles and lengths of daylight. Winnipeg, however, has a mean January temperature that is 20°C lower than Vancouver's. Conversely, Winnipeg's July mean is 2.6°C higher than Vancouver's. Although their latitudes are nearly the same, Winnipeg, which has no water influence, experiences much greater temperature extremes than does Vancouver. The key to Vancouver's moderate year-round climate is the Pacific Ocean.

On a different scale, the moderating influence of water may also be demonstrated when temperature variations in the Northern and Southern Hemispheres are compared. In the Northern Hemisphere, 61 percent is covered by water, and land accounts for the remaining 39 percent. However, in the Southern Hemisphere, 81 percent is covered by water and 19 percent by land. The Southern Hemisphere is correctly called the *water hemisphere* (see Figure 13.1, page 410). **TABLE 16.2** portrays the considerably smaller annual temperature variations in the water-dominated Southern Hemisphere as compared with the Northern Hemisphere.

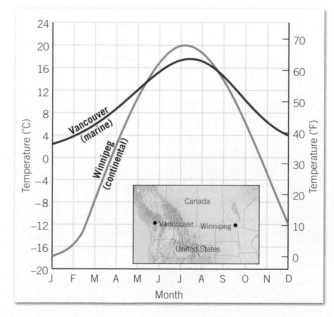

SmartFigure 16.28 Monthly Mean Temperatures for Vancouver, British Columbia, and Winnipeg, Manitoba Vancouver has a much smaller annual temperature range because of the strong marine influence of the Pacific Ocean. The curve for Winnipeg illustrates the greater extremes associated with an interior location.

---

[2]Realize that when a land surface is hot, the air above is cooler. For example, while the surface of a sandy beach can be painfully hot, the air temperature above the surface is more comfortable.

TABLE 16.2 Variation in Annual Mean Temperature Range (°C) with Latitude

Latitude	Northern Hemisphere	Southern Hemisphere
0	0	0
15	3	4
30	13	7
45	23	6
60	30	11
75	32	26
90	40	31

Altitude

The two cities in Ecuador mentioned earlier—Quito and Guayaquil—demonstrate the influence of altitude on mean temperatures. Although both cities are near the equator and not far apart, the annual mean temperature at Guayaquil is 25°C (77°F), as compared to Quito's mean of 13°C (55°F). The difference is explained largely by the difference in the cities' elevations: Guayaquil is only 12 meters (40 feet) above sea level, whereas Quito is high in the Andes Mountains, at 2800 meters (9200 feet). **FIGURE 16.29** provides another example.

Recall that temperatures drop an average of 6.5°C per kilometer in the troposphere; thus, cooler temperatures are to be expected at greater heights. Yet the magnitude of the difference is not explained completely by the normal lapse rate. If the normal lapse rate is used, we would expect Quito to be about 18°C cooler than Guayaquil, but the difference is only 12°C. The fact that high-altitude places such as Quito are warmer than the value calculated using the normal lapse rate results from the absorption and reradiation of solar energy by the ground surface.

Geographic Position

The geographic setting can greatly influence the temperatures experienced at a specific location. A coastal location where

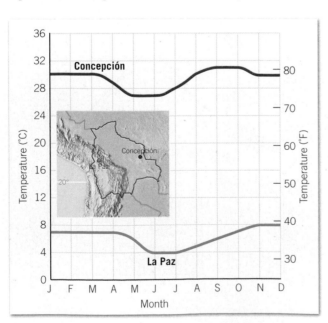

FIGURE 16.29 Monthly Mean Temperatures for Concepción and La Paz, Bolivia Both cities have nearly the same latitude (about 16° south). However, because La Paz is high in the Andes, at 4103 meters (13,461 feet), it experiences much cooler temperatures than Concepción, which is at an elevation of 490 meters (1608 feet).

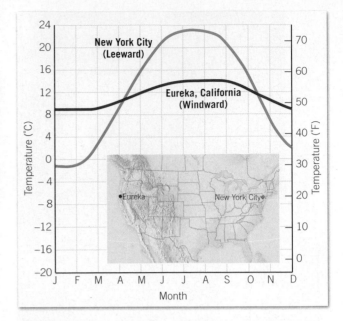

FIGURE 16.30 Monthly Mean Temperatures for Eureka, California, and New York City Both cities are coastal and located at about the same latitude. Because Eureka is strongly influenced by prevailing winds from the ocean and New York City is not, the annual temperature range at Eureka is much smaller.

prevailing winds blow from the ocean onto the shore (a **windward coast**) experiences considerably different temperatures than does a coastal location where the prevailing winds blow from the land toward the ocean (a **leeward coast**). In the first situation, the windward coast will experience the full moderating influence of the ocean—cool summers and mild winters—compared to an inland station at the same latitude.

A leeward coast, on the other hand, will have a more continental temperature pattern because the winds do not carry the ocean's influence onshore. Eureka, California, and New York City, the two cities mentioned earlier, illustrate this aspect of geographic position. The annual temperature range at New York City is 19°C (34°F) greater than Eureka's (**FIGURE 16.30**).

Seattle and Spokane, both in the state of Washington, illustrate a second aspect of geographic position: mountains that act as barriers. Although Spokane is only about 360 kilometers (220 miles) east of Seattle, the towering Cascade Range separates the cities. Consequently, Seattle's temperatures show a marked marine influence, but Spokane's are more typically continental (**FIGURE 16.31**). Spokane is 7°C (13°F) cooler than Seattle in January and 4°C (7°F) warmer than Seattle in July. The annual range at Spokane is 11°C (20°F) greater than at Seattle. The Cascade Range effectively cuts off Spokane from the moderating influence of the Pacific Ocean.

Cloud Cover and Albedo

You may have noticed that clear days are often warmer than cloudy ones and that clear nights are usually cooler than cloudy ones. This demonstrates that cloud cover is another factor that influences temperature in the lower atmosphere. Studies using satellite images show that at

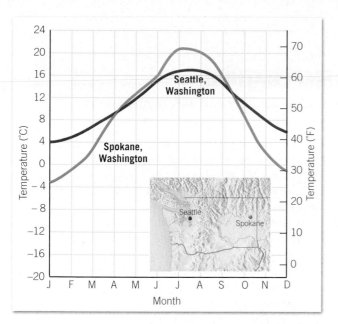

FIGURE 16.31 Monthly Mean Temperatures for Seattle and Spokane, Washington Because the Cascade Mountains cut off Spokane from the moderating influence of the Pacific Ocean, Spokane's annual temperature range is greater than Seattle's.

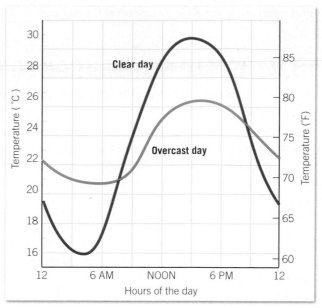

SmartFigure 16.32 The Daily Cycle of Temperature at Peoria, Illinois, for Two July Days Clouds reduce the daily temperature range. During daylight hours, clouds reflect solar radiation back to space. Therefore, the maximum temperature is lower than if the sky were clear. At night, the minimum temperature will not fall as low because clouds retard the loss of heat.

any particular time, about half of our planet is covered by clouds. Cloud cover is important because many clouds have a high albedo; therefore, clouds reflect a significant portion of the sunlight that strikes them back into space (see Figure 16.23). Cloud cover reduces the amount of incoming solar radiation, and daytime temperatures will be lower than if the clouds were not present and the sky were clear.

At night, clouds have the opposite effect as during daylight: They act as a blanket by absorbing radiation emitted by Earth's surface and reradiating a portion of it back to the surface. Consequently, some of the heat that otherwise would have been lost remains near the ground. Thus, nighttime air temperatures do not drop as low as they would on a clear night. The effect of cloud cover is to reduce the daily temperature range by lowering the daytime maximum and raising the nighttime minimum (**FIGURE 16.32**).

Clouds are not the only phenomenon that increase albedo and thereby reduces air temperatures. We also recognize that snow- and ice-covered surfaces have high albedos. This is one reason why mountain glaciers do not melt away in the summer and why snow may still be present on a mild spring day. In addition, during the winter, when snow covers the ground, daytime maximums on a sunny day are less than they otherwise would be because energy that the land would have absorbed and used to heat the air is reflected and lost.

16.8 CONCEPT CHECKS

1 List the factors that cause land and water to heat and cool differently.

2 Quito, Ecuador, is located on the equator and is *not* a coastal city. It has an average annual temperature of only 13°C (55°F). What is the likely cause for this low average temperature?

3 In what ways can geographic position be considered a control of temperature?

4 How does cloud cover influence the maximum temperature on an overcast day? How is the nighttime minimum influenced by clouds?

EYE ON EARTH

For more than a decade, scientists have used the Moderate Resolution Imaging Spectroradiometer (MODIS, for short) aboard NASA's *Aqua* and *Terra* satellites to gather surface temperature data from around the globe. This image shows average land-surface temperatures for the month of February over a 10-year span (2001–2010). (NASA)

QUESTION 1 *What are the approximate temperatures for southern Great Britain and northern Newfoundland (white arrows)?*

QUESTION 2 *Both southern Great Britain and northern Newfoundland are coastal areas at the same latitude, yet average February temperatures are quite different. Suggest a reason for this disparity.*

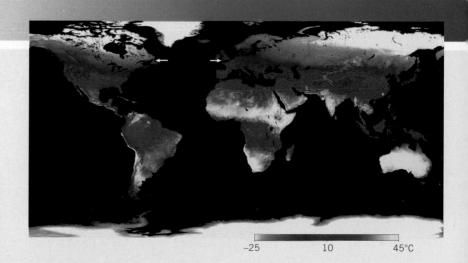

16.9 | WORLD DISTRIBUTION OF TEMPERATURE

Interpret the patterns depicted on world maps of January and July temperatures.

Take a moment to study the two world isothermal maps in **FIGURES 16.33** and **16.34**. From hot colors near the equator to cool colors toward the poles, these maps portray sea-level temperatures in the seasonally extreme months of January and July. Temperature distribution is shown by using isotherms. On these maps you can study global temperature patterns and the effects of the controlling factors of temperature, especially latitude, the distribution of land and water, and ocean currents. Like most other isothermal maps of large regions, all temperatures on these world maps have been reduced to sea level to eliminate the complications caused by differences in altitude.

On both maps, the isotherms generally trend east and west and show a decrease in temperatures poleward from the tropics. They illustrate one of the most fundamental aspects of world temperature distribution: that the effectiveness of incoming solar radiation in heating Earth's surface and the atmosphere above it is largely a function of latitude.

Moreover, there is a latitudinal shifting of temperatures caused by the seasonal migration of the Sun's vertical rays. To see this, compare the color bands by latitude on the two maps. On the January map, the "hot spots" of 30°C are *south* of the equator, but in July they have shifted *north* of the equator.

If latitude were the only control of temperature distribution, our analysis could end here, but that is not the case. The added effect of the differential heating of land and water is also reflected on the January and July temperature maps. The warmest and coldest temperatures are found over land; note the coldest area, a purple oval in Siberia, and the hottest areas, the deep orange ovals, all over land. Because temperatures do not fluctuate as much over water as over land, the north–south migration of isotherms is greater over the continents than over the oceans.

In addition, it is clear that the isotherms in the Southern Hemisphere, where there is little land and where the oceans predominate, are much more regular than in the Northern Hemisphere, where they bend sharply northward in July and southward in January over the continents.

Isotherms also show the presence of ocean currents. Warm currents cause isotherms to be deflected toward the poles, whereas cold currents cause an equatorward bending. The horizontal transport of water poleward warms the overlying air and results in air temperatures that are higher than would otherwise be expected for the latitude. Conversely, currents moving toward the equator produce cooler-than-expected air temperatures.

Because Figures 16.33 and 16.34 show the seasonal extremes of temperature, they can be used to evaluate variations in the annual range of temperature from place to place. A comparison of the two maps shows that a station near the equator has a very small annual range because it experiences little variation in the length of daylight, and it always has a relatively high Sun angle. A station in the middle latitudes, however, experiences wide variations in Sun angle and length of daylight and hence large variations in temperature. Therefore, we can state that the

FIGURE 16.33 World Mean Sea-Level Temperatures in January, in Celsius (°C) and Fahrenheit (°F)

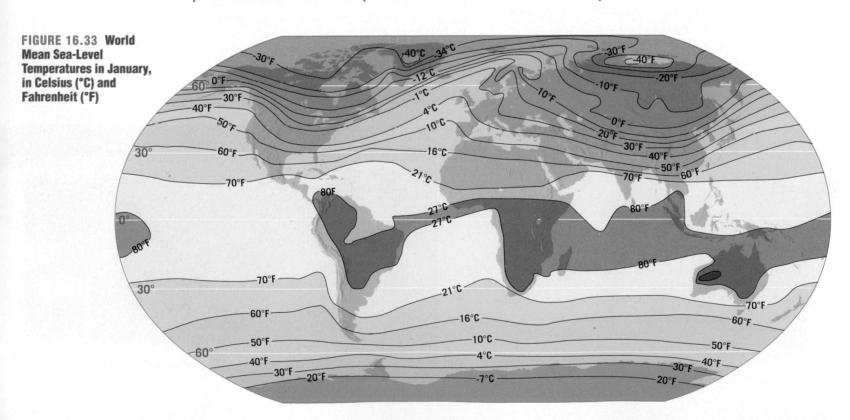

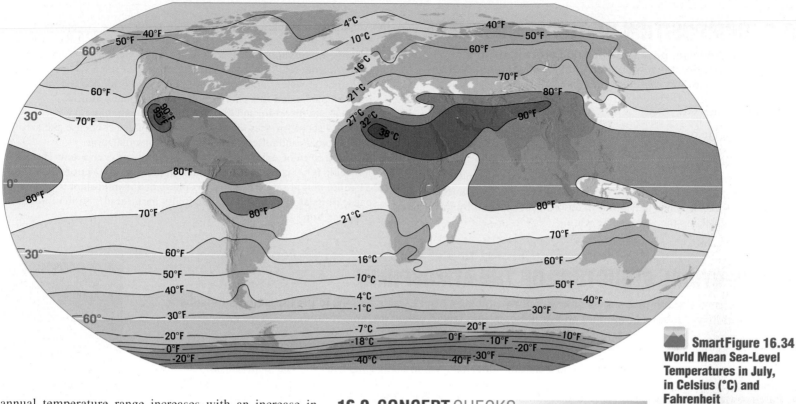

SmartFigure 16.34 World Mean Sea-Level Temperatures in July, in Celsius (°C) and Fahrenheit (°F)

annual temperature range increases with an increase in latitude.

Moreover, land and water also affect seasonal temperature variations, especially outside the tropics. A continental location must endure hotter summers and colder winters than a coastal location. Consequently, outside the tropics, the annual temperature range will increase with an increase in continentality.

16.9 CONCEPT CHECKS

1 Why do isotherms generally trend east–west?

2 Why do isotherms shift north and south from season to season?

3 Where do isotherms shift most, over land or water? Explain.

4 Which area on Earth experiences the highest annual temperature range?

16 CONCEPTS IN REVIEW | The Atmosphere: Composition, Structure, and Temperature

16.1 FOCUS ON THE ATMOSPHERE

Distinguish between weather and climate and name the basic elements of weather and climate.

KEY TERMS: weather, climate, elements (of weather and climate)

- Weather is the state of the atmosphere at a particular place for a short period of time. Climate, on the other hand, is a generalization of the weather conditions of a place over a long period of time.
- The most important elements—quantities or properties that are measured regularly—of weather and climate are (1) air temperature, (2) humidity, (3) type and amount of cloudiness, (4) type and amount of precipitation, (5) air pressure, and (6) the speed and direction of the wind.

Q This is a scene on a summer day in Antarctica, showing a joint British–American research team. Write two brief statements about the locale in this image—one that relates to weather and one that relates to climate.

David Vaughan/Science Source

16.2 COMPOSITION OF THE ATMOSPHERE

List the major gases composing Earth's atmosphere and identify the components that are most important to understanding weather and climate.

KEY TERMS: air, aerosols, ozone

- Air is a mixture of many discrete gases, and its composition varies from time to time and from place to place. If water vapor, dust, and other variable components of the atmosphere are removed, clean, dry air is composed almost entirely of nitrogen (N_2) and oxygen (O_2). Carbon dioxide (CO_2), although present only in minute amounts, is important because it has the ability to absorb heat radiated by Earth and thus helps keep the atmosphere warm.

- Among the variable components of air, water vapor is important because it is the source of all clouds and precipitation. Like carbon dioxide, water vapor can absorb heat emitted by Earth. When water changes from one state to another, it absorbs or releases heat. In the atmosphere, water vapor transports this latent ("hidden") heat from place to place, and this energy helps to drive many storms.

- Aerosols are tiny solid and liquid particles that are important because they may act as surfaces on which water vapor can condense and are also absorbers and reflectors of incoming solar radiation.

- Ozone, a form of oxygen that combines three oxygen atoms into each molecule (O_3), is a gas concentrated in the 10- to 50-kilometer (6- to 31-mile) height range in the atmosphere and is important to life because of its ability to absorb potentially harmful ultraviolet radiation from the Sun.

16.3 VERTICAL STRUCTURE OF THE ATMOSPHERE

Interpret a graph that shows changes in air pressure from Earth's surface to the top of the atmosphere. Sketch and label a graph that shows atmospheric layers based on temperature.

KEY TERMS: troposphere, environmental lapse rate, radiosonde, stratosphere, mesosphere, thermosphere

David R. Frazier/Science Source

- Because the atmosphere gradually thins with increasing altitude, it has no sharp upper boundary but simply blends into outer space.

- Based on temperature, the atmosphere is divided vertically into four layers. The *troposphere* is the lowermost layer. In the troposphere, temperature usually decreases with increasing altitude. This *environmental lapse rate* is variable but averages about 6.5°C per kilometer (3.5°F per 1000 feet). Essentially, all important weather phenomena occur in the troposphere.

- Beyond the troposphere is the *stratosphere*, which exhibits warming because of absorption of UV radiation by ozone. In the *mesosphere*, temperatures again decrease. Upward from the mesosphere is the *thermosphere*, a layer with only a tiny fraction of the atmosphere's mass and no well-defined upper limit.

- **Q** When the weather balloon in this photo was launched, the surface temperature was 17°C. The balloon is now at an altitude of 1 kilometer. What term is applied to the instrument package being carried aloft by the balloon? In what layer of the atmosphere is the balloon? If average conditions prevail, what is the air temperature at this altitude? How did you figure this out?

16.4 EARTH—SUN RELATIONSHIPS

Explain what causes the Sun angle and length of daylight to change during the year and describe how these changes produce the seasons.

KEY TERMS: rotation, circle of illumination, revolution, inclination of the axis, Tropic of Cancer, summer solstice, Tropic of Capricorn, winter solstice, autumnal (fall) equinox, spring equinox

- The two principal motions of Earth are (1) rotation, the spinning about its axis that produces the daily cycle of daylight and darkness, and (2) revolution, the movement in its orbit around the Sun.

- The seasons are caused by changes in the angle at which the Sun's rays strike the surface and the changes in the length of daylight at each latitude. These seasonal changes are the result of the tilt of Earth's axis as it revolves around the Sun.

- **Q** Assume that the date is December 22. At what latitude are the Sun's vertical rays striking? Is this date an equinox or a solstice?

16.5 ENERGY, HEAT, AND TEMPERATURE

Distinguish between heat and temperature. List and describe the three mechanisms of heat transfer.

KEY TERMS: heat, temperature, conduction, convection, radiation, electromagnetic radiation, visible light, infrared, ultraviolet (UV)

- Heat refers to the quantity of energy present in a material, whereas temperature refers to intensity, or the degree of "hotness."

- The three mechanisms of heat transfer are (1) conduction, the transfer of heat through matter by molecular activity; (2) convection, the transfer of heat by the movement of a mass or substance from one place to another; and (3) radiation, the transfer of heat by electromagnetic waves.

■ Electromagnetic radiation is energy emitted in the form of rays, or waves, called electromagnetic waves. All radiation is capable of transmitting energy through the vacuum of space. One of the most important differences between electromagnetic waves is their wavelengths, which range from very long radio waves to very short gamma rays. Visible light is the only portion of the electromagnetic spectrum we can see.

■ Some basic laws that relate to radiation are (1) all objects emit radiant energy; (2) hotter objects radiate more total energy than do colder objects; (3) the hotter the radiating body, the shorter the wavelengths of maximum radiation; and (4) objects that are good absorbers of radiation are good emitters as well.

Q **Describe how each of the three basic mechanisms of heat transfer are illustrated in this image.**

16.6 HEATING THE ATMOSPHERE

Sketch and label a diagram that shows the paths taken by incoming solar radiation. Summarize the greenhouse effect.

KEY TERMS: reflection, scattering, albedo, diffused light, selective absorbers, greenhouse effect

■ About 50 percent of the solar radiation that strikes the atmosphere reaches Earth's surface. About 30 percent is reflected back to space. The fraction of radiation reflected by a surface is called its albedo. The remaining 20 percent of the energy is absorbed by clouds and the atmosphere's gases.

■ Radiant energy absorbed at Earth's surface is eventually radiated skyward. Because Earth has a much lower surface temperature than the Sun, its radiation is in the form of long-wave infrared radiation. Because atmospheric gases, primarily water vapor and carbon dioxide, are more efficient absorbers of long-wave radiation, the atmosphere is heated from the ground up.

■ The selective absorption of Earth radiation by water vapor and carbon dioxide that results in Earth's average temperature being warmer than it would be otherwise is referred to as the greenhouse effect.

16.7 FOR THE RECORD: AIR TEMPERATURE DATA

Calculate five commonly used types of temperature data and interpret a map that depicts temperature data using isotherms.

KEY TERMS: daily mean temperature, daily range, monthly mean, annual mean, annual temperature range, isotherm, temperature gradient

■ Daily mean temperature is an average of the daily maximum and daily minimum temperatures, whereas the daily range is the difference between the daily maximum and daily minimum temperatures. The monthly mean is determined by averaging the daily means for a particular month. The annual mean is an average of the 12 monthly means, whereas the annual temperature range is the difference between the highest and lowest monthly means.

■ Temperature distribution is shown on a map by using isotherms, which are lines of equal temperature. Temperature gradient is the amount of temperature change per unit of distance. Closely spaced isotherms indicate a rapid rate of change.

16.8 WHY TEMPERATURES VARY: THE CONTROLS OF TEMPERATURE

Discuss the principal controls of temperature and use examples to describe their effects.

KEY TERMS: temperature control, specific heat, windward coast, leeward coast

■ Controls of temperature are factors that cause temperature to vary from place to place and from time to time. Latitude (Earth–Sun relationships) is one example. Ocean currents (discussed in Chapter 10) provide another example.

■ Unequal heating of land and water is a temperature control. Because land and water heat and cool differently, land areas experience greater temperature extremes than water-dominated areas.

■ Altitude is an easy-to-visualize control: The higher up you go, the colder it gets; therefore, mountains are cooler than adjacent lowlands.

■ Geographic position as a temperature control involves factors such as mountains acting as barriers to marine influence and a place being on a windward or a leeward coast.

Q **The graph shows monthly high temperatures for Urbana, Illinois, and San Francisco, California. Although the two cities are located at about the same latitude, the temperatures they experience are quite different. Which line on the graph represents Urbana, and which represents San Francisco? How did you figure this out?**

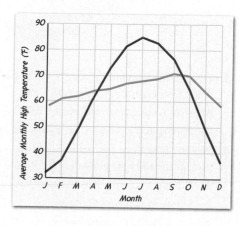

16.9 WORLD DISTRIBUTION OF TEMPERATURE

Interpret the patterns depicted on world maps of January and July temperatures.

- On world maps showing January and July mean temperatures, isotherms generally trend east–west and show a decrease in temperature moving poleward from the equator. When the two maps are compared, a latitudinal shifting of temperatures is seen. Bending isotherms reveal the locations of ocean currents.
- Annual temperature range is small near the equator and increases with an increase in latitude. Outside the tropics, annual temperature range also increases as marine influence diminishes.

Q Refer to Figure 16.33. What causes the bend or kink in the isotherms in the North Atlantic?

GIVE IT SOME **THOUGHT**

1. Determine which statements refer to weather and which refer to climate. (*Note:* One statement includes aspects of *both* weather and climate.)

   a. The baseball game was rained out today.

   b. January is Omaha's coldest month.

   c. North Africa is a desert.

   d. The high this afternoon was 25°C.

   e. Last evening a tornado ripped through central Oklahoma.

   f. I am moving to southern Arizona because it is warm and sunny.

   g. Thursday's low of −20°C is the coldest temperature ever recorded for that city.

   h. It is partly cloudy.

2. This map shows the mean percentage of possible sunshine received in the month of November across the 48 contiguous United States.

   a. Does this map relate more to climate or to weather?

   b. If you visited Yuma, Arizona, in November, would you *expect* to experience a sunny day or an overcast day?

   c. Might what you actually experience during your visit be different from what you expected? Explain.

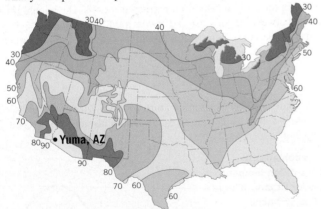

3. Refer to the graph in Figure 16.3 to answer the following questions about temperatures in New York City:

   a. What is the approximate average daily high temperature in January? In July?

   b. Approximately what are the highest and lowest temperatures ever recorded?

4. Which of the three mechanisms of heat transfer is clearly illustrated in each of the following situations?

   a. Driving a car with the seat heater turned on

   b. Sitting in an outdoor hot tub

   c. Lying inside a tanning bed

   d. Driving a car with the air conditioning turned on

5. The circumference of Earth at the equator is 24,900 miles. Calculate how fast someone at the equator is rotating in miles per hour. If the rotational speed of Earth were to slow down, how might this impact daytime highs and nighttime lows?

6. Rank the following according to the wavelengths of radiant energy each emits, from the shortest wavelengths to the longest:

   a. A light bulb with a filament glowing at 4000°C

   b. A rock at room temperature

   c. A car engine at 140°C

7. Imagine being at the beach in this photo on a sunny summer afternoon.

   a. Describe the temperatures you would expect if you measured the surface of the beach and at a depth of 12 inches in the sand.

   b. If you stood waist deep in the water and measured the water's surface temperature and its temperature at a depth of 12 inches, how would these measurements compare to those taken on the beach?

Tequilab/Shutterstock

8. On which summer day would you expect the *greatest* temperature range? Which would have the *smallest* range in temperature? Explain your choices.

   a. Cloudy skies during the day and clear skies at night

   b. Clear skies during the day and cloudy skies at night

   c. Clear skies during the day and clear skies at night

   d. Cloudy skies during the day and cloudy skies at night

9. The accompanying sketch map represents a hypothetical continent in the Northern Hemisphere. One isotherm has been placed on the map.

   a. Is the temperature higher at City A or City B?

   b. Is the season winter or summer? How are you able to determine this?

   c. Describe (or sketch) the position of this isotherm 6 months later.

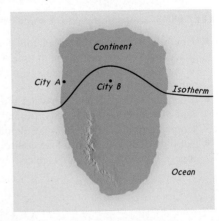

10. This photo shows a snow-covered area in the middle latitudes on a sunny day in late winter. Assume that 1 week after this photo was taken, conditions were essentially identical, except that the snow was gone. Would you expect the air temperatures to be different on the two days? If so, which day would be warmer? Suggest an explanation.

CoolR/Shutterstock

11. The Sun shines continually at the North Pole for 6 months, from the spring equinox until the fall equinox, yet temperatures never get very warm. Explain why this is the case.

12. The data below are mean monthly temperatures in degrees Celsius for an inland location that lacks any significant ocean influence. *Based on annual temperature range*, what is the approximate latitude of this place? Are these temperatures what you would normally expect for this latitude? If not, what control would explain these temperatures?

J	F	M	A	M	J	J	A	S	O	N	D
6.1	6.6	6.6	6.6	6.6	6.1	6.1	6.1	6.1	6.1	6.6	6.6

EXAMINING THE **EARTH SYSTEM**

1. Earth's axis is inclined 23½° to the plane of its orbit. What if the inclination of the axis changed? Answer the following questions that address this possibility:

  a. How would seasons be affected if Earth's axis were perpendicular to the plane of its orbit?

  b. Describe the seasons if Earth's axis were inclined 40°. Where would the Tropics of Cancer and Capricorn be located? How about the Arctic and Antarctic Circles?

3. The accompanying photo shows the explosive 1991 eruption of Mt. Pinatubo in the Philippines. How would you expect global temperatures to respond to the ash and debris that this volcano spewed high into the atmosphere? Speculate about how a change in temperature might impact one or more of the spheres in the Earth system.

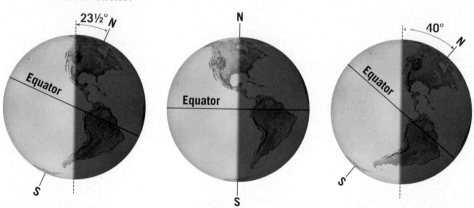

D. Harlow/USGS

2. Speculate on the changes in global temperatures that might occur if Earth had substantially more land area and less ocean area than at present. How might such changes influence the biosphere?

4. Figure 16.21 shows that about 30 percent of the Sun's energy is reflected and scattered back to space. If Earth's albedo were to increase to 50 percent, how would you expect average surface temperatures to change? Explain.

MasteringGeology™

Looking for additional review and test prep materials? Visit the Self Study area in **www.masteringgeology.com** to find practice quizzes, study tools, and multimedia that will aid in your understanding of this chapter's content. In **MasteringGeology™** you will find:

- GEODe: Earth Science: An interactive visual walkthrough of key concepts
- Geoscience Animation Library: More than 100 animations illuminating many difficult-to-understand Earth science concepts

- In The News RSS Feeds: Current Earth science events and news articles are pulled into the site with assessment
- Pearson eText
- Optional Self Study Quizzes
- Web Links
- Glossary
- Flashcards

17

Moisture, Clouds, and Precipitation

Cumulonimbus clouds are often associated with thunderstorms and severe weather. (Photo by Cusp/SuperStock)

Water vapor is an odorless, colorless gas that mixes freely with the other gases of the atmosphere. Unlike oxygen and nitrogen—the two most abundant components of the atmosphere—water can change from one state of matter to another (solid, liquid, or gas) at the temperatures and pressures experienced on Earth. Because of this unique property, water freely leaves the oceans as a gas and returns again as a liquid or solid.

As you observe day-to-day weather changes, you might ask: Why is it generally more humid in the summer than in the winter? Why do clouds form on some occasions but not on others? Why do some clouds look thin and harmless, whereas others form gray and ominous towers? Answers to these questions involve the role of water vapor in the atmosphere, the central theme of this chapter.

17.1 | WATER'S CHANGES OF STATE

List and describe the processes that cause water to change from one state of matter to another. Define *latent heat* and explain why it is important.

Water is the only substance that exists in the atmosphere as a solid, liquid, and gas (**FIGURE 17.1**). It is made of hydrogen and oxygen atoms that are bonded together to form water molecules (H_2O). In all three states of matter (even ice), these molecules are in constant motion; the higher the temperature, the more vigorous the movement. The chief difference among liquid water, ice, and water vapor is the arrangement of the water molecules.

Ice, Liquid Water, and Water Vapor

Ice is composed of water molecules that are held together by mutual molecular attractions. The molecules form a tight, orderly network, as shown in **FIGURE 17.2**. As a consequence, the water molecules in ice are not free to move relative to each other but rather vibrate about fixed sites. When ice is heated, the molecules oscillate more rapidly. When the rate of molecular movement increases sufficiently, the bonds between some of the water molecules are broken, resulting in melting.

In the liquid state, water molecules are still tightly packed but are moving fast enough that they are able to slide past one another. As a result, liquid water is fluid and will take the shape of its container.

As liquid water gains heat from its environment, some of the molecules will acquire enough energy to break the remaining molecular attractions and escape from the surface, becoming water vapor. Water-vapor molecules are widely spaced compared to liquid water and exhibit very energetic random motion. What distinguishes a gas from a liquid is its compressibility (and expandability). For example, you can easily put more and more air into a tire and increase its volume only slightly. However, you can't put 10 gallons of gasoline into a 5-gallon can.

To summarize, when water changes state, it does not turn into a different substance; only the distances and interactions among the water molecules change.

Latent Heat

Whenever water changes state, heat is exchanged between water and its surroundings. When water evaporates, heat is absorbed (see Figure 17.2). Meteorologists often measure heat energy in calories. One **calorie** is the amount of heat required to raise the temperature of 1 gram of water 1°C (1.8°F). Thus, when 10 calories of heat are absorbed by 1 gram of water, the molecules vibrate faster, and a 10°C (18°F) temperature rise occurs.

Under certain conditions, heat may be added to a substance without an accompanying rise in temperature. For example, when a glass of ice water is warmed, the temperature of the ice–water mixture remains a constant 0°C (32°F) until all the ice has melted. If adding heat does not raise the temperature, where does this energy go? In this case, the added energy goes into breaking the molecular attractions between the water molecules in the ice cubes.

Because the heat used to melt ice does not produce a temperature change, it is referred to as **latent heat**. (*Latent* means "hidden," like the latent fingerprints hidden at a crime

FIGURE 17.1 Caught in a Downpour (Photo by AP Photo/Keystone, Marcel Bier)

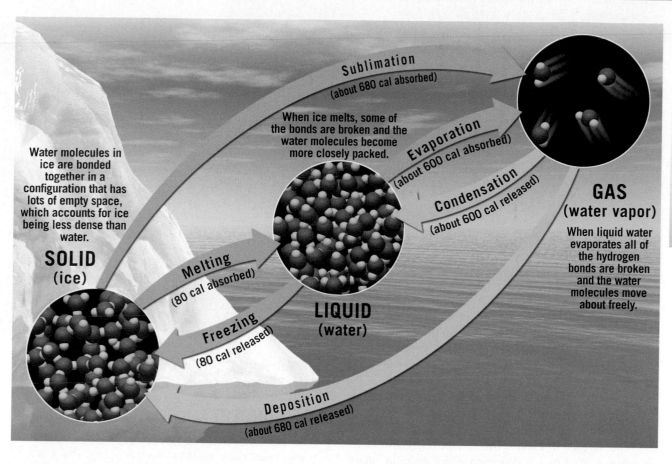

Water molecules in ice are bonded together in a configuration that has lots of empty space, which accounts for ice being less dense than water.

When ice melts, some of the bonds are broken and the water molecules become more closely packed.

Sublimation (about 680 cal absorbed)

Evaporation (about 600 cal absorbed)

Condensation (about 600 cal released)

Melting (80 cal absorbed)

Freezing (80 cal released)

Deposition (about 680 cal released)

SOLID (ice)

LIQUID (water)

GAS (water vapor)

When liquid water evaporates all of the hydrogen bonds are broken and the water molecules move about freely.

SmartFigure 17.2 Changes of State Involve an Exchange of Heat The numbers represent the approximate number of calories either absorbed or released when 1 gram of water changes from one state of matter to another.

scene.) This energy can be thought of as being stored in liquid water, and it is not released to its surroundings as heat until the liquid returns to the solid state.

Melting 1 gram of ice requires 80 calories, an amount referred to as *latent heat of melting. Freezing*, the reverse process, releases these 80 calories per gram to the environment as *latent heat of fusion.*

Evaporation and Condensation

We saw that heat is absorbed when ice is converted to liquid water. Heat is also absorbed during **evaporation**, the process of converting a liquid to a gas (vapor). The energy absorbed by water molecules during evaporation is used to give them the motion needed to escape the surface of the liquid and become a gas. This energy is referred to as the *latent heat of vaporization.* During the process of evaporation, it is the higher-temperature (faster-moving) molecules that escape the surface. As a result, the average molecular motion (temperature) of the remaining water is reduced—hence the common expression "evaporation is a cooling process." You have undoubtedly experienced this cooling effect when stepping dripping wet from a swimming pool or bathtub. In this situation, the energy used to evaporate water comes from your skin—hence, you feel cool.

The reverse process, **condensation**, occurs when water vapor changes to the liquid state. During condensation, water-vapor molecules release energy (*latent heat of condensation*) in an amount equivalent to what was absorbed during evaporation. When condensation occurs in the atmosphere, it results in the formation of such phenomena as fog and clouds.

As you will see, latent heat plays an important role in many atmospheric processes. In particular, when water vapor condenses to form cloud droplets, latent heat of condensation is released, warming the surrounding air and giving it buoyancy. When the moisture content of air is high, this process can spur the growth of towering storm clouds.

EYE ON EARTH

The Navajo Generating Station, located on the Navajo Indian Reservation near Page, Arizona, has three 236-meter (560-feet) stacks. (Photo by Michael Collier)

QUESTION 1 *What fuel is this plant burning to generate electricity? (Hint: Look directly behind the facility.)*

QUESTION 2 *Why do power-generating facilities such as this one have tall stacks?*

QUESTION 3 *Explain why the "smoke" changes color from bright white to pale yellow when it reaches a height of about 200 feet above the stacks.*

520

FIGURE 17.3 Examples of Condensation and Deposition (Top photo by NaturePL/SuperStock; bottom photo by elen_studio/Fotolia)

Condensation of water vapor generates phenomena such as dew, clouds, and fog.

Frost on a windowpane, an example of deposition

Sublimation and Deposition You are probably least familiar with the last two processes illustrated in Figure 17.2—sublimation and deposition. **Sublimation** is the conversion of a solid directly to a gas, without passing through the liquid state. Examples you may have observed include the gradual shrinking of unused ice cubes in the freezer and the rapid conversion of dry ice (frozen carbon dioxide) to wispy clouds that quickly disappear.

Deposition refers to the reverse process, the conversion of a vapor directly to a solid. This change occurs, for example, when water vapor is deposited as ice on solid objects such as grass or windows (**FIGURE 17.3**). These deposits are called *white frost* or *hoar frost* and are frequently referred to simply as *frost*. A household example of the process of deposition is the "frost" that accumulates in a freezer. As shown in Figure 17.2, deposition releases an amount of energy equal to the total amount released by condensation and freezing.

17.1 CONCEPT CHECKS

1 Summarize the processes by which water changes from one state of matter to another. Indicate whether energy is absorbed or released.

2 What is *latent heat*?

3 What is a common example of sublimation?

4 How does frost form?

17.2 | HUMIDITY: WATER VAPOR IN THE AIR

Distinguish between relative humidity and dew point. Write a generalization relating how temperature changes affect relative humidity.

Water vapor constitutes only a small fraction of the atmosphere, varying from as little as one-tenth of 1 percent up to about 4 percent by volume. But the importance of water in the air is far greater than these small percentages would indicate. Indeed, scientists agree that *water vapor* is the most important gas in the atmosphere when it comes to understanding atmospheric processes.

Humidity is the general term for the amount of water vapor in air. Meteorologists employ several methods to express the water-vapor content of the air; we examine three: mixing ratio, relative humidity, and dew-point temperature.

Saturation

Before we consider these humidity measures further, it is important to understand the concept of **saturation**. Imagine a closed jar that contains water overlain by dry air, both at the same temperature. As the water begins to evaporate from the water surface, a small increase in pressure can be detected in the air above. This increase is the result of the motion of the water-vapor molecules that were added to the air through evaporation. In the open atmosphere, this pressure is termed **vapor pressure** and is defined as the part of the total atmospheric pressure that can be attributed to the water-vapor content.

In the closed container, as more and more molecules escape from the water surface, the steadily increasing vapor pressure in the air above forces more and more of these molecules to return to the liquid. Eventually the number of vapor molecules returning to the surface will balance the number leaving. At that point, the air is *saturated*. If we add heat to the container, thereby increasing the temperature of the water and air, more water will evaporate before a balance is reached. Consequently, at higher temperatures, more moisture is required for saturation. The amount of water vapor required for saturation at various temperatures is shown in **TABLE 17.1**.

TABLE 17.1 Amount of Water Vapor Needed to Saturate 1 Kilogram of Air at Various Temperatures

Temperature °C (°F)	Water-Vapor Content at Saturation (grams)
–40 (–40)	0.1
–30 (–22)	0.3
–20 (–4)	0.75
–10 (14)	2
0 (32)	3.5
5 (41)	5
10 (50)	7
15 (59)	10
20 (68)	14
25 (77)	20
30 (86)	26.5
35 (95)	35
40 (104)	47

Mixing Ratio

Not all air is saturated, of course. Thus, we need ways to express how humid a parcel of air is. One method is to specify the amount of water vapor contained in a unit of air. The **mixing ratio** is the mass of water vapor in a unit of air compared to the remaining mass of dry air:

$$\text{mixing ratio} = \frac{\text{mass of water vapor (grams)}}{\text{mass of dry air (kilograms)}}$$

Table 17.1 shows the mixing ratios of saturated air at various temperatures. For example, at 25°C (77°F), a saturated parcel of air (1 kilogram) would contain 20 grams of water vapor.

Because the mixing ratio is expressed in units of mass (usually in grams per kilogram), it is not affected by changes in pressure or temperature. However, measuring the mixing ratio by direct sampling is time-consuming. Thus, other methods are employed to express the moisture content of the air. These include relative humidity and dew-point temperature.

Relative Humidity

The most familiar and, unfortunately, the most misunderstood term used to describe the moisture content of air is relative humidity.

Relative humidity *is a ratio of the air's actual water-vapor content compared with the amount of water vapor required for saturation at that temperature (and pressure).* Thus, relative humidity indicates how near the air is to saturation rather than the actual quantity of water vapor in the air.

To illustrate, we see from Table 17.1 that at 25°C (77°F), air is saturated when it contains 20 grams of water vapor per kilogram of air. Thus, if the air contains 10 grams per kilogram on a 25°C day, the relative humidity is expressed as 10/20, or 50 percent. If air with a temperature of 25°C had a water-vapor content of 20 grams per kilogram, the relative humidity would be expressed as 20/20, or 100 percent. When the relative humidity reaches 100 percent, the air is saturated.

Because relative humidity is based on the air's water-vapor content, as well as the amount of moisture required for saturation, it can be changed in either of two ways. First, relative humidity can be changed by the addition or removal of water vapor. Second, because the amount of moisture required for saturation is a function of air temperature, relative humidity varies with temperature. (Recall that the amount of water vapor required for saturation is temperature dependent, and at higher temperatures, it takes more water vapor to saturate air than at lower temperatures.)

Adding or Subtracting Moisture Notice in **FIGURE 17.4** that when water vapor is added to a parcel of air, its relative humidity increases until saturation occurs (100 percent relative humidity). What if even more moisture is added to this parcel of saturated air? Does the relative humidity exceed 100 percent? Normally, this situation does not occur. Instead, the excess water vapor condenses to form liquid water.

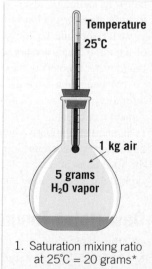
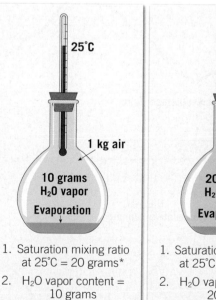
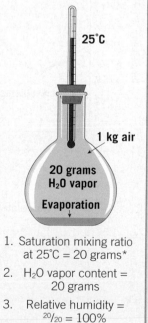

FIGURE 17.4 The Relationship Between Relative Humidity and Water Vapor Content At a constant temperature, the capacity (saturation mixing ratio) remains unchanged at 20 grams per kilogram, while the relative humidity rises from 25 to 100 percent as the water-vapor content increases.

Temperature 25°C
1 kg air
5 grams H₂O vapor

1. Saturation mixing ratio at 25°C = 20 grams*
2. H₂O vapor content = 5 grams
3. Relative humidity = 5/20 = 25%

*See Table 17.1

A. Initial condition: 5 grams of water vapor

25°C
1 kg air
10 grams H₂O vapor
Evaporation

1. Saturation mixing ratio at 25°C = 20 grams*
2. H₂O vapor content = 10 grams
3. Relative humidity = 10/20 = 50%

B. Addition of 5 grams of water vapor = 10 grams

25°C
1 kg air
20 grams H₂O vapor
Evaporation

1. Saturation mixing ratio at 25°C = 20 grams*
2. H₂O vapor content = 20 grams
3. Relative humidity = 20/20 = 100%

C. Addition of 10 grams of water vapor = 20 grams

FIGURE 17.5 How Relative Humidity Varies with Temperature When the water-vapor content (mixing ratio) is constant, a decrease in air temperature causes an increase in relative humidity. In this example, when the temperature of the air in the flask was lowered from 20° to 10°C, the relative humidity increased from 50 to 100 percent. Further cooling (from 10° to 0°C) causes one-half of the water vapor to condense. In nature, cooling of air below its saturation mixing ratio generally causes condensation in the form of clouds, dew, or fog.

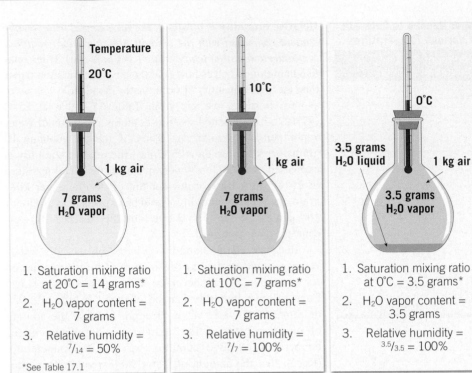

Temperature
20°C

1 kg air

7 grams H₂O vapor

1. Saturation mixing ratio at 20°C = 14 grams*
2. H₂O vapor content = 7 grams
3. Relative humidity = ⁷/₁₄ = 50%

*See Table 17.1

A. Initial condition: 20°C

10°C

1 kg air

7 grams H₂O vapor

1. Saturation mixing ratio at 10°C = 7 grams*
2. H₂O vapor content = 7 grams
3. Relative humidity = ⁷/₇ = 100%

B. Cooled to 10°C

0°C

3.5 grams H₂O liquid

1 kg air

3.5 grams H₂O vapor

1. Saturation mixing ratio at 0°C = 3.5 grams*
2. H₂O vapor content = 3.5 grams
3. Relative humidity = ³·⁵/₃·₅ = 100%

C. Cooled to 0°C

In nature, moisture is added to the air mainly via evaporation from the oceans. However, plants, soil, and smaller bodies of water also make substantial contributions.

Changes with Temperature The second condition that affects relative humidity is air temperature. Examine

FIGURE 17.6 Typical Daily Variations in Temperature and Relative Humidity This graph shows the daily variations in temperature and relative humidity during a spring day at Washington, DC. When temperature increases, relative humidity drops (see midafternoon) and vice versa.

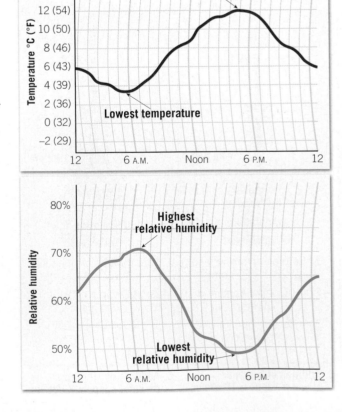

FIGURE 17.5 carefully. Note in Figure 17.5A that when air at 20°C contains 7 grams of water vapor per kilogram it has a relative humidity of 50 percent. This can be verified by referring to Table 17.1. Here we can see that at 20°C, air is saturated when it contains 14 grams of water vapor per kilogram of air. Because the air in Figure 17.5A contains 7 grams of water vapor, its relative humidity is 7/14, or 50 percent.

When the flask is cooled from 20° to 10°C, as shown in Figure 17.5B, the relative humidity increases from 50 to 100 percent. We can conclude that when the water-vapor content remains constant, *a decrease in temperature results in an increase in relative humidity.*

What happens when the air is cooled below the temperature at which saturation occurs? Figure 17.5C illustrates this situation. Notice from Table 17.1 that when the flask is cooled to 0°C, the air is saturated at 3.5 grams of water vapor per kilogram of air. Because this flask originally contained 7 grams of water vapor, 3.5 grams of water vapor will condense to form liquid droplets that collect on the walls of the container. The relative humidity of the air inside remains at 100 percent. This brings up an important concept. When air aloft is cooled below its saturation level, some of the water vapor condenses to form clouds. As clouds are made of liquid droplets (or ice crystals), they are no longer part of the *water-vapor* content of the air. (Clouds are not water vapor; they are composed of liquid water droplets or ice crystals too tiny to fall to Earth.)

We can summarize the effects of temperature on relative humidity as follows: When the water-vapor content of air remains at a constant level, a decrease in air temperature results in an increase in relative humidity, and an increase in temperature causes a decrease in relative humidity. In **FIGURE 17.6** the variations in temperature and relative humidity during a typical day demonstrate the relationship just described.

Dew-Point Temperature

Another important measure of humidity is the dew-point temperature. The **dew-point temperature**, or simply the **dew point**, is the temperature to which a parcel of air would need to be cooled to reach saturation. For example, in Figure 17.5, the unsaturated air in the flask had to be cooled to 10°C before saturation occurred. Therefore, 10°C is the dew-point temperature for this air. In nature, cooling below the dew point causes water vapor to condense, typically as dew, fog, or clouds. The term *dew point*

FIGURE 17.7 Condensation, or "Dew," on a Cold Drinking Glass The cold glass chills the surrounding layer of air below its dew-point temperature, causing condensation. (amana images inc./Alamy)

stems from the fact that during nighttime hours, objects near the ground often cool below the dew-point temperature and become coated with dew. A similar phenomenon is the condensation of water that occurs when the air adjacent to a cold drink reaches its dew point (**FIGURE 17.7**).

Unlike relative humidity, which is a measure of how near the air is to being saturated, dew-point temperature is a measure of its *actual moisture* content. Because the dew-point temperature is directly related to the amount of water vapor in the air, and because it is easy to determine, it is one of the most useful measures of humidity.

The amount of water vapor needed for saturation is temperature dependent: For every 10°C (18°F) increase in temperature, the amount of water vapor needed for saturation doubles (see Table 17.1). Therefore, relatively cold *saturated air* (0°C [32°F]) contains about half the water vapor of *saturated air* having a temperature of 10°C (50°F) and roughly one-fourth that of warm *saturated air* with a temperature of 20°C (68°F). Because the dew point is the temperature at which saturation occurs, we can conclude that high dew-point temperatures indicate moist air, and low dew-point temperatures indicate dry air. More precisely, we know that air over Fort Myers, Florida, with a dew point of 25°C (77°F), contains about twice the water vapor of air situated over St. Louis, Missouri, with a dew point of 15°C (59°F), and four times that of Tucson, Arizona, with a dew point of 5°C (41°F).

Because the dew-point temperature is a good measure of the amount of water vapor in the air, it is the measure of atmospheric moisture that appears on a variety of weather maps. Notice on the map in **FIGURE 17.8** that most of the places located near the warm Gulf of Mexico have dew-point temperatures that exceed 70°F (21°C). When the dew point exceeds 65°F (18°C), most people consider the air to be humid, and air with a dew point of 75°F (24°C) or higher is oppressive. Also notice in Figure 17.8 that although the Southeast is dominated by humid conditions (dew points above 65°F), most of the remainder of the country is experiencing drier air.

Measuring Humidity

Relative humidity is commonly measured using a **hygrometer** (*hygro* = moisture, *metron* = measuring instrument). One type of hygrometer, called a **psychrometer**, consists of two identical thermometers mounted side by side (**FIGURE 17.9**). One thermometer, the *dry-bulb* thermometer, gives the current air temperature. The other, called the *wet-bulb* thermometer, has a thin muslin wick tied around the end (see the ends of the thermometers in Figure 17.9).

To use the psychrometer, the cloth sleeve is saturated with water, and a continuous current of air is passed over the wick (see Figure 17.9). This is done either by swinging the instrument freely in the air or by fanning air past it. As a consequence, water evaporates from the wick, and the heat absorbed by the evaporating water makes the temperature of the wet bulb drop. The loss of heat that was required to evaporate water from the wet bulb lowers the thermometer reading.

The amount of cooling that takes place is directly proportional to the dryness of the air. The drier the air, the

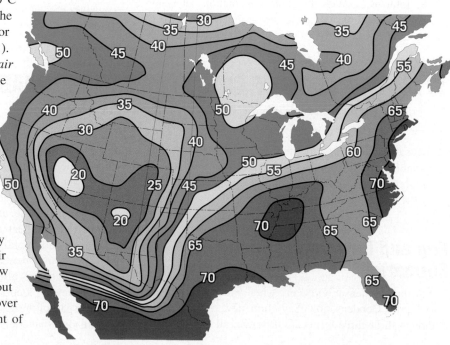

SmartFigure 17.8 Map Showing Dew-Point Temperatures on a Typical September Day Dew-point temperatures above 60°F dominate the southeastern United States, indicating that this region is blanketed with humid air, while much drier conditions prevail over the American Southwest.

FIGURE 17.9 Sling Psychrometer Used to Determine Both Relative Humidity and Dew Point

(Photo by E. J. Tarbuck)

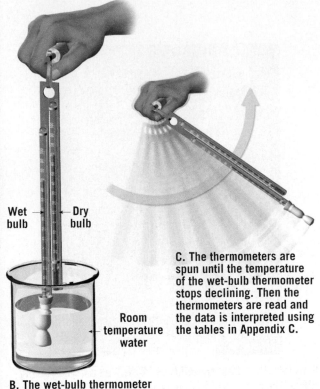

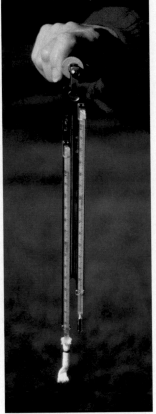

A. The dry-bulb thermometer gives the current air temperature.

B. The wet-bulb thermometer is covered with a cloth wick that is dipped in water.

C. The thermometers are spun until the temperature of the wet-bulb thermometer stops declining. Then the thermometers are read and the data is interpreted using the tables in Appendix C.

greater the amount of moisture that evaporates. The more heat the evaporating water absorbs, the greater the cooling. Therefore, the larger the difference that is observed between the thermometer readings, the lower the relative humidity; the smaller the difference, the higher the relative humidity. If the air is saturated, no evaporation will occur, and the two thermometers will have identical readings.

To determine the precise relative humidity from the thermometer readings, a standard table is used (refer to Appendix B, Table B.1). With the same information but using a different table (Table B.2), the dew-point temperature can also be calculated.

17.2 CONCEPT CHECKS

1. Describe how the water vapor content of air at saturation is related to air temperature.

2. List three measures that are used to express humidity.

3. How do relative humidity and mixing ratio differ?

4. If the amount of water vapor in the air remains unchanged, how does a decrease in temperature affect relative humidity?

5. On a warm summer day when the relative humidity is high, it may seem even warmer than the thermometer indicates. Why do we feel so uncomfortable on these muggy days?

17.3 | THE BASIS OF CLOUD FORMATION: ADIABATIC COOLING

Explain how adiabatic cooling results in cloud formation.

We have considered basic properties of water vapor and how its variability is measured. This section examines some of the important roles that water vapor plays in weather, especially the formation of clouds.

Fog and Dew Versus Cloud Formation

Recall that condensation occurs when water vapor changes to a liquid. Condensation may form dew, fog, or clouds. Although these three forms are different, all require that air reach saturation. As indicated earlier, saturation occurs either when sufficient water vapor is added to the air or, more commonly, when the air is cooled to its dew point.

Near Earth's surface, heat is readily exchanged between the ground and the air above. During evening hours, the surface radiates heat away, causing the surface and adjacent air to cool rapidly. This radiation cooling accounts for the formation of dew and some types of fog. Thus, surface cooling that occurs after sunset accounts for some condensation. However, cloud formation often takes place during the warmest part of the day. Clearly some other mechanism must operate aloft that cools air sufficiently to generate clouds.

Adiabatic Temperature Changes

The process that is responsible for most cloud formation is easily demonstrated if you have ever pumped up a bicycle tire and noticed that the pump barrel became quite warm. The heat you felt was a result of the work you did on the air to compress it. When energy is used to compress air, the motion of the gas molecules increases, and therefore the temperature of the air rises. Conversely, air that is allowed to escape from a bicycle tire *expands and cools*. This results because the expanding air pushes (does work on) the surrounding air and must cool by an amount equivalent to the energy expended.

You have probably felt the cooling effect of the propellant gas expanding as you have applied hairspray or spray deodorant. As the compressed gas in an aerosol spray can is released, it quickly expands and cools. This drop in temperature occurs *even though heat is neither added nor subtracted*. Such variations are known as **adiabatic temperature changes** and result when air is compressed or allowed to expand. In summary, *when air is allowed to expand, it cools, and when it is compressed, it warms*.

Adiabatic Cooling and Condensation

To simplify the following discussion, it helps to imagine a volume of air enclosed in a thin elastic cover. Meteorologists call this imaginary volume of air a **parcel**. Typically, we consider a parcel to be a few hundred cubic meters in volume, and we assume that it acts independently of the surrounding air. It is also assumed that no heat is transferred into or out of the parcel. Although this image is highly idealized, over short time spans, a parcel of air behaves in a manner much like an actual volume of air moving vertically in the atmosphere.

Dry Adiabatic Rate
As you travel from Earth's surface upward through the atmosphere, the atmospheric pressure rapidly diminishes because there are fewer and fewer gas molecules. Thus, any time a parcel of air moves upward, it passes through regions of successively lower pressure, and the ascending air expands. As it expands, it cools adiabatically. Unsaturated air cools at a constant rate of 10°C for every 1000 meters of ascent (5.5°F per 1000 feet).

Conversely, descending air comes under increasingly higher pressures, compresses, and is heated 10°C for every

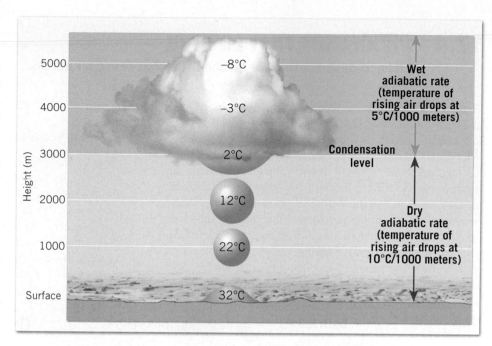

FIGURE 17.10 Dry Versus Wet Adiabatic Rates of Cooling Rising air cools at the dry adiabatic rate of 10° per 1000 meters, until the air reaches the dew point and condensation (cloud formation) begins. As air continues to rise, the latent heat released by condensation reduces the rate of cooling. The wet adiabatic rate is therefore always less than the dry adiabatic rate.

1000 meters of descent. This rate of cooling or heating applies only to *unsaturated air* and is known as the **dry adiabatic rate**.

Wet Adiabatic Rate
If a parcel of air rises high enough, it will eventually cool to its dew point, where the process of condensation begins. From this point on along its ascent, *latent heat of condensation* stored in the water vapor will be liberated. Although the air will continue to cool after condensation begins, the released latent heat works against the adiabatic process, thereby reducing the rate at which the air cools. This slower rate of cooling caused by the addition of latent heat is called the **wet adiabatic rate** of cooling. Because the amount of latent heat released depends on the quantity of moisture present in the air, the wet adiabatic rate varies from 5°C per 1000 meters for air with a high moisture content to 9°C per 1000 meters for dry air.

FIGURE 17.10 illustrates the role of adiabatic cooling in the formation of clouds. Note that from the surface up to the condensation level, the air cools at the dry adiabatic rate. The wet adiabatic rate begins at the condensation level.

17.3 CONCEPT CHECKS

1 How is the formation of dew different from cloud formation? How are they similar?

2 Why does air cool when it rises through the atmosphere?

3 What do meteorologists mean when they use the word *parcel*?

4 Why does the adiabatic rate of cooling change when condensation begins? Why is the wet adiabatic rate not a constant number?

5 The contents of an aerosol spray can are under very high pressure. Explain why the spray feels cold when it is allowed to escape the container.

17.4 | PROCESSES THAT LIFT AIR

List and describe the four mechanisms that cause air to rise.

To review, when air rises, it expands and cools adiabatically. If air is lifted sufficiently, it will eventually cool to its dew-point temperature, saturation will occur, and clouds will develop. Why does air rise on some occasions but not on others?

Generally, air tends to resist vertical movement; air located near the surface tends to stay near the surface, and air aloft tends to remain aloft. Exceptions to these patterns include conditions in the atmosphere that give air sufficient buoyancy to rise without the aid of outside forces. One example, called **convective lifting**, occurs when a mass of air is warmer and therefore less dense than the surrounding air. In most situations, however, when you see clouds forming, there is some mechanical

phenomenon at work that forces the air to rise (at least initially).

Here we will look at four mechanisms that cause air to rise:

1. *Orographic lifting*—Air is forced to rise over a mountainous barrier.
2. *Frontal wedging*—Warmer, less dense air is forced over cooler, denser air.
3. *Convergence*—A pileup of horizontal airflow results in upward movement.
4. *Localized convective lifting*—Unequal surface heating causes localized pockets of air to rise because of their buoyancy.

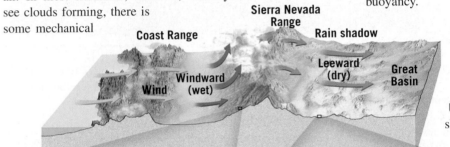

A. Orographic lifting leads to precipitation on windward slopes.

B. By the time air reaches the leeward side of the mountains, much of the moisture has been lost, resulting in a *rain shadow desert*.

Orographic Lifting

Orographic lifting occurs when elevated terrains, such as mountains, act as barriers to the flow of air (**FIGURE 17.11**). As air ascends a mountain slope, adiabatic cooling often generates clouds and copious precipitation. In fact, many of the rainiest places in the world are located on windward mountain slopes.

By the time air reaches the leeward side of a mountain, much of its moisture has been lost. If the air descends, it warms adiabatically, making condensation and precipitation even less likely. As shown in Figure 17.11, the result can be a **rain shadow desert**. The Great Basin Desert of the western United States lies only a few hundred kilometers from the Pacific Ocean, but it is effectively cut off from the ocean's moisture by the imposing Sierra Nevada. The Gobi Desert of Mongolia, the Takla Makan of China, and the Patagonia Desert of Argentina are other examples of deserts that exist because they are on the leeward sides of mountains. (For a map showing deserts, see Figure 6.30, page 193.)

Frontal Wedging

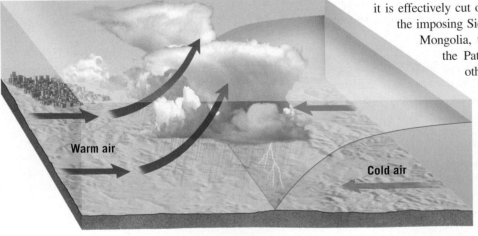

If orographic lifting were the only mechanism that forced air aloft, the relatively flat central portion of North America would be an expansive desert instead of the nation's breadbasket. Fortunately, this is not the case.

In central North America, masses of warm and cold air collide, producing a **front**. Here the cooler, denser air acts as a barrier over which the warmer, less dense air rises. This process, called **frontal wedging**, is illustrated in **FIGURE 17.12**.

It should be noted that weather-producing fronts are associated with storm systems called *middle-latitude cyclones*. Because these storms are responsible for producing a high percentage of the precipitation in the middle latitudes, we examine them closely in Chapter 19.

Convergence

We saw that the collision of contrasting air masses forces air to rise. In a more general sense, whenever air in the lower atmosphere flows together, lifting results. This phenomenon is called **convergence**. When air flows in from more than one direction, it must go somewhere. Because it cannot go down, it goes up. This leads to adiabatic cooling and possible cloud formation.

Convergence can also occur whenever an obstacle slows or restricts horizontal airflow (wind). When air moves from a relatively smooth surface, such as the ocean, onto an irregular landscape, its speed is reduced. The result is a pileup of air (convergence). This is similar to what happens when people leave a well-attended sporting event and a pileup results at the exits. When air converges, the air molecules do not simply squeeze closer together (like people); rather, there is a net upward flow.

The Florida peninsula provides an excellent example of the role that convergence can play in initiating cloud development and precipitation (**FIGURE 17.13**). On warm days, the airflow is from the ocean to the land along both coasts of Florida. This leads to a pileup of air along the coasts and general convergence over the peninsula. This pattern of air movement and the uplift that results is aided by intense solar heating of the land. This is why the peninsula of Florida experiences the greatest frequency of midafternoon thunderstorms in the United States.

Convergence as a mechanism of forceful lifting is a major contributor to the weather associated with middle-latitude cyclones and hurricanes. The low-level horizontal

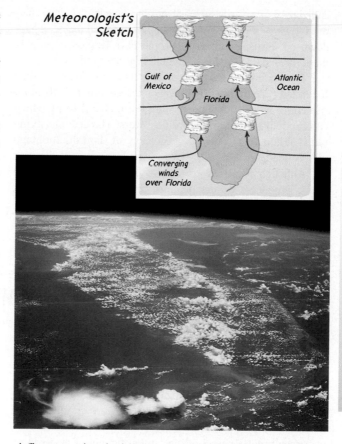

airflow associated with these systems is inward and upward around their centers. These important weather producers are covered in more detail later, but for now remember that convergence near the surface results in a general upward flow.

Localized Convective Lifting

On warm summer days, unequal heating of Earth's surface may cause pockets of air to be warmed more than the surrounding air. For instance, air above a paved parking lot will be warmed more than the air above an adjacent wooded park. Consequently, the parcel of air above the parking lot, which is warmer (less dense) than the surrounding air, will be buoyed upward (**FIGURE 17.14**). These rising parcels of warmer air are

SmartFigure 17.13 Surface Convergence Enhances Cloud Development When surface air converges, the column of air increases in height because the air is crammed into a smaller and smaller space. The result is forced lifting and adiabatic cooling that generates clouds. This photo shows southern Florida, viewed from the Space Shuttle. On warm days, airflow from the Atlantic Ocean and Gulf of Mexico onto the Florida peninsula generates many midafternoon thunderstorms. (Photo by NASA)

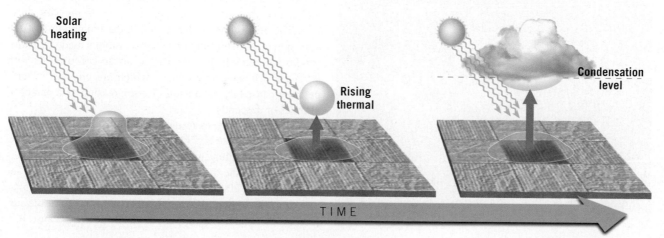

FIGURE 17.14 Localized Convective Lifting Unequal heating of Earth's surface causes pockets of air to be warmed more than the surrounding air. These buoyant parcels of hot air (thermals) rise, and if they reach the condensation level, clouds form.

called *thermals*. Birds such as hawks and eagles use these thermals to carry them to great heights, from which they can gaze down on unsuspecting prey. People have learned to employ these rising parcels using hang gliders as a way to "fly."

The phenomenon that produces rising thermals is called **localized convective lifting**. When these warm parcels of air rise above the lifting condensation level, clouds form, which can produce midafternoon rain showers. The accompanying rains, although occasionally heavy, are of short duration and widely scattered.

Although localized convective lifting by itself is not a major producer of precipitation, the added buoyancy that results from surface heating contributes significantly to the lifting initiated by the other mechanisms. In addition, even though the other mechanisms force air to rise, convective lifting occurs because the air is warmer (less dense) than the surrounding air and rises for the same reasons as does a hot-air balloon.

17.4 CONCEPT CHECKS

1 List four mechanisms that cause air to rise.

2 How do orographic lifting and frontal wedging force air to rise?

3 Explain why the Great Basin area of the western United States is dry. What term is applied to this situation?

4 What causes the Florida peninsula to experience the greatest frequency of midafternoon thunderstorms in the United States?

17.5 | THE WEATHERMAKER: ATMOSPHERIC STABILITY

Describe how atmospheric stability is determined and compare conditional instability with absolute instability.

When air rises, it cools and usually produces clouds. Why do clouds vary so much in size, and why does the resulting precipitation vary so much? The answers are closely related to the *stability* of the air.

Recall that a parcel of air can be thought of as having a thin, flexible cover that allows it to expand but prevents it from mixing with the surrounding air (picture a hot-air balloon). If this parcel were forced to rise, *its temperature would decrease*

FIGURE 17.15 Hot Air Rises Hot-air balloons rise through the atmosphere because the heated air inside the balloons is warmer (less dense) than the surrounding air. (Photo by Steve Vidler/SuperStock)

because of expansion. By comparing the parcel's temperature to that of the surrounding air, we can determine the stability of the bubble. If the parcel's temperature is lower than that of the surrounding environment, it will be denser; if it is allowed to move freely, it will sink to its original position. Air of this type, called **stable air**, resists vertical movement.

If, however, our imaginary rising parcel is *warmer* and hence less dense than the surrounding air, it will continue to rise until it reaches an altitude where its temperature equals that of its surroundings. This is exactly how a hot-air balloon works, rising as long as it is warmer and less dense than the surrounding air (**FIGURE 17.15**). This type of air is classified as **unstable air**. In summary, stability is a property of air that describes its tendency to remain in its original position (stable) or to rise (unstable).

Types of Stability

The stability of air is determined by measuring the temperature of the atmosphere at various heights. Recall from Chapter 16 that this measure is called the *environmental lapse rate*. The environmental lapse rate is the temperature of the atmosphere, as determined from observations made by radiosondes and aircraft. It is important not to confuse this with *adiabatic temperature changes*, which are changes in temperature caused by expansion or compression as a parcel of air rises or descends.

To illustrate, we examine a situation in which the environmental lapse rate is 5°C per 1000 meters (**FIGURE 17.16**). Under this condition, when air at the surface has a temperature of 25°C, the air at 1000 meters will be 5° cooler, or 20°C, the air at 2000 meters will have a temperature of 15°C, and so forth. At first glance, it appears that the air at the surface is less dense than the air at 1000 meters because it is

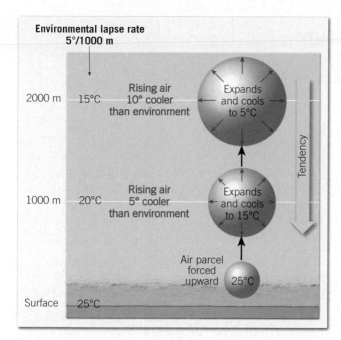

Environmental lapse rate
5°/1000 m

FIGURE 17.16 When an Unsaturated Parcel of Air Is Lifted, It Expands and Cools at the Dry Adiabatic Rate (10°C per 1000 meters) Because the temperature of the rising parcel of air is lower than the surrounding environment, it will be heavier, and sink to its original position if it is allowed to do so. Air of this type is referred to as *stable*.

5° warmer. However, if the air near the surface were unsaturated and were to rise to 1000 meters, it would expand and cool at the dry adiabatic rate of 10°C per 1000 meters. Therefore, upon reaching 1000 meters, its temperature would have dropped 10°C. Being 5° cooler than its environment, it would be denser and tend to sink to its original position. Hence, we

say that the air near the surface is potentially cooler than the air aloft and therefore will not rise on its own. The air just described is *stable* and resists vertical movement.

Absolute Stability Stated quantitatively, **absolute stability** prevails when the environmental lapse rate is less than the wet adiabatic rate. **FIGURE 17.17** depicts this situation using an environmental lapse rate of 5°C per 1000 meters and a wet adiabatic rate of 6°C per 1000 meters. Note that at 1000 meters, the temperature of the surrounding air is 15°C, while the rising parcel of air has cooled to 10°C and is therefore the denser air. Even if this stable air were to be forced above the condensation level, it would remain cooler and denser than its environment, and thus it would tend to return to the surface.

The most stable conditions occur when the temperature in a layer of air actually increases with altitude rather than decreases. When such a reversal occurs, a *temperature inversion* is said to exist. Temperature inversions frequently occur on clear nights as a result of radiation cooling of Earth's surface. Under these conditions, an inversion is created because the ground and the air immediately above will cool more rapidly than the air aloft. When warm air overlies cooler air, it acts as a lid and prevents appreciable vertical mixing. Because of this, temperature inversions are responsible for trapping pollutants in a narrow zone near Earth's surface.

Absolute Instability At the other extreme from absolute stability, air is said to exhibit **absolute instability** when the environmental lapse rate is greater than the dry adiabatic rate. As shown in **FIGURE 17.18**, the ascending parcel of air is always warmer than its environment and

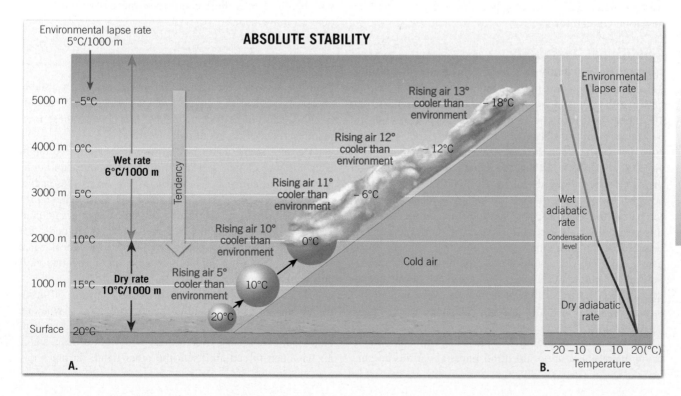

SmartFigure 17.17 Atmospheric Conditions That Result in Absolute Stability A. The rising parcel of air is always cooler and heavier than the surrounding air and is therefore stable. **B.** Graphic representation of the conditions shown in part A.

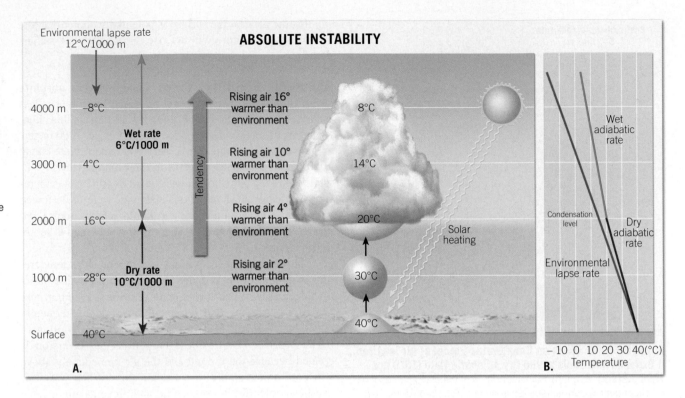

will continue to rise because of its own buoyancy. However, absolute instability is generally found near Earth's surface. On hot, sunny days the air above some surfaces, such as shopping center parking lots, is heated more than the air over adjacent surfaces. These invisible pockets of more intensely heated air, being less dense than the air aloft, will rise like a hot-air balloon. This phenomenon produces the small, fluffy clouds we associate with fair weather. Occasionally, when the surface air is considerably warmer than the air aloft, clouds with considerable vertical development can form.

Conditional Instability   A more common type of atmospheric instability is called **conditional instability**. This occurs when moist air has an environmental lapse rate between the dry and wet adiabatic rates (between 5°C and 10°C per 1000 meters). Simply, the atmosphere is said to be conditionally unstable when it is *stable* for an *unsaturated* parcel of air but *unstable* for a *saturated* parcel of air. Notice in **FIGURE 17.19** that the rising parcel of air is cooler than the surrounding air for nearly 3000 meters. With the addition of latent heat above the lifting condensation level, the parcel becomes warmer than the surrounding air. From this point along its ascent, the parcel will continue to rise because of its own buoyancy, without an outside force. Thus, conditional instability depends on whether the rising air is saturated. The word *conditional* is used because the air must be forced upward, such as over mountainous terrain, before it becomes unstable and rises because of its own buoyancy.

In summary, the stability of air is determined by measuring the temperature of the atmosphere at various heights.

In simple terms, a column of air is deemed unstable when the air near the bottom of this layer is significantly warmer (less dense) than the air aloft, indicating a steep environmental lapse rate. Under these conditions, the air actually turns over, as the warm air below rises and displaces the colder air aloft. Conversely, the air is considered to be stable when the temperature drops gradually with increasing altitude. The most stable conditions occur during a temperature inversion, when the temperature actually increases with height. Under these conditions, there is very little vertical air movement.

Stability and Daily Weather

From the previous discussion, we can conclude that stable air resists vertical movement, whereas unstable air ascends freely because of its own buoyancy. But how do these facts manifest themselves in our daily weather?

Because stable air resists upward movement, we might conclude that clouds will not form when stable conditions prevail in the atmosphere. Although this seems reasonable, recall that processes exist that *force* air aloft. These processes include orographic lifting, frontal wedging, and convergence. When stable air is forced aloft, the clouds that form are widespread and have little vertical thickness when compared to their horizontal dimension, and precipitation, if any, is light to moderate.

By contrast, clouds associated with the lifting of unstable air are towering and often generate thunderstorms and occasionally even tornadoes. For this reason, we can conclude that on a dreary, overcast day with light drizzle, stable air has been forced aloft. On the other hand, during a day

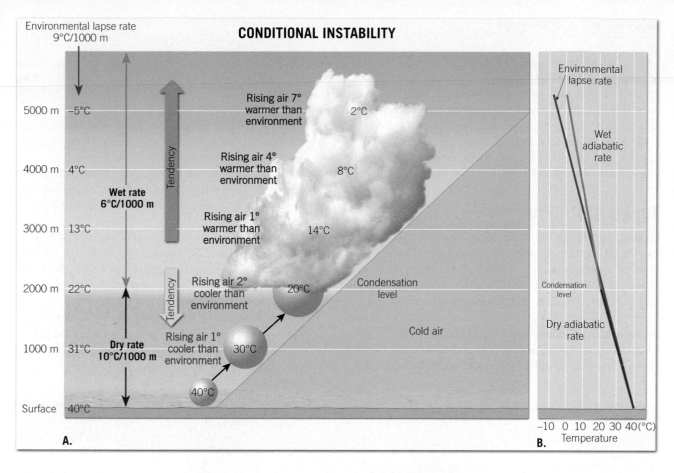

Environmental lapse rate
9°C/1000 m

5000 m   −5°C   Rising air 7°
                        warmer than   2°C
                        environment

Tendency

          Wet rate
          6°C/1000 m   Rising air 4°
4000 m   4°C   warmer than   8°C
                        environment

3000 m   13°C   Rising air 1°
                        warmer than   14°C
                        environment

          Dry rate   Condensation
          10°C/1000 m   Rising air 2°   level
2000 m   22°C   cooler than   20°C
                        environment
                                                       Cold air
                        Rising air 1°
1000 m   31°C   cooler than   30°C
                        environment

                                          40°C

Surface   40°C

A.

FIGURE 17.19
Conditional Instability May Result When Warm Air Is Forced to Rise Along a Frontal Boundary A. Note that the environmental lapse rate of 9°C per 1000 meters lies between the dry and wet adiabatic rates. The parcel of air is cooler than the surrounding air up to nearly 3000 meters, where its tendency is to sink toward the surface (stable). Above this level, however, the parcel is warmer than its environment and will rise because of its own buoyancy (unstable). Thus, when conditionally unstable air is forced to rise, the result can be towering cumulus clouds. **B.** Graphic representation of the conditions shown in part A.

Environmental lapse rate

Wet adiabatic rate

Condensation level

Dry adiabatic rate

−10 0 10 20 30 40(°C)
Temperature
B.

when cauliflower-shaped clouds appear to be growing as if bubbles of hot air are surging upward, we can be fairly certain that the ascending air is unstable.

In summary, stability plays an important role in determining our daily weather. To a large degree, stability determines the type of clouds that develop and whether precipitation will come as a gentle shower or a heavy downpour.

17.5 CONCEPT CHECKS

1 Explain the difference between the environmental lapse rate and adiabatic cooling.

2 Describe absolute stability in your own words.

3 Compare absolute instability and conditional instability.

4 What types of clouds and precipitation, if any, form when stable air is forced aloft?

5 Describe the weather associated with unstable air.

17.6 | CONDENSATION AND CLOUD FORMATION List the necessary conditions for condensation and briefly describe the two criteria used for cloud classification.

To review briefly, condensation occurs when water vapor in the air changes to a liquid. The result of this process may be dew, fog, or clouds. For any of these forms of condensation to occur, the air must be saturated. Saturation occurs most commonly when air is cooled to its dew point, or less often when water vapor is added to the air.

Generally, there must be a surface on which the water vapor can condense. When dew occurs, objects at or near the ground, such as grass and car windows, serve this purpose. But when condensation occurs in the air above the ground, tiny bits of particulate matter, known as **condensation nuclei**, serve as surfaces for water-vapor condensation. These nuclei are very important, for in their absence, a relative humidity well in excess of 100 percent is needed to produce clouds.

Condensation nuclei such as microscopic dust, smoke, and salt particles (from the ocean) are profuse in the lower atmosphere. Because of this abundance of particles, relative humidity rarely exceeds 101 percent. Some particles, such as ocean salt, are particularly good nuclei because they absorb water. These particles are termed **hygroscopic** (*hygro* = moisture, *scopic* = to seek) **nuclei**. When condensation takes place, the initial growth rate of cloud droplets is rapid. It diminishes quickly because the excess water vapor is quickly absorbed by the numerous competing particles. This results in the formation of a cloud consisting of millions upon millions of tiny water droplets, all so fine that they remain suspended in air. When cloud formation occurs at below-freezing temperatures, tiny ice crystals form.

Thus, a cloud might consist of water droplets, ice crystals, or both.

The slow growth of cloud droplets by additional condensation and the immense size difference between cloud droplets and raindrops suggest that condensation alone is not responsible for the formation of drops large enough to fall as rain. We first examine clouds and then return to the question of how precipitation forms.

Types of Clouds

SmartFigure 17.20 Classification of Clouds, Based on Height and Form

Clouds are among the most conspicuous and observable aspects of the atmosphere and its weather. **Clouds** are a form of condensation best described as *visible aggregates of minute droplets of water or tiny crystals of ice.* In addition to being prominent and sometimes spectacular features in the sky, clouds are of continual interest to meteorologists because they provide a visible indication of what is going on in the atmosphere. Anyone who observes clouds with the hope of recognizing different types often finds that there is a

bewildering variety of these familiar white and gray masses streaming across the sky. Once one comes to know the basic classification scheme for clouds, most of the confusion vanishes.

Clouds are classified on the basis of their *form* and *height* (**FIGURE 17.20**). Three basic forms are recognized:

- **Cirrus** (*cirrus* = a curl of hair) clouds are high, white, and thin. They can occur as patches or as delicate veil-like sheets or extended wispy fibers that often have a feathery appearance.
- **Cumulus** (*cumulus* = a pile) clouds consist of individual globular cloud masses. They normally exhibit a flat base and have the appearance of rising domes or towers. Such clouds are frequently described as having a cauliflower structure.
- **Stratus** (*stratum* = a layer) clouds are best described as sheets or layers that cover much or all of the sky. While there may be minor breaks, there are no distinct individual cloud units.

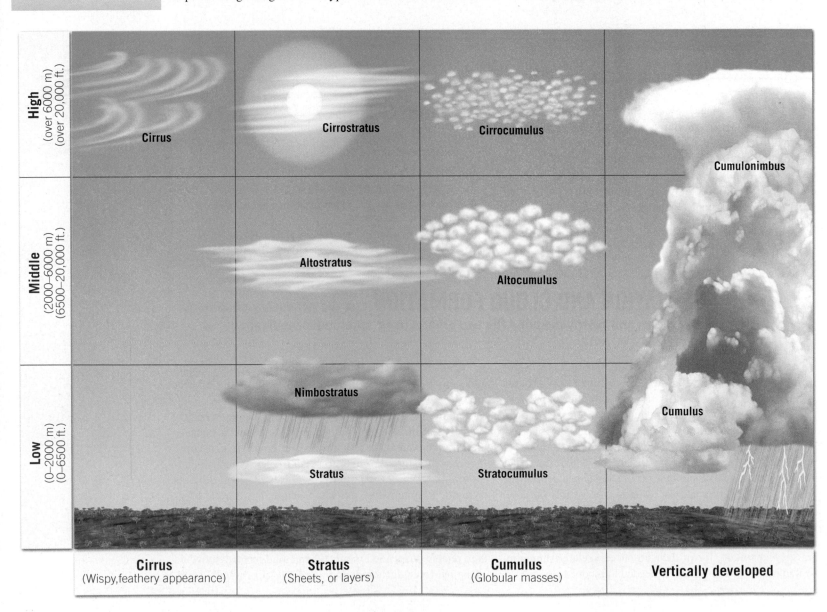

All other clouds reflect one of these three basic forms or are combinations or modifications of them.

Four levels of cloud heights are recognized: high, middle, low, and clouds of vertical displacement (see Figure 17.20). **High clouds** normally have bases above 6000 meters (20,000 feet), **middle clouds** generally occupy heights from 2000 to 6000 meters (6500 to 20,000 feet), and **low clouds** form below 2000 meters (6500 feet). The altitudes listed for each height category are not hard and fast. There is some seasonal as well as latitudinal variation. For example, at high latitudes or during cold winter months in the midlatitudes, high clouds are often found at lower altitudes.

High Clouds   Three cloud types make up the family of high clouds (above 6000 meters [20,000 feet]): cirrus, cirrostratus, and cirrocumulus. *Cirrus* clouds are thin and delicate and sometimes appear as hooked filaments called "mares' tails" (**FIGURE 17.21A**). As the names suggest, *cirrocumulus* clouds consist of fluffy masses (**FIGURE 17.21B**), whereas *cirrostratus* clouds are flat layers (**FIGURE 17.21C**). Because of the low temperatures and small quantities of water vapor present at high altitudes, all high clouds are thin and white and are made up of ice crystals. Furthermore, these clouds are not considered precipitation makers. However, when cirrus clouds are followed by cirrocumulus clouds and increased sky coverage, they may warn of impending stormy weather.

Middle Clouds   Clouds that appear in the middle range (2000–6000 meters [6500–20,000 feet]) have the prefix *alto* as part of their name. *Altocumulus* clouds are composed of globular masses that differ from cirrocumulus clouds in that they are larger and denser (**FIGURE 17.21D**). *Altostratus* clouds create a uniform white to grayish sheet covering the sky, with the Sun or Moon visible as a bright spot (**FIGURE 17.21E**). Infrequent light snow or drizzle may accompany these clouds.

Low Clouds   There are three members in the family of low clouds: stratus, stratocumulus, and nimbostratus. *Stratus* are a uniform foglike layer of clouds that frequently covers much of the sky. On occasion these clouds may produce light precipitation. When stratus clouds develop a scalloped bottom that appears as long parallel rolls or broken globular patches, they are called *stratocumulus* clouds.

Nimbostratus clouds derive their name from the Latin *nimbus*, which means "rainy cloud," and *stratus*, which means "to cover with a layer" (**FIGURE 17.21F**). As the name suggests, nimbostratus clouds are some of the chief precipitation producers. Nimbostratus clouds form in association with stable conditions. We might not expect clouds to grow or persist in stable air, but cloud growth of this type is common when air is forced to rise, as occurs along a mountain range, along a front, or near the center of a cyclone, where converging winds cause air to ascend. Such forced ascent of stable air leads to the formation of a stratified cloud layer that is large horizontally compared to its depth.

Clouds of Vertical Development   Some clouds do not fit into any one of the three height categories just mentioned. Such clouds have their bases in the low height range but often extend upward into the middle or high altitudes. Consequently, clouds in this category are called **clouds of vertical development**. They are all related to one another and are associated with unstable air. Although *cumulus* clouds are often connected with fair weather (**FIGURE 17.21G**), they may grow dramatically under the proper circumstances. Once

EYE ON EARTH

The cloud perched on the top of this volcano is called a *cap cloud*, and it may remain in place for hours. Cap clouds belong to a group of clouds called *orographic clouds*. (Photo by amana images inc./Alamy)

QUESTION 1 *Describe how cap clouds form, based on the fact that they are referred to as orographic clouds.*

QUESTION 2 *Explain why this group of clouds have relatively flat bases.*

QUESTION 3 *Cap clouds are related to another cloud type that has a very similar shape and form in mountainous regions. Do an internet search to find the name of the related cloud type.*

FIGURE 17.21 Common Forms of Several Different Cloud Types (Photos A, B, D, E, F, and G by E. J. Tarbuck; photo C by Jung-Pang Wu/Getty Images Inc; photo H by Doug Millar/Science Source)

A. Cirrus

B. Cirrocumulus

C. Cirrostratus

D. Altocumulus

TABLE 17.2   Cloud Types and Characteristics

Cloud Family and Height	Cloud Type	Characteristics
High clouds: above 6000 meters (20,000 feet)	Cirrus	Thin, delicate, fibrous, ice-crystal clouds. Sometimes appear as hooked filaments called "mares' tails" (see Figure 17.21A).
	Cirrocumulus	Thin, white, ice-crystal clouds in the form of ripples, waves, or globular masses, all in a row. May produce a "mackerel sky." Least common of the high clouds (see Figure 17.21B).
	Cirrostratus	Thin sheet of white, ice-crystal clouds that may give the sky a milky look. Sometimes produce halos around the Sun or Moon (see Figure 17.21C).
Middle clouds: 2000–6000 meters (6500–20,000 feet)	Altocumulus	White to gray clouds often composed of separate globules; "sheep-back" clouds (see Figure 17.21D).
	Altostratus	Stratified veil of clouds that are generally thin and may produce very light precipitation. When thin, the Sun or Moon may be visible as a bright spot, but no halos are produced (see Figure 17.21E).
Low clouds: below 2000 meters (6500 feet)	Stratocumulus	Soft gray clouds in globular patches or rolls. Rolls may join together to make a continuous cloud.
	Stratus	Low uniform layer resembling fog but not resting on the ground. May produce drizzle.
	Nimbostratus	Amorphous layer of dark gray clouds. Some of the chief precipitation-producing clouds (see Figure 17.21F).
Clouds of vertical development: 500–18,000 meters (1600–60,000 feet)	Cumulus	Dense, billowy clouds, often characterized by flat bases. May occur as isolated clouds or closely packed (see Figure 17.21G).
	Cumulonimbus	Towering cloud sometimes spreading out on top to form an "anvil head." Associated with heavy rainfall, thunder, lightning, hail, and tornadoes (see Figure 17.21H).

E. Altostratus

F. Nimbostratus

FIGURE 17.21 Common Forms of Several Different Cloud Types (continued)

G. Cumulus

H. Cumulonimbus

upward movement is triggered, acceleration is powerful, and clouds with great vertical extent form. The end result is often a towering cloud, called a *cumulonimbus*, that usually produces rain showers or a thunderstorm (**FIGURE 17.21H**).

Definite weather patterns can often be associated with particular clouds or certain combinations of cloud types, so it is important to become familiar with cloud descriptions and characteristics. **TABLE 17.2** lists the 10 basic cloud types that are recognized internationally and gives some characteristics of each.

17.6 CONCEPT CHECKS

1 As you drink an ice-cold beverage on a warm, humid day, the outside of the glass or bottle becomes wet. Explain.

2 What is the function of condensation nuclei in cloud formation?

3 What is the basis for the classification of clouds?

4 Why are high clouds always thin?

5 Which cloud types are associated with the following characteristics: thunder, halos, precipitation, hail, lightning, mares' tails?

17.7 | FOG

Define *fog* and explain how the various types of fog form.

Fog is generally considered to be an atmospheric hazard. When it is light, visibility is reduced to 2 or 3 kilometers (1 or 2 miles). However, when it is dense, visibility may be cut to a few dozen meters or less, making travel by any mode not only difficult but often dangerous. Officially, visibility must be reduced to 1 kilometer (0.6 mile) or less before fog is reported. While this figure is arbitrary, it does provide an objective criterion for comparing fog frequencies at different locations.

Fog is defined as *a cloud with its base at or very near the ground*. There is no basic physical difference between fog and a cloud; their appearance and structure are the same. The essential difference has to do with the method and place of formation. Whereas clouds result when air rises and cools adiabatically, most fogs result from radiation cooling or the movement of air over a cold surface. (The exception is upslope fog.) In other circumstances, fogs form when enough water vapor is added to the air to bring about saturation (evaporation fogs). We examine both of these types.

Meteorologist's Sketch

FIGURE 17.22 Advection Fog Rolling into San Francisco Bay Advection fog in coastal areas often occurs when warm, moist air flows over the cold waters of an ocean current and is cooled below its dew-point temperature. (Photo by Ed Pritchard/Getty Images)

Snow in Sierra Nevada

Fog

Pacific Ocean

A.

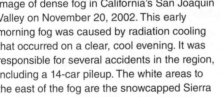

B.

FIGURE 17.23 Radiation Fog A. Satellite image of dense fog in California's San Joaquin Valley on November 20, 2002. This early morning fog was caused by radiation cooling that occurred on a clear, cool evening. It was responsible for several accidents in the region, including a 14-car pileup. The white areas to the east of the fog are the snowcapped Sierra Nevada. (Photo by NASA) **B.** Radiation fog can make the morning commute hazardous. (Photo by Tim Gainey/Alamy)

Fogs Caused by Cooling

Three common fogs form when air at Earth's surface is chilled below its dew point. Often a fog is a combination of these types.

Advection Fog When warm, moist air moves over a cool surface, the result might be a blanket of fog called **advection fog** (**FIGURE 17.22**). Examples of such fogs are very common. The foggiest location in the United States, and perhaps in the world, is Cape Disappointment, Washington. The name is indeed appropriate because this station averages 2552 hours of fog each year (there are about 8760 hours in a year). The fog experienced at Cape Disappointment, and at other West Coast locations, is produced when warm, moist air from the Pacific Ocean moves over the cold California Current and is then carried onshore by the prevailing winds. Advection fogs are also relatively common in the winter season, when warm air from the Gulf of Mexico moves across cold, often snow-covered surfaces of the Midwest and East.

Radiation Fog Radiation fog forms on cool, clear, calm nights, when Earth's surface cools rapidly by radiation. As the night progresses, a thin layer of air in contact with the ground is cooled below its dew point. As the air cools and becomes denser, it drains into low areas, resulting in "pockets" of fog. The largest pockets are often river valleys, where thick accumulations may occur (**FIGURE 17.23**).

Radiation fog may also form when the skies clear after a rainfall. In these situations, the air near the surface is close to saturation, and only a small amount of radiation cooling is needed to promote condensation. Radiation fog of this type often occurs around sunset, and it can make driving hazardous.

Upslope Fog As its name implies, **upslope fog** is created when relatively humid air moves up a gradually sloping plain or up the steep slopes of a mountain. As a result of this upward movement, air expands and cools adiabatically. If the dew point is reached, an extensive layer of fog may form. In the United States, the Great Plains offers an excellent example. When humid easterly or southeasterly winds blow westward from the Mississippi River upslope toward the Rocky Mountains, the air gradually rises, resulting in an adiabatic temperature decrease of about 13°C. When the difference between the air temperature and dew point of westward-moving air is less than 13°C, an extensive fog can result in the western plains.

Evaporation Fogs

When the saturation of air occurs primarily because of the addition of water vapor, the resulting fogs are called *evaporation fogs*. Two types of evaporation fogs are recognized: steam fog and frontal, or precipitation, fog.

Steam Fog When cool air moves over warm water, enough moisture may evaporate from the water surface to produce saturation. As the rising water vapor meets the cold air, it immediately re-condenses and rises with the air that is being warmed from below. Because the water has a steaming appearance, the phenomenon is called **steam fog**. Steam fog is fairly common over lakes and rivers in the fall and early winter, when the water may still be relatively warm and the air is rather crisp (**FIGURE 17.24**). Steam fog is often shallow because as the steam rises, it evaporates in the unsaturated air above.

Frontal Fog When frontal wedging occurs, warm air is lifted over colder air. If the resulting clouds yield rain, and the cold air below is near the dew point, enough rain will evaporate to produce fog. A fog formed in this manner is called **frontal fog**, or **precipitation fog**. The result is a more-or-less continuous zone of condensed water droplets reaching from the ground up through the clouds.

Both steam fog and frontal fog result from the addition of moisture to a layer of air. The air is usually cool or cold and already near saturation. Because air's capacity for water vapor at low temperatures is small, only a relatively modest amount of evaporation is necessary to produce saturated conditions and fog.

The frequency of dense fog varies considerably from place to place (**FIGURE 17.25**). As might be expected, fog incidence is highest in coastal areas, especially where cold currents prevail, as along the Pacific and New England

FIGURE 17.24 Steam Fog Rising from Sierra Lake, Blanca, Arizona (Photo by Michael Collier)

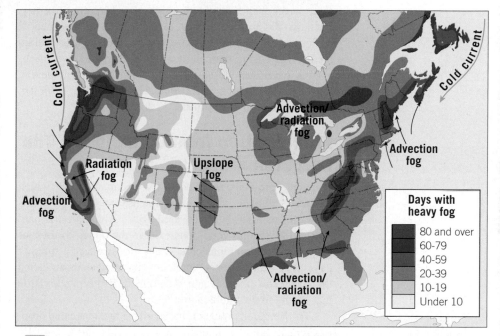

 SmartFigure 17.25 Map Showing the Average Number of Days per Year with Heavy Fog Notice that the frequency of dense fog varies considerably from place to place. Coastal areas, particularly the Pacific Northwest and New England, where cold currents prevail, have high occurrences of dense fog.

coasts. Relatively high frequencies are also found in the Great Lakes region and in the humid Appalachian Mountains of the East. By contrast, fogs are rare in the interior of the continent, especially in the arid and semiarid areas of the West.

17.7 CONCEPT CHECKS

1 Define *fog*.

2 List five main types of fog and describe how each type forms.

17.8 | HOW PRECIPITATION FORMS

Describe the two mechanisms that produce precipitation.

All clouds contain water. But why do some clouds produce precipitation while others drift placidly overhead? This simple question perplexed meteorologists for many years. First, cloud droplets are very small, averaging less than 10 micrometers in diameter; for comparison, a human hair is about 75 micrometers in diameter (**FIGURE 17.26**). Because of their small size, cloud droplets fall incredibly slowly. In addition, clouds are made up of many billions of these droplets, all competing for the available water vapor; thus, their continued growth via condensation is extremely slow. So, what causes precipitation?

A raindrop large enough to reach the ground without completely evaporating contains roughly 1 million times more water than a single cloud droplet. Therefore, for precipitation to form, millions of cloud droplets must somehow coalesce (join together) into drops large enough to sustain themselves during their descent to the surface. Two mechanisms have been proposed to explain this phenomenon: the *Bergeron process* in **FIGURE 17.27** and the *collision-coalescence process* in **FIGURE 17.28**.

Precipitation from Cold Clouds: The Bergeron Process

You have probably watched a TV documentary in which mountain climbers have braved intense cold and ferocious snow storms to scale ice-covered peaks. Although it is hard to imagine, similar conditions exist in the upper portions of towering cumulonimbus clouds, even on sweltering summer days. It is within these cold clouds that a mechanism called the **Bergeron process** generates much of the precipitation that occurs in the middle latitudes.

The Bergeron process is based on the fact that cloud droplets remain liquid at temperatures as low as –40°C (–40°F).

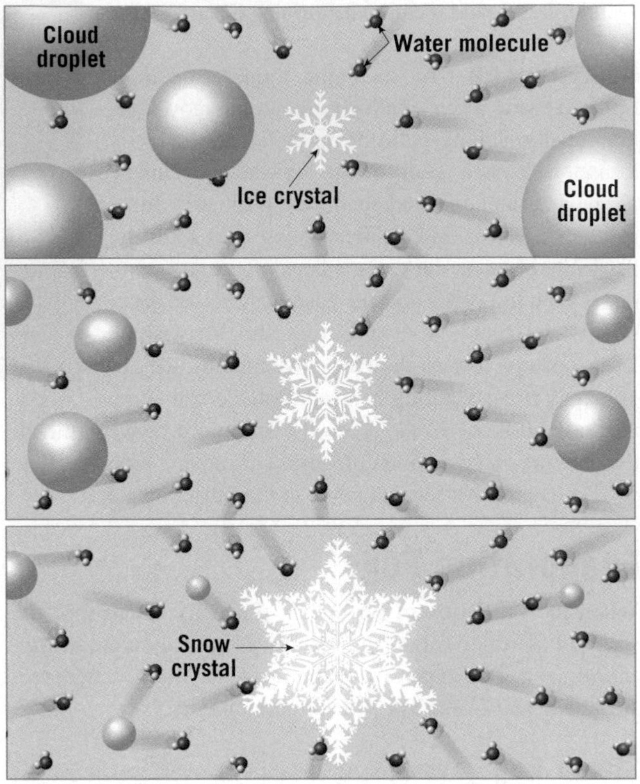

FIGURE 17.27 The Bergeron Process Ice crystals grow at the expense of cloud droplets until they are large enough to fall. The size of these particles has been greatly exaggerated.

Liquid water at temperatures below freezing is referred to as **supercooled**, and it becomes solid, or freezes, upon impact with objects. This explains why airplanes collect ice when they pass through a cloud composed of subzero droplets, a condition called *icing*. Supercooled water droplets also freeze upon contact with particles in the atmosphere known as **freezing nuclei**. Because freezing nuclei are relatively sparse, cold clouds primarily consist of supercooled droplets intermixed with a lesser amount of ice crystals.

When ice crystals and supercooled water droplets coexist in a cloud, the conditions are ideal for generating precipitation. Because ice crystals have a greater affinity for water vapor than does liquid water, they collect the available water vapor at a much faster rate. In turn, the water droplets evaporate to replenish the diminishing water vapor, thereby providing a continual source of moisture for the growth of ice crystals. As shown in Figure 17.27, the result is that ice crystals grow larger, at the expense of the water droplets, which shrink in size.

Eventually, this process generates ice crystals large enough to fall as snowflakes. During their descent, these ice crystals become larger as they intercept supercooled cloud droplets that freeze on them. When the surface temperature is about 4°C (39°F) or higher, snowflakes usually melt before they reach the ground and continue their descent as rain.

FIGURE 17.26 Comparative Diameters of Particles Involved in Condensation and Precipitation Processes

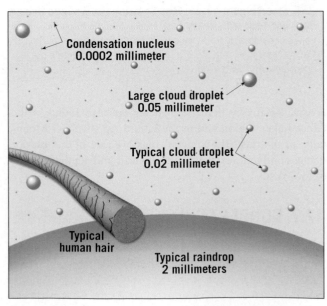

Condensation nucleus
0.0002 millimeter

Large cloud droplet
0.05 millimeter

Typical cloud droplet
0.02 millimeter

Typical human hair

Typical raindrop
2 millimeters

Precipitation from Warm Clouds: The Collision–Coalescence Process

A few decades ago, meteorologists believed that the Bergeron process was responsible for the formation of most precipitation. However, it was discovered that copious rainfall may be produced within clouds located well below the freezing level (*warm clouds*), particularly in the tropics. This led to the proposal of a second mechanism thought to produce precipitation—the **collision–coalescence process**.

Research has shown that clouds composed entirely of liquid droplets must contain some droplets larger than 20 micrometers (0.02 millimeters) for precipitation to form. These large droplets usually form when *hygroscopic particles* (particles that attract water), such as sea salt, are abundant in the atmosphere. Hygroscopic particles begin to remove water vapor from the air when the relative humidity is under 100 percent, and the cloud droplets that form on them can grow quite large. Because the rate at which drops fall is size dependent, these "giant" droplets fall most rapidly. As they plummet, they collide with smaller, slower droplets (see Figure 17.28). After many such collisions, these droplets may grow large enough to fall to the surface without evaporating. Updrafts also aid this process because they cause the droplets to traverse the cloud repeatedly.

Raindrops can grow to a maximum size of 5 millimeters, at which point they fall at a rate of 33 kilometers (20 miles) per hour. At this size and speed, the water's surface tension, which holds the drop together, is overcome by the drag imposed by the air, causing the drops to break apart. The resulting breakup of a large raindrop produces numerous smaller drops that begin anew the task of sweeping up cloud droplets. Drops that are less than 0.5 millimeter (0.02 inch) upon reaching the ground are termed *drizzle* and require about 10 minutes to fall from a cloud 1000 meters (3300 feet) overhead.

17.8 CONCEPT CHECKS

1 What is the difference between precipitation and condensation?

2 Describe the Bergeron process in your own words.

3 What conditions favor the collision–coalescence process?

FIGURE 17.28 The Collision–Coalescence Process Most cloud droplets are so small that the motion of the air keeps them suspended. The drops are not drawn to scale; a typical raindrop has a volume equal to roughly 1 million cloud droplets.

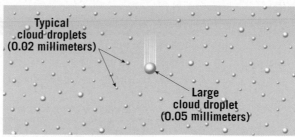

Typical cloud droplets (0.02 millimeters)
Large cloud droplet (0.05 millimeters)

A. Because large cloud droplets fall more rapidly than smaller droplets, they are able to sweep up the smaller ones in their path and grow.

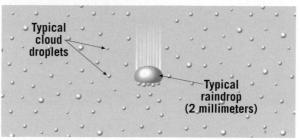

Typical cloud droplets
Typical raindrop (2 millimeters)

B. As drops increase in size, their fall velocity increases, resulting in increased air resistance, which causes the raindrop to flatten.

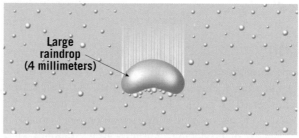

Large raindrop (4 millimeters)

C. As the raindrop approaches 4 millimeters in size, it develops a depression in the bottom.

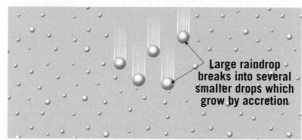

Large raindrop breaks into several smaller drops which grow by accretion

D. Finally, when the diameter exceeds about 5 millimeters, the depression grows upward almost explosively, forming a donut–like ring of water that immediately breaks into smaller drops.

17.9 | FORMS OF PRECIPITATION

List the different types of precipitation and explain how each type forms.

Much of the world's precipitation begins as snow crystals or other solid forms, such as hail or graupel (**TABLE 17.3**). Entering the warmer air below the cloud, these ice particles often melt and reach the ground as raindrops. In some parts of the world, particularly the subtropics, precipitation often forms in clouds that are warmer than 0°C (32°F). These rains frequently occur over the ocean, where cloud condensation nuclei are not plentiful and those that exist vary in size. Under such conditions, cloud droplets can grow rapidly by the collision–coalescence process to produce copious amounts of rain.

Because atmospheric conditions vary greatly from place to place as well as seasonally, several different forms of precipitation are possible. Rain and snow are the most common and familiar, but the other forms of precipitation listed in Table

TABLE 17.3   Forms of Precipitation

Type	Appropriate Size	State of Matter	Description
Mist	0.005–0.05 mm	Liquid	Droplets large enough to be felt on the face when air is moving 1 meter/second. Associated with stratus clouds.
Drizzle	0.05–0.5 mm	Liquid	Small uniform drops that fall from stratus clouds, generally for several hours.
Rain	0.5–5 mm	Liquid	Generally produced by nimbostratus or cumulonimbus clouds. When heavy, it can show high variability from one place to another.
Sleet	0.5–5 mm	Solid	Small, spherical to lumpy ice particles that form when raindrops freeze while falling through a layer of subfreezing air. Because the ice particles are small, damage, if any, is generally minor. Sleet can make travel hazardous.
Glaze	Layers 1 mm–2 cm thick	Solid	Produced when supercooled raindrops freeze on contact with solid objects. Glaze can form a thick coating of ice having sufficient weight to seriously damage trees and power lines.
Rime	Variable accumulations	Solid	Deposits usually consisting of ice feathers that point into the wind. These delicate, frostlike accumulations form as supercooled cloud or fog droplets encounter objects and freeze on contact.
Snow	1 mm–2 cm	Solid	The crystalline nature of snow allows it to assume many shapes, including six-sided crystals, plates, and needles. Produced in supercooled clouds, where water vapor is deposited as ice crystals that remain frozen during their descent.
Hail	5 mm–10 cm or larger	Solid	Precipitation in the form of hard, rounded pellets or irregular lumps of ice. Produced in large cumulonimbus clouds, where frozen ice particles and supercooled water coexist.
Graupel	2–5 mm	Solid	Sometimes called soft hail, graupel forms when rime collects on snow crystals to produce irregular masses of "soft" ice. Because these particles are softer than hailstones, they normally flatten out upon impact.

17.3 are important as well. The occurrence of sleet, glaze, or hail is often associated with important weather events. Although limited in occurrence and sporadic in both time and space, these forms, especially glaze and hail, can cause considerable damage.

Rain

In meteorology, the term **rain** is restricted to drops of water that fall from a cloud and have a diameter of at least 0.5 millimeter (0.02 inch). Most rain originates either in nimbostratus clouds or in towering cumulonimbus clouds that are capable of producing unusually heavy rainfalls known as *cloudbursts*. Raindrops rarely exceed about 5 millimeters (0.2 inch) in diameter. Larger drops do not survive because surface tension, which holds the drops together, is exceeded by the frictional drag of the air. Consequently, large raindrops regularly break apart into smaller ones.

Fine, uniform drops of water having a diameter less than 0.5 millimeter (0.02 inch) are called *drizzle*. Drizzle can be so fine that the tiny drops appear to float, and their impact is almost imperceptible. Drizzle and small raindrops generally are produced in stratus or nimbostratus clouds, where precipitation can be continuous for several hours or, on rare occasions, for days.

Snow

Snow is precipitation in the form of ice crystals (snowflakes) or, more often, aggregates of crystals. The size, shape, and concentration of snowflakes depend to a great extent on the temperature at which they form.

Recall that at very low temperatures, the moisture content of air is small. The result is the formation of very light, fluffy snow made up of individual six-sided ice crystals. This is the "powder" that downhill skiers talk so much about. By contrast, at temperatures warmer than about −5°C (23°F), the ice crystals join together into larger clumps consisting of tangled aggregates of crystals. Snowfalls composed of these composite snowflakes are generally heavy and have a high moisture content, which makes them ideal for making snowballs.

Sleet and Glaze

Sleet is a wintertime phenomenon and refers to the fall of small particles of ice that are clear to translucent. For sleet to be produced, a layer of air with temperatures above freezing must overlie a subfreezing layer near the ground. When raindrops, which are often melted snow, leave the warmer air and encounter the colder air below, they freeze and reach the ground as small pellets of ice the size of the raindrops from which they formed.

On some occasions, when the vertical distribution of temperatures is similar to that associated with the formation of sleet, **glaze** (**freezing rain**) results instead. In such situations, the subfreezing air near the ground is not thick enough to allow the raindrops to freeze. The raindrops, however, do become supercooled as they fall through the cold air and turn to ice upon colliding with solid objects. The result can be a thick coating of ice having sufficient weight to break tree limbs, down power lines, and make walking or driving extremely hazardous (**FIGURE 17.29**).

FIGURE 17.29 Glaze Forms When Supercooled Raindrops Freeze on Contact with Objects In January 1998, an ice storm of historic proportions caused enormous damage in New England and southeastern Canada. Nearly 5 days of freezing rain (glaze) left millions without electricity—some for as long as a month. (Photo by Dick Blume/Syracuse Newspapers/The Image Works)

Hail

Hail is precipitation in the form of hard, rounded pellets or irregular lumps of ice. Moreover, large hailstones often consist of a series of nearly concentric shells of differing densities and degrees of opaqueness (**FIGURE 17.30**). Most hailstones have diameters between 1 centimeter (0.4 inch, pea size) and 5 centimeters (2 inches, golf ball size), although some can be as big as an orange or larger. Occasionally, hailstones weighing a pound or more have been reported. Many of these were probably composites of several stones frozen together.

Hail is produced only in large cumulonimbus clouds, where updrafts can sometimes reach speeds approaching 160 kilometers (100 miles) per hour and where there is an abundant supply of supercooled water. Hailstones begin as small embryonic ice pellets that grow by collecting supercooled water droplets as they fall through the cloud (see Figure 17.30A). If they encounter a strong updraft, they may be carried upward again and begin the downward journey anew. Each trip through the supercooled portion of the cloud results in an additional layer of ice. Hailstones can also form from a single descent through an updraft. Either way, the process continues until the hailstone encounters a downdraft or grows too heavy to remain suspended by the thunderstorm's updraft.

The record for the largest hailstone ever found in the United States was set on July 23, 2010, in Vivian, South Dakota. The stone was over 20 centimeters (8 inches) in diameter and weighed nearly 900 grams (2 pounds). The stone that held the previous record of 0.75 kilograms (1.67 pounds) fell in Coffeyville, Kansas, in 1970 (see Figure 17.30B). The diameter of the stone found in South Dakota also surpassed the previous record of a 17.8-centimeter (7-inch) stone that fell in Aurora, Nebraska, in 2003. Even larger hailstones have reportedly been recorded in Bangladesh, where a 1987 hailstorm killed more than 90 people. It is estimated that large hailstones hit the ground at speeds exceeding 160 kilometers (100 miles) per hour.

The destructive effects of large hailstones are well known, especially to farmers whose crops have been devastated in a few minutes and to people whose windows and roofs have been damaged (**FIGURE 17.31**). In the United States, hail damage each year costs hundreds of millions of dollars.

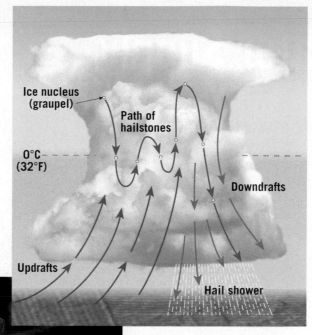

A. Hailstones begin as small ice pellets that grow by adding supercooled water droplets as they move through a cloud. Strong updrafts may carry stones upward in several cycles, increasing the size of the hail by adding a new layer with each cycle. Eventually, the hailstones encounter a downdraft or grow too large to be supported by the updraft.

Ice nucleus (graupel)
Path of hailstones
0°C (32°F)
Downdrafts
Updrafts
Hail shower

SmartFigure 17.30 Formation of Hailstones (Photo courtesy of University Corporation for Atmospheric Research / Science Source)

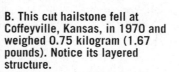

B. This cut hailstone fell at Coffeyville, Kansas, in 1970 and weighed 0.75 kilogram (1.67 pounds). Notice its layered structure.

FIGURE 17.31 Hail Damage to Auto Hail damage that occured northeast of Denver, Colorado. (Photo by Hyoung Chang/Getty Images)

Our Water Supply

FRESH WATER IN THE UNITED STATES

Water is basic to life. It has been called the "bloodstream" of both the biosphere and society. The needs of the world's rapidly growing population are putting Earth's water supply under unprecedented pressure.

RMBrown/Alamy

SOURCES OF FRESH WATER IN THE UNITED STATES

23% GROUNDWATER
79 BILLION GALLONS PER DAY

77% SURFACE WATER
270 BILLION GALLONS PER DAY

GROUNDWATER
Use by category

68.4% IRRIGATION

19.3% PUBLIC SUPPLY

THERMOELECTRIC **0.05%**
MINING **0.92%**
AQUACULTURE **1.3%**
LIVESTOCK **1.3%**
INDUSTRIAL **4.3%**
DOMESTIC **4.2%**

Irrigation is number one

There are nearly 60 million acres (nearly 243,000 square kilometers or about 93,700 square miles) of irrigated land in the United States. That is an area nearly the size of the state of Wyoming. About 42 percent of the water used for irrigation is groundwater.

DRILLING DOWN

Countries are increasingly meeting demand by withdrawing groundwater from nonrenewable sources. Groundwater withdrawal worldwide tripled between 1950 and 2010 according to the United Nations. India has had the largest growth and accounts for about one-quarter of the total.

THE WATER WE USE EACH DAY

According to the American Water Works Association, daily indoor per capita use in an average single family home is nearly 70 gallons.

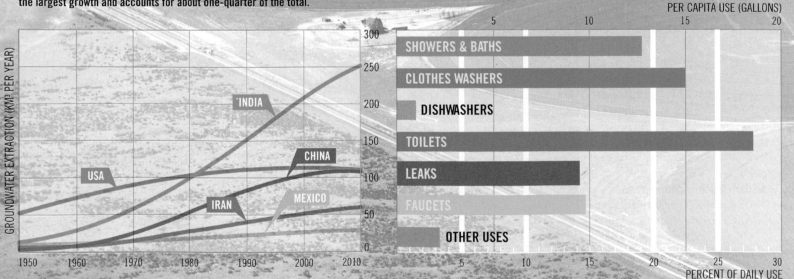

GROUNDWATER EXTRACTION (KM³ PER YEAR)

INDIA
CHINA
USA
IRAN
MEXICO

1950 1960 1970 1980 1990 2000 2010

300 250 200 150 100 50 0

PER CAPITA USE (GALLONS)
5 10 15 20

SHOWERS & BATHS
CLOTHES WASHERS
DISHWASHERS
TOILETS
LEAKS
FAUCETS
OTHER USES

5 10 15 20 25 30
PERCENT OF DAILY USE

igone/Shutterstock

13%
#1 INDIA

12%
#2 CHINA

TOP 7 GLOBAL WATER CONSUMERS

Three nations—India, China, and the United States--together use about one third of the roughly 4000 cubic kilometers (960 cubic miles) of water extracted worldwide each year.

8%
#3 UNITED STATES

4%
#4 RUSSIA

4%
#5 INDONESIA

3%
#6 NIGERIA

3%
#7 BRAZIL

52%
REST OF WORLD

Nils Z/Shutterstock

Adam Gilchrist

Dan Peretz/Shutterstock

6 gallons
It takes about 6 gallons of water to grow a single serving of lettuce.

49 gallons
It takes almost 49 gallons of water to produce just one eight ounce glass of milk. That includes the water consumed by the cow and to grow the food she eats, plus water to process the milk.

2,600 gallons
More than 2,600 gallons of water is required to produce a single serving of steak.

544

FIGURE 17.32 Rime Rime consists of delicate ice crystals that form when supercooled fog or cloud droplets freeze on contact with objects. (Photo by Siepman/Getty Images)

Rime

Rime is a deposit of ice crystals formed by the freezing of supercooled fog or cloud droplets on objects whose surface temperature is below freezing. When rime forms on trees, it adorns them with its characteristic ice feathers, which can be spectacular to behold (**FIGURE 17.32**). In these situations, objects such as pine needles act as freezing nuclei, causing the supercooled droplets to freeze on contact. When the wind is blowing, only the windward surfaces of objects will accumulate the layer of rime.

17.9 CONCEPT CHECKS

1 List the forms of precipitation and the circumstances of their formation.

2 Explain why snow can sometimes reach the ground as rain, but the reverse does not occur.

3 How is sleet different from glaze? Which is usually more hazardous?

17.10 | MEASURING PRECIPITATION

Explain how precipitation is measured.

The most common form of precipitation—rain—is the easiest to measure. Any open container having a consistent cross section throughout can be used as a rain gauge. In general practice, however, more sophisticated devices are used so that small amounts of rainfall can be measured accurately and losses from evaporation can be reduced. A *standard rain gauge* (**FIGURE 17.33**) has a diameter of about 20 centimeters (8 inches) at the top. When the water is caught, a funnel conducts the rain into a cylindrical measuring tube that has a cross-sectional area only one-tenth as large as the receiver. Consequently, rainfall depth is magnified 10 times, which allows for accurate measurements to the nearest 0.025 centimeter (0.01 inch). The narrow opening also minimizes evaporation. When the amount of rain is less than 0.025 centimeter, it is reported as a *trace* of precipitation.

Measuring Snowfall

Snow records are typically kept with two measurements: depth and water equivalent. The depth of snow is usually measured with a calibrated stick. The actual measurement is simple, but choosing a *representative* spot often poses a dilemma. Even when winds are light or moderate, snow drifts freely, making measurement of snowfall difficult. As a rule, it is best to take several measurements in an open place away from trees and obstructions and then average them. To obtain the water equivalent, samples are melted and then weighed or measured as rain.

The quantity of water in a given volume of snow is not constant. A general ratio of 10 units of snow to 1 unit of water is often used when exact information is not available, but the actual water content of snow may deviate widely from this figure. It may take as much as 30 centimeters of light, fluffy dry snow or as little as 4 centimeters of wet snow to produce 1 centimeter of water.

Precipitation Measurement by Weather Radar

Weather forecasts use maps like the one in **FIGURE 17.34** to depict precipitation patterns. The instrument that produces

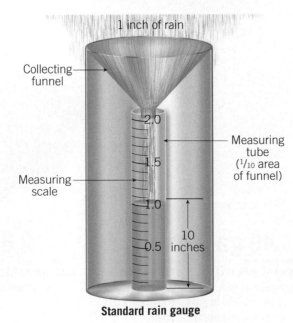

FIGURE 17.33 Precipitation Measurement Using a Standard Rain Gauge A standard rain gauge allows for accurate rainfall measurement to the nearest 0.025 centimeter (0.01 inch). Because the cross-sectional area of the measuring tube is only one-tenth as large as the collector, rainfall is magnified 10 times.

1 inch of rain

Collecting funnel

Measuring tube (¹/₁₀ area of funnel)

Measuring scale

10 inches

Standard rain gauge

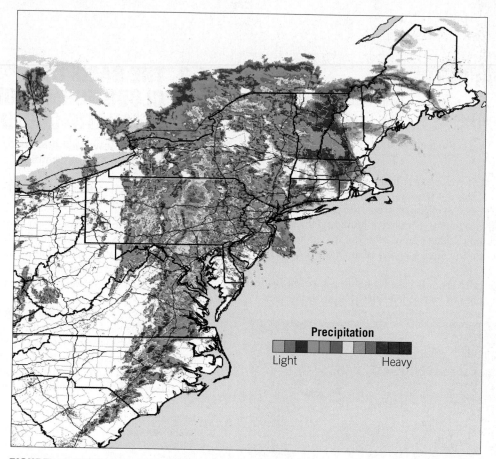

FIGURE 17.34 Doppler Radar Display Produced by the National Weather Service Colors indicate different intensities of precipitation. Note the band of heavy precipitation along the eastern seaboard. (Courtesy of NOAA)

these images, called *weather radar*, has given meteorologists an important tool to probe storm systems as far as a few hundred kilometers away.

Weather radar units have transmitters that send out short pulses of radio waves at wavelengths that can penetrate clouds consisting of small droplets but are reflected by larger raindrops, ice crystals, or hailstones. The reflected signal, called an *echo*, is received and displayed on a monitor. Because the echo is "brighter" when the precipitation is more intense, weather radar is able to depict not only the regional extent of the precipitation but also the rate of rainfall. Figure 17.34 is a typical radar display in which precipitation intensity is shown using colors. Weather radar is also an important tool for determining the rate and direction of storm movement.

17.10 CONCEPT CHECKS

1 Sometimes, when rainfall is light, it is reported as a *trace*. When this occurs, how much (or how little) rain has fallen?

2 Why is rainfall easier to measure than snowfall?

17 CONCEPTS IN REVIEW | Moisture, Clouds, and Precipitation

17.1 WATER'S CHANGES OF STATE

List and describe the processes that cause water to change from one state of matter to another. Define *latent heat* and explain why it is important.

KEY TERMS: calorie, latent heat, evaporation, condensation, sublimation, deposition

■ Water vapor, an odorless, colorless gas, changes from one state of matter (solid, liquid, or gas) to another at the temperatures and pressures experienced near Earth's surface.

■ The processes that result in a change of state are evaporation, condensation, melting, freezing, sublimation, and deposition. Evaporation, melting, and sublimation require heat, whereas condensation, freezing, and deposition result in latent (hidden) heat being released.

Q Label the accompanying diagram with the appropriate terms for the changes of state that are shown.

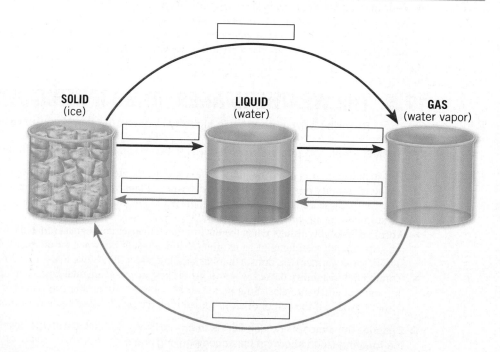

17.2 HUMIDITY: WATER VAPOR IN THE AIR

Distinguish between relative humidity and dew point. Write a generalization relating how temperature changes affect relative humidity.

KEY TERMS: humidity, saturation, vapor pressure, mixing ratio, relative humidity, dew-point temperature (dew point), hygrometer, psychrometer

- Humidity is a general term to describe the amount of water vapor in the air. The methods used to express humidity quantitatively include (1) mixing ratio, the mass of water vapor in a unit of air compared to the remaining mass of dry air; (2) vapor pressure, the part of the total atmospheric pressure attributable to its water-vapor content; (3) relative humidity, the ratio of the air's actual water-vapor content to the amount of water vapor required for saturation at that temperature; and (4) dew point, the temperature to which a parcel of air would need to be cooled to reach saturation.

- When air is saturated, the pressure exerted by the water vapor is called saturation vapor pressure. At saturation, a balance is reached between the number of water molecules leaving the surface of the water and the number returning. At higher temperatures, more water vapor is required to reach saturation because the saturation vapor pressure is temperature dependent.

- Relative humidity can be changed in two ways: by adding or subtracting water vapor or by changing the air's temperature.

Q Refer to the accompanying photo and explain how the relative humidity inside the house compares to the relative humidity outside the house on this particular day.

Photo by Clynt Garnham Housing/Alamy

17.3 THE BASIS OF CLOUD FORMATION: ADIABATIC COOLING

Explain how adiabatic cooling results in cloud formation.

KEY TERMS: adiabatic temperature change, parcel, dry adiabatic rate, wet adiabatic rate

- Cooling of air as it rises and expands, due to successively lower air pressure, is the basic cloud-forming process. Temperature changes that result when air is compressed or when air expands are called adiabatic temperature changes.

- Unsaturated air warms by compression and cools by expansion at the rather constant rate of 10°C per 1000 meters (5.5°F per 1000 feet) of altitude change, a quantity called the dry adiabatic rate. When air rises high enough, it will cool sufficiently to cause condensation and form clouds. Air that continues to rise above the condensation level will cool at the wet adiabatic rate, which varies from 5°C to 9°C per 1000 meters of ascent. The difference in the wet and dry adiabatic rates results because condensation releases latent heat, thereby reducing the rate at which air cools as it ascends.

17.4 PROCESSES THAT LIFT AIR

List and describe the four mechanisms that cause air to rise.

KEY TERMS: convective lifting, orographic lifting, rain shadow desert, front, frontal wedging, convergence, localized convective lifting

- Four mechanisms that can initiate the vertical movement of air are (1) orographic lifting, which occurs when elevated terrains, such as mountains, act as barriers to the flow of air; (2) frontal wedging, when cool air acts as a barrier over which warmer, less dense air rises; (3) convergence, which happens when air flows together, causing an upward movement of air; and (4) localized convective lifting, caused by unequal surface heating, which results in pockets of air rising because of their lower density (buoyancy).

17.5 THE WEATHERMAKER: ATMOSPHERIC STABILITY

Describe how atmospheric stability is determined and compare conditional instability with absolute instability.

KEY TERMS: stable air, unstable air, absolute stability, absolute instability, conditional instability

- The stability of air is determined by comparing the environmental lapse rate (change in temperature with altitude) to the wet and dry adiabatic rates.
- Absolute stability prevails when the environmental lapse rate is less than the wet adiabatic rate. Under these conditions, a parcel of air that is forced to rise will always be heavier than the surrounding air, producing stability.
- Absolute instability results when the environmental lapse rate is greater than the dry adiabatic rate. Specifically, a column of air exhibits absolute instability when the air near the ground is significantly warmer (less dense) than the air aloft and will rise like a hot-air balloon.
- Conditional instability occurs when moist air has an environmental lapse rate between the dry and wet adiabatic rates. Simply, the air is stable for an unsaturated parcel of air but becomes unstable if the parcel of air is forced high enough for it to become saturated.

Q Describe the atmospheric conditions that were likely associated with the development of the towering cloud shown in the accompanying photo.

Photo by Rolf Nussbaumer/Alamy

17.6 CONDENSATION AND CLOUD FORMATION

List the necessary conditions for condensation and briefly describe the two criteria used for cloud classification.

KEY TERMS: condensation nuclei, hygroscopic nuclei, cloud, cirrus, cumulus, stratus, high clouds, middle clouds, low clouds, clouds of vertical development

- For condensation to occur, air must reach saturation. Saturation occurs most commonly when air is cooled to its dew point, or, less often water, when vapor is added to the air. Condensation normally requires a surface on which water vapor can condense. In cloud and fog formation, tiny particles called condensation nuclei serve this purpose.
- Clouds are a form of condensation described as visible aggregates of minute droplets of water or tiny crystals of ice.
- Clouds are classified on the basis of their form and height. The three basic forms are cirrus (high, white, thin, wispy fibers), cumulus (globular, individual cloud masses), and stratus (sheets or layers that cover much or all of the sky). The four categories based on height are high clouds (bases normally above 6000 meters [20,000 feet]), middle clouds (from 2000 to 6000 meters [6500 to 20,000 feet]), low clouds (below 2000 meters [6500 feet]), and clouds of vertical development.

17.7 FOG

Define *fog* and explain how the various types of fog form.

KEY TERMS: fog, advection fog, radiation fog, upslope fog, steam fog, frontal fog (precipitation fog)

- Fog is a cloud with its base at or very near the ground. Fogs form when air is cooled below its dew point or when enough water vapor is added to the air to cause saturation.
- Advection fog forms when warm, moist air moves over a cool surface. Radiation fog forms on cool, clear, calm nights, when Earth's surface cools rapidly by the process of radiation cooling.
- When air becomes saturated through the addition of water vapor, the resulting fogs are called evaporation fogs. During a gentle rain when the air near Earth's surface is cool and near saturation, enough rain may evaporate to produce fog. A fog formed in this manner is called frontal fog, or precipitation fog.

Q Identify the type of fog shown in the accompanying photo.

(Photo by Pat and Chuck Blackley/Alamy)

17.8 HOW PRECIPITATION FORMS

Describe the two mechanisms that produce precipitation.

KEY TERMS: Bergeron process, supercooled, freezing nuclei, collision–coalescence process

- For precipitation to form, millions of cloud droplets must join together into drops that are large enough to reach the ground before evaporating. Two mechanisms have been proposed to explain this phenomenon: the Bergeron process and the collision–coalescence process.
- The Bergeron process operates in cold clouds where ice crystals and supercooled liquid droplets coexist. The ice crystals, which have a greater affinity for water vapor, grow larger at the expense of supercooled cloud droplets. When the ice crystals grow large enough, they reach the ground as snowflakes in winter, whereas in the summer they melt to form rain.
- The collision–coalescence process operates in warm clouds that contain large hygroscopic (water-seeking) nuclei, such as salt particles. The hydroscopic particles form large droplets, which collide and join with smaller water droplets as they descend. After many collisions, the droplets grow large enough so as not to evaporate before they reach the ground.

17.9 FORMS OF PRECIPITATION

List the different types of precipitation and explain how each type forms.

KEY TERMS: rain, snow, sleet, glaze (freezing rain), hail, rime

- The forms of precipitation include rain, snow, sleet, glaze, hail, and rime. The term rain is restricted to drops of water that fall from a cloud and have a diameter of at least 0.5 millimeter (0.02 inch). Snow is precipitation in the form of ice crystals (snowflakes) or, more often, aggregates of crystals. Sleet is a wintertime phenomenon and refers to the fall of small particles of ice that are clear to translucent, while glaze forms when supercooled raindrops turn to ice upon colliding with solid objects. Hail is precipitation in the form of hard, rounded pellets or irregular lumps of ice that usually have diameters between 1 centimeter (0.4 inch, pea size) and 5 centimeters (2 inches, golf ball size). Rime is a deposit of feathery-looking ice crystals formed by the freezing of supercooled fog droplets on objects.

17.10 MEASURING PRECIPITATION

Explain how precipitation is measured.

- Any open container having a consistent cross section throughout can be used as a rain gauge.
- Weather radar units have transmitters that emit short pulses of radio waves at wavelengths that can penetrate clouds consisting of small cloud droplets but are reflected by larger raindrops, ice crystals, or hailstones. Because the echo is "brighter" when the precipitation is more intense, weather radar is able to depict not only the regional extent of the precipitation but also the rate of rainfall.

GIVE IT SOME **THOUGHT**

1. Refer to Figure 17.2 to complete the following:
   a. In which state of matter is water densest?
   b. In which state of matter are water molecules most energetic?
   c. In which state of matter is water compressible?

2. The accompanying photo shows a cup of hot coffee. What state of matter is the "steam" rising from the liquid? Explain your answer.

Photo by Dmitry Kolmakov/Shutterstock

3. The primary mechanism by which the human body cools itself is perspiration.
   a. Explain how perspiring cools the skin.
   b. Refer to the data for Phoenix, Arizona, and Tampa, Florida, in Table A.
   c. In which city would it be easier to stay cool by perspiring? Explain your choice.

Table A

City	Temperature	Dew-Point Temperature
Phoenix, AZ	101°F	47°F
Tampa, FL	101°F	77°F

4. During hot summer weather, many people put "koozies" around their beverages to keep the drinks cold. In addition to preventing a warm hand from heating the container through conduction, what other mechanisms slow the process of warming beverages?

5. Refer to Table 17.1 to answer this question. How much more water is contained in saturated air at a tropical location with a temperature of 40°C compared to a polar location with a temperature of −10°C?

6. Refer to the data for Phoenix, Arizona, and Bismarck, North Dakota, in Table B, to complete the following:
   a. Which city has a higher relative humidity?
   b. Which city has the greatest quantity of water vapor in the air?
   c. In which city is the air closest to its saturation point with respect to water vapor?
   d. In which city does the air require the most water vapor in order to reach saturation?

Table B

City	Temperature	Dew-Point Temperature
Phoenix, AZ	101°F	47°F
Bismarck, ND	39°F	38°F

7. The accompanying graph shows how air temperature and relative humidity change on a typical summer day in the Midwest. Assuming that the dew-point temperature remained constant, what would be the best time of day to water a lawn to minimize evaporation of the water spread on the grass?

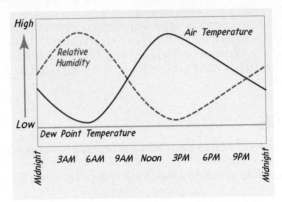

8. The accompanying diagram shows air flowing from the ocean over a coastal mountain range. Assume that the dew-point temperature remains constant in dry air (that is, relative humidity less than 100 percent). If the air parcel becomes saturated, the dew-point temperature will cool at the wet adiabatic rate as it ascends, but it will not change as the air parcel descends. Use this information to complete the following:
   a. Determine the air temperature and dew-point temperature for the air parcel at each location (B–G) shown on the diagram.
   b. At what elevation will clouds begin to form (with relative humidity = 100 percent)?
   c. Compare the air temperatures at points A and G. Why are they different?
   d. How did the water vapor content of the air change as the parcel of air traversed the mountain? (*Hint:* Compare dew-point temperatures.)
   e. On which side of the mountain might you expect lush vegetation, and on which side would you expect desert-like conditions?
   f. Where in the United States might you find a situation like what is pictured here?

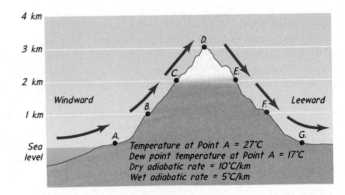

9. The cumulonimbus cloud pictured in Figure 17.21H is roughly 12 kilometers tall, 8 kilometers wide, and 8 kilometers long. Assume that the droplets in each cubic meter of the cloud total 0.5 cubic centimeter. How much liquid does the cloud contain? How many gallons is this? (*Note:* 3785 cm³ = 1 gallon.)

10. Cloud droplets form and grow as water vapor condenses onto hygroscopic condensation nuclei. Research has shown that the maximum radius for cloud droplets is about 0.05 millimeter. However, typical raindrops have volumes thousands of times greater. Explain how cloud droplets become raindrops.

11. Why does radiation fog form mainly on clear nights rather than on cloudy nights?

12. Which winter storm is likely to produce deeper snowfall: a low-pressure system that passes through the midwestern states of Nebraska, Iowa, and Illinois (26°F average temperature at the time of the storm) *or* exactly the same system passing through North Dakota, Minnesota, and Wisconsin (16°F average temperature at the time of the storm).

13. Weather radar provides information on the intensity of precipitation in addition to the total amount of precipitation that falls over a given time period. Table C shows the relationship between radar reflectivity values and rainfall rates. If radar measured a reflectivity value of 47 dBZ for 2½ hours over a location, how much rain will have fallen there?

14. What are the advantages and disadvantages of using rain gauges compared to weather radar in measuring rainfall?

Table C
Conversion of Radar Reflectivity to Rainfall Rate

Radar Reflectivity (dBZ)	Rainfall Rate (inches/hr)
65	16+
60	8.0
55	4.0
52	2.5
47	1.3
41	0.5
36	0.3
30	0.1
20	trace

EXAMINING THE **EARTH SYSTEM**

1. The interactions among Earth's spheres have produced the Great Basin area of the western United States, which includes some of the driest areas in the world. Examine the map of the region in Figure 6.30. Although the area is only a few hundred miles from the Pacific Ocean, it is a desert. Explain why. Did any geologic factor(s) contribute to the formation of this desert environment? Do any major rivers have their source in the Great Basin? Explain. (For information about deserts in the United States, visit the U.S. Geological Survey's [USGS's] "Deserts: Geology and Resources" Website, at http://pubs.usgs.gov/gip/deserts/.)

2. When Earth is viewed from space, the most striking feature of the planet is *water*. It is found as a liquid in the global oceans, as a solid in the polar ice caps, and as clouds and water vapor in the atmosphere. Although only one-thousandth of 1 percent of the water on Earth exists as water vapor, it has a huge influence on our planet's weather and climate. What role does water vapor play in heating Earth's surface? How does water vapor act to transfer heat from Earth's land–sea surface to the atmosphere?

(NASA)

3. The amount of precipitation that falls at any particular place and time is controlled by the quantity of moisture in the air and many other factors. How might each of the following alter the precipitation at a particular locale?

   a. An increase in the elevation of the land
   b. A decrease in the area covered by forests and other types of vegetation
   c. Lowering of average ocean-surface temperatures
   d. An increase in the percentage of time that the winds blow from an adjacent body of water
   e. A major episode of global volcanism lasting a decade

4. Phoenix and Flagstaff, Arizona, are both located in the southwestern United States, less than a 2-hour drive apart. Using the accompanying climate diagram (which gives the elevations of the two cities), describe the impact that elevation has on the precipitation and temperature of each city. Use the Internet to compare and explain the natural vegetation of these locations. Next, check the current weather conditions for Phoenix and Flagstaff, Arizona, by using The Weather Channel Website, at www.weather.com. Do the current conditions seem to fit climate data? Explain.

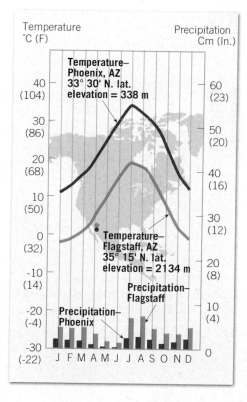

MasteringGeology™

18

Air Pressure and Wind

Horizontal differences in air pressure created the winds that are propelling these sailboats. (Photo by UpperCut Images/Alamy)

551

Of the various elements of weather and climate, changes in air pressure are the least noticeable. When listening to a weather report, we are generally interested in moisture conditions (humidity and precipitation), temperature, and perhaps wind. Rarely do people wonder about air pressure. Although the hour-to-hour and day-to-day variations in air pressure are generally not noticed by people, they are very important in producing changes in our weather. Variations in air pressure from place to place cause the movement of air we call wind and are a significant factor in weather forecasting. As we will see, air pressure is closely tied to the other elements of weather in a cause-and-effect relationship.

18.1 | UNDERSTANDING AIR PRESSURE

Define *air pressure* and describe the instruments used to measure this weather element.

In Chapter 16 we noted that **air pressure** is simply the pressure exerted by the weight of air above. Average air pressure at sea level is about 1 kilogram per square centimeter, or 14.7 pounds per square inch. Specifically, a column of air 1 square inch in cross section, measured from sea level to the top of the atmosphere, would weigh about 14.7 pounds (**FIGURE 18.1**). This is roughly the same pressure that is produced by a 1-square-inch column of water 10 meters (33 feet) in height. With some simple arithmetic, you can calculate that the air pressure exerted on the top of a small (50 centimeter-by-100 centimeter [20 inch-by-40 inch]) school desk exceeds 5000 kilograms (11,000 pounds), or about the weight of a 50-passenger school bus. Why doesn't the desk collapse under the weight of the ocean of air above? Simply, air pressure is exerted in all directions—down, up, and sideways. Thus, the air pressure pushing down on the desk exactly balances the air pressure pushing up on the desk.

Visualizing Air Pressure

Imagine a tall aquarium that has the same dimensions as the small desk mentioned in the preceding paragraph. When this aquarium is filled to a height of 10 meters (33 feet), the water pressure at the bottom equals 1 atmosphere (14.7 pounds per square inch). Now, imagine what will happen if this aquarium is placed on top of our student desk so that all the force is directed downward. Compare this to what results when the desk is placed inside the aquarium and allowed to sink to the bottom. In the latter example, the desk survives because the water pressure is exerted in all directions, not just downward, as in our earlier example. The desk, like your body, is "built" to withstand the pressure of 1 atmosphere. It is important to note that although we do not generally notice the pressure exerted by the ocean of air around us, except when ascending or descending in an elevator or airplane, it is nonetheless substantial. The pressurized suits that astronauts use on space walks are designed to duplicate the atmospheric pressure experienced at Earth's surface. Without these protective suits to keep body fluids from boiling away, astronauts would perish in minutes.

The concept of air pressure can also be understood if we examine the behavior of gas molecules. Gas molecules, unlike molecules of the liquid and solid phases, are not bound to one another but freely move about, filling all the space available to them. When two gas molecules collide, which happens frequently under normal conditions, they bounce off each other like elastic balls. If a gas is confined to a container, this motion is restricted by its sides, much as the walls of a handball court redirect the motion of the handball. The continuous bombardment of gas molecules against the sides of the container

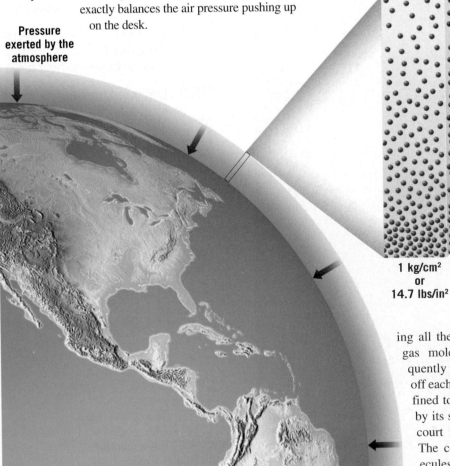

FIGURE 18.1 Sea-Level Pressure Air pressure can be thought of as the weight of the atmosphere above. A column of air 1-square inch in cross section extending from sea level to the top of the atmosphere would weigh about 14.7 pounds.

Pressure exerted by the atmosphere

1 kg/cm²
or
14.7 lbs/in²

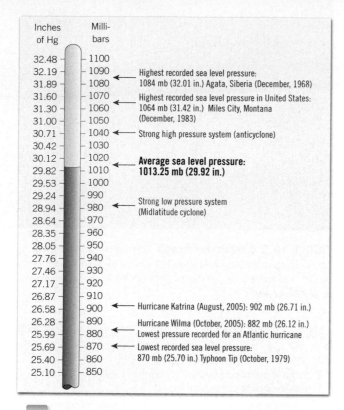

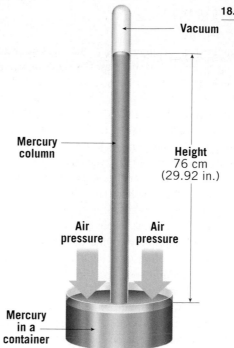

FIGURE 18.3 Simple Mercury Barometer The weight of the column of mercury is balanced by the pressure exerted on the dish of mercury by the air above. If the pressure drops, the column of mercury falls; if the pressure increases, the column rises.

SmartFigure 18.2
Inches and Millibars A comparison of two units commonly used to express air pressure

When air pressure increases, the mercury in the tube rises. Conversely, when air pressure decreases, so does the height of the mercury column. With some refinements, the mercury barometer invented by Torricelli is still the standard pressure-measuring instrument used today. Standard atmospheric pressure at sea level equals 29.92 inches of mercury.

The need for a smaller and more portable instrument for measuring air pressure led to the development of the **aneroid barometer** (*aneroid* means "without liquid"). Instead of having a mercury column held up by air pressure, an aneroid barometer uses a partially evacuated metal chamber (**FIGURE 18.4**). The chamber is extremely sensitive to variations in air pressure

exerts an outward push that we call air pressure. Although the atmosphere is without walls, it is confined from below by Earth's surface and effectively from above because the force of gravity prevents its escape. Here, we can define *air pressure* as the force exerted against a surface by the continuous collision of gas molecules.

Measuring Air Pressure

When meteorologists measure atmospheric pressure, they use a unit called the *millibar*. Standard sea-level pressure is 1013.2 millibars. Although the millibar has been the unit of measure on all U.S. weather maps since January 1940, the media use "inches of mercury" to describe atmospheric pressure. In the United States, the National Weather Service converts millibar values to inches of mercury for public and aviation use (**FIGURE 18.2**).

Inches of mercury is easy to understand. The use of mercury for measuring air pressure dates from 1643, when Torricelli, a student of the famous Italian scientist Galileo, invented the **mercury barometer**. Torricelli correctly described the atmosphere as a vast ocean of air that exerts pressure on us and all objects about us. To measure this force, he filled a glass tube, which was closed at one end, with mercury. The tube was then inverted into a dish of mercury (**FIGURE 18.3**). Torricelli found that the mercury flowed out of the tube until the weight of the column was balanced by the pressure that the atmosphere exerted on the surface of the mercury in the dish. In other words, the weight of mercury in the column equaled the weight of the same diameter column of air that extended from the ground to the top of the atmosphere.

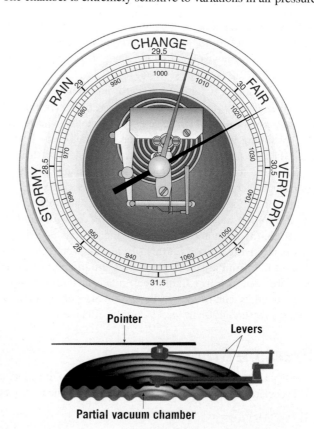

FIGURE 18.4 Aneroid Barometer The black pointer shows the current air pressure. When the barometer is read, the observer moves the other pointer to coincide with the current air pressure. Later, when the barometer is checked, the observer can see whether the air pressure has been rising, falling, or has remained steady. The bottom diagram is a cross section. An aneroid barometer has a partially evacuated chamber that changes shape, compressing as air pressure increases and expanding as pressure decreases.

and changes shape, compressing as the pressure increases and expanding as the pressure decreases. A series of levers transmit the movements of the chamber to a pointer on a dial that is calibrated to read in inches of mercury and/or millibars.

As shown in Figure 18.4, the face of an aneroid barometer intended for home use is inscribed with words such as *fair*, *change, rain*, and *stormy*. Notice that "fair" corresponds with high-pressure readings, whereas "rain" is associated with low pressures. Barometric readings, however, may not always indicate the weather. The dial may point to "fair" on a rainy day or to "rain" on a fair day. To "predict" the local weather, the change in air pressure over the past few hours is more important than the current pressure reading. Falling pressure is often associated with increasing cloudiness and the possibility of precipitation, whereas rising air pressure generally indicates clearing conditions. It is useful to remember, however, that particular barometer readings or trends do not always correspond to specific types of weather.

Another advantage of an aneroid barometer is that it can easily be connected to a recording mechanism. The resulting instrument is a **barograph**, which provides a continuous record of pressure changes with the passage of time (**FIGURE 18.5**). Another important adaptation of the aneroid barometer is its use to indicate altitude for aircraft, mountain climbers, and mapmakers.

FIGURE 18.5 Aneroid Barograph This instrument makes a continuous record of air pressure. The cylinder with the graph is a clock that turns once per day or once per week. (Photo courtesy of Stuart Aylmer/Alamy)

18.1 CONCEPT CHECKS

1 Describe air pressure in your own words.

2 What is standard sea-level pressure in millibars, in inches, and in pounds per square inch?

3 Describe the operating principles of the mercury barometer and the aneroid barometer.

4 List two advantages of the aneroid barometer.

18.2 | FACTORS AFFECTING WIND

Discuss the three forces that act on the atmosphere to either create or alter winds.

In Chapter 17, we examined the upward movement of air and its role in cloud formation. As important as vertical motion is, far more air moves horizontally, the phenomenon we call **wind**. What causes wind?

Simply stated, wind is the result of horizontal differences in air pressure. *Air flows from areas of higher pressure to areas of lower pressure.* You may have experienced this when opening something that is vacuum packed. The noise you hear is caused by air rushing from the higher pressure outside the can or jar to the lower pressure inside. Wind is nature's attempt to balance such inequalities in air pressure. Because unequal heating of Earth's surface generates these pressure differences, *solar radiation is the ultimate energy source for most wind.*

If Earth did not rotate, and if there were no friction between moving air and Earth's surface, air would flow in a straight line from areas of higher pressure to areas of lower pressure. But because Earth does rotate and friction does exist, wind is controlled by the following combination of forces: pressure gradient force, Coriolis effect, and friction.

Pressure Gradient Force

If an object experiences an unbalanced force in one direction, it will accelerate (experience a change in velocity). The force that generates winds results from horizontal pressure differences. When air is subjected to greater pressure on one side than on another, the imbalance produces a force that is directed from the region of higher pressure toward the area of lower pressure. Thus, pressure differences cause the wind to blow, and the greater these differences, the greater the wind speed.

Variations in air pressure over Earth's surface are determined from barometric readings taken at hundreds of weather stations. These pressure measurements are shown on surface weather maps using **isobars** (*iso* = equal, *bar* = pressure), or lines connecting places of equal air pressure (**FIGURE 18.6**). The *spacing* of the isobars indicates the amount of pressure change occurring over a given distance and is called the **pressure gradient force**. Pressure gradient is analogous to gravity acting on a ball rolling down a hill. A steep pressure gradient, like a steep hill, causes greater acceleration of a parcel of air than does a weak pressure gradient (a gentle hill). Thus, the relationship between wind speed and the pressure gradient is straightforward: *Closely spaced isobars indicate a steep pressure gradient and strong winds; widely spaced isobars indicate a weak pressure gradient and light winds.* Figure 18.6 illustrates the relationship between the spacing of isobars and wind speed. Note also that the pressure gradient force is always directed at *right angles* to the isobars.

In order to draw isobars on a weather map to show air pressure patterns, meteorologists must compensate for the *elevation* of each station. Otherwise, all high-elevation locations, such as Denver, Colorado, would always be mapped as having low pressure. This compensation is accomplished by converting all pressure measurements to sea-level equivalents.

FIGURE 18.7 is a surface weather map that shows isobars (representing corrected sea-level air pressure) and winds. Wind *direction* is shown as wind arrow shafts, and *speed* is shown as wind bars (see the accompanying key). Isobars, used to depict pressure patterns, are rarely straight or evenly spaced on surface maps. Consequently, wind generated by the pressure gradient force typically changes speed and direction as it flows.

The area of somewhat circular closed isobars in eastern North America represented by the red letter *L* is a *low-pressure system*. In western Canada, a *high-pressure system*, denoted by the blue letter *H*, can also be seen. We will discuss *highs* and *lows* in the next section.

In summary, the *horizontal pressure gradient is the driving force of wind*. The magnitude of the pressure gradient force is shown by the spacing of isobars. The direction of force is always from areas of higher pressure toward areas of lower pressure and at right angles to the isobars.

Coriolis Effect

Figure 18.7 shows the typical air movements associated with high- and low-pressure systems. As expected, the air moves out of the regions of higher pressure and into the regions of

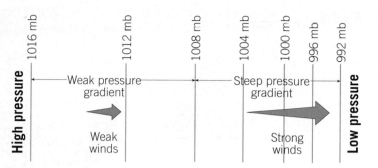

FIGURE 18.6 Isobars Show the Pressure Gradient Isobars are lines connecting places of equal air pressure. The spacing of isobars indicates the amount of pressure change occurring over a given distance—called the pressure gradient. Closely spaced isobars indicate a strong pressure gradient and high wind speeds, whereas widely spaced isobars indicate a weak pressure gradient and low wind speeds.

lower pressure. However, the wind does not cross the isobars at right angles as the pressure gradient force directs it to do. The direction deviates as a result of Earth's rotation. This has been named the **Coriolis effect**, after the French scientist who first thoroughly described it.

All free-moving objects or fluids, including the wind, are deflected to the *right* of their path of motion in the Northern Hemisphere and to the *left* in the Southern Hemisphere. The reason for this deflection can be illustrated by imagining the path of a rocket launched from the North Pole toward a target located on the equator (**FIGURE 18.8**). If the rocket took an hour to reach its target, during its flight, Earth would have rotated 15 degrees to the east. To someone standing on Earth, it would look as if the rocket had veered off its path and hit Earth 15 degrees west of its target. The true path of

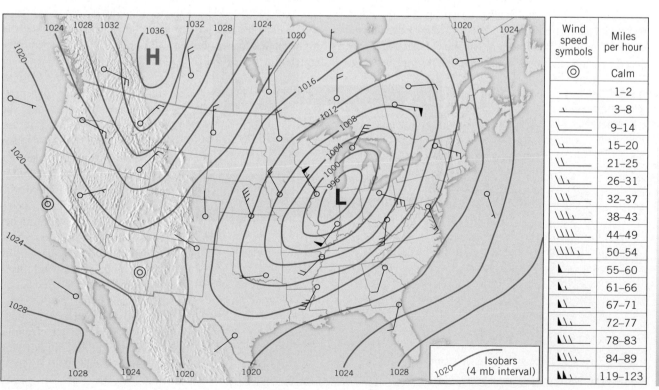

Wind speed symbols	Miles per hour
⊚	Calm
——	1–2
⊥	3–8
\\	9–14
\\\	15–20
\\\\	21–25
\\\\\	26–31
\\\\\\	32–37
\\\\\\\	38–43
\\\\\\\\	44–49
\\\\\\\\\	50–54
◣	55–60
◣\	61–66
◣\\	67–71
◣\\\	72–77
◣\\\\	78–83
◣\\\\\	84–89
◤◤	119–123

SmartFigure 18.7 Isobars on a Weather Map Isobars are used to show the distribution of pressure on daily weather maps. Isobars are seldom straight but usually form broad curves. Concentric isobars indicate cells of high and low pressure. The "wind flags" indicate the expected airflow surrounding pressure cells and are plotted as "flying" with the wind (that is, the wind blows toward the station circle). Notice on this map that the isobars are more closely spaced and the wind speed is faster around the low-pressure center than around the high.

SmartFigure 18.8 Coriolis Effect The Coriolis effect, illustrated using the 1-hour flight of a rocket traveling from the North Pole to a location on the equator. **A.** On a nonrotating Earth, the rocket would travel straight to its target. **B.** However, Earth rotates 15 degrees each hour. Thus, although the rocket travels in a straight line, when we plot the path of the rocket on Earth's surface, it follows a curved path that veers to the right of the target.

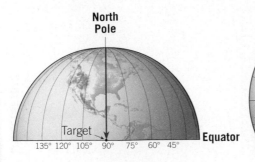

A. Nonrotating Earth

B. Rotating Earth

We attribute the apparent shift in wind direction to the Coriolis effect. This deflection (1) is always directed at right angles to the direction of airflow; (2) affects only wind direction, not wind speed; (3) is affected by wind speed (the stronger the wind, the greater the deflection); and (4) is strongest at the poles and weakens equatorward, becoming nonexistent at the equator.

Note that any free-moving object will experience a deflection caused by the Coriolis effect. The U.S. Navy dramatically discovered this fact in World War II. During target practice, long-range guns on battleships continually missed their targets by as much as several hundred yards until ballistic corrections were made for the changing position of a seemingly stationary target. Over a short distance, however, the Coriolis effect is relatively small.

Friction with Earth's Surface

The effect of friction on wind is important only within a few kilometers of Earth's surface. Friction acts to slow air movement and, as a consequence, alters wind direction. To illustrate friction's effect on wind direction, let us look at a situation in which friction has no role. Above the friction layer, the pressure gradient force and Coriolis effect work together to direct the flow of air. Under these conditions, the pressure gradient force causes air to start moving across the isobars. As soon as the air starts to move, the Coriolis effect acts at right angles to this motion. The faster the wind speed, the greater the deflection.

the rocket is straight and would appear so to someone out in space looking at Earth. It is Earth turning under the rocket that gives it its *apparent* deflection.

Note that the rocket is deflected to the right of its path of motion because of the counterclockwise rotation of the Northern Hemisphere. In the Southern Hemisphere, the effect is reversed. Clockwise rotation produces a similar deflection, but to the *left* of the path of motion. The same deflection is experienced by wind regardless of the direction it is moving.

Eventually, the Coriolis effect will balance the pressure gradient force, and the wind will blow parallel to the isobars (**FIGURE 18.9**). Upperair winds generally take this path and are called **geostrophic winds**. The lack of friction with Earth's surface allows geostrophic winds to travel at higher speeds than do surface winds. This can be observed in **FIGURE 18.10** by noting the wind flags, many of which indicate winds of 50 to 100 miles per hour.

The most prominent features of upper-level flow are **jet streams**. First encountered by high-flying bombers during World War II, these fast-moving "rivers" of air travel

FIGURE 18.9 The Geostrophic Wind The only force acting on a stationary parcel of air is the pressure gradient force. Once the air begins to accelerate, the Coriolis effect deflects it to the right in the Northern Hemisphere. Greater wind speeds result in a stronger Coriolis effect (deflection) until the flow is parallel to the isobars. At this point, the pressure gradient force and Coriolis effect are in balance, and the flow is called a *geostrophic wind*. It is important to note that in the "real" atmosphere, airflow is continually adjusting for variations in the pressure field. As a result, the adjustment to geostrophic equilibrium is much more irregular than shown.

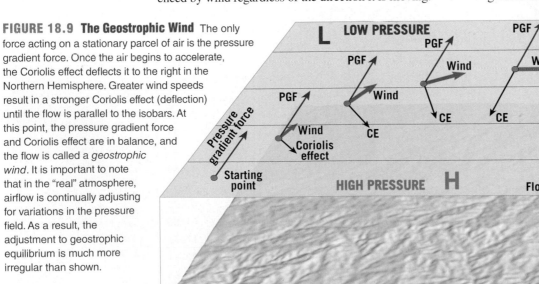

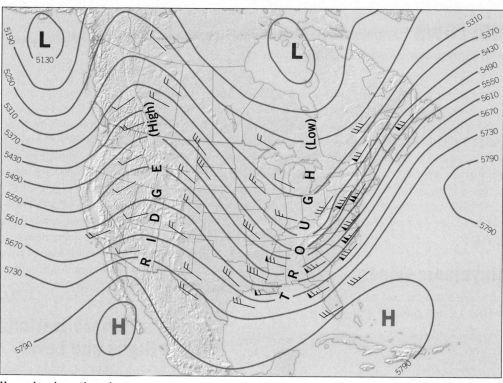

Wind speed symbols	Miles per hour
◎	Calm
—	1–2
⌐	3–8
⌐	9–14
⌐	15–20
⌐	21–25
⌐	26–31
⌐	32–37
⌐	38–43
⌐	44–49
⌐	50–54
⌐	55–60
⌐	61–66
⌐	67–71
⌐	72–77
⌐	78–83
⌐	84–89
⌐	119–123

Upper-level weather chart

FIGURE 18.10 Upper-Air Weather Chart This simplified weather chart shows the direction and speed of the upper-air winds. Note from the flags that the airflow is almost parallel to the contours. Like most other upper-air charts, this one shows variations in the height (in meters) at which a selected pressure (500 millibars) is found instead of showing variations in pressure at a fixed height, like surface maps. Do not let this confuse you because there is a simple relationship between height contours and pressure. Places experiencing 500-millibar pressure at higher altitudes (toward the south on this map) are experiencing higher pressures than are places where the height contours indicate lower altitudes. Thus, *higher-elevation* contours indicate *higher* surface pressures, and *lower-elevation* contours indicate *lower* surface pressures.

between 120 and 240 kilometers (75 and 150 miles) per hour in a west-to-east direction. One such stream is situated over the polar front, which is the zone separating cool polar air from warm subtropical air.

Below 600 meters (2000 feet), friction complicates the airflow just described. Recall that the Coriolis effect is proportional to wind speed. Friction lowers the wind

Representation of upper-level chart

speed, so it reduces the Coriolis effect. Because the pressure gradient force is not affected by wind speed, it wins the tug of war shown in **FIGURE 18.11**. The result is a movement of air at an angle across the isobars, toward the area of lower pressure.

The roughness of the terrain determines the angle of airflow across the isobars. Over the smooth ocean surface, friction is low, and the angle is small. Over rugged terrain, where friction is higher, the angle that air makes as it flows across the isobars can be as great as 45 degrees.

In summary, upper airflow is nearly parallel to the isobars, whereas the effect of friction causes the surface winds to move more slowly and cross the isobars at an angle.

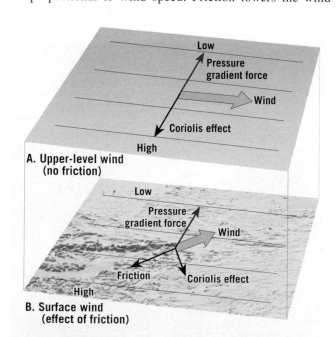

FIGURE 18.11 The Effects of Friction on Wind Friction has little effect on winds aloft, so airflow is parallel to the isobars. In contrast, friction slows surface winds, which weakens the Coriolis effect, causing winds to cross the isobars and move toward the lower pressure.

18.2 CONCEPT CHECKS

1 List three factors that combine to direct horizontal airflow (wind).

2 What force is responsible for generating wind?

3 Write a generalization relating the spacing of isobars to wind speed.

4 Briefly describe how the Coriolis effect influences air movement.

5 Unlike winds aloft, which blow nearly parallel to the isobars, surface winds generally cross the isobars. Explain what causes this difference.

18.3 | HIGHS AND LOWS Contrast the weather associated with low-pressure centers (cyclones) and high-pressure centers (anticyclones).

Among the most common features on a weather map are areas designated as pressure centers. **Cyclones,** or **lows,** are centers of low pressure, and **anticyclones,** or **highs,** are high-pressure centers. As **FIGURE 18.12** illustrates, the pressure decreases from the outer isobars toward the center in a cyclone. In an anticyclone, just the opposite is the case: The values of the isobars increase from the outside toward the center. Knowing just a few basic facts about centers of high and low pressure greatly increases your understanding of current and forthcoming weather.

Cyclonic and Anticyclonic Winds

In the preceding section, you learned that the two most significant factors that affect wind are the pressure gradient force and the Coriolis effect. Winds move from higher pressure toward lower pressure and are deflected to the right or left by Earth's rotation. When these controls of airflow are applied to pressure centers in the Northern Hemisphere, the result is that winds blow inward and counterclockwise around a low (**FIGURE 18.13A**). Around a high, they blow outward and clockwise (see left side of map in Figure 18.12).

In the Southern Hemisphere, the Coriolis effect deflects the winds to the left; therefore, winds around a low blow clockwise, and winds around a high move counterclockwise (**FIGURE 18.13B**). In either hemisphere, friction causes a net inflow (**convergence**) around a cyclone and a net outflow (**divergence**) around an anticyclone.

Weather Generalizations About Highs and Lows

Rising air is associated with cloud formation and precipitation, whereas subsidence produces clear skies. In this section we will discuss how the movement of air can itself create pressure change and generate winds. After doing so, we will examine the relationship between horizontal and vertical flow, and their effects on weather.

Let us first consider a surface low-pressure system where the air is spiraling inward. The net inward transport of air causes a shrinking of the

FIGURE 18.12 Cyclonic and Anticyclonic Winds in the Northern Hemisphere Arrows show the winds blowing inward and counterclockwise around a low and outward and clockwise around a high.

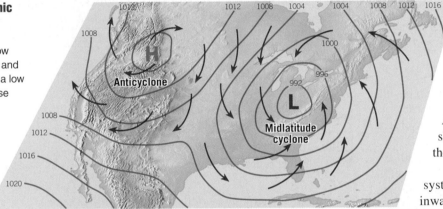

FIGURE 18.13 Cyclonic Circulation in the Northern and Southern Hemispheres The cloud patterns in these images allow us to "see" the circulation pattern in the lower atmosphere. (NASA images)

A. This satellite image shows a large low-pressure center in the Gulf of Alaska. The cloud pattern clearly shows an inward and counterclockwise spiral.

B. This satellite image shows a strong cyclonic storm in the South Atlantic near the coast of Brazil. The cloud pattern shows an inward and clockwise circulation.

area occupied by the air mass, a process that is termed *horizontal convergence.* Whenever air converges horizontally, it must pile up—that is, increase in height to allow for the decreased area it now occupies. This generates a "taller" and therefore heavier air column. Yet a surface low can exist only as long as the column of air exerts less pressure than that occurring in surrounding regions. We seem to have encountered a paradox: A low-pressure center causes a net accumulation of air, which increases its pressure. Consequently, a surface cyclone should quickly eradicate itself in a manner not unlike what happens when a vacuum-packed can is opened.

For a surface low to exist for very long, compensation must occur at some layer aloft. For example, surface convergence could be maintained if divergence (spreading out) aloft occurred at a rate equal to the inflow below. **FIGURE 18.14** shows the relationship between surface convergence and divergence aloft that is needed to maintain a low-pressure center.

Divergence aloft may even exceed surface convergence, thereby resulting in intensified surface inflow and accelerated vertical motion. Thus, divergence aloft can intensify storm centers as well as maintain them. On the other hand, inadequate divergence aloft permits surface flow to "fill" and weaken the accompanying cyclone.

Note that surface convergence about a cyclone causes a net upward movement. The rate of this vertical movement is slow, generally less than 1 kilometer (0.6 mile) per day. Nevertheless, because rising air often results in cloud formation and precipitation, a low-pressure center is generally related to unstable conditions and stormy weather (**FIGURE 18.15A**).

As often as not, it is divergence aloft that creates a surface low. Spreading out aloft initiates upflow in the atmosphere directly below, eventually working its way to the surface, where inflow is encouraged.

Like their cyclonic counterparts, anticyclones must be maintained from above. Outflow near the surface is accompanied by convergence aloft and general subsidence of the air column (see Figure 13.14). Because descending air is compressed and warmed, cloud formation and precipitation are unlikely in an anticyclone. Thus, fair weather can usually be expected with the approach of a high-pressure center (**FIGURE 18.15B**).

For reasons that should now be obvious, it has been common practice to print on household barometers the words "stormy" at the low-pressure end and "fair" on the high-pressure end. By noting whether the pressure is rising, falling, or steady, we have a good indication of what the forthcoming weather will be. Such a determination, called the **pressure,** or **barometric, tendency**, is a useful aid in short-range weather prediction.

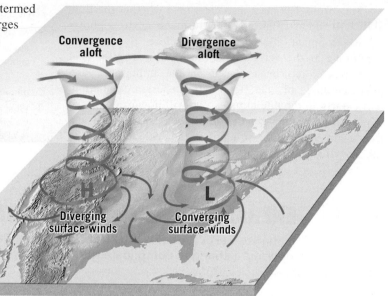

Flow aloft

Convergence aloft   **Divergence aloft**

H   L

Diverging surface winds   **Converging surface winds**

FIGURE 18.14 Airflow Associated with Surface Cyclones (L) and Anticyclones (H)
A low, or cyclone, has converging surface winds and rising air, resulting in cloudy conditions. A high, or anticyclone, has diverging surface winds and descending air, which leads to clear skies and fair weather.

You should now be better able to understand why television weather reporters emphasize the positions and projected paths of cyclones and anticyclones. The "villain" on these weather programs is always the low-pressure center, which produces "bad" weather in any season. Lows move in roughly a west-to-east direction across the United States and require a few days to more than a week for the journey. Because their paths can be somewhat erratic, accurate prediction of their migration is difficult, although essential, for short-range forecasting.

(A)

(B)

FIGURE 18.15 Weather Generalizations Related to Pressure Centers
A. A rainy day in London. Low-pressure systems are frequently associated with cloudy conditions and precipitation. (Photo by Lourens Smak/Alamy) **B.** Clear skies and "fair" weather may be expected when an area is under the influence of high pressure. (Photo by Prisma Bildagentur AG/Alamy)

Meteorologists must also determine whether the flow aloft will intensify an embryo storm or act to suppress its development. Because of the close tie between conditions at the surface and those aloft, a great deal of emphasis has been placed on the importance and understanding of the total atmospheric circulation, particularly in the midlatitudes. We will now examine the workings of Earth's general atmospheric circulation and then again consider the structure of the cyclone in light of this knowledge.

18.4 | GENERAL CIRCULATION OF THE ATMOSPHERE

Summarize Earth's idealized global circulation. Describe how continents and seasonal temperature changes complicate the idealized pattern.

The underlying cause of wind is unequal heating of Earth's surface. In tropical regions, more solar radiation is received than is radiated back to space. In polar regions, the opposite is true: Less solar energy is received than is lost. Attempting to balance these differences, the atmosphere acts as a giant heat-transfer system, moving warm air poleward and cool air equatorward. On a smaller scale, but for the same reason, ocean currents also contribute to this global heat transfer. The general circulation is complex, and a great deal has yet to be explained. We can, however, develop a general understanding by first considering the circulation that would occur on a nonrotating Earth having a uniform surface. We will then modify this system to fit observed patterns.

Circulation on a Nonrotating Earth

On a hypothetical nonrotating planet with a smooth surface of either all land or all water, two large thermally produced cells would form (**FIGURE 18.16**). The heated equatorial air would rise until it reached the tropopause, which acts like a lid and deflects the air poleward. Eventually, this upper-level airflow would reach the poles, sink, spread out in all directions at the surface, and move back toward the equator. Once there, it would be reheated and start its journey over again. This hypothetical circulation system has upper-level air flowing poleward and surface air flowing equatorward.

If we add the effect of rotation, this simple convection system will break down into smaller cells. **FIGURE 18.17** illustrates the three pairs of cells proposed to carry on the task of heat redistribution on a rotating planet. The polar and tropical cells retain the characteristics of the thermally generated convection described earlier. The nature of the midlatitude circulation is more complex and will be discussed in more detail in a later section.

Idealized Global Circulation

Near the equator, the rising air is associated with the pressure zone known as the **equatorial low**. This region of ascending moist, hot air is marked by abundant precipitation. Because this region of low pressure is a zone where winds converge, it is also referred to as the **intertropical convergence zone (ITCZ)**. As the upper-level flow from the equatorial low reaches 20° to 30° latitude, north or south, it sinks back toward the surface. This subsidence and associated adiabatic heating produce hot, arid conditions. The center of this zone of subsiding dry air is the **subtropical high**, which encircles the globe near 30° latitude, north and south (see Figure 18.17). The great deserts of Australia, Arabia, and Africa exist because of the stable, dry condition caused by the subtropical highs.

At the surface, airflow is outward from the center of the subtropical high. Some of the air travels equatorward and is deflected by the Coriolis effect, producing the reliable **trade winds**. The remainder travels poleward and is also deflected, generating the prevailing **westerlies** of the midlatitudes. As the westerlies move poleward, they encounter the cool

FIGURE 18.16 Global Circulation on a Nonrotating Earth A simple convection system is produced by unequal heating of the atmosphere.

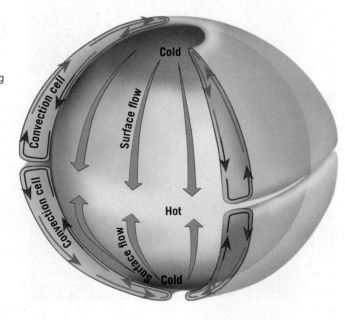

polar **easterlies** in the region of the **subpolar low**. The interaction of these warm and cool winds produces the stormy belt known as the **polar front**. The source region for the variable polar easterlies is the **polar high**. Here, cold polar air is subsiding and spreading equatorward.

In summary, this simplified global circulation is dominated by four pressure zones. The subtropical and polar highs are areas of dry subsiding air that flows outward at the surface, producing the prevailing winds. The low-pressure zones of the equatorial and subpolar regions are associated with inward and upward airflow accompanied by clouds and precipitation.

**SmartFigure 18.17
Idealized Global
Circulation Proposed
for the Three-Cell
Circulation Model of a
Rotating
Earth**

Influence of Continents

Up to this point, we have described the surface pressure and associated winds as continuous belts around Earth. However, the only truly continuous pressure belt is the subpolar low in the Southern Hemisphere, where the ocean is uninterrupted by landmasses. At other latitudes, particularly in the Northern Hemisphere, where landmasses break up the ocean surface, large seasonal temperature differences disrupt the pattern. **FIGURE 18.18** shows the resulting pressure and wind patterns for January and July. The circulation over the oceans is dominated by semipermanent cells of high pressure in the subtropics and cells of low pressure over the subpolar regions. The subtropical highs are responsible for the trade winds and westerlies, as mentioned earlier.

The large landmasses, on the other hand, particularly Asia, become cold in the winter and develop a seasonal high-pressure system from which surface flow is directed off the land (see Figure 18.18). In the summer, the opposite occurs: The landmasses are heated and develop a low-pressure cell, which permits air to flow onto the land. These seasonal changes in wind direction are known as the **monsoons**. During warm months, areas such as India experience a flow of warm, water-laden air from the Indian Ocean, which produces the rainy summer monsoon. The winter monsoon is dominated by dry continental air. A similar situation exists, but to a lesser extent, over North America.

In summary, the general circulation is produced by semipermanent cells of high and low pressure over the

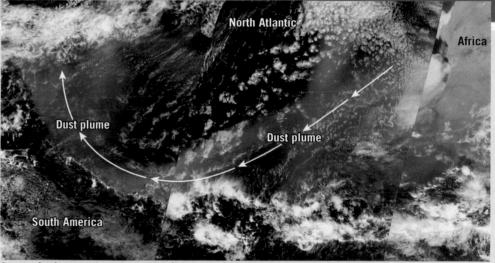

EYE ON EARTH

This image shows a January dust storm in Africa that produced a dust plume that reached all the way to the northeast coast of South America. It has been estimated that plumes such as this transport about 40 million tons of dust from the Sahara Desert to the Amazon basin each year. The minerals carried by these dust plumes help to replenish nutrients in rain forest soils that are continually being washed out of the soils by heavy tropical rains.

QUESTION 1 *Notice that the dust plume is following a curved path. Does the atmospheric circulation carrying this dust plume exhibit a clockwise or counterclockwise rotation?*

QUESTION 2 *What global pressure system was responsible for transporting this dust from Africa to South America? Referring to Figure 18.18 might be helpful.*

Composite image stitched together from a series of images collected by MODIS. (Courtesy of NASA)

FIGURE 18.18 Average Surface Air Pressure
These maps show the average surface air pressure, in millibars, for **A.** January and **B.** July, with associated winds.

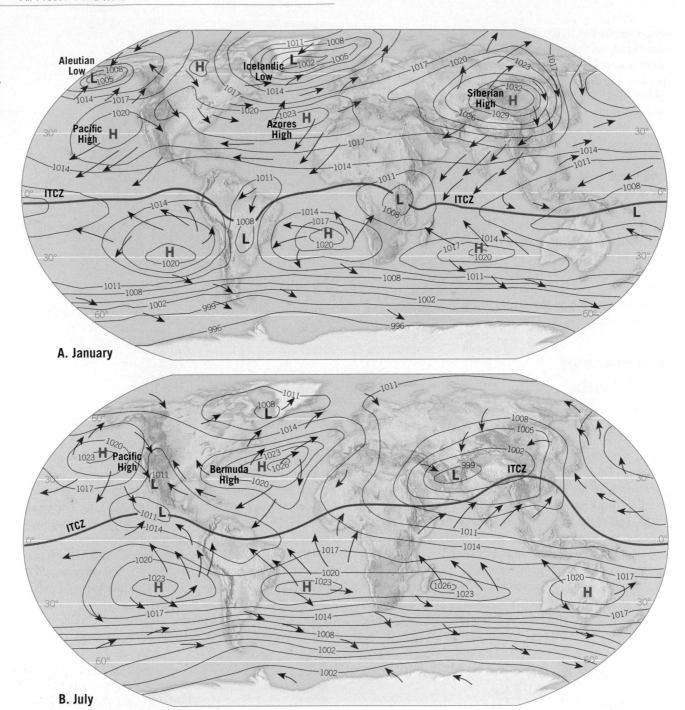

A. January

B. July

oceans and is complicated by seasonal pressure changes over land.

The Westerlies

The circulation in the midlatitudes, the zone of the westerlies, is complex and does not fit the convection system proposed for the tropics. Between 30° and 60° latitude, the general west-to-east flow is interrupted by the migration of cyclones and anticyclones. In the Northern Hemisphere, these cells move from west to east around the globe, creating an anticyclonic (clockwise) flow or a cyclonic (counterclockwise) flow in their area of influence. A close correlation exists between the paths taken by these surface pressure

systems and the position of the upper-level airflow, indicating that the upper air steers the movement of cyclonic and anticyclonic systems.

Among the most obvious features of the flow aloft are the seasonal changes. The steep temperature gradient across the middle latitudes in the winter months corresponds to a stronger flow aloft. In addition, the polar jet stream fluctuates seasonally such that its average position migrates southward with the approach of winter and northward as summer nears. By midwinter, the jet core may penetrate as far south as central Florida.

Because the paths of low-pressure centers are guided by the flow aloft, we can expect the southern tier of states to experience more of their stormy weather in the winter season. During the hot summer months, the storm track is across the northern states, and some cyclones never leave Canada. The northerly storm track associated with summer also applies to Pacific storms, which move toward Alaska during the warm months, thus producing an extended dry season for much of the West coast. The number of cyclones generated is seasonal as well, with the largest number occurring in the cooler months, when the temperature gradients are greatest. This fact is in agreement with the role of cyclonic storms in the distribution of heat across the midlatitudes.

18.5 | LOCAL WINDS

List three types of local winds and describe their formation.

Now that we have examined Earth's large-scale circulation, let us turn briefly to winds that influence much smaller areas. Remember that all winds are produced for the same reason: pressure differences that arise because of temperature differences caused by unequal heating of Earth's surface. **Local winds** are small-scale winds produced by a locally generated pressure gradient. Those described here are caused either by topographic effects or by variations in surface composition in the immediate area.

Land and Sea Breezes

In coastal areas during the warm summer months, the land is heated more intensely during the daylight hours than is the adjacent body of water. As a result, the air above the land surface heats, expands, and rises, creating an area of lower pressure. A **sea breeze** then develops because cooler air over the water (higher pressure) moves toward the warmer land (lower pressure) (**FIGURE 18.19A**). The sea breeze begins to develop shortly before noon and generally reaches its greatest intensity during the mid- to late afternoon. These relatively cool winds can be a significant moderating influence on afternoon temperatures in coastal areas. Small-scale sea breezes can also develop along the shores of large lakes. People who live in a city near the Great Lakes, such as Chicago, recognize this lake effect, especially in the summer. They are reminded daily by weather reports of the cooler temperatures near the lake as compared to warmer outlying areas.

At night, the reverse may take place. The land cools more rapidly than the sea, and the **land breeze** develops (**FIGURE 18.19B**).

Mountain and Valley Breezes

A daily wind similar to land and sea breezes occurs in many mountainous regions. During daylight hours, the air along the slopes of the mountains is heated more intensely than the air at the same elevation over the valley floor. Because this warmer air is less dense, it glides up along the slope and generates a **valley breeze** (**FIGURE 18.20A**). The occurrence of these daytime upslope breezes can often be identified by the cumulus clouds that develop on adjacent mountain peaks.

After sunset, the pattern may reverse. Rapid radiation cooling along the mountain slopes produces a layer of cooler air next to the ground. Because cool air is denser than warm air, it drains downslope into the valley. This movement of air is called a **mountain breeze** (**FIGURE 18.20B**). The same type of cool air drainage can occur in places that have very modest slopes. The result is that the coldest pockets of air are usually found in the lowest spots.

**SmartFigure 18.19
Sea and Land Breezes**

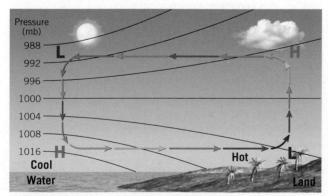

A. During daylight hours, cooler and denser air over the water moves onto the land, generating a sea breeze.

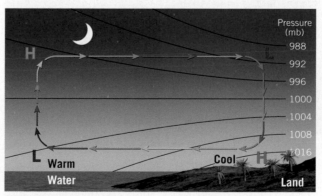

B. At night the land cools more rapidly than the sea, generating an offshore flow called a land breeze.

FIGURE 18.20 Valley and Mountain Breezes

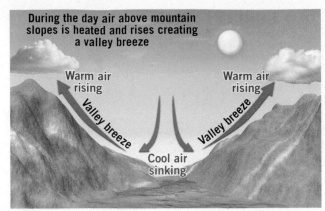

During the day air above mountain slopes is heated and rises creating a valley breeze

Warm air rising

Warm air rising

Valley breeze

Valley breeze

Cool air sinking

A. Valley breeze

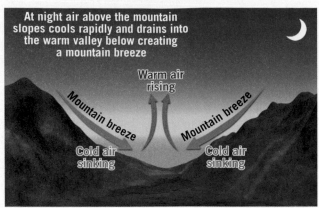

At night air above the mountain slopes cools rapidly and drains into the warm valley below creating a mountain breeze

Warm air rising

Mountain breeze

Mountain breeze

Cold air sinking

Cold air sinking

B. Mountain breeze

FIGURE 18.21 Santa Ana Winds Wildfires fanned by Santa Ana winds raged in southern California in October 2003. These fires scorched more than 740,000 acres and destroyed more than 3000 homes. The inset shows an idealized high-pressure center composed of cool, dry air that drives Santa Ana winds. Adiabatic heating causes the air temperature to increase and the relative humidity to decrease. (NASA)

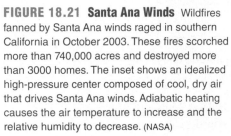

H

Cool dry air

Santa Ana winds

Los Angeles

San Diego

Like many other winds, mountain and valley breezes have seasonal preferences. Although valley breezes are most common during the warm season, when solar heating is most intense, mountain breezes tend to be more dominant in the cold season.

Chinook and Santa Ana Winds

Warm, dry winds sometimes move down the east slopes of the Rockies, where they are called **chinooks**. Such winds are often created when a strong pressure gradient develops in a mountainous region. As the air descends the leeward slopes of the mountains, it is heated adiabatically (by compression). Because condensation may have occurred as the air ascended the windward side, releasing latent heat, the air descending the leeward slope will be warmer and drier than it was at a similar elevation on the windward side. Although the temperature of these winds is generally less than 10°C (50°F), which is not particularly warm, the winds occur mostly in the winter and spring, when the affected areas may be experiencing below-freezing temperatures. Thus, by comparison, these dry, warm winds often bring a drastic change. When the ground has a snow cover, these winds are known to melt it in short order.

A chinooklike wind that occurs in southern California is the **Santa Ana**. This hot, desiccating wind greatly increases the threat of fire in this already dry area (**FIGURE 18.21**).

18.5 CONCEPT CHECKS

1 What is a local wind?

2 Describe the formation of a sea breeze.

3 Does a land breeze blow toward or away from the shore?

4 During what time of day would you expect to experience a well-developed valley breeze—midnight, late morning, or late afternoon?

18.6 | MEASURING WIND Describe the instruments used to measure wind. Explain how wind direction is expressed using compass directions.

Two basic wind measurements, direction and speed, are particularly significant to weather observers. One simple device for determining both measurements is a *wind sock*, which is a common sight at small airports and landing strips (**FIGURE 18.22A**). The cone-shaped bag is open at both ends and is free to change position with shifts in wind direction. The degree to which the sock is inflated is an indication of wind speed.

Winds are always labeled by the direction from which they blow. A north wind blows *from* the north *toward* the south, an east wind *from* the east *toward* the west. The instrument most commonly used to determine wind direction is the **wind vane** (**FIGURE 18.22B**, upper right). This instrument, a common sight on many buildings, always points *into* the wind. The wind direction is often shown on a dial that is connected to the wind vane. The dial indicates wind direction, either by points of the compass (N, NE, E, SE, etc.) or by a scale of 0 to 360 degrees. On the latter scale, 0 degrees and 360 degrees are both north, 90 degrees is east, 180 degrees is south, and 270 degrees is west.

Wind speed is commonly measured using a **cup anemometer** (see Figure 18.22B, upper left). The wind speed is read from a dial much like the speedometer of an automobile. Places where winds are steady and speeds are relatively high are potential sites for tapping wind energy.

When the wind consistently blows more often from one direction than from any other, it is called a **prevailing wind**. You may be familiar with the prevailing westerlies that dominate the circulation in the midlatitudes. In the United States, for example, these winds consistently move the "weather"

A.

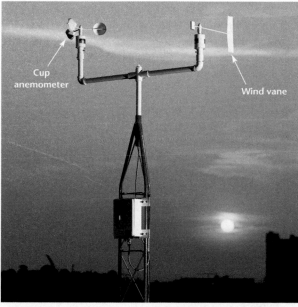

B.

FIGURE 18.22 **Measuring the Wind** The two basic wind measurements are speed and direction. (Photo A by Lourens Smak/ Alamy Images; photo B by Belfort Instrument Company)

Cup anemometer

Wind vane

EYE ON EARTH

This mountain area was cloud free as this summer day began. By afternoon these clouds had formed. The clouds were associated with a local wind. (Photo by Herbert Koeppel/Alamy Images)

QUESTION 1 With which local wind are the clouds in this photo most likely associated?

QUESTION 2 Describe the process that created the local wind that was associated with the formation of these clouds.

QUESTION 3 Would you expect clouds such as these to form at night?

FIGURE 18.23 Wind Roses These graphs show the percentage of time winds come from various directions.

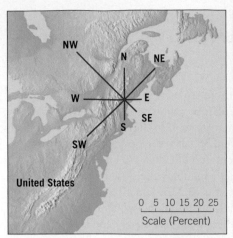

Wind frequency for winter in the northeastern United States.

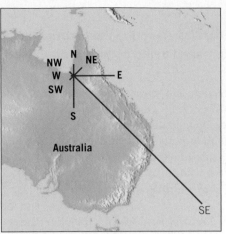

Wind frequency for winter in northeastern Australia. Note the reliability of the southeast trade winds in Australia as compared to the westerlies in the northeastern United States.

from west to east across the continent. Embedded within this general eastward flow are cells of high and low pressure, with their characteristic clockwise and counterclockwise flow. As a result, the winds associated with the westerlies, as measured at the surface, often vary considerably from day to day and from place to place. By contrast, the direction of airflow associated with the belt of trade winds is much more consistent, as can be seen in **FIGURE 18.23**.

By knowing the locations of cyclones and anticyclones in relation to where you are, you can predict the changes in wind direction that will occur as a pressure center moves past. Because changes in wind direction often bring changes in temperature and moisture conditions, the ability to predict the winds can be very useful. In the Midwest, for example, a north wind may bring cool, dry air from Canada, whereas a south wind may bring warm, humid air from the Gulf of Mexico. Sir Francis Bacon summed it up nicely when he wrote, "Every wind has its weather."

18.6 CONCEPT CHECKS

1 What are the two basic wind measurements? What instruments are used to make these measurements?

2 *From* what direction does a northeast wind blow? *Toward* what direction does a south wind blow?

18.7 | EL NIÑO AND LA NIÑA AND THE SOUTHERN OSCILLATION

Describe the Southern Oscillation and its relationship to El Niño and La Niña. List the climate impacts to North America of El Niño and La Niña.

El Niño was first recognized by fishermen from Ecuador and Peru, who noted a gradual warming of waters in the eastern Pacific in December or January. Because the warming usually occurred near the Christmas season, the event was named El Niño—"little boy," or "Christ child," in Spanish. These periods of abnormal warming happen at irregular intervals of 2 to 7 years and usually persist for spans of 9 months to 2 years. **La Niña**, which means "little girl," is the opposite of El Niño and refers to colder-than-normal sea-surface temperatures along the coastline of Ecuador and Peru.

As **FIGURE 18.24A** illustrates, the atmospheric circulation in the central Pacific during a La Niña event is dominated by strong trade winds. These wind systems, in turn, generate a strong equatorial current that flows westward from South America toward Australia and Indonesia. In addition, a cold ocean current is observed flowing equatorward along the coast of Ecuador and Peru. The latter flow, called the Peru Current, encourages upwelling of cold, nutrient-filled waters that serve as the primary food source for millions of small feeder fish, particularly anchovies. Therefore, fishing is particularly good during the periods of strong upwelling.

Every few years, however, the circulation associated with La Niña is replaced by an El Niño event (**FIGURE 18.24B**).

Impact of El Niño

El Niño is noted for its potentially catastrophic impact on the weather and economies of Peru, Chile, and Australia, among other countries. As shown in Figure 18.24B, during an El Niño, strong equatorial countercurrents amass large quantities of warm water that block the upwelling of colder, nutrient-filled water along the west coast of South America. As a result, the anchovies, which support the population of game fish, starve, devastating the fishing industry. At the same time, some inland areas of Peru and Chile that are normally arid receive above-average rainfall, which can cause major flooding. These climatic fluctuations have been known for years, but they were considered local phenomena.

Scientists now recognize that El Niño is part of the global atmospheric circulation pattern that affects the weather at great distances from Peru. One of the most severe El Niño events on record occurred in 1997–1998 and was responsible

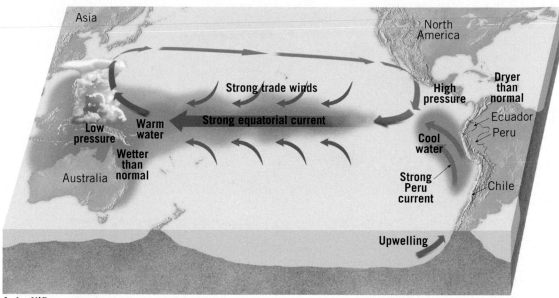

A. La Niña

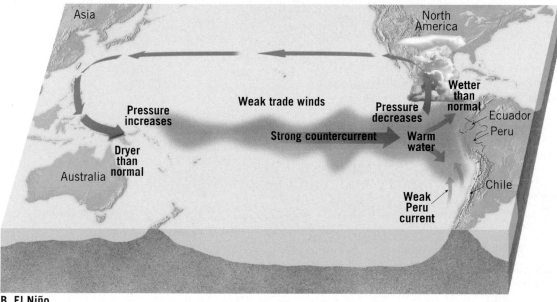

B. El Niño

FIGURE 18.24 The Relationship Between El Niño, La Niña, and the Southern Oscillation
A. During a La Niña event, strong trade winds drive the equatorial currents toward the west. At the same time, the strong Peru Current causes upwelling of cold water along the west coast of South America. **B.** When the Southern Oscillation occurs, the pressure over the eastern and western Pacific flip-flops. This causes the trade winds to diminish, leading to an eastward movement of warm water along the equator and the beginning of an El Niño. As a result, the surface waters of the central and eastern Pacific warm, with far-reaching consequences for weather patterns.

for a variety of weather extremes in many parts of the world. During this episode, ferocious winter storms struck the California coast, causing unprecedented beach erosion, landslides, and floods. In the southern United States, heavy rains also brought floods to Texas and the Gulf states.

Although the effects of El Niño are somewhat variable, some locales appear to be affected more consistently. In particular, during the winter, warmer-than-normal conditions prevail in the north-central United States and parts of Canada (**FIGURE 18.25A**). In addition, significantly wetter winters are experienced in the southwestern United States and northwestern Mexico, while the southeastern United States experiences wetter and cooler conditions. In the western Pacific, drought conditions are observed in parts of Indonesia, Australia, and the Philippines (see Figure 18.25A). One major benefit of El Niño is a lower-than-average number of Atlantic hurricanes. El Niño is credited with suppressing

hurricanes during the 2009 hurricane season, the least active in 12 years.

Impact of La Niña

La Niña was once thought to be the normal conditions that occur between two El Niño events, but meteorologists now consider La Niña an important atmospheric phenomenon in its own right. Researchers have come to recognize that when surface temperatures in the eastern Pacific are *colder than average*, a La Niña event is triggered and exhibits a distinctive set of weather patterns (see Figure 18.24A).

Typical La Niña winter weather includes cooler and wetter conditions over the northwestern United States and especially cold winter temperatures in the Northern Plains states (**FIGURE 18.25C**). In addition, unusually warm conditions occur in the Southwest and Southeast. In the western Pacific,

FIGURE 18.25 Climatic Impacts of El Niño and La Niña El Niño has the most significant impact on the climate of North America during the winter. In addition, El Niño affects the areas around the tropical Pacific in both winter and summer. Likewise, La Niña has its most significant impact on North America in the winter but influences other areas during all seasons.

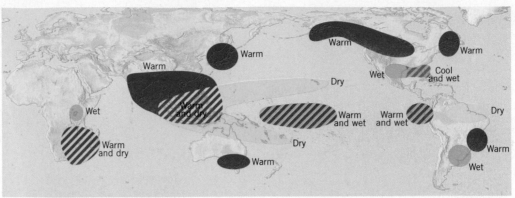

A. El Niño: December to February

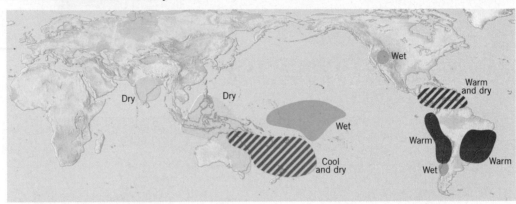

B. El Niño: June to August

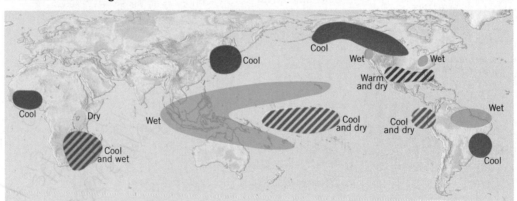

C. La Niña: December to February

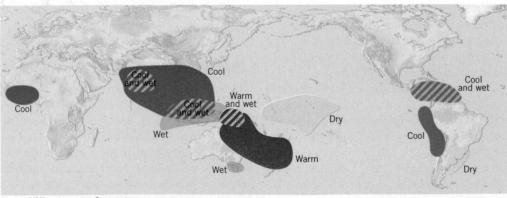

D. La Niña: June to August

 Cool   Warm   Dry   Wet   Cool and dry    Cool and wet   Warm and dry   Warm and wet

The 1930s Dust Bowl
An Environmental Disaster

During a span of dry years in the 1930s, large dust storms plagued the Great Plains. Topsoil was stripped from millions of acres. Because of the size and severity of these storms, the region came to be called the Dust Bowl and the time period the Dirty Thirties.

HEART OF THE
DUST BOWL

OTHER AREAS SEVERELY
AFFECTED BY DUST STORMS

WA

ND

MT

WY

SD

MN

OR

CA

NV

NE

IA

CO

The southern Great
Plains were most ►
severely affected.

KS

UT

OK

In places, dust drifted like snow, covering farm buildings, fences, and fields. Crop failure and economic hardship resulted in many farms being abandoned.

TX

NM

AZ

Arthur Rothstein/ Library of Congress

Dust blackens the sky near Elkhart, Kansas, on May 21, 1937. The transformation of semiarid grasslands into farms during an unusually wet period set the stage for this disastrous period of soil erosion. When drought struck, the unprotected soils were vulnerable to the wind.

LANDS

Budweiser

WALKERS

FIGURE 18.26
Queensland Floods A local resident rows through flood waters that turned this street in Rockhampton into a river. The Queensland floods of 2010–2011 have been attributed to one of the strongest La Niña events as far back as records have been kept. Unusually warm sea-surface temperatures around Australia contributed to the heavy rains. In other parts of Australia, this strong La Niña event brought relief from a decade-long drought. (Photo by Jonathan Wood/Getty Images)

La Niña events are associated with wetter-than-normal conditions. The 2010–2011 La Niña contributed to a deluge in Australia, which resulted in one of the country's worst natural disasters: Large portions of the state of Queensland were extensively flooded (**FIGURE 18.26**). Another La Niña impact is more frequent hurricane activity in the Atlantic. A recent study concluded that the cost of hurricane damages in the United States is 20 times greater in La Niña years than in El Niño years.

Southern Oscillation

Major El Niño and La Niña events are intimately related to the large-scale atmospheric circulation. Each time an El Niño occurs, the barometric pressure drops over large portions of the eastern Pacific and rises in the western Pacific (see Figure 18.24B). Then, as a major El Niño event comes to an end, the pressure difference between these two regions swings back in the opposite direction, triggering a La Niña event (see Figure 18.24A). This seesaw pattern of atmospheric pressure between the eastern and western Pacific is called the **Southern Oscillation** (**FIGURE 18.27**).

Winds are the link between the pressure change associated with the Southern Oscillation and the ocean warming and cooling associated with El Niño and La Niña. The start of an El Niño event begins with a rise in surface pressure over Australia and Indonesia and a decrease in pressure over the eastern Pacific (see Figure 18.24B). During a strong El Niño, this pressure change causes the trade winds to weaken and countercurrents to develop and move warm water eastward. The resulting change in atmospheric circulation takes the rain with it, causing drought in the western Pacific and increased rainfall in the normally dry regions of Peru and Chile. The opposite circulation develops during La Niña events, as shown in Figure 18.24A. When a strong La Niña develops, the trade winds strengthen, causing dryer-than-normal conditions in the eastern Pacific, while extreme flooding may occur in Indonesia and northeastern Australia.

FIGURE 18.27 **The Seesaw Pattern of the Southern Oscillation** Negative values (blue) represent the cold La Niña phase, whereas positive values (red) represent the warm El Niño phase. The graph was created by analyzing six variables, including sea-surface temperatures and sea-level pressures.

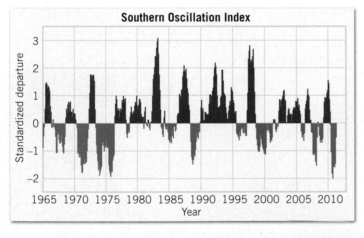

18.7 CONCEPT CHECKS

1 Describe how a major El Niño event tends to affect the weather in Peru and Chile as compared to Indonesia and Australia.

2 Describe the sea-surface temperatures on both sides of the tropical Pacific during a La Niña event.

3 How does a major La Niña event influence the hurricane season in the Atlantic Ocean?

4 Briefly describe the Southern Oscillation and how it is related to El Niño and La Niña.

5 Describe how an El Niño event might affect the climate in North America during the winter. Describe the same for a La Niña event.

18.8 | GLOBAL DISTRIBUTION OF PRECIPITATION
Discuss the major factors that influence the global distribution of precipitation.

A casual glance at **FIGURE 18.28** shows a relatively complex pattern for the distribution of precipitation. Although the map appears to be complicated, the general features of the map can be explained by applying our knowledge of global winds and pressure systems.

The Influence of Pressure and Wind Belts

In general, regions influenced by high pressure, with its associated subsidence and diverging winds, experience relatively dry conditions. On the other hand, regions under the influence of low pressure and its converging winds and ascending air receive ample precipitation. This pattern is illustrated by noting that the tropical regions dominated by the equatorial low are the rainiest regions on Earth. It is here that we find the rain forests of the Amazon basin in South America and the Congo basin in Africa. The warm, humid trade winds converge to yield abundant rainfall throughout the year. By contrast, areas dominated by the subtropical high-pressure cells clearly receive much smaller amounts of precipitation. These are the regions of

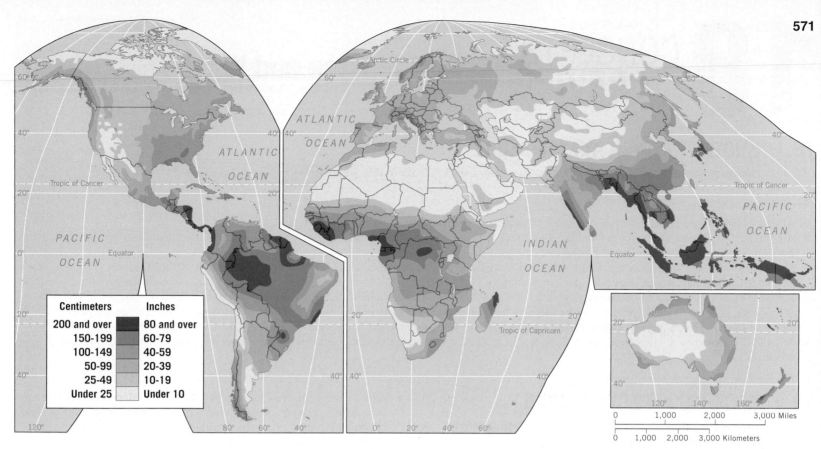

Centimeters / **Inches**

Centimeters		Inches
200 and over		80 and over
150-199		60-79
100-149		40-59
50-99		20-39
25-49		10-19
Under 25		Under 10

FIGURE 18.28 Global Distribution of Average Annual Precipitation

extensive subtropical deserts. In the Northern Hemisphere, the largest desert is the Sahara. Examples in the Southern Hemisphere include the Kalahari in southern Africa and the dry lands of Australia.

Other Factors

If Earth's pressure and wind belts were the only factors controlling precipitation distribution, the pattern shown in Figure 18.28 would be simpler. The inherent nature of the air is also an important factor in determining precipitation potential. Because cold air has a low capacity for moisture compared with warm air, we would expect a latitudinal variation in precipitation, with low latitudes receiving the greatest amounts of precipitation and high latitudes receiving the smallest amounts. Figure 18.28 indeed shows heavy rainfall in equatorial regions and meager precipitation in high-latitude areas. Recall that the dry region in the warm subtropics is explained by the presence of the subtropical high.

The distribution of land and water also complicates the precipitation pattern. Large landmasses in the middle latitudes commonly experience decreased precipitation toward their interiors. For example, central North America and central Eurasia receive considerably less precipitation than do coastal regions at the same latitude. Mountain barriers also alter precipitation patterns. Windward mountain slopes receive abundant precipitation, whereas leeward slopes and adjacent lowlands are often deficient in moisture.

18.8 CONCEPT CHECKS

1. With which global pressure belt are the rain forests of Africa's Congo basin associated? Which pressure system is linked to the Sahara Desert?

2. What factors, in addition to the distribution of wind and pressure, influence the global distribution of precipitation?

EYE ON EARTH

This satellite image was produced with data from the *Tropical Rainfall Measuring Mission* (*TRMM*). Notice the band of heavy rainfall shown in reds and yellows that extends east–west across the image.

QUESTION 1 With which pressure zone is this band of rainy weather associated?

QUESTION 2 Is it more likely that this image was acquired in July or January? Explain.

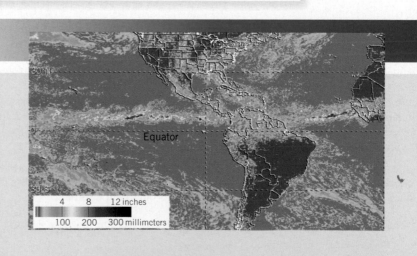

18.1 UNDERSTANDING AIR PRESSURE

Define *air pressure* and describe the instruments used to measure this weather element.

KEY TERMS: air pressure, mercury barometer, aneroid barometer, barograph

- Air has weight: At sea level, it exerts a pressure of 1 kilogram per square centimeter (14.7 pounds per square inch).
- Air pressure is the force exerted by the weight of air above. With increasing altitude, there is less air above to exert a force, and thus air pressure decreases with altitude—rapidly at first and then much more slowly.
- The unit meteorologists use to measure atmospheric pressure is the millibar. Standard sea-level pressure is expressed as 1013.2 millibars. Isobars are lines on a weather map that connect places of equal air pressures.
- A mercury barometer measures air pressure using a column of mercury in a glass tube sealed at one end and inverted in a dish of mercury. It measures atmospheric pressure in inches of mercury, the height of the column of mercury in the barometer. Standard atmospheric pressure at sea level equals 29.92 inches of mercury. As air pressure increases, the mercury in the tube rises, and when air pressure decreases, so does the height of the column of mercury.
- Aneroid ("without liquid") barometers consist of partially evacuated metal chambers that compress as air pressure increases and expand as pressure decreases.

18.2 FACTORS AFFECTING WIND

Discuss the three factors that act on the atmosphere to either create or alter winds.

KEY TERMS: wind, isobar, pressure gradient force, Coriolis effect, geostrophic wind, jet stream

- Wind is controlled by a combination of (1) the pressure gradient force, (2) the Coriolis effect, and (3) friction. The pressure gradient force is the primary driving force of wind that results from pressure differences that are depicted by the spacing of isobars on a map. Closely spaced isobars indicate a steep pressure gradient and strong winds; widely spaced isobars indicate a weak pressure gradient and light winds.
- The Coriolis effect produces deviation in the path of wind due to Earth's rotation (to the right in the Northern Hemisphere and to the left in the Southern Hemisphere). Friction, which significantly influences airflow near Earth's surface, is negligible above a height of a few kilometers.

- Above a height of a few kilometers, the Coriolis effect is equal to and opposite the pressure gradient force, which results in geostrophic winds. Geostrophic winds follow a path parallel to the isobars, with velocities proportional to the pressure gradient force.

Q These diagrams show surface winds at two locations. All factors in both situations are identical except that one surface is land and the other is water. Which diagram represents winds over the land? Explain your choice.

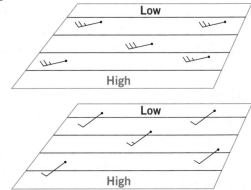

18.3 HIGHS AND LOWS

Contrast the weather associated with low-pressure centers (cyclones) and high-pressure centers (anticyclones).

KEY TERMS: cyclone (low), anticyclone (high), convergence, divergence, pressure (barometric) tendency

- The two types of pressure centers are (1) cyclones, or lows (centers of low pressure), and (2) anticyclones, or highs (high-pressure centers). In the Northern Hemisphere, winds around a low (cyclone) are counterclockwise and inward. Around a high (anticyclone), they are clockwise and outward. In the Southern Hemisphere, the Coriolis effect causes winds to be clockwise around a low and counterclockwise around a high.
- Because air rises and cools adiabatically in a low-pressure center, cloudy conditions and precipitation are often associated with their passage. In a high-pressure center, descending air is compressed and warmed; therefore, cloud formation and precipitation are unlikely in an anticyclone, and "fair" weather is usually expected.

Q Assume that you are an observer checking this aneroid barometer several hours after it was last checked. What is the pressure tendency? How did you figure this out? What does the tendency shown on the barometer indicate about forthcoming weather?

18.4 GENERAL CIRCULATION OF THE ATMOSPHERE

Summarize Earth's idealized global circulation. Describe how continents and seasonal temperature changes complicate the idealized pattern.

KEY TERMS: equatorial low, intertropical convergence zone (ITCZ), subtropical high, trade winds, westerlies, polar easterlies, subpolar low, polar front, polar high, monsoon

- If Earth's surface were uniform, four belts of pressure oriented east to west would exist in each hemisphere. Beginning at the equator, the four belts would be the (1) equatorial low, also referred to as the intertropical convergence zone (ITCZ), (2) subtropical high at about 25° to 35° on either side of the equator, (3) subpolar low, situated at about 50° to 60° latitude, and (4) polar high, near Earth's poles.
- Particularly in the Northern Hemisphere, large seasonal temperature differences over continents disrupt the idealized, or zonal, global patterns of pressure and wind. In winter, large, cold landmasses develop a seasonal high-pressure system from which surface airflow is directed off the land. In summer, landmasses are heated, and a low-pressure system develops over them, which permits air to flow onto the land. These seasonal changes in wind direction are known as monsoons.
- In the middle latitudes, between 30° and 60° latitude, the general west-to-east flow of the westerlies is interrupted by the migration of cyclones and anticyclones. The paths taken by these cyclonic and anticyclonic systems is closely correlated to upper-level airflow and the polar jet stream. The average position of the polar jet stream, and hence the paths followed by cyclones, migrates southward with the approach of winter and northward as summer nears.

18.5 LOCAL WINDS

List three types of local winds and describe their formation.

KEY TERMS: local wind, sea breeze, land breeze, valley breeze, mountain breeze, chinook, Santa Ana

- Local winds are small-scale winds produced by a locally generated pressure gradient. Sea and land breezes form along coasts and are brought about by temperature contrasts between land and water. Valley and mountain breezes occur in mountainous areas where the air along slopes heats differently than does the air at the same elevation over the valley floor. Chinook and Santa Ana winds are warm, dry winds created when air descends the leeward side of a mountain and warms by compression.

Q It is late afternoon on a warm summer day and you are enjoying some time at the beach. Until the last hour or two, winds were calm. Then a breeze began to develop. Is it more likely a cool breeze from the water or a warm breeze from the adjacent land area? Explain.

18.6 MEASURING WIND

Describe the instruments used to measure wind. Explain how wind direction is expressed using compass directions.

KEY TERMS: wind vane, cup anemometer, prevailing wind

- The two basic wind measurements are direction and speed. Winds are always labeled by the direction from which they blow. Wind direction is measured with a wind vane, and wind speed is measured using a cup anemometer.

Q When designing an airport, it is important to have planes take off *into* the wind. Refer to the accompanying wind rose and describe the orientation of the runway and the direction planes would usually travel when they took off. Where on Earth might you find a wind rose like this?

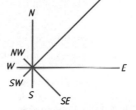

18.7 EL NIÑO AND LA NIÑA AND THE SOUTHERN OSCILLATION

Describe the Southern Oscillation and its relationship to El Niño and La Niña. List the climate impacts to North America of El Niño and La Niña.

KEY TERMS: El Niño, La Niña, Southern Oscillation

- El Niño refers to episodes of ocean warming in the eastern Pacific along the coasts of Ecuador and Peru. It is associated with weak trade winds, a strong eastward moving equatorial countercurrent, a weakened Peru Current, and diminished upwelling along the western margin of South America.
- A La Niña event is associated with colder-than-average surface temperatures in the eastern Pacific. La Niña is linked to strong trade winds, a strong westward-moving equatorial current, and a strong Peru Current with significant coastal upwelling.
- El Niño and La Niña events are part of the global circulation and are related to a seesaw pattern of atmospheric pressure between the eastern and western Pacific called the Southern Oscillation. El Niño and La Niña events influence weather on both sides of the tropical Pacific Ocean as well as weather in the United States.

Torsten Blackwood/AFP/Getty Images

Q This image shows floods occurring in eastern Australia. Is this region more likely being influenced by La Niña or El Niño?

18.8 GLOBAL DISTRIBUTION OF PRECIPITATION

Discuss the major factors that influence the global distribution of precipitation.

- The general features of the global distribution of precipitation can be explained by global winds and pressure systems. In general, regions influenced by high pressure, with its associated subsidence and divergent winds, experience dry conditions. Regions under the influence of low pressure and its converging winds and ascending air receive ample precipitation.
- Air temperature, the distribution of continents and oceans, and the location of mountains also influence the distribution of precipitation.

Q Why do most high-latitude regions receive relatively meager precipitation?

GIVE IT SOME **THOUGHT**

1. Mercury is 13.5 times denser (heavier) than water. If you built a barometer using water rather than mercury, how tall, in inches, would it have to be to record standard sea-level pressure?

2. This satellite image shows a tropical cyclone (hurricane).

   a. Examine the cloud pattern and determine whether the flow is clockwise or counterclockwise.

   b. In which hemisphere is the storm located?

   c. What factor determines whether the flow is clockwise or counterclockwise?

NASA

3. If divergence in the jet stream above a surface low-pressure center exceeds convergence at the surface, will surface winds likely get stronger or weaker? Explain.

4. The accompanying map is a simplified surface weather map for April 2, 2011, on which three pressure cells are numbered.

   a. Identify which of the pressure cells are anticyclones (highs) and which are cyclones (lows).

   b. Which pressure cell has the steepest pressure gradient and therefore the strongest winds?

   c. Refer to Figure 18.2 to determine whether pressure cell 3 should be considered strong or weak.

5. You and a friend are watching TV on a rainy day, when the weather reporter says, "The barometric pressure is 28.8 inches and rising." Hearing this, you say, "It looks like fair weather is on its way." Your friend responds with the following questions: "I thought air pressure had something to do with the weight of air. How does inches relate to weight? And why do you think the weather is going to improve?" How would you respond to your friend's queries?

6. If you live in the Northern Hemisphere and are directly west of the center of a cyclone, what is the probable wind direction at your location? What if you were west of an anticyclone?

7. If Earth did not rotate on its axis and if its surface were completely covered with water, what direction would a boat drift if it started its journey in the middle latitudes of the Northern Hemisphere? (*Hint:* What would the global circulation pattern be like for a nonrotating Earth?)

8. The accompanying sketch shows a cross section of the idealized circulation in the Northern Hemisphere. Match the appropriate number on the sketch to each of the following features:

   a. equatorial low

   b. polar front

   c. subtropical high

   d. polar high

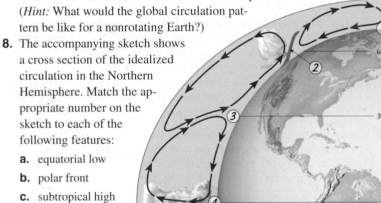

9. The accompanying maps of Africa show the distribution of precipitation for July and January. Which map represents July, and which represents January? How were you able to figure this out?

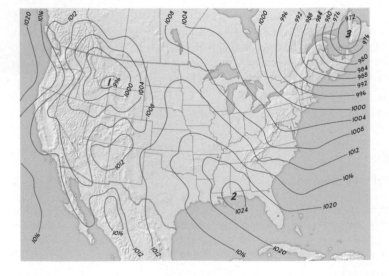

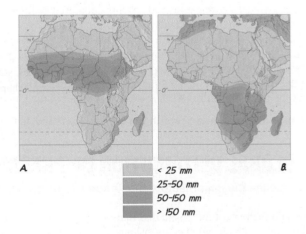

A

B

| < 25 mm |
| 25–50 mm |
| 50–150 mm |
| > 150 mm |

EXAMINING THE **EARTH SYSTEM**

1. This satellite image shows a portion of a wildfire that occurred in May and June 2011. Known as the Wallow Fire, it burned more than 538 thousand acres (840 square miles) in southeastern Arizona. It was the largest wildfire in Arizona history. Suggest two ways that wind contributed to this wildfire. How might this event influence the geosphere and the biosphere?

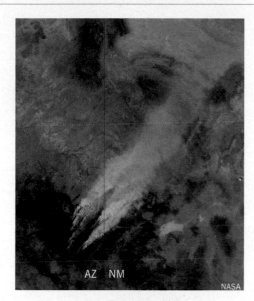

2. Examine this classic image of Africa from space and pick out the region dominated by the equatorial low and the areas influenced by the subtropical highs in each hemisphere. What clue(s) did you use? Speculate on the differences in the biosphere between the regions dominated by high pressure and the zone influenced by low pressure.

3. The accompanying map shows wintertime sea-surface temperature anomalies (differences from average) over the equatorial Pacific Ocean. Based on this map, answer the following questions:

   a. In what phase was the Southern Oscillation (El Niño or La Niña) when this image was made?
   b. Would the trade winds be strong or weak at this time?
   c. If you lived in Australia during this event, what weather conditions would you expect?
   d. If you were attending college in the southeastern United States during winter months, what type of weather conditions would you expect? (*Hint:* See Figure 18.25.)

4. How are global winds related to surface ocean currents? What is the *ultimate* source of energy that drives both of these circulations?

MasteringGeology™

Looking for additional review and test prep materials? Visit the Self Study area in **www.masteringgeology.com** to find practice quizzes, study tools, and multimedia that will aid in your understanding of this chapter's content. In **MasteringGeology™** you will find:

- GEODe: Earth Science: An interactive visual walkthrough of key concepts
- Geoscience Animation Library: More than 100 animations illuminating many difficult-to-understand Earth science concepts

- In The News RSS Feeds: Current Earth science events and news articles are pulled into the site with assessment
- Pearson eText
- Optional Self Study Quizzes
- Web Links
- Glossary
- Flashcards

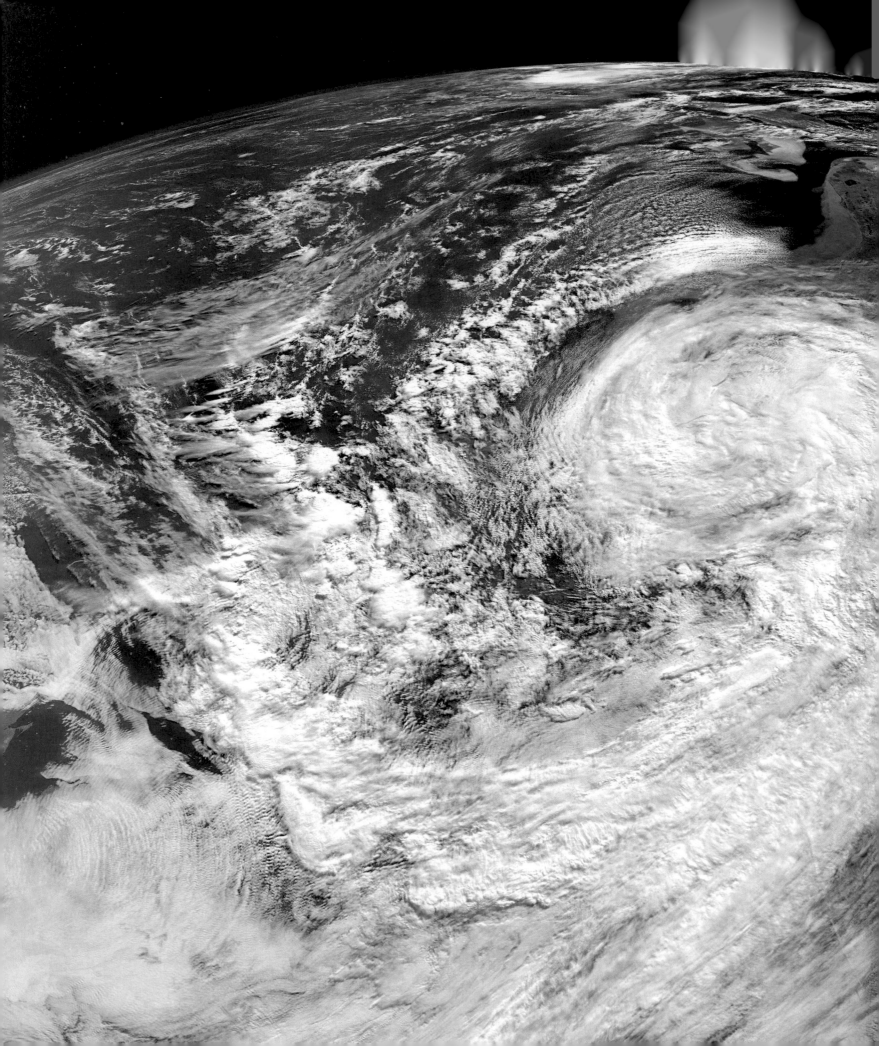

19

Weather Patterns and Severe Storms

This satellite image shows Hurricane Sandy, called Superstorm Sandy in the media, battering the East coast on October 30, 2012. This view of the storm is looking south from Canada. Florida is near the top of the image. (NASA)

Tornadoes and hurricanes rank among nature's most destructive forces. Each spring, newspapers report the death and destruction left in the wake of a band of tornadoes. During late summer and fall, we hear occasional news reports about hurricanes. Storms with names such as Katrina, Rita, Sandy, and Ike make front-page headlines. Thunderstorms, although less intense and far more common than tornadoes and hurricanes, are also part of our discussion on severe weather in this chapter. Before looking at violent weather, however, we will study the atmospheric phenomena that most often affect our day-to-day weather: air masses, fronts, and traveling midlatitude cyclones. We will see the interplay of the elements of weather discussed in Chapters 16, 17, and 18.

19.1 | AIR MASSES

Discuss air masses, their classification, and associated weather.

For many people who live in the middle latitudes, which include much of the United States, summer heat waves and winter cold spells are familiar experiences. In the first instance, several days of high temperatures and oppressive humidity may finally end when a series of thunderstorms pass through the area, followed by a few days of relatively cool relief. By contrast, the clear skies that often accompany a span of frigid subzero days may be replaced by thick gray clouds and a period of snow as temperatures rise to levels that seem mild compared to those that existed just a day earlier. In both examples, what was experienced was a period of generally constant weather conditions followed by a relatively short period of change and then the reestablishment of a new set of weather conditions that remained for perhaps several days before changing again.

What Is an Air Mass?

The weather patterns just described result from movements of large bodies of air, called air masses. An **air mass**, as the term implies, is an immense body of air, usually 1,600 kilometers (1,000 miles) or more across and perhaps

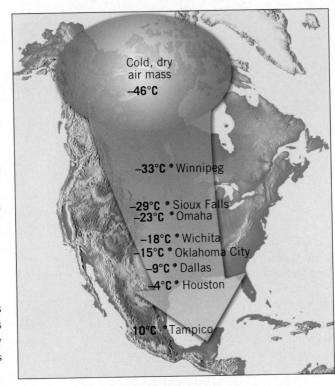

FIGURE 19.2 **An Invasion of Frigid Air** As this very cold air mass moved southward from Canada, it brought some of the coldest weather of the winter to the areas in its path. As it advanced into the United States, the air mass slowly got warmer. Thus, the air mass was gradually modified at the same time that it modified the weather in the areas over which it moved. (From *Physical Geography: A Landscape Appreciation*, 9th edition, by Tom L. McKnight and Darrell Hess, ©2008. Reprinted and electronically reproduced by permission of Pearson Education, Inc., Upper Saddle River, NJ.)

several kilometers thick, that is characterized by a similarity of temperature and moisture at any given altitude. When this air moves out of its region of origin, it will carry these temperatures and moisture conditions with it, eventually affecting a large portion of a continent (**FIGURE 19.1**).

An excellent example of the influence of an air mass is illustrated in **FIGURE 19.2**, which shows a cold, dry mass from northern Canada moving southward. With a beginning temperature of −46°C (−51°F), the air mass warms to −33°C (−27°F) by the time it reaches Winnipeg. It continues to warm as it moves

FIGURE 19.1 **Lake-Effect Snow Storm** This satellite image shows a cold, dry air mass moving from its source region in Canada across Lake Superior. It illustrates the process that leads to lake-effect snow storms. (NASA)

southward through the Great Plains and into Mexico. Throughout its southward journey, the air mass becomes warmer. But it also brings some of the coldest weather of the winter to the places in its path. Thus, the air mass is modified, but it also modifies the weather in the areas over which it moves.

The horizontal uniformity of an air mass is not perfect, of course. Because air masses extend over large areas, small differences occur in temperature and humidity from place to place. Still, the differences observed within an air mass are small compared to the rapid changes experienced across air-mass boundaries.

Since it may take several days for an air mass to move across an area, the region under its influence will probably experience fairly constant weather, a situation called **air-mass weather**. Certainly, there are usually some day-to-day variations, but the events will be very unlike those in an adjacent air mass.

The air-mass concept is an important one because it is closely related to the study of atmospheric disturbances. Most disturbances in the middle latitudes originate along the boundary zones that separate different air masses.

Source Regions

When a portion of the lower atmosphere moves slowly or stagnates over a relatively uniform surface, the air will assume the distinguishing features of that area, particularly with regard to temperature and moisture conditions.

The area where an air mass acquires its characteristic properties of temperature and moisture is called its **source region**. The source regions that produce air masses influencing North America are shown in **FIGURE 19.3**.

Air masses are classified according to their source region. **Polar (P)** and **arctic (A) air masses** originate in high latitudes toward Earth's poles, whereas those that form in low latitudes are called **tropical (T) air masses**. The designation *polar, arctic,* or *tropical* gives an indication of the temperature characteristics of an air mass. *Polar* and *arctic* indicate cold, and *tropical* indicates warm.

In addition, air masses are classified according to the nature of the surface in the source region. **Continental (c) air masses** form over land, and **maritime (m) air masses** originate over water. The designation *continental* or *maritime* thus suggests the moisture characteristics of the air mass. Continental air is likely to be dry, and maritime air is likely to be humid.

The basic types of air masses according to this scheme of classification are continental polar (cP), continental arctic (cA), continental tropical (cT), maritime polar (mP), and maritime tropical (mT).

Weather Associated with Air Masses

Continental polar and maritime tropical air masses influence the weather of North America most, especially east of the Rocky Mountains. Continental polar air masses originate in northern Canada, interior Alaska, and the Arctic—areas that are uniformly cold and dry in winter and cool and dry in summer. In winter, an invasion of continental polar air brings the clear skies and cold temperatures we associate with a cold wave as it moves southward from Canada into

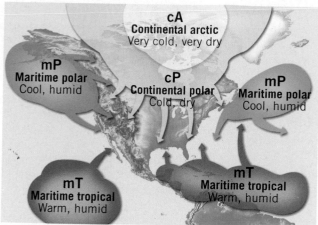

A. Winter pattern

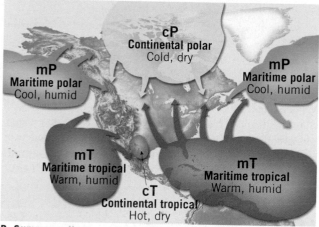

B. Summer pattern

FIGURE 19.3 Air-Mass Source Regions for North America Source regions are largely confined to subtropical and subpolar locations. The fact that the middle latitudes are where cold and warm air masses clash, often because the converging winds of a traveling cyclone draw them together, means that this zone lacks the conditions necessary to be a source region. The differences between polar and arctic are relatively small and serve to indicate the degree of coldness of the respective air masses. By comparing the winter (**A**) and summer (**B**) maps, it is clear that the extent and temperature characteristics fluctuate.

the United States. In summer, this air mass may bring a few days of cooling relief.

Although cP air masses are not, as a rule, associated with heavy precipitation, those that cross the Great Lakes during late autumn and winter sometimes bring snow to the leeward shores. These localized storms often form when the surface weather map indicates no apparent cause for a snowstorm. These are known as **lake-effect snows**, and they make Buffalo and Rochester, New York, among the snowiest cities in the United States (**FIGURE 19.4**).

What causes lake-effect snow? During late autumn and early winter, the temperature contrast between the lakes and adjacent land areas can be large.[1] The temperature contrast can be especially great when a very cold cP air mass pushes southward across the lakes. The satellite image in Figure 19.1 illustrates the process. Notice that as the cloud-free air moves across Lake Superior, clouds develop because the air acquires large quantities of heat and moisture from the relatively warm lake surface. By the time the cP air reaches the opposite shore, the air mass is humid and unstable, and heavy snow showers are occurring.

Maritime tropical air masses affecting North America most often originate over the warm waters of the Gulf of

---

[1]Recall that land cools more rapidly and to lower temperatures than water. See the discussion of land and water in the section "Why Temperatures Vary: The Controls of Temperature" in Chapter 16.

SmartFigure 19.4 Snowfall Map The snowbelts of the Great Lakes are easy to pick out on this snowfall map. (Data from NOAA) The photo was taken following a 6-day lake-effect snowstorm in November 1996 that dropped 175 centimeters (nearly 69 inches) of snow on Chardon, Ohio, setting a new state record. (Photo by Tony Dejak/ AP Photo)

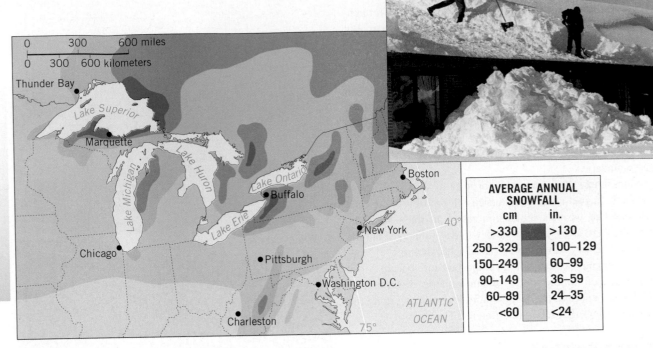

AVERAGE ANNUAL SNOWFALL	
cm	in.
>330	>130
250–329	100–129
150–249	60–99
90–149	36–59
60–89	24–35
<60	<24

Mexico, the Caribbean Sea, or the adjacent Atlantic Ocean. As you might expect, these air masses are warm, moisture laden, and usually unstable. Maritime tropical air is the source of much, if not most, of the precipitation in the eastern two-thirds of the United States. In summer, when an mT air mass invades the central and eastern United States, and occasionally southern Canada, it brings the high temperatures and oppressive humidity typically associated with its source region.

Of the two remaining air masses, maritime polar and continental tropical, the latter has the least influence on the weather of North America. Hot, dry continental tropical air, originating in the Southwest and Mexico during the summer, only occasionally affects the weather outside its source region.

During the winter, maritime polar air masses coming from the North Pacific often originate as continental polar air masses in Siberia. The cold, dry cP air is transformed into relatively mild, humid, unstable mP air during its long journey across the North Pacific (**FIGURE 19.5**). As this mP air arrives at the western shore of North America, it is often accompanied by low clouds and shower activity. When this air advances inland against the western mountains, orographic uplift produces heavy rain or snow on the windward slopes of the mountains. Maritime polar air also

FIGURE 19.5 Air-Mass Modification During winter, maritime polar (mP) air masses in the North Pacific usually begin as continental polar (cP) air masses in Siberia. The cP air is modified to mP as it slowly crosses the ocean.

cP Cold, dry, stable

Modified cP Cool, less dry, stable

mP Cool, moist, unstable

EYE ON EARTH

This satellite image from December 27, 2010, shows a strong winter storm off the East Coast of the United States. (NASA)

QUESTION 1 Can you identify the very center of the storm?

QUESTION 2 What air mass is being drawn into the storm to produce the dense clouds in the upper right?

QUESTION 3 What term is applied to a storm such as this?

QUESTION 4 Farther south, a cold air mass over the southeastern states is cloud free. What is its likely classification? Explain how it is being modified as it moves over the Atlantic.

FIGURE 19.6 Classic Nor'easter The satellite image shows a strong winter storm called a *nor'easter* along the coast of New England on January 12, 2011. In winter, a nor'easter exhibits a weather pattern in which strong northeast winds carry cold, humid mP air from the North Atlantic into New England and the middle Atlantic states. The ground-level view of the storm in Boston shows that the combination of ample moisture and strong convergence can result in heavy snow. (Satellite image by NASA; photo by Michael Dwyer/ Alamy Images)

originates in the North Atlantic off the coast of eastern Canada and occasionally influences the weather of the northeastern United States. In winter, when New England is on the northern or northwestern side of a passing low-pressure center, the counterclockwise cyclonic winds draw in maritime polar air. The result is a storm characterized by snow and cold temperatures, known locally as a **nor'easter** (**FIGURE 19.6**).

19.1 CONCEPT CHECKS

1 Define *air mass*. What is air-mass weather?

2 On what basis are air masses classified?

3 Compare the temperature and moisture characteristics of the following air masses: cP, mP, mT, and cT.

4 Which air mass is associated with lake-effect snow? What causes lake-effect snow?

19.2 | FRONTS Compare and contrast typical weather associated with a warm front and a cold front. Describe an occluded front and a stationary front.

Fronts are boundaries that separate different air masses, one warmer than the other and often having a higher moisture content. A front can form between any two contrasting air masses. Considering the vast size of the air masses involved, fronts are relatively narrow, being 15- to 200-kilometer-wide (9- to 120-mile-wide) bands of discontinuity. On the scale of a weather map, they are generally narrow enough to be represented by a broad line. Above Earth's surface, a front slopes at a low angle so that warmer air overlies cooler air (**FIGURE 19.7**). In the ideal case, the air masses on both sides of the front move in the same direction and at the same speed. Under this condition, the front acts simply as a barrier, moving along between the two contrasting air masses.

Generally, however, an air mass on one side of a front moves faster relative to the frontal boundary than the air mass on the other side. Thus, one air mass actively advances into another and "clashes" with it. In fact, during World War I,

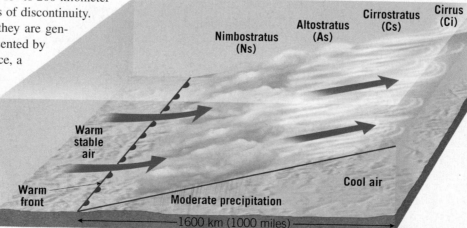

FIGURE 19.7 Warm Front This diagram shows the idealized clouds and weather associated with a warm front. During most of the year, warm fronts produce light to moderate precipitation over a wide area.

Norwegian meteorologists called these boundaries *fronts*, analogous to battle lines between two armies. Along these "battlegrounds," centers of low pressure develop and generate much of the precipitation and severe weather in the middle latitudes. As one air mass advances into another, limited mixing occurs along the frontal surface, but for the most part, the air masses retain their distinct identities as one is displaced upward over the other. No matter which air mass is advancing, *it is always the warmer (less dense) air that is forced aloft,* whereas *the cooler (denser) air acts as the wedge on which lifting takes place.* The term **overrunning** is generally applied to warmer air gliding up along a colder air mass. We will now take a look at different types of fronts.

Warm Fronts

When the surface position of a front moves so that warm air occupies territory formerly covered by cooler air, it is called a **warm front** (see Figure 19.7). On a weather map, the surface position of a warm front is shown by a red line with red semicircles protruding into the cooler air.

East of the Rockies, warm tropical air often enters the United States from the Gulf of Mexico and overruns receding cool air. As the cold air retreats, friction with the ground slows the advance of the surface position of the front more so than its position aloft. Stated another way, less dense, warm air has a hard time displacing denser, cold air. For this reason, the boundary separating these air masses acquires a very gradual slope. The average slope of a warm front is about 1:200, which means that if you are 200 kilometers (120 miles) ahead of the surface location of a warm front, you will find the frontal surface at a height of 1 kilometer (0.6 mile).

As warm air ascends the retreating wedge of cold air, it expands and cools adiabatically to produce clouds and, frequently, precipitation. The sequence of clouds shown in Figure 19.7 typically precedes a warm front. The first sign of the approach of a warm front is the appearance of cirrus clouds overhead. These high clouds form 1000 kilometers (600 miles) or more ahead of the surface front, where the overrunning warm air has ascended high up the wedge of cold air.

As the front nears, cirrus clouds grade into cirrostratus, which blend into denser sheets of altostratus. About 300 kilometers (180 miles) ahead of the front, thicker stratus and nimbostratus clouds appear, and rain or snow begins. Because of their slow rate of advance and very low slope, warm fronts usually produce light to moderate precipitation over a large area for an extended period. Warm fronts, however, are occasionally associated with cumulonimbus clouds and thunderstorms. This occurs when the overrunning air is unstable and the temperatures on opposite sides of the front contrast sharply. At the other extreme, a warm front associated with a dry air mass could pass unnoticed at the surface.

A gradual increase in temperature occurs with the passage of a warm front. The increase is most noticeable when there is a large temperature difference between the adjacent air masses. The moisture content and stability of the encroaching warm air mass largely determine when clear skies will return. During summer, cumulus, and occasionally cumulonimbus, clouds are embedded in the warm unstable air mass that follows the front. Precipitation from these clouds can be heavy but is usually scattered and of short duration.

Cold Fronts

When dense cold air is actively advancing into a region occupied by warmer air, the boundary is called a **cold front** (**FIGURE 19.8**). As with warm fronts, friction tends to slow the surface position of a cold front more so than its position aloft. However, because of the relative positions of the adjacent air masses, the cold front steepens as it moves. On average, cold fronts are about twice as steep as warm fronts, having a slope of perhaps 1:100. In addition, cold fronts advance at speeds around 35 to 50 kilometers (20 to 35 miles) per hour compared to 25 to 35 kilometers (15 to 20 miles) per hour for warm fronts. These two differences—rate of movement and steepness of slope—largely account for the more violent nature of cold-front weather compared to the weather generally accompanying a warm front (**FIGURE 19.9**).

As a cold front approaches, commonly from the west or northwest, towering clouds can often be seen in the distance. Near the front, a dark band of ominous clouds foretells the coming weather. The forceful lifting of air along a cold front is often so rapid that the latent heat released when water vapor condenses increases the air's buoyancy appreciably. The heavy downpours and vigorous wind gusts associated with mature cumulonimbus clouds frequently result. A cold front produces roughly the same amount of lifting as a warm front, but over a shorter distance. As a result,

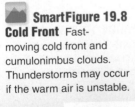

SmartFigure 19.8 Cold Front Fast-moving cold front and cumulonimbus clouds. Thunderstorms may occur if the warm air is unstable.

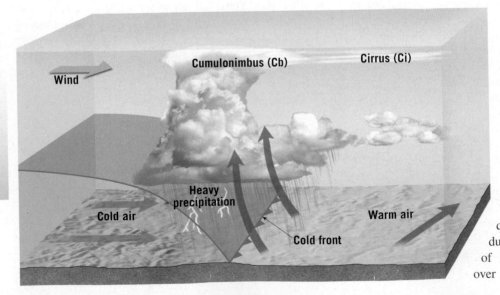

Wind

Cumulonimbus (Cb)

Cirrus (Ci)

Cold air

Heavy precipitation

Cold front

Warm air

FIGURE 19.10 Stages in the Formation of an Occluded Front

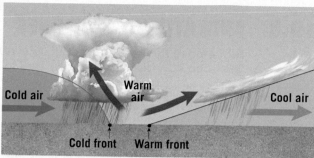

In this example, the air behind the cold front is colder and denser than the air ahead of the warm front.

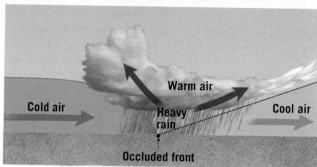

The surface cold front moves faster than the surface warm front and overtakes it to form an occluded front.

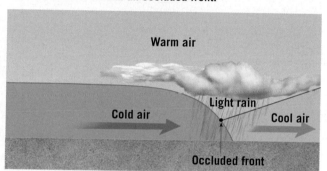

The denser cold air lifts the warm air and advances into and displaces the cool air.

FIGURE 19.9 Splashing Hailstones Cumulonimbus clouds along a cold front produced hail and heavy rain at this ballpark in Wichita, Kansas. Dozens of cars in the parking lot were damaged. (Photo by Fernando Salazar/AP Photo)

the intensity of precipitation is greater, but the duration is shorter. In addition, a marked temperature drop and a wind shift from the south to west or northwest accompany the passage of the front. The sometimes violent weather and sharp temperature contrast along the cold front are symbolized on a weather map by a blue line with blue triangle-shaped points that extend into the warmer air mass (see Figure 19.8).

The weather behind a cold front is dominated by a subsiding and relatively cold air mass. Thus, clearing usually begins soon after the front passes. Although the compression of air due to subsidence causes some adiabatic heating, the effect on surface temperatures is minor. In winter, the long, cloudless nights that often follow the passage of a cold front allow for abundant radiation cooling that reduces surface temperatures. When a cold front moves over a relatively warm area, surface heating can produce shallow convection. This, in turn, may generate low cumulus or stratocumulus clouds behind the front.

Stationary Fronts and Occluded Fronts

Occasionally, the flow on both sides of a front is neither toward the cold air mass nor toward the warm air mass but almost parallel to the line of the front. Thus, the surface position of the front does not move. This condition is called a **stationary front**. On a weather map, stationary fronts are shown with blue triangular points on one side of the front and red semicircles on the other. At times, some overrunning occurs along a stationary front, most likely causing gentle to moderate precipitation.

The fourth type of front is an **occluded front**, an active cold front that overtakes a warm front, as shown in **FIGURE 19.10**. As the advancing cold air wedges the warm front upward, a new front emerges between the advancing cold air and the air over which the warm front is gliding. The weather of an occluded front is generally complex. Most precipitation is

associated with the warm air being forced aloft. When conditions are suitable, however, the newly formed front is capable of initiating precipitation of its own.

A word of caution is in order concerning the weather associated with various fronts. Although the preceding discussion will help you recognize the weather patterns associated with fronts, remember that these descriptions are generalizations. The weather generated along any individual front may or may not conform fully to this idealized picture. Fronts, like all other aspects of nature, do not lend themselves to classification as easily as we would like.

19.2 CONCEPT CHECKS

1 Compare the weather of a typical warm front with that of a typical cold front.

2 Why is cold-front weather usually more severe than warm-front weather?

3 Describe a stationary front and an occluded front.

19.3 | MIDLATITUDE CYCLONES

Summarize the weather associated with the passage of a mature midlatitude cyclone. Describe how airflow aloft is related to cyclones and anticyclones at the surface.

So far, we have examined the basic elements of weather as well as the dynamics of atmospheric motions. We are now ready to apply our knowledge of these diverse phenomena to an understanding of day-to-day weather patterns in the middle latitudes. For our purposes, *middle latitudes* refers to the region between southern Florida and Alaska. The primary weather producers here are **midlatitude,** or **middle-latitude, cyclones**. On weather maps they are shown by an *L*, meaning *low-pressure system.* **FIGURE 19.11** shows two views of a large idealized midlatitude cyclone with probable air masses, fronts, and surface wind patterns.

Midlatitude cyclones are large centers of low pressure that generally travel from west to east. Lasting from a few days to more than a week, these weather systems have a counterclockwise circulation, with an airflow inward toward their centers. Most midlatitude cyclones also have a cold front extending from the central area of low pressure, and frequently a warm front as well. Convergence and forceful lifting initiate cloud development and frequently cause abundant precipitation.

As early as the 1800s, it was known that middle-latitude cyclones were the bearers of precipitation and severe weather. But it was not until the early part of the 1900s that a model was developed to explain how cyclones form. A group of Norwegian scientists formulated and published this model in 1918. The model was created primarily from near-surface observations.

Years later, as data from the middle and upper troposphere and from satellite images became available, modifications were necessary. However, this model is still a useful working tool for interpreting the weather. If you keep this model in mind when you observe changes in the weather, the changes will no longer come as a surprise. You should begin to see some order in what once appeared to be disorder, and you might even occasionally "predict" the impending weather.

Idealized Weather of a Midlatitude Cyclone

The midlatitude cyclone model provides a useful tool for examining the weather patterns of the middle latitudes. **FIGURE 19.12** illustrates the distribution of clouds and thus the regions of possible precipitation associated with a mature system. Compare this drawing to the satellite image shown in **FIGURE 19.13**. It is easy to see why we often refer to the cloud pattern of a midlatitude cyclone as having a "comma" shape.

Guided by the westerlies aloft, cyclones generally move eastward across the United States, so we can expect the first signs of their arrival in the west. However, often in the region of the Mississippi River Valley, cyclones begin a more northeasterly path and occasionally move directly northward. A midlatitude cyclone typically requires 2

SmartFigure 19.11 Idealized Structure of a Large, Mature Midlatitude Cyclone

A. This map view shows fronts, air masses, and surface winds. **B.** The three-dimensional view is a cross section through warm and cold fronts along a line from point A to point B.

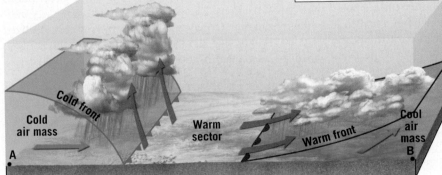

mP

cP

L

mP

A

B

mT

mT

Key

- ●●● Warm front
- ▲▲▲ Cold front
- ▲●▲● Stationary front
- ●▲●▲ Occluded front

A. Map view

Cold air mass

Cold front

Warm sector

Warm front

Cool air mass

A

B

B. Three-dimensional view from point A to point B

to 4 days to move completely across a region. During that brief period, abrupt changes in atmospheric conditions may be experienced. This is particularly true in the winter and spring, when the largest temperature contrasts occur across the middle latitudes.

Using Figure 19.12 as a guide, we will now consider these weather producers and what we should expect from them as they move over an area. To facilitate our discussion, Figure 19.12 includes two profiles along lines A–E and F–G:

- Imagine the change in weather as you move along profile A–E. At point A, the sighting of high cirrus clouds would be the first sign of the approaching cyclone. These high clouds can precede the surface front by 1000 kilometers (600 miles) or more, and they generally are accompanied by falling pressure. As the warm front advances, a lowering and thickening of the cloud deck is noticed.

- Usually within 12 to 24 hours after the first sighting of cirrus clouds, light precipitation begins (point B). As the front nears, the rate of precipitation increases, a rise in temperature is noticed, and winds begin to change from east or southeast to south or southwest.

- With the passage of the warm front, an area is under the influence of a maritime tropical air mass (point C). Generally, the region affected by this sector of the cyclone experiences warm to hot temperatures, southwesterly winds, fairly high humidity, and clear to partly cloudy skies containing cumulus clouds.

- The relatively warm, humid weather of the warm sector passes quickly and is replaced by gusty winds and precipitation generated along the cold front. The approach of a rapidly advancing cold front is marked by a wall of dark clouds (point D). Severe weather accompanied by heavy precipitation, hail, and an occasional tornado is a definite possibility, especially during spring and summer. The passage of the cold front is easily detected by a wind shift: The southwest winds are replaced by winds from the west to northwest and by a pronounced drop in temperature. Also, the rising pressure hints of the subsiding cool, dry air behind the front.

- Once the front passes, skies clear as cooler air invades the region (point E). Often a day or two of almost cloudless deep blue skies occurs, unless another cyclone is edging into the region.

A very different set of weather conditions prevails in the regions north of the storm's center along profile F–G of Figure 19.12. In this part of the storm, temperatures remain cool. The first hints of the approaching low-pressure center are a continual drop in air pressure and increasingly overcast conditions that bring varying amounts of precipitation. This section of the cyclone most often generates snow during the winter months.

Once the formation of an occluded front begins, the character of the storm changes. Because occluded fronts tend to move more slowly than other fronts, the entire wishbone-shaped frontal structure of the storm rotates counterclockwise. As a result, the occluded front appears to "bend over backward." This effect adds to the misery of the region

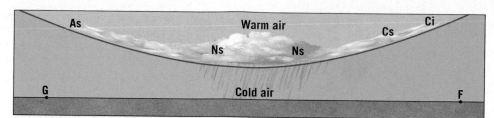

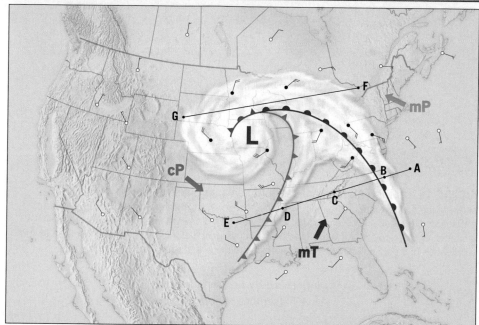

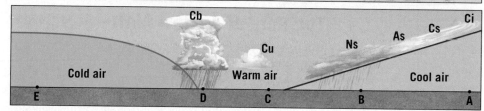

FIGURE 19.12 Cloud Patterns Typically Associated with a Mature Midlatitude Cyclone The middle section is a map view. Note the cross-sectional lines (F–G, A–E). Above the map is a vertical cross section along line F–G. Below the map is a section along A–E. For cloud abbreviations, refer to Figures 19.7 and 19.8.

FIGURE 19.13 Satellite View of a Mature Midlatitude Cyclone This storm swept across the central United States and produced strong wind gusts (up to 125 kilometers [78 miles] per hour), rain, hail, and snow. It also spawned 61 tornadoes on October 26, 2010. This cyclone set a record for the lowest pressure not associated with a hurricane ever recorded over land in the continental United States: 28.21 inches of mercury. It is easy to see why we often refer to the cloud pattern of a cyclone as having a "comma" shape. (NASA)

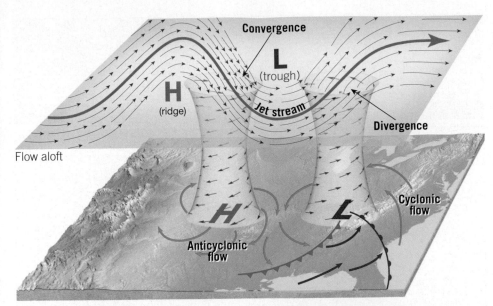

FIGURE 19.14 Flow Aloft Influences Surface Winds and Pressure Idealized depiction showing the support that divergence and convergence aloft provide to cyclonic and anticyclonic circulation at the surface. Divergence aloft initiates upward air movement, reduced surface pressure, and cyclonic flow. On the other hand, convergence along the jet stream results in general subsidence of the air column, increased surface pressure, and anticyclonic surface winds.

or coming together (**FIGURE 19.14**). The resulting accumulation of air must be accompanied by a corresponding increase in surface pressure. Consequently, we might expect a low-pressure system to "fill" rapidly and be eliminated. However, this does not occur. On the contrary, cyclones often exist for a week or longer. For this to happen, surface convergence must be offset by a mass outflow at some level aloft (see Figure 19.14). As long as divergence (spreading out) aloft is equal to or greater than surface inflow, the low pressure and its accompanying convergence can be sustained.

Because cyclones are bearers of stormy weather, they have received far more attention than anticyclones. Nevertheless, a close relationship exists, which makes it difficult to separate any discussion of these two types of pressure systems. The surface air that feeds a cyclone, for example, generally originates as air flowing out of an anticyclone. Consequently, cyclones and anticyclones typically are found adjacent to each other. Like a cyclone, an anticyclone depends on the flow far above to maintain its circulation. Divergence at the surface is balanced by convergence aloft and general subsidence of the air column (see Figure 19.14).

influenced by the occluded front because it lingers over the area longer than the other fronts.

The Role of Airflow Aloft

When the earliest studies of midlatitude cyclones were made, little was known about the nature of the airflow in the middle and upper troposphere. Since then, a close relationship has been established between surface disturbances and the flow aloft. Airflow aloft plays an important role in maintaining cyclonic and anticyclonic circulation. In fact, more often than not, these rotating surface wind systems are actually generated by upper-level flow.

Recall that the airflow around a cyclone (low-pressure system) is inward, a fact that leads to mass convergence,

19.3 CONCEPT CHECKS

1 Briefly describe the weather associated with the passage of a mature midlatitude cyclone when the center of low pressure is about 200 to 300 kilometers (125 to 200 miles) north of your location.

2 If the midlatitude cyclone described in Question 1 took 3 days to pass your location, on which day would temperatures likely be warmest? On which day would they likely be coldest?

3 What winter weather might be expected with the passage of a mature midlatitude cyclone when the center of low pressure is located about 100 to 200 kilometers (60 to 125 miles) south of your location?

4 Briefly explain how flow aloft aids the formation of cyclones at the surface.

EYE ON EARTH

This image of a line of clouds was taken by astronauts aboard the International Space Station. The dashed line on the image shows the approximate surface position of the front responsible for the cloud development. Assume that this front is located over the central United States as you answer the following questions. (NASA)

QUESTION 1 What is the cloud type of the tallest clouds in the image?

QUESTION 2 Are these clouds more typical of a cold front or a warm front?

QUESTION 3 Is the front most likely moving toward the southeast or toward the northwest?

QUESTION 4 Is the air mass located to the southeast of the front more likely continental polar (cP) or maritime tropical (mT)?

19.4 | THUNDERSTORMS

List the basic requirements for thunderstorm formation and locate places on a map that exhibit frequent thunderstorm activity. Describe the stages in the development of a thunderstorm.

Thunderstorms are the first of three severe weather types we will examine in this chapter. Sections on tornadoes and hurricanes follow. All these phenomena can be related to low-pressure systems (cyclones).

Severe weather is more fascinating than everyday weather phenomena. The lightning display and booming thunder generated by a severe thunderstorm can be a spectacular event that elicits both awe and fear (**FIGURE 19.15**). Of course, hurricanes and tornadoes also attract a great deal of much-deserved attention. A single tornado outbreak or hurricane can cause many deaths as well as billions of dollars in property damage. In a typical year, the United States experiences thousands of violent thunderstorms, hundreds of tornadoes, and several hurricanes.

What's in a Name?

Up to now we have examined midlatitude cyclones, which play an important role in causing day-to-day weather changes. Yet the use of the term *cyclone* is often confusing. To many people, the term implies only an intense storm, such as a tornado or a hurricane. When a hurricane unleashes its fury on India or Bangladesh, for example, it is usually

In southern Asia and Australia, the term *cyclone* is applied to storms that are called *hurricanes* in the United States. This image shows Cyclone Yasi, which struck eastern Australia in February 2011.

In parts of the Great Plains, *cyclone* is a synonym for *tornado*. The nickname for the athletic teams at Iowa State University is the *Cyclones.**

FIGURE 19.16 The Term Cyclone Sometimes the use of the term *cyclone* can be confusing. (Satellite image courtesy of NASA; logo courtesy of Iowa State University)

reported in the media as a *cyclone* (the term denoting a hurricane in that part of the world).

Similarly, tornadoes are referred to as *cyclones* in some places. This custom is particularly common in portions of the Great Plains of the United States. Recall that in *The Wizard of Oz*, Dorothy's house was carried from her Kansas farm to the land of Oz by a cyclone. Indeed, the nickname for the athletic teams at Iowa State University is the *Cyclones* (**FIGURE 19.16**). Although hurricanes and tornadoes are, in fact, cyclones, the vast majority of cyclones are *not* hurricanes or tornadoes. The term *cyclone* simply refers to the circulation around any low-pressure center, no matter how large or intense it is.

Tornadoes and hurricanes are both smaller and more violent than midlatitude cyclones. Midlatitude cyclones can have a diameter of 1600 kilometers (1000 miles) or more. By contrast, hurricanes average only 600 kilometers (375 miles) across, and tornadoes, with a typical diameter of just 0.25 kilometer (0.16 mile), are much too small to show up on a weather map.

FIGURE 19.15 Summertime Lightning Display A storm is classified as a thunderstorm only after thunder is heard. Because thunder is produced by lightning, lightning must also occur. (Photo by agefotostock/SuperStock)

* Iowa State University is the only Division I school to use Cyclones as its team name. The pictured logo was created to better communicate the school's image by combining the mascot, a cardinal bird named Cy, and the Cyclone team name.

The thunderstorm, a much more familiar weather event, hardly needs to be distinguished from tornadoes, hurricanes, and midlatitude cyclones. Unlike the flow of air about these latter storms, the circulation associated with thunderstorms is characterized by strong up-and-down movements. Winds in the vicinity of a thunderstorm do not follow the inward spiral of a cyclone, but they are typically variable and gusty.

Although thunderstorms form "on their own," away from cyclonic storms, they also form in conjunction with cyclones. For instance, thunderstorms are frequently spawned along the cold front of a midlatitude cyclone, where on rare occasions a tornado may descend from the thunderstorm's cumulonimbus tower. Hurricanes also generate widespread thunderstorm activity. Thus, thunderstorms are related in some manner to all three types of cyclones mentioned here.

Thunderstorm Occurrence

Almost everyone has observed various small-scale phenomena that result from the vertical movements of relatively warm, unstable air. Perhaps you have seen a dust devil over an open field on a hot day, whirling its dusty load to great heights. Or maybe you have noticed a bird glide effortlessly skyward on an invisible thermal of hot air. These examples illustrate the dynamic thermal instability that occurs during the development of a thunderstorm.

A **thunderstorm** is a storm that generates lightning and thunder. Thunderstorms frequently produce gusty winds, heavy rain, and hail. A thunderstorm may be produced by a single cumulonimbus cloud and influence only a small area, or it may be associated with clusters of cumulonimbus clouds covering a large area.

Thunderstorms form when warm, humid air rises in an unstable environment. Various mechanisms can trigger the upward air movement needed to create thunderstorm-producing cumulonimbus clouds. One mechanism, the unequal heating of Earth's surface, significantly contributes to the formation *of air-mass thunderstorms*. These storms are associated with the scattered puffy cumulonimbus clouds that commonly form

within maritime tropical air masses and produce scattered thunderstorms on summer days. Such storms are usually short-lived and seldom produce strong winds or hail.

Another type of thunderstorm not only benefits from uneven surface heating but is associated with the lifting of warm air, as occurs along a front or a mountain slope. Moreover, diverging winds aloft frequently contribute to the formation of these storms because they tend to draw air from lower levels upward beneath them. Some of the thunderstorms of this type may produce high winds, damaging hail, flash floods, and tornadoes. Such storms are described as *severe*.

At any given time, an estimated 2000 thunderstorms are in progress on Earth. As we would expect, the greatest number occur in the tropics, where warmth, plentiful moisture and instability are always present. About 45,000 thunderstorms take place each day, and more than 16 million occur annually around the world. The lightning from these storms strikes Earth 100 times each second (**FIGURE 19.17A**). Annually, the United States experiences about 100,000 thunderstorms and millions of lightning strikes. A glance at **FIGURE 19.17B** shows that thunderstorms are most frequent in Florida and the eastern Gulf coast region, where such activity is recorded between 70 and 100 days each year. The region on the eastern side of the Rockies in Colorado and New Mexico is next, with thunderstorms occurring on 60 to 70 days each year. Most of the rest of the nation experiences thunderstorms on 30 to 50 days annually. The western margin of the United States has little thunderstorm activity. The same is true for the northern tier of states and for Canada, where warm, moist, unstable mT air seldom penetrates.

Stages of Thunderstorm Development

All thunderstorms require warm, moist air, which, when lifted, releases sufficient latent heat to provide the buoyancy necessary to maintain its upward flight. This instability and associated buoyancy are triggered by a number of different processes, yet most thunderstorms have a similar life history.

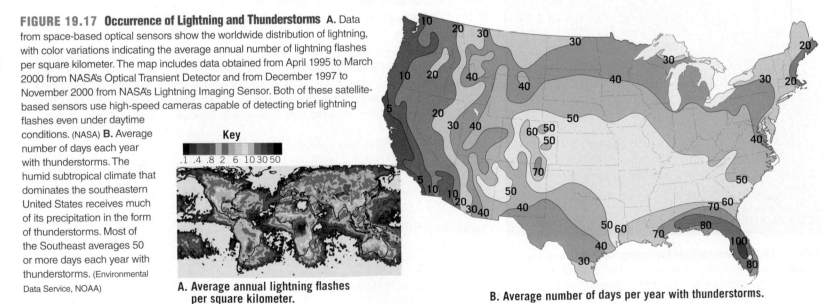

FIGURE 19.17 Occurrence of Lightning and Thunderstorms A. Data from space-based optical sensors show the worldwide distribution of lightning, with color variations indicating the average annual number of lightning flashes per square kilometer. The map includes data obtained from April 1995 to March 2000 from NASA's Optical Transient Detector and from December 1997 to November 2000 from NASA's Lightning Imaging Sensor. Both of these satellite-based sensors use high-speed cameras capable of detecting brief lightning flashes even under daytime conditions. (NASA) **B.** Average number of days each year with thunderstorms. The humid subtropical climate that dominates the southeastern United States receives much of its precipitation in the form of thunderstorms. Most of the Southeast averages 50 or more days each year with thunderstorms. (Environmental Data Service, NOAA)

Key
.1 .4 .8 2 6 10 30 50

A. Average annual lightning flashes per square kilometer.

B. Average number of days per year with thunderstorms.

A.   **B.**

FIGURE 19.18 Cumulus Development **A.** Buoyant thermals often produce fair-weather cumulus clouds that soon evaporate into the surrounding air, making it more humid. As this process of cumulus development and evaporation continues, the air eventually becomes sufficiently humid so that newly forming clouds do not evaporate but continue to grow. (Photo by Henry Lansford/ Science Source) **B.** This developing cumulonimbus cloud became a towering August thunderstorm over central Illinois. (Photo by E. J. Tarbuck)

Because instability and buoyancy are enhanced by high surface temperatures, thunderstorms are most common in the afternoon and early evening (**FIGURE 19.18A**). However, surface heating alone is not sufficient for the growth of towering cumulonimbus clouds. A solitary cell of rising hot air produced by surface heating could, at best, produce a small cumulus cloud, which would evaporate within 10 to 15 minutes.

The development of 12,000-meter (40,000-foot) (or, on rare occasions, 18,000-meter [60,000-foot]) cumulonimbus towers requires a continual supply of moist air (**FIGURE 19.18B**). Each new surge of warm air rises higher than the last, adding to the height of the cloud (**FIGURE 19.19**). These updrafts occasionally reach speeds greater than 100 kilometers (60 miles) per hour, based on the size of hailstones they are capable of carrying upward. Usually within an hour, the amount and size of precipitation that has accumulated is too much for the updrafts to support, and consequently downdrafts develop in one part of the cloud, releasing heavy precipitation. This is the most active stage of the thunderstorm. Gusty winds, lightning, heavy precipitation, and sometimes hail are experienced.

Eventually the warm, moist air supplied by updrafts ceases as downdrafts dominate throughout the cloud. The cooling effect of falling precipitation, coupled with the influx of colder air aloft, marks the end of the thunderstorm activity. The life span of a typical cumulonimbus cell within a thunderstorm complex is only about an hour, but as the storm moves, fresh supplies of warm, water-laden air generate new cells to replace those that are dissipating.

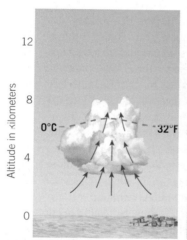 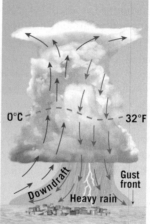 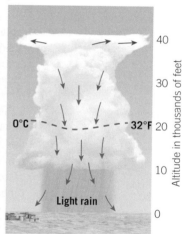

Altitude in kilometers: 12, 8, 4, 0
Altitude in thousands of feet: 40, 30, 20, 10, 0

0°C   32°F   0°C   32°F   0°C   32°F

Downdraft   Gust front   Heavy rain   Light rain

During the cumulus stage, strong updrafts provide moisture that condenses and builds the cloud.

The mature stage is marked by heavy precipitation. Updrafts exist side by side with downdrafts and continue to enlarge the cloud.

When updrafts disappear, precipitation becomes light and then stops. Without a supply of moisture from updrafts, the cloud evaporates.

SmartFigure 19.19 Thunderstorm Development Once a cloud passes beyond the freezing level, the Bergeron process begins producing precipitation. Eventually, the accumulation of precipitation in the cloud is too great for the updraft to support. The falling precipitation causes drag on the air and initiates a downdraft. Once downdrafts dominate, rainfall diminishes, and the cloud starts to dissipate.

19.4 CONCEPT CHECKS

1 Briefly compare and contrast midlatitude cyclones, hurricanes, and tornadoes. How are thunderstorms related to each?

2 What are the basic requirements for the formation of a thunderstorm?

3 Where are thunderstorms most common on Earth? In the United States?

4 Summarize the stages in the development of a thunderstorm.

19.5 | TORNADOES Summarize the atmospheric conditions and locations that are favorable to the formation of tornadoes. Discuss tornado destruction and tornado forecasting.

Tornadoes are local storms of short duration that rank high among nature's most destructive forces (**FIGURE 19.20**). Their sporadic occurrence and violent winds cause many deaths each year. The nearly total destruction in some stricken areas has led many to liken their passage to bombing raids during war (**FIGURE 19.21**).

During the very stormy spring of 2011, there were 753 confirmed tornadoes during April, setting a record for the number of tornadoes in a single month. The deadliest and most destructive outbreak occurred between April 25 and 28, when 326 confirmed tornadoes struck, mostly in the South. The loss of life and property were extraordinary. Estimated fatalities numbered between 350 and 400, and damages were in the billions of dollars. The tornado that struck Tuscaloosa, Alabama, was especially notable for the death and destruction it caused. A few weeks later, on May 22, another outbreak struck the Midwest. Joplin, Missouri, was in the direct path of a storm that took more than 150 lives.

Tornadoes, sometimes called *twisters* or *cyclones*, are violent windstorms that take the form of a rotating column of air, or *vortex*. Pressures within some tornadoes have been estimated to be as much as 10 percent lower than immediately outside the storm. Drawn by the much lower pressure in the center of the vortex, air near the ground rushes into the tornado from all directions. As the air streams inward, it spirals upward around the core until it eventually merges with the airflow of the parent thunderstorm deep in the cumulonimbus tower. Because of the tremendous pressure gradient associated with a strong tornado, maximum winds can sometimes approach 480 kilometers (300 miles) per hour.

A tornado may consist of a single vortex, but within many stronger tornadoes are smaller whirls called *suction vortices* that rotate within the main vortex (**FIGURE 19.22**). Suction vortices have diameters of only about 10 meters (33 feet)

FIGURE 19.21 Tornado Destruction at Moore, Oklahoma On May 20, 2013, central Oklahoma was devastated by an EF-5 tornado, the most severe category. It took 24 lives, injured 377, and caused damages in excess of $2 billion. At least 13,000 structures were destroyed or damaged. The tornado was on the ground for 39 minutes and had a path that extended for 27 kilometers (17 miles). At its peak, the tornado was 2.1 kilometers (1.3 miles) wide and had winds of 340 kilometers (210 miles) per hour. (Photo by Jewel Samad/Getty Images)

and rotate very rapidly. This structure accounts for occasional observations of virtually total destruction of one building while another one, just 10 meters (33 feet) away, suffers little damage.

Tornado Occurrence and Development

Tornadoes form in association with severe thunderstorms that produce high winds, heavy (sometimes torrential) rainfall, and often damaging hail. Fortunately, fewer than 1 percent of all thunderstorms produce tornadoes. Nevertheless, a much higher number of thunderstorms must be monitored as potential tornado producers. A tornado is the product of the interaction between strong updrafts in a thunderstorm and the winds in the troposphere.

Tornadoes can form in any situation that produces severe weather, including cold fronts and tropical cyclones (hurricanes). The most intense tornadoes are usually those that form in association with huge

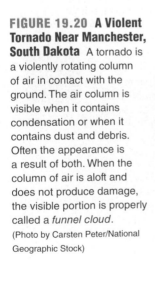

FIGURE 19.20 A Violent Tornado Near Manchester, South Dakota A tornado is a violently rotating column of air in contact with the ground. The air column is visible when it contains condensation or when it contains dust and debris. Often the appearance is a result of both. When the column of air is aloft and does not produce damage, the visible portion is properly called a *funnel cloud*. (Photo by Carsten Peter/National Geographic Stock)

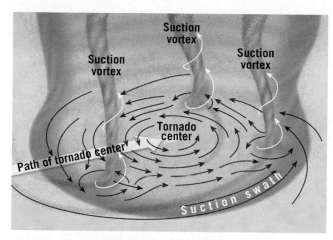

FIGURE 19.22 Multiple-Vortex Tornado Some tornadoes have multiple suction vortices. These small and very intense vortices are roughly 10 meters (30 feet) across and move in a counterclockwise path around the tornado center. Because of this multiple-vortex structure, one building might be heavily damaged and another one, just 10 meters away, might suffer little damage.

thunderstorms called *supercells*. An important precondition linked to tornado formation in severe thunderstorms is the development of a **mesocyclone**—a vertical cylinder of rotating air, typically about 3 to 10 kilometers (2 to 6 miles) across, that develops in the updraft of a severe thunderstorm (**FIGURE 19.23**). The formation of this large vortex often precedes tornado formation by 30 minutes or so.

The formation of a mesocyclone does not necessarily mean that tornado formation will follow. Only about half of all mesocyclones produce tornadoes. Forecasters cannot determine in advance which mesocyclones will spawn tornadoes.

General Atmospheric Conditions Severe thunderstorms—and, hence, tornadoes—are most often spawned along the cold front of a midlatitude cyclone or in association with a supercell thunderstorm such as the one pictured in Figure 19.23D. Throughout spring, air masses associated with midlatitude cyclones are most likely to have greatly contrasting conditions. Continental polar air from Canada may still be very cold and dry, whereas maritime tropical air from the Gulf of Mexico is warm, humid, and unstable. The greater the contrast when these air masses meet, the more intense the storm. The two contrasting air masses are most likely to meet in the central United States because there is no significant natural barrier separating the center of the country from the Arctic or the Gulf of Mexico. Consequently, this region generates more tornadoes than any other area of the country or, in fact, the world. The map in **FIGURE 19.24**, which depicts tornado incidence in the United States for a 27-year period, readily substantiates this fact.

Tornado Climatology An average of 1352 tornadoes were reported annually in the United States between 2003 and 2012. Still, the actual number that occurs from one year to the next varies greatly. During this 10-year span, for example, yearly totals ranged from a low of 1043 in 2012 to a high of 1820 in 2004.

Tornadoes occur during every month of the year. April through June is the period of greatest tornado frequency in

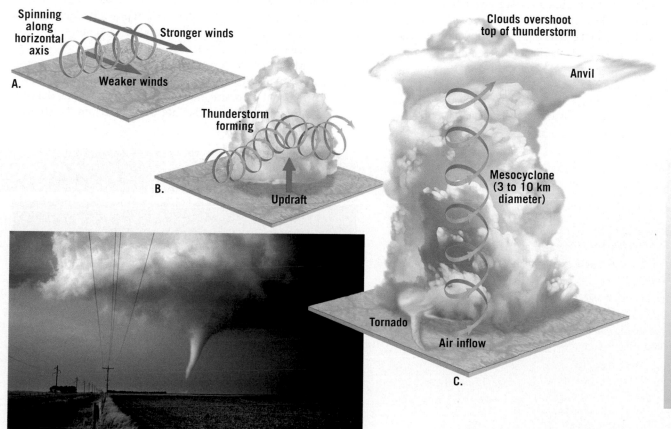

SmartFigure 19.23 The Formation of a Mesocyclone Often Precedes Tornado Formation A. Winds are stronger aloft than at the surface (called speed wind shear), producing a rolling motion about a horizontal axis. **B.** Strong thunderstorm updrafts tilt the horizontally rotating air to a nearly vertical alignment. **C.** The mesocyclone, a vertical cylinder of rotating air, is established. **D.** If a tornado develops, it will descend from a slowly rotating wall cloud in the lower portion of the mesocyclone. (Photo by Gene Rhoden/ Weatherpix/ Getty Images)

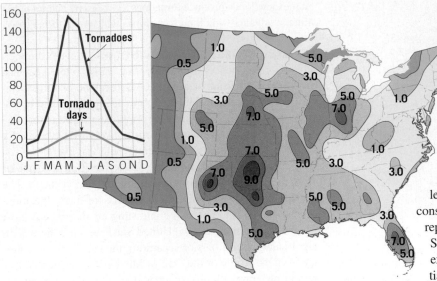

FIGURE 19.24 Tornado Occurrence The map shows average annual tornado incidence per 26,000 square kilometers (10,000 square miles) for a 27-year period. The graph shows the average number of tornadoes and tornado days each month in the United States for the same period.

the United States; the number is lowest during December and January (see Figure 19.24, graph inset). Of the 40,522 confirmed tornadoes reported over the contiguous 48 states during the 50-year period from 1950 through 1999, an average of almost 6 per day occurred during May. At the other extreme, a tornado was reported only about every other day in December and January.

Profile of a Tornado
An average tornado has a diameter of between 150 and 600 meters (500 and 2000 feet), travels across the landscape at approximately 45 kilometers (30 miles) per hour, and cuts a path about 10 kilometers (6 miles) long.[2] Because tornadoes usually occur slightly ahead of a cold front, in the zone of southwest winds, most move toward the northeast. Of the hundreds of tornadoes reported in the

[2]The 10-kilometer (6-mile) figure applies to documented tornadoes. Because many small tornadoes may go undocumented, the real average path of all tornadoes is unknown but is shorter than 10 kilometers.

United States annually, more than half are comparatively weak and short-lived. Most of these small tornadoes have lifetimes of 3 minutes or less and paths that seldom exceed 1 kilometer (0.6 mile) in length and 100 meters (330 feet) in width. Typical wind speeds are on the order of 150 kilometers (90 miles) per hour or less. On the other end of the tornado spectrum are the infrequent and often long-lived violent tornadoes. Although large tornadoes constitute only a small percentage of the total reported, their effects are often devastating. Such tornadoes may exist for periods in excess of 3 hours and produce an essentially continuous damage path more than 150 kilometers (90 miles) long and perhaps 1 kilometer (0.6 mile) or more wide. Maximum winds range beyond 500 kilometers (310 miles) per hour.

Tornado Destruction and Loss of Life
The potential for tornado destruction depends largely on the strength of the winds generated by the storm. Because tornadoes generate the strongest winds in nature, they have accomplished many seemingly impossible tasks, such as driving a piece of straw through a thick wooden plank and uprooting huge trees. Although it may seem impossible for winds to cause some of the extensive damage attributed to tornadoes, tests in engineering facilities have repeatedly demonstrated that winds in excess of 320 kilometers (200 miles) per hour are capable of incredible feats (**FIGURE 19.25**).

Most tornado losses are associated with a few storms that strike urban areas or devastate entire small communities. The amount of destruction caused by such storms depends to a significant degree (but not completely) on the strength

EYE ON EARTH

This satellite image shows a portion of the diagonal path left by a tornado as it moved across northern Wisconsin in 2007. (NASA)

QUESTION 1 *Toward what direction did the storm advance: the northeast or the southwest?*

QUESTION 2 *Did the tornado more likely occur ahead of or behind a cold front? Explain.*

QUESTION 3 *Is it more probable that the storm took place in March or June? Why is the date you selected more likely?*

A.

B.

FIGURE 19.25 Tornado Winds: The Strongest in Nature A. The force of the wind during a tornado near Wichita, Kansas, in April 1991 was enough to drive this piece of metal into a utility pole. (Photo by John Sokich/NOAA) **B.** The remains of a truck wrapped around a tree in Bridge Creek, Oklahoma, on May 4, 1999, following a major tornado outbreak. (LM Otero/AP Photo)

of the winds. A wide spectrum of tornado strengths, sizes, and lifetimes are observed. The commonly used guide to tornado intensity is the **Enhanced Fujita intensity scale**, or **EF-scale** for short (**TABLE 19.1**). Because tornado winds cannot be measured directly, a rating on the EF-scale is determined by assessing the worst damage produced by a storm. Although widely used, the EF-scale is not perfect. Estimating tornado intensity based on damage alone does not take into account the structural integrity of the objects hit by a tornado. A well-constructed building can withstand very high winds, whereas a poorly built structure can suffer devastating damage from the same or weaker winds.

Although the greatest part of tornado damage is caused by violent winds, most tornado injuries and deaths result from flying debris. The proportion of tornadoes that result in loss of life is small. In most years, slightly fewer than 2 percent of all reported tornadoes in the United States are "killers." Although the percentage of tornadoes resulting in death is small, each tornado is potentially lethal. When tornado fatalities and storm intensities are compared, the results are quite interesting: The majority (63 percent) of tornadoes are weak (EF-0 and EF-1), and the number of storms decreases as tornado intensity increases. The distribution of tornado fatalities, however, is just the opposite. Although only 2 percent of tornadoes are classified as violent (EF-4 and EF-5), they account for nearly 70 percent of tornado deaths.

Tornado Forecasting

Because severe thunderstorms and tornadoes are small and relatively short-lived phenomena, they are among the most difficult weather features to forecast precisely. Nevertheless, the prediction, detection, and monitoring of such storms are among the most important services provided by professional meteorologists. Both the timely issuance and dissemination of watches and warnings are critical to the protection of life and property.

The Storm Prediction Center (SPC) located in Norman, Oklahoma, is part of the National Weather Service (NWS) and the National Centers for Environmental Prediction (NCEP). The mission of the SPC is to provide timely and accurate forecasts and watches for severe thunderstorms and tornadoes.

Severe thunderstorm outlooks are issued several times daily. *Day 1* outlooks identify the areas that are likely to be affected by severe thunderstorms during the next 6 to 30 hours, and *day 2* outlooks extend the forecast through the following day. Both outlooks describe the type, coverage, and intensity of the severe weather expected. Many local NWS field offices also issue severe weather outlooks that provide more local descriptions of the severe weather potential for the next 12 to 24 hours.

TABLE 19.1 Enhanced Fujita Intensity Scale*

Scale	Wind Speed		Damage
	Km/Hr	Mi/Hr	
EF-0	105–137	65–85	*Light.* Some damage to siding and shingles.
EF-1	138–177	86–110	*Moderate.* Considerable roof damage. Winds can uproot trees and overturn single-wide mobile homes. Flagpoles bend.
EF-2	178–217	111–135	*Considerable.* Most single-wide homes destroyed. Permanent homes can shift off foundations. Flagpoles collapse. Softwood trees debarked.
EF-3	218–265	136–165	*Severe.* Hardwood trees debarked. All but small portions of houses destroyed.
EF-4	266–322	166–200	*Devastating.* Complete destruction of well-built residences, large sections of school buildings.
EF-5	>322	>200	*Incredible.* Significant structural deformation of mid- and high-rise buildings.

*The original Fujita scale was developed by T. Theodore Fujita in 1971 and put into use in 1973. The Enhanced Fujita intensity scale is a revision that was put into use in February 2007. Winds speeds are estimates (not measurements) based on damage, and represent 3-second gusts at the point of damage.

FIGURE 19.26 Doppler Radar This is a dual Doppler radar image of a violent (EF-5) tornado near Moore, Oklahoma, on May 3, 1999. The left image (reflectivity) shows precipitation in the supercell thunderstorm. The right image shows motion of the precipitation along the radar beam—that is, how fast rain or hail is moving toward or away from the radar. In this example, the radar was unusually close to the tornado—close enough to make out the signature of the tornado itself. Most of the time only the weaker and larger mesocyclone is detected. (NOAA)

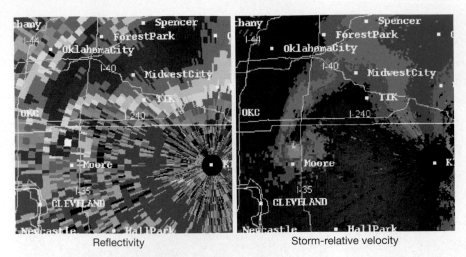

Reflectivity

Storm-relative velocity

Tornado Watches and Warnings Tornado watches alert the public to the possibility of tornadoes over a specified area for a particular time interval. Watches serve to fine-tune forecast areas already identified in severe weather outlooks. A typical watch covers an area of about 65,000 square kilometers (25,000 square miles) for a 4- to 6-hour period. A tornado watch is an important part of the tornado alert system because it sets in motion the procedures necessary to deal adequately with detection, tracking, warning, and response. Watches are generally reserved for organized severe weather events where the tornado threat will affect at least 26,000 square kilometers (10,000 square miles) and/or persist for at least 3 hours. Watches typically are not issued when the threat is thought to be isolated and/or short-lived.

Whereas a tornado watch is designed to alert people to the possibility of tornadoes, a **tornado warning** is issued by local offices of the NWS when a tornado has actually been sighted in an area or is indicated by weather radar. It warns of a high probability of imminent danger. Warnings are issued for much smaller areas than for watches, usually covering portions of a county or counties. In addition, they are in effect for much shorter periods, typically 30 to 60 minutes. Because a tornado warning may be based on an actual sighting, warnings are occasionally issued after a tornado has already developed. However, most warnings are issued prior to tornado formation, sometimes by several tens of minutes, based on Doppler radar data and/or spotter reports of funnel clouds.

If the direction and the approximate speed of the storm are known, an estimate of its most probable path can be made. Because tornadoes often move erratically, the warning area is fan-shaped downwind from the point where the tornado has been spotted. Improved forecasts and advances in technology have contributed to a significant decline in tornado deaths over the past 50 years.

Doppler Radar Many of the difficulties that once limited the accuracy of tornado warnings have been reduced or eliminated by an advancement in radar technology called **Doppler radar**. Doppler radar not only performs the same tasks as conventional radar but also has the ability to detect motion directly (**FIGURE 19.26**). Doppler radar can detect the initial formation and subsequent development of a mesocyclone, the intense rotating wind system in the lower part of a thunderstorm that frequently precedes tornado development. Almost all mesocyclones produce damaging hail, severe winds, or tornadoes. Those that produce tornadoes (about 50 percent) can sometimes be distinguished by their stronger wind speeds and their sharper gradients of wind speeds.

It should also be pointed out that not all tornado-bearing storms have clear-cut radar signatures and that other storms can give false signatures. Detection, therefore, is sometimes a subjective process, and a given display could be interpreted in several ways. Consequently, trained observers continue to form an important part of the warning system.

The benefits of Doppler radar are many. As a research tool, it is not only providing data on the formation of tornadoes but also helping meteorologists gain new insights into thunderstorm development, the structure and dynamics of hurricanes, and air-turbulence hazards that plague aircraft. As a practical tool for tornado detection, Doppler radar has significantly improved our ability to track thunderstorms and issue warnings.

19.5 CONCEPT CHECKS

1 Why do tornadoes have such high wind speeds?

2 What general atmospheric conditions are most conducive to the formation of tornadoes?

3 During what months is tornado activity most pronounced in the United States?

4 Name the scale commonly used to rate tornado intensity. How is a rating on this scale determined?

5 Distinguish between a tornado watch and a tornado warning.

19.6 | HURRICANES

Identify areas of hurricane formation on a world map and discuss the conditions that promote hurricane formation. List the three broad categories of hurricane destruction.

Most of us view the weather in the tropics with favor. Places such as the islands of the Caribbean are known for their lack of significant day-to-day variations. Warm breezes, steady temperatures, and rains that come as heavy but brief tropical showers are often the rule. It is ironic that these relatively tranquil regions produce some of the most violent storms on Earth.

Hurricanes are intense centers of low pressure that form over tropical oceans and are characterized by intense convective (thunderstorm) activity and strong cyclonic circulation (**FIGURE 19.27**). Sustained winds must equal or exceed 119 kilometers (74 miles) per hour. Unlike midlatitude cyclones, hurricanes lack contrasting air masses and fronts. Rather, the source of energy that produces and maintains hurricane-force winds is the huge quantity of latent heat liberated during the formation of the storm's cumulonimbus towers.

The vast majority of hurricane-related deaths and damage are caused by relatively infrequent, yet powerful, storms. Hurricane Sandy, shown in the chapter-opening photo, devastated coastal areas of New Jersey, New York, and Connecticut and caused billions of dollars in damages in late October 2012. The storm that pounded an unsuspecting Galveston, Texas, in 1900 was not just the deadliest U.S. hurricane ever but the deadliest natural disaster of *any kind* to affect the United States. The deadliest and most costly storm in recent memory occurred in August 2005, when Hurricane Katrina devastated the Gulf Coast of Louisiana, Mississippi, and Alabama and took an estimated 1800 lives. Although hundreds of thousands fled before the storm made landfall, thousands of others were caught by the storm. In addition to the human suffering and tragic loss of life that were left in the wake of Hurricane Katrina, the financial losses caused by the storm are practically incalculable.

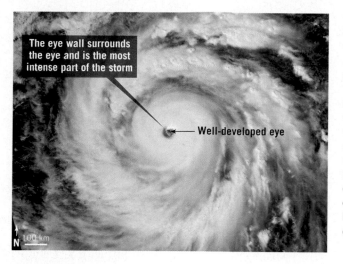

The eye wall surrounds the eye and is the most intense part of the storm

Well-developed eye

100 km

FIGURE 19.27 Super Typhoon Jangmi In the western Pacific, hurricanes are called typhoons. This storm struck portions of Taiwan, China, and Japan in late September 2008. It was the strongest storm worldwide that year, with sustained winds that reached 270 kilometers (165 miles) per hour. The counterclockwise spiral of the clouds indicates that it is a Northern Hemisphere storm. (NASA)

Profile of a Hurricane

Most hurricanes form between the latitudes of 5° and 20° over all the tropical oceans except the South Atlantic and the eastern South Pacific (**FIGURE 19.28**). The North Pacific has the greatest number of storms, averaging 20 each year. Fortunately for those living in the coastal regions of the southern and eastern United States, fewer than 5 hurricanes, on the average, develop annually in the warm sector of the North Atlantic.

These intense tropical storms are known in various parts of the world by different names. In the western Pacific, they are called *typhoons*, and in the Indian Ocean, including the Bay of Bengal and Arabian Sea, they are simply called *cyclones*. In the following discussion, these storms will be referred to as hurricanes. The term *hurricane* is derived from Huracan, a Carib god of evil.

Although many tropical disturbances develop each year, only a few reach hurricane status. By international

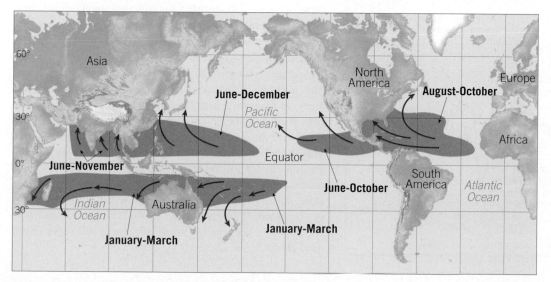

60°

Asia

June-December

Pacific Ocean

30°

North America

August-October

Europe

June-November

0°

Equator

Africa

June-October

South America

Atlantic Ocean

30°

Indian Ocean

Australia

January-March

January-March

FIGURE 19.28 Regions Where Hurricanes Form This world map shows the regions where most hurricanes form as well as their principal months of occurrence and the most common tracks they follow. Hurricanes do not develop within about 5° of the equator because the Coriolis effect is too weak. Because warm surface ocean temperatures are necessary for hurricane formation, they seldom form poleward of 20° latitude nor over the cool waters of the South Atlantic and the eastern South Pacific.

FIGURE 19.29 Hurricane Fran These weather maps show Hurricane Fran at 7:00 A.M. EST on two successive days, September 5 and 6, 1996. On September 5, winds exceeded 190 kilometers (118 miles) per hour. As the storm moved inland, heavy rains caused flash floods, killed 30 people, and caused more than $3 billion in damages. The station information plotted off the Gulf and Atlantic coasts is from data buoys, which are remote floating instrument packages. The small boxes extending southeast from the storm's center show the position of the eye at 6-hour intervals.

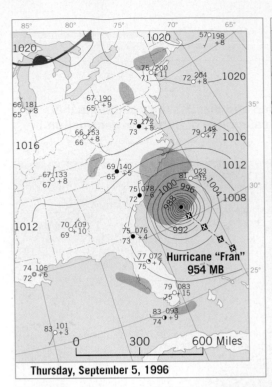

Thursday, September 5, 1996

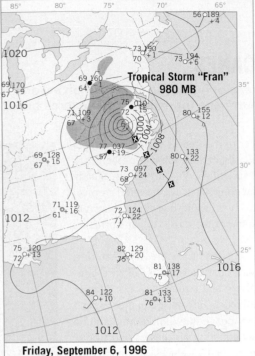

Friday, September 6, 1996

FIGURE 19.30 Cross Section of a Hurricane Note that the vertical dimension is greatly exaggerated. (After NOAA)

Outflow of air at the top of the hurricane is important because it prevents the convergent flow at lower levels from "filling in" the storm.

Sinking air in the eye warms by compression.

Eye

Eye wall, the zone where winds and rain are most intense.

Tropical moisture spiraling inward creates rain bands that pinwheel around the storm center.

Measurements of surface pressure and wind speed during the passage of Cyclone Monty at Mardie Station, Western Australia, between February 29 and March 2, 2004. (Hurricanes are called "cyclones" in this part of the world.)

Pressure

Mean speed

Minimum pressure 964 on 3 March

agreement, a hurricane has wind speeds in excess of 119 kilometers (74 miles) per hour and a rotary circulation. Mature hurricanes average 600 kilometers (375 miles) across, although they can range in diameter from 100 kilometers (60 miles) up to about 1500 kilometers (930 miles). From the outer edge to the center, the barometric pressure has on occasion dropped 60 millibars, from 1010 millibars to 950 millibars. The lowest pressures ever recorded in the Western Hemisphere are associated with these storms.

A steep pressure gradient generates the rapid, inward-spiraling winds of a hurricane (**FIGURE 19.29**). As the air rushes toward the center of the storm, its velocity increases. This occurs for the same reason that skaters with their arms extended spin faster as they pull their arms in close to their bodies.

As the inward rush of warm, moist surface air approaches the core of the storm, it turns upward and ascends in a ring of cumulonimbus towers (**FIGURE 19.30**). This doughnut-shaped wall of intense convective activity surrounding the center of the storm is called the **eye wall**. It is here that the greatest wind speeds and heaviest rainfall occur. Surrounding the eye wall are curved bands of clouds that trail away in a spiral fashion. Near the top of the hurricane, the airflow is outward, carrying the rising air away from the storm center, thereby providing room for more inward flow at the surface.

At the very center of the storm is the **eye** of the hurricane. This well-known feature is a zone about 20 kilometers (12.5 miles) in diameter where precipitation ceases and winds subside. It offers a brief but deceptive break from the extreme weather in the enormous curving wall clouds that surround it. The air within the eye gradually descends and heats by compression, making it the warmest part of the storm. Although many people believe that the eye is characterized by clear blue skies, this is usually not the case because the subsidence in the eye is seldom strong enough to produce cloudless conditions. Although the sky appears much brighter in this region, scattered clouds at various levels are common.

GEOGRAPHICS
Hurricane Katrina from Space

Satellites allow us to track the formation, movement, and growth of hurricanes. In addition, their specialized instruments provide data that can be transformed into images that allow scientists to analyze the internal structure and workings of these huge destructive storms.

KEY TO SYMBOLS

+ Location of storm center (at 8 A.M. EDT)
○ Tropical depression
𝟨 Tropical storm
𝟨 Hurricane

TOTAL RAINFALL

8	16	24	32 cm
3.2	6.4	9.6	12.8 in

NOAA

Storm track and rainfall values for Hurricane Katrina for the period August 23 to 31, 2005. The highest totals (dark red) exceeded 30 centimeters (12 inches) over parts of Cuba and the Florida Keys. Because the storm moved rapidly, rainfall totals (green to blue) in many areas were generally less than 13 centimeters (5 inches). Data is from the Tropical Rainfall Measuring Mission (TRMM) satellite.

A color-enhanced infrared image from the GOES-East satellite. Infrared wavelengths are invisible but can be detected as heat. Wavelengths of radiation emitted by an object depend on temperature. Longer infrared wavelengths indicate colder temperatures and shorter wavelengths are associated with warmer temperatures. The high tops of towering storm clouds are colder than the tops of clouds that produce less intense weather. The highest (coldest) cloud tops and thus the most intense storms are easily seen.

GOES-EAST AVNCOLOR IR CH 4 - AUG 29 05 00:45 UTC   NASA

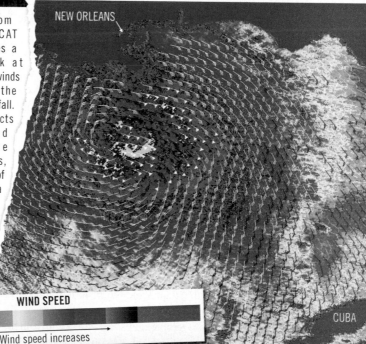

This image from NASA's QuikSCAT satellite provides a detailed look at Katrina's surface winds shortly before the storm made landfall. The image depicts relative wind speeds. The strongest winds, shown in shades of purple, circle a well-defined eye. The barbs fly with the wind and show the strong counterclockwise flow of the storm.

NEW ORLEANS

WIND SPEED

Wind speed increases

CUBA

YUCATAN PEN.

NASA

This eye wall tower rises 16 kilometers (10 miles) above the ocean surface. Towers this tall near the storm's center are often an indication that a storm is intensifying. Katrina grew from a category 3 to a category 4 storm soon after this image was received.

HEIGHT (km)
20
10
0

RAIN RATE (mm/hr)

0	10	20	30	40	50

NASA

Tropical Rainfall Measuring Mission (TRMM) satellite image of Katrina early on August 28, 2005. The cutaway view of the inner portion of the storm shows cloud height (vertical scale) and rainfall rates (horizontal scale). Two isolated towers (in red) are visible: one in an outer rain band and the other in the eye wall.

NASA

ALABAMA   FLORIDA

MISSISSIPPI

LOUISIANA

TEXAS

YUCATAN PENINSULA

200 KM

This relatively "traditional" image shows Katrina on August 28, 2005 as the massive storm approached the Gulf Coast. After passing over Florida as a category 1 hurricane, Katrina entered the Gulf and intensified to a category 5 storm with winds of 257 kilometers (160 miles) per hour, and even stronger gusts. When Katrina came ashore the next day it was a slightly less vigorous category 4 storm.

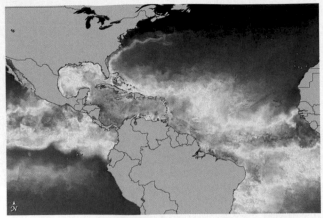

Sea Surface Temperature (°C)

-2   16.5   27.8   35

Hurricane Formation and Decay

A hurricane is a heat engine that is fueled by the latent heat liberated when huge quantities of water vapor condense. The amount of energy produced by a typical hurricane in just a single day is truly immense. The release of latent heat warms the air and provides buoyancy for its upward flight. The result is to reduce the pressure near the surface, which encourages a more rapid inward flow of air. To get this engine started, a large quantity of warm, moisture-laden air is required, and a continual supply is needed to keep it going.

Hurricane Formation Hurricanes develop most often in the late summer, when ocean waters have reached temperatures of 27°C (80°F) or higher and thus are able to provide the necessary heat and moisture to the air (**FIGURE 19.31**). This ocean-water temperature requirement accounts for the fact that hurricanes do not form over the relatively cool waters of the South Atlantic and the eastern South Pacific. For the same reason, few hurricanes form poleward of 20° latitude. Although water temperatures are sufficiently high, hurricanes do not form within 5° of the equator because the Coriolis effect is too weak to initiate the necessary rotary motion.

Many tropical storms begin as disorganized arrays of clouds and thunderstorms that develop weak pressure gradients but exhibit little or no rotation. Such areas of low-level convergence and lifting are called *tropical disturbances*. Most of the time, these zones of convective activity die out. However, tropical disturbances occasionally grow larger and develop a strong cyclonic rotation.

What happens on occasions when conditions favor hurricane development? As latent heat is released from the clusters of thunderstorms that make up the tropical disturbance, areas within the disturbance get warmer. As a result, air density

lowers and surface pressure drops, creating a region of weak low pressure and cyclonic circulation. As pressure drops at the storm center, the pressure gradient steepens. If you were watching an animated weather map of the storm, you would see the isobars get closer together. In response, surface wind speeds increase and bring additional supplies of moisture to nurture storm growth. The water vapor condenses, releasing latent heat, and the heated air rises. Adiabatic cooling of rising air triggers more condensation and the release of more latent heat, which causes a further increase in buoyancy. And so it goes.

Meanwhile, at the top of the storm, air is diverging. Without this outward flow up top, the inflow at lower levels would soon raise surface pressures (that is, fill in the low) and thwart storm development.

Other Tropical Storms Many tropical disturbances occur each year, but only a few develop into full-fledged hurricanes. By international agreement, lesser tropical cyclones are placed in different categories, based on wind strength. When a cyclone's strongest winds do not exceed 61 kilometers (38 miles) per hour, it is called a **tropical depression**. When winds are between 61 and 119 kilometers (38 and 74 miles) per hour, the cyclone is termed a **tropical storm**. It is during this phase that a name is given (Andrew, Katrina, Sandy, etc.). If the tropical storm becomes a hurricane, the name remains the same. Each year, between 80 and 100 tropical storms develop around the world. Of these, usually half or more eventually become hurricanes.

Hurricane Decay Hurricanes diminish in intensity whenever they (1) move over ocean waters that cannot supply warm, moist tropical air; (2) move onto land; or (3) reach a location where the large-scale flow aloft is unfavorable. When a hurricane moves onto land, it loses its punch rapidly. The most important reason for this rapid demise is the fact that the storm's source of warm, moist air is cut off. When an adequate supply of water vapor does not exist, condensation and the release of latent heat must diminish. In addition, friction from the increased roughness of the land surface rapidly slows surface wind speeds. This factor causes the winds to move more directly into the center of the low, thus helping to eliminate the large pressure differences.

Hurricane Destruction

A location only a few hundred kilometers from a hurricane—just 1 day's striking distance away—may experience clear skies and virtually no wind. Prior to the age of weather satellites, this situation made the task of warning people of impending storms very difficult.

The amount of damage caused by a hurricane depends on several factors, including the size and population density of the area affected and the shape of the ocean bottom near the shore. The most significant factor, of course, is the strength of the storm itself. By studying past storms, a scale has been established to rank the relative intensities of hurricanes. As **TABLE 19.2** indicates, a *category 5* storm is the worst possible, whereas a *category 1* hurricane is least severe.

TABLE 19.2 Saffir–Simpson Hurricane Scale

Scale Number (category)	Central Pressure (millibars)	Winds (km/hr)	Storm Surge (meters)	Damage
1	980	119–153	1.2–1.5	Minimal
2	965–979	154–177	1.6–2.4	Moderate
3	945–964	178–209	2.5–3.6	Extensive
4	920–944	210–250	3.7–5.4	Extreme
5	<920	>250	>5.4	Catastrophic

During hurricane season, it is common to hear scientists and reporters use the numbers from the **Saffir–Simpson hurricane scale**. When Hurricane Katrina made landfall, sustained winds were 225 kilometers (140 miles) per hour, making it a strong category 4 storm. Storms that fall into category 5 are rare. Damage caused by hurricanes can be divided into three categories: (1) storm surge, (2) wind damage, and (3) heavy rains and inland flooding.

Storm Surge   The most devastating damage in the coastal zone is usually caused by storm surge (**FIGURE 19.32**). It not only accounts for a large share of coastal property losses but also is responsible for a high percentage of all hurricane-caused deaths. A **storm surge** is a dome of water 65 to 80 kilometers (40 to 50 miles) wide that sweeps across the coast near the point where the eye makes landfall. If all wave activity were smoothed out, the storm surge would be the height of the water above normal tide level. In addition, tremendous wave activity is superimposed on the surge. The worst surges occur in places like the Gulf of Mexico, where the continental shelf is very shallow and gently sloping. In addition, local features such as bays and rivers can cause the surge height to double and increase in speed.

As a hurricane advances toward the coast in the Northern Hemisphere, storm surge is always most intense on the right side of the eye (viewed from the ocean), where winds are blowing *toward* the shore. In addition, on this side of the storm, the forward movement of the hurricane contributes to the storm surge. In **FIGURE 19.33**, assume that a hurricane

FIGURE 19.33 An Approaching Hurricane Winds associated with a Northern Hemisphere hurricane that is advancing toward the coast. This hypothetical storm, with peak winds of 175 kilometers (109 miles) per hour, is moving toward the coast at 50 kilometers (31 miles) per hour. On the right side of the advancing storm (as viewed from the ocean), the 175-kilometer-per-hour winds are in the same direction as the movement of the storm (50 kilometers per hour). Therefore, the *net* wind speed on the right side of the storm is 225 kilometers (140 miles) per hour. On the left side, the hurricane's winds are blowing opposite the direction of storm movement, so the *net* winds of 125 kilometers (78 miles) per hour are away from the coast. Storm surge will be greatest along the part of the coast hit by the right side of the advancing hurricane.

with peak winds of 175 kilometers (109 miles) per hour is moving toward the shore at 50 kilometers (31 miles) per hour. In this case, the net wind speed on the right side of the advancing storm is 225 kilometers (140 miles) per hour. On the left side, the hurricane's winds are blowing opposite the direction of storm movement, so the net winds are *away* from the coast at 125 kilometers (78 miles) per hour. Along the shore facing the left side of the oncoming hurricane, the water level may actually decrease as the storm makes landfall.

Wind Damage

Destruction caused by wind is perhaps the most obvious of the classes of hurricane damage. Debris such as signs, roofing materials, and small items left outside become dangerous flying missiles in hurricanes. For some structures, the force of the wind is sufficient to cause total ruin. Mobile homes are particularly vulnerable. High-rise buildings are also susceptible to hurricane-force winds. Upper floors are most vulnerable because wind speeds usually increase with height. Recent research suggests that people should stay below the 10th floor but remain above any floors at risk for flooding. In regions with good building codes, wind damage is usually not as catastrophic as storm-surge damage. However, hurricane-force winds affect a much larger area than storm surge and can cause huge economic losses. For example, in 1992 it was largely the winds associated with Hurricane Andrew that produced more than $25 billion of damage in southern Florida and Louisiana.

A hurricane may produce tornadoes that contribute to the storm's destructive power. Studies have shown that more than half of the hurricanes that make landfall produce at least one tornado. In 2004 the number of tornadoes associated with tropical storms and hurricanes was extraordinary. Tropical Storm Bonnie and five landfalling hurricanes—Charley, Frances, Gaston, Ivan, and Jeanne—produced nearly 300 tornadoes that affected the southeastern and mid-Atlantic states.

Heavy Rains and Inland Flooding

The torrential rains that accompany most hurricanes contribute to a third significant threat: flooding. Whereas the effects of storm surge and strong winds are concentrated in coastal areas, heavy rains may affect places hundreds of kilometers from the coast for up to several days after the storm has lost its hurricane-force winds.

In September 1999, Hurricane Floyd brought flooding rains, high winds, and rough seas to a large portion of the Atlantic seaboard. More than 2.5 million people evacuated their homes from Florida north to the Carolinas and beyond. It was the largest peacetime evacuation in U.S. history up to that time. Torrential rains falling on already saturated ground created devastating inland flooding. Altogether Floyd dumped more than 48 centimeters (19 inches) of rain on Wilmington, North Carolina, 33.98 centimeters (13.38 inches) in a single 24-hour span.

Tracking Hurricanes

Today we have the benefit of numerous observational tools for tracking tropical storms and hurricanes. Using input from satellites, aircraft reconnaissance, coastal radar, and remote data buoys in conjunction with sophisticated computer models, meteorologists monitor and forecast storm movements and intensity. The goal is to issue timely watches and warnings.

An important part of this process is the *track forecast*—the predicted path of the storm. The track forecast is probably the most basic information because accurate prediction of other storm characteristics (winds and rainfall) is of little value if there is significant uncertainty about where the storm is going. Accurate track forecasts are important because they can lead to timely evacuations from the surge zone, where the greatest number of deaths usually occur. Fortunately, track forecasts have been steadily improving. During the span

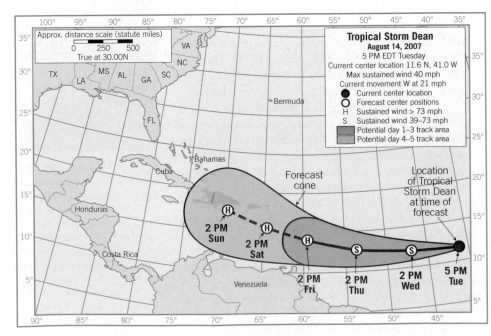

FIGURE 19.34 Five-Day Track Forecast for Tropical Storm Dean, Issued at 5 P.M. EDT, Tuesday, August 14, 2007 When a hurricane track forecast is issued by the National Hurricane Center, it is termed a *forecast cone*. The cone represents the probable track of the center of the storm and is formed by enclosing the area swept out by a set of circles along the forecast track (at 12 hours, 24 hours, 36 hours, etc.). The size of each circle gets larger with time. Based on statistics from 2003 to 2007, the entire track of an Atlantic tropical cyclone can be expected to remain entirely within the cone roughly 60 to 70 percent of the time. (National Weather Service/National Hurricane Center)

2001 to 2005, forecast errors were roughly half of what they were in 1990. During the very active 2004 and 2005 Atlantic hurricane seasons, 12- to 72-hour track forecast accuracy was at or near record levels. Consequently, the length of official track forecasts issued by the National Hurricane Center was extended from 3 days to 5 days (**FIGURE 19.34**). Current 5-day track forecasts are now as accurate as 3-day forecasts were 15 years ago.

Despite improvements in accuracy, forecast uncertainty still requires that hurricane warnings be issued for relatively large coastal areas. During the span 2000 to 2005, the average length of coastline under a hurricane warning in the United States was 510 kilometers (316 miles). This represents a significant improvement over the preceding decade, when the average was 730 kilometers (452 miles). Nevertheless, only about one-quarter of an average warning area experiences hurricane conditions.

19.6 CONCEPT CHECKS

1 Define *hurricane*. What other names are used for this storm?

2 In what latitude zone do hurricanes develop?

3 Distinguish between the eye and the eye wall of a hurricane. How do conditions differ in these zones?

4 What is the source of energy that drives a hurricane?

5 Why do hurricanes *not* form near the equator? Explain the lack of hurricanes in the South Atlantic and eastern South Pacific.

6 When do most hurricanes in the North Atlantic and Caribbean occur? Why are these months favored?

7 Why does the intensity of a hurricane diminish rapidly when it moves over land?

8 What are the three broad categories of hurricane damage?

19 CONCEPTS IN REVIEW | Weather Patterns and Severe Storms

19.1 AIR MASSES

Discuss air masses, their classification, and associated weather.

KEY TERMS: air mass, air-mass weather, source region, polar (P) air mass, arctic (A) air mass, tropical (T) air mass, continental (c) air mass, maritime (m) air mass, lake-effect snow, nor'easter

- An *air mass* is a large body of air, usually 1600 kilometers (1000 miles) or more across, that is characterized by a sameness of temperature and moisture at any given altitude. When this air moves out of its region of origin, called the source region, it carries these temperatures and moisture conditions elsewhere, perhaps eventually affecting a large portion of a continent.

- Air masses are classified according to the nature of the surface in the source region and the latitude of the source region. Continental (c) designates an air mass of land origin, with the air likely to be dry; a maritime (m) air mass originates over water and, therefore, will be relatively humid. Polar (P) and arctic (A) air masses originate in high latitudes and are cold. Tropical (T) air masses form in low latitudes and are warm. According to this classification scheme, the four basic types of air masses are continental polar (cP), continental tropical (cT), maritime polar (mP), and maritime tropical (mT).

- Continental polar (cP) and maritime tropical (mT) air masses influence the weather of North America most, especially east of the Rocky Mountains. Maritime tropical air is the source of much, if not most, of the precipitation received in the eastern two-thirds of the United States.

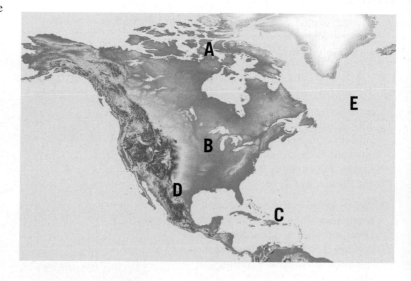

Q Identify the source region associated with each letter on this map. One letter *is not* associated with a source region. Which one is it?

19.2 FRONTS

Compare and contrast typical weather associated with a warm front and a cold front. Describe an occluded front and a stationary front.

KEY TERMS: front, overrunning, warm front, cold front, stationary front, occluded front

- Fronts are boundary surfaces that separate air masses of different densities, one usually warmer and more humid than the other. As one air mass moves into another, the warmer, less dense air mass is forced aloft in a process called overrunning.
- Along a warm front, a warm air mass overrides a retreating mass of cooler air. As the warm air ascends, it cools adiabatically to produce clouds and, frequently, light to moderate precipitation over a large area.
- A cold front forms where cold air is actively advancing into a region occupied by warmer air. Cold fronts are about twice as steep as and move more rapidly than warm fronts. Because of these two differences, precipitation along a cold front is generally more intense and of shorter duration than precipitation associated with a warm front.

Q Identify each of these symbols used to designate fronts. On which side of each symbol are the warm air and the cool air?

19.3 MIDLATITUDE CYCLONES

Summarize the weather associated with the passage of a mature midlatitude cyclone. Describe how airflow aloft is related to cyclones and anticyclones at the surface.

KEY TERMS: midlatitude (middle-latitude) cyclone

- The primary weather producers in the middle latitudes are large centers of low pressure that generally travel from west to east, called midlatitude cyclones. These bearers of stormy weather, which last from a few days to a week, have a counterclockwise circulation pattern in the Northern Hemisphere, with an inward flow of air toward their centers.
- Most midlatitude cyclones have a cold front and frequently a warm front extending from the central area of low pressure. Convergence and forceful lifting along the fronts initiate cloud development and frequently cause precipitation. The particular weather experienced by an area depends on the path of the cyclone.
- Guided by west-to-east-moving jet streams, cyclones generally move eastward across the United States. Airflow aloft (divergence and convergence) plays an important role in maintaining cyclonic and anticyclonic circulation. In cyclones, divergence aloft supports the inward flow at the surface.

19.4 THUNDERSTORMS

List the basic requirements for thunderstorm formation and locate places on a map that exhibit frequent thunderstorm activity. Describe the stages in the development of a thunderstorm.

KEY TERM: thunderstorm

- Thunderstorms are caused by the upward movement of warm, moist, unstable air. They are associated with cumulonimbus clouds that generate heavy rainfall, lightning, thunder, and occasionally hail and tornadoes.
- Air mass thunderstorms frequently occur in maritime tropical (mT) air during spring and summer in the middle latitudes. Generally, three stages are involved in the development of these storms: the cumulus stage, mature stage, and dissipating stage.

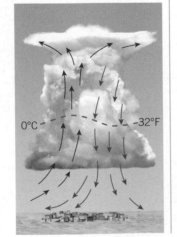

Q Which stage in the development of a thunderstorm is shown in this sketch? Describe what is occurring. Is there a stage that follows this one? If so, describe what occurs during that stage.

19.5 TORNADOES

Summarize the atmospheric conditions and locations that are favorable to the formation of tornadoes. Discuss tornado destruction and tornado forecasting.

KEY TERMS: tornado, mesocyclone, Enhanced Fujita intensity scale (EF-scale), tornado watch, tornado warning, Doppler radar

- A tornado is a violent windstorm that takes the form of a rotating column of air called a vortex that extends downward from a cumulonimbus cloud. Many strong tornadoes are multiple-vortex storms. Because of the tremendous pressure gradient associated with a strong tornado, maximum winds can approach 480 kilometers (300 miles) per hour.
- Tornadoes are most often spawned along the cold front of a midlatitude cyclone or in association with a supercell thunderstorm. Tornadoes also form in association with tropical cyclones (hurricanes). In the United States, April through June is the period of greatest tornado activity, but tornadoes can occur during any month of the year.
- Most tornado damage is caused by the tremendously strong winds. One commonly used guide to tornado intensity is the Enhanced Fujita intensity scale (EF-scale). A rating on the EF-scale is determined by assessing damage produced by the storm.
- Because severe thunderstorms and tornadoes are small and short-lived phenomena, they are among the most difficult weather features to forecast precisely. When weather conditions favor the formation of tornadoes, a tornado watch is issued. The National Weather Service issues a tornado warning when a tornado has been sighted in an area or is indicated by Doppler radar.

19.6 HURRICANES

Identify areas of hurricane formation on a world map and discuss the conditions that promote hurricane formation. List the three broad categories of hurricane destruction.

KEY TERMS: hurricane, eye wall, eye, tropical depression, tropical storm, Saffir–Simpson hurricane scale, storm surge

- Hurricanes, the greatest storms on Earth, are tropical cyclones with wind speeds in excess of 119 kilometers (74 miles) per hour. These complex tropical disturbances develop over tropical ocean waters and are fueled by the latent heat that is liberated when huge quantities of water vapor condense.

- Hurricanes form most often in late summer, when ocean-surface temperatures reach 27°C (80°F) or higher and thus are able to provide the necessary heat and moisture to the air. Hurricanes diminish in intensity when they move over cool ocean water that cannot supply adequate heat and moisture, move onto land, or reach a location where large-scale flow aloft is unfavorable.

- The Saffir–Simpson scale ranks the relative intensities of hurricanes. A 5 on the scale represents the strongest storm possible, and a 1 indicates the lowest severity. Damage caused by hurricanes is divided into three categories: (1) storm surge, (2) wind damage, and (3) heavy rains and inland flooding.

Q **This image taken from the International Space Station shows the inner portion of Hurricane Igor in September 2010. Identify the eye and the eye wall. In which of these zones are winds and rainfall most intense?**

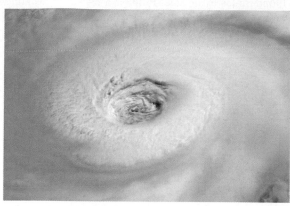

NASA

GIVE IT SOME **THOUGHT**

1. Refer to Figure 19.4 to answer these questions:
   a. Thunder Bay and Marquette are both on the shore of Lake Superior, yet Marquette gets much more snow than Thunder Bay. Why is this the case?
   b. Notice the narrow, north–south zone of relatively heavy snow east of Pittsburgh and Charleston. This region is too far from the Great Lakes to receive lake-effect snowfall. Speculate on a likely reason for the higher snowfalls here. Does your answer explain the shape of this snowy zone?

2. Refer to the accompanying weather map to answer the following questions:
   a. What is a likely wind direction at each city?
   b. Identify the likely air mass that is influencing each city.
   c. Identify the cold front, warm front, and occluded front.
   d. What is the barometric tendency at city A and city C?
   e. Which one of the three cities is probably coldest? Which one is probably warmest?

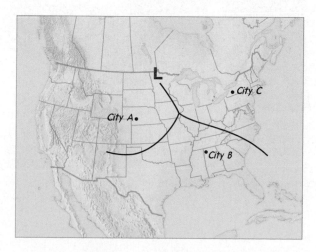

3. Apply your knowledge of fronts to explain the following weather proverb:
   Rain long foretold, long last;
   Short notice, soon past.

4. If you hear that a cyclone is approaching, should you immediately seek shelter? Why or why not?

5. The accompanying diagrams show surface temperatures with isotherms labeled in degrees Fahrenheit for noon and 6 P.M. on January 29, 2008. On this day, a powerful front moved through Missouri and Illinois.
   a. What type of front passed through the Midwest?
   b. Describe how the temperature changed in St. Louis, Missouri, over the 6-hour period.
   c. Describe the likely shift in wind direction in St. Louis during this time span.

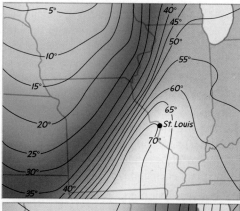

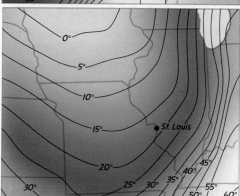

6. If you were located 400 kilometers ahead of the surface position of a typical warm front that had a slope of 1:200, how high would the frontal surface be above you?

7. Assume that after seeing a lightning bolt you heard thunder 10 seconds later. About how far away did the lightning occur?

8. The accompanying table lists the number of tornadoes reported in the United States by decade. Propose a reason to explain why the totals for the 1990s and 2000s are so much higher than for the 1950s and 1960s.

Number of U.S. Tornadoes Reported, by Decade	
Decade	Number of Tornadoes Reported
1950-1959	4796
1960-1969	6613
1970-1979	8579
1980-1989	8196
1990-1999	12,138
2000-2009	12,914

9. The number of tornado deaths in the United States in the 2000s was less than 40 percent the number that occurred in the 1950s, even though there was a significant increase in population. Suggest a likely reason for the decline in the death toll.

10. A television meteorologist is able to inform viewers about the intensity of an approaching hurricane. However, the meteorologist can report the intensity of a tornado only after it has occurred. Why is this the case?

11. Refer to the graph in Figure 19.30. Explain why wind speeds are greatest when the slope of the pressure curve is steepest.

12. Assume that it is late September 2016, and that the eye of Hurricane Gaston, a category 5 storm, is projected to follow the path shown on the accompanying map of Texas. Answer the following questions:

  a. Name the stages of development that Gaston must have gone through to become a hurricane. At what stage did the storm receive its name?

  b. If the storm follows the projected path, will the city of Houston experience Gaston's fastest winds and greatest storm surge? Explain why or why not.

  c. What is the greatest threat to life and property if this storm approaches the Dallas–Fort Worth area?

EXAMINING THE **EARTH SYSTEM**

1. This image shows the effects of a major snowstorm that dropped nearly 2 meters (7 feet) of snow on Buffalo, New York, in December 2001. This weather event was unrelated to a midlatitude cyclone. Places not far from Buffalo received only modest amounts of snow or no snow at all. Which spheres of the Earth system interacted in the Great Lakes region to produce this snowstorm? What term is applied to heavy snows such as this?

2. The situations described below involve interactions between the atmosphere and Earth's surface. In each case indicate whether the air mass is being made more stable or more unstable. Briefly explain each choice.

  a. An mT air mass moving northward from the Gulf of Mexico over the southeastern United States in winter.

  b. An mT air mass from the Gulf of Mexico moving northward over the southeastern United States in summer.

  c. A wintertime cP air mass from Siberia moving eastward from Asia across the North Pacific.

AP Photo/David Duprey

3. This world map shows the tracks and intensities of thousands of hurricanes and other tropical cyclones. It was put together by the National Hurricane Center and the Joint Typhoon Warning Center.

   a. What area has experienced the greatest number of category 4 and 5 storms?

   b. Why do hurricanes not form in the very heart of the tropics, astride the equator?

   c. Explain the absence of storms in the South Atlantic and the eastern South Pacific.

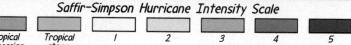

Saffir–Simpson Hurricane Intensity Scale

Tropical depression | Tropical storm | 1 | 2 | 3 | 4 | 5

4. This satellite image shows Tropical Cyclone Favia as it came ashore along the coast of Mozambique, Africa, on February 22, 2007. This powerful storm was moving from east to west. Portions of the storm had sustained winds of 203 kilometers (126 miles) per hour as it made landfall. Letters A–D relate to Question c.

   a. Identify the eye and the eye wall of the cyclone.

   b. Based on wind speed, classify the storm using the Saffir–Simpson hurricane scale.

   c. Which one of the lettered sites should experience the strongest storm surge? Explain.

   d. Describe the possible effects of the storm on coastal lands (geosphere), drainage networks (hydrosphere), and plant and animal life (biosphere).

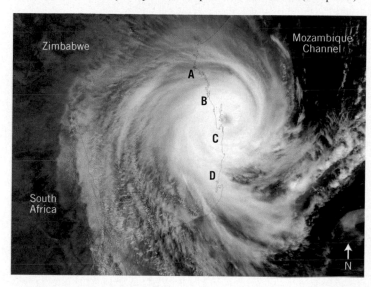

MasteringGeology™

Looking for additional review and test prep materials? Visit the Self Study area in **www.masteringgeology.com** to find practice quizzes, study tools, and multimedia that will aid in your understanding of this chapter's content. In **MasteringGeology™** you will find:

- GEODe: Earth Science: An interactive visual walkthrough of key concepts
- Geoscience Animation Library: More than 100 animations illuminating many difficult-to-understand Earth science concepts

- In The News RSS Feeds: Current Earth science events and news articles are pulled into the site with assessment
- Pearson eText
- Optional Self Study Quizzes
- Web Links
- Glossary
- Flashcards

20

World Climates and Global Climate Change

Mauna Loa Observatory is an important atmospheric research facility on Hawaii's Big Island. It has been collecting data and monitoring atmospheric change since the 1950s. (Photo by Forrest M. Mimms III)

The focus of this chapter is *climate*, the long-term aggregate of weather. Climate is more than just an expression of average atmospheric conditions. In order to accurately portray the character of a place or an area, variations and extremes must also be included. Climate strongly influences the nature of plant and animal life, the soil, and many external geologic processes. Climate influences people as well.

Although climate has a significant impact on people, we are learning that people also have a strong influence on climate. In fact, today global climate change caused by humans is an important global environmental issue. Unlike changes in the geologic past, which represented natural variations, modern climate change is dominated by human influences that are sufficiently large that they exceed the bounds of natural variability. Moreover, these changes are likely to continue for many centuries. The effects of this venture into the unknown with climate could be very disruptive not only to humans but to many other life-forms as well.

20.1 | THE CLIMATE SYSTEM

List the five parts of the climate system and provide examples of each.

Throughout this book, you have been reminded that Earth is a multidimensional system that consists of many interacting parts. A change in any one part can produce changes in any or all of the other parts—often in ways that are neither obvious nor immediately apparent. This fact is certainly true when it comes to the study of climate and climate change.

To understand and appreciate climate, it is important to realize that climate involves more than just the atmosphere. Indeed, we must recognize that there is a **climate system** that includes the atmosphere, hydrosphere, geosphere, biosphere, and cryosphere. (The *cryosphere* refers to the ice and snow that exist at Earth's surface.) The climate system *involves the exchanges of energy and moisture that occur among the five spheres.* These exchanges link the atmosphere to the other spheres so that the whole functions as an extremely complex, interactive unit. Changes to the climate system do not occur in isolation. Rather, when one part of it changes, the other components also react. The major components of the climate system are shown in **FIGURE 20.1**.

FIGURE 20.1 Earth's Climate System Schematic view showing several components of Earth's climate system. Many interactions occur among the various components on a wide range of space and time scales, making the system extremely complex.

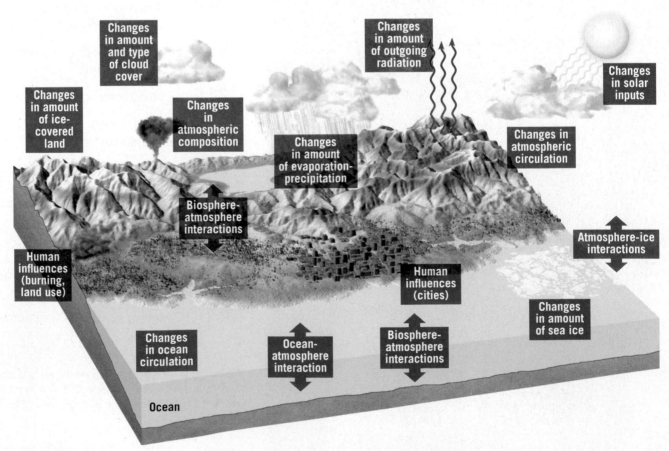

Climate has a profound impact on many of Earth's external processes. When climate changes, these processes respond. A glance back at the rock cycle in Chapter 3 (page 61) reminds us about many of the connections. Of course, rock weathering has an obvious climate connection, as do processes that operate in arid, tropical, and glacial landscapes. Phenomena such as debris flows and river flooding are often triggered by atmospheric events such as periods of extraordinary rainfall. Clearly, the atmosphere is a basic link in the hydrologic cycle. Other connections involve the impact of internal processes on the atmosphere. For example, the particles and gases emitted by volcanoes can change the composition of the atmosphere, and mountain building can have a significant impact on regional temperature, precipitation, and wind patterns.

The study of sediments, sedimentary rocks, and fossils clearly demonstrates that, through the ages, practically every place on our planet has experienced wide swings in climate, from ice ages to conditions associated with subtropical coal swamps or desert dunes. Chapter 12 reinforces this fact. Time scales for climate change vary from decades to millions of years.

20.1 CONCEPT CHECKS

1 What are the five major parts of the climate system?
2 List at least five connections between climate and Earth's external and internal processes.

20.2 | WORLD CLIMATES Explain why classification is a necessary process when studying world climates. Discuss the criteria used in the Köppen system of climate classification.

Previous chapters have already presented the spatial and seasonal variations of the major elements of weather and climate. Chapter 16 examined the controls of temperature and the world distribution of temperature. In Chapter 18, you studied the general circulation of the atmosphere and the global distribution of precipitation. You are now ready to investigate the *combined* effects of these variations in different parts of the world. The varied nature of Earth's surface and the many interactions that occur among atmospheric

processes give every location on our planet a distinctive, even unique, climate. However, we are not going to describe the unique climatic character of countless different locales. Instead, we introduce the major climate regions of the world. The discussion in this chapter examines large areas and uses particular places only to illustrate the characteristics of these major climate regions.

Temperature and precipitation are the most important elements in a climate description because they have the

EYE ON EARTH

This image shows a portion of Alaska's Delta Range, a mountainous area southeast of Fairbanks. (Photo by Michael Collier)

QUESTION 1 *What are the five major parts of the climate system?.*

QUESTION 2 *Which of the five parts are represented in this photo?*

QUESTION 3 *Speculate on how climate change in the coming decades might cause change to the area shown here.*

greatest influence on people and their activities, and they also have an important impact on the distribution of such phenomena as vegetation and soils. Nevertheless, other factors are also important for a complete climatic description. When possible, some of these factors are introduced into our discussion of world climates.

Climate Classification

The worldwide distribution of temperature, precipitation, pressure, and wind is, to say the least, complex. Because of the many differences from place to place and time to time, it is unlikely that any two places that are more than a very short distance apart can experience identical weather. The virtually infinite variety of places on Earth makes it apparent that the number of different climates must be extremely large. Having such a diversity of information to investigate is not unique to the study of the atmosphere. It is a problem basic to all science. (Consider astronomy, which deals with billions of stars, and biology, which studies millions of complex organisms.) To cope with such variety, we must devise some means of *classifying* the vast array of data to be studied. By establishing groups of items that have common characteristics, order and manageability are introduced. Bringing order to large quantities of information not only aids comprehension and understanding but also facilitates analysis and explanation.

One of the first attempts at climate classification was made by the ancient Greeks, who divided each hemisphere into three zones: *torrid*, *temperate*, and *frigid* (**FIGURE 20.2**).

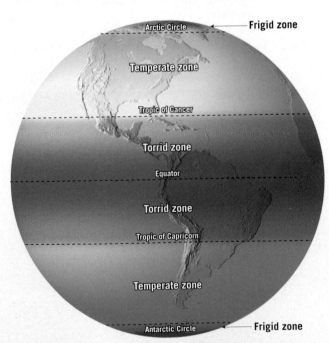

FIGURE 20.2 An Early Climate Classification Among the first attempts at climate classification was one made by the ancient Greeks. They divided each hemisphere into three zones. The winterless *torrid* zone was separated from the summerless *frigid* zone by the *temperate* zone, which had features of the other two.

The basis of this simple scheme was Earth–Sun relationships. The boundaries were the four astronomically important parallels of latitude: the Tropic of Cancer (23.5° north), the Tropic of Capricorn (23.5° south), the Arctic Circle (66.5° north), and the Antarctic Circle (66.5° south). Thus, the globe was divided into winterless climates and summerless climates and an intermediate type that had features of the other two.

Few other attempts were made until the beginning of the twentieth century. Since then, many climate-classification schemes have been devised. Remember that the classification of climates (or of anything else) is not a natural phenomenon but the product of human ingenuity. The value of any particular classification system is determined largely by its *intended use*. A system designed for one purpose may not work well for another.

The Köppen Classification

In this chapter, we use a classification devised by Russian-born German climatologist Wladimir Köppen (1846–1940). As a tool for presenting the general world pattern of climates, the **Köppen classification** has been the best-known and most used system for decades. It is widely accepted for many reasons. For example, it uses only easily obtained data: mean monthly and annual values of temperature and precipitation. Furthermore, the criteria are unambiguous, are relatively simple to apply, and divide the world into climate regions in a realistic way.

Köppen believed that the distribution of natural vegetation is an excellent expression of the totality of climate. Consequently, the boundaries he chose were largely based on the limits of certain plant associations. Five principal groups were recognized, and each group was designated by a capital letter, as follows:

A. **Humid tropical.** Winterless climates; all months have a mean temperature above 18°C (64°F).

B. **Dry.** Climates where evaporation exceeds precipitation; there is a constant water deficiency.

C. **Humid middle-latitude, mild winters.** The average temperature of the coldest month is below 18°C (64°F) but above –3°C (27°F).

D. **Humid middle-latitude, severe winters.** The average temperature of the coldest month is below –3°C (27°F), and the warmest monthly mean exceeds 10°C (50°F).

E. **Polar.** Summerless climates; the average temperature of the warmest month is below 10°C (50°F).

Notice that four of the major groups (A, C, D, and E) are defined on the basis of temperature characteristics, and the fifth, the B group, has precipitation as its primary criterion. Each of the five groups is further subdivided by using the criteria and symbols presented in **FIGURE 20.3**.

Letter Symbol 1st	2nd	3rd		
A			Average temperature of the coldest month is 18°C or higher.	
	f		Every month has 6 cm of precipitation or more.	
	m		Short dry season; precipitation in driest month less than 6 cm but equal to or greater than 10 − R/25 (R is annual rainfall in cm).	
	w		Well-defined winter dry season; precipitation in driest month less than 10 − R/25.	
	s		Well-defined summer dry season (rare).	
B			Potential evaporation exceeds precipitation. The dry–humid boundary is defined by the following formulas: (Note: R is the average annual precipitation in cm, and T is the average annual temperature in °C.) $R < 2T + 28$ when 70% or more of rain falls in warmer 6 months. $R < 2T$ when 70% or more of rain falls in cooler 6 months. $R < 2T + 14$ when neither half year has 70% or more of rain.	
	S		Steppe The BS–BW boundary is 1/2 the dry–humid boundary.	
	W		Desert	
		h	Average annual temperature is 18°C or greater.	
		k	Average annual temperature is less than 18°C.	
C			Average temperature of the coldest month is under 18°C and above −3°C.	
	w		At least 10 times as much precipitation in a summer month as in the driest winter month.	
	s		At least three times as much precipitation in a winter month as in the driest summer month; precipitation in driest summer month less than 4 cm.	
	f		Criteria for w and s cannot be met.	
		a	Warmest month is over 22°C; at least 4 months over 10°C.	
		b	No month above 22°C; at least 4 months over 10°C.	
		c	One to 3 months above 10°C.	
D			Average temperature of coldest month is −3°C or below; average temperature of warmest month is greater than 10°C.	
	w		Same as under C.	
	s		Same as under C.	
	f		Same as under C.	
		a	Same as under C.	
		b	Same as under C.	
		c	Same as under C.	
		d	Average temperature of the coldest month is −3°C or below.	
E			Average temperature of the warmest month is below 10°C.	
	T		Average temperature of the warmest month is greater than 0°C and less than 10°C.	
	F		Average temperature of the warmest month is 0°C or below.	

FIGURE 20.3 The Köppen System of Climate Classification This system uses easily obtained data: mean monthly and annual values of temperature and precipitation. When using this figure to classify climate data, first determine whether the data meet the criteria for the E climates. If the station is not a polar climate, proceed to the criteria for B climates. If the data do not fit into either the E or B groups, check the data against the criteria for A, C, and D climates, in that order. (Photos, in order from A–E, are by Michael Collier, Marek Zak/Alamy Images, Ed Reschke/Getty Images, Michael Collier, J.G. Paren/Science Source)

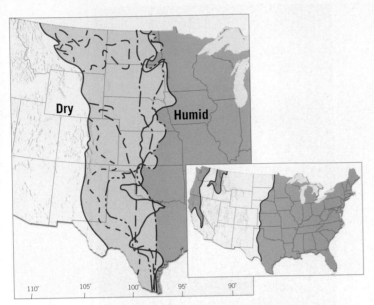

FIGURE 20.4 Conditions Change from Year to Year Yearly fluctuations in the dry–humid boundary during a 5-year period. The small inset shows the average position of the dry–humid boundary.

A strength of the Köppen system is the relative ease with which boundaries are determined. However, these boundaries cannot be viewed as fixed. On the contrary, all climate boundaries shift from year to year (**FIGURE 20.4**). The boundaries shown on climate maps are simply average locations based on data collected over many years. Thus, a climate boundary should be regarded as a broad transition zone and not a sharp line.

The world distribution of climates according to the Köppen classification is shown in **FIGURE 20.5**. You will refer to this map several times as Earth's climates are discussed in the following pages.

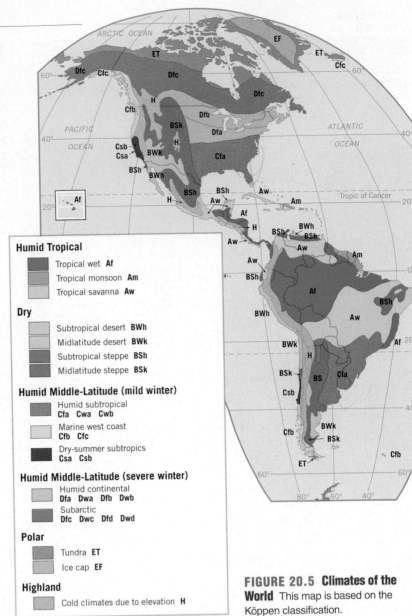

Humid Tropical

- Tropical wet **Af**
- Tropical monsoon **Am**
- Tropical savanna **Aw**

Dry

- Subtropical desert **BWh**
- Midlatitude desert **BWk**
- Subtropical steppe **BSh**
- Midlatitude steppe **BSk**

Humid Middle-Latitude (mild winter)

- Humid subtropical **Cfa Cwa Cwb**
- Marine west coast **Cfb Cfc**
- Dry-summer subtropics **Csa Csb**

Humid Middle-Latitude (severe winter)

- Humid continental **Dfa Dwa Dfb Dwb**
- Subarctic **Dfc Dwc Dfd Dwd**

Polar

- Tundra **ET**
- Ice cap **EF**

Highland

- Cold climates due to elevation **H**

FIGURE 20.5 Climates of the World This map is based on the Köppen classification.

20.2 CONCEPT CHECKS

1 Why is classification often a necessary task in science?

2 What climate data are needed to classify a climate using the Köppen system?

3 Should climate boundaries, such as those shown on the world map in Figure 20.5, be regarded as fixed? Explain.

20.3 | HUMID TROPICAL (A) CLIMATES

Compare the two broad categories of tropical climates.

Within the A group of climates, two main types are recognized: wet tropical climates (Af and Am) and tropical wet and dry (Aw).

The Wet Tropics

The constantly high temperatures and year-round rainfall in the wet tropics combine to produce the most luxuriant vegetation found in any climatic realm: the **tropical rain forest** (**FIGURE 20.6**).

The environment of the wet tropics characterizes almost 10 percent of Earth's land area. An examination of Figure 20.5 shows that Af and Am climates form a discontinuous belt astride the equator that typically extends 5° to 10° into each hemisphere. The poleward margins are most often marked by diminishing rainfall, but occasionally decreasing

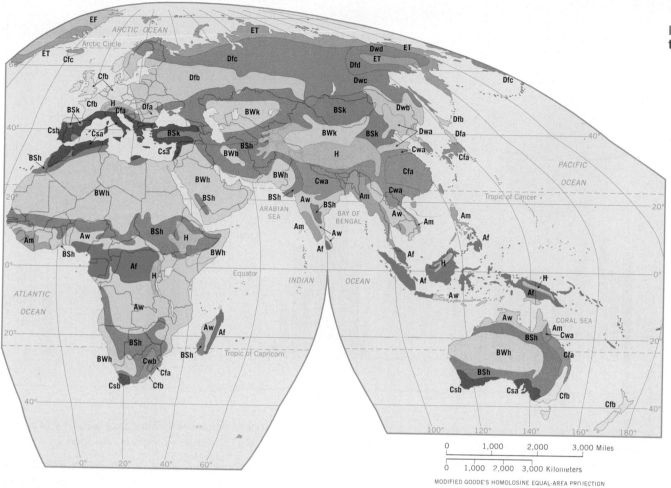

FIGURE 20.5 Climates of the World (continued)

FIGURE 20.6 Tropical Rain Forest Unexcelled in luxuriance and characterized by hundreds of different species per square kilometer, the tropical rain forest is a broadleaf evergreen forest that dominates the wet tropics. This image shows Borneo's Segama River passing through virgin tropical rain forest. (Photo by Peter Lilja/AGE Fotostock)

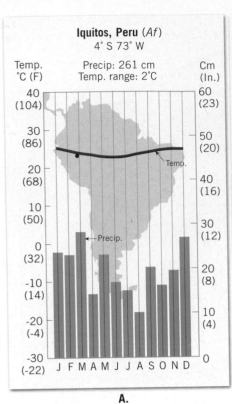

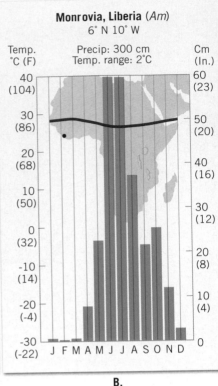

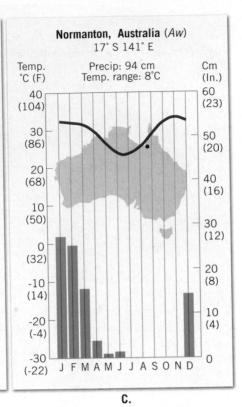

Iquitos, Peru (*Af*)
4° S 73° W
Precip: 261 cm
Temp. range: 2°C

A.

Monrovia, Liberia (*Am*)
6° N 10° W
Precip: 300 cm
Temp. range: 2°C

B.

Normanton, Australia (*Aw*)
17° S 141° E
Precip: 94 cm
Temp. range: 8°C

C.

temperatures mark the boundary. Because of the general decrease in temperature with height in the troposphere, this climate region is restricted to elevations below 1000 meters (nearly 3300 feet). Consequently, the major interruptions near the equator are principally cooler highland areas.

Data for some representative stations in the wet tropics are shown in **FIGURE 20.7A,B**. A brief examination reveals the most obvious features that characterize the climate in these areas:

■ Temperatures usually average 25°C (77°F) or more each month. Consequently, not only is the annual mean temperature high, but the annual temperature range is very small.

■ The total precipitation for the year is high, often exceeding 200 centimeters (80 inches).

■ Although rainfall is not evenly distributed throughout the year, tropical rain forest stations are generally wet in all months. If a dry season exists, it is very short.

Because places with an Af or Am designation lie near the equator, the reason for the uniform temperature rhythm experienced in such locales is clear: The intensity of solar radiation is consistently high. The vertical rays of the Sun are always relatively close, and changes in the length of daylight throughout the year are slight; therefore, seasonal temperature variations are minimal.

The region is strongly influenced by the equatorial low. Its converging trade winds and the accompanying ascent of warm, humid, unstable air produce conditions that are ideal for the formation of precipitation.

Tropical Wet and Dry

In the latitude zone poleward of the wet tropics and equatorward of the subtropical deserts lies the transitional **tropical wet and dry climate**. Here the rain forest gives way to the *savanna*, a tropical grassland with scattered drought-tolerant trees (**FIGURE 20.8**). Because temperature characteristics among all A climates are quite similar, the primary factor that distinguishes the Aw climate from Af and Am is

FIGURE 20.8 Tropical Savanna Grassland This tropical savanna in Tanzania's Serengeti National Park, with its stunted, drought-resistant trees, was probably strongly influenced by seasonal burnings carried out by native human populations. (Photo by Kondrachov Vladimir/Shutterstock)

precipitation. Although the overall amount of precipitation in the tropical wet and dry realm is often considerably less than in the wet tropics, the most distinctive feature of this climate is not the annual rainfall total but the markedly seasonal character of the rainfall. The climate diagram for Normanton, Australia (**FIGURE 20.7C**), clearly illustrates this trait. As the equatorial low advances poleward in summer, the rainy season commences and features weather patterns typical of the wet tropics. Later, with the retreat of the equatorial low, the subtropical high advances into the region and brings with it pronounced dryness. In some Aw regions, such as India, Southeast Asia, and portions of Australia, the alternating periods of rainfall and dryness are associated with a well-established monsoon circulation (see Chapter 18).

20.3 CONCEPT CHECKS

1 What is the main factor that distinguishes Aw climates from Af and Am? How is this difference reflected in the vegetation of these climate regions?

2 How do the equatorial low and the subtropical high influence the seasonal distribution of rainfall in the Aw climate?

20.4 | DRY (B) CLIMATES

Contrast low-latitude dry climates and middle-latitude dry climates.

It is important to realize that the concept of dryness is a relative one and refers to any situation in which a water deficiency exists. Climatologists define a dry climate as one in which the yearly precipitation is not as great as the potential loss of water by evaporation. Thus, dryness is not only related to annual rainfall totals but is also a function of evaporation, which in turn is closely dependent on temperature.

To establish the boundary between dry and humid climates, the Köppen classification uses formulas that involve three variables: average annual precipitation, average annual temperature, and seasonal distribution of precipitation. The use of average annual temperature reflects its importance as an index of evaporation. The amount of rainfall defining the humid–dry boundary increases as the annual mean temperature increases. The use of seasonal precipitation as a variable is also related to this idea. If rain is concentrated in the warmest months, loss to evaporation is greater than if the precipitation is concentrated in the cooler months.

Within the regions defined by a general water deficiency are two climatic types: **arid**, or **desert** (BW), and **semiarid**, or **steppe** (BS). These two groups have many features in common; their differences are primarily a matter of degree. The semiarid is a marginal and more humid variant of the arid climate type and represents a transition zone that surrounds the desert and separates it from the bordering humid climates (see Figure 6.30, page 193).

Low-Latitude Deserts and Steppes

The heart of low-latitude dry climates lies in the vicinities of the Tropics of Cancer and Capricorn. A glance at Figure 20.5 shows a virtually unbroken desert environment stretching for more than 9300 kilometers (nearly 6000 miles) from the Atlantic coast of North Africa to the dry lands of northwestern India. In addition to this single great expanse, the Northern Hemisphere contains another, much smaller area of subtropical desert and steppe in northern Mexico and the southwestern United States. In the Southern Hemisphere, dry climates dominate Australia. Almost 40 percent of the continent is desert, and much of the remainder is steppe. In addition, arid and semiarid areas are found in southern Africa and make a limited appearance in coastal Chile and Peru.

The existence of this dry subtropical realm is primarily the result of the prevailing global distribution of air pressure and winds. Earth's low-latitude deserts and steppes coincide with the subtropical high-pressure belts where air is subsiding (see Figure 18.17, page 561). When air sinks, it is compressed and warmed. Such conditions are opposite of what is needed for cloud formation and precipitation. Therefore, clear skies, a maximum of sunshine, and dryness are to be expected. The classic view of Africa and the Arabian Peninsula from space in **FIGURE 20.9** reinforces this idea. The climate diagrams for Cairo, Egypt, and Monterrey, Mexico (**FIGURE 20.10A,B**), illustrate the characteristics of low-latitude dry climates.

FIGURE 20.9 Deserts of Africa and the Arabian Peninsula In this view of Earth from space, North Africa's Sahara Desert, the adjacent Arabian Desert, and the Kalahari and Namib Deserts in southern Africa are clearly visible as tan-colored, cloud-free zones. These low-latitude deserts are dominated by the dry, subsiding air associated with pressure belts known as the *subtropical highs*. By contrast, the band of clouds that extends across central Africa and the adjacent oceans coincides with the equatorial low-pressure belt, the rainiest region on Earth. (NASA)

FIGURE 20.10
Climate Diagrams for Arid and Semiarid Stations Stations A. and B. are in the subtropics, whereas C. is in the middle latitudes. Cairo and Lovelock are classified as deserts; Monterrey is a steppe. Lovelock, Nevada, may also be called a rain shadow desert.

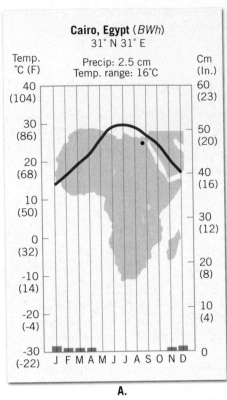

Cairo, Egypt (*BWh*)
31° N 31° E
Precip: 2.5 cm
Temp. range: 16°C

A.

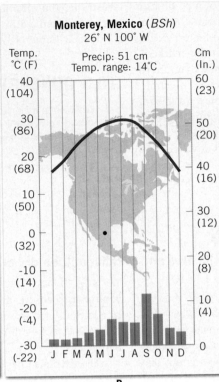

Monterey, Mexico (*BSh*)
26° N 100° W
Precip: 51 cm
Temp. range: 14°C

B.

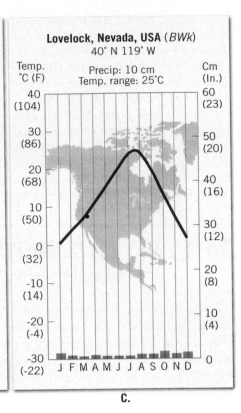

Lovelock, Nevada, USA (*BWk*)
40° N 119° W
Precip: 10 cm
Temp. range: 25°C

C.

Middle-Latitude Deserts and Steppes

Unlike their low-latitude counterparts, middle-latitude deserts and steppes are not controlled by the subsiding air masses associated with high pressure. Instead, these dry lands exist principally because of their positions in the deep interiors of large landmasses far removed from the oceans, which are the ultimate source of moisture for cloud formation and precipitation. In addition, the presence of high mountains across the paths of prevailing winds further acts to separate these areas from water-bearing maritime air masses.

Windward sides of mountains are often wet. As prevailing winds meet mountain barriers, the air is forced to ascend, producing clouds and precipitation. By contrast, the leeward sides of mountains are usually much dryer and are often arid enough to be referred to as rain shadow deserts (see Figure 17.11, page 526). Because

EYE ON EARTH

This classic view of Earth from space was taken in December 1968 by an *Apollo 8* astronaut. The image shows the Western Hemisphere. Dense clouds cover much of North America. A large portion of South America is also cloud covered. However, the western margin of South America is cloud free. As you answer the following questions, you will find it helpful to consult Figure 6.30 (page 193) and Figures 15.2 and 15.3 (pages 455–456) and the discussions that relate to these figures.

QUESTION 1 *What is the name of the desert that occupies the cloudless area along the western margin of South America?*

QUESTION 2 *An ocean current flows just offshore of this desert area. Is it warm or cold? Toward what direction must the current flow—toward or away from the equator?*

QUESTION 3 *How does this desert's close proximity to the Pacific Ocean influence its aridity?*

NASA

many middle-latitude deserts occupy sites on the leeward sides of the mountains, they can also be classified as rain shadow deserts (**FIGURE 20.10C**). In North America, the Coast Ranges, Sierra Nevada, and Cascades are the foremost mountain barriers. In Asia, the great Himalayan chain prevents the summertime monsoon flow of moist Indian Ocean air from reaching the interior. Because the Southern Hemisphere lacks extensive land areas in the middle latitudes, only a small area of desert and steppe is found in this latitude range, existing primarily in the rain shadow of the towering Andes.

In the case of middle-latitude deserts, we have an example of the impact of tectonic processes on climate.

Rain shadow deserts exist because of the mountains produced when plates collide. Without such mountain-building episodes, wetter climates would prevail where many dry regions exist today.

20.4 CONCEPT CHECKS

1 Why is the amount of precipitation that defines the boundary between humid and dry climates variable?

2 What is the primary reason (control) for the existence of the dry subtropical realm (BWh and BSh)?

3 What factors contribute to the existence of middle-latitude deserts and steppes?

20.5 | HUMID MIDDLE-LATITUDE CLIMATES (C AND D CLIMATES)

Distinguish among five different humid middle-latitude climates.

The humid middle-latitude climates dominate large portions of North America, Europe, and Asia. Humid middle-latitude climates are divided into two categories: C climates, which have mild winters, and D climates, which have severe winters.

Humid Middle-Latitude Climates with Mild Winters (C Climates)

Although the term *subtropical* is often used for the C climates, it can be misleading. Although many areas with C climates do indeed possess some near-tropical characteristics, other regions do not. For example, we would be stretching

the use of the term *subtropical* to describe the climates of coastal Alaska and Norway, which belong to the C group. Within the C group of climates, several subgroups are recognized.

Humid Subtropics Located on the eastern sides of the continents, in the 25° to 40° latitude range, the **humid subtropical climate** dominates the southeastern United States, as well as other similarly situated areas around the world (see Figure 20.5). The climate diagram for Guangzhou, China (**FIGURE 20.11A**), shows a typical example. In summer, the humid subtropics experience hot, sultry weather of the type one expects to find in the rainy tropics. Daytime temperatures

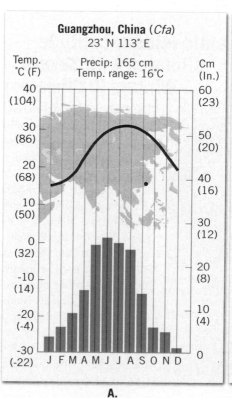

A.

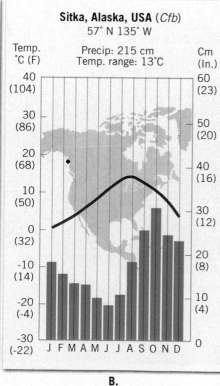

B.

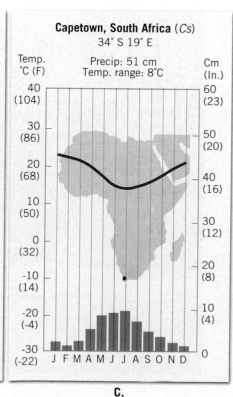

C.

FIGURE 20.11 Examples of C Climates A. humid subtropical, **B.** marine west coast, and **C.** dry-summer subtropical.

are generally high, and because both mixing ratio and relative humidity are high, the night brings little relief. An afternoon or evening thunderstorm is also possible, for these areas experience such storms on an average of 40–100 days each year, the majority during the summer months.

As summer turns to autumn, the humid subtropics lose their similarity to the rainy tropics. Although winters are mild, frosts are common in the higher-latitude Cfa areas and occasionally plague the tropical margins as well. The winter precipitation is also different in character from the summer. Some is in the form of snow, and most is generated along fronts of the frequent middle-latitude cyclones that sweep over these regions.

Marine West Coast

Situated on the western (windward) side of continents, from about 40° to 65° north and south latitude, is a climate region dominated by the onshore flow of oceanic air (**FIGURE 20.12**). In North America, the **marine west coast climate** extends from near the U.S.–Canadian border northward as a narrow belt into southern Alaska (**FIGURE 20.11B**). The largest area of Cfb climate is found in Europe because there are no mountain barriers blocking the movement of cool maritime air from the North Atlantic.

The prevalence of maritime air masses means that mild winters and cool summers are the rule, as is an ample amount of rainfall throughout the year. Although there is no pronounced dry period, there is a drop in monthly precipitation totals during the summer. The reason for the reduced summer rainfall is the poleward migration of the oceanic subtropical highs. Although the areas of marine west

coast climate are situated too far poleward to be dominated by these dry anticyclones, their influence is sufficient to cause a decrease in warm season rainfall.

Dry-Summer Subtropics

The **dry-summer subtropical climate** is typically located along the west sides of continents between latitudes 30° and 45°. Situated between the marine west coast climate on the poleward side and the subtropical steppes on the equatorward side, this climate is best described as transitional in character. It is unique because it is the only humid climate that has a strong winter rainfall maximum, a feature that reflects its intermediate position (**FIGURE 20.11C**). In summer, the region is dominated by stable conditions associated with the oceanic subtropical highs. In winter, as the wind and pressure systems follow the Sun equatorward, the region is within range of the cyclonic storms of the polar front. Thus, during the course of a year, these areas alternate between becoming a part of the dry subtropics and an extension of the humid middle latitudes. Whereas middle-latitude changeability characterizes the winter, subtropical constancy describes the summer.

As is the case for the marine west coast climate, mountain ranges limit the dry-summer subtropics to a relatively narrow coastal zone in both North and South America. Because Australia and southern Africa barely reach to the latitudes where dry-summer climates exist, the development of this climatic type is limited on those continents as well. Consequently, because of the arrangement of the continents and their mountain ranges, inland development occurs only in the Mediterranean basin. Here the zone of subsidence extends far to the east in summer; in winter, the sea is a major route of cyclonic disturbances. Because the dry-summer climate is particularly extensive in this region, the name *Mediterranean climate* is often used as a synonym.

FIGURE 20.12 Marine West Coast Climate Fog is common along the rocky Pacific coastline at Olympic National Park, Washington. As the name of this climate implies, the ocean exerts a strong influence. (Photo by Richard J. Green/Science Source)

Humid Middle-Latitude Climates with Severe Winters (D Climates)

The C climates that were just described characteristically have mild winters. By contrast, D climates experience severe winters. Two types of D climates are recognized: the humid continental and the subarctic climates. Climatic diagrams of representative locations are shown in **FIGURE 20.13**. The D climates are land-controlled climates, the result of broad continents in the middle latitudes. Because continentality is a basic feature, D climates are absent in the Southern

Hemisphere, where the middle-latitude zone is dominated by the oceans.

Humid Continental

The **humid continental climate** is confined to the central and eastern portions of North America and Eurasia in the latitude range between approximately 40° and 50° north latitude. It may at first seem unusual that a continental climate should extend eastward to the margins of the ocean. However, because the prevailing atmospheric circulation is from the west, deep and persistent incursions of maritime air from the east are not likely to occur.

Both winter and summer temperatures in the humid continental climate can be characterized as relatively severe. Consequently, annual temperature ranges are high throughout the climate.

Precipitation is generally greater in summer than in winter. Precipitation totals generally decrease toward the interior of the continents, as well as from south to north, primarily because of increasing distance from the sources of maritime tropical (mT) air. Furthermore, the more northerly stations are also influenced for a greater part of the year by drier polar air masses.

Wintertime precipitation in humid continental climates is chiefly associated with the passage of fronts connected with traveling midlatitude cyclones. Part of this precipitation is in the form of snow, and the proportion of snow increases with latitude. Although precipitation is often considerably less during the cold season, it is usually more conspicuous than the greater amounts that fall during summer. An obvious reason is that snow remains on the ground, often for extended periods.

Subarctic

Situated north of the humid continental climate and south of the polar tundra is an extensive **subarctic climate** region covering broad, uninterrupted

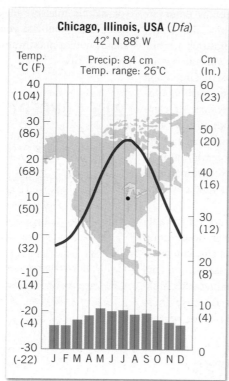

Chicago, Illinois, USA (*Dfa*)
42° N 88° W
Precip: 84 cm
Temp. range: 26°C

A.

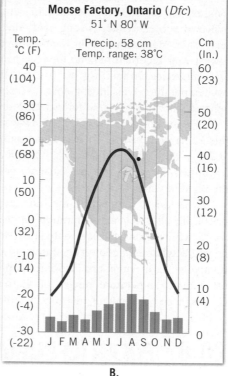

Moose Factory, Ontario (*Dfc*)
51° N 80° W
Precip: 58 cm
Temp. range: 38°C

B.

FIGURE 20.13 Examples of D Climates Climates in this category are associated with the interiors of large landmasses in the mid- to high latitudes of the Northern Hemisphere. Winters can be harsh in Chicago's humid continental (Dfa) climate, and the subarctic environment (Dfc) of Moose Factory is more extreme.

expanses from western Alaska to Newfoundland in North America and from Norway to the Pacific coast of Russia in Eurasia (see Figure 20.5). It is often referred to as the *taiga* climate, for its extent closely corresponds to the northern coniferous forest region of the same name (**FIGURE 20.14**). Although scrawny, the spruce, fir, larch, and birch trees in the taiga represent the largest stretch of continuous forest on the surface of Earth.

FIGURE 20.14 Subarctic Climate The northern coniferous forest, also called the taiga, is associated with subarctic climates. This climate typically experiences the highest annual temperature ranges on Earth. This scene is in Quebec's Gaspésie Peninsula. (Photo by Michael P. Gadomski/Science Source)

Here in the source regions of continental polar air masses, the outstanding feature is certainly the dominance of winter. Winter is long, and temperatures are bitterly cold. Winter minimum temperatures are among the lowest ever recorded outside the ice sheets of Greenland and Antarctica. In fact, for many years, the world's coldest temperature was attributed to Verkhoyansk in east-central Siberia, where the temperature dropped to –68°C (–90°F) on February 5 and 7, 1892. Over a 23-year period, this same station had an average monthly minimum of –62°C (–80°F) during January. Although exceptional temperatures, they illustrate the extreme cold that envelops the taiga in winter.

By contrast, summers in the subarctic are remarkably warm, despite their short duration. However, when compared with regions farther south, this short season must be characterized as cool. The extremely cold winters and relatively warm summers combine to produce the highest annual temperature ranges on Earth. Because these far northerly continental interiors are the source regions

for cP air masses, there is very limited moisture available throughout the year. Precipitation totals are therefore small, with a maximum occurring during the warmer summer months.

20.6 | POLAR (E) CLIMATES

Contrast ice cap and tundra climates.

Polar climates are those in which the mean temperature of the warmest month is below 10°C (50°F). Thus, just as the tropics are defined by their year-round warmth, the polar realm is known for its enduring cold. As winters are periods of perpetual night, or nearly so, temperatures at most polar locations are understandably bitter. During the summer months temperatures remain cool despite the long days, because the Sun is so low in the sky that its oblique rays are not effective in bringing about a genuine warming. Although polar climates are classified as humid, precipitation is generally meager. Evaporation, of course, is also limited. The

scanty precipitation totals are easily understood in view of the temperature characteristics of the region. The amount of water vapor in the air is always small because low mixing ratios must accompany low temperatures. Usually precipitation is most abundant during the warmer summer months, when the moisture content of the air is highest.

Two types of polar climates are recognized. The **tundra climate** (ET) is a treeless climate found almost exclusively in the Northern Hemisphere (**FIGURE 20.15**). Because of the combination of high latitude and continentality, winters are severe, summers are cool, and annual temperature ranges are

FIGURE 20.15 Ice Cap Climate Greenland and Antarctica are the major examples of this extreme climate. In this image, scientists are conducting research on the Greenland ice sheet. (Photo by Dr. Joel Harper)

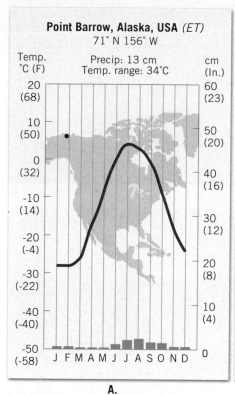

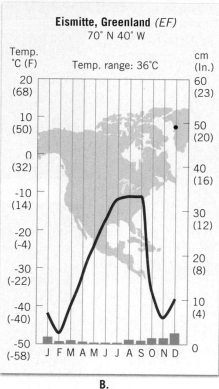

Examples of E Climates These climatic diagrams represent the two basic types of polar climates. **A.** Barrow, Alaska, exhibits a tundra (ET) climate. **B.** Eismitte, Greenland, a station located on a massive ice sheet, is classified as an ice cap (EF) climate.

high (**FIGURE 20.16A**). Furthermore, yearly precipitation is small, with a modest summer maximum.

The **ice cap climate** (EF) does not have a single monthly mean above 0°C (32°F) (**FIGURE 20.16B**). Consequently, because the average temperature for all months is below freezing, the growth of vegetation is prohibited, and the landscape is one of permanent ice and snow. This climate of perpetual frost covers a surprisingly large area of more than 15.5 million square kilometers (6 million square miles), or about 9 percent of Earth's land area. Aside from scattered occurrences in high mountain areas, it is confined

to the ice sheets of Greenland and Antarctica (see Figure 6.2, page 173).

20.6 CONCEPT CHECKS

1 Although polar regions experience extended periods of sunlight in the summer, temperatures remain cool. Explain.

2 Why are precipitation totals low in polar climates? Which season has the most precipitation? Why?

3 Where are EF climates most extensively developed?

20.7 | HIGHLAND CLIMATES

Summarize the characteristics associated with highland climates.

It is a well-known fact that mountains have climate conditions that are distinctly different from those found in adjacent lowlands. Compared to nearby places at lower elevations, sites with **highland climates** are cooler and usually wetter. Unlike the world climate types already discussed, which consist of large, relatively homogeneous regions, the outstanding characteristic of highland climates is the great diversity of climatic conditions that occur.

The best-known climatic effect of increased altitude is lower temperatures. In addition, an increase in precipitation due to orographic lifting usually occurs at higher elevations. Despite the fact that mountain stations are colder and often wetter than locations at lower elevations,

highland climates are often very similar to those in adjacent lowlands in terms of seasonal temperature cycles and precipitation distribution. **FIGURE 20.17** illustrates this relationship.

Phoenix, at an elevation of 338 meters (1109 feet), lies in the desert lowlands of southern Arizona. By contrast, Flagstaff is located at an altitude of 2100 meters (7000 feet) on the Colorado Plateau in northern Arizona. When summer averages climb to 34°C (93°F) in Phoenix, Flagstaff is experiencing a pleasant 19°C (66°F), which is a full 15°C (27°F) cooler. Although the temperatures at each city are quite different, the pattern of monthly temperature changes for each place is similar. Both experience their minimum and maximum monthly means in the same months. When

SmartFigure 20.17 Highland Climate Diagrams for two stations in Arizona illustrate the general influence of elevation on climate. Flagstaff is cooler and wetter because of its position on the Colorado Plateau, nearly 1800 meters (6000 feet) higher than Phoenix. Only scanty drought-tolerant natural vegetation can survive in the hot, dry climate of southern Arizona, near Phoenix. (Photo by Design Pics/SuperStock) The natural vegetation associated with the cooler, wetter highlands near Flagstaff, Arizona, is much different from the desert lowlands. (Photo by Jim Cole/Alamy)

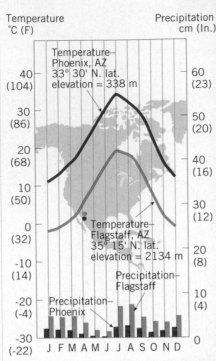

Phoenix, AZ

Flagstaff, AZ

precipitation data are examined, both places have a similar seasonal pattern, but the amounts at Flagstaff are higher in every month. In addition, because of its higher altitude, much of Flagstaff's winter precipitation is in the form of snow. By contrast, all of the precipitation at Phoenix is rain.

Because topographic variations are pronounced in mountains, every change in slope with respect to the Sun's rays produces a different microclimate. In the Northern Hemisphere, south-facing slopes are warmer and drier because they receive more direct sunlight than north-facing slopes and deep valleys. Wind direction and speed in mountains can be highly variable. Mountains create various obstacles to winds. Locally, winds may be funneled through valleys or forced over ridges and around mountain peaks. When weather conditions are fair, mountain and valley breezes are created by the topography itself.[1]

We know that climate strongly influences vegetation, which is the basis for the Köppen system. Thus, where there are vertical differences in climate, we should expect a vertical zonation of vegetation as well. By ascending a mountain, we can view dramatic vegetation changes that

otherwise might require a poleward journey of thousands of kilometers (see Figure 20.17). This occurs because altitude duplicates, in some respects, the influence of latitude on the distribution of vegetation.

Perhaps the terms *variety* and *changeability* best describe mountain climates. Because atmospheric conditions fluctuate rapidly with changes in altitude and exposure, a nearly limitless variety of local climates occur in mountainous regions. The climate in a protected valley is very different from that of an exposed peak. Conditions on windward slopes contrast sharply with those on the leeward sides, whereas slopes facing the Sun are unlike those that lie mainly in the shadows.

20.7 CONCEPT CHECKS

1 The Arizona cities of Flagstaff and Phoenix are relatively close to one another yet have contrasting climates. Briefly describe the differences and why they occur.

---
[1]Mountain and valley breezes are described in the section "Local Winds" in Chapter 18. Also see Figure 18.20, page 564.

20.8 | HUMAN IMPACT ON GLOBAL CLIMATE Summarize the nature and cause of the atmosphere's changing composition since about 1750. Describe the climate's response.

Proposals to explain global climate change are many and varied. In Chapter 6, we examined some possible causes for ice-age climates. These hypotheses, which involved plate tectonics and variations in Earth's orbit, were natural causes. Another natural cause, discussed in Chapter 9, is the possible role of explosive volcanic eruptions in modifying the atmosphere. It is important to remember that these mechanisms, as well as others, were not only responsible for climate change in the geologic past but will also contribute to future shifts in climate.

When relatively recent and future changes in our climate are considered, we must also examine the impact of human beings. In this section, we examine how humans contribute to global climate change. One impact largely results from the addition of carbon dioxide and other greenhouse gases to the atmosphere. A second impact is related to the addition of human-generated aerosols to the atmosphere.

Human influence on regional and global climate did not just begin with the onset of the modern industrial period. There is good evidence that people have been modifying the environment over extensive areas for thousands of years. The use of fire and the overgrazing of marginal lands by domesticated animals have both reduced the abundance and distribution of vegetation. By altering ground cover, humans have modified such important climate factors as surface albedo, evaporation rates, and surface winds.

Rising CO$_2$ Levels

In Chapter 16 you learned that carbon dioxide (CO$_2$) represents only about 0.0397 percent (397 parts per million) of the gases that make up clean, dry air. Nevertheless, it is a very significant component meteorologically. Carbon dioxide is influential because it is transparent to incoming short-wavelength solar radiation, but it is not transparent to some of the longer-wavelength outgoing Earth radiation. A portion of the energy leaving the ground is absorbed by atmospheric CO$_2$. This energy is subsequently re-emitted, part of it back toward the surface, thereby keeping the air near the ground warmer than it would be without CO$_2$. Thus, along with water vapor, carbon dioxide is largely responsible for the *greenhouse effect* of the atmosphere. Carbon dioxide is an important heat absorber, and it follows logically that any change in the air's CO$_2$ content could alter temperatures in the lower atmosphere.

Earth's tremendous industrialization of the past two centuries has been fueled—and still is fueled—by burning fossil fuels: coal, natural gas, and petroleum (**FIGURE 20.18**). Combustion of these fuels has added great quantities of carbon dioxide to the atmosphere. Figure 16.5 (page 489) shows changes in CO$_2$ concentrations at Hawaii's Mauna Loa Observatory, where measurements have been made since 1958. The graph shows an annual seasonal cycle and

EYE ON EARTH

Amundsen–Scott South Pole Station is an American research facility. Among the phenomena that are monitored there are variations in atmospheric composition. The accompanying graph shows changes in the air's CO$_2$ content at South Pole Station (90° south latitude) and at a similar facility at Barrow, Alaska (71° north latitude). (Photo by Vicky Beaver/Alamy)

QUESTION 1 *Describe how the two lines on the graph differ.*

QUESTION 2 *Which line on the graph represents the South Pole, and which represents Barrow, Alaska?*

QUESTION 3 *Explain how you were able to determine which line is which.*

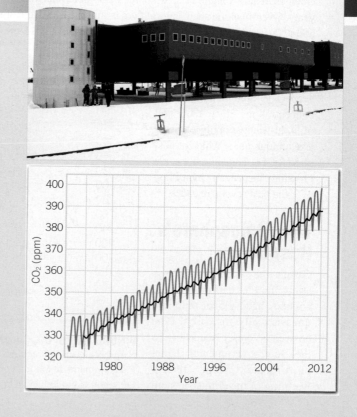

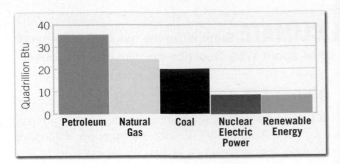

FIGURE 20.18 U.S. Energy Consumption The graph shows energy consumption in 2011. The total was 97.5 quadrillion Btu. A quadrillion is 10 raised to the 12th power, or a million million. The burning of fossil fuels represents slightly more than 82 percent of the total. (Data from U.S. Energy Information Administration)

a steady upward trend over the years. The up-and-down of the seasonal cycle is due to the vast land area of the Northern Hemisphere, which contains the majority of land-based vegetation. During spring and summer in the Northern Hemisphere, when plants are absorbing CO_2 as part of photosynthesis, concentrations decrease. The annual increase in CO_2 during the cold months occurs as vegetation dies and leaves fall and decompose, which releases CO_2 back into the air.

The use of coal and other fuels is the most prominent means by which humans add CO_2 to the atmosphere, but it is not the only way. The clearing of forests also contributes substantially because CO_2 is released as vegetation is burned or decays. Deforestation is particularly pronounced in the tropics, where vast tracts are cleared for ranching and agriculture or are subjected to inefficient commercial logging operations (**FIGURE 20.19**). According to United Nations estimates, nearly 10.2 million hectares (25.1 million acres) of tropical forest were permanently destroyed each year during the 1990s. Between the years 2000 and 2005, the average figure increased to 10.4 million hectares (25.7 million acres) per year.

Some of the excess CO_2 is taken up by plants or is dissolved in the ocean. It is estimated that about 45 percent remains in the atmosphere. **FIGURE 20.20** is a graphic record of changes in atmospheric CO_2 extending back more than 400,000 years. Over this long span, natural fluctuations have varied from about 180 to 300 ppm. As a result of human activities, the present CO_2 level is about 30 percent higher than its highest level over at least the past 650,000 years. The rapid increase in CO_2 concentrations since the onset of industrialization is obvious. The annual rate at which atmospheric CO_2 concentrations are growing has been increasing over the past several decades.

The Atmosphere's Response

Given the increase in the atmosphere's carbon dioxide content, have global temperatures actually increased? The answer is "yes." According to the Intergovernmental Panel on Climate Change (IPCC), warming of the climate system

FIGURE 20.19 Tropical Deforestation Clearing the tropical rain forest is a serious environmental issue. In addition to causing a loss of biodiversity, it is a significant source of carbon dioxide. Fires are frequently used to clear the land. This scene is in Brazil's Amazon basin. (Photo by Pete Oxford/Nature Picture Library)

is unequivocal, as is now evident from observations of increases in global average air and ocean temperatures, widespread melting of snow and ice, and rising global sea level.[2] Most of the observed increase in global average temperatures since the mid-twentieth century is *very likely* due to the observed increase in human-generated greenhouse gas concentrations. (As used by the IPCC, *very likely* indicates a probability of 90–99 percent.) Global warming since the mid-1970s is now about 0.6°C (1°F), and total warming in

[2]IPCC, "Summary for Policy Makers." In *Climate Change 2013: The Physical Science Basis*. The Intergovernmental Panel on Climate Change is an authoritative group that provides advice to the world community through periodic reports that assess the state of knowledge of causes of climate change. More than 259 authors and 1089 scientific reviewers from more than 55 countries contributed to the 2013 report.

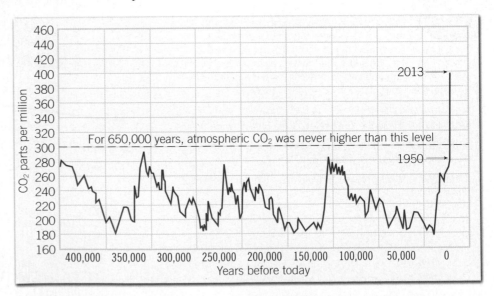

FIGURE 20.20 CO_2 Concentrations over the Past 400,000 Years Most of these data come from the analysis of air bubbles trapped in ice cores. The record since 1958 comes from direct measurements at Mauna Loa Observatory, Hawaii. The rapid increase in CO_2 concentrations since the onset of the Industrial Revolution is obvious. (NOAA)

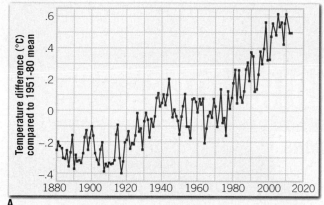

A.

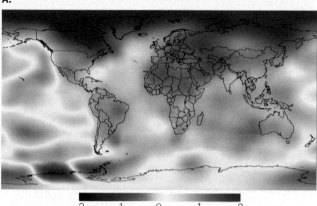

B.   **Temperature difference (°C) compared to 1951–80 mean**

SmartFigure 20.21 Global Temperatures The year 2012 was among the 10 warmest years on record. **A.** The graph depicts global temperature change in degrees Celsius since the year 1880. **B.** The world map shows global temperature differences averaged from 2008 through 2012 compared to the mean for the 1951–1980 base period. The high latitudes in the Northern Hemisphere clearly stand out. (NASA/ Goddard Institute for Space Studies)

the past century is about 0.8°C (1.4°F). The upward trend in surface temperatures is shown in **FIGURE 20.21A**. The world map in **FIGURE 20.21B** compares surface temperatures for 2012 to the base period (1951–1980). You can see that the greatest warming has been in the Arctic and neighboring high-latitude regions. Here are some related facts:

■ When we consider the 132-year span for which there are instrumental records (since 1880), the 10 warmest years have all occurred since 1998.

■ Global mean temperature is now higher than at any time in at least the past 500 to 1000 years.

■ The average temperature of the global ocean has increased to depths of at least 3000 meters (10,000 feet).

Are these temperature trends caused by human activities, or would they have occurred anyway? The scientific consensus of the IPCC is that human activities were *very likely* responsible for most of the temperature increase since 1950.

What about the future? Projections for the years ahead depend in part on the quantities of greenhouse gases that are emitted. **FIGURE 20.22** shows the best estimates of global warming for several different scenarios. The IPCC estimates that if there is a doubling of the preindustrial level of carbon dioxide (280 ppm) to 560 ppm, the "likely" temperature increase will be in the range of 2° to 4.5°C (3.5° to 8.1°F). The increase is "very

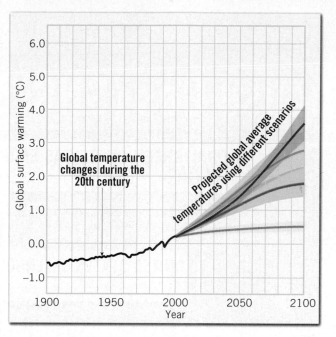

FIGURE 20.22 Temperature Projections to 2100 The right half of the graph shows projected global warming based on different emissions scenarios. The shaded zone adjacent to each colored line shows the uncertainty range for each scenario. The basis for comparison (0.0 on the vertical axis) is the global average for the period 1980 to 1999. The orange line represents the scenario in which CO_2 concentrations were held constant at values for the year 2000. (NOAA)

unlikely" (1 to 10 percent probability) to be less than 1.5°C (2.7°F), and values higher than 4.5°C (8.1°F) are possible.

The Role of Trace Gases

Carbon dioxide is not the only gas that contributes to the global increase in temperature. In recent years, atmospheric scientists have come to realize that the industrial and agricultural activities of people are causing a buildup of several trace gases that also play significant roles. The substances are called *trace gases* because their concentrations are much lower than the concentration of carbon dioxide. The trace gases that are most important are methane (CH_4), nitrous oxide (N_2O), and chlorofluorocarbons (CFCs). These gases absorb wavelengths of outgoing radiation from Earth that would otherwise escape into space (**FIGURE 20.23**). Although individually their impact is

FIGURE 20.23 Methane Methane is produced by anaerobic bacteria in wet places, where oxygen is scarce. (*Anaerobic* means "without air," specifically oxygen.) Such places include swamps, bogs, wetlands, and the guts of termites and grazing animals such as cattle and sheep. Methane is also generated in flooded paddy fields ("artificial swamps") used for growing rice. Coal mining and drilling for oil and natural gas are other sources of methane. (Photo by Robert Harding/ SuperStock)

Greenhouse Gas (GHG) Emissions

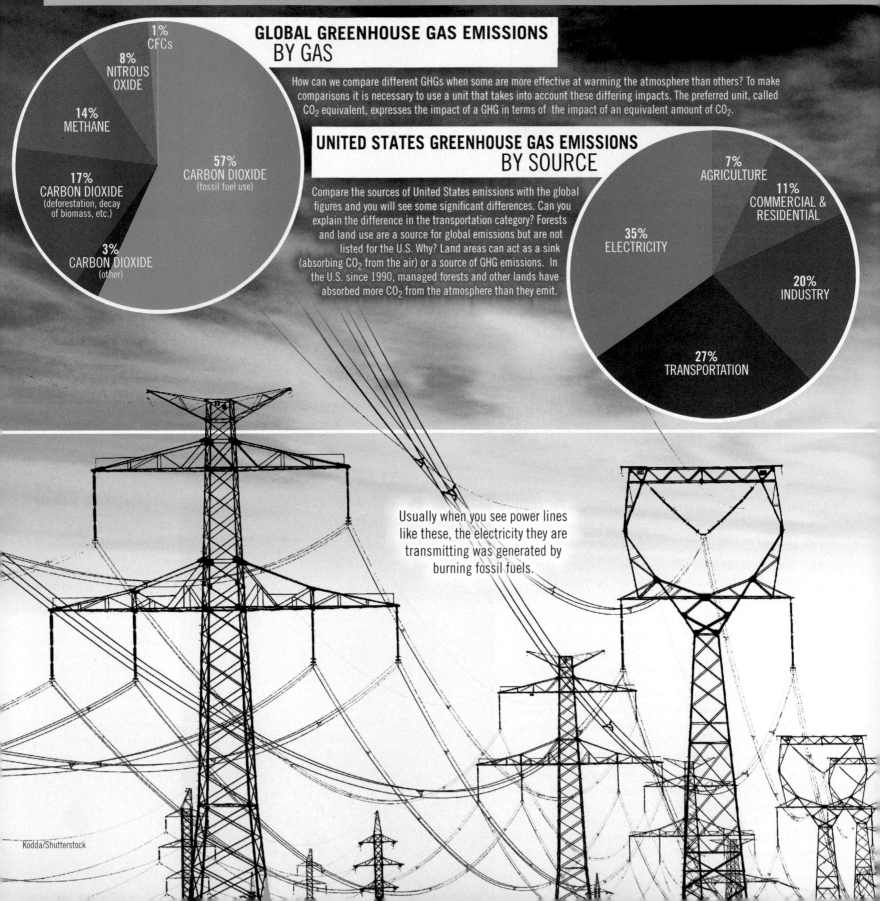

GLOBAL GREENHOUSE GAS EMISSIONS
BY GAS

How can we compare different GHGs when some are more effective at warming the atmosphere than others? To make comparisons it is necessary to use a unit that takes into account these differing impacts. The preferred unit, called CO_2 equivalent, expresses the impact of a GHG in terms of the impact of an equivalent amount of CO_2.

1%
CFCs

8%
NITROUS
OXIDE

14%
METHANE

17%
CARBON DIOXIDE
(deforestation, decay
of biomass, etc.)

3%
CARBON DIOXIDE
(other)

57%
CARBON DIOXIDE
(fossil fuel use)

UNITED STATES GREENHOUSE GAS EMISSIONS
BY SOURCE

Compare the sources of United States emissions with the global figures and you will see some significant differences. Can you explain the difference in the transportation category? Forests and land use are a source for global emissions but are not listed for the U.S. Why? Land areas can act as a sink (absorbing CO_2 from the air) or a source of GHG emissions. In the U.S. since 1990, managed forests and other lands have absorbed more CO_2 from the atmosphere than they emit.

7%
AGRICULTURE

11%
COMMERCIAL &
RESIDENTIAL

35%
ELECTRICITY

20%
INDUSTRY

27%
TRANSPORTATION

Usually when you see power lines like these, the electricity they are transmitting was generated by burning fossil fuels.

Kodda/Shutterstock

The ever-escalating emissions of greenhouse gases (GHG) can be traced to all parts of the planet and to all sections of society and the economy.

GLOBAL GREENHOUSE GAS EMISSIONS BY SOURCE

The burning of fossil fuels is a major part of emissions from energy supply, industry, transport, and residential and commercial buildings.

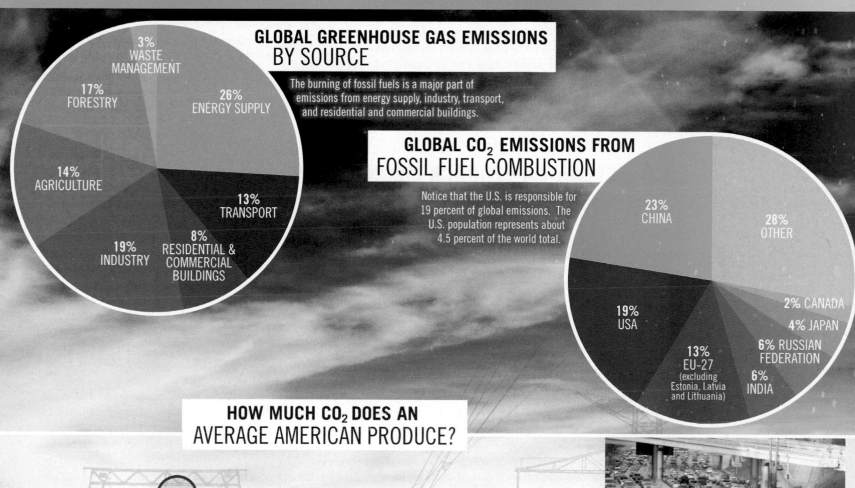

- 3% WASTE MANAGEMENT
- 17% FORESTRY
- 26% ENERGY SUPPLY
- 14% AGRICULTURE
- 13% TRANSPORT
- 19% INDUSTRY
- 8% RESIDENTIAL & COMMERCIAL BUILDINGS

GLOBAL CO_2 EMISSIONS FROM FOSSIL FUEL COMBUSTION

Notice that the U.S. is responsible for 19 percent of global emissions. The U.S. population represents about 4.5 percent of the world total.

- 23% CHINA
- 28% OTHER
- 2% CANADA
- 4% JAPAN
- 6% RUSSIAN FEDERATION
- 6% INDIA
- 13% EU-27 (excluding Estonia, Latvia and Lithuania)
- 19% USA

HOW MUCH CO_2 DOES AN AVERAGE AMERICAN PRODUCE?

17,000 POUNDS OF CO_2
by using **1,100 kilowatt-hours of electricity** per month

8,800 POUNDS OF CO_2
by using **6,300 cubic feet of natural gas** per month

1,000 POUNDS OF CO_2
by creating **4.5 pounds of trash** per day

1,000 POUNDS OF CO_2
by **driving 160 miles** per week

8,900 POUNDS OF CO_2
by **flying 1,900 miles** per year

modest, taken together, these trace gases play a significant role in warming the troposphere.

Sophisticated computer models show that the warming of the lower atmosphere caused by CO_2 and trace gases will not be the same everywhere. Rather, the temperature response in polar regions could be two to three times greater than the global average. One reason is that the polar troposphere is very stable, which suppresses vertical mixing and thus limits the amount of surface heat that is transferred upward. In addition, the expected reduction in sea ice will contribute to the greater temperature increase. This topic will be explored more fully in the next section.

20.9 | CLIMATE-FEEDBACK MECHANISMS

Contrast positive- and negative-feedback mechanisms and provide examples of each.

Climate is a very complex interactive physical system. Thus, when any component of the climate system is altered, scientists must consider many possible outcomes, some of which amplify the initial effect and some of which balance it out. These possible outcomes are called **climate-feedback mechanisms**. They complicate climate-modeling efforts and add greater uncertainty to climate predictions.

Types of Feedback Mechanisms

What climate-feedback mechanisms are related to carbon dioxide and other greenhouse gases? One important mechanism is that warmer surface temperatures increase evaporation rates. This in turn increases the atmosphere's water vapor content. Remember that water vapor is an even more powerful absorber of radiation emitted by Earth than is carbon dioxide. Therefore, with more water vapor in the air, the temperature increase caused by carbon dioxide and the trace gases is reinforced.

Recall that the temperature increase at high latitudes may be two to three times greater than the global average.

This projection is based in part on the likelihood that the area covered by sea ice will decrease as surface temperatures rise. Because ice reflects a much larger percentage of incoming solar radiation than does open water, the melting of the sea ice causes a surface having a high albedo to be replaced by a surface with a much lower albedo (**FIGURE 20.24**). The result is a substantial increase in the solar energy absorbed at the surface. This in turn feeds back to the atmosphere and magnifies the initial temperature increase created by higher levels of greenhouse gases.

So far, the climate-feedback mechanisms discussed have magnified the temperature rise caused by the buildup of carbon dioxide. Because these effects reinforce the initial change, they are called **positive-feedback mechanisms**. However, other effects must be classified as **negative-feedback mechanisms** because they produce results that are just the opposite of the initial change and tend to offset it.

One probable result of a global temperature rise would be an accompanying increase in cloud cover due to the higher moisture content of the atmosphere. Most clouds are good reflectors of solar radiation. At the same time, however, they are also good absorbers and emitters of radiation emitted by Earth. Consequently, clouds produce two opposite effects. They are a negative-feedback mechanism because they increase the reflection of solar radiation and thus diminish the amount of solar energy available to heat the atmosphere. On the other hand, clouds act as a positive-feedback mechanism by absorbing and emitting radiation that would otherwise be lost from the troposphere.

Which effect, if either, is stronger? Scientists still are not sure whether clouds will produce net positive or negative feedback. Although recent studies have not settled the question, they seem to lean toward the idea that clouds do not dampen global warming but rather produce a small positive feedback overall.[3]

Global warming caused by human-induced changes in atmospheric composition continues to be one of the most-studied aspects of climate change. Although no models yet incorporate

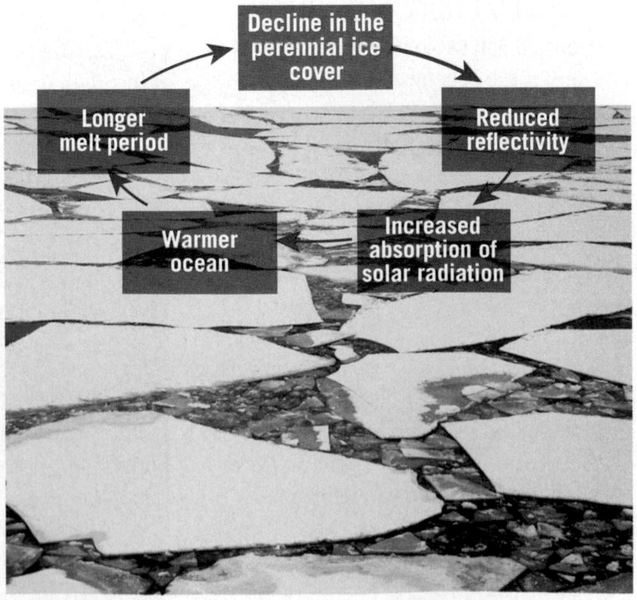

FIGURE 20.24 Sea Ice As a Feedback Mechanism The image shows the springtime breakup of sea ice near Antarctica. The diagram shows a likely feedback loop. A reduction in sea ice acts as a positive-feedback mechanism because surface albedo decreases, and the amount of energy absorbed at the surface increases. (Photo by Radius Images/Alamy)

Decline in the perennial ice cover

Longer melt period

Reduced reflectivity

Warmer ocean

Increased absorption of solar radiation

---

[3] A. E. Dessler, "A Determination of the Cloud Feedback from Climate Variations over the Past Decade," *Science* 330: 1523–1526, December 10, 2010.

the full range of potential factors and feedbacks, there is strong scientific consensus that the increasing levels of atmospheric carbon dioxide and trace gases have already warmed the planet and will continue to do so into the foreseeable future.

Computer Models of Climate: Important yet Imperfect Tools

Earth's climate system is amazingly complex. Comprehensive state-of-the-science climate simulation models are among the basic tools used to develop possible climate-change scenarios. Called *general circulation models* (*GCMs*), they are based on fundamental laws of physics and chemistry and incorporate human and biological interactions. The models simulate many variables, including temperature, rainfall, snow cover, soil moisture, winds, clouds, sea ice, and ocean circulation over the entire globe through the seasons and over spans of decades.

In many other fields of study, hypotheses can be tested by direct experimentation in the laboratory or by observations and measurements in the field. However, this is often not possible in the study of climate. Rather, scientists must construct computer models of how our planet's climate system works. If we understand the climate system correctly and construct a model appropriately, then the behavior of the model climate system should mimic the behavior of Earth's climate system (**FIGURE 20.25**).

What factors influence the accuracy of climate models? Clearly, mathematical models are *simplified* versions of the real Earth and cannot capture its full complexity, especially at smaller geographic scales. Moreover, when computer models are used to simulate future climate change, many assumptions have to be made that significantly influence

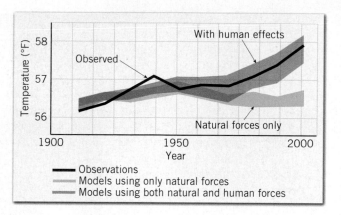

FIGURE 20.25 Computer Models The blue band shows how global average temperatures would have changed due to natural forces only, as simulated by climate models. The red band shows model projections of the effects of human and natural forces combined. The black line shows actual observed global average temperatures. As the blue band indicates, without human influences, temperatures over the past century would actually have first warmed and then cooled slightly over recent decades. Bands of color are used to express the range of uncertainty. (U.S. Global Change Research Program)

the outcome. They must consider a wide range of possibilities for future changes in population, economic growth, consumption of fossil fuels, technological development, improvements in energy efficiency, and more.

Despite many obstacles, our ability to use supercomputers to simulate climate continues to improve. Although today's models are far from infallible, they are powerful tools for understanding what Earth's future climate might be like.

20.9 CONCEPT CHECKS

1 Distinguish between positive and negative climate-feedback mechanisms.

2 Provide at least one example of each type of feedback mechanism.

3 List some factors that influence the accuracy of computer models of climate.

20.10 | HOW AEROSOLS INFLUENCE CLIMATE

Discuss the possible impacts of aerosols on climate change.

Increasing the levels of carbon dioxide and other greenhouse gases in the atmosphere is the most direct human influence on global climate. But it is not the only impact. Global climate is also affected by human activities that contribute to the atmosphere's aerosol content. **Aerosols** are the tiny, often microscopic, liquid and solid particles that are suspended in the air. Unlike cloud droplets, aerosols are present even in relatively dry air. Atmospheric aerosols are composed of many different materials, including soil, smoke, sea salt, and sulfuric acid. Natural sources are numerous and include such phenomena as dust storms and volcanoes.

Most human-generated aerosols come from the sulfur dioxide emitted during the combustion of fossil fuels and as a consequence of burning vegetation to clear agricultural land. Chemical reactions in the atmosphere convert the sulfur dioxide into sulfate aerosols, the same material that produces acid precipitation. The satellite images in **FIGURE 20.26** provide an example.

How do aerosols affect climate? Aerosols act directly by reflecting sunlight back to space and indirectly by making

clouds "brighter" reflectors. The second effect relates to the fact that many aerosols (such as those composed of salt or sulfuric acid) attract water and thus are especially effective as cloud condensation nuclei. The large quantity of aerosols produced by human activities (especially industrial emissions) trigger an increase in the number of cloud droplets that form within a cloud. A greater number of small droplets increases the cloud's brightness, causing more sunlight to be reflected back to space.

One category of aerosols, called **black carbon**, is soot generated by combustion processes and fires. Unlike most other aerosols, black carbon warms the atmosphere because it is an effective absorber of incoming solar radiation. In addition, when deposited on snow and ice, black carbon reduces surface albedo, thus increasing the amount of light absorbed. Nevertheless, despite the warming effect of black carbon, the overall effect of atmospheric aerosols is to cool Earth.

Studies indicate that the cooling effect of human-generated aerosols offsets a portion of the global warming caused by the growing quantities of greenhouse gases in the

FIGURE 20.26 Human-Generated Aerosols These satellite images show a serious air pollution episode that plagued China on October 8, 2010. (NASA)

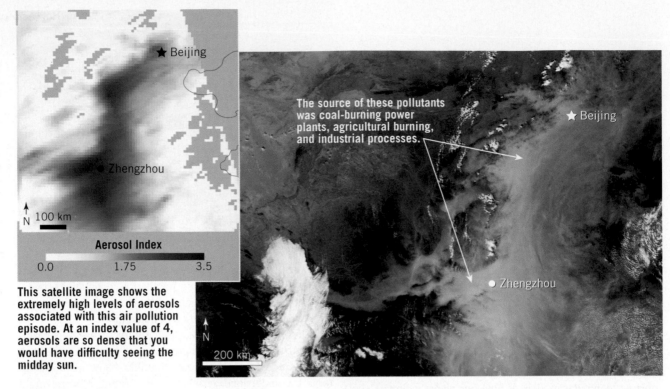

This satellite image shows the extremely high levels of aerosols associated with this air pollution episode. At an index value of 4, aerosols are so dense that you would have difficulty seeing the midday sun.

The source of these pollutants was coal-burning power plants, agricultural burning, and industrial processes.

atmosphere. The magnitude and extent of the cooling effect of aerosols is uncertain. This uncertainty is a significant hurdle in advancing our understanding of how humans alter Earth's climate.

It is important to point out some significant differences between global warming by greenhouse gases and aerosol cooling. After being emitted, greenhouse gases, such as carbon dioxide, remain in the atmosphere for many decades. By contrast, aerosols released into the troposphere remain there for only a few days or, at most, a few weeks before they are "washed out" by precipitation. Because of their short lifetime in the troposphere, aerosols are distributed unevenly over the globe. As expected, human-generated aerosols are concentrated near the areas that produce them—namely industrialized regions that burn fossil fuels and land areas where vegetation is burned.

The lifetime of aerosols in the atmosphere is short. Therefore, the effect of aerosols on today's climate is determined by the amount emitted during the preceding couple weeks. By contrast, the carbon dioxide and trace gases released into the atmosphere remain for much longer spans and thus influence climate for many decades.

20.10 CONCEPT CHECKS

1 What are the main sources of human-generated aerosols?

2 What effect does black carbon have on atmospheric temperatures?

3 What is the net effect of aerosols on temperatures in the troposphere?

4 How long do aerosols remain in the atmosphere before they are removed?

5 How does the residence time of aerosols compare to that of CO_2?

20.11 | SOME POSSIBLE CONSEQUENCES OF GLOBAL WARMING

Describe some possible consequences of global warming.

What consequences can be expected if the carbon dioxide content of the atmosphere reaches a level that is twice what it was early in the twentieth century? Because the climate system is complex, predicting the distribution of particular regional changes can be speculative. It is not yet possible to pinpoint specifics, such as where or when it will become drier or wetter. Nevertheless, plausible scenarios can be given for larger scales of space and time.

As noted, the magnitude of the temperature increase will not be the same everywhere. The temperature rise will probably be smallest in the tropics and increase toward the poles. As for precipitation, the models indicate that some regions will experience significantly more precipitation and runoff, whereas others will experience a decrease in runoff due to reduced precipitation or greater evaporation caused by higher temperatures.

TABLE 20.1 summarizes some of the most likely effects and their possible consequences. The table also provides the IPCC's estimate of the probability of each effect. Levels of confidence for these projections vary from "likely" (67 to 90 percent probability) to "very likely" (90 to 99 percent probability) to "virtually certain" (greater than 99 percent probability).

TABLE 20.1   Projected Changes and Effects of Global Warming in the Twenty-First Century

Projected Changes and Estimated Probability*	Examples of Projected Impacts
Higher maximum temperatures; more hot days and heat waves over nearly all land areas (*virtually certain*)	Increased incidence of death and serious illness in older age groups and urban poor. Increased heat stress in livestock and wildlife. Shift in tourist destinations. Increased risk of damage to a number of crops. Increased electric cooling demand and reduced energy supply reliability.
Higher minimum temperatures; fewer cold days, frost days, and cold waves over nearly all land areas (*virtually certain*)	Decreased cold-related human morbidity and mortality. Decreased risk of damage to a number of crops and increased risk to others. Extended range and activity of some pest and disease vectors. Reduced heating energy demand.
Increases in frequency of heavy precipitation events over most areas (*very likely*)	Increased flood, landslide, avalanche, and debris flow damage. Increased soil erosion. Increased flood runoff could increase recharge of some floodplain aquifers. Increased pressure on government and private flood insurance systems and disaster relief.
Increases in area affected by drought (*likely*)	Decreased crop yields. Increased damage to building foundations caused by ground shrinkage. Decreased water-resource quantity and quality. Increased risk of wildfires.
Increases in intense tropical cyclone activity (*likely*)	Increased risks to human life, risk of infectious-disease epidemics, and many other risks. Increased coastal erosion and damage to coastal buildings and infrastructure. Increased damage to coastal ecosystems, such as coral reefs and mangroves.

* *Virtually certain* indicates a probability greater than 99 percent, *very likely* indicates a probability of 90–99 percent, and *likely* indicates a probability of 67–90 percent.

Source: Adapted from Climate Change 2007: Synthesis Report. Contribution of Working Groups I, II and III to the Fourth Assessment Report of the Intergovernmental Panel on Climate Change, Figure SPM.3. IPCC, Geneva, Switzerland.

Sea-Level Rise

A significant impact of human-induced global warming is a rise in sea level. As this occurs, coastal cities, wetlands, and low-lying islands could be threatened with more frequent flooding, increased shoreline erosion, and saltwater encroachment into coastal rivers and aquifers.

Research indicates that sea level has risen about 25 centimeters (9.75 inches) since 1870. As **FIGURE 20.27** indicates, the rate of sea-level rise has been greater in recent years. Some models indicate that additional rise may approach or even exceed 50 centimeters (20 inches) by the end of the twenty-first century. Such a change may seem modest, but scientists realize that any rise in sea level along a *gently* sloping shoreline, such as the Atlantic and Gulf coasts of the United States, will lead to significant erosion and severe and permanent inland flooding (**FIGURE 20.28**). If this happens, many beaches and wetlands will be eliminated, and coastal civilization will be severely disrupted. Low-lying and densely populated places such as Bangladesh and the small island nation of the Maldives are especially vulnerable. The average elevation in the Maldives is 1.5 meters (less than 5 feet), and its highest point is just 2.4 meters (less than 8 feet) above sea level.

How is a warmer atmosphere related to a rise in sea level? One significant factor is thermal expansion. Higher air temperatures warm the adjacent upper layers of the ocean, which in turn causes the water to expand and sea level to rise.

Perhaps a more easily visualized contributor to global sea-level rise is melting glaciers. With few exceptions, glaciers around the world have been retreating at unprecedented rates over the past century. Some mountain glaciers have disappeared altogether. A recent 18-year satellite study showed that the mass of the Greenland and Antarctic Ice Sheets dropped an average of 475 gigatons per year. (A gigaton is 1 billion metric tons.) That is enough water to raise sea level 1.5 millimeters (0.05 inch) per year. The loss of ice was not steady but was occurring at an accelerating rate during the study period. Each year over the course of the study period, the two ice sheets lost a combined average of 36.3 gigatons more than they did the year before. During the same span, mountain

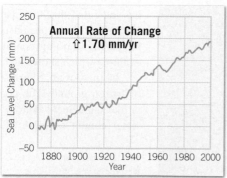

A. Sea Level Change 1870 to 2000

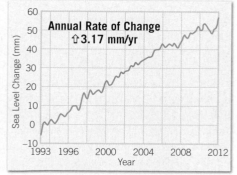

B. Sea Level Change 1993 to 2012

FIGURE 20.27 Rising Sea Level These graphs show sea-level change from 1870 until 2012. Note the difference in the rate of change between the two graphs. **A.** This graph shows historical sea-level data derived from coastal tide gauges. **B.** This graph shows the average sea level since 1993, derived from global satellite measurements. (NASA)

SmartFigure 20.28 Slope of the Shoreline The slope of the shoreline is critical to determining the degree to which sea-level changes will affect it. As sea level gradually rises, the shoreline retreats, and structures that were once thought to be safe from wave attack become vulnerable.

Where the slope is gentle a small rise in sea level causes a substantial shift

Original shoreline

Shoreline shift

Sea level rise

Where the slope is steep the same sea level rise causes a small shift

Original shoreline

Shoreline shift

Sea level rise

glaciers and ice caps lost an average of slightly more than 400 gigatons per year.

Because rising sea level is a gradual phenomenon, coastal residents may overlook it as an important contributor to shoreline erosion problems. Rather, the blame may be assigned to other forces, especially storm activity. Although a given storm may be the immediate cause, the magnitude of its destruction may result from the relatively small sea-level rise that allowed

FIGURE 20.29 Siberian Lakes This false-color image pair shows lakes dotting the tundra in 1973 and 2002. The tundra vegetation is colored a faded red, whereas lakes appear blue or blue-green. Many lakes disappeared or shrunk considerably between 1973 and 2002. After studying satellite imagery of about 10,000 large lakes in a 500,000-square-kilometer (195,000-square-mile) area in northern Siberia, scientists documented an 11 percent decline in the number of lakes, at least 125 of which disappeared completely. (NASA)

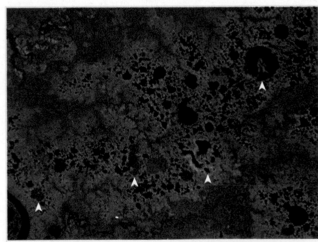

(a) June 27, 1973

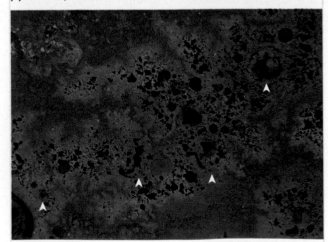

(b) July 2, 2002

the storm's power to cross a much greater land area.

The Changing Arctic

A 2005 study of climate change in the Arctic began with the following statement:

> For nearly 30 years, Arctic sea ice extent and thickness have been falling dramatically. Permafrost temperatures are rising and coverage is decreasing. Mountain glaciers and the Greenland ice sheet are shrinking. Evidence suggests we are witnessing the early stage of an anthropogenically induced global warming superimposed on natural cycles, reinforced by reductions in Arctic ice.[4]

Arctic Sea Ice Climate models are in general agreement that one of the strongest signals of global warming should be a loss of sea ice in the Arctic. This is indeed occurring. The area covered by sea ice naturally grows during the frigid Arctic winters and shrinks when temperatures climb in the spring and summer. Since 1979, satellites have observed a 13 percent decline per decade in the minimum summertime extent of sea ice in the Arctic. The thickness of the ice has also been declining.

The map in Figure 14.3A (page 435) compares the average sea ice extent for early September 2012 to the long-term average for the period 1979–2000. The extent of sea ice in September 2012 set a record. On that date the extent was less than 4 million square kilometers (1.54 million square miles)— 70,000 square kilometers (27,000 square miles) less than the previous record low set in September 2007. The trend is also clear when you examine the graph in Figure 14.3B. Is it possible that this trend may be part of a natural cycle? Yes, but it is more likely that the sea ice decline represents a combination of natural variability and human-induced global warming, with the latter becoming increasingly evident in coming decades. As was noted in the section "Climate-Feedback Mechanisms," a reduction in sea ice represents a positive-feedback mechanism that reinforces global warming.

Permafrost During the past decade, mounting evidence has indicated that the extent of permafrost in the Northern Hemisphere has decreased, as would be expected under long-term warming conditions. **FIGURE 20.29** presents one example which shows that such a decline is occurring.

In the Arctic, short summers thaw only the top layer of frozen ground. The permafrost beneath this *active layer* is like the cement bottom of a swimming pool. In summer, water cannot percolate downward, so it saturates the soil

---

[4]J. T. Overpeck, et al., "Arctic System on Trajectory to New, Seasonally Ice-Free States," *EOS, Transactions, American Geophysical Union*, 86(34): 309, August 23, 2005.

above the permafrost and collects on the surface in thousands of lakes. However, as Arctic temperatures climb, the bottom of the "pool" seems to be "cracking." Satellite imagery shows that over a 20-year span, a significant number of lakes have shrunk or disappeared altogether. As the permafrost thaws, lake water drains deeper into the ground.

Thawing permafrost represents a potentially significant positive-feedback mechanism that may reinforce global warming. When vegetation dies in the Arctic, cold temperatures inhibit its total decomposition. As a consequence, over thousands of years, a great deal of organic matter has become stored in the permafrost. When the permafrost thaws, organic matter that may have been frozen for millennia comes out of "cold storage" and decomposes. The result is the release of carbon dioxide and methane—greenhouse gases that contribute to global warming.

The Potential for "Surprises"

You have seen that climate in the twenty-first century, unlike in the preceding 1000 years, is not expected to be stable. Rather, a constant state of change is very likely. Many of the changes will probably be gradual environmental shifts, imperceptible from year to year. Nevertheless, the effects, accumulated over decades, will have powerful economic, social, and political consequences.

Despite our best efforts to understand future climate shifts, there is also the potential for "surprises." This simply means that, due to the complexity of Earth's climate system, we might experience relatively sudden, unexpected changes or see some aspects of climate shift in an unexpected manner. The report *Climate Change Impacts on the United States* describes the situation like this:

> Surprises challenge humans' ability to adapt, because of how quickly and unexpectedly they occur. For example, what if the Pacific Ocean warms in such a way that El Niño events become much more extreme? This could reduce the frequency, but perhaps not the strength, of hurricanes along the East Coast, while on the West Coast, more severe winter storms, ex-

treme precipitation events, and damaging winds could become common. What if large quantities of methane, a potent greenhouse gas currently frozen in icy Arctic tundra and sediments, began to be released to the atmosphere by warming, potentially creating an amplifying "feedback loop" that would cause even more warming? We simply do not know how far the climate system or other systems it affects can be pushed before they respond in unexpected ways.

There are many examples of potential surprises, each of which would have large consequences. Most of these potential outcomes are rarely reported, in this study or elsewhere. Even if the chance of any particular surprise happening is small, the chance that at least one such surprise will occur is much greater. In other words, while we can't know which of these events will occur, it is likely that one or more will eventually occur.[5]

The impact on climate of an increase in atmospheric carbon dioxide and trace gases is obscured by some uncertainties. Yet climate scientists continue to improve our understanding of the climate system and the potential impacts and effects of global climate change. Policymakers are confronted with responding to the risks posed by emissions of greenhouse gases, knowing that our understanding is imperfect. However, they are also faced with the fact that climate-induced environmental changes cannot be reversed quickly, if at all, due to the lengthy time scales associated with the climate system.

---

[5]National Assessment Synthesis Team, *Climate Change Impacts on the United States: The Potential Consequences of Climate Variability and Change* (Washington, DC: U.S. Global Research Program, 2000), p. 19.

20.11 CONCEPT CHECKS

1 List and describe the factors that are causing sea level to rise.

2 Is global warming greater near the equator or near the poles? Explain.

3 Based on Table 20.1, what projected changes relate to something other than temperature?

20 CONCEPTS IN REVIEW | World Climates and Global Climate Change

20.1 THE CLIMATE SYSTEM

List the five parts of the climate system and provide examples of each.

KEY TERM: climate system

- Climate is the aggregate of weather conditions for a place or region over a long period of time.
- Earth's climate system involves the exchanges of energy and moisture that occur among the atmosphere, hydrosphere, solid Earth, biosphere, and *cryosphere* (the ice and snow that exist at Earth's surface).

Q Which sphere of the climate system dominates this image? What other sphere or spheres are present?

James Balog/Getty Images

20.2 WORLD CLIMATES

Explain why classification is a necessary process when studying world climates. Discuss the criteria used in the Köppen system of climate classification.

KEY TERM: Köppen classification

- Climate classification brings order to large quantities of information, which aids comprehension and understanding and facilitates analysis and explanation.
- The most important elements in climate descriptions are temperature and precipitation because they have the greatest influence on people and their activities, and they also have an important impact on the distribution of vegetation and the development of soils.

- An early attempt at climate classification by the Greeks divided each hemisphere into three zones: torrid, temperate, and frigid. Many climate classifications have been devised, with the value of each determined by its intended use.
- The Köppen classification, which uses mean monthly and annual values of temperature and precipitation, is a widely used system. The boundaries Köppen chose were largely based on the limits of certain plant associations.
- Five principal climate groups, each with subdivisions, were recognized. Each group is designated by a capital letter. Four of the climate groups—A, C, D, and E—are defined on the basis of temperature characteristics, and the fifth, the B group, has precipitation as its primary criterion.

Q Why are three different formulas used to determine whether a climate is considered dry?

20.3 HUMID TROPICAL (A) CLIMATES

Compare the two broad categories of tropical climates.

KEY TERMS: tropical rain forest, tropical wet and dry

- Humid tropical (A) climates are winterless, with all months having a mean temperature above 18°C (64°F).
- Wet tropical climates (Af and Am), which lie near the equator, have constantly high temperatures and enough rainfall to support the most luxuriant vegetation (tropical rain forest) found in any climatic realm.
- Tropical wet and dry climates (Aw) are found poleward of the wet tropics and equatorward of the subtropical deserts, where the rain forest gives way to the tropical grasslands and scattered drought-tolerant trees of the savanna. The most distinctive feature of this climate is the seasonal character of the rainfall.

Q When does the rainy season occur in a tropical wet and dry climate: winter or summer? Explain.

20.4 DRY (B) CLIMATES

Contrast low-latitude dry climates and middle-latitude dry climates.

KEY TERMS: arid (desert), semiarid (steppe)

- Dry (B) climates, in which the yearly precipitation is less than the potential loss of water by evaporation, are subdivided into two types: arid or desert (BW) and semiarid or steppe (BS).
- Differences between desert and steppe are primarily a matter of degree, with semiarid being a marginal and more humid variant of arid.
- Low-latitude deserts and steppes coincide with the clear skies caused by subsiding air beneath the subtropical high-pressure belts.
- Middle-latitude deserts and steppes exist principally because of their position in the deep interiors of large landmasses far removed from the ocean. Because many middle-latitude deserts occupy sites on the leeward sides of mountains, they can also be classified as rain shadow deserts.

Dennis Tasa

Q This photo was taken in Nevada's Great Basin Desert, looking west toward the Sierra Nevada. What is the basic cause of the arid conditions in this region?

20.5 HUMID MIDDLE-LATITUDE CLIMATES (C AND D CLIMATES)

Distinguish among five different humid middle-latitude climates.

KEY TERMS: humid subtropical climate, marine west coast climate, dry-summer subtropical climate, humid continental climate, subarctic climate

- Middle-latitude climates with mild winters (C climates) occur where the average temperature of the coldest month is below 18°C (64°F) but above –3°C (27°F). Three C climate subgroups exist.
- Humid subtropical climates (Cfa) are located on the eastern sides of the continents, in the 25°–40° latitude range. Summer weather is hot and sultry, and winters are mild. In North America, the marine west coast climate (Cfb, Cfc) extends from near the U.S.–Canada border northward as a narrow belt into southern Alaska. The prevalence of maritime air masses means that mild winters and cool summers are the rule. Dry-summer subtropical climates (Csa, Csb) are typically located along the west sides of continents between latitudes 30° and 45°. In summer, the regions are dominated by stable, dry conditions associated with the oceanic subtropical highs. In winter, they are within range of the cyclonic storms of the polar front.
- Humid middle-latitude climates with severe winters (D climates) are land-controlled climates that are absent in the Southern Hemisphere. The D climates have severe winters. The average temperature of the coldest month is –3°C (27°F) or below, and the warmest monthly mean exceeds 10°C (50°F).
- Humid continental climates (Dfa, Dfb, Dwa, Dwb) are confined to the eastern portions of North America and Eurasia in the latitude range between approximately 40° and 50° north latitude. Both winter and summer temperatures can be characterized as relatively severe. Precipitation is generally greater in summer than in winter. Subarctic climates (Dfc, Dfd, Dwc, Dwd) are situated north of the humid continental climates and south of the polar tundras. The outstanding feature of subarctic climates is the dominance of winter. By contrast, summers in the subarctic are remarkably warm, despite their short duration. The highest annual temperature ranges on Earth occur here.

20.6 POLAR (E) CLIMATES

Contrast ice cap and tundra climates.

KEY TERMS: polar climate, tundra climate, ice cap climate

- Polar climates (ET, EF) are those in which the mean temperature of the warmest month is below 10°C (50°F). Annual temperature ranges are extreme, with the lowest annual means on the planet. Although polar climates are classified as humid, precipitation is generally meager, with many nonmarine stations receiving less than 25 centimeters (10 inches) annually. Two types of polar climates are recognized.

- The tundra climate (ET) is found almost exclusively in the Northern Hemisphere. The 10°C (50°F) summer isotherm represents its equatorward limit. It is a treeless region of grasses, sedges, mosses, and lichens with permanently frozen subsoil, called permafrost.

- The ice cap climate (EF) does not have a single monthly mean above 0°C (32°F). Consequently, the growth of vegetation is prohibited, and the landscape is one of permanent ice and snow. The ice sheets of Greenland and Antarctica are important examples.

20.7 HIGHLAND CLIMATES

Summarize the characteristics associated with highland climates.

KEY TERM: highland climates

- Highland climates are characterized by a great diversity of climatic conditions over a small area. Although the best-known climatic effect of increased altitude is lower temperatures, greater precipitation due to orographic lifting is also common. Variety and changeability best describe highland climates. Because atmospheric conditions fluctuate with altitude and exposure to the Sun's rays, a nearly limitless variety of local climates occur in mountainous regions.

Q Which of the local winds discussed in Chapter 18 are likely associated with highland climates?

20.8 HUMAN IMPACT ON GLOBAL CLIMATE

Summarize the nature and cause of the atmosphere's changing composition since about 1750. Describe the climate's response.

- Humans have been modifying the environment for thousands of years. By altering ground cover with the use of fire and the overgrazing of land, people have modified such important climatic factors as surface albedo, evaporation rates, and surface winds.

- Human activities produce climate change through the release of carbon dioxide (CO_2) and trace gases. Humans release CO_2 when they cut down forests and when they burn fossil fuels such as coal, oil, and natural gas. A steady rise in atmospheric CO_2 levels has been documented at Mauna Loa, Hawaii, and other locations around the world.

- More than half of the carbon released by humans is absorbed by new plant matter or dissolved in the oceans. About 45 percent remains in the atmosphere, where it can influence climate for decades. Air bubbles trapped in glacial ice reveal that there is currently about 30 percent more CO_2 than the atmosphere has contained in the past 650,000 years.

- As a result of the extra heat retention due to added CO_2, Earth's atmosphere has warmed by about 0.8°C (1.4°F) in the past 100 years, most of it since the 1970s. Temperatures are projected to increase by another 2° to 4.5°C (3.6° to 8.1°F) in the future.

- Trace gases such as methane, nitrous oxide, and CFCs also play a significant role in increasing global temperature.

20.9 CLIMATE-FEEDBACK MECHANISMS

Contrast positive- and negative-feedback mechanisms and provide examples of each.

KEY TERMS: climate-feedback mechanism, positive-feedback mechanism, negative-feedback mechanism

- A change in one part of the climate system may trigger changes in other parts of the climate system that amplify or diminish the initial effect. These climate-feedback mechanisms are called positive-feedback mechanisms if they reinforce the initial change and negative-feedback mechanisms if they counteract the initial effect.

- The melting of sea ice due to global warming (decreasing albedo and increasing the initial effect of warming) is one example of a positive-feedback mechanism. The production of more clouds (blotting out incoming solar radiation, leading to cooling) is an example of a negative-feedback mechanism.

- Computer models of climate give scientists a tool for testing hypotheses about climate change. Although these models are far simpler than the real climate system, they are useful tools for predicting the future climate.

Michael Collier

Q Changes in precipitation and temperature due to climate change can increase the risk of forest fires. Describe two ways that the event shown in this photo could contribute to global warming.

20.10 HOW AEROSOLS INFLUENCE CLIMATE

Discuss the possible impacts of aerosols on climate change.

KEY TERMS: aerosols, black carbon

- Aerosols are tiny liquid and solid particles that are suspended in the air. Global climate is affected by human activities that contribute to the atmosphere's aerosol content.
- Most aerosols reflect a portion of incoming solar radiation back to space and therefore have a cooling effect.
- Overall, aerosols have a cooling effect, yet some aerosols called black carbon (soot from combustion processes and fires) absorb incoming solar radiation and warm the atmosphere. When black carbon is deposited on snow and ice, it reduces surface albedo and increases the amount of light absorbed at the surface.

Q Do aerosols spend more or less time in the atmosphere than greenhouse gases such as carbon dioxide? What is the significance of this difference in residence time? Explain.

20.11 SOME POSSIBLE CONSEQUENCES OF GLOBAL WARMING

Describe some possible consequences of global warming.

- In the future, Earth's surface temperature is likely to continue to rise. The temperature increase will likely be greatest in the polar regions and least in the tropics. Some areas will get drier, and other areas will get wetter.
- Sea level is predicted to rise for several reasons, including the melting of glacial ice and thermal expansion (a given mass of seawater takes up more volume when it is warm than when it is cool). Low-lying, gently sloped, highly populated coastal areas are most at risk.
- Sea ice cover and thickness in the Arctic have been declining since satellite observations began in 1979.
- Because of the warming of the Arctic, permafrost is melting, releasing CO_2 and methane to the atmosphere in a positive-feedback mechanism.
- Because the climate system is complicated, dynamic, and imperfectly understood, it could produce sudden, unexpected changes with little warning.

GIVE IT SOME **THOUGHT**

1. Refer to Figure 20.1, which illustrates various components of Earth's climate system. Boxes represent interactions or changes that occur in the climate system. Select three boxes and provide an example of an interaction or change associated with each. Explain how these interactions may influence temperature.

2. Describe one way in which changes in the biosphere can cause changes in the climate system. Next, suggest one way in which the biosphere is affected by changes in some other part of the climate system. Finally, indicate one way in which the biosphere records changes in the climate system.

3. Refer to the monthly rainfall data (in millimeters) for three cities in Africa. Their locations are shown on the accompanying map. Match the data for each city to the correct location (1, 2, or 3) on the map. How were you able to figure this out? *Bonus:* Which figure in Chapter 18 would be especially useful in explaining or illustrating why these places have rainfall maximums and minimums when they do?

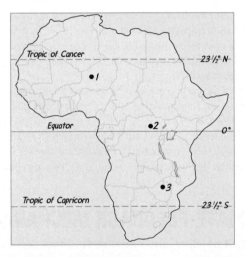

	J	F	M	A	M	J	J	A	S	O	N	D
CITY A	81	102	155	140	133	119	99	109	206	213	196	122
CITY B	0	2	0	0	1	15	88	249	163	49	5	6
CITY C	236	168	86	46	13	8	0	3	8	38	94	201

4. Refer to Figure 20.5, which shows climates of the world. Humid continental (Dfb and Dwb) and subarctic (Dfc) climates are usually described as being "land controlled"—that is, they lack marine influence. Nevertheless, these climates are found along the margins of the North Atlantic and the North Pacific oceans. Explain why this occurs.

5. It has been suggested that global warming over the past several decades likely would have been greater were it not for the effect of certain types of air pollution. Explain how this could be true.

6. Motor vehicles are a significant source of CO_2. Using electric cars, such as the one pictured here, is one way to reduce emissions from this source. Although these vehicles emit little or no CO_2 or other air pollutants directly into the air, can they still be connected to such emissions? If so, explain.

David Pearson/Alamy

7. If a fellow student who, unlike you, had not studied climate were to ask, "Isn't the greenhouse effect a bad thing because it's responsible for global warming?" how would you respond?

8. During a conversation, an acquaintance indicates that he is skeptical about global warming. When you ask why he feels that way, he says, "The past couple of years in this area have been among the coolest I can remember." While you assure this person that it is useful to question scientific findings, you suggest to him that his reasoning in this case may be flawed. Use your understanding of the definition of *climate* along with one or more graphs in the chapter to persuade this person to reevaluate his reasoning.

EXAMINING THE **EARTH SYSTEM**

1. The Köppen climate classification is based on the fact that there is an excellent association between natural vegetation (biosphere) and climate (atmosphere). Briefly describe the climate conditions (temperature and precipitation) and natural vegetation associated with each of the following Köppen climates: Af, BWh, Dfc, and ET.

☐	< 10 cm (4 in.)
☐	10–20 cm (4–8 in.)
☐	20–40 cm (8–16 in.)
■	> 40 cm (16 in.)

2. Examine the precipitation map for the state of Nevada. Notice that the areas receiving the most precipitation resemble long, slender "islands" scattered across the state. Provide an explanation for this pattern. Are average temperatures in these wetter areas likely different from those in nearby less rainy places? Why or why not? A look back at Section 6.9 and Figure 6.32 (page 195) might be helpful.

3. How might the burning of fossil fuels, such as the gasoline to run your car, influence global temperature? If such a temperature change occurs, how might sea level be affected? How might the intensity of hurricanes change? How might these changes impact people who live on a beach or barrier island along the Atlantic or Gulf coasts?

4. This satellite image from August 2007 shows the effects of tropical deforestation in a portion of the Amazon basin in western Brazil. Intact forest is dark green, whereas cleared areas are tan (bare ground) or light green (crops and pasture). Notice the relatively dense smoke in the left center of the image. How does deforestation of tropical forests change the composition of the atmosphere? Describe the effect that tropical deforestation has on global warming.

NASA

MasteringGeology™

Looking for additional review and test prep materials? Visit the Self Study area in **www.masteringgeology.com** to find practice quizzes, study tools, and multimedia that will aid in your understanding of this chapter's content. In **MasteringGeology™** you will find:

- GEODe: Earth Science: An interactive visual walkthrough of key concepts
- Geoscience Animation Library: More than 100 animations illuminating many difficult-to-understand Earth science concepts

- In The News RSS Feeds: Current Earth science events and news articles are pulled into the site with assessment
- Pearson eText
- Optional Self Study Quizzes
- Web Links
- Glossary
- Flashcards

UNIT SEVEN | EARTH'S PLACE IN THE UNIVERSE

21

Origins of Modern Astronomy*

FOCUS ON CONCEPTS

Each statement represents the primary **LEARNING OBJECTIVE** for the corresponding major heading within the chapter. After you complete the chapter, you should be able to:

21.1 Explain the geocentric view of the solar system and describe how it differs from the heliocentric view.

21.2 List and describe the contributions to modern astronomy of Nicolaus Copernicus, Tycho Brahe, Johannes Kepler, Galileo Galilei, and Isaac Newton.

21.3 Compare the equatorial system of coordinates used to establish the position of the stars with longitude and latitude. Explain how the positions of stars are described using declination and right ascension.

21.4 Describe the two primary motions of Earth and explain the difference between a solar day and a sidereal day.

21.5 Sketch the changing positions of the Earth–Moon system that produce the regular cycle we call the phases of the Moon.

21.6 Sketch the positions of the Earth–Moon system that produce a lunar eclipse, as well as a solar eclipse.

*This chapter was revised with the assistance of Professors Mark Watry and Teresa Tarbuck.

An observer at a very dark site with the Milky Way galaxy in the background. (Photo by Babak Tafreshi/Science Source)

The science of astronomy provides a rational way of knowing and understanding the origins of Earth, the solar system, and the universe. Earth was once thought to be unique, different in every way from everything else in the universe. However, through the science of astronomy, we have discovered that Earth and the Sun are similar to other objects in the universe and that the physical laws that apply on Earth seem to apply everywhere else in the universe.

How did our understanding of the universe change so drastically? In this chapter we examine the transformation from the ancient view of the universe, which focused on the positions and movements of celestial objects, to the modern perspective, which focuses on *understanding how* these objects came to be and *why* they move the way they do.

21.1 | ANCIENT ASTRONOMY Explain the geocentric view of the solar system and describe how it differs from the hellocentric view.

Long before recorded history, people were aware of the close relationship between events on Earth and the positions of heavenly bodies. They realized that changes in the seasons and floods of great rivers such as the Nile in Egypt occurred when certain celestial bodies, including the Sun, Moon, planets, and stars, reached particular places in the heavens. Early agrarian cultures, whose survival depended on seasonal change, believed that if these heavenly objects could control the seasons, they could also strongly influence all Earthly events. These beliefs undoubtedly encouraged early civilizations to begin keeping records of the positions of celestial objects.

The origin of astronomy began more than 5000 years ago, when humans began to track the motion of celestial objects so they knew when to plant their crops or prepare to hunt migrating herds. The ancient Chinese, Egyptians, and Babylonians are well known for their record keeping. These cultures recorded the locations of the Sun, the Moon, and the five visible planets as these objects moved slowly against the background of "fixed" stars. Eventually, it was not enough to track the motions of celestial objects; predicting their future positions (to avoid getting married at an unfavorable time, for example) became important.

A study of Chinese archives shows that the Chinese recorded every appearance of the famous Halley's Comet for at least 10 centuries. However, because this comet appears only once every 76 years, they were unable to link these appearances to establish that what they saw was the same object multiple times. Like most other ancients, the Chinese considered comets to be mystical. Generally, comets were seen as bad omens and were blamed for a variety of disasters, from wars to plagues (**FIGURE 21.1**). In addition, the Chinese kept quite accurate records of "guest stars." Today we know that a "guest star" is a normal star, usually too faint to be visible, which increases its brightness as it explosively ejects gases from its surface, a phenomenon we call a *nova* (*novus* = new) or *supernova* (**FIGURE 21.2**).

FIGURE 21.1 The Bayeux Tapestry that Hangs in Bayeux, France This tapestry shows the apprehension caused by Halley's Comet in A.D. 1066. This event preceded the defeat of King Harold by William the Conqueror. ("Sighting of a comet." Detail from Bayeux Tapestry. Musee de la Tapisserie, Bayeux. "With special authorization of the City of Bayeux." Bridgeman-Giraudon/Art Resource, NY)

The Golden Age of Astronomy

The "Golden Age" of early astronomy (600 B.C.–A.D. 150) was centered in Greece. Although the early Greeks have been criticized for using purely philosophical arguments to explain natural phenomena, they employed observational data as well. The basics of geometry and trigonometry, which they developed, were used to measure the sizes of and distances to the largest-appearing bodies in the heavens—the Sun and the Moon.

The early Greeks held the incorrect **geocentric** (*geo* = Earth, *centric* = centered) view of the universe—which professed that Earth was a sphere that remained motionless at the center of the universe. Orbiting Earth were the Moon, Sun, and known planets—Mercury, Venus, Mars, Jupiter, and Saturn. The Sun and Moon were thought to be perfect crystal spheres. Beyond the planets was a transparent, hollow **celestial sphere** to which the stars were attached and which traveled daily around Earth. (Although it appears that the stars and planets move across the sky, this effect is actually caused by Earth's rotation on its axis.) Some early Greeks realized that the motion of the stars could be explained just as easily by a rotating Earth, but they rejected that idea because people can't sense Earth's motion, and the planet seemed too large to be movable. In fact, proof of Earth's rotation was not demonstrated until 1851.

FIGURE 21.2 The Sudden Appearance of a "Guest Star" The Chinese recorded the sudden appearance of a "guest star" in 1054 A.D. The scattered remains of that supernova is the Crab Nebula in the constellation Taurus. This image comes from the Hubble Space Telescope. (NASA/Jet Propulsion Laboratory)

To the Greeks, all except seven of the heavenly bodies appeared to remain in the same position relative to one another. These seven wanderers (*planetai* in Greek) were the Sun, the Moon, Mercury, Venus, Mars, Jupiter, and Saturn. Each was thought to have a circular orbit around Earth. Although this system was incorrect, the Greeks refined it to the point that it explained the apparent movements of all celestial bodies.

The famous Greek philosopher Aristotle (384–322 B.C.) concluded that Earth is spherical because it always casts a curved shadow when it eclipses the moon. Although many considered most of Aristotle's teachings infallible, his belief in a spherical Earth was lost during the Middle Ages.

Measuring the Earth's Circumference

The first successful attempt to establish the size of Earth is credited to Eratosthenes (276–194 B.C.). Eratosthenes observed the angles of the noonday Sun in two Egyptian cities that were roughly north and south of each other—Syene (presently Aswan) and Alexandria (**FIGURE 21.3**). Finding that the angles of the noonday sun differed by 7 degrees, or 1/50 of a complete circle, he concluded that the circumference of Earth must be 50 times the distance between these two cities. The cities were 5000 *stadia* apart, giving him a measurement of 250,000 *stadia*.

Many historians believe the *stadia* was 157.6 meters (517 feet), which would make Eratosthenes's calculation of Earth's circumference—39,400 kilometers (24,428 miles)—very close to the modern value of 40,075 kilometers (24,902 miles).

A Sun-Centered Universe?

The first Greek to profess a *Sun-centered*, or **heliocentric** (*helios* = Sun, *centric* = centered), universe was Aristarchus (312–230 B.C.).

Aristarchus also used simple geometric relations to calculate the relative distances from Earth to the Sun and the Moon. He later used these data to calculate their sizes. As a result of an observational error beyond his control, he came up with measurements that were much too small. However, he did discover that the Sun was many times more distant than the Moon and many times larger than Earth. The latter fact may have prompted him to suggest a Sun-centered universe. Nevertheless, because of the strong influence of Aristotle's writings, the Earth-centered view dominated Western thought for nearly 2000 years.

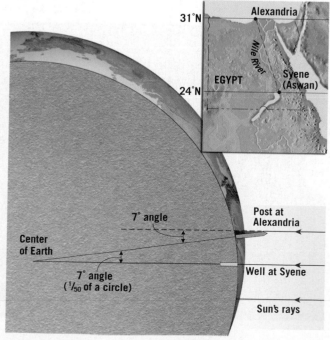

SmartFigure 21.3 Orientation of the Sun's Rays at Syene (Aswan) and Alexandria, Egypt on June 21 From these data, Eratosthenes calculated Earth's circumference.

FIGURE 21.4 The Universe According to Ptolemy, Second Century A.D. A. Ptolemy believed that the star-studded celestial sphere made a daily trip around a motionless Earth. In addition, he proposed that the Sun, Moon, and planets made trips of various lengths along individual orbits. **B.** A three-dimensional model of an Earth-centered system. Ptolemy likely utilized something similar to this to calculate the motions of the heavens. (Photo by Science Museum, London/The Bridgeman Art Library)

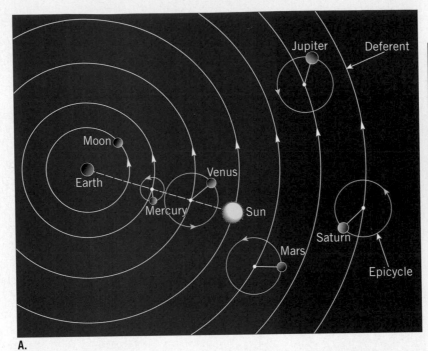

A.

B.

Mapping the Stars Probably the greatest of the early Greek astronomers was Hipparchus (second century B.C.), best known for his star catalogue. Hipparchus determined the location of almost 850 stars, which he divided into six groups according to their brightness. (This system is still used today.) He measured the length of the year to within minutes of the modern value and developed a method for predicting the times of lunar eclipses to within a few hours.

Although many of the Greek discoveries were lost during the Middle Ages, the Earth-centered view that the Greeks proposed became entrenched in Europe. Presented in its finest form by Claudius Ptolemy, this geocentric outlook became known as the **Ptolemaic System**.

Ptolemy's Model

Much of our knowledge of Greek astronomy comes from a 13-volume treatise, *Almagest* (meaning "the great work"), which was compiled by Ptolemy in A.D. 141. In addition to presenting a summary of Greek astronomical knowledge, Ptolemy is credited with developing a model of the universe that accounted for the observable motions of the celestial bodies (**FIGURE 21.4**).

In the Greek tradition, the Ptolemaic model had the planets moving in perfect circular orbits around a motionless Earth. (The Greeks considered the circle to be the pure and perfect shape.) However, the motion of the planets, as seen against the background of stars, is not so simple. Each planet, if watched night after night, moves slightly eastward among the stars. Periodically, each planet appears to stop, reverse direction for a period of time, and then resume an eastward motion. The apparent westward drift is called **retrograde** (*retro* = to go back, *gradus* = walking) **motion**. This rather odd apparent motion results from the combination of the motion of Earth and the planet's own motion around the Sun.

The retrograde motion of Mars is shown in **FIGURE 21.5**. Because Earth has a faster orbital speed than Mars, it overtakes its neighbor. While doing so, Mars *appears* to be moving backward, in

FIGURE 21.5 Retrograde Motion of Mars, as Seen Against the Background of Distant Stars When viewed from Earth, Mars moves eastward among the stars each day and then periodically appears to stop and reverse direction. This apparent westward drift is a result of the fact that Earth has a faster orbital speed than Mars and overtakes it. As this occurs, Mars appears to be moving backward—that is, it exhibits retrograde motion.

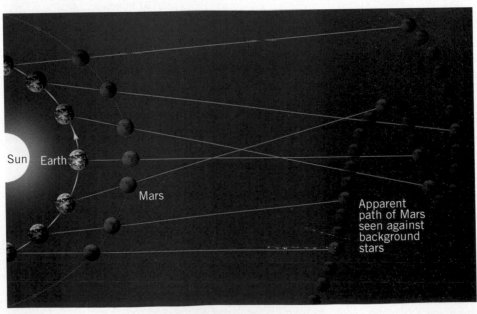

Apparent path of Mars seen against background stars

retrograde motion. This is analogous to what a driver sees out the side window when passing a slower car. The slower planet, like the slower car, appears to be going backward, although its actual motion is in the same direction as the faster-moving body.

Although it is difficult to accurately represent retrograde motion using the incorrect Earth-centered model, Ptolemy did so successfully (**FIGURE 21.6**). Rather than using a single circle for each planet's orbit, he proposed that the planets orbited on small circles (*epicycles*), revolving along large circles (*deferents*). By trial and error, he found the right combination of circles to produce the amount of retrograde motion observed for each planet. (An interesting note is that almost any closed curve can be produced by the combination of two circular motions, a fact that can be verified by anyone who has used the Spirograph™ design-drawing toy.)

It is a tribute to Ptolemy's genius that he was able to account for the planets' motions as well as he did, considering that he used an incorrect model. The precision with which his model was able to predict planetary motion is attested to by the fact that it went virtually unchallenged, in principle if not in detail, until the seventeenth century. When Ptolemy's predicted positions for the planets became out of step with the observed positions (which took 100 years or more), his model was simply recalibrated, using the new observed positions as a starting point.

With the decline of the Roman Empire around the fourth century, much of the accumulated knowledge disappeared as libraries were destroyed. After the decline of Greek and Roman civilizations, the center of astronomical study moved east to Baghdad where, fortunately, Ptolemy's work was translated into

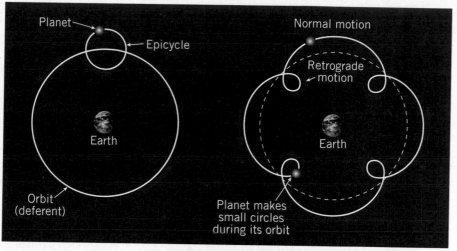

Arabic. Later, Arabic astronomers expanded Hipparchus's star catalog and divided the sky into 48 constellations—the foundation of our present-day constellation system. It wasn't until sometime after the tenth century that the ancient Greeks' contributions to astronomy were reintroduced to Europe through the Arabic community. The Ptolemaic model soon dominated European thought as the correct representation of the heavens, which created problems for anyone who found errors in it.

SmartFigure 21.6 Ptolemy's Explanation of Retrograde Motion
Retrograde motion is the apparent backward motion of planets against the background of fixed stars. In Ptolemy's model, the planets move on small circles (epicycles) while they orbit Earth on larger circles (deferents). Through trial and error, Ptolemy discovered the right combination of circles to produce the retrograde motion observed for each planet.

21.1 CONCEPT CHECKS

1 Why did the ancients believe that celestial objects had some influence over their lives?

2 What is the modern explanation of "guest stars" that suddenly appear in the night sky?

3 Explain the *geocentric* view of the universe.

4 In the Greek model of the universe, what were the seven wanderers, or *planetai*? How were they different from stars?

5 Describe what produces the retrograde motion of Mars. What geometric arrangements did Ptolemy use to explain this motion?

21.2 | THE BIRTH OF MODERN ASTRONOMY

List and describe the contributions to modern astronomy of Nicolaus Copernicus, Tycho Brahe, Johannes Kepler, Galileo Galilei, and Isaac Newton

Ptolemy's Earth-centered universe was not discarded overnight. Modern astronomy's development was more than a scientific endeavor; it required a break from deeply entrenched philosophical and religious views that had been a basic part of Western society for thousands of years. Its development was brought about by the discovery of a new and much larger universe, governed by discernible laws.

We examine the work of five noted scientists involved in this transition from an astronomy that merely describes what is observed to an astronomy that tries to explain what is observed and, more importantly, why the universe behaves the way it does. They are Nicolaus Copernicus,

Tycho Brahe, Johannes Kepler, Galileo Galilei, and Sir Isaac Newton.

Nicolaus Copernicus

For almost 13 centuries after the time of Ptolemy, very few astronomical advances were made in Europe; some were even lost, including the notion of a spherical Earth. The first great astronomer to emerge after the Middle Ages was Nicolaus Copernicus (1473–1543) from Poland (**FIGURE 21.7**). After discovering Aristarchus's writings, Copernicus became convinced that Earth is a planet, just

FIGURE 21.7 Painting of Polish Astronomer Nicolaus Copernicus (1473–1543) Copernicus's proclamation that Earth was just another planet was very controversial for more than 100 years after his death. (Detlev van Ravenswaay/Science Source)

FIGURE 21.8 Tycho Brahe (1546–1601) in His Observatory, in Uraniborg, on the Danish Island of Hveen Tycho (central figure) and the background are painted on the wall of the observatory within the arc of the sighting instrument called a quadrant. In the far right, Tycho can be seen "sighting" a celestial object through the "hole" in the wall. Tycho's accurate measurements of Mars enabled Johannes Kepler to formulate his three laws of planetary motion. (Courtesy of Royal Geographic Society, London/The Bridgeman Library International)

like the other five then-known planets. The daily motions of the heavens, he reasoned, could be more simply explained by a rotating Earth.

Having concluded that Earth is a planet, Copernicus constructed a *heliocentric* model for the solar system, with the Sun at the center and the planets Mercury, Venus, Earth, Mars, Jupiter, and Saturn orbiting it. This was a major break from the ancient and prevailing idea that a motionless Earth lies at the center of all movement in the universe. However, Copernicus retained a link to the past and used circles to represent the orbits of the planets. Because of this, Copernicus was unable to accurately predict the future locations of the planets. Copernicus found it necessary to add smaller circles (epicycles) like those that Ptolemy had used. The discovery that the planets actually have *elliptical* orbits occurred a century later and is credited to Johannes Kepler.

Like his predecessors, Copernicus also used philosophical justifications to support his point of view: "In the midst of all stands the Sun. For who could in this most beautiful temple place this lamp in another or better place than that from which it can at the same time illuminate the whole?"

Copernicus's monumental work *De Revolutionibus, Orbium Coelestium* (*On the Revolution of the Heavenly Spheres*), which set forth his controversial Sun-centered solar system, was published as he lay on his deathbed. Hence, he never suffered the criticisms that fell on many of his followers. Although Copernicus's model was a vast improvement over Ptolemy's, it did not attempt to explain how planetary motions occurred or why.

The greatest contribution of the Copernican system to modern science is its challenge of the primacy of Earth in the universe. At the time, many Europeans considered this heretical. Professing the Sun-centered model cost at least one person his life. Giordano Bruno was seized by the Inquisition, a Church tribunal, in 1600, and, refusing to denounce the Copernican theory, was burned at the stake.

Tycho Brahe

Tycho Brahe (1546–1601) was born of Danish nobility 3 years after the death of Copernicus. Reportedly, Tycho became interested in astronomy while viewing a solar eclipse that astronomers had predicted. He persuaded King Frederick II to establish an observatory near Copenhagen, which Tycho headed. There he designed and built pointers (the telescope would not be invented for a few more decades), which he used for 20 years to systematically measure the locations of the heavenly bodies in an effort to disprove the Copernican theory (**FIGURE 21.8**). His observations, particularly of Mars, were far more precise than any made previously and are his legacy to astronomy.

Tycho did not believe in the Copernican model because he was unable to observe an apparent shift in the position of stars that should result if Earth traveled around the Sun. His argument went like this: If Earth orbits the Sun, the position of a nearby star, when observed from two locations in Earth's orbit 6 months apart, should shift with respect to the more distant stars. Tycho was correct, but his measurements did not have great enough precision to show any displacement. The apparent shift of the stars is called *stellar parallax*, and today it is used to measure distances to the nearest stars. (Stellar parallax is discussed in Appendix C, page 745.)

FIGURE 21.9 German Astronomer Johannes Kepler (1571–1630) Kepler's contribution to modern astronomy was the derivation of his three laws of planetary motion. (Photo by Imagno/Getty Images)

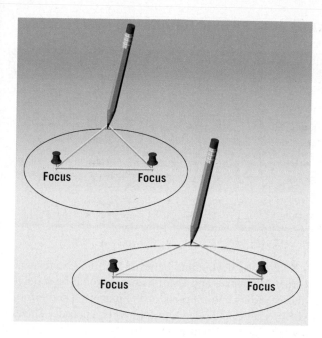

FIGURE 21.10 Drawing Ellipses with Various Eccentricities Using two straight pins for foci and a loop of string, trace out a curve while keeping the string taut, and you will have drawn an ellipse. The farther the pins (the foci) are moved apart, the more flattened (more eccentric) is the resulting ellipse.

The principle of parallax is easy to visualize: Close one eye, and with your index finger vertical, use your eye to line up your finger with some distant object. Now, without moving your finger, view the object with your other eye and notice that the object's position appears to change. The farther away you hold your finger, the less the object's position seems to shift. Herein lay the flaw in Tycho's argument. He was right about parallax, but the distance to even the nearest stars is enormous compared to the width of Earth's orbit. Consequently, the shift that Tycho was looking for is too small to be detected without the aid of a telescope—an instrument that had not yet been invented.

With the death of his patron, the king of Denmark, Tycho was forced to leave his observatory. Known for his arrogance and extravagant nature, Tycho was unable to continue his work under Denmark's new ruler. As a result, Tycho moved to Prague in the present-day Czech Republic, where, in the last year of his life, he acquired an able assistant, Johannes Kepler. Kepler retained most of the observations made by Tycho and put them to exceptional use. Ironically, the data Tycho collected to refute the Copernican view of the solar system would later be used by Kepler to support it.

Johannes Kepler

If Copernicus ushered out the old astronomy, Johannes Kepler (1571–1630) ushered in the new (**FIGURE 21.9**). Armed with Tycho's data, a good mathematical mind, and, of greater importance, a strong belief in the accuracy of

Tycho's work, Kepler derived three basic laws of planetary motion. The first two laws resulted from his inability to fit Tycho's observations of Mars to a circular orbit. Unwilling to concede that the discrepancies were a result of observational error, he searched for another solution. This endeavor led him to discover that the orbit of Mars is not a perfect circle but is slightly elliptical (**FIGURE 21.10**). About the same time, he realized that the orbital speed of Mars varies in a predictable way. As it approaches the Sun, it speeds up, and as it moves away, it slows down.

In 1609, after nearly a decade of work, Kepler proposed his first two laws of planetary motion:

1. The path of each planet around the Sun, while almost circular, is actually an ellipse, with the Sun at one focus (see Figure 21.10).

2. Each planet revolves so that an imaginary line connecting it to the Sun sweeps over equal areas in equal intervals of time (**FIGURE 21.11**). This *law of equal areas* geometrically expresses the variations in orbital speeds of the planets.

Figure 21.11 illustrates the second law. Note that in order for a planet to sweep equal areas in the same amount of time, it must travel more rapidly when it is nearer the Sun and more slowly when it is farther from the Sun.

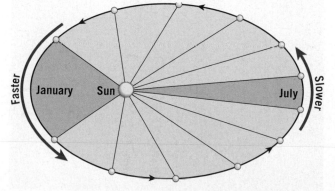

FIGURE 21.11 Kepler's Law of Equal Areas A line connecting a planet (Earth) to the Sun sweeps out an area in such a manner that equal areas are swept out in equal times. Thus, Earth revolves slower when it is farther from the Sun (aphelion) and faster when it is closest (perihelion). The eccentricity of Earth's orbit is greatly exaggerated in this diagram.

TABLE 21.1   Period of Revolution and Solar Distances of Planets

Planet	Solar Distance (AU) *	Period (years)	Ellipticity (0 = circle)
Mercury	0.39	0.24	0.205
Venus	0.72	0.62	0.007
Earth	1.00	1.00	0.017
Mars	1.52	1.88	0.094
Jupiter	5.20	11.86	0.049
Saturn	9.54	29.46	0.057
Uranus	19.18	84.01	0.046
Neptune	30.06	164.80	0.011

* AU = astronomical unit

Kepler was devout and believed that the Creator made an orderly universe and that this order would be reflected in the positions and motions of the planets. The uniformity he tried to find eluded him for nearly a decade. Then in 1619, Kepler published his third law in *The Harmony of the Worlds*:

3. The orbital periods of the planets and their distances to the Sun are proportional.

In its simplest form, the orbital period is measured in Earth years, and the planet's distance to the Sun is expressed in terms of Earth's mean distance to the Sun. The latter "yardstick" is called the **astronomical unit (AU)** and is equal to about 150 million kilometers (93 million miles). Using these units, Kepler's third law states that the planet's orbital period squared is equal to its mean solar distance cubed. Consequently, the solar distances of the planets can be calculated when their periods of revolution are known. For example, Mars has an orbital period of 1.88 years, and

1.88 squared equals 3.54. The cube root of 3.54 is 1.52, and that is the average distance from Mars to the Sun, in astronomical units (**TABLE 21.1**).

Kepler's laws assert that the planets revolve around the Sun and therefore support the Copernican theory. Kepler, however, did not determine the *forces* that act to produce the planetary motion he had so ably described. That task would remain for Galileo Galilei and Sir Isaac Newton.

Galileo Galilei

Galileo Galilei (1564–1642) was the greatest Italian scientist of the Renaissance (**FIGURE 21.12**). He was a contemporary of Kepler and, like Kepler, strongly supported the Copernican theory of a Sun-centered solar system. Galileo's greatest contributions to science were his descriptions of the behavior of moving objects, which he derived from experimentation. The method of using experiments to determine natural laws had essentially been lost since the time of the early Greeks.

All astronomical discoveries before Galileo's time were made without the aid of a telescope. In 1609, Galileo heard that a Dutch lens maker had devised a system of lenses that magnified objects. Apparently without ever having seen a telescope, Galileo constructed his own, which magnified distant objects three times the size seen by the unaided eye. He

FIGURE 21.12 Italian Scientist Galileo Galilei (1564–1642) Galileo was the first scientist to use a new invention, the telescope, to observe the Sun, Moon, and planets in more detail than ever before. (Nimatallah/Art Resource, NY)

FIGURE 21.13 One of Galileo's Many Telescopes Although Galileo did not invent the telescope, he built several—the largest of which had a magnification of 30. (Photo by Gianni Tortoli/Science Source)

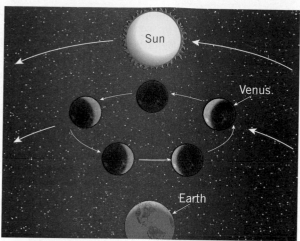

A. In the Ptolemaic (Earth-centered) system, the orbit of Venus lies between the Sun and Earth, as shown here. Thus, in an Earth-centered solar system, only the crescent phase of Venus would be visible from Earth.

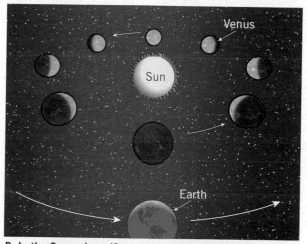

B. In the Copernican (Sun-centered) system, Venus orbits the Sun and hence all of the phases of Venus should be visible from Earth.

C. As Galileo observed, Venus goes through a series of Moonlike phases. Venus appears smallest during the full phase when it is farthest from Earth and largest in the crescent phase when it is closest to Earth. This led Galileo to conclude that the Sun was the center of the solar system.

SmartFigure 21.15 Using a Telescope, Galileo Discovered That Venus Has Phases Like Earth's Moon

A. In the Ptolemaic (Earth-centered) system, the orbit of Venus lies between the Sun and Earth, as shown in Figure 21.4A. Thus, in an Earth-centered solar system, only the crescent phase of Venus would be visible from Earth. **B.** In the Copernican (Sun-centered) system, Venus orbits the Sun, and hence all the phases of Venus should be visible from Earth. **C.** As Galileo observed, Venus goes through a series of Moonlike phases. Venus appears smallest during the full phase, when it is farthest from Earth, and largest in the crescent phase, when it is closest to Earth. This verified Galileo's belief that the Sun is the center of the solar system. (Photo courtesy of Lowell Observatory)

FIGURE 21.14 Sketch by Galileo of Jupiter and Its Four Largest Satellites Using a telescope, Galileo discovered Jupiter's four largest Moons (drawn as stars) and noted that their positions change nightly. You can observe these same changes with binoculars. (Yerkes Observatory Photograph/University of Chicago)

immediately made others, the best having a magnification of about 30 (**FIGURE 21.13**).

With the telescope, Galileo was able to view the universe in a new way. He made many important discoveries that supported the Copernican view of the universe, including the following:

1. *The discovery of Jupiter's four largest satellites, or moons* (**FIGURE 21.14**). This finding dispelled the old idea that Earth was the sole center of motion in the universe; for here, plainly visible, was another center of motion—Jupiter. It also countered the frequently used argument that the Moon would be left behind if Earth revolved around the Sun.

2. *The discovery that the planets are circular disks rather than just points of light, as was previously thought.* This indicated that the planets must be Earth-like as opposed to star-like.

3. *The discovery that Venus exhibits phases just as the Moon does and that Venus appears smallest when it is in full phase and thus is farthest from Earth* (**FIGURE 21.15B,C**). This observation demonstrates that Venus orbits its source of light—the Sun. In the Ptolemaic system, shown in **FIGURE 21.15A**, the orbit of Venus lies between Earth

and the Sun, which means that only the crescent phases of Venus should ever be seen from Earth.

4. *The discovery that the Moon's surface is not a smooth glass sphere, as the ancients had proclaimed.* Rather, Galileo saw mountains, craters, and plains, indicating that the Moon is Earth-like. He thought the plains might be bodies of water, and this idea was strongly promoted by others, as we can tell from the names given to these features (Sea of Tranquility, Sea of Storms, etc.).

5. *The discovery that the Sun (the viewing of which may have caused the eye damage that later blinded him) had sunspots—dark regions caused by slightly lower temperatures.* He tracked the movement of these spots and estimated the rotational period of the Sun as just under a month. Hence, another heavenly body was found to have both "blemishes" and rotational motion.

Each of these observations eroded a bedrock principle held by the prevailing view on the nature of the universe.

In 1616, the Church condemned the Copernican theory as contrary to Scripture because it did not put humans at their rightful place in the center of Creation, and Galileo was told to abandon this theory. Undeterred, Galileo began writing his most famous work, *Dialogue of the Great World Systems.* Despite poor health, he completed the project and in 1630 went to Rome, seeking permission from Pope Urban VIII to publish. Since the book was a dialogue that expounded both the Ptolemaic and Copernican systems, publication was allowed. However, Galileo's detractors were quick to realize that he was promoting the Copernican view at the expense of the Ptolemaic system. Sale of the book was quickly halted, and Galileo was called before the Inquisition. Tried and convicted of proclaiming doctrines contrary to religious teachings, he was sentenced to permanent house arrest, under which he remained for the last 10 years of his life.

Despite this restriction, and his grief following the death of his eldest daughter, Galileo continued to work. In 1637 he became totally blind, yet during the next few years, he completed his finest scientific work, a book on the study of motion in which he stated that the natural tendency of an object in motion is to remain in motion. Later, as more scientific evidence in support of the Copernican system was discovered, the Church allowed Galileo's works to be published.

Sir Isaac Newton

Sir Isaac Newton (1642–1727) was born in the year of Galileo's death (**FIGURE 21.16**). His many accomplishments in mathematics and physics led a successor to say, "Newton was the greatest genius that ever existed."

Although Kepler and those who followed attempted to explain the forces involved in planetary motion, their explanations were less than satisfactory. Kepler believed that some force pushed the planets along in their orbits. Galileo, however, correctly reasoned that no force is required to keep an object in motion. Instead, Galileo proposed that the natural tendency for a moving object that is unaffected by an outside force is to continue moving at a uniform speed and in a straight line. Newton later formalized this concept, **inertia**, as his first law of motion.

The problem, then, was not to explain the force that keeps the planets moving but rather to determine the force that *keeps them from going in a straight line out into space.* It was to this end that Newton conceptualized the force of gravity. At the early age of 23, he envisioned a force that extends from Earth into space and holds the Moon in orbit around Earth. Although others had theorized the existence of such a force, he was the first to formulate and test the **law of universal gravitation**. It states:

> Every body in the universe attracts every other body with a force that is directly proportional to their masses and inversely proportional to the square of the distance between them.

Thus, gravitational force decreases with distance, so that two objects 3 kilometers apart have 3^2, or 9, times less gravitational attraction than if the same objects were 1 kilometer apart.

The law of gravitation also states that the greater the mass of an object, the greater its gravitational force. For example, the large mass of the Moon has a gravitational force strong enough to cause ocean tides on Earth, whereas the tiny mass of a communications satellite has very little effect on Earth.

With his laws of motion, Newton proved that the force of gravity—combined with the tendency of a planet to remain in straight-line motion—would result in a planet having an elliptical orbit as established by Kepler. Earth, for example, moves forward in its orbit about 30 kilometers (18.5 miles)

FIGURE 21.16 Prominent English Scientist Sir Isaac Newton (1642–1727) Newton discovered that gravity is the force that holds planets in orbit around the Sun. (Photo by De Agostini/Getty Images)

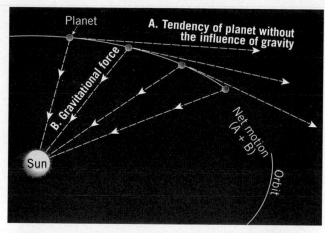

SmartFigure 21.17
Orbital Motion of Earth and Other Planets

each second, and during the same second, the force of gravity pulls it toward the Sun about 0.5 centimeter (1/8 inch). Therefore, as Newton concluded, it is the combination of Earth's forward motion and its "falling" motion that defines its orbit (**FIGURE 21.17**). If gravity were somehow eliminated, Earth would move in a straight line out into space. Conversely, if Earth's forward motion suddenly stopped, gravity would pull it until it crashed into the Sun.

Newton used the law of universal gravitation to express Kepler's third law, which defines the relationship between the orbital periods of the planets and their solar distances. In its new form, Kepler's third law takes into account the masses of the bodies involved and thereby provides a method for determining the mass of a body when the orbit of one of its satellites is known. For example, the mass of the Sun is known from Earth's orbit, and Earth's mass has been determined from the orbit of the Moon. In fact, the mass of any body with a satellite can be determined. The masses of bodies that do not have satellites can be determined only if the bodies noticeably affect the orbit of a neighboring body or of a nearby artificial satellite.

21.2 CONCEPT CHECKS

1 What major change did Copernicus make in the Ptolemaic system? Why was this change philosophically different?

2 What data did Tycho Brahe collect that was useful to Johannes Kepler in his quest to describe planetary motion?

3 Who discovered that planetary orbits are ellipses rather than circles?

4 Does Earth move faster in its orbit near perihelion (January) or near aphelion (July)?

5 Explain why Galileo's discovery of a rotating Sun supports the Copernican view of a Sun-centered universe.

6 Newton discovered that the orbits of the planets result from opposing forces. Briefly explain these forces.

21.3 | POSITIONS IN THE SKY
Compare the equatorial system of coordinates used to establish the position of the stars with longitude and latitude. Explain how the positions of stars are described using declination and right ascension.

If you gaze at the stars away from city lights, you will get the distinct impression that the stars produce a spherical shell surrounding Earth. This impression seems so real that it is easy to understand why the early Greeks regarded the stars as being fixed to a crystalline celestial sphere. Although we realize that no such sphere exists, it is convenient to use this concept to map the stars and other celestial objects. We describe two mapping systems that use the concept of celestial sphere: (1) the division of the sky into areas called *constellations* and (2) the extension of Earth's lines of longitude and latitude into space (the *equatorial system*).

Constellations

The natural fascination people have with the star-studded skies led them to name the patterns they saw. These configurations, called **constellations** (*con* = with, *stella* = star), were named in honor of mythological characters or great heroes, such as Orion the hunter (see GEOgraphics on pages 000–000). Sometimes it takes a bit of imagination to identify the intended subjects, as most constellations were probably not originally thought of as likenesses. Although many of the constellations originated from Greek mythology, the Greeks adopted most of them from the Babylonians, Egyptians, and Mesopotamians.

Although the stars that make up constellations all appear to be the same distance from Earth, this is not the case. Some are many times farther away than others. Thus, the stars in a particular constellation are not associated with each other in any important physical way. In addition, various cultural groups, including Native Americans and the Chinese, attached their own names, pictures, and stories to the constellations. For example, the stars that compose the constellation Leo the lion are said to represent a horse in the ancient Chinese zodiac.

Because the solar system is "flat," like a whirling Frisbee, the planets orbit the Sun along nearly the same plane. Therefore, the planets, Sun, and Moon all appear to move along a band around the sky known as the *zodiac*. Because Earth's Moon cycles through its phases about 12 times each year, the Babylonians divided the zodiac into

Orion the Hunter

Orion, a prominent constellation located on the celestial equator, is visible around the world. It is one of the most conspicuous constellations to decorate the winter skies in the Northern Hemisphere. Named after a hunter in Greek mythology, its brightest stars are Rigel, a luminous blue-white star and Betelgeuse, a red supergiant. Many of the other brightest stars in the constellation are hot, massive blue stars.

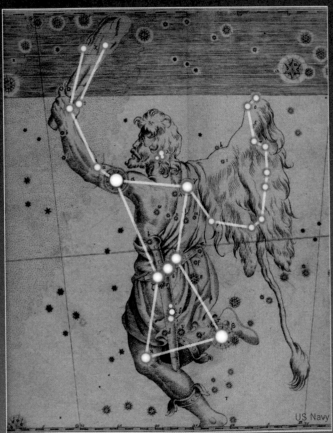

▲ Artist's depiction of Orion based on Greek mythology. Ancient Chinese observers visualized a different image and named this constellation the White Tiger.

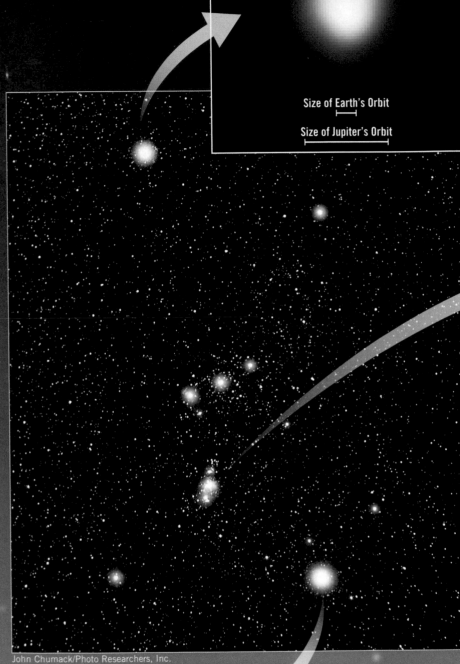

HST/NASA

Size of Betelgeuse

Size of Earth's Orbit

Size of Jupiter's Orbit

John Chumack/Photo Researchers, Inc.

Rigel, also known as Beta Orionis (β Orionis), is the brightest star in the constellation and the sixth brightest star in the night sky. As seen from Earth, Rigel is actually a triple star system. The primary star (Rigel A) of the three-star system is a blue-white massive star that is about 130,000 times more luminous than the Sun. Although Rigel has the designation "beta," it is almost always brighter than Alpha Orionis (Betelgeuse). Rigel is one of the "model stars" by which other stars are compared and classified.

Betelgeuse (Alpha Orionis) is a red supergiant that makes up the shoulder of the winter constellation Orion the hunter. Betelgeuse is so huge that if placed at the center of our solar system, its outer atmosphere would extend beyond the orbit of Jupiter.

Occurring together at the heart of the Orion Nebula is a cluster of four dazzling, young stars, known as the Trapezium. This quartet of stars is much hotter and brighter than the Sun, and collectively provides enough radiation to make the entire nebula glow.

HST/NASA

HST/NASA

The Orion Nebula is visible with the naked eye as the middle "star" in the sword of Orion, which consists of the three stars located south of Orion's Belt. The star appears fuzzy to sharp-eyed observers, and its cloud-like nature is obvious through binoculars or a small telescope. It is one of the brightest nebulae in the night sky and the closest region to Earth where massive stars are born. Within this nebula, astronomers have observed disk-shaped structures composed of dust and gases that orbit protostars. These structures provide insight into the processes of how stars and planetary systems develop from collapsing clouds of gas and dust.

FIGURE 21.18
Astronomical Coordinate System on the Celestial Sphere

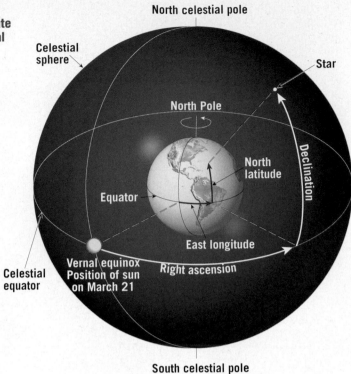

North celestial pole

Celestial sphere

Star

North Pole

Declination

North latitude

Equator

East longitude

Celestial equator

Vernal equinox
Position of sun
on March 21

Right ascension

South celestial pole

12 constellations. The dozen constellations of the zodiac ("Zone of Animals," so named because some constellations represent animals) are Aries, Taurus, Gemini, Cancer, Leo, Virgo, Libra, Scorpio, Sagittarius, Capricorn, Aquarius, and Pisces. These names may be familiar to you as the astrological signs of the zodiac.

Today, 88 constellations are recognized, and they are used to divide the sky into units, just as state boundaries divide the United States. Every star in the sky is within the boundaries of one of these constellations. Astronomers use

constellations when they want to roughly identify the area of the heavens they are observing. For a student, constellations provide a good way to become familiar with the night sky.

Some of the brightest stars in the heavens were given proper names, such as Sirius, Arcturus, and Betelgeuse. In addition, the brightest stars in a constellation are generally named in order of their brightness by the letters of the Greek alphabet—alpha (α), beta (β), and so on—followed by the name of the parent constellation. For example, Sirius, the brightest star in the constellation Canis Major (Larger Dog), is also called Alpha (α) Canis Majoris.

The Equatorial System

The **equatorial system** divides the celestial sphere into coordinates that are similar to the latitude and longitude system we use for establishing locations on Earth's surface (**FIGURE 21.18**). Because the celestial sphere (night sky) appears to rotate around an imaginary line extending from Earth's axis, the north and south celestial poles are aligned with the terrestrial North Pole and South Pole. The north celestial pole happens to be very near the bright star whose various names reflect its location: "pole star," Polaris, and North Star. To an observer in the Northern Hemisphere, the stars appear to circle Polaris, because it, like the North Pole, is in the center of motion (**FIGURE 21.19**). (**FIGURE 21.20** shows how to locate the North Star using two stars, called pointer stars, located on the outside of the "dipper" in the easily located constellation the Big Dipper.)

Now, imagine a plane through Earth's equator, a plane that extends outward from Earth and intersects the celestial sphere. The intersection of this plane with the celestial sphere is called the *celestial equator* (see Figure 21.18). In the equatorial system, the term *declination* is analogous to latitude, and the term *right ascension* is analogous to longitude (see Figure 21.18). **Declination** (*declinare* = to turn away), like latitude, is the angular distance north or south of the celestial equator. **Right ascension** (*ascendere* = to climb up) is the angular distance measured eastward along the celestial equator from the position of the vernal equinox. (The *vernal equinox* is at the point in the sky where the Sun crosses the celestial equator, at the onset of spring.) While declination (Dec) is expressed in

FIGURE 21.19 Star Trails in the Region of Polaris (North Celestial Pole) on a Time Exposure (Photo by Douglas Kirkland/CORBIS)

degrees, right ascension (RA) is expressed in hours, minutes, and seconds, where each hour is equivalent to 15 degrees. (Earth rotates 15 degrees each hour.) For example, the position of Betelgeuse, a bright star in the constellation Orion, is RA: 5h55m10.3s, Dec: +7°24'25.4".

To visualize distances in the night sky, it helps to remember that the Sun and full Moon have an apparent width of about 0.5 degree.

21.3 CONCEPT CHECKS

1 How do modern astronomers use constellations?
2 How many constellations are currently recognized?
3 How are the brightest stars in a constellation denoted?
4 Briefly describe the *equatorial system*.

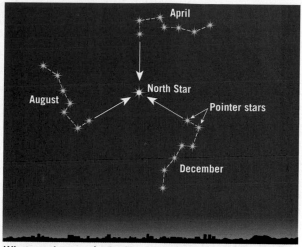

SmartFigure 21.20 Locating the North Star (Polaris) from the Pointer Stars in the Big Dipper The Big Dipper, which is part of the constellation Ursa Major, is shown soon after sunset in December (lower figure), April (upper figure), and August (left).

What an observer in the Northern Hemisphere sees looking north on a clear night three times a year.

21.4 | THE MOTIONS OF EARTH Describe the two primary motions of Earth and explain the difference between a solar day and a sidereal day.

The two primary motions of Earth are *rotation* and *revolution*. A lesser motion is *axial precession*. **Rotation** is the turning, or spinning, of a body on its axis. **Revolution** is the motion of a body, such as a planet or moon, along a path around some point in space. For example, Earth *revolves* around the Sun, and the Moon *revolves* around Earth. Earth also has another very slow motion, known as **axial precession**, which is the gradual change in the orientation of Earth's axis over a period of 26,000 years.

Rotation

The main consequences of Earth's rotation are day and night. Earth's rotation has become a standard method of measuring time because it is so dependable and easy to use. You may be surprised to learn that Earth's rotation is measured in two ways, making two kinds of days. Most familiar is the

mean solar day, the time interval from one noon to the next, which averages about 24 hours. Noon is when the Sun has reached its highest point in the sky.

The **sidereal** (*sider* = star, *at* = pertaining to) **day**, on the other hand, is the time it takes for Earth to make one complete rotation (360 degrees) with respect to a star other than our Sun. The sidereal day is measured by the time required for a star to reappear at the identical position in the sky. The sidereal day has a period of 23 hours, 56 minutes, and 4 seconds (measured in solar time), which is almost 4 minutes shorter than the mean solar day. This difference results because the direction to distant stars changes only infinitesimally, whereas the direction to the Sun changes by almost 1 degree each day. This difference is shown in **FIGURE 21.21**.

Why do we use the mean solar day rather than the sidereal day to measure time? Consider the fact that in sidereal time, "noon" occurs 4 minutes earlier each day. Therefore,

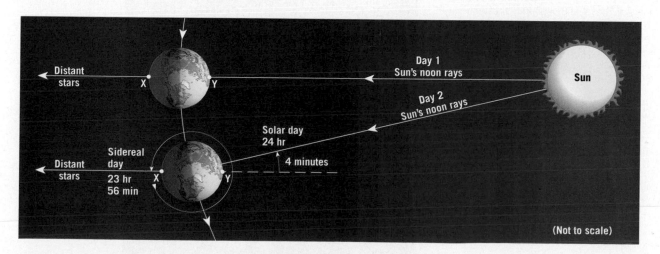

FIGURE 21.21 Illustration of the Difference Between a Solar Day and a Sidereal Day Locations X and Y are directly opposite each other. It takes Earth 23 hours and 56 minutes to make one rotation with respect to the stars (sidereal day). However, notice that after Earth has rotated once with respect to the stars, point Y is not yet returned to the "noon position" with respect to the Sun. Earth has to rotate another 4 minutes to complete the solar day.

FIGURE 21.22
Depiction of the Ecliptic and the Plane of the Ecliptic Earth's orbital motion causes the apparent position of the Sun to shift about 1 degree each day on the celestial sphere. The ecliptic is the path the Sun appears to trace through the stars. The plane of the ecliptic is an imaginary plane that connects the points on the ecliptic.

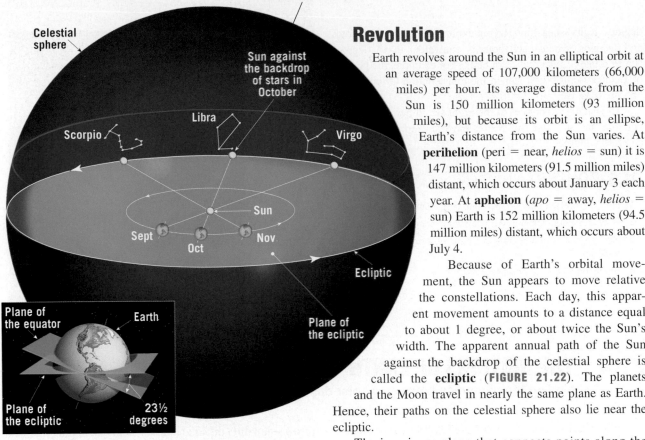

Revolution

Earth revolves around the Sun in an elliptical orbit at an average speed of 107,000 kilometers (66,000 miles) per hour. Its average distance from the Sun is 150 million kilometers (93 million miles), but because its orbit is an ellipse, Earth's distance from the Sun varies. At **perihelion** (peri = near, *helios* = sun) it is 147 million kilometers (91.5 million miles) distant, which occurs about January 3 each year. At **aphelion** (apo = away, *helios* = sun) Earth is 152 million kilometers (94.5 million miles) distant, which occurs about July 4.

Because of Earth's orbital movement, the Sun appears to move relative the constellations. Each day, this apparent movement amounts to a distance equal to about 1 degree, or about twice the Sun's width. The apparent annual path of the Sun against the backdrop of the celestial sphere is called the **ecliptic** (**FIGURE 21.22**). The planets and the Moon travel in nearly the same plane as Earth. Hence, their paths on the celestial sphere also lie near the ecliptic.

The imaginary plane that connects points along the ecliptic is called the **plane of the ecliptic**. As measured from this imaginary plane, Earth's axis is tilted about $23\frac{1}{2}$ degrees (see Figure 21.22). This angle is very important to Earth's inhabitants because the inclination of Earth's axis causes the yearly cycle of seasons, a topic discussed in detail in Chapter 16.

after a span of 6 months, "noon" would occur at "midnight." However, observatories use clocks that keep sidereal time because the stars appear to move through the sky in sidereal time. Simply, if a star is sighted directly south of an observatory at 9:00 P.M. (sidereal time), it will appear in the same direction at that time every (sidereal) day.

EYE ON THE UNIVERSE

Stonehenge, a prehistoric monument whose construction began about 5000 years ago, is located north of the modern city of Salisbury, England. It is the remains of a ring of massive stones; the largest stands 9 meters (30 feet) tall and weighs 25 tons. Some of the smaller stones appear to have been transported 250 kilometers (150 miles) from southwestern Wales. The arrangement of the stones indicates that Stonehenge must have an astronomical connection. Each year on June 21 or 22, an observer standing within the stone circle, looking though the entrance, would see the Sun rising above the heel stone, as shown in the inset image. (To answer the following questions, you may want to refer to the section "Earth–Sun Relationships" in Chapter 16)

QUESTION 1 *What is the significance of June 21–22? Does it mark the occurrence of the spring equinox, the summer equinox, the spring solstice, or the summer solstice?*

QUESTION 2 *On June 21–22, at what latitude do the vertical rays of the Sun strike Earth's surface?*

QUESTION 3 *Most world maps and globes label the latitude that is the answer to Question 2. What is this line of latitude called?*

Adam Woolfitt/Robert Harding

Robin Scagell/Science Source

Precession

A third and very slow movement of Earth is called *axial precession*. Although Earth's axis maintains approximately the same angle of tilt, the direction in which the axis points continually changes (**FIGURE 21.23A**). As a result, the axis traces a circle on the sky. This movement is similar to the "wobble" of a spinning top (**FIGURE 21.23B**). At the present time, the axis points toward the bright star Polaris. In A.D. 14000, it will point toward the bright star Vega, which will then be the North Star for about 1000 years or so (**FIGURE 21.23C**). The period of precession is 26,000 years. By the year 28000, Polaris will once again be the North Star. Precession has only a minor effect on the seasons because Earth's angle of tilt changes only slightly.

In addition to its own movements, Earth accompanies the Sun as it speeds in the direction of the bright star Vega at 20 kilometers (12 miles) per second. Also, the Sun, like other nearby stars, revolves around the galaxy, a trip that requires 230 million years to complete at speeds approaching 250 kilometers (150 miles) per second. In addition, the galaxies themselves are in motion. We are presently approaching one of our nearest galactic neighbors, the Great Galaxy in Andromeda.

21.4 CONCEPT CHECKS

1 Describe the three primary motions of Earth.

2 Explain the difference between the mean solar day and the sidereal day.

3 Define the *ecliptic*.

4 Why does axial precession have little effect on the seasons?

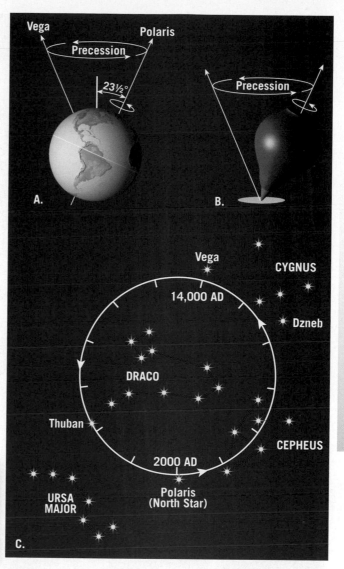

SmartFigure 21.23 Precession of Earth's Axis A. The precession of Earth's axis causes the North Pole to "trace" a circle through the sky during a 26,000-year cycle. Currently, the North Pole points toward Polaris (North Star). In about 12,000 years, Vega will be the North Star. Around 3000 B.C., the North Star was Thuban, a bright star in the constellation Draco. **B.** Precession illustrated by a spinning toy top. **C.** The circle shows the path of the North Pole among some prominent stars and constellations in the northern sky.

21.5 | MOTIONS OF THE EARTH–MOON SYSTEM

Sketch the changing positions of the Earth–Moon system that produce the regular cycle we call the phases of the Moon.

Earth has one natural satellite, the Moon. In addition to accompanying Earth in its annual trek around the Sun, our Moon orbits Earth about once each month. When viewed from a Northern Hemisphere perspective, the Moon moves counterclockwise (eastward) around Earth. The Moon's orbit is elliptical, causing the Earth–Moon distance to vary by about 6 percent, averaging 384,401 kilometers (238,329 miles).

The motions of the Earth–Moon system constantly change the relative positions of the Sun, Earth, and Moon. The results are some of the most noticeable astronomical phenomena: the *phases of the Moon* and the occasional *eclipses of the Sun and Moon*.

Lunar Motions

The cycle of the Moon through its phases requires $29\frac{1}{2}$ days—a time span called the **synodic month**. This cycle was the basis for the first Roman calendar. However, this is the *apparent period* of the Moon's revolution around Earth and not the true period, which takes only $27\frac{1}{3}$ days and is known as the **sidereal month**. The reason for the difference of nearly 2 days each cycle is shown in **FIGURE 21.24**. Notice that as the Moon orbits Earth, the Earth–Moon system also moves in an orbit around the Sun. Consequently, even after the Moon has made a complete revolution around Earth, it has not yet reached its starting position with respect to the Sun, which is directly between the Sun and Earth (*new-Moon phase*). This motion takes an additional 2 days.

An interesting fact concerning the motions of the Moon is that its period of rotation around its axis and its revolution around Earth are the same—$27\frac{1}{3}$ days. Because of this, the same lunar hemisphere always faces Earth. All of the landings of the manned *Apollo* missions were confined to the Earth-facing side. Only orbiting satellites and astronauts have seen the "back" side of the Moon.

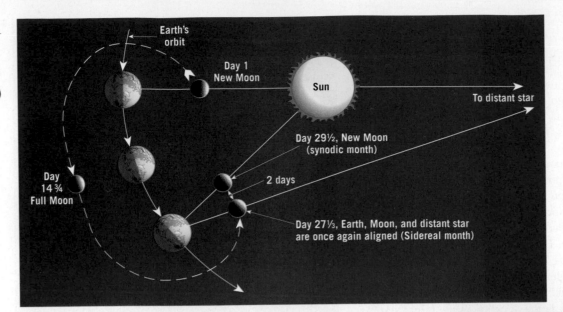

FIGURE 21.24 The Difference Between a Sidereal Month (27⅓ days) and a Synodic Month (29½ days) Distances and angles are not shown to scale.

SmartFigure 21.25 Phases of the Moon
A. The outer figures show the phases as seen from Earth. **B.** Compare these photographs with the diagram. (Photos © UC Regents/Lick Observatory)

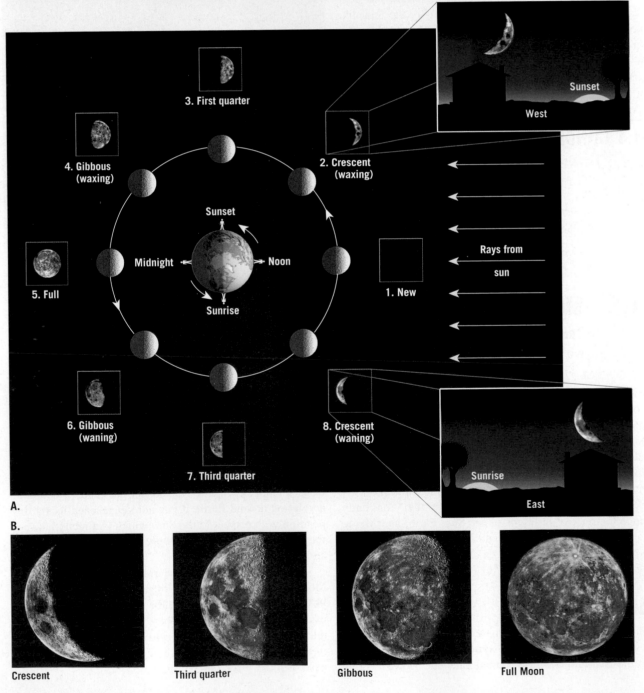

A.

B.

Crescent

Third quarter

Gibbous

Full Moon

Because the Moon rotates on its axis only once every $27\frac{1}{3}$ days, any location on its surface experiences periods of daylight and darkness lasting about 2 weeks. This, along with the absence of an atmosphere, accounts for the high surface temperature of 127°C (261°F) on the day side of the Moon and the low surface temperature of –173°C (–280°F) on its night side.

Phases of the Moon

The first astronomical phenomenon to be understood was the regular cycle of the **phases of the Moon**. On a monthly basis, we observe the phases as a systematic change in the amount of the Moon that appears illuminated (**FIGURE 21.25**). We will choose the "new-Moon" position in the cycle as a starting point. About 2 days after the new Moon, a thin sliver (*crescent phase*) can be seen with the naked eye low in the western sky just after sunset. During the following week, the illuminated portion of the Moon that is visible from Earth increases (*waxing*) to a half-circle (*first-quarter phase*) that can be seen from about noon to midnight. In another week, the complete disk (*full-Moon phase*) can be seen rising in the east as the Sun sinks in the west. During the next 2 weeks, the percentage of the Moon that can be seen steadily declines (*waning*), until the Moon disappears altogether (*new-Moon phase*). The cycle soon begins anew, with the reappearance of the crescent Moon.

The lunar phases are a result of the motion of the Moon and the sunlight that is reflected from its surface (see Figure 21.25B). Half of the Moon is illuminated at all times (note the inner group of Moon sketches in Figure 21.25A). But to an earthbound observer, the percentage of the bright side that is visible depends on the location of the Moon with respect to the Sun and Earth. When the Moon lies *between* the Sun and Earth, none of its bright side faces Earth, so we see the new-Moon ("no-Moon") phase. Conversely, when the Moon lies on the side of Earth opposite the Sun, all of its lighted side faces Earth, so we see the full Moon. At all positions between these extremes, an intermediate amount of the Moon's illuminated side is visible from Earth.

21.5 CONCEPT CHECKS

1 Compare the *synodic month* with the *sidereal month*.

2 What is the approximate length of the cycle of the phases of the Moon?

3 What phenomenon results from the fact that the Moon's periods of rotation and revolution are the same?

4 The Moon rotates very slowly (once in $27\frac{1}{3}$ days) on its axis. How does this affect the lunar surface temperature?

5 What is different about the crescent phase that precedes the new-Moon phase and that which follows the new-Moon phase?

6 What phase of the Moon occurs approximately 1 week after the new Moon? 2 weeks?

21.6 | ECLIPSES OF THE SUN AND MOON Sketch the positions of the Earth–Moon system that produce a lunar eclipse, as well as a solar eclipse.

Along with understanding the Moon's phases, the early Greeks also realized that eclipses are simply shadow effects. When the Moon moves in a line directly between Earth and the Sun, which can occur only during the new-Moon phase, it casts a dark shadow on Earth, producing a **solar eclipse** (*eclipsis* = failure to appear) (**FIGURE 21.26**). Conversely, the Moon is eclipsed (**lunar eclipse**) when it moves within Earth's shadow, a situation that is possible only during the full-Moon phase (**FIGURE 21.27**).

Why does a solar eclipse not occur with every new-Moon phase and a lunar eclipse with every full Moon? They would, if the orbit of the Moon lay exactly along the plane of Earth's orbit. However, the Moon's orbit is inclined about 5 degrees to the plane of the ecliptic. Thus, during most new-Moon

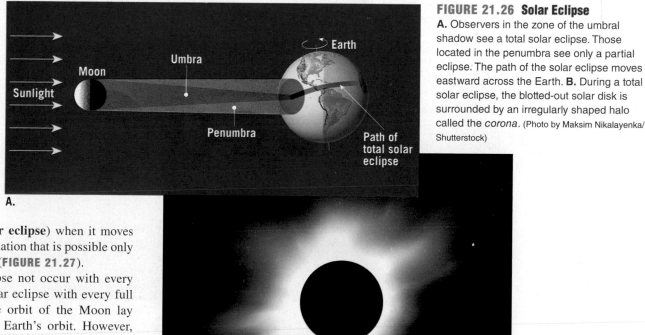

FIGURE 21.26 Solar Eclipse
A. Observers in the zone of the umbral shadow see a total solar eclipse. Those located in the penumbra see only a partial eclipse. The path of the solar eclipse moves eastward across the Earth. **B.** During a total solar eclipse, the blotted-out solar disk is surrounded by an irregularly shaped halo called the *corona*. (Photo by Maksim Nikalayenka/Shutterstock)

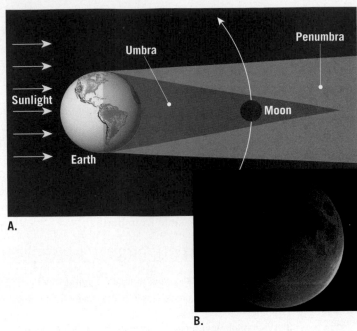

SmartFigure 21.27 Lunar Eclipse A. During a total lunar eclipse, the Moon's orbit carries it into the dark shadow of Earth (umbra). During a partial eclipse, only a portion of the Moon enters the umbra. B. During a total lunar eclipse, a dark, copper-colored Moon is observed. The color is a result of a small amount of sunlight that is reddened by Earth's atmosphere—for the same reason sunsets appear red. This light is refracted (bent) toward the Moon's surface. (Photo by Eckhard Slawik/Science Source)

beginning, middle, and end. These occur as a solar eclipse flanked by two lunar eclipses or vice versa. Furthermore, it occasionally happens that the first set of eclipses for the year occurs at the very beginning of a year, the second set in the middle, and a third set before the calendar year ends, resulting in six eclipses in that year. More rarely, if one of these sets consists of three eclipses, the total number of eclipses in a year can reach seven, which is the maximum.

During a total lunar eclipse, Earth's circular shadow moves slowly across the disk of the full Moon. When totally eclipsed, the Moon is completely within Earth's shadow but is still visible as a coppery disk because Earth's atmosphere bends some long-wavelength light (red) into its shadow. Some of this light reflects off the Moon and back to us. A total eclipse of the Moon can last up to 4 hours and is visible to anyone on the side of Earth facing the Moon.

During a total solar eclipse, the Moon casts a circular shadow that is never wider than 275 kilometers (170 miles), about the size of South Carolina. This shadow traces a stripe on Earth's surface. Anyone observing in this region will see the Moon slowly block the Sun from view and the sky darken (**FIGURE 21.28**). Near totality, a sharp drop in temperature of a few degrees is experienced. The solar disk is completely blocked for a maximum of only 7 minutes because the Moon's shadow is so small. At totality, the dark Moon is seen covering the complete solar disk, and

phases, the shadow of the Moon passes either above or below Earth; during most full-Moon phases, the shadow of Earth misses the Moon. An eclipse can take place only when a new- or full-Moon phase occurs while the Moon's orbit crosses the plane of the ecliptic.

Because these conditions are normally met only twice a year, the usual number of eclipses is four. These occur as a set of one solar and one lunar eclipse, followed 6 months later with another set. Occasionally the alignment is such that three eclipses can occur in a 1-month period—at the

FIGURE 21.28 Montage of Images Showing a Complete Solar Eclipse This sequence of photos moving from the upper left to the lower right shows the stages of a total solar eclipse. (Photo by Dr. Fred Espenak/ Science Source)

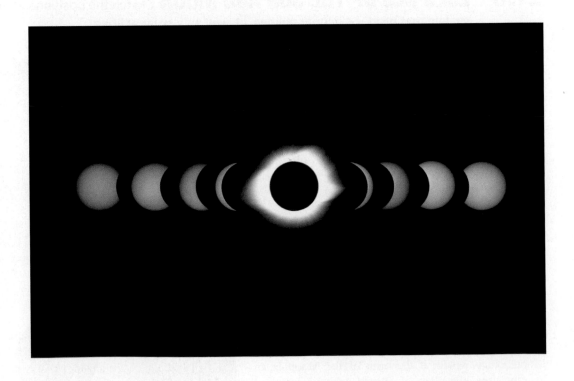

only the Sun's brilliant white outer atmosphere is visible (see Figure 21.26). Total solar eclipses are visible only to people in the dark part of the Moon's shadow (*umbra*), while partial eclipses are seen by those in the light portion (*penumbra*) (see Figure 21.26).

Partial solar eclipses are most common in the polar regions because it is these areas that the penumbra blankets when the dark umbra of the Moon's shadow just misses Earth. A total solar eclipse is a rare event at any given location. The next one that will be visible from the contiguous United States will occur on August 21, 2017.

21.6 CONCEPT CHECKS

1 Sketch the locations of the Sun, Moon, and Earth during a solar eclipse and during a lunar eclipse.

2 How many eclipses normally occur each year?

3 Solar eclipses are slightly more common than lunar eclipses. Why, then, is it more likely that your region of the country will experience a lunar eclipse than a solar eclipse?

4 How long can a total eclipse of the Moon last? How about a total eclipse of the Sun?

21 CONCEPTS IN REVIEW | Origins of Modern Astronomy

21.1 ANCIENT ASTRONOMY

Explain the geocentric view of the solar system and describe how it differs from the heliocentric view.

KEY TERMS: geocentric, celestial sphere, heliocentric, Ptolemaic system, retrograde motion

- Early agrarian cultures, whose survival depended on seasonal change, believed that if the heavenly objects could control the seasons, they must also strongly influence all other Earth events.
- The early Greeks held a geocentric ("Earth-centered") view of the universe, believing that Earth is a sphere that stays motionless at the center of the universe. They believed that the Moon, the Sun, and the known planets—Mercury, Venus, Mars, Jupiter, and Saturn—orbited Earth.
- The early Greeks believed that the stars traveled daily around Earth on a transparent, hollow celestial sphere. In A.D. 141, Claudius Ptolemy documented this geocentric view, called the Ptolemaic system, which became the dominant view of the solar system for over 15 centuries.

21.2 THE BIRTH OF MODERN ASTRONOMY

List and describe the contributions to modern astronomy of Nicolaus Copernicus, Tycho Brahe, Johannes Kepler, Galileo Galilei, and Isaac Newton.

KEY TERMS: astronomical unit (AU), inertia, law of universal gravitation,

- Modern astronomy evolved during the 1500s and 1600s. The scientists involved in the transition from an astronomy that merely describes what is observed to an astronomy that tries to explain why the universe behaves the way it does included Nicolaus Copernicus, Tycho Brahe, Johannes Kepler, Galileo Galilei, and Isaac Newton.
- Nicolaus Copernicus (1473–1543) reconstructed the solar system with the Sun at the center and the planets orbiting around it, but he erroneously continued to use circles to represent the orbits of planets. The establishment of his day rejected his Sun-centered view.
- Tycho Brahe's (1546–1601) observations of the planets were far more precise than any made previously and are his legacy to astronomy.
- Johannes Kepler (1571–1630) used Tycho Brahe's observations to usher in a new astronomy with the formulation of his three laws of planetary motion.
- After constructing his own telescope, Galileo Galilei (1564–1642) made many important discoveries that supported the Copernican view of a Sun-centered solar system. This included charting the movement of Jupiter's four largest moons, proving that Earth was not the center of all planetary motion.
- Sir Isaac Newton (1642–1727) demonstrated that the orbit of a planet is a result of a planet's inertia (the tendency of a planet to move in a straight line) and the Sun's gravitational attraction, which bends the planet's path into an elliptical orbit.

21.3 POSITIONS IN THE SKY

Compare the equatorial system of coordinates used to establish the position of the stars with longitude and latitude. Explain how the positions of stars are described using declination and right ascension.

KEY TERMS: constellations, equatorial system, declination, right ascension

- As early as 5000 years ago, people began naming the configurations of stars, called *constellations*, in honor of mythological characters or great heroes. Today, 88 constellations are recognized that divide the sky into units, just as state boundaries divide the United States.
- One method for locating stars, called the *equatorial system*, divides the celestial sphere into a coordinate system similar to the grid system of latitude–longitude used for locating places on Earth's surface. *Declination*, like latitude, is the angular distance north or south of the *celestial equator*. *Right ascension* is the angular distance measured eastward from the position of the *vernal equinox* (the point in the sky where the Sun crosses the celestial equator at the onset of spring).

21.4 THE MOTIONS OF EARTH

Describe the two primary motions of Earth and explain the difference between a solar day and a sidereal day.

KEY TERMS: rotation, revolution, axial precession, mean solar day, sidereal day, perihelion, aphelion, ecliptic, plane of the ecliptic

- The two primary motions of Earth are *rotation* (the turning, or spinning, of a body on its axis) and *revolution* (the motion of a body, such as a planet or moon, along a path around some point in space).
- Earth's rotation can be measured in two ways, making two kinds of days. The *mean solar day* is the time interval from one noon (the time of day when the Sun is highest in the sky) to the next, which averages about 24 hours. In contrast, the *sidereal day* is the time it takes for Earth to make one complete rotation with respect to a star other than the Sun, a period of 23 hours, 56 minutes, and 4 seconds.
- Earth revolves around the Sun in an elliptical orbit at an average distance from the Sun of 150 million kilometers (93 million miles). At *perihelion* (closest to the Sun), which occurs in January, Earth is 147 million kilometers (91.5 million miles) from the Sun. At *aphelion* (farthest from the Sun), which occurs in July, Earth is 152 million kilometers (94.5 million miles) distant. The imaginary plane that connects Earth's orbit with the celestial sphere is called the *plane of the ecliptic*.
- Another very slow motion of Earth is *precession*—the slow motion of Earth's axis that traces out a cone over a period of 26,000 years.

21.5 MOTIONS OF THE EARTH–MOON SYSTEM

Sketch the changing positions of the Earth–Moon system that produce the regular cycle we call the phases of the Moon.

KEY TERMS: synodic month, sidereal month, phases of the Moon

- One of the first astronomical phenomena to be understood was the regular cycle of the phases of the Moon. The phases of the Moon are a result of the motion of the Moon around Earth and the portion of the bright side of the Moon that is visible to an observer on Earth.
- The cycle of the Moon through its phases requires $29\frac{1}{2}$ days, a time span called the *synodic month*. However, the true period of the Moon's revolution around Earth takes $27\frac{1}{3}$ days and is known as the *sidereal month*. The difference of nearly 2 days is due to the fact that as the Moon orbits Earth, the Earth–Moon system also moves in an orbit around the Sun.

21.6 ECLIPSES OF THE SUN AND MOON

Sketch the positions of the Earth–Moon system that produce a lunar eclipse, as well as a solar eclipse

KEY TERMS: solar eclipse, lunar eclipse

- In addition to understanding the Moon's phases, the early Greeks also realized that eclipses are simply shadow effects. When the Moon moves in a line directly between Earth and the Sun, which can occur only during the new-Moon phase, it casts a dark shadow on Earth, producing a *solar eclipse*.
- A *lunar eclipse* takes place when the Moon moves within the shadow of Earth during the full-Moon phase.
- Because the Moon's orbit is inclined about 5 degrees to the plane that contains the Earth and Sun (the plane of the ecliptic), during most new- and full-Moon phases, no eclipse occurs. Only if a new- or full-Moon phase occurs as the Moon crosses the plane of the ecliptic can an eclipse take place. The usual number of eclipses is four per year.

GIVE IT SOME THOUGHT

1. Refer to Figure 21.3 and imagine that Eratosthenes had measured the difference in the angles of the noonday Sun between Syene and Alexandria to be 10 degrees instead of 7 degrees. Consider how this measurement would have affected his calculation of Earth's circumference as you answer the following questions.
   a. Would this new measurement lead to a more accurate calculation?
   b. Would this new measurement lead to an estimate for the circumference of Earth that is larger or smaller than Eratosthenes's original estimate?

2. Use Kepler's third law to answer the following questions:
   a. Determine the period of a planet with a solar distance of 10 AU.
   b. Determine the distance between the Sun and a planet with a rotational period of 5 years.
   c. Imagine two bodies, one twice as large as the other, orbiting the Sun at the same distance. Which of the bodies, if either, would move faster than the other?

3. Galileo used his telescope to observe the planets and moons in our solar system. These observations allowed him to determine the positions and relative motions of the Sun, Earth, and other objects in the solar system. Refer to Figure 21.15A, which shows an Earth-centered solar system, and Figure 21.15B, which shows a Sun-centered solar system, to complete the following:
   a. Describe the phases of Venus that an observer on Earth would see for the Earth-centered model of the solar system.
   b. Describe the phases of Venus that an observer on Earth would see for the Sun-centered model of the solar system.
   c. Explain how Galileo used observations of the phases of Venus to determine the correct positions of the Sun, Earth, and Venus.

4. Refer to the accompanying diagram, which shows three asteroids (A, B, and C) that are being pulled by the gravitational force exerted on them by their partner asteroid shown on the left. How will the strength of the

gravitational force felt by each asteroid (A, B, and C) compare? (Assume that all these asteroids are composed of the same material.)

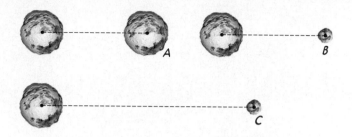

5. Refer to the accompanying diagram, which shows two pairs of asteroids, Pair D and Pair E. Is it possible for the asteroids in Pair D to be experiencing the same degree of gravitational force as the asteroids in Pair E? Explain your answer.

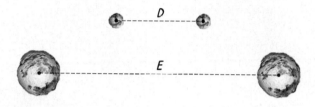

6. Imagine that Earth rotates on its axis at half its current rate. How much time would be required to capture the time-lapse photo shown in Figure 21.19?

7. If we were able to reverse the direction of Earth's rotation, would the solar day be longer, shorter, or the same?

8. Refer to the accompanying photo of the Moon to complete the following:

   a. When you observe the phase of the Moon shown, is the Moon waxing or waning?

   b. What time of day can this phase of the Moon be observed?

9. Imagine that you are looking up at a full Moon. At the same time, an astronaut on the Moon is viewing Earth. In what phase will Earth appear to be from the astronaut's vantage point? Sketch a diagram to illustrate your answer.

© UC Regents/Lick Oservatory

10. If the Moon's orbit were precisly aligned with the plane of Earth's orbit, how many eclipses (solar and lunar) would occur in a 6-month period of time? If the Moon's orbit were tilted 90 degrees with respect to the plane of Earth's orbit, how many eclipses (solar and lunar) would occur in a 6-month period?

EXAMINING THE **EARTH SYSTEM**

1. Currently, Earth is closest to the Sun (perihelion) in January (147 million kilometers [91.5 million miles]) and farthest from the Sun in July (152 million kilometers [94.5 million miles]). As a result of the precession of Earth's axis, 12,000 years from now perihelion (closest) will occur in July, and aphelion (farthest) will take place in January. Assuming no other changes, how might this change *average* summer temperatures for your location? What about *average* winter temperatures? What might the impact be on the biosphere and hydrosphere? (To aid your understanding of the effect of Earth's orbital parameters on the seasons, you may want to review the section "Variations in Earth's Orbit" in Chapter 6, page 191.)

2. In what ways do the interactions between Earth and its Moon influence the Earth system? If Earth did not have a moon, how might the atmosphere, hydrosphere, geosphere, and biosphere be different?

MasteringGeology™

Looking for additional review and test prep materials? Visit the Self Study area in **www.masteringgeology.com** to find practice quizzes, study tools, and multimedia that will aid in your understanding of this chapter's content. In **MasteringGeology™** you will find:

- GEODe: Earth Science: An interactive visual walkthrough of key concepts
- Geoscience Animation Library: More than 100 animations illuminating many difficult-to-understand Earth science concepts

- In The News RSS Feeds: Current Earth science events and news articles are pulled into the site with assessment
- Pearson eText
- Optional Self Study Quizzes
- Web Links
- Glossary
- Flashcards

22 Touring Our Solar System[1]

FOCUS ON CONCEPTS

Each statement represents the primary **LEARNING OBJECTIVE** for the corresponding major heading within the chapter. After you complete the chapter, you should be able to:

22.1 Describe the formation of the solar system according to the nebular theory. Compare and contrast the terrestrial and Jovian planets.

22.2 List and describe the major features of Earth's Moon and explain how maria basins were produced.

22.3 Outline the principal characteristics of Mercury, Venus, and Mars. Describe their similarities to and differences from Earth.

22.4 Compare and contrast the four Jovian planets.

22.5 List and describe the principal characteristics of the small bodies that inhabit the solar system.

View of the windblown Martian surface obtained by the NASA rover *Curiosity*. The black-colored rocks are volcanic and have a composition similar to that of the rock basalt found on the Hawaiian islands. (Photo courtesy of NASA/JPL-Caltech/MSSS)

---

[1]This chapter was revised with the assistance of Professors Teresa Tarbuck and Mark Watry.

lanetary geology is the study of the formation and evolution of the bodies in our solar system—including the eight planets and myriad smaller objects: moons, dwarf planets, asteroids, comets, and meteoroids. Studying these objects provides valuable insights into the dynamic processes that operate on Earth. Understanding how other atmospheres evolve helps scientists build better models for predicting climate change. Studying tectonic processes on other planets helps us appreciate how these complex interactions alter Earth. In addition, seeing how erosional forces work on other bodies allows us to observe the many ways landscapes are created. Finally, the uniqueness of Earth, a body that harbors life, is revealed through the exploration of other planetary bodies.

22.1 | OUR SOLAR SYSTEM: AN OVERVIEW

Describe the formation of the solar system according to the nebular theory. Compare and contrast the terrestrial and Jovian planets.

The Sun is at the center of a revolving system, trillions of miles wide, consisting of eight planets, their satellites, and numerous smaller asteroids, comets, and meteoroids. An estimated 99.85 percent of the mass of our solar system is contained within the Sun. Collectively, the planets account for most of the remaining 0.15 percent. Starting from the Sun, the planets are Mercury, Venus, Earth, Mars, Jupiter, Saturn, Uranus, and Neptune (**FIGURE 22.1**). Pluto was

 SmartFigure 22.1 Orbits of the Planets **A.** Artistic view of the solar system, in which planets are not drawn to scale. **B.** Positions of the planets shown to scale using astronomical units (AU), where 1 AU is equal to the average distance from Earth to the Sun—150 million kilometers (93 million miles).

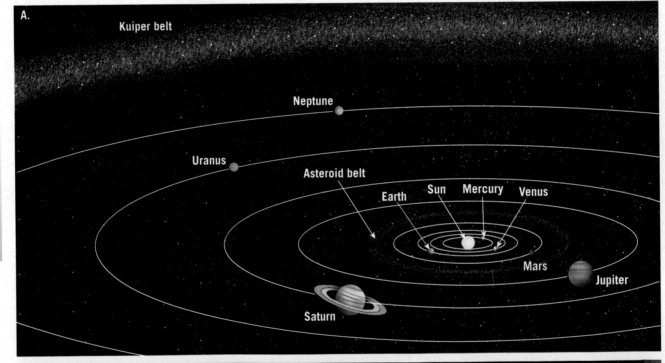

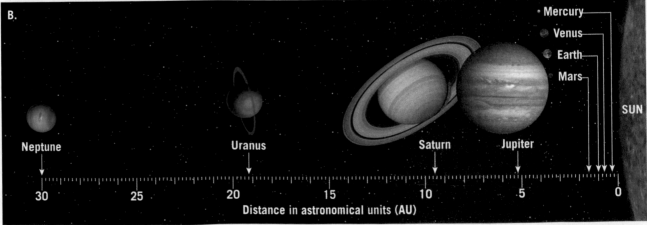

TABLE 22.1 Planetary Data

Planet	Symbol	AU*	Mean Distance from Sun		Period of Revolution	Inclination of Orbit	Orbital Velocity	
			Millions of Miles	Millions of Kilometers			mi/s	km/s
Mercury	☿	0.39	36	58	88^d	7°00′	29.5	47.5
Venus	♀	0.72	67	108	225^d	3°24′	21.8	35.0
Earth	⊕	1.00	93	150	365.25^d	0°00′	18.5	29.8
Mars	♂	1.52	142	228	687^d	1°51′	14.9	24.1
Jupiter	♃	5.20	483	778	12^{yr}	1°18′	8.1	13.1
Saturn	♄	9.54	886	1427	30^{yr}	2°29′	6.0	9.6
Uranus	♅	19.18	1783	2870	84^{yr}	0°46′	4.2	6.8
Neptune	♆	30.06	2794	4497	165^{yr}	1°46′	3.3	5.3

Planet	Period of Rotation	Diameter		Relative Mass (Earth = 1)	Average Density (g/cm^3)	Polar Flattening (%)	Eccentricity†	Number of Known Satellites††
		Miles	Kilometers					
Mercury	59^d	3015	4878	0.06	5.4	0.0	0.206	0
Venus	243^d	7526	12,104	0.82	5.2	0.0	0.007	0
Earth	23^{h}56^{m}04^s	7920	12,756	1.00	5.5	0.3	0.017	1
Mars	24^{h}37^{m}23^s	4216	6794	0.11	3.9	0.5	0.093	2
Jupiter	9^{h}56^m	88,700	143,884	317.87	1.3	6.7	0.048	67
Saturn	10^{h}30^m	75,000	120,536	95.14	0.7	10.4	0.056	62
Uranus	17^{h}14^m	29,000	51,118	14.56	1.2	2.3	0.047	27
Neptune	16^{h}07^m	28,900	50,530	17.21	1.7	1.8	0.009	13

*AU = astronomical unit, Earth's mean distance from the Sun.

†Eccentricity is a measure of the amount an orbit deviates from a circular shape. The larger the number, the less circular the orbit.

††Includes all satellites discovered as of December 2012.

recently reclassified as a member of a new class of solar system bodies called *dwarf planets*.

Tethered to the Sun by gravity, all the planets travel in the same direction, on slightly elliptical orbits (**TABLE 22.1**). Gravity causes objects nearest the Sun to travel fastest. Therefore, Mercury has the highest orbital velocity, 48 kilometers (30 miles) per second, and the shortest period of revolution around the Sun, 88 Earth-days. By contrast, the distant dwarf planet Pluto has an orbital speed of just 5 kilometers (3 miles) per second and requires 248 Earth-years to complete one revolution. Most large bodies orbit the Sun approximately in the same plane. The planets' inclination with respect to the Earth–Sun orbital plane, known as the *ecliptic*, is shown in Table 22.1.

Nebular Theory: Formation of the Solar System

The **nebular theory**, which explains the formation of the solar system, proposes that the Sun and planets formed from a rotating cloud of interstellar gases (mainly hydrogen and helium) and dust called the **solar nebula**. As the solar nebula contracted due to gravity, most of the material collected in the center to form the hot *protosun*. The remaining materials formed a thick, flattened, rotating disk, within which matter gradually cooled and condensed into grains and clumps of icy, rocky material. Repeated collisions resulted in most of the material clumping together into larger and larger chunks that eventually became asteroid-sized objects called **planetesimals**.

The composition of planetesimals was largely determined by their proximity to the protosun. As you might expect, temperatures were highest in the inner solar system and decreased toward the outer edge of the disk. Therefore, between the present orbits of Mercury and Mars, the planetesimals were composed of materials with high melting temperatures—metals and rocky substances. Then, through repeated collisions and accretion, these asteroid-sized rocky bodies combined to form the four **protoplanets** that eventually became Mercury, Venus, Earth, and Mars.

The planetesimals that formed beyond the orbit of Mars, where temperatures were low, contained high percentages of ices—water, carbon dioxide, ammonia, and methane—as well as small amounts of rocky and metallic debris. It was mainly from these planetesimals that the four outer planets eventually formed. The accumulation of ices accounts, in part, for the large sizes and low densities of the outer planets. The two most massive planets, Jupiter and Saturn, had surface gravities sufficient to attract and retain large quantities of hydrogen and helium, the lightest elements.

It took roughly 1 billion years after the protoplanets formed for the planets to gravitationally accumulate most of the interplanetary debris. This was a period of intense bombardment as the planets cleared their orbits by collecting much of the leftover material. The "scars" of this period are still evident on the Moon's surface. Because of the gravitational effect of the planets, particularly Jupiter, small bodies were flung into planet-crossing orbits or into interstellar space. The small fraction of interplanetary matter that escaped this violent period became either asteroids or comets. By comparison, the present-day solar system is a much quieter place, although many of these processes continue today at a reduced pace.

The Planets: Internal Structures and Atmospheres

The planets fall into two groups, based on location, size, and density: the **terrestrial (Earth-like) planets** (Mercury, Venus, Earth, and Mars) and the **Jovian (Jupiter-like) planets** (Jupiter, Saturn, Uranus, and Neptune). Because of their relative locations, the four terrestrial planets are also known as *inner planets*, and the four Jovian planets are known as *outer planets*. A correlation exists between planetary locations and sizes: The inner planets are substantially smaller than the outer planets, also known as *gas giants*. For example, the diameter of Neptune (the smallest Jovian planet) is nearly 4 times larger than the diameter of Earth. Furthermore, Neptune's mass is 17 times greater than that of Earth or Venus.

Other properties that differ among the planets include densities, chemical compositions, orbital periods, and numbers of satellites. Variations in the chemical composition of planets are largely responsible for their density differences. Specifically, the average density of the terrestrial planets is about 5 times the density of water, whereas the average density of the Jovian planets is only 1.5 times that of water. In fact, Saturn has a density only 0.7 times that of water, which means that it would float in a sufficiently large tank of water. The outer planets are also characterized by long orbital periods and numerous satellites.

Internal Structures Shortly after Earth formed, segregation of material resulted in the formation of three major layers, defined by their chemical composition—the crust, mantle, and core. This type of chemical separation occurred in the other planets as well. However, because the terrestrial planets are compositionally different from the Jovian planets, the nature of these layers is different as well (**FIGURE 22.2**).

The terrestrial planets are dense, having relatively large cores of iron and nickel. The outer cores of Earth and Mercury are liquid, whereas the cores of Venus and Mars are thought to be only partially molten. This difference is attributable to Venus and Mars having lower internal temperatures than those of Earth and Mercury. Silicate minerals and other lighter compounds make up the mantles of the terrestrial planets. Finally, the silicate crusts of terrestrial planets are relatively thin compared to their mantles.

The two largest Jovian planets, Jupiter and Saturn, likely have small, solid cores consisting of iron compounds, like the cores of the terrestrial planets, and rocky material similar to Earth's mantle. Progressing outward, the layer above the core consists of liquid hydrogen that is under extremely high temperatures and pressures. There is substantial evidence that under these conditions, hydrogen behaves like a metal in that its electrons move freely about and are efficient conductors of both heat and electricity. Jupiter's intense magnetic field is thought to be the result of electric currents flowing within a spinning layer of liquid metallic hydrogen. Saturn's magnetic field is much weaker than Jupiter's, due to its smaller shell of liquid metallic hydrogen. Above this metallic layer, both Jupiter and Saturn are thought to be composed of molecular liquid hydrogen that is intermixed with helium. The outermost layers are gases of hydrogen and helium, as well as ices of water, ammonia, and methane—which mainly account for the low densities of these giants.

Uranus and Neptune also have small iron-rich, rocky cores, but their mantles are likely hot, dense water and ammonia. Above their mantles, the amount of hydrogen and helium increases, but these gases exist in much lower concentrations than in Jupiter and Saturn.

All planets, except Venus and Mars, have significant magnetic fields generated by flow of metallic materials in their liquid cores, or mantles. Venus has a weak field due to the interaction between the solar wind and its uppermost atmosphere (ionosphere), while the weak Martian magnetic field is thought

FIGURE 22.2 Comparing Internal Structures of the Planets

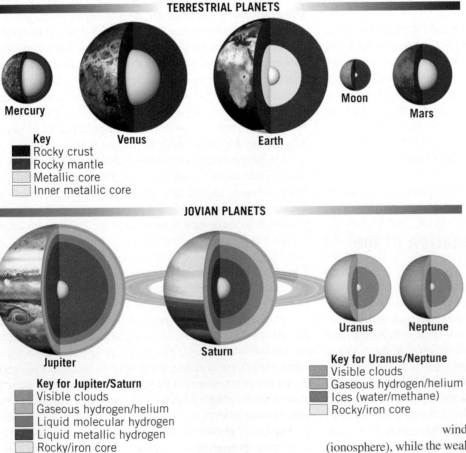

TERRESTRIAL PLANETS

Mercury

Venus

Earth

Moon

Mars

Key
- Rocky crust
- Rocky mantle
- Metallic core
- Inner metallic core

JOVIAN PLANETS

Jupiter

Saturn

Uranus

Neptune

Key for Jupiter/Saturn
- Visible clouds
- Gaseous hydrogen/helium
- Liquid molecular hydrogen
- Liquid metallic hydrogen
- Rocky/iron core

Key for Uranus/Neptune
- Visible clouds
- Gaseous hydrogen/helium
- Ices (water/methane)
- Rocky/iron core

to be a remnant from when its interior was hotter. Magnetic fields play an important role in protecting a planet's surface from bombardment by charged particles of the solar wind—a necessary condition for the survival of life-forms.

The Atmospheres of the Planets

The Jovian planets have very thick atmospheres composed mainly of hydrogen and helium, with lesser amounts of water, methane, ammonia, and other hydrocarbons. By contrast, the terrestrial planets, including Earth, have relatively meager atmospheres composed of carbon dioxide, nitrogen, and oxygen. Two factors explain these significant differences—solar heating (temperature) and gravity (**FIGURE 22.3**). These variables determine what planetary gases, if any, were captured by planets during the formation of the solar system and which were ultimately retained.

During planetary formation, the inner regions of the developing solar system were too hot for ices and gases to condense. By contrast, the Jovian planets formed where temperatures were low and solar heating of planetesimals was minimal. This allowed water vapor, ammonia, and methane to condense into ices. Hence, the gas giants contain large amounts of these volatiles. As the planets grew, the largest Jovian planets, Jupiter and Saturn, also attracted large quantities of the lightest gases, hydrogen and helium.

How did Earth acquire water and other volatile gases? It seems that early in the history of the solar system, gravitational tugs by the developing protoplanets sent planetesimals into very eccentric orbits. As a result, Earth was bombarded with icy objects that originated beyond the orbit of Mars. This was a fortuitous event for organisms that currently inhabit our planet. Mercury, our Moon, and numerous other small bodies lack significant atmospheres, even though they certainly would have been bombarded by icy bodies early in their development.

Airless bodies develop where solar heating is strong and/or gravities are weak. Simply stated, *small warm bodies* have a better chance of losing their atmospheres because gas molecules are more energetic and need less speed to escape their weak gravities. For example, warm bodies with small surface gravity, such as our Moon, are unable to hold even heavy gases such as carbon dioxide and nitrogen. Mercury is massive enough to hold trace amounts of hydrogen, helium, and oxygen gas.

The slightly larger terrestrial planets, Earth, Venus, and Mars, retain some heavy gases, including water vapor, nitrogen, and carbon dioxide. However, their atmospheres are miniscule compared to their total mass. Early in their development, the terrestrial planets probably had much thicker atmospheres. Over time, however, these primitive atmospheres gradually changed as light gases trickled away into space. For example, Earth's atmosphere continues to leak hydrogen and helium (the two lightest gases) into space. This phenomenon occurs near the top of Earth's atmosphere, where air is so tenuous that nothing stops the fastest-moving particles from flying off into

space. The speed required to escape a planet's gravity is called **escape velocity**. Because hydrogen is the lightest gas, it most easily reaches the speed needed to overcome Earth's gravity.

Billions of years in the future, the loss of hydrogen (one of the components of water) will eventually "dry out" Earth's oceans, ending its hydrologic cycle. Life, however, may remain sustainable in Earth's polar regions.

The massive Jovian planets have strong gravitational fields and thick atmospheres. Furthermore, because of their great distance from the Sun, solar heating is minimal. This explains why Saturn's moon Titan, which is small compared to Earth but much further from the Sun, retains an atmosphere. Because the molecular motion of a gas is temperature dependent, even hydrogen and helium move too slowly to escape the gravitational pull of the Jovian planets.

Planetary Impacts

Planetary impacts between solar system bodies have occurred throughout the history of the solar system. On bodies that have little or no atmosphere (like the Moon) and, therefore, no air resistance, even the smallest pieces of interplanetary debris (meteorites) can reach the surface. At high enough velocities, this debris can produce microscopic cavities on individual mineral grains. By contrast, large **impact craters** result from collisions with massive bodies, such as asteroids and comets.

Planetary impacts were considerably more common in the early history of the solar system than they are today.

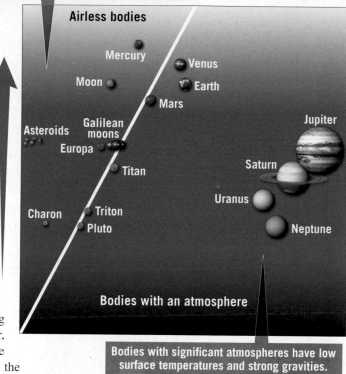

Airless worlds have relatively warm surface temperatures and/or weak gravities.

Bodies with significant atmospheres have low surface temperatures and strong gravities.

SmartFigure 22.3 Bodies with Atmospheres Versus Airless Bodies Two factors largely explain why some solar system bodies have thick atmospheres, whereas others are airless. Airless worlds have relatively warm surface temperatures and/or weak gravities. Bodies with significant atmospheres have low surface temperatures and strong gravities.

FIGURE 22.4 Formation of an Impact Crater

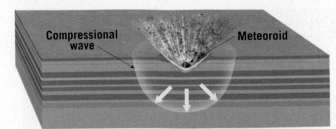

A. The energy of a rapidly moving body is transformed into heat and shock waves.

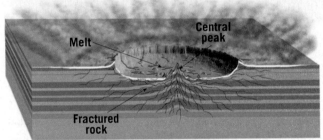

B. The rebound of over-compressed rock causes debris to be explosively ejected from the crater.

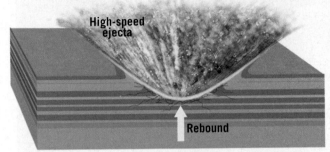

C. Heating melts some material that may be ejected from the crater as glass beads.

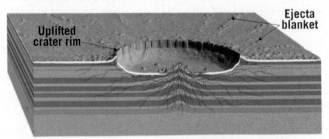

D. Small secondary craters often form when the material "splashed" from the impact crater strikes the surrounding landscape.

FIGURE 22.5 Lunar Crater Euler This 20-kilometer-wide (12-mile-wide) crater is located in the southwestern part of Mare Imbrium. Clearly visible are the bright rays, central peak, secondary craters, and large accumulation of ejecta near the crater rim. (Courtesy of NASA)

The formation of a large impact crater is illustrated in **FIGURE 22.4**. The meteoroid's high-speed impact compresses the material it strikes, causing an almost instantaneous rebound, which ejects material from the surface. On Earth, impacts can occur at speeds that exceed 50 kilometers (30 miles) per second. Impacts at such high speeds produce shock waves that compress both the impactor and the material being impacted. Almost instantaneously, the over-compressed material rebounds and explosively ejects material out of the newly formed crater. This process is analogous to the detonation of an explosive device that has been buried underground.

Craters excavated by objects that are several kilometers across often exhibit a central peak, such as the one in the large crater in **FIGURE 22.5**. Much of the material expelled, called *ejecta*, lands in or near the crater, where it accumulates to form a rim. Large meteoroids may generate sufficient heat to melt and eject some of the impacted rock as glass beads. Specimens of glass beads produced in this manner, as well as melt breccia consisting of broken fragments welded by the heat of impact, have been collected on Earth, as well as the Moon, allowing planetary geologists to more fully understand such events.

Following that early period of intense bombardment, the rate of cratering diminished dramatically and now remains essentially constant. Because weathering and erosion are almost nonexistent on the Moon and Mercury, evidence of their cratered past is clearly evident.

On larger bodies, thick atmospheres may cause the impacting objects to break up and/or decelerate. For example, Earth's atmosphere causes meteoroids with masses of less than 10 kilograms (22 pounds) to lose up to 90 percent of their speed as they penetrate the atmosphere. Therefore, impacts of low-mass bodies produce only small craters on Earth. Our atmosphere is much less effective in slowing large bodies; fortunately, they make very rare appearances.

22.1 CONCEPT CHECKS

1 Briefly outline the steps in the formation of our solar system, according to the nebular theory.

2 By what criteria are planets considered either terrestrial or Jovian?

3 What accounts for the large density differences between the terrestrial and Jovian planets?

4 Explain why the terrestrial planets have meager atmospheres, as compared to the Jovian planets.

5 Why are impact craters more common on the Moon than on Earth, even though the Moon is a much smaller target and has a weaker gravitational field?

6 When did the solar system experience the period of heaviest planetary impacts?

22.2 | EARTH'S MOON: A CHIP OFF THE OLD BLOCK List and describe the
major features of Earth's Moon and explain how maria basins were formed.

The Earth–Moon system is unique because the Moon is the largest satellite relative to its planet. Mars is the only other terrestrial planet that has moons, but its tiny satellites are likely captured asteroids. Most of the 150 or so satellites of the Jovian planets are composed of low-density rock–ice mixtures, none of which resemble the Moon. As we will see later, our unique planet–satellite system is closely related to the mechanism that created it.

The diameter of the Moon is 3475 kilometers (2160 miles), about one-fourth of Earth's 12,756 kilometers (7926 miles). The Moon's surface temperature averages about 107°C (225°F) during daylight hours and −153°C (−243°F) at night. Because its period of rotation on its axis equals its period of revolution around Earth, the same lunar hemisphere always faces Earth. All of the landings of staffed *Apollo* missions were confined to the side of the Moon that faces Earth.

The Moon's density is 3.3 times that of water, comparable to that of *mantle* rocks on Earth but considerably less than Earth's average density (5.5 times that of water). The Moon's relatively small iron core is thought to account for much of this difference.

The Moon's low mass relative to Earth results in a lunar gravitational attraction that is one-sixth that of Earth. A person who weighs 150 pounds on Earth weighs only 25 pounds on the Moon, although the person's mass remains the same. This difference allows an astronaut to carry a heavy life-support system with relative ease. If not burdened with such a load, an astronaut could jump six times higher on the Moon than on Earth. The Moon's small mass (and low gravity) is the primary reason it was not able to retain an atmosphere.

How Did the Moon Form?

Until recently, the origin of the Moon—our nearest planetary neighbor—was a topic of considerable debate among scientists. Current models show that Earth is too small to have formed with a moon, particularly one so large. Furthermore, a captured moon would likely have an eccentric orbit similar to the captured moons that orbit the Jovian planets.

The current consensus is that the Moon formed as a result of a collision between a Mars-sized body and a youthful, semimolten Earth about 4.5 billion years ago. During this collision, some of the ejected debris was thrown into orbit around Earth and gradually coalesced to form the Moon. Computer simulations show that most of the ejected material would have come from the rocky mantle of the impactor, while its core was assimilated into the growing Earth. This *impact model* is consistent with the Moon having a proportionately smaller core than Earth's and, hence, a lower density.

The Lunar Surface When Galileo first pointed his telescope toward the Moon, he observed two different types of terrain: dark lowlands and brighter, highly cratered highlands (**FIGURE 22.6**). Because the dark regions appeared to be smooth, resembling seas on Earth, they were called **maria** (*mar* = sea, singular *mare*). The *Apollo 11* mission showed conclusively that the maria are exceedingly smooth plains composed of basaltic lavas. These vast plains are strongly concentrated on the side of the Moon facing Earth and cover about 16 percent of the lunar surface. The lack of large volcanic cones on these surfaces is

EYE ON THE UNIVERSE

Mariner 9 obtained this image of Phobos, one of two tiny satellites of Mars. Phobos has a diameter of only 24 kilometers (15 miles). The two moons of Mars were not discovered until the late 1800s because their size made them nearly impossible to view telescopically. (Photo by NASA)

QUESTION 1 *In what way is Phobos similar to Earth's Moon?*

QUESTION 2 *List characteristics of Phobos that make it different from Earth's Moon.*

QUESTION 3 *Do an Internet search to learn how Phobos and its companion moon, Deimos, got their names.*

FIGURE 22.6 Telescopic View of the Lunar Surface The major features are the dark maria and the light, highly cratered highlands. (NASA)

SmartFigure 22.7 Formation and Filling of Large Impact Basins

(Photo © UC Regents/Lick Observatory)

evidence of high eruption rates of very fluid basaltic lavas similar to the Columbia Plateau flood basalts on Earth.

By contrast, the Moon's light-colored areas resemble Earth's continents, so the first observers dubbed them **terrae** (Latin for "lands"). These areas are now generally referred to as the **lunar highlands** because they are elevated several kilometers above the maria. Rocks retrieved from the highlands are mainly breccias, pulverized by massive bombardment early in the Moon's history. The arrangement of terrae and maria has resulted in the legendary "face" of the "man in the Moon."

Some of the most obvious lunar features are impact craters. A meteoroid 3 meters (10 feet) in diameter can blast out a crater 50 times larger, or about 150 meters (500 feet) in diameter. The larger craters shown in Figure 22.6, such as Kepler and Copernicus (32 and 93 kilometers [20 and 57 miles] in diameter, respectively), were created from bombardment by bodies 1 kilometer (0.62 mile) or more in diameter. These two craters are thought to be relatively young because of the bright *rays* (light-colored ejected material) that radiate from them for hundreds of kilometers.

History of the Lunar Surface

The evidence used to unravel the history of the lunar surface comes primarily from radiometric dating of rocks returned from *Apollo* missions and studies of crater densities—counting the number of craters per unit area. The greater the crater density, the older the feature is inferred to be. Such evidence suggests that, after the Moon coalesced, it passed through the following four phases: (1) formation of the original crust, (2) excavation of the large impact basins, (3) filling of maria basins, and (4) formation of rayed craters.

During the late stages of its accretion, the Moon's outer shell was most likely completely melted—literally a magma ocean. Then, about 4.4 billion years ago, the magma ocean began to cool and underwent magmatic differentiation (see Chapter 4). Most of the dense minerals, olivine and pyroxene, sank, while less-dense silicate minerals floated to form the Moon's crust. The highlands are made of these igneous

Impact of an asteroid-size body produced a huge crater hundreds of kilometers in diameter and disturbed the lunar crust far beyond the crater.

Filling of the impact crater with fluid basalts, perhaps derived from partial melting deep within the lunar mantle.

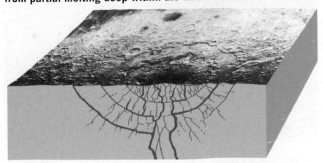

Today these basins make up the lunar maria and a few similar large structures on Mercury.

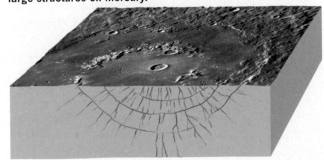

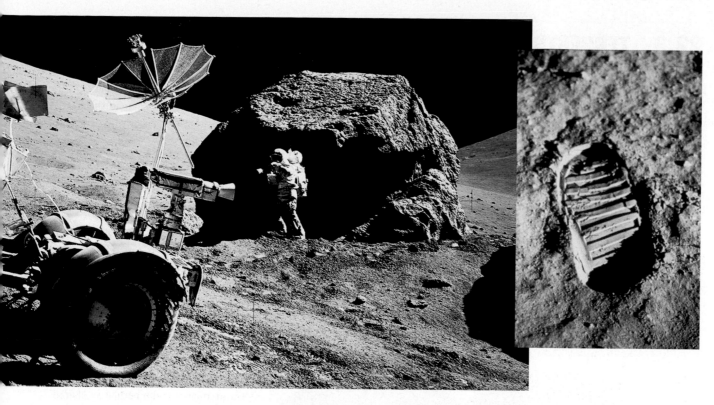

FIGURE 22.8 Astronaut Harrison Schmitt, Sampling the Lunar Surface Notice the footprint (inset) in the lunar "soil," called regolith, which lacks organic material and is therefore not a true soil. (Courtesy of NASA)

rocks, which rose buoyantly like "scum" from the crystallizing magma. The most common highland rock type is *anorthosite*, which is composed mainly of calcium-rich plagioclase feldspar.

Once formed, the lunar crust was continually impacted as the Moon swept up debris from the solar nebula. During this time, several large impact basins were created. Then, about 3.8 billion years ago, the Moon, as well as the rest of the solar system, experienced a sudden drop in the rate of meteoritic bombardment.

The Moon's next major event was the filling of the large impact basins, which were created at least 300 million years earlier (**FIGURE 22.7**). Radiometric dating of the maria basalts puts their age between 3.0 billion and 3.5 billion years, considerably younger than the initial lunar crust.

The maria basalts are thought to have originated at depths between 200 and 400 kilometers (125 and 250 miles). They were likely generated by a slow rise in temperature attributed to the decay of radioactive elements. Partial melting probably occurred in several isolated pockets, as indicated by the diverse chemical makeup of the rocks retrieved during the *Apollo* missions. Recent evidence suggests that some mare-forming eruptions may have occurred as recently as 1 billion years ago.

Other lunar surface features related to this period of volcanism include small shield volcanoes (8–12 kilometers [5–7.5 miles] in diameter), evidence of pyroclastic eruptions, rilles (narrow winding valleys thought to be lava channels), and grabens (down-faulted valleys).

The last prominent features to form were rayed craters, as exemplified by the 90-kilometer-wide (56-mile-wide) Copernicus crater shown in Figure 22.7. Material ejected from these craters blankets the maria surfaces and many older, rayless craters. The relatively young Copernicus crater is thought to be about 1 billion years old. Had it formed on Earth, weathering and erosion would have long since obliterated it.

Today's Lunar Surface The Moon's small mass and low gravity account for its lack of atmosphere and flowing water. The processes of weathering and erosion that continually modify Earth's surface are absent on the Moon. In addition, tectonic forces are no longer active on the Moon, so quakes and volcanic eruptions have ceased. Because the Moon is unprotected by an atmosphere, erosion is dominated by the impact of tiny particles from space (*micrometeorites*) that continually bombard its surface and gradually smooth the landscape. This activity has crushed and repeatedly mixed the upper portions of the lunar crust.

Both the maria and terrae are mantled with a layer of gray, unconsolidated debris derived from a few billion years of meteoric bombardment (**FIGURE 22.8**). This soil-like layer, properly called **lunar regolith** (*rhegos* = blanket, *lithos* = stone), is composed of igneous rocks, breccia, glass beads, and fine *lunar dust*. The lunar regolith is anywhere from 2 to 20 meters (6.5 to 65 feet) thick, depending on the age of the surface.

22.2 CONCEPT CHECKS

1 Briefly describe the origin of the Moon.

2 Compare and contrast the Moon's maria and highlands.

3 How are maria on the Moon similar to the Columbia Plateau in the Pacific Northwest?

4 How is crater density used in the relative dating of surface features on the Moon?

5 List the major stages in the development of the modern lunar surface.

6 Compare and contrast the processes of weathering and erosion on Earth with the same processes on the Moon.

22.3 | TERRESTRIAL PLANETS Outline the principal characteristics of Mercury, Venus, and Mars. Describe their similarities to and differences from Earth.

The terrestrial planets, in order from the Sun, are Mercury, Venus, Earth, and Mars. Because most of the book focuses on Earth, we consider the three other Earth-like planets next.

Mercury: The Innermost Planet

Mercury, the innermost and smallest planet, revolves around the Sun quickly (88 days) but rotates slowly on its axis. Mercury's day–night cycle, which lasts 176 Earth-days, is very long compared to Earth's 24-hour cycle. One "night" on Mercury is roughly equivalent to 3 months on Earth and is followed by the same duration of daylight. Mercury has the greatest temperature extremes, from as low as −173°C (−280°F) at night to noontime temperatures exceeding 427°C (800°F), hot enough to melt tin and lead. These extreme temperatures make life as we know it impossible on Mercury.

Mercury absorbs most of the sunlight that strikes it, reflecting only 6 percent into space, a characteristic of terrestrial bodies that have little or no atmosphere. The minuscule amount of gas that is present on Mercury may have originated from several sources, including ionized gas from the Sun, ices that vaporized during a recent comet impact, and/or outgassing of the planet's interior.

Although Mercury is small and scientists expected the planet's interior to have already cooled, *Mariner 10* measured Mercury's magnetic field in 2012. It found Mercury's magnetic field to be about 100 times less than Earth's, which suggests that Mercury has a large core that remains hot and fluid—a requirement for generating a magnetic field.

Mercury resembles Earth's Moon in that it has very low reflectivity, no sustained atmosphere, numerous volcanic features, and a heavily cratered terrain (**FIGURE 22.9**). The largest-known impact crater (1300 kilometers [800 miles] in diameter) on Mercury is Caloris Basin. Images and other data gathered by *Mariner 10* show evidence of volcanism in and around Caloris Basin and a few other smaller basins. Also like our Moon, Mercury has smooth plains that cover nearly 40 percent of the area imaged by *Mariner 10*. Most of these smooth areas are associated with large impact basins, including Caloris Basin, where lava partially filled the basins and the surrounding lowlands. Consequently, these smooth plains appear to be similar in origin to lunar maria. Recently, *Messenger* found evidence

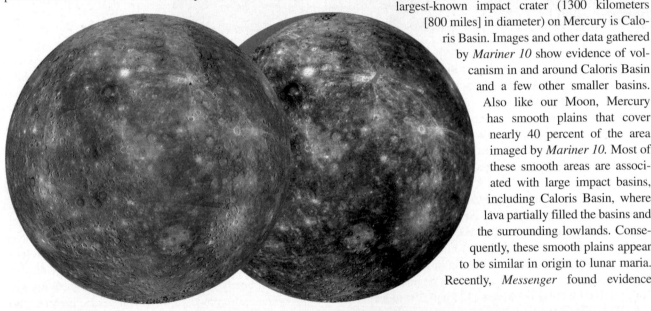

FIGURE 22.9 Two Views of Mercury On the left is a monochromatic image, while the image on the right is color enhanced. These are high-resolution mosaics constructed from thousands of images obtained by the *Messenger* orbiter. (Courtesy of NASA)

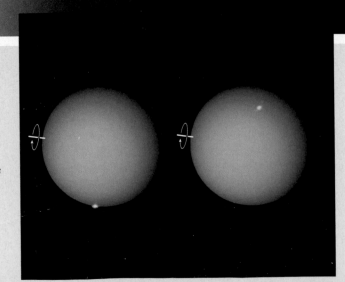

EYE ON THE UNIVERSE

In 2012, the Hubble Space Telescope captured these images of auroras above the giant planet Uranus. These light shows on Uranus appear to last for only a few minutes and consist of faint glowing dots. These are unlike auroras on Earth, which can color the sky shades of green, red, or purple for hours. (Photo by NASA)

QUESTION 1 *What is unusual about the location of the auroras on Uranus?*

QUESTION 2 *What does this indicate about the locations of Uranus's "north" and "south" magnetic poles?*

of volcanism by revealing thick volcanic deposits similar to those on Earth in the Columbia Basin. In addition, researchers were surprised by the recent detection of probable ice caps on Mercury.

Venus: The Veiled Planet

Venus, second only to the Moon in brilliance in the night sky, is named for the Roman goddess of love and beauty. It orbits the Sun in a nearly perfect circle once every 225 Earth-days. However, Venus rotates in the opposite direction of the other planets (*retrograde motion*) at an agonizingly slow pace: 1 Venus day is equivalent to about 244 Earth days. Venus has the densest atmosphere of the terrestrial planets, consisting mostly of carbon dioxide (97 percent)—the prototype for an extreme *greenhouse effect*. As a consequence, the surface temperature of Venus averages about 450°C (900°F) day and night. Temperature variations at the surface are generally minimal because of the intense mixing within the planet's dense atmosphere. Investigations of the extreme and uniform surface temperature led scientists to more fully understand how the greenhouse effect operates on Earth.

The composition of the Venusian interior is probably similar to Earth's. However, Venus's weak magnetic field means its internal dynamics must be very different from Earth's. Mantle convection is thought to operate on Venus, but the processes of plate tectonics, which recycle rigid lithosphere, do not appear to have contributed to the present Venusian topography.

The surface of Venus is completely hidden from view by a thick cloud layer composed mainly of tiny sulfuric acid droplets. In the 1970s, despite extreme temperatures and pressures, four Russian spacecraft landed successfully and obtained surface images. As expected, however, all the probes were crushed by the planet's immense atmospheric pressure, approximately 90 times that on Earth, within an hour of landing. Using radar imaging, the unstaffed spacecraft *Magellan* mapped Venus's surface in stunning detail (**FIGURE 22.10**).

A few thousand impact craters have been identified on Venus—far fewer than on Mercury and Mars but more than on Earth. Researchers expected that Venus would show evidence of extensive cratering from the heavy bombardment period but found instead that a period of extensive volcanism was responsible for resurfacing Venus. The planet's thick atmosphere also limits the number of impacts by breaking up large incoming meteoroids and incinerating most of the small debris.

About 80 percent of the Venusian surface consists of low-lying plains covered by lava flows, some of which

Aphrodite highlands

Elevation of surface

Low ⟶ High

FIGURE 22.10 **Global View of the Surface of Venus** This computer-generated image of Venus was constructed from years of investigations, culminating with the *Magellan* mission. The twisting bright features that cross the globe are highly fractured ridges and canyons of the eastern Aphrodite highland. (Courtesy of NASA)

traveled along lava channels that extend hundreds of kilometers (**FIGURE 22.11**). Venus's Baltis Vallis, the longest-known lava channel in the solar system, meanders 6800 kilometers (4255 miles) across the planet. More than 1000 volcanoes with diameters greater than 20 kilometers (12 miles) have been identified on Venus. However, high surface pressures keep the gaseous components in lava from escaping. This retards the production of pyroclastic material and lava fountaining, phenomena that tend to steepen volcanic cones. In addition, because of Venus's high temperature, lava remains mobile longer and thus flows far from the vent. Both of these factors result in volcanoes that tend to be shorter and wider than those on Earth or Mars (**FIGURE 22.12**). Maat Mons, the largest volcano on Venus, is about 8.5 kilometers high (5 miles) and 400 kilometers (250 miles) wide. By comparison, Mauna Loa, the largest volcano on Earth, is about 9 kilometers high (5.5 miles) and only 120 kilometers (75 miles) wide.

Venus also has major highlands that consist of plateaus, ridges, and topographic rises that stand above the plains. The rises are thought to have formed where hot mantle plumes encountered the base of the planet's crust, causing uplift. Much as with mantle plumes on Earth, abundant volcanism is associated with mantle upwelling on Venus. Recent data

FIGURE 22.11 Extensive Lava Flows on Venus This *Magellan* radar image shows a system of lava flows that originated from a volcano named Ammavaru, which lies approximately 300 kilometers (186 miles) west of the scene. The lava, which appears bright in this radar image, has rough surfaces, whereas the darker flows are smooth. Upon breaking through the ridge belt (left of center), the lava collected in a 100,000-square-kilometer pool. (Courtesy of NASA)

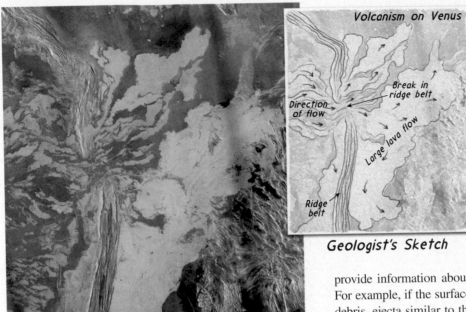

Geologist's Sketch

collected by the European Space Agency's *Venus Express* suggest that Venus's highlands contain silica-rich granitic rock. As such, these elevated landmasses resemble Earth's continents, albeit on a much smaller scale.

Mars: The Red Planet

Mars, approximately one-half the diameter of Earth, revolves around the Sun in 687 Earth-days. Mean surface temperatures range from lows of −140°C (−220°F) at the poles in the winter to highs of 20°C (68°F) at the equator in the summer. Although seasonal temperature variations are similar to Earth's, daily temperature variations are greater due to the very

thin atmosphere of Mars (only 1 percent as dense as Earth's). The tenuous Martian atmosphere consists primarily of carbon dioxide (95 percent), with small amounts of nitrogen, oxygen, and water vapor.

Topography Mars, like the Moon, is pitted with impact craters. The smaller craters are usually filled with wind-blown dust—confirming that Mars is a dry, desert world. The reddish color of the Martian landscape is due to iron oxide (rust). Large impact craters provide information about the nature of the Martian surface. For example, if the surface is composed of dry dust and rocky debris, ejecta similar to that surrounding lunar craters is to be expected. But the ejecta surrounding some Martian craters has a different appearance: It looks like a muddy slurry that was splashed from the crater. Planetary geologists infer that a layer of permafrost (frozen, icy soil) lies below portions of the Martian surface and that impacts heated and melted the ice to produce the fluid-like appearance of these ejecta.

About two-thirds of the surface of Mars consists of heavily cratered highlands, concentrated mostly in its southern hemisphere (**FIGURE 22.13**). The period of extreme cratering occurred early in the planet's history and ended about 3.8 billion years ago, as it did in the rest of the solar system. Thus, Martian highlands are similar in age to the lunar highlands.

Based on relatively low crater counts, the northern plains, which account for the remaining one-third of the planet, are younger than the highlands. If Mars once had abundant water, it would have flowed to the north, which is lower in elevation, forming an expansive ocean (shown in blue in Figure 22.13, indicating lower elevation). The relatively flat topography of the northern plains, possibly the smoothest surface in the solar system, is consistent with vast outpourings of fluid basaltic lavas. Visible on these plains are volcanic cones, some with summit pits (craters) and lava flows with wrinkled edges.

Located along the Martian equator is an enormous elevated region, about the size of North America, called the *Tharsis bulge*. This feature, about 10 kilometers (6 miles) high, appears to have been uplifted and capped with a massive accumulation of volcanic rock that includes the solar system's largest volcanoes.

The tectonic forces that created the Tharsis region also produced fractures that radiate from its center, like spokes on a bicycle wheel. Along the

FIGURE 22.12 Volcanoes on Venus Sapas Mons is a broad volcano, 400 kilometers (250 miles) wide. The bright areas in the foreground are lava flows. Another large volcano, Maat Mons, is in the background. (Courtesy of NASA)

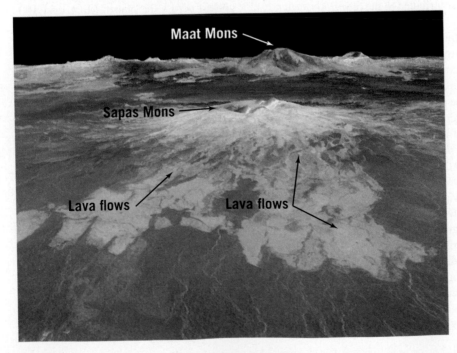

Two hemispheres of Mars

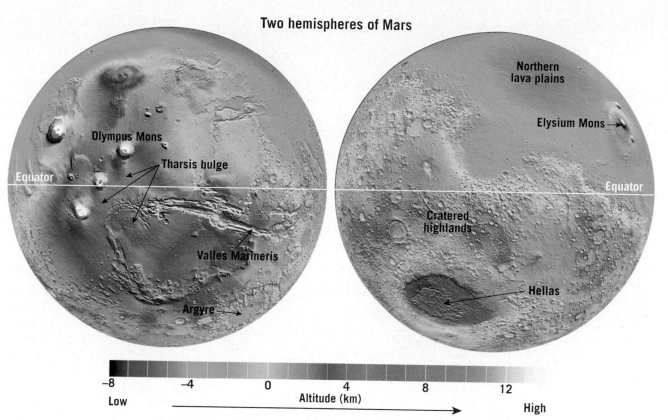

Northern
lava plains

Elysium Mons →

Olympus Mons

Tharsis bulge

Equator

Equator

Valles Marineris

Cratered
highlands

← Hellas

Argyre →

-8   -4   0   4   8   12
Low   Altitude (km)   High

FIGURE 22.13 Two Hemispheres of Mars Color represents height above (or below) the mean planetary radius: White is about 12 kilometers above average, and dark blue is 8 kilometers below average. (Courtesy of NASA)

eastern flanks of the bulge, a series of vast canyons called *Valles Marineris* (Mariner Valleys) developed. Valles Marineris is so vast that it can be seen in the image of Mars in Figure 22.13. This canyon network was largely created by down-faulting, not by stream erosion, as is the case for Arizona's Grand Canyon. Thus, it consists of grabenlike valleys similar to the East African Rift valleys. Once formed, Valles Marineris grew by water erosion and collapse of the rift walls. The main canyon is more than 5000 kilometers (3000 miles) long, 7 kilometers (4 miles) deep, and 100 kilometers (60 miles) wide.

Other prominent features on the Martian landscape are large impact basins. Hellas, the largest identifiable impact structure on the planet, is about 2300 kilometers (1400 miles) in diameter and has the planet's lowest elevation. Debris ejected from this basin contributed to the elevation of the adjacent highlands. Other buried crater basins that are even larger than Hellas probably exist.

Volcanoes on Mars

Volcanism has been prevalent on Mars during most of its history. The scarcity of impact craters on some volcanic surfaces suggests that the planet is still active. Mars has several of the largest-known volcanoes in the solar system, including the largest, Olympus Mons, which is about the size of Arizona and stands nearly three times higher than Mount Everest. This gigantic volcano was last active about 100 million years ago and resembles Earth's Hawaiian shield volcanoes (**FIGURE 22.14**).

How did the volcanoes on Mars grow so much larger than similar structures on Earth? The largest volcanoes on the terrestrial planets tend to form where plumes of hot rock rise from deep within their interiors. On Earth, moving plates keep the crust in constant motion. Consequently, mantle plumes tend to produce a chain of volcanic structures, like the Hawaiian islands. By contrast, plate tectonics on Mars is absent, so successive eruptions accumulate in the same location. As a result, enormous volcanoes such as Olympus Mons form rather than a string of smaller ones.

Wind Erosion on Mars

Currently, the dominant force shaping the Martian surface is wind erosion. Extensive dust storms, with winds up to 270 kilometers (170 miles) per hour, can persist for weeks. Dust devils have also been photographed.

Caldera of
Olympus Mons

Outline of the
state of Arizona

SmartFigure 22.14 Olympus Mons This massive inactive shield volcano on Mars covers an area about the size of the state of Arizona. (Courtesy of NASA)

Mars Exploration

Since the first close-up picture of Mars was obtained in 1965, spacecraft voyages to the fourth planet from the Sun have revealed a world that is strangely familiar. Mars has a thin atmosphere, polar ice caps, volcanoes, lava plains, sand dunes, and seasons. Unlike Earth, Mars appears to lack liquid water on its surface; however, many Martian landscapes suggest that, in the past, running water was an effective erosion agent. The defining question for Mars exploration is "has Mars ever harbored life?"

NASA's Phoenix lander dug into the Martian surface to uncover water ice in a northern region of the planet. Whether ice becomes available as liquid water to support microbial life remains unanswered.

PHOENIX

VIKING 2

VIKING 1 • • PATHFINDER

• OPPORTUNITY

CURIOSITY

SPIRIT

MARS LANDING SITES

NASA

The U.S. has successfully landed seven rovers on the surface of Mars. The most recent was NASA's Curiosity, which landed in Gale Crater in August, 2012.

NASA's Curiosity rover, the size of a car, gracefully landed on Mars after decelerating from 13,000 miles per hour to a complete stop. The landing, described as "seven minutes of terror," began a two-year mission in and around Gale Crater to discover signs of past or present microbial life.

NASA

The layers in the background, located at the base of Mount Sharp, are thought to be surviving remnants of extensive deposits laid down in a lake long ago, or possibly wind-delivered sediments subsequently cemented together by ground water. Curiosity will use 10 instruments to investigate whether Mars had ever provided a water-rich environment.

This image captured by Curiosity shows Mount Sharp, a central peak located in the 96-mile wide Gale Crater.

CAPE ST. VINCENT

VICTORIA CRATER

NASA

This false-color image obtained by Rover Opportunity shows Cape St. Vincent, one of many promontories that jut out from the walls of Victoria Crater. Below the loose, jumbled rocks, layering in the crater walls shows evidence of ancient wind-blown dunes.

	Spacecraft	Type	Landed on Mars	Years Active**
1	Curiosity	Rover	August 2012	Remains in operation
2	Phoenix	Lander	May 2008	Ran out of power during its first Martian winter
3	Mars Reconnaissance	Orbiter	March 2006	Planned 2-year mission, remains in operation
4	Spirit	Rover	January 2004	Planned 4-year mission, operated for more than 6 years
5	Opportunity	Rover	January 2004	Planned 90-day mission, remains in operation
6	Odyssey	Orbiter	October 2001	Remains active, longest active spacecraft in orbit around another planet
7	Viking I	Lander*	July 1976	Operational for more than 6 years
8	Viking II	Lander*	September 1976	Operational for more than 3 years

* Also had an orbiter; ** As of late 2012

NASA

This NASA image obtained by *Curiosity* rover shows rounded gravel fragments within a rock outcrop consistent with the sedimentary rock conglomerate. Weathered rock fragments can be seen below.

A typical sample of the sedimentary rock conglomerate that contains rounded gravel fragments deposited in a stream bed on Earth.

FIGURE 22.16 Earth-like Stream Channels Are Strong Evidence That Mars Once Had Flowing Water Inset shows a close-up of a streamlined island where running water encountered resistant material along its channel. (Courtesy of NASA)

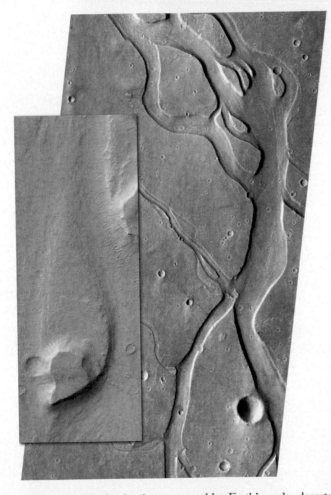

Most of the Martian landscape resembles Earth's rocky deserts, with abundant dunes and low areas partially filled with dust.

Water Ice on Mars Liquid water does not appear to exist anywhere on the Martian surface. However, poleward of about 30° latitude, ice can be found within 1 meter (3 feet) of the surface. In the polar regions, it forms small permanent ice caps. Current estimates place the maximum amount of water ice held by the Martian polar ice caps at about 1.5 times the amount covering Greenland.

Considerable evidence indicates that in the first 1 billion years of the planet's history, liquid water flowed on the surface, creating stream valleys and related features (**FIGURE 22.15**). One location where running water was involved in carving valleys can be seen in the *Mars Reconnaissance Orbiter* image in **FIGURE 22.16**. Notice the streamlike banks that contain numerous teardrop-shaped islands. These valleys appear to have been cut by catastrophic floods with discharge rates that were more than 1000 times greater than those of the Mississippi River. Most of these large flood channels emerge from areas of chaotic topography that appear to have formed when the surface collapsed. The most likely source of water for these flood-created valleys was the melting of subsurface ice. If the meltwater was trapped beneath a thick layer of permafrost, pressure could mount until a catastrophic release occurred. As the water escaped, the overlying surface would collapse, creating the chaotic terrain.

Not all Martian valleys appear to have resulted from water released in this manner. Some exhibit branching, tree-like patterns that resemble dendritic drainage networks on Earth. In addition, the *Opportunity* rover investigated structures similar to features created by water on Earth—including layered sedimentary rocks, playas (salt flats), and lake beds. Minerals that form only in the presence of water, such as hydrated sulfates, were also detected. Small spherical concretions of hematite, dubbed "blueberries," were found that probably precipitated from water to form lake sediments. Nevertheless, except in the polar regions, water does not appear to have significantly altered the topography of Mars for more than 1 billion years.

On August 6, 2012, the Mars rover *Curiosity* landed in Gale Crater, near what NASA calls Mount Sharp (officially known as Aeolis Mons). We can only imagine what *Curiosity* will uncover in its quest to study the habitability, climate, and geology of Mars.

22.3 CONCEPT CHECKS

1 What body in our solar system is most like Mercury?

2 Why are the surface temperatures so much higher on Venus than on Earth?

3 Venus was once referred to as "Earth's twin." How are these two planets similar? How do they differ from one another?

4 What surface features do Mars and Earth have in common?

5 Why are the largest volcanoes on Earth so much smaller than the largest ones on Mars?

6 What evidence suggests that Mars had an active hydrologic cycle in the past?

22.4 | JOVIAN PLANETS

Compare and contrast the four Jovian planets.

The four Jovian (Jupiter-like) planets, in order from the Sun, are Jupiter, Saturn, Uranus, and Neptune. Because of their location within the solar system and their size and composition, they are also commonly called the *outer planets* and the *gas giants*.

Jupiter: Lord of the Heavens

The giant among planets, Jupiter has a mass 2.5 times greater than the combined mass of all other planets, satellites, and asteroids in the solar system. However, it pales in comparison to the Sun, with only 1/800 of the Sun's mass.

Jupiter orbits the Sun once every 12 Earth-years, and it rotates more rapidly than any other planet, completing one rotation in slightly less than 10 hours. When viewed telescopically, the effect of this fast spin is noticeable. The bulge of the equatorial region and the slight flattening at the poles are evident (see the Polar Flattening column in Table 22.1).

Jupiter's appearance is mainly attributable to the colors of light reflected from its three main cloud layers (**FIGURE 22.17**). The warmest, and lowest, layer is composed mainly of water ice and appears blue-gray; it is generally not seen in visible-light images. The middle layer, where temperatures are lower, consists of brown to orange-brown clouds of ammonium hydrosulfide droplets. These colors are thought to be by-products of chemical reactions occurring in Jupiter's atmosphere. Near the top of its atmosphere lie white wispy clouds of ammonia ice.

Because of its immense gravity, Jupiter is shrinking a few centimeters each year. This contraction generates most of the heat that drives Jupiter's atmospheric circulation. Thus, unlike winds on Earth, which are driven by solar energy, the heat emanating from Jupiter's interior produces the huge convection currents observed in its atmosphere.

Jupiter's convective flow produces alternating dark-colored *belts* and light-colored *zones*, as shown in Figure 22.17. The light clouds (*zones*) are regions where warm material is ascending and cooling, whereas the dark belts represent cool material that is sinking and warming. This convective circulation, along with Jupiter's rapid rotation, generates the high-speed, east–west flow observed between the belts and zones.

The largest storm on the planet is the Great Red Spot. This enormous anticyclonic storm that is twice the size of Earth has been known for 300 years. In addition to the Great Red Spot, there are various white and brown oval-shaped storms. The white ovals are the cold cloud tops of huge storms many times larger than hurricanes on Earth. The brown storm clouds reside at lower levels in the atmosphere. Lightning in various white oval storms has been photographed by the *Cassini* spacecraft, but the strikes appear to be less frequent than on Earth.

Jupiter's magnetic field, the strongest in the solar system, is probably generated by a rapidly rotating, liquid metallic hydrogen layer surrounding its core. Bright auroras, associated with the magnetic field, have been photographed over Jupiter's poles (**FIGURE 22.18**). Unlike Earth's auroras, which occur only in conjunction with heightened solar activity, Jupiter's auroras are continuous.

Jupiter's Moons   Jupiter's satellite system, consisting of 67 moons discovered thus far, resembles a miniature solar system. Galileo discovered the four largest satellites, referred to as Galilean satellites, in 1610 (**FIGURE 22.19**). The two largest, Ganymede and Callisto, are roughly the size of Mercury, whereas the two smaller ones, Europa and Io, are about the size of Earth's Moon. The eight largest moons appear to have formed around Jupiter as the solar system condensed.

Jupiter also has many very small satellites (about 20 kilometers [12 miles] in diameter) that revolve in the opposite direction (*retrograde motion*) of the largest moons and have

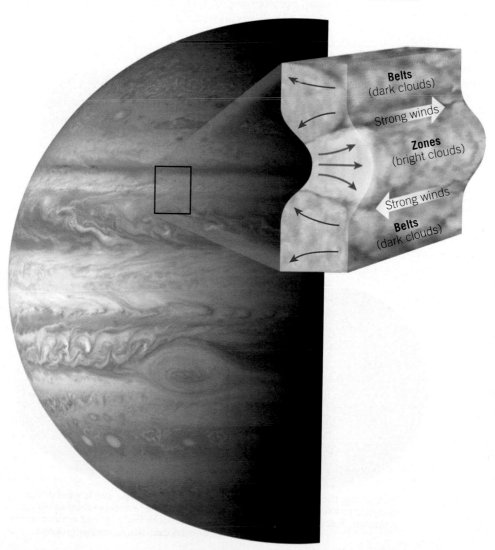

FIGURE 22.17 The Structure of Jupiter's Atmosphere The areas of light clouds (*zones*) are regions where gases are ascending and cooling. Sinking dominates the flow in the darker cloud layers (*belts*). This convective circulation, along with the rapid rotation of the planet, generates the high-speed winds observed between the belts and zones.

Belts (dark clouds)

Strong winds

Zones (bright clouds)

Strong winds

Belts (dark clouds)

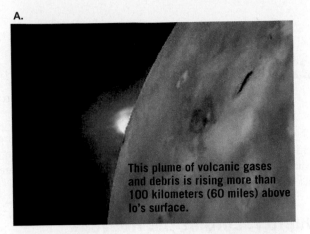

This plume of volcanic gases and debris is rising more than 100 kilometers (60 miles) above Io's surface.

FIGURE 22.18 View of Jupiter's Aurora, Taken by the Hubble Space Telescope This phenomenon is produced by high-energy electrons racing along Jupiter's magnetic field. The electrons excite atmospheric gases and make them glow. (Courtesy of NASA/John Clark)

B.

The bright red area on the left side of the image (see arrow) is newly erupted lava.

eccentric (elongated) orbits steeply inclined to the Jovian equator. These satellites appear to be asteroids or comets that passed near enough to be gravitationally captured by Jupiter or are remnants of the collisions of larger bodies.

The Galilean moons can be observed with binoculars or a small telescope and are interesting in their own right. Images from *Voyagers 1* and *2* revealed, to the surprise of most geoscientists, that each of the four Galilean satellites is a unique world (Figure 22.19). The *Galileo* mission also unexpectedly revealed that the composition of each satellite is strikingly different, implying a different evolution for each. For example, Ganymede has a dynamic core that generates a strong magnetic field not observed in other satellites.

FIGURE 22.20 A Volcanic Eruption on Jupiter's Moon Io (Courtesy of NASA; Jet Propulsion Laboratory/University of Arizona/NASA)

The innermost of the Galilean moons, Io, is perhaps the most volcanically active body in our solar system. In all, more than 80 active, sulfurous volcanic centers have been discovered. Umbrella-shaped plumes have been observed rising from Io's surface to heights approaching 200 kilometers (125 miles) (**FIGURE 22.20A**). The heat source for volcanic activity is tidal energy generated by a relentless "tug of war" between Jupiter and the other Galilean satellites—with Io as the rope. The gravitational field of Jupiter and the other nearby satellites pull and push on Io's tidal bulge as its slightly eccentric orbit takes it alternately closer to and farther from Jupiter. This gravitational flexing of Io is transformed into heat (similar to the back-and-forth bending of a piece of sheet metal) and results in Io's spectacular sulfurous volcanic eruptions. Moreover, lava, thought to be mainly composed of silicate minerals, regularly erupts on its surface (**FIGURE 22.20B**).

FIGURE 22.19 Jupiter's Four Largest Moons These moons are often referred to as the Galilean moons because Galileo discovered them. (Courtesy of NASA)

A. Io is one of only three volcanically active bodies other than Earth known to exist in the solar system.

B. Europa, the smallest of the Galilean moons, has an icy surface that is crisscrossed by many linear features.

D. Callisto, the outermost of the Galilean satellites, is densely cratered, much like Earth's Moon

C. Ganymede, the largest Jovian satellite, exhibits cratered areas, smooth regions, and areas covered by numerous parallel grooves.

The planets closer to the Sun than Earth are considered too warm to contain liquid water, and those farther from the Sun are generally too cold (although some features on Mars indicate that it probably had abundant liquid water at some point in its history). The best prospects of finding liquid water within our solar system lie beneath the icy surfaces of some of Jupiter's moons. For instance, an ocean of liquid water is possibly hidden under Europa's outer covering of ice. Detailed images from *Galileo* have revealed that Europa's icy surface is quite young and exhibits cracks apparently filled with dark fluid from below. This suggests that under its icy shell, Europa must have a warm, mobile interior—perhaps an ocean. Because

Encke gap   Cassini division

D C B A

Saturn

FIGURE 22.21 Saturn's Dynamic Ring System The two bright rings, called A ring (outer) and B ring (inner), are separated by the Cassini division. A second small gap (Encke gap) is also visible as a thin line in the outer portion of the A ring. (Courtesy of NASA)

liquid water is a necessity for life as we know it, there is considerable interest in sending an orbiter to Europa—and, eventually, a lander capable of launching a robotic submarine—to determine whether it harbors life.

Jupiter's Rings

One of the surprising aspects of the *Voyager 1* mission was the discovery of Jupiter's ring system. More recently, the ring system was thoroughly investigated by the *Galileo* mission. By analyzing how these rings scatter light, researchers determined that the rings are composed of fine, dark particles that are similar in size to smoke particles. Furthermore, the faint nature of the rings indicates that these minute particles are widely dispersed. The main ring is composed of particles believed to be fragments blasted from the surfaces of Metis and Adrastea, two small moons of Jupiter. Impacts on Jupiter's moons Amalthea and Thebe are believed to be the source of the debris from which the outer gossamer ring formed.

Saturn: The Elegant Planet

Requiring more than 29 Earth-years to make one revolution, Saturn is almost twice as far from the Sun as Jupiter, yet their atmospheres, compositions, and internal structures are remarkably similar. The most striking feature of Saturn is its system of rings, first observed by Galileo in 1610 (**FIGURE 22.21**). Through his primitive telescope, the rings appeared as two small bodies adjacent to the planet. Their ring nature was determined 50 years later by Dutch astronomer Christian Huygens.

Saturn's atmosphere, like Jupiter's, is dynamic. Although the bands of clouds are fainter and wider near the equator, rotating "storms" similar to Jupiter's Great Red Spot occur in Saturn's atmosphere, as does intense lightning. Although the atmosphere is nearly 75 percent hydrogen and 25 percent helium, the clouds (or condensed gases) are composed of ammonia, ammonia hydrosulfide, and water, each segregated by temperature. Like Jupiter, the atmosphere's dynamics are driven by the heat released by gravitational compression.

Saturn's Moons The Saturnian satellite system consists of 62 known moons, of which 53 have been named. The moons vary significantly in size, shape, surface age, and origin. Twenty-three of the moons are "original" satellites that formed in tandem with their parent planet. At least three (Rhea, Dione, and Tethys) show evidence of tectonic activity, where internal forces have ripped apart their icy surfaces. Others, like Hyperion, are so porous that impacts punch into their surfaces (**FIGURE 22.22**). Many of Saturn's smallest moons have irregular shapes and are only a few tens of kilometers in diameter.

FIGURE 22.22 Hyperion, Saturn's Impact-Pummeled Satellite Planetary geologists think Hyperion's surface is so weak and porous that impacts punch into its surface. (Courtesy of NASA)

FIGURE 22.23
Enceladus, Saturn's Tectonically Active, Icy Satellite The Northern Hemisphere contains a 1-kilometer-deep (0.6 mile) chasm, and linear features, called tiger stripes, are visible in the lower right. Inset image shows jets spurting ice particles, water vapor, and organic compounds from the area of the tiger stripes. (Courtesy of NASA)

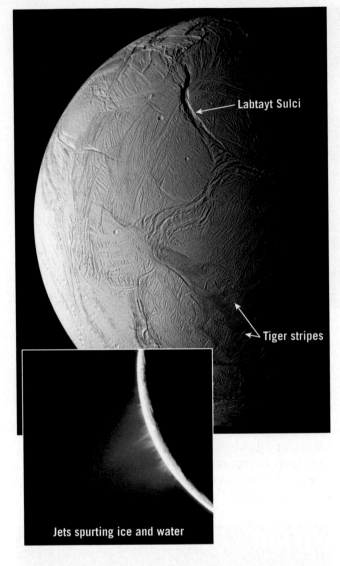

Labtayt Sulci

Tiger stripes

Jets spurting ice and water

Saturn's largest moon, Titan, is larger than Mercury and is the second-largest satellite in the solar system. Titan and Neptune's Triton are the only satellites in the solar system known to have substantial atmospheres. Titan was visited and photographed by the *Cassini-Huygens* probe in 2005. The atmospheric pressure at Titan's surface is about 1.5 times that at Earth's surface, and the atmospheric composition is about 98 percent nitrogen and 2 percent methane, with trace organic compounds. Titan has Earth-like geologic landforms and geologic processes, such as dune formation and stream-like erosion caused by methane "rain." In addition, the northern latitudes appear to have lakes of liquid methane.

Enceladus is another unique satellite of Saturn—one of the few where active eruptions have been observed (**FIGURE 22.23**). The outgassing, comprised mostly of water, is thought to be the source that replenishes the material in Saturn's E ring. The volcanic-like activity occurs in areas called "tiger stripes" that consists of four large fractures with ridges on either side.

Saturn's Ring System

In the early 1980s, the nuclear-powered *Voyagers 1* and *2* explored Saturn within 160,000 kilometers (100,000 miles) of its surface. More information

was collected about Saturn in that short time than had been acquired since Galileo first viewed this "elegant planet" in the early 1600s. More recently, observations from ground-based telescopes, the Hubble Space Telescope, and the *Cassini-Huygens* spacecraft, have added to our knowledge of Saturn's ring system. In 1995 and 1996, when the positions of Earth and Saturn allowed the rings to be viewed edge-on, Saturn's faintest rings and satellites became visible. (The rings were visible edge-on again in 2009.)

Saturn's ring system is more like a large rotating disk of varying density and brightness than a series of independent ringlets. Each ring is composed of individual particles—mainly water ice, with lesser amounts of rocky debris—that circle the planet while regularly impacting one another. There are only a few gaps; most of the areas that look like empty space either contain fine dust particles or coated ice particles that are inefficient reflectors of light.

Most of Saturn's rings fall into one of two categories, based on density. Saturn's main (bright) rings, designated A and B, are tightly packed and contain particles that range in size from a few centimeters (pebble-size) to tens of meters (house-size), with most of the particles being roughly the size of a large snowball (see Figure 22.21). In the dense rings, particles collide frequently as they orbit the planet. Although Saturn's main rings (A and B) are 40,000 kilometers (25,000 miles) wide, they are very thin, only 10–30 meters (30–100 feet) from top to bottom.

At the other extreme are Saturn's faint rings. Saturn's outermost ring (E ring), not visible in Figure 22.21, is composed of widely dispersed, tiny particles. Recall that volcanic-like activity on Saturn's satellite Enceladus is thought to be the source of material for the E ring.

Studies have shown that the gravitational tugs of nearby moons tend to shepherd the ring particles by gravitationally altering their orbits (**FIGURE 22.24**). For example, the F ring, which is very narrow, appears to be the work of satellites located on either side that confine the ring by pulling back particles that try to escape. On the other hand, the Cassini division, a clearly visible gap in Figure 22.21, arises from the gravitational pull of Mimas, one of Saturn's moons.

Some of the ring particles are believed to be debris ejected from the moons embedded in them. It is also possible that material is continually recycled between the rings and the ring moons. The ring moons gradually sweep up particles, which are subsequently ejected by collisions with large chunks of ring material, or perhaps by energetic collisions with other moons. It seems, then, that planetary rings are not the timeless features that we once thought; rather, they are continually recycled.

The origin of planetary ring systems is still being debated. Perhaps the rings formed simultaneously and from the same material as the planets and moons—condensing from a flattened cloud of dust and gases that encircled the parent planet. Or perhaps the rings formed later, when a moon or large asteroid was gravitationally pulled apart after straying too close to a planet. Yet another hypothesis suggests that a foreign body collided catastrophically with one of the planet's moons, the fragments of which would tend to jostle one another and form a flat, thin ring. Researchers expect more light to be shed on the origin of planetary rings as the *Cassini* spacecraft continues its tour of Saturn.

Uranus and Neptune: Twins

Although Earth and Venus have many similar traits, Uranus and Neptune are perhaps more deserving of being called "twins." They are nearly equal in diameter (both about four times the size of Earth), and they are both bluish in appearance, as a result of methane in their atmospheres. Their days are nearly the same length, and their cores are made of rocky silicates and iron—similar to the other gas giants. Their mantles, made mainly of water, ammonia, and methane, are thought to be very different from Jupiter and Saturn. One of the most pronounced differences between Uranus and Neptune is the time they take to complete one revolution around the Sun—84 and 165 Earth-years, respectively.

Uranus: The Sideways Planet
Unique to Uranus is the orientation of its axis of rotation. Whereas the other planets resemble spinning toy tops as they circle the Sun, Uranus is like a top that has been knocked on its side but remains spinning (**FIGURE 22.25**). This unusual characteristic of Uranus is likely due to one or more impacts essentially knocking the planet sideways from its original orientation early in its evolution.

Uranus shows evidence of huge storm systems equivalent in size to those in the United States. Recent photographs from the Hubble Space Telescope also reveal banded clouds composed mainly of ammonia and methane ice—similar to the cloud systems of the other gas giants.

Uranus's Moons
Spectacular views from *Voyager 2* showed that Uranus's five largest moons have varied terrains. Some have long, deep canyons and linear scars, whereas others possess large, smooth areas on otherwise crater-riddled surfaces. Studies conducted at California's Jet Propulsion Laboratory suggest that Miranda, the innermost of the five largest moons, was recently geologically active—most likely driven by gravitational heating, as occurs on Io.

Uranus's Rings
A surprise discovery in 1977 showed that Uranus has a ring system. The discovery was made as Uranus passed in front of a distant star and blocked its view, a process called *occultation* (*occult* = *hidden*). Observers saw

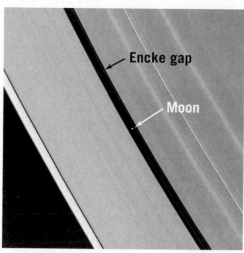

A. Pan is a small moon about 30 kilometers in diameter that orbits in the Encke gap, located in the A ring. It is responsible for keeping the Encke gap open by sweeping up any stray material that may enter.

Encke gap

Moon

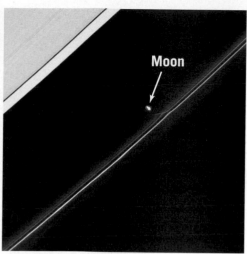

B. Prometheus, a potato-shaped moon, acts as a ring shepherd. Its gravity helps confine the moonlets in Saturn's thin Fring.

Moon

FIGURE 22.24 Two of Saturn's Ring Moons (Courtesy of NASA)

the star "wink" briefly five times (meaning five rings) before the primary occultation and again five times afterward. More recent ground- and space-based observations indicate that Uranus has at least 10 sharp-edged, distinct rings orbiting its equatorial region. Interspersed among these distinct structures are broad sheets of dust.

Neptune: The Windy Planet
Because of Neptune's great distance from Earth, astronomers knew very little about

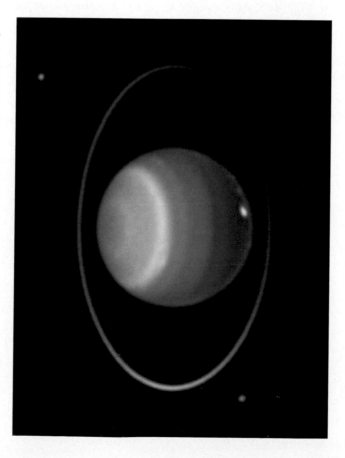

FIGURE 22.25 Uranus, Surrounded by Its Major Rings and a Few of Its Known Moons Also visible in this image are cloud patterns and several oval storm systems. This false-color image was generated from data obtained by Hubble's Near Infrared Camera. (Image by Hubble Space Telescope, courtesy of NASA)

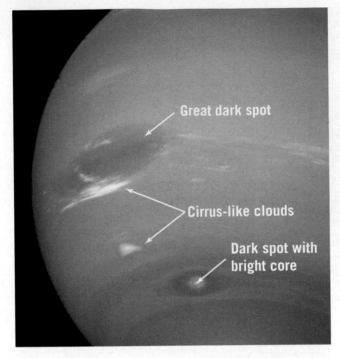

this planet until 1989. Twelve years and nearly 3 billion miles of *Voyager 2* travel provided investigators an amazing opportunity to view the outermost planet in the solar system.

Neptune has a dynamic atmosphere, much like that of the other Jovian planets (**FIGURE 22.26**). Record wind speeds exceeding 2400 kilometers (1500 miles) per hour encircle the planet, making Neptune one of the windiest places in the solar system. In addition, Neptune exhibits large dark spots thought to be rotating storms similar to Jupiter's Great Red Spot. However, Neptune's storms

FIGURE 22.27 Triton, Neptune's Largest Moon The bottom of the image shows Triton's wind and sublimation-eroded south polar cap. Sublimation is the process whereby a solid (ice) changes directly to a gas. (Courtesy of NASA)

appear to have comparatively short life spans—usually only a few years. Another feature that Neptune has in common with the other Jovian planets is layers of white, cirrus-like clouds (probably frozen methane) about 50 kilometers (30 miles) above the main cloud deck.

Neptune's Moons Neptune has 13 known satellites, the largest of which is the moon Triton; the remaining 12 are small, irregularly shaped bodies. Triton is the only large moon in the solar system that exhibits retrograde motion, indicating that it most likely formed independently and was later gravitationally captured by Neptune (**FIGURE 22.27**).

Triton and a few other icy moons erupt "fluid" ices—an amazing manifestation of volcanism. **Cryovolcanism** (from the Greek *kryos*, meaning "frost") describes the eruption of magmas derived from the partial melting of ice instead of silicate rocks. Triton's icy magma is a mixture of water ice, methane, and probably ammonia. When partially melted, this mixture behaves as molten rock does on Earth. In fact, upon reaching the surface, these magmas can generate quiet outpourings of ice lavas or occasionally produce explosive eruptions. An explosive eruptive column can generate the ice equivalent of volcanic ash. In 1989, *Voyager 2* detected active plumes on Triton that rose 8 kilometers (5 miles) above the surface and were blown downwind for more than 100 kilometers (60 miles). In other environments, ice lavas develop that can flow great distances from their source—similar to the fluid basaltic flows on Hawaii.

Neptune's Rings Neptune has five named rings; two of them are broad, and three are narrow, perhaps no more than 100 kilometers (60 miles) wide. The outermost ring appears to be partially confined by the satellite Galatea. Neptune's rings are most similar to Jupiter's in that they appear faint, which suggests that they are composed mostly of dust-size particles. Neptune's rings also display red colors, indicating that the dust is composed of organic compounds.

22.4 CONCEPT CHECKS

1 What is the nature of Jupiter's Great Red Spot?

2 Why are the Galilean satellites of Jupiter so named?

3 What is distinctive about Jupiter's satellite Io?

4 Why are many of Jupiter's small satellites thought to have been captured?

5 How are Jupiter and Saturn similar to one another?

6 What two roles do ring moons play in the nature of planetary ring systems?

7 How are Saturn's satellite Titan and Neptune's satellite Triton similar to one another?

8 Name three bodies in the solar system that exhibit active volcanism.

22.5 | SMALL SOLAR SYSTEM BODIES List and describe the principal characteristics of the small bodies that inhabit the solar system.

There are countless chunks of debris in the vast spaces separating the eight planets and in the outer reaches of the solar system. In 2006, the International Astronomical Union organized solar system objects not classified as planets or moons into two broad categories: (1) **small solar system bodies**, including *asteroids*, *comets*, and *meteoroids*, and (2) **dwarf planets**. The newest grouping, dwarf planets, includes Ceres, the largest known object in the asteroid belt, and Pluto, a former planet.

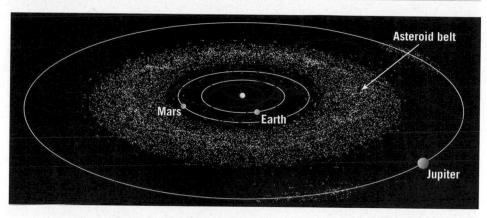

FIGURE 22.28 The Asteroid Belt The orbits of most asteroids lie between Mars and Jupiter. Also shown in red are the orbits of a few known near-Earth asteroids.

Asteroids and meteoroids are composed of rocky and/or metallic material with compositions somewhat like the terrestrial planets. They are distinguished according to size: Asteroids are larger than 100 meters (60 miles) in diameter, whereas meteoroids have diameters less than 100 meters. Comets, on the other hand, are loose collections of ices, dust, and small rocky particles that originate in the outer reaches of the solar system.

Asteroids: Leftover Planetesimals

Asteroids are small bodies (planetesimals) that remain from the formation of the solar system, which means they are about 4.6 billion years old. Most asteroids orbit the Sun between Mars and Jupiter, in the region known as the **asteroid belt** (**FIGURE 22.28**). Only five asteroids are more than 400 kilometers (250 miles) in diameter, but the solar system hosts an estimated 1 to 2 million asteroids larger than 1 kilometer (0.6 mile) and many millions that are smaller. Some travel along eccentric orbits that take them very near the Sun, and others regularly pass close to Earth and the Moon (Earth-crossing asteroids). Many of the recent large-impact craters on the Moon and Earth probably resulted from collisions with asteroids. There are an estimated 1000 to 2000 Earth-crossing asteroids that are more than 0.6 kilometer in diameter. Inevitably, Earth–asteroid collisions will occur again.

Because most asteroids have irregular shapes, planetary geologists initially speculated that they might be fragments of a broken planet that once orbited between Mars and Jupiter. However, the combined mass of all asteroids is now estimated to be only 1/1000 of the modest-sized Earth. Today, most researchers agree that asteroids are leftover debris from the solar nebula. Asteroids have lower densities than scientists originally thought, suggesting that they are porous bodies, like "piles of rubble," loosely bound together (**FIGURE 22.29**).

In February 2001, an American spacecraft became the first visitor to an asteroid. Although it was not designed for landing, *NEAR Shoemaker* landed successfully on Eros and collected information that has planetary geologists both intrigued and perplexed. Images obtained as the spacecraft drifted toward the surface of Eros revealed a barren, rocky surface composed of particles ranging in size from fine dust to boulders up to 10 meters (30 feet) across. Researchers unexpectedly discovered that fine debris tends to concentrate in the low areas, where it forms flat deposits resembling ponds. Surrounding the low areas, the landscape is marked by an abundance of large boulders.

One of several hypotheses to explain the boulder-strewn topography is seismic shaking, which would cause the boulders to move upward as the finer materials sink. This is analogous to what happens when a jar of sand and various-sized pebbles is shaken: The larger pebbles rise to the top, while the smaller sand grains settle to the bottom (sometimes referred to as the Brazil nut effect).

Indirect evidence from meteorites suggests that some asteroids might have been heated by a large impact event. A few large asteroids may have completely melted, causing

FIGURE 22.29 Giant Asteroid Vesta (Photo courtesy of NASA)

FIGURE 22.30 Changing Orientation of a Comet's Tail as It Orbits the Sun
(Photo by Dan Schechter/Science Source)

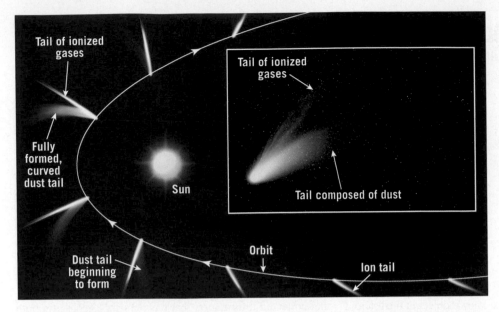

them to differentiate into a dense iron core and a rocky mantle. In November 2005, the Japanese probe *Hayabusa* landed on a small near-Earth asteroid named 25143 Itokawa; it returned to Earth in June 2010. Analyzed samples suggest that the surface of the asteroid was identical in composition to meteorites and was once part of a larger asteroid. *Hayabusa 2* is scheduled to launch in 2014, to eventually expose subsurface samples by blasting a crater in asteroid 1999 JU3.

Comets: Dirty Snowballs

Comets, like asteroids, are leftover material from the formation of the solar system. They are loose collections of rocky material, dust, water ice, and frozen gases (ammonia, methane,

and carbon dioxide), thus the nickname "dirty snowballs." Recent space missions to comets have shown their surfaces to be dry and dusty, which indicates that their ices are hidden beneath a rocky layer.

Most comets reside in the outer reaches of the solar system and take hundreds of thousands of years to complete a single orbit around the Sun. However, a smaller number of *short-period comets* (those having orbital periods of less than 200 years), such as the famous Halley's Comet, make regular encounters with the inner solar system (**FIGURE 22.30**). The shortest-period comet (Encke's Comet) orbits around the Sun once every 3 years.

Structure and Composition of Comets All the phenomena associated with comets come from a small central body called the **nucleus**. These structures are typically 1 to 10 kilometers in diameter, but nuclei 40 kilometers across have been observed. When comets reach the inner solar system, solar energy begins to vaporize their ices. The escaping gases carry dust from the comet's surface, producing a highly reflective halo called a **coma** (**FIGURE 22.31**). Within the coma, the small glowing nucleus with a diameter of only a few kilometers can sometimes be detected.

As comets approach the Sun, most develop tails that can extend for millions of kilometers. The tail of a comet points away from the Sun in a slightly curved manner (see Figure 22.30), which led early astronomers to believe that the Sun has a repulsive force that pushes away particles of the coma to form the tail. Scientists have identified two solar forces known to contribute to tail formation. One is *radiation pressure* caused by radiant energy (light) emitted by the Sun, and the second is the *solar wind*, a stream of charged particles ejected from the Sun. Sometimes a single tail composed of both dust and ionized gases is produced, but two tails are often observed (see Figure 22.30). The heavier dust particles produce a slightly curved tail that follows the comet's orbit, whereas the extremely light ionized gases are "pushed" directly away from the Sun, forming the second tail.

As a comet's orbit carries it away from the Sun, the gases forming the coma recondense, the tail disappears, and the comet returns to cold storage. Material that was blown from the coma to form the tail is lost forever. When all the gases are expelled, the inactive comet, which closely resembles an asteroid, continues its orbit without a coma or tail. It is believed that few comets remain active for more than a few hundred close orbits of the Sun.

The very first samples from a comet's coma (Comet Wild 2) were returned to Earth in January 2006 by NASA's

FIGURE 22.31 Coma of Comet Holmes The nucleus of the comet is within the bright spot in the center. Comet Holmes, which orbits the Sun every six years, was uncharacteristically active during its most recent entry into the inner solar system. (Courtesy of NASA)

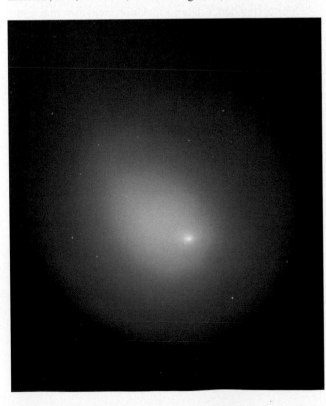

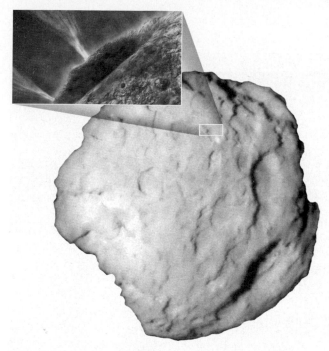

FIGURE 22.32 Comet Wild 2 This image shows Comet Wild 2, as seen by NASA's *Stardust* spacecraft. The inset shows an artist's depiction of jets of gas and dust erupting from Comet Wild 2. (Courtesy of NASA)

Stardust spacecraft (**FIGURE 22.32**). Images from *Stardust* show that the comet's surface was riddled with flat-bottomed depressions and appeared dry, although at least 10 gas jets were active. Laboratory studies revealed that the coma contained a wide range of organic compounds and substantial amounts of silicate crystals.

The Realm of Comets: The Kuiper Belt and Oort Cloud

Most comets originate in one of two regions: the *Kuiper belt* or the *Oort cloud*. Named in honor of astronomer Gerald Kuiper, who predicted its existence, the **Kuiper belt** hosts comets that orbit in the outer solar system, beyond Neptune (see Figure 22.1). This disc-shaped structure is thought to contain about a billion objects over 1 kilometer in size. However, most comets are too small and too distant to be observed from Earth, even using the Hubble Space Telescope. Like the asteroids in the inner solar system, most Kuiper belt comets move in slightly elliptical orbits that lie roughly in the same plane as the planets. A chance collision between two Kuiper belt comets or the gravitational influence of one of the Jovian planets occasionally alters their orbits sufficiently to send them into our view.

Halley's Comet originated in the Kuiper belt. Its orbital period averages 76 years, and every one of its 29 appearances since 240 B.C. has been recorded, thanks to ancient Chinese astronomers—testimony to their dedication as astronomical observers and the endurance of Chinese culture. In 1910, Halley's Comet made a very close approach to Earth, making for a spectacular display.

Named for Dutch astronomer Jan Oort, the **Oort cloud** consists of comets that are distributed in all directions from the Sun, forming a spherical shell around the solar system. Most Oort cloud comets orbit the Sun at distances greater than 10,000 times the Earth–Sun distance. The gravitational

effect of a distant passing star may send an occasional Oort cloud comet into a highly eccentric orbit that carries it toward the Sun. However, only a tiny fraction of Oort cloud comets have orbits that bring them into the inner solar system.

Meteoroids: Visitors to Earth

Nearly everyone has seen **meteors**, commonly (but inaccurately) called "shooting stars." These streaks of light can be observed in as little as the blink of an eye or can last as "long" as a few seconds. They occur when a small solid particle, a **meteoroid**, enters Earth's atmosphere from interplanetary space. Heat, created by friction between the meteoroid and the air, produces the light we see trailing across the sky. Most meteoroids originate from one of the following three sources: (1) interplanetary debris missed by the gravitational sweep of the planets during formation of the solar system, (2) material that is continually being ejected from the asteroid belt, or (3) the rocky and/or metallic remains of comets that once passed through Earth's orbit. A few meteoroids are probably fragments of the Moon, Mars, or possibly Mercury, ejected by a violent asteroid impact. Before *Apollo* astronauts brought Moon rocks back to Earth, meteorites were the only extraterrestrial materials that could be studied in the laboratory.

Meteoroids less than about 1 meter (3 feet) in diameter generally vaporize before reaching Earth's surface. Some, called *micrometeorites*, are so tiny and their rate of fall so slow that they drift to Earth continually as space dust. Researchers estimate that thousands of meteoroids enter Earth's atmosphere every day. After sunset on a clear, dark night, many are bright enough to be seen with the naked eye from Earth.

Meteor Showers Occasionally, meteor sightings increase dramatically to 60 or more per hour. These displays, called **meteor showers**, result when Earth encounters a swarm of meteoroids traveling in the same direction at nearly the same speed as Earth. The close association of these swarms to the orbits of some short-term comets strongly suggests that they represent material lost by these comets (**TABLE 22.2**). Some swarms, not associated with the orbits of known comets, are probably the scattered remains

TABLE 22.2 Major Meteor Showers

Shower	Approximate Dates	Associated Comet
Quadrantids	January 4–6	
Lyrids	April 20–23	Comet 1861 I
Eta Aquarids	May 3–5	Halley's Comet
Delta Aquarids	July 30	
Perseids	August 12	Comet 1862 III
Draconids	October 7–10	Comet Giacobini-Zinner
Orionids	October 20	Halley's Comet
Taurids	November 3–13	Comet Encke
Andromedids	November 14	Comet Biela
Leonids	November 18	Comet 1866 I
Geminids	December 4–16	

Meteor Crater
(cross-section)

Overturned rock units
Ejecta
Kaibab ls
Moenkopi
Coconino ss

Geologist's Sketch

Original rim profile
Moenkopi
Recent sediment
Fallback breccia
Kaibab ls
Coconino ss
Highly fractured bedrock

SmartFigure 22.33 Meteor Crater, Near Winslow, Arizona This cavity is about 1.2 kilometers (0.75 mile) across and 170 meters (560 feet) deep. The solar system is cluttered with asteroids and comets that can strike Earth with explosive force. (Photo by Michael Collier)

of the nucleus of a long-defunct comet. The notable *Perseid meteor shower* that occurs each year around August 12 is likely material ejected from the comet *Swift–Tuttle* on previous approaches to the Sun.

Most meteoroids large enough to survive passage through the atmosphere to impact Earth probably originate among the asteroids, where chance collisions or gravitational interactions with Jupiter modify their orbits and send them toward Earth. Earth's gravity does the rest.

A few very large meteoroids have blasted craters on Earth's surface that strongly resemble those on our Moon. At least 40 terrestrial craters exhibit features that could be produced only by an explosive impact of a large asteroid, or perhaps even a comet nucleus. More than 250 others may be of impact origin. Notable among them is Arizona's Meteor Crater, a huge cavity more than 1 kilometer (0.6 mile) wide and 170 meters (560 feet) deep, with an upturned rim that rises above the surrounding countryside (**FIGURE 22.33**). More than 30 tons of iron fragments have been found in the immediate area, but attempts to locate the main body have been unsuccessful. Based on the

amount of erosion observed on the crater rim, the impact likely occurred within the past 50,000 years.

Types of Meteorites The remains of meteoroids, when found on Earth, are referred to as **meteorites** (**FIGURE 22.34**). Classified by their composition, meteorites are either (1) *irons*, mostly aggregates of iron with 5–20 percent nickel; (2) *stony* (also called *chondrites*), silicate minerals with inclusions of other minerals; or (3) *stony–irons*, mixtures of the two. Although stony meteorites are the most common, irons are found in large numbers because metallic meteorites withstand impacts better, weather more slowly, and are easily distinguished from terrestrial rocks. Iron meteorites are probably fragments of once-molten cores of large asteroids or small planets.

One type of stony meteorite, called a *carbonaceous chondrite*, contains organic compounds and occasionally simple amino acids, which are some of the basic building blocks of life. This discovery confirms similar findings in observational astronomy, which indicate that numerous organic compounds exist in interstellar space.

Data from meteorites have been used to ascertain the internal structure of Earth and the age of the solar system. If meteorites represent the composition of the terrestrial planets, as some planetary geologists suggest, our planet must contain a much larger percentage of iron than is indicated by surface rocks. This is one reason that geologists think Earth's core is mostly iron and nickel. In addition, radiometric dating of meteorites indicates that the age of our solar system is about 4.6 billion years. This "old age" has been confirmed by data obtained from lunar samples.

FIGURE 22.34 Iron Meteorite Found Near Meteor Crater, Arizona (Courtesy of M2 Photography/Alamy)

Dwarf Planets

Since its discovery in 1930, Pluto has been a mystery to astronomers who were searching for another planet in order to explain irregularities in Neptune's orbit. At the time of its discovery, Pluto was thought to be the size of Earth—too small to significantly alter Neptune's orbit. Later, estimates of Pluto's diameter, adjusted because of improved satellite images, indicated that it was less than half Earth's diameter. Then, in 1978, astronomers realized that Pluto appeared much larger than it really is because of the brightness of its newly discovered satellite, Charon (**FIGURE 22.35**). Most recently, calculations based on images obtained by the Hubble Space Telescope show that Pluto's diameter is 2300 kilometers (1430 miles), about one-fifth the diameter of Earth and less than half that of Mercury (long considered the solar system's "runt"). In fact, seven moons in the solar system, including Earth's, are larger than Pluto.

Even more attention was given to Pluto's status as a planet when astronomers discovered another large icy body in orbit beyond Neptune. Soon, more than 1000 of these *Kuiper belt objects* were discovered forming a band of objects— a second "asteroid belt," but located at the outskirts of the solar system. The Kuiper belt objects are rich in ices and have physical properties similar to those of comets. Many other planetary objects, some perhaps larger than Pluto, are thought to exist in this belt of icy worlds beyond Neptune's orbit. Researchers soon recognized that Pluto was unique among the planets—completely different from the four rocky, innermost planets, as well as the four gaseous giants.

In 2006, the International Astronomical Union, the group responsible for naming and classifying celestial objects, voted to designate a new class of solar system objects called *dwarf planets*. These are celestial bodies that orbit the Sun and are essentially spherical due to their own gravity but are not large enough to sweep their orbits clear of other debris. By this definition, Pluto is recognized as a dwarf planet and the prototype of this new category of planetary objects. Other

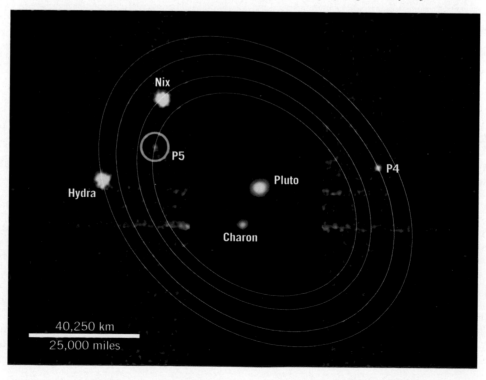

FIGURE 22.35 Pluto, with Its Five Known Moons (Courtesy of NASA)

Nix

P5

Hydra

Pluto

P4

Charon

40,250 km

25,000 miles

dwarf planets include Eris, a Kuiper belt object, and Ceres, the largest-known asteroid (**FIGURE 22.36**).

Pluto's reclassification was not the first such "demotion." In the mid-1800s, astronomy textbooks listed as many as 11 planets in our solar system, including the asteroids Vesta, Juno, Ceres, and Pallas. Astronomers continued to discover dozens of other "planets," a clear signal that these small bodies represent a class of objects separate from the planets.

The new classification will give a home to the hundreds of additional dwarf planets astronomers assume exist in the solar system. *New Horizons*, the first spacecraft designed to explore the outer solar system, was launched in January 2006. As of September 2012, *New Horizons* was halfway between the orbits of Uranus and Neptune. Scheduled to fly by Pluto in July 2015 and later explore the Kuiper belt, *New Horizons* carries tremendous potential for aiding researchers in further understanding the solar system.

FIGURE 22.36 An Artist's Drawing Showing the Relative Sizes of the Best-Known Dwarf Planets Compared to Earth and Its Moon Eris, the largest known dwarf planet, has a very eccentric orbit that takes it as far as 100 AU from the Sun. Both Eris and Pluto are composed mainly of ices of water, methane, and ammonia. Ceres is the only identified dwarf planet in the asteroid belt. (Courtesy of NASA)

22.5 CONCEPT CHECKS

1 Where are most asteroids found?

2 Compare and contrast asteroids and comets.

3 What do you think would happen if Earth passed through the tail of a comet?

4 Where are most comets thought to reside? What eventually becomes of comets that orbit close to the Sun?

5 Differentiate among the following solar system bodies: meteoroid, meteor, and meteorite.

6 What are the three main sources of meteoroids?

7 Why was Pluto demoted from the ranks of the officially recognized planets?

22 CONCEPTS IN REVIEW | Touring Our Solar System

22.1 OUR SOLAR SYSTEM: AN OVERVIEW

Describe the formation of the solar system according to the nebular theory. Compare and contrast the terrestrial and Jovian planets.

KEY TERMS: nebular theory, solar nebula, planetesimal, protoplanet, terrestrial (Earth-like) planet, Jovian (Jupiter-like) planet, escape velocity, impact crater

- Our Sun is the most massive body in a solar system, which includes planets, dwarf planets, moons, and other small bodies. The planets all orbit in the same direction and at speeds proportional to their distance from the Sun, with inner planets moving faster and outer planets moving more slowly.
- The solar system's formation is described by the nebular theory, which proposes that the system began as a solar nebula before condensing due to gravity. While most of the matter ended up in the Sun, some material formed a thick disc around the early Sun and later clumped together into larger and larger bodies. Planetesimals collided to form protoplanets, and protoplanets grew into planets.
- The four terrestrial planets are enriched in rocky materials, whereas the Jovian planets have a higher proportion of ice and gas. The terrestrial planets are relatively dense, with thin atmospheres, while the Jovian planets are less dense and have thick atmospheres.
- Smaller planets have less gravity to retain gases in their atmosphere. It's easier for lightweight gases such as hydrogen and helium to reach escape velocity in this situation, so the atmospheres of the terrestrial planets tend to be enriched in heavier gases, such as water vapor, carbon dioxide, and nitrogen.

22.2 EARTH'S MOON: A CHIP OFF THE OLD BLOCK

List and describe the major features of Earth's Moon and explain how maria basins were formed.

KEY TERMS: maria, lunar highlands (terrae), lunar regolith

- Earth's Moon is the largest moon relative to its planet, and its composition is unique in the solar system, approximately the same as the composition of Earth's mantle (density = 3.3 g/cm^3). The Moon likely formed due to a collision between a Mars-sized protoplanet and the early Earth. The bulk of the protoplanet's iron core material was incorporated into Earth, and its rocky mantle material spun off to make the Moon.
- Two types of topography dominate the lunar surface: (1) light-colored lunar highlands (or terrae) dominated by relatively old anorthosite breccia and (2) darker lowlands called maria, which are dominated by younger flood basalts. Both terrae and maria are partially covered by a layer called lunar regolith, which is produced by micrometeorite bombardment.

Q Briefly describe the formation of our Moon and how its formation accounts for its low density compared to that of Earth.

22.3 TERRESTRIAL PLANETS

Outline the principal characteristics of Mercury, Venus, and Mars. Describe their similarities to and differences from Earth.

- Mercury is the planet closest to the Sun. It has a very thin atmosphere and a weak magnetic field. Because it has a very thin atmosphere and its rate of rotation is extremely slow, the temperature on the surface varies from less than −173°C (−280°F) at night to 427°C (800°F) during daylight hours. The lobate scarps on Mercury's surface are likely the traces of thrust faults, which formed due to the planet's cooling and contraction.
- Venus, the second planet from the Sun, has a very dense atmosphere that is dominated by carbon dioxide. The resulting extreme greenhouse effect produces surface temperatures around 450°C (900°F). The topography of Venus has been resurfaced by active volcanism.
- Mars is the fourth planet from the Sun. It has about 1 percent as much atmosphere as Earth, so it is relatively cold (−140°C to 20°C [−220°F to 68°F]). Mars appears to be the closest planetary analogue to Earth, showing surface evidence of rifting, volcanism, and modification by flowing water. Volcanoes on Mars, such as Olympus Mons, are much bigger than volcanoes on Earth because of the lack of plate motion on Mars: The lava accumulates in a single cone rather than forming a long chain of cones, as exemplified by the Hawaiian islands.

Q As you can see from this graph, Mercury's temperature varies a lot from "day" to "night," but Venus's temperature is relatively constant "around the clock." Suggest a reason for this difference.

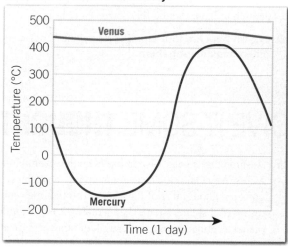

Idealized graph comparing the daily temperature variations on Venus and Mercury.

22.4 JOVIAN PLANETS

Compare and contrast the four Jovian planets.

KEY TERM: cryovolcanism

- Jupiter is the fifth planet from the Sun. It is very big—several times greater than the combined mass of everything else in the solar system except for the Sun. Convective flow among its three layers of clouds produces its characteristic banded appearance. Persistent, giant rotating storms exist between these bands. Many moons orbit Jupiter, including Io, which shows active volcanism, and Europa, which has an icy shell.
- Saturn is the sixth planet from the Sun. Like Jupiter, it is big, gaseous, and endowed with dozens of moons. Some of these moons show evidence of tectonics, while Titan has its own atmosphere. Saturn's well-developed rings are made of many particles of water ice and rocky debris.
- Uranus is the seventh planet from the Sun. Like its "twin" Neptune, it has a blue atmosphere dominated by methane, and its diameter is about four times greater than Earth's. Uranus rotates sideways relative to the plane of the solar system. It has a relatively thin ring system and at least five moons.
- Neptune, the eighth planet from the Sun, has an active atmosphere, with fierce wind speeds and giant storms. It has one large moon, Triton, which shows evidence of cryovolcanism, as well as a dozen smaller moons and a ring system.

Q Prepare and label a sketch comparing the typical characteristics of terrestrial planets and those of Jovian planets.

22.5 SMALL SOLAR SYSTEM BODIES

List and describe the principal characteristics of the small bodies that inhabit the solar system.

KEY TERMS: small solar system body, dwarf planet, asteroid, asteroid belt, comet, nucleus, coma, Kuiper belt, Oort cloud, meteor, meteoroid, meteor shower, meteorite

- Small solar system bodies include rocky asteroids and icy comets. Both are basically scraps left over from the formation of the solar system or fragments from later impacts.
- Most asteroids are concentrated in a wide belt between the orbits of Mars and Jupiter. Some are rocky, some are metallic, and some are basically "piles of rubble" loosely held together by their own weak gravity.
- Comets are dominated by ices, "dirtied" by rocky material and dust. They originate in either the Kuiper belt (a second "asteroid belt" beyond Neptune) or the Oort cloud (a spherical "shell" around the otherwise planar "disc" of the solar system). When the orbit of a comet brings it through the inner solar system, solar radiation (sunlight) causes its ices to begin to vaporize, generating the coma (gaseous envelope around the comet's nucleus) and its characteristic "tail."
- A meteoroid is debris that enters Earth's atmosphere, flaring briefly as a meteor before either burning up or striking Earth's surface to become a meteorite. Asteroids and material lost from comets as they travel through the inner solar system are the most common source of meteoroids.
- Dwarf planets include Ceres (located in the asteroid belt), the dwarf planet Pluto, and Eris, a Kuiper belt object. They are spherical bodies that orbit the Sun but are not massive enough to have cleared their orbits of debris.

Q **Shown here are four small solar system bodies. Identify each and explain the differences among them.**

Muellek Josef/Shutterstock National Science Foundation

Jerry Schad/Science Source NASA/JPL-Caltech/UCAL/MPS/DLR/IDA

GIVE IT SOME THOUGHT

1. Assume that a solar system has been discovered in a nearby region of the Milky Way Galaxy. The accompanying table shows data that have been gathered about three of the planets orbiting the central star of this newly discovered solar system. Using Table 22.1 as a guide, classify each planet as either Jovian, terrestrial, or neither. Explain your reasoning.

	Planet 1	Planet 2	Planet 3
Relative Mass (Earth = 1)	1.2	15	0.1
Diameter (km)	15,000	52,000	5000
Mean Distance from Star (AU)	1.4	17	35
Density (g/cm3)	4.8	1.22	5.3
Orbital Eccentricity	0.01	0.05	0.23

2. In order to conceptualize the size and scale of Earth and Moon as they relate to the solar system, complete the following:

   a. Approximately how many Moons (diameter 3475 kilometers [2160 miles]) would fit side-by-side across the diameter of Earth (diameter 12,756 kilometers [7926 miles])?

   b. Given that the Moon's orbital radius is 384,798 kilometers, approximately how many Earths would fit side-by-side between Earth and the Moon?

   c. Approximately how many Earths would fit side-by-side across the Sun, whose diameter is about 1,390,000 kilometers?

   d. Approximately how many Suns would fit side-by-side between Earth and the Sun, a distance of about 150,000,000 kilometers?

3. The accompanying graph shows the temperatures at various distances from the Sun during the formation of our solar system. Use it to complete the following:

   a. Which planets formed at locations where the temperature in the solar system was hotter than the boiling point of water?

   b. Which planets formed at locations where the temperature in the solar system was cooler than the freezing point of water?

Condition	Temperature (Farenheit)	Temperature (Kelvins)
Water freezes	32	199
Room temp.	72	296
Human body	98.6	310
Water boils	212	373

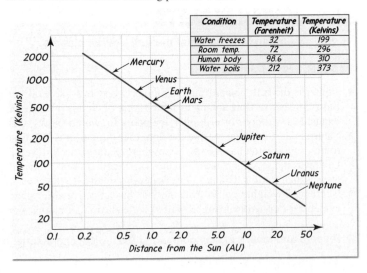

4. The accompanying sketch shows four primary craters (A, B, C, and D). The impact that produced crater A produced two secondary craters (labeled "a") and three rays. Crater D has one secondary crater (labeled "d"). Rank the four primary craters from oldest to youngest and explain your ranking.

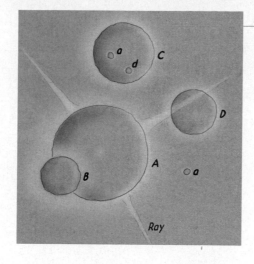

5. The accompanying diagram shows two of Uranus's moons, Ophelia and Cordelia, which act as shepherd moons for the Epsilon ring. Explain what would happen to the Epsilon ring if a large asteroid struck Ophelia, knocking it out of the Uranian system.

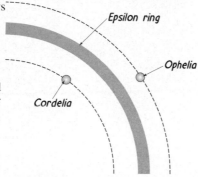

6. It has been estimated that Halley's Comet has a mass of 100 billion tons. Furthermore, it is estimated to lose about 100 million tons of material when its orbit brings it close to the Sun. With an orbital period of 76 years, calculate the maximum remaining life span of Halley's Comet.

7. The accompanying diagram shows a comet traveling toward the Sun at the first position where it has both an ion tail and a dust tail. Refer to this diagram to complete the following:

   a. For each of the three numbered sites, indicate whether the comet will have no tails, one tail, or two tails. If one tail or two tails are present, in what direction will they point?

   b. Would your answers to the preceding question change if the Sun's energy output were to increase significantly? If so, how would they change?

   c. If the solar wind suddenly ceased, how would this affect the comet and its tails?

8. Assume that three irregularly shaped planet-like objects, each smaller than our Moon, have just been discovered orbiting the Sun at a distance of 35 AU. One of your friends argues that the objects should be classified as planets because they are large and orbit the Sun. Another friend argues that the objects should be classified as dwarf planets, such as Pluto. State whether you agree or disagree with either or both of your friends. Explain your reasoning.

EXAMINING THE **EARTH SYSTEM**

1. On Earth the four major spheres (atmosphere, hydrosphere, geosphere, and biosphere) interact as a system with occasional influences from our near-space neighbors. Which of these spheres are absent, or nearly absent, on the Moon? Because the Moon lacks these spheres, list at least five processes that operate on Earth but are absent on the Moon.

2. Among the planets in our solar system, Earth is unique because water exists in all three states (solid, liquid, and gas) on and near its surface. In what state(s) of matter is water found on Mercury, Venus, and Mars?

   a. How would Earth's hydrologic cycle be different if its orbit were inside the orbit of Venus?

   b. How would Earth's hydrologic cycle be different if its orbit were outside the orbit of Mars?

3. If a large meteorite were to strike Earth in the near future, what effect might this event have on the atmosphere (in particular, on average temperatures and climate)? If these conditions persisted for several years, how might the changes influence the biosphere?

MasteringGeology™

Looking for additional review and test prep materials? Visit the Self Study area in **www.masteringgeology.com** to find practice quizzes, study tools, and multimedia that will aid in your understanding of this chapter's content. In **MasteringGeology™** you will find:

- GEODe: Earth Science: An interactive visual walkthrough of key concepts
- Geoscience Animation Library: More than 100 animations illuminating many difficult-to-understand Earth science concepts

- In The News RSS Feeds: Current Earth science events and news articles are pulled into the site with assessment
- Pearson eText
- Optional Self Study Quizzes
- Web Links
- Glossary
- Flashcards

23

Light, Astronomical Observations, and the Sun[1]

FOCUS ON CONCEPTS

Each statement represents the primary **LEARNING OBJECTIVE** for the corresponding major heading within the chapter. After you complete the chapter, you should be able to:

23.1 List and describe the various types of electro-magnetic radiation.

23.2 Explain how the three types of spectra are generated and what they tell astronomers about the radiating body that produced them.

23.3 Compare and contrast refracting and reflecting telescopes. Explain why modern telescopes are built on mountaintops.

23.4 Explain the advantages of radio telescopes and orbiting observatories over optical telescopes.

23.5 Write a statement explaining why the Sun is important to the study of astronomy. Sketch the Sun's structure and describe each of its four major layers.

23.6 List and describe the three types of explosive activity that occur at the Sun's surface.

23.7 Summarize the process called the proton–proton chain reaction.

Comet Hale-Bopp above the observatories on the summit of Mauna Kea, Hawaii. (Photo by David Nunuk/Science Source)

[1]This chapter was revised with the assistance of Mark Watry and Teresa Tarbuck.

Since astronomers cannot study the universe by bringing it into the laboratory, and because the vast majority of celestial objects are too far away to visit, astronomers collect and study those things that come to Earth from space. Overwhelmingly, this means collecting and studying light emitted or reflected by objects found in the universe. In fact, everything that is known about the universe beyond the solar system comes from the analysis of the light from distant sources. This chapter examines the properties and utility of light, some of the tools astronomers use to collect and study light, and what is known about the nearest source of light, the Sun.

23.1 | SIGNALS FROM SPACE

List and describe the various types of electromagnetic radiation.

Although visible light is most familiar to us, it constitutes only a tiny sliver of an array of energy referred to as **electromagnetic radiation** (**FIGURE 23.1**). Included in this array are *gamma rays*, *X-rays*, *ultraviolet light*, *visible light*, *infrared radiation* (heat), *microwaves*, and *radio waves* (**FIGURE 23.2**). All forms of radiant energy travel through the vacuum of space in a straight line at the rate of 300,000 kilometers (186,000 miles) per second.[2] Over 24 hours, this is a staggering 26 billion kilometers. The light that we collect tells us about the processes that created it and about the matter lying between us and the source of the light.

---
[2]Light rays are "bent" slightly when they pass near a very massive object such as the Sun.

Nature of Light

Experiments have demonstrated that light can be described in two ways. In some instances, light behaves like waves, and in others, it behaves like discrete particles. In the wave sense, light is analogous to swells in the ocean. This motion is characterized by *wavelength*—the distance from one wave crest to the next. Wavelengths vary from several kilometers for some radio waves to less than one-billionth of a centimeter for gamma rays (see Figure 23.2). Most of these waves are either too long or too short for our eyes to detect; however, the primary characteristics of all electromagnetic radiation can be described using visible light as an example.

The extremely narrow band of electromagnetic radiation we can see (which is labeled *visible light* in Figure 23.2) is

FIGURE 23.1 A Face-on View of Galaxy NGC 1232 Despite being 100 million light-years away, modern telescopes allow astronomers to study its intricate details. Older, reddish stars are located mainly in the galaxy's central region, while young, hot blue stars make up the spiral arms. (Photo by NASA)

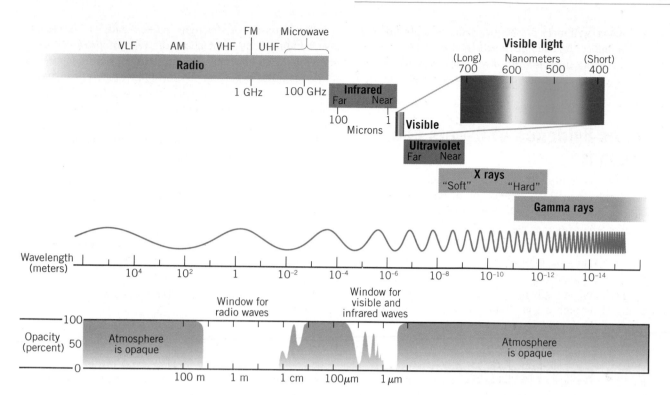

FIGURE 23.2
Electromagnetic Radiation
The electromagnetic spectrum ranges from long-wavelength radio waves to short-wavelength gamma radiation.

sometimes referred to as *white light*. White light consists of an array of waves having various wavelengths, a fact easily demonstrated with a prism (**FIGURE 23.3A**). As white light passes through a prism, the color with the shortest wavelength, violet, is bent more than blue, which is bent more than green, and so forth (**TABLE 23.1**). Thus, white light can be separated into its component colors, producing the familiar "rainbow of colors" (see Figure 23.3A).

Wave theory, however, cannot explain some of the observed characteristics of light. In these cases, light acts like a stream of particles, analogous to infinitesimally small bullets fired from a machine gun. These particles, called **photons**, can exert a pressure (push) on matter, which is called **radiation pressure**. Recall that photons from the Sun are responsible for pushing material away from a comet to produce its dust tail. Each photon has a specific amount of energy, which is related to its wavelength in a simple way: *Shorter wavelengths* correspond to *more energetic photons*. Thus, blue light has more energetic photons than red light.

Which theory of light—the wave theory or the particle theory—is correct? The answer is that both are correct

TABLE 23.1 Colors and Corresponding Wavelengths

Color	Wavelength (Nanometers*)
Violet	380–440
Blue	440–500
Green	500–560
Yellow	560–590
Orange	590–640
Red	640–750

*1 nanometer is 10^{-9} meter.

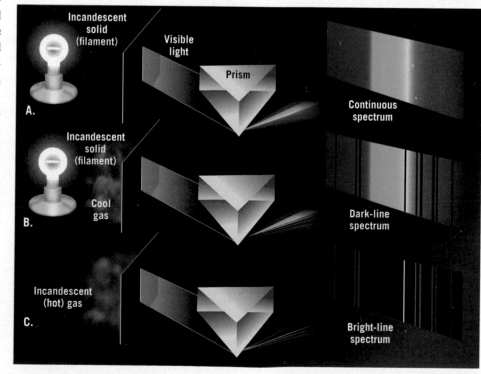

SmartFigure 23.3
Formation of the Three Types of Spectra
A. Continuous spectrum.
B. Dark-line spectrum.
C. Bright-line spectrum.

because each will predict the behavior of light for certain phenomena. As George Abell, a prominent astronomer, stated about all scientific laws, "The mistake is only to apply them to situations that are outside their range of validity."

Light As Evidence of Events and Processes

Light energy from all parts of the electromagnetic spectrum reaches Earth from stars, interstellar matter, and galaxies throughout the universe. This light provides evidence of events and processes that we could not otherwise observe. For example, when matter is engulfed by a black hole, the result is the emission of high-energy x-rays. By contrast, when less violent processes occur, small amounts of low-energy radiation are released. This occurs when a shock wave moves through a gas cloud, and it raises the cloud's temperature, causing infrared (heat) energy to be emitted.

The intensity of the light emitted and its wavelength distribution tell us a lot about the type of process that is occurring. This information can often be used to support or refute scientific hypotheses. For example, theoretical studies predicted the existence of black holes long before observational evidence existed. The concept of black holes gained considerable support when x-rays matching the wavelengths predicted by this hypothesis were detected around objects suspected of being black holes.

23.1 CONCEPT CHECKS

1 What term is used to describe the collection that includes gamma rays, x-rays, ultraviolet light, visible light, infrared radiation, microwaves, and radio waves?

2 Which color has the longest wavelength? The shortest?

3 How does the amount of energy contained in a photon relate to its wavelength?

23.2 | SPECTROSCOPY Explain how the three types of spectra are generated and what they tell astronomers about the radiating body that produced them.

When Sir Isaac Newton used a prism to disperse white light into its component colors, he unknowingly initiated the field of **spectroscopy**—the study of those properties of light that are wavelength dependent. The rainbow of colors Newton produced is called a *continuous spectrum* because all wavelengths of visible light are included. Later it was learned that two other types of spectra (*dark line* and *bright line*) exist and that each one is generated under somewhat different conditions (see Figure 23.3). Just as visible light can produce a spectrum, other regions of the electromagnetic spectrum can be dispersed to produce spectra.

Continuous Spectrum

A **continuous spectrum** is produced by an incandescent (glowing) solid, liquid, or gas under high pressure. (*Incandescent* means "to emit light when hot.") It consists of a continuous band of wavelengths like that generated by a common 100-watt incandescent light bulb (see Figure 23.3A). A continuous spectrum contains two important pieces of information about radiating bodies.

First, a continuous spectrum provides information about the total energy output of the radiating body. If the temperature of a radiating surface increases, the total amount of energy emitted increases. The rate of increase is stated in the Stefan–Boltzmann law: *The energy radiated by a body is directly proportional to the fourth power of its absolute temperature.* For example, if the temperature of a star is twice that of another star, the total radiation emitted by the hotter star is $2^4 = 2 \times 2 \times 2 \times 2$, or 16 times greater than that of the cooler star.

Second, a continuous spectrum contains information about the surface temperature of the radiating body. As

the surface temperature of an object increases, a larger proportion of its energy is radiated at shorter wavelengths (higher energy). To illustrate, imagine a metal rod that is heated slowly. Initially, the rod appears dull red (longer wavelengths), then yellow, and later bluish-white (shorter wavelengths). All incandescent bodies show this behavior, so it follows that blue stars are hotter than yellow stars (like the Sun), which are hotter than red stars (see Table 23.1).

Dark-Line Spectrum

When a telescope collects light radiating from a star and the light is passed through an instrument called a **spectroscope** (which spreads out the wavelengths in a manner similar to a prism), a "continuous spectrum" that contains a series of dark is produced. This type of spectrum, called a **dark-line** (or **absorption**) **spectrum**, is generated whenever visible light is passed through a comparatively cool gas at low pressure (**FIGURE 23.3B**). For example, when visible light is passed through a glass jar containing hydrogen gas, the hydrogen atoms absorb specific wavelengths of light, resulting in a unique set of dark lines. Each set of dark lines, like a set of fingerprints, identifies the matter that is present. Elements such as iron that exist in the gaseous state on the Sun have been identified in numerous other stars through study of their spectra. Even organic molecules have been discovered in distant interstellar clouds of dust and gases using this technique. Thus, when carefully analyzed, a dark-line spectrum reveals the composition of the radiating body.

The spectra of most stars are of the dark-line type. Imagine the light produced in the Sun's interior passing outward through its atmosphere. The gas in the solar atmosphere is cooler than that inside, and when it absorbs some of the sunlight (and re-emits it in a random direction), we do not see it, and a dark area (line) appears in the spectrum. Although the lines appear black, they just look that way next to the bright parts of the spectrum. The relative intensities of the light in the dark lines contain information about the relative amounts of each kind of matter present.

Bright-Line Spectrum

A **bright-line** (or **emission**) **spectrum** is a series of bright lines that appear in the same locations as the dark lines for the same gas. They are produced by hot (incandescent), gaseous materials, at low pressure (**FIGURE 23.3C**). These spectra contain information about the temperature of the gas and the matter in it. **FIGURE 23.4** shows the emission spectra of hydrogen and helium, the two most abundant elements in the universe.

Bright-line, or emission, spectra are produced by large interstellar clouds (nebula) consisting largely of hydrogen gas excited by extremely hot stars. Because the brightest emission line produced by hydrogen is red, these clouds tend to have a red glow that is characteristic of excited hydrogen gas. The Orion Nebula is a well-known emission nebula that is bright enough to be seen by the naked eye (**FIGURE 23.5**). It is located in the constellation Orion, in the sword of the hunter.

The Doppler Effect

The positions of the bright and dark lines in the spectra described earlier shift when the source of energy moves relative to the observer. This effect is observed for all types of

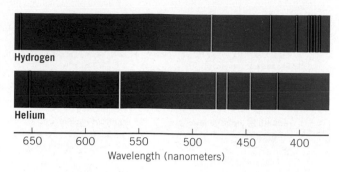

waves. You may have heard the change in pitch of a car horn or an ambulance siren as it passes by. When it is approaching, the sound seems to have a higher-than-normal pitch, and when it is moving away, the pitch sounds lower than normal. This effect, first explained by Christian Doppler in 1842, is called the **Doppler effect**. The reason for the difference in pitch is that it takes time for the wave to be emitted. If the source is moving away, the beginning of the wave is emitted nearer to you than the end, which stretches the wave—that is, gives it a longer wavelength (**FIGURE 23.6**). The opposite is true for an approaching source.

In the case of light, when a source is moving away, its light appears redder than it actually is because its waves are lengthened. Objects approaching have their light waves shifted toward the blue (shorter-wavelength) end of the spectrum. The same effect is produced if you are moving and the light remains stationary.

Doppler shifts are generally established by comparing the dark lines in the spectra of a star (or other celestial body) with the spectrum of a motionless body produced in

FIGURE 23.4 Bright-Line Spectra of Hydrogen and Helium These gases are the two most abundant elements in the universe.

FIGURE 23.5 The Orion Nebula, an Emission Nebula The red color of Orion Nebula is produced by hydrogen gas that is being heated by nearby hot stars. Bright enough to be seen by the naked eye, the Orion Nebula is located in the sword of the hunter in the constellation of the same name. (Courtesy of National Optical Astronomy Observatory/ Association of Universities for Research in Astronomy/National Science Foundation [NOAO/AURA/ NSF])

SmartFigure 23.6 The Doppler Effect Illustration showing the apparent lengthening and shortening of wavelengths caused by the relative motion between a source and an observer.

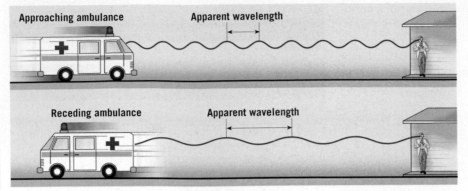

Approaching ambulance   Apparent wavelength

Receding ambulance   Apparent wavelength

the laboratory (**FIGURE 23.7**). When the spectral lines are shifted toward the long end of the electromagnetic spectrum, called a *red shift*, the star is moving away from the observer. By contrast, when the spectral lines are moving toward the short end (blue/violet end) of the spectrum, the object is approaching Earth.

In addition, the amount of shift allows us to calculate the rate at which the relative movement is occurring. Large

Doppler shifts indicate high velocities; small Doppler shifts indicate low velocities. For example, notice that the amount of the red shift in Figure 23.7B is less than that shown in Figure 23.7C.

Two types of Doppler shifts are important in astronomy: those caused by local motions and those caused by the expansion of the universe. Doppler shifts due to local motions are used to measure how fast one star orbits another in a binary (two-star) system or how fast a pulsing star expands and contracts. Those shifts caused by the expansion of the universe (where space is continually being created between the galaxies) can tell us how far away distant objects are. These measurements coupled with the speed of light tell us how long ago the light left these distant objects, and as we look farther out, we can get a sense of the age of the universe. (The expansion of the universe will be discussed in depth in Chapter 24.)

FIGURE 23.7 Determining the Relative Motion of Two Bodies We can determine whether Earth is approaching or receding from a celestial body by comparing the spectrum of a motionless light source to the spectrum of a moving body. **A.** Standard dark-line spectrum for sodium produced in the laboratory. **B.** and **C.** Sodium lines as they would appear when a light source is receding (red shift). **D.** Sodium lines produced by an approaching star (blue shift).

A. Standard sodium lines

B. Red-shifted sodium lines

C. Large red-shifted sodium lines

D. Blue-shifted sodium lines

23.2 CONCEPT CHECKS

1 What is *spectroscopy*?

2 Describe a continuous spectrum. Give an example of a natural phenomenon that exhibits a continuous spectrum.

3 What can a continuous spectrum tell astronomers about a star?

4 What can be learned about a star (or other celestial objects) from a dark-line (absorption) spectrum?

5 What produces emission lines (bright lines) in a spectrum?

6 Briefly describe the Doppler effect and describe how astronomers determine whether a star is moving toward or away from Earth.

23.3 | COLLECTING LIGHT USING OPTICAL TELESCOPES

Compare and contrast refracting and reflecting telescopes. Explain why modern telescopes are built on mountaintops.

The earliest tools used to observe the heavens were human eyes—the only mechanisms early astronomers like Tycho Brahe had available to them. However, the human eye is a poor instrument for astronomical observation because it cannot collect much light, is not very sensitive to faint colors, and collects only visible light. Early optical telescopes vastly improved over the naked eye, allowing for the collection of large amounts of light. Optical telescopes collect light with visible (or nearly visible) wavelengths and come in two basic types—*refracting* and *reflecting* telescopes.

Refracting Telescopes

Much like the one used by Galileo, **refracting telescopes** employ lenses to collect and focus light (**FIGURE 23.8**). The light coming from a distant object can be thought of as a ray or beam by the time it reaches Earth. Our eye or a telescope lens intercepts some portion of the incoming light. To collect more light, one simply uses a larger lens.

Two major problems prevent the manufacture of large refracting telescopes. First, the lens acts like a prism,

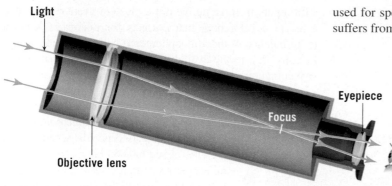

Light

Objective lens

Focus

Eyepiece

FIGURE 23.8 Refracting Telescope A refracting telescope uses a lens to collect and focus light.

spreading out the colors in the light, an effect called **chromatic aberration**. This is a problem that grows quickly as lenses get larger. Second, large lenses weigh so much that they sag under their own weight, changing their shape and, hence, changing their focusing properties.

The world's largest refracting telescope is the 1-meter (40-inch) telescope at Yerkes Observatory in Williams Bay, Wisconsin (**FIGURE 23.9**). This telescope was successfully

used for spectroscopic work (and other observations), but it suffers from the problems described above.

Reflecting Telescopes

Small refracting telescopes work very well for observing objects in the solar system and for observing any other bright source. However, through experimentation, Sir Isaac Newton discovered that a large lens would cause white light to separate into its constituent parts (chromatic aberration), causing a halo of colored light to form around the object being viewed. By designing a telescope that used a mirror rather than a lens, Newton avoided this problem because the light does not travel through glass but is reflected from a coated surface instead (**FIGURE 23.10**). As a result, refracting telescopes have mainly been replaced by **reflecting telescopes** that use a curved mirror to collect and focus the light (**FIGURE 23.11A**).

All large telescopes built today are of the reflecting type, having a mirror that is made of a special glass that is finely ground to a nearly perfect parabolic shape. A parabola is the geometric shape that takes parallel lines—or parallel light rays—and focuses them to a point. The Hubble Space Telescope has a 2.4-meter (94.5-inch) mirror that is ground to within about one-millionth of an inch of being a perfect paraboloid. (If you have the time and patience, you could grind your own 8-, 10-, or even 12-inch mirror.) Once ground, the surface of the mirror is coated with a highly reflective material.

Reflecting telescopes collect more light as the diameter of the mirror increases, just like refracting telescopes with larger lenses. However, there are difficulties in increasing the size of the mirror beyond several meters. These include

FIGURE 23.9 World's Largest Refracting Telescope This 1-meter (40-inch) refractor is located at Yerkes Observatory, Williams Bay, Wisconsin. (Photo by AP Photo/The Janesville Gazette, Lukas Keapproth)

FIGURE 23.10 Newton's Reflecting Telescope (Photo by Dave King © Dorling Kindersley, Courtesy of The Science Museum, London)

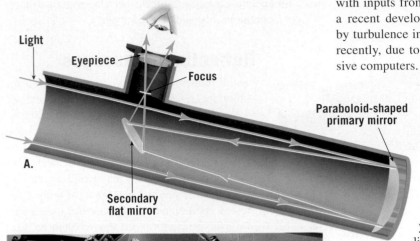

Light
Eyepiece
Focus
Paraboloid-shaped primary mirror
Secondary flat mirror
A.

B.

with inputs from an active optics system. Active optics are a recent development that corrects for distortions caused by turbulence in the atmosphere and became practical only recently, due to the availability of fast, relatively inexpensive computers.

Larger telescopes not only allow us to collect more light from faint nearby objects, they also allow us to collect more light from very distant objects. Since the speed of light is finite (about 300,000 kilometers [186,000 miles] per second), it takes time for the light to get to us. Even light from the Sun takes about $8\frac{1}{2}$ minutes to reach Earth, and light from the nearest large galaxy takes 2 million years to reach us. Larger telescopes allow us to literally look back in time. Our desire to understand the nature and evolution of the universe has motivated us to develop telescopes that look farther and farther back in time. Larger telescopes also generally provide better *resolution*, or clarity (**FIGURE 23.12**).

Light Collection

Telescopes simply collect light. The earliest light collectors were the astronomer's eyes. Astronomers would look through telescopes and draw what they saw (**FIGURE 23.13**). Each person's eyes perceive light intensity and faint color differently (and each person has a different amount of drawing talent), so under the same conditions, different images of the same object were produced. In addition, personal biases can influence what a person observes. For example, in the early twentieth century, noted astronomer Percival Lowell (1855–1916) was convinced that there were canals on the surface of Mars. Thus, he "saw" them in his telescope and drew them in his images. Subsequent studies did not support Lowell's observations.

Photographic film was a revolutionary improvement. It is not impeded by personal biases, it records reasonably accurate relative light intensities, and it records faint colors more

supporting such a large mass, moving that mass to realign the telescope, warping of the mirror surface under its own weight, and the time required to grind a nearly perfect surface over such a large area. For example, it took 14 years to construct the mirror for the Hale Telescope.

These difficulties have recently been overcome in two ways. First, we can use an array of several smaller deformable mirrors under computer control to give the effect of one large mirror. Second, we can use a single, very thin mirror mounted on actuators that control the mirror shape,

accurately than does the human eye. However, only about 2 percent of the light that strikes film is recorded. This means that long exposure times are required when recording faint images. Furthermore, photographic film, like the human eye, is not equally sensitive to all wavelengths. There are also differences between individual batches or even between pieces of film that need to be considered when making quantitative comparisons.

Advances in semiconductor technology have produced the *charge-coupled device* (*CCD*), which takes an electronic photograph and effectively uses the same piece of "film" over and over again. (Charge-coupled devices are used in digital cameras, like cell phone cameras, as the light-sensing component.) CCD cameras offer a tremendous improvement over photographic film for detection of visible and near-visible light. They typically detect 70 percent, or more, of all incoming light and are easily calibrated for variations in wavelength sensitivity. Using CCD cameras, astronomers can collect light from distant objects for hours, as long as the telescope is accurately steered. Light can also be collected over several nights and synthesized to make a single image.

Once collected, light emitted from distant sources is analyzed to determine the temperature, composition, relative motion, and distance to celestial objects. For faint or distant sources, as much light as possible must be collected, for the longest amount of time that is reasonable. This requires very large instruments with very sensitive detectors and

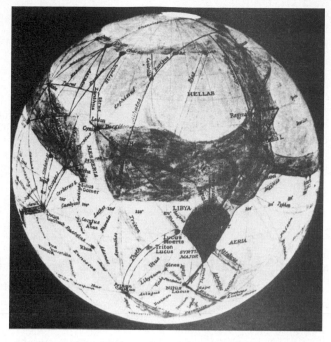

FIGURE 23.13 Percival Lowell's Drawing of Mars Percival Lowell believed that life existed on Mars and drew these canals, influenced perhaps by his personal biases. (Photo SPL/Science Source)

virtually no interference from other sources of electromagnetic energy. Further, Earth's atmosphere is very turbulent, which turns faint points of light into very faint smudges.

The largest telescopes were built on mountaintops away from urban areas to get above as much of the turbulent atmosphere as possible and to reduce the effects of light pollution (**FIGURE 23.14**). In addition, astronomers have

FIGURE 23.14 Kitt Peak National Observatory on a Starlit Night (Photo by Bryan Allen/CORBIS)

designed *adaptive optics* to overcome much of the blurring introduced by the constant motion of Earth's atmosphere. However, these instruments are limited to collecting light in the visible (or radio) regions of the electromagnetic spectrum because other wavelengths do not penetrate Earth's atmosphere (see Figure 23.2).

Finally, with the dawn of the space age, even these wavelength limitations have been overcome. It has become practical to put astronomical observatories in space, avoiding the turbulent atmosphere and allowing for the collection of electromagnetic radiation at all wavelengths.

23.3 CONCEPT CHECKS

1. What is the main difference between reflecting and refracting telescopes?

2. Why do astronomers seek to design telescopes with larger and larger mirrors?

3. Why do all large optical telescopes use mirrors rather than lenses to collect light?

4. What are the advantages of *charge-coupled devices* (*CCDs*) over photographic film?

5. Why is the human eye an ineffective tool for astronomical observation?

6. Provide two reasons why the largest telescopes are built on mountaintops away from large cities.

23.4 | RADIO- AND SPACE-BASED ASTRONOMY

Explain the advantages of radio telescopes and orbiting observatories over optical telescopes.

Sunlight consists of more than the visible portion of the electromagnetic spectrum. Gamma rays, x-rays, ultraviolet radiation, infrared radiation, and radio waves are also produced by stars and other celestial objects. CCD cameras that are sensitive to ultraviolet and infrared radiation have been developed to extend the limits of our vision. However, much of the radiation produced by celestial objects cannot penetrate our atmosphere or is not detectable by optical telescopes. As a result, astronomers have developed other observational techniques covering the remaining portions of the electromagnetic spectrum.

Radio Telescopes

Of great importance is a narrow band of radio waves that penetrates the atmosphere (see Figure 23.2). One particular wavelength is the 21-centimeter line produced by neutral hydrogen (hydrogen atoms that still hold their electron). Measurement of this radiation has permitted us to map the galactic distribution of hydrogen, the material from which stars are made.

The detection of radio waves is accomplished by "big dishes" called **radio telescopes** (**FIGURE 23.15A**). In principle,

Secondary reflector

Receiver

Primary reflector

A.

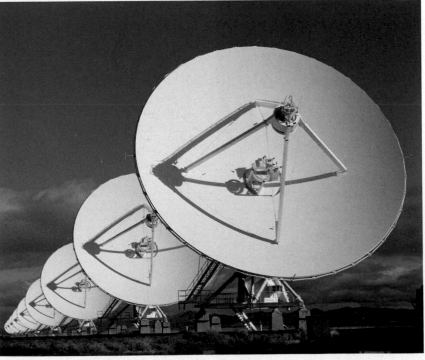

B.

FIGURE 23.15 Radio Telescopes A. The 100-meter (330-foot) steerable Robert C. Byrd radio telescope at Green Bank, West Virginia. The dish acts like the mirror of a reflecting optical telescope to focus radio waves onto the receiver. (Photo by National Radio Astronomy Observatory) **B.** Twenty-seven identical radio telescopes operate together to form the Very Large Array near Socorro, New Mexico. (Photo by Prisma/SuperStock)

the dish of a radio telescope operates in the same manner as the mirror of an optical telescope. It is parabolic in shape and focuses the incoming radio waves on an antenna, which collects and transmits these waves to an amplifier.

Because radio waves are about 100,000 times longer than visible radiation, the surface of a dish need not be as smooth as a mirror. In fact, except for the shortest radio waves, wire mesh is an adequate reflector. On the other hand, because radio signals from celestial sources are very weak, large dishes are necessary in order to intercept a signal that is strong enough to be detected. The largest radio telescope is a bowl-shaped antenna hung in a natural depression in Puerto Rico (**FIGURE 23.16**). It is 300 meters (1000 feet) in diameter and has some directional flexibility because of its movable antenna. The largest steerable types have about 100-meter (330-foot) dishes. The National Radio Astronomy Observatory in Green Bank, West Virginia, provides an example (see Figure 23.15A).

Radio telescopes have relatively poor resolution, making it difficult to pinpoint the radio source. Pairs or groups of telescopes are used to reduce this problem. When several radio telescopes are wired together, the resulting network is called a **radio interferometer** (**FIGURE 23.15B**).

Orbiting Observatories

Orbiting observatories circumvent all the problems caused by Earth's atmosphere and have led to many significant discoveries in astronomy. NASA's series of "four great observatories" provides a good illustration.

The Hubble Space Telescope
Launched in 1990, the *Hubble Space Telescope* (*HST*) is an optical reflecting telescope in orbit around Earth (see the GEOgraphics on page 712). Its images are not distorted by the atmosphere, and there is no atmospherically scattered light to drown out faint sources of light. In addition, it can collect ultraviolet light that is absorbed by Earth's ozone layer and is thus unavailable to ground-based telescopes. Hubble must be considered to be one of the most important instruments in the history of astronomy because of the large number of discoveries that have been made from its images. The 2.4-meter (94.5-inch) mirror has produced images with a sensitivity and resolution that are only now being matched by much larger (10-meter [33-foot]) ground-based telescopes.

Here are just a few of the many discoveries made with Hubble. HST provided visual proof that pancake-shaped disks of dust are common around young stars, providing support for the nebular hypothesis of solar system formation. Hubble provided decisive evidence that super-massive black holes reside in the center of many galaxies by imaging the

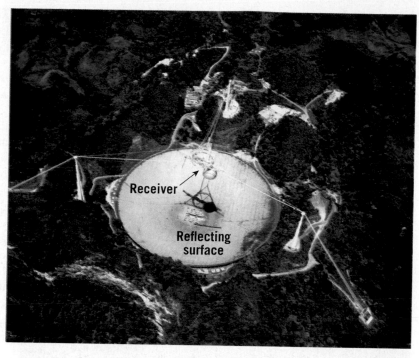

Receiver

Reflecting surface

FIGURE 23.16 The 300-Meter (1000-Foot) Radio Telescope at Arecibo, Puerto Rico (Courtesy of David Parker/Science Source)

movements of dust and gas in the interiors of galaxies. The HST has also allowed us to look farther out into the universe (and farther back in time) than ever before, in the process producing the most "elusive" astronomical image ever taken, the *Ultra-Deep Field* (see the GEOgraphics on page 713). This image was acquired by looking at a patch of "empty" sky for a total of 1 million seconds; the faintest objects put only one photon per minute into the exposure.

The scientific successor to the Hubble Telescope, called the *James Webb Space Telescope*, is scheduled for launch in 2018. The Webb telescope will have a large 6.6–meter-wide (21-foot-wide) mirror and a sunshield the size of a tennis court.

The Compton Gamma Ray Observatory
Designed to collect data on some of the most violent physical processes in the universe, the *Compton Gamma Ray Observatory* (*CGRO*) was launched in 1991. It had a sensitivity 10 times greater than any previous gamma ray instrument and collected an incredible range of high-energy radiation. One of the main scientific discoveries made by CGRO was the uniform distribution of *gamma ray bursts*, which suggest that they are common events associated with ordinary objects.

Gamma ray bursts are flashes of gamma rays that come from seemingly random places deep in the universe at random times. They are probably the most luminous and, therefore, the most energetic events occurring in the universe since the Big Bang. It is quite likely that many of them are caused by rapidly rotating massive stars as they collapse to form black holes (see Chapter 24).

The Chandra X-Ray Observatory
The *Chandra X-Ray Observatory* (*CXO*), launched in 1999, was designed to observe objects such as black holes, quasars, and high-temperature gases at x-ray wavelengths to better understand the

FIGURE 23.17 Supernova Remnant Captured by Chandra X-Ray Observatory This scattered glowing debris ejected from a massive star that is barely visible in the optical part of the spectrum is awash in brilliantly glowing gases emitting x-rays. (NASA/Marshall Space Flight Center)

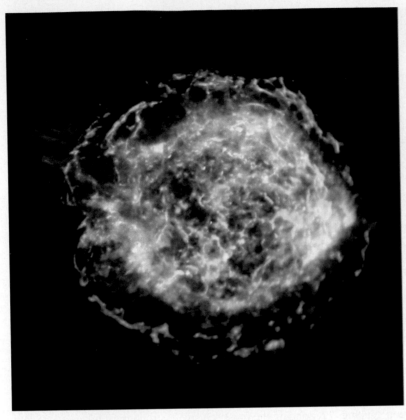

The Spitzer Space Telescope

Launched in 2003, the *Spitzer Space Telescope* (*SST*) was designed to collect infrared (heat) energy that is mostly blocked by Earth's atmosphere. Its instruments must be cooled to near zero Kelvin so that heat from nearby objects (and the satellite itself) does not interfere with the measurements. The telescope is actually in an orbit around the Sun to keep it away from the thermal energy radiated by Earth, and it is outfitted with a shield to deflect solar radiation.

Spitzer's highly sensitive instruments give us unique views of the universe and allow us to peer into regions of space that are hidden from optical telescopes by vast, dense clouds of gas and dust (nebula). Fortunately, infrared light can pass through these clouds, allowing us to peer into regions of star formation, the centers of galaxies, and newly forming planetary systems. Infrared light also brings us information about cool celestial objects, such as small stars that are too dim to be detected at visible wavelengths, planets that lay outside our solar system, and molecular clouds.

structure and evolution of the universe. With a resolution 25 times greater than any other x-ray observatory, it uses only as much power as an ordinary hair dryer (**FIGURE 23.17**). The CXO has observed a black hole pulling in matter and two black holes merging into one. In addition, it has provided an independent measurement of the age of the universe, reinforcing the estimated age of 12–14 billion years. The CXO has shown what galaxies were like when the universe was only a few billion years old.

23.4 CONCEPT CHECKS

1 Why are radio telescopes much larger than optical telescopes?

2 What are some of the advantages of radio telescopes over optical telescopes?

3 Explain why space makes a good site for an optical observatory.

4 What can astronomers learn about the universe by studying it at multiple wavelengths?

23.5 | THE SUN Write a statement explaining why the Sun is important to the study of astronomy. Sketch the Sun's structure and describe each of its four major layers.

The Sun is one of the 200 billion stars that make up the Milky Way Galaxy. Although the Sun is of little significance to the universe as a whole, to those of us who inhabit Earth, it is the primary source of energy. Everything from the food we eat to the fossil fuels we burn in our automobiles and power plants is ultimately derived from solar energy (**FIGURE 23.18**). The Sun is also important in astronomy, since it is the only star close enough to permit easy study of its surface. Even with the largest telescopes, most other stars appear only as points of light.

Because the Sun is so bright and emits eye-damaging radiation, observing it directly is unsafe. However, we can study it safely by using a telescope to project the Sun's image on a piece of cardboard placed behind

the telescope's eyepiece. Several telescopes around the world use this method to keep a constant vigil of the Sun. One of the finest is at the Kitt Peak National Observatory in southern Arizona (**FIGURE 23.19**). It consists of a 150-meter (500-foot) sloped enclosure that directs sunlight to a mirror situated below ground. From the mirror, an 85-centimeter (33-inch) image of the Sun is projected to an observing room, where it is studied.

Compared to other stars of the universe, many of which are larger, smaller, hotter, cooler, more red, or more blue, the Sun is an "average star." However, on the scale of our solar system, it is truly gigantic, having a diameter equal to 109 Earth diameters (1.35 million kilometers) and a volume 1.25 million times as great as that of Earth. Yet, because of its

FIGURE 23.18 The Sun Is the Source of 99.9 Percent of All Energy on Earth (Photo by Jerry and Marcy Monkman/Danita Delimont/Alamy)

gaseous nature, the density is only one-quarter that of Earth, a little greater than the density of water.

For convenience of discussion, we divide the Sun into four parts: the *solar interior*; the *visible surface*, or *photosphere*; and the two layers of its atmosphere, the *chromosphere* and the *corona* (**FIGURE 23.20**). Because the Sun is gaseous throughout, no sharp boundaries exist between these layers. The Sun's interior makes up all but a tiny fraction of the solar mass, and unlike the outer three layers, it is not accessible to direct observation. We discuss the visible layers first.

Photosphere

The **photosphere** (*photos* = light, *sphere* = ball) is aptly named because it radiates most of the sunlight we see and therefore appears as the bright disk of the Sun. Although it is

FIGURE 23.19 The Robert J. McMath Solar Telescope Located at Kitt Peak, near Tucson, Arizona, this telescope has movable mirrors at the top to follow the Sun and reflect the Sun's light down the sloping tunnel. (Photo by Kent Wood/ Science Source) Inset photo shows a view of the solar disk obtained by a solar telescope. (Photo by SOHO/ LASCO [ESA & NASA])

2002/06/09 13:19

SmartFigure 23.20 Diagram of the Sun's Structure Earth is shown for scale.

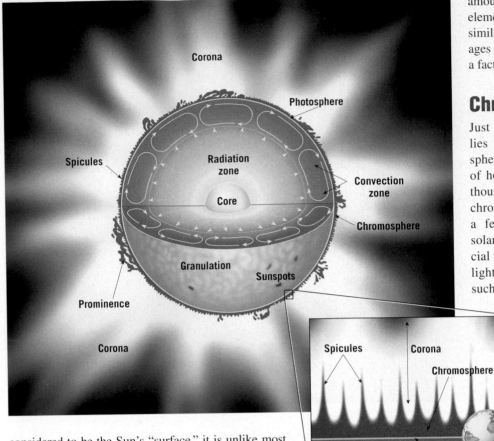

amounts of the other detectable elements. Other stars also show similar disproportionate percentages of these two lightest elements, a fact we consider later.

Chromosphere

Just above the photosphere lies the **chromosphere** (color sphere), a relatively thin layer of hot, incandescent gases a few thousand kilometers thick. The chromosphere is observable for a few moments during a total solar eclipse or by using a special instrument that blocks out the light from the photosphere. Under such conditions, it appears as a thin red rim around the Sun. Because the chromosphere consists of hot, incandescent gases under low pressure, it produces a bright-line spectrum that is nearly the reverse of the dark-line spectrum of the photosphere. One of the bright lines of hydrogen contributes a good portion of its total output and accounts for this sphere's red color.

In 1868, a study of the chromospheric spectrum revealed the existence of an element unknown on Earth. It was named helium, from *helios*, the Greek word for Sun. Originally, helium was thought to be an element unique to the stars, but 27 years later, it was discovered in a natural-gas well on Earth.

The top of the chromosphere contains numerous **spicules** (*spica* = point), flamelike structures that extend upward about 10,000 kilometers (6000 miles) into the lower corona, almost like trees that reach into our atmosphere (**FIGURE 23.22**). Spicules are produced by the turbulent motion of the granules below.

Corona

The outermost portion of the solar atmosphere, the **corona** (*corona* = crown) is very tenuous and, like the chromosphere, is visible only when the brilliant photosphere is blocked (**FIGURE 23.23**). This envelope of ionized gases normally extends 1 million kilometers or so from the Sun and produces a glow about half as bright as the full Moon.

At the outer fringe of the corona, the ionized gases have speeds great enough to escape the gravitational pull of the

FIGURE 23.21 Granules of the Solar Photosphere Granules appear as yellowish-orange patches. Each granule is about the size of Texas and lasts for only 10–20 minutes before being replaced by a new granule. (Courtesy of National Optical Astronomy Observatory/Association of Universities for Research in Astronomy/National Science Foundation)

considered to be the Sun's "surface," it is unlike most surfaces to which we are accustomed. The photosphere consists of a layer of incandescent gas less than 500 kilometers (300 miles) thick, having a pressure less than 1/100 of our atmosphere. Furthermore, it is neither smooth nor uniformly bright, as the ancients had imagined. It has numerous blemishes.

When viewed telescopically, the photosphere's grainy texture is apparent. This is the result of numerous comparatively small, bright markings called **granules** (*granum* = small grain) that are surrounded by narrow, dark regions (**FIGURE 23.21**). Granules are typically the size of Texas and owe their brightness to hotter gases that are rising from below. As this gas spreads laterally, cooling causes it to darken and sink back into the interior. Each granule survives for only 10–20 minutes, while the combined motion of old granules being replaced by new ones gives the photosphere the appearance of boiling. This up-and-down movement of gas, called *convection*, produces the grainy appearance of the photosphere and is responsible for the transfer of energy in the uppermost part of the Sun's interior (see Figure 23.20).

The composition of the photosphere has been determined from the dark lines of its absorption spectrum (see Figure 23.3). When these fingerprints are compared to the spectra of known elements, they indicate that most of the elements found on Earth also exist on the Sun. When the strengths of the absorption lines are analyzed, the relative abundance of the elements can be determined. These studies show that 90 percent of the Sun's surface atoms are hydrogen and almost 10 percent are helium. That leaves only minor

FIGURE 23.22 Spicules of the Chromosphere This view shows the outer edge of the solar disk. (National Solar Observatory/ Sacramento Peak)

Sun. The streams of charged particles (protons and electrons) that boil from the corona constitute the **solar wind**. The solar wind travels outward through the solar system at high speeds, about 250–800 kilometers per second. The solar wind and the Sun's magnetic field form a sort of bubble that reaches far beyond the orbit of Pluto. This immense region, called the **heliosphere**, extends to the outermost boundary of the Sun's influence, where interstellar space begins.

During its journey, the solar wind interacts with the bodies of the solar system, continually bombarding lunar rocks and altering their appearance. Although Earth's magnetic field prevents the solar winds from reaching the surface, these streams of charged particles interact with gases in our atmosphere—a topic we will discuss later.

Studies of the energy emitted from the Sun indicate that its temperature averages about 6000K (10,000°F) in the photosphere. Upward from the photosphere, the temperature unexpectedly increases, exceeding 1 million K at the top of the corona. It should be noted that although the coronal temperature exceeds that of the photosphere by many times, it radiates much less energy overall because of its very low density.

NASA

FIGURE 23.23 Solar Corona This photograph was obtained during a total eclipse.

Surprisingly, the high temperature of the corona is probably caused by sound waves generated by the convective motion of the photosphere. Just as boiling water makes noise, energetic sound waves generated in the photosphere are absorbed by the gases that compose the corona, thereby increasing its temperature.

23.5 CONCEPT CHECKS

1 Why is the Sun significant to the study of astronomy?
2 Describe the Sun in relationship to other stars in the universe.
3 Describe the photosphere, chromosphere, and corona.
4 Why are there no distinct boundaries between the layers of the Sun?
5 Why is the photosphere considered the Sun's "surface"?
6 Briefly describe the *solar wind*.

23.6 | THE ACTIVE SUN

List and describe the three types of explosive activity that occurs at the Sun's surface.

Most of the Sun's energy that reaches Earth is a result of a rather steady, continuous emission from the photosphere. In addition to this predictable aspect of our Sun's energy output, there is a much more irregular component, characterized by explosive surface activity that includes *sunspots*, *prominences*, and *solar flares*.

Sunspots

The most conspicuous features on the surface of the Sun are the dark blemishes called **sunspots** (**FIGURE 23.24A**). Although large sunspots were occasionally observed before the advent of the telescope, they were generally regarded as

FIGURE 23.24 Sunspots
A. Large sunspot group on the solar disk. (Solar & Heliospheric Observatory consortium (ESA & NASA)/Science Source) **B.** Sunspots having visible umbra (dark central area) and penumbra (lighter area surrounding umbra). (Courtesy of National Optical Astronomy Observatory/Association of Universities for Research in Astronomy/National Science Foundation [NOAO/AURA/NSF])

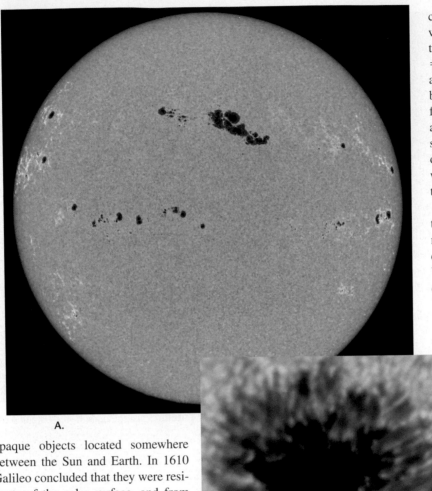

A.

B.

center, the *umbra* (*umbra* = shadow), which is rimmed by a lighter region, the *penumbra* (*paene* = almost, *umbra* = shadow) (**FIGURE 23.24B**). Sunspots appear dark only by contrast with the brilliant photosphere, a fact accounted for by their temperature, which is about 1500K less than that of the solar surface. If these dark spots could be observed away from the Sun, they would appear many times brighter than the full moon.

During the early nineteenth century, it was believed that a tiny planet named Vulcan orbited between Mercury and the Sun. In the search for Vulcan, an accurate record of sunspot occurrences was kept. Although the planet was never found, the sunspot data showed that the number of sunspots on the solar disk varies in an 11-year cycle (**FIGURE 23.25**).

First, the number of sunspots increases to a maximum, with perhaps 100 or more visible at a given time. Then, over a period of five to seven years, their numbers decline to a minimum, when only a few or even none are visible. At the beginning of each cycle, the first sunspots form about 30° from the solar equator, but as the cycle progresses and their numbers increase, they form nearer the equator. During the period when sunspots are most abundant, the majority form about 15° from the equator. They rarely occur more than 40° away from the Sun's equator or within 5° of it.

Another interesting characteristic of sunspots was discovered by astronomer George Hale, for whom the Hale Telescope is named. Hale deduced that the large spots are strongly magnetized, and when they occur in pairs, they have opposite magnetic poles. For instance, if one member of the pair is a north magnetic pole, then the other member is a south magnetic pole. Also, every pair located in the same hemisphere is magnetized in the same manner. However, all pairs in the *other* hemisphere are magnetized in the opposite manner. At the beginning of each sunspot cycle, the situation reverses, and the polarity of these sunspot pairs is opposite those of the previous cycle. The cause of this change in polarity—in fact, the cause of sunspots themselves—is not fully understood. However, other modes of solar activity vary in the same cyclic manner as sunspots, indicating the likelihood of a common origin.

opaque objects located somewhere between the Sun and Earth. In 1610 Galileo concluded that they were residents of the solar surface, and from their motion, he deduced that the Sun rotates on its axis about once a month.

Later observations indicated that the time required for one rotation varied by latitude. The Sun's equator rotates once in 25 days, whereas a place located 70° from the solar equator, either north or south, requires 33 days for one rotation. If Earth rotated in a similar disjointed manner, imagine the consequences! The Sun's nonuniform rotation is a testament to its gaseous nature.

Sunspots begin as small, dark pores about 1600 kilometers (1000 miles) in diameter. Although most sunspots last for only a few hours, some grow into blemishes many times larger than Earth and last for a month or more. The largest sunspots often occur in pairs surrounded by several smaller sunspots. An individual spot contains a black

FIGURE 23.25 Mean Annual Sunspot Numbers The number of sunspots reaches a maximum on roughly an 11-year cycle.

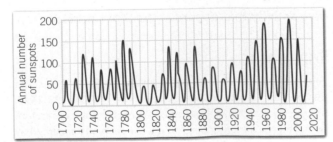

Prominences

Among the most spectacular features of the active Sun are **prominences** (*prominere* = to jut out). These huge cloud-like structures, consisting of concentrations of chromospheric gases, are best observed when they are on the edge, or limb, of the Sun, where they often appear as bright arches that extend well into the corona (**FIGURE 23.26**). *Quiescent prominences* have the appearance of a fine tapestry and seem to hang motionless for days at a time, but motion pictures reveal that the material within them is continually falling like luminescent rain.

By contrast, *eruptive prominences* rise almost explosively away from the Sun. These active prominences reach velocities up to 1000 kilometers (620 miles) per second and may leave the Sun entirely. Whether eruptive or quiescent, prominences are ionized chromospheric gases trapped by magnetic fields that extend from regions of intense solar activity.

Solar Flares

Solar flares are brief outbursts that normally last an hour or so and appear as a sudden brightening of the region above a sunspot cluster. During their existence, enormous quantities of energy are released across the entire electromagnetic spectrum, much of it in the form of ultraviolet, radio, and x-ray radiation. Simultaneously, fast-moving atomic particles are ejected, causing the *solar wind* to intensify noticeably. Although a major flare could conceivably endanger a staffed space flight, these are relatively rare events. Within hours after a large outburst, the ejected particles reach Earth

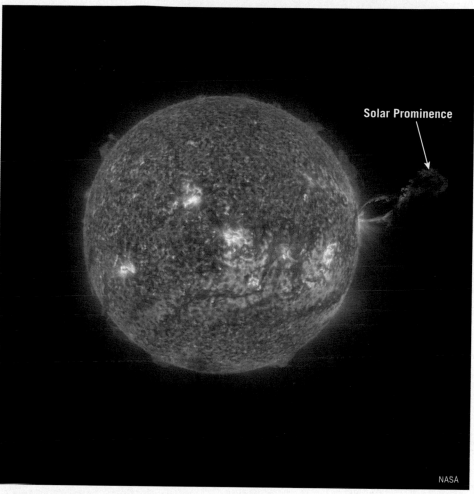

Solar Prominence

NASA

FIGURE 23.26 Solar Prominence

EYE ON THE UNIVERSE

The accompanying image shows one of two telescopes located at Keck Observatory atop Hawaii's Mauna Kea volcano, at an elevation of 4200 meters (13,800 feet). The 10-meter (33-foot) mirrors of these telescopes consist of 36 distinct pieces. These telescopes can work independently or in tandem to double their light-gathering capacity. (Photo by Enrico Sacchetti/Science Source)

QUESTION 1 *Examine the primary mirror located near the center of the image. Based on what you see, describe the shape of the 36 segments that make up this mirror.*

QUESTION 2 *Assume that these mirrors have a perfect circular shape and a diameter of 10 meters (33-foot) and that they are working in tandem (gathering light from exactly the same source). How much more light-gathering capacity do they have compared to the Hale Telescope, which has a mirror with a 5-meter (16-foot) diameter? (Hint: Compare the surface areas of these instruments.)*

QUESTION 3 *The Keck Observatory telescopes are located far out in the Pacific, on the Big Island of Hawaii, and the Hale Telescope is located 80 kilometers (50 miles) northeast of San Diego. Does the location of the Keck telescopes represent an advantage or a disadvantage as compared to the Hale Telescope? Explain.*

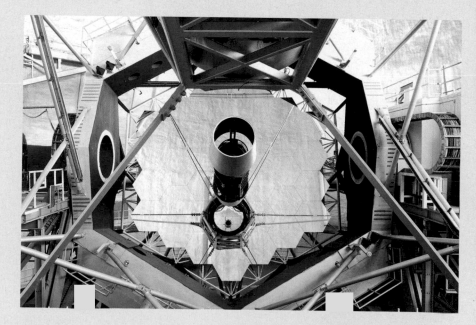

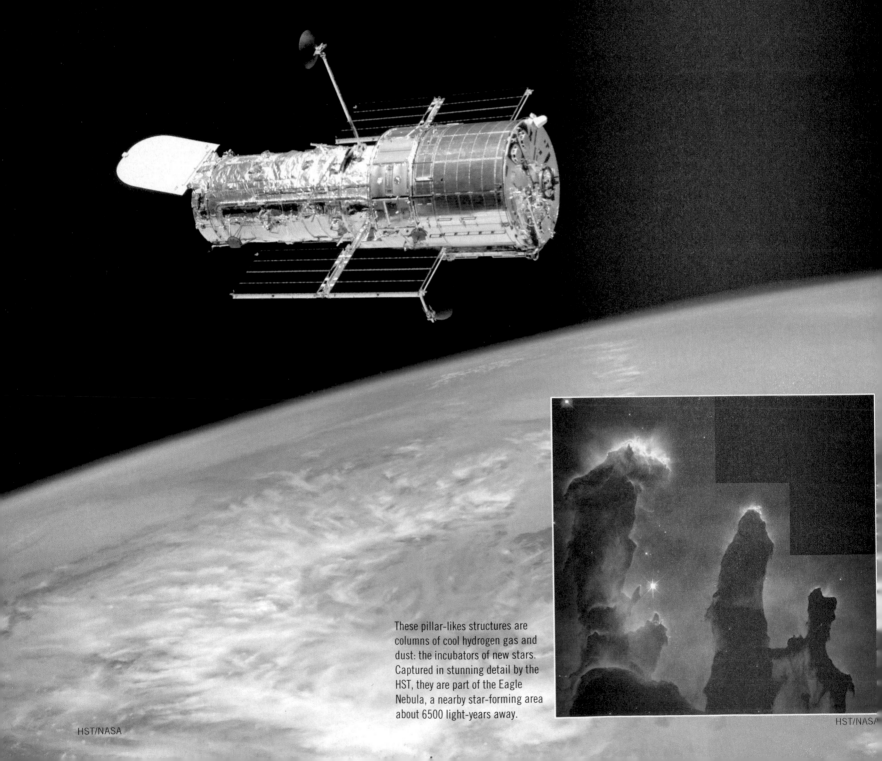

Hubble Space Telescope

In 1990, the Space Shuttle *Atlantis* carried the Hubble Space Telescope (HST) into an orbit 557 kilometers (347 miles) above Earth's surface. At this altitude, the HST avoids many of the problems experienced by Earth-bound telescopes, which are affected by the atmosphere that distorts and blocks much of the light reaching our planet. The thousands of images beamed back to Earth have helped scientists solve many great mysteries. For example, data from the HST allowed scientists to determine that the age of the universe is between 13 and 14 billion years, significantly narrowing the previous range of 10 to 20 billion years.

These pillar-likes structures are columns of cool hydrogen gas and dust: the incubators of new stars. Captured in stunning detail by the HST, they are part of the Eagle Nebula, a nearby star-forming area about 6500 light-years away.

HST/NASA

HST/NASA

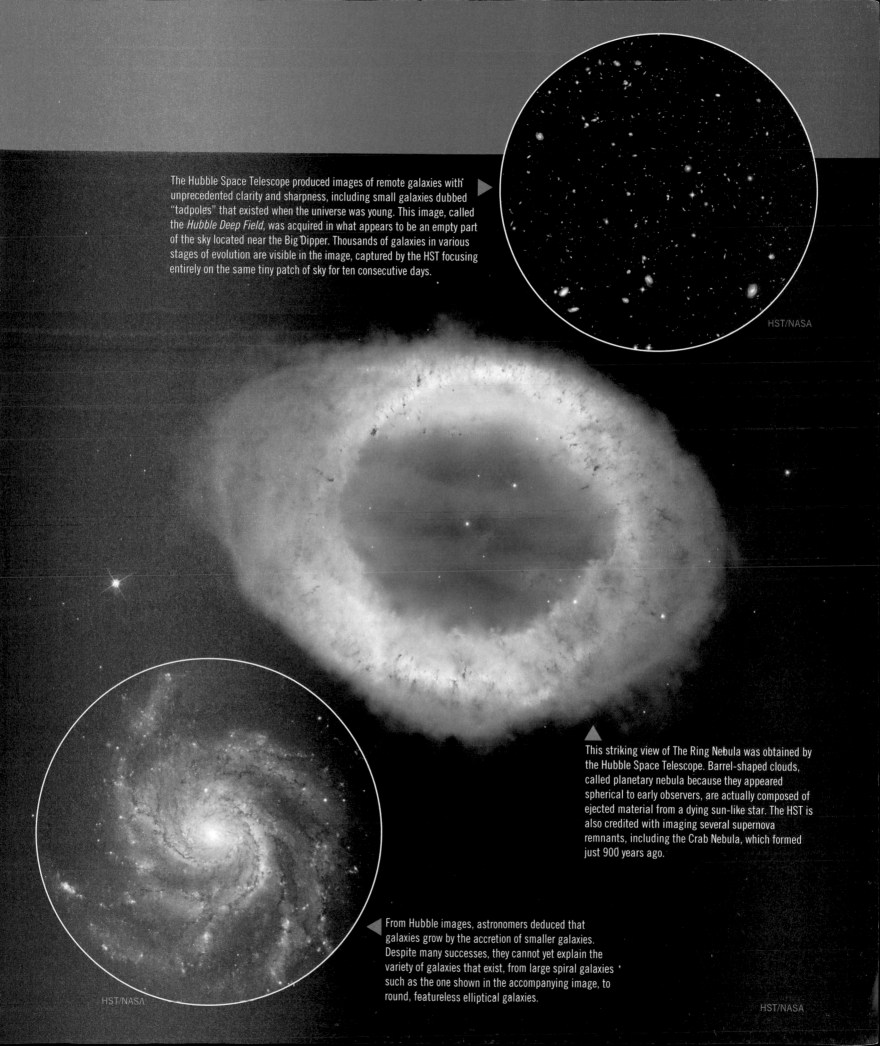

The Hubble Space Telescope produced images of remote galaxies with unprecedented clarity and sharpness, including small galaxies dubbed "tadpoles" that existed when the universe was young. This image, called the *Hubble Deep Field*, was acquired in what appears to be an empty part of the sky located near the Big Dipper. Thousands of galaxies in various stages of evolution are visible in the image, captured by the HST focusing entirely on the same tiny patch of sky for ten consecutive days.

HST/NASA

This striking view of The Ring Nebula was obtained by the Hubble Space Telescope. Barrel-shaped clouds, called planetary nebula because they appeared spherical to early observers, are actually composed of ejected material from a dying sun-like star. The HST is also credited with imaging several supernova remnants, including the Crab Nebula, which formed just 900 years ago.

From Hubble images, astronomers deduced that galaxies grow by the accretion of smaller galaxies. Despite many successes, they cannot yet explain the variety of galaxies that exist, from large spiral galaxies such as the one shown in the accompanying image, to round, featureless elliptical galaxies.

HST/NASA

HST/NASA

atmosphere above the magnetic poles is set aglow for several nights. The auroras appear in a wide variety of forms. Sometimes the display consists of vertical streamers with considerable movement. At other times, the auroras appear as a series of luminous, expanding arcs or as a quiet, almost foglike, glow. Auroral displays, like other solar activities, vary in intensity with the 11-year sunspot cycle.

and disturb the ionosphere,[3] affecting long-distance radio communications.

The most spectacular effects of solar flares are the **auroras**, also called the Northern and Southern Lights (**FIGURE 23.27**). Following a strong solar flare, Earth's upper

[3]The ionosphere is a complex atmospheric zone of ionized gases extending between about 80 and 400 kilometers (50 and 250 miles) above Earth's surface.

23.6 CONCEPT CHECKS

1 What did Galileo learn about the Sun from his observations of sunspots?

2 Briefly describe the 11-year sunspot cycle.

3 What are prominences?

4 How do solar flares affect the solar wind?

23.7 | THE SOURCE OF SOLAR ENERGY

Summarize the process called the proton–proton chain reaction.

The interior of the Sun cannot be observed directly. For that reason, what we know about it is based on information acquired from the energy radiated from the photosphere and solar atmosphere and from theoretical studies.

The source of the Sun's energy, **nuclear fusion** (*fusus* = to melt), was not discovered until the late 1930s. Deep in its interior, a nuclear reaction called the **proton–proton chain reaction** converts four hydrogen nuclei (protons) into the nucleus of a helium atom. The energy released from the proton–proton reaction results because some of the matter involved is actually converted to radiant energy. This can be illustrated by noting that four hydrogen atoms have a combined atomic mass of 4.032 (4×1.008), whereas the atomic mass of helium is 4.003, which is 0.029 less than the combined mass of the hydrogen. The tiny missing mass is emitted as energy according to Einstein's formula $E = mc^2$, where E equals energy, m equals mass, and c equals the speed of light. Because the speed of light is very great, the amount of energy released from even a small amount of mass is enormous.

The conversion of just one pinhead's worth of hydrogen to helium generates more energy than burning thousands of tons of coal. (This process is often referred to as "hydrogen burning," but it is nothing like the type of burning to which we are accustomed.) Most of this energy is in the form of very high-energy photons that work their way toward the solar surface,

being absorbed and re-emitted many times until they reach an opaque layer just below the photosphere. Here, convection currents transport this energy to the solar surface, where it radiates through the nearly transparent chromosphere and corona as mostly visible light (see Figure 23.20).

Only a small percentage (0.7%) of the hydrogen in the proton–proton reaction is actually converted to energy. Nevertheless, the Sun consumes an estimated 600 million tons of hydrogen each second, with about 4 million tons of it being converted to energy. The by-product of hydrogen burning is helium, which forms the solar core. Consequently, the core continually grows in size.

How long can the Sun produce energy at its present rate before all of its fuel (hydrogen) is consumed? Even at the enormous rate of consumption, the Sun has enough fuel to easily last another 100 billion years. However, evidence from other stars indicates that the Sun will grow dramatically and engulf Earth long before all of its hydrogen is gone. It is likely that a star the size of the Sun can remain in a stable state for about 10 billion years. Since the Sun is already 5 billion years old, it is "middle-aged."

To initiate the proton–proton reaction, the Sun's internal temperature must have reached several million degrees. What was the source of this heat? As previously noted, the solar system formed from an enormous cloud of dust and gases (mostly hydrogen) that gravitationally collapsed. When a gas is squeezed (compressed), its temperature increases. Although all the bodies in the solar system were heated in this manner, the Sun was the only one that, because of its mass, became hot enough to trigger the proton–proton reaction. Astronomers currently estimate its internal temperature at 15 million K.

The planet Jupiter is basically a hydrogen-rich gas ball. Why didn't it become a star? Although Jupiter is a huge planet, the stars with the lowest mass are between 75 and 80 times the size of Jupiter.

23.7 CONCEPT CHECKS

1 What "fuel" does the Sun consume?

2 What happens to the matter that is consumed in the proton–proton chain reaction?

23 CONCEPTS IN REVIEW | Light, Astronomical Observations, and the Sun

23.1 SIGNALS FROM SPACE

List and describe the various types of electromagnetic radiation.

KEY TERMS: electromagnetic radiation, photons, radiation pressure

- Electromagnetic radiation consists of an array of energy that includes gamma rays, x-rays, ultraviolet light, visible light, infrared radiation (heat), microwaves, and radio waves.
- Light, a type of electromagnetic radiation, can be described in two ways: (1) as waves and (2) as a stream of particles, called photons.
- The wavelengths of electromagnetic radiation vary from several kilometers for radio waves to less than one-billionth of a centimeter for gamma rays. Shorter wavelengths correspond to more energetic photons, and the longer wavelengths are less energetic.

Q List the following types of electromagnetic radiation in order from long-wavelength radiation to short-wavelength radiation: visible light, gamma rays, ultraviolet radiation, x-rays, infrared radiation, and radio waves.

23.2 SPECTROSCOPY

Explain how the three types of spectra are generated and what they tell astronomers about the radiating body that produced them.

KEY TERMS: spectroscopy, continuous spectrum, spectroscope, dark-line (absorption) spectrum, bright-line (emission) spectrum, Doppler effect

- Spectroscopy is the study of the properties of light that depend on wavelength. When a prism is used to disperse visible light into its component parts (wavelengths), one of three possible types of spectra is produced. (A *spectrum*, the singular form of *spectra*, is the light pattern produced by passing light through a prism.)
- The three types of spectra are (1) continuous spectrum, (2) dark-line (absorption) spectrum, and (3) bright-line (emission) spectrum. The spectra of most stars are of the dark-line type, which indicate the composition of the radiating body as well as the relative amount of each type of matter. Continuous spectra provide information about the total output of energy and the surface temperature of the radiating body. Finally, bright-line spectra are produced by gaseous bodies (nebula) at low pressure and contain information about the temperature of the gas and its composition.
- Spectroscopy can be used to determine the motion of an object, a phenomena called the Doppler effect.

Q Use the accompanying diagram, which shows two dark-line spectra, to determine whether the moving body is going toward the observer or receding from the observer. Explain.

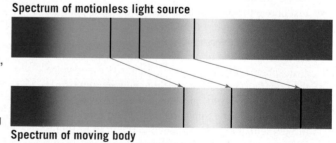

Spectrum of motionless light source

Spectrum of moving body

23.3 COLLECTING LIGHT USING OPTICAL TELESCOPES

Compare and contrast refracting and reflecting telescopes. Explain why modern telescopes are built on mountaintops.

KEY TERMS: refracting telescope, chromatic aberration, reflecting telescope

- There are two types of optical telescopes: (1) the refracting telescope, which uses a lens to bend or refract light, and (2) the reflecting telescope, which uses a concave mirror to focus (gather) the light. Most large modern telescopes use mirrors to collect light.
- Telescopes simply collect light. When correctly analyzed, the collected light can be used to determine the temperature, composition, relative motion, and distance to a celestial object.
- Historically, astronomers relied on their eyes to collect light. Then photographic film was developed, which was a revolutionary advancement. Presently, light is collected using a charge-coupled device (CCD). A CCD camera produces a digital image akin to that of a digital camera. Despite these advances, Earth-bound optical telescopes can only "view" a tiny portion of the electromagnetic spectrum and are hindered by "poor seeing" as a result of atmospheric turbulence.

23.4 RADIO- AND SPACE-BASED ASTRONOMY

Explain the advantages of radio telescopes and orbiting observatories over optical telescopes.

KEY TERMS: radio telescope, radio interferometer

- The detection of *radio waves* is accomplished by "big dishes" known as *radio telescopes*. A parabolic-shaped dish, often made of wire mesh, operates in a manner similar to the mirror of a reflecting telescope.
- Of great importance is a narrow band of radio waves that is able to penetrate Earth's atmosphere. Because this radiation is produced by neutral hydrogen, it has permitted us to map the galactic distribution of this gaseous material from which stars are made.
- Orbiting observatories, like the *Hubble Space Telescope*, circumvent all the problems caused by the Earth's atmosphere and have led to many significant discoveries in astronomy.

23.5 THE SUN

Write a statement explaining why the Sun is important to the study of astronomy. Sketch the Sun's structure and describe each of its four major layers.

KEY TERMS: photosphere, granule, chromosphere, spicule, corona, solar wind, heliosphere

- The *Sun* is an average star, one of 200 billion stars that make up the Milky Way Galaxy. The Sun can be divided into four parts: (1) the solar interior, (2) the photosphere (visible surface) and the two layers of its atmosphere, (3) the chromosphere, and (4) the corona.
- The photosphere radiates most of the light we see. Unlike most surfaces to which we are accustomed, it consists of a layer of incandescent gas less than 500 kilometers (300 miles) thick and has a grainy texture consisting of numerous relatively small, bright markings called granules.
- Just above the photosphere lies the chromosphere, a relatively thin layer of hot incandescent gases a few thousand kilometers thick.
- At the edge of the uppermost portion of the solar atmosphere, called the corona, ionized gases escape the gravitational pull of the Sun and stream toward Earth at high speeds, producing the solar wind. The solar wind and the Sun's magnetic field form a sort of bubble that extends far beyond the orbit of Pluto. This immense region, called the heliosphere, extends to the outermost boundary of the Sun's influence, where interstellar space begins.

23.6 THE ACTIVE SUN

List and describe the three types of explosive activity that occur at the Sun's surface.

KEY TERMS: sunspot, prominence, solar flare, aurora

- Numerous features have been identified on the active Sun. Sunspots are dark blemishes with a black center, the umbra, which is rimmed by a lighter region, the penumbra. The number of sunspots observable on the solar disk varies in an 11-year cycle.
- Prominences, huge cloudlike structures best observed when they are on the edge, or limb, of the Sun, are produced by ionized chromospheric gases trapped by magnetic fields that extend from regions of intense solar activity.
- The most explosive events associated with sunspots are solar flares. Flares are brief outbursts that release enormous quantities of energy that appear as a sudden brightening of the region above sunspot clusters. During the event, radiation and fast-moving atomic particles are ejected, causing the solar wind to intensify. When the ejected particles reach Earth and disturb the ionosphere, radio communication is disrupted, and the auroras, also called the Northern and Southern Lights, occur.

Q The accompanying image is a close-up view of the Sun's surface. What name is given to the Sun's surface? Name the labeled features.

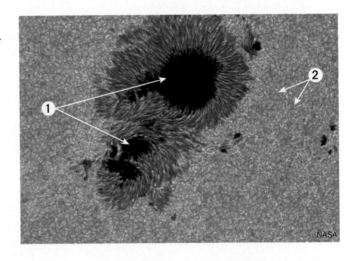

NASA

23.7 THE SOURCE OF SOLAR ENERGY

Summarize the process called the proton–proton chain reaction.

KEY TERMS: nuclear fusion, proton–proton chain reaction

- The source of the Sun's energy is nuclear fusion. Deep in the solar interior, at a temperature of 15 million K, a type of nuclear fusion called the proton–proton chain reaction converts four hydrogen nuclei (protons) into the nucleus of a helium atom. During the reaction, some of the matter is converted to the energy that the Sun radiates to space.
- The Sun can continue to operate its nuclear furnace and exist in its present stable state for 10 billion years. As the Sun is already 5 billion years old, it is a "middle-aged" star.

GIVE IT SOME **THOUGHT**

1. Refer to Figure 23.2 to answer the following questions:
   a. Is the atmosphere mostly transparent or opaque to visible light?
   b. Is the atmosphere mostly transparent or opaque to radio waves with a wavelength of 1 meter (3 feet)?
   c. Is the atmosphere mostly transparent or opaque to gamma rays?

2. Imagine that the composition of Earth's atmosphere were altered so that its ability to absorb visible and far infrared light were reversed.
   a. If you were outdoors when the Sun was at its highest point in the sky, how would the sky appear?
   b. Would there be an increase or a decrease in Earth's average surface temperature?

3. Suppose a well-known scientist claimed that stars consist primarily of helium rather than hydrogen.
   a. What type of object in the galaxy could you study to investigate whether stars consist primarily of helium or hydrogen?
   b. How could spectroscopy help you verify or disprove the scientist's claim? Explain your reasoning.

4. Imagine that you are responsible for funding the construction of observatories. After considering the four proposals listed below, state whether you would or would not recommend funding for each proposal and explain your reasoning.

   Proposal A: A ground-based x-ray telescope on top of a mountain in Arizona, designed to observe supernovae in distant galaxies.

   Proposal B: A space-based 3-meter reflecting infrared telescope designed to observe very distant galaxies.

   Proposal C: A ground-based 8-meter refracting optical telescope located on the top of Mauna Kea in Hawaii, designed to measure the spectra of binary stars in our galaxy.

   Proposal D: A ground-based 250-meter radio telescope array in New Mexico, designed to measure the distribution of hydrogen gas clouds in the disk of our galaxy.

5. An important absorption line in the spectrum of stars occurs at a wavelength of 656 nm for stars not moving toward or away from Earth.

Imagine that you observe four stars in our galaxy and discover that this absorption line is at the wavelength shown in the accompanying diagram. Using this data, complete the following questions. Explain the reasoning behind your answers. If you are unable to determine the answer to any of these questions from the given information, explain.

STAR	Wavelength of Absorption Line
A	649 nm
B	656 nm
C	658 nm
D	647 nm

   a. Which of these stars is moving toward Earth the fastest?
   b. Which of these stars is closest to Earth?
   c. Which of these stars is moving away from Earth?

6. Consider the following discussion among three of your classmates regarding why telescopes are put in space. Support or refute each statement.

   Student 1: "I think it is because the atmosphere distorts and magnifies light, which causes objects to look larger than they actually are."

   Student 2: "I thought it was because some of the wavelengths of light being sent out from the telescopes can be blocked by Earth's atmosphere, so the telescopes need to be above the atmosphere."

   Student 3: "Wait, I thought it was because by moving the telescope above the atmosphere, the telescope is closer to the objects, which makes them appear brighter."

7. Refer to the accompanying spectra, which represent four identical stars in our galaxy. One star is not moving, another is moving away from you, and two stars are moving toward you. Determine which star is which and explain how you reached your conclusion.

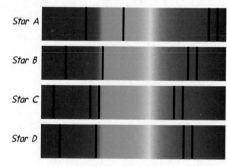

Star A
Star B
Star C
Star D

EXAMINING THE **EARTH SYSTEM**

1. Of the two sources of energy that power the Earth system, the Sun is the main driver of Earth's external processes. If the Sun increased its energy output by 10 percent, what would happen to global temperatures? What effect would this temperature change have on the percentage of water that exists as ice? What would be the impact on the position of the ocean shoreline? Speculate about whether the change in temperature might produce an increase or a decrease in the amount of surface vegetation. In turn, what impact might this change in vegetation have on the level of atmospheric carbon dioxide? How would such a change in the amount of carbon dioxide in the atmosphere affect global temperatures?

MasteringGeology™

Looking for additional review and test prep materials? Visit the Self Study area in **www.masteringgeology.com** to find practice quizzes, study tools, and multimedia that will aid in your understanding of this chapter's content. In **MasteringGeology™** you will find:

- GEODe: Earth Science: An interactive visual walkthrough of key concepts
- Geoscience Animation Library: More than 100 animations illuminating many difficult-to-understand Earth science concepts

- In The News RSS Feeds: Current Earth science events and news articles are pulled into the site with assessment
- Pearson eText
- Optional Self Study Quizzes
- Web Links
- Glossary
- Flashcards

24

Beyond Our Solar System[1]

FOCUS ON CONCEPTS

Each statement represents the primary **LEARNING OBJECTIVE** for the corresponding major heading within the chapter. After you complete the chapter, you should be able to:

24.1 Define *cosmology* and describe Edwin Hubble's most significant discovery about the universe.

24.2 Explain why interstellar matter is often referred to as a stellar nursery. Compare and contrast bright and dark nebulae.

24.3 Define *main-sequence star*. Explain the criteria used to classify stars as giants.

24.4 List and describe the stages in the evolution of a typical Sun-like star.

24.5 Compare and contrast the final state of Sun-like stars to the remnants of the most massive stars.

24.6 List the three major types of galaxies. Explain the formation of large elliptical galaxies.

24.7 Describe the Big Bang theory. Explain what it tells us about the universe.

The Whirlpool Galaxy (M51) has excited astronomers for nearly three centuries. Its majestic spiral arms are star-forming factories. (NASA)

[1]Revised with the assistance of Professors Teresa Tarbuck and Mark Watry.

Astronomers and cosmologists study the nature of our vast universe, attempting to answer questions such as these: Is our Sun a typical star? Do other stars have solar systems with planets similar to Earth? Are galaxies distributed randomly, or are they organized into groups? How do stars form? What happens when stars expend their fuel? If the early universe consisted mostly of hydrogen and helium, how did other elements come into existence? How large is the universe? Did it have a beginning? Will it have an end? This chapter explores these as well as other questions.

24.1 | THE UNIVERSE   Define *cosmology* and describe Edwin Hubble's most significant discovery about the universe.

The universe is more than a collection of dust clouds, stars, stellar remnants, and galaxies (**FIGURE 24.1**). It is an entity with its own properties.

Cosmology is the study of the universe, including its properties, structure, and evolution. Over the years, cosmologists have developed a comprehensive theory that describes the structure and evolution of the universe. Some of the questions cosmologists seek to answer with this theory include these: How did the universe evolve to its present state? How long has it existed, and how will it end? Modern cosmology addresses these important issues and helps us understand the universe we inhabit.

How Large Is It?

For most of human existence, our universe was thought to be Earth centered, containing only the Sun, Moon, 5 wandering stars, and about 6000 fixed stars. Even after the Copernican view of a Sun-centered universe became widely accepted, the entire universe was believed to consist of a single galaxy, the Milky Way, composed of innumerable stars, along with many faint "fuzzy patches," thought to be clouds of dust and gases.

In the mid-1700s, German philosopher Immanuel Kant proposed that many of the telescopically visible fuzzy patches of light scattered among the stars were actually distant galaxies similar to the Milky Way. Kant described them as "island universes." Each galaxy, he believed, contained billions of stars and was a universe in itself. In Kant's time, however, the weight of opinion favored the hypothesis that the faint patches of light occurred within our galaxy. Admitting otherwise would have implied a vastly larger universe, thereby diminishing the status of Earth and, likewise, humankind.

In 1919 Edwin Hubble arrived at the observatory at Mount Wilson, California, to conduct research using a 2.5-meter

FIGURE 24.1 The Trifid Nebula, in the Constellation Sagittarius This colorful nebula is a cloud consisting mostly of hydrogen and helium gases. These gases are excited by light emitted by hot, young stars within and produce a reddish glow. (National Optical Astronomy Observatory/Association of Universities for Research in Astronomy/National Science Foundation)

(100-inch) telescope, then the world's largest and most advanced astronomical instrument. Armed with this modern tool, Hubble set out to solve the mystery of the "fuzzy patches." At that time, the debate was still raging as to whether the fuzzy patches were "island universes," as Kant had proposed more than 150 years earlier, or clouds of dust and gases (nebulae). To accomplish this task, Hubble studied a group of pulsating stars known as *Cepheid variables*—extremely bright variable stars that increase and decrease in brightness in a repetitive cycle. This group of stars is significant because their "true" brightness, called **absolute magnitude**, can be determined by know-

A.

B.

FIGURE 24.2 Andromeda: A Nearby Galaxy That Is Larger Than Our Milky Way A. Photo of Andromeda Galaxy, taken by an optical telescope aboard the *GALEX* orbiter. (NASA) **B.** Andromeda Galaxy as it appears under low magnification. When viewed with the naked eye, Andromeda appears as a fuzzy patch surrounded by stars. (European Southern Observatory/ESO)

ing the rate at which they pulsate (see Appendix C). When the absolute magnitude of a star is compared to its observed brightness, a reliable approximation of distance can be established. (This is similar to how we judge the distance of an oncoming vehicle when driving at night.) Thus, Cepheid variables are important because they can be used to measure large astronomical distances.

Using the telescope at Mount Wilson, Hubble found several Cepheid variables embedded in one of the fuzzy patches. However, because these intrinsically bright stars appeared faint, Hubble concluded that they must lie outside the Milky Way. Indeed, one of the objects Hubble observed lies more than 2 million light-years away and is now known as the *Andromeda Galaxy* (**FIGURE 24.2**).

Based on his observations, Hubble determined that the universe extended far beyond the limits of our imagination. Today, we know that there are hundreds of billions of galaxies, each containing hundreds of billions of stars. For example, researchers estimate that a million galaxies exist in the portion of the sky bounded by the cup of the Big Dipper. There are literally more stars in the heavens than grains of sand in all the beaches on Earth.

A Brief History of the Universe

Large telescopes can actually "look back in time," which accounts for much of the knowledge astronomers have acquired regarding the history of the universe. Light from

celestial objects that are great distances from Earth require millions or even billions of years to reach Earth. The distance light travels in 1 year is called a **light-year** (slightly less than 10 trillion kilometers [6 trillion miles]). Therefore, the farther out telescopes can "see," the farther back in time astronomers are able to study. Even the closest large galaxy, the Andromeda Galaxy, is a staggering 2.5 million light-years away. Light that left Andromeda Galaxy 2.5 million years ago is just now reaching Earth, allowing scientists to observe Andromeda as it was 2.5 million years ago. Light from the farthest-known objects, about 13 billion light-years away, came from stars that have long since burned out.

The time line for the history of the universe, shown in **FIGURE 24.3**, highlights some of the major events in the evolution of matter and energy. The model that most accurately describes the birth and current state of the universe is the **Big Bang theory**. According to this theory, all of the energy and matter of the universe existed in an incomprehensibly hot and dense state. About 13.7 billion years ago, our universe began as a cataclysmic explosion, which continued to expand, cool, and evolve to its current state.

In the earliest moments of this expansion, only energy and quarks (subatomic particles that are the building blocks of protons and neutrons) existed. Not until 380,000 years after the initial expansion did the universe cool sufficiently for electrons and protons to combine to form hydrogen and helium atoms—the lightest elements in the universe. For the

FIGURE 24.3 Time Line for the Evolution of the Universe According to the Big Bang theory, the universe began 13.7 billion years ago and has been expanding ever since.

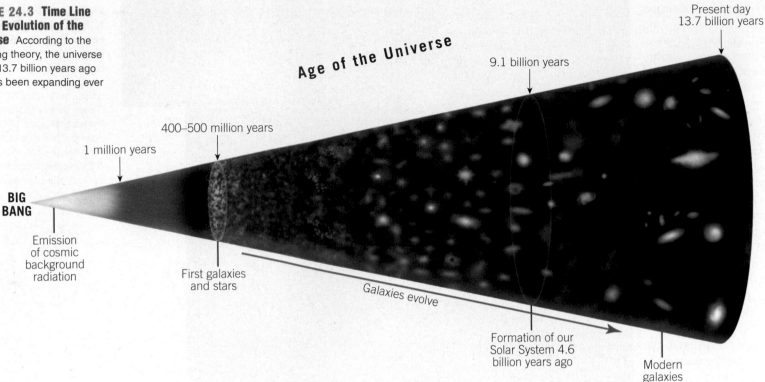

first time, light traveled through space. Eventually, temperatures decreased sufficiently to allow clumps of matter to collect. This material formed massive clouds of dust and gases (nebulae), which quickly evolved into the first stars and galaxies. Our Sun and planetary system, having formed about 5 billion years ago (nearly 9 billion years after the Big Bang), is a latecomer to the universe.

24.1 CONCEPT CHECKS

1 What is cosmology?
2 Explain how Edwin Hubble used Cepheid variables to change our view of the structure of the universe.
3 When do cosmologists think the universe began?
4 What two elements were the first to form?

24.2 | INTERSTELLAR MATTER: NURSERY OF THE STARS

Explain why interstellar matter is often referred to as a stellar nursery. Compare and contrast bright and dark nebulae.

As the universe expanded, gravity caused matter to accumulate into large "clumps" and "strands" of **interstellar matter** known as **nebulae** (meaning "clouds"; the singular of *nebulae* is *nebula*). In addition, considerable amounts of interstellar matter once resided in the interiors of stars and were subsequently returned to space. Some stars ejected matter as part of their normal life cycle, some exploded when they died, and some formed black holes that ejected streams of matter through structures called jets.

Interstellar matter resides between stars within the galaxies and consists of roughly 90 percent hydrogen and 9 percent helium. The remainder, *interstellar dust*, is composed of atoms, molecules, and larger dust grains of the heavier elements. These huge concentrations of interstellar dust and gases are extremely *diffuse*, similar to fog, with no distinct edges or boundaries. However, because nebulae are enormous, their total mass is many times that of the Sun. If a

nebula is dense enough, it will contract due to gravity, leading to processes that form stars and planets.

When nebulae are in close proximity to very hot (blue) stars, they glow and are called **bright nebulae**. By contrast, when clouds of interstellar material are too far from bright stars to be illuminated, they are referred to as **dark nebulae**.

Bright Nebulae

There are three main types of bright nebulae: *emission*, *reflection*, and *planetary nebulae*. Emission and reflection nebulae consist of the material from which new stars are born—stellar nurseries. By contrast, planetary nebulae form when stars die and expel material into space. Thus, all stars form from nebulae, and many ultimately return to the material of nebulae when their life cycle ends.

Emission Nebulae Glowing clouds of hydrogen gas, called **emission nebulae**, are produced in active star-forming regions of galaxies. Energetic ultraviolet light emitted from hot, young stars ionizes the hydrogen atoms in the nebulae. Because these gases exist under extremely low pressure, they radiate, or emit, their energy as less energetic *visible light*. The conversion of ultraviolet light to visible light is known as *fluorescence*—the same phenomenon that causes neon lights to glow. Hydrogen emits much of its energy in the red portion of the spectrum, which accounts for the red glow from emission nebulae (**FIGURE 24.4**).

When elements other than hydrogen are ionized, the glowing nebula may exhibit a broader range of colors. An emission nebula that is easily seen with binoculars is located in the constellation Orion, in the sword of the hunter.

Reflection Nebulae As the name implies, **reflection nebulae** merely reflect the light of nearby stars (**FIGURE 24.5**). Reflection nebulae are likely composed of significant amounts of comparatively large debris, including grains of carbon compounds. This view is supported by the fact that atomic gases with low densities could not reflect light sufficiently to produce the glow observed. Reflection nebulae are usually blue because blue light (shorter wavelength) is scattered more efficiently than red light (longer wavelength)—a process that also produces the blue color of the sky. The blue wisps in the Pleiades, shown in Figure 24.5, are reflection nebulae.

Planetary Nebulae **Planetary nebulae** are generally not as diffuse as other nebulae, and they originate from the remnants of dying Sun-like stars (**FIGURE 24.6**). Planetary

FIGURE 24.5 Reflection Nebulae This blue reflection nebula, located in the Pleiades star cluster, is a result of the scattering of starlight by relatively large molecules and interstellar dust. The Pleiades star cluster, barely visible to the naked eye in the constellation Taurus, is spectacular when viewed through binoculars or a small telescope. (David Malin/AAO/Royal Observatory, Edinburgh/Science Source)

FIGURE 24.4 Emission Nebulae Lagoon Nebula is a large emission nebula, composed mainly of hydrogen. Its red color is attributed to ionized gases, which are excited by the energetic light emitted from young, hot stars embedded in the nebula. (National Optical Astronomy Observatory/Association of Universities for Research in Astronomy/National Science Foundation [NOAO/AURA/NSF])

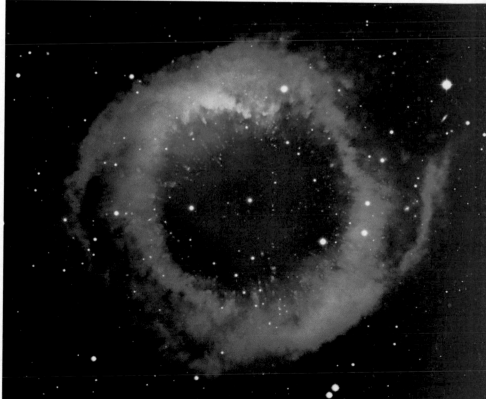

FIGURE 24.6 Planetary Nebulae The Helix Nebula is the nearest planetary nebula to our solar system. A planetary nebula is the ejected outer envelope of a Sun-like star that formed during the star's collapse from a red giant to a white dwarf. (© Anglo-Australian Observatory. Photography by David Malin)

FIGURE 24.7 Dark Nebulae The Horsehead Nebula is a dark nebula in the constellation Orion. (Courtesy of the European Southern Observatory)

name. A good example of a planetary nebula is the Helix Nebula in the constellation Aquarius (see Figure 24.6).

Dark Nebulae

Recall that when clouds of interstellar material are too distant from bright stars to be illuminated, they are referred to as *dark nebulae*. Exemplified by the Horsehead Nebula in Orion, dark nebulae appear as opaque objects silhouetted against bright backgrounds (**FIGURE 24.7**). In addition, dark nebulae can also easily be seen as starless regions— "holes in the heavens"—when viewing the Milky Way (see Figure 24.14). Although dark nebulae often appear dense, they are made of the same matter as bright nebulae and consist of thinly scattered matter.

nebulae consist of glowing clouds of dust and hot gases that have been expelled near the end of a star's life. When first viewed through primitive optical telescopes, planetary nebulae resembled giant planets such as Jupiter—hence their

24.2 CONCEPT CHECKS

1 Why is the phrase "nursery of stars" an appropriate way of describing interstellar matter (nebulae)?

2 Compare and contrast bright and dark nebulae.

3 Why are reflection nebulae generally blue?

4 How are planetary nebulae different from other types of bright nebulae?

24.3 | CLASSIFYING STARS: HERTZSPRUNG–RUSSELL DIAGRAMS (H-R DIAGRAMS) Define *main-sequence star*. Explain the criteria used to classify stars as giants.

Early in the twentieth century, Einar Hertzsprung and Henry Russell independently studied the relationship between the true brightness (absolute magnitude) of stars and their respective temperatures. Their work resulted in the development of a graph, called a **Hertzsprung–Russell diagram (H-R diagram)**. By studying H-R diagrams, we can learn a great deal of information about the relationships among the sizes, colors, and temperatures of stars (see Appendix D). For example, we learned that the hottest stars are blue in color and the coolest are red.

To produce an H-R diagram, astronomers survey a portion of the sky and plot each star according to its absolute magnitude (stellar brightness) and temperature (**FIGURE 24.8**). Notice that the stars in Figure 24.8 are not uniformly distributed. Rather, about 90 percent of all stars fall along a band that runs from the upper-left corner to the lower-right corner of the H-R diagram. These "ordinary" stars are called **main-sequence stars**. As you can see in Figure 24.8, the hottest main-sequence stars are intrinsically the brightest, and, conversely, the coolest are the dimmest.

The absolute magnitude of main-sequence stars is also related to their mass. The hottest (blue) stars are about 50 times more massive than the Sun, whereas the coolest (red) stars are only 1/10 as massive. Therefore, on the H-R diagram, the main-sequence stars appear in decreasing order, from hotter, more massive blue stars to cooler, less massive red stars.

Note the location of the Sun in Figure 24.8. The Sun is a yellow main-sequence star with an absolute magnitude, or "true" brightness, of about 5 (see Appendix D). Because the vast majority of main-sequence stars have magnitudes between –10 and 20, the Sun's midpoint position in this range results in its classification as an "average star." (Note that stellar magnitudes are measured so that the lower the number, the brighter the star and vice versa.)

Just as all humans do not fall into the normal size range, some stars differ significantly from main-sequence stars. Above and to the right of the main sequence stars lies a group of very luminous stars called *giants*, or, on the basis of their color, **red giants** (see Figure 24.8). The size of these giants can be estimated by comparing them with stars of known size that have the same surface temperature. Scientists have discovered that objects having equal surface temperatures radiate the same amount of energy per unit area. Any difference in the brightness of two stars having the same surface temperature can be attributed to their relative sizes. Therefore, if one red star is 100 times more luminous than another red star, it must have a surface area that is 100 times larger. Stars with large radiating surfaces appear in the upper-right position of an H-R diagram and are appropriately called *giants*.

Some stars are so immense that they are called **supergiants**. Betelgeuse, a bright red supergiant in the constellation Orion, has a radius about 800 times that of the Sun. If this star were at the center of our solar system, it would extend beyond the orbit of Mars, and Earth would find itself buried inside this supergiant!

In the lower portion of the H-R diagram, opposite conditions are observed. These stars are much fainter than main-sequence stars of the same temperature and likewise are much smaller. Some probably approximate Earth in size. These stars are called *white dwarfs*.

H-R diagrams have proved to be an important tool for interpreting stellar evolution: The stages in which stars, similar to living things, are born, age, and die. Considering that almost 90 percent of stars lie on the main sequence, we can be relatively certain that they spend most of their active years as main-sequence stars. Only a few percent are giants, and perhaps 10 percent are white dwarfs.

24.3 CONCEPT CHECKS

1 On an H-R diagram, where do stars spend most of their life span?

2 How does the Sun compare in size and brightness to other main-sequence stars?

3 Describe how the H-R diagram is used to determine which stars are "giants."

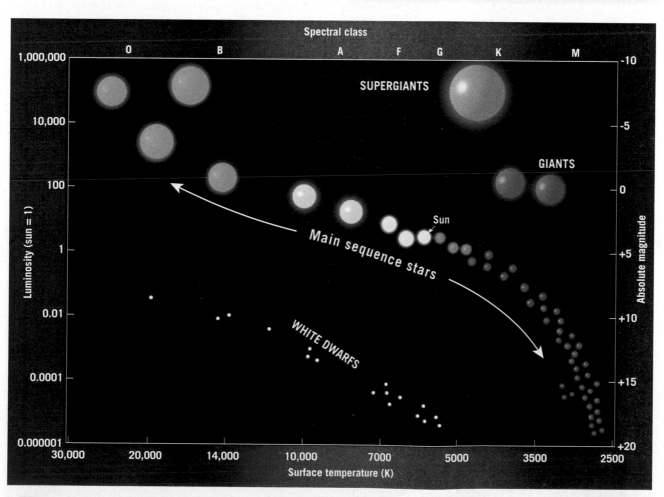

SmartFigure 24.8 Hertzsprung–Russell Diagram Astronomers use these diagrams to study stellar evolution by plotted stars according to their temperatures and luminosities (absolute magnitudes).

24.4 | STELLAR EVOLUTION

List and describe the stages in the evolution of a typical Sun-like star.

The idea of describing how a star is born, ages, and dies may seem a bit presumptuous, for most stars have life spans that exceed billions of years. However, by studying stars of different ages, at different points in their life cycles, astronomers have been able to assemble a model for stellar evolution.

The method used to create this model is analogous to how an alien, upon reaching Earth, might determine the developmental stages of human life. By observing large numbers of humans, this stranger would witness the onset of life, the progression of life in children and adults, and the death of the elderly. From this information, the alien could put the stages of human development into their natural sequence. Based on the relative abundance of humans in each stage of development, it would even be possible to conclude that humans spend more of their lives as adults than as toddlers. Similarly, astronomers have pieced together the life story of stars.

The first stars probably formed about 300 million years after the Big Bang. The most massive nebulae were the birthplaces of the first stars because their immense gravity caused them to collapse quickly. As a result, the stars that formed early in the history of the universe, often referred to as *first-generation stars*, were very massive. First-generation stars consisted mostly of hydrogen, with lesser amounts of helium, the primary elements formed during the Big Bang. Massive stars have relatively short lifetimes, followed by violent, explosive deaths. These explosions create the heavier elements and expel them into space. Some of this ejected matter is incorporated into subsequent generations of stars such as our Sun.

Every stage of a star's life is ruled by gravity. The mutual gravitational attraction of particles in a thin, gaseous nebula causes the cloud to collapse on itself. As the cloud is squeezed to unimaginable pressures, its temperature increases, eventually igniting its nuclear furnace, and a star is born. A star is a ball of very hot gases, caught between the opposing forces of gravity trying to contract it and thermal nuclear energy trying to expand it. Eventually, all of a star's nuclear fuel will be exhausted, and gravity will prevail, collapsing the stellar remnant into a small, dense body.

Stellar Birth

The birthplaces of stars are interstellar clouds, rich in dust and gases (see Figure 24.4). In the Milky Way, these gaseous clouds are about 92 percent hydrogen, 7 percent helium, and less than 1 percent heavier elements.

If these thin gaseous clouds become sufficiently concentrated, they begin to gravitationally contract (**FIGURE 24.9**). A mechanism that may trigger star formation is a shock wave from a catastrophic explosion (supernova) of a nearby star. Slow dissipation of thermal energy is also thought to cause nebulae to collapse. Regardless of how the process is initiated, once it begins, mutual gravitational attraction of the particles causes the cloud to contract, pulling every particle toward the center.

FIGURE 24.9 H-R Diagram Illustrating the Evolution of a Sun-Like Star

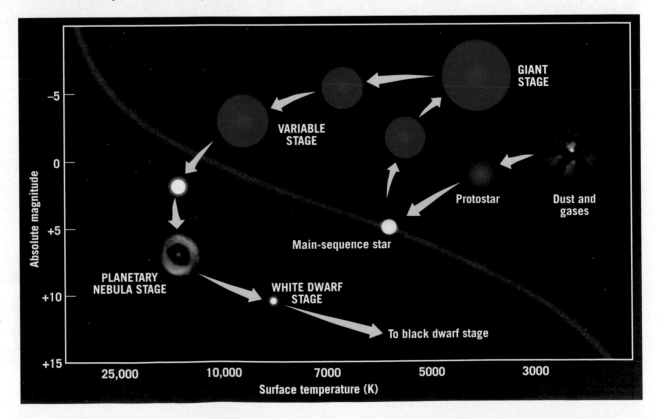

As the cloud collapses, gravitational energy is converted into energy of motion, or thermal energy, causing the contracting gases to gradually increase in temperature. When the temperature of these gaseous bodies increases sufficiently, they begin to radiate energy in the form of long-wavelength red light. Because these large red objects are not hot enough to engage in nuclear fusion, they are not yet stars. The name **protostar** is applied to these bodies (see Figure 24.9).

Protostar Stage

During the protostar stage, gravitational contraction continues, slowly at first and then much more rapidly. This collapse causes the core of the developing star to heat much faster than its outer envelope. (Stellar temperatures are expressed in kelvin [K]; see Appendix A.) When the core reaches a temperature of 10 million K, the pressure within is so intense that groups of four hydrogen nuclei (through a several-step process) fuse into a single helium nucleus. Astronomers refer to this nuclear reaction, in which hydrogen nuclei are fused into helium, as **hydrogen fusion**.

The immense amount of heat released by hydrogen fusion causes the gases inside stars to move with increased vigor, raising the internal gas pressure. At some point, the increased atomic motion produces an outward force (gas pressure) that balances the inward-directed force of gravity. Upon reaching this balance, the stars become *stable main-sequence stars* (see Figure 24.9). In other words, a main-sequence star is one in which the force of gravity, in an effort to squeeze the star into the smallest possible ball, is precisely balanced by gas pressure created by hydrogen fusion in the star's interior.

Main-Sequence Stage

During the main-sequence stage, stars experience minimal changes in size or energy output. Hydrogen is continually being converted into helium, and the energy released maintains the gas pressure sufficiently high to prevent gravitational collapse.

How long can stars maintain this balance? Hot, massive blue stars radiate energy at such an enormous rate that they substantially deplete their hydrogen fuel in only a few million years, approaching the end of their main-sequence stage rapidly. By contrast, the smallest (red) main-sequence stars may take hundreds of billions of years to burn their hydrogen; they live practically forever. A yellow star, such as the Sun, typically remains a main-sequence star for about 10 billion years. Because the solar system is about 5 billion years old, the Sun is expected to remain a stable main-sequence star for another 5 billion years.

An average star spends 90 percent of its life as a hydrogen-burning main-sequence star. Once the hydrogen fuel in a star's core is depleted, the star evolves rapidly and dies. However, with the exception of the least-massive stars, death is delayed when another type of nuclear reaction is triggered and the star becomes a *red giant* (see Figure 24.9).

Red Giant Stage

Evolution to the red giant stage begins when the usable hydrogen in a star's interior is consumed, leaving a helium-rich core. Although hydrogen fusion is still progressing in the star's outer shell, no fusion is taking place in the core. Without a source of energy, the core no longer has the gas pressure necessary to support itself against the inward force of gravity. As a result, the core begins to contract.

The collapse of a star's interior causes its temperature to rise rapidly as gravitational energy is converted into thermal energy. Some of this energy is radiated outward, initiating a more vigorous level of hydrogen fusion in the shell surrounding the star's core. The additional heat from the accelerated rate of hydrogen fusion expands the star's outer gaseous shell enormously. Sun-like stars become bloated **red giants**, while the most massive stars become

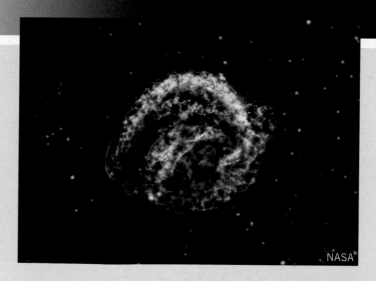

EYE ON THE UNIVERSE

The accompanying image shows the remnant of a supernova that Johannes Kepler observed in 1604. The red, green, and blue show low-, intermediate-, and high-energy x-rays observed by NASA's Chandra X-ray Observatory.

QUESTION 1 *If a supernova explosion were to occur within the immediate vicinity of our solar system, what might be some possible consequences of the intense x-ray and gamma radiation that would reach Earth?*

QUESTION 2 *Explain the difference between a nova and a supernova.*

NASA

supergiants, which are thousands of times larger than their main-sequence size.

As a star expands, its surface cools, which explains the star's color: Relatively cool objects radiate more energy as long-wavelength radiation (nearer the red end of the visible spectrum). Eventually the star's gravitational force stops this outward expansion, and the two opposing forces—gravity and gas pressure—once again achieve balance. The star enters a stable state but is much larger in size. Some red giants overshoot the equilibrium point and instead rebound like an overextended spring. These stars, which alternately expand and contract, and never reach equilibrium, are known as **variable stars**.

While the outer envelope of a red giant expands, the core continues to collapse, and the internal temperature eventually reaches 100 million K. This astonishing temperature triggers another nuclear reaction, in which helium is converted to carbon. At this point, a red giant consumes both hydrogen and helium to produce energy. In stars more massive than the Sun, other thermonuclear reactions occur that generate the elements on the periodic table up to and including number 26, iron.

Burnout and Death

What happens to stars after the red giant phase? We know that stars, regardless of their size, must eventually exhaust their usable nuclear fuel and collapse in response to their immense gravity. Because the gravitational field of a star is dependent on its mass, low-mass stars and high-mass stars have different fates.

Death of Low-Mass Stars   Stars less than one-half the mass of the Sun (0.5 solar mass) consume their fuel at relatively low rates (**FIGURE 24.10A**). Consequently, there are many small, *cool red stars* that may remain stable for as long as 100 billion years. Because the interiors of low-mass stars never attain sufficiently high temperatures and pressures to fuse helium, their only energy source is hydrogen fusion. Thus, a low-mass star never becomes a bloated red giant. Rather, it remains a stable main-sequence star until it consumes its usable hydrogen fuel and collapses into a hot, dense *white dwarf*.

**SmartFigure 24.10
Evolutionary Stages of Stars
Having Various Masses**

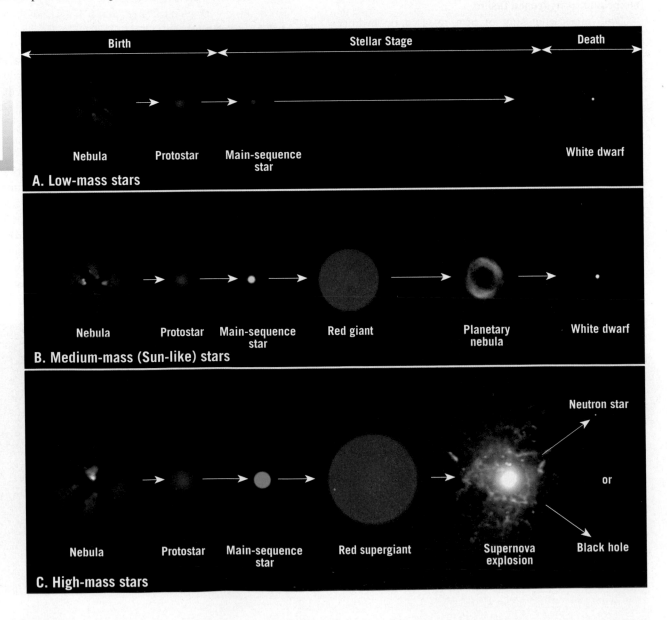

A. Low-mass stars

B. Medium-mass (Sun-like) stars

C. High-mass stars

Death of Medium-Mass (Sun-Like) Stars

Stars with masses ranging between one-half and eight times that of the Sun have a similar evolutionary history (**FIGURE 24.10B**). During their red giant phase, Sun-like stars fuse hydrogen and helium fuel at accelerated rates. Once this fuel is exhausted, these stars (like low-mass stars) collapse into Earth-size bodies of great density—white dwarfs. Without a source of nuclear energy, white dwarfs become cooler and dimmer as they continually radiate thermal energy into space.

During their collapse from red giants to white dwarfs, medium-mass stars cast off their bloated outer atmosphere, creating an expanding spherical cloud of gas. The remaining hot, central white dwarf heats the gas cloud, causing it to glow. Recall that these spectacular spherical clouds are called *planetary nebulae* (see Figure 24.6).

Death of Massive Stars

In contrast to Sun-like stars, which expire nonviolently, stars that are more than eight times the mass of the Sun have relatively short life spans and terminate in brilliant explosions called **supernovae** (**FIGURE 24.10C**). During supernova events, these stars become millions of times brighter than they were in prenova stages. If a star located near Earth produced such an outburst, its brilliance would surpass that of the Sun. Fortunately for us, supernovae are relatively rare events; none have been observed in our galaxy since the advent of the telescope, although Tycho Brahe and Johannes Kepler each recorded one, about 30 years apart, late in the sixteenth century. Chinese astronomers recorded an even brighter supernova in A.D. 1054. Today, the remnant of that great outburst is the Crab Nebula, shown in **FIGURE 24.11**.

A supernova event is triggered when a massive star has consumed most of its nuclear fuel. Without a heat engine to generate the gas pressure required to balance its immense gravitational field, it collapses. This implosion is enormous, resulting in a shock wave that rebounds out from the star's interior. This energetic shock wave blasts the star's outer shell into space, generating the supernova event.

Theoretical work predicts that during a supernova, the star's interior condenses into an incredibly hot object, possibly no larger than 20 kilometers (12 miles) in diameter. These incomprehensibly dense bodies have been named *neutron stars*. Some supernova events are thought to produce even smaller and more intriguing objects called *black holes*. We consider the nature of neutron stars and black holes in the following section.

FIGURE 24.11 Crab Nebula in the Constellation Taurus This spectacular nebula is thought to be the remains of the supernova of A.D. 1054. (Courtesy of NASA)

24.4 CONCEPT CHECKS

1 What element is the fuel for main-sequence stars?

2 Describe how main-sequence stars become giants.

3 Why are less massive stars thought to age more slowly than more massive stars, despite the fact they have much less "fuel"?

4 List the steps thought to be involved in the evolution of Sun-like stars.

24.5 | STELLAR REMNANTS Compare and contrast the final state of Sun-like stars to the remnants of the most massive stars.

Eventually, all stars consume their nuclear fuel and collapse into one of three celestial objects—*white dwarfs*, *neutron stars*, or *black holes*. How a star's life ends, and what final form it takes, depends largely on the star's mass during its main-sequence stage (**TABLE 24.1**).

White Dwarfs

After low- and medium-mass stars consume their remaining fuel, gravity causes them to collapse into **white dwarfs**. The density of these Earth-sized objects, having masses

TABLE 24.1 Summary of Evolution for Stars of Various Masses

Initial Mass of Main-Sequence Star (Sun = 1)*	Main-Sequence Stage	Giant Phase	Evolution After Giant Phase	Terminal State (Final Mass)
0.001	None (planet)	No	Not applicable	Planet (0.001)
0.1	Red	No	Not applicable	White dwarf (0.1)
1–3	Yellow	Yes	Planetary nebula	White dwarf (<1.4)
8	White	Yes	Supernova	Neutron star (1.4–3)
25	Blue	Yes (supergiant)	Supernova	Black hole (>3)

*These mass numbers are estimates.

FIGURE 24.12 Crab Pulsar: A Young Neutron Star Centered in the Crab Nebula This is the first pulsar to have been associated with a supernova. The energy emitted from this star illuminates the Crab Nebula. (Courtesy of NASA)

roughly equal to the Sun, may be 1 million times greater than water. A spoonful of such matter would weigh several tons on Earth. Densities of this magnitude are possible only when electrons are displaced inward from their regular orbits around an atom's nucleus. Material in this state is called **degenerate matter**.

The atoms in degenerate matter have been squeezed together so tightly that the electrons are pushed very close to the nucleus. However, the electrical repulsion that occurs between the negatively charged electrons supports the star against complete gravitational collapse.

As main-sequence stars contract into white dwarfs, their surfaces become extremely hot, sometimes exceeding 25,000K. Without sources of energy, main-sequence stars slowly cool and eventually become small, cold, burned-out embers called *black dwarfs*. However, our galaxy is not yet old enough for any black dwarfs to have formed.

Neutron Stars

A study of white dwarfs produced a surprising conclusion: The *smallest white dwarfs* are the *most massive*, and the *largest* are the *least massive*. Researchers have discovered that more massive stars, because of their greater gravitational fields, are squeezed into smaller, more densely packed objects than less massive stars. Thus, the smallest white dwarfs were produced from the collapse of larger, more massive main-sequence stars than are the largest white dwarfs.

This conclusion led to the prediction that stellar remnants even *smaller* and *more massive* than white dwarfs must exist. Named **neutron stars**, these objects are the remnants of explosive supernova events. In white dwarfs, the electrons are pushed close to the nucleus, whereas in neutron stars, the electrons are forced to combine with

protons in the nucleus to produce neutrons (hence the name *neutron star*). A pea-size sample of this matter would weigh 100 million tons. This is approximately the density of an atomic nucleus; thus, a neutron star can be thought of as a large atomic nucleus, composed entirely of neutrons.

During a supernova implosion, the outer envelope of the star is ejected, while the core collapses into a very hot star that is only about 20 to 30 kilometers (12 to 18 miles) in diameter. Although neutron stars have high surface temperatures, their small size greatly limits their luminosity, making them nearly impossible to locate visually.

Theoretical models predict that neutron stars have a very strong magnetic field and a high rate of rotation. As stars collapse, they rotate faster, for the same reason ice skaters rotate faster as they pull their arms in as they spin. Radio waves generated by the rotating magnetic fields of neutron stars are concentrated into two narrow zones that align with the star's magnetic poles. Consequently, these stars resemble rapidly rotating beacons emitting strong radio waves. If Earth happened to be in the path of these beacons, the star would appear to blink on and off, or pulsate, as the waves swept past.

In the early 1970s, a source that radiates short pulses of radio energy named a **pulsar** (*pulsating radio source*) was discovered in the Crab Nebula (**FIGURE 24.12**). Visual inspection of this radio source indicated that it was coming from a small star centered in the nebula. The pulsar found in the Crab Nebula is most likely the remains of the supernova of A.D. 1054 (see Figure 24.11).

Black Holes

Although neutron stars are extremely dense, they are not the densest objects in the universe. Stellar evolutionary theory predicts that neutron stars cannot exceed three times the mass of the Sun. Above this mass, not even tightly packed neutrons can withstand the star's gravitational pull. Following supernova explosions, if the core of a remaining star exceeds three solar masses, gravity prevails, and the stellar remnant collapses into an object that is denser than a neutron star. (The pre-supernova mass of such a star was likely 25 times that of our Sun.) The incredible objects, or celestial phenomena, created by such a collapse are called **black holes**.

Einstein's theory of general relativity predicts that even though black holes are extremely hot, their surface gravity is so immense that even light cannot escape. Consequently, they literally disappear from sight. Anything that moves too close to a black hole can be swept in and devoured by its immense gravitational field.

How do astronomers find objects whose gravitational field prevents the escape of all matter and energy? Theory predicts that as matter is pulled into a black hole, it should become extremely hot and emit a flood of x-rays before it is engulfed. Because *isolated* black holes do not have a source of matter to engulf, astronomers decided to look at binary-star

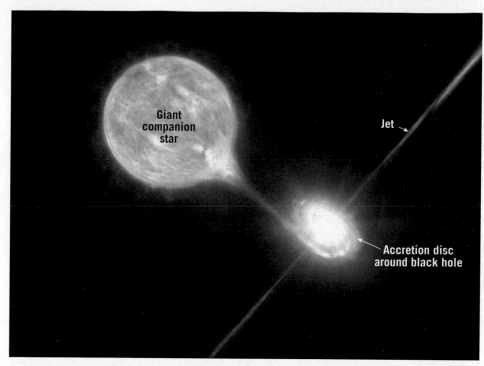

FIGURE 24.13 Artist's Depiction of a Black Hole and a Giant Companion Star Note the accretion disk surrounding the black hole. (Courtesy of European Southern Observatory/L. Calcada/M. Kornmesser)

systems for evidence of matter emitting x-rays while being rapidly swept into a region of apparent nothingness.

X-rays cannot penetrate our atmosphere; therefore, the existence of black holes was not confirmed until the advent of orbiting observatories. The first black hole to be identified, Cygnus X-1, orbits a massive supergiant companion once every 5.6 days. The gases that are pulled from the companion form an *accretion disk* that spirals around a "void" while emitting a steady stream of x-rays (**FIGURE 24.13**). Recent observations have determined that pairs of jets extend outward from these accretion disks and are thought to return some of this material back to space (see Figure 24.13).

Cygnus X-1, which is about 8 or 9 times as massive as our Sun, probably formed from a star of approximately 40 solar masses. Since the discovery of Cygnus X-1, many other x-ray sources have been discovered that are assumed to be black holes.

Astronomers have established that black holes are common objects in the universe and vary considerably in size. Small black holes have masses approximately 10 times that of our Sun but are only about 32 kilometers (20 miles) across, less than the distance of a marathon course. Intermediate black holes have masses 1000 times our Sun, and the largest black holes (*supermassive black holes*), found in the centers of galaxies, are estimated to be millions of solar masses. Because the earliest stars were thought to be massive, their deaths could have provided the seeds that eventually formed the supermassive black holes at the centers of galaxies.

24.5 CONCEPT CHECKS

1 Describe degenerate matter.

2 What is the final state of a medium-mass (Sun-like) star?

3 How do the "lives" of the most massive stars end? What are the two possible products of this event?

4 Explain how it is possible for the *smallest* white dwarfs to be the *most* massive.

5 Black holes are thought to be abundant, yet they are hard to find. Explain why.

24.6 | GALAXIES AND GALACTIC CLUSTERS

List the three major types of galaxies. Explain the formation of large elliptical galaxies.

On a clear and moonless night away from city lights, you can see a truly marvelous sight—a band of light stretching from horizon to horizon. With his telescope, Galileo discovered that this band of light was composed of countless individual stars. Today, we realize that the Sun is actually a part of this vast system of stars, the Milky Way Galaxy (**FIGURE 24.14**).

Galaxies (*galaxias* = milky) are collections of interstellar matter, stars, and stellar remnants that are gravitationally bound (see Figure 24.14). Recent observational data indicate that supermassive black holes may exist at the centers of most galaxies. In addition, spherical halos of very tenuous gas and numerous star clusters (*globular clusters*) appear to surround many of the largest galaxies.

FIGURE 24.14 View of Our Milky Way Galaxy at Sunset The dark patches in the "milky" band of light are caused by the presence of dark nebulae. (Courtesy of ESO/European Southern Observatory)

The Milky Way

The Milky Way derives it name from its appearance as a dim, "milky" glowing band that arches across the night sky. In this magnificent 360-degree panoramic image we see the plane of our galaxy, edge-on from our perspective on Earth. From this vantage point, the components of the Milky Way come into view. We can see the galaxy's bright central bulge, its disc that contains both dark and glowing nebulae that harbor bright, young stars. Also visible adjacent to the Milky Way are a few of its satellite galaxies.

ESO

Halo of old stars and hot gas

Galactic disc

Central bulge

Solar system

Globular clusters

This profile view shows that, in addition to its dense central bulge and galactic disc containing spiral arms, the Milky Way is surrounded by a spherical halo that extends far beyond the galactic disc. Recent evidence indicates that the halo contains a large amount of hot gas, but lacks star formation. The halo also contains old stars and numerous large stellar groups called globular clusters.

The age of the stars in globular cluster NGC 6397 are more than 13 billion years old, which confirms that the Milky Way is among the oldest of galaxies.

NGC 6397

ESO

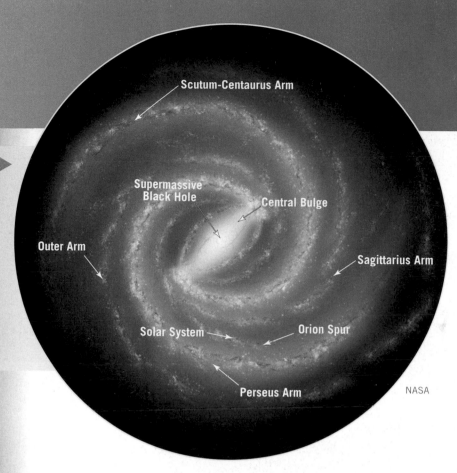

An artist's conception of the Milky Way Galaxy showing its dense central bulge and flat disc consisting of curved spiral arms. Our galaxy is a barred-spiral type that contains more than 200 billion stars. Its diameter exceeds 100,000 light-years and it rotates such that our solar system makes one complete trip around the galactic center every 250 million years. The galactic center of our galaxy houses a supermassive black hole with a mass of at least 40,000 Suns. Like the black holes found in the center of most large galaxies this behemoth tries to consume anything that happens to be nearby.

Scutum-Centaurus Arm

Supermassive Black Hole

Central Bulge

Outer Arm

Sagittarius Arm

Solar System

Orion Spur

Perseus Arm

NASA

Our galaxy is just one of an assemblage of more than 50 galaxies that collectively make up the Local Group. Members include Andromeda Galaxy, which is an even larger spiral galaxy than the Milky Way. For many years astronomers thought that the Magellanic Clouds, that can be seen as fuzzy patches in the southern sky, were our closest galactic neighbor. But, in fact, the closest galaxy so far discovered, named Canis Major Dwarf Galaxy, lies within our galaxy. Astronomers have recently concluded that the Milky Way grew to its current size by "eating up" dwarf galaxies like Canis Major.

Large Magellanic Cloud

Small Magellanic Cloud

ESO

FIGURE 24.15 **Dramatic Image of the Spiral Galaxy Messier 83** Although smaller, Messier 83 is thought to be very similar to the Milky Way Galaxy. (ESO/European Space Observatory)

Types of Galaxies

Among the hundreds of billions of galaxies, three basic types have been identified: *spiral*, *elliptical*, and *irregular*. Within each of these categories are many variations, the causes of which are still a mystery.

Spiral Galaxies Our Milky Way Galaxy is an example of a large **spiral galaxy** (FIGURE 24.15). Spiral galaxies are flat, disk-shaped objects that range from 20,000 to about 125,000 light-years in diameter. Typically, spiral galaxies have a greater concentration of stars near their centers, but there are numerous variations. As shown in FIGURE 24.16, spiral galaxies have arms (usually two) extending from the central nucleus. Spiral galaxies rotate rapidly in the center, while the outermost stars rotate more slowly, which gives these galaxies the appearance of a fireworks pinwheel. Generally, the central bulge contains older stars that give it a yellowish color, while younger hot stars are located in the arms. The young, hot stars in the arms are found in large groups that appear as bright patches of blue and violet light shown in Figure 24.15.

Many spiral galaxies have a band of stars extending outward from the central bulge that merges with the spiral arms. These are known as **barred spiral galaxies** (FIGURE 24.17). Recent investigations have found evidence that our galaxy probably has a bar structure. What produces these bar-shaped structures is a matter of ongoing research.

The first galaxies were small and composed mainly of massive stars and abundant interstellar matter. These galaxies grew quickly by accreting nearby interstellar matter and by colliding and merging with other galaxies. In fact, our galaxy is currently absorbing at least two tiny satellite galaxies.

A. Face-on view

B. Edge-on view

SmartFigure 24.16 Spiral Galaxies A. Spiral galaxies typically have a greater concentration of older stars near their center, which gives the central bulge its yellowish color. By contrast, the arms of spiral galaxies contain numerous hot, young stars that give these structures a bluish or violet tint. **B.** Edge-on view showing the central bulge. **C.** Surrounding most large galaxies are spherical halos of very tenuous gases and groups of stars called globular clusters. This large globular cluster contains an estimated 10 million stars. (Image A courtesy of NASA; images B and C courtesy of ESO/European Southern Observatory)

C. Globular cluster

FIGURE 24.17 Barred Spiral Galaxy (Courtesy of NASA, ESA, and the Hubble Heritage Team [STScI/AURA])

Elliptical Galaxies   As the name implies, **elliptical galaxies** have an ellipsoidal shape that can be nearly spherical, and they lack arms (**FIGURE 24.18**). Some of the largest and the smallest galaxies are elliptical. The smallest of these are known as **dwarf galaxies**. The two small companions of Andromeda shown in Figure 24.2 are dwarf galaxies.

The very largest known galaxies (1 million light-years in diameter) are also elliptical. For comparison, the Milky Way, a large spiral galaxy, is about one-half the diameter of a large elliptical galaxy. Most large elliptical galaxies are believed to result from the merger of two or more smaller galaxies.

Large elliptical galaxies tend to be composed of older, low-mass stars (red) and have minimal amounts of interstellar matter. Thus, unlike the arms of spiral galaxies, they have low rates of star formation. As a result, elliptical galaxies appear yellow to red in color, as compared to the bluish tint emanating from the young, hot stars in the arms of spiral galaxies.

Irregular Galaxies   Approximately 25 percent of known galaxies show no symmetry and are classified as **irregular galaxies**. Some were once spiral or elliptical galaxies that were subsequently distorted by the gravity of a larger neighbor. Two well-known irregular galaxies, the *Large* and *Small Magellanic Clouds*, are named for explorer Ferdinand Magellan, who observed them when he circumnavigated the globe in 1520. They are among our nearest galactic neighbors.

Recent images of the Large Magellanic Cloud revealed a central bar-shaped structure. Thus, the Large Magellanic Cloud was once a barred spiral galaxy that was subsequently distorted by gravity exerted by another galaxy as it passed by.

Galactic Clusters

Once astronomers discovered that stars occur in groups (galaxies), they set out to determine whether galaxies also occur in groups or whether they are just randomly distributed. They found that galaxies are grouped into gravitationally bound clusters (**FIGURE 24.19**). Some large **galactic clusters** contain thousands of galaxies. Our own galactic

FIGURE 24.18 Large Elliptical Galaxy This large elliptical galaxy belongs to the Fornax Cluster. Dark clouds of interstellar matter are visible within the central nucleus of this galaxy. Some of the star-like objects in this image are large groups of stars (globular clusters) that belong to the galaxy. (Courtesy of ESO/European Southern Observatory)

FIGURE 24.19 The Fornax Galaxy Cluster This is one of the nearest groupings of galaxies to our Local Group. Although many of the galaxies shown are elliptical, an elegant barred spiral galaxy is visible in the lower right. (Courtesy of ESO/European Southern Observatory/ J. Emerson/VISTA)

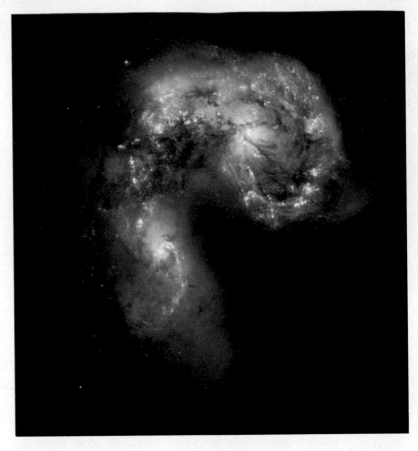

Galactic Collisions

Within galactic clusters, interactions between galaxies, often driven by one galaxy's gravity disturbing another, are common. For example, a large galaxy may engulf a dwarf satellite galaxy. In this case, the larger galaxy will retain its form, while the smaller galaxy will be torn apart and assimilated into the larger galaxy. Recall that two dwarf satellite galaxies are currently merging with the Milky Way.

Galactic interactions may also involve two galaxies of similar size that pass through one another without merging. It is unlikely that the individual stars within these galaxies will collide because they are so widely dispersed. However, the interstellar matter will likely interact, triggering an intense period of star formation.

In an extreme case, two large galaxies may collide and merge into a single large galaxy (**FIGURE 24.20**). Many of the largest elliptical galaxies are thought to have been produced by the merger of two large spiral galaxies. Some studies have predicted that in 2 to 4 billion years, there is a 50 percent probability that the Milky Way and Andromeda Galaxies will collide and merge.

cluster, called the **Local Group**, consists of more than 40 galaxies and may contain many undiscovered dwarf galaxies. Of the Local Group galaxies, three are large spiral galaxies, including the Milky Way and Andromeda Galaxies.

Galactic clusters also reside in huge groups called *superclusters*. There are possibly 10 million superclusters; our Local Group is found in the Virgo Supercluster. From visual observations, it appears that superclusters are the largest entities in the universe.

24.6 CONCEPT CHECKS

1 Compare the three main types of galaxies.

2 What type of galaxy is our Milky Way?

3 Describe a possible scenario for the formation of a large elliptical galaxy.

24.7 | THE BIG BANG THEORY

Describe the Big Bang theory. Explain what it tells us about the universe.

The *Big Bang theory* describes the birth, evolution, and fate of the universe. According to the Big Bang theory, the universe was originally in an extremely hot, supermassive state that expanded rapidly in all directions. Based on astronomers' best calculations, this expansion began about 13.7 billion years ago. What scientific evidence exists to support this theory?

Evidence for an Expanding Universe

In 1912 Vesto Slipher, while working at the Lowell Observatory in Flagstaff, Arizona, was the first to discover that galaxies exhibit motion. The motions he detected were twofold: Galaxies rotate, and galaxies move relative to each other. Slipher's efforts focused on the shifts in the spectra of the light emanating from galaxies. (See the section titled "The Doppler Effect" in Chapter 23.) When a source of light is moving away from an observer, the spectral lines shift toward the red end of the spectrum (longer wavelengths). Conversely, when celestial objects approach the observer, the spectral lines shift to the blue end of the spectrum (shorter wavelengths).

In 1929, a study of galaxies conducted by Edwin Hubble expanded the groundwork established by Slipher. Hubble noticed that most galaxies have spectral shifts toward the red end of the spectrum—which occurs when an object emitting light is receding from an observer (**FIGURE 24.21**). Therefore, all galaxies (except those in the Local Group) appear to be moving away from the Milky Way. These patterns were later named **cosmological red shifts** because the movement they reveal is a result of the expansion of the universe.

Recall that Hubble had also found a way to measure galactic distances. By comparing the distance to a galaxy with Vesto Slipher's measurements of its red shift, Hubble

Standard spectral lines (unshifted)

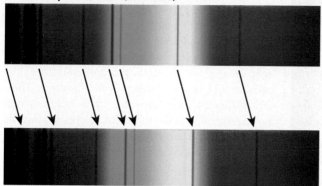

Red shift moves spectral lines to longer wavelengths

FIGURE 24.21 Red Shift Illustration of the shift in spectral lines toward the red end of the spectrum, which occurs when an object emitting light is receding from an observer.

made an unexpected discovery: He discovered that the red shifts of galaxies increase with distance and that the most distant galaxies are receding from the Milky Way at the fastest rate. This concept, now called **Hubble's law**, states that *galaxies recede at speeds proportional to their distances from the observer.*

This discovery surprised Hubble because conventional wisdom was that the universe was unchanging and would likely remain unchanged. What cosmological theory could explain Hubble's findings? Researchers concluded that an expanding universe accounts for the observed red shifts.

To help visualize why Hubble's law implies an expanding universe, imagine a batch of raisin bread dough that has been set out to rise for a few hours (**FIGURE 24.22**). In this analogy, the raisins represent galaxies, and the dough represents space. As the dough doubles in size, so does the distance between all the raisins. The distance between raisins that were originally 2 centimeters apart will become 4 centimeters, while the distance between raisins originally 6 centimeters apart will increase to 12 centimeters. The raisins that were originally farthest apart travel greater distances than those located closer together. Therefore, in an expanding universe, as in our analogy, more space is

created between two objects located farther apart than between two objects that are closer together.

Another feature of the expanding universe can be demonstrated using the same bread analogy. Regardless of which raisin you look at, it will move away from all the other raisins. Likewise, at any point in the universe, all other galaxies (except those in the same cluster) are receding from an observer at that location. Hubble's law implies a centerless universe that is expanding uniformly. The Hubble Space Telescope is named in honor of Edwin Hubble's invaluable contributions to the scientific understanding of the universe.

Predictions of the Big Bang Theory

Recall from the Introduction that in order for a hypothesis to become an accepted component of scientific knowledge (a theory), it must incorporate predictions that can be tested. One prediction of the Big Bang model is that if the universe was initially unimaginably hot, then researchers should be able to detect the remnant of that heat. The electromagnetic radiation (light) emitted by a white-hot universe would have extremely high energy and short wavelengths. However, according to the Big Bang theory, the continued expansion of the universe would have stretched the waves so that by now they should be detectable as long-wavelength radio waves called microwave radiation. Scientists began to search for this "missing" radiation, which they named *cosmic microwave background radiation*. As predicted, in 1965, this microwave radiation was detected and found to fill the entire visible universe.

Detailed observations of the cosmic microwave background radiation since its original discovery have confirmed many theoretical details of the Big Bang theory, including the order and timing of important events in the early history of the universe.

What Is the Fate of the Universe?

Cosmologists have developed different scenarios for the ultimate fate of the universe (**FIGURE 24.23**). In one scenario, the stars will slowly burn out and be replaced by invisible degenerate matter and black holes that travel outward through an endless, dark, cold universe. This scenario is sometimes called the *Big Chill* because the universe will slowly cool as it expands, to the point that it is unable to sustain life. Another possibility is that the outward flight of the galaxies will slow and eventually stop. Gravitational contraction would follow, causing all matter to eventually collide and coalesce into the high-energy, high-density state, from which the universe began. This fiery death of the universe, the Big Bang operating in reverse, has been called the *Big Crunch*.

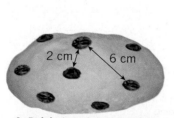

A. Raisin bread dough before it rises.

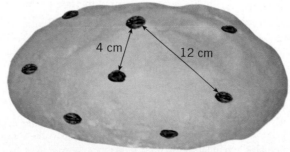

B. Raisin bread dough a few hours later.

SmartFigure 24.22 Raisin Bread Analogy for an Expanding Universe As the dough rises, raisins originally farthest apart travel greater distances than those located closer together. Thus, in an expanding universe (as with the raisins), more space is created between two objects that are farther apart than between two objects that are closer together.

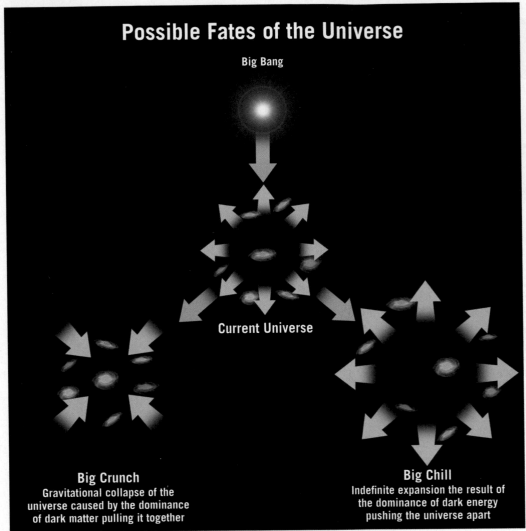

Possible Fates of the Universe

Big Bang

Current Universe

Big Crunch
Gravitational collapse of the universe caused by the dominance of dark matter pulling it together

Big Chill
Indefinite expansion the result of the dominance of dark energy pushing the universe apart

FIGURE 24.23 Cosmic Tug of War Illustration showing two possible fates of the universe. The gravity of dark matter tries to pull the universe together, while dark energy tries to push it apart. There is growing consensus that dark energy will prevail and produce an ever-expanding universe.

Whether the universe will expand forever or eventually collapse on itself is contingent on its density. If the average density of the universe is greater than an amount known as its *critical density* (about one atom for every cubic meter), gravitational attraction would be sufficient to stop the outward expansion and cause the universe to collapse. On the other hand, if the density of the universe is less than the critical value, the universe will continue to expand forever. Complicating the possibilities in the fate of the universe are two other types of matter that may exist. These are called *dark matter* and *dark energy*.

Dark Matter   The universe contains perhaps 100 billion galaxies, each with billions of stars, massive clouds of gas and dust, and large numbers of planets, moons, and other debris. Yet everything we see is like the tip of the cosmic iceberg; it accounts for less than 5 percent of the total mass of the universe. Astronomers came to this conclusion when studying the rotational periods of stars as they orbit the center of the Milky Way Galaxy. The law of gravity states that the stars closest to galactic center should travel faster than those near the galaxy's outer edge. (This is the reason Mercury travels around the Sun at a much faster speed than Pluto.) Yet these researchers found that all stars orbit the galactic center at roughly the

same speed. This implies that something surrounding our galaxy is tugging on the stars. This yet undetected material was named **dark matter**.

Approximately one-quarter of the universe consists of dark matter, which produces no detectable light energy but exerts a gravitational force that pulls on all "visible" matter in the universe. Thus, dark matter exerts a force that helps hold our galaxy together and at the same time works to slow the expansion of the universe as a whole.

Although the concept of dark matter may sound foreboding, it simply allows for the possibility that matter exists that does not interact with electromagnetic radiation. Recall that most of our knowledge about the universe comes to us via light. If there is a form of matter that does not interact with light, we will not be able to "see" it—hence the term *dark matter*.

Dark Energy   In the early 1990s, most cosmologists held the view that gravity was certain to slow the expansion of the universe over time, resulting eventually in the Big Crunch. However, in 1998 observations of very distant galaxies by the Hubble Space Telescope showed that the universe is actually expanding faster today than it was early in its history. Therefore, the expansion of the universe was not slowing due to gravity as scientists thought; rather, it was accelerating. To explain this unexpected result, researchers concluded that some unusual material, generally referred to as **dark energy**, must exist. Unlike dark matter, which works to slow the expansion of the universe, dark energy exerts a force that pushes matter outward, causing the expansion to speed up.

It has not been determined whether dark matter and dark energy are related—or exactly what they are. Most researchers think that dark matter consists of a type of subatomic particle that has not yet been detected. Dark energy may have its own particle, but there is no evidence of its existence.

There is growing consensus among cosmologists that dark energy, which is propelling the universe outward, is the dominant force. If dark energy is, in fact, the driving force behind the fate of the universe, the universe will expand forever (see Figure 24.23). Consider the following as astronomers search for dark energy: "Absence of evidence is not evidence of absence."

24.7 CONCEPT CHECKS

1 In your own words, explain how astronomers determined that the universe is expanding.

2 What did the Big Bang theory predict that was finally confirmed years after it was formulated?

3 Which view of the fate of the universe is currently favored: the Big Crunch or the Big Chill?

4 What property does the universe possess that will determine its final state?

24.1 THE UNIVERSE

Define *cosmology* and describe Edwin Hubble's most significant discovery about the universe.

KEY TERMS: cosmology, absolute magnitude, light-year, Big Bang theory

- Cosmology is the study of the universe, including its properties, structure, and evolution.
- The universe consists of hundreds of billions of galaxies, each containing billions of stars.
- The model that most accurately describes the birth and current state of the universe is the Big Bang theory. According to this model, the universe began about 13.7 billion years ago, in a cataclysmic explosion, and then it continued to expand, cool, and evolve to its current state.

24.2 INTERSTELLAR MATTER: NURSERY OF THE STARS

Explain why interstellar matter is often referred to as a stellar nursery. Compare and contrast bright and dark nebulae.

KEY TERMS: interstellar matter, nebula, bright nebula, dark nebula, emission nebula, reflection nebula, planetary nebula

- New stars are born out of enormous accumulations of dust and gases, called nebulae, which are scattered between existing stars.
- Emission nebulae derive their visible light from nearby hot stars or stars that are embedded in them. Reflection nebulae contain comparatively large debris, including grains of carbon compounds that reflect the light of nearby stars. Planetary nebulae consist of glowing clouds of dust and hot gases that have been expelled near the end of a star's life. Nebulae that are too distant from bright stars to be illuminated are referred to as dark nebulae.

Q Based on color, what type of nebulae is likely shown in the accompanying image?

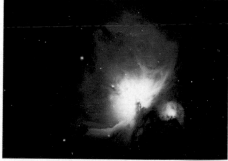

National Optical Astronomy Observatory (NOAO)

24.3 CLASSIFYING STARS: HERTZSPRUNG–RUSSELL DIAGRAMS (H-R DIAGRAMS)

Define *main-sequence star*. Explain the criteria used to classify stars as giants.

KEY TERMS: Hertzsprung–Russell diagram (H-R diagram), main-sequence star, red giant, supergiant

- Hertzsprung–Russell diagrams are constructed by plotting the absolute magnitudes and temperatures of stars on graphs. Considerable information about stars and stellar evolution has been discovered through the use of H-R diagrams.
- Stars are positioned within H-R diagrams as follows: (1) Main-sequence stars—90 percent of all stars—are in the band that runs from the upper-left corner (massive, hot blue stars) to the lower-right corner (low-mass, red stars); (2) red giants and supergiants—very luminous stars with large radii—are located in the upper-right position; and (3) white dwarfs—small, dense stars—are located in the lower portion.

Q Identify the type of stars located in the positions labeled A, B, and C on the accompanying H-R diagram.

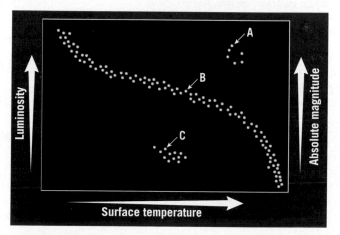

24.4 STELLAR EVOLUTION

List and describe the stages in the evolution of a typical Sun-like star.

KEY TERMS: protostar, hydrogen fusion, variable star, supernova

- Stars are born when their nuclear furnaces are ignited by the unimaginable pressures and temperatures generated during the collapse of nebulae.
- Red star-like objects not yet hot enough for nuclear fusion are called protostars. When the core of a protostar reaches a temperature of about 10 million K, a process called hydrogen fusion begins, marking the birth of the star. Hydrogen fusion involves the conversion of four hydrogen nuclei into a single helium nucleus and the release of thermal nuclear energy.
- Two opposing forces act on a star: gravity, which tries to contract it into the smallest possible ball, and gas pressure (created by thermal nuclear energy), which tries to blow it apart. When the forces are balanced, the star becomes a stable main-sequence star.
- Medium- and high-mass stars experience another type of nuclear fusion that causes their outer envelopes to expand enormously (hundreds to thousands of times larger), making them red giants or supergiants. When a star exhausts all its usable nuclear fuel, gravity takes over, and the stellar remnant collapses into a small, dense body.

24.5 STELLAR REMNANTS

Compare and contrast the final state of Sun-like stars to the remnants of the most massive stars.

KEY TERMS: white dwarf, degenerate matter, neutron star, pulsar, black hole

- The final fate of a star is determined by its mass.
- Stars with less than one-half the mass of the Sun collapse into hot, dense white dwarfs.
- Medium-mass stars, like the Sun, become red giants, collapse, and end up as white dwarfs, often surrounded by expanding spherical clouds of glowing gas called planetary nebulae.
- Massive stars terminate in a brilliant explosion called a supernova. Supernova events can produce small, extremely dense neutron stars, composed entirely of neutrons, or smaller, even denser black holes—objects that have such immense gravity that light cannot escape their surface.

24.6 GALAXIES AND GALACTIC CLUSTERS

List the three major types of galaxies. Explain the formation of large elliptical galaxies.

KEY TERMS: spiral galaxy, barred spiral galaxy, elliptical galaxy, dwarf galaxy, irregular galaxy, galactic cluster, Local Group

(NASA)

- The various types of galaxies include (1) irregular galaxies, which lack symmetry and account for about 25 percent of the known galaxies; (2) spiral galaxies, which are disk shaped and have a greater concentration of stars near their centers and arms extending from their central nucleus; and (3) elliptical galaxies, which have an ellipsoidal shape that may be nearly spherical.
- Galaxies are grouped in galactic clusters, some containing thousands of galaxies. Our own, called the Local Group, contains at least 40 galaxies.

Q What type of galaxy is shown in the accompanying image?

24.7 THE BIG BANG THEORY

Describe the Big Bang theory. Explain what it tells us about the universe.

KEY TERMS: cosmological red shift, Hubble's law, dark matter, dark energy

- Evidence for an expanding universe came from the study of red shifts in the spectra of galaxies. Edwin Hubble concluded that the observed red shifts, called cosmological red shifts, result from the expansion of space. This evidence strongly supports the Big Bang model of an expanded universe.
- One question that remains is whether the universe will expand forever in a Big Chill or gravitationally contract in a Big Crunch. Dark matter slows the expansion of the universe, while dark energy exerts a force that pushes matter outward, causing the expansion to speed up. Most cosmologists favor an endless, ever-expanding universe.

GIVE IT SOME **THOUGHT**

1. Assume that NASA is sending a space probe to each of the following locations:

   a. Polaris (the North Star)

   b. A comet near the outer edge of our solar system

   c. Jupiter

   d. The far edge of the Milky Way Galaxy

   e. The near side of the Andromeda Galaxy

   f. The Sun

   List the locations in order, *from nearest to farthest.*

2. Use the information provided below about three main-sequence stars (A, B, and C) to complete the following and explain your reasoning:

   ■ Star A has a main-sequence life span of 5 billion years.

   ■ Star B has the same luminosity (absolute magnitude) as the Sun.

   ■ Star C has a surface temperature of 5000K.

   a. Rank the mass of these stars from *greatest to least.*

   b. Rank the energy output of these stars from *greatest to least.*

   c. Rank the main-sequence life span of these stars from *longest to shortest.*

3. The masses of three clouds of gas and dust (nebulae) are provided below. Imagine that each cloud will collapse to form a single star. Use this information to complete the following and explain your reasoning.

   ■ Cloud A is 60 times the mass of the Sun.

   ■ Cloud B is 7 times the mass of the Sun.

   ■ Cloud C is 2 times the mass of the Sun.

   a. Which cloud or clouds, if any, will evolve into a red main-sequence star?

   b. Which of the stars that will form from these clouds, if any, will reach the giant stage?

   c. Which of the stars that will form from these clouds, if any, will go through the supernova stage?

4. The accompanying photo shows the Trifid Nebula, which can be easily observed with a small telescope. What unique properties does this nebula exhibit?

5. Refer to the accompanying images (A, B, C, and D) to complete the following:

   a. Which of these nebulae, if any, is an emission nebula?

   b. Which of these nebulae, if any, formed near the end of a star's lifetime?

   c. Which of these nebulae, if any, is a reflection nebula?

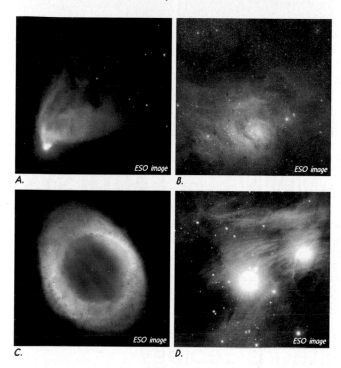

A.   B.

C.   D.

6. How a star evolves is closely related to its mass as a main-sequence star. Complete the accompanying diagram by labeling the evolutionary stages for the three groups of main-sequence stars shown.

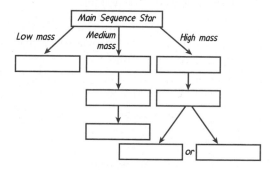

7. Refer to the accompanying photos of an elliptical galaxy and a spiral galaxy to complete the following:

   a. Which image (A or B) is an elliptical galaxy?

   b. Which of these galaxies appears to contain more young, hot massive stars? How did you determine your answer?

A.   B.

c. When stars are born from a cloud of dust and gases, large and small stars form at about the same time. Which group of stars, the large or the small stars, will die out first? Over time, how will this affect the color of the light we observe coming from this group of stars? Based on your response, which of these galaxies appears to be older? Explain.

8. Consider these three characteristics of the universe:

   ■ It does not have a center.

   ■ It does not have edges.

   ■ Its galaxies are all moving away from each other.

   a. Which of the three characteristics of the universe is/are depicted in the raisin bread dough analogy (see Figure 24.22)?

   b. Which of the three characteristics of the universe is/are not accurately depicted in Figure 24.22?

EXAMINING THE **EARTH SYSTEM**

1. Briefly describe how the atmosphere, hydrosphere, geosphere, and biosphere are each related to the death of stars that occurred billions of years ago.

2. Scientists are continuously searching the Milky Way Galaxy for other stars that may have planets. What types of stars would most likely have a planet or planets suitable for life as we know it?

3. Based on your knowledge of the Earth system, the planets in our solar system, and the universe in general, speculate about the likelihood that extra-solar planets exist with atmospheres, hydrospheres, geospheres, and biospheres similar to Earth's. Explain your speculation.

MasteringGeology™

Looking for additional review and test prep materials? Visit the Self Study area in **www.masteringgeology.com** to find practice quizzes, study tools, and multimedia that will aid in your understanding of this chapter's content. In **MasteringGeology**™ you will find:

■ GEODe: Earth Science: An interactive visual walkthrough of key concepts

■ Geoscience Animation Library: More than 100 animations illuminating many difficult-to-understand Earth science concepts

■ In The News RSS Feeds: Current Earth science events and news articles are pulled into the site with assessment

■ Pearson eText

■ Optional Self Study Quizzes

■ Web Links

■ Glossary

■ Flashcards

APPENDIX A | Metric and English Units Compared

Units

1 kilometer (km)	= 1000 meters (m)
1 meter (m)	= 100 centimeters
1 centimeter (cm)	= 0.39 inch (in.)
1 mile (mi)	= 5280 feet (ft)
1 foot (ft)	= 12 inches (in.)
1 inch (in.)	= 2.54 centimeters (cm)
1 square mile (mi^2)	= 640 acres (a)
1 kilogram (kg)	= 1000 grams (g)
1 pound (lb)	= 16 ounces (oz)
1 fathom	= 6 feet (ft)

Conversions

Length

When you want to convert:	multiply by:	to find:
inches	2.54	centimeters
centimeters	0.39	inches
feet	0.30	meters
meters	3.28	feet
yards	0.91	meters
meters	1.09	yards
miles	1.61	kilometers
kilometers	0.62	miles

Area

When you want to convert:	multiply by:	to find:
square inches	6.45	square centimeters
square centimeters	0.15	square inches
square feet	0.09	square meters
square meters	10.76	square feet
square miles	2.59	square kilometers
square kilometers	0.39	square miles

Volume

When you want to convert:	multiply by:	to find:
cubic inches	16.38	cubic centimeters
cubic centimeters	0.06	cubic inches
cubic feet	0.028	cubic meters
cubic meters	35.3	cubic feet
cubic miles	4.17	cubic kilometers
cubic kilometers	0.24	cubic miles
liters	1.06	quarts
liters	0.26	gallons
gallons	3.78	liters

Masses and Weights

When you want to convert:	multiply by:	to find:
ounces	28.35	grams
grams	0.035	ounces
pounds	0.45	kilograms
kilograms	2.205	pounds

Temperature

When you want to convert degrees Fahrenheit (°F) to degrees Celsius (°C), subtract 32 degrees and divide by 1.8.

When you want to convert degrees Celsius (°C) to degrees Fahrenheit (°F), multiply by 1.8 and add 32 degrees.

When you want to convert degrees Celsius (°C) to kelvin (K), delete the degree symbol and add 273. When you want to convert kelvin (K) to degrees Celsius (°C), add the degree symbol and subtract 273.

FIGURE A.1
Temperature Scales

APPENDIX B | Relative Humidity and Dew-Point Tables

TABLE B.1 Relative Humidity (Percent)

Dry bulb (°C)	Dry-Bulb Temperature Minus Wet-Bulb Temperature = Depression of the Wet Bulb																					
	1	2	3	4	5	6	7	8	9	10	11	12	13	14	15	16	17	18	19	20	21	22
−20	28																					
−18	40																					
−16	48	0																				
−14	55	11																				
−12	61	23																				
−10	66	33	0																			
−8	71	41	13																			
−6	73	48	20	0																		
−4	77	54	32	11																		
−2	79	58	37	20	1																	
0	81	63	45	28	11																	
2	83	67	51	36	20	6																
4	85	70	56	42	27	14																
6	86	72	59	46	35	22	10	0														
8	87	74	62	51	39	28	17	6														
10	88	76	65	54	43	33	24	13	4													
12	88	78	67	57	48	38	28	19	10	2												
14	89	79	69	60	50	41	33	25	16	8	1											
16	90	80	71	62	54	45	37	29	21	14	7	1										
18	91	81	72	64	56	48	40	33	26	19	12	6	0									
20	91	82	74	66	58	51	44	36	30	23	17	11	5									
22	92	83	75	68	60	53	46	40	33	27	21	15	10	4	0							
24	92	84	76	69	62	55	49	42	36	30	25	20	14	9	4	0						
26	92	85	77	70	64	57	51	45	39	34	28	23	18	13	9	5						
28	93	86	78	71	65	59	53	47	42	36	31	26	21	17	12	8	4					
30	93	86	79	72	66	61	55	49	44	39	34	29	25	20	16	12	8	4				
32	93	86	80	73	68	62	56	51	46	41	36	32	27	22	19	14	11	8	4			
34	93	86	81	74	69	63	58	52	48	43	38	34	30	26	22	18	14	11	8	5		
36	94	87	81	75	69	64	59	54	50	44	40	36	32	28	24	21	17	13	10	7	4	
38	94	87	82	76	70	66	60	55	51	46	42	38	34	30	26	23	20	16	13	10	7	5
40	94	89	82	76	71	67	61	57	52	48	44	40	36	33	29	25	22	19	16	13	10	7

Relative Humidity Values

To determine the relative humidity, find the air (dry-bulb) temperature on the vertical axis (far left) and the depression of the wet bulb on the horizontal axis (top). Where the two meet, the relative humidity is found. For example, when the dry-bulb temperature is 20°C and a wet-bulb temperature is 14°C, then the depression of the wet bulb is 6°C (20°C−14°C). From **TABLE B.1**, the relative humidity is 51 percent and from **TABLE B.2**, the dew point is 10°C.

TABLE B.2 Dew-Point Temperature (°C)

Dry bulb (°C)	Dry-Bulb Temperature Minus Wet-Bulb Temperature = Depression of the Wet Bulb																					
	1	2	3	4	5	6	7	8	9	10	11	12	13	14	15	16	17	18	19	20	21	22
−20	−33																					
−18	−28																					
−16	−24																					
−14	−21	−36																				
−12	−18	−28																				
−10	−14	−22																				
−8	−12	−18	−29																			
−6	−10	−14	−22																			
−4	−7	−12	−17	−29																		
−2	−5	−8	−13	−20																		
0	−3	−6	−9	−15	−24																	
2	−1	−3	−6	−11	−17																	
4	1	−1	−4	−7	−11	−19																
6	4	1	−1	−4	−7	−13	−21															
8	6	3	1	−2	−5	−9	−14															
10	8	6	4	1	−2	−5	−9	−14	−18													
12	10	8	6	4	1	−2	−5	−9	−16													
14	12	11	9	6	4	1	−2	−5	−10	−17												
16	14	13	11	9	7	4	1	−1	−6	−10	−17											
18	16	15	13	11	9	7	4	2	−2	−5	−10	−19										
20	19	17	15	14	12	10	7	4	2	−2	−5	−10	−19									
22	21	19	17	16	14	12	10	8	5	3	−1	−5	−10	−19								
24	23	21	20	18	16	14	12	10	8	6	2	−1	−5	−10	−18							
26	25	23	22	20	18	17	15	13	11	9	6	3	0	−4	−9	−18						
28	27	25	24	22	20	19	17	16	14	11	9	7	4	1	−3	−9	−16					
30	29	27	26	24	22	21	19	18	16	14	12	10	8	5	1	−2	−8	−15				
32	31	29	28	27	25	24	22	21	19	17	15	13	11	8	5	2	−2	−7	−14			
34	33	31	30	29	27	26	24	23	21	20	18	16	14	12	9	6	3	−1	−5	−12	−29	
36	35	33	32	31	29	28	27	25	24	22	20	19	17	15	13	10	7	4	0	−4	−10	
38	37	35	34	33	32	30	29	28	26	25	23	21	19	17	15	13	11	8	5	1	−3	9
40	39	37	36	35	34	32	31	30	28	27	25	24	22	20	18	16	14	12	9	6	2	−2

Dew-Point Values

APPENDIX C | Stellar Properties

Measuring Distances to the Closest Stars

Measuring the distance to stars is difficult. Nevertheless, astronomers have developed some direct as well as indirect methods to measure stellar distances. One simple measurement, called *stellar parallax*, is effective in determining the distances to only the closest stars.

Stellar parallax is the slight back-and-forth shift of the apparent position of a nearby star due to the orbital motion of Earth around the Sun. The principle of parallax is easy to visualize. Close one eye, and with your index finger in a vertical position, use your open eye to align your finger with some distant object. Without moving your finger, view the object with your other eye and notice that its position appears to have changed. Now repeat the exercise, holding your finger farther away, and notice that the farther away you hold your finger, the less its position seems to shift. In principle, this method of measuring stellar distances is elementary and was practiced by ancient Greek astronomers.

Modern cosmologists determine parallax by photographing a nearby star against the background of distant stars. Then, when Earth has moved halfway around its orbit six months later, the same star is photographed again. When these two photographs are compared, the position of the nearby star appears to have shifted with respect to the background stars. **FIGURE C.1** illustrates this shift, and the parallax angle that is determined from it. The nearest stars have the largest parallax angles, whereas those of distant stars are much too small to measure.

In practice, conducting parallax measurements is quite complex because of the miniscule angles being measured. The process is further complicated because both the Sun and the star being measured are moving relative to each other. The first accurate stellar parallax was not determined until 1838. Even today, parallax angles for only a few thousand of the nearest stars are known

with certainty—nearly all others have such small parallax shifts that accurate measurements are not possible. Fortunately, other methods have been developed to estimate distances to more distant stars. In addition, the Hubble Space Telescope, which is not hindered by Earth's light-distorting atmosphere, has obtained accurate parallax distances for many more stars.

Stellar Brightness

The oldest means of classifying stars is based on their *brightness*, also called *luminosity* or *magnitude*. Three factors control the brightness of a star as seen from Earth: *how large* it is, *how hot* it is, and its *distance* from Earth. The stars in the night sky come in a grand assortment of sizes, temperatures, and distances, so their apparent brightness varies widely.

Apparent Magnitude

Stars have been classified according to their apparent brightness since at least the second century B.C., when Hipparchus placed about 850 stars into six categories, based on his ability to see differences in brightness. Because he could only reliably see six different brightness levels, he created six categories. These categories were later called *magnitudes*, with first magnitude being the brightest and sixth magnitude the dimmest. Because some stars may appear dimmer than others only because they are farther away, a star's brightness, *as it appears when viewed from Earth*, is called its *apparent magnitude*. With the invention of the telescope, many stars fainter than the sixth magnitude were discovered.

In the mid-1800s, a method was developed to standardize the magnitude scale. An absolute comparison was made between the light coming from stars of the first magnitude and those of the sixth magnitude. It was determined that

FIGURE C.1 Geometry of Stellar Parallax The parallax angle shown here is enormously exaggerated to illustrate the principle. Because distances to even the nearest stars are thousands of times greater than the Earth–Sun distance, the triangles that astronomers work with are extremely long and narrow, making the angles that are measured very small.

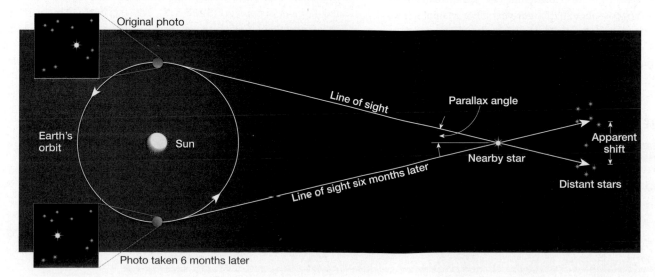

TABLE C.1 Ratios of Star Brightness

Difference in Magnitude	Brightness Ratio
0.5	1.6:1
1	2.5:1
2	6.3:1
3	16:1
4	40:1
5	100:1
10	10,000:1
20	100,000,000:1

*Calculations: 2.512 × 2.512 × 2.512 × 2.512 × 2.512, or 2.512 raised to the fifth power, equals 100.

a first-magnitude star was about 100 times brighter than a sixth-magnitude star. On the scale that was devised, any two stars that differ by five magnitudes have a ratio in brightness of 100 to 1. Hence, a third-magnitude star is 100 times brighter than an eighth-magnitude star. It follows, then, that the brightness ratio of two stars differing by only one magnitude is about 2.5.[1] A star of the first magnitude is about 2.5 times brighter than a star of the second magnitude. **TABLE C.1** shows how differences in magnitude correspond to brightness ratios.

Because some celestial bodies are brighter than first-magnitude stars, zero and negative magnitudes were introduced. On this scale, the Sun has an apparent magnitude of –26.7. At its brightest, Venus has a magnitude of –4.3. At the other end of the scale, the Hubble Space Telescope can view stars with an apparent magnitude of 30, more than 1 billion times dimmer than stars that are visible to the unaided eye.

Absolute Magnitude

Apparent magnitudes were good approximations of the true brightness of stars when astronomers thought that the universe was very small—containing no more than a few thousand stars that were all at very similar distances from Earth. However, we now know that the universe is unimaginably large and contains innumerable stars at wildly varying distances. Since astronomers are interested in the "true" brightness of stars, they devised a measure called *absolute magnitude*.

Stars of the same apparent magnitude usually do not have the same brightness because their distances from us are not equal. Astronomers correct for distance by determining what brightness (magnitude) the stars would have if they were at a standard distance—about 32.6 light-years. For example, if the Sun, which has an apparent magnitude of –26.7, were located 32.6 light-years

[1]The more negative, the brighter; the more positive, the dimmer.

from Earth, it would have an absolute magnitude of about +5. Thus, stars with absolute magnitudes greater than 5 (smaller numerical value) are intrinsically brighter than the Sun but appear much dimmer because of their distance from Earth. **TABLE C.2** lists the absolute and apparent magnitudes of some stars as well as their distances from Earth. Most stars have an absolute magnitude between −5 (very bright) and 15 (very dim). The Sun is near the midpoint of this range.

Stellar Color and Temperature

The next time you are outside on a clear night, look carefully at the stars and note their colors (**FIGURE C.2**). Because human eyes do not respond well to color in low-intensity light (when it is very dark, we see in only black and white), we tend to look at the brightest stars. Some that are quite colorful can be found in the constellation Orion. Of the two brightest stars in Orion, Rigel (β Orionis) appears blue, whereas Betelgeuse (α Orionis) is definitely red.

Very hot stars with surface temperatures above 30,000K emit most of their energy in the form of short-wavelength light and therefore appear blue. On the other hand, cooler red stars, with surface temperatures generally less than 3000K, emit most of their energy as longer-wavelength red light. Stars such as the Sun with surface temperatures between 5000 and 6000K appear yellow. Because color is primarily a manifestation of a star's surface temperature, this characteristic provides astronomers with useful information. Combining temperature data with stellar magnitude tells us a great deal about the size and mass of stars.

Binary Stars and Stellar Mass

One of the night sky's best-known constellations, the Big Dipper, appears to consist of seven stars. But those with good eyesight can recognize that the second star in the handle is actually two stars. In the early nineteenth century, careful examination of numerous star pairs by William Herschel showed that many stars found in pairs actually orbit one another. In such cases, the two stars are in fact united by their mutual gravitation. These pairs, in which the members are far enough apart to be telescopically identified as two stars, are called *visual binaries* (*binaries* = double). The idea of one star orbiting another may seem unusual, but many stars in the universe exist in pairs or multiples.

Binary stars can be used to determine the star property most difficult to calculate—its *mass*. The mass of a body can be established if it is gravitationally attached to a partner. Binary stars orbit each other around a common point called the *center of mass* (**FIGURE C.3**). For stars of equal mass, the center of mass lies exactly halfway between them. When one star is more massive than its partner, their common center will be located closer to the more massive one. Thus, if the sizes of their orbits can be observed, their individual masses

TABLE C.2 Distance, Apparent Magnitude, and Absolute Magnitude of Some Stars

Name	Distance (light-years)	Apparent Magnitude	Absolute Magnitude
Sun	NA	−26.7	5.0
Alpha Centauri	4.27	0.0	4.4
Sirius	8.70	−1.4	1.5
Arcturus	36	−0.1	−0.3
Betelgeuse	520	0.8	−5.5
Deneb	1600	1.3	−6.9

FIGURE C.2 Time-Lapse Photograph of Stars in the Constellation Orion These star trails show some of the various star colors. It is important to note that the human eye sees color somewhat differently than photographic film. (Courtesy of National Optical Astronomy Observatories)

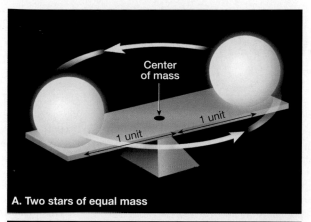

A. Two stars of equal mass

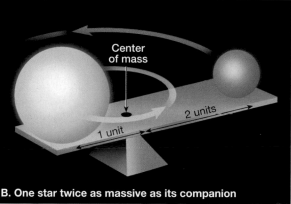

B. One star twice as massive as its companion

FIGURE C.3 Binary Stars Orbit Each Other Around Their Common Center of Mass A. For stars of equal mass, the center of mass lies exactly halfway between them. **B.** If one star is twice as massive as its companion, it is twice as close to their common center of mass. Therefore, more massive stars have proportionately smaller orbits than their less massive companions.

can be determined. You can experience this relationship on a seesaw by trying to balance a person who has a much greater (or smaller) mass.

For illustration, when one star has an orbit half the size (radius) of its companion, it is twice as massive as its companion. If their combined masses are equal to three solar masses, then the larger will be twice as massive as the Sun, and the smaller will have a mass equal to that of the Sun. Most stars have a mass that falls in a range between 1/10 and 50 times the mass of the Sun.

GLOSSARY

Aa flow A type of lava flow that has a jagged, blocky surface.

Abrasion The grinding and scraping of a rock surface by the friction and impact of rock particles carried by water, wind, or ice.

Absolute humidity The weight of water vapor in a given volume of air (usually expressed in grams/m^3).

Absolute instability Air that has a lapse rate greater than the dry adiabatic rate.

Absolute magnitude The apparent brightness of a star if it were viewed from a distance of 10 parsecs (32.6 light-years). Used to compare the true brightness of stars.

Absolute stability Air with a lapse rate less than the wet adiabatic rate.

Absorption spectrum A continuous spectrum with dark lines superimposed. Also known as *dark-line spectrum*.

Abyssal plain A very level area of the deep-ocean floor, usually lying at the foot of the continental rise.

Abyssal zone A subdivision of the benthic zone characterized by extremely high pressures, low temperatures, low oxygen, few nutrients, and no sunlight.

Accretionary wedge A large wedge-shaped mass of sediment that accumulates in subduction zones. Here, sediment is scraped from the subducting oceanic plate and accreted to the overriding crustal block.

Acid precipitation Rain or snow with a pH value that is less than the pH of unpolluted precipitation.

Active continental margin A portion of the seafloor adjacent to the continents that is usually narrow and consisting of highly deformed sediments. These margins occur where oceanic lithosphere is being subducted beneath the margin of a continent.

Adiabatic temperature change Cooling or warming of air caused when air is allowed to expand or is compressed, not because heat is added or subtracted.

Advection Horizontal convective motion, such as wind.

Advection fog A fog formed when warm, moist air is blown over a cool surface.

Aerosols Tiny solid and liquid particles suspended in the atmosphere.

Aftershocks Smaller earthquakes that follow the main earthquake.

Air A mixture of many discrete gases, of which nitrogen and oxygen are most abundant, in which varying quantities of tiny solid and liquid particles are suspended.

Air mass A large body of air that is characterized by a sameness of temperature and humidity.

Air pollutants Airborne particles and gases that occur in concentrations that endanger the health and well-being of organisms or disrupt the orderly functioning of the environment.

Air pressure The force exerted by the weight of a column of air above a given point.

Air-mass weather The conditions experienced in an area as an air mass passes over it. Because air masses are large and fairly homogenous, air-mass weather will be fairly constant and may last for several days.

Albedo The reflectivity of a substance, usually expressed as a percentage of the incident radiation reflected.

Alluvial fan A fan-shaped deposit of sediment formed when a stream's slope is abruptly reduced.

Alluvium Unconsolidated sediment deposited by a stream.

Alpine glacier A glacier confined to a mountain valley, which in most instances had previously been a stream valley. Also known as a *valley glacier*.

Altitude (of the Sun) The angle of the Sun above the horizon.

Andean-type plate margin A plate boundary that generates continental volcanic arcs.

Andesitic composition *See* Intermediate composition.

Anemometer An instrument used to determine wind speed. Also known as a *cup anemometer*.

Aneroid barometer An instrument for measuring air pressure that consists of evacuated metal chambers that are very sensitive to variations in air pressure.

Angiosperm A flowering plant in which fruits contain the seeds.

Angle of repose The steepest angle at which loose material remains stationary, without sliding downslope.

Angular unconformity An unconformity in which the strata below dip at an angle different from that of the beds above.

Annual mean temperature An average of the 12 monthly temperature means.

Annual temperature range The difference between the highest and lowest monthly temperature means.

Anthracite A hard, metamorphic form of coal that burns clean and hot.

Anticline A fold in sedimentary strata that resembles an arch; the opposite of syncline.

Anticyclone A high-pressure center characterized by a clockwise flow of air in the Northern Hemisphere.

Aphelion The place in the orbit of a planet where the planet is farthest from the Sun.

Aphotic zone The portion of the ocean where there is no sunlight.

Apparent magnitude The brightness of a star when viewed from Earth.

Aquifer Rock or soil through which groundwater moves easily.

Aquitard Impermeable beds that hinder or prevent groundwater movement.

Archean eon The second eon of Precambrian time, following the Hadean and preceding the Proterozoic. It extends between 3.8 billion and 2.5 billion years before the present.

Arctic (A) air mass A bitterly cold air mass that forms over the frozen Arctic Ocean.

Arête A narrow knifelike ridge separating two adjacent glaciated valleys.

Arid *See* Desert.

Arid climate *See* Dry climate.

Arkose A feldspar-rich sandstone.

Artesian well A well in which the water rises above the level where it was initially encountered.

Asteroid belt The region in which most asteroids orbit the Sun between Mars and Jupiter.

Asteroids Thousands of small planetlike bodies, ranging in size from a few hundred kilometers to less than a kilometer, whose orbits lie mainly between those of Mars and Jupiter.

Asthenosphere A subdivision of the mantle situated below the lithosphere. This zone of weak material exists below a depth of about 100 kilometers (60 miles) and in some regions extends as deep as 700 kilometers (430 miles). The rock within this zone is easily deformed. Also known as the *low-velocity zone*.

Astronomical theory A theory of climatic change first developed by Yugoslavian astronomer Milutin Milankovitch. It is based on changes in the shape of Earth's orbit, variations in the obliquity of Earth's axis, and the wobbling of Earth's axis.

Astronomical unit (AU) Average distance from Earth to the Sun; 1.5×10^8 kilometers (93×10^6 miles).

Astronomy The scientific study of the universe; it includes the observation and interpretation of celestial bodies and phenomena.

Atmosphere The gaseous portion of a planet; the planet's envelope of air. One of the traditional subdivisions of Earth's physical environment.

Atoll A continuous or broken ring of coral reef surrounding a central lagoon.

Atom The smallest particle that exists as an element.

Atomic number The number of protons in the nucleus of an atom.

Atomic weight The average of the atomic masses of isotopes for a given element.

Aurora A bright display of ever-changing light caused by solar radiation interacting with the upper atmosphere in the region of the poles.

Autumnal equinox The equinox that occurs on September 21–23 in the Northern Hemisphere and on March 21–22 in the Southern Hemisphere.

Axial precession A slow motion of Earth's axis that traces out a cone over a period of 26,000 years. Also known simply as *precession*.

Backshore The inner portion of the shore, lying landward of the high-tide shoreline. It is usually dry, being affected by waves only during storms.

Backswamp A poorly drained area on a floodplain that results when natural levees are present.

Bajada An apron of sediment along a mountain front created by the coalescence of alluvial fans.

Banded iron formations A finely layered iron and silica-rich (chert) layer deposited mainly during the Precambrian.

Bar Common term for sand and gravel deposits in a stream channel.

Barchan dune A solitary sand dune shaped like a crescent with its tips pointing downward.

Barchanoid dune Dunes forming scalloped rows of sand oriented at right angles to the wind. This form is intermediate between isolated barchans and extensive waves of transverse dunes.

Barograph A recording barometer.

Barometer An instrument that measures atmospheric pressure.

Barometric tendency *See* Pressure tendency.

Barred spiral galaxy A galaxy having straight arms extending from its nucleus.

Barrier island A low, elongate ridge of sand that parallels the coast.

Basalt A fine-grained igneous rock of mafic composition.

Basalt plateau The broad and extensive accumulation of lava from a succession of flows emanating from fissure eruptions.

Basaltic composition A compositional group of igneous rocks indicating that the rock contains substantial dark silicate minerals and calcium-rich plagioclase feldspar.

Base level The level below which a stream cannot erode.

Basin A circular downfolded structure.

Batholith A large mass of igneous rock that formed when magma was emplaced at depth, crystallized, and subsequently exposed by erosion.

Bathymetry The measurement of ocean depths and the charting of the shape or topography of the ocean floor.

Baymouth bar A sandbar that completely crosses a bay, sealing it off from the open ocean.

Beach An accumulation of sediment found along the landward margin of the ocean or a lake.

Beach drift The transport of sediment in a zigzag pattern along a beach caused by the uprush of water from obliquely breaking waves.

Beach face The wet, sloping surface that extends from the berm to the shoreline.

Beach nourishment The process by which large quantities of sand are added to the beach system to offset losses caused by wave erosion.

Bed load Sediment that is carried by a stream along the bottom of its channel.

Benioff zone The zone of inclined seismic activity that extends from a trench downward into the asthenosphere.

Benthic zone The marine life zone that includes *any* seafloor surface, regardless of its distance from shore.

Benthos The forms of marine life that live on or in the ocean bottom.

Bergeron process A theory that relates the formation of precipitation to supercooled clouds, freezing nuclei, and the different saturation levels of ice and liquid water.

Berm The dry, gently sloping zone on the backshore of a beach at the foot of the coastal cliffs or dunes.

Big Bang theory The theory which proposes that the universe originated as a single mass, which subsequently exploded.

Binary stars Two stars revolving around a common center of mass under their mutual gravitational attraction.

Biochemical sedimentary rock Sediment that forms when material dissolved in water is precipitated by water-dwelling organisms. Shells are common examples.

Biogenous sediment Seafloor sediments consisting of material of marine-organic origin.

Biomass The total mass of a defined organism or group of organisms in a particular area or ecosystem.

Biosphere The totality of life on Earth; the parts of the solid Earth, hydrosphere, and atmosphere in which living organisms can be found.

Bituminous coal The most common form of coal, often called soft, black coal.

Black carbon Soot generated by combustion processes and fires.

Black dwarf A final state of evolution for a star, in which all of its energy sources are exhausted and it no longer emits radiation.

Black hole A massive star that has collapsed to such a small volume that its gravity prevents the escape of all radiation.

Blowout (deflation hollow) A depression excavated by the wind in easily eroded deposits.

Bode's law A sequence of numbers that approximates the mean distances of the planets from the Sun.

Body waves Seismic waves that travel through Earth's interior.

Bowen's reaction series A concept proposed by N. L. Bowen that illustrates the relationships between magma and the minerals crystallizing from it during the formation of igneous rocks.

Braided stream A stream consisting of numerous intertwining channels.

Breakwater A structure protecting a nearshore area from breaking waves.

Breccia A sedimentary rock composed of angular fragments that were lithified.

Bright nebula A cloud of glowing gas excited by ultraviolet radiation from hot stars.

Bright-line spectrum The bright lines produced by an incandescent gas under low pressure. Also known as *emission spectrum*.

Brittle deformation Deformation that involves the fracturing of rock. Associated with rocks near the surface.

Cactolith A quasi-horizontal chonolith composed of anastomosing ductoliths, whose distal ends curl like a harpolith, thin like a sphenolith, or bulge discordantly like an akmolith or ethmolith.

Caldera A large depression typically caused by collapse or ejection of the summit area of a volcano.

Calorie The amount of heat required to raise the temperature of 1 gram of water 1°C.

Calving Wastage of a glacier that occurs when large pieces of ice break off into water.

Cambrian explosion The huge expansion in biodiversity that occurred at the beginning of the Paleozoic era.

Cap rock A necessary part of an oil trap. The cap rock is impermeable and hence keeps upwardly mobile oil and gas from escaping at the surface.

Capacity The total amount of sediment a stream is able to transport.

Carbonate group Mineral group whose members contain the carbonate ion (CO_2^{2-}) and one or more kinds of positive ions. Calcite is a common example.

Carbonic acid A weak acid formed when carbon dioxide is dissolved in water. It plays an important role in chemical weathering.

Cassini division A wide gap in the ring system of Saturn between the A ring and the B ring.

Catastrophism The concept that Earth was shaped by catastrophic events of a short-term nature.

Cavern A naturally formed underground chamber or series of chambers most commonly produced by solution activity in limestone.

Celestial sphere An imaginary hollow sphere upon which the ancients believed the stars were hung and carried around Earth.

Cementation One way in which sedimentary rocks are lithified. As material precipitates from water that percolates through the sediment, open spaces are filled, and particles are joined into a solid mass.

Cenozoic era A span on the geologic time scale beginning about 65 million years ago following the Mesozoic era.

Cepheid variable A star whose brightness varies periodically because it expands and contracts. A type of pulsating star.

Chemical bond A strong attractive force that exists between atoms in a substance. It involves the transfer or sharing of electrons that allows each atom to attain a full valence shell.

Chemical compound A substance formed by the chemical combination of two or more elements in definite proportions and usually having properties different from those of its constituent elements.

Chemical sedimentary rock Sedimentary rock consisting of material that was precipitated from water by either inorganic or organic means.

Chemical weathering The processes by which the internal structure of a mineral is altered by the removal and/or addition of elements.

Chinook A wind blowing down the leeward side of a mountain and warming by compression.

Chromatic aberration The property of a lens whereby light of different colors is focused at different places.

Chromosphere The first layer of the solar atmosphere found directly above the photosphere.

Cinder cone A rather small volcano built primarily of pyroclastics ejected from a single vent. Also known as a *scoria cone*.

Circle of illumination The great circle that separates daylight from darkness.

Circular orbital motion A reference to the movement of water in a wave. As a wave travels, energy is passed along by moving in a circle. The waveform advances but the water does not advance appreciably.

Circum-Pacific belt An area approximately 40,000 kilometers (24,000 miles) in length surrounding the basin of the Pacific Ocean where oceanic lithosphere is continually subducted beneath the surrounding continental plates causing most of Earth's largest earthquakes.

Cirque An amphitheater-shaped basin at the head of a glaciated valley produced by frost wedging and plucking.

Cirrus One of three basic cloud forms; also one of the three high cloud types. They are thin, delicate ice-crystal clouds often appearing as veil-like patches or thin, wispy fibers.

Clastic rock A sedimentary rock made of broken fragments of preexisting rock.

Cleavage The tendency of a mineral to break along planes of weak bonding.

Climate A description of aggregate weather conditions; the sum of all statistical weather information that helps describe a place or region.

Climate system The exchanges of energy and moisture that occur among the atmosphere, hydrosphere, solid Earth, biosphere, and cryosphere.

Climate-feedback mechanism Several different possible outcomes that may result when one of the atmosphere's elements is altered.

Climatology The scientific study of climate.

Closed system A system that is self-contained with regard to matter—that is, no matter enters or leaves.

Cloud A form of condensation best described as a dense concentration of suspended water droplets or tiny ice crystals.

Clouds of vertical development Clouds that have their bases in the low-height range but extend upward into the middle or high altitudes.

Cluster (star) A large group of stars.

Coarse-grained texture An igneous rock texture in which the crystals are roughly equal in size and large enough so that individual minerals can be identified with the unaided eye.

Coast A strip of land that extends inland from the coastline as far as ocean-related features can be found.

Coastline The coast's seaward edge. The landward limit of the effect of the highest storm waves on the shore.

Col A pass between mountain valleys where the headwalls of two cirques intersect.

Cold front A front along which a cold air mass thrusts beneath a warmer air mass.

Collision–coalescence process A theory of raindrop formation in warm clouds (above 0°C) in which large cloud droplets (giants) collide and join together with smaller droplets to form a raindrop. Opposite electrical charges may bind the cloud droplets together.

Color A phenomenon of light by which otherwise identical objects may be differentiated.

Column A feature found in caves that is formed when a stalactite and stalagmite join.

Columnar joints A pattern of cracks that form during cooling of molten rock to generate columns that are generally six sided.

Coma The fuzzy, gaseous component of a comet's head.

Comet A small body that generally revolves about the Sun in an elongated orbit.

Compaction A type of lithification in which the weight of overlying material compresses more deeply buried sediment. It is most important in the fine-grained sedimentary rocks such as shale.

Competence A measure of the largest particle a stream can transport; a factor that is dependent on velocity.

Composite cone A volcano composed of both lava flows and pyroclastic material. Also known as a *stratovolcano*.

Compound A substance formed by the chemical combination of two or more elements in definite proportions and usually having properties different from those of its constituent elements.

Compressional mountains Mountains in which great horizontal forces have shortened and thickened the crust. Most major mountain belts are of this type.

Compressional stress Differential stress that shortens a rock body.

Concordant A term used to describe intrusive igneous masses that form parallel to the bedding of the surrounding rock.

Condensation nuclei Tiny bits of particulate matter that serve as surfaces on which water vapor condenses.

Condensation The change of state from a gas to a liquid.

Conditional instability Moist air with a lapse rate between the dry and wet adiabatic rates.

Conduction The transfer of heat through matter by molecular activity. Energy is transferred through collisions from one molecule to another.

Conduit A pipelike opening through which magma moves toward Earth's surface. It terminates at a surface opening called a vent.

Cone of depression A cone-shaped depression in the water table immediately surrounding a well.

Confined aquifer An aquifer that has impermeable layers (aquitards) both above and below.

Confining pressure Stress that is applied uniformly in all directions.

Conformable Layers of rock that were deposited without interruption.

Conglomerate A sedimentary rock composed of rounded, gravel-size particles.

Constellation An apparent group of stars originally named for mythical characters. The sky is presently divided into 88 constellations.

Contact metamorphism Changes in rock caused by the heat from a nearby magma body. Also known as *thermal metamorphism*.

Continent Large, continuous areas of land that include the adjacent continental shelf and islands that are structurally connected to the mainland.

Continental (c) air mass An air mass that forms over land; it is normally relatively dry.

Continental drift A theory that originally proposed that the continents are rafted about. It has essentially been replaced by the plate tectonics theory.

Continental margin The portion of the seafloor adjacent to the continents. It may include the continental shelf, continental slope, and continental rise.

Continental rift A linear zone along which continental lithosphere stretches and pulls apart. Its creation may mark the beginning of a new ocean basin.

Continental rise The gently sloping surface at the base of the continental slope.

Continental shelf The gently sloping submerged portion of the continental margin, extending from the shoreline to the continental slope.

Continental slope The steep gradient that leads to the deep-ocean floor and marks the seaward edge of the continental shelf.

Continental volcanic arc Mountains formed in part by igneous activity associated with the subduction of oceanic lithosphere beneath a continent.

Continuous spectrum An uninterrupted band of light emitted by an incandescent solid, liquid, or gas under pressure.

Convection The transfer of heat by the movement of a mass or substance. It can take place only in fluids.

Convergence The condition that exists when the distribution of winds in a given area results in a net horizontal inflow of air into the area. Because convergence at lower levels is associated with an upward movement of air, areas of convergent winds are regions favorable to cloud formation and precipitation.

Convergent plate boundary A boundary in which two plates move together, causing one of the slabs of lithosphere to be consumed into the mantle as it descends beneath on an overriding plate.

Coral reef A structure formed in a warm, shallow, sunlit ocean environment that consists primarily of the calcite-rich remains of corals as well as the limy secretions of algae and the hard parts of many other small organisms.

Core The innermost layer of Earth, located beneath the mantle. The core is divided into an outer core and an inner core.

Coriolis force (effect) The deflective force of Earth's rotation on all free-moving objects, including the atmosphere and oceans. Deflection is to the right in the Northern Hemisphere and to the left in the Southern Hemisphere.

Corona The outer, tenuous layer of the solar atmosphere.

Correlation Establishing the equivalence of rocks of similar age in different areas.

Cosmological red shift Changes in the spectra of galaxies which indicate that they are moving away from the Milky Way as a result of the expansion of space.

Cosmology The study of the universe.

Country breeze A circulation pattern characterized by a light wind blowing into a city from the surrounding countryside. It is best developed on clear and otherwise calm nights when the urban heat island is most pronounced.

Covalent bond A chemical bond produced by the sharing of electrons.

Crater The depression at the summit of a volcano, or that which is produced by a meteorite impact.

Craton The part of the continental crust that has attained stability; that is, it has not been affected by significant tectonic activity during the Phanerozoic eon. It consists of the shield and stable platform.

Creep The slow downhill movement of soil and regolith.

Crevasse A deep crack in the brittle surface of a glacier.

Cross-bedding A structure in which relatively thin layers are inclined at an angle to the main bedding. Formed by currents of wind or water.

Cross-cutting A principle of relative dating which says that a rock or fault is younger than any rock (or fault) through which it cuts.

Crust The very thin outermost layer of Earth.

Cryovolcanism A type of volcanism that results from the eruption of magmas derived from the partial melting of ice.

Crystal An orderly arrangement of atoms.

Crystal form *See* Habit.

Crystal settling During the crystallization of magma, the settling of the earlier-formed minerals that are denser than the liquid portion to the bottom of the magma chamber.

Crystal shape *See* Habit.

Crystallization The formation and growth of a crystalline solid from a liquid or gas.

Cumulus One of three basic cloud forms; also the name given one of the clouds of vertical development. Cumulus are billowy individual cloud masses that often have flat bases.

Cup anemometer *See* Anemometer.

Curie point The temperature above which a material loses its magnetization.

Cut bank The area of active erosion on the outside of a meander.

Cutoff A short channel segment created when a river erodes through the narrow neck of land between meanders.

Cyclone A low-pressure center characterized by a counterclockwise flow of air in the Northern Hemisphere.

Daily mean temperature The mean temperature for a day, which is determined by averaging the hourly readings or, more commonly, by averaging the maximum and minimum temperatures for a day.

Daily temperature range The difference between the maximum and minimum temperatures for a day.

Dark matter Undetected matter that is thought to exist in great quantities in the universe.

Dark nebula A cloud of interstellar dust that obscures the light of more distant stars and appears as an opaque curtain.

Dark silicate mineral A silicate mineral that contains ions of iron and/or magnesium in its structure. Dark silicates are dark in color and have a higher specific gravity than nonferromagnesian silicates.

Dark-line spectrum *See* Absorption spectrum.

Daughter product An isotope that results from radioactive decay.

Debris flow A relatively rapid type of mass wasting that involves a flow of soil and regolith containing a large amount of water. Also called *mudflow*.

Declination (stellar) The angular distance north or south of the celestial equator denoting the position of a celestial body.

Decompression melting Melting that occurs as rock ascends due to a drop in confining pressure.

Deep-ocean basin The portion of seafloor that lies between the continental margin and the oceanic ridge system. This region comprises almost 30 percent of Earth's surface.

Deep-ocean trench *See* Trench.

Deep-sea fan A cone-shaped deposit at the base of the continental slope. The sediment is transported to the fan by turbidity currents that follow submarine canyons.

Deflation The lifting and removal of loose material by wind.

Deformation General term for the processes of folding, faulting, shearing, compression, or extension of rocks as the result of various natural forces.

Degenerate matter Extremely dense solar material created by electrons being displaced inward toward an atom's nucleus.

Delta An accumulation of sediment formed where a stream enters a lake or an ocean.

Dendritic pattern A stream system that resembles the pattern of a branching tree.

Density Mass per unit volume of a substance, usually expressed as grams per cubic centimeter (g/cm^3).

Deposition The process by which water vapor is changed directly to a solid without passing through the liquid state.

Desalination The removal of salts and other chemicals from seawater.

Desert One of the two types of dry climate; the driest of the dry climates. Also known as *arid*.

Desert pavement A layer of coarse pebbles and gravel created when wind removed the finer material.

Detachment fault A nearly horizontal fault that may extend hundreds of kilometers below the surface. Such a fault represents a boundary between rocks that exhibit ductile deformation and rocks that exhibit brittle deformation.

Detrital sedimentary rock Rock formed from the accumulation of material that originated and was transported in the form of solid particles derived from both mechanical and chemical weathering.

Dew-point temperature The temperature to which air has to be cooled in order to reach saturation.

Differential stress Forces that are unequal in different directions.

Differential weathering The variation in the rate and degree of weathering caused by such factors as mineral makeup, degree of jointing, and climate.

Diffused light Solar energy scattered and reflected in the atmosphere that reaches Earth's surface in the form of diffuse blue light from the sky.

Dike A tabular-shaped intrusive igneous feature that cuts through the surrounding rock.

Dip-slip fault A fault in which the movement is parallel to the dip of the fault.

Discharge The quantity of water in a stream that passes a given point in a period of time.

Disconformity A type of unconformity in which the beds above and below are parallel.

Discordant A term used to describe plutons that cut across existing rock structures, such as bedding planes.

Disseminated deposit Any economic mineral deposit in which the desired mineral occurs as scattered particles in the rock but in sufficient quantity to make the deposit an ore.

Dissolved load That portion of a stream's load that is carried in solution.

Distributary A section of a stream that leaves the main flow.

Diurnal tidal pattern A tidal pattern exhibiting one high tide and one low tide during a tidal day; a daily tide.

Divergence The condition that exists when the distribution of winds in a given area results in a net horizontal outflow of air from the region. In divergence at lower levels, the resulting deficit is compensated for by a downward movement of air from aloft; hence, areas of divergent winds are unfavorable to cloud formation and precipitation.

Divergent plate boundary A region where the rigid plates are moving apart, typified by the mid-ocean ridges. Also known as *spreading center*.

Divide An imaginary line that separates the drainage of two streams; often found along a ridge.

Dome A roughly circular upfolded structure similar to an anticline.

Doppler effect The apparent change in wavelength of radiation caused by the relative motions of the source and the observer.

Doppler radar In addition to performing the tasks of conventional radar, a new generation of weather radar that can detect motion directly and hence greatly improve tornado and severe storm warnings.

Drainage basin The land area that contributes water to a stream.

Drawdown The difference in height between the bottom of a cone of depression and the original height of the water table.

Drift *See* Glacial drift.

Drumlin A streamlined asymmetrical hill composed of glacial till. The steep side of the hill faces the direction from which the ice advanced.

Dry adiabatic rate The rate of adiabatic cooling or warming in unsaturated air. The rate of temperature change is 1°C per 100 meters.

Dry climate A climate in which yearly precipitation is not as great as the potential loss of water by evaporation. Also known as *arid climate*.

Dry-summer subtropical climate A climate located on the west sides of continents between latitudes 30° and 45°. It is the only humid climate with a strong winter precipitation maximum.

Ductile deformation A type of solid state flow that produces a change in the size and shape of a rock body without fracturing. Occurs at depths where temperatures and confining pressures are high.

Dune A hill or ridge of wind-deposited sand.

Dwarf galaxy Very small galaxies, usually elliptical and lacking spiral arms.

Dwarf planets Celestial bodies that orbit stars and are massive enough to be spherical but have not cleared their neighboring regions of planetesimals.

Earth science The name for all the sciences that collectively seek to understand Earth. It includes geology, oceanography, meteorology, and astronomy.

Earth system science An interdisciplinary study that seeks to examine Earth as a system composed of numerous interacting parts or subsystems.

Earthflow The downslope movement of water-saturated, clay-rich sediment. Most characteristic of humid regions.

Earthquake The vibration of Earth produced by the rapid release of energy.

Ebb current The movement of a tidal current away from the shore.

Eccentricity The variation of an ellipse from a circle.

Echo sounder An instrument used to determine the depth of water by measuring the time interval between emission of a sound signal and the return of its echo from the bottom.

Eclipse The cutting off of the light of one celestial body by another passing in front of it.

Ecliptic The yearly path of the Sun plotted against the background of stars.

Economic mineral A concentration of a mineral resource or reserve that can be profitably extracted from Earth.

El Niño The name given to the periodic warming of the ocean that occurs in the central and eastern Pacific. A major El Niño episode can cause extreme weather in many parts of the world.

Elastic deformation Rock deformation in which the rock will return to nearly its original size and shape when the stress is removed.

Elastic rebound The sudden release of stored strain in rocks that results in movement along a fault.

Electromagnetic radiation *See* Radiation.

Electromagnetic spectrum The distribution of electromagnetic radiation by wavelength.

Electron A negatively charged subatomic particle that has a negligible mass and is found outside an atom's nucleus.

Element A substance that cannot be decomposed into simpler substances by ordinary chemical or physical means.

Elements of weather and climate Quantities or properties of the atmosphere that are measured regularly and that are used to express the nature of weather and climate.

Elliptical galaxy A galaxy that is round or elliptical in outline. It contains little gas and dust, no disk or spiral arms, and few hot, bright stars.

Eluviation The washing out of fine soil components from the horizon by downward-percolating water.

Emergent coast A coast where land that was formerly below sea level has been exposed either because of crustal uplift or a drop in sea level or both.

Emission nebula A gaseous nebula that derives its visible light from the fluorescence of ultraviolet light from a star in or near the nebula.

Emission spectrum *See* Bright-line spectrum.

End moraine A ridge of till marking a former position of the front of a glacier.

Energy The capacity to do work.

Energy levels Spherically shaped, negatively charged zones that surround the nucleus of an atom. Also known as *principal shells*.

Enhanced Fujita intensity scale (EF-scale) A scale originally developed by T. Theodore Fujita for classifying the severity of a tornado, based on the correlation of wind speed with the degree of destruction.

Environment Everything that surrounds and influences an organism.

Environmental lapse rate The rate of temperature decrease with increasing height in the troposphere.

Eon The largest time unit on the geologic time scale, next in order of magnitude above era.

Ephemeral stream A stream that is usually dry because it carries water only in response to specific episodes of rainfall. Most desert streams are of this type.

Epicenter The location on Earth's surface that lies directly above the focus of an earthquake.

Epoch A unit of the geologic calendar that is a subdivision of a period.

Equatorial low A belt of low pressure lying near the equator and between the subtropical highs.

Equatorial system A method of locating stellar objects much like the coordinate system used on Earth's surface.

Equinox The time when the vertical rays of the Sun are striking the equator. The length of daylight and darkness is equal at all latitudes at equinox.

Era A major division on the geologic calendar; eras are divided into shorter units called periods.

Erosion The incorporation and transportation of material by a mobile agent, such as water, wind, or ice.

Eruption column Buoyant plumes of hot, ash-laden gases that can extend thousands of meters into the atmosphere.

Eruptive variable A star that varies in brightness.

Escape velocity The initial velocity an object needs to escape from the surface of a celestial body.

Esker A sinuous ridge composed largely of sand and gravel deposited by a stream flowing in a tunnel beneath a glacier near its terminus.

Estuary A partially enclosed coastal water body that is connected to the ocean. Salinity here is measurably reduced by the freshwater flow of rivers.

Eukaryotes An organism whose genetic material is enclosed in a nucleus; plants, animals, and fungi are eukaryotes.

Euphotic zone The portion of the photic zone near the surface where light is bright enough for photosynthesis to occur.

Evaporation The process of converting a liquid to a gas.

Evaporite deposits A sedimentary rock formed of material deposited from solution by evaporation of water.

Evapotranspiration The combined effect of evaporation and transpiration.

Evolution (theory of) A fundamental theory in biology and paleontology that sets forth the process by which members of a population of organisms come to differ from their ancestors. Organisms evolve by means of mutations, natural selection, and genetic factors. Modern species are descended from related but different species that lived in earlier times.

Exfoliation dome A large, dome-shaped structure, usually composed of granite, formed by sheeting.

Exotic stream A permanent stream that traverses a desert and has its source in well-watered areas outside the desert.

External process Process such as weathering, mass wasting, or erosion that is powered by the Sun and transforms solid rock into sediment.

Extrusive Igneous activity that occurs outside the crust.

Eye A zone of scattered clouds and calm averaging about 20 kilometers (12 miles) in diameter at the center of a hurricane.

Eye wall The doughnut-shaped area of intense cumulonimbus development and very strong winds that surrounds the eye of a hurricane.

Eyepiece A short-focal-length lens used to enlarge the image in a telescope. The lens nearest the eye.

Fall A type of movement common to mass-wasting processes that refers to the free falling of detached individual pieces of any size.

Fault A break in a rock mass along which movement has occurred.

Fault creep Displacement along a fault that is so slow and gradual that little seismic activity occurs.

Fault scarp A cliff created by movement along a fault. It represents the exposed surface of the fault prior to modification by weathering and erosion.

Fault-block mountain A mountain formed by the displacement of rock along a fault.

Felsic The group of igneous rocks composed primarily of feldspar and quartz.

Fetch The distance that wind has traveled across open water. It is one of three factors that influence the height, length, and period of a wave.

Filaments Dark, thin streaks that appear across the bright solar disk.

Fine-grained texture A texture of igneous rocks in which the crystals are too small for individual minerals to be distinguished with the unaided eye.

Fiord A steep-sided inlet of the sea formed when a glacial trough was partially submerged.

Fissure A crack in rock along which there is a distinct separation.

Fissure eruption An eruption in which lava is extruded from narrow fractures or cracks in the crust.

Flare A sudden brightening of an area on the Sun.

Flood basalts Flows of basaltic lava that issue from numerous cracks or fissures and commonly cover extensive areas to thicknesses of hundreds of meters.

Flood current The tidal current associated with the increase in the height of the tide.

Floodplain The flat, low-lying portion of a stream valley subject to periodic inundation.

Flow A type of movement common to mass-wasting processes in which water-saturated material moves downslope as a viscous fluid.

Fluorescence The absorption of ultraviolet light, which is reemitted as visible light.

Focal length The distance from a lens to the point where it focuses parallel rays of light.

Focus (earthquake) The zone within Earth where rock displacement produces an earthquake. Also known as the *hypocenter*.

Focus (light) The point where a lens or mirror causes light rays to converge.

Fog A cloud with its base at or very near Earth's surface.

Fold A bent rock layer or series of layers that were originally horizontal and subsequently deformed.

Foliation A texture of metamorphic rocks that gives the rock a layered appearance.

Food chain A succession of organisms in an ecological community through which food energy is transferred from producers through herbivores and on to one or more carnivores.

Food web A group of interrelated food chains.

Footwall block The rock surface below a fault.

Forearc basin The region located between a volcanic arc and an accretionary wedge where shallow-water marine sediments typically accumulate.

Foreshocks Small earthquakes that often precede a major earthquake.

Foreshore That portion of the shore lying between the normal high and low water marks; the intertidal zone.

Fossil The remains or traces of organisms preserved from the geologic past.

Fossil assemblage The overlapping ranges of a group of fossils (assemblage) collected from a layer. By examining such an assemblage, the age of the sedimentary layer can be established.

Fossil fuel General term for any hydrocarbon that may be used as a fuel, including coal, oil, natural gas, bitumen from tar sands, and shale oil.

Fossil magnetism See *Paleomagnetism.*

Fossil succession A principle in which fossil organisms succeed one another in a definite and determinable order, so any time period can be recognized by its fossil content.

Fracture zone Any break or rupture in rock along which no appreciable movement has taken place.

Fragmental texture See *Pyroclastic texture.*

Freezing The change of state from a liquid to a solid.

Freezing nuclei Solid particles that serve as cores for the formation of ice crystals.

Freezing rain See Glaze.

Front The boundary between two adjoining air masses having contrasting characteristics.

Frontal fog Fog formed when rain evaporates as it falls through a layer of cool air.

Frontal wedging Lifting of air resulting when cool air acts as a barrier over which warmer, lighter air will rise.

Frost wedging The mechanical breakup of rock caused by the expansion of freezing water in cracks and crevices.

Fumarole A vent in a volcanic area from which fumes or gases escape.

Galactic cluster Groups of gravitationally bound galaxies that sometimes contain thousands of galaxies.

Geocentric The concept of an Earth-centered universe.

Geologic structure See Rock structure.

Geologic time The span of time since the formation of Earth, about 4.6 billion years.

Geologic time scale The division of Earth history into blocks of time—eons, eras, periods, and epochs. The time scale was created using relative dating principles.

Geology The science that examines Earth, its form and composition, and the changes it has undergone and is undergoing.

Geosphere The solid Earth, the largest of Earth's four major spheres.

Geostrophic wind A wind, usually above a height of 600 meters (2000 feet), that blows parallel to the isobars.

Geothermal energy Natural steam used for power generation.

Geothermal gradient The gradual increase in temperature with depth in the crust. The average is 30°C per kilometer in the upper crust.

Geyser A fountain of hot water ejected periodically.

Giant (star) A luminous star of large radius.

Glacial budget The balance, or lack of balance, between ice formation at the upper end of a glacier and ice loss in the zone of wastage.

Glacial drift An all-embracing term for sediments of glacial origin, no matter how, where, or in what shape they were deposited. Also known simply as *drift*.

Glacial erratic An ice-transported boulder that was not derived from bedrock near its present site.

Glacial striations Scratches and grooves on bedrock caused by glacial abrasion.

Glacial trough A mountain valley that has been widened, deepened, and straightened by a glacier.

Glacier A thick mass of ice originating on land from the compaction and recrystallization of snow that shows evidence of past or present flow.

Glassy texture A term used to describe the texture of certain igneous rocks, such as obsidian, that contain no crystals.

Glaze A coating of ice on objects formed when supercooled rain freezes on contact.

Globular cluster A nearly spherically shaped group of densely packed stars.

Globule A dense, dark nebula thought to be the birthplace of stars. Also known as glaze.

Gondwanaland The southern portion of Pangaea, consisting of South America, Africa, Australia, India, and Antarctica.

Graben A valley formed by the downward displacement of a fault-bounded block.

Graded bed A sediment layer that is characterized by a decrease in sediment size from bottom to top.

Gradient The slope of a stream; generally measured in feet per mile.

Granitic composition A compositional group of igneous rocks that indicates a rock is composed almost entirely of light-colored silicates.

Granule A fine structure visible on the solar surface caused by convective cells below.

Gravitational collapse The gradual subsidence of mountains caused by lateral spreading of weak material located deep within these structures.

Great Oxygenation Event A time about 2.5 billion years ago, when a significant amount of oxygen appeared in the atmosphere.

Greenhouse effect The transmission of short-wave solar radiation by the atmosphere, coupled with the selective absorption of longer-wavelength terrestrial radiation, especially by water vapor and carbon dioxide.

Groin A short wall built at a right angle to the shore to trap moving sand.

Ground moraine An undulating layer of till deposited as the ice front retreats.

Groundmass The matrix of smaller crystals within an igneous rock that has porphyritic texture.

Groundwater Water in the zone of saturation.

Guyot A submerged flat-topped seamount.

Gymnosperm A group of seed-bearing plants that includes conifers and Ginkgo. The term means "naked seed," a reference to the unenclosed condition of the seeds.

Gyre The large circular surface current pattern found in each ocean.

Habit Refers to the common or characteristic shape of a crystal, or aggregate of crystals. Also known as *crystal form* and *crystal shape*.

Hadean eon A term found on some versions of the geologic time scale. It refers to the earliest interval (eon) of Earth history and ended 4 billion years ago.

Hail Nearly spherical ice pellets having concentric layers and formed by the successive freezing of layers of water.

Half graben A tilted fault block in which the higher side is associated with mountainous topography and the lower side is a basin that fills with sediment.

Half-life The time required for one-half of the atoms of a radioactive substance to decay.

Halocline A layer of water in which there is a high rate of change in salinity in the vertical dimension.

Hanging valley A tributary valley that enters a glacial trough at a considerable height above its floor.

Hanging wall block The rock surface immediately above a fault.

Hard stabilization Any form of artificial structure built to protect a coast or to prevent the movement of sand along a beach. Examples include groins, jetties, breakwaters, and seawalls.

Hardness The resistance a mineral offers to scratching.

Heat The kinetic energy of random molecular motion.

Heliocentric The view that the Sun is at the center of the solar system.

Heliosphere A large region of space that extends far beyond Pluto's orbit, marked by solar winds and the Sun's magnetic field.

Hertzsprung-Russell diagram *See* H-R diagram.

High A center of high pressure characterized by anticyclonic winds.

High cloud A cloud that normally has its base above 6000 meters (3,728 miles); the base may be lower in winter and at high-latitude locations.

Highland climate A complex pattern of climate conditions associated with mountains. Highland climates are characterized by large differences that occur over short distances.

Hogback A narrow, sharp-crested ridge formed by the upturned edge of a steeply dipping bed of resistant rock.

Horizon A layer in a soil profile. Also known as the soil horizon.

Horn A pyramid-like peak formed by glacial action in three or more cirques surrounding a mountain summit.

Horst An elongated, uplifted block of crust bounded by faults.

Hot spot A concentration of heat in the mantle capable of producing magma, which in turn extrudes onto Earth's surface. The intraplate volcanism that produced the Hawaiian islands is one example.

Hot spot track A chain of volcanic structures produced as a lithospheric plate moves over a mantle plume.

Hot spring A spring in which the water is 6–9°C (10–15°F) warmer than the mean annual air temperature of its locality.

H-R diagram A plot of stars according to their absolute magnitudes and spectral types. Stands for *Hertzsprung-Russell diagram*.

Hubble's law A law that relates the distance to a galaxy and its velocity.

Humid continental climate A relatively severe climate characteristic of broad continents in the middle latitudes between approximately 40° and 50° north latitude. This climate is not found in the Southern Hemisphere, where the middle latitudes are dominated by the oceans.

Humid subtropical climate A climate generally located on the eastern side of a continent and characterized by hot, sultry summers and cool winters.

Humidity A general term referring to water vapor in the air but not to liquid droplets of fog, cloud, or rain.

Humus Organic matter in soil produced by the decomposition of plants and animals.

Hurricane A tropical cyclonic storm having winds in excess of 119 kilometers (74 miles) per hour.

Hydraulic fracturing A method of opening up pore space in otherwise impermeable rocks, permitting natural gas to flow out into wells.

Hydrogen burning The conversion of hydrogen through fusion to form helium.

Hydrogen fusion The nuclear reaction in which hydrogen nuclei are fused into helium nuclei.

Hydrogenous sediment Seafloor sediments consisting of minerals that crystallize from seawater. An important example is manganese nodules.

Hydrosphere The water portion of our planet; one of the traditional subdivisions of Earth's physical environment.

Hydrothermal solution The hot, watery solution that escapes from a mass of magma during the later stages of crystallization. Such solutions may alter the surrounding country rock and are frequently the source of significant ore deposits.

Hygrometer An instrument designed to measure relative humidity.

Hygroscopic nuclei Condensation nuclei having a high affinity for water, such as salt particles.

Hypocenter *See* Focus (earthquake).

Hypothesis A tentative explanation that is tested to determine whether it is valid.

Ice cap A mass of glacial ice covering a high upland or plateau and spreading out radially.

Ice cap climate A climate that has no monthly means above freezing and supports no vegetative cover except in a few scattered high mountain areas. This climate, with its perpetual ice and snow, is confined largely to the ice sheets of Greenland and Antarctica.

Ice sheet A very large, thick mass of glacial ice flowing outward in all directions from one or more accumulation centers.

Ice shelf A large, relatively flat mass of floating ice that forms where glacial ice flows into bays and extends seaward from the coast but remains attached to the land along one or more sides.

Iceberg A mass of floating ice produced by a calving glacier. Usually 20 percent or less of the iceberg protrudes above the waterline.

Igneous rock A rock formed by the crystallization of molten magma.

Immature soil A soil lacking horizons.

Impact craters Depressions result from collisions with bodies such as asteroids and comets.

Incised meander A meandering channel that flows in a steep, narrow valley. Incised meanders form either when an area is uplifted or when base level drops.

Inclination of the axis The tilt of Earth's axis from the perpendicular to the plane of Earth's orbit.

Inclusion A piece of one rock unit contained within another. Inclusions are used in relative dating. The rock mass adjacent to the one containing the inclusion must have been there first in order to provide the fragment.

Index fossil A fossil that is associated with a particular span of geologic time.

Inertia A property of matter that resists a change in its motion.

Infiltration The movement of surface water into rock or soil through cracks and pore spaces.

Infrared Radiation with a wavelength from 0.7 to 200 micrometers.

Inner core The solid innermost layer of Earth, about 1300 kilometers (800 miles) in radius.

Inner planets *See* Terrestrial planets.

Inselberg An isolated mountain remnant characteristic of the late stage of erosion in an arid region.

Intensity (earthquake) A measure of the degree of earthquake shaking at a given locale, based on the amount of damage.

Interface A common boundary where different parts of a system interact.

Interior drainage A discontinuous pattern of intermittent streams that do not flow to the ocean.

Intermediate composition The composition of igneous rocks lying between felsic and mafic. Also known as andesitic composition.

Interstellar matter Dust and gases found between stars.

Intertidal zone The area where land and sea meet and overlap; the zone between high and low tides.

Intertropical convergence zone (ITCZ) The zone of general convergence between the Northern and Southern Hemisphere trade winds.

Intraplate volcanism Igneous activity that occurs within a tectonic plate away from plate boundaries.

Intrusion *See* Pluton.

Intrusive Igneous rock that formed below Earth's surface.

Ion An atom or a molecule that possesses an electrical charge.

Ionic bond A chemical bond between two oppositely charged ions formed by the transfer of valence electrons from one atom to the other.

Ionosphere A complex zone of ionized gases that coincides with the lower portion of the thermosphere.

Iron meteorite One of the three main categories of meteorites. This group is composed largely of iron with varying amounts of nickel (5–20 percent). Most meteorite finds are irons.

Irregular galaxy A galaxy that lacks symmetry.

Island arc *See* Volcanic island arc.

Isobar A line drawn on a map connecting points of equal atmospheric pressure, usually corrected to sea level.

Isostasy The concept that Earth's crust is floating in gravitational balance on the material of the mantle.

Isostatic adjustment Compensation of the lithosphere when weight is added or removed. When weight is added, the lithosphere responds by subsiding, and when weight is removed, there is uplift.

Isotherms Lines connecting points of equal temperature.

Isotopes Varieties of the same element that have different mass numbers; their nuclei contain the same number of protons but different numbers of neutrons.

Jet stream Swift (120–240 kilometers per hour), high-altitude winds.

Jetties A pair of structures extending into the ocean at the entrance to a harbor or river that are built for the purpose of protecting against storm waves and sediment deposition.

Joint A fracture in rock along which there has been no movement.

Jovian planets The Jupiter-like planets: Jupiter, Saturn, Uranus, and Neptune. These planets have relatively low densities. Also known as the *outer planets*.

Kame A steep-sided hill composed of sand and gravel that originates when sediment is collected in openings in stagnant glacial ice.

Karst A topography consisting of numerous depressions called *sinkholes*.

Kettle holes Depressions created when blocks of ice became lodged in glacial deposits and subsequently melted.

Köppen classification A system for classifying climates devised by Wladimir Köppen that is based on mean monthly and annual values of temperature and precipitation.

Kuiper belt A region outside the orbit of Neptune where most short-period comets are thought to originate.

La Niña An episode of strong trade winds and unusually low sea-surface temperatures in the central and eastern Pacific. The opposite of *El Niño*.

Laccolith A massive igneous body intruded between preexisting strata.

Lahar Mudflows on the slopes of volcanoes that result when unstable layers of ash and debris become saturated and flow downslope, usually following stream channels.

Lake-effect snow Snow showers associated with a cP air mass to which moisture and heat are added from below as the air mass traverses a large and relatively warm lake (such as one of the Great Lakes), rendering the air mass humid and unstable.

Laminar flow The movement of water particles in straight-line paths that are parallel to the channel. The water particles move downstream, without mixing.

Land breeze A local wind blowing from land toward the water during the night in coastal areas.

Lapse rate (normal) The average drop in temperature (6.5°C per kilometer [3.5°F per 1000 feet]) with increased altitude in the troposphere.

Latent heat The energy absorbed or released during a change in state.

Lateral continuity (principle of) A principle which states that sedimentary beds originate as continuous layers that extend in all directions until they grade into a different type of sediment or thin out at the edge of a sedimentary basin.

Lateral moraine A ridge of till along the sides of an alpine glacier composed primarily of debris that fell to the glacier from the valley walls.

Laurasia The northern portion of Pangaea, consisting of North America and Eurasia.

Lava Magma that reaches Earth's surface.

Lava tube A tunnel in hardened lava that acts as a horizontal conduit for lava flowing from a volcanic vent. Lava tubes allow fluid lavas to advance great distances.

Law of conservation of angular momentum The product of the velocity of an object around a center of rotation (axis), and the distance squared of the object from the axis is constant.

Leaching The depletion of soluble materials from the upper soil by downward-percolating water.

Light silicate mineral A silicate mineral that lacks iron and/or magnesium. Light silicates are generally lighter in color and have lower specific gravities than dark silicates.

Lightning A sudden flash of light generated by the flow of electrons between oppositely charged parts of a cumulonimbus cloud or between the cloud and the ground.

Light-year The distance light travels in a year; about 6 trillion miles.

Liquefaction A phenomenon, sometimes associated with earthquakes, in which soils and other unconsolidated materials containing abundant water are turned into a fluid-like mass that is not capable of supporting buildings.

Lithification The process, generally cementation and/or compaction, of converting sediments to solid rock.

Lithosphere The rigid outer layer of Earth, including the crust and upper mantle.

Lithospheric plate A coherent unit of Earth's rigid outer layer that includes the crust and upper unit. Also known simply as a *plate*.

Local Group The cluster of 20 or so galaxies to which our galaxy belongs.

Local wind A small-scale wind produced by a locally generated pressure gradient. Examples include land and sea breezes and mountain and valley breezes.

Localized convective lifting Unequal surface heating that causes localized pockets of air (thermals) to rise because of their buoyancy.

Loess Deposits of windblown silt, lacking visible layers, generally buff-colored, and capable of maintaining a nearly vertical cliff.

Longitudinal (seif) dunes Long ridges of sand oriented parallel to the prevailing wind; these dunes form where sand supplies are limited.

Longshore current A nearshore current that flows parallel to the shore.

Low A center of low pressure characterized by cyclonic winds.

Low cloud A cloud that forms below a height of 2000 meters (1,200 miles).

Lower mantle The part of the mantle that extends from the core–mantle boundary to a depth of 660 kilometers (400 miles).

Low-velocity zone *See* Asthenosphere.

Luminosity The brightness of a star. The amount of energy radiated by a star.

Lunar breccia A lunar rock formed when angular fragments and dust are welded together by the heat generated by the impact of a meteoroid.

Lunar eclipse An eclipse of the Moon.

Lunar highlands *See* Terrae.

Lunar regolith A thin, gray layer on the surface of the Moon, consisting of loosely compacted, fragmented material believed to have been formed by repeated meteoritic impacts.

Luster The appearance or quality of light reflected from the surface of a mineral.

Mafic Igneous rocks with a low silica content and a high iron–magnesium content.

Magma A body of molten rock found at depth, including any dissolved gases and crystals.

Magmatic differentiation The process of generating more than one rock type from a single magma.

Magnetic reversal A change in Earth's magnetic field from normal to reverse or vice versa.

Magnetic time scale A scale that shows the ages of magnetic reversals and is based on the polarity of lava flows of various ages.

Magnetometer A sensitive instrument used to measure the intensity of Earth's magnetic field at various points.

Magnitude (earthquake) The total amount of energy released during an earthquake.

Magnitude (stellar) A number given to a celestial object to express its relative brightness.

Main-sequence stars A sequence of stars on the Hertzsprung-Russell diagram, containing the majority of stars, that runs diagonally from the upper left to the lower right.

Manganese nodules Rounded lumps of hydrogenous sediment scattered on the ocean floor, consisting mainly of manganese and iron and usually containing small amounts of copper, nickel, and cobalt.

Mantle The 2900-kilometer- (1800-mile-) thick layer of Earth located below the crust.

Mantle plume A mass of hotter-than-normal mantle material that ascends toward the surface, where it may lead to igneous activity. These plumes of solid yet mobile material may originate as deep as the core–mantle boundary.

Maria The Latin name for the smooth areas of the Moon formerly thought to be seas.

Marine terrace A wave-cut platform that has been exposed above sea level.

Marine west coast climate A climate found on windward coasts from latitudes 40° to 65° and dominated by maritime air masses. Winters are mild and summers are cool.

Maritime (m) air mass An air mass that originates over the ocean. These air masses are relatively humid.

Mass extinction An event in which a large percentage of species become extinct.

Mass number The number of neutrons and protons in the nucleus of an atom.

Mass wasting The downslope movement of rock, regolith, and soil under the direct influence of gravity.

Massive An igneous pluton that is not tabular in shape.

Mean solar day The average time between two passages of the Sun across the local celestial meridian.

Meander A looplike bend in the course of a stream.

Mechanical weathering The physical disintegration of rock, resulting in smaller fragments.

Medial moraine A ridge of till formed when lateral moraines from two coalescing alpine glaciers join.

Megathrust fault The plate boundary separating a subducting slab of oceanic lithosphere and the overlying plate.

Melt The liquid portion of magma, excluding the solid crystals.

Melting The change of state from a solid to a liquid.

Mercalli intensity scale *See* Modified Mercalli intensity scale.

Mercury barometer A mercury-filled glass tube in which the height of the mercury column is a measure of air pressure.

Mesocyclone A vertical cylinder of cyclonically rotating air (3 to 10 kilometers in diameter) that develops in the updraft of a severe thunderstorm and that often precedes the development of damaging hail or tornadoes.

Mesopause The boundary between the mesosphere and the thermosphere.

Mesosphere The layer of the atmosphere immediately above the stratosphere and characterized by decreasing temperatures with height.

Mesozoic era A span on the geologic time scale between the Paleozoic and Cenozoic eras from about 248 million to 65 million years ago.

Metallic bond A chemical bond present in all metals that may be characterized as an extreme type of electron sharing in which the electrons move freely from atom to atom.

Metamorphic rock Rocks formed by the alteration of preexisting rock deep within Earth (but still in the solid state) by heat, pressure, and/or chemically active fluids.

Metamorphism The changes in mineral composition and texture of a rock subjected to high temperature and pressure within Earth.

Meteor The luminous phenomenon observed when a meteoroid enters Earth's atmosphere and burns up; popularly called a "shooting star."

Meteor shower Many meteors appearing in the sky, caused by Earth intercepting a swarm of meteoritic particles.

Meteorite Any portion of a meteoroid that survives its traverse through Earth's atmosphere and strikes Earth's surface.

Meteoroid Small solid particles that have orbits in the solar system.

Meteorology The scientific study of the atmosphere and atmospheric phenomena; the study of weather and climate.

Microcontinents Relatively small fragments of continental crust that may lie above sea level, such as the island of Madagascar, or may be submerged, as exemplified by the Campbell Plateau located near New Zealand.

Middle cloud A cloud occupying the height range from 2000 to 6000 meters.

Midlatitude (middle-latitude) cyclone A large center of low pressure with an associated cold front and often a warm front. Frequently accompanied by abundant precipitation.

Mid-ocean ridge A continuous mountainous ridge on the floor of all the major ocean basins and varying in width from 500 to 5000 kilometers (300 to 3000 miles). The rifts at the crests of these ridges represent divergent plate boundaries.

Mineral A naturally occurring, inorganic crystalline material with a unique chemical composition.

Mineral resource All discovered and undiscovered deposits of a useful mineral that can be extracted now or at some time in the future.

Mineralogy The study of minerals.

Mixed tidal pattern A tidal pattern exhibiting two high tides and two low tides per tidal day, with a large inequality in high water heights, low water heights, or both. Coastal locations that experience such a tidal pattern may also show alternating periods of diurnal and semidiurnal tidal patterns. Also called mixed semidiurnal.

Mixing depth The height to which convectional movements extend above Earth's surface. The greater the mixing depth, the better the air quality.

Mixing ratio The mass of water vapor in a unit mass of dry air; commonly expressed as grams of water vapor per kilogram of dry air.

Model A term often used synonymously with hypothesis but that is less precise because it is sometimes used to describe a theory as well.

Modified Mercalli intensity scale A 12-point scale developed to evaluate earthquake intensity based on the amount of damage to various structures.

Mohorovičić discontinuity (Moho) The boundary separating the crust from the mantle, discernible by an increase in seismic velocity.

Mohs scale A series of 10 minerals used as a standard in determining hardness.

Moment magnitude A more precise measure of earthquake magnitude than the Richter scale that is derived from the amount of displacement that occurs along a fault zone.

Monocline A one-limbed flexure in strata. The strata are unusually flat-lying or very gently dipping on both sides of the monocline.

Monsoon Seasonal reversal of wind direction associated with large continents, especially Asia. In winter, the wind blows from land to sea; in summer, from sea to land.

Monthly mean temperature The mean temperature for a month that is calculated by averaging the daily means.

Mountain belt A geographic area of roughly parallel and geologically connected mountain ranges developed as a result of plate tectonics.

Mountain breeze The nightly downslope winds commonly encountered in mountain valleys.

Natural levees The elevated landforms that parallel some streams and act to confine their waters, except during floodstage.

Neap tide The lowest tidal range, which occurs near the times of the first- and third-quarter phases of the Moon.

Nearshore zone The zone of beach that extends from the low-tide shoreline seaward to where waves break at low tide.

Nebula A cloud of interstellar gas and/or dust.

Nebular theory The basic idea that the Sun and planets formed from the same cloud of gas and dust in interstellar space.

Negative-feedback mechanism A feedback mechanism that tends to maintain a system as it is—that is, maintain the status quo.

Nekton Pelagic organisms that can move independently of ocean currents by swimming or other means of propulsion.

Neritic zone The marine-life zone that extends from the low tideline out to the shelf break.

Neutron A subatomic particle found in the nucleus of an atom. A neutron is electrically neutral and has a mass approximately that of a proton.

Neutron star A star of extremely high density, composed entirely of neutrons.

Nonconformity An unconformity in which older metamorphic or intrusive igneous rocks are overlain by younger sedimentary strata.

Nonfoliated texture Metamorphic rocks that do not exhibit foliation.

Nonmetallic mineral resource A mineral resource that is not a fuel or processed for the metals it contains.

Nonrenewable resource A resource that forms or accumulates over such long time spans that it must be considered as fixed in total quantity.

Nonsilicates Mineral groups that lack silicas in their structures and account for less than 10 percent of Earth's crust.

Nor'easter The term used to describe the weather associated with an incursion of *mP* air from the North Atlantic into the Northeast and Mid-Atlantic regions; strong northeast winds, freezing or near-freezing temperatures, and the possibility of precipitation make this an unwelcome weather event.

Normal fault A fault in which the rock above the fault plane has moved down relative to the rock below.

Normal polarity A magnetic field that is the same as that which exists at present.

Nova A star that explosively increases in brightness.

Nuclear fusion The source of the Sun's energy.

Nucleus The small heavy core of an atom that contains all of its positive charge and most of its mass.

Nuée ardente Incandescent volcanic debris buoyed up by hot gases that moves downslope in an avalanche fashion.

Numerical date A date that specifies the actual number of years that have passed since an event occurred.

Obliquity The angle between the planes of Earth's equator and orbit.

Obsidian A volcanic glass of felsic composition.

Occluded front A front formed when a cold front overtakes a warm front. It marks the beginning of the end of a middle-latitude cyclone.

Occlusion The overtaking of one front by another.

Occultation An eclipse of a star or planet by the Moon or a planet.

Ocean basin A deep submerged region that lies beyond the continental margins.

Oceanic plateau An extensive region on the ocean floor composed of thick accumulations of pillow basalts and other mafic rocks that in some cases exceed 30 kilometers (20 miles) in thickness.

Oceanic ridge system A continuous elevated zone on the floor of all the major ocean basins and varying in width from 500 to 5000 kilometers (300–3000 miles). The rifts at the crests of ridges represent divergent plate boundaries.

Oceanic rise *See* Mid-ocean ridge.

Oceanic zone The marine-life zone beyond the continental shelf.

Oceanography The scientific study of the oceans and oceanic phenomena.

Octet rule A rule which says that atoms combine in order that each may have the electron arrangement of a noble gas; that is, the outer energy level contains eight neutrons.

Offshore zone The relatively flat submerged zone that extends from the breaker line to the edge of the continental shelf.

Oil trap A geologic structure that allows for significant amounts of oil and gas to accumulate.

Oort cloud A spherical shell composed of comets that orbit the Sun at distances generally greater than 10,000 times the Earth–Sun distance.

Open cluster A loosely formed group of stars of similar origin.

Open system A system in which both matter and energy flow into and out of the system. Most natural systems are of this type.

Orbit The path of a body in revolution around a center of mass.

Ore Usually a useful metallic mineral that can be mined at a profit. The term is also applied to certain nonmetallic minerals such as fluorite and sulfur.

Ore deposit A naturally occurring concentration of one or more metallic minerals that can be extracted economically.

Organic matter Material composed of organic compounds consisting of the remains of once-living plants and animals and their waste products in the environment.

Original horizontality Layers of sediments are generally deposited in a horizontal or nearly horizontal position.

Orogenesis The processes that collectively result in the formation of mountains.

Orographic lifting Mountains acting as barriers to the flow of air, forcing the air to ascend. The air cools adiabatically, and clouds and precipitation may result.

Outer core A layer beneath the mantle about 2200 kilometers (1364 miles) thick that has the properties of a liquid.

Outer planets *See* Jovian planets.

Outgassing The escape of gases that had been dissolved in magma.

Outlet glacier A tongue of ice that normally flows rapidly outward from an ice cap or ice sheet, usually through mountainous terrain to the sea.

Outwash plain A relatively flat, gently sloping plain consisting of materials deposited by meltwater streams in front of the margin of an ice sheet.

Overrunning Warm air gliding up a retreating cold air mass.

Oxbow lake A curved lake produced when a stream cuts off a meander.

Ozone A molecule of oxygen that contains three oxygen atoms.

Pahoehoe flow A lava flow with a smooth-to-ropey surface.

Paleomagnetism The natural remnant magnetism in rock bodies. The permanent magnetization acquired by rock that can be used to determine the location of the magnetic poles and the latitude of the rock at the time it became magnetized. Also known as *fossil magnetism*.

Paleontology The systematic study of fossils and the history of life on Earth.

Paleozoic era A span on the geologic time scale between the eons of the Precambrian and Mesozoic era from about 540 million to 248 million years ago.

Pangaea The proposed supercontinent that 200 million years ago began to break apart and form the present landmasses.

Parabolic dunes Dunes that resemble barchans, except that their tips point into the wind; they often form along coasts that have strong onshore winds, abundant sand, and vegetation that partly covers the sand.

Paradigm A theory that is held with a very high degree of confidence and is comprehensive in scope.

Parallax The apparent shift of an object when viewed from two different locations.

Parasitic cone A volcanic cone that forms on the flank of a larger volcano.

Parcel An imaginary volume of air enclosed in a thin elastic cover. Typically it is considered to be a few hundred cubic meters in volume and is assumed to act independently of the surrounding air.

Parent material The material on which a soil develops.

Parsec The distance at which an object would have a parallax angle of 1 second of arc (3.26 light-years).

Partial melting The process by which most igneous rocks melt. Since individual minerals have different melting points, most igneous rocks melt over a temperature range of a few hundred degrees. If the liquid is squeezed out after some melting has occurred, a melt with a higher silica content results.

Passive continental margin A margin that consists of a continental shelf, continental slope, and continental rise. These margins are *not* associated with plate boundaries and therefore experience little volcanism and few earthquakes.

Pegmatite A very coarse-grained igneous rock (typically granite) commonly found as a dike associated with a large mass of plutonic rock that has smaller crystals. Crystallization in a water-rich environment is believed to be responsible for the very large crystals.

Pelagic zone Open ocean of *any* depth. Animals in this zone swim or float freely.

Penumbra The portion of a shadow from which only part of the light source is blocked by an opaque body.

Perched water table A localized zone of saturation above the main water table created by an impermeable layer (aquiclude).

Peridotite An igneous rock of ultramafic composition thought to be abundant in the upper mantle.

Perihelion The point in the orbit of a planet where it is closest to the Sun.

Period A basic unit of the geologic calendar that is a subdivision of an era. Periods may be divided into smaller units called epochs.

Periodic table The tabular arrangement of the elements according to atomic number.

Permeability A measure of a material's ability to transmit water.

Perturbation The gravitational disturbance of the orbit of one celestial body by another.

pH scale A common measure of the degree of acidity or alkalinity of a solution, it is a logarithmic scale ranging from 0 to 14. A value of 7 denotes a neutral solution, values below 7 indicate greater acidity, and numbers above 7 indicate greater alkalinity.

Phanerozoic eon That part of geologic time represented by rocks containing abundant fossil evidence. The eon extending from the end of the Proterozoic eon (about 540 million years ago) to the present.

Phases of the Moon The progression of changes in the Moon's appearance during the month.

Phenocryst In an igneous rock with a porphyritic texture, a conspicuously large crystal embedded in a matrix of finer-grained crystals called the groundmass.

Photic zone The upper part of the ocean into which any sunlight penetrates.

Photochemical reaction A chemical reaction in the atmosphere that is triggered by sunlight, often yielding a secondary pollutant.

Photon A discrete amount (quantum) of electromagnetic energy.

Photosphere The region of the Sun that radiates energy to space. The visible surface of the Sun.

Photosynthesis The process by which plants and algae produce carbohydrates from carbon dioxide and water in the presence of chlorophyll, using light energy and releasing oxygen.

Physical environment The part of the environment that encompasses water, air, soil, and rock, as well as conditions such as temperature, humidity, and sunlight.

Phytoplankton Algal plankton, which are the most important community of primary producers in the ocean.

Piedmont glacier A glacier that forms when one or more valley glaciers emerge from the confining walls of mountain valleys and spread out to create a broad sheet in the lowlands at the base of the mountains.

Pillow lava Basaltic lava that solidifies in an underwater environment and develops a structure that resembles a pile of pillows.

Pipe A vertical conduit through which magmatic materials have passed.

Placer A deposit formed when heavy minerals are mechanically concentrated by currents, most commonly streams and waves. Placers are sources of gold, tin, platinum, diamonds, and other valuable minerals.

Plane of the ecliptic The imaginary plane that connects Earth's orbit with the celestial sphere.

Planetary nebula A shell of incandescent gas expanding from a star.

Planetesimal A solid celestial body that accumulated during the first stages of planetary formation. Planetesimals aggregated into increasingly larger bodies, ultimately forming the planets.

Plankton Passively drifting or weakly swimming organisms that cannot move independently of ocean currents. Includes microscopic algae, protozoa, jellyfish, and larval forms of many animals.

Plate *See* Lithospheric plate.

Plate tectonics The theory which proposes that Earth's outer shell consists of individual plates that interact in various ways and thereby produce earthquakes, volcanoes, mountains, and the crust itself.

Playa A flat area on the floor of an undrained desert basin. Following heavy rain, the playa becomes a lake.

Playa lake A temporary lake in a playa.

Pleistocene epoch An epoch of the Quaternary period beginning about 1.8 million years ago and ending about 10,000 years ago. Best known as a time of extensive continental glaciation.

Plucking (quarrying) The process by which pieces of bedrock are lifted out of place by a glacier.

Plug *See* Volcanic neck.

Pluton A structure that results from the emplacement and crystallization of magma beneath the surface of Earth. Also known as an *intrusion*.

Pluvial lake A lake formed during a period of increased rainfall. During the Pleistocene epoch, this occurred in some nonglaciated regions during periods of ice advance elsewhere.

Point bar A crescent-shaped accumulation of sand and gravel deposited on the inside of a meander.

Polar (P) air mass A cold air mass that forms in a high-latitude source region.

Polar easterlies In the global pattern of prevailing winds, winds that blow from the polar high toward the subpolar low. These winds, however, should not be thought of as persistent winds, such as the trade winds.

Polar front The stormy frontal zone separating air masses of polar origin from air masses of tropical origin.

Polar high Anticyclones that are assumed to occupy the inner polar regions and are believed to be thermally induced, at least in part.

Polar wandering As a result of paleomagnetic studies in the 1950s, researchers proposed that either the magnetic poles migrated greatly through time or the continents had gradually shifted their positions.

Population I Stars rich in atoms heavier than helium. Nearly always relatively young stars found in the disk of the galaxy.

Population II Stars poor in atoms heavier than helium. Nearly always relatively old stars found in the halo, globular clusters, or nuclear bulge.

Porosity The volume of open spaces in rock or soil.

Porphyritic texture An igneous texture consisting of large crystals embedded in a matrix of much smaller crystals.

Positive-feedback mechanism A feedback mechanism that enhances or drives change.

Pothole A circular depression in a bedrock stream channel created by the abrasive action of particles swirling in fast-moving eddies.

Precambrian All geologic time prior to the Paleozoic era.

Precession *See* Axial precession.

Precipitation fog Fog formed when rain evaporates as it falls through a layer of cool air.

Pressure gradient The amount of pressure change occurring over a given distance.

Pressure tendency The nature of the change in atmospheric pressure over the past several hours. It can be a useful aid in short-range weather prediction. Also known as *barometric tendency*.

Prevailing wind A wind that consistently blows from one direction more than from another.

Primary (P) wave A type of seismic wave that involves alternating compression and expansion of the material through which it passes.

Primary pollutants Pollutants emitted directly from identifiable sources.

Primary productivity The amount of organic matter synthesized by organisms from inorganic substances through photosynthesis or chemosynthesis within a given volume of water or habitat in a unit of time.

Principal shells *See* Energy levels.

Proglacial lake A lake created when a glacier acts as a dam, blocking the flow of a river or trapping glacial meltwater. The term refers to the position of such lakes just beyond the outer limits of a glacier.

Prokaryotes Cells or organisms such as bacteria whose genetic material is not enclosed in a nucleus.

Prominence A concentration of material above the solar surface that appears as a bright archlike structure.

Proterozoic eon The eon following the Archean and preceding the Phanerozoic. It extends between about 2500 million (2.5 billion) and 540 million years ago.

Proton A positively charged subatomic particle found in the nucleus of an atom.

Proton–proton chain A chain of thermonuclear reactions by which nuclei of hydrogen are built up into nuclei of helium.

Protoplanet A developing planetary body that grows by the accumulation of planetesimals.

Protostar A collapsing cloud of gas and dust destined to become a star.

Psychrometer A device consisting of two thermometers (wet bulb and dry bulb) that is rapidly whirled and, with the use of tables, yields the relative humidity and dew point.

Ptolemaic system An Earth-centered system of the universe.

Pulsar A variable radio source of small size that emits radio pulses in very regular periods.

Pulsating variable A variable star that pulsates in size and luminosity.

Pycnocline A layer of water in which there is a rapid change of density with depth.

Pyroclastic flow A highly heated mixture, largely of ash and pumice fragments, traveling down the flanks of a volcano or along the surface of the ground.

Pyroclastic material The volcanic rock ejected during an eruption, including ash, bombs, and blocks.

Pyroclastic texture An igneous rock texture resulting from the consolidation of individual rock fragments that are ejected during a violent volcanic eruption. Also known as *fragmental texture*.

Quaternary period The most recent period on the geologic time scale. It began about 2.6 million years ago and extends to the present.

Radial pattern A system of streams running in all directions away from a central elevated structure, such as a volcano.

Radiation The transfer of energy (heat) through space by electromagnetic waves. Also known as *electromagnetic radiation*.

Radiation fog Fog resulting from radiation heat loss by Earth.

Radiation pressure The force exerted by electromagnetic radiation from an object such as the Sun.

Radio interferometer Two or more radio telescopes that combine their signals to achieve the resolving power of a larger telescope.

Radio telescope A telescope designed to make observations in radio wavelengths.

Radioactive decay The spontaneous decay of certain unstable atomic nuclei.

Radioactivity The spontaneous emission of certain unstable atomic nuclei.

Radiocarbon (carbon-14) The radioactive isotope of carbon, which is produced continuously in the atmosphere and is used in dating events from the very recent geologic past (the last few tens of thousands of years).

Radiometric dating The procedure of calculating the absolute ages of rocks and minerals that contain radioactive isotopes.

Radiosonde A lightweight package of weather instruments fitted with a radio transmitter and carried aloft by a balloon.

Rain Drops of water that fall from clouds that have a diameter of at least 0.5 millimeter (0.02 inch).

Rain shadow desert A dry area on the lee side of a mountain range. Many middle-latitude deserts are of this type.

Rapids A part of a stream channel in which the water suddenly begins flowing more swiftly and turbulently because of an abrupt steepening of the gradient.

Ray (lunar) Any of a system of bright elongated streaks, sometimes associated with a crater on the Moon.

Recessional moraine An end moraine formed as the ice front stagnated during glacial retreat.

Rectangular pattern A drainage pattern characterized by numerous right-angle bends that develops on jointed or fractured bedrock.

Red giant A large, cool star of high luminosity; a star occupying the upper-right portion of the Hertzsprung-Russell diagram.

Reflecting telescope A telescope that concentrates light from distant objects by using a concave mirror.

Reflection The process whereby light bounces back from an object at the same angle at which it encounters a surface and with the same intensity.

Reflection nebula A relatively dense dust cloud in interstellar space that is illuminated by starlight.

Refracting telescope A telescope that uses a lens to bend and concentrate the light from distant objects.

Refraction The process by which the portion of a wave in shallow water slows, causing the wave to bend and tend to align itself with the underwater contours. Also known as *wave refraction*.

Regional metamorphism Metamorphism associated with large-scale mountain-building processes.

Regolith The layer of rock and mineral fragments that nearly everywhere covers Earth's surface.

Relative dating Placing rocks in their proper sequence or order to determine the chronological order of events.

Relative humidity The ratio of the air's water-vapor content to its water-vapor capacity.

Renewable resource A resource that is virtually inexhaustible or that can be replenished over relatively short time spans.

Reserve An already identified deposit from which minerals can be extracted profitably.

Reservoir rock The porous, permeable portion of an oil trap that yields oil and gas.

Residual soil Soil developed directly from the weathering of the bedrock below.

Resolving power The ability of a telescope to separate objects that would otherwise appear as one.

Retrograde motion The apparent westward motion of the planets with respect to the stars.

Reverse fault A fault in which the material above the fault plane moves up in relation to the material below.

Reverse polarity A magnetic field opposite to that which exists at present.

Revolution The motion of one body about another, as Earth about the Sun.

Richter scale A scale of earthquake magnitude based on the motion of a seismograph.

Ridge push A mechanism that may contribute to plate motion. It involves the oceanic lithosphere sliding down the oceanic ridge under the pull of gravity.

Rift valley A long, narrow trough bounded by normal faults. It represents a region where divergence is taking place.

Rift zone A region of Earth's crust along which divergence is taking place.

Right ascension An angular distance measured eastward along the celestial equator from the vernal equinox. Used with declination in a coordinate system to describe the position of celestial bodies.

Rime A thin coating of ice on objects produced when supercooled fog droplets freeze on contact.

Ring of Fire The zone of active volcanoes surrounding the Pacific Ocean.

Rip current A strong narrow surface or near-surface current of short duration and high speed flowing seaward through the breaker zone at nearly right angles to the shore. It represents the return to the ocean of water that has been piled up on the shore by incoming waves.

Rock A consolidated mixture of minerals.

Rock avalanche Very rapid downslope movement of rock and debris. These rapid movements may be aided by a layer of air trapped beneath the debris, and they have been known to reach speeds of over 200 kilometers (125 miles) per hour.

Rock cycle A model that illustrates the origin of the three basic rock types and the interrelatedness of Earth materials and processes.

Rock flour Ground-up rock produced by the grinding effect of a glacier.

Rock structure All features created by the processes of deformation from minor fractures in bedrock to a major mountain chain. Also known as *geologic structure*.

Rock-forming minerals The minerals that make up most of the rocks of Earth's crust.

Rockslide The rapid slide of a mass of rock downslope along planes of weakness.

Rotation The spinning of a body, such as Earth, about its axis.

Runoff Water that flows over the land rather than infiltrating into the ground.

Saffir–Simpson hurricane scale A scale, from 1 to 5, used to rank the relative intensities of hurricanes.

Salinity The proportion of dissolved salts to pure water, usually expressed in parts per thousand (‰).

Saltation Transportation of sediment through a series of leaps or bounces.

Santa Ana The local name given a chinook wind in southern California.

Saturation The maximum quantity of water vapor that the air can hold at any given temperature and pressure.

Scattering The redirecting (in all directions) of light by small particles and gas molecules in the atmosphere. The result is diffused light.

Scoria Hardened lava that has retained the vesicles produced by escaping gases.

Scoria cone *See* Cinder cone.

Sea arch An arch formed by wave erosion when caves on opposite sides of a headland unite.

Sea breeze A local wind blowing from the sea during the afternoon in coastal areas.

Sea ice Frozen seawater that is associated with polar regions. The area covered by sea ice expands in winter and shrinks in summer.

Sea stack An isolated mass of rock standing just offshore, produced by wave erosion of a headland.

Seafloor spreading The process of producing new seafloor between two diverging plates.

Seamount An isolated volcanic peak that rises at least 1000 meters (3000 feet) above the deep-ocean floor.

Seawall A barrier constructed to prevent waves from reaching the area behind the wall. Its purpose is to defend property from the force of breaking waves.

Secondary enrichment The concentration of minor amounts of metals that are scattered through unweathered rock into economically valuable concentrations by weathering processes.

Secondary pollutants Pollutants that are produced in the atmosphere by chemical reactions that occur among primary pollutants.

Secondary (S) wave A seismic wave that involves oscillation perpendicular to the direction of propagation.

Sediment Unconsolidated particles created by the weathering and erosion of rock, by chemical precipitation from solution in water, or from the secretions of organisms and transported by water, wind, or glaciers.

Sedimentary rock Rock formed from the weathered products of preexisting rocks that have been transported, deposited, and lithified.

Seismic gap A segment of an active fault zone that has not experienced a major earthquake over a span when most other segments have. Such segments are probable sites for future major earthquakes.

Seismic waves A rapidly moving ocean wave generated by earthquake activity capable of inflicting heavy damage in coastal regions.

Seismogram The record made by a seismograph.

Seismograph An instrument that records earthquake waves. Also known as a *seismometer*.

Seismology The study of earthquakes and seismic waves.

Seismometer *See* Seismograph.

Selective absorbers Gases that absorb and emit radiation only in certain wavelengths.

Semiarid *See* Steppe.

Semidiurnal tidal pattern A tidal pattern exhibiting two high tides and two low tides per tidal day with small inequalities between successive highs and successive lows; a semi-daily tide.

Settling velocity The speed at which a particle falls through a still fluid. The size, shape, and specific gravity of particles influence settling velocity.

Shadow zone The zone between 104 and 143 degrees distance from an earthquake epicenter in which direct waves do not arrive because of refraction by Earth's core.

Shear Stress that causes two adjacent parts of a body to slide past one another.

Sheeting A mechanical weathering process characterized by the splitting-off of slablike sheets of rock.

Shelf break The point where a rapid steepening of the gradient occurs, marking the outer edge of the continental shelf and the beginning of the continental slope.

Shield A large, relatively flat expanse of ancient metamorphic rock within the stable continental interior.

Shield volcano A broad, gently sloping volcano built from fluid basaltic lavas.

Shore Seaward of the coast, a zone that extends from the highest level of wave action during storms to the lowest tide level.

Shoreline The line that marks the contact between land and sea. It migrates up and down as the tide rises and falls.

Sidereal day The period of Earth's rotation with respect to the stars.

Sidereal month A time period based on the revolution of the Moon around Earth with respect to the stars.

Silicate Any one of numerous minerals that have the oxygen and silicon tetrahedron as their basic structure.

Silicon-oxygen tetrahedron A structure composed of four oxygen atoms surrounding a silicon atom that constitutes the basic building block of silicate minerals.

Sill A tabular igneous body that was intruded parallel to the layering of preexisting rock.

Sinkhole A depression produced in a region where soluble rock has been removed by groundwater.

Slab pull A mechanism that contributes to plate motion in which cool, dense oceanic crust sinks into the mantle and "pulls" the trailing lithosphere along.

Sleet Frozen or semifrozen rain formed when raindrops freeze as they pass through a layer of cold air.

Slide A movement common to mass-wasting processes in which the material moving downslope remains fairly coherent and moves along a well-defined surface.

Slip face The steep, leeward slope of a sand dune; it maintains an angle of about 34 degrees.

Slump The downward slipping of a mass of rock or unconsolidated material moving as a unit along a curved surface.

Small solar system bodies Solar system objects not classified as planets or moons that include dwarf planets, asteroids, comets, and meteoroids.

Snow A solid form of precipitation produced by sublimination of water vapor.

Snowfield An area where snow persists year-round.

Snowline The lower limit of perennial snow.

Soil A combination of mineral and organic matter, water, and air; the portion of the regolith that supports plant growth.

Soil horizon A layer of soil that has identifiable characteristics produced by chemical weathering and other soil-forming processes.

Soil profile A vertical section through a soil, showing its succession of horizons and the underlying parent material.

Soil taxonomy A soil classification system consisting of six hierarchical categories based on observable soil characteristics. The system recognizes 12 soil orders.

Soil texture The relative proportions of clay, silt, and sand in a soil. A soil's texture strongly influences its ability to retain and transmit water and air.

Solar constant The rate at which solar radiation is received outside Earth's atmosphere on a surface perpendicular to the Sun's rays when Earth is at an average distance from the Sun.

Solar eclipse An eclipse of the Sun.

Solar flare A sudden and tremendous eruption in the solar chromosphere.

Solar nebula The cloud of interstellar gas and/or dust from which the bodies of our solar system formed.

Solar winds Subatomic particles ejected at high speed from the solar corona.

Solifluction A slow, downslope flow of water-saturated materials common to permafrost areas.

Solstice The time when the vertical rays of the Sun are striking either the Tropic of Cancer or the Tropic of Capricorn. Solstice represents the longest or shortest day (length of daylight) of the year.

Solum The O, A, and B horizons in a soil profile. Living roots and other plant and animal life are largely confined to this zone.

Sorting The process by which solid particles of various sizes are separated by moving water or wind. Also, the degree of similarity in particle size in sediment or sedimentary rock.

Source region The area where an air mass acquires its characteristic properties of temperature and moisture.

Specific gravity The ratio of a substance's weight to the weight of an equal volume of water.

Spectral class A classification of a star according to the characteristics of its spectrum.

Spectroscope An instrument for directly viewing the spectrum of a light source.

Spectroscopy The study of spectra.

Spheroidal weathering Any weathering process that tends to produce a spherical shape from an initially blocky shape.

Spicule A narrow jet of rising material in the solar chromosphere.

Spiral galaxy A flattened, rotating galaxy with pinwheel-like arms of interstellar material and young stars winding out from its nucleus.

Spit An elongated ridge of sand that projects from the land into the mouth of an adjacent bay.

Spreading center *See* Divergent plate boundary.

Spring A flow of groundwater that emerges naturally at the ground surface.

Spring equinox The equinox that occurs on March 21–22 in the Northern Hemisphere and on September 21–23 in the Southern Hemisphere.

Spring tide The highest tidal range, which occurs near the times of the new and full moons.

Stable air Air that resists vertical displacement. If it is lifted, adiabatic cooling will cause its temperature to be lower than the surrounding environment; if it is allowed, it will sink to its original position.

Stable platform The part of a craton that is mantled by relatively undeformed sedimentary rocks and underlain by a basement complex of igneous and metamorphic rocks.

Stalactite An icicle-like structure that hangs from the ceiling of a cavern.

Stalagmite A columnlike form that grows upward from the floor of a cavern.

Star dune An isolated hill of sand that exhibits a complex form and develops where wind directions are variable.

Stationary front A situation in which the surface position of a front does not move; the flow on either side of such a boundary is nearly parallel to the position of the front.

Steam fog Fog having the appearance of steam, produced by evaporation from a warm water surface into the cool air above.

Stellar parallax A measure of stellar distance.

Steppe One of the two types of dry climate. A marginal and more humid variant of the desert that separates it from bordering humid climates. Also known as *semiarid*.

Stock A pluton similar to but smaller than a batholith.

Stony meteorite One of the three main categories of meteorites. Such meteorites are composed largely of silicate minerals with inclusions of other minerals.

Stony-iron meteorite One of the three main categories of meteorites. This group, as the name implies, is a mixture of iron and silicate minerals.

Storm surge The abnormal rise of the sea along a shore as a result of strong winds.

Strain An irreversible change in the shape and size of a rock body that is caused by stress.

Strata Parallel layers of sedimentary rock.

Stratified drift Sediments deposited by glacial meltwater.

Stratopause The boundary between the stratosphere and the mesosphere.

Stratosphere The layer of the atmosphere immediately above the troposphere, characterized by increasing temperatures with height, due to the concentration of ozone.

Stratovolcano *See* Composite cone.

Stratus One of three basic cloud forms; also, the name given one of the flow clouds. They are sheets or layers that cover much or all of the sky.

Streak The color of a mineral in powdered form.

Stream valley The channel, valley floor, and sloping valley walls of a stream.

Stress The force per unit area acting on any surface within a solid.

Striations (glacial) Scratches or grooves in a bedrock surface caused by the grinding action of a glacier and its load of sediment.

Strike-slip fault A fault along which the movement is horizontal.

Stromatolite Structures that are deposited by algae and consist of layered mounds of calcium carbonate.

Subarctic climate A climate found north of the humid continental climate and south of the polar climate and characterized by bitterly cold winters and short, cool summers. Places within this climatic realm experience the highest annual temperature ranges on Earth.

Subduction The process of thrusting oceanic lithosphere into the mantle along a convergent boundary.

Subduction erosion A process in subduction zones in which sediment and rock are scraped off the bottom of the overriding plate and transported into the mantle.

Subduction zone A long, narrow zone where one lithospheric plate descends beneath another.

Sublimation The conversion of a solid directly to a gas without passing through the liquid state.

Submarine canyon A seaward extension of a valley that was cut on the continental shelf during a time when sea level was lower, or a canyon carved into the outer continental shelf, slope, and rise by turbidity currents.

Submergent coast A coast with a form that is largely the result of the partial drowning of a former land surface either because of a rise of sea level or subsidence of the crust or both.

Subpolar low Low pressure located at about the latitudes of the Arctic and Antarctic Circles. In the Northern Hemisphere the low takes the form of individual oceanic cells; in the Southern Hemisphere there is a deep and continuous trough of low pressure.

Subsoil A term applied to the B horizon of a soil profile.

Subtropical high Not a continuous belt of high pressure but rather several semipermanent, anticyclonic centers characterized by subsidence and divergence located roughly between latitudes 25° and 35°.

Summer solstice The solstice that occurs on June 21–22 in the Northern Hemisphere and on December 21–22 in the Southern Hemisphere.

Sunspot A dark spot on the Sun, which is cool in contrast to the surrounding photosphere.

Supercontinent A large landmass that contains all, or nearly all, of the existing continents.

Supercontinent cycle The idea that the rifting and dispersal of one supercontinent is followed by a long period during which the fragments gradually reassemble into a new supercontinent.

Supercooled The condition of water droplets that remain in the liquid state at temperatures well below 0°C (32°F).

Supergiant A very large star of high luminosity.

Supernova An exploding star that increases in brightness many thousands of times.

Superposition The principle that in any undeformed sequence of sedimentary rocks, each bed is older than the layers above and younger than the layers below.

Supersaturation The condition of being more highly concentrated than is normally possible under given temperature and pressure conditions. When describing humidity, it refers to a relative humidity that is greater than 100 percent.

Surf A collective term for breakers; also, the wave activity in the area between the shoreline and the outer limit of breakers.

Surface soil The uppermost layer in a soil profile: the A horizon.

Surface waves Seismic waves that travel along the outer layer of Earth.

Suspended load The fine sediment carried within the body of flowing water.

Suture A zone along which two crustal fragments are jointed together. For example, following a continental collision, the two continental blocks are sutured together.

Swells Wind-generated waves that have moved into an area of weaker winds or calm.

Syncline A linear downfold in sedimentary strata; the opposite of anticline.

Synodic month The period of revolution of the Moon with respect to the Sun, or its cycle of phases.

System Any size group of interacting parts that form a complex whole.

Talus An accumulation of rock debris at the base of a cliff.

Tarn A small lake in a cirque.

Tectonic plate A coherent unit of Earth's rigid outer layer that includes the crust and upper unit.

Tectonics The study of the large-scale processes that collectively deform Earth's crust.

Temperature A measure of the degree of hotness or coldness of a substance; a measure of the *average* kinetic energy of individual atoms or molecules in a substance.

Temperature gradient The amount of temperature change per unit of distance.

Temperature inversion A layer in the atmosphere of limited depth where the temperature increases rather than decreases with height.

Temporary (local) base level The level of a lake, resistant rock layer, or any other base level that stands above sea level.

Tenacity A mineral's toughness or resistance to breaking or deforming.

Tensional stress The type of stress that tends to pull apart a body.

Terminal moraine The end moraine marking the farthest advance of a glacier.

Terrace A flat, benchlike structure produced by a stream, which was left elevated as the stream cut downward.

Terrae The extensively cratered highland areas of the Moon. Also known as *lunar highlands*.

Terrane A crustal block bounded by faults, whose geologic history is distinct from the histories of adjoining crustal blocks.

Terrestrial planets Any of the Earth-like planets, including Mercury, Venus, Mars, and Earth. Also known as the *inner planets*.

Terrigenous sediment Seafloor sediments derived from terrestrial weathering and erosion.

Texture The size, shape, and distribution of the particles that collectively constitute a rock.

Theory A well-tested and widely accepted view that explains certain observable facts.

Thermal gradient The increase in temperature with depth. It averages 1°C per 30 meters (1–2°F per 100 feet) in the crust.

Thermal metamorphism *See* Contact metamorphism.

Thermocline A layer of water in which there is a rapid change in temperature in the vertical dimension.

Thermohaline circulation Movements of ocean water caused by density differences brought about by variations in temperature and salinity.

Thermosphere The region of the atmosphere immediately above the mesosphere and characterized by increasing temperatures due to absorption of very shortwave solar energy by oxygen.

Thrust fault A low-angle reverse fault.

Thunder The sound emitted by rapidly expanding gases along the channel of lightning discharge.

Thunderstorm A storm produced by a cumulonimbus cloud and always accompanied by lightning and thunder. It is of relatively short duration and usually accompanied by strong wind gusts, heavy rain, and sometimes hail.

Tidal current The alternating horizontal movement of water associated with the rise and fall of the tide.

Tidal delta A deltalike feature created when a rapidly moving tidal current emerges from a narrow inlet and slows, depositing its load of sediment.

Tidal flat A marshy or muddy area that is covered and uncovered by the rise and fall of the tide.

Tide Periodic change in the elevation of the ocean surface.

Till Unsorted sediment deposited directly by a glacier.

Tombolo A ridge of sand that connects an island to the mainland or to another island.

Tornado A small, very intense cyclonic storm with exceedingly high winds, most often produced along cold fronts in conjunction with severe thunderstorms.

Tornado warning A warning issued when a tornado has actually been sighted in an area or is indicated by radar.

Tornado watch A warning issued for areas of about 65,000 square kilometers (25,000 square miles), indicating that conditions are such that tornadoes may develop; it is intended to alert people to the possibility of tornadoes.

Trade winds Two belts of winds that blow almost constantly from easterly directions and are located on the equatorward sides of the subtropical highs.

Transform fault A major strike-slip fault that cuts through the lithosphere and accommodates motion between two plates.

Transform plate boundary A boundary in which two plates slide past one another without creating or destroying lithosphere.

Transpiration The release of water vapor to the atmosphere by plants.

Transported soil Soils that form on unconsolidated deposits.

Transverse dunes A series of long ridges oriented at right angles to the prevailing wind; these dunes form where vegetation is sparse and sand is very plentiful.

Travertine A form of limestone ($CaCO_3$) that is deposited by hot springs or as a cave deposit.

Trellis pattern A system of streams in which nearly parallel tributaries occupy valleys cut in folded strata.

Trench An elongated depression in the seafloor produced by bending of oceanic crust during subduction. Also known as *deep-ocean trench*.

Trophic level A nourishment level in a food chain. Plant and algae producers constitute the lowest level, followed by herbivores and a series of carnivores at progressively higher levels.

Tropic of Cancer The parallel of latitude, 23½° north latitude, marking the northern limit of the Sun's vertical rays.

Tropic of Capricorn The parallel of latitude, 23½° south latitude, marking the southern limit of the Sun's vertical rays.

Tropical depression By international agreement, a tropical cyclone with maximum winds that do not exceed 61 kilometers (38 miles) per hour.

Tropical rain forest A luxuriant broadleaf evergreen forest; also, the name given the climate associated with this vegetation.

Tropical storm By international agreement, a tropical cyclone with maximum winds between 61 and 119 kilometers (38 and 74 miles) per hour.

Tropical wet and dry A climate that is transitional between the wet tropics and the subtropical steppes.

Tropopause The boundary between the troposphere and the stratosphere.

Troposphere The lowermost layer of the atmosphere. It is generally characterized by a decrease in temperature with height.

Tsunami The Japanese word for a seismic sea wave.

Tundra climate A climate found almost exclusively in the Northern Hemisphere or at high altitudes in many mountainous regions. A treeless climatic realm of sedges, grasses, mosses, and lichens that is dominated by a long, bitterly cold winter.

Turbidite A turbidity current deposit characterized by graded bedding.

Turbidity current A downslope movement of dense, sediment-laden water created when sand and mud on the continental shelf and slope are dislodged and thrown into suspension.

Turbulent flow The movement of water in an erratic fashion, often characterized by swirling, whirlpool-like eddies. Most streamflow is of this type.

Ultimate base level Sea level; the lowest level to which stream erosion could lower the land.

Ultramafic composition Igneous rocks composed mainly of iron and magnesium-rich minerals.

Ultraviolet Radiation with a wavelength from 0.2 to 0.4 micrometer.

Umbra The central, completely dark part of a shadow produced during an eclipse.

Unconformity A surface that represents a break in the rock record, caused by erosion or nondeposition.

Uniformitarianism The concept that the processes that have shaped Earth in the geologic past are essentially the same as those operating today.

Unsaturated zone The area above the water table where openings in soil, sediment, and rock are not saturated but filled mainly with air.

Unstable air Air that does not resist vertical displacement. If it is lifted, its temperature will not cool as rapidly as the surrounding environment, so it will continue to rise on its own.

Upslope fog Fog created when air moves up a slope and cools adiabatically.

Upwelling The rising of cold water from deeper layers to replace warmer surface water that has been moved away.

Urban heat island The phenomenon of temperatures within a city being generally higher than in surrounding rural areas.

Valence electron The electrons involved in the bonding process; the electrons occupying the highest principal energy level of an atom.

Valley breeze The daily upslope winds commonly encountered in a mountain valley.

Valley glacier *See* Alpine glacier.

Valley train A relatively narrow body of stratified drift deposited on a valley floor by meltwater streams that issue from a valley glacier.

Vapor pressure The part of the total atmospheric pressure attributable to water-vapor content.

Variable stars Red giants that overshoot equilibrium and then alternately expand and contract.

Vein deposit A mineral filling a fracture or fault in a host rock. Such deposits have a sheetlike, or tabular, form.

Vent The surface opening of a conduit or pipe.

Ventifact A cobble or pebble polished and shaped by the sandblasting effect of wind.

Vesicular texture A term applied to igneous rocks that contain small cavities called vesicles, which are formed when gases escape from lava.

Viscosity A measure of a fluid's resistance to flow.

Visible light Radiation with a wavelength from 0.4 to 0.7 micrometer.

Volatiles Gaseous components of magma dissolved in melt. Volatiles readily vaporize (form a gas) at surface pressures.

Volcanic bomb A streamlined pyroclastic fragment ejected from a volcano while molten.

Volcanic cone A cone-shaped structure built by successive eruptions of lava and/or pyroclastic materials.

Volcanic island arc A chain of volcanic islands generally located a few hundred kilometers from a trench where active subduction of one oceanic slab beneath another is occurring. Also known simply as *island arc*.

Volcanic neck An isolated, steep-sided, erosional remnant consisting of lava that once occupied the vent of a volcano. Also known as a *plug*.

Warm front A front along which a warm air mass overrides a retreating mass of cooler air.

Wash A common term for a desert stream course that is typically dry except for brief periods immediately following a rain.

Water table The upper level of the saturated zone of groundwater.

Wave base A depth equal to one-half the wavelength measured from the still water level. Below this depth, water movement associated with a wave is negligible.

Wave height The vertical distance between the trough and crest of a wave.

Wave of oscillation A water wave in which the waveform advances as the water particles move in circular orbits.

Wave of translation The turbulent advance of water created by breaking waves.

Wave period The time interval between the passage of successive crests at a stationary point.

Wave refraction *See* Refraction.

Wave-cut cliff A seaward-facing cliff along a steep shoreline formed by wave erosion at its base and mass wasting.

Wave-cut platform A bench or shelf in the bedrock at sea level, cut by wave erosion.

Wavelength The horizontal distance separating successive crests or troughs.

Weather The state of the atmosphere at any given time.

Weathering The disintegration and decomposition of rock at or near Earth's surface.

Welded tuff A pyroclastic rock composed of particles that have been fused together by the combination of heat still contained in the deposit after it has come to rest and by the weight of overlying material.

Well An opening bored into the zone of saturation.

Westerlies The dominant west-to-east motion of the atmosphere that characterizes the regions on the poleward side of the subtropical highs.

Wet adiabatic rate The rate of adiabatic temperature change in saturated air. The rate of temperature change is variable, but it is always less than the dry adiabatic rate.

White dwarf A star that has exhausted most or all of its nuclear fuel and has collapsed to a very small size; believed to be near its final stage of evolution.

White frost Ice crystals instead of dew that form on surfaces when the dew point is below freezing.

Wind Air flowing horizontally with respect to Earth's surface.

Wind vane An instrument used to determine wind direction.

Winter solstice The solstice that occurs on December 21–22 in the Northern Hemisphere and on June 21–22 in the Southern Hemisphere.

Yazoo tributary A tributary that flows parallel to the main stream because a natural levee is present.

Zodiac A band along the ecliptic containing the 12 constellations of the zodiac.

Zone of accumulation The part of a glacier characterized by snow accumulation and ice formation. Its outer limit is the snowline.

Zone of fracture The upper portion of a glacier, consisting of brittle ice.

Zone of saturation The zone where all open spaces in sediment and rock are completely filled with water.

Zone of wastage The part of a glacier beyond the zone of accumulation where all of the snow from the previous winter melts, as does some of the glacial ice.

Zooplankton Animal plankton.

INDEX